27.50
2850

The world in figures

The world in figures

Facts On File, Inc.
119 West 57th Street,
New York, N.7. 10019

Editorial information compiled by The Economist

This edition by arrangement with the
Economist Newspaper Limited.
This edition published 1980 by
Fact On File, Inc.
119 W. 57 St.
New York, N.Y. 10019

ISBN 0-87196-436-8

Printed and bound in Hong Kong

Introduction

This book provides detailed figures on each country in the world, and also on the relative importance of countries in the world under various subject headings. It can be used to supplement and expand upon information provided in The Economist Diary.

The volume is divided into two main sections: a world section and a section of individual countries. Countries are grouped in the second section by main region, and a country name index is included at the back of the book for easy reference (pages 292 and 293); this index also includes alternative names and old names for various countries.

Points to note in using the figures include:
1. *Rates of exchange*. Conversions between currencies have in general been made by using the relevant average par or market ('free') rate of exchange for any particular year (see also notes on page 290).
2. *Growth rates*. Rates of growth are affected by the choice of starting and finishing point. Further, they are higher when starting from a relatively low point; for example, an increase of 5 units from 5 to 10 in one year is a growth rate of 100 % for the year, an increase of 5 units from 10 to 15 is a growth rate of 50 %, and from 15 to 20 is only 33½ %. The natural course of events is for a growth rate to be high in the early stages of a new development, slowing down as size increases.
3. *Accuracy and estimates*. Figures are measures which are in most cases approximate, the degree of accuracy varying according to the country and the item being measured. For example, full census figures for population can usually be regarded as reasonably accurate, but even here there can be wide variations – population estimates for Nigeria vary widely. Further, most figures for latest years are provisional and subject to revision even when provided officially by the country concerned; this applies, for example, to UK national income and investment figures.

In this book, estimates are marked with an asterisk or asterisks; a gradation has been introduced to give some indication of the possible degree of error, as follows:
*   provisional or estimate
**   rough estimate
*** very rough estimate ('guesstimate')
This is not meant to imply that other figures are completely accurate, since virtually all figures are subject to some official estimation during their compilation.

Other main symbols are (see the glossary on page 289 for a full list):
na   not available or not applicable
%pa percentage per annum (compound growth rate)
000   thousand
mn   million
bn   billion, where 1 billion means 1 thousand million
t   tonne or metric ton
km   kilometre
mi   mile

Every care has been taken in the compilation of the information in this book, but no responsibility can be accepted for the accuracy of the data presented.

It is impossible to mention here all the sources from which the figures are compiled, but all are gratefully acknowledged. A list of main sources which can be used for additional information is included on page 288. British official statistics are reprinted by kind permission of the Controller of HM Stationery Office.

Contents

World population

Number of people, and growth %pa 1970–76

Rank in 1976		People 000, 1976	Growth %pa
	World total	*4 040 000**	*1.9**
1	China	850 000**	1.7**
2	India	610 080*	2.1*
3	Soviet Union	256 670	0.9
4	United States	215 120	0.8
5	Indonesia	139 620*	2.6*
6	Japan	112 770	1.3
7	Brazil	109 180*	2.8*
8	Bangladesh	78 660*	2.4*
9	Pakistan	72 370*	3.3*
10	Nigeria	64 750**	2.7**
11	Mexico	62 330*	3.5*
12	Germany, West	61 510	0.2
13	Italy	56 170	0.8
14	United Kingdom	55 890	0.1
15	France	52 920	0.7
16	Vietnam	46 520*	2.9*
17	Philippines	43 750*	2.9*
18	Thailand	42 960*	2.8*
19	Turkey	41 090*	2.8*
20	Egypt	38 070*	2.2*
21	Spain	35 970*	1.1*
22	Korea, South	35 860*	1.8*
23	Poland	34 360	0.9
24	Iran	33 590*	2.7*
25	Burma	30 830*	2.2*
26	Ethiopia	28 680*	2.6*
27	South Africa	26 130	2.5
28	Argentina	25 720*	1.3*
29	Zaire	25 630*	2.8*
30	Colombia	24 370*	2.9*
31	Canada	23 140	1.4
32	Jugoslavia	21 560	1.0
33	Rumania	21 450	1.0
34	Afghanistan	19 800*	2.5*
35	Sudan	18 850**	3.1**
36	Morocco	17 830*	2.9**
37	Algeria	17 300*	3.2*
38	Germany, East	16 790	−0.3
39	Taiwan	16 330	1.8
40	Korea, North	16 250*	2.6*
41	Peru	16 090*	3.0*
42	Tanzania	15 610*	2.7*
43	Czechoslovakia	14 920	0.7
44	Australia	13 920	1.8
45	Kenya	13 850*	3.6*
46	Netherlands	13 770	0.9
47	Sri Lanka	13 730*	1.5*
48	Nepal	12 860*	2.3*
49	Venezuela	12 390*	2.9*
50	Malaysia	12 300*	2.9*
51	Uganda	11 940*	3.3*
52	Iraq	11 510*	3.4*
53	Hungary	10 600	0.4
54	Chile	10 450	1.8
55	Ghana	10 310*	3.0*
56	Belgium	9 820	0.3
57	Portugal	9 690*	1.2*
58	Cuba	9 460	1.7
59	Mozambique	9 440*	2.3*
60	Saudi Arabia	9 240**	3.0**
61	Greece	9 170	0.7
62	Bulgaria	8 760	0.5
63	Cambodia	8 350*	2.8*
64	Madagascar	8 270**	3.0**
65	Sweden	8 220	0.4
66	Syria	7 600*	3.3*
67	Austria	7 510	0.2
68	Ecuador	7 300*	3.4*
69	Yemen, North	6 870*	3.0*
70	Angola	6 560*	2.4*
71	Cameroon	6 531*	1.9*
72	Rhodesia	6 530*	3.5*

Rank in 1976		People 000, 1976	Growth %pa
73	Switzerland	6 350	0.4
74	Guatemala	6 260	2.9
75	Upper Volta	6 170*	2.3*
76	Mali	5 840*	2.4*
77	Bolivia	5 790*	2.7*
78	Tunisia	5 740*	1.9*
79	Malawi	5 180*	2.6*
80	Zambia	5 140*	3.5*
81	Senegal	5 090**	4.4**
82	Denmark	5 070	0.5
83	Ivory Coast	5 020*	2.6*
84	Dominican Rep	4 840*	3.0*
85=	Finland	4 730	0.4
85=	Niger	4 730*	2.8*
87	Haiti	4 670*	1.6*
88	Guinea	4 530*	2.4**
89	Hongkong	4 440*	1.9*
90	Rwanda	4 290*	2.6*
91	El Salvador	4 123*	3.1*
92	Chad	4 116*	2.1*
93	Norway	4 030	0.6
94	Burundi	3 860*	2.4*
95	Israel	3 530	2.9
96	Laos	3 380*	2.2*
97	Somalia	3 260*	2.6*
98	Puerto Rico	3 210	2.8
99	Benin	3 200*	2.7*
100	Ireland	3 160	1.2
101	New Zealand	3 130	1.8
102	Sierra Leone	3 110*	3.4*
103	Lebanon	2 960*	3.1*
104	Honduras	2 831*	3.0*
105	Papua New Guinea	2 829*	2.2*
106	Uruguay	2 800*	0.6**
107	Jordan	2 780*	3.2*
108	Paraguay	2 720*	2.8*
109	Central African Emp	2 690**	2.1**
110	Albania	2 550	3.0
111	Libya	2 510*	4.0*
112	Togo	2 283*	2.6*
113	Singapore	2 278	1.6
114	Nicaragua	2 230*	3.3*
115	Jamaica	2 060	1.6
116	Costa Rica	2 020	2.6
117	Liberia	1 751*	2.4*
118	Yemen, South	1 749*	3.3*
119	Panama	1 720*	3.1*
120	Mongolia	1 490*	3.0*
121	Congo	1 390*	2.6*
122	Mauritania	1 350*	2.5*
123	Bhutan	1 200*	2.3*
124=	Lesotho	1 070*	2.4*
124=	Trinidad & Tobago	1 070*	0.7*
126	Kuwait	1 060*	6.1*
127	Mauritius	914*	1.6*
128	South-West Africa	880**	2.7**
129	Oman	791**	3.1*
130	Guyana	783*	1.5*
131	Utd Arab Emirates	700**	15.2**
132	Botswana	690*	3.0*
133	Timor, East	688*	2.2*
134	Cyprus	670**	1.9**
135	Fiji	588	2.1
136	Gambia	540*	2.6*
137	Guinea-Bissau	534*	1.5*
138	Gabon	526*	0.9*
139	Reunion	511*	2.2*
140	Swaziland	500*	2.9*
141	Surinam	435*	2.6*
142	Martinique	369	1.5
143	Guadeloupe	360*	1.6*
144	Luxembourg	358	0.9
145	Malta	330*	0.2*

Rank in 1976		People 000, 1976	Growth %pa
146	Equatorial Guinea	319*	1.9*
147	Cape Verde	306*	1.9*
148	Macao	275*	1.7*
149	Comoros	274*	2.5*
150	Bahrain	259*	3.2*
151	Barbados	247	0.6
152	Netherlands Antilles	241*	1.4*
153	Djibouti	226**	7.0**
154	Iceland	220*	1.3*
155	Bahamas	211	3.2
156	Qatar	210**	11.4**
157	Solomon Islands	200*	3.5*
158	Brunei	177*	5.3*
159	Samoa, Western	151*	0.9*
160	Belize	144*	2.9*
161	Maldives	136*	3.6*
162	New Caledonia	135*	3.6*
163	French Polynesia	134*	2.6*
164	Sahara, Western	120**	8.0**
165	St Lucia	110*	1.4*
166	Pacific Islands, US	108*	5.2**
167	St Vincent	107**	3.1**
168	Guam	102*	3.1*
169	New Hebrides	97*	2.6*
170=	Grenada	96*	0.3*
170=	Virgin Islands, US	96*	7.4*
172	Tonga	90*	0.8*
173	Sao Tome & Principe	80*	1.4*
174	Dominica	76*	1.3*
175	Jersey	74*	0.6*
176	Antigua	71*	1.4*
177	French Guiana	62*	3.3*
178	Isle of Man	60	1.5*
179=	Gilbert Islands	59*	3.0*
179=	Seychelles	59*	2.1*
181=	Bermuda	57*	1.5*
181=	Guernsey	57*	1.1*
183	Greenland	50	1.3
184	St Kitts-Nevis	48*	0.8*
185	Panama Canal Zone	44*	0.0*
186	Faroes	41	1.1
187	Mayotte	40*	2.5**
188	Samoa, American	31*	2.2*
189	Gibraltar	30*	2.0*
190	Andorra	28*	5.0*
191	Monaco	25*	1.3*
192	Liechtenstein	24*	2.1*
193	San Marino	19*	1.3*
194	Cook Islands	18	−2.4*
195	Northern Marianas	15*	5.2**
196	Cayman Islands	14	4.7
197	Montserrat	13*	2.5*
198	Virgin Islands, British	10*	0.0*
199	Wallis & Futuna Is	9.2*	1.1*
200	Tuvalu	9.0*	2.4*
201	Nauru	7.5*	2.0*
202=	Anguilla	6.6*	0.8**
202=	St Helena	6.6**	1.3**
204=	St Pierre & Miquelon	6.0*	1.6*
204=	Turks & Caicos Is	6.0*	0.9*
206	Niue	3.8	−4.2*
207	Christmas Island	3.3*	1.4*
208	Midway Islands	2.3*	0.6*
209	Wake Island	2.0*	3.3**
210	Falkland Islands	1.9*	−1.2*
211	Norfolk Island	1.8*	2.9*
212	Tokelau	1.6	−0.7*
213	Br Indian Ocean Terr	1.5*	−4.5**
214	Johnston Island	1.4***	5.6***
215	Vatican	1.0*	0.0*
216	Cocos Islands	0.5*	−3.6*
217	Pitcairn	0.07*	−3.9*

World gross domestic product

Total value in 1976, and growth %pa at constant prices for 1970-76

Rank in 1976		$ mn 1976	Growth[d] %pa	Rank in 1976		$ mn 1976	Growth[d] %pa	Rank in 1976		$ mn 1976	Growth[d] %pa
	World	6 770 000**	3.9[a]	72	Dominican Rep	3 915	9.0	144	Rwanda	410**	na
1	United States	1 702 000	2.9	73=	Burma	3 700*	2.6	145	Mauritania	380**	na
2	Soviet Union	700 000***	5.6[b]	73=	Zaire	3 700**	na	146	Burundi	370**	na
3	Japan	555 157	5.5	75	Uruguay	3 693	1.0	147	Barbados	360**	na
4	Germany, West	445 910	2.5	76	Angola	3 500**	na	148=	Bermuda	350**	na
5	France	346 740	4.1	77	Kenya	3 411	4.8*[a]	148=	Somalia	350**	na
6	China	340 000***	6.5**[c]	78	Rhodesia	3 247	4.6	150	Jersey	330**	na
7	United Kingdom	220 295	2.0	79	Sri Lanka	3 132	4.1	151=	Laos	300***	na
8	Canada	194 606	5.0	80	Ethiopia	3 100**	3.6[e]	151=	Panama Canal Zone	300***	na
9	Italy	170 765	2.9	81	Jamaica	3 045	2.1[e]	153	Botswana	280**	na
10	Brazil	130 000**	na	82	Uganda	3 000**	0.5[e]	154	Liechtenstein	260*	na
11	Spain	104 619	4.9	83	Trinidad and Tobago	2 684	2.5[a]	155	Djibouti	240**	na
12	Poland	95 000***	9.0[b]	84=	Ghana	2 600**	na	156=	Greenland	230***	na
13	Australia	94 533	3.9	84=	Mozambique	2 600**	na	156=	Guernsey	230***	na
14	Netherlands	89 523	3.6	86	Zambia	2 514	−4.1[a]	156=	Swaziland	230**	na
15	India	85 000*	2.4	87	Tanzania	2 450**	na	159	Faroes	215***	na
16	Mexico	79 119	5.0	88	Oman	2 430	na	160=	Macao	210**	na
17	Sweden	74 214	2.2	89	Costa Rica	2 345	6.3[a]	160=	Monaco	210***	na
18	Germany, East	70 000***	5.1[b]	90=	Bolivia	2 300**	5.9[a]	162=	Guinea-Bissau	200**	na
19	Belgium	67 460	4.0	90=	Qatar	2 300**	na	162=	Lesotho	200**	na
20	Iran	62 800*	19.2[a]	92	Afghanistan	2 200**	na	164	Isle of Man	188*	na
21	Switzerland	56 284	0.3	93	Luxembourg	2 197	1.8[a]	165	Andorra	150***	na
22	Czechoslovakia	55 000***	5.2[b]	94	El Salvador	2 186	5.1	166	Cape Verde	130**	na
23	Argentina	50 000**	2.6	95	Cameroon	2 100**	na	167=	Belize	120**	na
24	Saudi Arabia	49 300*	12.3	96	Panama	2 028	4.1	167=	French Guiana	120**	na
25	Turkey	40 703	7.5[a]	97	Lebanon	2 000***	na	167=	Samoa, American	120***	na
26	Austria	40 619	4.1	98	Gabon	1 900**	na	170	Sahara, Western	110***	na
27	Rumania	40 000***	11.7[b, e]	99	Nicaragua	1 842	5.6	171	Gambia	105**	na
28	Denmark	38 527	2.5	100	Madagascar	1 750**	na	172=	Equatorial Guinea	100**	na
29	Indonesia	37 270	na	101	Paraguay	1 700	6.1[a]	172=	Nauru	100***	na
30	Jugoslavia	35 000**	5.7[b]	102	Nepal	1 600**	2.3[a]	172=	Pacific Islands, US	100***	na
31	South Africa	32 217*	3.6[f]	103	Yemen, North	1 500**	na	172=	Timor, East	100***	na
32	Norway	31 307	4.8	104	Senegal	1 476	na	176=	Bhutan	90**	na
33	Venezuela	31 019	5.2	105	Iceland	1 454	5.1	176=	Gibraltar	90**	na
34	Finland	28 145	3.5	106	Papua New Guinea	1 430*	8.4[e]	178	San Marino	85***	na
35	Korea, South	25 318	11.0	107	Albania	1 300***	na	179=	Christmas Island	60***	na
36	Nigeria	25 000**	na	108	Jordan	1 206	na	179=	Comoros	60**	na
37	Hungary	24 000***	5.8[b]	109	Honduras	1 201	3.5	179=	Gilbert Islands	60**	na
38	Greece	22 244	5.2	110=	Mongolia	1 200***	na	179=	Solomon Islands	60**	na
39	Bulgaria	20 000***	7.8[a, b]	110=	South-West Africa	1 200***	na	183	St Lucia	54**	na
40	Philippines	17 795	6.2	112	Haiti	1 040**	4.4[c]	184=	New Hebrides	50***	na
41	Taiwan	17 258	8.3	113=	Brunei	1 000**	na	184=	Samoa, Western	50**	na
42	Thailand	16 609	6.6	113=	Reunion	1 000**	na	186	Cayman Islands	48***	na
43	Portugal	16 000**	4.3[a]	115	Liberia	910**	2.9*[a]	187=	Antigua	40**	na
44=	Algeria	15 600**	na	116	Martinique	840**	na	187=	Sao Tome & Principe	40**	na
44=	Iraq	15 600*	na	117	Cambodia	800***	na	187=	Tonga	40**	na
46	Libya	15 000**	na	118	Cyprus	778	−1.0	190=	Grenada	38**	na
47	Colombia	14 900**	6.2[a]	119	Guinea	750**	na	190=	St Vincent	38**	na
48	Pakistan	13 840*	3.2	120	Malawi	707	6.4	192	St Pierre & Miquelon	31***	na
49	Israel	13 640	6.4	121	Sierra Leone	680*	na	193=	Dominica	30**	na
50	New Zealand	12 630*	3.4	122	Congo	670**	na	193=	St Kitts-Nevis	30**	na
51	Egypt	12 300**	na	123	Fiji	662	7.1[e]	195	Virgin Islands, British	28**	na
52=	Kuwait	12 000*	na	124=	Guam	650**	na	196	Seychelles	27**	na
52=	Peru	12 000**	5.5[a]	124=	Niger	650**	na	197=	Maldives	20***	na
54	Chile	11 459	0.3	126=	Bahrain	630**	na	197=	Northern Marianas	20***	na
55	Malaysia	10 993	6.6[a]	126=	French Polynesia	630**	na	199	Cook Islands	18**	na
56	Hongkong	9 726	7.7	126=	New Caledonia	630**	na	200	Mayotte	15***	na
57	Utd Arab Emirates	9 500**	na	129	Togo	590**	na	201	Midway Islands	14***	na
58	Puerto Rico	9 000**	3.1[a]	130	Mauritius	570**	na	202	Wake Island	13***	na
59	Cuba	8 500***	na	131	Surinam	560**	na	203	Montserrat	10**	na
60	Vietnam	8 000**	na	132=	Bahamas	550**	na	204	Norfolk Island	8.5***	na
61	Ireland	7 975	3.1	132=	Guadeloupe	550**	na	205	Turks & Caicos Islands	8.0***	na
62	Korea, North	7 600***	na	132=	Virgin Islands, US	550**	na	206	Falkland Islands	3.8*	na
63	Bangladesh	7 500*	na	135=	Chad	540**	na	207	Saint Helena	3.6***	na
64	Morocco	7 000**	na	135=	Mali	540**	na	208	Wallis & Futuna Islands	3.5***	na
65	Syria	5 944	10.3	137=	Netherlands Antilles	500**	na	209=	Anguilla	3.0**	na
66	Singapore	5 915	9.1	137=	Upper Volta	500**	na	209=	Tuvalu	3.0***	na
67	Ecuador	4 955	9.3	139	Malta	479	8.6*	211	Niue	2.5***	na
68	Sudan	4 900**	na	140	Guyana	470**	na	212	Tokelau	0.8***	na
69	Tunisia	4 442	9.0[a]	141	Yemen, South	430***	na	213	Cocos Islands	0.7***	na
70	Guatemala	4 363	5.9	142=	Benin	420**	na	214[c]	Pitcairn	0.2***	na
71	Ivory Coast	3 950**	na	142=	Central African Emp	420**	na				

[a] 1970–75 [b] Growth of net material product at constant prices [c] Not included in this table are: British Indian Ocean Territory, Johnston Island, Vatican [d] Growth of gross domestic product at constant prices, 1970-76 [e] 1970–74 [f] Including South-West Africa

World population and gross domestic product

Total population and total gross domestic product for each country[a], 1976

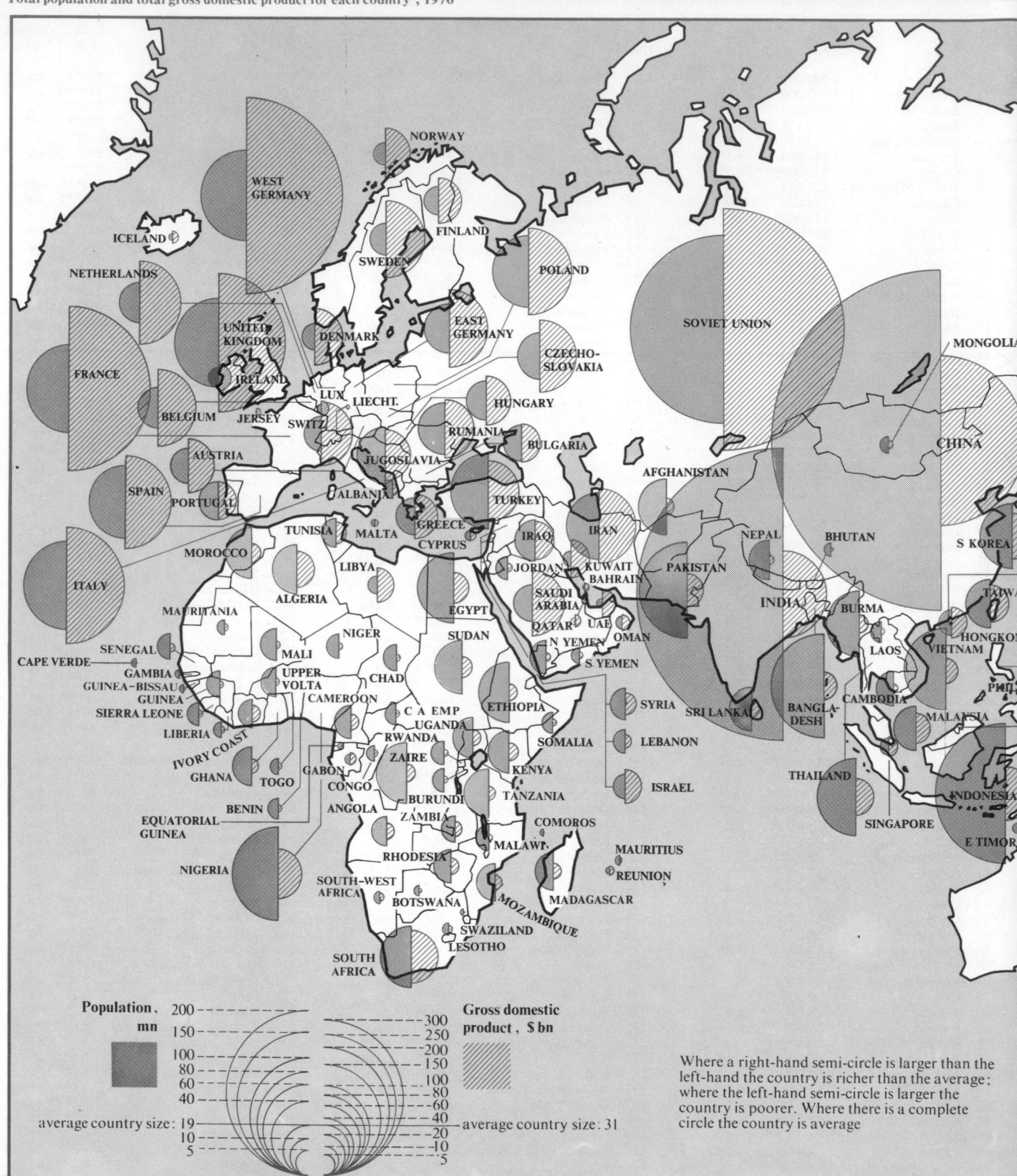

Population, mn

200 --------
150 --------
100 --------
80 --------
60 --------
40 --------

average country size: 19

10 --------
5 --------

-------- 300
-------- 250
-------- 200
-------- 150
-------- 100
-------- 80
-------- 60
-------- 40
-------- 20
-------- 10
-------- 5

average country size: 31

Gross domestic product, $ bn

Where a right-hand semi-circle is larger than the left-hand the country is richer than the average; where the left-hand semi-circle is larger the country is poorer. Where there is a complete circle the country is average

[a]Figures for population are ranked on page 8, and for gross domestic product on page 9; these figures are also available for each country in the country section.　　Countries shown he

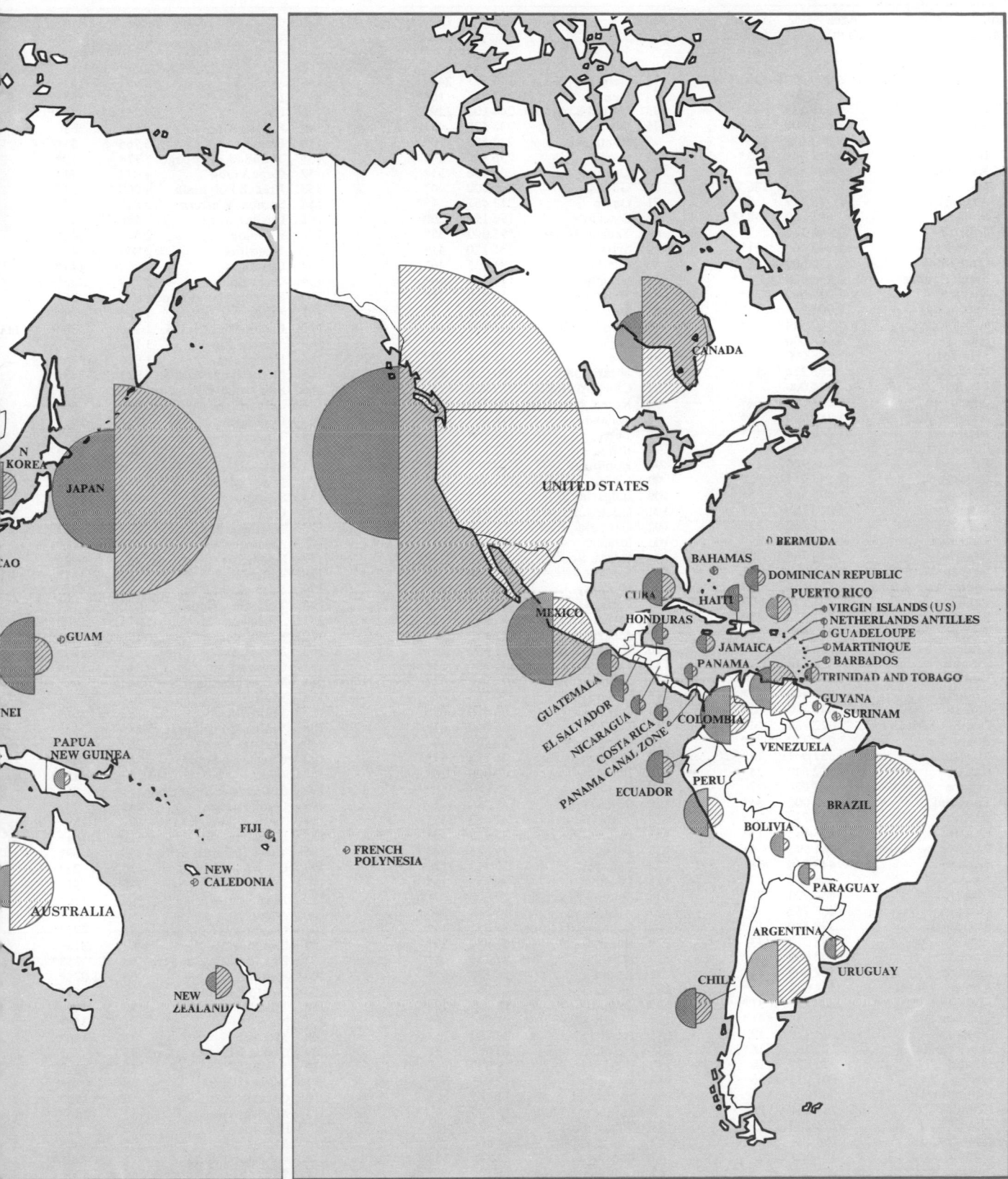

World area

Country area (square kilometres) and density of population (people per square kilometre), 1976

Rank by area		Area (km²)	Density (people/km²)
	World country area	136 000 000*	30*
1	Soviet Union	22 402 200	11
2	Canada	9 976 139	2
3	China	9 561 000	87**
4	United States	9 363 123	23
5	Brazil	8 511 965	13*
6	Australia	7 686 848	2
7	India	3 287 590	186*
8	Argentina	2 776 889	9*
9	Sudan	2 505 800	8**
10	Algeria	2 381 741	7*
11	Zaire	2 345 409	11*
12	Greenland	2 175 600	0.1
13	Saudi Arabia	2 150 000*	4**
14	Mexico	2 022 060	31*
15	Indonesia	1 904 345	73*
16	Libya	1 759 540	1*
17	Iran	1 648 000	20*
18	Mongolia	1 565 000	1*
19	Peru	1 285 216	13*
20	Chad	1 284 000	3*
21	Niger	1 267 000	4*
22	Angola	1 246 700	5*
23	Mali	1 240 000	5*
24	South Africa	1 222 161	21
25	Ethiopia	1 221 900	23*
26	Colombia	1 138 914	21*
27	Bolivia	1 098 581	5*
28	Mauritania	1 030 700	1*
29	Egypt	1 001 449	38*
30	Tanzania	942 000	17*
31	Nigeria	923 768	70**
32	Venezuela	912 050	14*
33	South-West Africa	823 168	1**
34	Pakistan	803 940	90*
35	Mozambique	783 030	12*
36	Turkey	780 576	53*
37	Chile	756 945	14
38	Zambia	752 614	7*
39	Burma	676 552	46*
40	Afghanistan	647 500	31*
41	Somalia	637 700	5*
42	Central African Emp	622 984	4**
43	Botswana	600 372	1*
44	Madagascar	587 041	14**
45	Kenya	582 640	24*
46	France	547 026	97
47	Thailand	514 000	84*
48	Spain	504 782	71*
49	Cameroon	475 442	14*
50	Papua New Guinea	461 691	6*
51	Sweden	449 964	18
52	Morocco	446 550	40*
53	Iraq	434 924	26*
54	Paraguay	406 752	7*
55	Rhodesia	390 580	17*
56	Japan	372 313	303
57	Congo	342 000	4*
58	Finland	337 009	14
59	Yemen, South	333 038	5*
60	Malaysia	329 749	37*
61	Vietnam	329 556	141*
62	Norway	324 219	12
63	Ivory Coast	322 462	16*
64	Poland	312 677	110
65	Italy	301 225	186
66	Philippines	300 000*	146*
67	Ecuador	283 561	26*
68	Upper Volta	274 200	23*
69	New Zealand	268 704	11
70	Gabon	267 667	2*
71	Sahara, Western	266 000	0.5**
72	Jugoslavia	255 804	84
73	Germany, West	248 577	247
74	Guinea	245 857	18**
75	United Kingdom	244 103	229
76	Ghana	238 500	43*
77	Rumania	237 500	90
78	Laos	236 800	14*
79	Uganda	236 036	51*
80	Guyana	215 000	4*
81	Oman	212 457	4**
82	Senegal	196 192	26*
83	Yemen, North	195 000*	35*
84	Syria	185 180	41*
85	Cambodia	181 035	46*
86	Uruguay	177 508	16*
87	Tunisia	163 610	35*
88	Surinam	163 265	3*
89	Bangladesh	143 998	546
90	Nepal	140 797	91*
91	Greece	131 944	69
92	Nicaragua	130 000	17*
93	Czechoslovakia	127 869	117
94	Korea, North	120 538	135*
95	Malawi	118 484	44*
96	Cuba	114 524	83
97	Benin	112 622	28*
98	Honduras	112 088	25*
99	Liberia	111 400	16*
100	Bulgaria	110 912	79
101	Guatemala	108 889	57
102	Germany, East	108 178	155
103	Iceland	103 000	2*
104	Korea, South	98 484	364*
105	Jordan	97 740	28*
106	Hungary	93 030	114
107	Portugal	92 082	105*
108	French Guiana	91 000	1*
109	Austria	83 849	90
110	Utd Arab Emirates	83 600	8**
111	Panama	75 650	23*
112	Sierra Leone	71 740	43*
113	Ireland	70 283	45
114	Sri Lanka	65 610	209*
115	Togo	56 000	41*
116	Costa Rica	50 700	40
117	Dominican Rep	48 734	99*
118	Bhutan	47 000	26*
119	Denmark	43 069	118
120	Switzerland	41 288	154
121	Netherlands	41 160	335
122	Guinea-Bissau	36 125	15*
123	Taiwan	35 981	454
124	Belgium	30 513	322
125	Lesotho	30 355	35*
126	Solomon Islands	29 800*	7*
127	Albania	28 748	89
128	Equatorial Guinea	28 051	11*
129	Burundi	27 834	139*
130	Haiti	27 750	168*
131	Rwanda	26 338	163*
132	Belize	22 963	6*
133	Djibouti	22 000	10**
134	El Salvador	21 041	196*
135	Israel	20 700	171
136	New Caledonia	19 058	7*
137	Fiji	18 272	32
138	Kuwait	17 818	59*
139	Swaziland	17 365	29*
140	Falkland Islands	16 260	0.1*
141	Timor, East	14 925	46*
142	New Hebrides	14 800	7*
143	Bahamas	13 935	15
144	Gambia	11 295	48*
145	Qatar	11 000*	19**
146	Jamaica	10 991	187
147	Lebanon	10 400	285*
148	Cyprus	9 251	72**
149	Puerto Rico	8 897	361
150	Brunei	5 765	31*
151	Trinidad & Tobago	5 128	210*
152	Cape Verde	4 033	73*
153	French Polynesia	4 000	33*
154	Samoa, Western	2 842	53*
155	Luxembourg	2 586	138
156	Reunion	2 512	204*
157	Mauritius	2 045	447*
158	Comoros	1 862	147*
159	Guadeloupe	1 780	202*
160	Panama Canal Zone	1 676	26*
161	Faroes	1 399	29
162	Pacific Islands, US	1 378	78*
163	Martinique	1 100	335
164	Hongkong	1 046	4 245*
165	Netherlands Antilles	993	243*
166	Sao Tome & Principe	964	83*
167	Gilbert Islands	860	69*
168	Dominica	751	101*
169	Tonga	700	129*
170	Bahrain	662	391*
171	St Lucia	616	179*
172	Isle of Man	588	103
173	Singapore	581	3 921
174	Guam	549	186*
175	Northern Marianas	479	31*
176	Andorra	453	62*
177	Seychelles	443	133*
178	Antigua	442	161*
179	Barbados	431	573
180	Turks & Caicos Is	430	14*
181	St Helena	413	16**
182	St Vincent	389	275*
183	Mayotte	374	107*
184 =	Grenada	344	279*
184 =	Virgin Islands, US	344	279*
186	Malta	316	1 040*
187	Maldives	298	456*
188	Wallis & Futuna Is	275	33*
189	St Kitts-Nevis	267	180
190 =	Cayman Islands	260	54
190 =	Niue	260	15
192	St Pierre & Miquelon	242	25*
193	Cook Islands	241	75
194	Samoa, American	197	157*
195	Liechtenstein	160	151*
196	Virgin Islands, British	153	69*
197	Christmas Island	135	24*
198	Jersey	116	641
199	Montserrat	102	130*
200	Anguilla	91	73*
201	Guernsey	78	723*
202	San Marino	61	318*
203	Br Indian Ocean Terr	60	25*
204	Bermuda	53	1 075*
205	Norfolk Island	36	51*
206	Tuvalu	26	346*
207	Nauru	21	357*
208	Macao	16	17 190*
209	Cocos Islands	14	36*
210	Tokelau	10	157
211	Wake Island	8	250**
212	Gibraltar	6	5 190*
213 =	Midway Islands	5	460*
213 =	Pitcairn	5	15
215	Monaco	2	13 300*
216	Johnston Island	1	1 400***
217	Vatican	0.4	2 300*

World standard of living

National income per person (in $) and energy consumption per person (in kilograms of coal equivalent)

Rank by income in 1976	Income $, 1976	Energy[a] kg, 1975
World	*1 520**	*2 028**
1 Nauru	12 000***	6 626
2 Kuwait	11 300*	8 718
3 Utd Arab Emirates	11 000***	13 699
4 Liechtenstein	10 800*	na
5 Qatar	9 000***	35 328
6 Switzerland	8 246	3 642
7 Sweden	8 043	6 178
8 Monaco	8 000***	na
9 Canada	7 341	9 880
10 United States	6 996	10 999
11 Denmark	6 803	5 268
12 Germany, West	6 451	5 345
13 Norway	6 400*	4 607
14 Belgium	6 371	5 584
15 Australia	6 288	6 485
16 Jersey	6 100**	na
17= Guam	6 000***	6 097
17= Midway Islands	6 000***	na
17= Wake Island	6 000***	24 025
20 Netherlands	5 892	5 784
21 France	5 859	3 944
22 Luxembourg	5 800*	15 504
23= Bermuda	5 500**	3 090
23= Iceland	5 500	4 720
25 Finland	5 351	4 766
26 Virgin Islands, US	5 100**	50 157
27= Andorra	5 000**	na
27= Libya	5 000**	1 299
27= Panama Canal Zone	5 000**	14 150
27= St Pierre & Miquelon	5 000***	4 122
31 Saudi Arabia	4 990*	1 398
32 Austria	4 823	3 700
33 Brunei	4 800**	9 628
34= Faroes	4 600***	4 325
34= French Polynesia	4 600**	877
34= New Caledonia	4 600***	9 933
37 Greenland	4 500***	5 465
38 Japan	4 465	3 622
39= Guernsey	4 300**	na
39= Norfolk Islands	4 300***	na
41 San Marino	4 200***	na
42 Germany, East	4 000***	6 835
43 New Zealand	3 663*	3 111
44 United Kingdom	3 550	5 265
45 Czechoslovakia	3 500***	7 151
46 Israel	3 288	2 806
47= Cayman Islands	3 000**	2 838
47= Christmas Island	3 000***	20 537
47= Samoa, American	3 000***	1 682
50 Gabon	2 900**	1 026
51 Isle of Man	2 830*	na
52 Italy	2 723	3 012
53= Gibraltar	2 700***	1 267
53= Pitcairn	2 700***	na
55 Spain	2 663	2 147
56 Soviet Union	2 600***	5 546
57= Bahamas	2 500**	6 279
57= Poland	2 500**	5 007
57= Virgin Islands, British	2 500**	1 121
60 Singapore	2 470*	2 151
61 Oman	2 400*	334
62 Ireland	2 367	3 097
63 Greece	2 322	2 090
64 Puerto Rico	2 200**	3 203
65= Bulgaria	2 100**	4 781
65= Hungary	2 100***	3 624
67= Hongkong	2 090*	1 119
67= Trinidad & Tobago	2 090**	3 132
69 Venezuela	2 070*	2 639
70 Bahrain	2 030**	12 079
71= Argentina	1 900**	1 754
71= Martinique	1 900**	1 062
71= Reunion	1 900**	434
74 Falkland Islands	1 770*	750*
75= Iran	1 700*	1 353
75= Netherlands Antilles	1 700**	12 231
77 Rumania	1 600**	3 803
78 French Guiana	1 570**	953
79 Jugoslavia	1 540**	1 930
80 Malta	1 533	1 032
81 Portugal	1 500*	983
82 Barbados	1 400**	1 078
83 Jamaica	1 310*	1 427
84 Iraq	1 280*	713
85 Guadeloupe	1 250**	564
86 Uruguay	1 240*	942
87= Cocos Islands	1 200***	na
87= Northern Marianas	1 200***	na
87= Turks & Caicos Is	1 200***	na
90 Cyprus	1 157	1 278
91 Mexico	1 130	1 221
92 Fiji	1 117*	582
93= Brazil	1 100**	670
93= South-West Africa	1 100**	na
95 South Africa	1 070*	2 953*
96 Costa Rica	1 064	544
97 Surinam	1 060**	2 063
98 Panama	1 055	865
99 Djibouti	1 020**	450
100 Gilbert Islands	1 000**	346
101 Turkey	980*	630
102 Taiwan	970	na
103 Chile	960*	765
104= Pacific Islands, US	900***	909[b]
104= Sahara, Western	900***	na
106= Algeria	840*	754
106= Malaysia	840*	588
108= Cook Islands	800**	na
108= Cuba	800***	1 157
110 Jordan	780*	408
111 Nicaragua	761	479
112= Macao	750***	264
112= Mongolia	750***	1 091
112= Syria	750*	477
115 Tunisia	735*	447
116= Belize	730**	520
116= Ivory Coast	730**	366
118= Lebanon	700***	928
118= Montserrat	700**	716
118= Peru	700**	682
121 Dominican Rep	694	458
122 Korea, South	641	1 038
123 Ecuador	621	442
124 Guyana	620**	1 114
125= Guatemala	600*	237
125= Mauritius	600**	279
127 Paraguay	574	153
128 St Kitts-Nevis	570**	na
129= Niue	550**	na
129= Saint Helena	550***	na
131= Antigua	540**	2 184
131= Colombia	540**	671
133 El Salvador	503	248
134= Angola	500**	174
134= New Hebrides	500***	561
134= Tokelau	500***	na
137 Albania	490**	741
138 St Lucia	460**	na
139 Papua New Guinea	453*	278
140= Korea, North	450***	2 808**
140= Rhodesia	450**	764
140= Sao Tome & Principe	450**	102
143= Congo	430**	209
143= Seychelles	430**	481
143= Tonga	430**	na
146 Anguilla	420**	na
147= Cape Verde	400**	61
147= Wallis & Futuna Is	400***	na
149 Honduras	394	232
150= Botswana	390**	na
150= Zambia	390**	504
152 China	380***	660**
153= Dominica	370**	na
153= Swaziland	370**	na
155= Liberia	360**	404
155= Morocco	360**	274
155= Philippines	360*	326
158 Thailand	358	284
159= Bolivia	350**	303
159= Nigeria	350**	90
161= Grenada	340**	323
161= St Vincent	340**	152
163= Guinea-Bissau	330**	82
163= Tuvalu	330***	na
165 Samoa, Western	320**	160
166 Egypt	310**	405
167 Cameroon	305**	104
168= Equatorial Guinea	300**	101
168= Mayotte	300***	na
168= Solomon Islands	300**	241
171 Senegal	270*	195
172 Mozambique	260**	186
173 Indonesia	242	178
174= Lesotho	240**	na
174= Uganda	240**	55
176= Ghana	230**	182
176= Sudan	230**	140
176= Togo	230**	65
176= Yemen, South	230**	328
180= Haiti	220**	30
180= Kenya	220*	174
180= Mauritania	220**	108
183 Sri Lanka	214	127
184 Madagascar	210**	71
185= Comoros	200***	na
185= Sierra Leone	200**	116
185= Yemen, North	200**	49
188 Gambia	190**	66
189 Pakistan	180*	183
190 Vietnam	160***	200**
191= Central African Emp	150**	34
191= Guinea	150**	92
191= Maldives	150***	na
191= Tanzania	150**	69
195 Malawi	137*	56
196 India	132*	221
197 Timor, East	130**	16
198= Chad	125**	39
198= Niger	125**	35
200 Zaire	124**	78
201 Benin	121**	52
202 Nepal	120**	10
203= Afghanistan	110**	52
203= Burma	110*	51
205 Ethiopia	108**	29
206 Somalia	105**	36
207= Cambodia	90***	16
207= Mali	90**	25
209 Burundi	87**	13
210= Bangladesh	85*	28
210= Laos	85***	63
210= Rwanda	85**	14
213 Upper Volta	75**	20
214 Bhutan	70**	na

[a]Although a guide to the standard of living, this total level of energy consumption is also affected by climate and by the nature of the economy; for example, small countries involved in fuelling aircraft and ships can be large energy users [b]Including Northern Marianas

World standard of living

National income per person[a], $, 1976

GREENLAND

WEST
GERMANY

NORWAY

FINLAND

ICELAND

NETHERLANDS

SWEDEN

EAST
GERMANY

FAROES

BELGIUM

SOVIET UNION

LUXEMBURG

POLAND

UNITED
KINGDOM

DENMARK

CZECHOSLOVAKIA

IRELAND

HUNGARY

MONGOLIA

AUSTRIA

FRANCE

RUMANIA

N KOREA

SWITZERLAND

JUGO

BULGARIA

ALBANIA

S KOREA

PORTUGAL

SPAIN

ITALY

TURKEY

CHINA

GREECE

MALTA

IRAN

MACAO

TUNISIA

CYPRUS

SYRIA
IRAQ

TAIWAN

MOROCCO

JORDAN

AFGHANISTAN

BAHRAIN

PAKISTAN

NEPAL

BHUTAN

HONGKONG

WESTERN SAHARA

ALGERIA

LIBYA

EGYPT

SAUDI
ARABIA

BANGLADESH

BURMA

LAOS

VIETNAM

CAPE VERDE

MAURITANIA

MALI

NIGER

CHAD

N YEMEN

OMAN

INDIA

THAILAND

CAMBODIA

PHILIPPIN

SENEGAL

GUINEA

SUDAN

S
YEMEN

DJIBOUTI

SRI LANKA

BRUNEI

GAMBIA

NIGERIA

MALDIVES

GUINEA-BISSAU

CAMEROON

C A EMP

ETHIOPIA

ISRAEL

SINGAPORE

MALAYSIA

SIERRA LEONE

UGANDA

SOMALIA

LEBANON

LIBERIA

IVORY COAST

CONGO

RWANDA

KENYA

SEYCHELLES

INDONESIA

GHANA

ZAIRE

BURUNDI

KUWAIT

E TIMOR

TOGO

TANZANIA

COMOROS

MADAGASCAR

COCOS IS

BENIN

GABON

ANGOLA

ZAMBIA

MALAWI

MAURITIUS

QATAR

CHRISTMAS I

SAO TOME & PRINCIPE

RHODESIA

EQUAT GUINEA

REUNION

ST HELENA

SOUTH-WEST
AFRICA

BOTSWANA

MOZAMBIQUE

UAE

SWAZILAND

SOUTH AFRICA

LESOTHO

$ per person

12 000
10 000
8 000
6 000

4 000

2 000
1 000
500

world average income
per person: $1 520

[a]Figures are ranked on page 13, and figures in local currency, US dollars and UK pounds are provided for each country in the country section.   The chart excludes some small countr

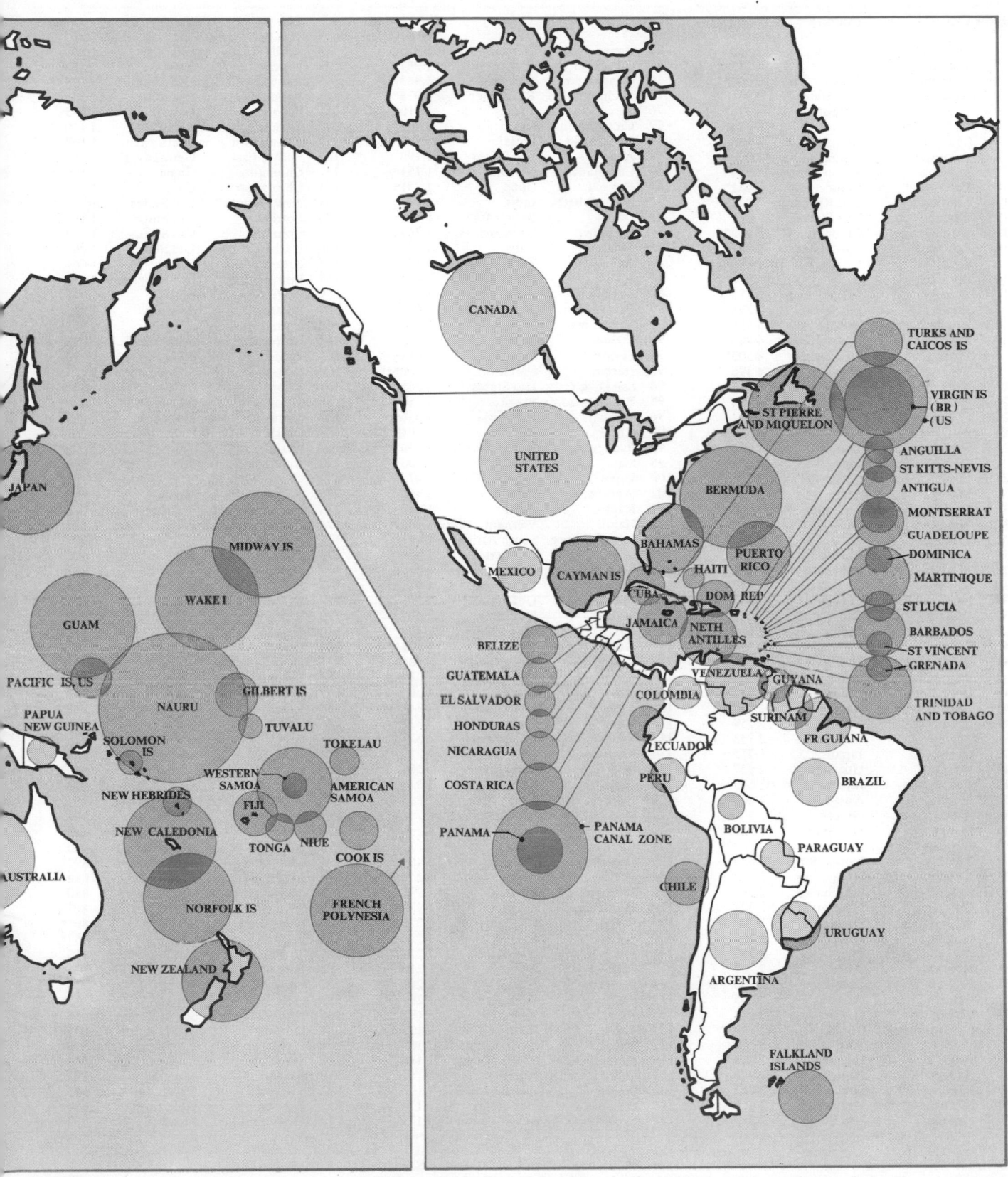

CANADA

UNITED
STATES

TURKS AND
CAICOS IS

VIRGIN IS
(BR)
(US

ST PIERRE
AND MIQUELON

ANGUILLA
ST KITTS-NEVIS
ANTIGUA

BERMUDA

MONTSERRAT
GUADELOUPE
DOMINICA

BAHAMAS

PUERTO
RICO

MARTINIQUE

MEXICO

CAYMAN IS

HAITI

ST LUCIA

CUBA

DOM REP

BARBADOS

JAMAICA

NETH
ANTILLES

ST VINCENT
GRENADA

BELIZE

TRINIDAD
AND TOBAGO

GUATEMALA

VENEZUELA

GUYANA

EL SALVADOR

COLOMBIA

SURINAM

HONDURAS

NICARAGUA

ECUADOR

FR GUIANA

COSTA RICA

PERU

BRAZIL

PANAMA

PANAMA
CANAL ZONE

BOLIVIA

PARAGUAY

CHILE

URUGUAY

JAPAN

MIDWAY IS

ARGENTINA

WAKE I

GUAM

NAURU

GILBERT IS

PACIFIC IS, US

PAPUA
NEW GUINEA

TUVALU

SOLOMON
IS

TOKELAU

WESTERN
SAMOA

AMERICAN
SAMOA

NEW HEBRIDES

FIJI

NIUE

NEW CALEDONIA

TONGA

COOK IS

AUSTRALIA

NORFOLK IS

FRENCH
POLYNESIA

NEW ZEALAND

FALKLAND
ISLANDS

World cities

Main cities, country and population[a]

Rank	City	Country	Population 000	Rank	City	Country	Population 000	Rank	City	Country	Population 000
1	Mexico City	Mexico	11 340	72	Berlin, West	Germany, W	1 985	143	Addis Ababa	Ethiopia	1 161
2	Tokyo	Japan	11 282	73	Cleveland	Utd States	1 967	144	Lyons	France	1 153
3	Shanghai	China	10 000**	74	Guadalajara	Mexico	1 963	145	Indianapolis	Utd States	1 139
4	New York	Utd States	9 561	75	Bucharest	Rumania	1 934	146	Vancouver	Canada	1 136
5	Buenos Aires	Argentina	8 925[b]	76	Barcelona	Spain	1 810[b]	147	Poona	India	1 135[c]
6	Paris	France	8 424	77	Ch'ang-ch'un	China	1 800**[c]	148	Fortaleza	Brazil	1 110
7	São Paulo	Brazil	8 100	78	Hyderabad	India	1 796[c]	149	Cape Town	South Africa	1 108[c]
8	Peking	China	8 000**	79	Atlanta	Utd States	1 790	150	Berlin, East	Germany, E	1 106
9	Moscow	Soviet Union	7 734	80	Casablanca	Morocco	1 753[b]	151 =	Ch'eng-tu	China	1 100**[c]
10	Calcutta	India	7 031[c]	81	Havana	Cuba	1 751[c]	151 =	Ch'ing-tao	China	1 100**[c]
11	London	Utd Kingdom	7 028	82	Ahmedabad	India	1 742[c]	153	New Orleans	Utd States	1 094
12	Chicago	Utd States	7 015	83	Dacca	Bangladesh	1 730[b]	154	Portland	Utd States	1 083
13	Tientsin	China	7 000**	84	Hamburg	Germany, W	1 717	155	Columbus	Utd States	1 069
14	Los Angeles[d]	Utd States	6 987	85	Milan	Italy	1 705	156	Hartford[q]	Utd States	1 063
15	Seoul	Korea, Sth	6 889	86	Ankara	Turkey	1 701	157	Kita-Kyushu	Japan	1 061
16	Bombay	India	5 971[c]	87	Anaheim[f]	Utd States	1 700	158	Birmingham	Utd Kingdom	1 059
17	Cairo	Egypt	5 715[b]	88	Bangalore	India	1 654[c]	159	Pôrto Alegre	Brazil	1 044
18	Jakarta	Indonesia	5 193	89	Tashkent	Soviet Union	1 643	160 =	Brussels	Belgium	1 042
19	Rio de Janeiro	Brazil	4 858	90	Monterrey	Mexico	1 638	160 =	Damascus	Syria	1 042
20	Philadelphia	Utd States	4 807	91	Vienna	Austria	1 615[c]	162	Rotterdam	Netherlands	1 032
21	Manila	Philippines	4 500*	92	Lisbon	Portugal	1 612[c]	163	Tbilisi	Soviet Union	1 030
22	Teheran	Iran	4 496	93	Harbin	China	1 600**[c]	164	Odessa	Soviet Union	1 023
23	Karachi	Pakistan	4 465*	94	San Diego	Utd States	1 585	165	Kaohsiung	Taiwan	1 020
24	Detroit	Utd States	4 424	95	Belo Horizonte	Brazil	1 557	166	Cologne	Germany, W	1 014
25	Leningrad	Soviet Union	4 372	96	Surabaja	Indonesia	1 556[c]	167	Marseilles	France	1 005
26	Bangkok	Thailand	4 340	97	Algiers	Algeria	1 504[b]	168	Omsk	Soviet Union	1 002
27	Shenyang	China	4 000**[c]	98 =	Pyongyang	Korea, North	1 500**	169	Fushun	China	1 000**[c]
28	Delhi[k]	India	3 647[c]	98 =	Sian	China	1 500**[c]	170 =	Amsterdam	Netherlands	989
29	Lüta	China	3 600**[c]	100	Lagos	Nigeria	1 477[c]	170 =	Chelyabinsk	Soviet Union	989
30	Madrid	Spain	3 520[b]	101	Warsaw	Poland	1 463	170 =	Kawasaki	Japan	989
31	Ho Chi Minh Cy	Vietnam	3 460	102	Kyoto	Japan	1 459	173	San Antonio	Utd States	982
32	Lima	Peru	3 303[b]	103	Hanoi	Vietnam	1 443	174	Dnepropetrovsk	Soviet Union	976
33	Santiago	Chile	3 263	104	Johannesburg	South Africa	1 441[c]	175	Rochester	Utd States	971
34	Rangoon	Burma	3 187*[b]	105	Miami	Utd States	1 439	176	Tunis	Tunisia	970*
35	Madras	India	3 170[c]	106	Medellín	Colombia	1 417[b]	177	Donetsk	Soviet Union	967
36	San Francisco[e]	Utd States	3 140	107	Denver[n]	Utd States	1 413	178	Sofia	Bulgaria	966
37	Istanbul	Turkey	3 135[b]	108	Milwaukee	Utd States	1 409	179	Fukuoka	Japan	965
38	Washington	Utd States	3 022	109	Seattle[o]	Utd States	1 407	180 =	Brisbane	Australia	958
39	Canton	China	3 000**[c]	110	Baku	Soviet Union	1 406	180 =	Kazan	Soviet Union	958
40	Sydney	Australia	2 936	111	Nan-ch'ing	China	1 400**[c]	182	Perm	Soviet Union	957
41	Boston	Utd States	2 890	112	Kharkov	Soviet Union	1 385	183	Nova Iguaçu	Brazil	932
42	Rome	Italy	2 884	113	Cincinnati	Utd States	1 381	184	Nagpur	India	930[c]
43	Bogotá	Colombia	2 855[b]	114	Stockholm	Sweden	1 364	185	Lille	France	929
44	Baghdad	Iraq	2 800*	115	Tampa[p]	Utd States	1 348	186	Yerevan	Soviet Union	928
45	Montreal	Canada	2 759	116	Kobe	Japan	1 338	187 =	Cali	Colombia	923[b]
46	Toronto	Canada	2 753	117	Buffalo	Utd States	1 327	187 =	Santo Domingo	Dominican Rep	923
47	Osaka	Japan	2 715	118 =	Munich	Germany, W	1 315	187 =	Ufa	Soviet Union	923
48	Nassau[s]	Utd States	2 657	118 =	Oporto	Portugal	1 315[c]	190	Mandalay	Burma	920*[b]
49	Yokohama	Japan	2 610	120	Taegu	Korea, South	1 311	191	Volgograd	Soviet Union	918
50	Melbourne	Australia	2 604	121	Gorky	Soviet Union	1 305	192	Rostov-na-Donu	Soviet Union	907
51	Athens	Greece	2 540[c]	122	Kansas City	Utd States	1 290	193	Providence[i]	Utd States	904
52	Dallas[m]	Utd States	2 527	123	Novosibirsk	Soviet Union	1 286	194 =	Abidjan	Ivory Coast	900*
53	Caracas	Venezuela	2 500*[b]	124	Kanpur	India	1 275[c]	194 =	Adelaide	Australia	900
54	Pusan	Korea, Sth	2 454	125	Copenhagen	Denmark	1 251	196	Chittagong	Bangladesh	890[b]
55	Kowloon	Hongkong	2 378	126	Recife	Brazil	1 250	197	Louisville	Utd States	888
56	St Louis	Utd States	2 367	127	Salvador	Brazil	1 237	198	Sacramento	Utd States	880
57	Pittsburgh	Utd States	2 322	128	Montevideo	Uruguay	1 230	199	Memphis	Utd States	867
58	Singapore	Singapore	2 308	129	Riverside[h]	Utd States	1 226	200	Guayaquil	Ecuador	861[b]
59	Houston	Utd States	2 286	130	Naples	Italy	1 224	201	Mosul	Iraq	857*
60	Alexandria	Egypt	2 259[b]	131	Phoenix	Utd States	1 221	202	Glasgow	Utd Kingdom	856
61	Kinshasa	Zaire	2 202	132	Sapporo	Japan	1 216	203 =	Basra	Iraq	854*
62	Wuhan	China	2 200**[c]	133	Belgrade	Jugoslavia	1 209[c]	203 =	Giza	Egypt	854[b]
63	Lahore	Pakistan	2 165[b]	134	Bandung	Indonesia	1 202[c]	205	Helsinki	Finland	853[b]
64	Baltimore	Utd States	2 148	135 =	Haiphong	Vietnam	1 191	206 =	Alma-Ata	Soviet Union	851
65	Ch'ung-ch'ing	China	2 100**[c]	135 =	Turin	Italy	1 191	206 =	Durban	South Africa	851[c]
66	Taipei	Taiwan	2 089	137	Minsk	Soviet Union	1 189	208 =	Hollywood[r]	Utd States	848
67	Nagoya	Japan	2 083	138	Kuibyshev	Soviet Union	1 186	208 =	Sarotov	Soviet Union	848
68	Budapest	Hungary	2 082	139	Tel Aviv-Jaffa[l]	Israel	1 181	210	Ibadan	Nigeria	847
69	Kiev	Soviet Union	2 013	140	Prague	Czechoslovakia	1 176	211	Victoria	Hongkong	845
70	Minneapolis[j]	Utd States	2 011	141	San Jose	Utd States	1 174	212	Dayton	Utd States	836
71	Newark	Utd States	1 999	142	Sverdlovsk	Soviet Union	1 171	213	Hiroshima	Japan	832

[a]1975, 1976 or 1977 [b]1972, 1973 or 1974 [c]Latest available [d]Los Angeles-Long Beach [e]San Francisco-Oakland [f]Anaheim-Santa Ana-Garden Grove [s]Nassau-Suffolk [h]Riverside-San Bernardino-Ontario [i]Providence-Warwick-Pawtucket [j]Minneapolis-St Paul [k]Including New Delhi [l]Including Ramat Gan [m]Dallas-Fort Worth [n]Denver-Boulder [o]Seattle-Everett [p]Tampa-St Petersburg [q]Hartford-New Britain-Bristol [r]Fort Lauderdale-Hollywood

World population ages

People in each main age group, as % of total population[a]; expectation of life[a]

Region and country	Age (years) Under 15 %	15–64 %	65 and over %	Expectation of life (no of years) Male	Female
World[e]	**36***	**58***	**6***	**na**	**na**
Africa[e]	**44***	**53***	**3***	**na**	**na**
Algeria[b]	47	49	4	52*	55*
Benin	46	50	4	39*	43*
Botswana	48	46	6	42	45
Burundi	44	54	2	40*	43*
Central African Emp[b]	42	55	3	33[d]	36[d]
Chad	41	56**	3*	29[d]	35[d]
Ethiopia[b]	45*	52**	3**	37*	40*
Gabon[b]	25	68	7	25*	45*
Gambia	41	57	2	39*	42*
Ghana	47	49	4	42*	45*
Guinea[b]	44	49	7	39*	42*
Kenya[b]	48	48	4	47	51
Lesotho	39	57	4	44*	48*
Liberia	42	55	3	46	44
Libya	49	47	4	51*	54*
Madagascar[b]	47	49	4	37[d]	38[d]
Mali	49	49	2	37*	40*
Mauritania	42	52	6	37*	40*
Mauritius	38	58	4	61	65
Morocco	46	51	3	51*	54*
Niger	43	54*	3*	37*	40*
Reunion[b]	46	51	3	56[d]	62[d]
Rhodesia[b]	47	51	2	50*	53*
Rwanda	44	53*	3*	39*	43*
Seychelles	43	51	6	62	68
South Africa	42	54	4	50	53
Swaziland	48	49	3	42	45
Tanzania[b]	44	53	3	39*	42*
Tunisia	45	51	4	52*	56*
Uganda[b]	46	50	4	48*	52*
Zambia	46	51	3	43*	46*
America[e]	**35***	**58***	**7***	**na**	**na**
Argentina	29	63	8	65	71
Bahamas	44	53	3	64	67
Bermuda	30	64	6	66[d]	72[d]
Bolivia	42	54	4	46*	48*
Brazil	42	55	3	58[d]	61[d]
Canada	26	66	8	69	76
Chile	35	60	5	60	66
Colombia	44	53	3	59*	63*
Costa Rica	44	52	4	62[d]	65[d]
Cuba	37	56	7	69	72
Dominican Republic	48	49	3	57[d]	59[d]
Ecuador	44	52	4	51[d]	54[d]
El Salvador	46	51	3	57[d]	60[d]
Guatemala	45	52	3	18[d]	50[d]
Guyana	44	53	3	59[d]	63[d]
Haiti	42	54	4	49	51
Honduras	47	51	2	52*	55*
Jamaica	46	48	6	63[d]	67[d]
Martinique[b]	43	52	5	63[d]	67[d]
Mexico	46	51	3	63	67
Nicaragua	48	49	3	51*	55*
Panama	43	53	4	64	67
Paraguay	45	52	3	60*	64*
Peru	45	52	3	53[d]	55[d]
Puerto Rico	37	56	7	69	76
St Lucia	50	45	5	55[d]	58[d]
Trinidad and Tobago	39	57	4	64	68
United States	24	65	11	69	77
Uruguay	28	63	9	66[d]	72[d]
Venezuela	45	52	3	63*[d]	69*[d]
Asia[e]	**38***	**58***	**4***	**na**	**na**
Bahrain	44	53	3	na	na
Bangladesh	43	54	3	36	36
Brunei	43	54	3	na	na
China	33*	61*	6*	60	63

Region and country	Age (years) Under 15 %	15–64 %	65 and over %	Expectation of life (no of years) Male	Female
Hongkong	30	64	6	67	75
India	40	57	3	42[d]	41[d]
Indonesia	44	53	3	47[d]	47[d]
Iran	47	50	3	51*	51*
Iraq	48	49	3	51*	54*
Israel	33	59	8	70	74
Japan	24	68	8	71	76
Jordan	48	49	3	53[d]	52[d]
Korea, South	39	58	3	63	67
Kuwait	44	54	2	66	72
Lebanon	43	52	5	61*	65*
Malaysia	45	52	3	63*	67*
Nepal	40	57	3	42	45
Pakistan[b]	43	53	4	na	na
Philippines	43	54	3	57*	60*
Singapore	32	64	4	65	70
Syria	48	49	3	54	59
Thailand	45	52	3	54[d]	59[d]
Yemen, South	48	48	4	44	46
Europe[e]	**24***	**64***	**12***	**na**	**na**
Austria	23	62	15	68	75
Belgium	23	63	14	68	74
Bulgaria	22	67	11	69	74
Cyprus	28	62	10	70	73
Czechoslovakia	23	65	12	67	73
Denmark	23	64	13	71	76
Faroes	32	59	9	na	na
Finland	22	68	10	67	75
France	24	62	14	69	76
Germany, East	21	63	16	69	74
Germany, West	21	65	14	68	75
Gibraltar	26	65	9	na	na
Greece	24	64	12	67[d]	71[d]
Greenland	39	58	3	59	66
Guernsey	23	62	15	na	na
Hungary	20	67	13	67	72
Iceland	30	61	9	72	77
Ireland	31	58	11	69[d]	73[d]
Isle of Man	20	60	20	na	na
Italy	24	64	12	69	75
Jersey	21	65	14	na	na
Jugoslavia	26	66	8	65	70
Liechtenstein	28	64	8	na	na
Luxembourg	20	67	13	67	74
Malta	26	65	9	68	72
Monaco[b]	13	65	22	na	na
Netherlands	25	64	11	71	77
Norway	24	62	14	72	78
Poland	24	66	10	67	74
Portugal	27	63	10	65	72
Rumania	25	65	10	67	72
San Marino	25	65	10	na	na
Soviet Union	26	65	9	64	74
Spain	28	62	10	70	75
Sweden	21	64	15	72	78
Switzerland	22	65	13	70	76
Turkey	40	55	5	52*[d]	56*[d]
United Kingdom	23	63	14	68	74
Oceania[e]	**31***	**62***	**7***	**na**	**na**
Australia	28	64	8	68[d]	74[d]
Fiji	39	58	3	68*	72*
French Polynesia	46	52	2	na	na
New Hebrides[b]	46	51	3	na	na
New Zealand	30	61	9	69	75
Pacific Islands, US[c]	46	50	4	na	na
Papua New Guinea	44	52	4	48	48
Solomon Islands	44	53	3	na	na
Samoa, American	47	51	2	65	69
Samoa, Western	50	47	3	61[d]	65[d]
Tonga	45	52	3	na	na

[a]Mainly 1970 to 1975   [b]Age group figures are the latest available   [c]Including Northern Marianas   [d]1960s   [e]1975

World education

Number of pupils, and % of population who are literate

Rank in 1975/76		Pupils[d] 000, 1975/76	Literacy[e] %[f]
	World	*750 000*[*][b]	na
1	China	200 000*	40**
2	India	93 200*	33
3	United States	57 346*	99
4	Soviet Union	53 540	100
5	Japan	21 281[a]	99*
6	Brazil	20 300*[a]	66
7	Indonesia	16 000*	57
8	Mexico	14 700*	74
9	Germany, West	12 730	100
10	United Kingdom	11 600[a]	97*
11	France	11 100*	100
12	Italy	10 500*	94
13	Vietnam	10 200**	60**
14	Philippines	10 092[a]	83
15	Bangladesh	10 000*	22[g]
16	Korea, South	9 072	88
17	Thailand	7 870*	79
18	Poland	7 466	98
19	Turkey	7 397	51
20	Spain	7 100*	91
21	Pakistan	7 080*[a]	15[g]
22	Egypt	6 536[a]	26[g]
23	Iran	6 244[a]	37
24	Canada	6 023	95[g]
25	Rumania	5 410	100
26	South Africa	5 320*[a]	57[g]
27	Colombia	5 280*[a]	81
28	Argentina	5 200*[a]	92
29	Nigeria	4 938*[a]	15[g]
30	Burma	4 819	68
31	Taiwan	4 180	81
32	Peru	4 160	72
33	Jugoslavia	4 000*	83
34	Zaire	3 892[a]	31[g]
35	Germany, East	3 360*	100
36	Australia	3 153*[a]	na
37	Kenya	3 132*	20[g]
38	Korea, North	3 000**[a]	90**
39	Algeria	2 950*[a]	26
40	Malaysia	2 930*	53
41	Netherlands	2 923[a]	100
42	Chile	2 900*[a]	88
43	Venezuela	2 830*[a]	77
44	Cuba	2 616	78[g]
45	Sri Lanka	2 534[a]	78
46	Czechoslovakia	2 458	100
47	Iraq	2 083[a]	24[g]
48	Morocco	2 048*	21
49	Belgium	1 923	100
50	Syria	1 793*	40
51	Portugal	1 751	71
52	Ecuador	1 710*	67[g]
53	Tanzania	1 658	28
54	Ghana	1 591[a]	43
55	Hungary	1 556	98
56	Greece	1 547[a]	84
57	Bulgaria	1 502	91
58	Austria	1 422	100
59	Sudan	1 409[a]	19
60	Sweden	1 380	100
61	Madagascar	1 171*[a]	39
62	Tunisia	1 135*	32
63	Cameroon	1 128[a]	19[g]
64	Switzerland	1 124[a]	100
65	Bolivia	1 120*	39[g]
66	Hongkong	1 100*	77
67	Ethiopia	1 047[a]	10*
68	Uganda	1 026[a]	35[g]
69	Dominican Rep	1 009[a]	67

Rank in 1975/76		Pupils[d] 000, 1975/76	Literacy[e] %[f]
70	Finland	988[a]	100
71	Rhodesia	948	39[g]
72	Zambia	934[a]	47
73	New Zealand	932	100
74	Puerto Rico	918	88
75	Afghanistan	891	6[g]
76	Denmark	836	100
77	Saudi Arabia	825[a]	15**
78	Ireland	824	100
79	Norway	804	99
80	El Salvador	794[a]	57
81	Israel	769*[a]	94[g]
82	Ivory Coast	751*[a]	20*
83	Guatemala	698[a]	46
84	Lebanon	665*[a]	70*
85	Libya	664[a]	22[g]
86	Nepal	659[a]	12
87	Mozambique	635[a]	11[g]
88	Albania	632[a]	71[g]
89	Malawi	630[a]	22
90	Angola	598[a]	30*
91	Cambodia	595*[a]	42[g]
92	Jamaica	575*[a]	82[g]
93	Uruguay	564*[a]	90[g]
94	Haiti	547*	23
95	Costa Rica	546*	88
96	Paraguay	545	80
97	Honduras	538	45[g]
98	Jordan	524[a]	46
99	Singapore	516	69
100	Panama	488*	78
101	Togo	430	16
102	Nicaragua	419[a]	58
103	Rwanda	399[a]	16[g]
104	Congo	398[a]	35*
105	Mongolia	367	99*
106	Senegal	339*[a]	6[g]
107	Laos	295*[a]	28[g]
108	Mali	290[a]	2[g]
109 =	Benin	287[a]	5
109 =	Trinidad & Tobago	287*[a]	92
111	Papua New Guinea	285*	32
112	Yemen, North	249	10*
113	Guinea	243*[a]	10
114 =	Lesotho	239*	59
114 =	Sierra Leone	239*[a]	7[g]
116	Yemen, South	237[a]	27
117	Chad	221*[a]	14
118 =	Central African Emp	218*[a]	7[g]
118 =	Mauritius	218*	62
120	Kuwait	205[a]	55
121	Guyana	202*[a]	85[g]
122	Liberia	191	24
123	South-West Africa	182	38[g]
124	Reunion	175[a]	63
125	Fiji	168	81
126	Upper Volta	150*[a]	11
127 =	Burundi	143[a]	14[g]
127 =	Gabon	143[a]	20
129	Cyprus	137	76[g]
130	Niger	134[a]	1[g]
131	Botswana	131*	33
132	Surinam	123*[a]	84[g]
133	Somalia	111*[a]	15
134	Swaziland	108*	29
135	Martinique	105*[a]	88
136	Guadeloupe	103[a]	83
137	Cape Verde	68	37
138 =	Oman	65	25
138 =	Utd Arab Emirates	65[a]	21

Rank in 1975/76		Pupils[d] 000, 1975/76	Literacy[e] %[f]
140	Malta	64	87
141	Barbados	62*[a]	96
142	Bahamas	60	90[g]
143 =	Bahrain	59	40
143 =	Guinea-Bissau	59	5[g]
145	Luxembourg	58*	99
146	Iceland	57	99
147	Mauritania	54*[a]	11
148	Netherlands Antilles	50*[a]	92
149	Samoa, Western	49*	98
150	Brunei	47	64
151	French Polynesia	43[a]	95[g]
152	Equatorial Guinea	42*[a]	na
153	Belize	41	87[g]
154	Pacific Islands, US	39[h]	na
155	New Caledonia	38*[a]	84[g]
156 =	Grenada	35*[a]	93
156 =	Timor, East	35[a]	na
158 =	Macao	33	79
158 =	St Vincent	33*[a]	na
160 =	Gambia	31	10
160 =	Guam	31*[a]	na
160 =	St Lucia	31[a]	na
163	Qatar	30[a]	na
164 =	Solomon Islands	28[a]	na
164 =	Tonga	28*[a]	na
166	Comoros	26[a],[c]	58
167	Virgin Islands, US	25*[a]	na
168	Dominica	24[a]	59[g]
169	New Hebrides	23	na
170	Antigua	20[a]	89[g]
171	Bhutan	19	5*
172	Gilbert Islands	18***[a]	90[g]
173 =	St Kitts-Nevis	15*[a]	88*[g]
173 =	Sao Tome & Principe	15*[a]	na
175	Seychelles	14*	58
176 =	Greenland	13*	na
176 =	Jersey	13	100*
176 =	Samoa, American	13[a]	na
179 =	Bermuda	12*[a]	98
179 =	French Guiana	12*[a]	74
179 =	Isle of Man	12	100*
182	Panama Canal Zone	11	na
183	Djibouti	10[a]	na
184	Faroes	9.7*	100*
185	Guernsey	8.1	100*
186	Cook Islands	6.6	na
187	Sahara, Western	5.3[a]	na
188	Monaco	5.0	98
189	Gibraltar	4.6	na
190	Maldives	3.9	na
191	Liechtenstein	3.8	100
192	Cayman Islands	3.5[a]	93[g]
193	Montserrat	3.2*	80[g]
194	San Marino	2.8[a]	95
195 =	Andorra	2.7[a]	na
195 =	Virgin Islands, Br	2.7	93[g]
195 =	Wallis and Futuna Is	2.7	na
198	Turks & Caicos Is	2.3	91[g]
199 =	Nauru	2.0	na
199 =	Tuvalu	2.0	na
201	Anguilla	1.8*[a]	88*[g]
202	St Pierre & Miquelon	1.7[a]	99
203	Niue	1.4*	94[g]
204	Saint Helena	1.3*[a]	98
205	Christmas Island	1.1	na
206	Tokelau	0.6**	97[g]
207	Falkland Islands	0.3[a]	98[g]

[a]Latest available information; mainly 1970 to 1974   [b]1974/75   [c]Including Mayotte   [d]Includes students in higher education   [e]Ability both to read and write   [f]Generally percentage of population aged 15 years and over; mainly around 1970   [g]Around 1960   [h]Including Northern Marianas

World labour force

Number in the labour force, and as % of total population, 1975 or 1976

Rank		Labour force[b] 000	% of total population	Rank		Labour force[b] 000	% of total population	Rank		Labour force[b] 000	% of total population
	World	*1 670 000**	*41**	69	Chile	2 607[a]	28[a]	138	Oman	150**[a]	21**[a]
1	China	400 000**	47**	70	Ivory Coast	2 571*	51*	139	Luxembourg	148	42
2	India	180 373[a]	33[a]	71	Denmark	2 486	49	140	Reunion	144*	28*
3	Soviet Union	129 154*	50*	72	Saudi Arabia	2 436	26	141	Guadeloupe	125	35
4	United States	96 917	45	73	Haiti	2 361	51	142	Martinique	123	33
5	Japan	53 780	48	74	Rwanda	2 315*	54*	143	Comoros	116*	42*
6	Indonesia	41 261[a]	35[a]	75	Malawi	2 291*	44*	144	Malta	113	35
7	Brazil	29 557[a]	32[a]	76	Finland	2 279	48	145	Surinam	108*	25*
8	Germany, West	26 878	43	77	Rhodesia	2 197*	34*	146	Barbados	100*	40*
9	United Kingdom	26 258	47	78	Guinea	2 072*	46*	147	Equatorial Guinea	96*	31*
10	Bangladesh	25 933	33	79	Hongkong	1 952	44	148	Iceland	93	42
11	Nigeria	25 193*	39*	80	Ecuador	1 941[a]	28[a]	149	Cape Verde	90*	29*
12	France	21 834	41	81	Yemen, North	1 931	28	150	Qatar	87	41
13	Pakistan	21 035	29	82	Zambia	1 928*	38*	151	Bahamas	84	41
14	Vietnam	20 749	45	83	Senegal	1 923*	38*	152	Netherlands Antilles	73[a]	34[a]
15	Italy	20 379	36	84	Burundi	1 898*	49*	153	Macao	72**	26**
16	Poland	17 507[a]	52[a]	85	Syria	1 839	25	154	Bahrain	60[a]	28[a]
17	Mexico	16 597	28	86	Norway	1 821	45	155	Maldives	52*	40*
18	Turkey	16 349	41	87	Angola	1 763*	27*	156	New Caledonia	50	38
19	Philippines	15 161	36	88	Laos	1 648	49	157	Brunei	41[a]	30[a]
20	Thailand	13 945	34	89	Chad	1 551*	38*	158	Virgin Islands, US	40**	43**
21	Spain	13 281	37	90	Guatemala	1 546[a]	28[a]	159	Jersey	39	53
22	Korea, South	13 061	36	91	Bolivia	1 511	26	160	Samoa, Western	38[a]	26[a]
23	Ethiopia	12 138*	42*	92	Benin	1 487*	46*	161 =	French Polynesia	35[a]	30[a]
24	Burma	11 933	39	93	Niger	1 486*	31*	161 =	New Hebrides	35[a]	45[a]
25	Zaire	10 881*	42*	94	Tunisia	1 414*	25*	163 =	Belize	33[a]	27[a]
26	Egypt	10 780*	28*	95	Papua New Guinea	1 410	50	163 =	Guam	33[a]	38[a]
27	Canada	10 645	46	96	Somalia	1 285*	39*	165	St Vincent	32*	30*
28	Rumania	10 227[e]	48[e]	97	New Zealand	1 276	41	166	St Lucia	30[a]	29[a]
29	South Africa	9 350*	36*	98	Dominican Rep	1 241[a]	31[a]	167	Grenada	29[a]	30[a]
30	Argentina	9 011[a]	38[a]	99	Sierra Leone	1 175*	38*	168 =	Bermuda	27[a]	50[a]
31	Jugoslavia	8 890[a]	43[a]	100	Israel	1 170	33	168 =	Guernsey	27	47
32	Germany, East	8 214[a]	48[a]	101	El Salvador	1 166[a]	33[a]	170	Antigua	25	35
33	Iran	7 725[a]	26[a]	102	Ireland	1 140	37	171	Isle of Man	23	38
34	Czechoslovakia	7 476	50	103	Uruguay	1 094	39	172	Dominica	21[a]	30[a]
35	Korea, North	7 168*	44*	104	Albania	1 086*	43*	173	Seychelles	20[a]	38[a]
36	Taiwan	6 836	42	105	Cent African Emp	1 002*	37*	174 =	Greenland	19[a]	40[a]
37	Afghanistan	6 728	34	106	Togo	970*	43*	174 =	Tonga	19[a]	24[a]
38	Tanzania	6 637*	43*	107	Honduras	931	33	176	Panama Canal Zone	18[a]	42[a]
39	Nepal	6 199	48	108	Puerto Rico	916	33	177	French Guiana	17[a]	38[a]
40	Colombia	5 975[a]	27[a]	109	Singapore	911	40	178	Faroes	15[a]	39[a]
41	Sudan	5 922*	31*	110	Jamaica	895	43	179	Pacific Islands, US	14[a,c]	16[a,c]
42	Australia	5 822[a]	44[a]	111	Paraguay	752[a]	32[a]	180 =	Gilbert Islands	13[a,f]	25[a,f]
43	Kenya	5 315*	38*	112	Nicaragua	714*	32*	180 =	St Kitts-Nevis	13[a]	29[a]
44	Hungary	5 093	48	113	Jordan	670	24	182	Gibraltar	12	42
45	Uganda	4 909*	41*	114	Liberia	665*	38*	183 =	Liechtenstein	10	43
46	Sri Lanka	4 869	35	115	Libya	599*	24*	183 =	Monaco	10[a]	45[a]
47	Morocco	4 733*	27*	116	Bhutan	588	49	185	Samoa, American	9.7[a]	33[a]
48	Netherlands	4 542	33	117	Costa Rica	585[a]	31[a]	186	San Marino	9.6	49
49	Bulgaria	4 448	51	118	Panama	580*	34*	187	Cook Islands	5.8[a]	30[a]
50	Malaysia	4 202	34	119	Lebanon	572[a]	23[a]	188	Montserrat	4.7[a]	37[a]
51	Sweden	4 154	50	120	Mongolia	550*	38*	189	Virgin Islands, Br	4.0[a]	38[a]
52	Madagascar	4 114*	50*	121	Lesotho	500**	47**	190	Cayman Islands	3.5[a]	33[a]
53	Belgium	4 031	41	122	Congo	483*	35*	191	Nauru	3.3[a]	48[a]
54	Algeria	3 891*	22*	123	Yemen, South	450	26	192	St Pierre & Miquelon	2.1[a]	37[a]
55	Peru	3 872[a]	29[a]	124	Mauritania	405*	30*	193	St Helena	1.7[d]	25[d]
56	Ghana	3 798*	37*	125	Trinidad & Tobago	391	37	194 =	Christmas Island	1.4[a]	51[a]
57	Venezuela	3 712	31	126	Botswana	340*	49*	194 =	Niue	1.4[a]	28[a]
58	Mozambique	3 697*	39*	127	Mauritius	310*	34*	196	Falkland Islands	1.0[a]	51[a]
59	Upper Volta	3 334*	54*	128	Kuwait	305	31	197	Norfolk Island	0.9[a]	51[a]
60	Cambodia	3 280	39	129	South-West Africa	299*	34*	198	Tokelau	0.4[a]	22[a]
61	Greece	3 235[a]	37[a]	130	Cyprus	278[a]	44[a]	199	Cocos Islands	0.3[a]	41[a]
62	Mali	3 198*	55*	131	Gambia	260*	48*				
63	Switzerland	3 128*	49*	132	Gabon	256*	49*				
64	Cameroon	3 127*	48*	133	Guyana	253	32				
65	Portugal	2 997[e]	31[e]	134	Swaziland	225*	45*				
66	Austria	2 969	39	135	Timor, East	212	31				
67	Iraq	2 868	25	136	Fiji	176	30				
68	Cuba	2 633[a]	31[a]	137	Guinea-Bissau	168*	31*				

[a] Latest available; mainly 1970 to 1974   [b] People economically active; includes those unemployed   [c] Including Northern Marianas   [d] Employees only   [e] Excluding armed forces
[f] Including Tuvalu

World health

People per physician and per hospital bed, 1970s

Rank by people per physician[a]		People per: physician	hospital bed	Rank by people per physician[a]		People per: physician	hospital bed	Rank by people per physician[a]		People per: physician	hospital bed
	World	*1 390******	*280******	68	Singapore	1 387	271[d]	136	Oman	5 050	900*
1	Albania	159	164[b]	69	Mexico	1 430	863	137	Gabon	5 200	98
2	Panama Canal Zone	290	76[d]	70	Gibraltar	1 450	110	138	Tunisia	5 220	420
3	Soviet Union	347	85	71	Barbados	1 476	114	139	St Vincent	5 420	170
4	Israel	351	177	72	Macao	1 492	209	140	Burma	5 440	1 500
5	Czechoslovakia	420	99	73	Virgin Islands, Br	1 500	250	141	Dominica	5 480	234[d]
6	Monaco	430	80	74	Costa Rica	1 503	264	142	Rhodesia	5 700	316
7	Argentina	460	176	75=	Egypt	1 520	465	143	Sao Tome & Pr'pe	6 700	36[b]
8	Bulgaria	466	116	75=	Utd Arab Emirates	1 520	na	144	Malaysia	7 490	290
9	Austria	479	88	77	Hongkong	1 528	245	145	Congo	7 900	220
10	Greece	491	155	78	Martinique	1 550	101[b]	146	Zambia	8 160	313[b]
11	Hungary	500	117	79	Sahara, Western	1 570	344	147	Algeria	8 190[b]	356[b]
12	Italy	502	95	80	Guam	1 575	376[d]	148	Thailand	8 370*	800
13	Germany, West	516	87	81	Nicaragua	1 600*	461	149	Swaziland	8 890	270
14	Belgium	530	112	82	Guadeloupe	1 613	93	150	Tuvalu	9 000	na
15	Mongolia	538	103	83	French Guinea	1 650	63	151	Bangladesh	9 554	9 420
16	Germany, East	549	92	84	Netherlands Antilles	1 783[b]	109[b]	152	Botswana	10 476	330
17	Norway	582	71	85	Peru	1 802	497	153	Papua New Guinea	10 800	169
18	Poland	584	129	86	Turkey	1 858	470	154	Madagascar	11 020	383
19	Canada	588	110	87	Dominican Rep	1 866	356	155	Ghana	11 230	695
20	Switzerland	591	87	88=	French Polynesia	1 900	121	156	Equatorial Guinea	11 600	170
21	Falkland Islands	606	71	88=	Turks & Caicos Is	1 900	290[d]	157	Haiti	11 700*	1 160
22	Sweden	615	66	90	Trinidad & Tobago	1 940	222	158=	Liberia	12 600	656
23	United States	622	152	91	Qatar	1 980	270	158=	Sudan	12 600	1 120
24	Denmark	624	103	92	Reunion	1 983	121	160	Yemen, South	13 600[b]	670
25	Netherlands	625	99	93	Montserrat	2 000	133	161	Ivory Coast	13 640	496
26	Iceland	640	70	94	South Africa	2 016	156	162=	Comoros	13 800[j]	460[j]
27	Spain	670	193	95	Brazil	2 025	266	162=	Morocco	13 800	732
28	France	680*	98	96	Surinam	2 030	180	164	Angola	15 404	322
29	Nauru	700	33	97	Fiji	2 070	350	165	Somalia	15 600	569
30	Finland	704	66	98	Bolivia	2 150	522	166	Kenya	16 290	760
31	Australia	721	81	99=	Ecuador	2 160	478	167	Guinea-Bissau	16 300[b]	550[b]
32	Jersey	730*	117	99=	St Helena	2 160	121	168	Mozambique	16 390	549[b]
33	United Kingdom	742	112[d]	101	Pacific Islands, US	2 180[i]	220[i]	169	Cambodia	16 575	968
34	New Zealand	754	93	102	Colombia	2 184	538	170	Senegal	16 600	816
35	Niue	780	131	103	Gilbert Islands	2 270	99	171	Sierra Leone	17 100	927
36	Kuwait	800	234	104	Chile	2 345[f]	270	172	Mauritania	17 750	2 730
37	Rumania	805	110	105	Vietnam	2 400**	600**	173	Indonesia	18 860*	1 470
38	Virgin Islands, US	824[b]	252	106	Jordan	2 438	937	174	Cape Verde	20 300[b]	710[b]
39	Ireland	830	93	107	Iraq	2 470	485	175	Lesotho	20 400	482
40	Isle of Man	840	83	108	Antigua	2 500	154	176	Uganda	20 700	710
41	Puerto Rico	848	220	109	Philippines	2 632	639	177	Togo	21 200	680
42	Jugoslavia	849	167	110	Korea, South	2 670	1 570	178	Guinea	22 390	600
43	Malta	850	95	111	Samoa, Western	2 700	230	179	Gambia	23 700[b]	850[b]
44	Japan	869	96	112	Iran	2 752	650	180	Maldives	24 000	2 870
45	Korea, North	889[b]	193[b]	113	Syria	3 060	1 071	181	Nigeria	25 460	1 378
46	Portugal	894	184	114	Honduras	3 100	606	182	Afghanistan	26 090	7 910
47	Cyprus	896[h]	149[h]	115	Belize	3 170	221	183	Cameroon	26 400	305
48	Venezuela	897	340	116	New Hebrides	3 200[b]	102	184	Yemen, North	26 450	1 443
49	Cook Islands	909	111	117	Tonga	3 240	297[d]	185	Tanzania	26 600*	750*
50	Uruguay	911	193	118	Guyana	3 290	190	186	Zaire	28 800	327
51	Guernsey	920*	101	119=	Seychelles	3 300	147[b]	187	Benin	36 071	826
52	Bermuda	950[c]	110[d]	119=	Wallis & Futuna Is	3 300	80	188	Nepal	36 450	6 630
53	Taiwan	962	503*[d]	121	Jamaica	3 509	257[d]	189	Malawi	37 980[b]	640
54	Luxembourg	970	93	122	Brunei	3 660	300	190	Mali	38 960	1 380
55	Cayman Islands	1 000*	300*	123	Mauritius	3 720[c]	271	191	Niger	41 100	1 460
56	St Pierre & Miquelon	1 020[b]	76	124	Djibouti	3 800	210	192	Cent African Emp	43 000	790
57	New Caledonia	1 040	87	125	Pakistan	3 804[g]	1 871	193	Chad	44 400	1 140
58	Greenland	1 100	70	126=	Grenada	4 000	137	194	Bhutan	49 000	390
59	Samoa, American	1 120	166	126=	St Kitts-Nevis	4 000	130*	195	Burundi	49 200	840
60	Libya	1 140	240	128	Sri Lanka	4 007	333	196	Rwanda	53 500	671
61	Cuba	1 150[b]	236	129	El Salvador	4 030	570	197	Upper Volta	59 600	1 170
62	Lebanon	1 170	230	130	India	4 162	1 571*[b]	198	Laos	73 500	1 021
63	Paraguay	1 190*	694	131	Saudi Arabia	4 200*	857	199	Ethiopia	74 000	3 080
64	Faroes	1 200[b]	160[b]	132	Guatemala	4 340	480				
65	Bahamas	1 267	194	133	Solomon Islands	4 470	120				
66	Panama	1 333	284	134	St Lucia	4 580	202				
67	Bahrain	1 350	248	135	China	4 800**	1 020**[b]				

[a]Rank showing fewest people per physician first (greatest number of physicians for the number of people) [b]Latest available figure [c]People per government employed physician only [d]People per hospital bed in government establishments only [e]People per physician in hospitals only [f]People per physician in government establishments only [g]People per registered physician [h]South only [i]Including Northern Marianas [j]Including Mayotte

World communications

Newspaper circulation (number per 1 000 people), **1970 to 1975**

Rank		No per 1 000	Rank		No per 1 000	Rank		No per 1 000	Rank		No per 1 000	Rank		No per 1 000
	World[b]	*110**[b]	27	Hungary	232	53=	Mauritius	92	81	Brazil	40	106=	Saudi Arabia	11
1	Sweden	536	28	Bulgaria	227	53=	Panama	92	82	Guatemala	39	109=	Ivory Coast	10
2	Japan	526	29	France	220	53=	Trinidad & Tobago	92	83	Honduras	37	109=	Yemen, North	10
3	Germany, East	452	30	Bermuda	214	57	Jamaica	90	84	Fiji	36	111=	Sierra Leone	9
4	Luxembourg	450	31	Guam	210	58	Malaysia	89	85	St Kitts-Nevis	30*	111=	Zaire	9*
5	United Kingdom	443*	32	Neth Antilles	197	59=	Jugoslavia	87	86	Belize	29	113=	Kenya	8
6	Finland	440	33	Singapore	190	59=	Portugal	87	87	Tunisia	28	113=	Madagascar	8
7	Iceland	436	34	Korea, South	170	61	Kuwait	86	88	Afghanistan	27	113=	Nepal	8
8	Norway	402	35	Bahamas	159	62	Martinique	83	89=	French Guiana	26	113=	Sudan	8
9	Switzerland	391	36	Guyana	155	63	Mongolia	80	89=	Nicaragua	26	117=	Liberia	7
10	Soviet Union	388	37	Argentina	147	64	Guadeloupe	70	91	Bolivia	25	117=	Papua New Guinea	7
11	Australia	386	38	Andorra	141	65	Colombia	63	92=	Egypt	22	117=	Senegal	7
12	New Zealand	376	39	Puerto Rico	134	66	New Caledonia	61	92=	Jordan	22	120	Togo	6
13	Denmark	355	40	Rumania	129	67	Brunei	60[a]	92=	Zambia	22	121=	Mozambique	5
14	Hongkong	327	41	Italy	126*	68	Surinam	59	95	Haiti	21	121=	Tanzania	5[a]
15	Netherlands	311*	42	Cyprus	124	69	Antigua	56	96	Botswana	20	121=	Uganda	5
16	Austria	308	43	Samoa, American	113	70	El Salvador	51	97	China	19[a]	124	Cameroon	3
17	Virgin Islands, US	300	44	Taiwan	110**	71=	Albania	48	98	Philippines	18	125	Ethiopia	2
18	United States	293	45	Greece	107	71=	Cook Islands	48	99	Algeria	17	126=	Guinea	1
19	Germany, West	289	46	Barbados	98	73	Reunion	46	100=	India	16	126=	Somalia	1
20	Czechoslovakia	288	47	Costa Rica	97	74	Turkey	45[a]	100=	Rhodesia	16	126=	Yemen, South	1
21	Liechtenstein	280	48	Spain	96	75	Dominican Rep	43	102	Iran	15	129	Mali	0.5
22	Uruguay	267	49=	Cuba	95	76	Sri Lanka	42	103	Morocco	14	130	Chad	0.4
23	Belgium	247	49=	Peru	95	77=	Ecuador	41	104	Angola	13	131=	Benin	0.3
24	Poland	237	51	French Polynesia	94	77=	Ghana	41	105	Guinea-Bissau	12	131=	Burundi	0.3
25	Ireland	236	52	Venezuela	93	77=	Paraguay	41	106=	Burma	11	131=	Upper Volta	0.3
26	Canada	235	53=	Faroes	92	77=	Seychelles	41	106=	Nigeria	11	134	Cent African Emp	0.2

Telephones (number per 1 000 people), **1975 or 1976**

Rank		No per 1 000	Rank		No per 1 000	Rank		No per 1 000	Rank		No per 1 000	Rank		No per 1 000
	World	97*	41=	Norfolk Island	220	82	Brunei	61	123	Samoa, Western	22	162=	Mozambique	5
1	Monaco	930	41=	Virgin Islands, Br	220	83=	Pacific Is, US	60[a,e]	124	Gabon	21[a]	162=	Sri Lanka	5
2	United States	718	43	Neth Antilles	200	83=	St Lucia	60	125	Libya	20[a]	166=	Cameroon	4[a]
3	Sweden	689	44	Nauru	190	83=	Wake Island	60	126	Peru	18	166=	Haiti	4
4	Liechtenstein	670	45	Malta	188	86	Venezuela	59	127	Saudi Arabia	17	166=	Lesotho	4
5	Bermuda	661	46	Czechoslovakia	183	87	Niue	58	128=	Djibouti	16	166=	Liberia	4
6	Switzerland	638	47	Cocos Islands	180	88=	Rumania	56	128=	Jordan	16	166=	Madagascar	4
7=	Canada	600	48	Barbados	170	88=	Seychelles	56	130=	Algeria	15	166=	Malawi	4
7=	Jersey	600	49	Germany, East	164	90	Reunion	55	130=	Paraguay	15	166=	Maldives	4
9	Midway Islands	580	50	Singapore	163	91=	Cook Islands	53	130=	Swaziland	15	166=	Sierra Leone	4
10	Guernsey	540	51	Puerto Rico	158	91=	Jamaica	53	130=	Zambia	15	166=	Tanzania	4
11	New Zealand	517	52	Ireland	151	91=	S-West Africa	53	134	Egypt	14[a]	166=	Togo	4[a]
12	Denmark	489[c]	53	New Caledonia	149	94=	Colombia	52	135=	El Salvador	13	166=	Uganda	4
13	Luxembourg	442	54	Montserrat	141	94=	Fiji	52	135=	Papua New Guinea	13	177=	Benin	3[a]
14	Japan	427	55	French Guiana	140	94=	Grenada	52	137=	Botswana	12	177=	India	3
15	Iceland	424	56	Samoa, American	138	94=	Mexico	52	137=	Ivory Coast	12	177=	Pakistan	3
16	Finland	409	57	Andorra	129[a]	98=	Dominica	47	137=	Philippines	12	177=	Sudan	3[a]
17	Australia	397	58	Kuwait	128	98=	Korea, South	47	140	Morocco	11	181=	Cent African Emp	2[a]
18	United Kingdom	394	59	Turks & Caicos Is	124	100=	Chile	45	141	Guatemala	10[a]	181=	Ethiopia	2
19	Netherlands	392	60	Bahrain	116	100=	St Kitts-Nevis	45	142=	Bolivia	9[a]	181=	Guinea	2[a]
20	Pitcairn	390	61	Portugal	115	100=	St Vincent	45	142=	Cambodia	9	181=	Indonesia	2
21	Guam	370	62=	Cyprus	113	103	Antigua	43[a]	142=	Kenya	9	181=	Laos	2
22	Norway	366	62=	Qatar	113	104=	Macao	42	142=	Oman	9	181=	Nigeria	2[a]
23	Germany, West	344	64	French Polynesia	108	104=	Surinam	42	142=	S Tome & Princ'pe	9	181=	Somalia	2
24=	Panama Canal Z'n	340**	65	Christmas Island	107	106	Belize	38	142=	Solomon Islands	9	181=	Zaire	2
24=	Virgin Islands, US	340	66	Hungary	101	107	Brazil	36	142=	Wallis & Futuna Is	9	189=	Afghanistan	1
26	Cayman Islands	336	67	Argentina	98	108	Cuba	31[a]	149=	Congo	8[a]	189=	Bangladesh	1
27	Austria	303	68	Bulgaria	97	109	Guyana	29	149=	Gilbert Islands	8[d]	189=	Bhutan	1
28	Belgium	300	69	Martinique	93	110=	Mauritius	28	149=	Senegal	8	189=	Burma	1
29	France	293	70	Uruguay	92	110=	Rhodesia	28	149=	Thailand	8	189=	Burundi	1
30	San Marino	290	71	Utd Arab Emirates	90	112=	Iraq	27	153=	Albania	7[a]	189=	Chad	1
31	Bahamas	275	72	Lebanon	87[a]	112=	Turkey	27	153=	Cape Verde	7	189=	Mali	1[a]
32	Gibraltar	272	73	Taiwan	85	114=	Dominican Rep	26	153=	Honduras	7	189=	Mauritania	1[a]
33	Italy	271	74=	Panama	80*	114=	Malaysia	26	156=	Angola	6[a]	189=	Nepal	1[a]
34	Falkland Islands	260**	74=	Poland	80	116=	Mongolia	25	156=	China	6**	189=	Niger	1
35	Isle of Man	254	76	South Africa	75	116=	Nicaragua	25	156=	Ghana	6	189=	Rwanda	1
36	Hongkong	253	77	Guadeloupe	74	118=	Ecuador	23	156=	Guinea-Bissau	6[a]	189=	Upper Volta	1
37=	Israel	245	78	Soviet Union	70	118=	Iran	23	156=	Tonga	6	189=	Yemen, North	1[a]
37=	St Pierre & Miq'n	245[a]	79	Jugoslavia	66	118=	New Hebrides	23	156=	Yemen, South	6[a]			
39=	Greece	238	80	Trinidad & Tobago	65	118=	Syria	23	162=	Comoros	5[a]			
39=	Spain	238	81	Costa Rica	62	118=	Tunisia	23	162=	Gambia	5			

[a]Latest available [b]Excluding China [c]Including Faroes and Greenland [d]Including Tuvalu [e]Including Northern Marianas

World equipment

Radio set ownership (number per 1 000 people), 1975 or 1976

Rank		No per 1 000
	World	245*
1	United States	1 875
2	Bermuda	909c
3	Canada	894
4	New Zealand	876
5	Argentina	838c
6	Guam	817
7	Virgin Is, US	815
8	Utd Kingdom	770*
9	Virgin Is, Br	750*
10	St Lucia	740
11	Pacific Is, US	660b
12	Norfolk Is	615
13	Belize	588
14	Hongkong	574
15	Peurto Rico	570
16	Faroes	563c
17	French Polynesia	560c
18	Neth Antilles	545
19	Uruguay	538
20	Fiji	524
21=	Falkland Is	500
21=	Turks & Caicos Is	500*
23=	Luxembourg	490
23=	Nauru	490
25	Barbados	477c
26	Finland	465
27	Soviet Union	461c
28=	Japan	460
28=	Lebanon	460
30	Bahamas	457c
31	New Caledonia	450
32	Belgium	435
33	Malta	401c
34=	Bahrain	390*
34=	Cook Islands	390
36	Sweden	389
37	Syria	374c
38	Germany, East	366*
39	Isle of Man	360*
40	St Pierre & Miq	350
41	Jersey	348*
42=	France	346
42=	Guyana	346c
42=	Monaco	346*
45	Denmark	345
46	Switzerland	335
47	Gibraltar	330*
48=	Germany, West	329
48=	Samoa, Western	329c
50	Norway	327
51	Cayman Is	321c
52=	Cyprus	320
52=	Jamaica	320c
52=	Yemen, South	320**
55	Guernsey	316*
56	Chile	313*
57	Mexico	301c
58	Austria	291
59	St Vincent	290c
60	Netherlands	289
61	Iceland	285
62	Ireland	280*
63	Greece	279*c
64	Jordan	271
65	Ecuador	270*c
66	Czechoslovakia	265
67	Bulgaria	260
68	Surinam	256
69	Iran	246c
70	Andorra	244
71=	Greenland	241
71=	Hungary	241
73=	El Salvador	238*c
73=	Poland	238
75	Trinidad & Tob	235c
76=	Italy	231
76=	Kuwait	231c
78	Spain	224*
79	Liechtenstein	223
80	Israel	221c
81	Grenada	220
82	Jugoslavia	209
83	Australia	208c
84	Antigua	201c
85=	Algeria	195c
85=	Cuba	195c
85=	Utd Arab Emirates	195**
88	Reunion	184c
89	San Marino	181c
90	Gabon	172c
91	Venezuela	168
92	Seychelles	161c
93	Niue	160
94	Panama	159
95	Singapore	158
96	Portugal	157
97	Liberia	154c
98=	Rumania	144
98=	Sahara, Western	144c
100=	Gilbert Islands	140
100=	Samoa, American	140c
102	Egypt	139c
103	Korea, South	138c
104=	Peru	131
104=	Thailand	131
106=	Brunei	125
106=	Mauritius	125c
108	Comoros	119c
109	Gambia	118c
110	New Hebrides	116
111	Saint Helena	115c
112	Colombia	114
113	Iraq	111
114	Swaziland	110c
115	Ghana	109c
116=	Madagascar	108c
116=	Tonga	108
118	Turkey	102
119=	Vietnam	100**
119=	Zaire	100c
121	Cameroon	96c
122	Sao Tome & P'pe	95c
123	South Africa	93c
124=	Martinique	90c
124=	Taiwan	90
126=	Botswana	81c
126=	Nigeria	81*c
128	Sudan	80c
129	Mongolia	79
130	Morocco	76
131	Bolivia	73*
132=	Albania	71
132=	Costa Rica	71
134	Korea, North	70**c
135	Paraguay	68
136	Mauritania	64c
137	Macao	61
138=	Brazil	60c
138=	Congo	60c
138=	Guadeloupe	60c
138=	Nicaragua	60c
138=	Senegal	60c
143=	Djibouti	58*c
143=	Honduras	58c
145	Solomon Islands	53
146	Benin	52c
147	Tunisia	50c
148	French Guiana	48c
149	Libya	44c
150	Philippines	43
151	Guatemala	41*
152	Dominican Rep	40c
153=	Kenya	39c
153=	Papua New Guinea	39*
155=	Laos	38*c
155=	Sri Lanka	38
157=	Indonesia	37*
157=	Malaysia	37
159=	Niger	36c
159=	Rhodesia	36c
161	Rwanda	32c
162	Saudi Arabia	28
163=	Burundi	27*c
163=	Cent African Emp	27c
165	Equatorial Guinea	26
166	Malawi	25c
167=	Guinea	24c
167=	India	24
169	Togo	23c
170=	Burma	22c
170=	Uganda	22c
172=	Sierra Leone	21c
172=	Somalia	21c
172=	Zambia	21c
175	Haiti	20c
176=	Cape Verde	19c
176=	Maldives	19
176=	Mozambique	19c
179=	Angola	18c
179=	Chad	18c
179=	Ivory Coast	18c
182=	Guinea-Bissau	17c
182=	Upper Volta	17c
184=	China	16**c
184=	Pakistan	16
186	Tanzania	15c
187	Cambodia	14
188=	Mali	13c
188=	Yemen, North	13
190	Lesotho	11c
191	Bangladesh	8c
192	Ethiopia	7c
193=	Aghanistan	6
193=	Nepal	6
193=	Timor, East	6c
196	Bhutan	3c
197	Oman	2**c

Television set ownership (number per 1 000 people), 1975 or 1976

Rank		No per 1 000
	World	94*
1	United States	589
2	Guam	442c
3	Bermuda	364c
4=	Canada	363
4=	Sweden	363
6	Isle of Man	338*
7	Monaco	334*
8	Jersey	328
9	Virgin Is, US	326
10=	Denmark	322
10=	Utd Kingdom	322
12	Guernsey	316*
13	Germany, East	307
14	Germany, West	301
15	Finland	300
16	Belgium	291
17	St Pierre & Miq	290
18	Switzerland	287
19	Netherlands	273
20	Norway	269
21	France	268
22	Austria	263
23	New Zealand	261
24	Czechoslovakia	253
25	Iceland	249
26	Luxembourg	245c
27	Hungary	233
28	Japan	232
29	Soviet Union	222
30	Australia	221c
31	Italy	220
32	Gibraltar	210
33	Spain	205
34	Puerto Rico	204
35	Poland	198
36	Israel	191c
37	Malta	190
38=	Kuwait	188c
38=	Liechtenstein	188
40	Ireland	186
41	Hongkong	185c
42	Argentina	180c
43	San Marino	174c
44=	Antigua	173c
44=	Bulgaria	173
46	Barbados	164c
47=	Jugoslavia	160
47=	Samoa, American	160
49=	Lebanon	143
49=	Neth Antilles	143c
51	Rumania	138
52	Cyprus	133c
53=	Greece	126
53=	Uruguay	126
55	Singapore	120
56	Bahrain	117
57	New Caledonia	114
58	Panama	111
59=	French Polynesia	105c
59=	Venezuela	105
61	Ut Arab Emirates	104**
62=	Portugal	94
62=	Trinidad & Tobago	94c
64	Andorra	85c
65	Mexico	84c
66	Brazil	83c
67	Surinam	79
68=	Chile	76*
68=	Costa Rica	76*
70	Colombia	73*
71	Cuba	64c
72	Reunion	61c
73	Taiwan	55
74	French Guiana	52c
75	Iran	51
76	Iraq	50c
77	Jamaica	49c
78=	Korea, South	46c
78=	Martinique	46c
80	Jordan	45c
81	Mauritius	44c
82	Turkey	43
83	Vietnam	40**
84	Guadeloupe	37c
85=	Ecuador	36c
85=	Nicaragua	36c
87	Dominican Rep	34c
88	Malaysia	33c
89	Peru	32
90	Syria	30
91=	El Salvador	28c
91=	Pacific Is, US	28b
93	Tunisia	27c
94	Morocco	26
95	Algeria	25c
96	Paraguay	20
97	South Africa	19**
98=	Sahara, Western	18c
98=	Yemen, South	18
100=	Egypt	17c
100=	Guatemala	17*
100=	Honduras	17c
100=	Philippines	17c
100=	Thailand	17c
105	St Lucia	16c
106	Saudi Arabia	14
107	Djibouti	12*c
108=	Gabon	10c
108=	Rhodesia	10c
110	Ivory Coast	9c
111	Bolivia	8*
112	Senegal	7c
113=	St Vincent	6c
113=	Sudan	6c
115=	Liberia	5c
115=	Zambia	5c
117=	Cambodia	3c
117=	Congo	3c
117=	Ghana	3c
117=	Haiti	3c
117=	Indonesia	3*
117=	Kenya	3c
117=	Libya	3c
124=	Albania	2
124=	Mongolia	2
124=	Nigeria	2c
124=	Pakistan	2c
124=	Sierra Leone	2c
129=	Madagascar	1c
129=	Uganda	1c
129=	Upper Volta	1c
132	Ethiopia	0.7c
133	India	0.5c
134	China	0.4**
135=	Samoa, Western	0.3c
135=	Tanzania	0.3**c
135=	Zaire	0.3c
138=	Bangladesh	0.2c
138=	Korea, North	0.2**c

aOther (no per 1 000): 140 Mozambique 0.1c, 141 Benin 0.03c bIncluding Northern Marianas cLatest available

World equipment

Passenger car ownership (number per 1 000 people), 1975 or 1976

Rank	No per 1 000	Rank	No per 1 000	Rank	No per 1 000	Rank	No per 1 000	Rank	No per 1 000
World	*64**	37 Japan	163	75= Djibouti	50*c	113= Colombia	14	148= Solomon Is	5c
1 Jersey	522	38 Malta	160	75= Dominica	50***	113= Congo	14c	151= Bolivia	4c
2 Andorra	500***	39= Fr Polynesia	148c	77= Jamaica	45c	113= Dominican Rep	14	151= Cambodia	4c
3 United States	498	39= Spain	148	77= Sahara, Western	45c	113= Swaziland	14	151= Central African Empire	4c
4 San Marino	480***c	41 Brunei	133	79 Seychelles	43	117= Guatemala	13c	151= Laos	4c
5 Guam	468	42 French Guiana	128c	80= Mexico	42	117= Samoa, Western	13	151= Mauritania	4c
6 Liechtenstein	409	43 Virgin Is, US	126	80= Panama	42c	119= Algeria	12c	151= Sierra Leone	4
7 Panama Canal Zone	400c	44 Martinique	115c	82 St Kitts-Nevis	38c, d	119= Turkey	12	157= Haiti	3c
8 Canada	386	45= Germany, East	112	83= Fiji	37	121 Tonga	11*	157= Indonesia	3
9 Isle of Man	380**	45= Guadeloupe	112c	83= Poland	37	122= El Salvador	10c	157= Korea, South	3
10 New Zealand	379	47 St Helena	110***c	85 Malaysia	36c	122= Iraq	10	157= Lesotho	3
11 Australia	378	48 Czechoslovakia	104	86= Belize	33*c	122= Mozambique	10c	157= Mali	3
12= Guernsey	370	49 Cyprus	101	86= St Lucia	33	122= Senegal	10c	157= Niger	3
12= Monaco	370***c	50 Argentina	100	88= Guyana	32	122= Togo	10	157= Pakistan	3c
14 Sweden	350	51 Libya	97c	88= St Vincent	32	127= Cape Verde	9c	157= Somalia	3c
15 Luxembourg	326	52 Antigua	95***	90 Rhodesia	29*c	127= Comoros	9***c	157= Tanzania	3c
16 Germany, West	312	53 Trinidad & Tob	94	91 Costa Rica	28c	127= Philippines	9c	157= Zaire	3c
17 France	306	54= Bahrain	91	92= Hongkong	27	130= Cuba	8*c	167= Albania	2*c
18 Cayman Islands	300**	54= Reunion	91c	92= New Hebrides	27c	130= Paraguay	8	167= Ethiopia	2
19 Switzerland	296	56 Portugal	90	94 Chile	25*	130= Thailand	8	167= Guinea	2c
20 Iceland	290	57= Barbados	84c	95= Macao	24c	133= Liberia	7c	167= Malawi	2c
21 New Caledonia	285c	57= Israel	84	95= Saudi Arabia	24	133= Madagascar	7c	167= Rwanda	2
22 Italy	283	59 South Africa	82	97 Bulgaria	23*	133= Sri Lanka	7	167= Sudan	2c
23 Belgium	275	60 Samoa, American	81c	98 Angola	21c	133= Syria	7	167= Uganda	2c
24 Netherlands	272	61 Lebanon	76c	99= Jordan	20	133= Yemen, South	7c	167= Upper Volta	2*
25 Denmark	264	62= Montserrat	75	99= Mauritius	20	138= Botswana	6	167= Vietnam	2***
26 United Kingdom	257	62= St Pierre & Miq	75**	99= S Tome & P'pe	20c	138= Cameroon	6c	176= Afghanistan	1
27 Norway	254	64 Jugoslavia	72	102 Gabon	19*c	138= Ecuador	6c	176= Burma	1c
28 Austria	243	65 Venezuela	71c	103= Iran	18c	138= Egypt	6	176= Burundi	1c
29 Bermuda	219	66 Pacific Is, US	67a	103= Morocco	18	138= Gambia	6c	176= Chad	1
30 Finland	218	67 Singapore	66	103= Soviet Union	18**	138= Ghana	6c	176= India	1
31 Bahamas	203c	68 Hungary	62	103= Tunisia	18	138= Guinea-Bissau	6c	176= Nigeria	1c
32= Kuwait	198	69 Brazil	57	103= Zambia	18*c	138= Kenya	6	182 Yemen, North	0.9c
32= Puerto Rico	198	70= Grenada	55*c	108= Oman	17	138= Papua New Guinea	6	183 Nepal	0.5*
34 Gibraltar	179	70= Uruguay	55*c	108= Peru	17c	138= Rumania	6*	184= Bangladesh	0.3
35 Ireland	173	72= Greece	54	110= Nicaragua	16*c	148= Benin	5c	184= Bhutan	0.3***c
36 Neth Antilles	167*c	72= Surinam	54c	110= Taiwan	16	148= Honduras	5c	186 China	0.1**
		74 Niue	53	112 Ivory Coast	15				

Commercial vehicle ownership (number per 1 000 people[b]), 1975 or 1976

Rank	No per 1 000	Rank	No per 1 000	Rank	No per 1 000	Rank	No per 1 000	Rank	No per 1 000
World	*16**	28= South Africa	34	56= Singapore	20	87= Ivory Coast	10*	114= Kenya	6
1 United States	122	28= Utd Kingdom	34	56= Soviet Union	20**	87= Nicaragua	10*c	114= Liberia	6c
2 New Caledonia	116c	32= Luxembourg	31	62= Costa Rica	19c	87= Niue	10*	114= Madagascar	6c
3 Japan	102	32= Uruguay	31*c	62= Fiji	19	87= St Helena	10***c	114= Philippines	6c
4 Guam	98	34= Andorra	30***	64 Ireland	18	87= St Pierre & Miquelon	10**	114= Turkey	6
5 Canada	94	34= Finland	30	65= Chile	16*	87= Samoa, American	10**	121= Albania	5*c
6 Australia	92	34= Spain	30	65= Guyana	16	87= Seychelles	10	121= Bolivia	5c
7 New Zealand	73	37 Israel	29	65= Mexico	16	96= New Hebrides	9c	121= Bulgaria	5*
8 Jersey	72	38= Bahamas	28c	65= St Lucia	16	96= Pacific, Is, US	9a	121= El Salvador	5c
9 Guernsey	69	38= Switzerland	28	69= Germany, East	15	96= Peru	9c	121= Gambia	5c
10 French Guiana	67c	40= Belgium	26	69= Poland	15	99= Congo	8c	121= Ghana	5c
11 Kuwait	66	40= Venezuela	26c	71= Gabon	14*c	99= Honduras	8c	121= Grenada	5***c
12= Denmark	50	42= Cyprus	25	71= Surinam	14c	99= Iraq	8	121= Portugal	5
12= Fr Polynesia	50c	42= Greece	25	73= Hungary	13	99= Jugoslavia	8	121= Sahara, Western	5c
12= Isle of Man	50**	42= Oman	25	73= Panama	13c	99= Lebanon	8c	121= St Kitts-Nevis	5c, d
12= Martinique	50c	45= Brunei	24	73= Samoa, Western	13	99= St Vincent	8	121= Sao Tome & Principe	5c
12= San Marino	50**c	45= Netherlands	24	73= Zambia	13*c	105= Brazil	7	121= Senegal	5c
17 Guadeloupe	47c	45= Trinidad & Tob	24	77= Barbados	12c	105= Colombia	7	121= Syria	5
18 Libya	46c	48= Belize	22*c	77= Malaysia	12c	105= Dominican Rep	7	121= Yemen, South	5
19 Argentina	43	48= Germany, West	22	77= Tunisia	12	105= Guatemala	7c	135= Angola	4c
20 France	41	48= Italy	22	80= Botswana	11	105= Jordan	7	135= Cuba	4*c
21= Malta	40	48= Montserrat	22	80= Hongkong	11	105= Morocco	7	135= Dominica	4***
21= Puerto Rico	40	48= Sweden	22	80= Jamaica	11c	105= Papua New Guinea	7	135= Guinea-Bissau	4c
23 Bermuda	39	48= Virgin Is, US	22	80= Mauritius	11	105= Paraguay	7	135= Lesotho	4
24= Bahrain	37	54= Austria	21	80= Panama Canal Zone	11c	105= Thailand	7	135= Macao	4c
24= Norway	37	54= Saudi Arabia	21	80= Rhodesia	11*c	114= Algeria	6c	135= Mauritania	4c
24= Reunion	37c	56= Antigua	20***	80= Swaziland	11	114= Cameroon	6c	135= Sri Lanka	4
27 Liechtenstein	36	56= Czechoslovakia	20	87= Djibouti	10*c			135= Tonga	4*
28= Iceland	34	56= Gibraltar	20	87= Ecuador	10c			135= Vietnam	4***
28= Neth Antilles	34*c	56= Monaco	20***c						

[a]Including Northern Marianas [b]Showing only countries with 4 or more commercial vehicles per 1 000 people [c]Latest available [d]Including Anguilla

World production

Agricultural production growth

Rank for 1970–76	Growth %pa 1960–70	1970–76
World	2.3	2.5
1 Libya	4.3	10.9
2 Syria	3.2	10.3
3 Rumania	2.0	8.7
4 Senegal	1.3	8.1
5 Zambia	2.3	6.8
6 Tunisia	0.1	6.3
7 Yemen, North	0.5	6.1
8 Philippines	3.2	6.0
9 Botswana	1.1	5.7
10 Israel	5.9	5.6
11 Malaysia[f]	5.3	5.4
12 Malawi	3.0	5.3
13= Bolivia	2.9	5.1
13= Iran	2.9	5.1
15= Swaziland	na	4.9
15= Turkey	2.4	4.9
17 South-West Africa	−1.0[e]	4.7
18= Jugoslavia	2.0	4.6
18= Rhodesia	2.1	4.6
20 Lebanon	5.4	4.5
21= Canada	1.8	4.4
21= Costa Rica	5.5	4.4
21= Guatemala	3.3	4.4
24 Korea, North	2.4*	4.2*
25= Albania	4.9	4.1
25= Nicaragua	5.2	4.1
27 El Salvador	3.0	4.0
28= Afghanistan	1.6	3.9
28= Hungary	1.8	3.9
28= Thailand	5.4	3.9
31 Colombia	3.0	3.8
32= Brazil	3.0	3.7
32= Honduras	4.0	3.7
32= Ivory Coast	4.1	3.7
32= Korea, South	5.7	3.7
36 Mauritius	1.2[h]	3.6
37= Finland	1.8	3.4
37= Gambia	1.4	3.4
37= United States	1.3	3.4
40 Sudan	4.9	3.3
41= Iceland	1.2	3.2
41= Reunion	−3.5[e]	3.2
41= Spain	3.3	3.2
44= Indonesia	1.7	3.1
44= Tanzania	2.6	3.1
46= Greece	4.6	3.0
46= Mongolia	0.0**	3.0
46= Rwanda	2.8	3.0
46= Taiwan	4.5	3.0
50= Argentina	2.9	2.9
50= Paraguay	3.9	2.9
50= Saudi Arabia	na	2.9
50= Yemen, South	−0.1	2.9
54= Czechoslovakia	2.4	2.8
54= Dominican Republic	0.5	2.8
54= Netherlands	2.0	2.8
54= Venezuela	5.3	2.8
58= Congo	0.6	2.6
58= Ireland	1.8	2.6
58= Morocco	2.4	2.6
61= Algeria	−1.5	2.5
61= Panama	4.2	2.5
63= Australia	3.0	2.4
63= China	3.1**	2.4**
63= Mexico	4.3	2.4
66 Niger	2.7	2.3
67= Papua New Guinea	5.0*	2.2
67= Poland	1.7	2.2
67= Sierra Leone	2.9	2.2
67= Switzerland	1.4	2.2
71= Ecuador	4.5	2.1
71= Guyana	1.1	2.1
71= South Africa	3.2	2.1
74= Bulgaria	3.6	2.0
74= Burma	2.3	2.0
76= Burundi	1.8	1.9*
76= Haiti	0.4	1.9
78= Egypt	2.5	1.8
78= New Zealand	2.5	1.8
78= Sweden	0.6	1.8
78= Surinam	9.7[e]	1.8
78= Trinidad and Tobago	3.1[e]	1.8
83= Austria	1.4	1.7
83= Pakistan	6.2*	1.7
85= Cameroon	5.3	1.6
85= Liberia	2.0	1.6
85= Soviet Union	3.3	1.6
88= Gabon	2.6	1.5
88= India	2.1	1.5
88= Iraq	3.7	1.5
91= Germany, East	0.6	1.3
91= Lesotho	−1.3[e]	1.3
93= Japan	3.6	1.2
93= Madagascar	3.2	1.2
93= Vietnam	na	1.2
96= France	2.1	1.1
96= Jamaica	0.2	1.1
96= Upper Volta	2.5	1.1
99= Chad	0.8	1.0
99= Mali	1.2	1.0
99= Nepal	0.4	1.0
102= Denmark	0.6	0.9
102= Kenya	2.0	0.9
104= Central African Empire	2.0	0.8
104= Laos	5.0	0.8
104= Norway	0.9	0.8
107= Bangladesh	1.7[c]	0.7
107= Italy	2.9	0.7
107= Peru	2.3	0.7
107= Zaire	1.0	0.7
111= Somalia	1.0	0.5
111= Uganda	4.0	0.5
113= Belgium-Luxembourg	1.4	0.4
113= Guinea	0.2[e]	0.4
113= United Kingdom	2.0	0.4
116 Puerto Rico	−2.4[e]	0.2
117 Germany, West	1.8	0.1
118= Ghana	1.8	−0.1
118= Uruguay	1.7	−0.1
120 Sri Lanka	2.7	−0.4
121 Chile	1.6	−0.6
122 Malta	4.8	−0.8
123= Ethiopia	2.8	−0.9
123= Nigeria	1.1	−0.9
123= Mozambique	2.9	−0.9
126 Benin	1.7	−1.1
127 Portugal	1.9	−1.7
128 Cyprus	7.9	−1.9
129 Cuba	2.9	−2.1
130 Mauritania	2.1	−3.1
131 Angola	2.4	−3.7
132 Barbados	0.1	−4.3
133 Togo	3.1	−5.1
134 Equatorial Guinea	0.0[e]	−7.0
135 Cambodia	4.4	−8.6

Industrial production growth

Rank for 1970–76	Growth %pa 1960–70	1970–76
World[j]	6.6	5.1
1 Korea, South	17.6	23.6
2 Bolivia[a]	13.2	15.4[b]
3 Taiwan	16.4	15.1
4 Jordan	na	14.8
5 Ecuador[a]	11.1	14.1[b]
6 Rumania	12.8	13.0[b]
7= Nigeria[a]	na	11.6
7= Syria	8.1	11.6[b]
9 Singapore[a]	na	10.8
10= Brazil	6.8	10.4[b]
10= Poland	8.6	10.4
12 Malawi[a]	na	10.1
13 Dominican Republic[a]	4.9	9.4[d]
14 Mongolia	10.0	9.2[b]
15 Greece[a]	8.6	8.9
16= Bulgaria	11.4	8.6
16= Malaysia	na	8.6
18 Kenya[a]	7.0	8.2[d]
19= Peru[a]	8.1	7.7[b]
19= Senegal	3.6	7.7
21 Colombia	5.6	7.4[d]
22 Jugoslavia	8.3	7.3
23 New Zealand[g]	6.8	7.2
24 Soviet Union	8.6	7.0
25 Spain	11.7	6.8
26= Czechoslovakia	6.0	6.5
26= Israel	11.8	6.5
26= Panama[a]	12.1	6.5[b]
29 Germany, East	6.2	6.4
30 Mexico	8.0	6.1
31 Morocco	3.8	6.0
32= Hungary	7.0	5.9
32= Philippines	6.2	5.9
34 Tunisia	na	5.5
35 Portugal	8.6	5.4
36= Egypt	10.5	5.1[d]
36= El Salvador[a]	10.8*	5.1[b]
38 India	6.1	4.7
39 Rhodesia	na	4.5
40 Canada	6.2	4.1
41 Netherlands	7.2	3.9
42 Austria	5.4	3.8
43= Finland	7.4	3.7
43= France	5.8	3.7
43= Japan	13.9	3.7
46= Ireland	6.7	3.6
46= South Africa[a]	8.5	3.6
48= Argentina	6.0	3.4
48= Italy	7.2	3.4
48= Venezuela[a]	6.2	3.4[b]
51 Paraguay	5.9[c]	3.2[b]
52 United States	5.0	3.1
53= Belgium	5.0	2.9
53= Norway	5.2	2.9
55 Australia[a]	5.0	2.5
56 Zambia	4.4[h]	2.4
57= Denmark	5.7	2.2
57= Sweden	6.2	2.2
59 Germany, West	5.7	1.9
60= Pakistan	12.4	1.4
60= Uruguay[a]	1.5	1.4[b]
62 United Kingdom[i]	2.8	0.2
63 Luxembourg	2.4	0.0
64= Cyprus	na	−0.7
64= Switzerland	5.4	−0.7
66 Chile[a]	4.2	−3.3
67 Mozambique	7.0	−3.7[b]

[a] Manufacturing only [b] 1970–75 [c] 1963–70 [d] 1970–74 [e] 1961–70 [f] Peninsular Malaysia only [g] Years ending March 31st [h] 1962–70 [i] Including construction [j] Excluding Albania, China, Mongolia, North Korea and Vietnam

World production

Manufacturing/agriculture ratio, 1970s[a]

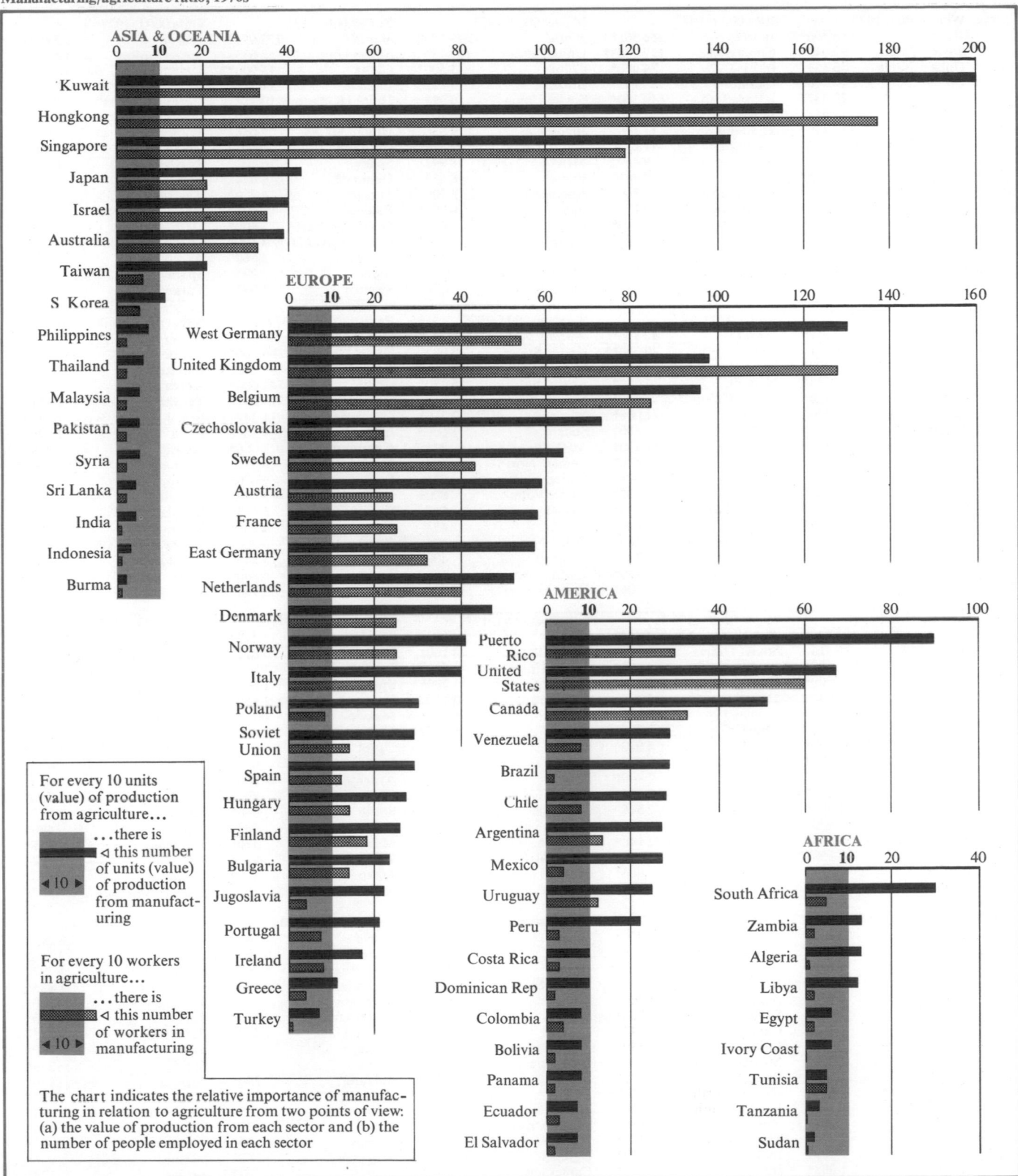

The chart indicates the relative importance of manufacturing in relation to agriculture from two points of view: (a) the value of production from each sector and (b) the number of people employed in each sector

For every 10 units (value) of production from agriculture... ...there is ◁ this number of units (value) of production from manufacturing

For every 10 workers in agriculture... ...there is ◁ this number of workers in manufacturing

[a]Showing the ratio of manufacturing to agriculture (degree of industrialisation) for countries with gross domestic product over $ 2 bn for which suitable information is available

World agriculture

Main products and main producers

Rank	Wheat 000 t, 1977		Rice 000 t, 1977		Maize[a] 000 t, 1977		Barley 000 t, 1977		Oats 000 t, 1977	
	World	*386 596**	*World*	*366 505**	*World*	*349 676**	*World*	*173 094**	*World*	*52 472**
1	Soviet Union	92 042	China	129 000**	United States	161 485	Soviet Union	52 653	Soviet Union	18 379
2	United States	55 134	India	74 000*	China	33 500**	China	15 000**	United States	10 856
3	China	40 000**	Indonesia	23 235	Brazil	19 122	Canada	11 515	Canada	4 303
4	India	29 082	Bangladesh	19 300*	Soviet Union	10 993	United Kingdom	10 784	Germany, West	2 723
5	Canada	19 651	Japan	17 000	Rumania	10 103	France	10 290	Poland	2 600*
6	France	17 450	Thailand	13 590	Jugoslavia	9 856	United States	9 056	France	1 928
7	Turkey	16 715	Vietnam	11 250	South Africa	9 714	Germany, West	7 497	China	1 900**
8	Australia	9 350	Burma	9 460	Mexico	8 991	Spain	6 707	Sweden	1 399*
9	Pakistan	9 155	Brazil	8 941	France	8 614	Denmark	6 084*	Finland	1 022
10	Germany, West	7 181	Korea, South	8 340	Argentina	8 300	Turkey	4 750*	Australia	962
11	Rumania	6 540*	Philippines	7 150	India	6 800*	Poland	3 404	United Kingdom	771
12	Italy	6 329	Korea, North	4 610*	Italy	6 456	Germany, East	3 400*	Germany, East	650*
13	Iran	6 200	United States	4 501	Hungary	6 150*	Czechoslovakia	3 100*	Czechoslovakia	600*
14	Jugoslavia	5 622	Pakistan	4 356	Canada	4 303	Australia	2 560	Argentina	570
15	Hungary	5 312*	Taiwan	2 700**	Philippines	3 037	India	2 296	Spain	428
16	Poland	5 310	Nepal	2 285	Indonesia	3 030	Sweden	1 992*	Turkey	380

Rank	Rye 000 t, 1977		Millet 000 t, 1977		Sorghum 000 t, 1977		Potatoes 000 t, 1977		Sweet potatoes 000 t, 1976	
	World	*23 767**	*World*	*42 886**	*World*	*55 413**	*World*	*306 552**	*World*	*135 855**
1	Soviet Union	8 471	China	20 000**	United States	20 083	Soviet Union	95 000*	China	112 000**
2	Poland	6 200*	India	10 000*	India	10 400*	Poland	49 000	Indonesia	2 478*
3	Germany, West	2 538	Nigeria	2 600*	Argentina	6 730*	China	41 600*	Korea, South	1 950*
4	Germany, East	1 500*	Soviet Union	2 000*	Nigeria	3 750*	United States	15 941	Taiwan	1 851
5	Czechoslovakia	870*	Niger	1 000*	Mexico	3 350*	Germany, West	11 347	Brazil	1 730*
6	Turkey	715	Mali	850*	Sudan	1 600*	India	7 287	India	1 672
7	United States	432	Egypt	830*	Australia	956	Germany, East	7 000*	Japan	1 279
8	Canada	392	Chad	574*	Yemen, North	800*	United Kingdom	6 571	Vietnam	1 200**
9	France	376	Uganda	471	Ethiopia	671	France	6 470*	Philippines	986*
10	Sweden	364*	Senegal	432*	Upper Volta	600*	Netherlands	5 752	Burundi	842*
11	Austria	351	Korea, North	418**	Uganda	516	Spain	5 674	Bangladesh	790
12	Denmark	320*	Sudan	410*	Brazil	453	Czechoslovakia	4 100*	Rwanda	675*
13	Spain	218	Cameroon	360*	Colombia	406	Rumania	4 000*	Uganda	664*
14	Argentina	170	Argentina	340	Pakistan	400*	Italy	3 310	United States	622
15	Hungary	147*	Upper Volta	330*	South Africa	386	Japan	3 200*	Kenya	555*
16	Portugal	90	Pakistan	305*	France	356	Turkey	2 900*	Tanzania	441*

Rank	Cassava 000 t, 1976		Sugar[b] 000 t, 1977		Tomatoes 000 t, 1977		Oranges 000 t, 1977		Apples 000 t, 1977	
	World	*104 952**	*World*	*91 353**	*World*	*44 886**	*World*	*32 611**	*World*	*21 517**
1	Brazil	26 816	Soviet Union	9 065*	United States	8 160*	United States	9 612	United States	3 053
2	Indonesia	12 500*	Brazil	8 500*	Soviet Union	4 733*	Brazil	7 058	France	2 190*
3	Nigeria	10 800*	Cuba	6 485*	China	3 200**	Spain	1 662	Italy	1 810
4	Zaire	9 832	United States	5 429	Italy	3 120	Italy	1 650	Germany, West	1 175
5	Thailand	7 850	India	5 239	Turkey	2 800*	Mexico	1 142*	Turkey	1 060*
6	India	6 307	China	4 200**	Egypt	2 400*	India	1 001*	Hungary	1 000*
7	Tanzania	5 100*	France	4 020	Spain	2 179	Egypt	990*	Poland	1 000*
8	Mozambique	2 400*	Australia	3 300	Greece	1 560*	Israel	897	Japan	904*
9	Colombia	1 900*	Germany, West	2 940*	Brazil	1 271	Argentina	800	Argentina	820
10	Ghana	1 800*	Philippines	2 685	Japan	1 243*	China	800**	India	719*
11	Angola	1 600*	Mexico	2 545	Rumania	1 150*	South Africa	630*	Spain	718
12	Paraguay	1 450*	Thailand	2 294*	Mexico	964*	Greece	582	China	490**
13	Madagascar	1 348	South Africa	2 140*	Bulgaria	800*	Morocco	570*	Rumania	455*
14	Vietnam	1 150*	Poland	1 900*	Portugal	790*	Turkey	513*	Canada	402
15	Burundi	896*	Argentina	1 666*	India	711*	Japan	406*	Bulgaria	400*
16	Cent African Emp	850*	Dominican Rep	1 361*	France	670*	Algeria	360*	Jugoslavia	367

Rank	Grapes 000 t, 1977		Soyabeans 000 t, 1977		Groundnuts[c] 000 t, 1977		Sunflowerseed 000 t, 1977		Olives 000 t, 1977	
	World	*57 116**	*World*	*77 100**	*World*	*18 383**	*World*	*11 762**	*World*	*8 226**
1	Italy	10 900	United States	45 796	India	5 500*	Soviet Union	6 200*	Italy	2 550
2	France	8 100*	China	13 000**	China	2 800**	United States	1 000*	Spain	1 787
3	Soviet Union	5 649*	Brazil	12 513	United States	1 627	Argentina	900	Greece	1 485*
4	United States	3 855	Argentina	1 400	Sudan	990*	Rumania	750*	Tunisia	500*
5	Spain	3 494	Soviet Union	700*	Senegal	890*	South Africa	476	Turkey	500*
6	Turkey	3 450*	Canada	517	Nigeria	850*	Jugoslavia	460*	Portugal	265
7	Argentina	3 400	Indonesia	489*	Argentina	600	Turkey	457*	Morocco	260*
8	Greece	1 585*	Mexico	400*	Indonesia	556*	Spain	373	Syria	238*
9	Rumania	1 500*	Paraguay	375*	Burma	530*	Bulgaria	370*	Algeria	140*
10	Germany, West	1 330*	Korea, North	310**	Brazil	324	Hungary	160*	Libya	100*
11	Jugoslavia	1 291	Korea, South	300*	Zaire	312	Canada	79	Argentina	87
12	Bulgaria	1 232*	Rumania	185*	South Africa	241	China	70**	United States	74*
13	Portugal	1 200*	Thailand	129*	Mali	230*	France	70*	Albania	47*
14	South Africa	1 180*	India	120*	Uganda	230*	Australia	68	Lebanon	46*
15	Chile	1 008*	Iran	103*	Cameroon	193*	Italy	59*	Jordan	23*
16	Iran	917*	Japan	100*	Thailand	176*	Uruguay	34*	Iran	17*

[a]Corn [b]Raw value [c]Peanuts

World agriculture

Main products and main producers

Rank	Coffee 000 t, 1977		Cocoa 000 t, 1977		Tea 000 t, 1977		Tobacco 000 t, 1977		Rubber 000 t, 1977	
	World	*4 407**	*World*	*1 467**	*World*	*1 761**	*World*	*5 626**	*World*	*3 602**
1	Brazil	943*	Ghana	320*	India	560*	China	1 000**	Malaysia	1 600*
2	Colombia	540*	Nigeria	250*	China	310**	United States	877	Indonesia	850*
3	Ivory Coast	318*	Brazil	237	Sri Lanka	213*	India	414	Thailand	436
4	Mexico	270*	Ivory Coast	235*	Japan	105*	Brazil	357	India	150
5	Uganda	202	Cameroon	90*	Soviet Union	99	Soviet Union	318*	Sri Lanka	148
6	El Salvador	180*	Ecuador	70	Kenya	90*	Turkey	223*	Nigeria	90*
7	Indonesia	180*	Dominican Rep	37*	Indonesia	75*	Japan	169*	Liberia	80*
8	Ethiopia	175*	Mexico	34*	Turkey	58*	Bulgaria	167*	Philippines	58*
9	Guatemala	147*	Papua New Guinea	34*	Bangladesh	34*	Korea, South	138*	Zaire	27*
10	India	103	Colombia	27*	Argentina	32*	Greece	112	China	25**
11	Madagascar	95	Togo	20*	Malawi	30*	Italy	110*	Brazil	20*
12	Cameroon	90*	Venezuela	17*	Taiwan	25*	Canada	103*	Ivory Coast	20*
13	Zaire	90*	Sao Tome & Principe	9*	Iran	24*	Poland	100*	Vietnam	20*
14	Kenya	87*	Costa Rica	8*	Uganda	22	Rhodesia	88*	Burma	16
15	Costa Rica	79	Equatorial Guinea	7*	Tanzania	15*	Indonesia	84*	Cameroon	16*
16	Ecuador	77	Sierra Leone	5*	Mozambique	14*	Argentina	80	Cambodia	15**

Rank	Cotton 000 t, 1977		Jute[a] 000 t, 1977		Wool 000 t, 1977		Milk 000 t, 1977		Meat 000 t, 1977	
	World	*14 254**	*World*	*4 314**	*World*	*1 545**	*World*	*405 271**	*World*	*125 697**
1	United States	3 156	China	1 400**	Australia	442*	Soviet Union	94 300	United States	25 651
2	Soviet Union	2 716	India	1 262*	Soviet Union	275	United States	55 824	China	16 000**
3	China	2 300**	Bangladesh	1 026	New Zealand	221*	France	30 100*	Soviet Union	14 800*
4	India	1 150*	Thailand	237	Argentina	90*	Germany, West	22 500*	France	4 850*
5	Brazil	570*	Brazil	96	South Africa	52*	Poland	17 800*	Germany, West	4 290
6	Turkey	570*	Nepal	56	Uruguay	38*	United Kingdom	15 041	Brazil	3 665
7	Pakistan	500*	Soviet Union	50*	China	37**	Brazil	11 428	Argentina	3 523
8	Egypt	435	Burma	43*	United Kingdom	31*	Netherlands	10 300*	Italy	3 031
9	Mexico	327	Vietnam	31*	Turkey	30*	Italy	9 700*	Australia	2 884
10	Sudan	208	Indonesia	14*	United States	24*	India	8 400*	United Kingdom	2 749
11	Iran	180*	Zaire	13*	Brazil	20*	Germany, East	8 000*	Poland	2 538*
12	Colombia	174	Cuba	6*	India	20*	Canada	7 500*	Japan	2 383*
13	Argentina	166	Bhutan	5*	Rumania	20*	New Zealand	6 635	Canada	2 305
14	Syria	160*	Peru	5*	Pakistan	19	Australia	5 897	Spain	2 204
15	Greece	152*	Mozambique	4*	Bulgaria	18*	Czechoslovakia	5 550*	Netherlands	1 724
16	Nicaragua	138	El Salvador	2*	Afghanistan	14*	Japan	5 534*	Germany, East	1 671*

Fish catch 000 t, 1976

Rank	Country		Rank	Country		Rank	Country		Rank	Country		Rank	Country	
	World	*73 467**	17	Vietnam	1 010**	34	Faroes	342	51	Angola	154*	67=	Malawi	75
1	Japan	10 620	18	Iceland	986	35	Portugal	339	52	Uganda	152	69	Cameroon	72*
2	Soviet Union	10 134	19	Brazil	950*	36	Netherlands	284	53	Venezuela	146	70	Greece	71*
3	China	6 800**	20	Taiwan	811	37	Argentina	282	54	Sri Lanka	136	71	New Zealand	70
4	Peru	4 343	21	France	806	38	Morocco	281	55=	Rumania	127	72=	Sierra Leone	68*
5	Norway	3 433	22	Korea, North	800**	39	Germany, East	279	55=	Yemen, South	127*	72=	UAE	68*
6	United States	3 004	23	Poland	750	40	Ghana	238	57	Finland	120	74	Papua New Guinea	63
7	Korea, South	2 407	24	Bangladesh	640*	41	Ecuador	223*	58	Zaire	118	75	Jugoslavia	59
8	India	2 400	25	South Africa	638*[b]	42	Sweden	209	59	Chad	115*	76	Madagascar	55*
9	Denmark	1 912	26	S-W Africa	574[c]	43	Pakistan	206	60	Australia	114	77	Zambia	54
10	Thailand	1 640	27	Mexico	572	44	Cuba	204	61	Egypt	107	78	Greenland	45
11	Spain	1 483	28	Malaysia	517	45	Oman	198	62	Mali	100*	79	Belgium	44
12	Indonesia	1 448	29	Burma	502	46	Tanzania	181	63	Ireland	94	80	Tunisia	43*
13	Philippines	1 430	30	Nigeria	495	47	Panama	172	64	Cambodia	85*	81	Kenya	41
14	Chile	1 264	31	Germany, W	454	48	Bulgaria	167	65	Puerto Rico	81	82	Algeria	35
15	Canada	1 136	32	Italy	420	49	Hongkong	158	66	Ivory Coast	77	83	Uruguay	34
16	United Kingdom	1 051	33	Senegal	361	50	Turkey	155*	67=	Colombia	75			

Timber 000 m³, 1976

Rank	Country		Rank	Country		Rank	Country		Rank	Country		Rank	Country	
	World	*2 524 219**	17	Ethiopia	24 220*	34	Ghana	13 058	51	Angola	7 836*	67	Ecuador	4 030
1	Soviet Union	384 534	18	Colombia	23 002*	35	Kenya	12 399*	52	Peru	7 300	68	Zambia	3 997
2	Utd States	341 397	19	Sudan	22 371	36	Argentina	12 242	53	Afghanistan	6 961	69	Haiti	3 989*
3	China	194 000**	20	Burma	21 655	37	Spain	12 122	54	Italy	6 560	70	Rwanda	3 930*
4	Brazil	163 995*	21	Poland	21 596	38	South Africa	11 568*	55	Madagascar	6 366	71	Chad	3 902
5	Canada	132 393	22	Thailand	21 119	39	Ivory Coast	10 450	56	Iran	6 239*	72	Honduras	3 868*
6	India	130 947*	23	Rumania	20 587	40	New Zealand	10 019	57	Rhodesia	5 902*	73	Bolivia	3 830
7	Indonesia	129 831	24	Vietnam	18 832	41	Korea, South	9 418*	58	Papua New Guinea	5 892	74	Costa Rica	3 528
8	Nigeria	68 883	25	Czechoslovakia	16 891	42	Nepal	9 260*	59	Guatemala	5 666	75	Switzerland	3 510
9	Sweden	55 660	26	Turkey	16 844	43	Mozambique	9 075*	60	Hungary	5 488	76	Utd Kingdom	3 343
10	Japan	38 134*	27	Mexico	14 783	44	Norway	8 968	61	Korea, Nth	5 240**	77	Malawi	3 316
11	Tanzania	37 526	28	Bangladesh	14 776	45	Pakistan	8 963	62	Sri Lanka	4 745	78	El Salvador	3 305
12	Malaysia	36 361	29	Uganda	14 611	46	Chile	8 816	63	Cambodia	4 570*	79	Somalia	3 230*
13	Philippines	33 527	30	Jugoslavia	14 036	47	Germany, E	8 706	64	Bulgaria	4 415	80	Laos	3 154*
14	Finland	32 950	31	Zaire	13 690*	48	Cameroon	8 252*	65	Upper Volta	4 370*	81	Guinea	3 139*
15	Germany, W	30 025	32	Australia	13 450	49	Venezuela	7 964*	66	Paraguay	4 295*	82	Morocco	3 093
16	France	29 127	33	Austria	13 175	50	Portugal	7 887				83	Mali	3 031*

[a]Including allied fibres [b]Excluding landings at Walvis Bay [c]Including landings at Walvis Bay

World agriculture

Main external trade products[a], 000 tonnes, 1976

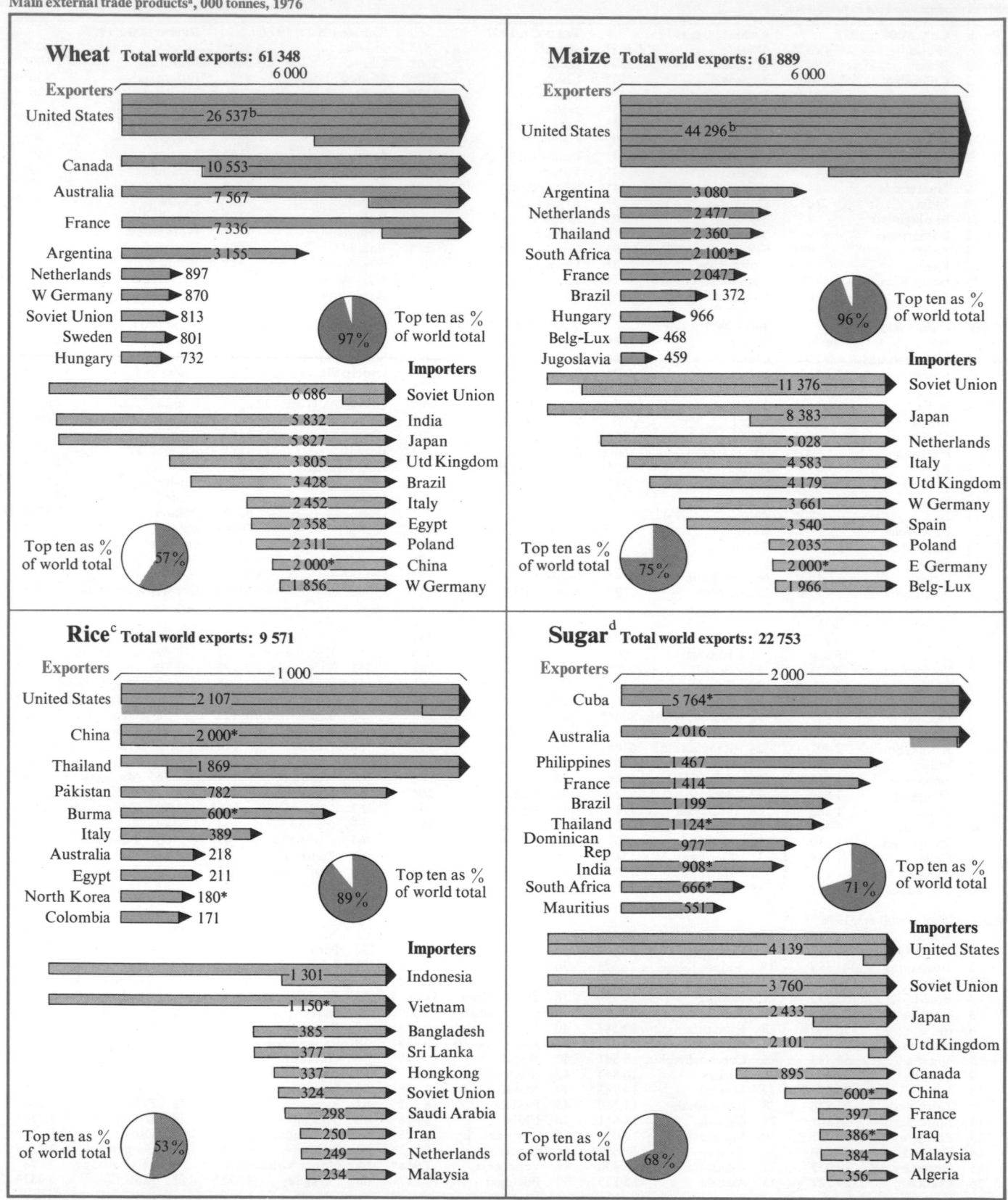

Wheat Total world exports: 61 348
6 000

Exporters
United States 26 537[b]
Canada 10 553
Australia 7 567
France 7 336
Argentina 3 155
Netherlands 897
W Germany 870
Soviet Union 813
Sweden 801
Hungary 732

Top ten as % of world total 97%

Importers
Soviet Union 6 686
India 5 832
Japan 5 827
Utd Kingdom 3 805
Brazil 3 428
Italy 2 452
Egypt 2 358
Poland 2 311
China 2 000*
W Germany 1 856

Top ten as % of world total 57%

Maize Total world exports: 61 889
6 000

Exporters
United States 44 296[b]
Argentina 3 080
Netherlands 2 477
Thailand 2 360
South Africa 2 100*
France 2 047
Brazil 1 372
Hungary 966
Belg-Lux 468
Jugoslavia 459

Top ten as % of world total 96%

Importers
Soviet Union 11 376
Japan 8 383
Netherlands 5 028
Italy 4 583
Utd Kingdom 4 179
W Germany 3 661
Spain 3 540
Poland 2 035
E Germany 2 000*
Belg-Lux 1 966

Top ten as % of world total 75%

Rice[c] Total world exports: 9 571
1 000

Exporters
United States 2 107
China 2 000*
Thailand 1 869
Pakistan 782
Burma 600*
Italy 389
Australia 218
Egypt 211
North Korea 180*
Colombia 171

Top ten as % of world total 89%

Importers
Indonesia 1 301
Vietnam 1 150*
Bangladesh 385
Sri Lanka 377
Hongkong 337
Soviet Union 324
Saudi Arabia 298
Iran 250
Netherlands 249
Malaysia 234

Top ten as % of world total 53%

Sugar[d] Total world exports: 22 753
2 000

Exporters
Cuba 5 764*
Australia 2 016
Philippines 1 467
France 1 414
Brazil 1 199
Thailand 1 124*
Dominican Rep 977
India 908*
South Africa 666*
Mauritius 551

Top ten as % of world total 71%

Importers
United States 4 139
Soviet Union 3 760
Japan 2 433
Utd Kingdom 2 101
Canada 895
China 600*
France 397
Iraq 386*
Malaysia 384
Algeria 356

Top ten as % of world total 68%

[a]Showing top ten exporters and top ten importers for each product [b]Including economic aid flows [c]Milled rice (equal to 65*% of quantity of paddy rice) [d]Raw value

World energy

Coal production

Top 20

Rank in 1976	000 tonnes			Growth %pa	
	1960	1970	1976	1960–70	1970–76
	World	*1 963 000**	*2 142 000**	*2 379 000**	*0.9**	*1.8**
1 United States	391 526	550 388	585 680	3.5	1.0
2 Soviet Union	355 918	432 715	494 000	2.0	2.2
3 China[c]	250 000**	360 000**	450 000**	3.7**	3.8**
4 Poland	104 438	140 101	179 303	3.0	4.2
5 Utd Kingdom	196 711	147 109	123 809	−2.9	−2.8
6 India	52 593	73 698	100 991	3.4	5.4
7 Germany, West	143 255	116 341	95 890	−2.0	−3.2
8 South Africa	38 173	54 612	75 730	3.6	5.6
9 Australia	20 400*	45 214	74 854	8.3*	8.8
10 Korea, North	6 800**	21 800**	35 000**[a]	12.4**	10.0**[b]
11 Czechoslovakia	26 400	28 195	28 266	0.7	0.1
12 France	55 960	37 838	23 300	−3.8	−7.8
13 Canada	8 020	11 598	20 798	3.8	10.2
14 Japan	51 067	39 694	18 396	−2.5	−12.0
15 Korea, South	5 340*	12 394	16 428	8.8*	4.8
16 Spain	13 783	10 751	10 483	−2.5	−0.4
17 Belgium	22 469	11 362	7 237	−6.6	−7.2
18 Rumania	3 405	6 402	7 120*	6.5	1.8*
19 Mexico	1 074	2 959	5 128[a]	10.7	11.6[b]
20 Turkey	3 653	4 574	4 639	2.3	0.2

Rest of the world

Rank in 1976	000 t 1976	Rank in 1976	000 t 1976
21 Vietnam	4 250**	36 Germany, East	456
22 Rhodesia	3 500*[a]	37 Mozambique	370
23 Taiwan	3 236	38 Bulgaria	288
24 Colombia	3 200[a]	39 Nigeria	237[a]
25 Brazil	3 130*	40 Botswana	223
26 Hungary	2 934	41= Indonesia	193
27 Chile	1 200	41= Portugal	193
28 Iran	1 150[b,e]	43 Afghanistan	189[a,c]
29 Pakistan	1 100*[c,f]	44 Mongolia	171[a]
30 Zambia	789	45 Philippines	149
31 Morocco	702	46 Swaziland	127[a]
32 Argentina	615	47 Zaire	90[a]
33 Jugoslavia	587	48 Venezuela	86
34 Norway	545	49 Peru	85*[a]
35 New Zealand	460	50[h] Ireland	51

Lignite production

Top 20

Rank in 1976	000 tonnes			Growth %pa	
	1960	1970	1976	1960–70	1970–76
	World	*635 000**	*794 000**	*893 000**	*2.3**	*2.0**
1 Germany, East	225 465	261 482	246 889	1.5	−1.0
2 Soviet Union	134 206	144 745	160 000	0.8	1.7
3 Germany, West	97 999	108 437	134 536	1.0	3.7
4 Czechoslovakia	58 403	81 783	89 467	3.4	1.5
5 Poland	9 327	32 766	39 302	13.4	3.1
6 Jugoslavia	21 430	27 779	35 694	2.6	4.3
7 Australia	14 200*	24 175	31 240	5.5*	4.4
8 Bulgaria	15 416	28 854	25 173	6.5	−2.3
9 United States	2 491	5 409	23 260	8.1	27.5
10 Hungary	23 676	23 679	22 323	0.0	−1.0
11 Greece	2 550	7 680	22 241	11.7	19.4
12 Rumania	3 363	14 129	18 730	15.4	4.8
13 Korea, North	3 800**	5 700**	9 000**[a]	4.1**	9.6**[b]
14 Turkey	1 911	4 437	8 250	8.8	10.9
15 Canada	1 969	3 465	4 678	5.8	5.1
16 Spain	1 762	2 831	4 140	4.9	6.5
17 India	47	3 545	3 895	54.0	1.6
18 Austria	5 973	3 670	3 184	−4.7	−2.3
19 France	2 276	2 785	3 135	2.0	2.0
20 Mongolia	619	1 915	2 549[a]	12.0	5.9[b]

Rest of the world

Rank in 1976	000 t 1976	Rank in 1976	000 t 1976
21 New Zealand	1 986	23 Albania	850*[a]
22 Italy	1 222	24[i] Thailand	681

Crude oil production

Rank in 1976	000 tonnes			Growth %pa	
	1960	1970	1976	1960–70	1970–76
	World	*1 054 000**	*2 274 000**	*2 860 000**	*8.0**	*3.9**
1 Soviet Union	147 859	353 039	520 000	9.1	6.7
2 Saudi Arabia[g]	65 712	188 408	424 232	11.1	14.5
3 United States	347 975	475 289	401 594	3.2	−2.8
4 Iran	52 392	191 470	296 500	13.8	7.5
5 Venezuela	149 372	194 306	119 756	2.7	−7.7
6 Iraq	47 467	76 457	112 415	4.9	6.6
7 Kuwait[g]	85 511	150 636	108 562	5.8	−5.3
8 Nigeria	850	54 203	102 655	51.5	11.2
9 Utd Arab Emirates	—	37 685	93 300	na	16.3
10 Libya	—	159 709	92 770	na	−8.7
11 China	5 500**	20 000**	85 000**	13.8**	27.3**
12 Indonesia	20 844	42 103	74 029	7.3	9.8
13 Canada	25 630	61 868	64 100	9.2	0.6
14 Algeria	8 632	47 202	50 090	18.5	1.0
15 Mexico	14 171	21 501	40 843	4.3	11.3
16 Qatar	8 212	17 516	23 534	7.9	5.0
17 Argentina	8 898	20 026	20 920	8.5	0.7
18 Australia[f]	—	4 125	20 515	na	30.6
19 Oman	—	16 583	18 293	na	1.6
20 Egypt	3 319	16 404	16 709	17.3	0.3
21 Rumania	11 500	13 377	14 700	1.5	1.6
22 Norway	—	—	13 691	na	na
23 United Kingdom	148	83	12 036	−5.6	129.2
24 Gabon	800	5 423	11 392	21.1	13.2
25 Trinidad & Tobago	5 994	7 223	10 992	1.9	7.2
26 Syria	—	4 243	9 976	na	15.3
27 Ecuador	360	193	9 480	−6.0	91.4
28 Brunei	4 583	6 685	8 639[a]	3.8	5.3[b]
29 India	454	6 809	8 623	31.1	4.0
30 Brazil	3 870	7 962	8 470	7.5	1.0
31 Malaysia	60	859	8 024	30.5	45.1
32 Colombia	7 584	11 327	7 347	4.1	−6.5
33 Angola	67	5 065	6 281	54.1	3.7
34 Germany, West	5 530	7 535	5 526	3.1	−5.0
35 Jugoslavia	944	2 854	3 880	11.7	5.2
36 Tunisia	—	4 151	3 712	na	−1.8
37 Peru	2 572	3 550	3 708	3.3	0.7
38 Bahrain	2 256	3 847	2 916	5.5	−4.5
39 Turkey	375	3 542	2 568	25.2	−5.2
40 Albania	728	1 487	2 300*[a]	7.4	9.1*[b]
41 Hungary	1 217	1 937	2 142	4.8	1.7
42 Austria	2 448	2 798	2 031	1.3	−5.2
43 Congo	52	19	2 002	−9.6	117.3
44 Spain	64	151	1 982	9.0	53.6
45 Bolivia	466	1 122	1 964	9.2	9.8
46 Netherlands	1 918	1 919	1 371	0.0	−5.4
47 Burma	545	801	1 160	3.9	6.4
48 Italy	1 998	1 405	1 108	−3.5	−3.9
49 Chile	943	1 601	1 092	5.4	−6.2
50 France	1 983	2 309	1 052	1.5	−12.3
51 Japan	526	770	580	3.9	−4.6
52 Poland	194	424	460	8.1	1.4
53 Pakistan	357	501	341	3.4	−6.2
54 Taiwan	2	101	247	46.0	16.1
55 Denmark	—	—	167[a]	na	na
56 Cuba	14*	159	150*[a]	27.5*	−1.2*[b]
57 New Zealand	1	58	136[e]	50.0	18.6[b]
58 Czechoslovakia	137	203	131	4.0	−7.0
59 Bulgaria	200	334	120	5.3	−15.7
60 Zaire	—	—	96[a]	na	na
61[d] Germany, East	na	90*	80[a]	na	−2.3*[b]

[a]1975 [b]1970–75 [c]Including lignite [d]Other (000 t): 62 Israel 36, 63 Morocco 25[a], 64 Thailand 6[a] [e]Year beginning March 21st [f]Year(s) ending June 30th [g]Including share of the Neutral Zone [h]Other (000 t): 51 Burma 15[a], 52 Sweden 11[a], 53 Algeria 8[a], 54 Tanzania 1[a] [i]Other (000 t): 25 Japan 52, 26 Chile 46[a]

World energy

Crude oil importers[a] and main suppliers, mn tonnes, 1975

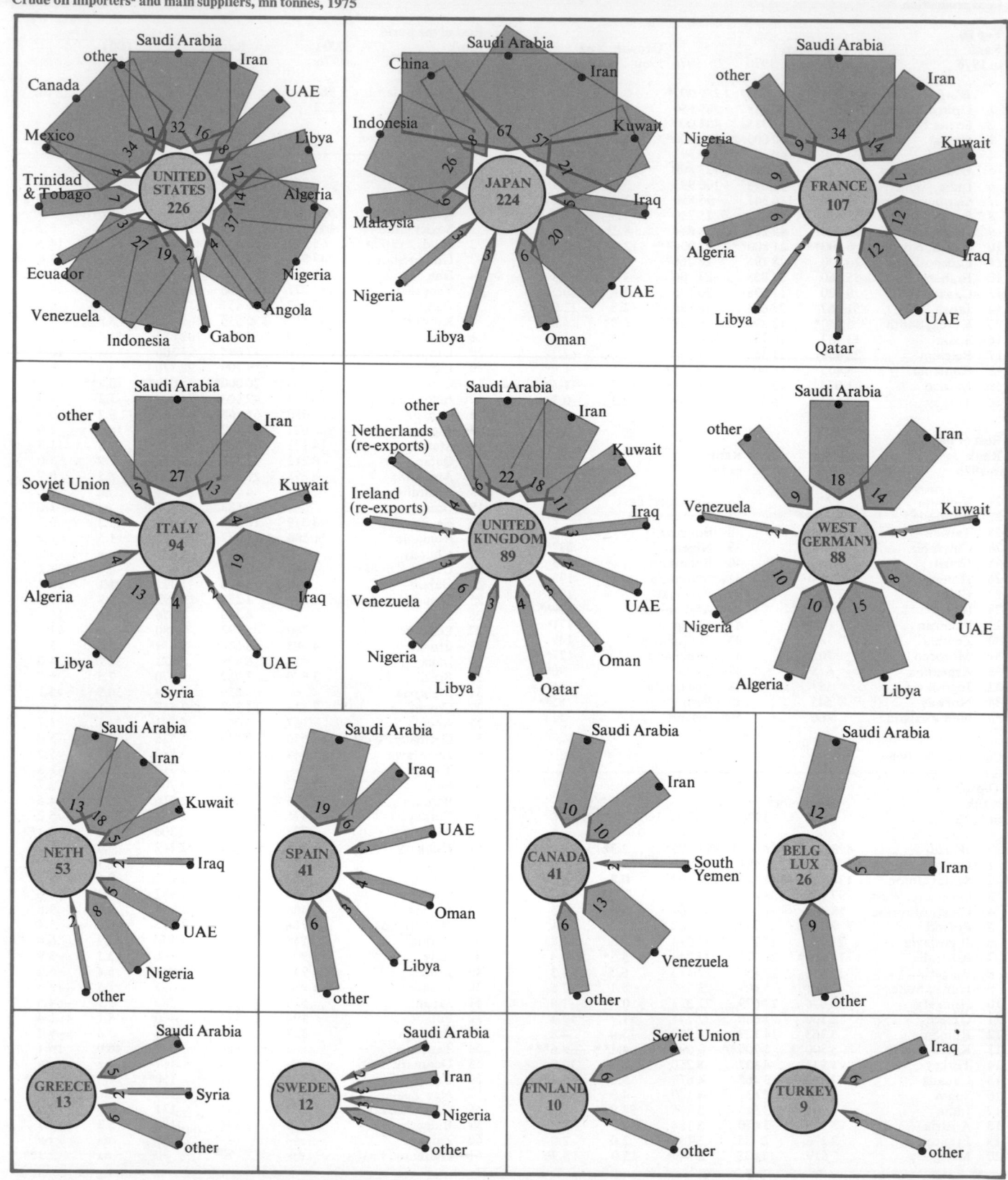

[a]Main OECD importers, showing total imports of crude oil (circled); imports from suppliers of at least 2mn tonnes are indicated by arrows

World energy

Petroleum products production

Top 20

Rank in 1975		000 tonnes 1960	1970	1975	Growth %pa 1960–70	1970–75
	World[a]	*860 000**	*1 910 000**	*2 163 000**	*8.3**	*2.5**
1	United States	369 331	530 603	601 209	3.7	2.5
2	Japan	25 744	163 255	205 300*	20.3	4.7*
3	France	29 400**	92 619	102 479	12.2**	2.0
4	Italy	28 278	111 325	94 553	14.7	−3.2
5	Germany, West	26 892	100 984	91 784	14.2	−1.9
6	United Kingdom	42 651	99 942	86 647	8.9	−2.8
7	Canada	34 332	58 431	77 950	5.5	5.9
8	China	5 000**	20 000**	60 000**	15.0**	20.0**
9	Netherlands	33 196	57 245	52 249	5.6	−1.8
10	Venezuela	43 761	66 700	43 570*	4.3	−8.2*
11	Brazil	8 416	24 101	41 930*	11.1	11.7*
12	Spain	6 005	30 590	41 500*	17.7	6.3*
13	Iran	16 715	24 400*	33 380*	3.8*	6.5*
14	Mexico	13 430	22 053	30 780*	5.1	6.9*
15	Belgium	6 328	26 145	27 401	15.2	1.0
16	Australia	11 313	22 987	24 604	7.3	1.4
17	Saudi Arabia[d]	11 510	29 506	23 798	9.9	−4.2
18	Virgin Is, US	—	11 600*	23 700*	na	15.4*
19	Neth Antilles	33 000*	44 418	22 110*	3.0*	−13.0*
20	Argentina	11 513	21 139	20 600*	6.3	−0.5*

Rest of the world

Rank in 1975		000 t 1975	Rank in 1975		000 t 1975
21	India	20 280*	64	Ireland	2 340*
22	Singapore	20 000**	65	Syria	2 284
23	Germany, East	17 800*	66	Lebanon	1 980*
24	Rumania	17 680*	67	Ecuador	1 977
25	Korea, South	16 100	68	Libya	1 850*
26	Czechoslovakia	15 223	69	Albania	1 800*
27	South Africa	13 920*	70	Uruguay	1 786
28	Kuwait[d]	13 700*	71	Sri Lanka	1 560*
29	Indonesia	12 750*	72	Jamaica	1 480*
30	Poland	12 489	73	Ivory Coast	1 250*
31	Turkey	12 110*	74	Yemen, South	1 200*
32	Sweden	11 980*	75	Ghana	1 183*
33	Trinidad & Tobago	11 682	76	Dominican Rep	1 159
34	Bahrain	10 490*	77	Tunisia	1 135
35	Taiwan	10 470	78	Guam	1 100*
36	Greece	10 340*	79	Sudan	1 072*
37	Bulgaria	10 220*	80	Burma	980*
38	Jugoslavia	10 068	81	Guatemala	955
39	Hungary	9 579[c]	82	Malaysia	930*
40	Bahamas	9 440*	83	Bolivia	900*
41	Puerto Rico	9 143	84	Gabon	810*
42	Philippines	8 790*	85	Angola	800*
43	Egypt	8 533	86	Jordan	756
44	Finland	7 763	87	Mozambique	750*
45	Austria	7 710*	88	Bangladesh	701
46	Denmark	7 662	89	Senegal	690*
47	Colombia	7 400*	90	Tanzania	669
48	Thailand	7 356	91	El Salvador	630*
49	Norway	6 600*	92	Nicaragua	626
50	Israel	6 523*	93	Honduras	598
51	Peru	5 470*	94	Madagascar	590*
52	Cuba	5 410*	95	Ethiopia	585*
53	Algeria	5 091	96	Zaire	564
54	Iraq	5 000*	97	Liberia	505*
55	Portugal	4 980*	98	Martinique	400**
56	Switzerland	4 440*	99	Cyprus	315
57=	Chile	3 770*	100	Sierra Leone	294
57=	Panama	3 770*	101	Costa Rica	257
59	Pakistan	3 340*	102	Paraguay	200*
60	New Zealand	2 970*	103	Qatar	171
61	Kenya	2 750*	104	Barbados	143
62	Nigeria	2 430*	105	Brunei	54
63	Morocco	2 415*			

Natural gas production

Top 10

Rank in 1976		Million cubic metres 1960	1970	1976	Growth %pa 1960–70	1970–76
	World	*468 300**	*1 038 000**	*1 326 000**	*8.3**	*4.2**
1	United States	344 800*	595 057	546 340*	5.6*	−1.4*
2	Soviet Union	45 303	197 945	321 000	15.9	8.4
3	Netherlands	330	31 617	97 302	57.8	20.6
4	Canada	12 770*	56 712	74 470*	16.1*	4.6*
5	United Kingdom	1	11 270	39 410	154.0	23.2
6	Rumania	10 143	23 629	32 180	8.8	5.3
7	Iran	950	11 223	22 476	28.0	12.3
8	Germany, West	950**	13 011	19 030	30.0**	6.5
9	Italy	6 447	13 171	15 370	7.4	2.6
10	Mexico	6 260*	11 700*	14 030*	6.5*	3.1*

Rest of the world[g]

Rank in 1976		mn m³ 1976	Rank in 1976		mn m³ 1976
11	Venezuela	11 660	30	Bolivia	2 000*
12	Algeria	9 532[b]	31	Taiwan	1 836
13	Indonesia	8 840	32	Jugoslavia	1 730
14	Argentina	7 710	33	Trinidad & Tobago	1 697
15	Germany, East	7 270[b]	34	Utd Arab Emirates	1 690*[b]
16	France	7 092	35	Iraq	1 654[h]
17	Poland	6 699	36	Colombia	1 625*[b]
18	Hungary	6 082	37	China	1 500**
19	Brunei	6 000[b]	38	India	1 208
20	Australia[f]	5 906	39	Czechoslovakia	937
21	Saudi Arabia[d]	5 900*	40	New Zealand	876
22	Kuwait[d]	5 581	41	Brazil	550*
23	Pakistan	5 140	42	Bangladesh	510*
24	Libya	3 250*[b]	43	Peru	450*[b]
25	Afghanistan	2 964[h]	44	Nigeria	402[h]
26	Japan	2 813	45	Tunisia	214
27	Qatar	2 209	46	Syria	210[b]
28	Bahrain	2 150[b]	47	Albania	155*[b]
29	Austria	2 144	48	Bulgaria	111[b]

Manufactured gas production

Top 10

Rank in 1976		Million cubic metres 1960	1970	1976	Growth %pa 1960–70	1970–76
	World	*140 000**	*158 000**	*146 000**	*1.2**	*−1.3**
1	Soviet Union	24 568	32 899	36 370	3.0	1.7
2	United States	24 003	29 083	26 750*	1.9	−1.4*
3	Germany, West	24 800**	20 119	15 385	−2.1**	−4.2
4	Czechoslovakia	4 331	7 093	7 936	5.0	1.9
5	Poland	5 163	6 682	7 548	2.6	2.0
6	Australia[f]	2 512	4 130*	7 100*[b]	5.1*	11.4**[c]
7	Japan	1 280*	3 180	6 500*	9.5*	12.6*
8	France	10 000**	8 847	6 097	−1.2**	−6.0
9	Germany, East	3 045	4 269	5 500	3.4	4.3
10	United Kingdom	22 727	20 580	3 680*	−1.0	−24.9

Rest of the world[g]

Rank in 1976		mn m³ 1976	Rank in 1976		mn m³ 1976
11	Italy	3 508	22	Turkey	688[b]
12	Belgium	2 668	23	Sweden	587[b]
13	Spain	2 656	24	Brazil	516*
14	Canada	2 588[b]	25	Denmark	330
15	South Africa	1 865[b]	26	Bulgaria	282[b]
16	Chile	1 150*[b]	27	Ireland	270
17	Austria	1 086	28	Portugal	220[b]
18	Hungary	1 029[b]	29	Mexico	200*[b]
19	Netherlands	1 007	30	Greece	163[b]
20	Rumania	950*[b]	31	Colombia	150*
21	Jugoslavia	740	32	Hongkong	112

[a]Excluding Soviet Union [b]1975 [c]1970–75 [d]Includes share of the Neutral Zone shared by Kuwait and Saudi Arabia [e]1976 [f]Year(s) ending June 30th [g]Showing only countries producing more than 100 mn m³

World energy

Electricity production

Top 70

Rank in 1976		Kilowatt-hours (mn) 1960	1970	1976	Growth %pa 1960–70	%pa 1970–76
	World	2 320 000*	4 915 000*	6 900 000*	7.8*	5.8*
1	United States	844 188	1 639 771	2 117 624	6.9	4.4
2	Soviet Union	292 274	740 926	1 111 000	9.7	7.0
3	Japan[a]	115 900*	361 200*	511 780	12.0*	6.0*
4	Germany, West	118 986	242 612	333 652	7.4	5.5
5	Canada	114 457	204 723	293 410	6.0	6.2
6	Utd Kingdom	137 300*	249 193	276 970	6.1*	1.8
7	France	72 118	140 708	191 199	6.9	5.2
8	Italy	56 240	117 423	163 550	7.6	5.7
9	China	58 000**	75 000**	120 000**	2.6**	8.1**
10	Poland	29 307	64 532	104 095	8.2	8.3
11	Spain	18 614	56 490	90 595	11.7	8.2
12	Germany, East	40 305	67 650	89 148	5.3	4.7
13	Sweden	34 718	60 645	86 416	5.7	6.1
14	India[f]	17 794	56 543	85 613*	12.3	7.2*
15	Norway	31 121	57 606	82 198	6.4	6.1
16	Brazil	22 865	45 460	78 068[a]	7.1	11.4[b]
17	South Africa	22 561[q]	50 791	78 056	8.0	7.4
18	Australia[q]	22 300*	53 892	76 597	9.2*	6.0
19	Czechoslovakia	24 450	45 163	62 629	6.3	5.6
20	Rumania	7 650	35 088	58 266	16.4	8.8
21	Netherlands	16 516	40 859	58 059	9.5	6.0
22	Belgium	15 152	30 523	47 350	7.2	7.6
23	Mexico	10 813	28 608	46 238	10.2	8.3
24	Jugoslavia	8 928	26 024	43 574	11.3	9.0
25	Switzerland[t]	19 072	33 173	42 282[a]	5.7	5.0[b]
26	Austria	15 965	30 036	35 037	6.5	2.6
27	Argentina	10 459	21 727	29 468[a]	7.6	6.3[b]
28	Finland	8 628	21 158	29 319	9.4	5.6
29	Bulgaria	4 657	19 513	27 741	15.4	6.0
30	Taiwan	3 628	13 213	26 877	13.8	12.6
31	Korea, South	1 758	9 597	24 240*	18.5	16.7*
32	Korea, North	9 139	16 500**	23 000**[a]	6.1**	6.8**[b]
33	Hungary	7 617	14 542	22 049	6.7	7.2
34	Venezuela	4 651	12 631	21 179[a]	10.5	10.9[b]
35	New Zealand[n,q]	6 361	12 926	20 064	7.3	7.6
36	Denmark	5 179	18 864	19 700*	13.8	0.7*
37	Turkey	2 815	8 623	18 270	11.8	13.3
38	Puerto Rico	2 151	8 027	16 203*[a]	14.1	15.1*[b]
39	Greece	2 277	9 399	15 151[a]	15.2	10.0[b]
40	Iran[u]	2 446[i]	7 004	15 000*[a]	19.3[j]	16.4*[b]
41	Colombia	3 750	8 750	14 500*	8.8	8.8*
42	Philippines	2 731	8 666	12 360*[a]	12.2	7.3*[b]
43	Egypt	2 639	7 591	10 421[a]	11.1	6.5[b]
44	Israel	2 313	6 838	10 344	11.4	7.1
45	Portugal	3 300**	7 488	9 594	8.5**	4.2
46	Chile	4 592	7 550	9 432	5.1	3.8
47	Pakistan	1 000**	6 513	8 800*[a]	20.6**	6.2*[b]
48	Ireland[f]	2 180*	5 399	8 600	9.5*	8.1
49	Hongkong	2 061[d]	5 097	8 340	13.8[e]	8.5
50	Peru	2 656	5 529	8 300*[a]	7.6	8.5*[b]
51	Thailand	594	4 545	7 910*[a]	22.6	11.7*[b]
52	Zambia	836	949	7 040	1.3	39.6
53	Rhodesia	2 388	6 410	6 738	10.4	0.8
54	Malaysia	1 242	3 545	6 510*	11.0	10.7*
55	Cuba	2 981	4 888	6 150*[a]	5.1	4.7*[b]
56	Kuwait	249	2 213	5 210	24.4	15.4
57	Singapore	659	2 205	4 604	12.8	13.1
58	Ghana	374	2 920	4 050*[a]	22.8	6.8*[b]
59	Algeria	1 387[g]	1 979	3 744*[a]	4.0[h]	13.6*[b]
60	Zaire	2 456	3 230	3 440*[a]	2.8	3.3[b]
61	Iraq	850*	1 909	3 400*[a]	8.4*	12.2*[b]
62	Indonesia	1 161	2 100	3 345*[a]	6.1	9.7*[b]
63	Morocco	1 012	1 982	3 220*	7.0	8.4*
64	Nigeria	528	1 550	3 211[a]	11.4	15.7[b]
65	Uruguay	1 310*	2 200	2 596*[a]	5.3*	3.4*[b]
66	Vietnam	562	2 060**	2 500**	13.9**	3.3**
67	Iceland	551	1 470	2 426	10.3	8.7

Rank in 1976		Kilowatt-hours (mn) 1960	1970	1976	Growth %pa 1960–70	%pa 1970–76
68	Jamaica	514	1 541	2 331[a]	11.6	8.6[b]
69	Saudi Arabia	50**	724	1 988[a]	30.6**	22.4[b]
70	Albania	194	944	1 800*[a]	17.1	13.8*[b]

Rest of the world

Rank in 1976		kW h (mn) 1976	Rank in 1976		kW h (mn) 1976
71	New Caledonia	1 791[a]	129	Sierra Leone	193[a]
72	Syria	1 780	130	Martinique	182[a]
73	Costa Rica	1 640	131	Yemen, South	180*
74	Dominican Rep	1 632*[a]	132	Guadeloupe	170*[a]
75	Surinam	1 600*[a]	133	Haiti	158[a]
76	Luxembourg	1 542	134	Cambodia	150*[a]
77	Netherlands Antilles	1 400*[a]	135	Macao	143[a]
			136	Rwanda	140*[a]
78	Trinidad & Tob	1 390*	137	Greenland	124*
79	Bangladesh	1 380*	138	Nepal	122*[a]
80	Tunisia	1 346[a]	139	Togo	118[a]
81	Cameroon	1 335	140	Pacific Islands, US	115*[c]
82	Panama	1 310*	141	Swaziland	112[a]
83	Angola	1 305*[a]	142	Congo	100[a]
84	Guam	1 300*[a]	143=	French Polynesia	95[a]
85	Ecuador	1 290*[a]	143=	Mauritania[n]	95*[a]
86	Lebanon	1 250*	145	Faroes	93*[f]
87	El Salvador	1 193	146=	Mali	70*[a]
88	Sri Lanka	1 149*[a]	146=	Niger	70*[a]
89	Kenya	1 110*	148	Sahara, Western	65*[a]
90	Guatemala	1 100*[a]	149	Samoa, American	64
91	Papua New Guinea	1 048[q]	150	Djibouti[o]	58*[a]
92	Bolivia	1 000*[a]	151	Benin	57[a]
93	Burma	990*	152=	Chad	56[a]
94	Mongolia	930	152=	French Guiana	56[a]
95	Libya[m]	900*	154	Upper Volta	53[a]
96	Liberia	870*[a]	155	Central African Empire	52[a]
97	Ivory Coast	860*[a]	156	Gibraltar	50[a]
98	Nicaragua	835*[a]	157=	Liechtenstein	49
99	Uganda	830*[a]	157=	Yemen, North	49[a]
100	Cyprus	802	159	Somalia[p]	42*[a]
101	Virgin Islands, US	768*[a]	160	St Lucia	36[a]
102	Afghanistan	748*[a]	161=	Belize	30*[a]
103	Mozambique	717*[a]	161=	Gambia	30*
104	Panama Canal Zone	692[a]	163	Christmas Island	28[a]
105	Ethiopia	665*[a,1]	164	Seychelles	27[a]
106	Bahamas	650*[a]	165	Nauru	26[a]
107	Tanzania[r]	636*[a]	166	Grenada	25[a]
108	Paraguay	510*[a]	167=	Burundi[k]	23*[a]
109	Guinea	500*[a]	167=	St Kitts-Nevis	23[a]
110	Honduras	480[a]	169=	Guinea-Bissau	20*[a]
111	Jordan	443[a]	169=	Samoa, Western	20[a]
112	Oman	413	171	Cayman Islands	17*[a]
113=	Bahrain	400*[a]	172=	St Vincent	16[a]
113=	Guyana	400	172=	Solomon Islands[n,o]	16[a]
115	Senegal	389*[a]	174	New Hebrides	15[a]
116	Sudan	350*[a]	175	Dominica	14[a]
117	Malta	338[f]	176	Virgin Islands, Br	12*[a]
118	Madagascar	335*[a]	177=	Cape Verde	8*[a]
119	Bermuda	300*[a]	177=	Cook Islands	8[a]
120	Botswana	274[a]	177=	Sao Tome & Principe[o]	8*[a]
121	Mauritius	272	180	Montserrat	7*
122	Laos[n]	255*[a]	181=	Gilbert Islands	5*[a]
123	Malawi[n]	254	181=	Tonga	5[a]
124	Fiji	241[a]	183=	Comoros	3*[a]
125	Gabon	235*[a]	183=	Timor, East	3*[a]
126	Brunei	230[a]	185	Falkland Islands	2[a]
127	Reunion	220*	186	St Helena	1*[a]
128	Barbados	214[a]			

[a]1975 [b]1970–75 [c]Including Northern Marianas [d]1963 [e]1963–70 [f]Year(s) ending March 31st [g]1962 [h]1962–70 [i]1964 [j]1964–70 [k]Includes hydro-electricity provided by Zaire [l]Year ending September 10th [m]Tripolitania only [n]Public supply only [o]Consumption [p]Main cities only [q]Year(s) ending June 30th [r]Mainland only [s]Years beginning April 1st [t]Years ending September 30th [u]Years beginning March 21st

World mining

Main products and main producers

Iron ore[a] 000 t, 1976 · Bauxite 000 t, 1976 · Copper ore[a] 000 t, 1976 · Lead ore[a] 000 t, 1976 · Manganese ore[a] 000 t, 1975

Rank	Iron ore		Bauxite		Copper ore		Lead ore		Manganese ore	
	World	536 000*	World	75 000*	World	7 600*	World	3 350*	World	9 320*
1	Soviet Union	130 700	Australia	23 540	United States	1 461	United States	553	Soviet Union	2 951
2	Australia	59 300[j]	Jamaica	10 310	Soviet Union	1 100**[b]	Soviet Union	500**	South Africa	2 006
3	United States	48 398*	Guinea	7 620	Chile	1 011	Australia	391[j]	Gabon	1 115
4	Brazil	46 621*[b]	Surinam	4 751[b]	Zambia	850	Canada	247	Brazil	820*
5	Canada	34 167**[g]	Soviet Union	4 500*	Canada	723	Mexico	166	Australia	673[j]
6	China	34 000**	Guyana	3 198[b]	Zaire	444	Peru	166[b]	India	575
7	India	26 868*	Hungary	2 919	Philippines	238*	Jugoslavia	123	China	300**
8	Sweden	18 370	Greece	2 455	Poland	230*[b]	Bulgaria	110*[b]	Ghana	199
9	Liberia	16 923[b]	United States	2 434	Australia	217[j]	China	100**	Zaire	160
10	Venezuela	15 425[b]	France	2 289	South Africa	179[b]	Korea, North	100**[b]	Mexico	154
11	France	13 555	Jugoslavia	2 032	Peru	176*[b]	Morocco	83	Morocco	105
12	South Africa	9 887	India	1 437	Papua New Guinea	172[b]	Sweden	70[b]	Japan	42
13	Chile	6 300*	Brazil	1 277*[b]	Jugoslavia	120	Poland	65*[b]	Rumania	31*
14	Mauritania	5 646[b]	China	1 000**	China	100**	Spain	62	Cuba	28[h]
15	Peru	5 067*[b]	Indonesia	938	Mexico	87	Japan	52	Hungary	28
16	Spain	3 800	Dominican Rep	785[b]	Japan	82	South-West Africa	48[b]	New Hebrides	19[k]
17	Korea, North	3 760**[b]	Rumania	779*[b]	Bulgaria	55*[b]	Rumania	45*	United States	17
18	Mexico	3 645	Haiti	733[k]	Turkey	43*[b]	Ireland	36[b]	Iran	14
19	Angola	3 388[b]	Malaysia	660	Finland	42	Germany, West	32	Turkey	13
20	Norway	2 550	Sierra Leone	645[b]	Sweden	41[b]	Brazil	30*[b]	Bulgaria	10*
21	Turkey	1 840*	Turkey	570[b]	Norway	31	Iran	30*[b]	Thailand	9
22	Algeria	1 728[b]	Ghana	325[b]	Rhodesia	30*	France	28	Argentina	8
23	Jugoslavia	1 490	Italy	24	South-West Africa	25[b]	Argentina	27[b]	Chile	8
24	Swaziland	1 417	Spain	9[b]	India	24[h]	Italy	26	Indonesia	6*
25	United Kingdom	1 190	Mozambique	2*[b]	Spain	19	Greenland	24	Jugoslavia	5

Nickel ore[a] 000 t, 1975 · Tin conc[d] 000 t, 1976 · Tungsten conc[e] 000 t, 1975 · Uranium[a] 000 t, 1976 · Zinc ore[a] 000 t, 1976

Rank	Nickel ore		Tin conc		Tungsten conc		Uranium		Zinc ore	
	World	744.7*	World[l]	182.0*	World	46.60*	World[f]	22.29	World[m]	5 500*
1	Canada	245.0	Malaysia	63.0	China	11.30**	United States	9.80	Canada	1 157
2	Soviet Union	160.0**	Bolivia	28.0	Soviet Union	10.00**	Canada	4.80	Soviet Union	720**
3	New Caledonia	133.3	Thailand	27.9	Korea, South	3.34	South Africa	3.41	Australia	468[j]
4	Australia	49.1[j]	Indonesia	22.2	United States	3.14	France	2.06	United States	433
5	Cuba	37.0	Australia	9.1	Bolivia	3.10*	Niger	1.60	Peru	360*[b]
6	Dominican Rep	26.9	Brazil	5.0*[b]	Korea, North	2.70**	Gabon	0.80[b]	Japan	260
7	Indonesia	21.0*	Zaire	4.0	Thailand	2.00	Australia	0.36	Mexico	259
8	South Africa	20.8	Nigeria	3.7	Portugal	1.77	Spain	0.17	Poland	190*[h]
9	United States	15.4	United Kingdom	3.3	Australia	1.58[j]	Portugal	0.09	Korea, North	162**[b]
10	Greece	14.8	South Africa	2.7	Brazil	1.43*	Argentina	0.05	Germany, West	115
11	Botswana	13.2	Rwanda	1.2*[h]	Canada	1.35	Germany, West	0.04	Sweden	111[c]
12	Rhodesia	10.0*	Rhodesia	0.8	Japan	0.97	Japan	0.002	Jugoslavia	107
13	Philippines	9.5	Burma	0.7[h]	France	0.78[h]			China	100**
14	Albania	6.0*	South-West Africa	0.7[h]	Peru	0.73*			Italy	87[b]
15	Finland	5.3[b]	Argentina	0.6*	Rwanda	0.50			Greenland	85[b]
16	Brazil	2.6*	Japan	0.6	Spain	0.49			Spain	81
17	Germany, East	2.4*	Laos	0.5[b]	Burma	0.44			Bulgaria	80*[b]
18	Poland	2.0*	Mexico	0.4[b]	Zaire	0.31			Zaire	70
19	Morocco	0.3*	Spain	0.4	Mexico	0.28			Ireland	67*[b]
20	Mexico	0.05	Portugal	0.3	Sweden	0.14			Iran	66[b]

Silver 000 kg, 1976 · Gold[i] 000 kg, 1976 · Diamonds 000 CM, 1975 · Phosphate rock 000 t, 1975 · Salt 000 t, 1975

Rank	Silver		Gold		Diamonds		Phosphate rock		Salt	
	World	9 300**	World	1 230.0*	World	42 700*	World	118 500*	World	159 200*
1	Soviet Union	1 370**	South Africa	709.0	Zaire	12 810	United States	44 285	United States	37 222
2	Mexico	1 330*	Soviet Union	240.0**	Soviet Union	9 900**	Soviet Union	24 120**	China	29 000**
3	Canada	1 270	Canada	52.0*	South Africa	7 023	Morocco	14 119	Soviet Union	14 300
4	Peru	1 240*	United States	33.0	Botswana	2 410	Tunisia	3 512	Germany, West	9 500
5	United States	1 067	Papua New Guinea	19.8*	Ghana	2 328	China	3 400**	United Kingdom	7 630
6	Australia	760	Rhodesia	17.0*	South-West Africa	1 740	Sahara, Western	3 300*	India	5 918[c]
7	Japan	270[b]	Ghana	16.0	Sierra Leone	1 650*	Togo	2 553[c]	France	5 575
8	Poland	230[b]	Philippines	15.6[b]	Tanzania	896	South Africa	1 824[c]	Canada	5 156
9	Chile	192[b]	Australia	15.5	Venezuela	819[c]	Senegal	1 600	Australia	5 057[j]
10	Jugoslavia	168[b]	Colombia	11.9	Angola	460	Nauru	1 535[j]	Rumania	4 210
11	Bolivia	148[c]	Germany, West	10.9	Liberia	406[k]	Christmas Island	1 487	Poland	3 818
12	Sweden	140[b]	Brazil	9.3	Central African Emp	338*	Vietnam	1 400**	Mexico	3 803
13	France	107[b]	Jugoslavia	5.5[b]	Brazil	270	Jordan	1 353	Netherlands	3 387[c]
14	Spain	104[c]	Mexico	4.8*	Ivory Coast	209	Israel	882	Italy	3 191
15	Honduras	99[c]	Chile	4.6*	Guinea	80	Syria	860	Germany, East	2 430
16	South Africa	96[b]	Japan	4.5*[h]	Guyana	50[c]	Algeria	802[e]	Spain	2 257[e]
17	Argentina	90[b]	Zaire	3.2[b]	India	20	Gilbert Islands	529	Brazil	1 500*
18	Zaire	71[b]	Peru	3.0*	Indonesia	15*	India	459	Argentina	1 150
19	Germany, East	62[b]	Sweden	2.4[b]	Lesotho	3	Korea, North	450**	Japan	1 068
20	Philippines	50[b]	Fiji	1.9[b]			Egypt	404	Colombia	926

[a]Metal content [b]1975 [c]1974 [d]Metal content of concentrates [e]Oxide (WO_3) content of concentrates [f]Excluding China, Soviet Union and other eastern Europe, and other countries for which information is not available [g]Shipments [h]1968 [i]Includes some gold refined in countries concerned from imported gold or ores [j]Year ending June 30th [k]Exports [l]Excluding China, Vietnam, North Korea, Mongolia, Soviet Union and East Germany [m]Excluding Czechoslovakia, Rumania and Vietnam

World manufacturing

Cement production

Top 50

Rank in 1976		000 tonnes 1960	1970	1976	Growth %pa 1960–70	1970–76
	World	317 000*	568 000*	742 000*	6.0*	4.6*
1	Soviet Union	45 520	95 248	124 000	7.7	4.5
2	Japan	22 537	57 189	68 712	9.8	3.1
3	United States	56 063	67 682	64 898*	1.9	−0.7*
4	Italy	16 014	33 076	36 323	7.5	1.6
5	Germany, West	24 905	38 325	34 097	4.4	−1.9
6	China	12 500*	15 000**	30 000**	1.8**	12.3**
7	France	14 349	29 009	29 500	7.3	0.3
8	Spain	5 234	16 702	25 291	12.3	7.1
9	Poland	6 599	12 180	19 808	6.3	8.4
10	India	7 845	13 956	18 499	5.9	4.8
11	Brazil	4 474	9 002	17 873	7.2	12.1
12	United Kingdom	13 501	17 171	15 781	2.4	−1.4
13	Rumania	3 054	8 127	13 088	10.3	8.3
14	Mexico	3 089	7 267	12 477	8.9	9.4
15	Turkey	2 038	6 374	12 392	12.1	11.7
16	Korea, South	464	5 782	11 872	28.7	12.7
17	Germany, East	5 032	7 984	11 345	4.7	6.0
18	Canada	5 338	7 283	9 898	3.2	5.2
19	Czechoslovakia	5 051	7 402	9 551	3.9	4.3
20	Greece	1 649	4 848	8 754	11.4	10.3
21	Taiwan	1 183	4 541	8 749	14.4	11.5
22	Jugoslavia	2 398	4 399	7 620	6.2	9.6
23	Belgium	4 388	6 729	7 506	4.4	1.8
24	South Africa	2 705	5 752	7 048	7.8	3.4
25	Korea, North	2 285	4 000**	6 000**a	5.8**	8.5**b
26	Austria	2 830	4 806	5 880	5.4	3.4
27	Argentina	2 641	4 770	5 603	6.1	2.7
28	Australia	2 799	4 499	5 039	4.9	1.9
29	Bulgaria	1 586	3 668	4 362	8.7	2.9
30	Hungary	1 571	2 771	4 298	5.8	7.6
31	Philippines	795	2 447	4 220	11.9	9.5
32	Thailand	526	2 627	4 106	17.4	7.7
33	Iran	797	2 575	3 900*a	12.4	8.7*b
34	Colombia	1 589	2 757	3 612	5.7	4.6
35	Switzerland	3 036	4 797	3 546	4.7	−4.9
36	Portugal	1 202	2 332	3 496	6.8	7.0
37	Netherlands	1 798	3 830	3 481	7.8	−1.6
38	Venezuela	1 501	2 318	3 455a	4.4	8.3b
39	Egypt	2 047	3 684	3 294	6.1	−1.8
40	Pakistan	981	2 656	3 180	10.5	3.0
41	Sweden	2 919	4 061	2 798	3.3	−6.0
42	Norway	1 151	2 635	2 680	8.6	0.3
43	Denmark	1 442	2 604	2 355	6.1	−1.7
44	Morocco	580	1 421	2 108	9.4	6.8
45	Cuba	813	742	2 083a	−0.9	22.9b
46	Israel	806	1 384	2 042	5.6	6.7
47	Peru	600	1 144	1 936a	6.7	11.1b
48	Finland	1 257	1 838	1 820	3.0	−0.2
49	Iraq	813	1 542	1 800*a	6.6	3.1*b
50	Lebanon	854	1 339	1 744c	4.6	6.8d

Rest of the world

Rank in 1976		000 t 1976	Rank in 1976		000 t 1976
51	Malaysia	1 733	64	Ivory Coast	720a
52	Ireland	1 570	65	Vietnam	700**
53	Puerto Rico	1 398	66	Angola	699a
54	Nigeria	1 383a	67	Ghana	688a
55	Syria	1 110	68	Rhodesia	677a
56	Saudi Arabia	1 056c	69	Uruguay	632a
57	Cyprus	1 026	70	Oman	626e
58	New Zealand	999	71	Libya	615a
59	Chile	964	72	Ecuador	604a
60	Algeria	940c	73	Zaire	577c
61	Kenya	897a	74	Dominican Republic	555a
62	Indonesia	881a	75	Jordan	533
63	Hongkong	764	76	Albania	517*a

Rank in 1976		000 t 1976	Rank in 1976		000 t 1976
77	Tunisia	478	88	Honduras	271a
78	Bahamas	463*	89	Mozambique	258a
79	Zambia	452a	90	Tanzania	246
80	Sri Lanka	421	91	Trinidad & Tobago	241
81	Senegal	380*	92	Haiti	232
82	Jamaica	365	93	Bolivia	226a
83	Luxembourg	343a	94	Nicaragua	193a
84	Guatemala	341a	95	Burma	184a
85	El Salvador	340a	96	Iceland	164a
86	Costa Rica	330a	97	Mongolia	159a
87	Panama	277a	98j	Togo	150a

Crude steel production

Rank in 1976[i]		000 tonnes 1960	1970	1976	Growth %pa 1960–70	1970–76
	World	346 000*	594 000*	674 000*	5.6*	2.1*
1	Soviet Union	65 294	115 889	144 650	5.9	3.7
2	United States	90 067	119 309	116 311	2.9	−0.4
3	Japan	22 138	93 322	107 399	15.5	2.4
4	Germany, West	34 100	45 040	42 413	2.8	−1.0
5	Italy	8 229	17 277	23 450	7.7	5.2
6	France	17 281	23 773	23 235	3.2	−0.4
7	United Kingdom	24 695	28 315	22 274	1.4	−3.9
8	China	18 000**	18 000**	22 000**	0.0**	3.4**
9	Poland	6 681	11 795	15 641	5.8	4.8
10	Czechoslovakia	6 768	11 480	14 693	5.4	4.2
11	Canada	5 270	11 198	13 135	7.8	2.7
12	Belgium	7 188	12 611	12 149	5.8	−0.6
13	Spain	1 919	7 394	10 980	14.4	6.7
14	Rumania	1 806	6 517	10 733	13.7	8.7
15	Brazil	2 260	5 390	9 090	9.1	9.1
16	India	3 200**	6 286	9 144	7.0**	6.4
17	Australiaf	3 600*	6 874	7 760	6.7*	2.0
18	South Africa	2 120**	4 674	7 155	8.2**	7.4
19	Germany, East	3 750	5 053	6 739	3.0	4.9
20	Netherlands	1 942	5 042	5 190	10.0	0.5
21	Sweden	3 218	5 494	5 140	5.5	−1.1
22	Mexico	1 500	3 846	5 124	9.9	4.9
23	Austria	3 163	4 079	4 910	2.6	3.1
24	Luxembourg	4 084	5 462	4 565	2.9	−3.0
25	Hungary	1 887	3 108	3 652	5.1	2.7
26	Korea, North	640**	2 200**	2 800**a	13.1**	4.9**b
27=	Jugoslavia	1 442	2 228	2 698	4.4	3.2
27=	Korea, South	50	481	2 698	25.4	33.3
29	Bulgaria	253	1 800	2 459	21.7	5.3
30	Argentina	277	1 859	2 410	21.0	4.4
31	Finland	254	1 167	1 643	16.5	5.9
32	Turkey	266	1 312	1 456	17.3	1.7
33	Norway	490	869	898	5.9	0.5
34	Venezuela	47	927	882	34.7	−0.8
35	Denmark	317	473	722	4.1	7.3
36	Greece	125	435	612*a	13.3	7.1*b
37	Taiwan	201*	370	597	6.3*	8.3
38	Chile	422	547	450	2.6	−3.2
39	Peru	60	94	443a	4.6	36.3b
40	Switzerland	275*	524*	420*a	6.7*	−4.3*b
41	Portugal	20**	385	391	34.4**	0.2
42	Rhodesia	86	150*	350*a	5.7*	18.5*b
43	Egypt	136	300	348a	8.2	3.0b
44	Cuba	63g	140	298a	12.1h	16.3b
45	Colombia	157	239	255	4.3	1.1
46	Algeria	31	31	181c	0.0	55.4d
47	Israel	40	120*	130*a	11.6*	1.6*b
48	Tunisia	—	100	129a	na	5.2b
49	Ireland	40*	80*	82*a	7.2*	0.5*b
50	Bangladesh	na	108**	76	na	−5.7**
51	Uruguay	10	16	16a	4.8	0.0b
52	Uganda	14	20	8a	3.6	−16.7b

a1975 b1970–75 c1974 d1970–74 e1972 fYears ending June 30th g1963 h1963–70 iExcludes countries for which information later than 1970 is not available jOther (000 t): 99 Afghanistan 147a, 100 Bangladesh 141, 101 Sudan 140*a, 102 Paraguay 138a, 103 Ethiopia 117a, 104 Malawi 104a, 105 Gabon 93a, 106 Uganda 88, 107 Liberia 86c, 108 Fiji 73a, 109 Madagascar 58a, 110 Congo 55a, 111 Cambodia 50*a, 112 Mali 49a, 113 Surinam 35a, 114 Niger 18*a, 115 Cape Verde 4c

World manufacturing

Beer production by main producers

Rank in 1976		000 hl 1976	Rank in 1976		000 hl 1976
	World[n]	787 000*,a	32	Argentina	3 943a
1	United States	188 800*	33	Nigeria	2 968a
2	Germany, West	88 426a	34	Portugal	2 934
3	United Kingdom	65 500	35	Finland	2 706b
4	Soviet Union	57 050a	36	Zambia	2 670c
5	Japan	38 966l	37	Cuba	2 111a
6	France	22 660a	38	Korea, South	1 896
7	Czechoslovakia	22 629	39	Norway	1 869b
8	Brazil	22 238a,h	40	Kenya	1 538a
9	Germany, East	21 202	41	Turkey	1 527a
10	Canada	20 100*	42	Greece	1 286a
11	Mexico	19 684a	43	Angola	1 196c
12	Australia	19 573a,m	44	Cameroon	1 150a
13	Spain	16 620a	45	Taiwan	1 141
14	Belgium	13 797a	46	Mozambique	965a
15	Netherlands	12 430a	47	Ecuador	889b
16	Poland	12 300	48	Chile	833a
17	Denmark	8 881a,d	49	Luxembourg	803a
18	Jugoslavia	8 685	50	Algeria	705c
19	Austria	7 757a	51	Ivory Coast	680a
20	Colombia	7 649c	52	Jamaica	663a
21	Rumania	7 449a	53	Thailand	613a
22	Hungary	6 619a	54	Puerto Rico	594
23	Italy	6 493a	55	Tanzania	589a
24	Venezuela	6 135a	56	India	583a
25	Zaire	5 723b	57	Bolivia	563b
26	South Africa	5 496a	58	Guatemala	526a
27	Ireland	5 050*,b	59	Ghana	500a
28	Bulgaria	4 516a	60	Indonesia	492a
29	Sweden	4 386c	61	Malawi	472a
30	Switzerland	4 325a	62	Singapore	462a
31	New Zealand	4 096	63	Dominican Rep	445a

Radio set production by main producers

Top ten

Rank in 1976		000 units 1960	1970	1976	Growth %pa 1960–70	1970–76
	World[n],[o]	52 000*	85 700*	65 700*,a	5.1*	−5.2*,e
1	Japan	12 851	32 618	16 770	9.7	−10.5
2	Soviet Union	4 165	7 815	8 443	6.5	1.3
3	Taiwan	—	6 248	6 849	na	1.5
4	Korea, South	na	1 088	6 578	na	35.0
5	United States	10 695	8 261	6 100**	−2.5	−4.9**
6	Germany, West	4 313	6 729	4 415a	4.5	−8.1e
7	France	2 214	2 921	3 458	2.8	2.8
8	Poland	627	987	2 038	4.6	12.8
9	Italy	935*	3 300*	1 800*,b	13.4*	−14.1*,f
10	Belgium	1 011	1 943	1 796a	6.8	−1.6e

Other main producers

Rank in 1976		000 units 1976	Rank in 1976		000 units 1976
11	India	1 677	26	Bulgaria	228a
12	Germany, East	1 068a	27	Sweden	194b
13	Mexico	1 030a	28	Czechoslovakia	183a
14	Australia	941b,m	29	Tanzania	177a
15	Indonesia	900c	30	Finland	174b
16	Canada	753	31	New Zealand	158
17	Rumania	712a	32=	Egypt	157b
18	United Kingdom	696a	32=	Morocco	157a,g
19	Brazil	640a	34	Philippines	151a
20	Portugal	610a	35	Denmark	146a
21	Spain	487a	36	Chile	139b
22	South Africa	313k	37	Sri Lanka	117c
23	Iran	281c,d	38	Cuba	113a
24	Turkey	272b	39	Norway	111c
25	Hungary	255a	40	Jugoslavia	109

Cotton fabric production[i] by main producers

Rank in 1976		mn m² 1976	Rank in 1976		mn m² 1976
	World	55 000***	32	Nigeria	276b
1	China	13 000****,k	33	Germany, East	274
2	India	8 034*,a	34	Philippines	205*
3	Soviet Union	7 400	35	Netherlands	195*,a
4	United States	4 158*	36	Greece	182*,b
5	Japan	2 237	37	Switzerland	146*,c
6	Germany, West	1 293*	38	Cuba	117c
7	France	1 196*	39	Austria	112*
8	Mexico	1 096*	40	South Africa	104*,a
9	Poland	948*	41=	Morocco	103*,k
10	Brazil	864*,a	41=	Sudan	103a
11	Hongkong	823	43	Israel	99*,a
12	Italy	820*,a	44	Venezuela	94*,j
13	Taiwan	811*	45	Canada	93a
14	Turkey	730**	46	Peru	90*,p
15	Rumania	677	47	Tanzania	86b
16	Egypt	647	48	Ethiopia	82c
17	Pakistan	608*,a,m	49=	Finland	78*
18	Argentina	595*,a	49=	Madagascar	78
19	Czechoslovakia	563*	51	Iraq	74*,b
20	Iran	482*,j	52	Zaire	69b
21	Thailand	443b	53	Bangladesh	68m
22	Spain	416*	54	Afghanistan	60*
23	Belgium	410*	55	Sweden	57*,b
24	Portugal	396*	56	Australia	55m
25	Jugoslavia	384	57	Chile	53*,c
26	United Kingdom	374*	58	Algeria	46*,b
27	Bulgaria	361*	59	Ghana	43j
28	Korea, South	340	60	Sri Lanka	38*,b
29	Hungary	319	61	El Salvador	37*,b
30	Syria	306*,a	62=	Mozambique	30*,b
31	Colombia	300*,p	62=	Uganda	30*,b

Television set production by main producers

Top ten

Rank in 1976		000 units 1960	1970	1976	Growth %pa 1960–70	1970–76
	World[n],[o]	19 800*	45 400*	50 900*,a	8.7*	2.3*,e
1	Japan	3 578	13 782	17 000*	14.4	3.6*
2	United States	5 828	9 483	7 800**	5.0	−3.2**
3	Soviet Union	1 726	6 682	7 060	14.5	0.9
4	Taiwan	—	1 254	3 850	na	20.6
5	Germany, West	2 164	2 936	3 356a	3.1	2.7e
6	Italy	728*	2 030*	2 330*,b	10.8*	3.5*,f
7	Korea, South	—	114	2 291	na	64.9
8	United Kingdom	2 141	2 214	2 100	0.3	−0.9
9	France	655	1 511	1 774	8.7	2.7
10	Brazil	183*	726	1 451a	14.8*	14.8e

Other main producers

Rank in 1976		000 units 1976	Rank in 1976		000 units 1976
11	Poland	963	26	Finland	248a,q
12	Belgium	579a	27	Iran	242c
13	Turkey	571a	28	Portugal	233a
14	Mexico	569a	29	Chile	223b
15	Rumania	548	30	Greece	196b
16	Australia	533m	31	New Zealand	147a
17	Spain	528a	32	Bulgaria	124a
18	Germany, East	509a	33	Norway	108b
19	Czechoslovakia	456	34	Ireland	106c
20	Canada	440*	35	Philippines	104a
21	Hungary	412	36	Malaysia	102a
22	Austria	404a	37	Venezuela	86j
23	Jugoslavia	402	38	Colombia	71c
24	Sweden	370b	39	Indonesia	70c
25	Argentina	290a	40	Egypt	68b

a1975 b1974 c1973 dSales or shipments e1970–75 f1970–74 gAssembly only hMain establishments only iFor the purpose of ranking in this table, 1 linear metre has been taken as equal to 1 square metre, and the following conversions for weight to square metres have been used: Europe 1 t = 6 500 m², India 1 t = 8 300 m², other 1 t = 8 500 m²; also see *The Economist Measurement Guide and Reckoner*, page 122 j1972 k1970 lYear ending March 31st mYear ending June 30th nExcluding China and Vietnam oExcluding Netherlands p1969 qIncluding sets assembled from imported parts

World manufacturing

Passenger cars

Production

Rank in 1976		000 units			Growth %pa	
		1960	1970	1976	1960–70	1970–76
	World[o]	12 810*	22 540*	29 050*	5.8*	4.3*
1	United States[h]	6 675	6 547	8 498	−0.2	4.4
2	Japan	165	3 179	5 028	34.4	7.9
3	Germany, West	1 817	3 528	3 546	6.8	0.1
4	France	1 136	2 458	3 388	8.0	5.5
5	Italy	596	1 720	1 469	11.2	−2.6
6	United Kingdom	1 353	1 641	1 333	2.0	−3.4
7	Soviet Union	139	344	1 239	9.5	23.8
8	Canada	326	923	1 137	11.0	3.5
9	Spain	42	455	755	26.8	8.8
10	Brazil[e]	57	255	556	16.2	13.9
11	Australia[e,k]	220**	390	370	6.0**	−0.9
12	Sweden	109	272	310*	9.6	2.2*
13	Mexico[e]	25	136	232	18.5	9.3
14	Poland	13	65	216	17.6	22.1
15	Czechoslovakia	56	143	179	9.8	3.8
16	Germany, East	64	127	164	7.0	4.3
17	Argentina[e]	49	169	141	13.2	−3.0
18	Jugoslavia	10*	62	139	20.0*	14.4
19	Netherlands	15	67	72	16.0	1.2
20[l]	Rumania	1	24	71	34.7	20.2

Assembly[a]

Rank in 1976		000 units 1976	Rank in 1976		000 units 1976
1	Belgium	1 047.0	12	Korea, South	26.6
2	South Africa	206.0	13	Morocco	25.0[e]
3	Turkey	74.6[e]	14	Colombia	23.1[e]
4	Iran	73.0[d]	15	Peru	21.2[e]
5	New Zealand	69.5	16	Switzerland	17.0[s]
6	Venezuela	56.0[t]	17	Netherlands	12.3
7	Jugoslavia	48.0[e]	18	Egypt	10.0[d]
8	Malaysia[j]	44.2	19=	Chile	5.0
9	Ireland	42.9	19=	Ivory Coast	5.0[e]
10	Portugal	39.4	21	Zambia	4.0[s]
11	Philippines	33.8	22=[m]	Greece	2.0[d]

Commercial vehicles

Production

Rank in 1976		000 units			Growth %pa	
		1960	1970	1976	1960–70	1970–76
	World[o]	3 690*	6 780*	9 980*	6.3*	6.7*
1	United States[h]	1 194	1 692	2 978	3.5	9.9
2	Japan	595	2 126	2 814	13.6	4.8
3	Soviet Union	385	572	786	4.0	5.4
4	Canada	72	236	503	12.6	13.4
5	France	234	292	471	2.2	8.3
6	Brazil[e]	76	161	416	7.8	17.1
7	United Kingdom	458	458	372	0.0	−3.4
8	Germany, West	239	318	330	2.9	0.6
9	Italy	49	134	120	10.6	−1.8
10	Spain	17	77	110	16.2	6.1
11	Mexico[e]	20	53	100[e]	10.2	13.5[f]
12	Australia[e,k]	70**	88	86	2.3**	−0.4
13=	Czechoslovakia	18	27	75	4.2	18.5
13=	Poland	24	53	75*	8.2	6.0*
15	Sweden	20	32	49[e]	4.7	9.0[f]
16	India	27	41	42	4.0	0.7
17	Rumania	9	37	39	14.8	0.9
18=	Argentina[e]	39	50	36	2.5	−5.3
18=	Germany, East	12	27	36	8.1	4.9
20[n]	Jugoslavia	5	29	21	18.0	−5.3

Assembly[a]

Rank in 1976		000 units 1976	Rank in 1976		000 units 1976
1	South Africa	104.5	13=	Colombia	6.5[d]
2	Belgium	70.6	15	Algeria	6.3[d]
3	Portugal	46.8	16	Morocco	4.5[e]
4	Iran	35.0[d]	17=	Greece	4.2[d]
5	Korea, South	23.0	17=	Ireland	4.2
6	Turkey	22.4	19	Israel	4.1[e]
7	Philippines	16.8	20	Venezuela	3.0[t]
8	Peru	12.8[e]	21	Tunisia	1.7[d]
9	Nigeria	12.2[e]	22	Egypt	1.2[d]
10	New Zealand	11.5	23	Ghana	1.0[t]
11	Netherlands	11.0	24	Zambia	0.8[s]
12	Malaysia[j]	8.1	25	Denmark	0.5[e]
13=	Chile	6.5			

Merchant vessels launched

Rank in 1976		000 gross registered tons			Growth %pa	
		1960	1970	1976	1960–70	1970–76
	World[o, p]	8 356*	21 690*	31 180*	10.0*	6.2*
1	Japan	1 732	10 476	14 524	19.7	5.6
2	Sweden	711	1 711	2 367	9.2	5.6
3	Germany, West	1 092	1 687	1 786	4.4	1.0
4	Spain	161	926	1 624	19.1	9.8
5	United Kingdom	1 331	1 237	1 347	−0.7	1.4
6	France	594	960	1 208	4.9	3.9
7	United States	485	338	1 047	−3.5	20.7
8	Denmark	219	514	948	8.9	10.7
9	Korea, South	—	—	814	na	na
10	Norway	198	639	756	12.4	2.8
11	Italy	434	598	664	3.3	1.8
12	Soviet Union	na	na	616[r]	na	na
13	Jugoslavia	161	393	583	9.3	6.8
14	Netherlands	567	461	578	−2.0	3.8
15	Poland	227	463	518	7.4	1.9
16	Brazil	—	100	407	na	26.4
17	Finland	77	222	389	11.2	9.8
18	Germany, East	na	334	355	na	1.0
19	Portugal	24	16	252	−4.0	58.3
20	Canada	116	33	244	−11.8	40.0
21	Rumania	na	na	217[r]	na	na
22	Belgium	130	155	184	1.8	2.9
23	Singapore	—	6	143	na	69.6
24	Bulgaria	na	na	133	na	na
25	Taiwan	23	91	83	14.7	−1.5
26	Greece	17	73	76	15.7	0.7
27=	Australia	28	54	46	6.8	−2.6
27=	India	13	29	46	8.3	8.0
29[q]	Argentina	3	18	39	19.6	13.8

Dwellings completed[b]

Rank in 1976		000 units 1976	Rank in 1976		000 units 1976
1	Soviet Union	2 200	11	Italy	184
2	Japan	1 719[i]	12	Rumania	165[a]
3	United States	1 549[i]	13	Germany, East	151
4	France	530	14	Jugoslavia	146[e]
5	Germany, West	392	15	Australia	142
6	Spain	370[a]	16	Czechoslovakia	140*
7	United Kingdom	325	17	Greece	129
8	Poland	273	18	Netherlands	107
9	Canada	236	19	Turkey	102*
10	Brazil	200[g]	20	Hungary	98[e]

[a] Assembly wholly or mainly from imported parts; this is not additional to the amount of production indicated, since the vehicles assembled in importing countries have already been counted in the production figures of exporting countries [b] New dwellings intended for habitation by one household; only the top 20 with suitable information are shown, owing to the lack of completeness in world building figures [c] 1975 [d] 1974 [e] Including assembly [f] 1970–75 [g] Permits issued, urban only [h] Factory sales [i] Dwellings started [j] Peninsular Malaysia only [k] Years ending June 30th [l] Rest (000 units, 1976): 21 India 39, 22 Taiwan 29[e], 23 Bulgaria 15[e], 24 Austria 1 [m] Rest (000 units): 22=Israel 2.0[e], 24 Tunisia 1.7[e], 25 Ghana 1.3[t], 26=Angola 1.0[d], 26=Denmark 1.0[e] [n] Rest (000 units, 1976): 21 Hungary 13, 22 Netherlands 11, 23 Austria 8, 24 Bulgaria 6[e] [o] Excluding China [p] Excluding Soviet Union and East Germany [q] Rest (000 grt, 1976): 30 Ireland 29, 31 Turkey 23, 32 Peru 20, 33 South Africa 16, 34 Pakistan 10, 35 Egypt 7, 36 Hongkong 6, 37 Philippines 4, 38 Indonesia 2 [r] Information is not complete [s] 1973 [t] 1972

World manufacturing

Other main products and main producers

Rank	Cotton yarn[m] 000 t, 1976		Wool yarn[m] 000 t, 1976		Man-made fibres 000 t, 1976		Wood pulp 000 t, 1975		Newsprint 000 t, 1976	
	World	9 800*[a]	World[o]	1 700*	World[o]	10 200*	World	102 800*	World	22 200*
1	Soviet Union	1 573[a]	Soviet Union	417[a]	United States	3 380*	United States	41 900	Canada	8 070
2	China	1 500**[g]	Italy	200[r]	Japan	1 621	Canada	14 707	United States	2 846
3	United States	1 480	United Kingdom	190	Soviet Union	955[a]	Japan	8 613	Japan	2 340
4	India	1 028	Japan	159	Germany, West	874	Sweden	8 344	Soviet Union	1 400*
5	Japan	467*	France	146	United Kingdom	618	Soviet Union	8 180*	Sweden	1 136
6	Pakistan	325	Poland	106	Italy	520*	Finland	5 174	Finland	1 010
7	France	245	Belgium	84	France	343	France	1 824	China	950**[a]
8	Poland	219	Germany, East	62[a]	Taiwan	341	Norway	1 734	Germany, West	501
9	Germany, West	208	Germany, West	60	Korea, South	338	China	1 600**	Norway	470*
10	Hongkong	196	Czechoslovakia	54[a]	Germany, East	283[a]	Germany, West	1 531	United Kingdom	321
11	Egypt	193	Rumania	51[a]	Poland	228	Brazil	1 301	France	255*
12	Korea, South	175	Jugoslavia	44	Spain	201	Austria	976	Italy	250
13	Italy	164	United States	40	Mexico	181[a]	Australia	967[i]	New Zealand	219[i]
14	Mexico	158[a]	Mexico	37[a]	Brazil	175*[a]	New Zealand	913[k,l]	South Africa	214
15	Taiwan	147	Bulgaria	33	India	153*[a]	Spain	891	Australia	205
16	Rumania	145[a]	Iran	29[n]	Netherlands	149[n]	Germany, East	864	Korea, South	155
17	Turkey	127[a]	Spain	29	Rumania	148*[a]	South Africa	767[c]	Switzerland	143[a]
18	Czechoslovakia	125	Turkey	26[g]	Czechoslovakia	147	Poland	763	Austria	141
19	Jugoslavia	117	Australia	23[l]	Canada	123[a]	Italy	758	Chile	135
20	United Kingdom	107	Switzerland	16[b]	Austria	101[a]	Rumania	668	Brazil	127
21	Argentina	89[a]	India	13[b]	Jugoslavia	82[a]	Jugoslavia	595	Netherlands	123
22	Portugal	84*	Canada	12[b]	Switzerland	76[a]	Portugal	587	Spain	106

Rank	Synthetic rubber 000 t, 1976		Sulphuric acid 000 t, 1976		Nitric acid 000 t, 1975		Caustic soda 000 t, 1976		Plastics & resins 000 t, 1976	
	World	5 200*[a]	World[o]	100 200*[a]	World	22 100*	World	24 920*[a]	World[o]	45 000**
1	United States	2 313	United States	29 954	United States	6 418	United States	9 214	United States	13 100**
2	Japan	941	Soviet Union	20 014	France	3 287	Germany, West	3 088	Germany, West	6 443
3	France	437	Japan	6 103	Germany, West	3 035	Soviet Union	2 604	Japan	4 954
4	Germany, West	378	Germany, West	4 682	Poland	2 185	Japan	2 589	Soviet Union	3 061
5	United Kingdom	320	France	3 958	Italy	967	France	1 273	Italy	2 650*
6	Netherlands	247	United Kingdom	3 272	Spain	895	Italy	1 134	United Kingdom	2 510
7	Italy	240*	Poland	3 187	Bulgaria	840	Canada	922	France	2 076[a]
8	Canada	210	Italy	2 887	Belgium	794	India	504	Netherlands	1 722
9	Brazil	164	Canada	2 840	Hungary	766	Sweden	470[b]	Spain	847
10	Rumania	147	Spain	2 414	Jugoslavia	635	Germany, East	441	Canada	722[a,c]
11	Germany, East	145	Belgium	1 890	Japan	553	Spain	410	Germany, East	694[c]
12	Poland	117	Mexico	1 805	Canada	436[b]	Poland	389	Belgium	662[a]
13	Belgium	90	India	1 689	Finland	398	Rumania	388	Poland	560
14	Spain	78	Rumania	1 555	Sweden	344[b]	Czechoslovakia	293	Sweden	442[a]
15	Mexico	75	Netherlands	1 462	Mexico	186	Brazil	258	Czechoslovakia	428[a]
16	China	70**	Australia	1 300[i]	Portugal	183	Mexico	217	Australia	407[i]
17	Czechoslovakia	57	Czechoslovakia	1 241	United Kingdom	160[r]	Australia	137[l]	Taiwan	405
18	Argentina	45	Finland	1 016	Greece	159[b]	Argentina	109	Austria	404
19	Australia	42	Germany, East	966	Australia	148[i]	Finland	95	Rumania	283[b]
20	Korea, South	35	Greece	911	Israel	85	Taiwan	92	Brazil	230*[a]
21	South Africa	35	Bulgaria	852	Colombia	65[r]	Hungary	89	Jugoslavia	205
22	India	22	Jugoslavia	848	Taiwan	52	Jugoslavia	89	Mexico	200*[a]

Rank	Fertilisers, nitro[p] 000 t, 1976[i]		Fertilisers, phos[q] 000 t, 1976[i]		Aluminium[f] 000 t, 1976		Copper[d] 000 t, 1976		Tin 000 t, 1976	
	World	43 880*	World	24 870*	World	11 960*	World	8 220*[a]	World[o]	220**
1	United States	9 262	United States	6 655	United States	3 858	United States	1 735	Malaysia	78
2	Soviet Union	8 535[a]	Soviet Union	4 103[a]	Soviet Union	1 600**	Soviet Union	1 400**[a]	Soviet Union	40**
3	China	3 000**	France	1 259	Japan	919	Japan	864	Indonesia	23
4	Japan	1 557	China	1 200**	Canada	878[a]	Zambia	706	Thailand	20
5	Poland	1 548[k]	Poland	928[a]	Germany, West	697	Chile	581	United Kingdom	11
6	India	1 508	Canada	653*	Norway	608	Canada	510	Bolivia	9.5
7	France	1 354	Germany, West	649	France	385	Belgium	480[h]	Brazil	6.6
8	Rumania	1 292[a]	Japan	585	United Kingdom	334	Germany, West	446	Belgium	6.1
9	Germany, West	1 259	Brazil	510*[a]	Netherlands	255	Poland	270	United States	5.7
10	Netherlands	1 153	Australia	474	Australia	232	Zaire	256[b]	Australia	5.6
11	United Kingdom	1 055	Belgium	470*	Italy	213	Spain	170*	Spain	5.4
12	Italy	1 000	United Kingdom	464	India	210	Australia	161	Nigeria	3.7
13	Canada	916[a]	Spain	458[a]	Spain	210	China	150**[a]	South Africa	2.4*
14	Spain	825[a]	Czechoslovakia	425[a]	Rumania	204[a]	United Kingdom	137	Germany, West	1.4
15	Germany, East	776	Germany, East	423	Jugoslavia	197[j]	Jugoslavia	136	Japan	1.1
16	Bulgaria	672[a]	Rumania	404[a]	China	160**[a]	Peru	135	Rhodesia	0.6
17	Belgium	610[a]	South Africa	375*[a]	Ghana	143[a]	South Africa	86[a]	Zaire	0.6
18	Mexico	581*	Italy	369	Greece	133	Mexico	74[a]	Mexico	0.4*
19	Korea, South	541[a]	New Zealand	350*	Brazil	122	Sweden	61	Portugal	0.3
20	Czechoslovakia	525[a]	Turkey	322[a]	Bahrain	116[a]	Bulgaria	52[a]	Argentina	0.1*
21	Hungary	453[a]	India	320	New Zealand	110*[a]	Germany, East	43*[a]		
22	Jugoslavia	357[a]	Jugoslavia	237[a]	Poland	103	France	39		

[a]1975 [b]1974 [c]1975/76 [d]Refined; includes secondary copper refined from scrap [e]Coverage is not complete [f]Primary [g]1970 [h]Includes processing of refined copper imported from Zaire [i]Year ending June 30th [j]Including secondary aluminium [k]1976 [l]Year ending March 31st [m]Mainly pure yarn; including some mixed yarn [n]1972 [o]Excluding China [p]N content of nitrogenous fertilisers [q]Phosphoric acid (P_2O_5) content of phosphate fertilisers. [r]1973 [s]1971

World transport

Roads, length in kilometres, 1975 or 1976

Rank		km	Rank		km	Rank		km	Rank		km	Rank		km	Rank		km
	World	*22 000 000**	40	Denmark	66 515	80	Afghanistan	17 973ª	119=	Korea, North	5 600**ª,ᵇ	160	Antigua	1 000*
1	Utd States	6 176 897	41	Venezuela	65 718ª	81	Taiwan	17 172	121	New		'161	Grenada	970*
2	Brazil	1 489 064	42	Switzerland	62 158	82	Somalia	17 100ª		Caledonia	5 214ª	162	Samoa, W	930*
3	France	1 486 000*ᶜ	43	Vietnam	60 000**	83	Puerto Rico	16 827ª	122	Upper Volta	5 250ª	163=	Isle of Man	800*
4	Soviet		44	Colombia	56 667	84	Paraguay	15 956ª	123	Libya	5 173	163=	St Lucia	800*
	Union	1 403 000	45	Peru	56 416	85	Cambodia	15 029ª	124	Albania	5 000*ª	165	Comoros	750*
5	India	1 232 300ª	46	S-W Africa	54 000*	86	Mali	14 704	125	Luxembourg	4 465	166	Virgin Is, US	730*
6	Japan	1 078 357	47	Iran	52 000	87	Syria	13 575ª	126	Jordan	4 152	167	St Vincent	665
7	Australia	837 866	48	Kenya	50 091	88	Guatemala	13 450*	127	Haiti	4 000*	168=	Gilbert Is	640*
8	Canada	834 152ᵇ	49	Pakistan	49 926	89	Senegal	13 271ª	128	Yemen, Nth	3 952	168=	New Hebrides	640*
9	China	808 000**	50	Uruguay	49 634ª	90	Nicaragua	12 500*	129	Guinea-		170	Fr Polynesia	583
10	Germany, W	469 568	51	Portugal	46 241	91	Iraq	11 859		Bissau	3 570ª	171	Fr Guiana	503
11	Utd Kingdom	368 370	52	Korea, South	45 514	92	Dominican		130=	Bahamas	3 000*	172	UAE	500***ª
12	Poland	300 822	53	Ivory Coast	45 214		Republic	11 844	130=	Nepal	3 000*	173	Tonga	433
13	Italy	291 081	54	Cameroon	43 500*	93	Jamaica	11 700*	132	Burundi	2 987ª	174	Guam	370*
14	Argentina	207 262	55	Thailand	39 721	94	Iceland	11 525ᶠ	133	Fiji	2 960	175	Sao Tome &	
15	Turkey	195 982	56	Mozambique	39 173ª	95	Malawi	11 025	134	Guyana	2 910*		Principe	288ª
16	Mexico	193 390	57	Bolivia	37 075	96	Congo	11 000*	135	Timor, East	2 896ª	176	Liechtenstein	250*
17	South Africa	185 326ᵇ	58	Greece	36 574	97	El Salvador	10 973ª	136	Swaziland	2 750*	177	Bermuda	240*
18	Sweden	179 654	59	Bulgaria	36 091	98	Israel	10 657ª	137	Lesotho	2 736	178	Panama	
19	Czecho-		60	Zambia	34 671ª	99	Yemen, Sth	10 494	138	Singapore	2 218		Canal Zone	232ᵈ
	slovakia	145 455ª	61	Ghana	32 000*	100	Botswana	10 219	139	Belize	2 197	179	Niue	229
20	Spain	145 328	62	Sri Lanka	31 150	101=	Guinea	10 000*	140	Reunion	2 052	180	San Marino	220ª
21	Zaire	145 000	63	Chad	30 725ª	101=	Liberia	10 000*	141	Surinam	2 000*	181	Seychelles	213
22	Germany, E	126 933	64	Uganda	27 536	103	Cyprus	9 838	142	Guadeloupe	1 975ª	182	St Kitts-Nevis	200*
23	Belgium	114 814	65	Madagascar	27 507	104	Oman	9 772	143	Cape Verde	1 946ª	183=	Cayman Is	180*
24	Jugoslavia	112 000*	66	Cuba	27 074ª	105	Mongolia	8 600ª,ᵇ	144	Kuwait	1 920*ª	183=	Montserrat	180*
25	Austria	102 858	67	Saudi Arabia	26 267	106	Togo	7 450	145	Mauritius	1 775	185	Samoa,	
26	Hungary	99 595	68	Egypt	25 976ª	107	Laos	7 395ª	146	Djibouti	1 600**		American	150*
27	Philippines	99 132ª	69	Morocco	25 286ª	108	Lebanon	7 100*	147	Martinique	1 510*	186	St Helena	107
28	Nigeria	97 000*ª	70	Costa Rica	24 700*	109	Panama	7 090*	148	Bhutan	1 500*	187	Turks &	
29	Indonesia	95 544ª	71	Ethiopia	23 000	110	Sierra Leone	7 064	149	Barbados	1 350*		Caicos Is	105
30	New Zealand	95 026	72	Burma	21 956	111	Niger	6 985	150	Gambia	1 270*	188	Andorra	100**
31	Ireland	89 006	73	Central		112	Benin	6 937ª	151	Malta	1 267ª	189=	Norfolk Island	80*
32	Netherlands	86 354		African Emp	21 950	113	Mauritania	6 904ª	152	Brunei	1 260*	189=	Virgin Is, Br	80*
33	Rhodesia	78 930	74	Tunisia	21 595	114	Gabon	6 878	153	Solomon Is	1 220	191	Anguilla	56
34	Algeria	78 408ª	75	Malaysia	21 324	115=	Rwanda	6 500*	154	Dominica	1 208ª	192	Gibraltar	50
35	Norway	78 116	76	Ecuador	21 300*ª	115=	Sahara, W	6 500*ª	155	Equat Guinea	1 180*	193	Monaco	46
36	Rumania	77 949	77	Sudan	19 535ª	115=	Trinidad		156	Neth Antilles	1 150*ᶜ	194	Macao	33ª
37	Chile	75 197	78	Tanzania	19 200*		& Tobago	6 500*	157	Qatar	1 100*	195	Bahrain	30ª,ᶜ
38	Finland	73 763	79	Papua		118	Bangladesh	6 300*	158	Hongkong	1 085	196	Nauru	24**
39	Angola	72 323ª		New Guinea	18 200*	119=	Honduras	5 600*	159	Pac Is, USⁱ	1 035	197	Falkland Is	21

Railways, length in kilometres, 1975 or 1976

Rank		km	Rank		km	Rank		km	Rank		km	Rank		km
	World	*1 410 000**	25	Pakistan	8 810	50	Tanzania	3 545	75=	Costa Rica	1 300*	100	Togo	445
1	Utd States	331 311	26	Hungary	8 243	51	Nigeria	3 524	75=	Uganda	1 300*	101	Lebanon	425*
2	Soviet Union	266 200	27	Turkey	8 138	52	Colombia	3 431	77	Senegal	1 186ª	102	Venezuela	419
3	Canada	70 715	28	Indonesia	7 610	53	Peru	3 400*ª	78	Ecuador	1 151ª	103=	Jamaica	373
4	India	60 301	29	Austria	6 494	54	Rhodesia	3 367	79	Philippines	1 150ª	103=	Nicaragua	373
5	China	48 000**	30	Finland	5 957	55	Angola	3 000*	80	Ethiopia	988*	105	Jordan	371
6	Australia	40 753ᵍ	31	Korea, South	5 653	56	Uruguay	2 987	81	Ghana	953	106	Panama	350*ª
7	Argentina	40 113	32	Zaire	5 257ª	57	Bangladesh	2 874	82	Guatemala	904*	107	Albania	300*ª
8	France	34 297	33	Switzerland	4 969	58	Netherlands	2 832	83=	Guinea	902	108	Luxembourg	275
9	Brazil	33 000	34	Iran	4 944	59	Kenya	2 729	83=	Israel	902	109	Haiti	240*
10	Germany, W	32 006	35	Egypt	4 855	60	Denmark	2 493	85	Madagascar	884	110	Swaziland	225
11	Japan	28 024ª	36	New Zealand	4 797	61	Greece	2 476	86	Ivory Coast	816	111	Surinam	150*
12	Mexico	24 700*	37	Sudan	4 556	62	S-W Africa	2 340	87	Congo	800	112	Guyana	130*
13	Poland	23 766	38	Korea, North	4 500**	63	Malaysia	2 290*	88	Mauritania	652	113	Nepal	106
14	Italy	20 171ª	39	Burma	4 328	64	Ireland	2 006	89	Cambodia	650**	114	Djibouti	100
15	South Africa	20 090	40	Bulgaria	4 290	65	Iraq	1 955	90	Fiji	644ʰ	115	Puerto Rico	96
16	Utd Kingdom	18 354	41	Norway	4 241	66	Zambia	1 930ª	91	Mali	642	116	Sierra Leone	84
17	Spain	15 839	42	Vietnam	4 230*	67	Tunisia	1 928	92	Botswana	634	117	Panama Canal	
18	Cuba	14 872ª	43	Taiwan	4 200	68	Honduras	1 780*	93	Saudi Arabia	612		Zone	76*
19	Germany, E	14 298	44	Mozambique	4 161	69	Morocco	1 756ª	94	El Salvador	602	118	St Kitts-Nevis	58
20	Czechoslovakia	13 215	45	Algeria	4 074ª	70	Dominican Rep	1 700**	95	Benin	579ª	119	Isle of Man	56
21	Sweden	12 070	46	Belgium	3 998	71	Syria	1 577	96	Malawi	566	120	Singapore	38
22	Rumania	11 039	47	Bolivia	3 787	72	Sri Lanka	1 498	97	Upper Volta	517	121	Hongkong	34
23	Jugoslavia	10 068	48	Thailand	3 765	73	Mongolia	1 425ª	98	Paraguay	498	122	Liechtenstein	18
24	Chile	9 960*	49	Portugal	3 563	74	Cameroon	1 320*	99	Liberia	493	123ʲ	Brunei	10*

ªLatest available information ᵇMain only ᶜSurfaced only ᵈPublic only ᵉIncludes rural roads ᶠExcluding urban roads ᵍGovernment only ʰOnly on sugar estates
ⁱIncluding Northern Marianas ʲOther: 124 Nauru 5 km, 125 = Lesotho and Monaco 2 km each

World railway traffic

Passenger[h]

Top 20

Rank in 1976		Passenger-km (mn) 1960	1970	1976	Growth %pa 1960–70	1970–76
	World[h]	860 000*	1 130 000*	1 280 000*	2.8*	2.1*
1	Japan	180 893	288 133	321 100	4.8	1.8
2	Soviet Union	170 800	265 406	312 517[a]	4.5	3.3[b]
3	India[i]	74 190	113 738	134 750	4.4	2.9
4	France	32 000	41 080	51 170	2.5	3.7
5	Poland	30 942	36 891	42 800	1.8	2.5
6	Italy	30 723	32 457	39 630	0.5	3.4
7	Germany, West	39 300**	38 129	38 349	−0.3**	0.1
8	Utd Kingdom[n]	34 676	35 708	36 840[a]	0.3	0.6[b]
9	Rumania	10 737	17 793	23 077	5.2	4.4
10	Germany, East	21 288	17 666	22 339	−1.8	4.0
11	Czechoslovakia	19 335	20 492	17 920	0.6	−2.2
12	Spain	7 341	13 293	16 684	6.1	3.9
13	United States[m]	34 211	17 284	15 680	−6.6	−1.6
14	Argentina	15 684	12 828	14 481	−2.0	2.0
15	Korea, South	4 935	9 819	13 890	7.1	6.0
16	Hungary	11 916	13 916	13 367	1.5	−0.7
17	Pakistan[e]	9 800*	9 566	12 360	−0.2*	4.4
18	Brazil	15 395	12 351	10 649[c]	−2.2	−3.6[d]
19	Jugoslavia	10 261	10 939	9 884	0.6	−1.7
20	Egypt	3 634[e]	6 529[e]	8 671[c]	6.0	6.5[d]

Rest of the world

Rank in 1976		Pass-km (mn) 1976		Rank in 1976		Pass-km (mn) 1976
21	Taiwan	8 480		61	Bolivia	310[a]
22	Netherlands	8 306		62	Luxembourg	294
23	Belgium	8 203		63	Albania	291[r]
24	Switzerland	8 130		64	Israel	280
25	Bulgaria	7 499		65	Peru	270[g]
26	Austria	6 504		66	Cameroon	260*
27	Thailand	5 530		67	Hongkong	251
28	Sweden	5 363		68	Madagascar	248[a]
29	Portugal	4 856[a]		69	Congo	246
30	Turkey	4 660		70	Senegal	220[c]
31	Mexico[o]	4 198[a]		71	Mongolia	213[a]
32	Denmark	3 420[i]		72	Mozambique	210[a]
33	Bangladesh	3 331[c,e]		73	Syria	166
34	Indonesia	3 258		74	Ethiopia	108[a,f]
35	Finland	3 046		75	Mali	100[a]
36	Canada	2 970*		76	Benin	97[a]
37	Burma	2 912[u]		77	Costa Rica	81[c]
38	Sri Lanka	2 900[a,u]		78	Saudi Arabia	72[c]
39	Chile	2 460		79=	Ecuador	65[a]
40	Iran	2 126[c]		79=	Togo	65[c]
41	Norway	1 990*		81	Jamaica	64[g]
42	Greece	1 471		82	Malawi	61
43	Algeria	1 058[c]		83	Cambodia	54[a]
44	Malaysia[w]	1 042[a]		84=	Guinea	42[p]
45	Vietnam	1 023[z]		84=	Venezuela	42[r]
46	Ivory Coast	946[a,q]		86	Nicaragua	28[a]
47	Morocco[o]	863		87	Paraguay	26[a]
48	Nigeria	785[a,i]		88	Jordan	18
49	Philippines	768		89	Guyana	6[c]
50	Ireland	739		90	Lebanon	2[c]
51	Cuba	695[a]				
52	Iraq	645[a,i]				
53	Tunisia	641				
54	New Zealand	589[i]				
55	Colombia	511				
56	Zaire	447[g]				
57	Ghana	431[g]				
58	Angola	418[c]				
59	Uruguay	358[a]				
60	Zambia	320[g]				

Cargo

Top 20

Rank in 1976		Tonne-km (mn) 1960	1970	1976	Growth %pa 1960–70	1970–76
	World	3 338 000*	5 020 000*	6 101 000*	4.2*	3.3*
1	Soviet Union	1 504 400	2 494 721	3 295 400	5.2	4.7
2	United States[m]	835 554	1 116 602	1 146 490	2.9	0.4
3	China	265 260[j]	272 000*	301 000*[r]	0.2*[k]	10.7*[y]
4	Canada	95 548	160 749	199 000*	5.3	3.6*
5	India[i]	69 120	128 304	143 100	6.4	1.8
6	Poland	66 547	99 262	130 956	4.1	4.7
7	Czechoslovakia	44 407	60 995	70 747	3.2	2.5
8	France	56 930	70 403	68 518	2.1	−0.5
9	South Africa[l]	28 400[i]	54 200[i]	68 114	6.7	3.4
10	Rumania	19 821	48 045	67 556	9.2	5.8
11	Germany, West	53 000**	71 287	59 202	3.0**	−3.0
12	Brazil	12 688	17 531	55 220[c]	3.3	33.2[d]
13	Germany, East	32 860	41 513	51 801	2.4	3.8
14	Japan	53 445	62 652	47 851	1.6	−4.4
15	Mexico[o]	14 004	22 863	32 542[a]	5.0	7.3[b]
16	Australia	13 090	25 403	31 509	6.9	3.7
17	Hungary	13 147	19 143	22 553	3.8	2.8
18	Jugoslavia	13 895	19 253	21 006	3.3	1.5
19	Utd Kingdom[n]	30 500*	26 807	20 448	−1.3*	−4.4
20	Bulgaria	6 981	13 858	17 055	7.1	3.5

Rest of the world

Rank in 1976		Tonne-km (mn) 1976		Rank in 1976		Tonne-km (mn) 1976
21	Italy	16 673		61	Malaysia[w]	829[a]
22	Sweden	15 458		62	Vietnam	809[z]
23	Argentina	11 039		63	Peru	735[g]
24	Spain	10 767		64	Indonesia	718
25	Korea, North	10 600[v]		65	Bangladesh	639[c,e]
26	Austria	10 540		66	Luxembourg	626
27	Korea, South	9 486		67	Cameroon	530*
28	Pakistan	8 300[e]		68	Ireland	523
29	Turkey	7 288		69	Congo	508
30	Mauritania	6 808[g]		70	Bolivia	465[a]
31	Belgium	6 638		71	Israel	449
32	Finland	6 546		72	Ivory Coast	443[a,q]
33	Rhodesia[l]	6 141[a,e]		73	Burma	382[r]
34	Switzerland	5 659		74	Guinea	341[r]
35	Angola	5 461[c]		75=	Ghana	305[g]
36	Iran	4 917[c]		75=	Syria	305
37	Liberia	4 396[x]		77	Sri Lanka	282[u]
38	New Zealand	3 648[i]		78	Uruguay	281[a]
39	Morocco[o]	3 131		79	Ethiopia	244[a,f]
40	Zaire	3 017[g]		80	Malawi	228
41	Taiwan	2 886		81	Madagascar	207[a]
42	Norway	2 774*		82	Albania	188[r]
43	Egypt	2 767[c]		83	Senegal	182[c]
44	Mongolia	2 701		84=	Jamaica	156[g]
45	Netherlands	2 695		84=	Mali	156[a]
46	Thailand	2 630		86=	Benin	127[a]
47	Sudan	2 288[g]		86=	Guatemala	127[a]
48	Mozambique	2 180[c]		88	Saudi Arabia	66[c]
49	Chile	2 160		89=	Ecuador	46[a]
50	Kenya	2 120[a]		89=	Hongkong	46
51	Algeria	1 901[c]		91	Lebanon	42[c]
52	Iraq	1 871[a,i]		92	Philippines	41
53	Cuba	1 825[a]		93	Paraguay	30[g]
54	Denmark	1 800[i]		94	Togo	22[c]
55	Tunisia	1 277		95	Venezuela	15[r]
56	Colombia	1 247		96=	Costa Rica	14[c]
57	Nigeria	972[a,i]		96=	Nicaragua	14[g]
58	Zambia	897[g]		98	Cambodia	10[g]
59	Portugal	856[a]		99	Jordan	6
60	Greece	845				

[a]1975 [b]1970–75 [c]1974 [d]1970–74 [e]Year(s) ending June 30th [f]Including traffic of Djibouti portion of Addis Ababa-Djibouti line; excluding Eritrea [g]1973 [h]Excludes, among other countries, Australia, China, North Korea and South Africa, for which information is not available [i]Year(s) ending March 31st [j]1959 [k]1959–70 [l]Including traffic in South-West Africa [m]Class I only [n]Excluding Northern Ireland [o]Main railways only [p]1967 [q]Including traffic in Upper Volta on the Abidjan-Ouagadougou line [r]1971 [s]1972 [t]Including traffic in Botswana [u]Year ending September 30th [v]1964 [w]Including traffic in Singapore [x]1970 [y]1970–71 [z]1960

World air traffic[a]

Passenger

Top 20

Rank in 1976		Passenger-km (mn) 1960	1970	1976	Growth %pa 1960–70	1970–76
	World[c]	*121 000*	*460 000*	*763 000*	*14.3*	*8.8*
1	United States	62 542	210 327	288 027	12.9	5.4
2	Soviet Union	12 111	78 226	130 529	20.5	8.9
3	Japan	1 051	14 954	32 334	30.4	13.7
4	United Kingdom	6 372	17 432	30 948	10.6	10.0
5	Canada	4 267	15 397	26 031	13.7	9.1
6	France	5 229	13 587	25 192	10.0	10.8
7	Australia	3 008	9 268	19 384	11.9	13.1
8	Germany, West	1 284	8 255	14 982	20.5	10.4
9	Spain	782	5 874	11 130	22.3	11.2
10	Italy	1 339	8 395	10 780	20.0	4.3
11	Netherlands	2 672	5 769	10 613	8.0	10.7
12	Brazil	2 679	4 385	10 366	5.1	15.4
13	Switzerland	1 138	4 420	8 493	14.5	11.5
14	Mexico	1 309	2 939	7 833	8.4	17.7
15	India	1 115	3 555	7 196	12.3	12.5
16	Singapore	190[h,j]	664[j]	6 362	36.7[i]	45.7
17	South Africa	513	2 872	6 050	18.8	13.2
18	Thailand	63	783	4 662	28.6	34.6
19	Greece	289	2 126	4 623	22.1	13.8
20	Korea, South	22	445	4 519	34.9	47.2

Rest of the world[d]

Rank in 1976		Pass-km (mn) 1976	Rank in 1976		Pass-km (mn) 1976
21	Israel	4 368	61	Rumania	769
22	New Zealand	4 324	62	Syria	712
23	Argentina	4 222	63	Libya	700
24	Sweden[k]	4 041	64	Zaire	690
25	Belgium	3 893	65	Cuba	663
26	Pakistan	3 410	66	Ethiopia	523
27	Norway	3 180	67	Hungary	510
28	Saudi Arabia	3 122	68	Bulgaria	500
29	Indonesia	3 112	69	Dominican Rep	487
30	Iran	3 059	70	Nigeria	470
31	Philippines	3 050	71	Bolivia	444
32	Colombia	2 976	72	Panama	437
33	Taiwan	2 908	73	Bangladesh	426
34	Portugal	2 841	74	Bahamas	402
35	Hongkong	2 632[f]	75	Zambia	356
36	Denmark[k]	2 603	76	Sudan	345
37	Venezuela	2 538	77	Malta	340
38	Jugoslavia	2 150	78	Costa Rica	326
39	Turkey	2 019	79 =	Bahrain[g]	320
40	Iceland	1 914	79 =	Oman[g]	320
41	Algeria	1 881	79 =	Qatar[g]	320
42	Malaysia	1 814	79 =	Utd Arab Emirates[g]	320
43	Lebanon	1 800	83	Ecuador	318
44	Egypt	1 739	84	Mozambique	312
45	Ireland	1 528	85	Sri Lanka	305
46	Germany, East	1 448	86	Cyprus	304
47	Poland	1 425	87	Madagascar	276
48	Finland	1 380	88	Afghanistan	258
49	Jamaica	1 379	89	Honduras	257
50	Peru	1 367	90	Barbados	251
51	Czechoslovakia	1 364	91	Yemen, North	237
52	Morocco	1 264	92	Cameroon	225
53	Chile	1 228	93	Papua New Guinea	220
54	Kuwait	1 135	94	Uganda[m]	197
55	Trinidad & Tobago	1 040	95 =	El Salvador	195
56	Tunisia	968	95 =	Tanzania[m]	195
57	Kenya[m]	879	97	Ghana	194
58	Iraq	863	98	Gabon[l]	190*
59	Austria	823	99	Burma	168
60	Jordan	805	100	Luxembourg	165

Cargo[b]

Top 20

Rank in 1976		Tonne-km, net (mn) 1960	1970	1976	Growth %pa 1960–70	1970–76
	World[c]	*3 350*	*15 230*	*24 470*	*16.3*	*8.2*
1	United States	1 454.9	7 440.0	9 083.8	17.7	3.4
2	Soviet Union	563.0	1 876.7	2 697.8	12.8	6.2
3	France	152.3	542.5	1 379.6	13.5	16.9
4	Japan	24.1	436.0	1 102.9	33.6	16.7
5	Germany, West	39.5	523.3	1 098.5	29.5	13.2
6	United Kingdom	180.7	604.5	914.0	12.9	7.1
7	Canada	63.7	424.9	681.6	21.0	8.2
8	Netherlands	117.0	393.0	663.4	13.3	9.1
9	Lebanon	18.5	145.2	525.2	22.9	23.9
10	Brazil	89.7	173.4	488.0	6.8	18.8
11	Italy	26.5	294.6	468.2	27.2	8.0
12	Australia	105.0	267.2	399.1	9.8	6.9
13	Korea, South	0.17	6.0	355.1	42.7	97.4
14	Switzerland	35.6	187.6	345.7	18.1	10.7
15	Belgium	39.6	192.1	325.8	17.1	9.2
16	Spain	8.4	107.2	289.7	29.0	18.0
17	India	46.5	115.2	279.3	9.5	15.9
18	Singapore	2.0[h,j]	10.8[j]	193.4	52.4[i]	61.7
19	Taiwan	2.0*	24.6	181.6	28.5*	39.5
20	Sweden[k]	26.9	111.0	172.5	15.2	7.6

Rest of the world[e]

Rank in 1976		Tonne-km, net (mn) 1976	Rank in 1976		Tonne-km, net (mn) 1976
21	South Africa	160.7	59 =	Mauritania[l]	14.4
22	Colombia	154.2	62 =	Congo[l]	14.2
23	Pakistan	148.3	62 =	Cuba	14.2
24	New Zealand	143.7	62 =	Poland	14.2
25	Israel	138.6	65 =	El Salvador	14.0
26	Thailand	126.8	65 =	Ivory Coast[l]	14.0
27	Norway[k]	118.9	67 =	Cent African Emp[l]	13.9
28	Denmark[k]	112.9	67 =	Niger[l]	13.9
29	Philippines	111.9	67 =	Senegal[l]	13.9
30	Argentina	104.0	70 =	Benin[l]	13.8
31	Saudi Arabia	85.2	70 =	Togo[l]	13.8
32	Mexico	83.4	70 =	Upper Volta[l]	13.8
33	Hongkong	78.0[f]	73	Costa Rica	13.5
34	Ireland	77.7	74	Afghanistan	13.0
35	Venezuela	76.8	75	Jamaica	11.5
36	Chile	75.7	76	Rumania	10.6
37	Iran	74.8	77	Austria	9.8
38	Portugal	73.5	78	Bangladesh	9.7
39	Greece	57.9	79	Algeria	8.8
40	Zaire	54.8	80	Cameroon	8.7
41	Germany, East	50.5	81	Nigeria	8.4
42	Indonesia	47.6	82	Uganda[m]	8.2
43	Malaysia	36.5	83	Madagascar	7.8
44	Finland	33.1	84 =	Bulgaria	7.6
45	Iceland	30.6	84 =	Syria	7.6
46	Kuwait	27.1	86 =	Sudan	7.5
47	Peru	25.9	86 =	Tunisia	7.5
48	Trinidad & Tobago	24.2	88	Ecuador	7.3
49	Egypt	21.9	89	Guatemala	7.1
50	Jugoslavia	20.6	90 =	Bahrain[g]	7.0
51	Ethiopia	20.1	90 =	Oman[g]	7.0
52	Zambia	19.8	90 =	Qatar[g]	7.0
53	Kenya[m]	19.7	90 =	Utd Arab Emirates[g]	7.0
54	Morocco	18.2	94	Libya	6.9
55	Czechoslovakia	17.4	95	Cyprus	6.7
56	Turkey	17.3	96	Mozambique	5.8
57	Iraq	15.7	97	Papua New Guinea	5.5
58	Chad[l]	14.8	98	Hungary	5.2
59 =	Gabon[l]	14.4	99	Panama	5.0
59 =	Jordan	14.4	100	Malawi	4.7

[a]Revenue scheduled traffic (traffic on flights according to a published timetable) of airlines registered in each country. Includes both domestic and international services
[b]Includes goods, livestock and mail   [c]Icao members only; excluding China, North Korea and Vietnam   [d]Countries with over 160 mn passenger-km traffic only   [e]Countries with over 4.5 mn tonne-km traffic only   [f]1975   [g]Apportionment of Gulf Air   [h]1966   [i]1966–70   [j]Apportionment of Malaysia-Singapore Airlines Ltd   [k]Including apportionment of SAS   [l]Including apportionment of Air Afrique (for Gabon until December 2, 1976)   [m]Including apportionment of East African Airways Corp (which was disbanded in 1977)

World sea traffic

Goods loaded[a,j]

Rank in 1976		Tonnes mn, 1976	Rank in 1976		Tonnes mn, 1976
	World	3 350*	71=	Germany, East	3.2
1	Saudi Arabia	390*c	71=	Ivory Coast	3.2b
2	United States	258	73	Guyana	2.9*c
3	Iran	191c	74=	Albania	2.7*c
4	Venezuela	190d	74=	Bulgaria	2.7b
5	Australia	159n	74=	Congo	2.7
6	Soviet Union	120b	77=	Dominican Rep	2.6d
7	Canada	114	77=	Senegal	2.6c
8	Kuwait	107b	77=	Togo	2.6c
9	Brazil	89.7	80	Pakistan	2.4n
10	Nigeria	87.6m	81=	Colombia	2.3
11	Indonesia	83.7	81=	Ghana	2.3d
12	Utd Arab Emirates	82.9*c	81=	Yemen, South	2.3c
13	Netherlands	82.5	84	Nauru	2.2*m
14	Japan	76.5	85	Sierra Leone	2.1b
15	Libya	73.0b	86=	Cyprus	1.8
16	United Kingdom	55.6	86=	Kenya	1.8b
17	Algeria	44.8c	88=	Christmas Island	1.7c
18	Netherlands Antilles	37.3*c	88=	Panama	1.7b
19	Poland	35.8	90	Jordan	1.6
20	Norway	34.2	91	Costa Rica	1.4
21	Belgium	32.9	92=	Guinea	1.3*l
22	Sweden	32.4	92=	Honduras	1.3b
23	Italy	30.9	94	Papua New Guinea	1.2d,n
24	India	30.7b	95	Sri Lanka	1.1
25	France	30.5c	96=	Korea, North	1.0**l
26	Germany, West	29.0	96=	Sudan	1.0b
27=	Iraq	26.4c	98	Madagascar	0.97
27=	South Africa	26.4	99	Uruguay	0.95c
29	Lebanon	25.8c	100	Vietnam	0.91*c
30	Qatar	25.4*c	101	Mauritius	0.90d
31	Liberia	25.2c	102=	Cameroon	0.84
32	Spain	23.8	102=	Tanzania	0.84b
33	Singapore	20.6	104	Guatemala	0.81*c
34	Trinidad and Tobago	20.3	105	Bangladesh	0.76c,n
35	Taiwan	19.9	106	Burma	0.71
36	Syria	17.0	107	Nicaragua	0.70c
37	Malaysia	16.5b	108	Haiti	0.63b
38	Brunei	16.2b	109=	Sahara, Western	0.62c
39	Morocco	15.4b	109=	Virgin Islands, US	0.62*c,l
40	Argentina	15.3	111	El Salvador	0.60
41=	Korea, South	14.3	112	Zaire	0.56c
41=	Mexico	14.3	113	Ethiopia	0.54b
41=	Oman	14.3*c	114	Gilbert Islands	0.53b,h
44	Greece	13.2	115	Somalia	0.47d
45	Thailand	12.7	116	Fiji	0.44
46	Finland	12.0	117	Iceland	0.42
47	Philippines	11.2	118	Martinique	0.40c
48	Ecuador	10.4b	119	Equatorial Guinea	0.31*c
49	Angola	10.0c	120	Reunion	0.30*c
50	Chile	9.9d	121	Guadeloupe	0.25d
51=	Bahrain	9.0b	122=	Macao	0.15c
51=	Mozambique	9.0b	122=	Solomon Islands	0.15d
51=	Peru	9.0b	124	Djibouti	0.14d
54	Mauritania	8.7m	125=	Benin	0.13c
55	New Zealand	8.2	125=	Malta	0.13
56	Jamaica	7.6	127=	Belize	0.11*l
57	Denmark	7.2	127=	Gambia	0.11b
58	Rumania	7.0*c	129=	Faroes	0.10*c
59	Gabon	6.6c	129=	Guam	0.10b,n
60	Hongkong	6.0	129=	New Hebrides	0.10c
61=	Cuba	5.9b	132	Greenland	0.080*c
61=	Egypt	5.9	133	Samoa, American	0.065c
63	Tunisia	5.5*b	134	Cambodia	0.050d
64	Surinam	4.9*c	135	Guinea-Bissau	0.040d
65	Portugal	4.3	136	Isle of Man	0.039*
66	Jugoslavia	4.0	137	St Vincent	0.037c
67	Turkey	3.8b	138	French Polynesia	0.034d
68	Israel	3.7	139	St Kitts-Nevis	0.027c,g
69	New Caledonia	3.6c	140	S Tome & Principe	0.026d
70	Ireland	3.4*c	141	St Lucia	0.025b

Goods unloaded[a,k]

Rank in 1976		Tonnes mn, 1976	Rank in 1976		Tonnes mn, 1976
	World	3 350*	71=	Tanzania	3.0b
1	Japan	576	72	Ghana	2.7d
2	United States	448	73=	Jamaica	2.6
3	Netherlands	256	73=	Uruguay	2.6b
4	Italy	217	75	Kuwait	2.5b
5	France	191c	76=	Colombia	2.2
6	United Kingdom	181	76=	Sudan	2.2b
7	Germany, West	111	78	Senegal	2.0c
8	Spain	84.2	79	Cyprus	1.7
9	Brazil	61.5	80	El Salvador	1.6
10	Belgium	57.5	81	Iraq	1.5d
11	Canada	56.5	82=	Cameroon	1.4
12	Sweden	55.1	82=	Jordan	1.4
13	Netherlands Antilles	45.9*c	82=	Sri Lanka	1.4
14	Korea, South	41.5	82=	Surinam	1.4*c
15	Taiwan	41.0	86=	Guatemala	1.3*c
16	Singapore	38.3	86=	New Caledonia	1.3c
17	Soviet Union	35.5b	86=	Nicaragua	1.3c
18	India	31.5b	89=	Costa Rica	1.2
19	Denmark	30.9	89=	Honduras	1.2b
20	Australia	26.9n	91=	Guyana	1.1*c
21	Greece	25.9	91=	Iceland	1.1
22	Virgin Islands, US	25.6*c,l	91=	Madagascar	1.1
23	Poland	23.6	91=	Papua New Guinea	1.1d,n
24	Finland	23.3	91=	Zaire	1.1c
25	Norway	22.0	96=	Malta	1.0
26	Bulgaria	20.0b	96=	United Arab Emirates	1.0*c
27	Turkey	17.7b	98	Oman	0.88c
28	Hongkong	17.4	99	Ethiopia	0.84b
29	Philippines	16.4	100	Reunion	0.81*c
30	Jugoslavia	16.2	101=	Djibouti	0.73d
31	Portugal	14.6	101=	Mauritius	0.73d
32	Iran	13.6c	103=	Congo	0.72
33	Cuba	13.3b	103=	Fiji	0.72
34	Thailand	13.1	105=	Albania	0.70*c
35	Trinidad & Tobago	13.0	105=	Bahrain	0.70b
36	Indonesia	12.1	107	Benin	0.61c
37	Rumania	12.0*c	108=	Cambodia	0.58d
38	Germany, East	11.6	108=	Qatar	0.58c
39=	Ireland	10.9*c	108=	Yemen, North	0.58c
39=	Malaysia	10.9b	111	Guinea	0.55*l
41	New Zealand	10.0	112	Guam	0.48b,n
42	Libya	9.6b	113	Haiti	0.46b
43=	Algeria	9.4c	114=	Brunei	0.44b
43=	Egypt	9.4	114=	Burma	0.44
45	Argentina	9.2	114=	French Polynesia	0.44d
46	South Africa	8.6f	114=	Somalia	0.44d
47	Syria	7.8	118	Korea, North	0.40**l
48	Morocco	7.3b	119	Sahara, Western	0.39c
49=	Mexico	7.2	120	Sierra Leone	0.38c
49=	Pakistan	7.2n	121	Martinique	0.36c
51	Chile	6.2m	122=	Gabon	0.35c
52	Venezuela	5.5d	123=	Gibraltar	0.34b
53	Bangladesh	5.4c,n	123=	Togo	0.34c
54	Israel	5.1	125	Samoa, American	0.32c
55=	Nigeria	4.7m	126	Faroes	0.31c
55=	Panama	4.7b	127=	Isle of Man	0.30*
57=	Bahamas	4.4e	127=	Macao	0.30c
57=	Lebanon	4.4c	129	Greenland	0.28*c
59	Kenya	4.2b	130	Guadeloupe	0.18d
60	Vietnam	4.1*c	131	Guinea-Bissau	0.17d
61=	Angola	4.0c	132	Belize	0.16*l
61=	Saudi Arabia	4.0*	133	Mauritania	0.15d
63	Tunisia	3.9b	134=	Cape Verde	0.14b
64=	Peru	3.8*c	134=	New Hebrides	0.14c
64=	Yemen, South	3.8c	136=	Equatorial Guinea	0.13*c
66	Ivory Coast	3.5c	136=	Gambia	0.13b
67	Ecuador	3.4c	138=	French Guiana	0.12
68	Mozambique	3.3b	138=	St Lucia	0.12b
69=	Dominican Rep	3.2d	140=	Bermuda	0.10b
69=	Liberia	3.2c	140=	Seychelles	0.10b

[a]In external trade; excluding China [b]1975 [c]1974 [d]1973 [e]1970 [f]Excluding crude oil (12.6* mn t in 1974) [g]Including Anguilla [h]Including Tuvalu [i]Excluding traffic with the United States [j]Countries loading less than 0.025 mn t in 1975 or 1976 are not specified [k]Countries unloading less than 0.10 mn t in 1975 or 1976 are not specified [l]1971 [m]1972 [n]Year ending June 30th [o]Year beginning March 31st

World tourism

Number of visitors[a], 1976

Rank	000	Rank	000	Rank	000	Rank	000	Rank	000	Rank	000
	World	270 000**k	31	Lebanon	1 555b	62	South Africa	460f	93	Senegal	136	123=	Ethiopia	31b
1	Spain	30 014j	32	Singapore	1 321	63	Bermuda	450	94	Sri Lanka	127	123=	Sudan	31b
2	United States	17 523	33	Puerto Rico	1 299u	64	Venezuela	426c	95	Gibraltar	125	126	Guadeloupe	30**e
3	Germany, East	17 313r	34	Malaysia	1 225	65	Guatemala	408	96	Martinique	120	127	Malawi	28c
4	Denmark	16 232	35	Argentina	1 200b	66	Kenya	407b	97	French Polynesia	117	128	Grenada	25
5	Czechoslovakia	14 078b	36	Norway	1 191h	67	New Zealand	384	98	Swaziland	110c	129=	Benin	19
6	Italy	13 930i	37	Jersey	1 120*	68	Botswana	353r	99	Ivory Coast	109b	129=	Mali	19h
7	France	13 470w	38	Morocco	1 108	69	Malta	340	100	Nepal	105	129=	Norfolk Island	19*
8	Canada	13 002	39	Thailand	1 098	70	Indonesia	313c	101	Paraguay	95d	129=	Zaire	19
9	Austria	11 598i	40	Taiwan	1 008	71	Costa Rica	300*	102	Afghanistan	91	133=	Burma	18
10	United Kingdom	10 089	41	Tunisia	978	72	Guernsey	280*c	103	Haiti	86	133=	New Hebrides	18
11	Germany, West	7 890j	42	Bahamas	940	73=	Chile	279	104	Liechtenstein	74	133=	St Vincent	18
12	Andorra	6 450	43	Egypt	913	73=	Panama	279	105	Mauritius	73c	133=	Yemen, South	18
13	Switzerland	5 879h	44	Korea, South	834	75	El Salvador	278	106=	Iceland	70	137	Dominica	16
14	Jugoslavia	5 572	45	Japan	795	76	Neth Antilles	276	106=	Tonga	70*	138=	Nigeria	13d
15	Hungary	5 551	46	Syria	793i	77=	Finland	264g	108=	Surinam	65*	138=	Timor, East	13e
16	Poland	4 428	47	Israel	733	77=	Peru	264	108=	Virgin Islands, Br	65b	140	St Kitts-Nevis	12
17	Bulgaria	4 033	48	Jordan	708b	79	Dominican Rep	260	110	Antigua	63	141=	Montserrat	11
18	Soviet Union	3 879	49	Iraq	630	80	Barbados	224	111	Bangladesh	61	141=	St Pierre & Miquelon	11e
19	Netherlands	3 848i	50	Iran	628	81	Cyprus	215	112	Togo	60i	141=	Upper Volta	11c
20	Greece	3 845	51	Kuwait	618p	82	Tanzania	210*b	113=	Ghana	56	144=	Madagascar	10*
21	Belgium	3 400**v	52	Philippines	615	83	Nicaragua	207	113=	Zambia	56	144=	Mauritania	10e
22	Mexico	3 218b	53	Brazil	556	84	Guam	205	115	Cayman Islands	54b	144=	Uganda	10c
23	Rumania	3 169	54	India	545	85	Pakistan	199	116	Gabon	52d	147=	Maldives	9
24	Macao	2 500*	55	Virgin Is, US	539	86	Algeria	185	117	Cuba	50*b	147=	Turks & Caicos Is	9b
25	San Marino	2 435	56	Australia	532	87	Monaco	181	118	Seychelles	47*	149	Sierra Leone	8c
26	Portugal	2 175	57	Colombia	522	88	Ecuador	172	119	St Lucia	46d	150=	Anguilla	2*
27	Sweden	1 781i	58=	Isle of Man	497	89	Fiji	169	120=	New Caledonia	35	150=	Bhutan	2*
28	Ireland	1 690	58=	Luxembourg	497	90	Trinidad & Tobago	157	120=	Samoa, American	35c,u	150=	Falkland Is	2
29	Turkey	1 676	60	Uruguay	492	91	Libya	145	122	Samoa, Western	34			
30	Hongkong	1 560	61	Jamaica	471	92	Honduras	140e	123=	Burundi	31			

Nationality of main non-resident visitors to top ten tourist countries, 000, 1976

Visitors to:	Spain[j]	United States	Germany, East[r]	Denmark	Czecho-slovakia[b]	Italy[i]	France[w]	Canada	Austria[i]	United Kingdom
Total visitors	30 014	17 523	17 313	16 232	14 078	13 930	13 470	13 002	11 598	10 089
from:										
Australia	89	168	na	62	na	213	85	48	72t	352
Austria	130	na	na	81	215	817	185	9	na	84
Belgium	901	45	na	95n	na	447	1 640	18	310n	683n
Brazil	51	105	na	na	na	na	55	9	na	na
Canada	140	11 164	na	92	na	173	170	na	67	477
Czechoslovakia	na	na	1 719	na	na	na	na	na	na	na
Denmark	409	na	na	na	17	171	610o	13	148	193
Finland	170	na	na	90	15	67	nas	9	19	47
France	9 476	217	na	148	32	1 678	na	92	385	1 171
Germany, East	na	na	na	na	6 006	na	na	na	na	na
Germany, West	3 885	366	3 069i	12 350	328	3 595	2 344	134	7 369	1 104
Greece	36	33	na	19	na	114	na	15	38	61
Hungary	na	na	288	na	3 625	na	na	na	na	na
Ireland	109	32	na	12	na	37	na	9	3	721
Italy	357	131	na	96	59	na	1 130	48	189	281
Japan	65	772	na	151	na	382	280	84	48	119
Jugoslavia	na	na	na	36	124	284	na	11	104	34
Mexico	50	1 921	na	na	na	na	45	30	na	na
Netherlands	1 045	95	na	272	19	471	1 320	65	857	832
Norway	159	39	na	208	na	57	nas	9	21	217
Poland	na	na	6 869	na	2 948	na	na	8	na	na
Portugal	4 930	na	na	14	na	39	60	15	6	47
Soviet Union	66	na	235	na	211	na	na	na	na	na
Spain	1 631m	57	na	51	na	205	615	11	46	254
Sweden	429	75	na	1 428	32	173	nas	19	217	311
Switzerland	542	91	na	102	20	751	620	28	336	249
United Kingdom	2 982	538	na	325	19	873	1 600	369	384	na
United States	793	na	na	379	38	1 845	1 010	11 642	514	1 490
Other[q]	1 569	1 674	5 106	221	370	1 538	1 701	307	465	1 362

[a]Generally arrivals at frontiers; excluding excursionists and cruise passengers [b]1975 [c]1974 [d]1973 [e]1972 [f]1971 [g]Excluding visitors from other Scandinavian countries (which include Denmark, Finland, Iceland, Norway and Sweden) [h]Arrivals at hotels and similar establishments [i]Arrivals at all forms of accommodation [j]Includes 12 014 thousand excursionists and 1 479 thousand cruise passengers [k]Includes some excursionists [l]West Berlin only [m]Own nationals resident abroad [n]Including from Luxembourg [o]Scandinavian total (includes also Finland, Iceland, Norway and Sweden) [p]1970 [q]Including for countries specified where figures are not available [r]Including excursionists [s]See Denmark for Scandinavian total [t]Including from New Zealand [u]Year ending June 30 [v]Estimate based on number of accommodation nights [w]Tourists making at least one over-night stay; estimates

World finance

Inflation (growth in consumer prices) and growth in money stock, %pa, 1970–76

Rank (consumer prices)		Growth %pa Consumer prices	Money stock
	World	*10.2****d**	*13.2****d**
1	Chile	208.3	204.3
2	Argentina	100.6	107.7
3	Uruguay	65.9	60.9ᵍ
4	Laos	35.5ᵃ	na
5	Zaire	27.9	23.5
6	Uganda	27.7	na
7	Yemen, North	27.1ᶜ	na
8	Iceland	26.3	28.3
9	Bangladesh	26.0ᶜ	na
10	Israel	25.3	25.9
11	Brazil	24.3	38.5
12	Ghana	22.8	29.3
13	Indonesia	19.6	36.3
14	Burma	19.1	na
15	Colombia	18.8	23.5
16	Turkey	18.4	28.1
17	Saudi Arabia	18.2	47.0
18	Jugoslavia	17.9	35.0
19	Bolivia	17.8	27.2
20	Barbados	16.2	13.2
21	Portugal	16.1	18.0
22	Peru	15.8	25.0
23	Korea, South	15.1	30.9
24	Nigeria	15.0	34.2
25	St Lucia	14.9	na
26	Pakistan	14.8	19.8**
27	Rwanda	14.1ᵃ	22.3
28	Ireland	14.0	12.4
29	Philippines	13.7	18.7
30=	Dominica	13.6	na
30=	United Kingdom	13.6	11.5*
32=	Bahrain	13.3	22.5
32=	Gibraltar	13.3	na
34	Sudan	13.2	18.7
35	Jamaica	13.1	17.3
36=	Ecuador	13.0	26.7
36=	Spain	13.0	21.6
38=	Mexico	12.7	19.7
38=	Trinidad & Tobago	12.7	25.4
40	Greece	12.5	19.6
41	Mauritius	12.4	29.8
42	Tanzania	12.3	20.9
43=	Haiti	12.2	18.8
43=	Italy	12.2	16.9
45	Finland	12.1	18.6
46	Costa Rica	11.9	23:4
47	Syria	11.8	24.2
48	Gambia	11.6	30.9
49	Jordan	11.5	16.5
50	Kenya	11.4	15.3
51	New Zealand	11.3	13.4
52	Japan	11.1	17.5
53	Papua New Guinea	11.0ᵇ	na
54	Fiji	10.9	13.7
55	Australia	10.8	11.8
56=	Guadeloupe	10.7ᵃ	na
56=	Martinique	10.7	na
56=	Yemen, South	10.7	18.8
59	Taiwan	10.5	24.8
60=	Dominican Republic	10.4	14.6
60=	Gabon	10.4ᵃ	42.0
62=	Mauritania	10.3	22.1
62=	Netherlands Antilles	10.3ᵇ	na
62=	Paraguay	10.3	19.8
62=	Reunion	10.3	na
62=	Samoa, Western	10.3	15.0*
67=	Liberia	10.1	na
67=	Niger	10.1	18.8
69=	Senegal	10.0	18.4

Rank (consumer prices)		Growth %pa Consumer prices	Money stock
69=	Togo	10.0	21.7
69=	Tonga	10.0	na
72=	French Guiana	9.9	na
72=	Sierra Leone	9.9	16.5
74	Iran	9.8	31.7
75=	Central African Emp	9.6	14.9
75=	South Africa	9.6	11.9
77	Cameroon	9.5	15.8
78=	French Polynesia	9.4	na
78=	Mozambique	9.4	na
80	Denmark	9.2	12.1
81	Zambia	9.1	11.0*
82	Kuwait	9.0ᶜ	26.7
83	France	8.9	12.0
84=	Guatemala	8.8	19.3
84=	New Caledonia	8.8	na
86=	Netherlands	8.7	11.7
86=	Somalia	8.7	17.6
88=	Madagascar	8.6	9.6
88=	Malawi	8.6	14.3
90=	Belgium	8.5	8.5
90=	Norway	8.5	14.0
90=	Solomon Islands	8.5ᵇ	na
90=	Surinam	8.5	na
94=	El Salvador	8.4	21.1
94=	Singapore	8.4	16.1
96	Sweden	8.3	11.1
97=	India	8.2	13.8
97=	Ivory Coast	8.2	20.8
99=	Guyana	7.9	23.9
99=	Hongkong	7.9	17.0
99=	Thailand	7.9	13.2
102	Congo	7.8	15.9
103=	Luxembourg	7.6	na
103=	Morocco	7.6	18.3
105=	Bahamas	7.4	–0.6
105=	Nepal	7.4	15.2
105=	Puerto Rico	7.4	na
108=	Austria	7.3	12.9
108=	Canada	7.3	8.7
108=	Ethiopia	7.3	11.2
111	Iraq	6.9	23.0
112	Cyprus	6.8	11.8
113=	Egypt	6.7	19.1
113=	Switzerland	6.7	5.5
115	United States	6.6	6.0
116	Malaysia	6.5	16.8
117=	Panama	6.4	na
117=	Sri Lanka	6.4	13.4
119=	Honduras	6.1	15.2
119=	Rhodesia	6.1	na
121	Venezuela	6.0	27.8
122=	Germany, West	5.9	9.5
122=	Libya	5.9	29.5
124	Lebanon	5.8ᵍ	19.6
125	Tunisia	5.2	17.8
126	Malta	5.0	16.2
127	Upper Volta	3.9	20.2
128	Afghanistan	3.3	13.8
129	Hungary	3.1	na
130	Poland	2.8	na
131	Rumania	0.6ᵃ	na
132	Czechoslovakia	0.3	na
133	Bulgaria	0.2ᵍ	na
134	Soviet Union	–0.1	na
135	Germany, East	–0.6	na

Share prices (industrial)

Rank for 1970–76		Growth %pa 1960–70	1970–76
1	Israel	2.0	19.2
2	Venezuela	2.5	15.3
3	Denmark	0.4	13.6
4	Japan	5.3	13.4
5	Finland	5.1	10.7
6	Sweden	2.8	9.0
7	Mexico	–1.3	8.5
8	Austria	2.4	4.7
9	Spain	10.4	4.0
10	United States	4.4	3.8
11	Ireland	9.5	2.6
12	United Kingdom	3.8	2.3
13	Canada	6.4	1.9
14	Belgium	–0.8	1.4
15	Norway	0.7	0.3
16	Germany, West	–0.1	0.1
17	India	1.1	0.0
18	South Africa	12.9	–0.2
19	Peru	–1.7	–0.3
20	Philippines	1.4	–0.8
21	New Zealand	5.8	–0.9
22	France	0.3	–1.0
23	Netherlands	3.6	–1.4
24	Pakistan	4.6*	–2.7
25	Colombia	4.2	–2.9
26	Australia	5.9	–3.5
27	Switzerland	1.0	–4.2
28	Italy	–2.9	–9.9

International reservesᶠ, end-1976

Rankᵉ		$ mn
	*World*ᵈ	*258 300*
1	Germany, West	34 801
2	Saudi Arabia	27 025
3	United States	18 320
4	Japan	16 605
5	Switzerland	12 993
6	France	10 194
7	Iran	8 833
8	Venezuela	8 578
9	Netherlands	7 387
10	Italy	6 654
11	Brazil	6 541
12	Canada	5 843
13	Spain	5 284
14	Belgium-Luxembourg	5 206
15	Nigeria	5 203
16	Iraq	4 601
17	Austria	4 410
18	United Kingdom	4 230
19	Singapore	3 364
20	Libya	3 206
21	Australia	3 170
22	India	3 074
23	Korea, South	2 960
24	Sweden	2 491
25	Malaysia	2 472
26	Norway	2 229
27	Jugoslavia	2 049
28	Algeria	1 987
29	Kuwait	1 929
30	Thailand	1 893
31	Ireland	1 837
32	Lebanon	1 677
33	Philippines	1 640
34	Argentina	1 608
35	Taiwan	1 607
36	Indonesia	1 499
37	Israel	1 373

ᵃ1970–75 ᵇ1971–76 ᶜ1972–76 ᵈExcluding communist countries ᵉTop 37 countries only ᶠGold, foreign exchange, SDRs and reserve position with the International Monetary Fund ᵍ1970–74

World economic aid

Economic aid[a] given by non-communist developed countries Average per year (1973-75)

	$ mn
Total flow of resources	28 182
less Private capital flows	14 248
less Official general flows	2 563
equals Official economic aid	**11 371**
of which,	
bilateral	*8 334*
to multilateral institutions	*3 037*

Outflow of economic aid by country

	$ mn	% of gdp[b]
Total[d]	**11 371**	**0.32**
Australia	408	0.53
Austria	54	0.17
Belgium	291	0.54
Canada	698	0.48
Denmark	168	0.54
Finland	38	0.17
France	1 734	0.61
Germany, West	1 294	0.34
Italy	192	0.12
Japan	1 078	0.24
Netherlands	449	0.64
New Zealand	45	0.37
Norway	134	0.57
Sweden	414	0.71
Switzerland	78	0.16
United Kingdom	721	0.36
United States	3 503	0.25

Inflow of economic aid by region and country

	$ mn	$ per person
Total[d]	**10 647[c]**	**5**
Africa[d]	*3 028*	*8*
Algeria	120	7
Benin	37	12
Botswana	40	61
Burundi	35	10
Cameroon	76	12
Cape Verde	1	4
Central African Empire	38	22
Chad	55	14
Comoros	22	75
Congo	40	29
Djibouti	28	274
Egypt	188	5
Ethiopia	103	4
Gabon	33	63
Gambia	8	13
Ghana	67	7
Guinea	10	2
Guinea-Bissau	4	9
Ivory Coast	81	17
Kenya	113	9
Lesotho	20	20
Liberia	15	9
Libya	10	4
Madagascar	67	9
Malawi	45	9
Mali	93	17
Mauritania	38	30
Mauritius	21	25
Morocco	128	8
Mozambique	4	1
Niger	99	22
Nigeria	89	1
Reunion	217	442
Rwanda	55	13
Senegal	103	24

	$ mn	$ per person
Seychelles	8	133
Sierra Leone	13	5
Somalia	46	15
Sudan	78	4
Swaziland	15	32
Tanzania	180	12
Togo	34	16
Tunisia	141	25
Uganda	14	1
Upper Volta	77	13
Zaire	172	7
Zambia	60	13
America[d]	*1 465*	*4*
Argentina	35	1
Bahamas	1	3
Barbados	4	18
Belize	6	48
Bolivia	49	9
Brazil	148	1
Chile	67	7
Colombia	116	5
Costa Rica	25	13
Cuba	14	1
Dominican Republic	26	6
Ecuador	50	7
El Salvador	29	7
French Guiana	43	748
Guadeloupe	112	320
Guatemala	31	5
Guyana	10	13
Haiti	25	6
Honduras	30	10
Jamaica	23	12
Martinique	124	347
Mexico	55	1
Netherlands Antilles	28	118
Nicaragua	40	19
Panama	28	17
Paraguay	32	12
Peru	83	5
St Pierre & Miquelon	10	1 722
Surinam	44	106
Trinidad & Tobago	5	5
Uruguay	14	4
Venezuela	15	1
Asia[d]	*4 790*	*4*
Afghanistan	45	2
Bangladesh	610	8
Burma	76	3
Cambodia	179	23
India	1 097	2
Indonesia	652	5
Iraq	17	2
Israel	257	76
Jordan	103	39
Korea, South	258	8
Laos	58	18
Lebanon	14	5
Malaysia	70	6
Maldives	1	10
Nepal	37	3
Pakistan	353	5
Philippines	188	5
Singapore	20	9
Sri Lanka	92	7
Syria	21	3
Thailand	72	2
Yemen, North	34	5
Yemen, South	14	8

	$ mn	$ per person
Oceania[d]	*522*	*123*
Fiji	16	29
French Polynesia	53	429
Gilbert Islands	6	90
New Caledonia	67	505
New Hebrides	12	129
Pacific Islands, US[f]	74	650
Papua New Guinea	255	96
Samoa, Western	7	48
Solomon Islands	15	82
Tonga	2	25

Economic aid[e] given by communist developed countries Average per year (1973-75)

Outflow of economic aid by country

	$ mn	% of gdp[b]
Total	**2 828***	**0.28***
Bulgaria	59	0.39***
China	394	0.15***
Czechoslovakia	193	0.42***
Germany, East	115*	0.21***
Hungary	136	0.68***
Poland	136	0.18***
Rumania	418	1.55***
Soviet Union	1 377	0.25***

Inflow of economic aid by region and country

	$ mn	$ per person
Africa[d]	*715*	*2*
Algeria	163	10
Cameroon	25	4
Chad	22	6
Egypt	158	4
Ghana	22	2
Guinea	28	6
Mauritania	13	10
Mozambique	20	2
Niger	18	4
Senegal	17	4
Somalia	21	7
Sudan	45	3
Tanzania	27	2
Tunisia	18	3
Upper Volta	37	6
Zaire	38	2
Zambia	20	4
America[d]	*393*	*1*
Argentina	290	12
Brazil	60	1
Chile	7	1
Mexico	22	0.4
Asia[d]	*1 423*	*1*
Afghanistan	235	12
Bangladesh	94	1
Cambodia	10	1
India	177	0.3
Indonesia	33	0.3
Iran	169	5
Iraq	141	13
Jordan	58	22
Nepal	27	2
Pakistan	87	1
Sri Lanka	32	2
Syria	383	54
Europe	*289*	*1*
Turkey	289	7

[a]Official development assistance, including both bilateral assistance and flows to and from multilateral institutions [b]Gross domestic product [c]This total of inflows differs from the total of outflows mainly due to 'pipe-line effects' following from the difference in the time of inflows to multilateral institutions and the outflow from them to developing countries [d]Including flows not shown separately [e]Bilateral commitments of capital to developing countries [f]Including Northern Marianas

World external services[a]

Imports of services[b], 1976

Rank	$ mn	Rank	$ mn	Rank	$ mn	Rank	$ mn	Rank	$ mn	Rank	$ mn
World	*283 000**	21 South Africa	3 217	42 Malaysia	991c	63 Dominican Rep	368	84 Sri Lanka	127		
1 United States	38 144	22 Denmark	3 093	43 Thailand	961	64 Guatemala	321	85 Mali	121		
2 Germany, West	30 028	23 Indonesia	2 881	44 Portugal	913	65 Congo	307	86 Guyana	113		
3 Japan	20 327	24 Libya	2 497	45 Peru	907	66 Bahamas	297	87= Chad	107		
4 France	18 664	25 Venezuela	2 306	46 Chile	869	67 Ghana	293	87= Mauritius	107		
5 United Kingdom	18 226	26 Israel	2 270	47 Ivory Coast	837	68 Cameroon	290	89 Yemen, North	104		
6 Canada	12 438	27 Jugoslavia	2 211	48 Ireland	801	69 Uruguay	248	90= Malta	99		
7 Saudi Arabia	11 425	28 Algeria	2 107	49 Neth Antilles	723	70 Costa Rica	237	90= Tanzania	99		
8 Netherlands	10 291	29 Finland	1 989	50 Morocco	716	71= Senegal	220c	92= Malawi	94		
9 Italy	9 968	30 Korea, South	1 718	51 Gabon	687	71= Sudan	220	92= Upper Volta	94e		
10 Belgium-Lux	8 830	31 Iraq	1 712c	52 Zaire	668c	73 Nicaragua	212	94 Barbados	92		
11 Iran	6 384	32 Taiwan	1 576	53 Trinidad & Tobago	637	74 Bolivia	210	95 Paraguay	89		
12 Australia	5 343	33 India	1 441c	54 Panama	620	75 El Salvador	198	96 Fiji	85		
13 Sweden	5 170	34 Turkey	1 360	55 Pakistan	617	76 Iceland	174	97 Uganda	84		
14 Brazil	5 098	35 Singapore	1 303	56 Syria	505	77 Bangladesh	171	98 Somalia	69		
15 Norway	5 084	36 Egypt	1 260	57 Tunisia	495	78- Ethiopia	147	99 Togo	68c		
16 Mexico	4 761	37 Argentina	1 247	58 Jamaica	463	78- Honduras	147	100 Cen African Emp	67		
17 Nigeria	3 884	38 New Zealand	1 244	59 Zambia	432	80 Madagascar	144	101 Haiti	62		
18 Austria	3 696	39 Philippines	1 128	60 Kenya	430	81 Mauritania	142	102 Rwanda	60		
19 Spain	3 685	40 Greece	1 023	61= Ecuador	389	82 Cyprus	141	103 Sierra Leone	56		
20 Switzerland	3 423	41 Colombia	994	61= Jordan	389	83 Surinam	128	104 Burma	39		

Exports of services[b], 1976

Rank	$ mn	Rank	$ mn	Rank	$ mn	Rank	$ mn	Rank	$ mn	Rank	$ mn
World	*260 000**	21 Australia	2 326	42 Bahamas	630	63 Cameroon	190	84 Congo	79		
1 United States	56 311	22 Egypt	1 975	43 Colombia	561	64 Iceland	175	85 Sri Lanka	76		
2 Germany, West	23 807	23 Greece	1 820	44 Tunisia	548	65 Indonesia	160	86 Bolivia	73		
3 United Kingdom	23 215	24 Israel	1 740	45 Iraq	543c	66 Tanzania	158	87 Surinam	69		
4 France	19 742	25 Korea, South	1 641	46 New Zealand	491	67 Gabon	149	88 Madagascar	51		
5 Japan	14 457	26 South Africa	1 608	47 Morocco	475	68 Zaire	145c	89 Honduras	49		
6 Italy	11 628	27 Finland	1 427	48 Kenya	419	69 Fiji	143	90 Paraguay	39		
7 Netherlands	11 561	28 Brazil	1 178	49 Malaysia	416c	70 Barbados	142	91= Cen African Emp	37		
8 Belgium-Lux	9 900	29 Taiwan	1 176	50 Peru	397	71 Uruguay	138	91= Guyana	37		
9 Switzerland	7 088	30 Venezuela	1 142	51 Libya	349	72= Costa Rica	127	93 Malawi	34		
10 Canada	6 291	31 India	973c	52 Trinidad & Tobago	343	72= Sudan	127	94 Chad	32		
11 Spain	5 559	32 Panama	946	53 Syria	314	74 Dominican Rep	126	95 Somalia	31		
12 Norway	5 066	33 Turkey	895	54 Pakistan	307	75 Ethiopia	115	96= Burma	30		
13 Austria	4 820	34 Philippines	872	55 Ivory Coast	293	76 Ecuador	111	96= Haiti	30		
14 Saudi Arabia	4 462	35 Ireland	834	56 Jamaica	286	77 Mauritius	108	98 Togo	27c		
15 Jugoslavia	3 850	36 Portugal	809	57 Algeria	253	78 Nicaragua	92	99 Mauritania	24		
16 Denmark	3 841	37= Jordan	794	58 Senegal	237c	79 Zambia	90	100 Upper Volta	23c		
17 Sweden	3 085	37= Nigeria	794	59 Malta	228	80 Ghana	86	101 Sierra Leone	22		
18 Mexico	3 084	39 Argentina	727	60 Chile	225	81= Bangladesh	85	102 Mali	21		
19 Iran	2 844	40 Neth Antilles	720	61 Cyprus	222	81= El Salvador	85	103 Uganda	12		
20 Singapore	2 763	41 Thailand	692	62 Guatemala	220	83 Yemen, North	83	104 Rwanda	9		

Balance of services[b], 1976

Rank	$ mn	Rank	$ mn	Rank	$ mn	Rank	$ mn	Rank	$ mn	Rank	$ mn
World	*−23 000**	21 Fiji	+58	42 Surinam	−59	63 Jamaica	−177	84 Ivory Coast	−544		
1 United States	+18 167	22 Tunisia	+53	43 Malawi	−60	64 Syria	−191	85 Finland	−562		
2 Utd Kingdom	+4 989	23 Barbados	+50	44 Upper Volta	−71c	65 Ghana	−207	86 Malaysia	−575c		
3 Switzerland	+3 665	24 Ireland	+33	45 Uganda	−72	66 Congo	−228	87 Chile	−644		
4 Spain	+1 874	25 Senegal	+17c	46 Chad	−75	67 Morocco	−241	88 New Zealand	−753		
5 Italy	+1 660	26= Iceland	+1	47 Guyana	−76	68 Dominican Rep	−242	89 Venezuela	−1 164		
6 Jugoslavia	+1 639	26= Mauritius	+1	48 Korea, South	−77	69 Philippines	−256	90 Iraq	−1 169c		
7 Singapore	+1 460	28 Neth Antilles	−3	49 Bangladesh	−86	70 Thailand	−269	91 South Africa	−1 609		
8 Netherlands	+1 270	29 Burma	−9	50= Madagascar	−93	71 Ecuador	−278	92 Mexico	−1 677		
9 Austria	+1 124	30 Kenya	−11	50= Sudan	−93	72 Trinidad & Tob	−293	93 Algeria	−1 854		
10 France	+1 078	31 Norway	−18	52 Honduras	−98	73 Pakistan	−310	94 Sweden	−2 085		
11 Belgium-Lux	+1 070	32 Yemen, North	−21	53= Cameroon	−100	74 Zambia	−342	95 Libya	−2 148		
12 Greece	+797	33 Cen African Emp	−30	53= Mali	−100	75 Taiwan	−400	96 Indonesia	−2 721		
13 Denmark	+748	34= Ethiopia	−32	55 Guatemala	−101	76 Colombia	−433	97 Australia	−3 017		
14 Egypt	+715	34= Haiti	−32	56 Portugal	−104	77 Turkey	−465	98 Nigeria	−3 090		
15 Jordan	+405	36 Sierra Leone	−34	57= Costa Rica	−110	78 India	−468c	99 Iran	−3 540		
16 Bahamas	+333	37 Somalia	−38	57= Uruguay	−110	79 Peru	−510	100 Brazil	−3 920		
17 Panama	+326	38 Togo	−41c	59 El Salvador	−113	80 Argentina	−520	101 Japan	−5 870		
18 Malta	+129	39 Paraguay	−50	60 Mauritania	−118	81 Zaire	−523c	102 Canada	−6 147		
19 Cyprus	+81	40= Rwanda	−51	61 Nicaragua	−120	82 Israel	−530	103 Germany, W	−6 221		
20 Tanzania	+59	40= Sri Lanka	−51	62 Bolivia	−137	83 Gabon	−538	104 Saudi Arabia	−6 963		

[a]Including tourism, transport, banking, etc; sometimes referred to as 'invisible trade' [b]Main countries for which detailed recent information is available; excludes, among other countries, China, North Korea, Vietnam, Soviet Union and Eastern Europe [c]1975

World trade

World markets (countries ranked by amount of imports), exports and balance of trade, 1976

Rank by imports		Imports $ mn	Exports $ mn	Balance of trade[d] surplus (+) or deficit (−) $ mn	Rank by imports		Imports $ mn	Exports $ mn	Balance of trade[d] surplus (+) or deficit (−) $ mn
	World	*1 038 000**	*1 013 000**	*−25 000**	68	Bahrain	1 672	1 348	−324
1	United States[n]	129 565	114 997	−14 568	69	Tunisia	1 531	789	−742
2	Germany, West	89 131	103 485	+14 354	70	Ivory Coast	1 304	1 642	+338
3	Japan	64 853	67 218	+2 365	71	Zambia	1 138[a]	810[a]	−328[a]
4	France[u]	64 460	57 163	−7 297	72	Jordan	1 022	207	−815
5	United Kingdom[t]	56 614	46 704	−9 910	73	Ecuador	1 010	1 127	+117
6	Italy[v]	43 626	37 130	−6 496	74	Guatemala	982	782	−200
7	Canada	40 640	40 235	−405	75	Sudan	980	554	−426
8	Netherlands	40 429	40 097	−332	76	Kenya	973	790	−183
9	Soviet Union	38 105	37 164	−941	77	Jamaica	913	633	−280
10	Belgium-Luxembourg	35 492	32 811	−2 681	78	Dominican Republic	878	716	−162
11	Sweden	19 284	18 415	−869	79	Bangladesh	859	401	−458
12	Spain	17 495	8 722	−8 773	80	Vietnam	850**	250**	−600**
13	Switzerland[l]	14 750	14 806	+56	81	Panama[o]	838	226	−612
14	Poland	13 877	11 057	−2 820	82	Qatar	830	2 209	+1 379
15	Brazil	13 258	9 945	−3 313	83	Ghana	788	804	+16
16	Germany, East	13 196	11 361	−1 835	84	Costa Rica	774	557	−217
17	Iran	12 890	23 516	+10 626	85=	Korea, North	750***	800***	+50***
18	Australia	12 476	13 160	+684	85=	Lebanon	750*	800*	+50*
19	Denmark	12 409	9 104	−3 305	87	Oman	725	1 575	+850
20	Austria	11 481	8 479	−3 002	88	El Salvador	705	721	+16
21	Norway	11 095	7 942	−3 153	89	Zaire	677	926	+249
22	Czechoslovakia	9 380	8 733	−647	90	Tanzania	639	491	−148
23	Singapore	9 067	6 583	−2 484	91	Angola	625[c]	1 224[c]	+599[c]
24	Hongkong	8 882	8 526	−356	92	Cameroon	609	511	−98
25	Korea, South	8 755	7 699	−1 056	93	Uruguay	599*	536*	−63*
26	Saudi Arabia	8 694	36 125	+27 431	94	Sri Lanka	585	572	−13
27	Nigeria	8 203	10 569	+2 366	95	Senegal	581	462	−119
28	Taiwan	7 609	8 156	+547	96	South-West Africa	570**	800**	+230**
29	Finland	7 391	6 342	−1 049	97	Bolivia	555	513	−42
30	Jugoslavia	7 369	4 880	−2 489	98	Rhodesia	553[f]	664[f]	+111[f]
31	South Africa[p]	7 285	7 939	+654	99	Mongolia	550*[a]	220*[a]	−330*[a]
32	Venezuela	6 832	9 149	+2 317	100	Gabon	540*	897	+357*
33	China	6 100**	7 200**	+1 100**	101	Nicaragua	534	544	+10
34	Rumania	6 095	6 138	+43	102	Cyprus	505	275	−230
35	Greece	6 064	2 565	−3 499	103	Papua New Guinea	499[h,k]	657[h,k]	+158[h,k]
36	Mexico	5 897	3 442	−2 455	104	Iceland	470	403	−67
37	Israel	5 715	2 415	−3 300	105	Afghanistan	454[i]	291[i]	−163[i]
38	Indonesia	5 673	8 547	+2 874	106	Honduras	453	392	−61
39	India	5 611[b,h]	5 573[b,h]	−38[b,h]	107	Reunion	450	94	−356
40	Bulgaria	5 604	5 361	−243	108	Malta	423	229	−194
41	Hungary	5 544	4 936	−608	109	Yemen, South	414	249	−165
42	Puerto Rico	5 432[k]	3 346[k]	−2 086[k]	110	Yemen, North	412	8	−404
43	Algeria	5 311	5 162	−149	111	Mozambique	410[a]	210[a]	−200[a]
44	Turkey	5 168	1 917	−3 251	112	Liberia	399	460	+61
45	Portugal	4 230	1 809	−2 421	113	Martinique	382	124	−258
46	Ireland	4 212	3 354	−858	114	Guyana	358	275	−83
47	Philippines	3 950	2 511	−1 439	115	Mauritius	356	265	−91
48	Malaysia	3 921	5 288	+1 367	116	Ethiopia	350	278	−72
49	Cuba	3 883[a]	3 680[a]	−203[a]	117	Guadeloupe	317	90	−227
50	Egypt	3 808	1 521	−2 287	118	French Polynesia	296	22	−274
51	Netherlands Antilles	3 661	2 519	−1 142	119	Madagascar	280[c]	243[c]	−37[c]
52	Thailand	3 644	3 040	−604	120	New Caledonia	277	301	+24
53	Libya	3 610*	8401	+4 791*	121	Fiji	263	139	−124
54	Bahamas	3 560	2 879	−681	122	Surinam	262[a]	277[a]	+15[a]
55	Iraq	3 461	9 248	+5 787	123	Guam	259[c]	20[c]	−239[c]
56	Utd Arab Emirates	3 331	8 555	+5 224	124	Brunei	255	1 295	+1 040
57	Kuwait	3 318	9 827	+6 509	125	Albania	250***	200***	−50***
58	New Zealand	3 258	2 804	−454	126	Liechtenstein	239***	239	0***
59	Argentina	3 034	3 916	+882	127	Barbados	237	87	−150
60	Virgin Islands, US	2 679	2 010	−669	128	Jersey	222	88	−134
61	Morocco	2 615	1 263	−1 352	129	Paraguay	219	178	−41
62	Syria	2 383	1 073	−1 310	130	Botswana	218[a]	144[a]	−74[a]
63	Pakistan	2 134	1 167	−967	131	Malawi	206	160	−46
64	Peru	2 061	1 289	−772	132	Togo	203*	134*	−69*
65	Trinidad & Tobago	1 969	2 200	+231	133	Burma	194	186	−8
66	Colombia	1 710	1 882	+172	134	Haiti	190*	124	−66*
67	Chile	1 684*	2 083*	+399*	135	Swaziland	184[a]	181[a]	−3[a]

[a]1975 [b]Year ending March 31st [c]1974 [d]Crude balance of exports of goods (fob) less imports of goods (cif) [e]Including Mayotte [f]1973 [g]Including Anguilla [h]1977
[i]Year ending March 20th [j]Trade with France and Spain only [k]Year ending June 30th [l]Customs union which includes also Liechtenstein [m]1972 [n]Includes Puerto Rico
[o]Excluding free zone of Cólon [p]South Africa Customs Union, which includes also Botswana, Lesotho, South-West Africa and Swaziland [q]Indicating in part the potential

Rank by imports	Imports $ mn	Exports $ mn	Balance of trade[d] surplus (+) or deficit (−) $ mn
136= Guinea	180**a	160**a	−20**a
136= Mauritania	180	178	−2
138 Nepal	163	98	−65
139= Macao	162	190	+28
139= Somalia	162a	91a	−71a
141 Uganda	160	359	+199
142 Congo	158a	243a	+85a
143 Andorra	156j	6j	−150j
144 Sierra Leone	154	112	−42
145= Mali	150	99	−51
145= Panama Canal Zone	150***	20***	−130***
147 Benin	146c	34c	−112c
148 Bermuda	143	42	−101
149 Upper Volta	136	39	−97
150 Faroes	131	104	−27
151 Greenland	129	86	−43
152 Guernsey	126a	93a	−33a
153 Lesotho	125c	14c	−111c
154= Belize	103a	72a	−31a
154= Rwanda	103	81	−22
156 Niger	102a	91a	−11a
157 Sahara, Western	100**	100**	0**
158 Chad	87e	38e	−49e
159 French Guiana	86	4	−82
160= Djibouti	74f	20f	−54f
160= Gambia	74	34	−40
162= Antigua	70c	32c	−38c
162= Cambodia	70*f	13*f	−57*f
162= Isle of Man	70***	1***r	−69***
165 Central African Empire	68a	47a	−21a
166 Laos	65c	11c	−54c
167 Gibraltar	59	25	−34
168 Burundi	58	63	+5
169 Samoa, American	51	65	+14
170 St Lucia	44c	16c	−28c
171 Cape Verde	42	14	−28
172 Seychelles	39	8	−31
173= Guinea-Bissau	38a	6a	−32a
173= Pacific Islands, USs	38*	5	−33*
175 Cayman Islands	36	1	−35
176 New Hebrides	34	17	−17
177 Samoa, Western	30	7	−23
178 Solomon Islands	27	24	−3
179= Comoros	26c,e	9c,e	−17c,e
179= St Pierre & Miquelon	26c	12c	−14c
181= Equatorial Guinea	25**	15**	−10**
181= Grenada	25	13	−12
181= St Vincent	25c	7c	−18c
184 Dominica	21a	11a	−10a
185 St Kitts-Neviss	16m	6m	−10m
186 Tonga	15	4	−11
187 Nauru	14*c	50*c	+36*c
188= Gilbert Islands	12	22	+10
188= Timor, East	12c	6c	−6c
188= Virgin Islands, Br	12c	0c	−12c
191 Sao Tome & Principe	11a	7a	−4a
192 Maldives	10*a	4a	−6*a
193 Montserrat	8.0	0.4	−7.6
194 Norfolk Island	7.4h,k	0.9h,k	−6.5h,k
195 Cook Islands	6.7f	3.9f	−2.8f
196 Turks & Caicos Is	6.0c	0.5c	−5.5c
197 Falkland Islands	3.4a	2.6a	−0.8a
198= Niue	2.5a	0.2a	−2.3a
198= Saint Helena	2.5a	0.0a	−2.5a
198= Wallis & Futuna Is	2.5	0.02**	−2.5**
201 Bhutan	1.5***	1.0***	−0.5***
202 Tokelau	1.0***b	0.01***b	−1.0***b

World balance of trade (top 140 importers ranked[q]), 1976

Rank		Balance of trade, $ mn	Rank		Balance of trade, $ mn
1	Saudi Arabia	+27 431	71	Algeria	−149
2	Germany, West	+14 354	72	Barbados	−150
3	Iran	+10 626	73	Dominican Rep	−162
4	Kuwait	+6 509	74	Afghanistan	−163i
5	Iraq	+5 787	75	Yemen, South	−165
6	Utd Arab Emirates	+5 224	76	Kenya	−183
7	Libya	+4 791*	77	Malta	−194
8	Indonesia	+2 874	78=	Guatemala	−200
9	Nigeria	+2 366	78=	Mozambique	−200a
10	Japan	+2 365	80	Cuba	−203a
11	Venezuela	+2 317	81	Costa Rica	−217
12	Qatar	+1 379	82	Guadeloupe	−227
13	Malaysia	+1 367	83	Cyprus	−230
14	China	+1 100**	84	Guam	−239c
15	Brunei	+1 040	85	Bulgaria	−243
16	Argentina	+882	86	Martinique	−258
17	Oman	+850	87	French Polynesia	−274
18	Australia	+684	88	Jamaica	−280
19	South Africap	+654	89	Bahrain	−324
20	Angola	+599c	90	Zambia	−328a
21	Taiwan	+547	91	Mongolia	−330*a
22	Chile	+399*	92	Netherlands	−332
23	Gabon	+357*	93=	Hongkong	−356
24	Ivory Coast	+338	93=	Reunion	−356
25	Zaire	+249	95	Yemen, North	−404
26	Trinidad & Tobago	+231	96	Canada	−405
27	South-West Africa	+230**	97	Sudan	−426
28	Colombia	+172	98	New Zealand	−454
29	Papua New Guinea	+158h,k	99	Bangladesh	−458
30	Ecuador	+117	100	Vietnam	−600**
31	Rhodesia	+111f	101	Thailand	−604
32	Liberia	+61	102	Hungary	−608
33	Switzerlandw	+56	103	Panamao	−612
34=	Korea, North	+50***	104	Czechoslovakia	−647
34=	Lebanon	+50*	105	Virgin Islands, US	−669
36	Rumania	+43	106	Bahamas	−681
37	Macao	+28	107	Tunisia	−742
38	New Caledonia	+24	108	Peru	−772
39=	El Salvador	+16	109	Jordan	−815
39=	Ghana	+16	110	Ireland	−858
41	Surinam	+15a	111	Sweden	−869
42	Nicaragua	+10	112	Soviet Union	−941
43	Liechtenstein	0***	113	Pakistan	−967
44	Mauritania	−2	114	Finland	−1 049
45	Swaziland	−3a	115	Korea, South	−1 056
46	Burma	−8	116	Neth Antilles	−1 142
47	Sri Lanka	−13	117	Syria	−1 310
48	Guinea	−20**a	118	Morocco	−1 352
49	Madagascar	−37c	119	Philippines	−1 439
50	India	−38b,h	120	Germany, East	−1 835
51	Paraguay	−41	121	Puerto Rico	−2 086k
52	Bolivia	−42	122	Egypt	−2 287
53	Malawi	−46	123	Portugal	−2 421
54	Albania	−50***	124	Mexico	−2 455
55	Honduras	−61	125	Singapore	−2 484
56	Uruguay	−63*	126	Jugoslavia	−2 489
57	Nepal	−65	127	Belgium-Lux	−2 681
58	Haiti	−66*	128	Poland	−2 820
59	Iceland	−67	129	Austria	−3 002
60	Togo	−69*	130	Norway	−3 153
61	Somalia	−71a	131	Turkey	−3 251
62	Ethiopia	−72	132	Israel	−3 300
63	Botswana	−74a	133	Denmark	−3 305
64	Guyana	−83	134	Brazil	−3 313
65	Mauritius	−91	135	Greece	−3 499
66	Cameroon	−98	136	Italyv	−6 496
67	Senegal	−119	137	Franceu	−7 297
68	Fiji	−124	138	Spain	−8 773
69	Jersey	−134	139	Utd Kingdomt	−9 910
70	Tanzania	−148	140	United Statesn	−14 568

for higher (+) or lower (−) imports, or investment in other countries; for less-developed countries, a high deficit on the balance of trade often indicates a high level of foreign investment in or support for the country (eg North Yemen) rMain exports only sIncluding Northern Marianas tCustoms union which includes also Guernsey, Isle of Man and Jersey uCustoms union which includes also Monaco vCustoms union which includes also San Marino

World trade

Main imports and exports[d], $ bn, 1975

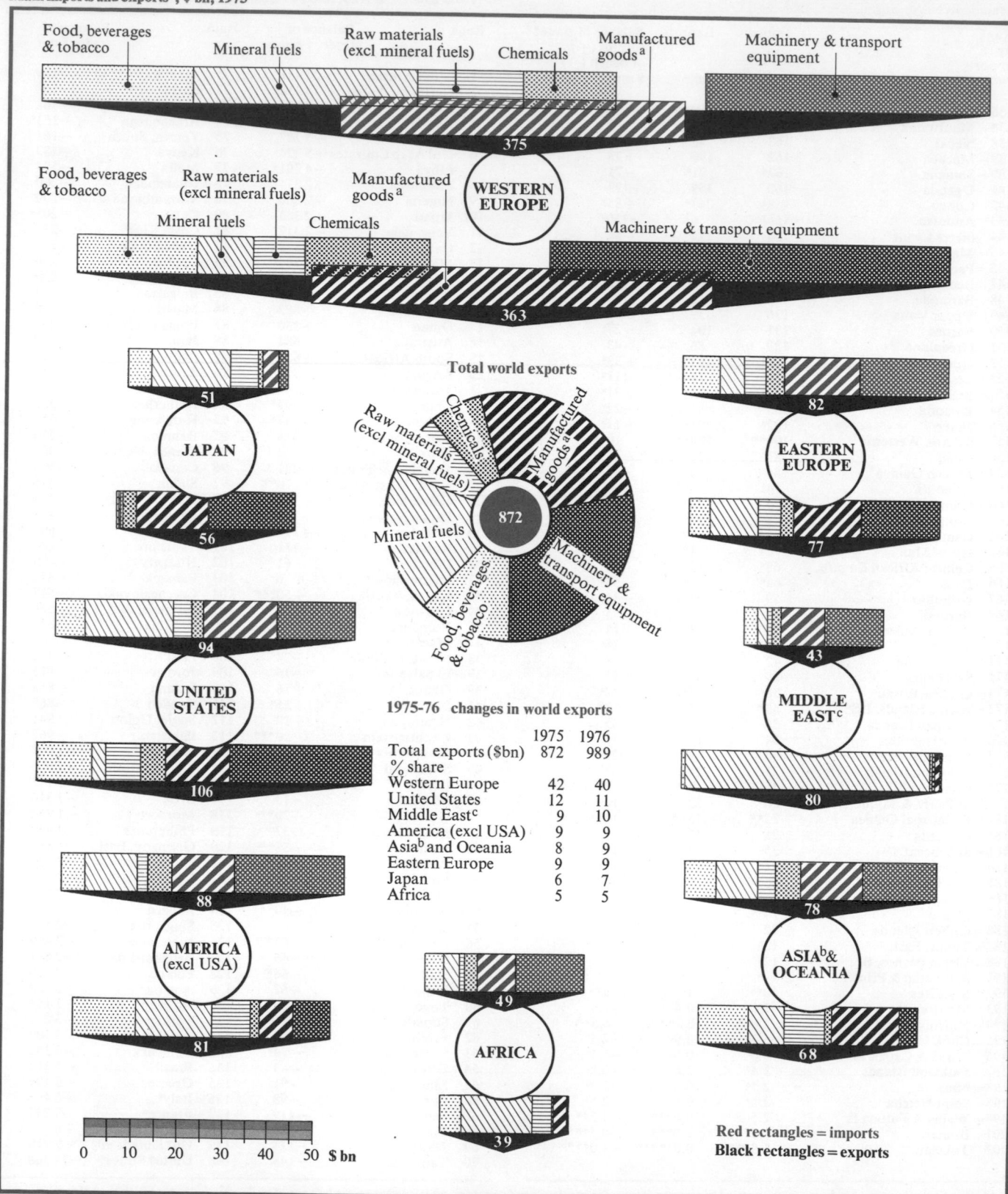

Food, beverages & tobacco

Mineral fuels

Raw materials (excl mineral fuels)

Chemicals

Manufactured goods[a]

Machinery & transport equipment

375

Food, beverages & tobacco

Raw materials (excl mineral fuels)

Mineral fuels

Manufactured goods[a]

Chemicals

Machinery & transport equipment

WESTERN EUROPE

363

51

JAPAN

56

94

UNITED STATES

106

88

AMERICA (excl USA)

81

Total world exports

Chemicals

Raw materials (excl mineral fuels)

Manufactured goods[a]

Mineral fuels

872

Machinery & transport equipment

Food, beverages & tobacco

82

EASTERN EUROPE

77

43

MIDDLE EAST[c]

80

78

ASIA[b] & OCEANIA

68

1975-76 changes in world exports

	1975	1976
Total exports ($bn)	872	989
% share		
Western Europe	42	40
United States	12	11
Middle East[c]	9	10
America (excl USA)	9	9
Asia[b] and Oceania	8	9
Eastern Europe	9	9
Japan	6	7
Africa	5	5

49

AFRICA

39

0 10 20 30 40 50 $ bn

Red rectangles = imports
Black rectangles = exports

[a]Excluding machinery and transport equipment; including trade not separately distinguished [b]Excluding Japan and Middle East [c]Middle East here includes also Cyprus and Egypt
[d]Excluding exports of Rhodesia, trade between East and West Germany, and inter-trade of communist countries of Asia

World trade

Main sources and destinations[d], $ bn, 1976

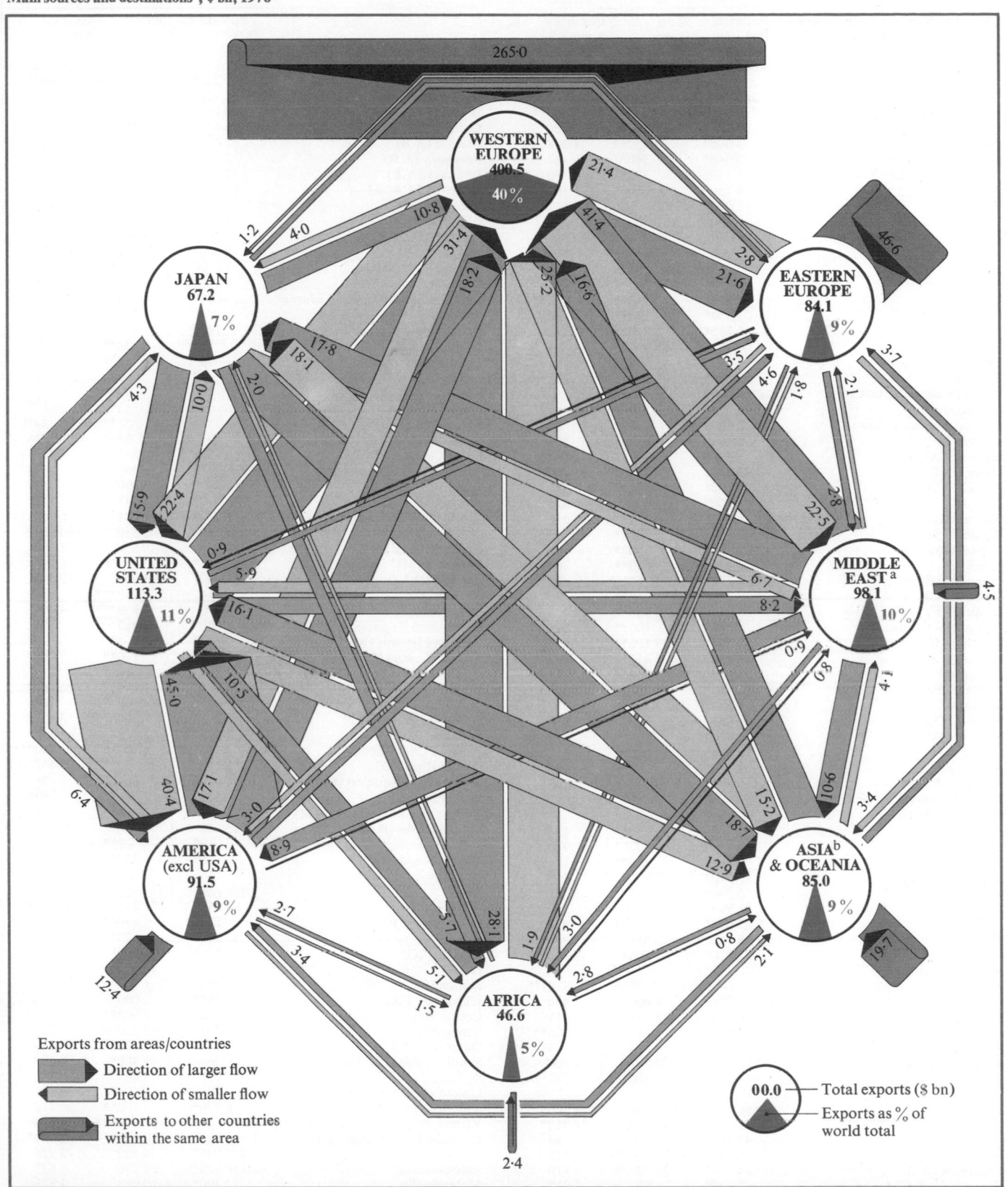

WESTERN EUROPE 400.5 40%

265.0

21.4

10.8

4.0

1.2

31.4 18.2 25.2 16.6

41.4 2.8 21.6

JAPAN 67.2 7%

EASTERN EUROPE 84.1 9%

46.6

3.7

17.8 18.1 2.0 0.0 4.3

15.9 22.4

3.5 4.6 1.8 2.1

2.8 22.5

UNITED STATES 113.3 11%

0.9 5.9 16.1

6.7 8.2 0.9

MIDDLE EAST[a] 98.1 10%

4.5

45.0 10.5 17.1 3.0 40.4 6.4 8.9

0.9 6.8 4.1 10.6 3.4

15.2 18.7 12.9

ASIA[b] & OCEANIA 85.0 9%

19.7

AMERICA (excl USA) 91.5 9%

12.4

2.7 3.4 5.1 1.5

5.7 28.1 1.9 3.0

2.8 0.8 2.1

AFRICA 46.6 5%

2.4

Exports from areas/countries

Direction of larger flow

Direction of smaller flow

Exports to other countries within the same area

00.0 — Total exports ($ bn)

Exports as % of world total

[a]Middle East here includes also Cyprus and Egypt   [b]Excluding Japan and Middle East   [c]Total includes special categories, etc, whose destinations could not be determined
[d]Excluding exports of Rhodesia, trade between East and West Germany, and inter-trade of communist countries of Asia

World groups

Population, gross domestic product and imports, 1976

Classified by main regional group[u] (for other groups see end of table)[n]	Population 000	Gdp $ mn	Imports[v] $ mn
World	4 040 000*	6 770 000**	1 038 000*
Africa	415 140*	171 170**	48 100*
OAU	380 920*	133 370**	38 500*
Ecowas ('West Africa')	119 210*	38 760**	12 660*
CEAO	28 200*	7 500**	2 460*
Ivory Coast (E)	5 020*	3 950**	1 304
Mali (E)	5 840*	540**	150
Mauritania (C,E)	1 350*	380**	180
Niger (E)	4 730*	650**	102[a]
Senegal (E)	5 090*	1 476	581[a]
Upper Volta (E)	6 170*	500**	136
Other Ecowas	91 010*	31 260**	10 200*
Benin (E)	3 200*	420**	146[c]
Gambia (D,E)	540*	105**	74
Ghana (D,E)	10 310*	2 600**	788[a]
Guinea (E)	4 530**	750**	180**[a]
Guinea-Bissau (E)	534*	200**	38[a]
Liberia (E)	1 750*	910**	399
Nigeria (D,E,G)	64 750**	25 000**	8 203
Sierra Leone (D,E)	3 110*	680**	154
Togo (E)	2 280*	590**	203[a]
Udeac ('Central Africa')	11 140*	5 090**	1 380*
Cameroon (E)	6 530*	2 100**	609
Central African Empire (E)	2 690**	420**	54
Congo (E)	1 390*	670**	177
Gabon (E,G)	526*	1 900**	540*
Other OAU	250 570*	89 520**	24 460*
Algeria (C,F,G)	17 300*	15 600**	5 311
Angola	6 560*	3 500**	625[c]
Botswana (D,E)	690*	280**	218[a]
Burundi (E)	3 860*	370**	58
Cape Verde (E)	306*	130**	42
Chad (E)	4 120*	540**	87[c]
Comoros (E)	274*	60***	20**[c]
Djibouti (C)	226**	240**	74[f]
Egypt (C,F)	38 070*	12 300**	3 808
Equatorial Guinea (E)	319*	100**	25**
Ethiopia (E)	28 680*	3 100**	350
Kenya (D,E)	13 850*	3 411	973
Lesotho (D,E)	1 070*	200**	125[c]
Libya (C,F,G)	2 510*	15 000**	3 610*
Madagascar (E)	8 270*	1 750**	287
Malawi (D,E)	5 180*	707	206
Mauritius (D,E)	914*	570**	356
Morocco (C)	17 830*	7 000**	2 615
Mozambique	9 440*	2 600**	300
Rwanda (E)	4 290*	410**	103
Sao Tome and Principe (E)	80*	40**	11[a]
Seychelles (D,E)	59*	27**	39
Somalia (C,E)	3 260*	350**	162[a]
Sudan (C,E)	18 850*	4 900**	980
Swaziland (D,E)	500*	230**	184[a]
Tanzania (D,E)	15 610*	2 450**	639
Tunisia (C)	5 740*	4 442	1 531
Uganda (D,E)	11 940*	3 000**	160
Zaire (E)	25 630*	3 700**	677
Zambia (D,E)	5 140*	2 514	800
Other Africa	34 220*	37 800*	9 600*
British Indian Ocean Territory (M)	2**	—	—
Mayotte (L)	40*	15***	6**[c]
Reunion (L)	511*	1 000**	450
Rhodesia	6 530*	3 247	553[f]
Sahara, Western	120**	110***	100**
Saint Helena (M)	7**	4***	3[a]
South Africa	26 130	32 217*	7 900*[p]
South-West Africa	880**	1 200***	570**
America	565 520*	2 285 550*	238 000*
Lafta	279 140*	341 150*	36 860*
Andean Group	65 940*	65 180*	12 168
Bolivia (A,B)	5 790*	2 300**	555
Colombia (A,B)	24 320*	14 900**	1 710
Ecuador (A,B,G)	7 300*	4 955	1 010
Peru (A,B)	16 090*	12 000**	2 061

	Population 000	Gdp $ mn	Imports[v] $ mn
Venezuela (A,B,G)	12 390*	31 019	6 832
Other Lafta	213 200*	275 970**	24 691*
Argentina (A,B)	25 720*	50 000**	3 034
Brazil (A,B)	109 180*	130 000**	13 258
Chile (A,B)	10 450	11 459	1 684*
Mexico (A,B)	62 330*	79 119	5 897
Paraguay (A,B)	2 720*	1 700	219
Uruguay (A,B)	2 800*	3 693	599*
Odeca and CACM	17 460*	11 937	3 448
Costa Rica (A,B)	2 020	2 345	774
El Salvador (A,B)	4 120*	2 186	705
Guatemala (A,B)	6 260	4 363	982
Honduras[g] (A,B)	2 830*	1 201	453
Nicaragua (A,B)	2 230*	1 842	534
Caricom	4 840*	6 930**	3 790*
East Caribbean Common Market	680*	370**	312*
Antigua [hh]	71*	40**	70[c]
Belize (M)	144*	120**	103[a]
Dominica [hh]	76*	30**	21[a]
Grenada (A,B,D,E)	96*	38**	25
Montserrat (M)	13*	10**	8
St Kitts-Nevis [hh]	48	30**	16[c]
Anguilla (M)	7*	3**	
St Lucia [hh]	110*	54**	44[c]
St Vincent [hh]	107**	38**	25[c]
Other Caricom	4 160*	6 560**	3 477
Barbados (A,B,D,E)	247	360**	237
Guyana (A,D,E)	783*	470**	358
Jamaica (A,B,D,E)	2 060	3 045	913
Trinidad & Tobago (A,B,D,E)	1 070*	2 684	1 969
Other America	264 080*	1 925 530*	193 900*
Bahamas (D,E)	211	550**	3 560
Bermuda (M)	57*	350**	143
Canada (D,I,J)	23 140	194 606	40 640
Cayman Islands (M)	14	48***	36
Cuba (A,O,P)	9 460	8 500***	4 066
Dominican Republic (A,B)	4 840*	3 915	878
Falkland Islands (M)	2*	4*	3[a]
French Guiana (L)	62*	120**	86
Guadeloupe (L)	360*	550**	317
Haiti (A,B)	4 670*	1 040**	190*
Martinique (L)	369	840**	382
Netherlands Antilles[ii]	241*	500**	3 661
Panama (A,B)	1 720*	2 028	838[f]
Panama Canal Zone (N)	44*	300***	150***
Puerto Rico (N)	3 210	9 000**	5 432[b]
St Pierre & Miquelon (L)	6*	31***	26[c]
Surinam (B,E)	435*	560**	262[a]
Turks and Caicos Islands (M)	6*	8***	6[c]
United States (B,I,J)	215 120	1 702 000	130 500*[q]
Virgin Islands, British (M)	10*	28**	12[c]
Virgin Islands, US (N)	96*	550**	2 679
Asia	2 260 000*	1 351 720**	180 900*
'Middle east'	82 850*	179 280**	45 620*
Bahrain (C,F)	259*	630**	1 672
Iran (G)	33 590*	62 800*	12 890
Iraq (C,F,G)	11 510*	15 600**	3 461
Israel	3 530	13 640	5 715
Jordan (C)	2 780*	1 206	1 022
Kuwait (C,F,G)	1 060*	12 000**	3 318
Lebanon (C)	2 960*	2 000***	750*
Oman (C)	791**	2 430	725
Qatar (C,F,G)	210**	2 300**	830
Saudi Arabia (C,F,G)	9 240*	49 300*	8 694
Syria (C,F)	7 600*	5 944	2 383
United Arab Emirates (C,F,G)	700*	9 500**	3 331
Yemen, North (C)	6 870*	1 500**	412
Yemen, South (C)	1 750*	430***	414
Colombo Plan in Asia[l] (K)	1 116 440*	230 990**	45 100*
Asean	240 910*	88 582	26 255
Indonesia (G)	139 620*	37 270	5 673
Malaysia (D)	12 300*	10 993	3 921
Philippines	43 750*	17 795	3 950

	Population 000	Gdp $ mn	Imports[v] $ mn
Singapore (D)	2 280	5 915	9 067
Thailand	42 960*	16 609	3 644
Other Colombo Plan in Asia[kk]	*875 530**	*142 400***	*18 840**
Afghanistan	19 800*	2 200**	454[s]
Bangladesh (D)	78 660*	7 500*	859
Bhutan	1 200*	90**	2***
Burma	30 830*	3 700*	194
India (D)	610 080*	85 000*	5 611[o,t]
Korea, South	35 860*	25 318	8 755
Maldives	136*	20***	10*[a]
Nepal	12 860*	1 600**	163
Pakistan	72 370*	13 840*	2 134
Sri Lanka (D)	13 730*	3 132	585
Communist countries in Asia (O)	**926 000***	**358 000***	**8 400***
Cambodia	8 350*	800***	70**
China	850 000**	340 000***	6 100**
Korea, North	16 250*	7 600***	750***
Laos	3 380*	300***	65[e]
Mongolia (P)	1 490*	1 200***	550*[a]
Vietnam (P)	46 520*	8 000***	850**
Other Asia	**134 680***	**583 450***	**81 780***
Brunei[jj]	177*	1 000**	255
Hongkong (M)	4 440*	9 726	8 882
Japan (I,J)	112 770	555 157	64 853
Macao[l]	275*	210***	162
Taiwan	16 330	17 258	7 609
Timor, East	688*	100***	12[e]
Europe	**774 730***	**2 848 640***	**553 200***
EEC	**258 670**	**1 389 495[A]**	**354 336**
Belgium (H,I)	9 820	67 460	⎫ 35 492[w]
Luxembourg (H,I)	358	2 300*	⎭
Denmark (H,I)	5 070	38 527	12 409
France (H,I,J)	52 920	346 740	64 460[aa]
Germany, West (H,I,J)	61 510	445 910	89 131
Ireland (H,I)	3 160	7 975	4 212
Italy (H,I,J)	56 170	170 765	43 626[bb]
Netherlands (H,I)	13 770	89 523	40 429
United Kingdom (D,H,I,J)	55 890	220 295	64 577[cc]
EEC linked territories	**240[v]**	**980***	**550***
Greenland[j,x]	50	230**	129
Guernsey[k]	57*	230**	126[a,ee]
Isle of Man[k]	60	188*	70***[ee]
Jersey[k]	74*	330**	222[cc]
Efta	**40 800***	**248 240***	**68 832**
Austria (H,I)	7 510	40 619	11 481
Faroes[v]	41	215***	131
Finland[h] (I)	4 730	28 145	7 391
Iceland (H,I)	220*	1 454	470
Norway (H,I)	4 030	31 307	11 095
Portugal (H,I)	9 690*	16 000**	4 230
Sweden (H,I)	8 220	74 214	19 284
Switzerland (H,I)	6 350	56 284	14 750[dd]
Communist countries in Europe (O)	**387 660**	**1 040 300***	**99 420***
Comecon in Europe	*363 550*	*1 004 000***	*91 801*
Bulgaria (P)	8 760	20 000***	5 604
Czechoslovakia (P)	14 920	55 000***	9 380
Germany, East (P)	16 790	70 000***	13 196
Hungary (P)	10 600	24 000***	5 544
Poland (P)	34 360	95 000***	13 877
Rumania (P)	21 450	40 000***	6 095
Soviet Union (P)	256 670	700 000***	38 105
Other communist countries in Europe	*24 110*	*36 300***	*7 619**

	Population 000	Gdp $ mn	Imports[v] $ mn
Albania	2 550	1 300***	250***
Jugoslavia (I[h])	21 560	35 000**	7 369
Other Europe	**87 360***	**169 620***	**30 060***
Andorra	28*	150***	99[d]
Cyprus (D,H)	670**	778	505
Gibraltar (M)	30*	90**	59
Greece (H,I)	9 170	22 244	6 064
Liechtenstein (H[ll])	24*	260*	239***[dd]
Malta (D,H)	330*	479	423
Monaco	25*	210***	na[aa]
San Marino	19*	85***	na[bb]
Spain (H,I)	35 970*	104 619	17 495
Turkey (H,I)	41 090*	40 703	5 168
Vatican	1*	na	na[bb]
Oceania	**21 660***	**111 900***	**17 600***
South Pacific Commission[i]	**4 600***	**4 640***	**1 840***
Cook Islands[y]	18	18**	7[f]
Fiji (D,E,K)	588	662	263
French Polynesia (L)	134*	630**	296
Gilbert Islands (M)	59*	60**	12
Guam (N)	102*	650**	259[e]
Nauru (D[h])	7*	100***	14*[e]
New Caledonia (L)	135*	630**	277
New Hebrides[m]	97*	50***	34
Niue[y]	4	3***	2[a]
Norfolk Island[z]	2*	9***	7[b,t]
Pacific Islands, US (N)[nn]	123*	120***	38
Papua New Guinea (D,E,K)	2 830*	1 430*	499[b,t]
Pitcairn (M)	0.1*	0.2***	—
Samoa, American (N)	31*	120***	51
Samoa, Western (D,E)	151*	50**	30
Solomon Islands (D)	200*	60**	27
Tokelau[y]	2	1***	1***
Tonga[h] (D,E)	90*	40**	15
Tuvalu (M)	9*	3***	na
Wallis and Futuna Islands (L)	9*	3***	2
Other Oceania	**17 060***	**107 260***	**15 760***
Australia (D,I)	13 920	94 533	12 476
Christmas Island[z]	3*	60***	na
Cocos Islands[z]	1*	1***	na
Johnston Island (N)	1***	na	na
Midway Islands (N)	2*	14***	na
New Zealand (D,I)	3 130	12 630*	3 258[ee]
Wake Island (N)	2**	13***	na
Other main group totals			
A Sela	321 550*	375 170**	49 790*
B OAS	526 870*	2 068 800*	176 120*
C Arab League	150 870*	163 060**	45 290*
D Commonwealth[mm]	956 130*	687 130*	162 760*
E EEC ACP states	291 800*	82 570*	28 660*
F Oapec	88 460*	138 180*	36 420*
G Opec	300 710*	282 250*	51 200*
H Council of Europe	381 950*	1 778 460*	445 540*
I OECD	775 290*	4 399 010*	710 860*
J 'Summit' group[gg]	577 520	3 635 473	497 790*
K Colombo Plan[i]	1 119 860*	233 090*	45 870*
L French territories[ff]	1 630*	3 820**	1 850*
M UK territories	4 800*	10 460*	9 270*
N US territories	3 610*	10 780**	8 650*
O Communist countries	1 323 120*	1 406 800***	111 890*
P Comecon	421 020*	1 021 700***	97 270*

[a] 1975   [b] Year ending June 30th   [c] 1972   [d] From France and Spain only   [e] 1974   [f] 1973   [g] Active participation in CACM suspended since 1970   [h] Associate or member with special status; included in the total   [i] Excluding developed country members   [j] Linked via Denmark   [k] Linked via United Kingdom (dependency of the Crown)   [l] Under Portuguese administration   [m] France/United Kingdom condominium   [n] Other main groups of which countries are members (at mid-1978) are indicated by capital letters in brackets after the name; the totals for these other groups are shown at the end of the table   [o] Year ending March 31st   [p] Estimate to exclude Botswana, Lesotho, South-West Africa and Swaziland   [q] Estimate to exclude Puerto Rico   [r] Excluding free zone of Colón   [s] Year ending March 20th   [t] 1977   [u] As at mid-1978   [v] Totals include, where figures are not available, estimates for 1976   [w] Belgium-Luxembourg economic union   [x] Danish territory   [y] New Zealand territory   [z] Australian territory   [aa] The customs area for France includes Monaco   [bb] The customs area for Italy includes San Marino and Vatican   [cc] The international trade figures for United Kingdom include Guernsey, Isle of Man and Jersey   [dd] The customs area for Switzerland includes Liechtenstein   [ee] Trade with Cook Islands, Niue and Tokelau is excluded   [ff] Including overseas departments   [gg] Large western industrial countries ('Rambouillet' group)   [hh] Associated state of the United Kingdom   [ii] Netherlands territory   [jj] United Kingdom is responsible for external affairs   [kk] Excluding members which became communist countries   [ll] Proposed member; included in total   [mm] Independent states only   [nn] Including Northern Marianas

World summary

World country area 136 000 000 km² = 52 500 000 mi²

People, resources and equipment

Population 1960 2 982* mn, 1970 3 610* mn, 1976 4 040* mn
Growth: 1960–70 1.9* %pa, 1970–76 1.9* %pa
Density (1976): 30* people per km²
Education Pupils (1975) 750*mn, teachers (1974) 23.8*[q]mn
Labour force (1976) 1 670* mn; in agriculture 790* mn (47* %)
Personnel (1975) Physicians: 2.9** mn, 1 per 1 390** people
Standard of living
National income per person (1976): $ 1 520* = £ 840*
Consumption per person (1975): energy 2 028*kg coal equivalent, electricity (1976) 1 710* kW h, newsprint 5* kg, steel 161* kg
Newspapers (1975): circulation 110**[b] per 1 000 people
Telephones (Dec 1976) 398* mn, 97* per 1 000 people
Livestock (mn, 1976) Cattle 1 212*, buffaloes 131*, sheep 1 036*, goats 419*, pigs 645*, horses 66*, asses 42*, mules 14*, camels 14*, chickens 6 140*, ducks 161*, turkeys 81*
Hospital beds (1975) 14.2** mn, 1 per 280** people
Roads (1976) 22.0* mn km = 13.7* mn mi, density 0.16* km per km²
Railways (1976) 1.41* mn km = 0.88* mn mi, density 0.010* km per km²
Ships (registered, 1977) 67 945, total of 394 mn gross tons

Durable equipment (Dec 1975)	mn	no per 1 000 people	no per km of road
Radio sets	990*	245*	
Television sets	380*	94*	
Passenger cars	258*	64*	12*
Commercial vehicles	65*	16*	3*

Production

Gross domestic product 1976 est: $ 6 770** bn = £ 3 750** bn
Growth in real terms: 1960–70 5.4* %pa, 1970–75 3.9* %pa

Production indices (1970 = 100)	1960	1970	1976	Growth %pa 1960–70	1970–76
Agricultural	80	100	115	2.3	2.4
Industrial	53	100	135	6.6	5.1
Main products (mn t)					
Agriculture					
Wheat	244.0*	319.1*	418.0*	2.7*	4.6*
Rice	236.5*	308.7*	350.2*	2.7*	2.1*
Maize	204.6*	261.2*	333.1*	2.5*	4.1*
Barley	91.0*	139.4*	184.4*	4.4*	4.8*
Potatoes	284.9*	312.6*	289.9*	0.9*	−1.2*
Sweet potatoes	94.0*	129.3*	135.9*	3.2*	0.8*
Cassava	73.5*	96.6*	105.0*	2.8*	1.4*
Sugar, raw value	55.0*	74.2*	85.3*	3.0*	2.4*
Tomatoes	15.2*	30.3*	42.3*	7.1*	5.7*
Oranges	15.0**	25.6*	32.4*	5.5**	4.0*
Grapes	44.0*	55.5*	59.6*	2.3*	1.2*
Wine	24.4*	30.2*	31.7*	2.2*	0.8*
Bananas	19.3*	32.5*	37.6*	5.3*	2.5*
Soyabeans	27.2*	46.5*	62.1*	5.5*	4.9*
Coffee	4.78*	3.84*	3.68*	−2.2*	−0.7*
Cocoa	1.17*	1.51*	1.34*	2.6*	−2.0*
Tea	0.97*	1.30*	1.63*	3.0*	3.8*
Tobacco	3.76*	4.67*	5.64*	2.2*	3.2*
Rubber	2.02*	3.10*	3.59*	4.4*	2.5*
Cotton	10.5*	11.8*	12.3*	1.2*	0.7*
Wool	1.48*	1.62*	1.54*	0.9*	−0.8*
Milk	313.7*	362.2*	394.8*	1.4*	1.4*
Butter	5.45**	5.82*	6.56*	0.7**	2.0*
Cheese	6.00**	8.50**	9.99*	3.5**	2.7**
Eggs	15.2*	21.2*	24.1*	3.4*	2.2*
Beef and buffalo meat	28.1*	40.2*	47.2*	3.6*	2.7*
Sheep and goat meat	5.70*	7.18*	7.18*	2.3*	0.0*
Pigmeat	27.9*	37.1*	41.7*	2.9*	2.0*
Poultrymeat	7.0**	17.7*	23.2*	9.7**	4.6*
Fish catch	40.2*	70.4*	73.5*	5.8*	0.7*
Timber (mn m³)	2 050*	2 388*	2 524*	1.5*	0.9*

Main products (mn t)	1960	1970	1976	Growth %pa 1960–70	1970–76
Energy					
Total energy (mn tce)	4 297*	6 989*	8 555*[c]	5.0*	4.1*[d]
Coal	1 963*	2 142*	2 379*	0.9*	1.8*
Lignite	635*	794*	893*	2.3*	2.0*
Crude oil	1 054*	2 274*	2 860*	8.0*	3.9*
Petroleum products[e]	860*	1 910*	2 163*[c]	8.3*	2.6*[d]
Natural gas (bn m³)	468*	1 038*	1 326*	8.3*	4.2*
Manufactured gas (bn m³)	140*	158*	146*	1.2*	−1.3*
Electricity (bn kW h)	2 320*	4 915*	6 900*	7.8*	5.8*
of which, nuclear (bn kW h)	2*	78*	343*[c]	44.2*	34.6*[d]
Mining					
Iron ore (Fe content)	257*	424*	536*	5.1*	4.0*
Antimony ore (Sb content)	55.2*	68.4*	72.6*[c]	2.2*	1.2*[d]
Bauxite	29.0*	57.6*	75.0*	7.1*	4.5*
Chromium ore (oxide content)[r]	1.85*	2.70*	3.66*[c]	3.9*	6.3*[d]
Copper ore (Cu content)	4.27*	6.47*	7.60*	4.2*	2.7*
Lead ore (Pb content)	2.43*	3.41*	3.35*	3.4*	−0.3*
Manganese ore (Mn content)	4.90*	7.78*	9.32*[c]	4.7*	3.7*[d]
Nickel ore (Ni content)	0.337*	0.657*	0.745*[c]	6.9*	2.5*[d]
Tin conc (Sn content)[g]	0.139*	0.184*	0.182*	2.8*	−0.2*
Tungsten conc (oxide content)	0.039*	0.040*	0.047*[c]	0.3*	3.0*[d]
Uranium (U content)[h]	na	0.0182*	0.0223*	na	3.4*[d]
Zinc ore (Zn content)[f]	3.38*	5.53*	5.50*	5.0*	−0.1*
Silver (mn kg)	7.50*	9.41*	9.30**	2.3*	−0.2**
Gold (mn kg)	1.14*	1.38*	1.23*	1.9*	−1.9*
Diamonds (mn CM)	28.4*	43.6*	42.7*[c]	4.4*	−0.4*[d]
Phosphate rock	38.0**	81.8*	118.5*[c]	8.0**	7.7*[d]
Manufacturing					
Beer (mn hl)	404*	634*	787*[c]	4.6*	4.4*[d]
Cigarettes (bn units)	1 889*	2 683*	3 112*[c]	3.6*	3.0*[d]
Man-made fibres[b]	3.30*	8.33*	10.20*	9.7*	3.4*
Wood pulp	59.2*	102.5*	102.8*[c]	5.6*	0.1*[d]
Newsprint	14.0*	21.6*	22.2*	4.4*	0.5*
Other paper	60.0**	106.5*	110.8*[c]	5.9**	0.8*[d]
Synthetic rubber	2.02*	5.05*	5.20*[c]	9.6*	0.6*[d]
Sulphuric acid[b]	56.7*[j]	85.9*	100.2*[c]	6.1*[k]	3.1*[d]
Nitric acid	11.8*[j]	21.2*	22.1*[c]	8.7*[k]	0.8*[d]
Caustic soda	12.3*[j]	22.4*	24.9*[c]	8.9*[k]	2.1*[d]
Plastics and resins[b]	6.77*	29.6*	45.0**	15.9*	7.2**
Fertilisers, nitrogenous[a]	9.9*	30.2*	43.9*	11.8*	6.4*
Fertilisers, phosphate[a]	13.0*[l]	19.2*	24.9*	5.7*[m]	4.4*
Fertilisers, potash[a]	11.2*[l]	16.9*	23.5*	6.1*[m]	5.6*
Cement	317*	568*	742*	6.0*	4.6*
Coke	280*	351*	369*[c]	2.3*	1.0*[d]
Pig iron	259*	439*	488*	5.4*	1.8*
Crude steel	346*	594*	674*	5.6*	2.1*
Aluminium	4.80*	9.58*	11.96*	7.2*	3.8*
Radio sets (mn units)[b,p]	52.0*	85.7*	65.7*[c]	5.1*	−5.2*[d]
Television sets (mn units)[b,p]	19.8*	45.4*	50.9*[c]	8.7*	2.3*[d]
Passenger cars (mn units)[b]	12.8*	22.5*	29.0*	5.8*	4.3*
Commercial vehicles (mn units)[b]	3.69*	6.78*	9.98*	6.3*	6.7*
Merchant vessels (mn grt)[b,i]	8.36*	21.7*	31.2*	10.0*	6.2*

Transport traffic	1960	1970	1976	Growth %pa 1960–70	1970–76
Passenger-km (bn)					
Rail[o]	860*	1 130*	1 280*	2.8*	2.1*
Air[n]	121	460	763	14.3	8.8
Cargo: tonne-kilometres (bn)					
Rail	3 338*	5 020*	6 101*	4.2*	3.3*
Air[n]	3.3	15.2	24.5	16.3	8.2
Sea: tonnes (bn)					
Goods loaded	1.08*	2.61*	3.35*	9.2*	4.2*

[a]Years ending June 30th [b]Excluding China [c]1975 [d]1970–75 [e]Excluding Soviet Union [f]Excluding Czechoslovakia, Rumania and Vietnam [g]Excluding China, Vietnam, North Korea, Mongolia, Soviet Union and East Germany [h]Excluding China, Soviet Union and eastern Europe, and other countries for which information is not available. [i]Excluding Soviet Union and East Germany [j]1963 [k]1963–70 [l]1963–65 average [m]1961–65 to 1970 [n]Icao members only; excluding, among others, China, North Korea and Vietnam [o]Excluding, among others, Australia, China, North Korea and South Africa [p]Excluding Netherlands [q]Excluding China, North Korea and Vietnam [r]Excluding Bulgaria, Rumania and Vietnam

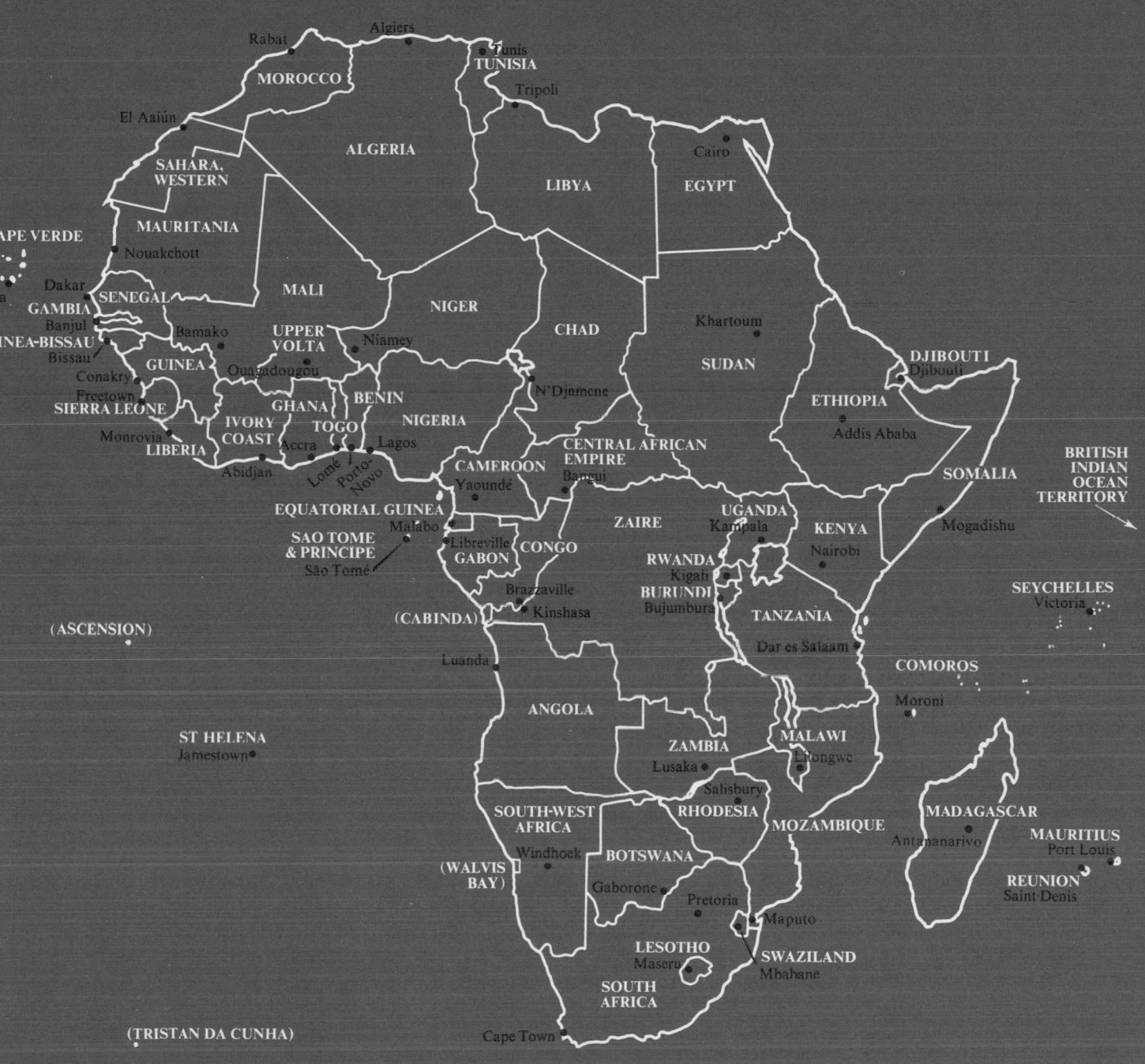

Africa

Rabat
Algiers
Tunis
TUNISIA
Tripoli
MOROCCO
El Aaiún
**SAHARA,
WESTERN**
ALGERIA
LIBYA
Cairo
EGYPT
MAURITANIA
CAPE VERDE
Nouakchott
Dakar
aia
SENEGAL
GAMBIA
MALI
Banjul
NIGER
Khartoum
GUINEA-BISSAU
Bamako
**UPPER
VOLTA**
Niamey
CHAD
SUDAN
DJIBOUTI
Bissau
Ouagadougou
N'Djamene
Djibouti
Conakry
GUINEA
Freetown
GHANA
BENIN
ETHIOPIA
SIERRA LEONE
TOGO
NIGERIA
Addis Ababa
**IVORY
COAST**
Accra
Lagos
**CENTRAL AFRICAN
EMPIRE**
Monrovia
**BRITISH
INDIAN
OCEAN
TERRITORY**
LIBERIA
Lomé
Porto-
Novo
CAMEROON
Bangui
SOMALIA
Abidjan
Yaoundé
Mogadishu
EQUATORIAL GUINEA
UGANDA
KENYA
**SAO TOME
& PRINCIPE**
Malabo
ZAIRE
Kampala
SEYCHELLES
Libreville
CONGO
RWANDA
Nairobi
Victoria
São Tomé
GABON
Kigali
(ASCENSION)
Brazzaville
BURUNDI
Kinshasa
Bujumbura
TANZANIA
(CABINDA)
Luanda
Dar es Salaam
COMOROS
Moroni
ST HELENA
Jamestown
ANGOLA
ZAMBIA
MALAWI
MADAGASCAR
Lusaka
Lilongwe
MAURITIUS
Salisbury
Antananarivo
Port Louis
**SOUTH-WEST
AFRICA**
RHODESIA
MOZAMBIQUE
REUNION
Saint Denis
(WALVIS
BAY)
Windhoek
BOTSWANA
Gaborone
Pretoria
Maputo
Maseru
SWAZILAND
Mbabane
LESOTHO
**SOUTH
AFRICA**
(TRISTAN DA CUNHA)
Cape Town

Algeria

The Democratic and Popular Republic of Algeria
Al Jumhuriya al Jaza'iriya ad Dimukratiya hash Sha'biya

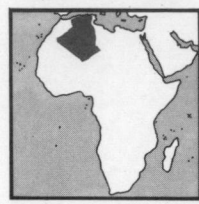

Location North-west Africa
The Mediterranean Sea forms the northern border, with Tunisia and Libya to the east, Morocco and Western Sahara to the west, and Mauritania, Mali and Niger to the south
Land Area 2 381 741 km² = 919 595 mi²
Usage (1975): agricultural 455 020* km² (19* %), of which, arable 64 500* km² (3* %), cropland 6 000* km² (0.2* %), pastures 384 520* km² (16* %); forests 24 240* km² (1* %)
Climate Temperate on the coast, varying to hot and dry in the south
Weather at Algiers, 59 m altitude
Temperature: hottest month August 22–29 °C, coldest January 9–15 °C
Rainfall (av monthly): driest month July 1 mm, wettest December 137 mm
Time 1 hour ahead of GMT (summer time, 2 hours ahead)
Measures Metric system
Monetary unit Algerian dinar (A D) = 100 centimes
Rate of exchange (1976 av): free A D 4.164 = $ 1, A D 7.521 = £ 1

Summary

Political Socialist republic, which became independent on July 3, 1962; the republic comprises mainly the former department of France (first annexed in 1842) and the former French colony of the Sahara. Member of UN, OAU, Arab League, Opec, Oapec and the Maghreb Organisation
Economic Earnings from exports are virtually all from crude oil, but one-half of the population depends on agriculture for a livelihood. The natural gas industry is being further developed, with plans for an under-sea pipeline to Italy

People

Population 1960 10.79* mn, 1970 14.33* mn, 1976 17.30* mn
Growth: 1960–70 2.9* %pa, 1970–76 3.2* %pa
Density (1976): 7* people per km²
Vital statistics (rate per 1 000 people, 1970–75): births 48.7*, deaths 15.4*
Regions (departments, population in 000, 1973; total of 14.43 mn)

Region		Region		Region	
Alger	2 179	Médéa	967	Saoura	260
Annaba	1212	Mostaganem	967	Sétif	1 328
Aurès	880	Oasis	606	Tiaret	418
Constantine	1 789	Oran	1 256	Tizi Ouzou	866
Al Asnam	895	Saïda	274	Tlemcen	534

Cities (population in 000, 1974)

Alger[a] (capital)	1 504	Annaba	313	Sétif	157
Oran	485	Tizi-Ouzou	224	Sidi-bel-Abbès	151
Constantine	350	Blida	159	Tlemcen	96[b]

[a]Algiers [b]1966
Race Arab-Berber
Language Arabic; Berber and French are also used
People speaking (1966): Arabic 80 %, Berber 19 %, French 1 %
Religion (1970) Moslem 98* %
Education (1974/75) Pupils: primary 2 499 605, secondary 327 039, vocational 84 735, teacher-training 7 955, higher 31 000*.
Teachers: primary 60 179, secondary 11 965, vocational 3 964, teacher-training 727, higher 2 900*
Labour force (1976) 3 891 000*; in agriculture 2 102 000* (54* %)
Personnel Scientists and engineers engaged in research (1972): 242
Physicians (1969): 1 698, 1 per 8 190 people
Standard of living
National income per person (1976): A D 3 500** = $ 840** = £ 460**
Consumption per person (1975): energy 754 kg coal equivalent, electricity (production) 223* kW h, newsprint 0.5 kg, steel 73 kg
Newspapers (1974): number 4; circulation 275 000, 17 per 1 000 people
Telephones (Dec 1975): 250 000, 15 per 1 000 people

Resources and equipment

Livestock (000, 1976) Cattle 1 281*, sheep 8 886*, goats 2 400*, camels 157*
Mineral reserves Coal (1957) 20 mn tonnes
Crude oil (1975) 1 280 mn tonnes
Natural gas (1975) 3 271 bn cubic metres
Uranium (1974) 28 000 tonnes
Petroleum refinery capacity (1975) 5.4 mn tonnes

Electrical capacity (1975) 1 110* megawatts, of which, hydro 286* megawatts
Hospital beds (1969) 39 053, 1 per 356 people
Roads (1974) 78 408 km = 48 720 mi, density 0.03 km per km²
Railways (1974) 4 074 km = 2 531 mi, density 0.002 km per km²
Ships (registered, 1977) 112, total of 1 055 962 gross tons
Ports (goods traffic, 000 tonnes, 1973)

	loaded	unloaded		loaded	unloaded
Arzew[a]	18 621	161	Annaba	2 220	1 286
Bejaia[a]	9 680	510	Oran	580	1 068
Algiers	1 329	2 392			

[a]Crude oil port
Airports Dar al Beïda (Algiers), Annaba, Constantine, Oran and 14 other airports with scheduled flights

Durable equipment (Dec 1974)	000	no per 1 000 people	
Radio sets	3 220	195	no per km of road
Television sets	410	25	
Passenger cars	204	12	2.6
Commercial vehicles	103	6	1.3

Production

Gross domestic product 1973: A D 29 700 mn = $ 7 100 mn = £ 2 897 mn
1976 est: A D 65 000** mn = $ 15 600** mn = £ 8 600** mn
Structure of gross domestic product

By origin (1972)	A D bn	% of gdp
Agriculture	3.0	11
Mining and quarrying	2.8	10
Manufacturing	3.8	14
Electricity, gas and water	0.4	1
Construction	2.5	9
Distribution and hotels	8.6	31
Transport and communication	2.3	8
Public administration	3.1	11
Other services	1.1	4
Total	27.5	100

By type of expenditure (1973)		
Government final consumption	4.7	16
Private final consumption	14.5	49
Stock investment	0.7	2
Gross fixed capital formation	12.0	40
Exports of goods and services	7.8	26
less Imports of goods and services	−10.0	−34
Total	29.7	100

Production index (1970 = 100)	1960	1970	1976	Growth %pa 1960–70	1970–76
Agricultural	116	100	116	−1.5	2.5

Main products (000 t)	1960	1970	1976	Growth %pa 1960–70	1970–76
Agriculture					
Wheat	1 505	1 435	2 000*	−0.5	5.7*
Barley	843	571	600*	−3.8	0.8*
Potatoes	259	262	576*	0.1	14.0*
Tomatoes	154	119	135*	−2.5	2.1*
Onions	83	38	107*	−7.5	18.8*
Watermelons	160**	190	300*	1.7**	7.9*
Oranges and mandarins	376	487	506	2.6	0.6
Olives	131	155	150*	1.7	−0.5*
Dates	89	144	136*	4.9	−0.9*
Grapes	2 035*	1 138	680*	−5.6*	−8.2*
Wine	1 585	869	500*	−5.8	−8.8*
Milk	97	313*	380*	12.4*	3.3*
Beef and veal	21	21*	29*	0.0*	5.5*
Sheep and goat meat	42	43*	55*	0.2*	4.2*
Fish catch	26	26	37[f]	0.0	7.3[g]
Energy					
Total energy (000 tce)	11 440	66 530	83 350[f]	19.3	4.6[g]
Crude oil	8 632	47 202	50 090	18.5	1.0
Petroleum products	1 490[d]	2 275	5 091[f]	7.3[e]	17.5[g]
Natural gas (mn m³)	231[b]	2 838	9 532[f]	32.1[e]	27.4[g]
Electricity (mn kW h)	1 387[b]	1 979	3 744*[f]	4.0[e]	13.6*[g]
Mining					
Iron ore (Fe content)	1 788	1 546	1 728[f]	−1.4	2.2[g]
Antimony ore (Sb content)	804	60*	60*[f]	−22.9	0.0*[g]
Zinc ore (Zn content)	40	18	11[f]	−7.6	−9.2[g]
Phosphate rock	563	493	802[h]	−1.3	12.9[i]

Algeria

Main products (000 t)	1960	1970	1976	Growth %pa	
Manufacturing				1960–70	1970–76
Cement	1 062	928	940[h]	−1.3	0.3[i]
Cotton fabrics (mn m²)	–	44	46*[h]	na	1.1*[i]
Crude steel	31	31	181[h]	0.0	55.4[i]
Fertilisers, nitrogenous[a]	–	15	61	na	26.3
Fertilisers, phophate[a]	14	16	64	1.3	26.2
Motor vehicles (000 units)[j]	1.2	3.0	6.3[h]	9.6	20.4[i]

[a]Years ending June 30th [b]1961 [c]1961–70 [d]1964 [e]1964–70 [f]1975 [g]1970–75
[h]1974 [i]1970–74 [j]Assembly only

Transport traffic	1960	1970	1976	Growth %pa	
Passenger-kilometres (mn)				1960–70	1970–76
Road[a]	na	2,510	4 617[d]	na	22.5[e]
Rail	626	1 013	1 058[f]	4.9	1.1[g]
Air	238[b]	515	1 881	10.1[c]	24.1
Cargo: tonne-kilometres (mn)					
Road	na	1 936	3 136[d]	na	17.4[e]
Rail	1 728	1 404	1 901[f]	−2.1	7.9[g]
Air	2.8[b]	3.7	8.8	3.4[c]	15.5
Sea: tonnes (mn)					
Goods loaded	14.2	44.2	44.8[f]	12.0	0.3[g]
Goods unloaded	7.4	7.8	9.4[f]	0.5	4.8[g]

[a]Public transport only [b]1962 [c]1962–70 [d]1973 [e]1970–73 [f]1974 [g]1970–74
Tourism (1975) Number of visitors 296 000, gross receipts $ 51 mn

Finance and trade

Money stock (end-year) 1964 A D 4 639 mn, 1970 A D 11 625 mn,
1976 A D 39 587 mn; growth: 1964–70 16.5 %pa, 1970–76 22.7 %pa
Budget (1976) Balanced at A D 14 600 mn = $ 3 500 mn = £ 1 900 mn

Balance of payments ($ mn)	1972	1973	1974	1975	1976
Balance of goods (fob)	−97	−309	+934	−1 011	na
Balance of services	−329	−515	−744	−1 076	na
Balance of transfers	+302	+380	−30	+425	na
Current balance	*−124*	*−443*	*+160*	*−1 661*	*na*
Long-term capital flow	+201	+1 038	+524	+1 371	na
Reserves and debt (end-year, $ mn)					
International reserves	493	1 143	1 689	1 353	1 987
External public debt	2 697	4 919	6 005	9 590	11 717

External trade (1976)
Imports: A D 22 123 mn = $ 5 311 mn = £ 2 940 mn
Exports: A D 21 499 mn = $ 5 162 mn = £ 2 858 mn

Main imports (1975)	% of total	Main exports (1976)	% of total
Machinery, non-electric	22	Crude oil	86
Iron and steel	13	Petroleum products	5
Chemicals	9*	Wine	3*
Motor vehicles	8		
Sugar	7		
Electrical machinery	7		
Cereals	6		
Textiles	3*		

Main sources (1975)		Main destinations (1975)	
France	34	United States	27
West Germany	12	West Germany	19
United States	11	France	15
Italy	8	Italy	11
Brazil	4	United Kingdom	4
Japan	4	Netherlands	3
Spain	4	Spain	3
United Kingdom	3	Belgium-Luxembourg	2
Belgium-Luxembourg	3	Soviet Union	2

Special focus

Government investment program 1976

	A D mn	% of total
Agriculture	1 856	7
Industry and tourism	12 405	48
Water supplies	747	3
Housing	1 578	6
Transport and communication	2 425	9
Education	1 855	7
Regional development	1 750	7
Other	3 404	13
Total	26 020	100

Angola
The People's Republic of Angola

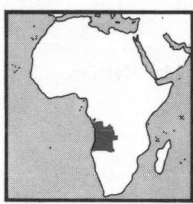

Location West south Africa
On the west coast of Africa, with South-West
Africa to the south, Zambia to the east, and
Zaire and Congo to the north. Cabinda, an
enclave in the north, is separated from the
rest of Angola by a Zaire corridor to the sea,
which includes the Zaire (Congo) estuary
Land Area 1 246 700 km² = 481 354 mi²
Climate Tropical in the north, sub-tropical
in the south, temperate on the high plateau
Weather at Lobito, 1m altitude
Temperature: hottest month March 24–31 °C, coldest Aug 17–23 °C
Rainfall (av monthly): driest months June, July 0 mm, wettest March
119 mm
Time 1 hour ahead of GMT
Measures Metric system
Monetary unit Kwanza (Kw) = 100 lwei; the kwanza replaced the
Angolan escudo at par from January 10, 1977
Rate of exchange (1976 av): free Kw 30.22 = $ 1, Kw 54.59 = £ 1

Summary

Political One-party socialist republic, which became independent on
November 11, 1975; formerly a Portuguese overseas province. Member of
UN and OAU
Economic Rich in oil, diamonds and iron ore; crude oil (mainly from
Cabinda) overtook coffee, the main export of the 1960s, in 1973.
Industry is being developed, but about two-thirds of the labour force
remains in agriculture

People

Population 1960 4.84* mn, 1970 5.67* mn, 1976 6.56* mn
Growth: 1960–70 1.6* %pa, 1970–76 2.4* %pa
Density (1976): 5* people per km²
Vital statistics (rate per 1 000 people, 1970–75): births 47.2*, deaths 24.5*
Regions (districts[a], population in 000, 1970; total of 5.67 mn)

Benguela	475	Cuanza Sul	459	Malanje	559
Bié	650	Huambo	838	Moçâmedes	53
Cabinda	81	Huila	645	Moxico	213
Cuando Cubango	112	Luanda	561	Uige	386
Cuanza Norte	298	Lunda	303	Zaire	42

[a] Reorganised since independence, with a new district, Cunenex
Cities (population in 000, 1970)

Luanda (capital)	481[b]	Benguela	41	Cabinda	21
Huambo[a]	62	Sá da Bandeira	32		
Lobito	60	Malanje	32		

[a]Formerly Nova Lisboa [b]1972
Race (1976) African 95** %, Portuguese 5** %
Language Portuguese and numerous Bantu African languages
Religion Mainly Animist, also Roman Catholic and Protestant
Education (1972/73) Pupils: primary 516 131, secondary and vocational
75 667, teacher-training 3 388, higher 2 660. Teachers: primary 12 622,
secondary and vocational 4 394, teacher-training 330, higher 324
Labour force (1976) 1 763 000*; in agriculture 1 061 000* (60* %)
Personnel (1973) Physicians: 383, 1 per 15 404 people
Standard of living
National income per person (1976): Kw 15 000** = $ 500** = £ 270**
Consumption per person (1975): energy 174 kg coal equivalent,
electricity (production) 204* kW h, newsprint 0.1 kg, steel 9 kg
Newspapers (1974): number 4; circulation 78 000, 13 per 1 000 people
Telephones (Dec 1973): 38 000, 6 per 1 000 people

Resources and equipment

Livestock (000, 1976) Cattle 3 000*, sheep 205*, goats 910*, pigs 360*
Mineral reserves (1975) Crude oil 184 mn tonnes
Natural gas 49 bn cubic metres
Petroleum refinery capacity (1975) 1.75 mn tonnes
Electrical capacity (1975) 510* megawatts, of which, hydro 368*
megawatts
Hospital beds (1972) 18 011, 1 per 322 people
Roads (1974) 72 323 km = 44 939 mi, density 0.06 km per km²
Railways (1975) 3 000* km = 1 860* mi, density 0.002 km per km²
Ships (registered, 1977) 22, total of 22 043 gross tons

Angola

Ports (goods traffic, 000 tonnes, 1973)

	loaded	unloaded		loaded	unloaded
Cabinda[a]	7 412	19	Lobito	1 322	1 009
Moçâmedes	6 168	155	Luanda	863	880

also Novo Redondo, Porto Amboim, Ambriz
[a]Crude oil port

Airports Luanda, Moçâmedes, Sá da Bandeira, Cabinda, Malanje, Benguela

Durable equipment (Dec 1974)	000	no per 1 000 people	no per km of road
Radio sets	116	18	
Passenger cars	134	21	1.9
Commercial vehicles	27	4	0.4

Production

Gross domestic product 1970: Kw 47 290 mn = $ 1 645 mn = £ 685 mn
1976 est: Kw 105 000**mn = $ 3 500**mn = £ 1 900**mn

Production index (1970 = 100)	1960	1970	1976	Growth %pa	
				1960–70	1970–76
Agricultural	79	100	80	2.4	−3.7

Main products (000 t)

Agriculture					
Maize	376	456	450*	1.9	−0.2*
Cassava	1 250	1 600*	1 600*	2.5*	0.0*
Bananas	na	300	300*	na	0.0*
Palm oil	36	38*	40*	0.5*	0.9*
Coffee	161	204	72*	2.4	−15.9*
Cotton	6	30	13*	17.6	−13.0*
Sisal	58	62	65*	0.6	0.8*
Milk	110	134*	140*	2.0*	0.7*
Beef and veal	17	41	44*	9.2	1.2*
Fish catch	252	368	184[c]	3.9	−13.0[d]
Timber (000 m³)	5 630	7 280	7 836[c]	2.6	1.5[d]
Energy					
Total energy (000 tce)	100	6 650	11 920[c]	52.2	12.4[d]
Crude oil	67	5 065	6 281	54.1	3.7
Petroleum products	180	657	800*[c]	13.8	4.0*[d]
Electricity (mn kW h)	143	644	1 305*[c]	16.2	15.1*[d]
Mining					
Iron ore (Fe content)	409	3 752	3 388[c]	24.8	−2.0[d]
Salt	58	88	100*[c]	4.3	2.6*[d]
Diamonds (000 CM)	1 057	2 396	460[c]	8.5	−28.1[d]
Manufacturing					
Beer (000 hl)	99	708	1 196[g]	21.7	19.1[h]
Fishmeal	45	63	63[c]	3.4	0.0[f]
Cigarettes (mn units)	1 065	2 016	2 400[c]	6.6	3.5[d]
Cement	161	446	699[c]	10.7	9.4[d]
Motor vehicles[a] (000 units)	na	1.24	1.00[e]	na	−5.2[f]
Motor cycles[b] (000 units)	na	2.0	6.1[i]	na	46.3[j]

[a]Assembly only [b]Including assembly [c]1975 [d]1970–75 [e]1974 [f]1970–74 [g]1973 [h]1970–73 [i]1972 [j]1970–72

Transport traffic	1960	1970	1976	Growth %pa	
				1960–70	1970–76
Passenger-kilometres (mn)					
Rail	99	204	418[a]	7.5	19.6[b]
Air	19	72	100	14.3	5.6
Cargo: tonne-kilometres (mn)					
Road	na	2 477	3 127[c]	na	12.4[d]
Rail	1 625	5 230	5 461[a]	12.4	1.1[b]
Water	na	180	220[c]	na	10.6[d]
Air	0.4	1.8	1.7	16.1	−0.9
Sea: tonnes (mn)					
Goods loaded	1.9	12.2	10.0[a]	20.8	−4.7[b]
Goods unloaded	0.65	1.37	3.98[a]	10.2	30.5[b]

[a]1974 [b]1970–74 [c]1972 [d]1970–72

Finance and trade

Budget (1974) Balanced at Kw 19 475 mn = $ 766 mn = £ 328 mn
External trade (1974) Imports: Kw 15 853 mn = $ 625 mn = £ 267 mn
Exports: Kw 31 215 mn = $ 1 227 mn = £ 525 mn

Main imports (1973)	% of total	Main exports (1973)	% of total
Machinery	25	Crude oil	30
Chemicals	13	Coffee	27
Motor vehicles	10	Diamonds	10
Food	9	Iron ore	6
Textile yarns and fabrics	7	Fishmeal	4
Iron and steel	6	Cotton	3
Petroleum and products	5	Sisal	3

Main sources (1973)		Main destinations (1973)	
Portugal	26	United States	28
West Germany	13	Portugal	25
United States	10	Canada	10
United Kingdom	8	Japan	9
France	7	West Germany	5

Special focus

Main exports (% of total exports)

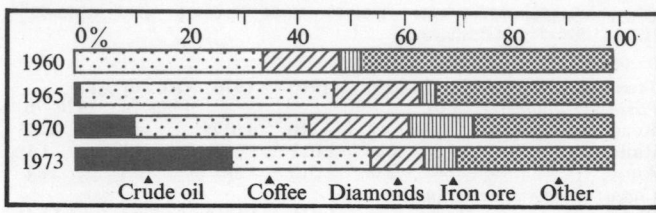

	0%	20	40	60	80	100
1960						
1965						
1970						
1973						

Crude oil   Coffee   Diamonds   Iron ore   Other

Benin

People's Republic of Benin
République Populaire du Bénin

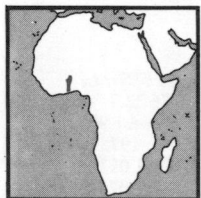

Location West Africa
The hinterland runs north from a coastal strip on the Atlantic Ocean; Togo is to the west, Nigeria to the east, and Upper Volta and Niger to the north
Land Area 112 622 km² = 43 484 mi²
Climate Tropical, hot and dry in the north
Weather at Cotonou, 7 m altitude
Temperature: hottest month March 26–28 °C, coldest August 23–25 °C
Rainfall (av monthly): driest month Dec 13 mm, wettest June 366 mm
Time 1 hour ahead of GMT
Measures Metric system
Monetary unit CFA franc (CFA Fr) = 100 centimes
Rate of exchange (1976 av): par CFA Fr 50 = Fr 1, free CFA Fr 239.0 = $ 1, CFA Fr 431.6 = £ 1

Summary

Political Socialist republic with military government, which became independent on August 1, 1960; formerly one of the provinces of French West Africa. Member of UN, OAU, Ocam, Ecowas, franc zone and an EEC ACP state. Before November 30, 1975 the name of this country was Dahomey
Economic An agricultural economy, with cotton, cocoa and palm products the main export earners. There is little mining or industrial activity, although there are oil prospects off-shore

People, resources and equipment

Population 1960 2.11* mn, 1970 2.72* mn, 1976 3.20* mn
Growth: 1960–70 2.6* %pa, 1970–76 2.7* %pa
Density (1976): 28* people per km²
Vital statistics (rate per 1 000 people, 1970–75): births 49.9*, deaths 23.0*
Cities (population in 000, 1975) Porto Novo (capital) 104, Cotonou 178, Natitingou 49*[a], Abomey 38*[a], Ouidah 25*[a], Parakou 21*[a]
[a]1973
Race (1969) Fon 32 %, Adja 8 %, Bariba 7 %, Yoruba 6 %
Language French; Fon, Yoruba and other African languages are also used
Religion (1976) Animist 70** %, Christian 15** %, Moslem 13** %
Education (1973/74) Pupils 286 711, teachers 6 090*
Labour force (1976) 1 487 000*; in agriculture 706 000* (47* %)
Personnel (1974) Physicians: 84, 1 per 36 071 people
Standard of living
National income per person (1976): CFA Fr 29 000** = $ 121** = £ 67**
Consumption per person (1975): energy 52 kg coal equivalent, electricity (production) 18 kW h
Newspapers (1974): number 1; circulation 1 000, 0.3 per 1 000 people
Telephones (Dec 1974): 8 000, 3 per 1 000 people
Livestock (000, 1976) Cattle 800*, sheep 850*, goats 840*, pigs 365*
Electrical capacity (1975) 15 megawatts
Hospital beds (1974) 3 667, 1 per 826 people
Roads (1974) 6 937 km = 4 310 mi, density 0.06 km per km²
Railways (1973) 579 km = 360 mi, density 0.005 km per km²

Benin

Ships (registered, 1977) 7, total of 912 gross tons
Port (1973) Cotonou (goods traffic, 000 tonnes) loaded 142, unloaded 546
Airport (1975) Cotonou: passenger departures and arrivals 45 113

Durable equipment	000	no per	
(at end-year)		1 000 people	
Radio sets (1972)	150	52	no per
Television sets (1972)	0.1	0.03	km of road
Passenger cars (1974)	14	5	2.0
Commercial vehicles (1974)	9	3	1.2

Production, finance and trade

Gross domestic product 1974/75 (year ending June 30th):
CFA Fr 80 800 mn = $ 363 mn = £ 155 mn
1976 est: CFA Fr 100 000**mn = $ 420**mn = £230**mn

Production index	1960	1970	1976	Growth %pa	
(1970 = 100)				1960–70	1970–76
Agricultural	84	100	94	1.7	−1.1
Main products (000 t)					
Maize	220*	229	221	0.4*	−0.6
Cassava	1 235ᵃ	736	381	−5.6ᵇ	−10.4
Yams	550**	534	450**	−0.3**	−2.8**
Groundnuts	18	57	46	12.2	−3.5
Palm kernels	50*	61	70*	2.0*	2.3*
Palm oil	24*	32	36*	2.9*	2.0*
Coffee	2.8	1.3	2.5*	−7.2	11.5*
Cottonseed	2	18	30*	24.8	8.5*
Cotton	2*	14	15*	21.5*	1.2*
Fish catch	28	32	29*ᶜ	1.2	−2.0*ᵈ
Timber (000 m³)	1 335	2 055	2 320ᶜ	4.4	2.4ᵈ
Electricity (mn kW h)	10*	33	57ᶜ	12.7ᵃ	11.5ᵈ

ᵃ1961 ᵇ1961–70 ᶜ1975 ᵈ1970–75

Transport traffic	1960	1970	1976	Growth %pa	
Passenger-kilometres (mn)				1960–70	1970–76
Rail	71	72	97ᵈ	0.1	6.1ᵉ
Airᵃ	41ᵇ	65	125	7.0ᶜ	11.5
Cargo: tonne-kilometres (mn)					
Rail	46	95	127ᵈ	7.5	6.0ᵉ
Airᵃ	2.5ʰ	6.3	13.8	14.0ᶜ	14.0
Sea: tonnes (mn)					
Goods loaded	0.13	0.17	0.13ᶠ	3.2	−7.3ᵍ
Goods unloaded	0.18	0.39	0.61ᶠ	8.0	12.1ᵍ

ᵃApportionment of Air Afrique ᵇ1963 ᶜ1963–70 ᵈ1975 ᵉ1970–75 ᶠ1974 ᵍ1970–74

Budget (1976) Balanced at CFA Fr 16 080 mn = $ 67 mn = £ 37 mn
External public debt (Dec 1975) $ 161 mn
External trade (1974) Imports: CFA Fr 35 174 mn = $ 146 mn = £ 62 mn
Exports: CFA Fr 8 185 mn = $ 34 mn = £ 15 mn

Main imports (1972)	% of total	*Main exports* (1972)	% of total
Textile yarns and fabrics	14	Cotton	28
Machinery	13	Cocoaᵃ	19
Food	13	Palm products	17
Chemicals	9	Cottonseed	5
Motor vehicles	7		
Main sources (1973)		*Main destinations* (1973)	
France	36	France	36
West Germany	7	West Germany	12
China	7	Netherlands	10

ᵃProduce grown mainly in Nigeria

Botswana

Republic of Botswana

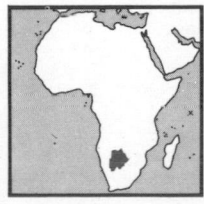

Location South central Africa
Rhodesia and Zambia are to the north-east,
South Africa to the south, and South-West
Africa to the west and north. A large part
forms the Kalahari Desert. Land-locked
Land Area 600 372 km² = 231 805 mi²
Climate Sub-tropical
Weather at Francistown, 1 004 m altitude
Temperature: hottest months Dec, Jan 18–31 °C,
coldest June 5–23 °C
Rainfall (av monthly): driest months July, Aug 1 mm, wettest Jan 107 mm
Time 2 hours ahead of GMT

Measures Metric system, which replaced the UK (imperial) system from December 1, 1974
Monetary unit Pula (Pu) = 100 thebe; the pula replaced the South African rand at par from August 23, 1976
Rate of exchange (1976 av): par $ 1.15 = Pu 1, free Pu 1.571 = £ 1

Summary

Political Republic, which became independent on September 30, 1966; formerly a UK protectorate and known as Bechuanaland. Member of UN, OAU, Commonwealth and an EEC ACP state
Economic Mainly an agricultural economy based on cattle, but with an important mining industry (especially for diamonds and nickel). About one-half of the non-agricultural labour force works in South African mines

People, resources and equipment

Population 1960 450 000**, 1970 579 000*, 1976 690 000*
Growth: 1960–70 2.5** %pa, 1970–76 3.0* %pa
Density (1976): 1* person per km²
Vital statistics (rate per 1 000 people, 1970–75): births 45.6*, deaths 23.0*
Cities (population in 000, 1976)
Gaborone (capital) 33* Selebi-Pikwe 21* Kanye 17*
Francistown 23* Serowe 20* Lobatsi 15*
Race (1964) African 98.6 %, European 0.7 %, Asian 0.1 %
Language Tswana and English; also various Tswana dialects
Religion (1970) Christian 60* %, Animist 40* %
Education (1975) Pupils 130 900*, teachers 4 410*
Labour force (1976) 340 000*; in agriculture 282 000* (83* %)
Working in South African mines (1973) 28 953, remittances and deferred pay Pu 2 085 000 = $ 3 011 000 = £ 1 228 000
Personnel Scientists and engineers (1973): 786
Physicians (1974): 63, 1 per 10 476 people
Standard of living
National income per person (1976): Pu 340** = $ 390** = £ 220**
Production per person (1975): electricity 400 kW h
Newspapers (1974): number 1; circulation 13 000, 20 per 1 000 people
Telephones (Dec 1975): 8,000, 12 per 1 000 people
Livestock (000, 1976) Cattle 2 200*, sheep 425*, goats 1 050*
Mineral reserves (known recoverable, 1961) Coal 506 mn tonnes
Electrical capacity (1975) 92 megawatts
Hospital beds (1975) 2 074, 1 per 330 people
Roads (1976) 10 219 km = 6 350 mi, density 0.02 km per km²
Railways (1975) 634 km = 394 mi, density 0.001 km per km²
Airports Gaborone, Francistown, Selebi-Pikwe

Durable equipment	000	no per	
(at end-year)		1 000 people	no per
Radio sets (1974)	55	81	km of road
Passenger cars (1976)	4.1	6	0.4
Commercial vehicles (1976)	7.8	11	0.8

Production, finance and trade

Gross domestic product
1974/75 (year ending June 30th): Pu 203.9 mn = $ 298 mn = £ 127 mn
1976 est: Pu 240** mn = $ 280** mn = £ 150** mn
Agricultural production index (1970 = 100) 1960 90, 1970 100, 1976 139
Growth: 1960–70 1.1 %pa, 1970–76 5.7 %pa
Main products (1976) *Agriculture* (000 t) Sorghum 56, maize 62, millet 5*, groundnuts 7*, milk 75*, beef and veal 46*, hides and skins 4.4, timber (000 m³, 1975) 739
Other Electricity (1975) 274 mn kW h, coal (1976) 223 000 t, diamonds 2.41 mn metric carats, beer (1973) 130 000 hl
Transport traffic Rail traffic is included with the total for Rhodesia
Budget (1975/76; year ending March 31st)
Revenue: Pu 79.3 mn = $ 91 mn = £ 43 mn
Expenditure: Pu 73.3 mn = $ 84 mn = £ 40 mn
External public debt (Dec 1976) $ 212 mn
External trade (1975)
Imports: Pu 159 mn = $ 218 mn = £ 98 mn
Exports: Pu 105 mn = $ 144 mn = £ 65 mn
Included in South Africa Customs Union

Main imports	% of total	*Main exports*	% of total
Food	18	Meat and products	35
Motor vehicles	12	Diamonds	31
Machinery	10	Nickel	21
Petroleum products	10		
Iron and steel	8		
Main source		*Main destinations*	
South Africa	80	United Kingdom	47
		South Africa	24
		United States	22

British Indian Ocean Territory

Chagos Archipelago

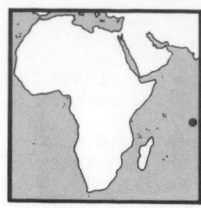

Location Western Indian Ocean
The archipelago is about 1 800 km east of
Seychelles
Land Area 60 km² = 23 mi² of which,
Diego Garcia 44 km²
Climate Tropical maritime
Weather at Diego Garcia
Temperature (annual av): max 29°C, min 25 °C
Rainfall (monthly av): 200 mm
Time 5 hours ahead of GMT
Measures US, UK (imperial) and metric systems
Monetary unit US dollar and UK pound

Summary

Political UK territory set up November 8, 1965; consists of the Chagos
Archipelago, formerly dependency of Mauritius. The islands of Aldabra,
Desroches and Farquhar, dependencies of Seychelles before 1965, and
part of the Territory from 1965, returned to Seychelles on June 28, 1976.
The Chagos Archipelago includes Diego Garcia, where there is a United
States naval support facility developed in terms of a UK–US agreement
in 1976
Economic Used as a defence base and communications centre. Some
1 200 residents were transferred to Mauritius on the development of a
defence base

People

Population 1960 1 130*, 1976 1 500**. There is no permanent population;
inhabitants are mainly service personnel and construction workers
Density (1976): 25** people per km²
The archipelago is administered from London, United Kingdom

Burundi

Republic of Burundi
République du Burundi
Republika y'u Kirundi

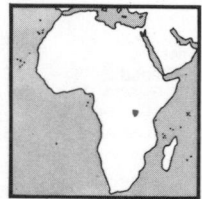

Location Central Africa
Lake Tanganyika is the south-west border, with
Zaire to the west, Rwanda to the north, and
Tanzania to the east and south. Land-locked
Land Area 27 834 km² = 10 747 mi²
Climate Tropical
Weather at Bujumbura
Temperature: average 24 °C
Rainfall (av monthly): 65 mm; driest months
June–September, wettest February–May
Time 2 hours ahead of GMT
Measures Metric system
Monetary unit Burundi franc (Bu Fr) = 100 centimes
Rate of exchange (1976 av): par Bu Fr 86.25 = $ 1, free Bu Fr 155.8 = £ 1

Summary

Political One-party republic, which became independent on July 1, 1962;
a customs union was formed with Rwanda until September 30, 1964.
Before 1962 Rwanda-Urundi was administered by Belgium as a trust
territory. Member of UN, OAU and an EEC ACP state
Economic Mainly dependent on coffee as an export crop; there are also
some earnings from cotton, hides and tea. There is some mining, and
exploitation of important nickel and uranium deposits is planned

People, resources and equipment

Population 1960 2.90* mn, 1970 3.35*mn, 1976 3.86* mn
Population includes 50 000* Tutsi refugees from Rwanda (1976), and
excludes 150 000* Hutu refugees from Burundi (of whom, 110 000* in
Tanzania, 25 000* in Zaire, 8 000* in Rwanda)
Growth: 1960–70 1.5* %pa, 1970–76 2.4* %pa
Density (1976): 139* people per km²
Vital statistics (rate per 1 000 people, 1970–71): births 42.0, deaths 20.4

Regions (provinces, population in 000, 1965; total of 3.21 mn)

| Bubanza | 395 | Bururi | 385 | Muramvya | 351 | Ngozi | 663 |
| Bujumbura | 97 | Gitega | 535 | Muyinga | 471 | Ruyigi | 314 |

Cities (population in 000, 1976) Bujumbura (capital) 100*, Muyinga 25*,
Gitega 10*
Race (1976) African 99½ % (Hutu 84* %, Tutsi 15* %), European 0.1* %
Language Kirundi and French; Swahili is also used
Religion (1965) Roman Catholic 50** %, Animist 45** %,
Protestant 4** %
Education Pupils (1974/75) 143 260, teachers (1971/72) 5 964
Labour force (1976) 1 898 000*; in agriculture 1 608 000* (85* %)
Personnel (1973) Physicians: 74, 1 per 49 200 people
Standard of living
National income per person (1976): Bu Fr 7 500** = $ 87** = £ 48**
Consumption per person (1975): energy 13 kg coal equivalent,
electricity 6* kW h
Newspapers (1974): number 1; circulation 1 200, 0.3 per 1 000 people
Telephones (Dec 1975): 4 000, 1 per 1 000 people
Livestock (000, Dec 1975) Cattle 800, sheep 311, goats 653, pigs 46
Electrical capacity (1975) 6 megawatts
Hospital beds (1972) 4 221, 1 per 840 people
Roads (1974) 2 987 km = 1 856 mi, density 0.11 km per km²
There are no railways
Inland waterway Lake Tanganyika
Port (on Lake Tanganyika): Bujumbura
Airport (1975) Bujumbura: passenger departures and arrivals 27 525
Durable equipment (Dec 1974) Radio sets: 100 000*, 27* per 1 000 people
Passenger cars: 4 200, 1.1 per 1 000 people, 1.4 per km of road
Commercial vehicles: 1 700, 0.5 per 1 000 people, 0.6 per km of road

Production, finance and trade

Gross domestic product 1970: Bu Fr 20 100 mn = $ 230 mn = £ 96 mn
1976 est: Bu Fr 32 000** mn = $ 370** mn = £ 205** mn

Production index (1970 = 100)	1960	1970	1976	Growth %pa 1960–70	1970–76
Agricultural	83	100	112*	1.8	1.9*
Main products (000 t)					
Cassava	1 361	1 000**	896*	−3.0**	−1.8**
Maize	94	182	160*	6.8	−2.1*
Sweet potatoes	648	800**	842*	2.1**	0.9**
Dry beans	152	170	150*	1.1	−2.1*
Bananas	na	987	915*	na	−1.3*
Coffee	10	23	21*	8.7	−1.5*
Cotton	3.0*	3.3	2.0*	1.0*	−8.0*
Beef and veal	6	9	11*	4.0	3.4*
Hides	1.7*	2.8	3.4*	5.0*	3.3*
Fish catch	9	13	15ᶠ	3.8	1.8ᵍ
Timber (000 m³)	690*	842*	912ᶠ	2.0*	1.6*ᵍ
Electricity (mn kW h)ᵃ	15ᵇ	22	23*ᶠ	6.6ᶜ	0.9*ᵍ
Tin conc (Sn content)	0.1ᵈ	0.1*	0.1*	0.0*ᵉ	0.0*
Beer (000 hl)	156	215	380ʰ	3.3	15.3ⁱ

ᵃIncludes hydro-electricity provided by Zaire ᵇ1964 ᶜ1964–70 ᵈ1966 ᵉ1966–70
ᶠ1975 ᵍ1970–75 ʰ1974 ⁱ1970–74
Transport traffic (1976) *Air* 1 mn passenger-km
Budget (1975) Revenue: Bu Fr 3 225 mn = $ 41 mn = £ 18 mn
Expenditure: Bu Fr 3 283 mn = $ 42 mn = £ 19mn
International reserves (Dec 1976) $ 49 mn
External public debt (Dec 1976) $ 74 mn
External trade (1976) Imports: Bu Fr 5 027 mn = $ 58 mn = £ 32 mn
Exports: Bu Fr 5 404 mn = $ 63 mn = £ 35 mn

Main imports (1974)	% of total	*Main exports* (1976)	% of total
Food	14	Coffee	89
Textile fabrics	13	Cotton	4
Machinery	7	Hides and skins	2*
Petroleum products	7	Tea	2*
Main sources (1976)		*Main destinations* (1976)	
Belgium-Luxembourg	18	United States	43
France	13	West Germany	17
West Germany	11	Belgium-Luxembourg	5
Iran	8	United Kingdom	5
Netherlands	6	France	4

Cameroon

United Republic of Cameroon
République Unie du Cameroun

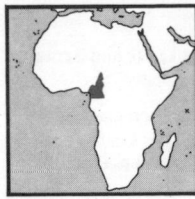

Location West central Africa
On the west coast of Africa, with Nigeria to the north-west, Chad to the north-east, Central African Empire to the east, and Congo, Gabon and Equatorial Guinea to the south
Land Area 475 442 km² = 183 569 mi²
Climate Tropical
Weather at Yaoundé, 770 m altitude
Temperature: hottest months Jan–Mar 19–29 °C, coldest Aug 18–27 °C
Rainfall (av monthly): driest month Jan 23 mm, wettest October 295 mm
Time 1 hour ahead of GMT
Measures Metric system
Monetary unit CFA franc (CFA Fr) = 100 centimes
Rate of exchange (1976 av): par CFA Fr 50 = Fr 1, free CFA Fr 239.0 = $ 1, CFA Fr 431.6 = £ 1

Summary

Political One-party republic, formed by the union of East and West Cameroon. East Cameroon, formerly under French administration, became independent on January 1, 1960, and West Cameroon, formerly the southern part of the Cameroons under UK administration, joined to form the Federal Republic of Cameroon on October 1, 1961. A completely unified state was formed from June 2, 1972, as the United Republic of Cameroon. Member of UN, OAU, Udeac, franc zone and an EEC ACP state
Economic A diversified agricultural economy, with main export earners cocoa, coffee and timber. Industrial activity is centred on the aluminium industry, with hydro-electricity being used in the smelter of alumina mainly imported from Guinea. The fourth development plan 1976-81 places emphasis on agricultural improvement, transport and industrial development

People, resources and equipment

Population 1960 4.70* mn, 1970 5.84* mn, 1976 6.53* mn
Growth: 1960–70 2.2* %pa, 1970–76 1.9* %pa
Density (1976): 14* people per km²
Vital statistics (rate per 1 000 people, 1970–75): births 40.4*, deaths 22.0*
Cities (population in 000, 1975)

Yaoundé (capital)	295*	Victoria	87*	Balfoussam	46*
Douala	345*	Kumba	67*	Foumban	46*
N'Kongsamba	96*	Maroua	47*	Bamenda	40*

Race (1970) Bamileke 26* %, Fulani 6* %, Bassa 3* %
Language French, English and several African languages
Religion (1976) Animist 45* %, Christian 35* %, Moslem 20* %
Education (1973/74) Pupils: primary 1 014 135, secondary 82 205, vocational 24 352, teacher-training 1 452, higher 5 533.
Teachers: primary 19 719, secondary 3 322, vocational 1 203, teacher-training (1972/73) 328, higher 158
Labour force (1976) 3 127 000*; in agriculture 2 574 000* (82* %)
Personnel (1971) Physicians: 225, 1 per 26 400 people
Standard of living National income per person (1976):
CFA Fr 73 000** = $ 305** = £ 170**
Consumption per person: energy (1975) 104 kg coal equivalent, electricity (production, 1976) 204 kW h
Newspapers (1974): number 1; circulation 20 000, 3 per 1 000 people
Telephones (June 1973): 22 000, 4 per 1 000 people
Livestock (000, Dec 1975) Cattle 2 655*, sheep 2 105*, goats 1 633*, pigs 412*
Electrical capacity (1975) 225* megawatts, of which, hydro 197* megawatts
Hospital beds (1970) 19 141, 1 per 305 people
Roads (1975) 43 500* km = 27 000* mi, density 0.09 km per km²
Railways (1975) 1 320* km = 820* mi, density 0.003 km per km²
Ships (registered, 1977) 29, total of 78 180 gross tons
Ports (goods traffic, 000 tonnes, 1973)

	loaded	unloaded
Douala	730*	1 147
Kribi	131	13
Victoria (1972)	28	15

Airports Douala, Yaoundé, Maroua, Balfoussam, and 9 other airports with scheduled flights
Durable equipment Radio sets (Dec 1974): 603 000, 96 per 1 000 people
Passenger cars (Dec 1972): 39 100, 6 per 1 000 people, 0.9 per km of road
Commercial vehicles (Dec 1972): 37 300, 6 per 1 000 people, 0.9 per km of road

Production

Gross domestic product 1973/74 (year ending June 30th):
CFA Fr 416 000 mn = $ 1 808 mn = £ 759 mn
1976 est: CFA Fr 500 000** mn = $ 2 100** mn = £ 1 200** mn
Structure of gross domestic product (1973/74) *By origin* Agriculture 32 %, manufacturing 12 %, distribution and hotels 29 %, other 27 %
By type Final consumption expenditure 81 %, (of which, government 13 %), stock investment 1 %, gross fixed capital formation 12 %, exports of goods and services 26 %, less imports of goods and services —19 %

Production index (1970 = 100)	1960	1970	1976	Growth %pa 1960–70	1970–76
Agricultural	60	100	110	5.3	1.6

Main products (000 t)
Agriculture

	1960	1970	1976	1960–70	1970–76
Maize	281	263	355*	−0.7	5.1*
Millet	310	312	390*	−0.1	3.8*
Sweet potatoes	176	328	160*	6.4	−11.3*
Cassava	750	930	800*	2.2	−2.5*
Groundnuts	78	178	179*	8.6	0.1*
Bananas	160	125	96*	−2.4	−4.3*
Plantains	530*	1 000*	1 000*	6.6*	0.0*
Palm kernels	27	40	40*	4.0	0.0*
Palm oil	25	54*	60*	8.0*	1.8*
Cocoa	82	112	82*	3.2	−5.1*
Coffee	41	82	80*	7.2	−0.4*
Cotton	10	15*	18*	4.1*	3.1*
Rubber	5**	12*	16*	9.1**	4.9*
Fish catch	49	71	72*b	3.8	0.3*c
Timber (000 m³)	5 600*	7 595*	8 252b	3.1*	1.7*c

Other

	1960	1970	1976	1960–70	1970–76
Electricity (mn kW h)	911	1 175	1 335	2.6	2.1
of which, hydro (mn kW h)	898	1 145	1 135b	2.5	−0.2c
Tin conc (Sn content)	0.06	0.04*	0.02*	−4.8*	−6.0*
Beer (000 hl)	234	702	1 150b	11.6	10.4c
Cigarettes (mn units)	876	975	1 367d	1.1	18.4e
Aluminium[a]	44	52	54	1.8	0.6
Radio sets (000 units)	—	59	60f	na	1.7g

[a]Produced from alumina imported from Guinea [b]1975 [c]1970–75 [d]1972
[e]1970–72 [f]1971 [g]1970–71

Transport traffic	1960	1970	1976	Growth %pa 1960–70	1970–76
Passenger-kilometres (mn)					
Rail	90	210	260*	8.8	3.6*
Air	54a,b	80a	225d	5.8c	18.8
Cargo: tonne-kilometres (mn)					
Rail	122	271	530*	8.3	11.8*
Air	2.7a,b	6.4a	8.7d	13.1c	5.3
Sea: tonnes (mn)					
Goods loaded	0.39	0.79	0.84	7.4	0.9
Goods unloaded	0.42	1.14	1.36	10.5	3.0

[a]Includes apportionment of traffic of Air Afrique; Cameroon withdrew from this group November 1971 [b]1963 [c]1963–70 [d]Cameroon Airlines
Tourism (1974) Number of visitors 96 000; gross receipts $ 17 mn

Finance and trade

Price index (1970 = 100)	1960	1970	1976	Growth %pa 1960–70	1970–76
Consumer prices	74.5a	100.0	172.6	3.8b	9.5
Money stock					
(end-year, CFA Fr bn)	14.1	38.4	92.4	10.6	15.8

[a]1962 [b]1962–70
Budget (1976/77; year ending June 30th)
Balanced at CFA Fr 128 000 mn = $ 517 mn = £ 302 mn

Balance of payments ($ mn)	1972	1973	1974	1975	1976
Balance of goods (fob)	−18	+99	+103	−28	−29
Balance of services	−71	−95	−109	−131	−100
Balance of transfers	+35	−21	−9	+9	+21
Current balance	−55	−17	−14	−151	−107
Long-term capital flow	+17	+52	+48	+59	+91
Reserves and debt (end-year, $ mn)					
International reserves	44	51	79	29	44
External public debt	276	424	570	688	na

External trade (1976)
Imports: CFA Fr 145 963 mn = $ 609 mn = £ 337 mn
Exports: CFA Fr 122 028 mn = $ 511 mn = £ 283 mn

Cameroon

Main imports (1975)	% of total	Main exports (1975)	% of total
Chemicals	14	Cocoa	34
Machinery, non-electric	13	Coffee	24
Motor vehicles	10	Timber	9
Petroleum products	10	Aluminium	7
Food	9		
Textile yarns & fabrics	6		
Electrical machinery	5		
Iron and steel	5		
Main sources (1975)		Main destinations (1975)	
France	46	France	27
West Germany	8	Netherlands	22
United States	7	Soviet Union	10
Italy	6	West Germany	7
Gabon	5	Japan	5

Special focus

Fourth development plan, 1976-81

	CFA Fr mn	% of total
Agriculture	125 236	17
Industry	240 785	33
Transport and communication	167 823	23
Education	36 721	5
Housing and town planning	89 180	12
Other	65 487	9
Total	725 232	100

Cape Verde

Republic of Cape Verde
República de Cabo Verde

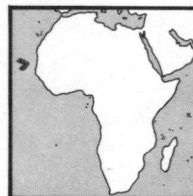

Location Eastern Atlantic Ocean
A group of 10 islands and 5 islets about 600 km
(400 mi) west of Dakar in Senegal
Land Area 4 033 km² = 1 557 mi²
Climate Tropical maritime
Weather at Praia, 34 m altitude
Temperature: hottest month Oct 24–29 °C,
coldest Feb 19–25 °C
Rainfall (av monthly): driest month May 0 mm,
wettest Sept 114 mm
Time 1 hour behind GMT
Measures Metric system
Monetary unit Escudo Caboverdiano (CV Esc) = 100 centavos; new
monetary unit, issued by the national bank from July 1, 1977,
at par with former Portuguese issued escudo
Rate of exchange (1976 av): par CV Esc 1 = Esc 1,
free CV Esc 30.22 = $ 1, CV Esc 54.59 = £ 1

Summary

Political One-party republic, which became independent July 5, 1975;
formerly a Portuguese overseas province. A federation with
Guinea-Bissau is being considered. Member of UN, OAU and an EEC
ACP state
Economic An agricultural economy, with some earnings from trading
activities including the supply of fuel to ships. Water conservation and
irrigation is important. There is substantial fishing potential

People, resources and equipment

Population 1960 199 000*, 1970 273 000*, 1976 306 000*
Growth: 1960–70 3.0* %pa, 1970–76 1.9* %pa
Density (1976): 73* people per km²
Vital statistics (rate per 1 000 people, 1974): births 29.2, deaths 8.8
Cities (population in 000, 1974) Praia (capital) 21*, Mindelo 29*
Race (1970) Mixed 70** %, African 30** %
Language Creole Portuguese
Religion Mainly Roman Catholic
Education (1975/76) Pupils 67 650, teachers 1 667
Labour force (1976) 90 000*; in agriculture 53 000* (59* %)
Personnel (1969) Physicians: 13, 1 per 20 300 people
Standard of living
National income per person (1976): CV Esc 12 000** = $ 400** = £ 220**
Consumption per person: energy (1975) 61 kg coal equivalent,
electricity (1975) 27 kW h

Telephones (Dec 1975): 2 000, 7 per 1 000 people
Livestock (000, 1976) Cattle 15*, sheep 2*, goats 20*, pigs 18*
Electrical capacity (1975) 6* megawatts
Hospital beds (1969) 376, 1 per 710 people
Roads (1972) 1 946 km = 1 209 mi, density 0.5 km per km²
Railways are being built (300* km in 1976)
Ships (registered, 1977) 9, total of 3 966 gross tons
Ports Porto Grande and Mindelo (São Vicente), Pedra Lume and Santa
Maria (Sal), Praia (San Tiago)
Airports Iiha do Sal, Praia and 6 other airports
Durable equipment Radio sets (Dec 1972): 5 200, 19 per 1 000 people
Passenger cars (Dec 1974): 2 700, 9 per 1 000 people, 1.4 per km of road
Commercial vehicles (Dec 1974): 800, 3 per 1 000 people, 0.4 per km of
road

Production, finance and trade

Gross domestic product 1974: CV Esc 3 300* mn = $ 130* mn = £ 55* mn
1976 est: CV Esc 4 000** mn = $ 130** mn = £ 73** mn
Main products (000 t, 1976) *Agriculture* Maize 17, bananas 7*, fish catch 4*
Other (1974) Electricity (mn kW h) 8*, salt 14, cement 4
Transport traffic (1975) *Sea* Goods loaded 20 000 t, unloaded 145 000 t
Budget (1976) Revenue CV Esc 684 mn = $ 23 mn = £ 13 mn
Expenditure CV Esc 903 mn = $ 30 mn = £ 17 mn
External trade (1976) Imports: CV Esc 1 272 mn = $ 42 mn = £ 23 mn
Exports: CV Esc 408 mn = $ 14 mn = £ 7 mn

Main imports[a]	% of total	Main exports[a]	% of total
Food	49	Fish	29
(of which, maize 9)		Bananas	19
Petroleum products	9	Salt	9
Main sources[a]		Main destinations[a]	
Portugal	58	Portugal	63
Netherlands	5	Angola	14
United Kingdom	3	Zaire	5
Angola	2	United Kingdom	5

[a] Excluding transit

Central African Empire

Empire Centrafricain

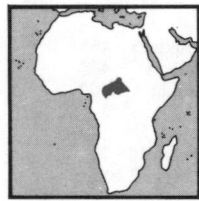

Location Central Africa
Chad is to the north, Sudan to the east, Zaire
and Congo to the south, and Cameroon to the
west; the Oubangui and Mbomou rivers form
the border with Zaire
Land Area 622 984 km² = 240 535 mi²
Climate Tropical
Weather at Bangui, 387 m altitude
Temperature: hottest month Feb 21–34 °C,
coldest July, Aug 21–29 °C
Rainfall (av monthly): driest month Dec 5 mm, wettest July 226 mm
Time 1 hour ahead of GMT
Measures Metric system
Monetary unit CFA franc (CFA Fr) = 100 centimes
Rate of exchange (1976 av): par CFA Fr 50 = Fr 1,
free CFA Fr 239.0 = $ 1, CFA Fr 431.6 = £ 1

Summary

Political Monarchy since December 4, 1976 (previously a republic)
which became independent in August 1960; formerly one of the four
territories included in French Equatorial Africa. Member of UN, OAU,
Ocam, Udeac, franc zone and an EEC ACP state
Economic An agricultural economy, with coffee, cotton and timber large
export earners; diamond production is important, but has fallen after
reaching a peak in 1968. There are uranium prospects

People, resources and equipment

Population 1960 1.80** mn, 1970 2.37** mn, 1976 2.69** mn
(Note: population estimates include refugees numbering 40 000* in 1960;
Alternate estimates show 1.23** mn in 1960, 1.61** mn in 1970 and
1.83** mn in 1976)
Growth: 1960–70 2.8** %pa, 1970–76 2.1** %pa
Density (1976): 4** people per km²
Vital statistics (rate per 1 000 people, 1970–75): births 43.4*, deaths 22.5*

Central African Empire

Cities (population in 000, 1971)
Bangui (capital) 187   Bossangoa 36   Bouar   29
Berbérati   40   Bambari 31   Bangassou 28
Race (1961) Banda 47 %, Baya 27 %, Azanda 10 %
Language French (official) and Sango (national)
Religion (1976) Animist 50** %, Christian 45** %, Moslem 5** %
Education (1973/74) Pupils 218 000*, teachers 3 700*
Labour force (1976) 1 002 000*; in agriculture 892 000* (89* %)
Personnel Scientists and engineers engaged in research (1969): 26
Physicians (1973): 59, 1 per 43 000 people
Standard of living
National income per person (1976): CFA Fr 37 000** = $ 150** = £ 85**
Consumption per person (1975): energy 34 kg coal equivalent,
electricity (production) 20 kW h
Newspapers (1972): number 1; circulation 500, 0.2 per 1 000 people
Telephones (Dec 1973): 5 000, 2 per 1 000 people
Livestock (000, 1976) Cattle 610*, sheep 76*, goats 566*, pigs 62*
Mineral reserves (1974) Uranium 8 000 tonnes
Electrical capacity (1975) 17 megawatts, of which, hydro 11 megawatts
Hospital beds (1972) 3 161, 1 per 790 people
Roads (1976) 21 950 km = 13 640 mi, density 0.04 km per km²
There are no railways, but they are planned (to link with Cameroon and
Sudan)
Inland waterways Oubangui river flowing into the Zaire (Congo) river
Port Bangui (on Oubangui river)
Airport Mpoko (Bangui)
Durable equipment (Dec 1974): Radio sets 70 000, 27 per 1 000 people
Passenger cars: 11 400, 4 per 1 000 people, 0.5 per km of road
Commercial vehicles: 3 000, 1 per 1 000 people, 0.1 per km of road

Production

Gross domestic product 1971: CFA Fr 57 100 mn = $ 206 mn = £ 85 mn
1976 est: CFA Fr 100 000** mn = $ 420** mn = £ 230** mn
Structure of gross domestic product (1971) *By origin* Agriculture 32 %,
mining and quarrying 6 %, manufacturing 9 %, other 53 %
By type Final consumption expenditure 88 % (of which, government
26 %), stock investment 1 %, gross fixed capital formation 17 %, exports
of goods and services 22 %, less imports of goods and services −27 %

Production index (1970 = 100)	1960	1970	1976	Growth % pa 1960–70	1970–76
Agricultural	82	100	105	2.0	0.8
Main products (000 t)					
Agriculture					
Maize	38	48	38	2.4	−3.8
Millet	26	40*	43*	4.4*	1.2*
Sweet potatoes	20	47*	60*	8.9*	4.2*
Cassava	1 200	1 000*	850*	−1.8*	−2.7*
Groundnuts	63	50*	36*	−2.3*	−5.3*
Sesameseed	5	10*	13*	7.2*	4.5*
Oranges	12	11*	11*	−0.9*	0.0*
Bananas	50*	60*	70*	1.8*	2.6*
Plantains	40*	55*	60*	3.0*	1.5*
Coffee	8	10*	10*	1.8*	0.0*
Cotton	11	20	16*	6.1	−3.6*
Timber (000 m³)	1 732	2 346*	2 200*ᵇ	3.1*	−1.3*ᶜ
Other					
Electricity (mn kW h)	8	47	52ᵇ	19.2	2.0ᶜ
Diamonds (000 CM)	70	482	338*ᵇ	21.3	−6.8*ᶜ
Beer (000 hl)	29	110	132ᵇ	14.3	3.7ᶜ
Cotton fabrics (mn m)	3	8	6ᵈ	10.0	−7.0ᵉ
Radio sets (000 units)	—	9	13ᵇ	na	7.6ᶜ
Motor cyclesª	—	6	4ᵇ	na	−7.8ᶜ

ª Assembly only ᵇ1975 ᶜ1970–75 ᵈ1974 ᵉ1970–74
Transport traffic (1976) *Air* (including apportionment of Air Afrique)
131 mn passenger-km, cargo 13.9 mn t-km

Finance and trade

Price indices (1970 = 100)	1960	1970	1976	Growth %pa 1960–70	1970–76
Consumer prices	72.6ª	100.0	173.1	4.8ᵇ	9.6
Wholesale prices	67.3	100.0	159.4	4.0	8.1
Money stock					
(end-year, CFA Fr bn)	2.8	7.3	16.9	10.1	14.9

ª 1963 ᵇ1963–70
Budget (1974) Revenue: CFA Fr 15 706 mn = $ 65 mn = £ 28 mn
Expenditure: CFA Fr 17 200 mn = $ 72 mn = £ 31 mn

Balance of payments ($ mn)	1972	1973	1974	1975	1976
Balance of goods (fob)	−4	+4	−5	−30	na
Balance of services	−16	−29	−41	−46	na
Balance of transfers	+22	+21	+32	+39	na
Current balance	*+1*	*−3*	*−15*	*−37*	*na*
Long-term capital flow	+6	−10	na	na	na
Reserves and debt (end-year, $ mn)					
International reserves	1.7	1.8	1.7	3.8	18.8
External public debt	62	72	86	110	na

External trade (1975) Imports: CFA Fr 14 614 mn = $ 68 mn = £ 31 mn
Exports: CFA Fr 10 112 mn = $ 47 mn = £ 21 mn

Main imports (1971)	% of total	Main exports (1975)	% of total
Food	14	Coffee	23
(of which, cereals 6)		Diamonds	20
Machinery, non-electric	14	Cotton	18
Motor vehicles	12	Timber	10ª
Chemicals	9	Tobacco	3ª
Textile yarns & fabrics	9		
Electrical machinery	8		
Main sources (1975)		*Main destinations* (1975)	
France	58	France	42
Jugoslavia	9	Belgium-Luxembourg	9
West Germany	7	Italy	8
United States	3	United States	8

ª1971

Special focus

Diamond production (000 metric carats)

1960	70	1964	442	1968	609	1972	524
1961	111	1965	537	1969	535	1973	524
1962	263	1966	541	1970	482	1974	338*
1963	402	1967	521	1971	468	1975	338*

Chad

Republic of Chad
République du Tchad

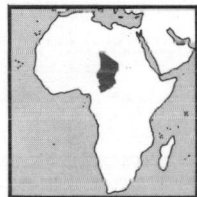

Location Central Africa
Libya is to the north, Sudan to the east,
Central African Empire to the south, and
Cameroon, Nigeria and Niger to the west.
Includes part of the Sahara desert. Land-locked
Land Area 1 284 000 km² = 496 000 mi²
Climate Tropical
Weather at N'Djamene, 295 m altitude
Temperature: hottest month April 23–42 °C,
coldest Dec 14–33 °C
Rainfall (av monthly): driest months Nov–March 0 mm, wettest Aug
320 mm
Time 1 hour ahead of GMT
Measures Metric system
Monetary unit CFA franc (CFA Fr) = 100 centimes
Rate of exchange (1976 av): par CFA Fr 50 = Fr 1,
free CFA Fr 239.0 = $ 1, CFA Fr 431.6 = £ 1

Summary

Political Republic with military government, which became independent
on August 11, 1960; formerly one of the four territories included in French
Equatorial Africa. Member of UN, OAU, franc zone and an EEC ACP
state; has observer status with Udeac
Economic An agricultural economy, with export earnings mainly from
cotton and cattle; affected in recent years by drought in the Sahel region.
Industry is mainly in food and textiles; there are some mineral prospects,
including crude oil

People, resources and equipment

Population 1960 2.98* mn, 1970 3.64* mn, 1976 4.12* mn
Growth: 1960–70 2.0* %pa, 1970–76 2.1* %pa
Density (1976): 3* people per km²
Vital statistics (rate per 1 000 people, 1970–75): births 44.0*, deaths 24.0*
Cities (population in 000, 1972)
N'Djameneª (capital) 179, Sarhᵇ 44, Moundou 40, Abéché 28
ªCalled Fort-Lamy before November 1973 ᵇCalled Fort-Archambault before July
1972

Chad

Race (1961) Sudanese Arab 30 %, Bagirmi, Sara and Kreish 25 %, Teda 7 %, Mbum 7 %, Maba and Masalit 7 %, Tama 6 %, Mubi and Sokaro 4 %, Kanuri 3 %
Language French; Arabic and African languages are also used
Religion (1976) Moslem 52** %, Animist 43** %, Christian 5** %
Education (1974/75) Pupils 221 500*, teachers 3 330*
Labour force (1976) 1 551 000*; in agriculture 1 342 000* (87* %)
Personnel Scientists and engineers engaged in research (1971): 85
Physicians (1974): 89, 1 per 44 400 people
Standard of living
National income per person (1976): CFA Fr 30 000** = $ 125** = £ 70**
Consumption per person (1975): energy 39 kg coal equivalent, electricity (production) 14 kW h
Newspapers (1974): number 4; circulation 1 500, 0.4 per 1 000 people
Telephones (Dec 1975): 5 000, 1 per 1 000 people
Livestock (000, 1976) Cattle 3 658*, sheep 2 424*, goats 2 424*, horses 145*, asses 300*, camels 310*
Electrical capacity (1975) 22 megawatts
Hospital beds (1974) 3 464, 1 per 1 140 people
Roads (1974) 30 725 km = 19 092 mi, density 0.02 km per km²
There are no railways, but they are planned
Inland waterways The Chari river is navigable from N'Djamene near Lake Chad to Sarh near the south border with Central African Empire; Logone river is also navigable in part
Airports N'Djamene, Sarh, Moundou, Abéché
Durable equipment Radio sets (Dec 1974): 70 000, 18 per 1 000 people
Passenger cars (Dec 1975): 6 060, 1.5 per 1 000 people, 0.2 per km of road
Commercial vehicles (Dec 1975): 8 380, 2.1 per 1 000 people, 0.3 per km of road

Production, finance and trade

Gross domestic product
1970: CFA Fr 74 900 mn = $ 270 mn = £ 112 mn
1976 est: CFA Fr 130 000** mn = $ 540** mn = £ 300** mn
Structure of gross domestic product (1970) *By origin* Agriculture 49 %, manufacturing 7 %, distribution and hotels 16 %, other 28 %

Production index (1970 = 100)	1960	1970	1976	Growth %pa 1960–70	1970–76
Agricultural	92	100	106	0.8	1.0

Main products (000 t)
Agriculture

	1960	1970	1976	1960–70	1970–76
Millet	900	610	533	−3.8	−2.2
Cassava	50	55*	56*	1.0*	0.3*
Groundnuts	120	100*	70*	−1.8*	−5.8*
Dates	25	25*	25*	0.0*	0.0*
Cotton	34	37*	55*	0.8*	6.8*
Milk	140	158*	120*	1.2*	−4.5*
Beef and veal	47	50*	50*	0.6*	0.0*
Sheep and goat meat	11	12*	18*	0.9*	7.0*
Cattle hides	4.0*	5.4	5.1*	3.0*	−0.9*
Fish catch	65	120	115*ᵃ	6.4	−0.8*ᵇ
Timber (000 m³)	2 370	3 330*	3 620*ᵃ	3.5*	1.7*ᵇ

Other

	1960	1970	1976	1960–70	1970–76
Electricity (mn kW h)	8	42	56ᵃ	18.0	5.9ᵇ
Beer (000 hl)	na	57	149ᵃ	na	21.2ᵇ
Cigarettes (mn units)	na	33	313ᵃ	na	56.8ᵇ
Cotton fabrics (mn m²)	na	5	6ᶜ	na	9.5ᵈ

ᵃ1975   ᵇ1970–75   ᶜ1972   ᵈ1970–72

Transport traffic (1976) *Air* (including apportionment of Air Afrique) 146 mn passenger-km, cargo 14.8 mn t-km
Budget (1976) Balanced at CFA Fr 15 785 mn = $ 66 mn − £ 37 mn
International reserves (Dec 1976) $ 23 mn
External public debt (Dec 1975) $ 163 mn
External trade (1974) Imports: CFA Fr 20 859 mn = $ 87 mn = £ 37 mn
Exports: CFA Fr 9 053 mn = $ 38 mn = £ 16 mn

Main imports (1973)	% of total	Main exports (1974)	% of total
Petroleum products	16	Cotton	67
Sugar	11	Beef and veal	7ᵇ
Transport equipment	8	Cattle	5ᵇ
Metals and minerals	7	Hides and skins	3ᵇ
Main sources (1974)		*Main destinations*ᵃ (1974)	
France	37	Zaire	5
Nigeria	12	Nigeria	4
United States	10	Congo	4
Cameroon	4	France	3

ᵃThe destination of 74 % of exports is not distinguished   ᵇ1973

Comoros

State of the Comoros
État Comorien

Location Western Indian Ocean
The archipelago comprises 4 islands off the east African coast of Mozambique
Land Area 2 236 km² = 863 mi², of which Anjouan 424 km², Grande Comore 1 148 km², Moheli 290 km²; includes Mayotte with 374 km²
Climate Tropical
Weather at Moroni, 59 m altitude
Temperature: hottest month March 24–31 °C, coldest August 19–27 °C
Rainfall (av monthly): driest month Oct 84 mm, wettest Jan 424 mm
Time 3 hours ahead of GMT
Measures Metric system
Monetary unit CFA franc (CFA Fr) = 100 centimes
Rate of exchange (1976 av): par CFA Fr 50 = Fr 1, free CFA Fr 239.0 = $ 1, CFA Fr 431.6 = £ 1

Summary

Political Republic, which declared unilateral independence July 6, 1975; formerly a French overseas territory. Mayotte, formerly part of the Comoros, seceded to remain a French territory. Information shown here in general includes Mayotte. Member of UN, OAU, franc zone and an EEC ACP state
Economic Foodstuffs, especially rice, are major imports, and exports are mainly essential oils for perfumes, vanilla and cloves

People, resources and equipment

Population 1960 190 000*, 1970 270 000*, 1976 314 000*
Of whom (1976): Comoros (excluding Mayotte) 274 000*, Mayotte 40 000*
Growth: 1960–70 3.6* %pa, 1970–76 2.5* %pa
Density (1976): 140* people per km²
Vital statistics (rate per 1 000 people, 1970–75): births 46.6*, deaths 21.7*
Regions (prefectures, population in 000, 1973; total of 290 000)
Anjouan 105, Grande Comore 135, Mayotte 38, Moheli 12
City (1974) Moroni (capital, on Grande Comore) 12 000*
Race Mixed Arab, Malagasy, Malay and African
Language French and Comorian; also Arabic and Swahili
Religion (1976) Comoros (excluding Mayotte): Moslem 99* %, Christian 1* %; Mayotte: mainly Christian
Education (1973/74) Pupils 26 391, teachers 682
Labour force (1976) 116 000*; in agriculture 76 000* (65* %)
Personnel (1973) Physicians: 21, 1 per 13 800 people
Standard of living National income per person (1976): CFA Fr 54 000** = $ 230** = £ 125**
Consumption per person (1975): energy 51 kg coal equivalent, electricity (production) 10* kW h
Telephones (Dec 1974): 2 000, 5 per 1 000 people
Livestock (000, 1976) Cattle 74*, goats 81*, sheep 8*
Electrical capacity (1975) 1 megawatt
Hospital beds (1972) 612, 1 per 460 people
Roads (1975) 750* km = 466* mi, density 0.34* km per km²
Port Mutsamudu
Airports Moroni, Anjouan, Moheli, Dzaoudzi (Mayotte)
Durable equipment Radio sets (Dec 1974): 36 000, 119 per 1 000 people
Motor vehicles (Dec 1973): 3 600, 12 per 1 000 people, 4.8 per km of road

Production, finance and trade

Gross domestic product 1975: CFA Fr 15 000* mn = $ 70* mn = £ 30* mn
1976 est: CFA Fr 18 000** mn = $ 75** mn = £ 42** mn
Main products *Agriculture* (000 t, 1976) Rice 15*, maize 4*, sweet potatoes 13*, cassava 82*, coconuts 57*, copra 4*
Other (1975) Electricity 3* mn kW h
Transport traffic (1973) *Sea* Goods loaded 9 000 t, unloaded 54 000 t
Budget (1975) Balanced at CFA Fr 2 949 mn = $ 14 mn = £ 6.2 mn
External trade (1974) Imports: CFA Fr 6 203 mn = $ 26 mn = £ 11mn
Exports: CFA Fr 2 138 mn = $ 9 mn = £ 4 mn

Main imports (1973)	% of total	Main exports (1972)	% of total
Rice	30**	Vanilla	41
Petroleum products	15**	Essential oils	33
Cement	15**	Cloves	11
Main sources (1973)		*Main destinations* (1973)	
France	50**	France	75**
Madagascar	15**	Madagascar	10**

Congo

People's Republic of the Congo
République Populaire du Congo

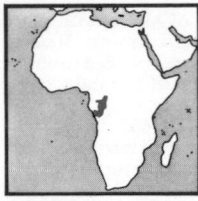

Location West central Africa
There is a short coast-line on the Atlantic Ocean
to the south-west, with Gabon to the west,
Angola (Cabinda) to the south, Zaire to the
south and east, and Cameroon and Central
African Empire to the north
Land Area 342 000 km² = 132 000 mi²
Climate Tropical
Weather at Brazzaville, 318 m altitude
Temperature: hottest month April 22–33 °C,
coldest July 17–28 °C
Rainfall (av monthly): driest months July, Aug 1 mm, wettest Nov 292 mm
Time 1 hour ahead of GMT
Measures Metric system
Monetary unit CFA franc (CFA Fr) = 100 centimes
Rate of exchange (1976 av): par CFA Fr 50 = Fr 1,
free CFA Fr 239.0 = $ 1, CFA Fr 431.6 = £ 1

Summary

Political Socialist republic, which became independent August 15, 1960;
formerly one of the four territories included in French Equatorial Africa.
Formerly often called Congo-Brazzaville. Member of UN, OAU, Udeac,
franc zone and an EEC ACP state
Economic Before 1973 the economy was based mainly on timber; since
then crude oil has become the main export product. An oil refinery is being
built at Pointe-Noire. Potash fertilisers are also important.

People, resources and equipment

Population 1960 0.80* mn, 1970 1.19* mn, 1976 1.39* mn
Growth: 1960–70 4.1* %pa, 1970–76 2.6* %pa
Density (1976): 4* people per km²
Vital statistics (rate per 1 000 people, 1970–75): births 45.1*, deaths 20.8*
Cities (population in 000, 1974)
Brazzaville (capital) 290, Pointe Noire 142, N'Kayi[a] 31, Loubomo[b] 30
[a]Formerly Jacob [b]Formerly Dolisie
Race (1974) Kongo 27 %, Téké 12 %, M'Bochi 7 %
Language French, Lingala, Likongo and other Bantu languages
Religion (1976) Animist 54** %, Christian 45** %, Moslem 1** %
Education (1974/75) Pupils 397 800*, teachers 7 400*
Labour force (1976) 483 000*; in agriculture 180 000* (37* %)
Personnel (1973) Physicians: 162, 1 per 7 900 people
Standard of living National income per person (1976):
CFA Fr 103 000** = $ 430** = £ 240**
Consumption per person (1975): energy 209 kg coal equivalent,
electricity (production) 74 kW h, steel 24 kg
Newspapers (1974): number 3
Telephones (Dec 1974): 10 000, 8 per 1 000 people
Livestock (000, 1976) Cattle 50, sheep 52, goats 101, pigs 44
Mineral reserves (1975) Crude oil 66 mn tonnes, natural gas 28 bn m³
Petroleum refinery capacity Under construction at Pointe-Noire
(capacity 1* mn tonnes)
Electrical capacity (1975) 32* megawatts, of which, hydro 15* megawatts
Hospital beds (1972) 5 541, 1 per 220 people
Roads (1975) 11 000* km = 6 800* mi, density 0.03 km per km²
Railways (1975) 800 km = 497 mi, density 0.002 km per km²
Ships (registered, 1977) 14, total of 4 172 gross tons
Ports (1973) Pointe Noire (goods traffic, 000 tonnes) loaded 2 659,
unloaded 595; Brazzaville, on the Zaire (Congo) river, acts as inland port
for Cameroon, Central African Empire and Chad
Airports Maya-Maya (Brazzaville): passenger arrivals and
departures (1975) 97 257; also (1977) Pointe-Noire and 8 other airports
with scheduled flights

Durable equipment	000	no per	
(at end-year)		1 000 people	
Radio sets (1974)	80	60	no per
Television sets (1973)	3.8	3	km of road
Passenger cars (1974)	19.0	14	1.7*
Commercial vehicles (1974)	10.5	8	1.0*

Production

Gross domestic product
1975: CFA Fr 140 000* mn = $ 650* mn = £ 290* mn
1976 est: CFA Fr 160 000** mn = $ 670** mn = £ 370** mn

Production index	1960	1970	1976	Growth %pa	
(1970 = 100)				1960–70	1970–76
Agricultural	94	100	117	0.6	2.6

Main products (000 t)
Agriculture

	1960	1970	1976	1960–70	1970–76
Cassava	800	450*	761	−5.6*	9.1*
Sweet potatoes	50	70*	96*	3.4*	5.4*
Groundnuts	12	20*	23	5.2*	2.4*
Sugar, raw value	15	31*	46*	7.5*	6.8*
Bananas	4	15*	24	14.0*	8.1*
Plantains	20*	28*	32*	3.4*	2.2*
Palm oil	5.2	3.3	2.2*	−4.4	−6.5*
Coffee	0.5	1.0*	2.0*	7.2*	12.3*
Cocoa	0.6	2.0	2.0*	12.8	0.0*
Tobacco	0.5	0.7*	2.0*	3.3*	19.1*
Timber (000 m³)	1 558	2 151	2 471[b]	3.3	2.8[c]
Energy					
Total energy (000 tce)	70	40	2 660[b]	−5.4	131.5[c]
Crude oil	52	19	2 002	−9.6	117.3
Natural gas (mn m³)	na	10	15	na	7.0
Electricity (mn kW h)	27	76	100[b]	11.0	5.6[c]
Mining					
Potash (oxide content)	na	206	462[b]	na	17.5[c]
Zinc ore (Zn content)	—	0.1	2.4[b]	na	88.8[c]
Gold (kg)	82	82	16[b]	0.0	−27.9[c]
Manufacturing					
Beer (000 hl)	33	66	342[b]	7.2	39.0[c]
Cigarettes (mn units)	825	989	1 193[b]	1.8	3.8[c]
Cement	—	92	55[b]	na	−9.8[c]
Fertilisers, potash[a]	na	67	277	na	26.7
Construction					
All buildings (000 m²)[f]	20	100	110[d]	17.5	9.9[e]

[a]Years ending June 30th [b]1975 [c]1970–75 [d]1971 [e]1970–71 [f]Permits

Transport traffic	1960	1970	1976	Growth %pa	
Passenger-kilometres (mn)				1960–70	1970–76
Rail	62	144	246	8.8	9.3
Air[a]	52[b]	73	137	5.0[c]	11.1
Cargo: tonne-kilometres (mn)					
Rail	209	512	508	9.4	−0.1
Air[a]	2.7[b]	6.4	14.2	13.1[c]	14.2
Sea: tonnes (mn)					
Goods loaded	0.42	2.35	2.67	18.7	2.1
Goods unloaded	0.34	0.52	0.72	4.2	5.6

[a]Includes apportionment of traffic of Air Afrique [b]1963 [c]1963–70

Finance and trade

Price index	1960	1970	1976	Growth %pa	
(1970 = 100)				1960–70	1970–76
Consumer prices	68.5	100.0	156.9	3.9	7.8
Money stock					
(end-year, CFA Fr bn)	7.3	12.4	30.0	5.3	15.9

Budget (1976) Balanced at CFA Fr 52 042 mn = $ 218 mn = £ 121 mn
International reserves (Dec 1976) $ 10 mn
External public debt (Dec 1975) $ 616 mn
External trade[a] (1975)
Imports: CFA Fr 33 890 mn = $ 158 mn = £ 71 mn
Exports: CFA Fr 52 070 mn = $ 243 mn = £ 109 mn

Main imports (1973)	% of total	Main exports (1975)	% of total
Machinery, non-electric	15	Crude oil	54
Food	10	Timber	19[d]
Chemicals	9	Veneers and	
Motor vehicles	8	plywood	16[c]
Metal small		Fertilisers, potash	13[c]
manufactures	7	Zinc ore	2[c]
Railway equipment	7	Diamonds[b]	2[c]
Main sources (1975)		*Main destinations* (1975)	
France	57*	France	22*
Netherlands	6*	Italy	15*
West Germany	6*	United States	8*
United States	5*	West Germany	7*

[a]Excluding trade with other Udeac countries [b]Re-exports of diamonds from Zaire,
not recorded in imports [c]1973 [d]1974

Special focus

Main exports % of total exports

	1970	1971	1972	1973	1974	1975
Crude oil	1	1	6	33	62	54
Timber	53	61	40	30	19	na

Djibouti

Republic of Djibouti
République de Djibouti

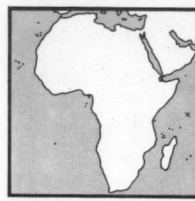

Location North-east Africa
On the Gulf of Aden, with Somalia to the south-east, and Ethiopia to the south and west
Land Area 22 000 km² = 8 500 mi²
Climate Tropical
Weather at Djibouti City, 7 m altitude
Temperature: hottest month July 31–41°C, coldest January 23–29°C
Rainfall (av monthly): driest month June 1 mm, wettest March 25 mm
Time 3 hours ahead of GMT
Measures Metric system
Monetary unit Djibouti franc (Dj Fr) = 100 centimes
Rate of exchange (1976 av): free Dj Fr 166 = $ 1, Dj Fr 300 = £ 1

Summary

Political Republic, which became independent on June 27, 1977; formerly the French Territory of the Afars and Issas. Previously known as French Somaliland. Member of UN, OAU and Arab League
Economic Djibouti City port is important for transit trade to Ethiopia (via the Djibouti-Addis Ababa railway). Saudi Arabia promised aid of $ 200 mn on independence

People, resources and equipment

Population 1960 80 000**, 1970 150 000 **, 1976 226 000**
Growth: 1960–70 6.5**%pa, 1970–76 7.0**%pa
Density (1976): 10** people per km²
Vital statistics (rate per 1 000 people, 1970): births 42.0, deaths 7.6
City (population, 1976) Djibouti City (capital) 120 000**
Race (1976) Issa and Somali 36 %, Afar 32 %, European 7 %, Arab 5 %
Language French; Arabic is also used
Religion (1970) Moslem 90 %, Roman Catholic 7 %
Education (1973/74) Pupils 10 444, teachers 335
Personnel (1974) Physicians: 52, 1 per 3 800 people
Standard of living
National income per person (1976): Dj Fr 170 000** = $ 1 020** = £ 570**
Consumption per person (1975): energy 450 kg coal equivalent, electricity 272 kW h
Telephones (Dec 1975): 4 000, 16 per 1 000 people
Livestock (000, 1976) Cattle 18*, sheep 98*, goats 580*, camels 25*
Electrical capacity (1975) 24 megawatts
Hospital beds (1974) 937 ,1 per 210 people
Roads (1976) 1 600** km = 1 000** mi, density 0.07** km per km²
Railways (1976) 100 km = 62 mi, density 0.005 km per km²
Port (goods traffic, 000 tonnes, 1973) Djibouti: goods loaded 142, unloaded 728
Airports, Djibouti, Obock, Tadjoura

Durable equipment (at end-year)	000	no per 1 000 people	
Radio sets (1974)	12	58	no per
Television sets (1973)	2.3	12	km of road
Passenger cars (1969)	7.4	50	5.0*
Commercial vehicles (1969)	1.5	10	1.0*

Production, finance and trade

Gross domestic product 1970: Dj Fr 22 300 mn = $ 104 mn = £ 43 mn 1976 est: Dj Fr 40 000**mn = $ 240**mn = £ 130**mn
Main products (1976) Mutton and goat meat 3 000*t, hides and skins 600*t, timber 23 000*m³, electricity (consumption, 1975) 58 000 000 kW h
Transport traffic Rail is included with the total for Ethiopia
Shipping entered (1971): 5 788 000 net reg tons; goods: see port
Budget (1976) Balanced at Dj Fr 5 713 mn = $ 34 mn = £ 19 mn
External trade (1973) Imports: Dj Fr 12 675 mn = $ 74 mn = £ 30 mn
Exports: Dj Fr 3 499 mn = $ 20 mn = £ 8 mn

Main imports	% of total	Main exports	% of total
Food	22	Special transactions[a]	54
Machinery	10	Ships	16
Motor vehicles	9	Leather and footwear	7
Main sources		*Main destinations*	
France	49	France	84
Ethiopia	12	Ethiopia	4

[a]Personal effects, second-hand goods, collectors items, war materials

Egypt

Arab Republic of Egypt
Jumhuriya Misr el Arabiya

Location North-east Africa
Forms the north-east corner of Africa, with an extension into Sinai; Libya is to the west, Sudan to the south, with the Mediterranean Sea forming the northern border, and the Red Sea and Israel the eastern border
Land Area 1 001 449 km² = 386 662 mi² of which, inhabited and cultivated territory 35 580 km² (4 %), including agricultural 28 620 km² (3 %)
Climate Hot and dry, with mild winters
Weather at Cairo, 116 m altitude
Temperature: hottest month July 21–36°C, coldest January 8–18°C
Rainfall (av monthly): driest months July, Aug 0 mm, wettest Dec 5 mm
Time 2 hours ahead of GMT
Measures Metric system; also: *area* 24 sahms = 1 kirat, 24 kirats = 1 feddân, 1 feddân = 0.42 hectare = 1.038 acres *capacity* 768 kadahs = 96 keilas = 8 ardebs = 1 dariba 1 dariba = 15.84 hectolitres = 43.55 UK bushels *weight* (*mass*) 14 400 dirhems = 100 rotls = 36 okes = 1 qantâr 1 qantâr = 44.928 kilograms = 99.049 pounds
Monetary unit Egyptian pound (E £) = 100 piastres = 1 000 millièmes
Rate of exchange (1976 av):
Official par $ 2.5556 = E £ 1, free £ 1.415 = E £ 1
Tourist (*incentive*) free E £ 0.67 = $ 1, E £ 1.21 = £ 1

Summary

Political Republic after June 1953 following the removal from power of King Farouk in July 1952. Egypt and Syria united to form the United Arab Republic in February 1958, and North Yemen also joined in March 1958. In 1961 the union was terminated, but Egypt retained the name United Arab Republic (UAR) until 1971. In September 1971 the Federation of Arab Republics was formed with Libya and Syria. In 1972–73 an agreement was reached with Libya to merge the two countries in due course. There have been two main conflicts with Israel: in June 1967 and October 1973. Member of UN, OAU, Arab League and Oapec
Economic Mainly an agricultural economy, with raw cotton accounting for one-third of exports and agriculture making up about 30 % of gross domestic product; manufacturing, including cotton yarn production, petroleum products and petrochemicals, accounts for one-sixth of gross domestic product. Industry is being developed and broadened, new projects including steelworks and a nuclear reactor; production of aluminium began in April 1977, with an output of about 100 000 tonnes per year. Earnings from the Suez Canal, reopened June 1975, were E £ 230 mn in 1976

People

Population 1960 25.92*mn, 1970 33.33*mn, 1976 38.07*mn
Growth: 1960–70 2.5*%pa, 1970–76 2.2*%pa
Density (1976): 38* people per km²; for inhabited and cultivated territory: 1 070* people per km²
Vital statistics (rate per 1 000 people, 1974): births 35.5, deaths 12.4
Regions (governates, population in 000, 1966; total of 30.08 mn)

Cities		Provinces: Lower Egypt		Provinces: Upper Egypt	
El Iskandarîya[b]	1 801	Beheira	1 979	Aswân	521
El Qâhira[a]	4 220	Dagahliya	2 279	Asyût	1 418
Ismâ'ilîya[e]	345	Dumyât[f]	432	Beni Suef	928
Bûr Sa'id[d]	283	Gharbiya	1 905	Faiyûm	941
El Suweis[c]	264	Kafr el Sheik	1 118	Gîza	1 650
Frontier districts		Minûfiya	1 458	Minya	1 706
El Wâdi el Gedid	59	Qalyûbîya	1 214	Qena	1 471
Matrûh	124	Sharqîya	2 102	Sohâg	1 696
Red Sea	38				
Sinai	131				

Cities (population in 000, 1974)

El Qâhira[a] (capital)	5 715	El Mahalla el Kûbra	288	Ismâ'ilîya[e]	190
El Iskandarîya[b]	2 259	Tanta	278	Damanhûr	176
Gîza	854	Aswân	246	El Faiyûm	168
El Suweis[c]	368	El Mansûra	232	El Minyâ	131
Shubrâ el Kheima	346	Asyût	197	Dumyât[f]	110
Bûr Sa'id[d]	342	Zagazig	195	Beni Suef	107

[a]Cairo [b]Alexandria [c]Suez [d]Port Said [e]Ismailia [f]Damietta

Egypt

Race (1966) Arab 99.6 %
Language Arabic
Religion (1966) Moslem 93 %, Christian 7 % (of whom Coptic 3* %)
Education (1974/75) Pupils: primary 4 145 454, secondary 1 601 171, vocational 348 306, teacher-training 33 275, higher 408 235.
Teachers: primary 103 600, secondary 46 080, vocational 20 717, teacher-training 2 673, higher 19 119
Labour force (1976) 10 780 000*; in agriculture 5 605 000* (52* %)
Personnel Scientists and engineers (1973): 593 254
Physicians (1973): 23 501, 1 per 1 520 people
Standard of living
National income per person (1976): E £ 120** = $ 310** = £ 170**
Consumption per person (1975): energy 405 kg coal equivalent, electricity (production) 280 kW h, newsprint 1.1 kg, steel 42 kg
Newspapers (1972): numbers 14; circulation 773 000, 22 per 1 000 people
Telephones (Dec 1974): 503 000, 14 per 1 000 people

Resources and equipment

Livestock (000, 1976) Cattle 2 392*, buffaloes 2 358*, sheep 2 000*, goats 1 372*, horses 29*, asses 1 539*, camels 113*, chickens 26 375*, ducks 3 295*, turkeys 705*
Mineral reserves Coal (known economic, 1965) 25 mn tonnes
Crude oil (1975) 210 mn tonnes
Natural gas (1975) 131 bn cubic metres
Petroleum refinery capacity (1975) 8.75* mn tonnes
Electrical capacity (1975) 3 893 megawatts, of which, hydro 2 445 megawatts
Hospital beds (1973) 76 611, 1 per 465 people
Roads (1972) 25 976 km = 16 141 mi, density 0.03 km per km²
Railways (1975) 4 855 km = 3 017 mi, density 0.005 km per km²
Waterways (1976) Nile river 1 550 km = 960 mi;
Suez canal (closed from June 1967 to June 1975) 162 km = 101 mi; vessels in transit (1975/76) 12 000*
Ships (registered, 1977) 176, total of 407 818 gross tons
Ports (goods traffic, 000 tonnes, 1972)

	loaded	unloaded		loaded	unloaded
Alexandria	2 990	9 253	Suez	193	4
Ras Shukheir[a]	4 000	21	Port Said	9[b]	na

[a]Crude oil port   [b]1971
Airports Cairo, Luxor, Aswân, Hurghada, Abu Simbel

Durable equipment	000	no per	
(at end-year)		1 000 people	
Radio sets (1974)	5 115	139	
Television sets (1974)	610	17	km of road
Passenger cars (1975)	215	6	8.3*
Commercial vehicles (1975)	46	1	1.8*

Production

Gross domestic product

1974: E £ 3 956 mn = $ 10 110 mn = £ 4 322 mn
1976 est: E £ 4 800** mn = $ 12 300** mn = £ 6 800** mn
Growth in real terms: 1968–72 5.6 %pa
Structure of gross domestic product (1974)

By origin	E £ mn	% of gdp
Agriculture	1 225	31
Mining and quarrying	87	2
Manufacturing	688	17
Electricity, gas and water	48	1
Construction	135	3
Distribution and hotels	361	9
Transport and communication	167	4
Other services	1 245	32
Total	3 956	100
By type of expenditure		
Government final consumption	1 097	28
Private final consumption	2 589	65
Stock investment	90	2
Gross fixed capital formation	640	16
Exports of goods and services	864	22
less Imports of goods and services	−1 324	−33
Total	3 956	100

Production indices	1960	1970	1976	Growth %pa	
(1970 = 100)				1960–70	1970–76
Agricultural	78	100	111	2.5	1.8
Industrial	37	100	122[e]	10.5	5.1[f]

Main products (000,t)

Agriculture	1960	1970	1976	1960–70	1970–76
Wheat	1 499	1 519	1 960	0.1	4.3
Maize	1 691	2 397	3 047	3.6	4.1
Rice	1 486	2 605	2 300	5.8	−2.0
Millet	603	874	800*	3.8	−1.5*
Potatoes	290	549	923*	6.6	9.0*
Dry broad beans	290	278	237*	−0.4	−2.6*
Cabbages	240*	279	330*	1.5*	2.8*
Pumpkins	220*	331	375*	4.2*	2.1*
Onions	640*	455	620*	−3.4*	5.3*
Tomatoes	1 100*	1 555	2 230*	3.5*	6.2*
Watermelons	820*	898	1 244*	0.9*	5.6*
Sugar, raw value	355	547	626*	4.4	2.3*
Oranges and mandarines	266	639	1 008*	9.2	7.9*
Dates	424	339	409*	−2.2	3.2*
Cotton	478	509	410*	0.6	−3.5*
Milk, cow	375	572	970*	4.3	9.2*
Milk, buffalo	750*	1 005	1 210*	3.0*	3.1*
Cheese	150*	207*	228*	3.3*	1.6*
Beef and buffalo meat	164	210	242*	2.5	2.4*
Fish catch	88	72	107[c]	−2.0	8.2[d]
Energy					
Total energy (000 tce)	4 360	22 040	13 300[c]	17.6	−9.6[d]
Crude oil	3 319	16 404	16 709	17.3	0.3
Petroleum products	4 160	3 236	8 533[c]	−2.5	21.4[d]
Natural gas (mn m³)	na	85*	55*[c]	na	−8.3*[d]
Electricity (mn kW h)	2 639	7 591	10 421[c]	11.1	6.5[d]
Mining					
Iron ore (Fe content)	120	226	621	6.5	18.4
Phosphate rock	566	716	404[c]	2.4	−10.8[d]
Manufacturing					
Cement	2 047	3 684	3 294	6.1	−1.8
Cigarettes (mn units)	10 609	12 153	23 769[c]	1.4	−14.4[d]
Cotton yarn	105	164	193	4.6	2.7
Cotton fabrics (mn m)	500*	730	647	3.9*	−2.0
Paper	39*	120	137[c]	11.9*	2.7[d]
Tyres (000 units)	280	430	468	4.4	1.4
Sulphuric acid	103	30	232[c]	−11.6	50.5[d]
Fertilisers, nitrogenous	55	118	150[c]	7.9	4.9[d]
Fertilisers, phosphate	36	74	77[c]	7.5	0.7[d]
Coke	30	318	365*[c]	26.6	2.8*[d]
Pig-iron	143	300	420[c]	7.7	7.0[d]
Crude steel	136	300	348[c]	8.2	3.0[d]
Radio sets (000 units)	64	148	157[e]	8.7	1.5[f]
Television sets (000 units)	—	64	68[e]	na	1.5[f]
Motor vehicles[g] (000 units)	1.9	6.1	11.2[e]	12.4	16.4[f]
Merchant vessels (000 grt)	7[a]	9	7	4.3[b]	−4.1

[a]1964 [b]1964–70 [c]1975 [d]1970–75 [e]1974 [f]1970–74 [g]Assembly only

Transport traffic	1960	1970	1976	Growth %pa	
Passenger-kilometres (mn)				1960–70	1970–76
Rail	3 634[a]	6 529[a]	8 671[d]	6.0	6.5[e]
Air	366[b]	1 010	1 739	13.5[c]	9.5
Cargo: tonne-kilometres (mn)					
Rail	2 096[a]	3 333[a]	2 767[d]	4.7	−4.1[e]
Air	5.7[b]	16.3	21.9	14.1[c]	5.0
Sea: tonnes (mn)					
Goods loaded	7.1[b]	12.7	5.9	7.5[c]	−11.9
Goods unloaded	9.3[b]	7.0	9.4	−3.5[c]	5.1

[a]Year ending June 30th [b]1962 [c]1962–70 [d]1974 [e]1970–74
Tourism Number of visitors (1975) 539 000, gross receipts (1972) $ 90 mn

Finance and trade

Price indices	1960	1970	1976	Growth %pa	
(1970 = 100)				1960–70	1970–76
Consumer prices	69.7	100.0	147.3	3.7	6.7
Wholesale prices	74.9	100.0	143.6	2.9	6.2
Money stock					
(end-year, E £ mn)	405	783	2 239	6.8	19.1

Budget (total, 1976)
Balanced at E £ 5 976 mn = $ 15 272 mn = £ 8 456 mn

Egypt

Balance of payments ($ mn)	1972	1973	1974	1975	1976
Balance of goods (fob)	−357	−429	−1 242	−2 374	−2 233
Balance of services	−109	−135	−121	−100	+715
Balance of transfers	+295	+641	+1 035	+1 076	+710
Current balance	*−170*	*+77*	*−327*	*−1 397*	*−807*
Long-term capital flow	+117	−62	−156	+588	+541

Reserves and debt (end-year, $ mn)

International reserves	139	363	356	294	339
External public debt	2 166	2 312	3 240	5 687	na

External trade (1976) Imports: E £ 1 490 mn = $ 3 808 mn = £ 2 108 mn
Exports: E £ 595 mn = $ 1 521 mn = £ 842 mn

Main imports (1974)	% of total	Main exports (1975)	% of total
Wheat	25	Cotton	37
Chemicals	14	Cotton yarn	11
Transport equipment	9	Crude oil and products	8
(of which, motor		Clothing	5
vehicles 6, aircraft 3)		Rice	4
Machinery	9	Chemicals	4
Iron and steel	5	Oranges	3
Fats and oils	5		
Timber	3		
Sugar	3		

Main sources (1975)		Main destinations (1975)	
United States	19	Soviet Union	43
France	11	Czechoslovakia	7
West Germany	8	East Germany	6
Soviet Union	6	Rumania	5
Italy	6	Italy	4
United Kingdom	5	China	4
Australia	4	Poland	3
Rumania	4		
Saudi Arabia	3		
Japan	3		

Special focus

Military expenditure

	Gdp	Military expenditure	
	$ mn	$ mn	as % of gdp
1972	7 859	1 500***	19***
1973	9 255	2 800***	30***
1974	10 110	3 600***	36***
1975	11 400**	5 700***	50***
1976	12 300**	5 300***	43***

Equatorial Guinea

Republic of Equatorial Guinea
República de Guinea Ecuatorial

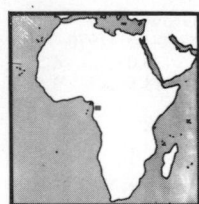

Location West central Africa
Comprises: Río Muni, on the west coast of
Africa, with Gabon to the south and east, and
Cameroon to the north; the island of Macias
Nguéma Biyogo (formerly Fernando Poo) in the
Bight of Biafra; the island of Pigalu (formerly
Annobon) in the Atlantic Ocean off the coast of
Gabon
Land Area 28 051 km² = 10 831 mi²
of which, Río Muni 26 017 km²

Climate Tropical
Weather at Malabo
Temperature: hottest month April 21–32 °C, coldest July 21–29 °C
Rainfall (av monthly): driest month January 5 mm, wettest June 302 mm
Time 1 hour ahead of GMT
Measures Metric system
Monetary unit Ekuele (E) = 100 céntimos
Rate of exchange (1976 av): par E 1 = Pa 1, free E 66.90 = $ 1,
E 120.8 = £ 1

Summary

Political One-party republic, which became independent October 12,
1968; formerly the two Spanish provinces of Fernando Poo and Río Muni.
Member of UN, OAU and an EEC ACP state

Economic An agricultural economy with cocoa, coffee and timber the
main export earners. After independence most Spanish professional
expatriates departed; a trade agreement with Soviet Union was signed in
1970. 20 000* Nigerian cocoa workers departed in 1976

People, resources and equipment

Population 1960 246 000*, 1970 285 000*, 1976 319 000*
Growth: 1960–70 1.5* %pa, 1970–76 1.9* %pa
Density (1976): 11* people per km²
Vital statistics (rate per 1 000 people, 1970–75): births 36.8*, deaths 19.7*
Refugees outside the country (1977): 100 000** (60 000** in Gabon,
30 000** in Cameroon)
Regions (population in 000, 1960) Río Muni 183, Macias Nguéma Biyogo
61, Pigalu 1.4
Cities (population in 000, 1974) Malaboª (capital) 25*, Bata 10*
ªFormerly called Santa Isabel
Race (1976) Fang 50** %, Bubi 5** %
Language Spanish; African languages (especially Fang) are also used
Religion (1966) Roman Catholic 85** %, Animist 15** %
Education (1973/74) Pupils 41 600*, teachers 830*
Labour force (1976) 96 000*; in agriculture 73 000* (77 %)
Personnel (1971) Physicians: 25, 1 per 11 600 people
Standard of living
National income per person (1976): E 20 000** = $ 300** = £ 160**
Consumption per person (1975): energy 101 kg coal equivalent
Livestock (000, 1976) Cattle 4*, sheep 31*, goats 7*, pigs 8*
Hospital beds (1967) 1 637, 1 per 170 people
Roads (1975) 1 180* km = 730* mi, density 0.04 km per km²
There are no railways
Ship (registered, 1977) 1, of 3 070 gross tons
Ports Malabo, San Carlos, Bata, Río Benito, Puerto Iradier
Airports Bata, Malabo
Durable equipment Radio sets (1970): 7 500, 26 per 1 000 people

Production, finance and trade

Gross domestic product 1970: E 5 300 mn = $ 76 mn = £ 32 mn
1976 est: E 6 700** mn = $ 100** mn = £ 55** mn
Agricultural production index (1970 = 100) 1976 65;
growth 1970–76 −7.0 %pa
Main products (000 t, 1976) Sweet potatoes 30*, cassava 49*,
bananas 15*, coconuts 7*, palm kernels 2.3*, palm oil 4.5*, coffee 5*,
cocoa 12*, timber (000 m³, 1975) 934
Transport traffic *Air* (1976) Passenger-km 5 mn. *Sea* (1970) Goods loaded
310 000* t, unloaded 135 000* t
Budget (1970) Revenue: E 709 mn = $ 10 mn = £ 4.2 mn
Expenditure: E 589 mn = $ 8 mn = £ 3.5 mn
External trade (1976) Imports: E 1 700** mn = $ 25** mn = £ 14** mn
Exports: E 1 000** mn = $ 15** mn = £ 8** mn

Main imports (1970)	% of total	Main exports (1970)	% of total
Food	17	Cocoa	66
Chemicals	14	Coffee	24
Beverages and tobacco	13	Timber	9
Transport equipment	8		
Machinery	6		

Main sources (1976) Spain, West Germany, Soviet Union, Cuba, China
Main destinations (1976) Spain, Soviet Union, China, Cuba

Ethiopia

Socialist Ethiopia

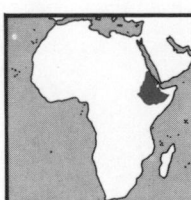

Location North-east Africa
Sudan is to the west, Kenya to the south,
Somalia to the south and east, and Djibouti
to the north-east; the northern province of
Eritrea is on the Red Sea coast
Land Area 1 221 900 km² = 471 800 mi²
of which, Eritrea 117 600 km²
Climate Mainly temperate on plateau, hot in
lowlands
Weather at Addis Ababa, 2 450 m altitude
Temperature: hottest months April, May 10–25 °C, coldest Dec 5–23 °C
Rainfall (av monthly): driest month Dec 5 mm, wettest Aug 300 mm
Time 3 hours ahead of GMT
The Ethiopian year, using the Coptic calendar, ends September 10th;
Ethiopian year 1970 = Gregorian 1977/78,
Gregorian 1978 = Ethiopian 1970/71

Ethiopia

Measures Metric system; also: *length* 1 kend = 0.5 metre = 1.64 feet
area 1 gasha = 40 hectares = 99 acres
weight (*mass*) 1 neter = 450 grams = 0.99 pound,
1 frasoulla = 17 kilograms = 37.48 pounds
Monetary unit Birr (Br) = 100 cents; the birr replaced the Ethiopian
dollar at par from October 14, 1976.
Rate of exchange (1976 av): par Br 2.0855 = $ 1, free Br 3.767 = £ 1

Summary

Political Socialist republic with military government; formerly ruled by
the late Emperor Haile Selassie, deposed September 12, 1974. Previously
also known as Abyssinia. Eritrea was fully incorporated into Ethiopia in
1962, after 10 years of formal autonomy; there is a separatist movement in
Eritrea. Italy occupied Ethiopia 1935–41; part of eastern Ethiopia
(Ogaden) has been claimed by Somalia. Member of UN, OAU and an
EEC ACP state
Economic An agricultural economy with a spread of agricultural export
earners, mainly coffee but also beans, oilseeds and hides and skins. There
is some industrial capacity, especially in food and textiles. Drought
affected the country very seriously 1973–76; the railway to
Djibouti was damaged in 1976–77 by military action

People, resources and equipment

Population 1960 20.70* mn, 1970 24.63* mn, 1976 28.68* mn
Growth: 1960–70 1.8* %pa, 1970–76 2.6* %pa
Density (1976): 23* people per km²
Vital statistics (rate per 1 000 people, 1970–75): births 49.4* , deaths 25.8*
Regions (provinces, population in 000, 1974; total of 27.8* mn)

Arusi	893*	Gwejam	1 830*	Sidamo	2 596*
Bale	740*	Harar	3 510*	Tigre	1 917*
Begemdir	1 419*	Illubabor	719*	Welega	1 327*
Eritrea	2 070*	Kefa	1 769*	Welo	2 570*
Gemu-Gwefa	731*	Shewa	5 712*		

Cities (population in 000, 1975)

Addis Ababa (capital)	1 161	Harar	54ª	Mak'ale	34ª
Asmara	318	Jima	52ª	Debre Zeyt	34ª
Diredawa	73ª	Nazaret	51ª	Dabra-Mark'os	34ª
Dese	55ª	Gonder	43ª	Bahir Dar	29ª

ª1974
Race (1970) Galla 33* %, Amhara 25* %, Tigréan 12* %, Walamo 8* %,
Somali 6* %, Gurage 4* %, Sidamo 3* %, Afar-Saho 3* %
Language Amharic; also Galla, Sidamo and Arabic, mainly in Eritrea
Religion (1976) Ethiopian Orthodox (Coptic) 50 %, Moslem 40 %,
Animist 10* %
Education (1973/74) Pupils: primary 849 831, secondary 182 263,
vocational 5 533, teacher-training 3 126, higher 6 474. Teachers:
primary 18 646, secondary 6 181, vocational 554, teacher-training 194,
higher 434
Labour force (1976) 12 138 000*; in agriculture 9 862 000* (81* %)
Personnel (1972) Physicians: 350, 1 per 74 000 people
Standard of living
National income per person (1976): Br 225 ** = $ 108** = £ 60**
Consumption per person (1975): energy 29 kg coal equivalent,
electricity (production) 24* kW h, newsprint 0.04 kg, steel 1 kg
Newspapers (1974): number 7; circulation 51 000, 2 per 1 000 people
Telephones (Dec 1975): 69 000, 2 per 1 000 people
Livestock (000, 1976) Cattle 25 963, sheep 23 065, goats 17 064,
horses 1 510*, mules 1 420*, asses 3 860, camels 960, chickens 51 300
Petroleum refinery capacity (1975) 0.74* mn tonnes
Electrical capacity (1975) 320* megawatts,
of which, hydro 206 megawatts
Hospital beds (1972) 8 415, 1 per 3 080 people
Roads (1976) 23 000 km = 14 300 mi, density 0.02 km per km²
Railways (1976) 988* km = 614* mi, density 0.001 km per km²; includes
682 km of the 782 km railway linking Addis Ababa with Djibouti
Ships (registered, 1977) 18, total of 23 989 gross tons
Ports (goods traffic, 000 tonnes, 1972) Assab: loaded 477, unloaded 652;
Massawa: loaded 206, unloaded 234.
Both ports are in Eritrea
Airports Passenger departures and arrivals (1975): Addis Ababa 157 081,
Asmara 44 123; also (1977) 36 other airports with scheduled flights

Durable equipment	000	no per	
(at end-year)		1 000 people	
Radio sets (1974)	200	7	no per
Television sets (1974)	20	0.7	km of road
Passenger cars (1975)	43	1.6	1.9
Commercial vehicles (1975)	9	0.3	0.4

Production

Gross domestic product
1973/74 (year ending July 7th): Br 5 586 mn = $ 2 678 mn = £ 1 123 mn
1976 est: Br 6 500** mn = $ 3 100** mn = £ 1 700** mn
Growth in real terms: 1960–70 4.0 %pa, 1970–74 3.6 %pa
Structure of gross domestic product (1973/74) *By origin* Agriculture 48 %,
manufacturing 8 %, construction 4 %, other 40 %
By type Final consumption expenditure 87 % (of which, government 11 %),
gross fixed capital formation and stock investment 10 %, exports of goods
and services 15 %, less imports of goods and services −12 %

Production index	1960	1970	1976	Growth % pa	
(1970 = 100)				1960–70	1970–76
Agricultural	76	100	95	2.8	−0.9
Main products (000 t)					
Agriculture					
Wheat	600	808	694	3.0	−2.5
Barley	1 200	1 525*	800*	2.4*	−10.2*
Maize	670	909	1 200*	3.1	4.7*
Sorghum and millet	1 950	1 197	1 192	−4.8	−0.1
Yams	220*	254*	270*	1.4*	1.0*
Chickpeas	158	185	109	1.6	−8.4
Dry peas	110*	127*	52*	1.4*	−13.8*
Dry broad beans	100	138	200*	3.3	6.4*
Sugar, raw value	43	73	136	5.4	10.9
Sesameseed	30	80*	70*	10.3*	−2.2*
Linseed	50	60	50*	1.8	−3.0*
Nigerseed	na	na	278ʲ	na	na
Coffee	101	175	170	5.7	−0.5
Cotton	3*	14	20*	16.7*	6.1*
Milk	451	516	531*	1.4	0.5*
Cattle hides	43*	49*	40*	1.3*	−3.3*
Beef and veal	217	267	210	2.1	−3.9
Timber (000 m³)	22 100	23 137*	24 220*ᵈ	4.7*	0.9*ᵉ
Other					
Petroleum products	—	594	585*ᵈ	na	−0.3*ᵉ
Electricity (mn kW h)ª	102	520	665*ᵈ	17.7	5.0*ᵉ
Saltª	125	218	68ᶠ	5.7	−25.3ᵍ
Gold (kg)	480	849	657ᵈ	5.9	−5.0ᵉ
Beer (000 hl)ª	157ᵇ	280	422ᵈ	12.3ᶜ	8.5ᵉ
Cigarettes (mn units)ª	336	870	1 193ᵈ	10.0	6.5ᵉ
Cotton yarnª	5.6ᵇ	10.5	11.7ʰ	13.4ᶜ	3.7ⁱ
Cotton fabrics (mn m²)ª	34ᵇ	70	82ʰ	15.5ᶜ	5.4ⁱ
Cementª	73ᵇ	175	117ᵈ	19.1ᶜ	−7.7ᵉ

ªYears ending September 10th ᵇ1965 ᶜ1965–70 ᵈ1975 ᵉ1970–75 ᶠ1974
ᵍ1970–74 ʰ1973 ⁱ1970–73 ʲ1972

Transport traffic	1960	1970	1976	Growth %pa	
Passenger-kilometres (mn)				1960–70	1970–76
Railª	65ᵇ	92	108ᵈ	5.1ᶜ	3.3ᵉ
Air	105	314	523	11.6	8.9
Cargo: tonne-kilometres (mn)					
Railª	202ᵇ	220	244ᵈ	1.2ᶜ	2.1ᵉ
Air	4.6	16.1	20.1	13.4	3.8
Sea: tonnes (mn)					
Goods loaded	0.24	0.73	0.54ᵈ	11.8	−5.7ᵉ
Goods unloaded	0.24	1.00	0.84ᵈ	15.4	−3.5ᵉ

ªIncluding traffic of Djibouti portion of Addis Ababa–Djibouti City line;
excluding Eritrea ᵇ1963 ᶜ1963–70 ᵈ1975 ᵉ1970–75
Tourism (1975) Number of visitors 31 000, gross receipts $ 7 mn

Finance and trade

Price index	1960	1970	1976	Growth %pa	
(1970 = 100)				1960–70	1970–76
Consumer prices	69.9ª	100.0	152.9	5.2ᵇ	7.3
Money stock					
(end-year, Br mn)	218	428	810	7.0	11.2

ª1963 ᵇ1963–70
Budget (1975/76; year ending July 7th)
Revenue: Br 1 175 mn = $ 563 mn = £ 282 mn
Expenditure: Br 1 331 mn = $ 638 mn = £ 320 mn

Balance of payments ($ mn)	1972	1973	1974	1975	1976
Balance of goods (fob)	+8	+60	+17	−44	−64
Balance of services	−17	−10	−13	−43	−32
Balance of transfers	+18	+25	+52	+40	+63
Current balance	+8	+75	+55	−46	−33
Long-term capital flow	+32	+53	+51	+70	+67
Reserves and debt (end-year, $ mn)					
International reserves	93	177	275	288	306
External public debt	404	480	566	674	na

Ethiopia

External trade (1976) Imports: Br 730 mn = $ 350 mn = £ 194 mn
Exports: Br 581 mn = $ 278 mn = £ 154 mn

Main imports (1975)	% of total	*Main exports* (1976)	% of total
Chemicals	18	Coffee	56
Machinery	16	Beans	10
Crude oil	15	Hides and skins	10
Motor vehicles	11	Oilseeds	5
Textile yarns and fabrics	5		
Iron and steel	4		
Main sources (1975)		*Main destinations* (1975)	
Saudi Arabia	15	United States	20
Japan	12	Saudi Arabia	14
Italy	11	Djibouti	10
West Germany	10	Egypt	9
United Kingdom	8	Japan	9
United States	8	West Germany	8
France	5	Italy	5

Special focus

The largest five countries in Africa, by population, 1976

	Population (mn)	Gross domestic product ($ mn)	National income per person ($)
Nigeria	64.75**	25 000**	350**
Egypt	38.07*	12 300**	310**
Ethiopia	28.68*	3 100**	108**
South Africa	26.13	32 217*	1 070*
Zaire	25.63*	3 700**	124**

Gabon

Gabonese Republic
République Gabonaise

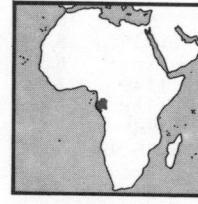

Location West central Africa
On the Atlantic coast, with Congo to the east
and south, and Equatorial Guinea and
Cameroon to the north
Land Area 267 667 km² = 103 347 mi²
Climate Tropical
Weather at Libreville, 35 m altitude
Temperature: hottest months March, April
23–32 °C, coldest July 20–28 °C
Rainfall (av monthly): driest month July 3 mm,
wettest November 373 mm
Time 1 hour ahead of GMT
Measures Metric system
Monetary unit CFA franc (CFA Fr) = 100 centimes
Rate of exchange (1976 av): par CFA Fr 50 = Fr 1,
free CFA Fr 239.0 = $ 1, CFA Fr 431.6 = £ 1

Summary

Political One-party republic, which became independent August 17, 1960;
formerly one of the four territories included in French Equatorial Africa.
Member of UN, OAU, Udeac, Opec, franc zone and an EEC ACP state
Economic Mineral production, especially crude oil and its products, is the
main basis of the economy, replacing timber which was important in the
1960s. Crude oil exploration continues, and the planned trans-Gabon
railway will open the way to iron ore deposits, adding to the important
mineral production of manganese and uranium

People, resources and equipment

Population[a] 1960 450 000*, 1970 498 000*[b], 1976 526 000*[b]
[a]UN estimates [b]Official census estimate for 1970 was 950,000 and for 1976 over
1 000 000 (including 60,000* refugees from Equatorial Guinea)
Growth: 1960–70 1.0* %pa, 1970–76 0.9* %pa
Density (1976): 2* people per km²
Vital statistics (rate per 1 000 people, 1970–75): births 32.2*, deaths 22.2*
Cities (population in 000, 1975) Libreville (capital) 130**,
Port Gentil 78, Lambaréné 23
Race (1970) Fang 45 %, Puno 19 %, Njawi 17 %, Kanda 10 %
Language French; Bantu languages are also used
Religion (1976) Christian 60** % (Roman Catholic 42** %),
Animist 39** %, Moslem 1** %
Education (1974/75) Pupils 143 392, teachers 3 300*
Labour force (1976) 256 000*; in agriculture 201 000* (78* %)

Personnel Scientists and engineers engaged in research (1970): 8
Physicians (1971): 96, 1 per 5 200 people
Standard of living National income per person (1976):
CFA Fr 700 000** = $ 2 900** = £ 1 600**
Consumption per person (1975): energy 1 026 kg coal equivalent,
electricity (production) 448* kW h, steel 208 kg
Newspapers (1974): number 1
Telephones (Dec 1973): 11 000, 21 per 1 000 people
Livestock (000, 1976) Cattle 5*, sheep 59*, goats 64*, pigs 5*
Mineral reserves (1975) Crude oil 108 mn tonnes
Natural gas 51 bn cubic metres
Uranium 20 000 tonnes
Petroleum refinery capacity (1975) 1.2* mn tonnes
Electrical capacity (1975) 45* megawatts
Hospital beds (1969) 4 995, 1 per 98 people
Roads (1976) 6 878 km = 4 274 mi, density 0.03 km per km²
Railways (cableway link to Congo rail system, 1976) 76 km = 47 mi,
density 0.0003 km per km²
Construction of a 970 km Trans-Gabon railway was begun in 1974
Inland waterway Ogoué river 320 km = 200 mi
Ships (registered, 1977) 15, total of 98 645 gross tons
Ports (goods traffic, 000 tonnes, 1973)

	loaded	unloaded
Port Gentil and Gamba[a]	5 337	140
Libreville-Owendo	360	516

A new port for minerals is planned for Santa Clara
[a]Crude oil port
Airports Port Gentil, Libreville, Lambaréné and 23 other airports with
scheduled services

Durable equipment (Dec 1974)	000	no per 1 000 people	
Radio sets	90	172	no per
Television sets	5.1	10	km of road
Passenger cars	10.1*	19*	1.5*
Commercial vehicles	7.3*	14*	1.1*

Production

Gross domestic product
1974: CFA Fr 371 700 mn = $ 1 546 mn = £ 661 mn
1976 est: CFA Fr 450 000** mn = $ 1 900** mn = £ 1 040** mn
Structure of gross domestic product *By origin* (1972) Agriculture 12 %,
mining and quarrying 32 %, manufacturing 8 %, construction 11 %,
other 37 %
By type (1974) Final consumption expenditure 34 % (of which
government 9 %), stock investment 10 %, gross fixed capital formation
42 %, exports of goods and services 58 %, less imports of goods and
services −43 %

Production index (1970 = 100)	1960	1970	1976	Growth %pa 1960–70	1970–76
Agricultural	77	100	109	2.6	1.5
Main products (000 t)					
Agriculture					
Cassava	120**	167*	180*	3.4**	1.2*
Bananas and plantains	90*	90*	90*	0.0*	0.0*
Palm kernels	na	0.23*	0.25*	na	1.4*
Palm oil	na	2.4*	2.7*	na	2.0*
Coffee	0.3	0.9*	0.9*	11.6*	0.0*
Cocoa	3.8	5.0	5.0*	2.8	0.0*
Timber (000 m³)	3 100	2 940	2 600*[a]	−0.5	−2.4*[b]
Energy					
Total energy (000 tce)	1 050	7 090	16 790[a]	21.0	18.8[b]
Crude oil	800	5 423	11 392	21.1	13.2
Petroleum products	—	876	810*[a]	na	−1.6*[b]
Natural gas (mn m³)	7	32	49	16.4	7.3
Electricity (mn kW h)	20	97	235*[a]	17.1	19.4*[b]
Mining					
Manganese ore (Mn content)	—	729	1 115[a]	na	8.9[b]
Uranium (U content)	—	0.40	0.80[a]	na	14.9[b]
Gold (kg)	506	501	100[a]	−0.1	−27.6[b]
Manufacturing					
Beer (000 hl)	45	110	327[a]	9.3	24.3[b]
Cigarettes (mn units)	—	—	332[a]	na	na
Cement	—	23	93[a]	na	32.2[b]
Plywood (000 m³)	na	72	71[c]	na	−0.3[d]
Construction					
Buildings (permits, 000 m²)	49	123	206[c]	9.6	13.8[d]

[a]1975 [b]1970–75 [c]1974 [d]1970–74
Transport traffic *Air* (including apportionment of Air Afrique, 1976)
passenger-km 190* mn, cargo 14.4 mn t-km *Sea* (1974) Goods loaded
6.6 mn t, unloaded 0.4 mn t

Gabon

Finance and trade

Price indices (1970 = 100)	1960	1970	1976	Growth %pa 1960–70	1970–76
Consumer prices	76.2[a]	100.0	164.1[c]	3.5[b]	10.4[d]
Wholesale prices	58.2	100.0	164.7[c]	5.6	10.5[d]
Money stock (end-year, CFA Fr bn)	5.5	12.5	102.6	8.5	42.0

[a]1962 [b]1962–70 [c]1975 [d]1970–75

Budget (1976) Balanced at CFA Fr 193 113 mn = $ 808 mn = £ 447 mn

Balance of payments ($ mn)	1972	1973	1974	1975	1976
Balance of goods (fob)	+112	+150	+450	+386	na
Balance of services	−131	−203	−335	−486	na
Balance of transfers	+14	+1	−9	−9	na
Current balance	*−4*	*−52*	*+105*	*−108*	*na*
Long-term capital flow	−6	+32	+114	+249	na
Reserves and debt (end-year, $ mn)					
International reserves	23	48	103	146	116
External public debt	215	383	505	515	na

External trade (1976)
Imports: CFA Fr 130 000* mn = $ 540* mn = £ 300* mn
Exports: CFA Fr 214 480 mn = $ 897 mn = £ 497 mn

Main imports (1971)	% of total	Main exports (1976)	% of total
Machinery, non-electric	17	Crude oil	82
Motor vehicles	10	Timber	5**
Food	9	Manganese	3**
Iron and steel	8	Uranium	1**
Chemicals	8		
Electrical machinery	7		
Aircraft	6		

Main sources (1975)		Main destinations (1975)	
France	63*	France	21*
United States	11*	United States	19*
Belgium-Luxembourg	5*	Spain	13*
West Germany	4*	Bahamas	11*

Special focus

Main exports	% of total exports			
	1960	1965	1970	1976
Crude oil	21	14	41	82
Timber	51	27	27	5**
Manganese	—	26	10	3**

Gambia

Republic of The Gambia

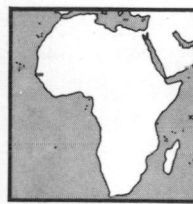

Location West Africa
With a coastal strip on the Atlantic Ocean, the country runs east on each side of the Gambia river; it is surrounded on all land boundaries by Senegal
Land Area 11 295 km² = 4 361 mi²
Climate Tropical
Weather at Banjul, 27 m altitude
Temperature: hottest month June 23–32 °C, coldest January 15–31 °C
Rainfall (av monthly): driest months March, April 1 mm, wettest Aug 500 mm
Time GMT
Measures UK (imperial) system, changing to the metric system
Monetary unit Dalasi (Di) = 100 bututs; the dalasi was introduced as a decimal currency July 1, 1971, replacing the Gambia pound at Di 5 = G £ 1
Rate of exchange (1976 av): par Di 4 = £ 1, free Di 2.215 = $ 1

Summary

Political Republic, which became independent February 18, 1965; formerly a UK colony, but with internal self-government from October 4, 1963. Member of UN, OAU, Ecowas, Commonwealth and an EEC ACP state
Economic A one-product agricultural economy based on groundnuts. Other agricultural production, especially rice, is being expanded in a program of diversification. There are some mineral prospects, especially china clay, and tourism is being developed

People, resources and equipment

Population 1960 300 000*, 1970 463 000*, 1976 540 000*
Growth: 1960–70 4.4* %pa, 1970–76 2.6* %pa
Density (1976): 48* people per km²
Vital statistics (rate per 1 000 people, 1970–75): births 43.3*, deaths 24.1*
Regions (population in 000, 1973; total of 493 000) Banjul City 39;
divisions:
Upper River 87; Lower River 137 (of which, North Bank 94);
MacCarthy Island 101; Western 130 (of which, Combo St Mary 39)
Cities (population in 000, 1975) Banjul[a] 43[b], Brikama 4*, Kuntaur 3*, Yundum 3*
[a]Formerly called Bathurst [b]Urban area only; total for Banjul-Combo St Mary area 78
Race (1973) Malinke 38 %, Fulani 16 %, Wolof 14 %, Jola 9 %, Serahuli 8 %
Language English; Malinke and other African languages are also used
Religion (1965) Moslem 58 %, Animist 38* %, Christian 4* %
Education (1975/76) Pupils 31 235, teachers 1 295
Labour force (1976) 260 000*; in agriculture 207 000* (80* %)
Personnel (1969) Physicians: 19, 1 per 23 700 people
Standard of living
National income per person (1976): Di 420** = $ 190** = £ 105**
Consumption per person: energy (1975) 66 kg coal equivalent, electricity (production, 1976) 56* kW h
Telephones (Dec 1975): 2 500, 5 per 1 000 people
Livestock (000, 1976) Cattle 310*, sheep 95*, goats 94*, pigs 8*
Electrical capacity (1975) 10 megawatts
Hospital beds (1966) 488, 1 per 850 people
Roads (1975) 1 270* km = 790* mi, density 0.11* km per km²
There are no railways
Inland waterway (1976) Gambia river 500* km = 300* mi
Ships (registered, 1977) 5, total of 1 608 gross tons
Ports Banjul, Basse Santa Su (on the Gambia river)
Airport Yundum (Banjul)
Durable equipment Radio sets (Dec 1974): 60 000, 118 per 1 000 people
Passenger cars (Dec 1972): 3 000, 6 per 1 000 people, 2.4* per km of road
Commercial vehicles (Dec 1972): 2 500, 5 per 1 000 people, 2.0* per km of road

Production, finance and trade

Gross domestic product
1975/76 (year ending June 30th): Di 210.7 mn = $ 105 mn = £ 53 mn
1976 est: Di 230** mn = $ 105** mn = £ 57** mn
Agricultural production index (1970 = 100) 1960 87, 1976 122
Growth: 1960–70 1.4 %pa, 1970–76 3.4 %pa
Main products (000 t, 1976) Rice 50*, millet 22, groundnuts 142*, palm kernels 1.7*, palm oil 2.4*, milk 5*, beef and veal 4*, cattle hides 0.5*, fish catch 11*, timber (000 m³) 250*, electricity (mn kW h) 30*
Transport traffic (1975) *Sea* Goods loaded 110 000 t, unloaded 126 000 t
Tourism (1975/76) Number of visitors 24 500, gross receipts $ 6* mn
Consumer price index (1970 = 100) 1961 89.2, 1976 192.8;
growth 1961–70 1.3 %pa, 1970–76 11.6 %pa
Money stock (end-year, Di mn) 1976 82.9; growth 1970–76 30.9 %pa
Budget (1976/77; year ending June 30th)
Revenue: Di 43.0 mn = $ 18 mn = £ 10.7 mn
Expenditure: Di 44.3 mn = $ 19 mn = £ 11.1 mn
International reserves (Dec 1976) $ 21 mn
External public debt (Dec 1976) $ 49 mn
External trade (1976) Imports: Di 164 mn = $ 74 mn = £ 41 mn
Exports: Di 76 mn = $ 34 mn = £ 19 mn

Main imports (1974/75[a])	% of total	Main exports (1974/75[a])	% of total
Textile yarns and fabrics	24	Groundnuts	47
Food	20	Groundnut oil	27
(of which, cereals 11*)		Groundnut meal and cake	8
Machinery and transport equipment	14	Fish	2
Chemicals	9		
Petroleum products	9		

Main sources (1976)		Main destinations (1976)	
United Kingdom	25	United Kingdom	30
China	13	Netherlands	22
Netherlands	6	France	10
Japan	6	Italy	7
France	5	Switzerland	6
Burma	5	Portugal	5
West Germany	5		

[a]Year ending June 30th

Ghana
Republic of Ghana

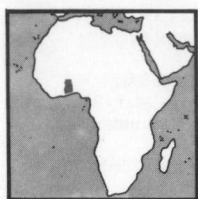

Location West Africa
With a south-facing coastline on the Atlantic
Ocean, Ivory Coast is to the west, Upper
Volta to the north, and Togo to the east
Land Area 238 500 km² = 92 100 mi²
Climate Tropical
Weather at Accra, 27 m altitude
Temperature: hottest months March, April
24–31 °C, coldest August 22–27 °C
Rainfall (av monthly): driest month January 15 mm, wettest June 178 mm
Time GMT
Measures Metric system; before 1975 the UK (imperial) system
Monetary unit Cedi (₵) = 100 pesewas; the name was changed from
'new cedi' to 'cedi' from February 7, 1972. The cedi was first introduced
July 19, 1965 to replace the Ghanaian pound at ₵ 2.4 = G £ 1
Rate of exchange (1976 av): par ₵ 1.15385 = $ 1, free ₵ 2.084 = £ 1

Summary

Political Republic with military government, which became independent
March 6, 1957; formerly the UK colony of the Gold Coast, joined by
British Togoland. A new constitution is planned for 1979. Member of
UN, OAU, Ecowas, Commonwealth and an EEC ACP state
Economic Mainly an agricultural economy, with cocoa accounting for
some two-thirds of export earnings; timber is also important. Mining
production includes bauxite, manganese, diamonds and gold; imported
bauxite is smelted into aluminium using hydro-electricity. A wide range of
light industrial products is manufactured; work began in 1977 on
a steelworks

People, resources and equipment

Population 1960 6.78* mn, 1970 8.63* mn, 1976 10.31* mn
Growth: 1960–70 2.5* %pa, 1970–76 3.0* %pa
Density (1976): 43* people per km²
Vital statistics (rate per 1 000 people, 1970–75): births 48.8*, deaths 21.9*
Regions (population in 000, 1970; census total of 8.55 mn)

Region		Region		Region	
Ashanti	1 477	Eastern	1 263	Upper	857
Brong-Ahafo	763	Greater Accra	849	Volta	947
Central	893	Northern	729	Western	768

Cities (population in 000, 1970)

City		City		City		City	
Accra (capital)	738ᵃ	Tamale	84	Obuasi	31	Sunyani	24
Kumasi	345	Koforidua	52	Winneba	31		
Sekondi-Takoradi	161	Cape Coast	52	Ho	24		

ᵃ Accra-Tema area, of which, Tema 61 thousand
Race (1960) Akan 44 %, Mossi-Dagomba 16 %, Ewe 13 %,
Ga-Adangme 8 %, Guan 4 %, Gurma 3 %, Yoruba 2 %
Language English; Ashanti, Fanti and other languages are also used
Religion (1976) Christian 40* %, Animist 40* %, Moslem 12* %
Education (1974/75) Pupils 1 591 000, teachers 59 100
Labour force (1976) 3 798 000*; in agriculture 2 042 000* (54* %)
Personnel Scientists and engineers (1970): 6 897
Physicians (1974): 856, 1 per 11 230 people
Standard of living
National income per person (1976): ₵ 270** = $ 230** = £ 130**
Consumption per person (1975): energy 182 kg coal equivalent,
electricity (production) 410* kW h, newsprint 1.1 kg, steel 9 kg
Newspapers (1973): number 6; circulation 381 000, 41 per 1 000 people
Telephones (Dec 1975): 60 000, 6 per 1 000 people
Livestock (000, 1976) Cattle 1 100*, sheep 1 800*, goats 2 000*, pigs 400*
Petroleum refinery capacity (1975) 1.45 mn tonnes
Electrical capacity (1975) 995* megawatts, of which, hydro 925*
megawatts
Hospital beds (1973) 13 461, 1 per 695 people
Roads (1975) 32 000* km = 20 000* mi, density 0.13* km per km²
Railways (1976) 953 km = 592 mi, density 0.004 km per km²
Ships (registered, 1977) 79, total of 182 696 gross tons
Ports (goods traffic, 000 tonnes, 1972) Tema: loaded 847, unloaded 2 031
Takoradi: loaded 1 963, unloaded 450
Airports (1975) Kotoka (Accra): passenger departures and arrivals
257 932; also Kumasi, Tamale, Sunyani, Takoradi

Durable equipment	000	no per	
(Dec 1974)		1 000 people.	
Radio sets	1 060	109	no per
Television sets	33	3	km of road
Passenger cars	55	6	1.7*
Commercial vehicles	44	5	1.4*

Production

Gross domestic product 1974: ₵ 2 196 mn = $ 1 910 mn = £ 816 mn
1976 est: ₵ 3 000** mn = $ 2 600** mn = £ 1 400** mn
Growth in real terms: 1967–71 5.2 %pa
Structure of gross domestic product (1972) *By origin* Agriculture 48 %,
mining and quarrying 2 %, manufacturing 10 %, other 40 %

Production indices (1970 = 100)	1960	1970	1976	Growth %pa 1960–70	1970–76
Agricultural	84	100	99	1.8	−0.1
Manufacturing	51ᵃ	100	129ᵍ	8.8ᵇ	13.6ʰ
Main products (000 t)					
Agriculture					
Maize	234	442	395*	6.6	−1.8*
Millet	110	93	71*	−1.7	−4.4*
Sorghum	100	90	81*	−1.0	−1.7*
Cassava	1 000	1 596	1 800*	4.8	2.0*
Taro	700**	1 050*	1 400*	4.1**	4.9*
Yams	800	1 617	800*	7.3	−11.1*
Tomatoes	15**	37*	97*	9.4**	17.4*
Groundnuts	49	60	60*	2.0	0.0*
Palm kernels	12	37*	32*	11.9*	−2.4*
Palm oil	7*	20*	32*	11.1*	8.1*
Bananas and plantains	650**	750*	900*	1.4**	3.1*
Oranges	25*	71	170*	11.0*	15.7*
Cocoa	439	406	320ᶜ	−0.8	−3.9*
Fish catch	32	171	255ᶜ	18.4	8.3ᵈ
Timber (000 m³)	8 100	10 133	11 973ᵉ	2.3	4.3ᶠ
Energy					
Total energy (000 tce)	–	360	490ᶜ	na	6.4ᵈ
Petroleum products	–	810	1 183*ᶜ	na	7.9*ᵈ
Electricity (mn kW h)	374	2 920	4 050*ᶜ	22.8	6.8*ᵈ
Mining					
Bauxite	194	342	325ᶜ	5.8	−1.0ᵈ
Manganese ore (Mn content)	266	191	199ᶜ	−2.9	0.8ᵈ
Gold (000 kg)	27	22	16	−2.2	−5.2
Diamonds (000 CM)	3 273	2 550	2 328ᶜ	−2.5	−1.8ᵈ
Manufacturing					
Beer (000 hl)	54	391	500ᶜ	21.9	5.0ᵈ
Cigarettes (mn units)	1 190ᵃ	1 536	2 339ᶜ	3.2ᵇ	8.8ᵈ
Cotton fabrics (mn m)	na	36	43ᵍ	na	9.3ʰ
Cement	–	442	688ᶜ	na	9.2ᵈ
Aluminium	–	113	143ᶜ	na	4.9ᵈ
Radio sets (000 units)	–	104	90ᶜ	na	−2.9ᵈ

ᵃ1962 ᵇ1962–70 ᶜ1975 ᵈ1970–75 ᵉ1974 ᶠ1970–74 ᵍ1972 ʰ1970–72

Transport traffic	1960	1970	1976	Growth % pa 1960–70	1970–76
Passenger-kilometres (mn)					
Rail	276	542	431ᶜ	7.0	−10.8ᵈ
Air	60	135	194	8.4	6.2
Cargo: tonne-kilometres (mn)					
Rail	357	311	305ᶜ	−1.4	−1.0ᵈ
Air	0.3	3.3	4.4	26.1	4.9
Sea: tonnes (mn)					
Goods loaded	2.07	2.19	2.26ᵃ	0.6	1.1ᵇ
Goods unloaded	1.87	4.23	2.74ᵃ	8.5	−13.4ᵈ

ᵃ1973 ᵇ1970–73 ᶜ1972 ᵈ1970–72

Finance and trade

Price index (1970 = 100)	1960	1970	1976	Growth %pa 1960–70	1970–76
Consumer prices	49.5	100.0	343.6	7.3	22.8
Money stock (end-year, ₵ mn)	134	306	1 429	8.6	29.3

Budget (total, 1975/76; year ending June 30th)
Revenue: ₵ 852 mn = $ 741mn = £ 370 mn
Expenditure: ₵ 1 199 mn = $ 1 043 mn = £ 521 mn

Balance of payments ($ mn)	1972	1973	1974	1975	1976
Balance of goods (fob)	+163	+213	−29	+150	+85
Balance of services	−67	−99	−166	−197	−207
Balance of transfers	+13	+12	+24	+45	+33
Current balance	*+109*	*+127*	*−171*	*−2*	*−89*
Long-term capital flow	+42	+31	+8	+92	+16
Reserves and debt (end-year, $ mn)					
International reserves	107	189	94	150	104
External public debt	610	668	701	797	na

Ghana

External trade (1975) Imports: ₵ 909 mn = $ 788 mn = £ 436 mn
Exports: ₵ 928 mn = $ 804 mn = £ 445 mn

Main imports	% of total	Main exports	% of total
Machinery	16	Cocoa	59
Chemicals	14	Timber	8
Crude oil	13	Aluminium	4
Food	12	Petroleum products	3
Motor vehicles	9*	Manganese ore	2
Textile yarns and fabrics	6		
Main sources		Main destinations	
United States	16	United Kingdom	15
United Kingdom	15	United States	11
West Germany	11	Netherlands	10
Nigeria	7	Switzerland	8
Japan	6	West Germany	8
Libya	5	Japan	7

Special focus

Cocoa prices (London, wholesale price)

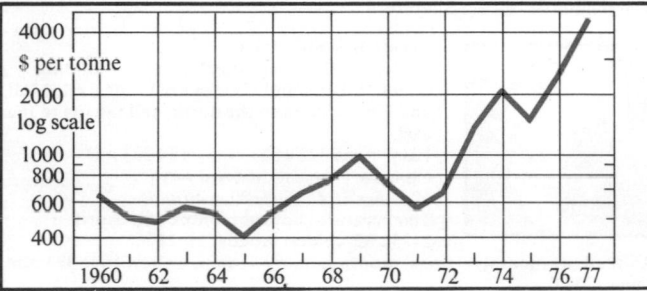

Guinea

Republic of Guinea
République de Guinée

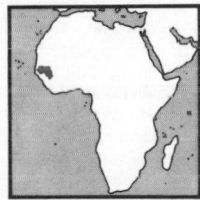

Location West Africa
With a coastline on the Atlantic Ocean, Guinea-Bissau, Senegal and Mali are to the north, Ivory Coast to the east, and Sierra Leone and Liberia to the south
Land Area 245 857 km² = 94 926 mi²
Climate Tropical, hot and wet on the coast
Weather at Conakry, 7 m altitude
Temperature: hottest month May 24–32 °C, coldest August 22–28 °C
Rainfall (av monthly): driest month January 3 mm, wettest July 1 298 mm
Time GMT
Measures Metric system
Monetary unit Syli (Sy) = 100 cauris; the Syli was introduced October 1972, replacing the Guinean franc (G Fr) at Sy 1 = G Fr 10.
Rate of exchange (1976 av): par Sy 24.6853 = SDR 1, free Sy 21.38 = $ 1, Sy 38.62 = £ 1

Summary

Political One-party people's republic, which became independent October 2, 1958; formerly a French colony. Member of UN, OAU, Ecowas and an EEC ACP state
Economic A broadly based agricultural economy, with pineapples, coffee and palm kernels important export products; however, the main export earner is aluminium in the form of bauxite and alumina—accounting for some three-quarters of export earnings. Iron ore deposits near Liberia are being further developed

People, resources and equipment

Population (UN est) 1960 3.07** mn, 1970 3.92** mn, 1976 4.53** mn
Official Guinea population estimate (1972): 5.14** mn, of whom 1.5** mn living outside the country
Growth: 1960–70 2.5** %pa, 1970–76 2.4** %pa
Density (1976): 18** people per km²
Vital statistics (rate per 1 000 people, 1970–75): births 46.6*, deaths 22.9*

Cities (population in 000, 1972)
Conakry (capital) 526 Kindia 45ᵃ Labe 26ᵃ Siguiri 18ᵃ
Kankan 176 N'Zérékoré 26ᵃ Mamou 18ᵃ
ᵃ1967
Race (1976) Malinke 34** %, Fulani 34** %, Susu 10** %, Kissi 6** %, Kpelle 5** %
Language French; local languages are also used, especially Malinke and Susu
Religion (1976) Moslem 85** %, Animist 14** %, Christian 1** %
Education (1971/72) Pupils 243 300*, teachers 7 800*
Labour force (1976) 2 072 000*; in agriculture 1 700 000* (82* %)
Personnel (1973) Physicians: 188, 1 per 22 390 people
Standard of living
National income per person (1976): Sy 3 300** = $ 150** = £ 85**
Consumption per person (1975): energy 92 kg coal equivalent, electricity (production) 113* kW h, steel 4 kg
Newspapers (1974): number 1; circulation 5 000, 1 per 1 000 people
Telephones (Dec 1974): 10 000, 2 per 1 000 people
Livestock (000, 1976) Cattle 1 550*, sheep 420*, goats, 385*, pigs 35*
Hospital beds (1972) 6 858, 1 per 600 people
Roads (1975) 10 000* km = 6 000* mi, density 0.04* km per km²
Railways (1975) 902 km = 560 mi, density 0.004 km per km²
A 1 200 km line is planned to link Conakry with east Guinea iron ore mines near Liberia
Ships (registered, 1977) 11, total of 12 597 gross tons
Ports Conakry, Kamsar
Airports Conakry, Kankan and 8 other airports with scheduled services
Durable equipment Radio sets (Dec 1974): 105 000, 24 per 1 000 people
Passenger cars (Dec 1972): 10 200, 2.5 per 1 000 people, 1.0* per km of road
Commercial vehicles (Dec 1972): 10 800, 2.6 per 1 000 people, 1.1* per km of road

Production, finance and trade

Gross domestic product
1976 est: Sy 16 000** mn = $ 750** mn = £ 410** mn
Agricultural production index (1970 = 100) 1976 103; growth 1970–76 0.4 %pa
Main products (000 t) *Agriculture* (1976) Rice 375*, maize 320*, cassava 480*, bananas 70*, plantains 215*, palm kernels 15*, palm oil 35*, coffee 5*, pineapples 15*, timber (000 m³ 1975) 3 139. *Other* (1975) Electricity (mn kW h) 500*, bauxite 7 620, alumina (1972) 700, iron ore (Fe content, 1970) 1 040, diamonds (000 CM) 80
Transport traffic *Passenger-kilometres* Rail (1967) 42 mn, air (1976) 26 mn
Cargo: tonne-kilometres Rail (1971) 341 mn, air (1976) 2.5 mn
Sea (1971) Goods loaded 1 300 000* t, unloaded 550 000* t
Budget (1972/73; year ending Sept 30th)
Balanced at Sy 4 500 mn = $ 210 mn = £ 90 mn
External trade (1975) Imports: Sy 3 800** mn = $ 180** mn = £ 80** mn
Exports: Sy 3 300** mn = $ 160** mn = £ 70** mn

Main imports (1970)	% of total	Main exports (1971)	% of total
Petroleum and products	14	Alumina and bauxite	72
Textiles and clothing	14	Pineapples	10
Rice	10	Coffee	6
Machinery and metals	6	Palm kernels	6
		Bananas	3*
Main sources (1975)		Main destinations (1975)	
France	24*	United States	16*
Soviet Union	15*	Spain	14*
United States	15*	Soviet Union	12*
Morocco	8*	Canada	9*
Belgium-Luxembourg	7*	West Germany	8*
United Kingdom	5*	France	7*
Spain	3*	Jugoslavia	6*

Special focus

Main non-agricultural products (000 tonnes)

	Bauxite	Alumina	Iron ore (Fe content)
1960	1 378	185	388
1965	1 870	522	378
1966	1 609	525	305
1967	1 639	530	356
1968	2 118	531	935
1969	2 459	572	1 040
1970	2 490	610	1 040
1971	2 630	665	na
1972	2 650	700	na
1973	3 660	na	na
1974	7 605	na	na
1975	7 620	na	na

Guinea-Bissau

Republic of Guinea-Bissau

Location West Africa
With a coastline on the Atlantic Ocean, Senegal is to the north, and Guinea to the east and south
Land Area 36 125 km² = 13 948 mi²
Climate Tropical
Weather at Bolama 19 m altitude
Temperature: hottest month May 24–32 °C, coldest January 19–31 °C
Rainfall (av monthly): driest month April 1 mm, wettest August 701mm

Time GMT
Measures Metric system
Monetary unit Guinea-Bissau peso (GB P) = 100 centavos; the peso replaced the escudo at par from March 2, 1976
Rate of exchange (1976 av): par GB P 1 = Esc 1, free GB P 30.22 = $ 1, GB P 54.59 = £ 1

Summary

Political One-party people's republic, which became independent September 10, 1974; formerly a Portuguese overseas province. Union with Cape Verde is being considered. Member of UN, OAU, Ecowas and an EEC ACP state
Economic An agricultural economy, with groundnuts accounting for about one-half of exports; palm kernels, timber, cashew nuts and hides and skins are other main export items. There is virtually no industry, but there are large deposits of bauxite. Rice production and fishing are being improved

People, resources and equipment

Population[a] 1960 521 000*, 1970 487 000*, 1976 534 000*[b]
[a] UN estimates [b] official estimate: 910 000*
Growth: 1960–70 −0.7* %pa, 1970–76 1.5* %pa
Density (1976): 15* people per km²
Vital statistics (rate per 1 000 people, 1970–75): births 40.1*, deaths 25.1*
Cities (population of municipalities in 000, 1970)
Bissau (capital) 119, Cacheu 71, Mansôa 35, Bissora 21
Race (1970) Balante 30** %, Fulani 20** %, Mandyako 14** %, Malinke 12** %
Language Guinean creole and Portuguese
Religion (1965) Animist 66** %, Moslem 30** %, Roman Catholic 4** %
Education (1975/76) Pupils 58 577, teachers 2 500*
Labour force (1976) 168 000*; in agriculture 141 000* (84* %)
Personnel (1969) Physicians: 30, 1 per 16 300 people
Standard of living
National income per person (1976): GB P 10 000** = $ 330** = £ 180**
Consumption per person (1975): energy 82 kg coal equivalent, electricity 38* kW h, newsprint 0.2 kg
Newspapers (1974): number 1; circulation 6 000, 12 per 1 000 people
Telephones (Dec 1973): 3 000, 6 per 1 000 people
Livestock (000, 1976) Cattle 258*, sheep 70*, goats 180*, pigs 176*
Electrical capacity (1975) 8* megawatts
Hospital beds (1969) 889, 1 per 550 people
Roads (1972) 3 570 km = 2 220 mi, density 0.10 km per km²
There are no railways
Ports Bissau, Bolama, Cacheu
Airport Bissau
Durable equipment Radio sets (Dec 1974): 9 000, 17 per 1 000 people
Passenger cars (Dec 1973): 3 300, 6 per 1 000 people, 0.9 per km of road
Commercial vehicles (Dec 1973): 2 000, 4 per 1 000 people, 0.6 per km of road

Production, finance and trade

Gross domestic product
1976 est: GB P 6 000** mn = $ 200** mn = £ 110** mn
Main products (000 t, 1976) Rice 35*, cassava 40*, groundnuts 28*, palm kernels 8.0*, palm oil 4.6*, cashew nuts 2.7*, fish catch 2*, timber (000 m³) 530*, electricity (mn kW h, 1975) 20*
Transport traffic (1973) *Sea* Goods loaded 40 000 t, unloaded 172 000 t
Budget (1972) Revenue: GB P 577 mm = $ 21 mn = £ 8.5 mn
Expenditure: GB P 552 mn = $ 20 mn = £ 8.1 mn
External trade (1975) Imports: GB P 965 mn = $ 38 mn = £ 17 mn
Exports: GB P 157 mn = $ 6 mn = £ 3 mn

Main imports (1973)	% of total	Main exports (1973)	% of total
Rice	16	Groundnuts	46
Textile yarns and		Transport equipment[a]	21
fabrics	13	Crude oil and	
Chemicals	10	products[a]	8
Crude oil and		Palm kernels	7
products	8	Metals[a]	6
Machinery	8	Timber	5
Metals	7	Cashew nuts	2*
Wine	3		

Main sources (1973)		Main destinations (1973)	
Portugal	56	Portugal	90
Spain	7	Netherlands	3
United Kingdom	5	Cape Verde	2
Japan	5		

[a] Mainly in transit

Ivory Coast

Republic of the Ivory Coast
République de la Côte d'Ivoire

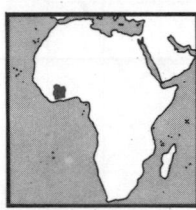

Location West Africa
With a south-facing coastline on the Atlantic Ocean, Liberia and Guinea are to the west, Mali and Upper Volta to the north, and Ghana to the east
Land Area 322 462 km² = 124 503 mi²
Climate Tropical, hot and wet
Weather at Abidjan, 20 m altitude
Temperature: hottest months Feb–April 24–32 °C, coldest August 22–28 °C
Rainfall (av monthly): driest month January 41 mm, wettest June 495 mm
Time GMT
Measures Metric system
Monetary unit CFA franc (CFA Fr) = 100 centimes
Rate of exchange (1976): par CFA Fr 50 = Fr 1, free CFA Fr 239.0 = $ 1, CFA Fr 431.6 = £ 1

Summary

Political One-party republic, which became independent August 7, 1960; formerly one of the territories of French West Africa. Member of UN, OAU, Ocam, CEAO, Ecowas, franc zone and an EEC ACP state
Economic A well-diversified agricultural economy, with main export earners coffee, cocoa and timber, and a wide spread of other agricultural products including especially palm oil. Nigeria has been overtaken as a producer of cocoa, and there is an aim to also overtake Ghana (the African cocoa leader) in the next decade. Industrial development has included hydro-electricity and crude oil refining, and there is some iron ore potential–an extension from the Liberian deposits. Light industry includes textiles, bottling plants, canneries, etc., and has expanded considerably in the 1970s. Foreign investment has aided these developments

People, resources and equipment

Population 1960 3.23* mn, 1970 4.31* mn, 1976 5.02* mn
Growth: 1960–70 2.9* %pa, 1970–76 2.6* %pa
Density (1976): 16* people per km²
Vital statistics (rate per 1 000 people, 1970–75): births 45.6*, deaths 20.6*
Cities (population in 000, 1970)
Abidjan (capital) 900* Daloa 60* Gagnoa 50*
Bouaké 173 Man 60* Korhogo 40*
Race (1961) Baule 23 %, Bete 18 %, Senuf 15 %, Malinke 11 %, Dan and Guro 10 %, Lobi 6 %, Lagoon 5 %, Ngere 5 %, Bakwe 5 %
Language French; Baule and other languages are also used
Religion (1976) Animist 65** %, Moslem 23** %, Christian 12** %
Education (1974/75) Pupils 751 200*, teachers 19 400*
Labour force (1976) 2 571 000*; in agriculture 2 096 000* (82* %)
Personnel Scientists and engineers engaged in research (1970): 319
Physicians (1971): 324, 1 per 13 640 people
Standard of living National income per person (1976):
CFA Fr 174 000** = $ 730** = £ 400**
Consumption per person (1975): energy 366 kg coal equivalent, electricity (production) 176* kW h, newsprint 0.3 kg, steel 23 kg
Newspapers: number (1974) 3; circulation (1972) 44 000, 10 per 1 000 people
Telephones (Dec 1975): 59 000, 12 per 1 000 people

Ivory Coast

Livestock (000, 1976) Cattle 600*, sheep 1 000*, goats 1 000*, pigs 210*
Petroleum refinery capacity (1975) 1.85 mn tonnes
Electrical capacity (1975) 350* megawatts, of which, hydro 224* megawatts
Hospital beds (1970) 8 682, 1 per 496 people
Roads (1976) 45 214 km = 28 095 mi, density 0.14 km per km²
Railways (1975) 816 km = 507 mi, density 0.003 km per km² (including part of the Abidjan-Ouagadougou line which totals 1 147 km)
Ships (registered 1977) 59, total of 115 717 gross tons
Ports (goods traffic, 000 tonnes, 1973)

	loaded	unloaded
Abidjan	3 041	3 468
Sassandra	991	27

Also San-Pedro and Tabou
Airports Abidjan and 10 other airports with scheduled services

Durable equipment
(at end-year)

	000	no per 1 000 people	
Radio sets (1971)	80	18	no per
Television sets (1972)	40	9	km of road
Passenger cars (1975)	76	15	1.7
Commercial vehicles (1975)	50*	10*	1.1*

Production

Gross domestic product
1975: CFA Fr 834 500 mn = $ 3 894 mn = £ 1 752 mn
1976 est: CFA Fr 950 000**mn = $ 3 950**mn = £ 2 200**mn
Structure of gross domestic product *By origin* (1974) Agriculture 26 %, manufacturing 15 %, construction 5 %, other 54 %
By type (1975) Final consumption expenditure 79 %, gross fixed capital formation 22 %, exports of goods and services 38 %, less imports of goods and services −39 %

Production index (1970 = 100)	1960	1970	1976	Growth %pa 1960–70	1970–76
Agricultural	67	100	124	4.1	3.7

Main products (000 t)

Agriculture	1960	1970	1976	1960–70	1970–76
Rice	160	316	420	7.0	4.9
Cassava	450	540	680*	1.8	3.9*
Taro	150*	182	200*	2.0*	1.6*
Yams	1 140*	1 551	1 700*	3.1*	1.5*
Bananas	85	179	170*	7.7	−0.8*
Pineapples	23	111	257*	17.0	15.0*
Palm kernels	16	21	38	2.8	10.4
Palm oil	18	50	158	10.5	21.1
Coffee	186	240	305*	2.6	4.1*
Cocoa	94	179	225*	6.7	3.9*
Cotton	2	13	19*	20.3	6.5*
Rubber	1*	11	19*	27.2*	9.5*
Fish catch	44	58	63e	2.8	1.7f
Timber (000 m³)	6 088	8 901	9 350e	3.9	1.0f
Other					
Petroleum products	—	696	1 250*e	na	12.4*f
Electricity (mn kW h)	67	517	860*e	22.7	10.7*f
Diamonds (000 CM)	199	213	209e	0.7	−0.4f
Beer (000 hl)	124a	170	680e	6.5b	32.0f
Cigarettes (mn units)	1 400a	2 000	2 620e	7.4b	5.6f
Cotton yarn	na	na	6.0e	na	na
Fertilisers, nitrogenous	—	1.8	4.5e	na	20.1f
Cement	109c	400	720e	38.4d	12.5f
Radio sets (000 units)	na	na	90h	na	na
Motor vehicles (000 units)g	1.9a	2.8	5.0e	8.8b	12.3f

a1965 b1965–70. c1966 d1966–70 e1975 f1970–75 gAssembly only h1974

Transport traffic	1960	1970	1976	Growth % pa 1960–70	1970–76
Passenger-kilometres (mn)					
Road	na	2 309	2 710d	na	3.3e
Raila	220	621	946d	10.9	8.8e
Airb	31	71	151	8.6	13.4
Cargo: tonne-kilometres (mn)					
Road	na	1 175	1 410d	na	3.7e
Raila	213	421	443d	7.1	1.0e
Airb	2.5	6.3	14.0	9.5	14.2
Seac: tonnes (mn)					
Goods loaded	1.1	3.6	3.2d	12.3	2.0e
Goods unloaded	0.8	2.3	3.5d	11.8	8.1e

aIncluding traffic in Upper Volta on the Abidjan–Ouagadougou line bIncluding apportionment of traffic of Air Afrique cIncluding coastwise traffic d1975 e1970–75
Tourism Number of visitors (1975) 109 000, gross receipts (1974) $ 23 mn

Finance and trade

Price index (1970 = 100)	1960	1970	1976	Growth %pa 1960–70	1970–76
Consumer prices	67.9	100.0	160.8	3.9	8.2
Money stock (end-year, CFA Fr bn)	29.6a	83.5	260.1	13.9b	20.8

a1962 b1962–70
Budget (1976) Balanced at CFA Fr 140 200 mn = $ 587 mn = £ 325 mn

Balance of payments ($ mn)	1972	1973	1974	1975	1976
Balance of goods (fob)	+135	+161	+359	+230	+506
Balance of services	−204	−311	−327	−479	−544
Balance of transfers	−28	−69	−93	−142	−167
Current balance	−97	−219	−61	−391	−205
Long-term capital flow	+29	+224	+119	+286	+263

Reserves and debt (end-year, $ mn)

International reserves		87	88	66	103	76
External public debt		703	923	1 196	1 526	2 215

External trade (1976)
Imports: CFA Fr 311 608 mn = $ 1 304 mn = £ 722 mn
Exports: CFA Fr 392 500 mn = $ 1 642 mn = £ 909 mn

Main imports (1974)	% of total	Main exports (1976)	% of total
Crude oil	13	Coffee	34
Machinery, non-electric	12	Cocoa	24
Chemicals	9	Timber	20
Motor vehicles	9	Petroleum products	4
Cereals	5		
Textile yarns and fabrics	5		
Electrical machinery	5		
Iron and steel	5		
Main sources (1976)		Main destinations (1976)	
France	38	France	25
United States	7	Netherlands	13
West Germany	7	United States	10
Japan	5	Italy	9
Iran	4	West Germany	7
Netherlands	4	United Kingdom	4

Special focus

Largest African producers of coffee, cocoa, palm oil, 1976 (000 tonnes)

	Coffee		Cocoa		Palm oil
Ivory Coast	305*	Ghana	320	Nigeria	510*
Uganda	211	*Ivory Coast*	225*	*Ivory Coast*	158
Ethiopia	170	Nigeria	210*	Zaire	155*

Kenya

Republic of Kenya
Jamhuri ya Kenya

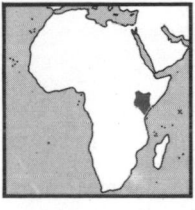

Location East central Africa
With a coastline on the Indian Ocean, Somalia is to the north-east, Ethiopia and Sudan to the north, Uganda to the west and Tanzania to the south
Land Area 582 640 km² = 224 960 mi² of which, 3 831 km² for Lake Victoria, 6 405 km² for Lake Rudolf
Climate Tropical, hot and humid on the coast, temperate inland, dry to the north

Weather at Nairobi, 1 820 m altitude
Temperature: hottest month Feb 13–26 °C, coldest July 11–21 °C
Rainfall (av monthly): driest month July 15 mm, wettest April 211 mm
Time 3 hours ahead of GMT
Measures Metric system, introduced from 1969 to replace the UK (imperial) system
Monetary unit Kenya shilling (K Sh) = 100 cents; also used is Kenya pound (K £) = 20 K Sh. The decimal currency was introduced September 14, 1966 to replace the East African pound (EA £) at K Sh 20 = EA £ 1
Rate of exchange (1976 av): par K Sh 9.66 = SDR 1, free K Sh 8.367 = $ 1, K Sh 15.11 = £ 1

Summary

Political One-party republic, which became independent December 12, 1963; formerly the UK colony and protectorate of Kenya. Member of UN, OAU, Commonwealth and an EEC ACP state. Member formerly of the East African Community during its existence from 1967 to July 1977.

Kenya

Economic Mainly an agricultural economy, with an expanding industrial sector. Agricultural exports are mainly coffee, tea and sisal, and development of irrigation is regarded as important. Hydro-electric capacity is being increased and tourism is becoming important. Industry includes especially oil refining

People, resources and equipment

Population 1960 8.11* mn, 1970 11.23* mn, 1976 13.85* mn
Growth: 1960–70 3.3* %pa, 1970–76 3.6* %pa
Density (1976): 24* people per km²
Vital statistics (rate per 1 000 people, 1970–75): births 48.7*, deaths 16.0*
Regions (population in 000, 1969; total of 10.94 mn)

Nairobi	509	Coast	944	Nyanza	2 122
Provinces		Eastern	1 907	Rift Valley	2 210
Central	1 676	North Eastern	246	Western	1 328

Cities (population in 000, 1974)

Nairobi (capital)	723	Kisumu	147	Eldoret	36
Mombasa	339	Nakuru	65	Thika	34

Race (1969) African 98 %, Asian 1¼ %, European ½ %, Arab ¼ %; of African: Kikuyu 20 %, Luo 14 %, Luhya 13 %, Kamba 11 %, Gusii 6 %
Language Swahili; English, Kikuyu, Luo and other languages are also used
Religion (1976) Christian 50** %, Animist 35** %, Moslem 6** %
Education (1975) Pupils 3 132 000*, teachers 97 000**
Labour force (1976) 5 315 000*; in agriculture 4 224 000* (79* %)
Personnel Scientists and engineers (1972): 3 955
Physicians (1973): 766, 1 per 16 290 people

Standard of living
National income per person (1976): K Sh 1 840* = $ 220* = £122*
Consumption per person (1975): energy 174 kg coal equivalent, electricity (production, 1976) 80* kW h, newsprint 0.5 kg, steel 9 kg
Newspapers (1974): number 3; circulation 97 000, 8 per 1 000 people
Telephones (Dec 1975): 122 000, 9 per 1 000 people
Livestock (000, 1976) Cattle 7 500*, sheep 3 611*, goats 4 100*, pigs 67*, camels 564*, chickens 15 428*
Petroleum refinery capacity (1975) 4.3 mn tonnes
Electrical capacity (1975) 282* megawatts, of which, hydro 172* megawatts
Hospital beds (1972) 15 904, 1 per 760 people
Roads (1976) 50 091 km = 31 125 mi, density 0.09 km per km²
Railways (1975) 2 729 km = 1 696 mi, density 0.005 km per km²
Oil pipeline (1978) 500* km = 310* mi
Ships (registered, 1977) 19, total of 15 192 gross tons
Port (goods traffic, 000 tonnes, 1973)
Mombasa: loaded 2 551, unloaded 4 173
Airports (1975) Nairobi: passenger departures and arrivals 931 782; also Mombasa, Kisumu, Lamu, Malindi, Mumias

Durable equipment (at end-year)	000	no per 1 000 people	
Radio sets (1974)	510	39	no per
Television sets (1974)	37	3	km of road
Passenger cars (1975)	84ᵃ	6	1.7
Commercial vehicles (1975)	84	6	1.7

ᵃExcludes light commercial vehicles

Production

Gross domestic product 1976: K Sh 28 536 mn = $ 3 411 mn = £ 1 889 mn
Growth in real terms: 1967–70 6.9 %pa, 1970–75 4.8* %pa
Structure of gross domestic product *By origin* (1975) Agriculture 27 %, manufacturing 12 %, construction 5 %, other 56 %
By type (1976) Final consumption expenditure 80 % (of which government 17 %), stock investment −1 %, gross fixed capital formation 21 %, exports of goods and services 34 %, less imports of goods and services −33 %

Production indices (1970 = 100)	1960	1970	1976	Growth %pa 1960–70	1970–76
Agricultural	82	100	105	2.0	0.9
Manufacturing	51	100	137ᶜ	7.0	8.2ᵈ
Main products (000 t)					
Agriculture					
Maize	1 200	1 524*	1 360*	2.4*	−1.9*
Millet and sorghum	379	330*	360*	−1.4*	1.5*
Potatoes	193	218*	250*	1.2*	2.3*
Sweet potatoes	450	460*	555*	0.2*	3.2*
Cassava	550	620*	677*	1.2*	1.5*

Main products (000 t)	1960	1970	1976	Growth %pa 1960–70	1970–76
Agriculture					
Sugar, raw value	33	136	173*	15.2	4.1*
Coffee	34	58	75*	5.5	4.4*
Tea	14	41	62*	11.3	7.1*
Sisal	63	44	33	−3.5	−4.7
Cotton	3.0	4.4	5.5*	3.8	3.8*
Milk	650*	820	763*	2.4*	−1.2*
Beef and veal	114	131*	115*	1.4*	−2.1*
Fish catch	13	34	27*	10.3	−4.5*
Timber (000 m³)	8 500*	11 160*	11 795ᵃ	2.8*	1.1*ᵇ
Other					
Petroleum products	—	2 210	2 750*ᵃ	na	4.5*ᵇ
Electricity (mn kW h)	250*	583*	1 110*	8.8*	11.3*
Beer (000 hl)	388	795	1 538ᵃ	7.4	14.1ᵇ
Cigarettes (mn units)	2 380	2 426	3 562ᵃ	0.2	8.0ᵇ
Soda ash	127	160	95ᵃ	2.3	−9.9ᵇ
Cement	341	792	897ᵃ	8.8	2.5ᵇ

ᵃ1975 ᵇ1970–75 ᶜ1974 ᵈ1970–74

Transport traffic	1960	1970	1976	Growth %pa 1960–70	1970–76
Passenger-kilometres (mn)					
Airᵇ	119ᶜ	267	879	14.4ᵈ	22.0
Cargo: tonne-kilometres (mn)					
Rail	2 925ᵃ	4 140ᵃ	2 120ᶜ,ᵍ	3.5ᵃ	na
Airᵇ	3.5ᶜ	10.4	19.7	19.9ᵈ	11.2
Sea: tonnes (mn)					
Goods loaded	1.0	2.3	1.8ᵉ	8.3	−4.5ᶠ
Goods unloaded	1.7	3.6	4.2ᵉ	8.0	3.0ᶠ
Tourism					
Number of visitors (000)	65ᶜ	343	407ᵉ	31.9ᵈ	3.5ᶠ
Gross receipts ($ mn)	26ᶜ	52	98ᵉ	12.2ᵈ	13.5ᶠ

ᵃTotal East African Railways Corporation (also including Tanzania and Uganda)
ᵇIncluding apportionment of traffic of East African Airways Corporation
ᶜ1964 ᵈ1964–70 ᵉ1975 ᶠ1970–75 ᵍKenya only

Finance and trade

Price index (1970 = 100)	1960	1970	1976	Growth %pa 1960–70	1970–76
Consumer prices	83.4	100.0	191.2	1.8	11.4
Money stock					
(end-year, K Sh mn)	1 333ᵃ	2 410	5 674	16.0ᵇ	15.3

ᵃ1966 ᵇ1966–70

Budget (actual, 1975/76; year ending June 30th)
Revenue: K Sh 5 037 mn = $ 628 mn = £ 315 mn
Expenditure: K Sh 6 614 mn = $ 825 mn = £ 413 mn

Balance of payments ($ mn)	1972	1973	1974	1975	1976
Balance of goods (fob)	−117	−75	−317	−242	−104
Balance of services	+11	−81	−22	−10	−11
Balance of transfers	+38	+30	+31	+38	+32
Current balance	*−68*	*−126*	*−308*	*−214*	*−83*
Long-term capital flow	+86	+130	+187	+141	+152
Reserves and debt (end-year, $ mn)					
International reserves	202	233	193	173	275
External public debt	513	608	770	1 071	na

External trade (1976) Imports: K Sh 8 142 mn = $ 973 mn = £ 539 mn
Exports: K Sh 6 610 mn = $ 790 mn = £ 437 mn

Main importsᵃ	% of total	Main exportsᵃ	% of total
Crude oil	24	Coffee	37
Machinery	20	Petroleum products	12
Chemicals	11	Tea	12
Motor vehicles	8	Fruit and vegetables	6
Iron and steel	7	Chemicals	4
Food	3	Hides and skins	3
Main sources		*Main destinations*	
United Kingdom	19	West Germany	13
Iran	17	United Kingdom	11
Japan	11	Tanzania	10
West Germany	10	Uganda	10
Saudi Arabia	6	United States	6
United States	6	Netherlands	5

ᵃDomestic (excluding Tanzania and Uganda)

Special focus

Structure of gross domestic product

	% of total		
	1964	1970	1975
Agriculture	38	31	27
Manufacturing	10	11	12
Other	52	58	61

Lesotho

Kingdom of Lesotho

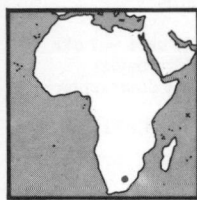

Location South Africa
Completely surrounded by South Africa, with
Cape Province and Natal between Lesotho and
the Indian Ocean, and Orange Free State
inland to the north-west. Land-locked
Land Area 30 355 km² = 11 720 mi²
Climate Continental: extreme temperatures
Weather at Maseru:
Temperature: hottest month January 15–33 °C,
coldest July minus 3–17 °C
Rainfall (av monthly): driest month June 7 mm, wettest February 141 mm
Time 2 hours ahead of GMT
Measures Metric system, introduced from 1970 to replace the UK
(imperial) system; the changeover was mainly completed in 1977
Monetary unit South African rand (R) = 100 cents
Rate of exchange (1976 av): par $ 1.15 = R 1, R 1.571 = £ 1

Summary

Political Parliamentary monarchy, which became independent October 4,
1966; formerly the UK territory of Basutoland. Member of UN, OAU,
Commonwealth and an EEC ACP state
Economic Mainly an agricultural economy based on livestock – wool,
mohair and live animals make up two-thirds of exports. Diamonds are
also an important export

People, resources and equipment

Population 1960 0.72*mn, 1970 0.93* mn, 1976 1.07* mn
excluding absentee workers in South Africa, numbering (1976) 110 000*
Growth: 1960–70 2.6* %pa, 1970–76 2.4* %pa
Density (1976): 35* people per km²
Vital statistics (rate per 1 000 people, 1970–75): births 39.0*, deaths 19.7*
City (population in 000, 1976) Maseru (capital) 30*
Race (1966) African 99.7 %, (of which Basotho 80 %), European 0.2 %
Language Sesotho (southern Sotho) and English
Religion (1966) Christian 72 % (Roman Catholic 39 %)
Education (1975) Pupils 239 000*, teachers 5 100*
Labour force (1976) 500 000**; in agriculture 450 000** (90** %)
Excludes Lesotho workers in South Africa (131 749 registered in 1972,
of whom 99 786 in mining and quarrying); remittances (1974): $ 60* mn
Personnel (1974) Physicians: 50, 1 per 20 400 people
Standard of living
National income per person (1976): R 210** = $ 240** = £ 130**
Telephones (Dec 1974): 4 000, 4 per 1 000 people
Livestock (000, 1976) Cattle 580*, sheep 1 640*, goats 915*, pigs 85*
Hospital beds (1974) 2 114, 1 per 482 people
Roads (1976) 2 736 km = 1 700 mi, density 0.09 km per km²
Railways (1976) 2 km = 1 mi, linking Maseru with South African railways
Airports Maseru, and 5 other airports with scheduled flights
Durable equipment Radio sets (Dec 1974): 11 000, 11 per 1 000 people
Passenger cars (Dec 1975): 3 500, 3 per 1 000 people, 1.3 per km of road
Commercial vehicles (Dec 1975): 4 300, 4 per 1 000 people,
1.6 per km of road

Production, finance and trade

Gross domestic product
1973/74 (year ending March 31st): R 84.1 mn = $ 124 mn = £ 51 mn
1976 est: R 170** mn = $ 200** mn = £ 110** mn
Agricultural production index (1970 = 100) 1976 108;
growth 1970–76 1.3 %pa
Main products (000 t, 1976) Wheat 50*, maize 130*, sorghum 70*,
milk 17*, wool 2.0*, beef and veal 13*, diamonds (000 CM, 1975) 3
Budget (1974/75; year ending March 31st)
Revenue: R 17.3 mn = $ 25 mn = £ 11 mn
Expenditure: R 16.0 mn = $ 23 mn = £ 10 mn
External public debt (Dec 1975) $ 23 mn
External trade (1974) Imports: R 85 mn = $ 125 mn = £ 53 mn
Exports: R 10 mn = $ 14 mn = £ 6 mn
Included in South Africa Customs Union

Main imports	% of total	Main exports	% of total
Food	23	Wool	35
Clothing	16	Mohair	16
Blankets	8	Livestock	16
Petroleum products	6	Diamonds	9

Most trade is with South Africa

Liberia

Republic of Liberia

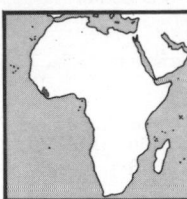

Location West Africa
With a coastline on the Atlantic Ocean
at the western end of the Gulf of Guinea, Sierra
Leone is to the north-west, Guinea to the north
and Ivory Coast to the east
Land Area 111 400 km² = 43 000 mi²
Climate Tropical, hot and wet
Weather at Monrovia, 23 m altitude
Temperature: hottest month March 23–31 °C,
coldest July 22–27 °C
Rainfall (av monthly): driest month January 30 mm, wettest July 996 mm
Time GMT
Measures UK (imperial) and US systems
Monetary unit Liberian dollar (L $) = 100 cents; the US dollar is also in
use
Rate of exchange (1976 av): par L $ 1 = $ 1, free L $ 1.806 = £1

Summary

Political Republic which was established July 26 1847 as an
independent state; originally formed as a country for freed US slaves,
the constitution is based on that of the United States. Under the 1973
Mano River Agreement, a customs union with Sierra Leone is planned.
Member of UN, OAU, Ecowas and an EEC ACP state
Economic Iron ore accounts for two-thirds of exports and is the
mainstay of the economy; in the mining sector, diamonds are also
important. Mining and agriculture each account roughly for one-quarter
of gross domestic product. The main agricultural export product is
rubber; timber is also important. Maritime fees from vessel registration,
etc. make up about 10 % of Budget revenue

People, resources and equipment

Population 1960 1.00*mn, 1970 1.52*mn, 1976 1.75*mn
Growth: 1960–70 4.3* %pa, 1970–76 2.4* %pa
Density (1976): 16* people per km²
Vital statistics (rate per 1 000 people, 1971): births 49.8, deaths 20.9
Cities (population in 000, 1974) Monrovia (capital) 172*, Greenville 18*,
Buchanan 16*, Robertsport 15*, Harper 10*
Race (1962) Kpelle 21 %, Bassa 16 %, Gia 8 %, Kru 8 %, Grebe 8 %,
Mano 7 %
Language English; local languages and dialects are also used
Religion (1965) Animist 90 %, Christian 7 %, Moslem 3 %
Education (1975) Pupils 190 799, teachers 5 705
Labour force (1976) 665 000*; in agriculture 480 000* (72* %)
Personnel (1973) Physicians: 132, 1 per 12 600 people
Standard of living
National income per person (1976): L $ 360** = $ 360** = £ 200**
Consumption per person (1975): energy 404 kg coal equivalent,
electricity (production) 509* kW h, newsprint 0.1* kg, steel 10 kg
Newspapers (1974): number 3; circulation 11 000, 7 per 1 000 people
Telephones (Dec 1974): 7 000, 4 per 1 000 people
Livestock (000, 1976) Cattle 35*, sheep 176*, goats 175*, pigs 93*,
chickens 2 000*
Petroleum refinery capacity (1975) 0.75 mn tonnes
Electrical capacity (1975) 300* megawatts, of which, hydro 75* megawatts
Hospital beds (1967) 2 181, 1 per 656 people
Roads (1975) 10 000* km = 6 200* mi, density 0.09* km per km²
Railways (1975) 493 km = 306 mi, density 0.004 km per km²
Ships (registered, 1977) 2 617, total of 79 982 968 gross tons;
of which oil tankers 910, total of 50 772 231 gross tons (63 %).
Mainly based in other countries and registered in Liberia for convenience
Ports (goods traffic, 000 tonnes, 1973)

	loaded	unloaded
Monrovia	12 574	881
Buchanan	12 783	212
Harper	23	—

Also Greenville and Robertsport
Airports Roberts International (50 km from Monrovia),
James Spriggs Payne (Sinkor, Monrovia), and 13 other airports with
scheduled flights

Durable equipment (at end-year)	000	no per 1 000 people	
Radio sets (1974)	261	154	no per
Television sets (1973)	8.5	5	km of road
Passenger cars (1974)	12.1	7	1.2*
Commercial vehicles (1974)	10.0	6	1.0*

Liberia

Production

Gross domestic product 1975: L $ 855 mn = $ 855 mn = £ 385 mn
1976 est: L $ 910**mn = $ 910**mn = £ 500**mn
Growth in real terms: 1970–75 2.9* %pa
Structure of gross domestic product (1974) *By origin* Agriculture 25 %, mining and quarrying 28 %, manufacturing 4 %, other 43 %

Production index (1970 = 100)	1960	1970	1976	Growth %pa 1960–70	1970–76
Agricultural	82	100	110	2.0	1.6

Main products (000 t)
Agriculture

	1960	1970	1976	1960–70	1970–76
Rice	148	138	230*	−0.7	8.9*
Cassava	400	235*	310*	−5.2*	4.7*
Bananas	55*	60	64*	0.9*	1.1*
Palm kernels	15	13*	15*	−1.6*	2.4*
Palm oil	40	14*	20*	−10.0*	6.1*
Coffee	1	4*	4*	15.0*	0.0*
Cocoa	1.0	2.2	3.0*	8.2	5.3*
Rubber	48	83	73*	5.6	−2.1*
Timber (000 m³)	1 600*	1 600*	1 973ᶜ	0.0*	4.3*ᵈ

Other

	1960	1970	1976	1960–70	1970–76
Petroleum products	—	403	505*ᶜ	na	4.6*ᵈ
Electricity (mn kW h)	115ᵃ	502	870*ᶜ	17.8ᵇ	11.6*ᵈ
of which, hydro (mn kW h)	17ᵃ	242	320*ᶜ	34.3ᵇ	5.7*ᵈ
Iron ore (Fe content)	2 192	15 813	16 923ᶜ	21.8	1.4ᵈ
Gold (kg)	32	20	140ᶜ	−4.6	47.6ᵈ
Diamonds (000 CM)ᵍ	976	811*	406ᶜ	−1.8*	−12.9*ᵈ
Soft drinks (000 hl)	—	80	122ᵉ	na	11.1ᶠ
Cement	—	88	86ᵉ	na	−0.6ᶠ

ᵃ1961 ᵇ1961–70 ᶜ1975 ᵈ1970–75 ᵉ1974 ᶠ1970–74 ᵍExports (mainly re-exports from Guinea and Sierra Leone)
Transport traffic: *Rail* (1970) Cargo 4 396 mn t-km
Air (1976) 3 mn passenger-km
Sea (1974) Goods loaded 25.2 mn t, unloaded 3.2 mn t

Finance and trade

Consumer price index (1970 = 100) 1976 178.2; growth rate 1970–76 10.1 %pa
Budget (1975) Revenue: L $ 154 mn = $ 154 mn = £ 69 mn
of which, maritime fees L $ 16 mn = $ 16 mn = £ 7 mn
Expenditure: L $ 147 mn = $ 147 mn = £ 66 mn
International reserves (Dec 1976) $ 17 mn
External public debt (Dec 1975) $ 275 mn
External trade (1976) Imports: L $ 399 mn = $ 399 mn = £ 221 mn
Exports: L $ 460 mn = $ 460 mn = £ 255 mn

Main imports	% of total	Main exports	% of total
Machinery, non-electric	22	Iron ore	72
Crude oil	13	Rubber	12
Food	10	Timber	8
Motor vehicles	9	Diamonds	4
Main sources		*Main destinations*	
United States	30	West Germany	28
Saudi Arabia	18	United States	19
West Germany	12	Italy	14
United Kingdom	8	France	8

Libya

Socialist People's Libyan Arab Jamahiriya

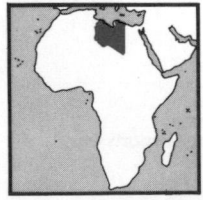

Location North Africa
With a long coastline on the Mediterranean Sea, Egypt and Sudan are to the east, Chad and Niger to the south, and Algeria and Tunisia to the west. Mainly desert inland
Land Area 1 759 540 km² = 679 360 mi²
Usage (1975): agricultural 93 440* km² (5.3* %), forests 5 340* km² (0.3* %)
Climate Hot and dry with mild winters

Weather at Tripoli, 22 m altitude
Temperature: hottest month August 22–30 °C, coldest January 8–16 °C
Rainfall (av monthly): driest month July 1 mm, wettest December 94 mm

Time 2 hours ahead of GMT
Measures Metric system, also:
length 1 draa milki = 50 centimetres = 19.7 inches
1 passo = 2 draa milki = 1 metre; 1 habl = 70 draa milki
area 1 dönüm = 919 square metres = 1 099 square yards
capacity 1 kilé = 36 litres = 7.92 UK gallons
weight (mass) 1 ukia = 32.05 grams = 1.131 ounces; 40 ukia = 1 oke
40 oke = 1 kantar (Tripolitania), 50 oke = 1 kantar (Cyrenaica)
Monetary unit Libyan dinar (L D) = 1 000 dirhams; the dinar replaced the Libyan pound at par from September 1, 1971
Rate of exchange (1976 av): par $ 3.377 8 = L D 1, free £ 1.870 = L D 1

Summary

Political Republic with military government, which became independent December 24, 1951; formerly an Italian colony, administered from 1942–51 by France and the United Kingdom. The former monarchy became a republic after the 1969 revolution. In September 1971 the Federation of Arab Republics was formed with Egypt and Syria. In 1972–73 an agreement was reached with Egypt to merge the two countries in due course. In January 1974 a proposed merger with Tunisia was announced. None of these various projects was carried out. Member of UN, OAU, Arab League, Oapec and Opec
Economic Crude oil and its products account for virtually all exports. Production of oil reached a peak in 1970, and a limitation policy was followed even before the high price rise in 1974; after a low in 1975, there has been some increase in production. The main development emphasis is on agriculture and housing; industrial developments include factories for chemicals, cement, etc.

People, resources and equipment

Population 1960 1.35*mn, 1970 1.99*mn, 1976 2.51*mn
Growth: 1960–70 3.9* %pa, 1970–76 4.0* %pa
Density (1976): 1* person per km²
Vital statistics (rate per 1 000 people, 1970–75): births 45.0*, deaths 14.7*
Regions (districts, population in 000, 1973; census total of 2.29 mn)

Tripoli	735	Khumsᵇ	162	Sabhah	113
Benghazi	337	Gharyan	156	Kalig	107
Zawiyah	248	Jabal al Akhdar	132		
Misratahᵃ	178	Darnahᵈ	123		

Cities (population in 000, 1973)

Tripoli (capital)	551	Az Zawiyah	50*ᵍ	Tubruqᶜ	25*ᵍ
Banghāziᶠ	282	Al Baydaᵉ	36*ᵍ	Darnahᵈ	25*ᵍ
Misratahᵃ	103	Al Khumsᵇ	30*ᵍ		

ᵃMisurata ᵇHoms ᶜTobruk ᵈDerna ᵉBeida ᶠBenghazi ᵍ1972
Race (1964) Arab-Berber 97 %, Italian 1 %
Italians mainly left after the 1969 revolution; there were (1977) 250 000* Egyptian workers in Libya
Language Arabic
Religion (1964) Moslem 97 %, Christian 2 %
Education (1974/75) Pupils 664 345, teachers 33 528
Labour force (1976) 599 000*; in agriculture 128 000* (21* %)
Personnel Scientists and engineers (1973): 8 319 (of whom, 79 % non-nationals). Physicians (1974): 2 063, 1 per 1 140 people
Standard of living
National income per person (1976): L D 1 500** = $ 5 000** = £ 2 800**
Consumption per person (1975): energy 1 299 kg coal equivalent, newsprint 0.3 kg, steel 321 kg
Newspapers (1974): number 5
Telephones (Dec 1971): 42 000, 20 per 1 000 people
Livestock (000, 1976) Cattle 123*, sheep 3 360*, goats 1 125*, horses 15*, asses 73*, camels 120*
Mineral reserves (1975) Crude oil 3 816 mn tonnes
Natural gas 806 bn cubic metres
Petroleum refinery capacity (1975) 3.3* mn tonnes
Electrical capacity (1975) 300* megawatts
Hospital beds (1974) 9 741, 1 per 240 people
Roads (1975) 5 173 km = 3 214 mi, density 0.003 km per km²
There are no railways in use
Ships (registered, 1977) 53, total of 673 969 gross tons
Ports (goods traffic, 000 tonnes, 1973)

	loaded	unloaded		loaded	unloaded
Es Sidrahᵃ	36 860	—	Al Harega		
Zweitinaᵃ	24 359	—	(Tobruk)ᵃ	11 336	—
Ras Lanufᵃ	16 863	—	Tripoli	11	3 146
Mersa Bregaᵃ	14 611	—	Benghazi Derna	11	1 956

ᵃCrude oil port
Airports International (Ben Gashir, Tripoli), Benina (Benghazi), Sebha, Misurata, Tobruk, Ghât, Ghudâmis, Al Kufra

Libya

Durable equipment (Dec 1974)	000	no per 1 000 people	
Radio sets	105	44	no per
Television sets	6	3	km of road
Passenger cars	234	97	45
Commercial vehicles	107	46	21

Production

Gross domestic product 1974: L D 4 092 mn = $ 13 822 mn = £ 5 909 mn
1976 est: L D 4 500**mn = $ 15 000**mn = £ 8 400**mn
Structure of gross domestic product (1974) *By origin* Agriculture 2 %, mining and quarrying 58 %, construction 10 %, other 30 %

Production index (1970 = 100)	1960	1970	1976	Growth %pa 1960–70	1970–76
Agricultural	66	100	186	4.3	10.9

Main products (000 t)
Agriculture

	1960	1970	1976	1960–70	1970–76
Wheat	34	21	70*	−4.7	22.2*
Barley	118	53	200*	−7.7	24.8*
Potatoes	4	10	66*	9.6	37.0*
Olives	29	71	99*	9.4	5.7*
Olive oil	6	13	19*	8.4	6.5*
Onions	5	24	42*	17.0	9.8*
Tomatoes	57	136	192*	9.1	5.9*
Watermelons	19*	21	100*	1.0*	29.7*
Oranges	19	17	23*	−1.1	5.2*
Dates	30	49	68*	5.0	5.6*
Timber (000 m³)	310*	431	462ᵃ	3.4*	1.4ᵇ

Other

	1960	1970	1976	1960–70	1970–76
Total energy (000 tce)	—	210 220	111 630ᵃ	na	−11.9ᵇ
Crude oil	—	159 709	92 770	na	−8.7
Petroleum products	—	436	1 850*ᵃ	na	33.5*ᵇ
Natural gas (mn m³)	—	—	3 250*ᵃ	na	na
Electricityᶜ (mn kW h)	105	426	900*	15.0	13.3*
Cigarettes (mn units)	618	1 639	2 710ᵈ	10.2	13.4*
Cement	—	95	615ᵇ	na	45.3ᵇ

ᵃ1975 ᵇ1970–75 ᶜTripolitania only ᵈ1974 ᵉ1970 74

Transport traffic	1960	1970	1976	Growth %pa 1960–70	1970–76
Passenger-kilometres (mn)					
Air	27*ᵃ	173	700	22.9*ᵇ	26.2
Cargo: tonne-kilometres (mn)					
Air	0.16*ᵃ	1.4	6.9	27.3*ᵇ	30.5
Sea: tonnes (mn)					
Goods loaded	0.086	161.1	73.0ᵉ	112.5	−14.3ᶠ
Goods unloaded	0.825	2.6	9.6ᵉ	12.2	29.9ᶠ
Tourism					
Number of visitors (000)	60ᶜ	77	238ᵉ	5.1ᵈ	25.3ᶠ
Gross receipts ($ mn)	9ᶜ	12	38ᵉ	5.5ᵈ	25.9ᶠ

ᵃ1961 ᵇ1961–70 ᶜ1965 ᵈ1965–70 ᵉ1975 ᶠ1970–75

Finance and trade

Price index (1970 = 100)	1960	1970	1976	Growth %pa 1960–70	1970–76
Consumer prices	72.0ᵃ	100.0	141.0	5.6ᵇ	5.9
Money stock (end-year, L D mn)	22.0	241.1	1 139.4	27.1	29.5

ᵃ1964 ᵇ1964–70

Budget (1974) Revenue: L D 2 067 mn = $ 6 982 mn = £ 2 985 mn
of which, petroleum revenue L D 1 442 mn = $ 4 871 mn = £ 2 082 mn
Expenditure: L D 2 062 mn = $ 6 965 mn = £ 2 978 mn

Balance of payments ($ mn)	1972	1973	1974	1975	1976
Balance of trade (fob)	+1 179	+1 517	+3 467	+1 816	+3 978
Balance of services	−787	−1 250	−1 430	−1 688	−2 148
Balance of transfers	−154	−200	−206	−196	−132
Current balance	*+238*	*+66*	*+1 832*	*−69*	*+1 698*
Long-term capital flow	−43	−510	−422	−1 557	−1 279
International reserves (end-year, $ mn)	2 925	2 127	3 616	2 195	3 206

External trade (1976) Imports: L D 1 070*mn = $ 3 610*mn = £ 2 000*mn
Exports: L D 2 487 mn = $ 8 401 mn = £ 4 651 mn

Main imports (1975)	% of total	Main exports (1975)	% of total
Food	15	Crude oil	95
Machinery, non-electric	14	Natural liquefied gas	3
Textiles	12	Petroleum products	2
Motor vehicles	10		
Electrical machinery	9		
Iron and steel	8		

Main sources (1975)	% of total	Main destinations (1975)	% of total
Italy	26	United States	22
West Germany	12	Italy	22
France	9	West Germany	19
Japan	8	Spain	5
United Kingdom	5	Bahamas	4
Greece	4	United Kingdom	4
Argentina	3	France	4

Special focus

Crude oil	Production million tonnes	Prices $ per tonne	$ per barrel
1968	126	17	2.23
1969	150	17	2.23
1970	160	17	2.23
1971	133	25	3.23
1972	107	28	3.62
1973	105	39	5.15
1974	73	120	15.77
1975	72	116	15.32
1976	93	123	16.20

Madagascar

Democratic Republic of Madagascar
Repoblika Demokratika Malagasy

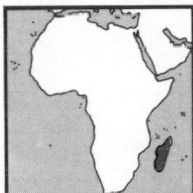

Location Western Indian Ocean
The republic occupies the island of Madagascar in the Indian Ocean, off the east African coast of Mozambique and separated from Africa by the Mozambique channel
Land Area 587 041 km² = 226 658 mi²
Climate Tropical; cooler in the highlands
Weather at Antananarivo, 1 370 m altitude
Temperature: hottest month December 16–27 °C, coldest July 9–20 °C
Rainfall (av monthly): driest month June 8 mm, wettest January 300 mm
Time 3 hours ahead of GMT
Measures Metric system
Monetary unit Madagascar franc (Mg Fr) = 100 centimes.
A new currency, the ariary (A), has been planned: A 1 = Mg Fr 5
Rate of exchange (1976 av): par Mg Fr 50 = Fr 1, Mg Fr 1 = CFA Fr 1, free Mg Fr 239.0 = $ 1, Mg Fr 431.6 = £ 1

Summary

Political One-party republic, which became independent on June 26, 1960; formerly a French colony. Member of UN, OAU and an EEC ACP state
Economic Mainly an agricultural economy with a wide spread of products and of export earners, with coffee, cloves, vanilla and essential oils making up nearly one-half of exports. There are varied mineral deposits including especially chromium ore; manufacturing accounts for some 12 % of gross domestic product, with food processing and oil refining important

People, resources and equipment

Population 1960 5.39**, 1970 6.93**, 1976 8.27**
Growth: 1960–70 2.5** %pa, 1970–76 3.0** %pa
Density (1976): 14** people per km²
Vital statistics (rate per 1 000 people, 1971): births 36.5, deaths 11.1
Regions (provinces, population in 000, 1971; total of 7.65 mn)
Diégo-Suarez 622 Majunga 918 Antananarivoᵃ 1 860
Fianarantsoa 1 861 Toamasinaᵇ 1 223 Toliaryᶜ 1 169
Cities (population in 000, 1972)
Antananarivoᵃ (capital) 439ᵈ Toamasinaᵇ 60 Diégo-Suarez 45
Majunga 67 Fianarantsoa 59 Toliaryᶜ 39
ᵃFormerly called Tananarive ᵇFormerly called Tamatave ᶜFormerly called Tuléar
ᵈ1975
Race (1972) Malagasy 99 % (Hova 24 %, Betsimisaraka 13 %, Betsileo 11 % in 1974), Comorians ½ %, French ½ %, Indian ¼ %, Chinese ⅛ %
Of 60 000* Comorians in 1976, 15 000* were repatriated in February 1977
Language Malagasy; French and local dialects are also used
Religion (1976) Animist 50** %, Christian 40** %, Moslem 10** %
Education (1973) Pupils 1 171 000*, teachers 22 000*

Madagascar

Labour force (1976) 4 114 000*; in agriculture 3 536 000* (86* %)
Personnel Scientists engaged in research (1971): 201
Physicians (1973): 687, 1 per 11 020 people
Standard of living
National income per person (1976): Mg Fr 50 000** = $ 210** = £ 120**
Consumption per person (1975): energy 71 kg coal equivalent,
electricity (production) 42* kW h, newsprint 0.1 kg, steel 5 kg
Newspapers (1974): number 9; circulation 59 000, 8 per 1 000 people
Telephones (Dec 1975): 31 000, 4 per 1 000 people
Livestock (000, 1976) Cattle 9 842*, sheep 700*, goats 1 300*, pigs 680*,
chickens 13 128*, ducks 2 352*, turkeys 1 362*
Mineral reserves (known economic, 1963) Coal 60 mn tonnes
Lignite 18 mn tonnes
Petroleum refinery capacity (1975) 0.75 mn tonnes
Electricity capacity (1975) 95* megawatts, of which, hydro 40* megawatts
Hospital beds (1973) 19 781, 1 per 383 people
Roads (1976) 27 507 km = 17 092 mi, density 0.05 km per km²
Railways (1975) 884 km = 549 mi, density 0.002 km per km²
Inland waterways (1975) Pangalanese Canal 700* km = 430* mi
Ships (registered, 1977) 44, total of 39 850 gross tons
Ports (goods traffic, 000 tonnes, 1973)

	loaded	unloaded
Toamasina	430	948
Majunga	55	93
Nossi-Bé/Saint Louis	72	16
Diégo-Suarez	27	28
Toliary	28	19

Airports Antananarivo, Majunga, Toamasina, Diégo-Suarez, Toliary
and 46 other airports

Durable equipment	000	no per	
(Dec 1974)		1 000 people	
			no per
Radio sets	855	108	km of road
Television sets	7.5	1	2.1
Passenger cars	57	7	1.6
Commercial vehicles	44	6	

Production

Gross domestic product 1973: Mg Fr 293 900 mn = $ 1 320 mn = £ 538 mn
1976 est: Mg Fr 420 000** mn = $ 1 750** mn = £ 970** mn
Structure of gross domestic product *By origin* (1971) Agriculture 29 %,
mining and quarrying 2 %, manufacturing 12 %, other 57 %
By type (1973) Final consumption expenditure 88 % (of which, government
17 %), stock investment 1 %, gross fixed capital formation 14 %, exports
of goods and services 14 %, less imports of goods and services −16 %

Production index	1960	1970	1976	Growth %pa	
(1970 = 100)				1960–70	1970–76
Agricultural	73	100	108	3.2	1.2
Main products (000 t)					
Agriculture					
Rice	1 212	1 865	1 814	4.4	−0.5
Maize	119*	109	123	−0.9*	2.0
Sweet potatoes	173	350	280*	7.3	−3.6*
Cassava	827	1 218	1 348	3.9	1.7
Sugar, raw value	87	101	117*	1.5	2.5*
Oranges	14*	57	84	15.1*	6.7
Bananas	130*	262	395	7.3*	7.1
Pineapples	7*	35	46	17.5*	4.7
Coffee	51	67	93	2.8	5.6
Sisal	13	25	21	6.8	−2.9
Cotton	1.4*	7.2	13.0	21.8*	10.3
Vanilla	0.3*	1.2	1.8ᵇ	14.9*	8.4ᶜ
Cloves	2.0*	12	20ᵈ	19.6*	13.6ᵉ
Pepper	1.1*	2.2*	3.0ᵇ	7.2*	6.4*ᶜ
Beef and veal	113	115*	114*	0.2*	−0.1*
Poultrymeat	24	32	42*	2.9	4.6*
Fish catch	28	45	56ᵇ	5.0	4.5ᶜ
Timber (000 m³)	3 705	5 291	6 366ᵇ	2.6	3.8ᶜ
Other					
Petroleum products	—	580*	590*ᵇ	na	0.3*ᶜ
Electricity (mn kW h)	107	246	335*ᵇ	8.7	6.4*ᶜ
Chromium ore (oxide content)	—	43	81ᵇ	na	13.2ᶜ
Beer (000 hl)	10	93	212ᵇ	25.0	17.9ᶜ
Cigarettes (mn units)	469	951	1 248ᵇ	7.3	5.6ᶜ
Cotton fabrics (mn m)	8*	48	78	19.5*	8.6
Cement	18	75	58ᵇ	15.3	−5.0ᶜ
Motor vehicles (000 units)ᵍ	na	3.0	na	na	na

ᵃAssembly only   ᵇ1975   ᶜ1970–75   ᵈ1974   ᵉ1970–74

Transport traffic	1960	1970	1976	Growth %pa	
Passenger-kilometres (mn)				1960–70	1970–76
Rail	136	182	248ᶜ	3.0	6.4ᵈ
Air	36ᵃ	242	276	26.8ᵇ	2.2
Cargo: tonne-kilometres (mn)					
Rail	131	292	207ᶜ	8.3	−6.6ᵈ
Air	2.1ᵃ	10.0	7.8	21.6ᵇ	−4.1
Sea: tonnes (mn)					
Goods loaded	0.23	0.72	0.97	11.8	5.1
Goods unloaded	0.45	1.08	1.14	9.2	0.9

ᵃ1962   ᵇ1962–70   ᶜ1975   ᵈ1970–75

Finance and trade

Consumer price index (1970 = 100) 1976 163.9;
growth 1970–76 8.6 %pa
Money stock (end-year, Mg Fr mn) 1976 79 880;
growth 1970–76 9.6 %pa
Budget (1974) Revenue: Mg Fr 71 000 mn = $ 295 mn = £ 126 mn
Expenditure: Mg Fr 93 400 mn = $ 388 mn = £ 166 mn

Balance of payments ($ mn)	1972	1973	1974	1975	1976
Balance of goods (fob)	−2	+22	+10	na	na
Balance of services	−25	−67	−73	na	na
Balance of transfers	+59	+34	+39	na	na
Current balance	*+33*	*−11*	*−24*	*na*	*na*
Long term capital flow	−7	+26	+18	na	na
Reserves and debt (end-year, $ mn)					
International reserves	52	68	49	36	42
External public debt	157	208	240	293	326

External trade (1974) Imports: Mg Fr 67 257 mn = $ 280 mn = £ 120 mn
Exports: Mg Fr 58 504 mn = $ 243 mn = £ 104 mn

Main imports	% of total	*Main exports*	% of total
Food	18	Coffee	27
(of which, rice 14)		Petroleum products	9
Crude oil	16	Vanilla	8
Chemicals	13	Meat and products	7
Machinery, non-electric	9	Cloves	7
Iron and steel	7	Fish	6
Electrical machinery	6	Sisal	4
Motor vehicles	5	Essential oils	4
Textile yarns and fabrics	4	Sugar	4
		Chromium ore	3
Main sources		*Main destinations*	
France	36	France	34
West Germany	9	United States	21
United States	7	Reunion	8
Japan	5	Japan	6
Italy	2	West Germany	5
Netherlands	2	United Kingdom	3

Malawi
Republic of Malawi

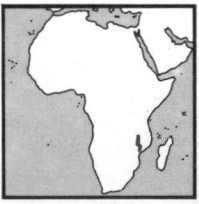

Location Central Africa
Lake Malawi forms the main part of the
eastern border, the larger part of the
Lake being territory of Malawi, with a
smaller area under Mozambique; Mozambique
surrounds the borders of the southern half
of Malawi, with Zambia to the west and
Tanzania to the north and east. Land-locked
Land Area 118 484 km² = 45 747 mi²
of which, Lake Malawi 22 970 km²
Climate Tropical, cooler in highlands
Weather at Lilongwe, 1 100 m altitude
Temperature: hottest month November 17–29 °C, coldest July 7–23 °C
Rainfall (av monthly): driest months June, July 1 mm, wettest
February 218 mm
Time 2 hours ahead of GMT
Measures Mainly metric system, which is being gradually introduced as
a change from the UK (imperial) system which is also in use
Monetary unit Malawi kwacha (M K) = 100 tambala; the Malawi
kwacha was introduced from February 15, 1971 to replace the Malawi
pound at the equivalent of M K 2 = M £ 1
Rate of exchange (1976 av): par M K 1.05407 = SDR 1,
free $ 1.095 = M K 1, M K 1.650 = £ 1

Malawi

Summary

Political One-party republic, which became independent July 6, 1964; formerly the UK protectorate of Nyasaland (from 1953–63 part of the Federation of Rhodesia and Nyasaland). Member of UN, OAU, Commonwealth and an EEC ACP state

Economic An agricultural economy, with tobacco and tea making up one-half of exports; other main products are sugar, groundnuts, cotton, maize and rice. There are no known mineral resources; light industry includes beer and cement production

People, resources and equipment

Population 1960 3.50**mn, 1970 4.44*mn, 1976 5.18*mn
Growth: 1960–70 2.4* %pa, 1970–76 2.6* %pa
Density (1976): 44* people per km²
Vital statistics (rate per 1 000 people, 1970–72): births 50.5, deaths 26.5
Cities (population in 000, 1975) Lilongwe (capitalª) 102, Blantyre 193, Zomba 20, Mzuzu 15
ªFrom January 1, 1975; formerly Zomba
Race (1966) African 99.5 %, Asian 0.3 %, European 0.2 %
Language Chichewa; English and local dialects are also used
Religion (1972) Animist 40** %, Christian 40** %, Moslem 15** %
Education (1974/75) Pupils 629 898, teachers 11 600*
Labour force (1976) 2 291 000*; in agriculture 1 972 000* (86* %)
Personnel (1969) Physicians: 114, 1 per 37 980 people

Standard of living

National income per person (1976): M K 125* = $ 137* = £ 76*
Consumption per person (1975): energy 56 kg coal equivalent, electricity (production) 58* kW h, newsprint 0.06 kg, steel 3 kg
Newspapers (1974): number 2; circulation 12 000, 2 per 1 000 people
Telephones (Dec 1975): 20 000, 4 per 1 000 people
Livestock (000, 1976) Cattle 700*, sheep 88, goats 739, pigs 189, chickens 8 092*
Mineral reserves Coal (1960) 38* mn tonnes
Electrical capacity (1975) 79* megawatts, of which, public supply 65* megawatts
Hospital beds (1970) 6 951, 1 per 640 people
Roads (1975) 11 025 km = 6 851 mi, density 0.09 km per km²
Railways (1975) 566 km = 352 mi, density 0.005 km per km²
Inland waterways Lake Malawi
Airports Chileka (Blantyre): passenger departures and arrivals (1975) 280 636; also Lilongwe, Mzuzu, Karonga
Durable equipment (at end-year)
Radio sets (1974): 125 000, 25 per 1 000 people
Passenger cars (1975): 10 500, 2 per 1 000 people, 1.0 per km of road
Commercial vehicles (1975): 8 800, 2 per 1 000 people, 0.8 per km of road

Production

Gross domestic product (1976) M K 646 mn = $ 707 mn = £ 392 mn
Growth in real terms: 1967–70 0.6 %pa, 1970–76 6.4 %pa
Structure of gross domestic product By origin (1973) Agriculture 49 %, manufacturing 9 %, distribution and hotels 12 %, other 30 %

Production indices (1970 = 100)	1960	1970	1976	Growth %pa 1960–70	1970–76
Agricultural	75	100	136	3.0	5.3
Manufacturing	na	100	178	na	10.1

Main products (000 t)	1960	1970	1976	Growth %pa 1960–70	1970–76
Agriculture					
Maize	650	900*	1 200*	3.3*	4.9*
Rice	10**	22	32*	8.2**	6.4*
Cassava	170	144*	80*	−1.6*	−9.3*
Groundnuts	120	190*	169*	4.7*	−1.9*
Sugar, raw value	—	34*	90*	na	17.6*
Tea	12	19	28	4.7	6.7
Tobacco	16	19	35	1.7	10.7
Cotton	3.0	7.6	6.0*	9.7	−3.9*
Fish catch	20*ª	63*	71ᶠ	15.0*ᵇ	2.4*ᵍ
Timber (000 m³)	3 407ª	4 222*	4 713ᶠ	2.7*ᵇ	2.2*ᵍ
Other					
Electricityᵉ (mn kW h)	28	133	254	16.9	11.4
Beer (000 hl)	na	146	472ᶠ	na	26.4ᵍ
Cigarettes (mn units)	317ᶜ	444	670ᶠ	7.0ᵈ	8.6ᵍ
Cement	30*	69	104ᶠ	8.7*	8.5ᵍ
Radio sets (000 units)	—	15	32ʰ	na	46.0ⁱ

ª1962 ᵇ1962–70 ᶜ1965 ᵈ1965–70 ᵉPublic supply only ᶠ1975 ᵍ1970–75 ʰ1972 ⁱ1970–72

Transport traffic	1960	1970	1976	Growth %pa 1960–70	1970–76
Passenger-kilometres (mn)					
Rail	50ª	58	61	1.9ᵇ	0.8
Air	14ᶜ	28	122	12.3ᵈ	27.8
Cargo: tonne-kilometres (mn)					
Rail	88ª	193	228	10.3ᵇ	2.8
Waterways	na	8.2	11.0ᵉ	na	6.1ᶠ
Air	0.2ᶜ	0.5	4.7	17.4ᵈ	45.3

ª1962 ᵇ1962–70 ᶜ1964 ᵈ1964–70 ᵉ1975 ᶠ1970–75
Tourism (1974) Number of visitors 28 000, gross receipts $ 4 mn

Finance and trade

Price index (1970 = 100)	1960	1970	1976	Growth %pa 1960–70	1970–76
Consumer prices	90.5ª	100.0	163.8	5.1ᵇ	8.6
Money stock (end-year, M K mn)	19.02ᶜ	32.69	72.80	11.4ᵈ	14.3

ª1968 ᵇ1968–70 ᶜ1965 ᵈ1965–70

Budget (1976) Revenue: M K 94 mn = $ 103 mn = £ 57 mn
Expenditure: M K 124 mn = $ 136 mn = £ 75 mn

Balance of payments ($ mn)	1972	1973	1974	1975	1976
Balance of goods (fob)	−33	−25	−47	−85	na
Balance of services	−30	−19	−1	−8	na
Balance of transfers	+13	+17	+13	+14	na
Current balance	−49	−28	−35	−78	*na*
Long-term capital flow	+38	+48	+70	+65	na

Reserves and debt (end-year, $ mn)					
International reserves	36	67	82	61	26
External public debts	221	267	322	331	343

External trade (1976) Imports: M K 188 mn = $ 206 mn = £ 114 mn
Exports: M K 146 mn = $ 160 mn = £ 89 mn

Main imports (1974)	% of total	Main exports (1975)	% of total
Machinery, non-electric	11	Tobacco	42
Petroleum products	10	Tea	17
Chemicals	10	Sugar	12
Textile yarns & fabrics	8	Groundnuts	5
Motor vehicles	7	Cotton	2
Food	7	Rice	2
Iron and steel	6		
Electrical machinery	5		
Metal small manufactures	4		

Main sources (1976)		Main destinations (1976)	
South Africa	29	United Kingdom	42
United Kingdom	22	United States	11
Japan	8	Netherlands	6
Rhodesia	5	South Africa	5
Canada	5	West Germany	3
Netherlands	5	Zambia	2
West Germany	5	Rhodesia	2

Mali

Republic of Mali
République du Mali

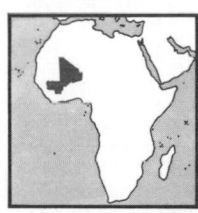

Location North-west Africa
Algeria is to the north, Niger to the east, Upper Volta, Ivory Coast and Guinea to the south, and Senegal and Mauritania to the west. The northern area is part of the Sahara desert. There is an outlet to the sea via the Senegal and Niger rivers
Land Area 1 240 000 km² = 479 000 mi²
Climate Hot and dry
Weather at Bamako, 340 m altitude
Temperature: hottest month April 24–39 °C, coldest January 16–33 °C
Rainfall (av monthly): driest months Dec, Jan, Feb 1 mm, wettest August 348 mm
Time GMT
Measures Metric system
Monetary unit Mali franc (M Fr) = 100 centimes; the Mali franc replaced the CFA franc from July 1962 at the equivalent M Fr 1 = CFA Fr 1, and was devalued May 1967 to M Fr 2 = CFA Fr 1
Rate of exchange (1976 av): par M Fr 100 = Fr 1, M Fr 2 = CFA Fr 1, free M Fr 478.0 = $ 1, M Fr 863.3 = £ 1

Mali

Summary

Political Republic with military government, which became independent September 22, 1960; formerly the territory of the French Soudan and, from January 1959–September 1960, linked with Senegal in the Federation of Mali. Member of UN, OAU, Ecowas, CEAO, franc zone and an EEC ACP state

Economic An agricultural economy, dependent on cotton and groundnuts as main exports. The Sahel drought of 1971–74 severely affected agriculture and especially the mainly nomad peoples in the north. There was some recovery by 1976. There is little industry, but a major dam project began in 1976, to provide hydro-electricity and irrigation resources

People, resources and equipment

Population[a] 1960 3.70* mn, 1970 5.05* mn, 1976 5.84* mn
[a]Including nomadic population estimated in 1960 at 200 000*
Growth: 1960–70 3.2* %pa, 1970–76 2.4* %pa
Density (1976): 5* people per km²
Vital statistics (rate per 1 000 people, 1970–75): births 50.1*, deaths 25.9*
Regions (population in 000, 1968; total of 4.83 mn)

Bamako	900	Kayes 731	Ségou 743
Gao	597	Mopti 973	Sikasso 888

Cities (population in 000, 1972)

Bamako (capital)	237	Ségou 40	Sikasso 29	Gao 17
Mopti	43	Kayes 37	San 18	Koutiala 16

Race (1963) Bambara 25 %, Fulani 16 %, Senufo 8 %, Marka 7 %, Tuareg 7 %, Sarakole 6 %, Dinla 5 %, Songhai 5 %, Malinke 5 %, Minianka 4 %
Language French; local languages, especially Bambara and Songhai, are also used
Religion (1976) Moslem 65* %, Animist 30** %, Christian 5** %
Education (1974/75) Pupils 290 081, teachers 9 200*
Labour force (1976) 3 198 000*; in agriculture 2 838 000* (89* %)
Personnel (1972) Physicians: 135, 1 per 38 960 people
Standard of living
National income per person (1976): M Fr 43 000** = $ 90** = £ 50**
Consumption per person (1975): energy 25 kg coal equivalent, electricity (production) 12* kW h
Newspapers (1972): number 1; circulation 3 000, 0.5 per 1 000 people
Telephones (Dec 1971): 5 000, 1 per 1 000 people
Livestock (000, 1976) Cattle 4 080*, sheep 4 219*, goats 3 929*, horses 150*, asses 400*, camels 178*
Electrical capacity (1975) 27* megawatts
Hospital beds (1971) 3 718, 1 per 1 380 people
Roads (1976) 14 704 km = 9 137 mi, density 0.01 km per km²
Railways (1975) 642 km = 399 mi, density 0.001 km per km²
Inland waterways Niger river 1 782 km = 1 107 mi; Senegal river
Airports Bamako, Mopti, and 7 other airports with scheduled flights
Durable equipment (at end-year)
Radio sets (1974): 75 000, 13 per 1 000 people
Passenger cars (1976): 15 200, 3 per 1 000 people, 1.0 per km of road
Commercial vehicles (1976): 3 600, 0.6 per 1 000 people, 0.2 per km of road

Production

Gross domestic product 1974: M Fr 203 300 mn = $ 423 mn = £ 181 mn
1976 est: M Fr 260 000**mn = $ 540**mn = £ 300**mn

Production index (1970 = 100)	1960	1970	1976	Growth %pa 1960–70	1970–76
Agricultural	89	100	106	1.2	1.0

Main products (000 t)

Agriculture					
Rice	143	138	237	−0.4	9.4
Maize	50	80*	81	4.8*	0.2*
Millet and sorghum	836	600	804	−3.3	5.0
Cassava	40*	40*	40*	0.0*	0.0*
Groundnuts	125	158	258	2.4	8.5
Cotton	4	22	41*	18.6	11.0*
Milk	70	53*	68*	−2.7*	4.2*
Beef and veal	31	51*	38*	5.1*	−4.8*
Sheep and goat meat	25	37*	30*	4.0*	−3.4*
Fish catch	80**	90*	100*	1.2**	1.8*
Timber (000 m³)	2 160	2 710*	3 040[a]	2.3*	2.3*[b]

[a]1975 [b]1970–75
Other (1975) Salt 5 000 t, electricity 70* mn kW h, beer 9 000 hl, cement 49 000 t

Transport traffic *Passenger-kilometres* (mn)	1960	1970	1976	Growth %pa 1960–70	1970–76
Rail	47[a]	75	100[c]	6.0[b]	5.9[d]
Air	29[a]	46	90	5.9[b]	11.8
Cargo: tonne-kilometres (mn)					
Rail	19[a]	128	156[c]	26.9[b]	4.0[d]
Waterways	na	25	27[c]	na	1.6[d]
Air	0.24[a]	2.5	1.5	34.2[b]	−8.2

[a]1962 [b]1962–70 [c]1975 [d]1970–75

Finance and trade

Money stock (end-year, M Fr bn) 1962 13.03, 1970 26.83, 1976 70.75
Growth rates: 1962–70 9.4 %pa, 1970–76 17.5 %pa
Budget (1976) Balanced at M Fr 49 272 mn = $ 103 mn = £ 57 mn

Balance of payments ($ mn)	1972	1973	1974	1975	1976
Balance of goods (fob)	−18	−48	−65	−66	−18
Balance of services	−26	−41	−81	−99	−100
Balance of transfers	+33	+61	+110	+113	+85
Current balance	*−12*	*−28*	*−36*	*−52*	*−33*
Long-term capital flow	+20	+10	+12	+18	+21

Reserves and debt (end-year, $ mn)

International reserves	3.8	4.2	6.1	4.2	6.9
External public debt	318	398	441	471	na

External trade (1976) Imports: M Fr 71 510 mn = $ 150 mn = £ 83 mn
Exports: M Fr 47 120 mn = $ 99 mn = £ 55 mn

Main imports (1972)	% of total	*Main exports* (1976)	% of total
Machinery	23	Cotton	44
Food	20	Groundnuts	11
(of which, cereals 8,		Textile yarns and fabrics	5**
sugar 7)		Fish	2
Petroleum products	11		
Chemicals	10		
Motor vehicles	8		
Main sources (1974)		*Main destinations* (1974)	
France	24	France	27
United States	13	China	17
Ivory Coast	9	Ivory Coast	15
China	9	Senegal	11
Soviet Union	7	West Germany	5

Mauritania

Islamic Republic of Mauritania
République Islamique de Mauritanie
Al Jumhuriya al Islamiya al Muritaniya

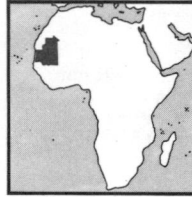

Location North-west Africa
With a coastline on the Atlantic Ocean, Western Sahara and Algeria are to the north, Mali to the east and south, and Senegal to the south. The north is part of the Sahara desert
Land Area 1 030 700 km² = 398 000 mi²
Excludes Tiris al Gharbia, southern part of Western Sahara (104 000* km² = 40 000* mi²)
Climate Hot and dry
Weather at Nouakchott, 21 m altitude
Temperature: hottest month Sept 24–34 °C, coldest Dec 13–28 °C
Rainfall (av monthly): driest month Jan 1 mm, wettest August 104 mm
Time GMT
Measures Metric system
Monetary unit Ouguiya (U) = 5 khoums; the Ouguiya replaced the CFA Franc on June 29, 1973 at U 1 = CFA Fr 5 (U 10 = Fr 1)
Rate of exchange (1976 av): free U 44.96 = $ 1, U 81.20 = £ 1

Summary

Political One-party republic, which became independent November 28, 1960; formerly a French colony. Controls southern part of Western Sahara, renamed Tiris al Gharbia (from February 1976); information shown here excludes that region. Member of UN, OAU, Arab League, Ecowas, CEAO, the Maghreb Organisation and an EEC ACP state
Economic Iron ore is the main export earner, and has been an important feature since 1961 and especially since 1964; a high level of production was reached in the world steel boom of 1973. Iron ore mining was nationalised in November 1974. Copper is also produced. Agricultural production is low, and livestock are important; fishing is being developed. Industrial development plans include oil refining and steel production, but have been affected by the cost of administering the Mauritanian part of Western Sahara

Mauritania

People, resources and equipment

Population 1960 0.95*mn, 1970 1.16*mn, 1976 1.35*mn
Growth: 1960–70 2.0* %pa, 1970–76 2.5* %pa
Density (1976): 1* person per km²
Vital statistics (rate per 1 000 people, 1970–75): births 44.8*, deaths 24.9*
Cities (population in 000, 1975)

Nouakchott (capital)	104	Zouerate	21	Atar	19
Nouadhibouª	23	Kaédi	20	Rosso	18

ªFormerly Port-Etienne
Race (1962) Arab 80 %, Fulani 14 %, Soninke 3 %, Zenaga 1 %
Language Arabic and French
Religion (1976) Moslem 99* %, Roman Catholic 1* %
Education (1974/75) Pupils 54 250*, teachers 2 100*
Labour force (1976) 405 000*; in agriculture 343 000* (85* %)
Personnel (1973) Physicians: 71, 1 per 17 750 people
Standard of living
National income per person (1976): U 10 000** = $ 220** = £ 120**
Consumption per person (1975): energy 108 kg coal equivalent,
electricity (production) 72* kW h, newsprint 1.2 kg
Newspapers (1972): number 1; circulation 300, 0.2 per 1 000 people
Telephones (Dec 1969): number 1 300, 1 per 1 000 people
Livestock (000, 1976) Cattle 2 000*, sheep 3 100* goats 2 500*,
horses 29*, asses 264*, camels 748*
Electrical capacity (1975) 39* megawatts
Hospital beds (1971) 440, 1 per 2 730 people
Roads (1972) 6 904 km = 4 290 mi, density 0.007 km per km²
Railways (1975) 652 km = 405 mi, density 0.0006 km per km²
Inland waterway Senegal river
Ships (registered, 1977) 4, total of 1 113 gross tons
Ports (goods traffic, 000 tonnes, 1972) Point-Central (iron ore port),
loaded 8 626; Nouadhibou, loaded 42, unloaded 89;
Nouakchott, loaded 23, unloaded 100
Airports Nouakchott, Nouadhibou and 14 other airports with
scheduled flights
Durable equipment (at end-year)
Radio sets (1974): 82 000, 64 per 1 000 people
Passenger cars (1972): 4 400, 4 per 1 000 people, 0.6 per km of road,
Commercial vehicles (1972): 5 000, 4 per 1 000 people, 0.7 per km of road

Production

Gross domestic product 1973: U 13 043 mn = $ 293 mn = £ 119 mn
1976 est: U 17 000** mn = $ 380** mn = £ 210** mn
Structure of gross domestic product (1973) *By origin* Agriculture 22 %,
mining and quarrying 33 %, manufacturing 5 %, other 40 %

Production index (1970 = 100)	1960	1970	1976	Growth %pa 1960–70	1970–76
Agricultural	82	100	83	2.1	−3.1
Main products (000 t)					
Agriculture					
Millet	60	81	60*	3.0	−4.9*
Dates	15	15	13*	0.0	−2.3*
Milk, cow	69*	84*	69*	2.0*	−3.2*
Milk, goat	68*	69*	63*	0.1*	−1.5*
Beef and veal	22*	28*	18*	2.4*	−7.1*
Sheep and goat meat	12	15	13*	2.3	−2.3*
Fish catch	20**	75*	34*ᵇ	14.0**	−14.6*ᶜ
Timber (000 m³)	430**	516*	565*ᵇ	1.8**	1.8*ᶜ
Other					
Electricityª (mn kW h)	na	73	95*ᵇ	na	5.4*ᶜ
Iron ore (Fe content)	—	5 923	5 646ᵇ	na	−1.0ᶜ
Copper ore (Cu content)	—	—	16ᵇ	na	na
Oxygen (000 m³)	na	75	70ᵈ	na	−1.9ᵉ
Acetylene (000 m³)	na	22	24ᵈ	na	2.4ᵉ

ªPublic supply only ᵇ1975 ᶜ1970–75 ᵈ1974 ᵉ1970–74

Transport traffic	1960	1970	1976	Growth %pa 1960–70	1970–76
Passenger-kilometres (mn)					
Air	41ᵇ	80	148	10.0ᶜ	10.8
Cargo: tonne-kilometres (mn)					
Rail	3 500ᵈ	6 208	6 808ᶠ	10.0ᵉ	3.1ᵍ
Airª	2.5ᵇ	6.3	14.4	14.1ᶜ	14.8
Sea: tonnes (mn)					
Goods loaded	0.015	9.19	8.69ʰ	90.0	−2.8ⁱ
Goods unloaded	0.036	0.090	0.153ᶠ	9.6	19.3ᵍ

ªIncludes apportionment of traffic of Air Afrique ᵇ1963 ᶜ1963–70 ᵈ1964
ᵉ1964–70 ᶠ1973 ᵍ1970–73 ʰ1972 ⁱ1970–72
Tourism Number of visitors (1972) 10 300; gross receipts (1975) $ 6 mn

Finance and trade

Consumer price index (1970 = 100) 1976 180.5; growth 1970–76 10.3 %pa
Money stock (end-year, U mn) 1970 1 110, 1976 3 684;
growth 1970–76 22.1 %pa
Budget (1976) Balanced at U 6 125 mn = $ 136 mn = £ 75 mn
International reserves (Dec 1976) $ 82 mn
External public debt (Dec 1976) $ 632 mn
External trade (1976) Imports: U 8 072 mn = $ 180 mn = £ 99 mn
Exports: U 8 013 mn = $ 178 mn = £ 99 mn

Main imports (1972)	% of total	Main exports (1972)	% of total
Food	20	Iron ore	74ª
(of which, sugar 6,		Fish	11
cereals 5)		Copper	10
Transport equipment	18	Natural gums	3
(of which, motor vehicles 9)			
Machinery, non-electric	17		
Electrical machinery	7		
Petroleum products	6		
Chemicals	5		

Main sources (1975)		Main destinations (1975)	
France	56*	France	20*
United States	8*	United Kingdom	16*
West Germany	8*	Italy	13*
United Kingdom	5*	Spain	11*
Senegal	5*	Japan	11*
		Belgium-Luxembourg	8*

ª82% in 1975

Special focus

Iron ore production (Fe content, 000 tonnes)

1960	—	1964	3 239ª	1968	5 006ª	1972	6 017
1961	198ª	1965	3 875ª	1969	5 641	1973	6 773
1962	656ª	1966	4 638ª	1970	5 923	1974	7 582
1963	841ª	1967	4 846ª	1971	5 497	1975	5 646

ªExports

Mauritius

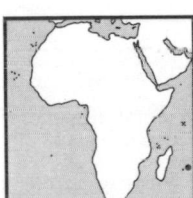

Location Western Indian Ocean
A group of islands, the main one of which, the
island of Mauritius, lies 900 km east of the
island of Madagascar; the other main islands are
Rodrigues, Agalega and St Brandon
Land Area 2 045 km² = 790 mi²
of which, island of Mauritius 1 865 km²,
Rodrigues 109 km²
Climate Sub-tropical
Weather at Mauritius, 55 m altitude
Temperature: hottest month January 23–30 °C, coldest July, Aug 17–24 °C
Rainfall (av monthly): driest month Sept 36 mm, wettest March 221 mm
Time 4 hours ahead of GMT
Measures Metric system: also *length* 12 lignes = 1 pouce,
12 pouces = 1 pied = 0.324484 metre = 1.07 feet
area 40 000 pied² = 1 arpent = 0.422 hectare = 1.043 acres
Monetary unit Mauritius rupee (M R) = 100 cents
Rate of exchange (1976 av): par M Rs 7.71376 = SDR 1,
free M Rs 6.682 = $ 1, M Rs 12.07 = £ 1

Summary

Political Parliamentary monarchy, which became independent March 12,
1968; formerly a UK colony (before 1810 a French colony). Member of
UN, OAU, Ocam, Commonwealth and an EEC ACP state
Economic Sugar is the main export product; although Mauritius is an
agricultural economy, cereals and other food are large imports and crop
diversification is under way. Industry is being developed, notably in
electronic components and textiles; tourism is of increasing importance.
There is a special Export Processing Zone (MEPZ)

People, resources and equipment

Population 1960 657 000, 1970 829 000, 1976 914 000*
Growth: 1960–70 2.4 %pa, 1970–76 1.6* %pa
Density (1976): 447* people per km²
Vital statistics (rate per 1 000 people, island of Mauritius only, 1976):
births 25.6, deaths 7.8

Mauritius

Regions (population in 000, 1975) Mauritius island 857, **Rodrigues 26***, Agalega and St Brandon 1*
Cities (population in 000, 1975) Port Louis (capital) 140, Beau Bassin-Rose Hill 83, Curepipe 54, Quatre Bornes 53, Vacoas-Phoenix 50
Race (1972) Indo-Mauritian 69 %, European, African and mixed 28 %
Language English and French; Creole is also used
Mother tongue (1972): Hindi 39 %, Creole 33 %, Urdu 9 %, Tamil 7 %, French 4 %, Telegu 3 %, Chinese 2 %
Religion (1975) Hindu 52 %, Roman Catholic 25** %, Moslem 17 %
Education Pupils (1975) 218 300*, teachers (1974) 8 600*
Labour force (1976) 310 000*; in agriculture 94 000* (30* %)
Personnel Scientists and engineers engaged in research (1973):91
Physicians (government employed only, 1974): 235, 1 per 3 720 people
Standard of living
National income per person (1976): M Rs 4 000** = $ 600** = £ 330**
Consumption per person (1975): energy 279 kg coal equivalent, electricity (1976) 298 kW h, newsprint 1.1 kg
Newspapers (1974): number 12; circulation 80 000, 92 per 1 000 people
Telephones (Dec 1975): 25 000, 28 per 1 000 people
Livestock (000, 1976) Cattle 53*, goats 67*, chickens 1 100*
Electrical capacity (1975) 121* megawatts
Hospital beds (1974) 3 209, 1 per 271 people
Roads (1976) 1 775 km = 1 103 mi, density 0.87 km per km²
There are no railways
Ships (registered, 1977) 17, total of 37 288 gross tons
Port (goods traffic, 000 tonnes, 1973) Port Louis: loaded 900, unloaded 731
Airports Plaisance, Rodrigues, Plaines des Roches (planned)

Durable equipment	000	no per	
(at end-year)		1 000 people	
Radio sets (1972)	107	125	no per
Television sets (1974)	38	44	km of road
Passenger cars (1975)	18	20	10.0
Commercial vehicles (1975)	10	11	5.4

Production, finance and trade

Gross domestic product 1975: M Rs 3 416 mn = $ 567 mn = £ 255 mn
1976 est: M Rs 3 800** mn = $ 570** mn = £ 310** mn
Structure of gross domestic product (1975) *By origin* Agriculture 30 %, manufacturing 16 %, construction 6 %, other 48 %
Agricultural production index (1970 = 100) 1976 124; growth 1970–76 3.6 % pa
Main products (000 t) *Agriculture* (1976) Sugar, raw value 731, potatoes 10*, tomatoes 7*, bananas 5*, tea 4*, tobacco 1*, milk 22*, fish catch (1975) 8, timber (000 m³, 1975) 49*
Other (1975) Electricity (mn kW h, 1976) 272, beer (000 hl) 115, cigarettes (mn units) 891, ethyl alcohol (000 hl) 27, nitrogenous fertilisers (1975/76) 4*, molasses 129
Transport traffic (1973) *Sea* Ships entered 2.47 mn grt; goods: see port
Tourism Number of visitors (1974) 73 340, gross receipts (1975) $ 22 mn
Consumer price index (1970 = 100) 1976 201.5; growth 1970–76 12.4 % pa
Money stock (end-year, M Rs mn) 1976 1 099; growth 1970–76 29.8 % pa
Budget (1975/76; year ending June 30th)
Revenue: M Rs 1 076 mn = $ 165 mn = £ 83 mn
Expenditure: M Rs 1 071 mn = $ 164 mn = £ 82 mn
International reserves (Dec 1976) $ 89 mn
External public debt (Dec 1976) $ 133 mn
External trade (1976) Imports: M Rs 2 381 = $ 356 mn = £ 197 mn
Exports: M Rs 1 768 mn = $ 265 mn = £ 146 mn

Main imports (1975)	% of total	*Main exports* (1975)	% of total
Machinery	18	Sugar	84
Cereals	13	Clothing	6
Petroleum products	10	Electronic components	3
Chemicals	9	Molasses	1
Textile yarns and fabrics	8		
Main sources (1975)		*Main destinations* (1975)	
United Kingdom	17	United Kingdom	77
South Africa	10	United States	6
Iran	9	France	6
France	9	Canada	4

Special focus

Sugar prices (in the United Kingdom)

	$ per tonne	£ per tonne		$ per tonne	£ per tonne
1971	110	45.2	1974	703	300.4
1972	179	71.6	1975	479	215.4
1973	239	97.6	1976	273	151.4

Morocco

Kingdom of Morocco
Al Mamlaka al Maghribiya

Location North-west Africa
On the north-west corner of Africa, with a coastline on the Atlantic Ocean and the Mediterranean Sea; Algeria is to the east and Western Sahara to the south
Land Area 446 550 km² = 172 410 mi²
Excludes northern part of Western Sahara (162 000* km² = 63 000* mi²)
Climate Warm on the coast, hot inland
Weather at Rabat, 65 m altitude
Temperature: hottest month August 18–28 °C, coldest January 8–17 °C
Rainfall (av monthly): driest month July 1 mm, wettest December 86 mm
Time GMT
Measures Metric system; also *length* 1 kala = 0.5 metre = 1.64 feet
capacity 1 fanega = 0.555 hectolitre = 1.53 UK bushels
weight (*mass*) 1 kantar = 100 kilograms = 220.5 pounds
Monetary unit Dirham (Dh) = 100 centimes
Rate of exchange (1976 av): free Dh 4.419 = $ 1, Dh 7.982 = £ 1

Summary

Political Monarchy, which became independent in 1956; formerly French and Spanish protectorates and the International Zone of Tangier. Ifni, a former Spanish province, was returned to Morocco June 30, 1969. Controls northern part of Western Sahara (from February 1976); information shown here excludes that region. Member of UN, OAU, Arab League and the Maghreb Organisation
Economic The main export is phosphates used mainly for fertilisers; in other respects predominantly an agricultural economy, with some exports of fruit, vegetables and fish. Livestock is important locally. Industry is being developed, especially textiles and chemicals. Guerilla warfare in Western Sahara has caused increased military expenditure (to 5* % of gross domestic product in 1977)

People, resources and equipment

Population 1960 11.63* mn, 1970 15.00** mn, 1976 17.83* mn
Growth: 1960–70 2.6** % pa, 1970–76 2.9** %pa
Density (1976): 40* people per km²
Vital statistics (rate per 1 000 people, 1970–75): births 46.2*, deaths 15.7*
Regions (population in 000, 1973; total of 16.31 mn)

Prefectures		Fès	1 138	Oujda	679
Casablanca	1 894	Kenitra	1 416	Safi	943
Rabat-Salé	703	Khouribga	351	Settat	702
Provinces		Ksar es Souk	491	Tanger	236
Agadir	1 221	Marrakech	1 642	Tarfaya	26
Al Hoceima	257	Meknès	806	Taza	604
Beni Mellal	695	Nador	501	Tetouan	844
Al Jadida	618	Ouarzazate	542		

Cities (population in 000, 1973)

Rabat (capital)	597	Meknès	403	Safi	215
Casablanca	1 753	Oujda	349	Tanger[b]	208
Marrakech[a]	436	Kenitra	342	Agadir	189
Fès[c]	426	Tetouan	309	Khouribga	159

[a]Marrakesh [b]Tangier [c]Fez
Race (1971) Moroccan 99 %
Language Arabic; Berber, Spanish and French are also used
Religion (1976) Moslem 98* %, Christian 2* %
Education (1975/76) Pupils: primary 1 547 647, secondary 451 575, vocational (1970/71) 6 986, teacher-training 4 030, higher (1974/75) 34 092.
Teachers: primary 37 226, secondary 19 613, vocational (1970/71) 572, teacher-training 486, higher (1974/75) 1 921
Labour force (1976) 4 733 000*; in agriculture 2 527 000* (53* %)
Personnel Physicians: 1 223, 1 per 13 800 people
Standard of living
National income per person (1976): Dh 1600** = $ 360** = £ 200**
Consumption per person (1975): energy 274 kg coal equivalent, electricity (production, 1976) 181* kW h, newsprint 0.2 kg, steel 28 kg
Newspapers (1974): number 6; circulation 235 000, 14 per 1 000 people
Telephones (Dec 1975): 198 000, 11 per 1 000 people
Livestock (000, 1976) Cattle 3 400*, sheep 16 800*, goats 7 200*, horses 300*, mules 394*, asses 1 200*, camels 200*, chickens 23 712*
Mineral reserves Coal (1960) 96 mn tonnes
Petroleum refinery capacity (1975) 3.9* mn tonnes

Morocco

Electrical capacity (1975) 730* megawatts
Hospital beds (1974) 23 056, 1 per 732 people
Roads (1974) 25 286 km = 15 712 mi, density 0.06 km per km²
Railways (1974) 1 756 km = 1 091 mi, density 0.004 km per km²
Ships (registered, 1977) 91, total of 270 295 gross tons
Ports (goods traffic, 000 tonnes, 1973)

	loaded	unloaded
Casablanca	13 956	3 367
Safi	4 144	331
Kenitra	400	201

Also: Mohammedia, Tangier (free zone), Agadir
Airports Casablanca, Tangier, Rabat, Marrakesh, and 9 other airports with scheduled flights

Durable equipment (at end-year)	000	no per 1 000 people	no per km of road
Radio sets (1974)	1 300	76	
Television sets (1976)	474	26	
Passenger cars (1975)	320	18	12.7
Commercial vehicles (1975)	127	7	5.0

Production

Gross domestic product 1975: Dh 29 890 mn = $ 7 376 mn = £ 3 320 mn
1976 est: Dh 31 000** mn = $ 7 000** mn = £ 3 900** mn
Growth in real terms: 1970–73 3.8 %pa
Structure of gross domestic product By origin (at 1960 prices, 1974)
Agriculture 28 %, mining, quarrying and utilities 11 %, manufacturing 14 %, construction 6 %, other 41 %
By use (1975) Final consumption expenditure 86 % (of which, government 17 %), gross fixed capital formation 26 %, exports of goods and services 27 %, less imports of goods and services −39 %

Production indices (1970 = 100)	1960	1970	1976	Growth %pa 1960–70	1970–76
Agricultural	79	100	117	2.4	2.6
Industrial	69	100	142	3.8	6.0
of which, mining	87	100	120	1.4	3.1

Main products (000 t)
Agriculture

	1960	1970	1976	1960–70	1970–76
Wheat	1 280	1 801	2 135	3.5	2.9
Barley	1 392	1 955	2 862	3.5	6.6
Maize	470	320	493	−3.8	7.5
Tomatoes	270	460	317*	5.5	−6.0*
Sugar, raw value	—	163	310*	na	11.3*
Dry broad beans	112	190	230	5.4	3.2
Oranges	430	527	566	2.1	1.2
Grapes	364	203	250*	−5.7	3.5*
Wine	231	125	70*	−6.0	−9.2*
Olives	181	280	312*	4.5	1.8*
Olive oil	25*	25*	42	0.0*	9.0*
Dates	49	90*	70*	6.3*	−4.1*
Figs	78	65*	67*	−1.8*	0.5*
Cotton	2	6	7*	11.6	2.6*
Milk	380	450*	497*	1.7*	1.7*
Meat	161	203*	214*	2.3*	0.9*
Fish catch	163	249	210b	4.3	−3.3c
Timber (000 m³)	1 950**	2 733	3 154*b	3.4**	2.9*c

Energy

	1960	1970	1976	1960–70	1970–76
Total energy (000 tce)	660	720	910b	0.9	4.8c
Coal	412	433	702	0.5	8.4
Crude oil	92	46	25b	−7.5	−11.5c
Petroleum products	205	1 490	2 415*b	21.9	10.1*c
Natural gas (mn m³)	9	44	67b	17.2	8.8c
Electricity (mn kW h)	1 012	1 982	3 220*	7.0	8.4*

Mining

	1960	1970	1976	1960–70	1970–76
Iron ore (Fe content)	874	522	210*	−5.0	−14.1*
Lead ore (Pb content)	94	84	83	−1.1	−0.2
Manganese ore (Mn content)	225	60	105b	−12.4	12.1c
Phosphate rock	7 473	11 424	14 119b	4.3	4.3c

Manufacturing

	1960	1970	1976	1960–70	1970–76
Cigarettes (mn units)	3 490	4 977	9 000b	3.6	12.6c
Wood pulp	17	45	55b	10.2	4.1c
Sulphuric acid	32	28	45d	−1.3	12.6e
Fertilisers, phosphate	30**	99	116b	12.7**	3.2c
Cement	580	1 421	2 108	9.4	6.8
Radio sets (000 units)a	na	na	157b	na	na
Motor vehicles (000 units)a	—	25.4	32.5	na	4.2

aAssembly only b1975 c1970–75 d1974 e1970–74

Transport traffic	1960	1970	1976	Growth %pa 1960–70	1970–76
Passenger-kilometres (mn)					
Railb	477	521	863	0.9	8.8
Air	158	392	1 264	9.5	21.5
Cargo: tonne-kilometres (mn)					
Roada	na	687	720c	na	1.6d
Railb	1 757	2 647	3 131	4.2	2.8
Air	5.0	5.4	18.2	0.8	22.4
Sea: tonnes (mn)					
Goods loaded	10.5	14.0	15.4e	2.9	1.9f
Goods unloaded	2.8	4.5	7.3e	4.9	9.9f

aPublic transport only bPrincipal railways only c1973 d1970–73 e1975 f1970–75
Tourism (1975) Number of visitors 1 245 000, gross receipts $ 296 mn

Finance and trade

Price index (1970 = 100)	1960	1970	1976	Growth %pa 1960–70	1970–76
Consumer prices	79.9	100.0	155.1	2.3	7.6
Money stock (end-year, Dh mn)	2 643	5 540	15 168	7.7	18.3

Budget (1974) Revenue: Dh 7 326 mn = $ 1 677 mn = £ 717 mn
Expenditure: Dh 8 748 mn = $ 2 002 mn = £ 856 mn

Balance of payments ($ mn)	1972	1973	1974	1975	1976
Balance of goods (fob)	−67	−124	+14	−899	−1 660
Balance of services	−23	−23	−115	−146	−241
Balance of transfers	+138	+242	+330	+507	+504
Current balance	*+48*	*+97*	*+229*	*−538*	*−1 397*
Long-term capital flow	+47	+8	+43	+391	+1 298
Reserves and debt (end-year, $ mn)					
International reserves	237	266	417	377	491
External public debt	1 189	1 342	1 898	2 462	na

External trade (1976) Imports: Dh 11 555 mn = $ 2 615 mn = £ 1 448 mn
Exports: Dh 5 579 mn = $ 1 263 mn = £ 699 mn

Main imports	% of total	Main exports	% of total
Machinery, non-electric	17	Phosphates	39
Crude oil	9	Citrus fruit	11
Motor vehicles	7	Textile yarns and fabrics	5
Chemicals	7	Clothing	5
Iron and steel	7	Fish, preserved	4
Wheat	6	Non-ferrous ores	3
Sugar	6	Chemicals	3
Electrical machinery	5	Tomatoes	2

Main sources		Main destinations	
France	29	France	24
United States	9	West Germany	10
West Germany	8	Italy	7
Spain	6	United Kingdom	6
Italy	6	Belgium-Luxembourg	6
Iraq	5	Spain	5
United Kingdom	4	Poland	5
Soviet Union	3	Netherlands	4
Belgium-Luxembourg	2	Soviet Union	4
Cuba	2	Jugoslavia	2
Poland	2	Brazil	2

Special focus

Phosphate rock production (000 tonnes)

1960	7 473	1966	9 428	1972	15 105
1961	7 950	1967	9 922	1973	17 077
1962	8 162	1968	10 324	1974	19 750
1963	8 549	1969	11 294	1975	14 119
1964	10 098	1970	11 424		
1965	9 825	1971	12 030		

Mozambique
People's Republic of Mozambique

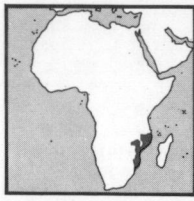

Location South-east Africa
With a coastline on the Indian Ocean, Tanzania, Malawi and Zambia are to the north, Rhodesia, Swaziland and South Africa to the west
Land Area 783 030 km² = 302 330 mi²
Climate Tropical in the north, sub-tropical in the south
Weather at Maputo, 59 m altitude
Temperature: hottest month February 22–31 °C, coldest July 13–24 °C
Rainfall (av monthly): driest months July, August 13 mm, wettest January 130 mm
Time 2 hours ahead of GMT
Measures Metric system
Monetary unit Mozambique escudo (M Esc) = 100 centavos
Rate of exchange (1976 av): par M Esc 1 = Esc 1, free M Esc 30.22 = $ 1, M Esc 54.59 = £ 1

Summary

Political One-party people's republic, which became independent June 25, 1975; formerly a Portuguese overseas province. Member of UN and OAU
Economic An agricultural economy, with main exports sugar, cotton, preserves, copra and cashew nuts. There has been little development of minerals but there are large resources notably of coal, bauxite, iron ore and natural gas. There is little industry, and the government is developing mainly the agriculture and mining sectors. Closure of the border with Rhodesia in 1976 affected trade considerably

People, resources and equipment

Population 1960 6.47**mn, 1970 8.23*mn, 1976 9.44*mn
Growth: 1960–70 2.4** %pa, 1970–76 2.3* %pa
Density (1976): 12* people per km²
Vital statistics (rate per 1 000 people, 1970–75): births 43.1*, deaths 20.1*
Cities (population in 000, 1975)
Maputoª (capital) 384ᶜ Sofalaᵇ 70* Quelimane 70*
Tete 150* Inhambane 70* Mozambique 50*
ªFormerly Lourenço Marques ᵇFormerly Beira
Race (1976) African 99* %
250 000* Portuguese left after independence in 1975
Language Portuguese and African languages and dialects
Religion (1976) Animist 60** %, Christian 24** %, Moslem 12** %
Education (1972/73) Pupils: primary 577 997, secondary 36 155, vocational 17 216, teacher-training 1 279, higher 2 621.
Teachers: primary 8 345, secondary 1 682, vocational 984, teacher-training 122, higher 326
Labour force (1976) 3 697 000*; in agriculture 2 522 000* (68* %)
Personnel (1971) Physicians: 510, 1 per 16 390 people
Standard of living
National income per person (1976): M Esc 8 000** = $ 260** = £ 150**
Consumption per person (1975): energy 186 kg coal equivalent, electricity (production) 78* kW h, newsprint 0.1 kg, steel 1 kg
Newspapers (1974): number 3; circulation 47 000, 5 per 1 000 people
Telephones (Dec 1975): 50 000, 5 per 1 000 people
Livestock (000, 1976) Cattle 1 420*, goats 570*, sheep 132*, pigs 183*
Mineral reserves Coal (1963) 700* mn tonnes
Lignite (1969) 400 mn tonnes
Natural gas (1965) 72 bn cubic metres
Petroleum refinery capacity (1975) 0.8 mn tonnes
Electrical capacity (1975) 393* megawatts
Hospital beds (1967) 13 102, 1 per 549 people
Roads (1974) 39 173 km = 24 341 mi, density 0.05 km per km²
Railways (1974) 4 161 km = 2 586 mi, density 0.005 km per km²
Ships (registered, 1977) 59, of 27 618 gross tons
Ports (goods traffic, 000 tonnes, 1973) Maputo: loaded 10 323, unloaded 3 182; Sofala: loaded 1 078, unloaded 702. Also: Nacala, Quelimane, Mozambique
Airports Passenger departures and arrivals (1975): Maputo 250 534, Sofala 215 833; also (1977) 6 other airports with scheduled flights

Durable equipment (at end-year)	000	no per 1 000 people	
Radio sets (1974)	176	19	no per
Television sets (1974)	1*	0.1*	km of road
Passenger cars (1972)	88	10	2.2
Commercial vehicles (1972)	14	2	0.3

Production

Gross domestic product 1970: M Esc 53 820 mn = $ 1 872 mn = £ 780
1976 est: M Esc 80 000** mn = $ 2 600** mn = £ 1 500** mn

Production indices (1970 = 100)	1960	1970	1976	Growth %pa 1960-70	1970-76
Agricultural	75	100	95	2.9	−0.9
Industrial	51	100	83ᶜ	7.0	−3.7ᵈ

Main products (000 t)	1960	1970	1976	1960-70	1970-76
Agriculture					
Rice	84	102	79*	2.0	−4.2*
Maize	340	310	450*	−0.9	6.4*
Sorghum	150	180*	250*	1.8*	5.6*
Cassava	2 100*	2 100*	2 400*	0.0*	2.2*
Groundnuts	51*	140*	100*	10.6*	−5.5*
Sugar, raw value	157	289	248*	6.3	−2.5*
Bananas	12	34	65*	11.0	11.4*
Cashew nuts	131ª	184	200*	5.0ᵇ	1.4*
Copra	44	60*	83*	3.2*	5.6*
Tea	7	17	13*	9.3	−4.4*
Sisal	30	29	19*	−0.3	−6.8*
Cotton	48	46	28*	−0.4	−7.9*
Milk	45	52*	60*	1.5*	2.4*
Beef and veal	24	31*	35*	2.6*	2.0*
Timber (000 m²)	8 000**	8 504*	9 075*ᶜ	0.6**	1.3*ᵈ
Other					
Total energy (000 tce)	280	380	540ᶜ	3.1	7.3ᵈ
Coal	270	351	370	2.7	0.9
Petroleum products	—	729	750*ᶜ	na	−0.6*ᵈ
Electricity (mn kW h)	122	683	717*ᶜ	18.8	1.0*ᵈ
Bauxite	4.8	7.2	2.0*ᶜ	4.1	−22.6*ᵈ
Salt	29	50	31ᵉ	5.6	−11.3ᶠ
Beer (000 hl)	79	439	965ᶜ	18.7	17.1ᵈ
Cigarettes (mn units)	1 280	2 570	2 696ᵉ	7.2	1.2ᶠ
Cement	222	394	258ᶜ	5.9	−8.1ᵈ
Sulphuric acid	—	14	23ᵉ	na	13.4ᶠ
Radio sets (000 units)	na	23	24ᵉ	na	1.0ᶠ

ª1961–65 average ᵇ1961–65 to 1970 ᶜ1975 ᵈ1970–75 ᵉ1974 ᶠ1970–74

Transport traffic	1960	1970	1976	Growth %pa 1960-70	1970-76
Passenger-kilometres (mn)					
Rail	199	315	210ª	4.7	−7.8ᵇ
Air	27	104	312	14.4	20.1
Cargo: tonne-kilometres (mn)					
Rail	2 020	2 957	2 180ᶜ	3.9	−7.3ᵈ
Air	0.61	3.8	5.8	20.1	7.3
Sea: tonnes (mn)					
Goods loaded	3.84	9.71	8.99ª	9.7	−1.5ᵇ
Goods unloaded	2.77	4.16	3.27ª	4.2	−4.7ᵇ

ª1975 ᵇ1970–75 ᶜ1974 ᵈ1970–74

Finance and trade

Consumer price index (Maputo, 1970 = 100) 1976 171.6; growth 1970–76 9.4 %pa
Budget (1976) Revenue: M Esc 7 100 mn = $ 235 mn = £ 130 mn
Expenditure: M Esc 9 300 mn = $ 308 mn = £ 170 mn
External trade (1975) Imports: M Esc 10 472 mn = $ 410 mn = £ 184 mn
Exports: M Esc 5 357 mn = $ 210 mn = £ 94 mn

Main imports (1973)	% of total	Main exports (1974)	% of total
Machinery	27	Sugar	22
Transport equipment	13	Preserves	14
(of which, motor vehicles 8)		Cotton	11
Metals	12	Copra	8
Chemicals	11	Cashew nuts	6
Textiles	9	Petroleum products	5
Food	7		
Minerals	7		
Main sources (1975)		Main destinations (1975)	
South Africa	18	Portugal	24
Portugal	15	United States	14
West Germany	11	South Africa	8
United Kingdom	8	India	5
Iraq	6	Japan	5
United States	5	United Kingdom	4
Japan	5	Netherlands	4
France	5	Brazil	4
Sweden	3	Egypt	3

Niger
Republic of Niger
République du Niger

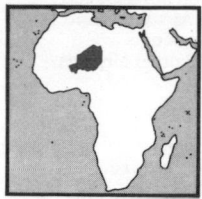

Location North central Africa
Algeria and Libya are to the north, Chad to the
east, Nigeria and Benin to the south, and
Upper Volta and Mali to the west. There is an
outlet to the sea via the Niger river
Land Area 1 267 000 km² = 489 000 mi²
Climate Hot and dry
Weather at Niamey, 216 m altitude
Temperature: hottest month May 27–41 °C,
coldest January 14–34 °C
Rainfall (av monthly): driest month Dec 0 mm, wettest August 188 mm
Time 1 hour ahead of GMT
Measures Metric system
Monetary unit CFA franc (CFA Fr) = 100 centimes
Rate of exchange (1976 av): par CFA Fr 50 = Fr 1,
free CFA Fr 239.0 = $ 1, CFA Fr 431.6 = £ 1

Summary

Political Republic with military government, which became independent
August 3, 1960; formerly a territory of French West Africa. Member of
UN, OAU, Ecowas, CEAO, Ocam, franc zone and an EEC ACP state
Economic An agricultural economy based largely on livestock, but with
mining, especially of uranium, of increasing importance; uranium
accounted for two-thirds of exports in 1975. France has developed and
received the supplies of uranium; phosphates and small quantities of oil
have also been discovered. Industry is limited. By 1976 there was
substantial recovery in the agricultural sector from the 1973–74 Sahel
drought

People, resources and equipment

Population 1960 3.01**mn, 1970 4.02*mn, 1976 4.73*mn
Growth: 1960–70 2.9** %pa, 1970–76 2.8* %pa
Density (1976): 4* people per km²
Vital statistics (rate per 1 000 people, 1970–75): births 52.2*, deaths 25.5*
Cities (population in 000, 1975) Niamey (capital) 150*, Maradi 42*,
Zinder 40*, Tahoua 30*
Race (1972) Hausa 54 %, Djerma-Songhai 24 %, Fulani 11 %,
Beriberi-Manga 9 %, Tuareg 3 %
Language French and local languages (especially Hausa)
Religion (1976) Moslem 85** %, Animist 14** %, Christian 1** %
Education (1974/75) Pupils 133 674, teachers 3 607
Labour force (1976) 1 486 000*; in agriculture 1 338 000* (90* %)
Personnel (1974) Physicians: 109, 1 per 41 100 people
Standard of living
National income per person (1976): CFA Fr 30 000** = $ 125** = £ 70**
Consumption per person (1975): energy 35 kg coal equivalent,
electricity (production) 15* kW h
Newspapers (1974): number 2
Telephones (Dec 1975): 5 000, 1 per 1 000 people
Livestock (000, 1976) Cattle 2 700*, goats 5 100*, sheep 2 300*,
horses 200*, asses 350*, camels 260*
Mineral reserves Uranium (1974) 40 000 tonnes
Electrical capacity (1975) 20* megawatts
Hospital beds (1973) 2 935, 1 per 1 460 people
Roads (1976) 6 985 km = 4 340 mi, density 0.006 km per km²
There are no railways
Inland waterways Niger river: 300* km = 190* mi
Airports Niamey: passenger departures and arrivals (1975) 67 178;
also Zinder, Maradi, Tahoua, Agades, Arlit
Durable equipment (at end-year)
Radio sets (1971): 150 000, 36 per 1 000 people
Passenger cars (1976): 16 600, 3 per 1 000 people, 2.4 per km of road
Commercial vehicles (1976): 3 500, 1 per 1 000 people, 0.5 per km of road

Production

Gross domestic product
1970: CFA Fr 111 100 mn = $ 400 mn = £ 167 mn
1976 est: CFA Fr 155 000** mn = $ 650** mn = £ 360** mn
Structure of gross domestic product *By origin* (1969) Agriculture 51 %,
manufacturing 6 %, construction 3 %, other 40 %
By type (1969) Final consumption expenditure 100 % (of which,
government 13 %), gross fixed capital formation 6 %, exports of goods
and services 11 %, less imports of goods and services −18 %

Production index (1970 = 100)	1960	1970	1976	Growth %pa 1960–70	1970–76
Agricultural	77	100	115	2.7	2.3
Main products (000 t)					
Agriculture					
Rice	10**	37	29	14.0**	−4.0
Millet	500**	610*	1 195*	2.0**	11.9*
Sorghum	222	337	308	4.3	−1.5
Cassava	101	154*	295*	4.3*	11.4*
Dry beans	70**	84	175*	1.8**	13.0*
Onions	19	33	42*	5.7	4.1*
Groundnuts	150	205	95	3.2	−12.0
Sugar cane	30**	36	75*	1.8**	13.0*
Cotton	1.0*	3.7*	5.0*	14.0*	5.1*
Milk, cow	88	108	57*	2.1	−10.1*
Milk, goat	95	119*	108*	2.3*	−1.6*
Beef and veal	39	48*	36*	2.1*	−4.7*
Cattle hides	3.8[a]	4.4**	3.8*	2.1**[b]	−2.4**
Goatskins	3.2[a]	3.8*	3.0*	2.5*[b]	−3.9*
Timber (000 m³)	1 660	2 275*	2 490*[c]	3.2*	1.8*[d]
Other					
Electricity (mn kW h)	8	39	70*[c]	17.2	12.4*[d]
Uranium (U content)	—	—	1.6	na	na
Tin conc (Sn content)	0.054	0.067	0.084*[c]	2.2	4.6*[d]
Beer (000 hl)	na	17	38*[c]	na	30.8[f]
Cement	—	33	18*[c]	na	−11.4*[d]

[a]1961–65 [b]1961–65 to 1970 [c]1975 [d]1970–75 [e]1973 [f]1970–73
Transport traffic (1976) *Air* (including apportionment of traffic of Air
Afrique) Passenger-km 137 mn, cargo 13.9 mn t-km

Finance and trade

Price index (1970 = 100)	1960	1970	1976	Growth %pa 1960–70	1970–76
Consumer prices	79.4[a]	100.0	178.1	3.9[b]	10.1
Money stock					
(end-year, CFA Fr mn)	5 742[a]	8 830	24 774	7.4[b]	18.8

[a]1964 [b]1964–70
Budget (1976/77; year ending September 30th)
Revenue: CFA Fr 34 100 mn = $ 138 mn = £ 81 mn
Expenditure: CFA Fr 35 300 mn = $ 143 mn = £ 84 mn
International reserves (Dec 1976) $ 82 mn
External public debt (Dec 1975) $ 132 mn
External trade (1975) Imports: CFA Fr 21 889 = $ 102 mn = £ 46 mn
Exports: CFA Fr 19 556 mn = $ 91 mn = £ 41 mn

Main imports (1974)	% of total	*Main exports* (1974)	% of total
Petroleum products	14	Uranium	50[a]
Machinery, non-electric	11	Cattle	12
Motor vehicles	10	Groundnut oil	9
Textile yarns & fabrics	7	Textile yarns & fabrics	5
Cereals	7	Sheep and goats	4
Chemicals	7		
Sugar	6		
Electrical machinery	6		
Main sources (1974)		*Main destinations* (1974)	
France	37	France	54
United States	13	Nigeria	27
Nigeria	9	West Germany	7
West Germany	8	Benin	2
United Kingdom	3		
Ivory Coast	3		

[a]65 % in 1975

Nigeria
Federal Republic of Nigeria

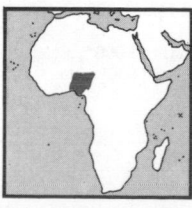

Location West Africa
With a south-facing coastline on the Atlantic
Ocean in the Gulf of Guinea, Benin is to the
west, Niger to the north and Chad and
Cameroon to the east
Land Area 923 768 km² = 356 669 mi²
Climate Tropical
Weather at Lagos, 3 m altitude
Temperature: hottest month March 26–32 °C,
coldest August 23–28 °C
Rainfall (av monthly): driest month Dec 25 mm, wettest June 460 mm
Time 1 hour ahead of GMT

Nigeria

Measures Metric system, which replaced the UK (imperial) system from January 1, 1973; also: 1 mudu (of rice) = 1.13* kilograms = 2.5* pounds, 1 tiya (of rice) = 2.27* kilograms = 5* pounds, 2 mudu = 1 tiya, 1 load (of cocoa) = 27.22 kilograms = 60 pounds
Monetary unit Naira (N) = 100 kobo; the naira was introduced as a decimal currency on January 1, 1973 to replace the Nigerian pound (N £) at the rate N 2 = N £ 1
Rate of exchange (1976 av): free $ 1.596 = N 1, N 1.132 = £ 1

Summary

Political Republic with military government, which became independent October 1, 1960; formerly the UK colony and protectorate of Nigeria. In 1961 the northern part of the UK Cameroons joined Nigeria, the southern portion becoming part of Cameroon. From 1967 to 1970 there was an attempt by the Eastern States to break away and form a new state of Biafra; the attempt was unsuccessful. Member of UN, OAU, Opec, Ecowas, Commonwealth, and an EEC ACP state
Economic Oil has become the most important feature of the economy, especially since the price rise of 1974, replacing agricultural produce as the main export. Agriculture remains important for the local economy. Manufacturing industry is developing rapidly, and new projects include petro-chemicals, fertilisers, steel, cement and paper

People, resources and equipment

Population[a] 1960 42.95**mn, 1970 55.07**mn, 1976 64.75**mn
[a]UN estimates; official Nigerian censuses indicate a higher population: 1963 55.67 mn, 1973 79.76 mn (growth 1963–73 3.7 %pa). The 1973 census results were officially abandoned in 1975
Growth: 1960–70 2.5** %pa, 1970–76 2.7** %pa
Density (1976): 70** people per km²
Vital statistics (rate per 1 000 people, 1970–75): births 49.3*, deaths 22.7*
Regions (States, as revised 1976; population in 000, February 1976)

Anambra	2 469	Imo	5 000	Ondo	2 728
Bauchi	3 240	Kaduna	4 098	Oyo	5 209
Bendel	3 536	Kano	5 775	Plateau	1 421
Benue	2 641	Kwara	2 399	Rivers	1 544
Borno	2 991	Lagos	1 100	Sokoto	2 873
Cross River	4 626	Niger	2 900		
Gongola	1 585	Ogun	1 449		

Cities (population in 000, 1975)

Lagos (capital[a])	1 477[b]	Ilesha	224	Aba	177
Ibadan	847	Onitsha	220	Ife	176
Ogbomosho	432	Iwo	214	Ila	155
Kano	399	Ado Ekiti	213	Oyo	152
Oshogbo	282	Kaduna	202	Ikerre Ekiti	145
Ilorin	282	Mushin	197	Benin City	136
Abeokuta	253	Maiduguri	189	Iseyin	115[b]
Port Harcourt	242	Enugu	187	Katsina	109[b]
Zaria	224	Ede	182	Calabar	103

[a]The capital is to be moved near to Abuja in a new Federal Capital Territory by 1991
[b]1971
Race (1976) African 99.9** % (in 1961: Hausa 21 %, Ibo 18 %, Yoruba 18 %, Fulani 10 %, Tiu 6 %, Kanuri 5 %, Ibibio 5 %)
Language English; also Hausa, Ibo, Yoruba and other local languages
Religion (1976) Moslem 50** %, Christian 25** %, Animist 20** %
Education Pupils (1974/75): primary 4 368 778, secondary 476 507, vocational 20 423, teacher-training 47 590, higher (1973/74) 23 228. Teachers (1973/74): primary 136 142, secondary 17 215, vocational 1 111, teacher-training 2 122, higher 3 459
Labour force (1976) 25 193 000*; in agriculture 14 326 000* (57* %)
Personnel Scientists and engineers (1970/71): 19 885
Physicians (1973): 2 343, 1 per 25 460 people
Standard of living
National income per person (1976) N 220** = $ 350** = £ 200**
Consumption per person (1975): energy 90 kg coal equivalent, electricity (production) 51 kW h, newsprint 0.3 kg, steel 22 kg
Newspapers (1974): number 12; circulation 660 000, 11 per 1 000 people
Telephones (Dec 1974): 111 000, 2 per 1 000 people
Livestock (000, 1976) Cattle 11 300*, sheep 7 900*, goats 23 000*, pigs 900*, horses 250*, asses 700*, camels 17*, chickens 90 000*
Mineral reserves (1975) Coal 135 mn tonnes
Crude oil 1 765 mn tonnes
Natural gas 1 422 bn cubic metres
Petroleum refinery capacity (1975) 3.0 mn tonnes
Electrical capacity (1975) 860* megawatts

Hospital beds (1972) 42 101, 1 per 1 378 people
Roads (1974) 97 000* km = 60 000* mi, density 0.10* km per km²
Railways (1975) 3 524 km = 2 190 mi, density 0.004 km per km²
Inland waterways Niger river 20 000* km = 12 500* mi
Ships (registered, 1977) 94, total of 335 540 gross tons
Ports (goods traffic, 000 tonnes, 1971) Bonny (crude oil port) loaded 54 880, Burutu (crude oil port) loaded 13 348, Lagos (Apada) loaded 1 048, unloaded 3 878. Also: Port Harcourt, Sapele, Calabar, Koko
Airports Murtala Mohammed (Lagos), Ibadan and 10 other airports with scheduled flights

Durable equipment (at end-year)	000	no per 1 000 people	
Radio sets (1974)	5 000*	81*	no per km of road
Television sets (1974)	110	2	
Passenger cars (1973)	82	1	0.8*
Commercial vehicles (1973)	58	1	0.6*

Production

Gross domestic product 1973/74 (year ending March 31st):
N 9 120 mn = $ 13 862 mn = £ 5 736 mn
1976 est: N 16 000** mn = $ 25 000** mn = £ 14 000** mn
Structure of gross domestic product (1973/74) *By origin*
Agriculture 36 %, mining and quarrying 22 %, manufacturing 7 %, construction 6 %, other 29 %.
By type Final consumption expenditure 76 % (of which, government 10 %), gross fixed capital formation 17 %, exports of goods and services 27 %, less imports of goods and services −20 %

Production indices (1970 = 100)	1960	1970	1976	Growth %pa 1960–70	Growth %pa 1970–76
Agricultural	90	100	95	1.1	−0.9
Manufacturing	na	100.0	193.7	na	11.6

Main products (000 t)

Agriculture					
Maize	1 143*	1 046*	1 050*	−0.9*	0.1*
Millet	2 605	3 284	3 200*	2.3	−0.4*,
Sorghum	2 557	4 080*	3 680*	4.8*	−1.7*
Cassava	6 900	9 084*	10 800*	2.8*	2.9*
Taro	1 497[a]	1 680*	1 850*	1.7*[b]	1.6*
Yams	12 400*	10 400*	15 500*	−1.7*	6.9*
Cowpeas	688	820*	870*	1.8*	1.0*
Groundnuts	1 150	1 581	700*	3.2	−12.7*
Sesameseed	57[a]	60*	70*	0.7*[b]	2.6*
Chillies and green peppers	409[a]	484*	570*	2.4*[b]	2.8*
Cocoa	189	305	210*	4.9	−6.0*
Tobacco	14	11	15*	−2.4	5.3*
Palm kernels	430	315	290*	−3.1*	−1.4*
Palm oil	552	488*	510*	−1.2*	0.7*
Cotton	117[a]	92	57	−3.4[b]	−7.7
Rubber	59	61*	85*	0.3*	5.7*
Milk	369	313*	316*	−1.6*	0.2*
Eggs	76**[a]	101*	113*	4.1*[b]	1.9*
Beef and veal	126	172*	169*	3.2*	−0.3*
Sheep and goat meat	89	100*	103*	1.2*	0.5*
Fish catch	na	543	507[c]	na	−1.4[d]
Timber (000 m³)	43 000**	56 860*	65 460*[c]	2.8**	2.8*[d]
Energy					
Total energy (000 tce)	1 690	70 660	131 070[c]	45.3	13.2[d]
Coal	600	59	237[c]	−20.7	32.1[d]
Crude oil	850	54 203	102 655	51.5	11.2
Petroleum products	—	837	2 430*[c]	na	23.8*[d]
Natural gas (mn m³)	—	111	402[c]	na	29.4[d]
Electricity (mn kW h)	528	1 550	3 211[c]	11.4	15.7[d]
Mining					
Tin conc (Sn content)	na	8.0	3.7	na	−11.9
Limestone	na	688	1 801[e]	na	27.2[f]
Manufacturing					
Beer (000 hl)	219	1 052	2 968[c]	17.0	23.1[d]
Cigarettes (mn units)	2 871	8 502	10 170[c]	11.5	3.7[d]
Cotton fabrics (mn m²)	na	275	276[e]	na	0.1[f]
Tyres (000 units)	na	266	307[e]	na	3.6[f]
Cement	167	596	1 383[c]	13.6	18.3[d]
Radio sets (000 units)	na	215	103[c]	na	−13.7[d]
Television sets (000 units)	—	6	7[c]	na	3.1[d]
Commercial vehicles (000 units)[g]	na	7.1	12.2[c]	na	11.4[d]

[a]1961–65 [b]1961–65 to 1970 [c]1975 [d]1970–75 [e]1974 [f]1970–74 [g]Assembly only

Nigeria

Transport traffic	1960	1970	1976	Growth %pa	
Passenger-kilometres (mn)				1960–70	1970–76
Rail[a]	576	728	785[d]	2.4	1.5[e]
Air	38	214	470	18.9	14.0
Cargo: tonne-kilometres (mn)					
Rail[a]	2 044	1 616	972[d]	−2.3	−5.0[e]
Air	1.1	6.4	8.4	19.3	4.6
Sea: tonnes (mn)					
Goods loaded	4.6[b]	53.6	87.6[f]	31.4[c]	27.9[g]
Goods unloaded	3.1[b]	3.7	4.7[f]	2.0[c]	13.1[g]

[a]Years ending March 31st [b]1961 [c]1961–70 [d]1975 [e]1970–75 [f]1972 [g]1970–72

Finance and trade ·

Price index	1960	1970	1976	Growth %pa	
(1970 = 100)				1960–70	1970–76
Consumer prices	66.4	100.0	231.8	4.2	15.0
Money stock					
(end-year, N mn)	240.8	642.5	3 752.6	10.3	34.2

Budget (Federal, 1976/77; year ending March 31st)
Revenue: N 5 756 mn = $ 9 158 mn = £ 5 276 mn
Expenditure (total): N 9 792 mn = $ 15 579 mn = £ 8 975 mn

Balance of payments ($ mn)	1972	1973	1974	1975	1976
Balance of goods (fob)	+858	+1 943	+7 292	+3 004	+2 882
Balance of services	−1 179	−1 898	−2 296	−2 835	−3 090
Balance of transfers	−22	−54	−98	−126	−142
Current balance	*−342*	*−8*	*+4 897*	*+42*	*−350*
Long-term capital flow	+367	+306	+170	+209	+27
Reserves and debt (end-year, $ mn)					
International reserves	376	583	5 626	5 803	5 203
External public debt	1 103	1 649	1 695	1 598	1 420

External trade (1976) Imports: N 5 139.7 mn = $ 8 203 mn = £ 5 818 mn
Exports: N 6 622.4 mn = $ 10 569 mn = £ 7 497 mn

Main imports (1975)	% of total	Main exports (1976)	% of total
Machinery	21	Crude oil	94
Motor vehicles	15	Cocoa	3
Iron and steel	10		
Chemicals	9		
Food	8		
Textile yarns and fabrics	6		

Main sources (1975)		Main destinations (1975)	
United Kingdom	23	United States	29
West Germany	15	United Kingdom	14
United States	11	Netherlands	11
Japan	10	France	11
France	8	Neth. Antilles	7

Special focus

Main exports (% of total exports)

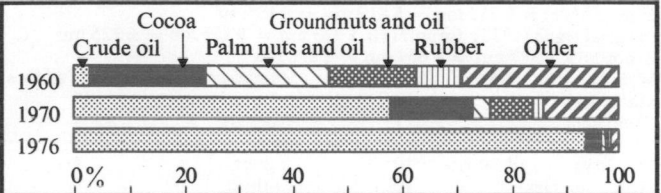

| | Crude oil | Cocoa | Palm nuts and oil | Groundnuts and oil | Rubber | Other |

Reunion

Department of Reunion
Départment de la Réunion

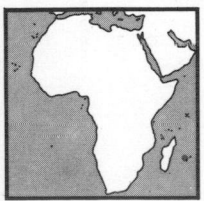

Location Western Indian Ocean
An island about 640 km (400 mi) east of the island of Madagascar
Land Area 2 512 km² = 970 mi²
Climate Sub-tropical
Weather at Hell-Bourg, 936 m altitude
Temperature: hottest month February
15–24 °C, coldest August 8–19 °C
Rainfall (av monthly): driest month September
51 mm, wettest January 569 mm

Time 4 hours ahead of GMT
Measures Metric system

Monetary unit French (metropolitan) franc (Fr) = 100 centimes
(before January 1, 1975 the CFA franc, CFA Fr 50 = Fr 1, was in use)
Rate of exchange (1976 av): free Fr 4.780 = $ 1, Fr 8.633 = £1

Summary

Political French overseas department. A French possession since 1643, it became a department in 1946. Member of franc zone
Economic The economy is based on sugar, accounting for about 80% of exports, and on rum; essential oils are also produced. There is substantial French aid, especially for development of tourism, housing and manufacturing industry

People, resources and equipment

Population 1960 340 000*, 1970 447 000*, 1976 511 000*
Growth: 1960–70 2.8* %pa, 1970–76 2.2* %pa
Density (1976): 204* people per km²
Vital statistics (rate per 1 000 people, 1973): births 28.1, deaths 7.1
Cities (population in 000, 1974)
Saint-Denis (capital) 104 Saint-Pierre 45* Saint-Louis 30*
Saint-Paul 50* Le Tampon 35*
Race Mixed, mainly of European, African, Indian and Chinese origin
Language French
Religion Mainly Roman Catholic
Education (1974/75) Pupils 175 480, teachers 6 450*
Labour force (1976) 144 000*; in agriculture 46 000* (32* %)
Personnel (1973) Physicians: 237, 1 per 1 983 people
Standard of living
National income per person (1976): Fr 9 300** = $ 1 900** = £ 1 100**
Consumption per person (1975): energy 434 kg coal equivalent, electricity (production, 1976) 430* kW h, newsprint 1.0 kg
Newspapers (1974): number 1; circulation 23 000, 46 per 1 000 people
Telephones (Dec 1975): 28 000, 55 per 1 000 people
Livestock (000, 1976) Cattle 21*, goats 40*, pigs 85*, chickens 1 918*
Electrical capacity (1975) 55* megawatts
Hospital beds (1972) 3 886, 1 per 121 people
Roads (1975) 2 052 km = 1 275 mi, density 0.82 km per km²
There are no railways
Port (goods traffic, 000 tonnes, 1973)
Pointe des Galets: loaded 316, unloaded 735
Airport Saint Denis-Gillot

Durable equipment	000	no per	
(Dec 1974)		1 000 people	
Radio sets	91	184	no per
Television sets	30	61	km of road
Passenger cars	45	91	22
Commercial vehicles	18	37	9

Production, finance and trade

Gross domestic product 1970: Fr 1 920 mn = $ 346 mn = £ 144 mn
1976 est: Fr 4 800** mn = $ 1 000** mn = £ 560** mn
Agricultural production index (1970 = 100) 1976 121; growth 1970–76 3.2 %pa
Main products (000 t, 1976) *Agriculture* Maize 14*, sweet potatoes 5*, cassava 4*, tomatoes 3*, sugar 243*, bananas 2*, pigmeat 4, poultrymeat 3*, timber (000 m³) 30*
Other Electricity (mn kWh) 220*
Transport traffic (000 t, 1974) *Sea* Goods loaded 297*, unloaded 814*
Consumer price index (Saint-Denis, 1970 = 100) 1976 180.0; growth 1970–76 10.3 %pa
Budget (1975) Balanced at Fr 2 471 mn = $ 576 mn = £ 259 mn
External trade (1976) Imports: Fr 2 152 mn = $ 450 mn = £ 249 mn
Exports: Fr 450 mn = $ 94 mn = £52 mn

Main imports (1975)	% of total	Main exports (1975)	% of total
Food	26	Sugar	82
(of which, cereals 8)		Rum	7
Chemicals	10	Essential oils	5
Motor vehicles	9		
Machinery, non-electric	6		
Petroleum products	5		
Electrical machinery	5		
Clothing	5		

Main sources (1975)		Main destinations (1975)	
France	63	France	94
Madagascar	8	United States	2
Italy	6	Mauritius	1
South Africa	5		

Rhodesia

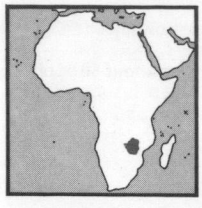

Location South central Africa
Mozambique is to the east, between Rhodesia and the Indian Ocean, Zambia to the north, Botswana to the west, and South Africa to the south. Land-locked
Land Area 390 580 km² = 150 800 mi²
Climate Sub-tropical
Weather at Salisbury, 1 472 *m altitude*
Temperature: hottest month November 16–27 °C, coldest June, July 7–21 °C
Rainfall (av monthly): driest month July 1 mm, wettest January 196 mm
Time 2 hours ahead of GMT
Measures UK (imperial) system; also:
US short ton = 2 000 pounds = 907.2 kilograms
Monetary unit Rhodesian dollar (R $) = 100 cents
Rate of exchange (1976 av): free R 1.34 = R $ 1, $ 1.54 = R $ 1, R $ 1.17 = £ 1

Summary

Political Officially a UK colony, which declared itself independent November 11, 1965 and a republic March 2, 1970; formerly, as Southern Rhodesia from 1953–63, a part of the Federation of Rhodesia and Nyasaland. A state of emergency has existed since the unilateral declaration of independence. The African name for this country is Zimbabwe.
Economic A mixed economy, with manufacturing accounting for one-quarter of gross domestic product and agriculture one-sixth. Mining is also important. The main agricultural export product, tobacco, was affected by UN trade sanctions, but the widespread export of minerals has helped obtain foreign currency. Chrome and nickel have been main exports since independence.

People, resources and equipment

Population 1960 3.84*mn, 1970 5.31*mn, 1976 6.53*mn
Growth: 1960–70 3.3* %pa, 1970–76 3.5* %pa
Density (1976): 17* people per km²
Vital statistics (rate per 1 000 people, 1970–75): births 47.9*, deaths 14.4*
Cities (population in 000, end-year, 1976)
Salisbury (capital) 566 Gwelo 64 Que Que 50 Wankie 28
Bulawayo 340 Umtali 61 Gatooma 33 Marandellas 20
Race (1976) African 95 %, European 4.2 %
Language English; also local languages
Usage (1969): Shona 67 %, Nguni 15 %, English 12 %
Religion (1976) Animist 80** %, Christian 20** %
Education (1975) Pupils 947 861, teachers 26 556
Labour force (1976) 2 197 000*; in agriculture 1 336 000* (61* %)
Personnel (1973) Physicians: 1 035, 1 per 5 700 people
Standard of living
National income per person (1976): R $ 290** = $ 450** = £250**
Consumption per person (1975): energy 764 kg coal equivalent, electricity (1976) 1 032 kW h, newsprint 2.1 kg, steel 45 kg
Newspapers (1974): number 2; circulation 100 000, 16 per 1 000 people
Telephones (Dec 1975) 183 000, 28 per 1 000 people
Livestock (000, 1976) Cattle 6 100*, sheep 770*, goats 2 050*, pigs 200*, chickens 8 404*
Mineral reserves Coal (1960) 6 613 mn tonnes
Petroleum refinery capacity (1975) 1.0* mn tonnes
Electrical capacity (1975) 1 192 megawatts, of which, hydro 705 megawatts
Hospital beds (1974) 19 285, 1 per 316 people
Roads (1975) 78 930 km = 49 040 mi, density 0.20 km per km²
Railways (1975) 3 367 km = 2 092 mi, density 0.009 km per km²
Airports Salisbury, Bulawayo and 6 other airports with scheduled flights

Durable equipment (at end-year)	000	no per 1 000 people	
Radio sets (1974)	225	36	no per km of road
Television sets (1972)	57	10	km of road
Passenger cars (1974)	180*	29*	2.3*
Commercial vehicles (1974)	70*	11*	0.9*

Production

Gross domestic product (1976) R $ 2 108 mn = $ 3 247 mn = £ 1 802 mn
Growth in real terms: 1967–70 6.5 %pa, 1970–76 4.6 %pa

Structure of gross domestic product (1975) *By origin* Agriculture 16 %, mining and quarrying 6 %, manufacturing 23 %, other 56 %

Production indices (1970 = 100)	1960	1970	1976	Growth %pa 1960–70	1970–76
Agricultural	81	100	131	2.1	4.6
Industrial	73.0ᵇ	100	130.1	6.5ᶜ	4.5

Main products (000 t)

Agriculture					
Wheat	—	40*	90*	na	14.5*
Maize	498	700*	1 400*	3.5*	12.3*
Millet	109	220*	220*	7.3*	0.0*
Groundnuts	74	132*	120*	6.0*	−1.6*
Sugar, raw value	31	145*	240*	16.7*	8.8*
Tea	1.1	2.3*	3.0*	7.5*	4.5*
Tobacco	94	62*	85*	−4.1*	5.4*
Cotton	—	43*	39*	na	−1.6*
Milk	210	237*	255*	1.2*	1.2*
Beef and veal	87	93*	142*	0.7*	7.3*
Timber (000 m³)	4 200**	5 532*	5 900*ᵈ	2.8**	1.3*ᵉ
Energy					
Total energy (000 tce)	3 680*	3 830	4 150ᵈ	0.4*	1.6ᵉ
Coal	3 559	3 171	3 500*ᵈ	−1.1	2.0*ᵉ
Electricity (mn kW h)	2 388	6 410	6 738	10.4	0.8
of which, hydro (mn kW h)	1 046	5 247	5 321ᵈ	17.5	0.3ᵉ
Mining					
Iron ore (Fe content)	98	325*	384*ᵈ	12.7*	3.4*ᵉ
Chromium ore (oxide content)	291	181	295ᵈ	−4.6	10.2ᵉ
Copper ore (Cu content)	14	23	30*	5.1	4.5*
Magnesite	7	20*	20*ᵈ	10.6*	0.0*ᵉ
Nickel ore (Ni content)	—	11*	10*ᵈ	na	−1.9*ᵉ
Gold (000 kg)	18	16*	17*	−1.2*	1.5*
Asbestos	122	160**	165**	2.7**	0.5*ᵉ
Phosphate rock	3	95	130ᵈ	41.3	6.5ᵉ
Manufacturing					
Cigarettes (mn units)	na	3 625	4 000ᶠ	na	2.5ᵍ
Fertilisers, nitrogenousᵃ	—	23*	65*	na	18.9*
Fertilisers, phosphateᵃ	na	28*	45*	na	8.2*
Cement	401	475	677ᵈ	1.7	7.3*
Coke	146	245*	255*ᵈ	5.3*	0.8*ᵉ
Pig-iron	86	250*	310*ᵈ	11.3*	4.4*ᵉ
Crude steel	86	150*	350*ᵈ	5.7*	18.5*ᵉ

ᵃYears ending June 30th ᵇ1965 ᶜ1965–70 ᵈ1975 ᵉ1970–75 ᶠ1974 ᵍ1970–74
Transport traffic (1974/75) Rail (including Botswana traffic) Cargo 6 141 mn t-km

Finance and trade

Consumer price index (1970 = 100) 1965 91.2, 1976 143.0; growth 1965–70 1.9 %pa, 1970–76 6.1 %pa
Budget (1976) Revenue: R $ 492 mn = $ 758 mn = £ 420 mn
Expenditure: R $ 532 mn = $ 819 mn = £ 455mn
External trade (1976) Imports: R $ 323 mn = $ 553 mn = £ 225 mn
Exports: R $ 387 mn = $ 664 mn = £ 270 mn

Main imports (1965)ᵃ	% of total	*Main exports* (1965)ᵃ	% of total
Machinery	18	Tobacco	33
Chemicals	11	Asbestos	8
(of which, fertilisers 4)		Meat	5
Textile yarns and fabrics	10	Copper	4
Motor vehicles	10	Clothing	4
Food	8	Chemicals	3
Iron and steel	4	Chrome ore	3
Main sources (1965)ᵃ		*Main destinations* (1965)ᵃ	
United Kingdom	30	Zambia	25
South Africa	23	United Kingdom	22
United States	7	South Africa	10
Japan	6	West Germany	9
West Germany	4	Malawi	5
Zambia	4	Japan	5

ᵃDetailed figures are not available since declaration of independence

Special focus

Migration of Europeans (000)

	Inward	Outward	Net inflow		Inward	Outward	Net inflow
1965	11.1	8.8	2.3	1971	14.7	5.3	9.4
1966	6.4	8.5	−2.1	1972	14.0	5.1	8.8
1967	9.6	7.6	2.0	1973	9.4	7.8	1.7
1968	11.9	5.6	6.2	1974	9.6	9.1	0.6
1969	10.9	5.9	5.0	1975	12.4	10.5	1.9
1970	12.2	5.9	6.3	1976	7.8	14.9	−7.1

Rwanda

Rwanda Republic
République Rwandaise
Republika ylu Rwanda

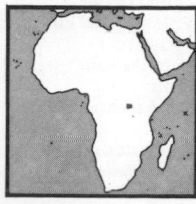

Location Central Africa
Uganda is to the north, Tanzania to the east, Burundi to the south, and Zaire to the west. Land-locked
Land Area 26 338 km² = 10 169 mi²
Climate Tropical, moderated by altitude
Weather at Kigali
Temperature: average annual 19°C
Rainfall (av monthly): 85 mm
Time 2 hours ahead of GMT

Measures Metric system
Monetary unit Rwanda franc (Rw Fr) = 100 centimes
Rate of exchange (1976 av): par Rw Fr 92.84 = $ 1,
free Rw Fr 167.7 = £ 1

Summary

Political Republic with military government, which became independent July, 1962; a customs union was formed with Burundi until September 30, 1964. Before 1962 Rwanda-Urundi was administered by Belgium as a trust territory. Member of UN, OAU, Ocam and an EEC ACP state
Economic Mainly an agricultural economy, with coffee accounting for some three-quarters of exports; there are also minerals, notably tin and tungsten. Plans include diversification of agriculture; industrial activity is limited and mainly involves processing of agricultural products

People, resources and equipment

Population 1960 2.60**mn, 1970 3.68* mn, 1976 4.29* mn
Population excludes 50 000* Tutsi refugees living in Burundi (1976)
Growth: 1960–70 3.5** %pa, 1970–76 2.6* %pa
Density (1976): 163* people per km²
Vital statistics (rate per 1 000 people ,1970–75): births 50.0*, deaths 23.6*
Cities (population in 000, 1975) Kigali (capital) 55*, Butare 20*, Nyanza 20*
Race (1970) Hutu 90 %, Tutsi 9 %, Twa 1 %
Language Kinyarwanda and French
Religion (1976) Animist 45** %, Roman Catholic 45** %, Protestant 9** %, Moslem 1** %
Education (1974/75) Pupils 398 969, teachers 8 900*
Labour force (1976) 2 315 000*; in agriculture 2 110 000* (91* %)
Personnel (1974) Physicians: 77, 1 per 53 500 people
Standard of living
National income per person (1976): Rw Fr 8 000** = $ 85** = £ 48**
Consumption per person (1975): energy 14 kg coal equivalent, electricity 33* kW h, steel 1* kg
Newspapers (1974): number 1; circulation 200, 0.05 per 1 000 people
Telephones (Dec 1975): 3 400, 1 per 1 000 people
Livestock (000, July 1976) Cattle 717*, sheep 252*, goats 570*, pigs 75*
Mineral reserves There are natural gas reserves
Electrical capacity (1975) 35*megawatts, of which, hydro 34* megawatts
Hospital beds (1974) 6 142, 1 per 671 people
Roads (1974) 6 500 km = 4 040 mi, density 0.25 km per km²
There are no railways
Inland waterway Lake Kivu
Airports Kigali, Kamembe
Durable equipment (at end-year)
Radio sets (1974): 133 000, 32 per 1 000 people
Passenger cars (1975): 6 500, 1.5 per 1 000 people, 1.0 per km of road
Commercial vehicles (1975): 4 800, 1.1 per 1 000 people, 0.7 per km of road

Production

Gross domestic product 1974: Rw Fr 28 680 mn = $ 309 mn = £ 132 mn
1976 est: Rw Fr 38 000** mn = $ 410** mn = £ 230** mn
Structure of gross domestic product (1974) *By origin* Agriculture 59 %, mining, quarrying and utilities 2 %, manufacturing 4 %, other 35 %
By type Final consumption expenditure 99 % (of which, government 12 %), gross fixed capital formation 10 %, exports of goods and services 12 %, less imports of goods and services −21 %

Production index (1970 = 100)	1960	1970	1976	Growth %pa 1960–70	Growth %pa 1970–76
Agricultural	76	100	120	2.8	3.0
Main products (000 t)					
Agriculture					
Maize	55	64	60*	1.5	−1.1*
Sorghum	137	156	140*	1.3	−1.8*
Potatoes	90	126	157*	3.4	3.7*
Sweet potatoes	450*	413	675*	−0.9*	8.5*
Cassava	125	345	416*	10.7	3.1*
Dry beans	98	144	162*	3.9	2.0*
Dry peas	60**	67	62*	1.1**	−1.3*
Pumpkins	45**	50*	60*	1.1**	3.1*
Bananas and plantains	800**	1 651	1 784*	7.5**	1.3*
Coffee	11	14	14*	2.6	0.0*
Tea	0.2*	1.2	4.0*	19.6*	22.2*
Pyrethrum	0.4*	0.6	0.2*g	3.4*	−16.7*h
Cattle hides	0.8a	1.5*	1.9*	10.0*b	4.0*
Milk	na	20*	22*	na	1.6*
Timber (000 m³)	2 700**	3 709*	3 930*e	3.2**	1.2*f
Other					
Electricity (mn kW h)	48c	81	140*e	14.0d	11.6*f
Tungsten conc (oxide content)	0.27	0.41	0.50e	4.3	4.0*f
Tin conc (Sn content)	1.30	1.32*	1.25*e	0.2*	−1.1f
Beer (000 hl)	–	142	266e	na	13.4f
Radio sets (000 units)	–	3	12e	na	32.0f

a1961–65 b1961–65 to 1970 c1966 d1966–70 e1975 f1970–75
g1974 h1970–74

Transport traffic *Road* (public transport only: 1974) Passenger-km: 605 mn
Waterway (Lake Kivu, 1967) Freight 70 000* t

Finance and trade

Consumer price index (1970 = 100) 1975 193.1; growth 1970–75 14.1 %pa
Money stock (end-year, Rw Fr mn) 1976 7 215; growth 1970–76 22.3 %pa
Budget (1975) Revenue: Rw Fr 4 374 mn = $ 47 mn = £ 21 mn
Expenditure: Rw Fr 4 388 mn = $ 47 mn = £ 21 mn

Balance of payments ($ mn)	1972	1973	1974	1975	1976
Balance of goods (fob)	−6	+18	−2	−24	+9
Balance of services	−19	−25	−28	−44	−51
Balance of transfers	+20	+29	+32	+58	+58
Current balance	−5	+22	+1	−10	+16
Long-term capital flow	+1	+3	+9	+17	+23
Reserves and debt (end-year, mn)					
International reserves	6	16	13	29	64
External public debt	38	46	63	82	na

External trade (1976) Imports: Rw Fr 9 607 mn = $103 mn = £ 57 mn
Exports: Rw Fr 7 535 mn = $ 81 mn = £ 45 mn

Main imports (1974)	% of total	Main exports (1975)	% of total
Textiles and clothing	18	Coffee	63b
Transport equipment	12	Cassiterite	13
(of which,		Tea	9
motor vehicles 8)		Tungsten	5
Machinery	11	Quinine	3
Food	11	Pyrethrum extract	2
(of which, cereals 4)		Hides and skins	2
Petroleum products	8		
Chemicals	8		
Iron and steel	6		
Tobacco	2		

Main sources (1975)		Main destinationsa (1975)	
Belgium-Luxembourg	16	Belgium-Luxembourg	17
Kenya	10	United Kingdom	4
West Germany	10	United States	2
France	7	West Germany	1
Japan	7		
Iran	6		
United States	6		
Italy	6		
United Kingdom	4		
China	4		

aThe destinations of 68 % of exports, sent via Mombasa, are not distinguished
b77 % in 1976

Sahara, Western

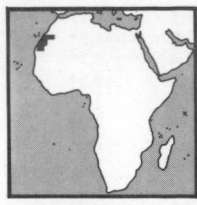

Location North-west Africa
With a coastline on the Atlantic Ocean,
Morocco is to the north, Algeria has a small
common border to the north-east, and
Mauritania is to the east and south
Land Area 266 000 km² = 103 000 mi²
Climate Hot and dry
Weather at Dakhlah, 11 m altitude
Temperature: hottest month September
19–27 °C, coldest January 13–22 °C
Rainfall (av monthly): driest month June 0 mm, wettest September 36 mm
Time GMT
Measures Metric system
Monetary unit Moroccan dirham and Mauritanian ouguiya

Summary

Political Territory administered by Morocco and Mauritania under
an agreement made November 14, 1975 with Spain; formerly a Spanish
province, Spain handed over full administration on February 26, 1976.
There is a guerrilla movement fighting for independence as the Sahraoui
Republic
Economic Phosphates are the most important feature, production
having begun in 1972; the mainly nomadic population subsists by
raising livestock. Guerrilla activity reduced the production of phosphates
from 1976

People, resources and equipment

Population[a] 1960 25 000**, 1970 76 000**, 1976 120 000**
Growth: 1960–70 11.8** %pa, 1970–76 8.0** %pa
Density ((1976): 0.5** people per km²
Vital statistics (rate per 1 000 people, 1972): births 20.9*, deaths 4.5*
[a] Estimates are very approximate due to the number of nomads in the population;
excludes refugees outside of the territory
Cities (population in 000, 1974) El Aaiún 28, Smara 7*, Dakhlah[a] 5*
[a] Called Villa Cisneros before January 12, 1976
Race (1970) Arab 78 %, European 22 %
Language (1970) Arabic and Hassania 78 %, Spanish 22 %
Religion (1970) Moslem 78 %, Roman Catholic 22 %
Education (1972/73) Pupils 5 270, teachers 260*
Personnel (1971) Physicians: 53, 1 per 1 570 people
National income per person (1976) Dh 4 000*** = $ 900*** = £ 500 ***
Consumption per person (1975) Electricity 570* kW h
Livestock (000, 1976) Goats 152*, sheep 17*, camels 83*
Mineral reserves (1974) Phosphates 1 700** mn tonnes
Electrical capacity (1975) 52* megawatts
Hospital beds (1972) 262, 1 per 344 people
Roads (1974) 6 500* km = 4 000* mi, density 0.02* km per km²
There are no railways
Port Dakhlah
Airports El Aaiún, Dakhlah

Durable equipment (Dec 1974)	000	no per 1 000 people	
Radio sets	16	144	no per km of road
Television sets	2.0	18	
Passenger cars	5.0	45	0.8*
Commercial vehicles	0.5	5	0.1*

Production, finance and trade

Gross domestic product
1976 est: Dh 500*** mn = $ 110*** mn = £ 60*** mn
Main products (000 t, 1975) Barley 1*, goat milk 41*, fish catch 4*,
electricity (mn kW h) 65*, phosphates 3 300*
Transport traffic (1974) *Sea* Goods loaded 624 000 t,
unloaded 387 000 t
Budget (1973) Balanced at Pa 1 553 mn = $ 26 mn = £ 11 mn
External trade (1976)
Imports: Dh 450*** mn = $ 100*** mn = £ 60*** mn
Exports: Dh 450*** mn = $ 100*** mn = £ 60*** mn
Main imports (% of total, 1973) Food 54, manufactures 44
Main export: Phosphates

Saint Helena

Colony of Saint Helena and dependencies

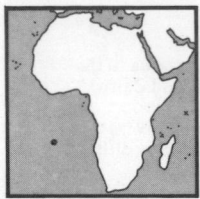

Location South Atlantic
The island of St Helena is 2 000 km west
of Angola with Ascension island 1 100 km
to the north-west and the island of
Tristan da Cunha 2 400 km to the south-west
Land Area 413 km² = 160 mi²
of which, St Helena 122 km², Ascension 88 km²,
Tristan da Cunha 98 km²
Climate Mild, temperate in Tristan da Cunha
Weather at Jamestown, 12 m altitude
Temperature: hottest month March 22–28 °C, coldest July–Sept 17–22 °C
Rainfall (av monthly): driest month Nov 0 mm, wettest March 20 mm
Time GMT
Measures UK (imperial) system, changing to the metric system in
line with the United Kingdom
Monetary unit UK pound (£) = 100 new pence
Rate of exchange (1976 av): free $ 1.806 = £1

Summary

Political St Helena is a UK colony, and Ascension, Tristan da Cunha,
Gough, Inaccessible and Nightingale islands are dependencies. Ascension
Island is also sometimes known as Wideawake Island. Tristan da Cunha
was evacuated from 1961–63 due to a volcanic eruption
Economic There is no industry nor are there minerals; agricultural
production is very limited. There are telecommunications and
meteorological stations, and a US air and missile base

People, resources and equipment

Population 1960 5 300**, 1970 6 100**, 1976 6 600**
Growth: 1960–70 1.4** %pa, 1970–76 1.3** %pa
Density (1976): 16** people per km²
Vital statistics (rate per 1 000 people, St Helena only, 1975)
births 25.0, deaths 8.1
Regions (population in 000, 1975) St Helena 5.1, Ascension 1.2*,
Tristan da Cunha 0.3*; Gough, Inaccessible and Nightingale islands:
uninhabited
Town (population in 000, 1976) Jamestown (capital) 1.6*
Race (1976) St Helenians 85* %
Language English
Religion (1966) Christian 99.6 % (mainly Anglican Communion)
Education (1974) Pupils 1 250*, teachers 71*
Labour force (employees, 1975) 1 667
Personnel (1973) Physicians: 3, 1 per 2 160 people
National income per person (1976) £ 300*** = $ 550***
Consumption per person (1975) Electricity 150* kW h
Livestock (000, Dec 1975) Cattle 1, sheep 3, goats 2, poultry 11
Electrical capacity (1972) 0.45 megawatt
Hospital beds (1975) 54, 1 per 121 people
Roads (1975) 107 km = 67 mi, density 0.26 km per km²
Port Jamestown
Airport There is an airfield on Ascension
Durable equipment (Dec 1974) Radio sets: 750, 115 per 1 000 people
Motor vehicles: 791, 121 per 1 000 people, 7.4 per km of road

Production, finance and trade

Gross domestic product 1976 est: £2.0*** mn = $ 3.6*** mn
Main products (1975) Potatoes, sweet potatoes, vegetables, fish 162 t,
electricity 1* mn kW h
Transport traffic *Sea* (1974) Goods unloaded 11 000 t
Budget (1975/76: year ending March 31st)
Revenue: £ 1 482 000 = $ 3 148 000
Expenditure: £ 1 544 000 = $ 3 279 000
External trade (1975) Imports: £ 1.1 mn = $ 2.5 mn. Exports: nil

Main imports (1974)	% of total	*Main exports* (1974)
Manufactures	45*	None (in 1968 there
(of which, motor vehicles 6)		were exports of
Food	30*	wool and fish)
Petroleum products	16	
Beer, wines and spirits	7	
Main sources (1968)		*Main destinations* (1974)
United Kingdom	61	None (in 1968: United
South Africa	28	Kingdom 78 %,
United States	4	South Africa 22 %)

Sao Tome and Principe

Democratic Republic of São Tomé and Príncipe

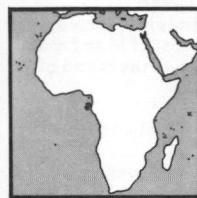

Location West Africa
The archipelago, consisting of the
main islands of Sao Tome and of
Principe, lies in the Gulf of Guinea
200 km west of the coast of Gabon
Land Area 964 km² = 372 mi²
Climate Warm and humid
Weather at Sao Tome, 5 m altitude
Temperature: hottest month March 23–31 °C,
coldest July, Aug 21–28 °C
Rainfall (av monthly): driest month July 1 mm, wettest March 150 mm
Time GMT
Measures Metric system
Monetary unit STP escudo (STP Esc) = 100 centavos
Rate of exchange (1976 av): par STP Esc 1 = Esc 1,
free STP Esc 30.22 = $ 1, STP Esc 54.59 = £ 1

Summary

Political Republic which became independent July 12, 1975;
formerly a Portuguese overseas province. Member of UN, OAU
and an EEC ACP State
Economic Agricultural economy, with cocoa the major export crop.
There is virtually no industry. Economic plans include the diversification
of agriculture to prevent high imports of food, and development
of fishing and tourism

People, resources and equipment

Population 1960 64 000, 1970 73 600 1976 80 000*
Growth: 1960–70 1.5* %pa, 1970–76 1.4* %pa
Density (1976): 83* people per km²
Vital statistics (rate per 1 000 people, 1972): births 45.0, deaths 11.2
City (population in 000, 1976) Sao Tome (capital) 20*
Race Mainly African
Language Portuguese and local languages
Religion Mainly Roman Catholic
Education (1972/73) Pupils 14 900*, teachers 470*
Personnel (1973) Physicians: 12, 1 per 6 700 people
Standard of living National income per person (1976):
STP Esc 14 000** = $ 450** = £ 250**
Consumption per person (1975): energy 102 kg coal equivalent,
electricity 100* kW h
Telephones (Dec 1975): 730, 9 per 1 000 people
Livestock (000, 1976) Cattle 2*, sheep 1*, goats 1*, pigs 3*
Electrical capacity (1975) 3* megawatts
Hospital beds (1969) 1 997, 1 per 36 people
Roads (1973) 288 km = 179 mi, density 0.30 km per km²
Port Sao Tome
Airports Sao Tome, Principe
Durable equipment (at end-year)
Radio sets (1974): 7 500, 95 per 1 000 people
Passenger cars (1973): 1 600, 20 per 1 000 people, 5.6 per km of road
Commercial vehicles (1973): 400, 5 per 1 000 people, 1.4 per km of road

Production, finance and trade

Gross domestic product
1976 est: STP Esc 1 200** mn = $ 40** mn = £ 22** mn
Main products (000 t, 1976) Cassava 3*, bananas 1*, coconuts 43*
copra 5*, palm kernels 2.4*, palm oil 1.0*, cocoa 8*,
fish catch 0.8*, timber (000 m³) 5*, electricity (mn kW h, 1975) 8*
Transport traffic *Sea* (1973) Goods loaded 26 000 t, unloaded 42 000 t
Budget (1974) Balanced at STP Esc 150 mn = $ 5.9 mn = £ 2.5 mn
External trade (1975) Imports: STP Esc 288 mn = $ 11.3 mn = £ 5.1 mn
Exports: STP Esc 180 mn = $ 7 mn = £ 3.2 mn

Main imports (1973)	% of total	Main exports (1973)	% of total
Food	39	Cocoa	87
Machinery	11	Copra	8
Textiles	11	Coconut oil and seeds	3
Chemicals	10		
Main sources (1975)		*Main destinations* (1975)	
Portugal	61	Netherlands	52
Angola	13	Portugal	33
Netherlands	4	West Germany	8
France	3	Belgium-Luxembourg	2

Senegal

Republic of Senegal
République du Sénégal

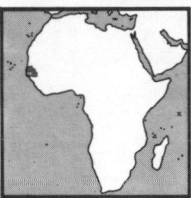

Location West Africa
With a coastline on the Atlantic Ocean,
Mauritania is to the north, Mali to the east,
and Guinea and Guinea-Bissau to the south;
Gambia forms an enclave within Senegal
Land Area 196 192 km² = 75 750 mi²
Climate Tropical
Weather at Dakar, 40 m altitude
Temperature: hottest months Sept, Oct
24–32 °C, coldest January 18–26 °C
Rainfall (av monthly): driest months April, May 1 mm,
wettest Aug 254 mm
Time GMT
Measures Metric system
Monetary unit CFA franc (CFA Fr) = 100 centimes
Rate of exchange (1976 av): par CFA Fr 50 = Fr 1,
free CFA Fr 239.0 = $ 1, CFA Fr 431.6 = £ 1

Summary

Political Republic which became independent August 20, 1960;
formerly a French territory, and from January 1959 until independence
linked with French Soudan (Mali) in the Federation of Mali. Member of
UN, OAU, Ecowas, CEAO, Ocam, franc zone and an EEC ACP state
Economic Mainly an agricultural economy, with groundnuts and
products the main export, accounting for about one-third of the total.
Phosphates are important and make up about one-quarter of exports;
there are iron ore deposits. There is a wide spread of light industry, and
Dakar is a large international port; the fishing industry has been
growing at a high rate and tourism is being developed

People, resources and equipment

Population 1960 3.11* mn, 1970 3.93** mn, 1976 5.09** mn
Growth: 1960–70 2.4** %pa, 1970–76 4.4** %pa
Density (1976): 26** people per km²
Vital statistics (rate per 1 000 people 1970–75): births 47.6*, deaths 23.9*
Cities (population in 000, 1976)
Dakar (capital) 800* Kaolack 106* Ziguinchor 73*
Thiès 117* Saint Louis 88* Diourbel 51*
Race (1971) African 99 % (Wolof 34 %, Serer 18 %, Tukulor 13* %,
Fulani 7* %, Dyola 7 %, Malinke and Bambara 6 %), European 1 %
Language French; local languages, especially Wolof and Tukulor,
are also used
Religion (1976) Moslem 80** %, Christian 10** %, Animist 10** %
Education Pupils (1972/73) 339 000*, teachers (1971/72) 8 800*
Labour force (1976) 1 923 000*; in agriculture 1 472 000* (77* %)
Personnel Scientists and engineers engaged in research (1972) : 392
Physicians (1974): 281, 1 per 16 600 people
Standard of living National income per person (1976):
CFA Fr 64 300* = $ 270* = £ 150*
Consumption per person (1975): energy 195 kg coal equivalent,
electricity 80* kW h, newsprint 0.2 kg, steel 14 kg
Newspapers (1974): number 1; circulation 30 000, 7 per 1 000 people
Telephones (Dec 1975): 38 000, 8 per 1 000 people
Livestock (000, 1976) Cattle 2 380, sheep 1 740, goats 873,
horses 226*, asses 196, camels 25*
Petroleum refinery capacity (1975) 0.7* mn tonnes
Electrical capacity (1975) 120* megawatts
Hospital beds (1974) 5 722, 1 per 816 people
Roads (1973) 13 271 km = 8 246 mi, density 0.07 km per km²
Railways (1973) 1 186 km = 737 mi, density 0.006 km per km²
Inland waterways Senegal river; also Saloun and Pasamance rivers
Ships (registered, 1977) 75, total of 28 044 gross tons
Port (goods traffic, 000 tonnes, 1973)
Dakar: loaded 2 386, unloaded 2 378
Airports Yoff (Dakar): passenger departures and arrivals (1975) 407 612;
also Saint Louis, Ziguinchor and (1977) 9 other airports with
scheduled flights

Durable equipment	000	no per	
(Dec 1974)		1 000 people	
Radio sets	286	60	no per
Television sets	35	7	km of road
Passenger cars	48	10	3.6
Commercial vehicles	25	5	1.9

Senegal

Production

Gross domestic product (1976) CFA Fr 353 bn = $ 1 476 mn = £ 817 mn

Production indices (1970 = 100)	1960	1970	1976	Growth %pa 1960–70	1970–76
Agricultural	88	100	159	1.3	8.1
Industrial	70	100	156	3.6	7.7

Main products (000 t)

Agriculture

	1960	1970	1976	1960–70	1970–76
Rice	83	91	112	0.9	3.5
Millet	396	405	555	0.2	5.4
Cassava	168	159	114	−0.5	−5.4
Groundnuts	892	583	1192	−4.2	12.7
Palm kernels	4.0**	7.0**	5.0*	5.8**	−5.5**
Cotton	1.0*	4.4	15.0*	16.0*	22.7*
Milk	98	120*	94*	2.0*	−4.0*
Beef and veal	19	31*	37*	5.0*	3.0*
Cattle hides	2.2	6.6*	6.5*	11.6*	−0.2*
Fish catch	122	189	362c	4.5	13.9d
Timber (000 m³)	2 049	2 465*	2 745c	1.9*	2.2*d

Other

	1960	1970	1976	1960–70	1970–76
Petroleum products	—	546	690*c	na	4.8*d
Electricityb (mn kW h)	127	287	389*c	8.5	6.3*d
Phosphate rock	198	998	1 600c	17.6	9.9d
Beer (000 hl)	82	106	113e	2.6	1.6f
Soft drinks (000 hl)	89	134	171g	4.2	8.5h
Cigarettes (mn units)	1 283	1 647	2 041e	2.5	5.5f
Fertilisers, phosphatea	—	12	20*	na	8.4*
Cement	168	241	380*	3.7	7.9*
Footwear (000 pairs)	3 756	5 292	3 588g	3.5	−12.1h

a Years ending June 30th b Consumption c 1975 d 1970–75 e 1974 f 1970–74
g 1973 h 1970–73

Transport traffic	1960	1970	1976	Growth %pa 1960–70	1970–76
Passenger-kilometres (mn)					
Rail	197	245	220d	2.2	−2.7e
Aira	41b	74	137	8.8c	10.8
Cargo: tonne-kilometres (mn)					
Rail	140	179	182d	2.5	0.4e
Aira	2.5b	6.3	13.9	14.1c	14.1
Sea: tonnes (mn)					
Goods loaded	0.97	2.80	2.60d	11.2	−1.9e
Goods unloaded	2.19	2.45	2.01d	1.1	−4.8e

a Includes apportionment of traffic of Air Afrique b 1963 c 1963–70 d 1974
e 1970–74

Tourism (1975) Number of visitors 129 000, gross receipts $ 7 mn

Finance and trade

Price index (1970 = 100)	1960	1970	1976	Growth %pa 1960–70	1970–76
Consumer prices	78	100	177	2.5	10.0
Money stock (end-year, CFA Fr bn)	35.3a	34.5	94.8	−0.3b	18.4

a 1962 b 1962–70

Budget (1976/77; year ending June 30th)
Balanced at CFA Fr 119 000 mn = $ 480 mn = £ 280 mn

Balance of payments ($ mn)	1972	1973	1974	1975	1976
Balance of goods (fob)		−59	−160	−136	−139	na
Balance of services		+5	+2	+7	+17	na
Balance of transfers		+64	+57	+63	+67	na
Current balance		*+11*	*−101*	*−66*	*−55*	*na*
Long-term capital flow		−2	54	40	54	na
Reserves and debt (end-year, $ mn)						
International reserves		38	12	6	31	25
External public debt		218	363	422	503	na

External trade (1975)
Imports: CFA Fr 124 610 mn = $ 581 mn = £ 262 mn
Exports: CFA Fr 99 100 mn = $ 462 mn = £ 208 mn

Main imports (1973)	% of total	*Main exports* (1974)	% of total
Cereals	20	Phosphates	28
Machinery	15	Groundnut oil	22
Chemicals	7	Fish	7
Sugar	7	Groundnut oilseed	
Crude oil	5	cake	7
Main sources (1974)		*Main destinations* (1974)	
France	41	France	50
West Germany	6	Netherlands	7
United States	6	United Kingdom	6
Nigeria	6	Ivory Coast	6
Iraq	4	Mauritania	

Seychelles
Republic of Seychelles

Location Western Indian Ocean
The group of about 90 islands is scattered, with the main, Mahé, 1 800 km east of Mombasa (Kenya) and 1 090 km north of the island of Madagascar.
Land Area 443 km² = 171 mi²
of which, Aldabra 155 km², Mahé 148 km²
Climate Tropical
Weather at Port Victoria, 5 m altitude
Temperature: hottest month April, 25–30 °C, coldest July, August 24–27 °C
Rainfall (av monthly): driest month August 69 mm, wettest Jan 386 mm
Time 4 hours ahead of GMT
Measures UK (imperial) system, with a gradual change to the metric system. Also: *length* 12 lignes = 1 pouce, 12 pouces = 1 pied = 0.324 84 metre (in general 1 pied is taken to mean 1 UK foot), 1 ell = 4 feet = 1.219 metre, 1 brasse = 1 fathom = 6 feet = 1.829 metres *area* 40 000 pied² = 1 arpent = 0.422 hectare = 1. 043 acres *capacity* 1 toise (masonry) = 40 cubic feet = 1.13 cubic metres
Monetary unit Seychelles rupee (S R) = 100 cents
Rate of exchange (1976 av): par S Rs 13.3333 = £ 1, free S Rs 7.382 = $ 1

Summary

Political Republic, which became independent on June 29, 1976; formerly a UK colony. Aldabra, Desroches and Farquhar, former dependencies of Seychelles, became part of the British Indian Ocean Territory in 1965; they returned to Seychelles on independence. Member of OAU and Commonwealth and an EEC ACP state
Economic Mainly based on agriculture and fishing industries with copra and cinnamon the main export products; bunkering of ships and aircraft is important. Agricultural diversification is under way to replace imported food; tourism has grown rapidly since 1971

People, resources and equipment

Population 1960 42 000*, 1970 52 000*, 1976 59 000*
Growth: 1960–70 2.1* %pa, 1970–76 2.1* %pa
Density (1976): 133* people per km²
Vital statistics (rate per 1 000 people, 1974): births 32.8, deaths 8.8
Regions (districts, population in 000, 1971)

Frigate Island	0.03	North Island	0.03	Victoria Wards	13.7
La Digue Island	2.0	Praslin Island	4.2	Outlying	
Mahé Island	31.7	Silhouette Island	0.4	islands	0.5

Town (population in 000, 1976) Victoria (capital) 15*
Race (1971) Creole 94 %, English 3 %, French 2 %
Language English and Creole; French is also used
Religion (1971) Christian 98 % (Roman Catholic 90* %), Hindu 1 %
Education (1975) Pupils 14 170*, teachers 613*
Labour force (1971) 19 827
Personnel Scientists and engineers (1973): 300* (of whom, 50* % non-nationals)
Physicians (1972): 16, 1 per 3 300 people
Standard of living
National income per person (1976): S Rs 3 200** = $ 430** = £ 240**
Consumption per person (1975): energy 481 kg coal equivalent, electricity (production) 474 kW h
Newspapers (1974): number 2; circulation 2 400, 41 per 1 000 people
Telephones (Dec 1975): 3 300, 56 per 1 000 people
Livestock (000, 1976) Cattle 5*, pigs 20*, goats 3*
Electrical capacity (1975) 11 megawatts
Hospital beds (1969) 348, 1 per 147 people
Roads (1975) 213 km = 132 mi, density 0.48 km per km²
Ships (registered, 1977) 10, total of 59 140 gross tons
Port Victoria
Airports Mahé international ,Praslin
Durable equipment (at end-year)
Radio sets (1973): 9 000, 161 per 1 000 people
Passenger cars (1975): 2 500, 43 per 1 000 people, 11.7 per km of road
Commercial vehicles (1975): 600, 10 per 1 000 people, 2.8 per km of road

Production, finance and trade

Gross domestic product 1976 est: S Rs 200** mn = $ 27** mn = £ 15**mn

Seychelles

Main products (000 t, 1976) Cassava 1*, coconuts 11*, copra 2*, electricity (mn kW h, 1975) 27, beer (000 hl, 1974) 28
Transport traffic *Sea* (1975) Goods loaded 11 000 t, unloaded 103 000 t
Tourism Number of visitors (1975) 37 300; growth 1971–75 87.4 %pa
Budget (1976) Revenue: S Rs 98.4 mn = $ 13 mn = £ 7.4 mn
Expenditure: S Rs 99.9 mn = $ 14 mn = £ 7.5 mn
External trade (1976) Imports: S Rs 291 mn = $ 39 mn = £22 mn
Exports: S Rs 57.6 mn = $ 8 mn = £4 mn

Main imports	% of total	Main exports	% of total
Food	20	Petroleum products	58
Petroleum products	19	(of which, jet fuel 36)	
Machinery	14	Copra	15
Motor vehicles	5	Cinnamon bark	6
Chemicals	5	Frozen fish	6
Main sources		*Main destinations*	
United Kingdom	30	Bunkers	66
Kenya	16	Pakistan	15
South Africa	7	Mauritius	5
South Yemen	7	Reunion	4
Japan	5	United States	3

Sierra Leone
Republic of Sierra Leone

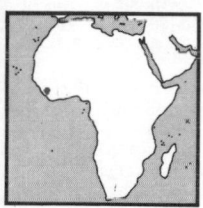

Location West Africa
With a coastline on the Atlantic Ocean, Guinea is to the north, Liberia to the south
Land Area 71 740 km² = 27 700 mi²
Climate Tropical
Weather at Freetown, 11 m altitude
Temperature: hottest month April 25–31 °C, coldest August 23–28 °C
Rainfall (av monthly): driest month February 3 mm, wettest August 902 mm

Time GMT
Measures Metric system, which replaced the UK (imperial) system in general from 1973
Monetary unit Leone (Le) = 100 cents
Rate of exchange (1976 av): par Le 2 = £1, free Le 1.107 = $ 1

Summary

Political Republic, which became independent April 27, 1961; formerly a UK colony and protectorate. Member of UN, OAU, Ecowas, Commonwealth and an EEC ACP state. Under the 1973 Mano River Agreement, a customs union with Liberia is planned
Economic Mining accounts for some three-quarters of exports (mainly diamonds), and agriculture for the remainder, principally palm kernels, cocoa and coffee. Iron ore production was halted at the Delco mine at the end of 1975. There is little manufacturing industry

People, resources and equipment

Population 1960 2.23* mn, 1970 2.55* mn, 1976 3.11 *mn
Growth: 1960–70 1.3* %pa, 1970–76 3.4* %pa
Density (1976): 43* people per km²
Vital statistics (rate per 1 000 people, 1970–75): births 44.7*, deaths 20.7*
Cities (population in 000, 1974)
Freetown (capital) 274, Bo 26, Kenema 13, Makeni 12
Race (1974) Mende 30** %, Temne 25** %, Limba 8** %
Language English; Mende, Temne, Krio (Creole) and local languages are also used
Religion (1976) Animist 65** %, Moslem 30** %, Christian 5** %
Education (1974/75) Pupils 239 000*, teachers 7 970*
Labour force (1976) 1 175 000*; in agriculture 796 000* (68* %)
Personnel (1970) Physicians: 149, 1 per 17 100 people
Standard of living
National income per person (1976): Le 220** — $ 200** = £ 110**
Consumption per person (1975): energy 116 kg coal equivalent, electricity (production) 64 kW h, newsprint 0.03 kg, steel 4 kg
Newspapers (1974): number 2; circulation 25 000, 9 per 1 000 people
Telephones (Dec 1975): 11 000, 4 per 1 000 people
Livestock (000, 1976) Cattle 305*, sheep 68*, goats 179*, pigs 36*, chickens 3 300*

Petroleum refinery capacity (1975) 0.5 mn tonnes
Hospital beds (1972) 2 837, 1 per 927 people
Roads (1976) 7 064 km = 4 389 mi, density 0.10 km per km²
Railways (1976) 84 km = 52 mi, density 0.001 km per km²
Inland waterways (1976) 800* km = 500* mi
Ships (registered, 1977) 12, total of 7 298 gross tons
Ports (goods traffic, 000 tonnes, 1970) Freetown: loaded 1 899, unloaded 850; also Bonthe
Airports Lungi (Freetown): passenger departures and arrivals (1975) 61 895; also (1977) 7 other airports with scheduled flights

Durable equipment	000	no per	
(at end-year)		1 000 people	
Radio sets (1974)	61	21	no per
Television sets (1974)	6	2	km of road
Passenger cars (1976)	12	4	1.7
Commercial vehicles (1976)	5	2	0.8

Production

Gross domestic product
1974/75 (year ending June 30th): Le 585 mn = $ 687 mn = £ 293 mn
1976 est: Le 750** mn = $ 680** mn = £ 370** mn
Structure of gross domestic product (1974/75) *By origin* Agriculture 33 %, mining and quarrying 14 %, manufacturing 7 %, other 46 %

Production index (1970 = 100)	1960	1970	1976	Growth %pa 1960–70	1970–76
Agricultural	75	100	114	2.9	2.2
Main products (000 t)					
Agriculture					
Rice	264	458	530*	5.7	2.5*
Cassava	56	83	87*	4.0	0.8*
Groundnuts	23*	21	18*	−0.9*	−2.5*
Mangoes	44**	50*	52*	1.3**	0.6*
Citrus fruit	84	104*	110*	2.2*	1.0*
Palm kernels	55	65*	54*	1.7*	−3.0*
Palm oil	35	53*	56*	4.2*	0.9*
Coffee	5.2*	7.5*	5.0*	3.7*	−6.5*
Cocoa	3.7	5.3	6.0*	3.7	2.1*
Fish catch	23[a]	31	68[c]	3.3[b]	17.2[d]
Timber (000 m³)	2 648	2 695	2 637[c]	0.2	−0.4[d]
Other					
Petroleum products	—	277	294[c]	na	1.2[d]
Electricity (mn kW h)	41	197	193[c]	17.0	−0.4[d]
Iron ore (Fe content)	881	1 377	916[c]	4.6	−7.8[d]
Bauxite	—	443	645[c]	na	7.8[d]
Diamonds (000 CM)	1 909*	1 955	1 650*[c]	0.2*	−3.3*[d]
Beer	na	101	181[c]	na	12.4[d]
Cigarettes (mn units)	38	492	5 000[c]	29.2	59.0[d]
Oxygen (000 m³)	na	80**	80*[c]	na	0.0**[f]
Acetylene (000 m³)	na	32**	82*[c]	na	26.5**[f]

[a]1961 [b]1961–70 [c]1975 [d]1970–75 [e]1974 [f]1970–74
Transport traffic *Air* 61 mn passenger-km, cargo 1.0 mn t-km.
Sea (1975) Goods loaded 2.11 mn t, unloaded 0.38 mn t
Tourism (1974) Number of visitors 7 800, gross receipts $ 3 mn

Finance and trade

Consumer price index (1970 = 100) 1976 176.3; growth 1970–76 9.9 %pa
Money stock (end-year, Le mn) 1976 72.1; growth 1970–76 16.5 %pa
Budget (1976/77; year ending June 30th)
Revenue: Le 104 mn = $ 89 mn = £ 52 mn
Expenditure: Le 156 mn = $ 134 mn = £ 78 mn
International reserves (Dec 1976) $ 25 mn
External public debt (Dec 1975) $ 198 mn
External trade (1976) Imports: Le 171 mn = $ 154 mn = £ 85 mn
Exports: Le 124 mn = $ 112 mn = £ 62 mn

Main imports (1974)	% of total	Main exports (1974)	% of total
Food	22	Diamonds	63[a]
(of which, cereals 12)		Iron ore	10
Machinery	13	Palm kernels	7
Textile yarns and fabrics	11	Cocoa	6
Crude oil	10	Palm kernel oil	4
Motor vehicles	7	Bauxite	3
Chemicals	6	Coffee	2
Main sources (1974)		*Main destinations* (1974)	
United Kingdom	21	United Kingdom	61
Japan	10	Netherlands	15
Nigeria	8	United States	6
West Germany	7	Japan	5
Pakistan	5	West Germany	4

[a]54 % in 1975

Somalia

Somali Democratic Republic
Jamhuuriyadda Dimuqraadiga Soomaaliya

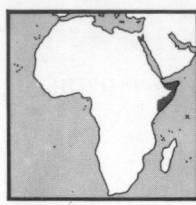

Location North-east Africa
The coastline extends from the Indian Ocean around Cape Guardafui into the Gulf of Aden, with Djibouti to the north-west, Ethiopia to the west and Kenya to the south-west
Land Area 637 700 km² = 246 200 mi²
Climate Hot and dry
Weather at Mogadishu, 12 m altitude
Temperature: hottest month April 26–32 °C, coldest July, Aug 23–28 °C
Rainfall (av monthly): driest months Jan, Feb 1 mm, wettest June 97 mm
Time 3 hours ahead of GMT
Measures Metric system
Monetary unit Somali shilling (So Sh) = 100 centesimi
Rate of exchange (1976 av): par So Sh 6.23270 = $ 1, free So Sh 11.26 = £ 1

Summary

Political One-party republic, which became independent July 1, 1960; formed from the British Somaliland protectorate, and the trust territory of Italian Somaliland. Parts of Ethiopia (Ogaden) and Kenya are claimed as belonging to Greater Somalia. A 1974 treaty of friendship with the Soviet Union was ended in November 1977. Member of UN, OAU, Arab League and an EEC ACP state
Economic An agricultural economy, with livestock and their products and bananas the main export earners; fishing is of growing importance. The nomadic dependence on livestock is planned to change gradually as irrigation and farming is developed

People, resources and equipment

Population 1960 2.23* mn, 1970 2.79*mn, 1976 3.26* mn
70** % of the population are nomadic
Growth: 1960–70 2.3* %pa, 1970–76 2.6* %pa
Density (1976): 5* people per km²
Vital statistics (rate per 1 000 people, 1970–75): births 47.2*, deaths 21.7*
Cities (population in 000, 1975)
Mogadishu (capital) 400** Chisimaio 70** Merca 60**
Hargeisa 70** Berbera 65** Giamame 30**
Race (1970) Somali 76** %, Sab 19** %, Bantu 4** %, Arab 1** %
Language Somali; Arabic, English and Italian are also used
Religion Mainly Moslem (Sunni)
Education (1973/74) Pupils 111 200*, teachers 3 720*
Labour force (1976) 1 285 000*; in agriculture 1 054 000* (82* %)
Personnel (1973) Physicians: 193, 1 per 15 600 people
Standard of living
National income per person (1976): So Sh 650** = $ 105** = £ 58**
Consumption per person (1975): energy 36 kg coal equivalent, electricity (production) 13* kW h, newsprint 0.2 kg
Newspapers (1973): number 2; circulation 4 000, 1 per 1 000 people
Telephones (Dec 1970): 5 000*, 2* per 1 000 people
Livestock (000, 1976) Cattle 2 600*, sheep 7 000*, goats 8 000*, camels 2 000*, chickens 2 500*
Electrical capacity (Mogadishu only, 1975) 18* megawatts
Hospital beds (1972) 5 163, 1 per 569 people
Roads (1973) 17 100 km = 10 625 mi, density 0.03 km per km²
There are no railways
Ships (registered, 1977) 31, total of 158 166 gross tons
Ports (goods traffic, 000 tonnes, 1972)

	loaded	unloaded
Berbera	220	72
Mogadishu	22	187
Chisimaio	84	80
Merca	64	50

Airports Mogadishu, Hargeisa and 8 other airports with scheduled flights
Durable equipment (at end-year)
Radio sets (1974): 67 000, 21 per 1 000 people
Passenger cars (1972): 8 000, 3 per 1 000 people, 0.5 per km of road
Commercial vehicles (1972): 8 000, 3 per 1 000 people, 0.5 per km of road

Production

Gross domestic product 1970: So Sh 1 779 mn = $ 249 mn = £ 104 mn
1976 est: So Sh 2 200** mn = $ 350** mn = £ 195** mn

Production index (1970 = 100)	1960	1970	1976	Growth %pa	
				1960–70	1970–76
Agricultural	91	100	103	1.0	0.5
Main products (000 t)					
Agriculture					
Maize	90**	95	120*	0.5**	4.0*
Millet and sorghum	180**	130**	120*	−3.2**	−1.3**
Cassava	15	25*	29*	5.2*	2.5*
Sesameseed	9	15**	22*	5.2**	6.6**
Sugar, raw value	13	52	30*	14.9	−8.8*
Bananas	91	150*	150*	5.1*	0.0*
Milk, cow	78	75	130*	−0.4	9.6*
Milk, sheep	45**	60**	75*	2.9**	3.8**
Milk, goat	90**	115**	140*	2.5**	3.3**
Beef and veal	18	25**	32*	3.3**	4.2**
Sheep and goat meat	30**	35**	57*	1.6**	8.5**
Hides and skins	9**	10**	11*	1.0**	1.6*
Fish catch	na	30*	33*b	na	1.9*c
Timber (000 m³)	2 300**	2 850	3 230b	2.2**	2.5c
Other					
Electricity (mn kW h)a	5*	28	42*b	18.8*	8.4*c
Ethyl alcohol (000 hl)	na	41	39d	na	−4.9c
Textiles (mn m)	—	—	18d	na	na

aMain cities only b1975 c1970–75 d1971 e1970–71

Transport traffic *Air* (1976) Passenger-km 23 mn, cargo 0.2 mn t-km
Sea (1973) Goods loaded 0.47 mn t, unloaded 0.44 mn t

Finance and trade

Price index (1970 = 100)	1960	1970	1976	Growth %pa	
				1960–70	1970–76
Consumer prices	65.3	100.0	165.4	4.4	8.7
Money stock					
(end-year, So Sh mn)	94.6	376.1	994.9	14.8	17.6

Budget (1975) Revenue: So Sh 667 mn = $ 106 mn = £ 48 mn
Expenditure: So Sh 583 mn = $ 93 mn = £ 42 mn

Balance of payments ($ mn)	1972	1973	1974	1975	1976
Balance of goods (fob)	−6	−40	−70	−52	−72
Balance of services	−18	−27	−34	−50	−38
Balance of transfers	+17	+29	+52	+102	+41
Current balance	−7	−39	−52	0	−69
Long-term capital flow	+18	+26	+61	+53	+74
Reserves and debt (end-year, $ mn)					
International reserves	31	35	42	68	85
External public debt	251	286	368	439	595

External trade (1975) Imports: So Sh 1 021 mn = $ 162 mn = £ 73 mn
Exports: So Sh 573 mn = $ 91 mn = £ 41 mn

Main imports (1974)	% of total	Main exports (1974)	% of total
Textile yarns & fabrics	11	Sheep and lambs	24
Machinery, non-electric	10	Bananas & plantains	20
Cereals	8	Goats	20
(of which, rice 6)		Meat	9
Iron and steel	8	Camels	8
Chemicals	8	Cattle	6
Sugar	7	Fish	4
Motor vehicles	7	Hides and skins	4
Petroleum products	7		
Main sources (1974)		*Main destinations* (1974)	
Italy	38	Saudi Arabia	65
Soviet Union	17	Italy	12
China	9	Iran	7
Kenya	7	Soviet Union	6
France	5	Kuwait	5
Thailand	5	China	4
Japan	5	South Yemen	4
United Kingdom	5		
West Germany	5		

Special focus

Impact of the 1973–74 drought

Number of livestock (000)	1971	1972	1973	1974	1975	1976
Cattle	2 850*	2 850*	2 500*	2 000*	2 300*	2 600*
Sheep	7 000*	7 500*	6 500*	5 000*	6 000*	7 000*

South Africa

Republic of South Africa
Republiek van Suid-Afrika

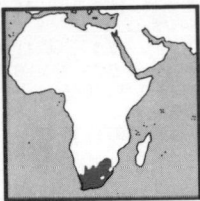

Location South Africa
Forms the southern part of Africa, with
a coastline on the Atlantic and Indian
Oceans. The northern border is formed by
Mozambique, Swaziland, Rhodesia, Botswana
and South-West Africa; Lesotho is
completely encircled within South Africa.
Walvis Bay is an enclave in South-West Africa
Land Area 1 222 161 km² = 471 879 mi²
(includes 1 124 km² for Walvis Bay)
Usage (1975): agricultural 963 000* km² (79* %), of which,
pastures 818 000* km² (67* %); forests 46 000* km² (4* %)
Climate Temperate: warm and sunny
Weather at Cape Town, 17 m altitude
Temperature: hottest month February 16–26 °C, coldest July 7–17 °C
Rainfall (av monthly): driest month Feb 8 mm, wettest July 89 mm
Time 2 hours ahead of GMT
Measures Metric system, which replaced the UK (imperial) system
in general from 1973
Monetary unit Rand (R) = 100 cents; the rand was introduced as a
decimal currency on February 14, 1961 to replace the South African
pound at R 2 = SA £ 1
Rate of exchange (1976 av): par $ 1.15 = R 1, free R 1.571 = £ 1

Summary

Political Republic, established as such May 31, 1961; previously
a parliamentary monarchy. The Union of South Africa was first
established in 1910. Since the creation of the republic, Bantu homelands
have been established to provide limited self-government; of them,
Transkei was declared independent by South Africa from October 26,
1976, and Bophuthatswana from December 6, 1977. Member of UN
Economic An industrial and mixed economy, with 24 % of gross
domestic product derived from manufacturing, 13 % from mining and
quarrying, and 8 % from agriculture. Exports are obtained from a
wide range of minerals, especially gold (about one-third of total exports)
and diamonds, and agricultural produce, especially cereals and sugar.
Among manufactures exported iron and steel is the most important.
Having no crude oil reserves, oil and energy from coal is being expanded;
a 1 843 megawatts nuclear power plant is under construction.

People

Population[a] 1960 17.12 mn, 1970 22.47 mn, 1976 26.13 mn
[a]Excluding Walvis Bay, administered before September 1, 1977 as part of
South-West Africa, with an estimated population of 27 000* in 1976
Growth: 1960–70 2.8 %pa, 1970–76 2.5 %pa
Density (1976): 21 people per km²
Vital statistics (rate per 1 000 people, 1970–75): births 42.9*, deaths 15.5*
Regions (population in 000, 1970; census total of 21.79 mn)

Bantu homelands	7 175	Lebowa	1 108	Provinces	14 619
Bophuthatswana[a]	904	Qwaqwa	25	Cape	4 294
Ciskei	538	Swazi	120	Natal	2 164
Gazankulu	272	Transkei	1 783	Orange Free State	1 682
Kwazulu	2 151	Venda	272	Transvaal	6 479

[a]Formerly called Tswanaland
Cities (population in 000, 1970) Cape Town (legislative capital) 1 108
Pretoria (administrative capital) 563

Johannesburg	1 441	Vereeniging	170	Umhlali	121
Durban	851	Pietermaritzburg	161	Roodepoort	114
Port Elizabeth	476	Benoni	150	Boksburg	105
Germiston	210	Springs	142	Kimberley	105
Bloemfontein	182	East London	125		

Race (1974): Bantu 71 %, European 17 %, Coloured 9 %, Asian 3 %
Main Bantu groups: Zulu 19 %, Xhosa 19 %, Tswana 8 %, Sepedi 8 %,
Seshoeshoe 6 %, Shangaan 3 %, Swazi 2 %, Venda 2 %
In Bantu homelands (1970)

Bophuthatswana	Tswana 67 %, Sepedi 8 %, Shangaan 6 %, Xhosa 3 %, Zulu 3 %, Seshoeshoe 3 %
Ciskei	Xhosa 97 %, Seshoeshoe 3 %
Gazankulu	Shangaan 87 %, Sepedi 10 %
Kwazulu	Zulu 97 %, Seshoeshoe 1 %, Xhosa 1 %
Lebowa	Sepedi 83 %, Shangaan 7 %
Qwaqwa	Seshoeshoe 96 %, Zulu 1 %
Swazi	Swazi 69 %, Zulu 18 %, Shangaan 9 %
Transkei	Xhosa 94 %, Seshoeshoe 3 %, Zulu 1 %
Venda	Venda 90 %, Shangaan 7 %, Sepedi 3 %

Language Afrikaans and English; African languages, especially Xhosa,
Zulu and Sesotho are also used
Usage (Europeans only, 1970): Afrikaans 48 %, English 30 %,
Afrikaans and English 18 %, German 1 %, Portuguese 1 %
Religion (1970) Christian 61 %, Bantu churches 13 %, Hindu 2 %,
Moslem 1 %, Jew 1 %
Education (1974) *European* Pupils 1 010 757, teachers 57 000*
African and other non-European Pupils 4 310 000*, teachers 91 200*
Labour force 1970: 7 986 200; in agriculture 2 239 200 (28 %),
mining and quarrying 676 100 (8 %), manufacturing 1 023 700 (13 %),
electricity, gas and water 49 700 (1 %), construction 446 400 (6 %),
distribution and hotels 716 100 (9 %), transport and communications
338 300 (4 %), other services 1 764 400 (22 %), not specified 732 400 (9 %)
1976: 9 350 000*
Personnel (1973) Physicians: 12 060, 1 per 2 016 people
Standard of living
National income per person (1976): R 930* = $ 1 070* = £ 592*
Consumption per person (1975): energy 2 953* kg coal equivalent,
electricity (production, 1976) 2 987 kW h, newsprint 6.6 kg, steel 263 kg
Newspapers (1974): number 24
Telephones (Dec 1975): 1 936 000, 75 per 1 000 people

Resources and equipment

Livestock (000, 1976) Cattle 12 700*, sheep 31 001, goats 5 200*,
pigs 1 380*, horses 230*, asses 210*, chickens 25 400*
Mineral reserves Coal (1969) 44 339 mn tonnes
Natural gas (1970) 11 bn cubic metres
Uranium (1974) 186 000 tonnes
Petroleum refinery capacity (1975) 22* mn tonnes
Electrical capacity (1975) 13 990 megawatts
Hospital beds (1973) 156 245, 1 per 156 people
Roads (main only, 1976) 185 326 km = 115 156 mi, density 0.15 km
per km²
Railways (1976) 20 090 km = 12 483 mi, density 0.02 km per km²
Ships (registered, 1977) 297, total of 476 324 gross tons
Ports (goods traffic, 000 tonnes, 1973)

	loaded	unloaded (excluding crude oil)
Durban	7 593	4 783
Port Elizabeth	6 716	1 712
Cape Town	3 181	1 551

Also East London, Walvis Bay, Richards Bay, Mossel Bay,
Port Nolloth, Lüderitz, Saldanha Bay
Airports Passenger departures and arrivals (1975): Jan Smuts
(Johannesburg) 3 287 389, Durban 1 102 049; also (1977) 28 other
airports with scheduled services

Durable equipment (at end-year)	000	no per 1 000 people	
Radio sets (1974)	2 335	93	no per km of road
Television sets (1976)	500**	19**	
Passenger cars (1976)	2 169	82	11.7
Commercial vehicles (1976)	900	34	4.9

Production

Gross domestic product[a] (1976)
R 28 015* mn = $ 32 217* mn = £ 17 833* mn
Growth in real terms (including South-West Africa): 1960–70 6.0 %pa,
1970–76 3.6 %pa
[a]Excludes South-West Africa, with a roughly estimated gdp of R 1 000*** mn
Structure of gross domestic product (including South-West Africa)

By origin (1975)	R mn	% of gdp
Agriculture	1 943	8
Mining and quarrying	3 171	13
Manufacturing	5 773	24
Electricity, gas and water	647	3
Construction	1 158	5
Distribution and hotels	3 287	14
Transport and communication	2 175	9
Other services	6 162	23
Total	24 316	100

By type of expenditure (1976)		
Government final consumption	4 521	16
Private final consumption	16 636	57
Stock investment	−440	−2
Gross fixed capital formation	8 736	30
Exports of goods and services	8 387	29
less Imports of goods and services	−8 750	−30
Total	29 015	100

Production indices (1970 = 100)	1960	1970	1976	Growth %pa 1960–70	1970–76
Agricultural	73	100	113	3.2	2.1
Manufacturing	44.1	100.0	123.5	8.5	3.6

South Africa

Main products (000 t)	1960	1970	1976	Growth %pa	
Agriculture				1960–70	1970–76
Wheat	740	1 396	2 060	6.6	6.7
Maize	3 801	6 132	7 312	4.9	3.0
Sorghum	226	445	260	7.0	−8.6
Total cereals	4 950*	8 152	9 808	5.1*	3.1
Potatoes	364	631	750*	5.7	2.9*
Groundnuts	214	326	153	4.3	−11.8
Sunflowerseed	90	96	255	0.6	17.7
Tomatoes	110	160**	260**	3.8**	8.4**
Sugar, raw value	946	1 399	2 042	4.0	6.5
Grapes	575	760*	1 150*	2.8*	7.1*
Wine	318	419*	630*	2.8*	7.0*
Apples	107	226*	310*	7.8*	5.4*
Pears	40	91*	110*	8.6*	3.2*
Peaches	66	155*	150*	8.9*	−0.5*
Oranges	289	543*	640*	6.5*	2.8*
Grapefruit	19	93*	85*	17.2*	−1.5*
Pineapples	115	123*	190*	0.7*	7.5*
Tobacco	30	36	34*	1.8	−0.9*
Sisal	1.2	5**	8*	15.3**	8.1**
Cotton	5	17	27*	13.0	8.0*
Beef and veal	370*	370*	400*	0.0*	1.3*
Sheep and goat meat	113*	170*	173*	4.2*	0.3*
Pigmeat	50	106*	90*	7.8*	−2.7*
Milk	2 570	2 912	2 560*	1.3	−2.1*
Eggs	66	112	162*	5.4	6.3*
Wool	65	57*	52*	−1.3*	−1.5*
Fish catch[a]	1 150**	1 562	1 315[c]	3.1**	−3.4[d]
Timber (000 m³)	4 333	9 893*	10 455*[c]	8.6*	1.1*[d]
Energy					
Total energy (000 tce)	38 210	54 760	76 110*	3.7	5.6*
Coal	38 173	54 612	75 730	3.6	5.6
Petroleum products	1 285	8 033	13 920*[c]	20.1	11.6*[d]
Manufactured gas (mn m³)	769	1 659	1 865[c]	8.0	2.4[c]
Electricity (mn k W h)	22 561[f]	50 791	78 056	8.0	7.4
Mining					
Iron ore (Fe content)	1 965	5 887	9 887	11.6	9.0
Antimony ore (Sb content)	12.3	17.3	16.3[c]	3.5	−1.2[d]
Chromium ore (oxide content)	340	642	906[c]	6.6	7.1[d]
Copper ore (Cu content)	46	148	179[c]	12.5	3.8[d]
Magnesite	61	84	61[c]	3.3	−6.2[d]
Manganese ore (Mn content)	455	1 182	2 006[c]	10.0	11.1[d]
Nickel ore (Ni content)	3*	12	21[c]	15.0*	12.4[d]
Uranium (U content)	6.0	3.2	2.6[c]	−6.1	−3.9[d]
Silver	0.07	0.110	0.096[c]	4.8	−2.7[d]
Gold (000 kg)	665	1 002	709	4.2	−5.6
Diamonds (000 CM)	3 140	8 112	7 023	10.0	−2.4
Fluorspar	103	173	291*	5.3	9.1*
Limestone and lime	7 273	15 076	18 543	7.6	3.5
Asbestos	160	287	370	6.0	4.3
Phosphate rock	268	1 685	1 824[k]	20.2	2.0[l]
Manufacturing					
Fishmeal[b]	148	340	250[c]	8.7	−6.0[d]
Beer (000 hl)	800	2 903	5 496[c]	13.8	13.6[d]
Cigarettes (mn units)	9 992	16 430	22 485[c]	5.1	6.5[d]
Cotton yarn	20[i]	60	73[k]	13.0[j]	5.0[l]
Cotton fabrics (mn m)	72[i]	118	104[c]	5.6[j]	−2.5[d]
Wood pulp[g]	74	630	767	23.9	3.3
Newsprint	—	160	214	na	5.0
Other paper[g]	146**	422	624	11.2**	6.7
Fertilisers, nitrogenous	22	200*	300*[c]	24.7*	8.4*[d]
Fertilisers, phosphate	128*	324*	375*[c]	9.7*	3.0*[d]
Cement	2 705	5 752	7 048	7.8	3.4
Coke oven coke	1 915	3 463	3 650*[c]	6.1	1.1*[d]
Pig iron	2 000	4 328	6 631	8.0	7.4
Crude steel[h]	2 120**	4 674	7 155	8.2**	7.4
Aluminium	—	—	78[c]	na	na
Tyres (000 units)	1 779	3 600	4 697[c]	7.3	5.5[d]
Passenger cars[e] (000 units)	87	195	206[c]	8.4	1.1[d]
Commercial vehicles[e] (000 units)	16	76	104[c]	16.9	6.6[d]

[a]Includes landings in Walvis Bay [b]Including production in South-West Africa [c]1975 [d]1970–75 [e]Assembly only [f]Year ending June 30th [g]Twelve months ending March 31st [h]Including concast steel billets [i]1961 [j]1961–70 [k]1974 [l]1970–74

Transport traffic	1960	1970	1976	Growth %pa	
Passenger-kilometres (mn)				1960–70	1970–76
Air	513	2 872	6 050	18.8	13.2
Cargo: tonne-kilometres (mn)					
Rail[a]	28 400[b]	54 200[b]	68 114	6.7	3.4
Air	14.7	66.0	160.7	16.2	16.0
Sea: tonnes (mn)					
Goods loaded	6.65	13.45	26.41	7.3	11.9
Goods unloaded[c]	7.17[d]	8.83	8.59	7.2[e]	−0.5

[a]Including traffic in South-West Africa [b]Year ending March 31st [c]Excluding crude oil (which amounted to 12.6* mn t in 1974) [d]1967 [e]1967–70

Tourism Number of visitors (1971) 460 000, gross receipts (1975) $ 179 mn

Finance and trade

Price indices	1960	1970	1976	Growth %pa	
(1970 = 100)				1960–70	1970–76
Consumer prices	76.1	100.0	173.4	2.8	9.6
Share prices	30	100	99	12.9	−0.2
Money stock					
(end-year, R mn)	992	2 258	4 437	8.6	11.9

Budget (1976/77; year ending March 31st)
Revenue: R 6 052 mn = $ 6 960 mn = £ 4 011 mn
Expenditure: R 7 832 mn = $ 9 007 mn = £ 5 190 mn,
of which (1974/75): education 6 %, defence 13 %

Balance of payments[a] ($ mn)	1972	1973	1974	1975	1976
Balance of goods (fob)	+688	+1 054	−61	−759	−283
Balance of services	−872	−1 164	−1 532	−1 906	−1 609
Balance of transfers	+62	+20	+123	+189	+151
Current balance	−122	−90	−1 471	−2 476	−1 741
Long-term capital flow	+726	+205	+1 646	+2 175	+1 280
Reserves and debt (end-year, $ mn)					
International reserves	1 291	1 234	1 159	1 216	940
Public debt: domestic	8 261	10 086	10 799	10 425	12 375
external	690	568	780	1 056	1 547

[a]Including South-West Africa

External trade (for South Africa Customs Union, including Botswana, Lesotho, South-West Africa and Swaziland, 1976)
Imports: R 6 335 mn = $ 7 285 mn = £ 4 032 mn
Exports (incl gold bullion): R 6 903 mn = $ 7 939 mn = £ 4 394 mn

Main imports[a]	% of total	Main exports[a]	% of total
Machinery	33	Gold	34
Motor vehicles	15	Diamonds	8
Chemicals	12	Iron and steel	7
Textiles and clothing	7	Gold coins	5
Food	5	Metal ores	4
Aircraft	5	Fruit and vegetables	4
Iron and steel	4	Cereals	4
Instruments	4	Sugar	3
Main sources[a]		Main destinations[a,b]	
United States	22	United Kingdom	15
West Germany	18	Japan	8
United Kingdom	18	West Germany	7
Japan	10	United States	7
France	4	Belgium-Luxembourg	3
Italy	4	Switzerland	2
Netherlands	3	France	2

[a]Excludes trade in crude oil and petroleum products, amounting to 7 % of imports and 3 % of exports in 1972 [b]Excluding gold bullion

Special focus

Population divisions, 1970 census (home populations, 000)

	Bantu	European	Coloured	Asian	Total
Provinces	*8 202*	*3 753*	*2 038*	*627*	*14 619*
Cape	1 385	1 109	1 778	21	4 294
Natal	1 133	440	68	524	2 164
Orange Free State	1 346	300	36	—	1 682
Transvaal	4 338	1 904	155	82	6 479
Bantu homelands	*7 138*	*20*	*13*	*4*	*7 175*
Bophuthatswana[b]	901	2	1	—	904
Ciskei	535	1	2	—	538
Gazankulu	271	—	—	—	272
Kwazulu	2 142	4	2	4	2 151
Lebowa	1 106	3	—	—	1 108
Qwaqwa	25	—	—	—	25
Swazi	120	—	—	—	120
Transkei[a]	1 767	9	8	—	1 783
Venda	272	1	—	—	272
Total	*15 340*	*3 773*	*2 051*	*630*	*21 794*

[a]Declared independent October 26, 1976 [b]Declared independent December 6, 1977

South-West Africa

Suidwes-Afrika
Namibia

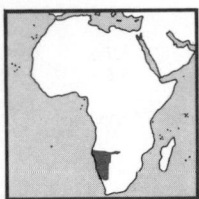

Location South-west Africa
With a coastline on the Atlantic Ocean, Angola is to the north, Botswana to the east and South Africa to the south; the Caprivi Strip, at the north east corner, runs east between Botswana and Angola and forms a small border with Zambia. Walvis Bay, part of South Africa, forms an enclave about 650 km (400 mi) north of Cape Province
Land Area 823 168 km² = 317 827 mi²
(excludes 1 124 km² for Walvis Bay)
Climate Temperate and sub-tropical
Weather at Windhoek, 1 728 m altitude
Temperature: hottest month January 17–29 °C, coldest July 6–20°C
Rainfall (av monthly): driest month August 1 mm, wettest March 79 mm
Time 2 hours ahead of GMT
Measures Metric system
Monetary unit South African rand (R) = 100 cents
Rate of exchange (1976 av): par $ 1.15 = R 1, free R 1.571 = £ 1

Summary

Political Territory under the administration of South Africa which was first given a mandate in 1920; the UN voted to end that mandate in October 1966. The UN changed the name of the territory to Namibia in June 1968, and there are proposals for Namibia to become independent. Member of ILO
Economic Mining (mainly diamonds), fishing and livestock are the main industries; a wide range of minerals are available and being developed, including uranium, vanadium and tin. Manufacturing industry is small

People, resources and equipment

Population[a] 1960 510 000**, 1970 750 000**, 1976 880 000**
[a]Excludes the population of Walvis Bay, part of South Africa
Growth: 1960–70 3.9** %pa, 1970–76 2.7** %pa
Density (1976): 1** person per km²
Vital statistics (rate per 1 000 people, 1970–75): births 45.5*, deaths 23.2*
Cities (population in 000, 1974)
Windhoek (capital) 76, Swakopmund (summer capital) 10**
Race (1974) African 84 %, European 12 %, Coloured 4 %
Main African tribes: Ovambo 46 %, Damara 9 %, Kavango 7 %, Herero 7 %, Nama 4 %, East Caprivians 3 %, Bushmen 3 %, Rehobothers 2 %
Language (1974) Afrikaans (67** %), German (23** %) and English (10** %); African languages are also in use
Religion (1974) Christian 50** % (Lutheran 40** %)
Education (1975) Pupils 181 616, teachers 5 457
Labour force (1976) 299 000*; in agriculture 153 000* (51* %)
Standard of living
National income per person (1976): R 950*** = $ 1 100*** = £ 600***
Telephones (Dec 1975): 46 000, 53 per 1 000 people
Livestock (000, 1976) Cattle 2 850*, sheep 5 000*, goats 2 000*
Hospital beds (1973) 6 905, 1 per 120 people
Roads (1975) 54 000* km = 34 000* mi, density 0.07 km per km²
Railways (1975) 2 340 km = 1 454 mi, density 0.003 km per km²
Ports Luderitz; Walvis Bay is the main port used
Airports Windhoek, Luderitz, Keetmanshoop. Walvis Bay and Alexander Bay airports (South Africa) are also used

Production, finance and trade

Gross domestic product
1976 est: R 1 000*** mn = $ 1 200*** mn = £ 700*** mn
Agricultural production index (1970 = 100) 1976 132; growth 1970–76 4.7 %pa
Main products (000 t) *Agriculture* (1976) Maize 15*, millet 20*, sorghum 3*, other cereals 90*, beef and veal 142*, sheep and goat meat 24*, milk 65*, wool 2.8*, hides and skins 7.5*, fish catch (excludes fish landed at Walvis Bay, 1975) 87
Mining (1976) Copper ore (Cu content) 25, lead ore (Pb content) 48, tin concentrates (Sb content) 0.7, zinc ore (Zn content) 46, diamonds (000 CM) 1 740, silver 0.05*, salt 210*
Manufacturing (1975) Canned fish 143; fishmeal 147

Rail traffic See South Africa
Budget (1975/76; year ending March 31st)
Balanced at R 104 mn = $ 120 mn = £ 56 mn
External trade (1976) Imports: R 500** mn = $ 570** mn = £ 300** mn
Exports: R 700** mn = $ 800** mn = £ 450** mn
Included in South Africa Customs Union

Main exports	% of total
Diamonds	45**
Minerals (excluding diamonds)	20**
Fish and products	7**
Livestock	7**
Karakul	7**

Main destination Most trade is with South Africa

Sudan

Democratic Republic of the Sudan
Jamhuriyat es Sudan el Dimukratiya

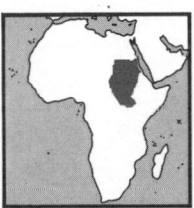

Location North-east Africa
With a coastline on the Red Sea, Egypt is to the north, Libya, Chad and Central African Empire to the west, Zaire, Uganda and Kenya to the south and Ethiopia to the east. The north, except for the Nile valley, is part of the Sahara desert
Land Area 2 505 800 km² = 967 500 mi²
Usage (1975): agricultural 315 000* km² (13* %); forests 915 000 km² (37 %)
Climate Tropical, dry in the north
Weather at Khartoum, 390 m altitude
Temperature: hottest month June 26–41 °C, coldest Jan 15–32°C
Rainfall (av monthly): driest months Jan, Feb 0 mm, wettest Aug 71 mm
Time 2 hours ahead of GMT
Measures Metric system; also:
length 1 diraa = 58 centimetres = 22.8 inches
area 1 feddan = 0.42 hectare = 1.04 acres
capacity 12 keilas = 1 ardeb = 1.98 hectolitres = 5.444 UK bushels
weight (*mass*)
100 rotl = 1 kantar (small) = 44.93 kilograms = 99.05 pounds
315 rotl = 1 kantar (large) = 141.5 kilograms = 312 pounds
Monetary unit Sudanese pound (S £) = 100 piastres = 1 000 millièmes
Rate of exchange (1976 av): par $ 2.8716 = S £ 1 (S £ 0.348 242 = $ 1), free £ 1.590 = S £ 1

Summary

Political One-party republic, which became independent January 1, 1956; formerly a condominium of the United Kingdom and Egypt. There is a separate regional constitution for the Southern Sudan. Member of UN, OAU, Arab League and an EEC ACP state
Economic Agricultural economy, with a wide spread of products and main exports of cotton, groundnuts and sesameseed. There is some co-operation with Egypt concerning especially irrigation schemes; new dams are being constructed. Industry includes processing of agricultural produce and refining of crude oil. The 1977/78 – 1982/83 plan envisages continued strong expansion of the agricultural sector, which will absorb 27 % of the plan's total of S £ 2 670 mn, and some expansion for industry, mining and tourism, which will absorb 20 %

People, resources and equipment

Population 1960 11.85*mn, 1970 15.70* mn, 1976 18.85** mn[a]
[a]There is an alternate UN estimate of 16.13* mn; includes 250 000** refugees from Eritrea (Ethiopia)
Growth: 1960–70 2.9* %pa, 1970–76 3.1** %pa
Density (1976): 8** people per km²
Vital statistics (rate per 1 000 people, 1970–75): births 47.8*, deaths 17.5*
Regions (provinces, population in 000, 1973; total of 16.90 mn)

Northern	12 473			Southern	4 428
Blue Nile	3 483	Khartoum	968	Bahr el Ghazal	1 575
Darfur	1 869	Kordofan	3 103	Equatoria	1 438
Kassala	1 798	Northern	1 252	Upper Nile	1 415

Cities (population in 000, 1973)

Khartoum (capital)	322	Khartoum North	161	Wad Medani	82
Omdurman	305	Port Sudan	123	El Obeid	74

Race (1970) Arab 50** %, Dinka 11** %, Nubian 8** %, Beja 6** %
Language Arabic; local languages, especially Nubian, are also used
Religion (1976) Moslem 70** %, Animist 28** %, Christian 2** %

Sudan

Education (1974/75) Pupils: primary 1 130 856, secondary 244 331, vocational 6 516, teacher-training 4 814, higher 22 204.
Teachers: primary 28 689, secondary 10 946, vocational 465, teacher-training 391, higher 1 320
Labour force (1976) 5 922 000*; in agriculture 4 678 000* (79* %)
Personnel Scientists and engineers (1971/72): 13 792*
Physicians (1974): 1 400, 1 per 12 600 people
Standard of living
National income per person (1976): S £ 80** = $ 230** = £ 130**
Consumption per person (1975): energy 140 kg coal equivalent, electricity (production) 19* kW h, newsprint 0.2 kg, steel 5 kg
Newspapers (1973): number 3; circulation 140 000, 8 per 1 000 people
Telephones (Dec 1974): 56 000, 3 per 1 000 people
Livestock (000, 1976) Cattle 15 400*, sheep 15 262*, goats 10 105*, camels 2 830, chickens 23 072*
Petroleum refinery capacity (1975) 1.2 mn tonnes
Electrical capacity (1975) 120* megawatts
Hospital beds (1974) 15 792, 1 per 1 120 people
Roads (1971) 19 535 km = 12 138 mi, density 0.01 km per km²
Railways (1976) 4 556 km = 2 831 mi, density 0.002 km per km²
Inland waterways (including White Nile, 1975) 4 068 km = 2 528 m
Ships (registered 1977) 13, total of 43 375 gross tons
Port (goods traffic, 000 tonnes, 1972)
Port Sudan: goods loaded 1 167, unloaded 2 006
Airports Khartoum, Juba, Malakal, Port Sudan and 11 other airports with scheduled flights

Durable equipment (at end-year)	000	no per 1 000 people	
Radio sets (1972)	1 310	80	no per
Television sets (1974)	100	6	km of road
Passenger cars (1972)	29	2	1.5
Commercial vehicles (1972)	21	1	1.1

Production

Gross domestic product
1974/75 (year ending June 30th): S £ 1511 mn = $ 4 338 mn = £ 1 846 mn
1976 est: S £ 1 700** mn = $ 4 900** mn = £ 2 700** mn
Structure of gross domestic product (1974/75) *By origin* Agriculture 39 %, manufacturing 9 %, construction 4 %, other 48 % *By type* Final consumption expenditure 91 % (of which, government 14 %), stock investment 3 %, gross fixed capital formation 14 %, exports of goods and services 12 %, less imports of goods and services −21 %

Production index (1970 = 100)	1960	1970	1976	Growth %pa 1960–70	1970–76
Agricultural	62	100	122	4.9	3.3

Main products (000 t)	1960	1970	1976	1960–70	1970–76
Agriculture					
Wheat	25	115	264	16.5	14.8
Millet	238	460	450*	6.8	−0.4*
Sorghum (dura)	1 051	1 529	1 800*	3.8	2.7*
Groundnuts	192	337	980	5.8	19.5
Sesameseed	127	297	265	8.9	−1.9
Dates	30	72*	105*	9.1*	6.5*
Sugar, raw value	10**	82*	185*	23.4**	14.5*
Cotton	128	246	114*	6.8	−12.0*
Milk, cow	600**	760**	866*	2.4**	2.2**
Milk, goat	200**	300**	362*	4.1**	3.2**
Milk, sheep	60**	90**	118*	4.1**	4.6**
Beef and veal	84	150**	158*	6.0**	0.9**
Sheep and goat meat	55**	94*	107*	5.5**	2.2*
Wool	2.8**	5.0**	6.0*	6.0**	3.1**
Hides and skins	30**	30*	36*	0.0**	3.1*
Fish catch	16	22	23ᵃ	3.3	1.0ᵇ
Timber (000 m³)	13 986	20 970	22 306ᵃ	4.1	1.2ᵇ
Other					
Petroleum products	—	668*	1 072*ᵃ	na	9.9*ᵇ
Electricity (mn kW h)	74	392	350*ᵃ	18.1	−2.2*ᵇ
Chromium ore (oxide content)	—	13.9	7.8ᵃ	na	−10.9ᵇ
Salt	54	63	66ᵃ	1.6	0.9ᵇ
Beer (000 hl)	45	57	93ᶜ	2.4	13.0ᵈ
Cigarettes (mn units)	na	734	680*ᵃ	na	−1.5*ᵇ
Cotton fabrics (mn m²)	na	82	103ᵃ	na	4.7ᵇ
Cement	91	156	140*ᵃ	5.5	−2.1*ᵇ

ᵃ1975 ᵇ1970–75 ᶜ1974 ᵈ1970–74

Transport traffic	1960	1970	1976	Growth %pa 1960–70	1970–76
Passenger-kilometres (mn)					
Air	46	156	345	13.0	14.1
Cargo: tonne-kilometres (mn)					
Rail	1 608	2 684	2 288ᶜ	5.3	−5.2ᵈ
Air	1.1	2.0	7.5	6.2	24.6
Sea: tonnes (mn)					
Goods loaded	0.69	0.99	1.01ᵃ	3.7	0.4ᵇ
Goods unloaded	0.97	1.85	2.22ᵃ	6.6	3.1ᵇ

ᵃ1975 ᵇ1970–75 ᶜ1973 ᵈ1970–73
Tourism (1975) Number of visitors 30 700, gross receipts $ 5 mn

Finance and trade

Price index (1970 = 100)	1960	1970	1976	Growth %pa 1960–70	1970–76
Consumer prices	71.7	100.0	210.8	3.4	13.2
Money stock (end-year, S £ mn)	37.1	109.6	306.7	11.4	18.7

Budget (1976/77; year ending June 30th)
Revenue: S £ 448 mn = $ 1 286 mn = £ 751 mn
Expenditure: S £ 378 mn = $ 1 085 mn = £ 634 mn

Balance of payments ($ mn)	1972	1973	1974	1975	1976
Balance of goods (fob)	+8	+107	−157	−321	−83
Balance of services	−67	−84	−139	−140	−93
Balance of transfers	+7	−1	+19	+45	+7
Current balance	−52	+22	−277	−416	−169
Long-term capital flow	+35	+6	+268	+241	+183
Reserves and debt (end-year, $ mn)					
International reserves	36	61	124	36	24
External public debt	475	667	1 264	1 535	na

External trade (1976) Imports: S £ 341 mn = $ 980 mn = £ 543 mn
Exports: S £ 193 mn = $ 554 mn = £ 307 mn

Main imports	% of total	Main exports	% of total
Machinery	33	Cotton	50
Chemicals	11	Groundnuts	20
Motor vehicles	10	Sesameseed	9
Textile yarns and fabrics	8	Gum arabic	5
Sugar	7		
Iron and steel	3		
Main sources		*Main destinations*	
United Kingdom	20	Italy	20
United States	9	Japan	7
West Germany	8	France	7
Italy	8	West Germany	7
Iraq	8	Jugoslavia	5
Japan	7	Netherlands	5
India	6	China	4

Special focus

Petroleum products (production began in 1964; 000 tonnes)

	1965	1970	1973	1974	1975
Naphtha	28	26	33*	35*	42*
Motor spirit	55	85*	105*	110*	114*
Jet fuel and kerosene	54	82	115*	120*	83*
Distillate fuel oils	148	234	265*	275*	447*
Residual fuel oil	303	239	185*	190*	383*
Liquefied petroleum gas	1	2	2*	3*	3*
Total	589	668*	705*	733*	1 072*

Swaziland

Kingdom of Swaziland
Umbuso weSwatini

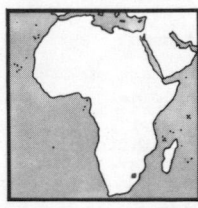

Location South-east Africa Mozambique is to the east and South Africa is on other borders. Land-locked
Land Area 17 365 km² = 6 705 mi²
Climate Sub-tropical
Weather at Mbabane, 1 163 *m altitude*
Temperature: hottest months Jan, Feb 15–25°C coldest June 6–19 °C
Rainfall (av monthly): driest month June 20 mm, wettest Jan 254 mm
Time 2 hours ahead of GMT
Measures Metric system, which replaced the UK (imperial) system in general from 1974

Swaziland

Monetary unit Lilangeni (Li) = 100 cents, lilangeni becoming emalangeni (Ei) for 2 or more; new currency introduced September 1974 to replace the South African rand at Li 1 = R 1
Rate of exchange (1976 av): par Li 1 = R 1, $ 1.15 = Li 1, free Li 1.571 = £ 1

Summary

Political Parliamentary monarchy, which became independent September 6, 1968; formerly a UK protected state. Member of UN, OAU, Commonwealth and an EEC ACP state
Economic The main exports are agricultural products, especially sugar, citrus fruit, and wood pulp, and mineral products, especially iron ore and asbestos. Livestock and products are important locally. Industry includes processing of agricultural produce, fertilisers and television set assembly (for the South African market). Tourism is being developed.

People, resources and equipment

Population 1960 320 000*, 1970 420 000*, 1976 500 000*
Growth: 1960–70 2.8* %pa, 1970–76 2.9* %pa
Density (1976): 29* people per km²
Vital statistics (rate per 1 000 people, 1970–75): births 49.0*, deaths 21.8*
Cities (population in 000, 1975) Mbabane (capital[a]) 24*, Manzini 10*
[a]Lobamba is to become the legislative capital
Race (1975) African 97* % (Swazi 90* %), European 2* %
Language English and siSwati (Swazi)
Religion (1976) Christian 60** %, Animist 40** %
Education (1975) Pupils 107 700*, teachers 3 300*
Labour force (1976) 225 000*; in agriculture 172 000* (76* %)
Employed in South Africa (1974): 9 051
Personnel (1974) Physicians 54, 1 per 8 890 people
Standard of living
National income per person (1976): Ei 320** = $ 370** = £ 200**
Consumption per person (1975): electricity 230* kW h
Telephones (Dec 1975): 7 400, 15 per 1 000 people
Livestock (000, 1976) Cattle 620*, sheep 35*, goats 260*, pigs 38*
Mineral reserves Coal (1961) 5 022 mn tonnes
Electrical capacity (1975) 72 megawatts
Hospital beds (1973) 1 719, 1 per 270 people
Roads (1975) 2 750* km = 1 700* mi, density 0.16* km per km²
Railways (1975) 225 km = 140 mi, density 0.013 km per km²
Airport Matsapa (Manzini)
Durable equipment Radio sets (Dec 1974): 53 000, 110 per 1 000 people
Passenger cars (1975): 7 100, 14 per 1 000 people, 2.6* per km of road
Commercial vehicles (1975): 5 600, 11 per 1 000 people, 2.0* per km of road

Production, finance and trade

Gross domestic product
1973/74 (year ending June 30th): Ei 152 mn = $ 227 mn = £ 95 mn
1976 est: Ei 200** mn = $ 230** mn = £130** mn
Agricultural production index (1970 = 100) 1976 133; growth 1970–76 4.9 %pa
Main products (000 t, 1976) *Agriculture* Maize 110, rice 4*, potatoes 7*, tomatoes 4*, sugar, raw value 208*, citrus fruit 75*, pineapples 21*, cotton 6*, milk 33*, beef and veal 14*, timber (000 m³, 1975) 1 850* *Other* (1975) Coal 127, electricity (mn kW h) 112, iron ore (Fe content) 1 417, asbestos 38, wood pulp 143
Transport traffic (1976) *Air* 5 mn passenger-km
Rail Mainly iron ore transported to Maputo in Mozambique
Tourism Number of visitors (1974) 110 000
Budget (1976/77; year ending March 31st)
Revenue: Ei 70 mn = $ 80 mn = £ 46 mn
Expenditure: Ei 75 mn = $ 86 mn = £ 50 mn
External public debt (Dec 1976) $ 67 mn
External trade (1975) Imports: Ei 135 mn = $ 184 mn = £ 83 mn
Exports: Ei 132 mn = $ 181 mn = £ 81 mn
Included in the South Africa Customs Union

Main imports (1973)	% of total	Main exports (1975)	% of total
Machinery	14*	Sugar	54
Chemicals	10	Wood pulp	9
Motor vehicles	10*	Iron ore	9
Petroleum products	8*	Asbestos	7
Food	8	Citrus fruit	3
Beverages and tobacco	6		
Main sources (1975)		Main destinations (1970)	
South Africa and		United Kingdom	25
United Kingdom		Japan	24
		South Africa	21

Tanzania

The United Republic of Tanzania
Jamhuri ya Muungano wa Tanzania

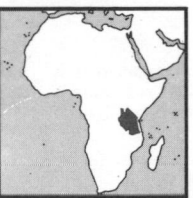

Location East central Africa
With a coastline on the Indian Ocean, Kenya and Uganda are to the north, Rwanda, Burundi and Zaire to the west, and Zambia, Malawi and Mozambique to the south. Territory includes the islands of Zanzibar and Pemba in the Indian Ocean, about 40 km off the coast
Land Area 942 000 km² = 363 710 mi² of which, mainland 939 360 km² = 362 690 mi², Zanzibar and Pemba 2 640 km² = 1 020 mi²
Climate Tropical on the coast, semi-temperate inland
Weather at Dar es Salaam, 14 m altitude
Temperature: hottest month Feb 25–31 °C, coldest July, Aug 19–28 °C
Rainfall (av monthly): driest month Aug 25 mm, wettest April 290 mm
Time 3 hours ahead of GMT
Measures Metric system, introduced from 1969 to replace the UK (imperial) system; also:
weight (*mass*) 36 ratili = 1 frasila = 36 pounds = 16.33 kilograms
Monetary unit Tanzanian shilling (T Sh) = 100 cents; also used is Tanzanian pound (T £) = 20 T Sh. The decimal currency was introduced June 14, 1966 to replace the East African pound (EA £) at T Sh 20 = EA £ 1
Rate of exchange (1976 av): par T Sh 9.66 = SDR1, free T Sh 8.367 = $ 1, T Sh 15.11 = £ 1

Summary

Political One-party republic, which became independent December 9, 1961; formerly, as Tanganyika, a UK trust territory. Zanzibar, formerly a UK protectorate, joined with Tanganyika on April 26, 1964, having become independent December 9, 1963. Member of UN, OAU, Commonwealth and an EEC ACP state. Member formerly of the East African Community during its existence from 1967 to July 1977
Economic Mainly an agricultural economy, with large exports of cotton, coffee, sisal, cloves (from Zanzibar), tobacco and cashew nuts. There are some minerals, including diamonds, coal and iron ore. Refining of imported crude oil is important. Some manufacturing industry has been developed, especially brewing, chemicals, textiles and cement

People, resources and equipment

Population 1960 10.33* mn, 1970 13.27* mn, 1976 15.61* mn
Growth: 1960–70 2.5* %pa, 1970–76 2.7* %pa
Density (1976): 17* people per km²
Vital statistics (rate per 1 000 people, 1967): births 47*, deaths 22*
Regions (population in 000, 1974; total of 14.76 mn)
Tanganyika (mainland) 14 346, Zanzibar 412
Cities (population in 000, 1975) Dar es Salaam (capital[b]) 517
Tanga 70[a] Arusha 47[a] Moshi 33[a] Dodoma 28[a] Mtwara 22[a] Zanzibar 68[a] Mwanza 42[a] Morogoro 30[a] Tabora 23[a] Kigoma 21*[a]
[a]1970 [b]The capital is to be moved to Dodoma
Race (1970) African 98 % (of whom, Sukuma 12 %, Makonde 4 %)
Language Swahili and English; local languages are also used
Religion (1976) Animist 35** %, Christian 30** %, Moslem 30** %
Education (1975/76) Pupils 1 658 085, teachers 33 435
Labour force (1976) 6 637 000*; in agriculture 5 516 000* (83* %)
Personnel (1973) Physicians: 540*, 1 per 26 600* people
Standard of living
National income per person (1976): T Sh 1 250** = $ 150** = £ 83**
Consumption per person (1975): energy 69 kg coal equivalent, electricity (production) 42* kW h, newsprint 0.1 kg, steel 5 kg
Newspapers: number (1974) 3; circulation (1969) 61 000, 5 per 1 000 people
Telephones (Dec 1975): 63 000, 4 per 1 000 people
Livestock (000, 1976) Cattle 14 362*, goats 4 600*, sheep 2 900*, chickens 20 354*, ducks 2 385*
Mineral reserves Coal (1967) 370 mn tonnes
Iron ore (1974) 90** mn tonnes
Petroleum refinery capacity (1975) 0.85* mn tonnes
Electrical capacity (1975) 160* megawatts
Hospital beds (1970) 17 600*, 1 per 750* people
Roads (1975) 19 200* km = 11 900* mi, density 0.02* km per km²
Railways (1976) 3 545 km = 2 203 mi, density 0.004 km per km²
Oil pipeline Dar es Salaam to Zambia 1 700 km = 1 060 mi
Inland waterways Lakes Tanganyika, Victoria and Malawi

Tanzania

Ships (registered, 1977) 22, total of 35 613 gross tons
Ports (goods traffic, 000 tonnes, 1973)

	loaded	unloaded
Dar es Salaam	815	2 391
Tanga	131	139
Mtwara	25	92

Also Mwanza (on Lake Victoria), Kigoma (on Lake Tanganyika)
Airports Passenger departures and arrivals (1975): Dar es Salaam 291 685, Arusha 39 412; also (1977) 11 other airports with scheduled flights

Durable equipment

(Dec 1974)	000	no per 1 000 people	
Radio sets	231	15	no per
Television sets	5**	0.3**	km of road
Passenger cars	39	3	2.0*
Commercial vehicles	42	3	2.2*

Production

Gross domestic product 1975[a]: T Sh 18 583 mn = $ 2 507 mn = £ 1 128 mn
1976 est[b]: T Sh 20 500** mn = $ 2 450** mn = £ 1 360** mn
[a]Mainland only [b]Includes estimate for Zanzibar of T Sh 500*** mn
Structure of gross domestic product (mainland only, 1975) *By origin*
Agriculture 37 %, mining and quarrying 1 %, manufacturing 9 %, construction 4 %, transport and communication 7 %, other 42 %

Production index	1960	1970	1976	Growth %pa	
(1970 = 100)				1960–70	1970–76
Agricultural	78	100	120	2.6	3.1
Main products (000 t)					
Agriculture					
Rice	112	184	430*	5.1	15.2*
Maize	559	767	1 619	3.2	13.3
Millet and sorghum	346	245	590*	−3.4	15.8*
Sweet potatoes	230	310	441*	3.0	6.0*
Cassava	1 796	6 261	5 100*	13.3	−3.3*
Dry beans	85	129[c]	146	4.3*	2.1*
Sugar, raw value	32	95	103*	11.5	1.3*
Mangoes	100**	145	172*	3.8**	2.9*
Bananas	430	679*	770*	4.7*	2.1*
Cashew nuts	60**	117	83	6.9**	−5.5
Coffee	27	47	55*	5.7	2.7*
Tea	4	8	13	7.8	8.4
Tobacco	2	11	19	18.6	9.5
Sisal	208	202	119*	−0.3	−8.4*
Cotton	33	77	68*	8.8	−2.0*
Milk	460	709	678*	4.4	−0.7*
Beef and veal	120	128*	121*	0.6*	−0.9*
Hides and skins	6*	27*	33*	16.2*	3.4*
Fish catch	69	185	181[b]	10.4	−0.4[c]
Timber (000 m³)	24 200*	31 595*	33 862*[b]	2.7*	1.4*[c]
Other					
Petroleum products	—	682	669[b]	na	−0.4[c]
Electricity[a] (mn kW h)	210*	462	636*[b]	8.2*	6.6*[c]
Diamonds (000 CM)	548	708	896[b]	2.6	4.8[c]
Salt	35	42	44[b]	1.8	0.9[c]
Beer (000 hl)	60	386	589[b]	20.5	8.8[c]
Cigarettes (mn units)	na	2 599	3 608[b]	na	6.8[c]
Cement	—	177	246	na	5.6
Cotton fabrics (mn m³)	na	39	86[d]	na	21.9[e]
Rayon fabrics (mn m³)	na	14	na	na	na
Plywood (000 m³)	na	1 122	1 147[d]	na	0.6[e]
Fertilisers, phosphate	—	—	14*[h]	na	na

[a]Mainland only (excludes Zanzibar) [b]1975 [c]1970–75 [d]1974 [e]1970–74

Transport traffic[a]	1960	1970	1976	Growth %pa	
Air[b] (mn)				1960–70	1970–76
Passenger-kilometres	119[d]	267	195	14.4[e]	−5.1
Cargo: tonne kilometres	3.4[d]	10.4	3.7	20.5[e]	−15.8
Sea[c]: tonnes (mn)					
Goods loaded	0.82	1.40	0.84[f]	5.5	−9.7[g]
Goods unloaded	0.69	2.29	2.97[f]	12.7	5.3[g]

[a]For rail traffic, see Kenya [b]Apportionment of traffic of East African Airways Corporation. Before October 1972 apportionment was one-third of the total; from that time (until the Corporation was disbanded in 1977) on the basis of ticket origin and destination [c]Mainland only. In 1962 Zanzibar accounted for 7 % of total goods loaded, and 9 % of goods unloaded. [d]1964 [e]1964–70 [f]1975 [g]1970–75
Tourism Number of visitors (1973) 120 000; gross receipts (1975) $ 11 mn

Finance and trade

Consumer price index (1970 = 100) 1976 200.6; growth 1970–76 12.3 %pa
Money stock (end-year, T Sh mn) 1976 5 486; growth 1970–76 20.9 %pa
Budget (1975/76; year ending June 30th)
Revenue: T Sh 4 063 mn = $ 507 mn = £ 254 mn
Expenditure: T Sh 6 325 mn = $ 789 mn = £ 395 mn

Balance of payments ($ mn)	1972	1973	1974	1975	1976
Balance of goods (fob)	−44	−74	−247	−296	−149
Balance of services	−18	−38	−71	−35	+59
Balance of transfers	−4	+5	+45	+93	+87
Current balance	−66	−107	−273	−237	−3
Long-term capital flow	+108	+155	+129	+172	+86
Reserves and debt (end-year, $ mn)					
International reserves	120	145	50	65	112
External public debt	701	834	1 122	1 210	1 414

External trade (1976) Imports: T Sh 5 350 mn = $ 639 mn = £ 354 mn
Exports: T Sh 4 108 mn = $ 491 mn = £ 272 mn

Main imports[a]	% of total	Main exports[a]	% of total
Machinery	21	Cotton	16
Crude oil	13	Coffee	13
Chemicals	9	Cloves	8
Iron and steel	5	Sisal	6
Railway vehicles	5	Tobacco	5
Motor vehicles	4	Cashew nuts	5
Textile yarns & fabrics	4	Petroleum products	5
Main sources		*Main destinations*	
United Kingdom	12	West Germany	14
Kenya	11	United Kingdom	13
Iran	9	United States	9
West Germany	8	Singapore	7
Japan	7	Italy	6
China	6	Kenya	6

[a]Domestic (excluding Kenya and Uganda)

Togo

Togolese Republic
République Togolaise

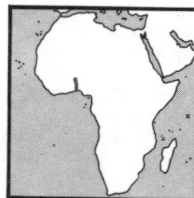

Location West Africa
The hinterland runs north from a coastal strip on the Atlantic Ocean; Benin is to the east, Ghana to the west and Upper Volta to the north
Land Area 56 000 km² = 21 600 mi²
Climate Tropical, drier in the north
Weather at Lomé
Temperature: av monthly 27 °C
Rainfall (av monthly): 65 mm
Time GMT

Measures Metric system
Monetary unit CFA franc (CFA Fr) = 100 centimes
Rate of exchange (1976 av): par CFA Fr 50 = Fr 1, free CFA Fr 239.0 = $ 1, CFA Fr 431.6 = £ 1

Summary

Political One-party republic with military government, which became independent April 27, 1960; formerly a trust territory administered by France (and before the 1914–18 war part of the German colony of Togoland). Member of UN, OAU, Ecowas, Ocam, franc zone and an EEC ACP state
Economic A mining and agricultural economy, with phosphates accounting for some two-thirds of exports and cocoa and coffee for one-quarter. There is some light industry. The 1976–80 development plan envisaged expenditure of CFA Fr 250 bn, expansion of transport facilities and telecommunications being important

People, resources and equipment

Population 1960 1.50** mn, 1970 1.96* mn, 1976 2.28* mn
Growth: 1960–70 2.7** %pa, 1970–76 2.6* %pa
Density (1976): 41* people per km²
Vital statistics (rate per 1 000 people, 1970–75): births 50.6*, deaths 23.3*
Cities (population in 000, 1975)

Lomé (capital)	230*	Palimé	24*	Bassari	21*
Sokodé	35*	Atakpamé	22*	Tsévié	17*

Race (1960) Ewe 21 %, Kabre 14 %, Watyi 12 %, Nandeba 6 %
Language French; also local languages, especially Ewe and Kabre
Religion (1975) Animist 65** %, Christian 24 %, Moslem 10 %
Education (1975/76) Pupils 429 838, teachers 7 950

Togo

Labour force (1976) 970 000*; in agriculture 681 000* (70* %)
Personnel Scientists and engineers (1971): 461
Physicians (1973): 100, 1 per 21 200 people
Standard of living National income per person (1976):
CFA Fr 56 000** = $ 230** = £ 130**
Consumption per person (1975): energy 65 kg coal equivalent,
electricity (production) 53 kW h, steel 9 kg
Newspapers (1973): number 3; circulation 13 000, 6 per 1 000 people
Telephones (Dec 1974): 8 000, 4 per 1 000 people
Livestock (000, 1976) Cattle 235*, sheep 750*, goats 630*, pigs 270*
Electrical capacity (1975) 25* megawatts
Hospital beds (1972) 3 075, 1 per 680 people
Roads (1976) 7 450 km = 4 629 mi, density 0.13 km per km²
Railways (1975) 445 km = 277 mi, density 0.008 km per km²
Ports (goods traffic, 000 tonnes, 1973) Port Kpémé (phosphates port):
loaded 2 197; Port Lomé: loaded 56, unloaded 385
Airport Tokoin (Lomé): passenger departures and arrivals (1975) 73 119
Durable equipment (at end-year)
Radio sets (1974): 50 000, 23 per 1 000 people
Passenger cars (1976): 22 500, 10 per 1 000 people, 3.0 per km of road
Commercial vehicles (1976): 3 400, 1 per 1 000 people, 0.5 per km of road

Production, finance and trade

Gross domestic product 1975: CFA Fr 123 600 mn = $ 577 mn = £ 260 mn
1976 est: CFA Fr 140 000** mn = $ 590** mn = £ 320** mn

Production index	1960	1970	1976	Growth %pa	
(1970 = 100)				1960–70	1970–76
Agricultural	74	100	73	3.1	−5.1

Main products (000 t)

Agriculture					
Cassava	574	700**	448	2.0**	−7.2**
Yams	800*	1 000**	750*	2.3**	−4.7**
Palm kernels	14*	17*	12*	2.0*	−5.6*
Palm oil	5*	6*	6*	1.8*	0.0*
Coffee	9	14	12*	4.5	−2.5*
Cocoa	13	28	16*	8.0	−8.9*
Tobacco	2.0**	2.0*	2.0*	0.0**	0.0*
Cotton	2.0**	2.0*	2.0*	0.0**	0.0*
Timber (000 m³)	778	1 197*	1 363*c	4.4*	2.6*d
Other					
Electricity (mn kW h)	14a	64	118e	20.6b	13.0d
Phosphate rock	—	1 508	2 553e	na	14.1f
Beer (000 hl)	—	94	200e	na	16.3d
Soft drinks (000 hl)	—	35	62g	na	21.0h
Cotton fabrics (mn m)	—	7	10g	na	12.6h
Cement	—	—	150c	na	na

a1962 b1962–70 c1975 d1970–75 e1974 f1970–74 g1973 h1970–73

Transport traffic	1960	1970	1976	Growth %pa	
Passenger-kilometres (mn)				1960–70	1970–76
Rail	79	84	65b	0.6	−6.2c
Aira	na	65	126	na	11.7
Cargo: tonne-kilometres (mn)					
Rail	8	12	22b	4.1	21.8c
Aira	na	6.3	13.8	na	14.0
Sea: tonnes (mn)					
Goods loaded	0.048	1.576	2.643b	41.8	13.8c
Goods unloaded	0.086	0.255	0.345b	11.5	7.8c

aIncludes apportionment of traffic of Air Afrique b1974 c1970–74
Tourism (1974) Number of visitors 24 100
Consumer price index (1970 = 100) 1976 176.9; growth 1970–76 10.0 %pa
Money stock (end-year, CFA Fr bn) 1976 32.9; growth 1970–76 21.7 %pa
Budget (1976) Balanced at CFA Fr 50 019 mn = $ 209 mn = £ 116 mn
International reserves (Dec 1976) $67 mn
External public debt (Dec 1975) $ 157 mn
External trade (1976)
Imports: CFA Fr 48 505* mn = $ 203* mn = £ 112* mn
Exports: CFA Fr 32 006* mn = $ 134* mn = £ 74* mn

Main imports (1974)	% of total	*Main exports* (1975)	% of total
Textile yarns and fabrics	17	Phosphates	65
Machinery	14	Cocoa	17
Food	11	Coffee	7
Petroleum products	10		
Motor vehicles	9		
Main sources (1975)		*Main destinations* (1975)	
France	35	France	39
United Kingdom	11	Netherlands	32
West Germany	11	West Germany	10

Tunisia

Republic of Tunisia
République Tunisienne
Al Jumhuriya al Tunisiya

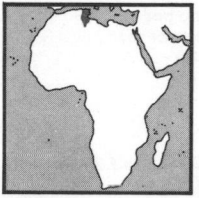

Location North Africa
The Mediterranean Sea forms the northern
border, with Algeria to the west and Libya to
the south-east
Land Area 163 610 km² = 63 170 mi²
Climate Temperate on the coast, hot and dry
inland
Weather at Tunis, 66 m altitude
Temperature: hottest month August 21–33 °C,
coldest January 6–14 °C
Rainfall (av monthly): driest month July 3 mm, wettest Jan 63 mm
Time 1 hour ahead of GMT
Measures Metric system
Monetary unit Tunisian dinar (T D) = 1 000 millimes
Rate of exchange (1976 av): free $ 2.332 = T D 1, £ 1.291 = T D 1

Summary

Political Republic, which became independent March 20, 1956; formerly
a French protectorate. Member of UN, OAU, Arab League, and the
Maghreb Organisation
Economic A mining and agricultural economy, with crude oil accounting
for over one-third of exports, phosphate fertilisers for about one-fifth and
olive oil for one-tenth. Tourism is also important. There is a wide range of
industry including especially oil refining, phosphate processing, cement,
sulphuric acid, steel, television sets, and assembly plants for motor
vehicles

People, resources and equipment

Population 1960 4.23*mn, 1970 5.13*mn, 1976 5.74*mn
Growth: 1960–70 1.9* %pa, 1970–76 1.9* %pa
Density (1976): 35* people per km²
Vital statistics (rate per 1 000 people): births (1976) 36.4,
deaths (1970–75) 13.8*
Regions (gouvernorates, population in 000, 1969; total of 5.03 mn)

Tunis	944	Jendouba	273	Nabeul	362
Béja	343	Kairouan	298	Sfax	471
Bizerte	359	Kasserine	226	Sousse	576
Gabès	221	Al Kef	335		
Gafsa	357	Medenine	262		

Cities (population in 000, 1975)

Tunis (capital)	970*	Bizerte	70*	Gabès	40*
Sfax	250*	Jerba	70**	Monastir	40*
Sousse	100*	Kairouan	55*	Béja	35*

Race (1961) Arab 94 %, Berber 2 %, French 2 %
Language Arabic; French is also used
Religion (1976) Moslem 99** %
Education (1975/76) Pupils: primary 932 787, other 202 000*
Teachers: primary 23 320, other 11 000*
Labour force (1976) 1 414 000*; in agriculture 624 000* (44* %)
Personnel Scientists and engineers engaged in research (1972): 818
Physicians (1971): 1 004, 1 per 5 220 people
Standard of living
National income per person (1976): T D 315* = $ 735* = £ 407*
Consumption per person (1975): energy 447 kg coal equivalent,
electricity (production) 240 kW h, newsprint 0.6 kg, steel 56 kg
Newspapers (1974): number 4; circulation 156 000, 28 per 1 000 people
Telephones (Dec 1975): 129 000, 23 per 1 000 people
Livestock (000, 1976) Cattle 880*, sheep 3 526*, goats 900*, camels 195*
Mineral reserves (1975) Crude oil 303 mn tonnes
Natural gas 81 bn cubic metres
Petroleum refinery capacity (1975) 1.2* mn tonnes
Electricity capacity (1975) 426 megawatts
Hospital beds (1972) 12 721, 1 per 420 people
Roads (1976) 21 595 km = 13 419 mi, density 0.13 km per km²
Railways (1975) 1 928 km = 1 198 mi, density 0.01 km per km²
Ships (registered, 1977) 39, total of 100 128 gross tons
Ports (goods traffic, 000 tonnes, 1973)

	loaded	unloaded
La Skhirraa	12 113	—
Sfax	2 698	482
Tunis-La Goulette	947	1 065
Bizerte	346	1 395
Sousse	133	182

aCrude oil port

Tunisia

Airports (1975) Passenger departures and arrivals: Tunis (Carthage)
1 500 569, Monastir 533 805, Jerba 305 853

Durable equipment (at end-year)	000	no per 1 000 people	
Radio sets (1974)	277	50	no per
Television sets (1973)	147	27	km of road
Passenger cars (1976)	105	18	4.9
Commercial vehicles (1976)	68	12	3.1

Production

Gross domestic product 1976: T D 1 905 mn = $ 4 442 mn = £ 2 459 mn
Growth in real terms: 1960–70 5.1 %pa, 1970–75 9.0 %pa
Structure of gross domestic product (1975) *By origin* Agriculture 17 %,
mining and quarrying 10 %, manufacturing 9 %, electricity, gas and water
1 %, construction 8 %, distribution and hotels 15* %, transport and
communication 5 %, other 35 %

Production indices (1970 = 100)	1960	1970	1976	Growth %pa 1960–70	1970–76
Agricultural	99	100	145	0.1	6.3
Industrial	88.0[b]	100.0	137.5	4.4[c]	5.5
Main products (000 t)					
Agriculture					
Wheat	439	519	919	1.7	10.0
Barley	136	151	231	1.1	7.3
Tomatoes	65	127	270*	6.9	13.4*
Grapes	230	102	160*	−7.8	7.8*
Oranges	51*	75	86	3.9*	2.3
Olives	550*	420	530*	−2.7*	4.0*
Olive oil	142	90	102	−4.5	2.1
Dates	48	39	48	−2.1	3.5
Milk	105	142	191*	3.1	5.1*
Fish catch	16	24	43[d]	4.1	12.4[e]
Timber (000 m³)	1 265	1 587	1 818[d]	2.3	2.7[e]
Energy					
Total energy (000 tce)	20	5 410	7 110[d]	75.1	5.6[e]
Crude oil	—	4 151	3 712	na	−1.8
Petroleum products	—	1 106	1 135[d]	na	0.5[e]
Natural gas (mn m³)	na	5	214	na	87.0
Manufactured gas (mn m³)	16	19	22	1.7	2.5
Electricity (mn kW h)	316	794	1 346[d]	9.7	11.1[e]
Mining					
Iron ore (Fe content)	563	422	262	−2.8	−7.6
Lead conc (Pb content)	17.9	22.0	10.4	2.1	−11.8
Salt	114	300	420[d]	10.2	7.0[e]
Phosphate rock	2 101	2 969	3 512[d]	3.5	3.4[e]
Manufacturing					
Beer (000 hl)	143	204	309[d]	3.6	8.7[e]
Cigarettes (mn units)	1 739	3 286	4 863[d]	6.6	8.1[e]
Sulphuric acid	na	424	634[d]	na	8.4[e]
Fertilisers, phosphate	58*	177	145[d]	12.0*	−3.9[e]
Tyres (000 units)	—	104	156	na	7.1
Cement	405	547	478	3.1	−2.2
Pig iron	—	125	103	na	−3.2
Crude steel	—	100	129[d]	na	5.2[e]
Radio sets (000 units)	—	37	91[d]	na	19.7[e]
Television sets (000 units)	—	24	35[d]	na	7.8[e]
Motor vehicles[a] (000 units)	—	0.94	4.18	na	28.0

[a]Assembly only [b]1967 [c]1967–70 [d]1975 [e]1970–75

Transport traffic	1960	1970	1976	Growth %pa 1960–70	1970–76
Passenger-kilometres (mn)					
Road[a]	na	1 130	1 540[b]	na	6.4[c]
Rail	351	442	641	2.3	6.4
Air	73	256	968	13.4	24.8
Cargo: tonne-kilometres (mn)					
Road	na	500	690[b]	na	6.7[c]
Rail	971	1 328	1 277	3.2	−0.7
Air	0.9	1.8	7.5	7.2	26.9
Sea: tonnes (mn)					
Goods loaded	3.9	7.2	5.5*[b]	6.3	−5.2*[c]
Goods unloaded	1.3	1.9	3.9[b]	3.9	15.2[c]
Tourism					
Number of visitors (000)	53	411	1 014[b]	22.7	19.8[c]
Gross receipts ($ mn)	3**	65	190[d]	36.0**	30.8[e]

[a]Public transport only [b]1975 [c]1970–75 [d]1974 [e]1970–74

Finance and trade

Price index (1970 = 100)	1960	1970	1976	Growth %pa 1960–70	1970–76
Consumer prices	76.0	100.0	135.8	2.8	5.2
Money stock					
(end-year, T D mn)	87.8	192.7	513.8	8.2	17.8

Budget (total, 1976) Balanced at T D 641 mn = $ 1 495 mn = £ 828 mn

Balance of payments ($ mn)	1972	1973	1974	1975	1976
Balance of goods (fob)	−140	−206	−119	−453	na
Balance of services	+50	+15	+16	+96	na
Balance of transfers	+86	+131	+132	+174	na
Current balance	*−4*	*−1*	*+29*	*−183*	na
Long-term capital flow	+90	+134	+160	+165	na
Reserves and debt (end-year, $ mn)					
International reserves	223	307	418	385	371
External public debt	1 118	1 309	1 472	1 738	na

External trade (1976) Imports: T D 657 mn = $ 1 531 mn = £848 mn
Exports: T D 338 mn = $ 789 mn = £ 437 mn

Main imports (1975)	% of total	Main exports (1975)	% of total
Machinery, non-electric	17	Crude oil	42
Food	13	Phosphate fertilisers	19
(of which, sugar 4, cereals 4)		Olive oil	9
Transport equipment	10	Clothing	5
(of which, motor vehicles 7)		Phosphoric acid	4
Chemicals	8		
Crude oil	6		
Main sources (1976)		*Main destinations* (1976)	
France	32	Italy	21
West Germany	10	France	17
Italy	9	Greece	15
United States	6	United States	14
Saudi Arabia	4	West Germany	7

Special focus

Structure of gross domestic product

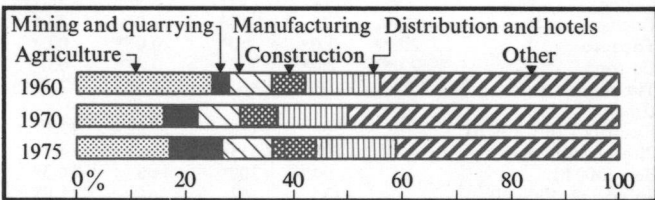

Uganda
Republic of Uganda

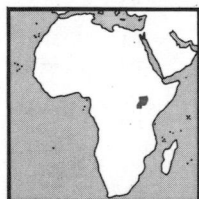

Location East Africa
Kenya is to the east, Tanzania and
Rwanda to the south, Zaire to the
west, and Sudan to the north. Lake
Victoria is part of the southern
border, and is shared with Kenya
and Tanzania. Land-locked
Land Area 236 036 km² = 91 134 mi²
Climate Tropical
Weather at Kampala, 1 312 m altitude
Temperature: hottest month Jan 18–28 °C, coldest July 17–25°C
Rainfall (av monthly): driest month July 46 mm, wettest April 175 mm
Time 3 hours ahead of GMT
Measures Metric system, introduced from 1970 to replace the UK
(imperial) system
Monetary unit Ugandan shilling (U Sh) = 100 cents; also used is
Ugandan pound (U £) = 20 U Sh. The decimal currency was introduced
in 1966 to replace the East African pound (EA £) at U Sh 20 = EA £ 1
Rate of exchange (1976 av): par U Sh 9.66 = SDR 1, free U Sh 8.367 = $ 1,
U Sh 15.11 = £1

Summary

Political One-party republic with military government, which became
independent October 9, 1962. Non-citizen Asians were expelled in 1972.
Member of UN, OAU, Commonwealth and an EEC ACP state. Member
formerly of the East African Community during its existence from 1967 to
to July 1977

Uganda

Economic An agricultural economy, with coffee accounting for four-fifths of exports, and cotton and tea also significant; there are some minerals, including especially copper. There is a range of light manufacturing industry

People, resources and equipment

Population 1960 6.68*mn, 1970 9.81*mn, 1976 11.94*mn
Growth 1960–70 3.9* %pa, 1970–76 3.3* %pa
Density (1976): 51* people per km²
Vital statistics (rate per 1 000 people, 1970–75): births 45.2*, deaths 15.9*
Regions (population in 000, 1969; total of 9.55 mn)
Buganda 2 667 Eastern 2 817 Northern 1 632 Western 2 433
Cities (population in 000, 1975) Kampala (capital) 332*, Jinja-Bugembe 100*, Mbale 24*, Gulu 22*, Entebbe 21*
Race (1976) African 99.8** %
Language English; local languages, especially Luganda, are also used
Religion (1976) Christian 60** %, Animist 30** %, Moslem 10** %
Education (1974) Pupils 1 025 894, teachers 31 537
Labour force (1976) 4 909 000*; in agriculture 4 075 000* (83* %)
Personnel (1974) Physicians: 540 1 per 20 700 people
Standard of living
National income per person (1976): U Sh 2 000** = $ 240** = £ 130**
Consumption per person (1975): energy 55 kg coal equivalent, electricity (production) 72* kWh, newsprint 0.2 kg, steel 1 kg
Newspapers (1974): number 4; circulation 58 000, 5 per 1 000 people
Telephones (Dec 1975): 45 000, 4 per 1 000 people
Livestock (000, Dec 1975) Cattle 4 900*, sheep 1 100*, goats 2 150*
Electrical capacity (1975) 163* megawatts
Hospital beds (1974) 15 723, 1 per 710 people
Roads (1975) 27 536 km = 17 110 mi, density 0.12 km per km²
Railways (1976) 1 300* km = 810* mi, density 0.006* km per km²
Inland waterway Lake Victoria
Ship (registered, 1977) 1, of 5 510 gross tons
Airport Entebbe (40 km from Kampala): passenger departures and arrivals (1975) 108 927

Durable equipment (at end-year)	000	no per 1 000 people	no per
Radio sets (1974)	250	22	km of road
Television sets (1972)	15	1	1.0
Passenger cars (1974)	27	2	1.0
Commercial vehicles (1974)	15	1	0.6

Production

Gross domestic product 1971: U Sh 10 367 mn = $ 1 451 mn = £ 597 mn
1976 est: U Sh 25 000** mn = $ 3 000** mn = £ 1 700** mn
Growth in real terms: 1963–70 4.7 %pa, 1970–74 0.5 %pa
Structure of gross domestic product (1971) *By origin* Agriculture 48 %, mining and quarrying 1 %, manufacturing 8 %, other 43 %

Production index (1970 = 100)	1960	1970	1976	Growth %pa 1960–70	1970–76
Agricultural	68	100	103	4.0	0.5
Main products (000 t)					
Agriculture					
Maize	155	391	623	9.7	8.1
Millet	380	630**	650*	5.2**	0.5**
Sorghum	208	500**	538	9.2**	1.2**
Sweet potatoes	440	700*	664*	4.8*	−0.9*
Cassava	1 000**	1 100**	1 000*	1.0**	−1.6**
Coffee	114	221	211	6.8	−0.8
Tea	5	18	21	13.7	2.6
Cotton	67	85	41	2.4	−11.5
Milk	200	275*	327*	3.2*	2.9*
Beef and veal	50**	65*	70*	2.7**	1.2*
Fish catch	63	129	170ª	7.4	5.7ᵇ
Timber (000 m³)	10 689	13 910	14 675*ª	2.7	1.1*ᵇ
Other					
Electricity (mn kW h)	420*	778	830*ª	6.4*	1.3*ᵇ
Copper ore (Cu content)	19	18	11ª	−0.8	−8.8ᵇ
Phosphate rock	4*	16*	15*	15.0*	−1.0*
Beer (000 hl)	78	278	389ª	13.6	7.0ᵇ
Cement	—	191	88	na	−12.1
Fertilisers, phosphate	na	5.9	4.0*ª	na	−7.5*ᵇ
Cotton fabrics (mn m)	12*	42	30ᶜ	13.3*	−8.1ᵈ
Crude steel	14	20	8ª	3.6	−16.7ᵇ

ª1975   ᵇ1970–75   ᶜ1974   ᵈ1970–74

Transport traffic *Rail* See Kenya *Air* (apportionment of traffic of East African Airways Corporation, 1976) Passenger-km 197 mn, cargo 8.2 mn t-km

Finance and trade

Consumer price index (1970 = 100) 1976 434.2; growth 1970–76 27.7 %pa
Money stock (end-year, U Sh mn) 1973 2 126; growth 1970–73 23.7 %pa
Budget (1976/77; year ending June 30th)
Revenue: U Sh 3 111 mn = $ 372 mn = £ 217 mn
Expenditure: U Sh 2 180 mn = $ 261 mn = £ 152 mn

Balance of payments ($ mn)	1972	1973	1974	1975	1976
Balance of goods (fob)	+92	+100	+35	−2	+115
Balance of services	−70	−55	−76	−73	−72
Balance of transfers	−6	−1	−1	+13	+2
Current balance	*+16*	*+43*	*−42*	*−63*	*+46*
Long-term capital flow	+32	−15	+16	+16	−10
External public debt					
(end-year, $ mn)	206	229	251	237	241

External trade (1976) Imports: U Sh 1 342 mn = $ 160 mn = £ 89 mn
Exports: U Sh 3 006 mn = $ 359 mn = £ 199 mn

Main importsª	% of total	Main exportsª	% of total
Machinery	29	Coffee	83
Motor vehicles	14	Cotton	6
Chemicals	13	Tea	3
Cereals	7	Copper	2
Main sources		*Main destinations*	
Kenya	50	United States	33
United Kingdom	15	United Kingdom	20
West Germany	9	France	6
Japan	4	Italy	6
United States	4	Japan	6

ªDomestic (excluding Kenya and Tanzania)

Upper Volta

Republic of Upper Volta
République de Haute-Volta

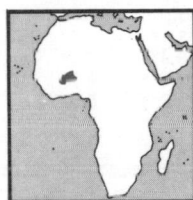

Location West Africa
Mali is to the north and west, Niger to the east, and Benin, Togo, Ghana and Ivory Coast to the south. Land-locked, but is the source of tributaries of the Volta River
Land Area 274 200 km² = 105 870 mi²
Climate Tropical
Weather at Ouagadougou, 302 m altitude
Temperature: hottest month April 26–39 °C, coldest Jan 16–33 °C
Rainfall (av monthly): driest month Dec 0 mm, wettest Aug 277 mm
Time GMT
Measures Metric system
Monetary unit CFA franc (CFA Fr) = 100 centimes
Rate of exchange (1976 av): par CFA Fr 50 = Fr 1, free CFA Fr 239.0 = $ 1, CFA Fr 431.6 = £ 1

Summary

Political One-party republic with military government, which became independent August 5, 1960; formerly a province of French West Africa. Member of UN, OAU, Ecowas, CEAO, Ocam, franc zone and an EEC ACP state
Economic An agricultural economy, with livestock and their products especially important locally and as exports; cotton, groundnuts and sesameseed are the main crop exports. The manufacturing sector is small

People, resources and equipment

Population 1960 4.40*mn, 1970 5.38*mn, 1976 6.17*mn
Growth: 1960–70 2.0* %pa, 1970–76 2.3* %pa
Density (1976): 23* people per km²
Vital statistics (rate per 1 000 people, 1970–75): births 48.4*, deaths 25.8*
Cities (population in 000, 1975)
Ouagadougou (capital) 169* Koudougou 36* Kaya 18*
Bobo Dioulasso 113* Ouahigouya 25* Banfora 12*
Race (1970) Mossi 48 %, Fulani 10 %, Lobi 7 %, Malinke 7 %, Bobo 7 %
Language French; local languages are also used

Upper Volta

Religion (1976) Animist 75** %, Moslem 20** %, Christian 5** %
Education (1974/75) Pupils 149 500*, teachers 3 600*
Labour force (1976) 3 334 000*; in agriculture 2 792 000* (84* %)
Personnel (1974) Physicians: 99, 1 per 59 600 people
Standard of living
National income per person (1976): CFA Fr 18 000** = $ 75** = £ 42**
Consumption per person (1975): energy 20 kg coal equivalent,
electricity (production) 9 kW h
Newspapers (1974): number 1; circulation 2 000, 0.3 per 1 000 people
Telephones (Dec 1975): 6 000, 1 per 1 000 people
Livestock (000, 1976) Cattle 1 900*, sheep 1 300*, goats 2 300*
Electrical capacity (1975) 18* megawatts
Hospital beds (1971) 4 675, 1 per 1 170 people
Roads (1974) 5 250 km = 3 262 mi, density 0.02 km per km²
Railways (1975) 517 km = 321 mi, density 0.002 km per km²
The railway links Ouagadougou with Abidjan, the Ivory Coast sea port
Airports (1975) Passenger departures and arrivals: Ouagadougou 44 740,
Bobo Dioulasso 8 339

Durable equipment (at end-year)	000	no per 1 000 people	
Radio sets (1974)	100	17	no per
Television sets (1971)	6	1	km of road
Passenger cars (1975)	10*	2*	2*
Commercial vehicles (1975)	10*	2*	2*

Production

Gross domestic product 1974: CFA Fr 109 600 mn = $ 456 mn = £ 195 mn
1976 est: CFA Fr 120 000** mn = $ 500** mn = £ 280** mn
Structure of gross domestic product *By origin* (1974) Agriculture 41 %,
manufacturing 10 %, construction 5 %, transport and communication
8 %, distribution and hotels 15 %, other 21 %

Production index (1970 = 100)	1960	1970	1976	Growth %pa 1960–70	1970–76
Agricultural	78	100	107	2.5	1.1

Main products (000 t)
Agriculture

	1960	1970	1976	1960–70	1970–76
Millet	326	378	370*	1.5	−0.3*
Sorghum	411	563	717*	3.2	4.1*
Maize	92	55	46*	−5.0	−2.9*
Cowpeas	96	70*	75*	−3.1*	1.2*
Groundnuts	95	68	87*	−3.3*	4.2*
Sesameseed	3.0**	6.3	8.0*	7.7**	4.1*
Cotton	1**	12	17*	28.0**	6.0*
Milk	50	69*	51*	3.3*	−4.9*
Meat	38	44*	37*	1.5*	−2.8*
Hides and skins	4.5**	5.7*	4.0*	2.4*	−5.7*
Timber (000 m³)	3 300**	4 102*	4 370*ª	2.2**	1.3*ᵇ

Other

	1960	1970	1976	1960–70	1970–76
Electricity (mn kW h)	8	27	53ª	12.9	14.4ᵇ
Beer (000 hl)	na	59	133ᶜ	na	22.5ᵈ
Cigarettes (mn units)	na	na	18ᶜ	na	na
Soap	na	2.3	3.6ª	na	9.4ᵇ

ª1975 ᵇ1970–75 ᶜ1974 ᵈ1970–74
Transport traffic *Rail* See Ivory Coast *Air* (including apportionment of
traffic of Air Afrique, 1976) Passenger-km 125 mn, cargo 13.8 mn t km
Tourism Number of visitors (1974) 10 747, gross receipts (1973) $ 2 mn

Finance and trade

Consumer price index (1970 = 100) 1976 126.1; growth 1970–76 3.9 %pa
Money stock (end-year, CFA Fr bn) 1976 27.6; growth 1970–76 20.2 %pa
Budget (total, 1976) Balanced at CFA Fr 21 122 mn = $ 88 mn = £ 49 mn
International reserves (Dec 1976) $ 71 mn
External public debt (Dec 1976) $ 248 mn
External trade (1976) Imports: CFA Fr 32 386 mn = $ 136 mn = £75 mn
Exports: CFA Fr 9 369 mn = $ 39 mn = £ 22 mn

Main imports (1972)	% of total	Main exports (1975)	% of total
Food	19	Livestock	36
(of which, sugar 6)		Hides and skins	18
Machinery	13	Cotton	16
Transport equipment	10	Groundnuts	15
Chemicals	8	Karité nuts	7
Petroleum products	8	Sesameseed	6
Textile yarns and fabrics	8		
Main sources (1975)		*Main destinations* (1975)	
France	43	Ivory Coast	48
Ivory Coast	12	France	19
United States	7*	Italy	7
West Germany	4	United Kingdom	6
		Ghana	3

Zaire

Republic of Zaire
République du Zaïre

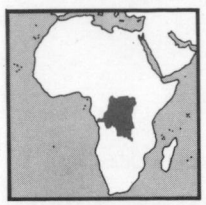

Location Central Africa
Forms the centre of Africa, reaching from
an eastern border with Uganda, Rwanda,
Burundi and Tanzania, to the Atlantic Ocean,
where there is a short coastline at the outlet of
the Zaire (Congo) river which provides a
corridor through Angolan territory. The main
border to the west is with Congo, to the north
with Central African Empire and Sudan, and
to the south with Angola and Zambia
Land Area 2 345 409 km² = 905 567 mi²
Climate Tropical
Weather at Kinshasa, 325 m altitude
Temperature: hottest months March, April 22–32 °C, coldest
July 18–27 °C
Rainfall (av monthly): driest months July, Aug 3 mm, wettest Nov 221 mm
Time Kinshasa, Mbandaka: 1 hour ahead of GMT
Kasai, Kivu, Shaba: 2 hours ahead of GMT
Measures Metric system
Monetary unit Zaire (Z) = 100 makuta = 10 000 sengi; the zaire was
introduced in June 1967, replacing the Congolese franc (C Fr) at
Z1 = C Fr 1 000
Rate of exchange (1976 av): free $ 1.239 = Z 1, Z 1.458 = £ 1

Summary

Political One-party republic, which became independent June 30, 1960;
formerly the Belgian Congo, administered by Belgium. From 1964 to
1971 known as the Democratic Republic of the Congo and commonly then
known as Congo-Kinshasa. Member of UN, OAU and an EEC ACP state
Economic A mining and agricultural economy, with copper, cobalt, crude
oil, diamonds and manganese main mineral exports, and coffee the main
agricultural export. There is some industry, including brewing, cement and
oil refining; hydro-electricity is important. Tourism is being developed in
addition to the wealth of mineral reserves. Off-shore oil production began
in 1975

People, resources and equipment

Population 1960 14.20** mn, 1970 21.69* mn, 1976 25.63* mn
Growth: 1960–70 4.3** %pa, 1970–76 2.8* %pa
Density (1976): 11* people per km²
Vital statistics (rate per 1 000 people, 1970–75): births 45.2*, deaths 20.5*
Regions (population in 000, 1974; total of 24.17 mn)

Bandundu	2 851	Haut-Zaïre	3 542	Kivu	3 721
Bas-Zaïre	1 658	Kasai Occidental	2 687	Shabaª	3 073
Equateur	2 633	Kasai Oriental	2 010	Kinshasa	1 991

ªFormerly Katanga
Cities (population in 000, 1974)

Kinshasa (capital)ᵉ	2 202ᵍ	Kisanganiᶜ	311	Matadi	144
Kanangaª	601	Bukavu	182	Mbandakaᶠ	134
Lubumbashiᵇ	404	Likasiᵈ	160**		
Mbuji Mayi	337	Kikwit	150		

ªFormerly Luluabourg ᵇFormerly Elisabethville ᶜFormerly Stanleyville
ᵈFormerly Jadotville ᵉFormerly Léopoldville ᶠFormerly Coquilhatville ᵍ1975
Race (1961) Luba 18 %, Mongo and Totela 17 %, Kongo 12 %, Rwanda
10 %
Language French; local langugages, including especially Lingala,
Swahili, Kiluba and Kikongo, are also used
Religion (1976) Christian 60** % (Roman Catholic 48** %),
Animist 39** %, Moslem 1** %
Education (1973/74) Pupils: primary 3 538 257, secondary 225 606,
vocational 47 579, teacher-training 62 018, higher 18 368. Teachers:
primary (1972/73) 80 481, secondary, vocational and teacher-training
14 483, higher 2 083
Labour force (1976) 10 881 000*; in agriculture 8 307 000* (76* %)
Personnel (1973) Physicians: 818, 1 per 28 800 people
Standard of living
National income per person (1976): Z 100** = $ 124** = £ 69**
Consumption per person (1975): energy 78 kg coal equivalent,
electricity (production) 138* kW h, newsprint 0.05 kg, steel 4 kg
Newspapers: number (1974) 11; circulation (1970) 200 000*, 9* per
1 000 people
Telephones (Dec 1975): 48 000, 2 per 1 000 people

Zaire

Livestock (000, Dec 1975) Cattle 1 144, sheep 711*, goats 2 256*
pigs 627, chickens 10 992*
Mineral reserves Coal (known economic, 1920) 720 mn tonnes
Crude oil (1975) 20 mn tonnes
Natural gas (1975) 2 bn cubic metres
Uranium (1974) 1 800 tonnes
Petroleum refinery capacity (1975) 0.85 mn tonnes
Electrical capacity (1975) 1 217* megawatts
Hospital beds (1973) 72 090, 1 per 327 people
Roads (1976) 145 000 km = 90 100 mi, density 0.06 km per km²
Railways (1974) 5 257 km = 3 267 mi, density 0.002 km per km²
Inland waterways (1976) 16 000* km = 10 000* mi (including Zaire river)
Ships (registered, 1977) 34, total of 109 785 gross tons
Ports (goods traffic, 000 tonnes, 1973)

	loaded	unloaded
Matadi	447	849
Boma	71	37

Also Banana
Airports N'Djili (Kinshasa): passenger departures and arrivals (1975)
314 636; also (1977) Kananga, Lubumbashi, Mbuji Mayi, Kisangani,
Bukaru and 23 other airports with scheduled flights

Durable equipment	000	no per	
(Dec 1974)		1 000 people	
Radio sets	2 448	100	no per
Television sets	7	0.3	km of road
Passenger cars	84	3	0.6
Commercial vehicles	71	3	0.5

Production

Gross domestic product 1974: Z 1 766 mn = $ 3 533 mn = £ 1 510 mn
1976 est: Z 3 000** mn = $ 3 700** mn = £ 2 100** mn
Structure of gross domestic product (1974) *By origin* Agriculture 15 %,
mining and quarrying 24 %, manufacturing 9 %, electricity, gas and water
1 %, construction 5 %, transport and communication 5 %, other 41 %.
By type Final consumption expenditure 77 % (of which, government 27 %),
gross fixed capital formation 23 %, exports of goods and services 46 %,
less imports of goods and services −46 %

Production index	1960	1970	1976	Growth %pa	
(1970 = 100)				1960–70	1970–76
Agricultural	91	100	104	1.0	0.7

Main products (000 t)					
Agriculture					
Rice	124	188	190*	4.2	0.2*
Maize	330	330	410*	0.0	3.7*
Sweet potatoes	374	350*	300*	−0.7*	−2.5*
Cassava	6 045	10 000*	9 832	5.2*	−0.3*
Oranges	63	90*	102*	3.6*	2.1*
Bananas	51	80*	81*	4.6*	0.2*
Groundnuts	175	240	289	3.2	3.1
Palm kernels	149	132	70	−1.2	−10.0
Palm oil	234	211	155*	−1.0	−5.0*
Coffee	54	81*	86*	4.1*	1.0*
Tea	4	5*	8*	2.3*	8.1*
Cocoa	5.5*	5.6	5.0*	0.2*	−1.9*
Jute	8**	7	12*	−1.3**	9.4*
Cotton	46	17	12*	−9.5	−5.6*
Rubber	36	32	36*	−1.2	2.0*
Meat	150	170*	189	1.3*	1.8*
Fish catch	77	137	125*ᶜ	5.9	−1.8*ᵈ
Timber (000 m³)	11 650ᵃ	13 940	13 690*ᶜ	2.0ᵇ	−0.4*ᵈ
Energy					
Total energy (000 tce)	450	500	650ᶜ	1.1	5.4ᵈ
Coal	163	102	90ᶜ	−4.6	−2.5ᵈ
Crude oil	—	—	96ᶜ	na	na
Petroleum products	—	634	564ᶜ	na	−2.3ᵈ
Electricity (mn kW h)	2 456	3 230	3 440*ᶜ	2.8	1.3*ᵈ
of which,					
hydro (mn kW h)	2 425	3 152	3 370*ᶜ	2.7	1.3*ᵈ
Mining					
Cobalt ore					
(Co content)	8	12	12*ᶜ	4.1	0.0*ᵈ
Copper ore					
(Cu content)	302	387	444	2.5	2.3
Manganese ore					
(Mn content)	207	156	160ᶜ	−2.8	0.6ᵈ
Tin conc (Sn content)	9.4	6.5	4.0	−3.6	−7.7
Zinc ore (Zn content)	109	104	70	−0.5	−6.4
Gold (000 kg)	9.9	5.6	3.2ᶜ	−5.5	−10.6ᵈ
Diamonds (000 CM)	13 453	14 087	12 810ᶜ	0.5	−1.9ᵈ

Main products (000 t)	1960	1970	1976	Growth %pa	
Manufacturing				1960–70	1970–76
Beer (000 hl)	na	3 394	5 723ᵉ	na	14.0ᶠ
Cigarettes (mn units)	na	3 510	4 910ᵉ	na	8.8ᶠ
Cotton fabrics (mn m²)	na	59	69ᵉ	na	4.0ᶠ
Sulphuric acid	na	136	110ᵉ	na	−6.8ʰ
Cement	200	419	577ᵉ	7.7	8.3ᶠ

ᵃ1961 ᵇ1961–70 ᶜ1975 ᵈ1970–75 ᵉ1974 ᶠ1970–74 ᵍ1973 ʰ1970–73

Transport traffic	1960	1970	1976	Growth %pa	
Passenger-kilometres (mn)				1960–70	1970–76
Road	na	149	124ᶠ	na	−5.9ᵍ
Rail	344	933	447ᶠ	10.5	−21.8ᵍ
Air	128ᵇ	465	690	17.5ᶜ	6.8
Cargo: tonne-kilometres (mn)					
Road	na	3*	3*ᵈ	na	0.0*ᵉ
Rail	1 725	2 610	3 017ᶠ	4.2	4.9ᵍ
Water	na	1.1	1.1ᶠ	na	0.0ᵍ
Air	11.7ᵇ	15.0	54.8	3.2ᶜ	24.1
Sea²: tonnes (mn)					
Goods loaded	na	0.62	0.56ᵈ	na	−2.4ᵉ
Goods unloaded	na	0.67	1.08ᵈ	na	12.4ᵉ

ᵃThrough Matadi and Boma only ᵇ1962 ᶜ1962–70 ᵈ1974 ᵉ1970–74 ᶠ1973
ᵍ1970–73

Tourism (1975) Number of visitors 41 000, gross receipts $ 11 mn

Finance and trade

Price index	1960	1970	1976	Growth %pa	
(1970 = 100)				1960–70	1970–76
Consumer prices	27.2ᵃ	100.0	436.8	20.4ᵇ	27.9
Money stock					
(end-year, Z mn)	43.9ᵃ	186.5	661.0	23.0ᵇ	23.5

ᵃ1963 ᵇ1963–70

Budget (actual, 1975) Revenue: Z 432 mn = $ 864 mn = £ 389 mn
Expenditure: Z 710 mn = $ 1 420 mn = £ 639 mn

Balance of payments ($ mn)	1972	1973	1974	1975	1976
Balance of goods (fob)	−62	+61	+81	−130	na
Balance of services	−309	−367	−556	−523	na
Balance of transfers	+6	+27	+3	+54	na
Current balance	−365	−279	−472	−599	na
Long-term capital flow	+317	+256	+259	+241	na
Reserves and debt (end-year, $ mn)					
International reserves	178	235	140	59	105
External public debt	761	1 679	2 561	2 738	na

External trade (1976) Imports: Z 547 mn = $ 677 mn = £375 mn
Exports: Z 747 mn = $ 926 mn = £513 mn

Main imports (1976)	% of total	*Main exports* (1976)	% of total
Food	20*	Copper	42*
(of which, cereals 3*, fish 3*)		Coffee	14*
Crude oil and products	12*	Cobalt	13*
Electrical machinery	12*	Crude oil	10*
Chemicals	10*	Diamonds	6*
Machinery, non-electric	9*	Manganese ore	5*
Iron and steel	9*		
Textile yarns and fabrics	7*		
Motor vehicles	6*		
Main sources (1975)		*Main destinations* (1975)	
United States	18*	Belgium-Luxembourg	38*
Belgium-Luxembourg	15*	Italy	14*
West Germany	13*	France	6*
France	13*	West Germany	6*
Italy	6*	United States	6*
United Kingdom	5*	United Kingdom	6*
Netherlands	4*	Japan	5*
Japan	4*		

Special focus

Copper prices (London wholesale prices)

	$ per tonne	£ per tonne		$ per tonne	£ per tonne
1965	1 295	462	1971	1 081	444
1966	1 526	545	1972	1 069	427
1967	1 140	414	1973	1 782	727
1968	1 240	517	1974	2 052	877
1969	1 466	611	1975	1 231	554
1970	1 415	589	1976	1 403	777

Zambia

Republic of Zambia

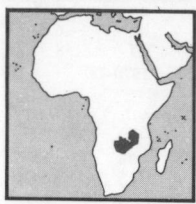

Location South central Africa
Zaire and Tanzania are to the north,
Malawi and Mozambique to the east,
Rhodesia, Botswana and South-West
Africa to the south, and Angola to the
west. The Zambesi river forms the
boundary with Rhodesia. Land-locked
Land Area 752 614 km² = 290 586 mi²
Climate Tropical, cool on high plateaux
Weather at Lusaka, 1 277 m altitude
Temperature: hottest month Oct 18–31 °C, coldest July 9–23 °C
Rainfall (av monthly): driest month Aug 0 mm, wettest Jan 231 mm
Time 2 hours ahead of GMT
Measures Metric system, which replaced the UK (imperial) system
in general from 1972
Monetary unit Kwacha (K) = 100 ngwee; the kwacha was introduced in
January 1968, replacing the Zambian pound (Z £) at K 2 = Z £ 1
Rate of exchange (1976 av): free $ 1.4023 = K 1, K 1.288 = £ 1

Summary

Political One-party republic, which became independent October 24,
1964; formerly the UK protectorate of Northern Rhodesia (from
1953–63 part of the Federation of Rhodesia and Nyasaland). Member of
UN, OAU, Commonwealth and an EEC ACP state
Economic The economy is dependent on copper, accounting for over
90 % of exports; the agricultural sector is important locally and is being
expanded to reduce dependence on copper. Completion in 1975 of the
Tanzam railway to Dar es Salaam eased the transport problem caused
by closure of the Rhodesian border in 1973. There is some industry in
addition to copper refining, mainly for import substitution

People, resources and equipment

Population 1960 3.21*mn, 1970 4.18*mn, 1976 5.14*mn
Growth: 1960–70 2.7* %pa, 1970–76 3.5* %pa
Density (1976): 7* people per km²
Vital statistics (rate per 1 000 people, 1970–75): births 51.5*, deaths 20.3*
Cities (population in 000, 1975)
Lusaka (capital) 415* Ndola 224* Mufulira 138*
Kitwe 314* Chingola 204* Luanshya 121*
Race (1969) African 99 %, European 1 %
Language English; local languages, especially Nyanja, Bemba,
Tonga, Lozi, Lunda and Luvale, are also used
Religion (1976) Christian 80** %; also Animist and Moslem
Education (1974) Pupils 933 846, teachers 21 400*
Labour force (1976) 1 928 000*; in agriculture 1 334 000* (69* %)
Personnel Scientists and engineers (1970): 5 900*
Physicians (1971): 527, 1 per 8 160 people

Standard of living
National income per person (1976): K 280** = $ 390** = £ 220**
Consumption per person (1975): energy 504 kg coal equivalent,
electricity (production, 1976) 1 370 kW h, newsprint 0.4 kg, steel 14 kg
Newspapers (1974): number 2; circulation 105 000, 22 per 1 000 people
Telephones (Dec 1975): 77 000, 15 per 1 000 people
Livestock (000, 1976) Cattle 2 300*, goats 283*, pigs 106*
Mineral reserves Coal (1973) 154 mn tonnes
Petroleum refinery capacity (1975) 1.1 mn tonnes
Electrical capacity (1975) 1 031 megawatts, of which, hydro 759 megawatts
Hospital beds (1969) 13 242, 1 per 313 people
Roads (1973) 34 671 km = 21 544 mi, density 0.05 km per km²
Railways (1976) 1 930* km = 1 200* mi, density 0.003* km per km²
Oil pipeline There is a pipeline link from Dar es Salaam in Tanzania
Ship (registered, 1977) 1, of 5 513 gross tons
Airports Lusaka: passenger departures and arrivals (1975) 356 123; also
(1977) Ndola, Kitwe and 13 other airports with scheduled flights

Durable equipment (Dec 1974)	000	no per 1 000 people	no per
Radio sets	100	21	
Television sets	22	5	km of road
Passenger cars	86*	18*	2.5*
Commercial vehicles	62*	13*	1.8*

Production

Gross domestic product (1976) K 1 793 mn = $ 2 514 mn = £ 1 392 mn
Growth in real terms: 1970–75 −4.1 % pa

Structure of gross domestic product (1975) *By origin* Agriculture 11 %,
mining and quarrying 10 %, manufacturing 18 %, electricity, gas and
water 2 %, construction 10 %, distribution and hotels 14 %,
transport and communication 5 %, other 30 %

Production indices (1970 = 100)	1960	1970	1976	Growth %pa 1960–70	1970–76
Agricultural	80	100	148	2.3	6.8
Industrial	71[b]	100.0	115.2	4.4[c]	2.4

Main products (000 t)

	1960	1970	1976	1960–70	1970–76
Agriculture					
Maize	600**	700*	1 070*	1.6**	7.3*
Cassava	145	143*	163*	−0.1*	2.2*
Sugar, raw value	na	40	85*	na	13.4*
Tobacco	6.0	5.1	7.0*	−1.6	5.4*
Cotton	—	2.4	4.0*	na	8.9*
Milk	35**	44*	50*	2.3**	2.2*
Beef and veal	18	27*	37*	4.1*	5.4*
Fish catch	20	48*	50*[d]	9.1*	0.8*[e]
Timber (000 m³)	3 771*	4 697*	4 715*[d]	2.2*	0.1*[e]
Energy					
Total energy (000 tce)	na	700	1 550[d]	na	17.2[e]
Coal	—	623	789	na	4.0
Electricity (mn kW h)	836	949	7 040	1.3	39.6
Mining					
Copper ore (Cu content)	680*	818	850	1.9*	0.6
Lead ore (Pb content)	15	33	16	8.2	−11.5
Zinc ore (Zn content)	40	65	49	5.0	−4.6
Silver (000 kg)	29	48	23[f]	5.2	−21.7[e]
Gold (000 kg)	0.20	0.36	0.15[d]	6.1	−16.2[e]
Manufacturing					
Copper	402	580	706	3.7	3.3
Beer (000 hl)	na	2 600	2 670[f]	na	0.9[e]
Cement	128	179	452[d]	3.4	20.3[e]
Sulphuric acid	na	na	283[d]	na	na
Motor vehicles (000 units)[a]	na	1.4	4.8[f]	na	50.8[g]

[a]Assembly only [b]1962 [c]1962–70 [d]1975 [e]1970–75 [f]1973 [g]1970–73

Transport traffic	1960	1970	1976	Growth %pa 1960–70	1970–76
Passenger-kilometres (mn)					
Rail	na[a]	308**	320[d]	na	1.9[e]
Air	63[b]	228	356	23.9[c]	7.7
Cargo: tonne-kilometres (mn)					
Rail	na[a]	485	897[d]	na	36.0[e]
Air	0.9[b]	5.3	19.8	34.4[c]	24.6

[a]Before July 1, 1967, Zambia Railways were part of Rhodesia Railways [b]1964
[c]1964–70 [d]1972 [e]1970–72

Tourism (1975) Number of visitors 52 000, gross receipts $ 10 mn

Finance and trade

Consumer price index (1970 = 100) 1976 168.6; growth 1970–76 9.1 %pa
Money stock (end-year, K mn) 1976 377; growth 1970–76 11.0 * %pa
Budget (1976) Revenue: K 457 mn = $ 641 mn = £ 355 mn
Expenditure: K 664 mn = $ 931 mn = £ 516 mn

Balance of payments ($ mn)	1972	1973	1974	1975	1976
Balance of goods (fob)	+194	+591	+606	−120	na
Balance of services	−269	−323	−403	−368	na
Balance of transfers	−135	−125	−126	−123	na
Current balance	−208	+143	+77	−611	na
Long-term capital flow	+101	−2	+53	+222	na

Reserves and debt (end-year, $ mn)

International reserves	165	193	172	149	100
External public debt	840	995	1 192	1 502	1 605

External trade (1975) Imports: K 732 mn = $ 1 138 mn = £ 512 mn
Exports: K 521 mn = $ 810 mn = £ 364 mn

Main imports (1974)	% of total	Main exports (1974)	% of total
Machinery	22	Copper	92[a]
Chemicals	10	Zinc	3
Crude oil	9	Maize	1
Food	9	Lead	1
Motor vehicles	8	Tobacco	1
Iron and steel	8		
Textile yarns and fabrics	7		

Main sources (1975)		Main destinations (1975)	
United Kingdom	20	United Kingdom	22
United States	12	Japan	17
Japan	9	West Germany	14
West Germany	7	Italy	13
South Africa	7	France	8

[a]Also 92 % in 1976

America

(ALASKA, USA)

CANADA

ST PIERRE AND MIQUELON
St Pierre

Ottawa

UNITED STATES

Washington DC

BERMUDA
Hamilton

(HAWAII, USA)

MEXICO
Mexico City

GUYANA
Georgetown

SURINAM
Paramaribo

Caracas

VENEZUELA

FRENCH GUIANA
Cayenne

COLOMBIA
Bogotá

Quito

ECUADOR

BRAZIL

PERU
Lima

La Paz

Brasilia

BOLIVIA
Sucre

PARAGUAY
Asunción

CHILE

Santiago

URUGUAY
Montevideo

Buenos Aires

ARGENTINA

FALKLAND ISLANDS
Stanley

Nassau

BAHAMAS

TURKS AND CAICOS IS
Cockburn Town

Havana

CUBA

DOMINICAN
REPUBLIC

9
10
11
12
13
14
15
16
17
18

CAYMAN IS
George Town

Port-au-Prince

JAMAICA
Kingston

HAITI

Santo
Domingo

Belmopan

5 6 7 8

BELIZE

1

HONDURAS
Tegucigalpa

2

NICARAGUA

4

TRINIDAD
AND TOBAGO
Port of Spain

Managua

PANAMA

COSTA RICA
San José

Panama City

Caracas

1
3

VENEZUELA

1 GUATEMALA
Guatemala City

2 EL SALVADOR
San Salvador

3 PANAMA CANAL ZONE
Balboa

4 NETHERLANDS ANTILLES
Willemstad, Curaçao
also Aruba, Bonaire, Saba,
St Eustatius, St Maarten (south part)

5 PUERTO RICO
San Juan

6 VIRGIN ISLANDS, US
Charlotte Amalie

7 VIRGIN IS, BR
Road Town

8 ST KITTS-NEVIS
Basseterre

9 ANGUILLA

10 ANTIGUA
St John's

11 MONTSERRAT
Plymouth

12 GUADELOUPE
also St Martin
(north part)
Basse-Terre

13 DOMINICA
Roseau

14 MARTINIQUE
Fort-de-France

15 ST LUCIA
Castries

16 ST VINCENT
Kingstown

17 BARBADOS
Bridgetown

18 GRENADA
St George's

Antigua
State of Antigua

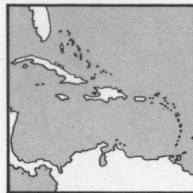

Location Eastern Caribbean Sea
Part of the Leeward Islands, the
island of Antigua has St Kitts and Nevis
islands to the west, and Montserrat
and Guadeloupe to the south.
Land Area 442 km² = 170 mi²
Climate Sub-tropical
Weather at Antigua Temperature: 21–30°C
Rainfall (monthly av): 85 mm
Time 4 hours behind GMT

Measures UK (imperial) system
Monetary unit East Caribbean dollar (EC $) = 100 cents
Rate of exchange (1976 av): free EC $ 2.61 = $ 1, EC $ 4.71 = £ 1
A link with the pound at EC $ 4.80 = £ 1 was changed from July 7, 1976
to a link with the US dollar at EC $ 2.70 = $ 1

Summary

Political An Associated State of the United Kingdom from
February 27, 1967; formerly a UK colony. Member of Caricom
Economic Cotton production and tourism are main industries;
petroleum refining was discontinued in 1975, although there have been
proposals to reopen the refinery. Sugar production was discontinued in
1972; sugar cane planting began again in early 1978.

People, resources and equipment

Population 1960 60 000*, 1970 65 500, 1976 71 000*
Growth: 1960–70 0.9* %pa, 1970–76 1.4* %pa
Density (1976): 161* people per km²
Vital statistics (rate per 1 000 people, 1975): births 19.3, deaths 6.6
Regions (population in 000, 1976) Antigua 70*, Barbuda 1*,
Redonda uninhabited
Town (population in 000, 1976) St John's (capital) 24*
Race (1960) African 92 %, Mixed 3 %, European 1 %
Language English
Religion Mainly Christian
Education (1973/74) Pupils 19 626, teachers 766
Labour force (1975) 24 533
Personnel (1974) Physicians: 28, 1 per 2 500 people
Standard of living
National income per person (1976): EC $ 1 400** = $ 540** = £ 300**
Consumption per person (1975): energy 2 184 kg coal equivalent
Newspapers (1973): number 1; circulation 4 000, 56 per 1 000 people
Telephones (Dec 1974): 3 000, 43 per 1 000 people
Livestock (000, 1976) Cattle 6*, sheep 11*, goats 7*, chickens 62*
Petroleum refinery capacity (1974) 900 000 tonnes (not operational from
1975)
Hospital beds (1975) 454, 1 per 154 people
Roads (1976) 1000* km = 600* mi, density 2.3* km per km²
Port St John's
Airport Coolidge (7 km from St John's)
Durable equipment (at end-year)
Radio sets (1974): 14 000, 201 per 1 000 people
Television sets (1974): 12 000, 173 per 1 000 people
Motor vehicles (1975): 8 100, 115 per 1 000 people, 8.1 per km of road

Production, finance and trade

Gross domestic product 1973: EC $ 92mn = $47mn = £ 19mn
1976 est: EC $ 105**mn = $ 40**mn = £ 22**mn
Main products (000 t, 1976) Milk 6*, sea island cotton 0.08,
fish catch 0.8*
Tourism (1976) Number of visitors 62 970
Budget (1976) Revenue: EC $ 27.0 mn = $ 10.4 mn = £ 5.7 mn
Expenditure: EC $ 36.1 mn = $ 13.8 mn = £ 7.7 mn
External trade (1974) Imports: EC $ 144 mn = $ 70 mn = £ 30 mn
Exports: EC $ 66 mn = $ 32 mn = £ 14 mn

Main imports (1973)	% of total	Main exports (1973)	% of total
Crude oil	34	Gas oil and diesel oil	58
Food	16	Motor spirit	26
Aircraft and engines	7	Aircraft and engines	6
Main sources (1973)		Main destinations (1973)	
Venezuela	31	Bunkers	37
United Kingdom	22	United States	21
United States	16	Switzerland	11

Argentina
Argentine Republic
República Argentina

Location South-east of South America
With a long coastline on the South
Atlantic, Bolivia, Brazil, Paraguay
and Uruguay are to the north and Chile
is to the west
Land Area 2 776 889 km² = 1 072 163 mi²
Usage (1975): agricultural 1 782 500* km²
(64* %), of which, arable 246 500* km² (9* %),
cropland 99 000* km² (4* %), pastures
1 437 000* (52* %); forests 607 000* (22* %)
Climate Varies from sub-tropical in the north to sub-arctic in the south
Weather at Buenos Aires, 27 m altitude
Temperature: hottest month Jan 17–29 °C, coldest June 5–14 °C
Rainfall (av monthly): driest month July 56 mm, wettest March 109 mm
Time 3 hours behind GMT
Measures Metric system
Monetary unit Argentine peso (Arg $) = 100 centavos; a new peso was
introduced from January 1, 1970 to replace the old peso at the rate
of 1 new peso = 100 old pesos
Rate of exchange (1976 av): free Arg $ 149.67 = $ 1, Arg $ 270.3 = £ 1

Summary

Political Republic with military government, which proclaimed
independence from Spain in 1816. The Malvinas (Falkland Islands) and
some Antarctic territory are claimed as part of the republic. Member of
UN, OAS, Sela and Lafta
Economic Exports are mainly of agricultural produce, with cereals
and meat products alone accounting for about one-half; however the
agricultural sector makes up only 12 % of gross domestic product
and absorbs only about 15 % of the labour force. There is a wide
spread of manufacturing industry making up one-third of gross domestic
product and absorbing 20 % of the labour force. Inflation is high

People

Population 1960 20.85* mn, 1970 23.75* mn, 1976 25.72* mn
Growth: 1960–70 1.3* %pa, 1970–76 1.3* %pa
Density (1976): 9* people per km²
Vital statistics (rate per 1 000 people, 1970): births 22.9, deaths 9.4
Regions (regions and provinces, population in 000, 1975; total of 25.38mn)

Litoral	18 022*	Norte	2 194*	Andina	1 929*
Federal district[a]	2 977*	Jujuy	312*	Catamarca	175*
Buenos Aires	9 948*	Salta	527*	La Rioja	137*
Chaco	586*	Santiago		Mendoza	1 058*
Corrientes	599*	del Estero	527*	Neuquén	161*
Entre Ríos	874*	Tucumán	828*	San Juan	398*
Formosa	238*	Centro	2 676*	Patagonia	563*
Misiones	460*	Córdoba	2 314*	Chubut	194*
Santa Fé	2 340*	La Pampa	175*	Río Negro	267*
		San Luis	187*	Santa Cruz	88*
				Tierra del	
				Fuego[b]	14*

[a] Federal capital [b] National territory
Cities (population in 000, 1970)

Buenos Aires (capital)	8 925[a]	Mar del Plata	300	Corrientes	131
Rosario	811	Santa Fé	245	Paraná	128
Córdoba	799	San Juan	224	Santiago	
La Plata	506	Salta	176	del Estero	105
Mendoza	471	Bahía Blanca	175		
San Miguel de Tucumán	366	Resistencia	143		

[a] 1974
Race (1976) European (mainly Spanish and Italian) 99** %, Indian 0.1* %
Language Spanish
Religion (1974) Roman Catholic 90** %, Protestant 2** %
Education (1974) Pupils: primary 3 570 615, secondary 440 304,
vocational 743 589, higher (1973) 423 824. Teachers: primary 191 958,
secondary 61 376, vocational 91 697, higher (1973) 38 964
Labour force (1970) 9 011 450; in agriculture 1 331 100 (15 %),
mining and quarrying 44 600 (½ %), manufacturing 1 771 250 (20 %),
electricity, gas and water 96 550 (1 %), construction 711 300 (8 %),
distribution and hotels 1 324 800 (15 %), transport and communication
593 250 (7 %), finance and real estate 252 650 (3 %), other services
2 098 750 (23 %), unknown 787 200 (9 %)
Personnel Scientists and engineers (1972): 333 000*
Physicians (in hospitals only, 1973): 53 684, 1 per 460 people

Argentina

Standard of living National income per person (1976):
Arg $ 280 000** = $ 1 900** = £ 1 000**
Consumption per person (1975): energy 1 754 kg coal equivalent,
electricity (production) 1 152 kW h, newsprint 7.4 kg, steel 172 kg
Newspapers (1974): number 167; circulation (149 dailies only) 3 683 000,
147 per 1 000 people
Telephones (Dec 1976): 2 539 535, 98 per 1 000 people

Resources and equipment

Livestock (000, 1976) Cattle 59 000*, sheep 36 500*, goats 5 600*.
pigs 5 000*, horses 3 500*, chickens 35 000*
Mineral reserves Coal (1974) 505 mn tonnes
Crude oil (1975) 360 mn tonnes
Natural gas (1975) 201 000 mn cubic metres
Uranium (1974) 9 300 tonnes
Petroleum refinery capacity (1975) 31 mn tonnes
Electrical capacity (1975) 9 259 megawatts
Hospital beds (1971) 133 847, 1 per 176 people
Roads (1976) 207 262 km = 128 787 mi, density 0.075 km per km²
Railways (1976) 40 113 km = 24 925 mi, density 0.014 km per km²
Inland waterways Rivers Paraguay, Paraná, Plate and Uruguay
Ships (registered, 1977) 401, total of 1 677 169 gross tons
Ports (goods traffic, 000 tonnes, 1976)

	loaded	unloaded		loaded	unloaded
Buenos Aires	6 066	2 996	Rosario	2 921	100
Bahía Blanca	2 996	617	La Plata	180	1 995
San Nicolás	366	2 817	Campana	157	388

Airports (passenger departures and arrivals, 000, 1976) Buenos Aires:
Aeroparque 2 968, Ezeiza 1 133; Córdoba 327; also (1977) 61 other
airports with scheduled flights

Durable equipment	000	no per	
(at end-year)		1 000 people	
Radio sets (1973)	21 000	838	no per
Television sets (1974)	4 500	180	km of road
Passenger cars (1976)	2 588	100	12.5
Commercial vehicles (1976)	1 101	43	5.3

Production

Gross domestic product
1975: Arg $ 1 345 000 mn = $ 49 100 mn = £ 22 100 mn
1976 est: Arg $ 7 500 000** mn = $ 50 000** mn = £ 28 000** mn
Growth in real terms: 1960–70 4.4 %pa, 1970–76 2.6 %pa
Structure of gross domestic product (1975)

By origin	Arg $ bn	% of gdp
Agriculture	164.3	12
Mining and quarrying	20.1	1
Manufacturing	448.1	33
Electricity, gas and water	20.8	2
Construction	54.7	4
Distribution and hotels	136.0	10
Transport and communication	112.0	8
Other	389.0	29
Total	1 345.0	100
By type of expenditure		
Government final consumption	136.8	10
Private final consumption	953.1	71
Stock investment	3.2	–
Gross fixed capital formation	286.8	21
Exports of goods and services	99.1	7
less Imports of goods and services	−134.0	−10
Total	1 345.0	100

Production indices	1960	1970	1976	Growth %pa	
(1970 = 100)				1960–70	1970–76
Agricultural	75	100	119	2.9	2.9
Industrial	56	100	122	6.0	3.4

Main products (000 t)

Agriculture	1960	1970	1976	1960–70	1970–76
Wheat	4 200	4 920	11 000	1.6	14.3
Maize	4 108	9 360	5 855	8.6	−7.5
Sorghum	831	4 068	5 158	17.2	4.0
Potatoes	1 860	2 336	1 528	2.3	−6.8
Sugar, raw value	837	979	1 559	1.6	8.1
Grapes	2 008	2 462	3 760	2.1	7.3
Wine	1 583	1 836	2 491*	1.5	5.2*
Apples	431	445	577	0.3	4.4
Oranges	460**	865	691	6.5**	−3.7
Lemons and limes	88	202	224	8.7	1.7

Main products (000 t)

Agriculture	1960	1970	1976	1960–70	1970–76
Soyabeans	1	27	695	39.0	71.8
Sunflowerseed	802	1 140	1 085	3.6	−0.8
Linseed	562	680	617	1.9	−1.6
Tobacco	41	66	93	4.9	5.9
Cotton	89	145	140	5.0	−0.6
Milk	4 511	4 196	5 526*	−0.7	4.7*
Cheese	119	167	245*	3.4	6.6*
Beef and veal	1 883	2 645	2 792	3.5	0.9
Pigmeat	184	210	248	1.3	2.8
Wool	109	92	90*	−1.7	−0.4*
Cattle hides	290*	388*	412*	3.0*	−1.0*
Fish catch	105	215	224e	7.4	0.8f
Timber (000 m³)	11 571**	11 782	11 428e	0.2**	−0.6f
Energy					
Total energy (000 tce)	14 020	38 800*	41 590e	10.7*	1.4*f
Coal	175	616	615	13.4	0.0
Crude oil	8 898	20 026	20 920	8.5	0.7
Petroleum products	11 513	21 139	20 600*e	6.3	−0.5*f
Natural gas (mn m³)	1 383	6 015	7 710	15.8	4.2
Electricity (mn kW h)	10 459	21 727	29 468e	7.6	6.3f
Mining					
Iron ore (Fe content)	58	107	212e	6.3	14.7f
Lead ore (Pb content)	27	36	27e	2.9	−5.2f
Zinc ore (Zn content)	35	39	38*	1.0	−0.7f
Uranium (U content)	0.029c	0.045	0.060e	7.6d	5.9f
Manufacturing					
Beer (000 hl)	2 430	3 565	3 943e	3.9	2.0f
Cigarettes (mn units)	23 662	30 220	38 621e	2.5	5.0f
Cotton yarn	95	90	89e	−0.6	−0.1f
Cotton fabrics	82	76	70e	−0.8	−1.6f
Man-made fibres	12	34	61e	11.0	12.4f
Wood pulp	46	166	276e	13.7	10.7f
Paper	235**	643	761e	10.6**	3.4f
Synthetic rubber	—	39	45	na	2.4
Sulphuric acid	132	180	234	3.2	4.5
Caustic soda	47	95	109	7.3	2.3
Plastics and resins	na	97	130e	na	6.0f
Fertilisers, nitrogenousª	2*	20	18	23.6*	−2.1
Cement	2 641	4 770	5 603	6.1	2.7
Coke	160	491	600*e	11.9	4.1*f
Pig iron	181	815	1 318	16.2	8.3
Crude steel	277	1 859	2 410	21.0	4.4
Television sets (000 units)	125	194	290e	4.5	8.4f
Passenger cars (000 units)ᵇ	49	169	141	13.2	−3.0
Commercial vehicles					
(000 units)ᵇ	39	50	36	2.5	−5.3
Merchant vessels (000 grt)	14e	18	39	4.3d	13.8
Construction					
All buildings (000 m³)	4 447	7 017	7 055g	4.7	0.5h

ªYears ending June 30th ᵇIncluding assembly c1964 d1964–70 e1975 f1970–75
g1971 h1970–71

Transport traffic	1960	1970	1976	Growth %pa	
Passenger-kilometres (mn)				1960–70	1970–76
Road	na	2 431	2 044a	na	−8.3b
Rail	15 684	12 828	14 481	−2.0	2.0
Air	990	2 395	4 222	9.2	9.9
Cargo: tonne-kilometres (mn)					
Rail	15 188	13 357	11 039	−1.3	−3.1
Air	16	54	104	13.0	11.5
Sea: tonnes (mn)					
Goods loaded	9.3	15.2	15.3	5.0	0.1
Goods unloaded	10.3	10.9	9.2	0.6	−2.8
Tourism					
Number of visitors (000)	na	695	1 200c	na	11.5d
Gross receipts ($ mn)	na	74	154c	na	15.8d

a1972 b1970–72 c1975 d1970–75

Finance and trade

Price indices	1960	1970	1976	Growth %pa	
(1970 = 100)				1960–70	1970–76
Consumer prices	15	100	6 517	21.1	100.6
Wholesale prices	18	100	7 788	19.0	106.7
Money stock					
(end-year, Arg $ bn)	2.2	17.8	1 428.7	23.5	107.7

Budget (1976) Revenue: Arg $ 421 390 mn = $ 2 815 mn = £ 1 559 mn
Expenditure: Arg $ 1 029 970 mn = $ 6 882 mn = £ 3 810 mn
of which (est), economic development 34 %, defence 11 %

Argentina

Balance of payments ($ mn)	1972	1973	1974	1975	1976
Balance of goods (fob)	+256	+1 288	+ 714	−549	+1 111
Balance of services	−476	−584	−590	−738	−520
Balance of transfers	−4	+11	—	+5	+18
Current balance	*−224*	*+715*	*+125*	*−1 282*	*+611*
Long-term capital flow	+215	+54	−5	−161	+886
Reserves and debt (end-year, $ mn)					
International reserves	465	1 318	1 315	452	1 608
External public debt	3 443	3 517	4 788	4 985	6 190

External trade (1976)
Imports: Arg $ 454 030 mn = $ 3 034 mn = £ 1 680 mn
Exports: Arg $ 741 953 mn = $ 3 916 mn[a] = £ 2 168 mn
[a]Valued at an average rate of Arg $ 189.5 = $1

Main imports (1975)	% of total	*Main exports* (1975)	% of total
Iron and steel	19	Cereals	37
Chemicals	19	(of which, maize 17,	
Machinery, non-electric	13	wheat 10)	
Crude oil	6	Meat and products	10
Non-ferrous metals	4	(of which, beef and	
Food	4	veal 3)	
Electrical machinery	4	Machinery	9
Natural gas	3	Fruit and vegetables	6
Paper	3	Animal feeding stuffs	5
Timber	2	Sugar	4
Motor vehicles	2	Chemicals	4
Instruments	2	Wool	4
		Oils and fats	3
		Leather	2
Main sources (1976)		*Main destinations* (1976)	
United States	18	Brazil	11
Brazil	12	Italy	9
West Germany	11	Netherlands	9
Japan	8	United States	7
Italy	5	Soviet Union	6
Chile	5	Japan	5
United Kingdom	4	West Germany	5
France	3	Chile	5
Bolívia	3	Spain	4

Special focus

Rate of inflation (consumer prices; % change over previous year)

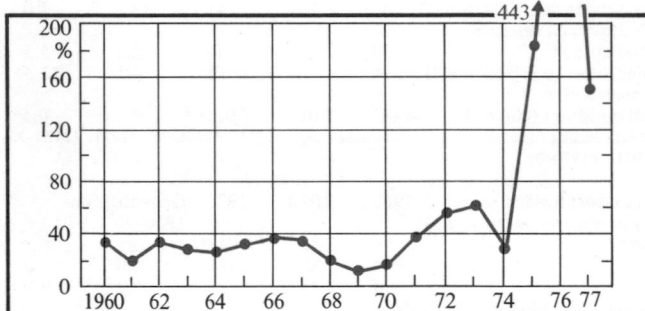

Bahamas

Commonwealth of the Bahamas

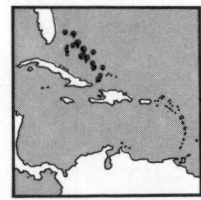

Location Western Atlantic Ocean
An archipelago of islands stretching from the Straits of Florida to end with the Turks and Caicos Islands; United States is to the north-west across the Straits of Florida, and Cuba and Haiti are to the south
Land Area 13 935 km² = 5 380 mi²
Climate Sub-tropical
Weather at Nassau, 4 m altitude
Temperature: hottest month August 24–32 °C, coldest February 18–25 °C
Rainfall (av monthly): driest month Dec 33 mm, wettest Sept 175 mm
Time 5 hours behind GMT (summer time, 4 hours behind)
Measures UK (imperial) system; also US gallon
Monetary unit Bahamian dollar (Ba $) = 100 cents
Rate of exchange (1976 av): par Ba $ 1 = $ 1, free Ba $ 1.806 = £ 1

Summary

Political Parliamentary monarchy, which became independent July 10, 1973; formerly there was internal self-government from January 7, 1964, with dependence on the United Kingdom. Member of UN, Commonwealth and an EEC ACP State
Economic Petroleum products, refined from imported crude oil, and re-exports of oil account for virtually all exports; tourism is important with gross receipts in 1975 of $ 317 mn accounting for about two-thirds of gross domestic product. There is some development of industry, especially chemicals, and financial business is significant

People, resources and equipment

Population 1960 113 000*, 1970 175 000, 1976 211 000
Growth: 1960–70 4.5* %pa, 1970–76 3.2 %pa
Density (1976): 15 people per km²
Vital statistics (rate per 1 000 people, 1975): births 19.8, deaths 5.4
Regions (main islands, population in 000, 1970)
New Providence 102 Eleuthera 9.5 Abaco 6.5
Grand Bahama 26 Andros 8.8 Long Island 3.9
Cities (population in 000, 1976) Nassau (capital)[a] 130*, Freeport[b] 16*
[a]On New Providence [b]On Grand Bahama
Race (1976) African 80* %
Language English
Religion Mainly Christian (Baptist, Anglican and Roman Catholic)
Education Pupils (1975/76) 60 010, teachers (1971/72) 1 343
Labour force (1975) 84 288
Personnel Scientists and engineers (1970): 3 000*
Physicians (1975): 161, 1 per 1 267 people
Standard of living
National income per person (1976): Ba $ 2 500** = $ 2 500** = £ 1 400**
Consumption per person (1975): energy 6 279 kg coal equivalent, electricity (production) 3 190* kW h, newsprint 4.9 kg
Newspapers (1974): number 2; circulation 31 000, 159 per 1 000 people
Telephones (Dec 1976): 58 000, 275 per 1 000 people
Livestock (000, 1976) Cattle 4*, sheep 31*, pigs 17*, goats 16*, chickens 750*
Petroleum refinery capacity (1975) 26 mn tonnes
Electrical capacity (1975) 250* megawatts
Hospital beds (1973) 996, 1 per 194 people
Roads (1976) 3 000* km = 1 900* mi, density 0.22* km per km²
There are no railways
Ships (registered, 1977) 109, total of 106 317 gross tons
Ports Nassau (New Providence), Freeport (Grand Bahama), Matthew Town (Inagua)
Airports Nassau, Freeport, West End (Grand Bahama), Rock Sound (Eleuthera); also 22 other airports with scheduled flights

Durable equipment[a]	000	no per	
(at end-year)		1 000 people	no per
Radio sets (1974)	90	457	km of road
Passenger cars (1974)	40	203	13.3
Commercial vehicles (1974)	5.5	28	1.8

[a]A television service started in 1977; some Florida TV stations can also be received

Production, finance and trade

Gross domestic product 1975: Ba $ 500* mn = $ 500* mn = £ 220* mn
1976 est: Ba $ 550** mn = $ 550** mn = £ 300** mn
Main products (000 t) *Agriculture* (1976) Cow milk 3*, goat milk 4*, poultry meat 5*, fish catch (1975) 2.8
Other (1975) Petroleum products 9 440*, electricity (mn kW h) 650*, salt (1974) 938, cement 463*
Transport traffic *Air* (1976) Passenger-km 402 mn, cargo 2.0 mn t-km
Sea Ships entered (1972) 8.54 mn nrt, goods unloaded (1970) 4.36 mn t
Tourism Number of visitors (including cruise passengers, 1975) 1 324 300, gross receipts $ 317 mn
Consumer price index (1970 = 100) 1976 153.3; growth 1970–76 7.4 %pa
Money stock (end-year, Ba $ mn) 1970 80.5, 1976 77.5; growth 1970–76 −0.6 %pa
Budget (1976) Revenue: Ba $ 134 mn = $ 134 mn = £ 74 mn
Expenditure: Ba $ 141 mn = $ 141 mn = £ 78 mn
International reserves (Dec 1976) $ 47 mn
External public debt (Dec 1976) $ 25 mn

Bahamas

External trade (1976) Imports: Ba $ 3 560 mn = $ 3 560 mn = £ 1 971 mn
Exports: Ba $ 2 879 mn = $ 2 879 mn = £ 1 594 mn

Main imports (1975)	% of total	Main exports (1975)	% of total
Crude oil	88	Crude oil (re-exports)	56
Petroleum products	3	Petroleum products	40
Food	2	Chemicals	3
Main sources (1975)		*Main destinations* (1975)	
Saudi Arabia	38	United States	76
Nigeria	18	Liberia	5
Libya	14	Canada	4
Indonesia	7	Puerto Rico	3

Special focus

Tourism, 1975

Origin of visitors	Number	%		Number	%
United States	662 000	50	Other	52 050	4
Canada	150 700	11	Total tourists	903 040	68
West Germany	23 310	2	Cruise passengers	421 280	32
United Kingdom	14 980	1	Total visitors	1 324 320	100

Barbados

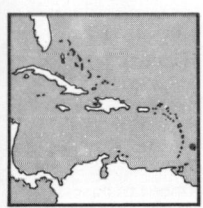

Location Eastern Caribbean Sea
The most easterly of the Caribbean islands, about 300 km north-east of Trinidad
Land Area 431 km² = 166 mi²
Climate Sub-tropical
Weather at Bridgetown, 55 m altitude
Temperature: hottest months June, Aug, Sept 23–31 °C, coldest February 21–28 °C
Rainfall (av monthly): driest month Feb 28 mm, wettest Nov 206 mm

Time 4 hours behind GMT
Measures UK (imperial) system; plans to convert gradually to the metric system came into operation August 1, 1977
Monetary unit Barbados dollar (Bds $) = 100 cents; the Barbados dollar was introduced in November 1973 to replace the East Caribbean dollar at par
Rate of exchange (1976 av): par Bds $ 2.00 = $ 1, free Bds $ 3.61 = £1

Summary

Political Parliamentary monarchy, which became independent November 30, 1966; formerly a UK colony, and from 1958-62 a member of the Federation of the West Indies. Member of UN, OAS, Sela, Caricom, Commonwealth and an EEC ACP state
Economic Tourism and sugar are the most important features of the economy; gross receipts from tourism are about equal to total exports, and sugar accounts for about one-third of exports. Agriculture is to be diversified, and there are plans to expand the financial sector

People, resources and equipment

Population 1960 233 000, 1970 238 000, 1976 247 000
Growth: 1960–70 0.2 %pa, 1970–76 0.6 %pa
Density (1976): 573 people per km²
Vital statistics (rate per 1 000 people, 1976): births 18.6, deaths 9.2
Town (population, 1976) Bridgetown (capital) 95 000*
Race (1970) African 90 %, European 4 %, Mixed 4 %
Language English
Religion (1970) Anglican 52 %, Methodist 9 %, Roman Catholic 4 %
Education Pupils (1973/74) 62 400*, teachers (1969/70) 2 088
Labour force (1976) 100 000*; in agriculture 18 000* (18* %)
Personnel (1975) Physicians: 166, 1 per 1 476 people
Standard of living
National income per person (1976): Bds $ 2 800** = $ 1 400** = £ 780**
Consumption per person (1975): energy 1 078 kg coal equivalent, electricity (production) 873 kW h, newsprint 2.0 kg
Newspapers (1974): number 1; circulation 24 000, 98 per 1 000 people
Telephones (Dec 1976): 44 050, 170 per 1 000 people

Livestock (000, 1976) Cattle 23*, sheep 49*, pigs 37*, goats 25*, chickens 375*
Mineral reserves (1975) Natural gas 1 000* mn cubic metres
Petroleum refinery capacity (1975) 150 000 tonnes
Electrical capacity (1975) 67 megawatts
Hospital beds (1975) 2 161, 1 per 114 people
Roads (1976) 1 350* km = 840* mi, density 3.1* km per km²
There are no railways
Ships (registered, 1977) 33, total of 4 448 gross tons
Port Bridgetown
Airport Grantley Adams (Seawell, 20 km from Bridgetown): passenger departures and arrivals (1976) 708 485

Durable equipment (at end-year)	000	no per 1 000 people	
Radio sets (1973)	116	477	no per
Television sets (1974)	40	164	km of road
Passenger cars (1973)	21	84	15.2
Commercial vehicles (1973)	3	12	2.2

Production

Gross domestic product 1975: Bds $ 700 mn = $ 350 mn = £ 158 mn
1976 est: Bds $ 730** mn = $ 360** mn = £ 200** mn

Production indices (1970 = 100)	1960	1970	1976	Growth %pa 1960–70	1970–76
Agricultural	99	100	77	0.1	−4.3
Main products (000 t)					
Agriculture					
Sugar, raw value	156	157	106*	0.1	−6.3*
Sweet potatoes	7*	5*	3*	−3.3*	−8.2*
Milk	6*	7*	8*	1.6*	2.0*
Other					
Petroleum products	11	119	143c	26.9	3.7d
Natural gas (mn m³)	2	3	2c	4.1	−7.5d
Electricity (mn kW h)	38	149	214c	14.6	7.5d
Beer (000 hl)	17a	48	59c	16.0b	4.2d
Cigarettes (mn units)	148	162	207c	0.9	5.0d
Margarine	1.1	1.6	1.8c	3.8	2.4d

a1963 b1963–70 c1975 d1970–75
Transport traffic *Air* (1976) Passenger-km 251 mn
Sea (1974) Ships entered 3.92 mn nrt
Tourism (1975) Number of visitors 222 000, gross receipts $ 78 mn

Finance and trade

Consumer price index (1970 = 100) 1976 246.6; growth 1970–76 16.2 %pa
Money stock (end-year, Bds $ mn) 1970 55.7, 1976 117.1; growth 1970–76 13.2 %pa
Budget (1976/77; year ending March 31st)
Revenue: Bds $ 182 mn = $ 91 mn = £ 52 mn
Expenditure: Bds $ 276 mn = $ 138 mn = £ 80 mn

Balance of payments ($ mn)	1972	1973	1974	1975	1976
Balance of goods (fob)	−90	−105	−118	−103	−121
Balance of services	+41	+45	+63	+54	+50
Balance of transfers	+6	+7	+7	+7	+13
Current balance	−43	−52	−48	−41	−57
Long-term capital flow	+21	+21	+4	+25	+22
International reserves (end-year, $ mn)	28	32	39	40	28

External trade (1976) Imports: Bds $ 473 mn = $ 237 mn = £ 131 mn
Exports: Bds $ 174 mn = $ 87 mn = £ 48 mn

Main imports	% of total	Main exports	% of total
Food	21	Sugar	27
(of which, meat 6, cereals 5)		Clothing	18
Crude oil & products	12	Petroleum products	13
Chemicals	9	Electrical parts	7
Electrical machinery	6	Chemicals	6
Machinery, non-electric	6	Molasses	5
Textile yarns and fabrics	5	Rum	3
Motor vehicles	5		
Main sources		*Main destinations*	
United States	24	United States	31
United Kingdom	19	Ireland	14
Trinidad and Tobago	11	Trinidad and Tobago	11
Canada	9	Canada	7
Venezuela	6	Windward Islands	6
Japan	3	Jamaica	4
Jamaica	3	United Kingdom	4

Belize
Colony of Belize

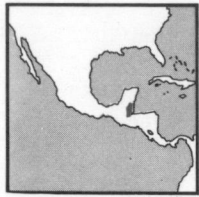

Location Central America
With a Caribbean Sea coastline, Mexico is to
the north and Guatemala to the west
Land Area 22 963 km² = 8 866 mi²
Climate Sub-tropical
Weather at Belize City, 5 m altitude
Temperature: hottest month Aug 24–31 °C,
coldest January 19–27 °C
Rainfall (av monthly): driest month
March 38 mm, wettest October 305 mm
Time 6 hours behind GMT
Measures UK (imperial) system, converting to the metric system
Monetary unit Belizean dollar (Bz $) = 100 cents
Rate of exchange (1976 av): free Bz $ 2.06 = $ 1, Bz $ 3.72 = £ 1
A link with the UK pound at Bz $ 4.00 = £ 1 was changed from May 1976
to a link with the US dollar at Bz $ 2.00 = $ 1

Summary

Political UK colony, with full internal self-government from January 6,
1964; formerly called British Honduras. Member of Caricom
Economic An agricultural economy, with sugar and citrus fruit
the main exports; fishing is also important and is being developed

People, resources and equipment

Population 1960 91 000*, 1970 121 000*, 1976 144 000*
Growth: 1960–70 2.9* %pa, 1970–76 2.9* %pa
Density (1976): 6* people per km²
Vital statistics (rate per 1 000 people, 1973): births 38.7, deaths 6.1
Cities (population in 000, 1975) Belmopan (capital) 4*, Belize City 45*
Race (1970) Creole (African and Mulatto) 52* %, Mestizo 22* %, Maya
13* %, Carib 6* %, Asian and Chinese 5* %, Caucasian (European) 2* %
Language English and Spanish; Creole, Carib and Maya are also used
Religion (1976) Roman Catholic 60** %, Protestant 40** %
Education (1975/76) Pupils 41 282, teachers (1973/74) 1 546
Labour force (1970) 33 121
Personnel Scientists and engineers (1970): 201
Physicians (1973): 41, 1 per 3 170 people
Standard of living
National income per person (1976): Bz $ 1 500** = $ 730** = £ 400**
Consumption per person (1975): energy 520 kg coal equivalent,
electricity (production) 214* kW h, newsprint 1.5 kg
Newspapers (1974): number 1; circulation 4 000, 29 per 1 000 people
Telephones (Dec 1976): 5 560, 38 per 1 000 people
Livestock (000, 1976) Cattle 46*, pigs 18*, sheep 3*, chickens 321*
Electrical capacity (1975) 7* megawatts
Hospital beds (1975) 642, 1 per 221 people
Roads (1976) 2 197 km = 1 365 mi, density 0.096 km per km²
Ships (registered, 1977) 3, total of 620 gross tons
Ports Belize City, Stann Creek, Punta Gorda
Airport International (Stanley Field, 15 km from Belize City)
Durable equipment (at end-year)
Radio sets (1974): 80 000, 588 per 1 000 people
Motor vehicles (1974): 7 500*, 55* per 1 000 people, 3.4* per km of road

Production, finance and trade

Gross domestic product 1974: Bz $ 150* mn = $ 90* mn = £ 37* mn
1976 est: Bz $ 250** mn = $ 120** mn = £ 70** mn
Main products *Agriculture* (000 t, 1976) Sugar, raw value 63*, maize 11*,
rice 6*, oranges 29*, grapefruit 15*, bananas 3*, milk 4*, fish catch (1975)
1.9, timber (000 m³, 1975) 116 *Other* (1975) Electricity 30* mn kW h,
cigarettes 88 mn units
Transport traffic (1971) *Sea* Goods loaded 110 000* t, unloaded 155 000* t
Budget (1975) Revenue: Bz $ 38.1 mn = $ 21.2 mn = £ 9.5 mn
Expenditure: Bz $ 28.7 mn = $ 15.9 mn = £ 7.2 mn
External trade (1975) Imports: Bz $ 185 mn = $ 103 mn = £ 46 mn
Exports: Bz $ 130 mn = $ 72 mn = £ 32 mn

Main imports (1973)	% of total	Main exports (1975)	% of total
Food	26	Sugar	61
Chemicals	10	Clothing	6*
Textile yarns and fabrics	7	Citrus and products	6
Machinery, non-electric	6	Fish	3
Main sources (1970)		*Main destinations* (1970)	
United States	34	United States	30
United Kingdom	25	United Kingdom	24

Bermuda
Colony of Bermuda

Location Western Atlantic Ocean
A group of about 150 islands and islets, about
900 km east of North Carolina (United States)
Land Area 53 km² = 21 mi²
Climate Mild and humid
Weather at Hamilton, 46 m altitude
Temperature: hottest month Aug 23–30 °C,
coldest Feb-Mar 14–20 °C
Rainfall (av monthly): driest month April
104 mm, wettest October 147 mm
Time 4 hours behind GMT (summer time, 3 hours behind)
Measures UK (imperial) system; also US units
Monetary unit Bermuda dollar (Bda $) = 100 cents
Rate of exchange (1976 av): par Bda $ 1 = $ 1, free Bda $ 1.806 = £ 1

Summary

Political UK colony, with internal self-government from June 8, 1968.
United States naval and military bases occupy 6 km² of the islands
Economic Tourism is the most important industry, and covers the deficit
in visible trade. Ship and aircraft bunkering and repairs are also important

People, resources and equipment

Population 1960 44 000, 1970 52 000, 1976 57 000*
Growth 1960–70 1.7 %pa, 1970–76 1.5* %pa
Density (1976): 1 075* people per km²
Vital statistics (rate per 1 000 people, 1975): births 15.0, deaths 7.1
Towns (population in 000, 1976) Hamilton (capital) 3*, St George 2*
Race (1970) African and Mixed 59 %, European 40 %
Language English
Religion (1970) Christian 91 %
Education (1973/74) Pupils 12 000*, teachers 710*
Labour force (1974) 27 300 (excluding US citizens at American bases)
Personnel (in hospitals only, 1973) Physicians: 58, 1 per 950 people
Standard of living National income per person (1976):
Bda $ 5 500** = $ 5 500** = £ 3 000**
Consumption per person (1975): energy 3 090 kg coal equivalent,
electricity (production) 5 300* kW h
Newspapers (1974): number 1; circulation 12 000, 214 per 1 000 people
Telephones (Dec 1976): 38 000, 661 per 1 000 people
Livestock (000, 1976) Chickens 74*, cattle 1*, pigs 1*
Electrical capacity (1975) 85* megawatts
Hospital beds (government only, 1973) 498, 1 per 110 people
Roads (1976) 240* km = 150* mi, density 4.5* km per km²
Ships (registered, 1977) 88, total of 1 751 515 gross tons
Ports Hamilton, St George; there is a free port at Ireland Island
Airport Kindley Field (20 km from Hamilton)

Durable equipment	000	no per	
(at end-year)		1 000 people	
Radio sets (1974)	50	909	no per
Television sets (1974)	20	364	km of road
Passenger cars (1975)	12.4	219	52
Commercial vehicles (1975)	2.2	39	9

Production, finance and trade

Gross domestic product 1974: Bda $ 300* mn = $ 300* mn = £ 130* mn
1976 est: Bda $ 350**mn = $ 350**mn = £ 190**mn
Main products (000 t, 1976) Potatoes 1*, bananas 1*, milk 1*,
fish catch (1975) 0.5, electricity (1975) 300* mn kW h
Transport traffic *Sea* Goods unloaded (1975) 0.1 mn t
Tourism (1975) Number of visitors 531 600, gross receipts $ 145 mn
Budget (1976/77; year ending March 31st)
Revenue: Bda $ 69.6 mn = $ 69.6 mn = £ 40 mn
Expenditure: Bda $ 69.5 mn = $ 69.5 mn = £ 40 mn
External trade (1976) Imports: Bda $ 143 mn = $ 143 mn = £ 79 mn
Exports: Bda $ 42 mn = $ 42 mn = £ 23 mn

Main imports	% of total	Main exports	% of total
Food	23	Drugs and medicines	39
Petroleum products	11	Petroleum products	34
Machinery	10	Aircraft supplies	7
Main sources		*Main destinations*	
United States	52	Bunkers	36
United Kingdom	13	United States	14
Netherlands Antilles	10	United Kingdom	6

Bolivia

Republic of Bolivia
República de Bolivia

Location West central South America
Brazil is to the north and east, Paraguay and
Argentina to the south, and Chile and Peru to
the west. Land-locked but with some access to
the sea via the Paraguay and Paraná rivers
which lead to the River Plate
Land Area 1 098 581 km² = 424 164 mi²
Climate Tropical below about 1 500 m,
cool above about 3 500 m
Weather at La Paz, 3 658 m altitude
Temperature: hottest month Nov 6–19 °C, coldest July 1–17 °C
Rainfall (av monthly): driest month June 8 mm, wettest Jan 14 mm
Time 4 hours behind GMT
Measures Metric system: also local units, including:
Weight (mass) 1 libra or arratel = 0.46 kilogram = 1.014 pounds
Monetary unit Bolivian peso (B $) = 100 centavos
Rate of exchange (1976 av): par B $ 20.00 = $ 1, free B $ 36.12 = £ 1

Summary

Political Republic, which became independent from Spain in 1825;
with a military government from 1971. Member of UN, OAS, Sela, Lafta
and Andean Group. Bolivia has claimed a strip of land with a coast line
on the Pacific, part of Chile, to give direct access to the sea
Economic There is a wide range of mining products, and tin accounts for
nearly one-half of exports. Oil and natural gas make up about one-third
of exports, but oil reserves are not large and new fields are being sought.
Smelting of local ores is of increasing importance, and other industry is
also being developed

People, resources and equipment

Population 1960 3.82*mn, 1970 4.93*mn, 1976 5.79*mn
Growth: 1960–70 2.6* %pa, 1970–76 2.7* %pa
Density (1976): 5* people per km²
Vital statistics (rate per 1 000 people, 1977): births 46.6*, deaths 18.0*
Regions (departments, population in 000, 1975; total of 5.63 mn)

Beni	224	La Paz	1 770	Potosí	997
Chuquisaca	528	Oruro	392	Santa Cruz	534
Cochabamba	915	Pando	37	Tarija	237

Cities (population in 000, 1975)

La Paz (administrative capital)	661	Cochabamba	184	Oruro	110
Sucre (legal capital)	107	Santa Cruz	149	Potosí	77

Race (1976) American Indian 55** % (Quechua 34** %, Aymará 21** %),
Mestizo 30** %, European (mainly Spanish) 14** %
Language Spanish; also Quechua and Aymará
Religion Mainly Roman Catholic
Education (1975) Pupils 1 120 000*, teachers 51 300*
Labour force (1976) 1 510 611; in agriculture 675 512 (45 %)
Personnel Scientists and engineers (1967): 10 925
Physicians (1974): 2 583, 1 per 2 150 people
Standard of living
National income per person (1976): B $ 7 000** = $ 350** = £ 194**
Consumption per person (1975): energy 303 kg coal equivalent,
electricity (production) 177* kW h, newsprint 0.9 kg, steel 21 kg
Newspapers (1974): number 14; circulation 135 000, 25 per 1 000 people
Telephones (Dec 1973): 49 000, 9 per 1 000 people
Livestock (000, 1976) Cattle 2 926*, sheep 7 767*, goats 2 848*, pigs 1 186*
Mineral reserves (1975) Crude oil 19 mn tonnes
Natural gas 133 000 mn cubic metres
Petroleum refinery capacity (1975) 1.3 mn tonnes
Electrical capacity (1975) 346* megawatts
Hospital beds (1970) 9 451, 1 per 522 people
Roads (1975) 37 075 km = 23 037 mi, density 0.034 km per km²
Railways (1975) 3 787 km = 2 353 mi, density 0.003 km per km²
Inland waterways (mainly rivers, 1976) 19 000*km = 12 000*mi
Ports Ports in neighbouring countries are used, especially Rosario
(Argentina) and Santos (Brazil)
Airports (passenger departures and arrivals, 000, 1976) La Paz 385,
Santa Cruz 280; also (1977) 23 other airports with scheduled flights

Durable equipment (at end-year)	000	no per 1 000 people	
Radio sets (1976)	426*	73*	no per
Television sets (1976)	45*	8*	km of road
Passenger cars (1974)	20	4	0.5
Commercial vehicles (1974)	29	5	0.8

Production

Gross domestic product 1975: B $ 43 079 mn = $ 2 154 mn = £ 969 mn
1976: B $ 46 000**mn = $ 2 300**mn = £ 1 270**mn
Growth in real terms: 1960–70 5.1 %pa, 1970–75 5.9 %pa
Structure of gross domestic product (1974) *By origin* Agriculture 15 %,
mining and quarrying 21 %, manufacturing 12 %, electricity, gas and
water 2 %, construction 5 %, transport & communication 8 %, other 38 %

Production indices (1970 = 100)	1960	1970	1976	Growth %pa 1960–70	1970–76
Agricultural	75	100	135	2.9	5.1
Manufacturing	29	100	205ᵇ	13.2	15.4ᶜ

Main products (000 t)

Agriculture					
Maize	248	286	300*	1.4	0.8*
Potatoes	605	655	870*	0.8	4.8*
Cassava	63	221	300*	13.4	5.2*
Sugar, raw value	27	123	213	16.4	9.6
Bananas	100**	212	260*	7.8**	3.5*
Coffee	3.0	11.2	16.0*	14.1	6.1*
Cotton	1.4**	5.0	12.0*	13.6**	15.7*
Beef and veal	33*	53*	78*	4.9*	6.7*
Timber (000 m³)	4 218	4 231	3 850ᵇ	0.0	−1.9ᶜ
Energy					
Total energy (000 tce)	650	1 580	4 950ᵇ	9.3	25.7ᶜ
Crude oil	466	1 122	1 964	9.2	9.8
Petroleum products	294	564	900*ᵇ	6.7	9.8*ᶜ
Natural gas (mn m³)	na	106	2 000*	na	63.2*
Electricity (mn kW h)	447	787	1 000*ᵇ	5.8	4.9*ᶜ
Mining					
Antimony ore (Sb content)	4.8*	10.6	14.4ᵇ	8.1*	6.3ᶜ
Copper ore (Cu content)	2.0*	8.2	5.9ᵇ	14.9*	−6.4ᶜ
Lead ore (Pb content)	18*	22	16ᵇ	2.2*	−5.6ᶜ
Tin conc (Sn content)	21*	30	28ᵇ	3.6*	−1.2ᶜ
Tungsten conc (oxide content)	1.1*	2.0	3.1*ᵇ	6.3*	9.5*ᶜ
Zinc ore (Zn content)ᵃ	4	7	47ᵇ	6.0	44.5ᶜ
Silverᵃ	0.15	0.15	0.15ᵈ	0.0	0.0ᵉ
Gold (000 kg)	1.04	0.86	1.66ᵇ	−1.9	13.9ᶜ
Sulphur	1	16	22*ᵇ	32.0	6.6*ᶜ
Manufacturing					
Beer (000 hl)	176	381	563ᵈ	8.0	10.3ᵉ
Cigarettes (mn units)	449	730	720ᵇ	5.0	−0.3ᶜ
Cotton fabrics (mn m)	9	12	11*ᵇ	2.9	−1.7*ᵉ
Cement	39	116	226ᵇ	11.5	14.3ᶜ

ᵃExports ᵇ1975 ᶜ1970–75 ᵈ1974 ᵉ1970–74

Transport traffic
Passenger-kilometres Rail (1975) 310 mn, air (1976) 444 mn
Cargo: tonne-kilometres Rail (1975) 465 mn, air (1976) 4.3 mn

Finance and trade

Price index (1970 = 100)	1960	1970	1976	Growth %pa 1960–70	1970–76
Consumer prices	58.5	100.0	266.6	5.5	17.8
Money stock (end-year, B $ mn)	419	1 532	6 497	13.8	27.2

Budget (1975) Revenue: B $ 5 689 mn = $ 284 mn = £ 128 mn
Expenditure: B $ 6 293 mn = $ 315 mn = £ 142 mn

Balance of payments ($ mn)	1972	1973	1974	1975	1976
Balance of goods (fob)	+6	+34	+212	−54	+3
Balance of services	−65	−70	−109	−118	−137
Balance of transfers	+13	+15	+14	+13	+14
Current balance	−45	−21	+117	−159	−119
Long-term capital flow	+93	+37	+100	+159	+235

Reserves and debt (end-year, $ mn)

		1972	1973	1974	1975	1976
International reserves		60	72	194	156	168
External public debt		742	768	889	1 207	1 576

External trade (1976)
Imports: B $ 11 090 mn = $ 555 mn = £ 307 mn
Exports: B $ 10 260 mn = $ 513 mn = £ 284 mn

Main imports (1972)	% of total	*Main exports* (1976)	% of total
Machinery	22	Tin	44
Chemicals	11	Crude oil	22
Motor vehicles	10	Natural gas	11
Cereals	9	Zinc ore	8
Iron and steel	7	Tungsten ore	7
Main sources (1976)		*Main destinations* (1976)	
United States	26	United States	34
Argentina	15	Argentina	23
Brazil	14	United Kingdom	9
Japan	11	Switzerland	4

Brazil
Brasil

Location Central and east South America
Uruguay, Argentina and Paraguay are to the
south-west, Bolivia and Peru to the west, and to
the north are Colombia, Venezuela, Guyana,
Surinam and French Guiana
Land Area 8 511 965 km² = 3 286 488 mi²
Usage (1975): agricultural 2 066 000* km²
(24* %), of which, arable 285 000* km² (3* %),
cropland 81 000* km² (1* %), pastures 1 700 000*
km² (20* %); forests 5 100 000* km² (60* %)
Climate Mainly tropical and sub-tropical; mild on the southern coast
and on the higher lands
Weather at Rio de Janeiro, 61 m altitude
Temperature: hottest month Feb 23–29 °C, coldest July 17–24 °C
Rainfall (av monthly): driest month July 41 mm, wettest Dec 137 mm
Time East, including all coast and Brasilia: 3 hours behind GMT
West: 4 hours behind GMT
Territory of Acre: 5 hours behind GMT
Fernando de Noronha: 2 hours behind GMT
Measures Metric system; also some local units, including
weight (mass) 1 metric arroba = 15 kilograms = 33.07 pounds
Monetary unit Cruzeiro (Cr) = 100 centavos; a new cruzeiro was
introduced from February 13, 1967, to replace the old cruzeiro at the
rate of 1 new cruzeiro = 1 000 old cruzeiros
Rate of exchange (1976 av): free Cr 10.77 = $ 1, Cr 19.45 = £ 1

Summary

Political Republic with military government, which became independent
from Portugal in 1822. Member of UN, OAS, Sela and Lafta
Economic A mixed economy with agriculture accounting for about 12 %
of gross domestic product and manufacturing 19 %; agricultural products
are the main exports, with coffee and sugar making up one-quarter to
one-third of the total. Large natural resources have helped to provide a
high rate of economic expansion—real growth of the economy was about
11 %pa for 1970–75; there has also been a high rate of inflation

People

Population[a] 1960 69.73*mn, 1970 92.52*mn, 1976 109.18*mn
[a]Excluding Indian jungle population (150 000 in 1956)
Growth: 1960–70 2.9 %pa, 1970–76 2.8 %pa
Density (1976): 13* people per km²
Vital statistics (rate per 1 000 people, 1970–75): births 37.1*, deaths 8.8*
Regions (population in 000, 1976; total of 110* mn)

Federal district	820*	Paraná	8 791
States		Pernambuco	5 995
Acre	256	Piauí	2 048
Alagoas	1 829	Rio de Janeiro	10 704[a]
Amazonas	1 120	Rio Grande do Norte	1 913
Bahia	8 641	Rio Grande do Sul	7 623
Ceará	5 258	Santa Catarina	3 451
Espírito Santo	1 750	São Paulo	21 268
Goiás	3 684*	Sergipe	1 012
Maranhão	3 399	*Territories*	
Mato Grosso	2 097	Amapá	148
Minas Gerais	12 764	Fernando de Noronha	1
Pará	2 626	Rondônia	147
Paraíba	2 729	Roraima	50

[a]Includes Guanabara united with Rio de Janeiro in 1975
Cities (population in 000, 1975)

Brasília (capital)	763	Nova Iguaçu	932	Manaus	389
São Paulo	8 100[a]	Belém	772	Osasco	377
Rio de Janeiro	4 858	Curitiba	766	Niterói	376
Belo Horizonte	1 557	Duque de Caxias	537	Natal	344
Recife	1 250	Goiânia	518	São Luís	330
Salvador	1 237	Santo André	515	Maceió	324
Fortaleza	1 110	Campinas	473	Guarulhos	311
Pôrto Alegre	1 044	Santos	396	Teresina	290

[a]1976 estimate
Race Mainly of European origin (Portuguese, Spanish, German, Italian),
but with large African and Indian elements and mixed groups
Language Portuguese
Religion (1970) Christian 98 % (Roman Catholic 92 %, Protestant 5 %)

Education (1972) Pupils: primary 14 082 098, secondary 4 459 522,
vocational 819 169, teacher-training 309 892, higher (1973) 785 159.
Teachers: primary 525 658, secondary 266 328, vocational 67 468,
teacher-training 37 468, higher (1973) 59 760
Labour force (1970) 29 557 224; in agriculture 13 090 358 (44 %),
mining and quarrying 175 424 (1 %), manufacturing 3 241 861 (11 %),
electricity, gas and water 158 428 ($\frac{1}{2}$ %), construction 1 719 714 (6 %),
distribution and hotels 2 263 539 (8 %), transport and communication
1 244 395 (4 %), finance and real estate 434 040 (1$\frac{1}{2}$ %),
other 7 229 465 (24 %)
Personnel (1972) Physicians: 48 726, 1 per 2 025 people
Standard of living
National income per person (1976): Cr 12 000** = $ 1 100** = £ 620**
Consumption per person (1975): energy 670 kg coal equivalent,
electricity (production) 735 kW h, newsprint 2.8 kg, steel 105 kg
Newspapers (1973): number 280; circulation 4 050 000, 40 per 1 000 people
Telephones (Dec 1976): 3 987 000, 36 per 1 000 people

Resources and equipment

Livestock (000, 1976) Cattle 95 000*, sheep 25 000*, goats 16 200*,
pigs 35 000*, horses 9 600*, chickens 280 000*
Mineral reserves Coal (1974) 3 256 mn tonnes
Crude oil (1975) 105 mn tonnes
Natural gas (1975) 26 000 mn cubic metres
Uranium (1974) 9 700 tonnes
Petroleum refinery capacity (1975) 51* mn tonnes
Electrical capacity (1975) 19 588 megawatts
Hospital beds (1973) 382 952, 1 per 266 people
Roads (1976) 1 489 064 km = 925 261 mi, density 0.17 km per km²
Railways (1975) 33 000 km = 20 505 mi, density 0.004 km per km²
Inland waterways (1976) Amazon and other rivers and tributaries:
43 000* km = 27 000* mi, of which, Amazon river 3 680 km = 2 290 mi
Ships (registered, 1977) 538, total of 3 329 951 gross tons
Ports (goods traffic, 000 tonnes, 1974)

	loaded	unloaded		loaded	unloaded
Vitória & Tubarão	51 929	970	Paranagua	2 924	374
Rio de Janeiro	3 805	12 443	Salvador	1 010	604
Santos	4 333	10 728	Macapá	1 520	2

Airports Congonhas (São Paulo), Galeao (Rio de Janeiro), Brasília,
Belo Horizonte, Recife and 92 other airports with scheduled flights

Durable equipment (at end-year)	000	no per 1 000 people	
Radio sets (1974)	6 275	60	no per
Television sets (1974)	8 650	83	km of road
Passenger cars (1976)	6 349	57	4.3
Commercial vehicles (1976)	825	7	0.6

Production

Gross domestic product
1975: Cr 895 600 mn = $ 109 166 mn = £ 49 134 mn
1976 est: Cr 1 400 000** mn = $ 130 000** mn = £ 72 000** mn
Growth in real terms: 1963–70 6.2 %pa, 1970–75 10.8 %pa
Structure of gross domestic product (1973)

By origin	Cr mn	% of gdp
Agriculture	58 413	12
Mining and quarrying	2 427	1
Manufacturing	91 599	19
Electricity, gas and water	8 446	2
Construction	24 801	5
Distribution	65 870	14
Transport and communication	19 984	4
Other services	205 623	43
Total	477 163	100
By type of expenditure		
Government final consumption	46 190	10
Private final consumption	327 060	69
Gross fixed capital formation	108 054	23
Exports of goods and services	40 030	8
less Imports of goods and services	−44 172	−9
Total	477 163	100

Production indices (1970 = 100)	1960	1970	1976	Growth %pa 1960–70	1970–76
Agricultural	74	100	124	3.0	3.7
Industrial	52	100	164[m]	6.8	10.4[n]

Brazil

Main products (000 t)	1960	1970	1976	Growth %pa	
				1960–70	1970–76
Agriculture					
Wheat	713	1 844	3 215	10.0	9.7
Rice	4 795	7 553	9 560	4.6	4.0
Maize	8 672	14 216	17 845	5.1	3.9
Potatoes	1 113	1 583	1 814	3.6	2.3
Sweet potatoes	1 283	2 134	1 730*	5.2	−3.4*
Cassava	17 613	29 464	26 816	5.3	−1.6
Dry beans	1 731	2 211	1 923	2.5	−2.3
Sugar, raw value	3 380	5 447	7 600*	4.9	5.7*
Tomatoes	397	764	1 171	6.8	7.4
Soyabeans	206	1 509	10 845	22.0	38.9
Groundnuts	408	928	528	8.6	−9.0
Oranges	1 821	3 099	7 373	5.5	15.5
Bananas	3 332*	6 408*	8 121	6.8*	4.0*
Coffee	2 085	1 551[i]	943[k]	−2.7[j]	−8.0[l]
Cocoa	163	197	233*	1.9	2.8*
Tobacco	161	244	298	4.2	3.4
Cotton	370	673	390	6.2	−8.7
Sisal	135	205	167	4.3	−3.3
Milk	5 052	7 353	13 850*	3.8	11.1*
Butter	25	45*	67*	6.1*	6.8*
Eggs	338	454	522*	3.0	2.4*
Wool	23	19	28*	−1.7	6.7*
Cattle hides	240**	315*	356*	2.8**	2.0*
Beef and veal	1 359	1 853*	2 226*	3.1*	3.1*
Pigmeat	474	767	750*	4.9	−0.4*
Poultrymeat	160**	336*	442*	7.7**	4.7*
Fish catch	251	518*	674*[m]	7.5*	5.4*[n]
Timber (000 m³)	115 000**	158 760*	164 000*[m]	3.3**	0.6*[n]
Energy					
Total energy (000 tce)	9 320	19 380	25 410[m]	7.6	5.6[n]
Coal	1 277	2 361	3 130*	6.3	4.8*
Crude oil	3 870	7 962	8 470	7.5	1.0
Petroleum products	8 416	24 101	41 930*[m]	11.1	11.7*[n]
Natural gas (mn m³)	50**	120**	550*	9.0**	29.0*[n]
Manufactured gas (mn m³)[o]	440	458	516[h]	0.4	2.0*
Electricity (mn kW h)	22 865	45 460	78 068[m]	7.1	11.4[n]
of which, hydro (mn kW h)	18 384	39 863	71 991[m]	8.0	12.5[n]
Mining					
Iron ore (Fe content)	6 355	24 739	46 621*[m]	14.6	13.5*[n]
Bauxite	121	510	1 277*[m]	15.5	20.1*[n]
Chromium ore (oxide content)	3.6	28	124[o]	22.8	64.6[p]
Lead ore (Pb content)	10	28	30*[m]	10.8	1.4*[n]
Magnesite	63	216	375*[m]	13.1	11.7*[n]
Manganese ore (Mn content)	438	1 202	820*[m]	10.6	−7.4*[n]
Nickel ore (Ni content)	0.1	3.0	2.6*[m]	41.2	−2.8*[n]
Tin conc (Sn content)	1.6	3.6	5.0*[m]	8.6	6.7*[n]
Zinc ore (Zn content)	—	19	30*[m]	na	9.6*[n]
Gold (000 kg)	3.7	5.6	9.3	4.2	8.8
Diamonds (000 CM)	52	41	270*[m]	−2.3	45.8[n]
Salt	923	1 826	1 500*[m]	7.1	−3.9*[n]
Asbestos	98	376	819[o]	14.4	29.6[p]
Manufacturing					
Beer (000 hl)[c]	7 360[d]	9 132	22 238[m]	2.7[e]	19.5[n]
Cigarettes (mn units)	59 992[d]	70 703	101 741[m]	2.1[e]	7.5[n]
Cotton yarn	40*	44*	60[m]	1.1*	6.4*[n]
Cotton fabrics (mn m)	680*[f]	784	864[m]	3.0*[g]	2.0[n]
Man-made fibres	45*	92*	175[m]	7.4*	13.7*[n]
Wood pulp	270*	811	1 301[m]	11.6*	9.9[n]
Newsprint	66	103	127	4.6	3.6
Other paper	480**	1 116	1 518[m]	8.8**	6.3[n]
Synthetic rubber	—	75	164	na	13.9
Tyres (000 units)	3 047	7 847	14 556	9.9	10.8
Fertilisers, nitrogenous	16	22	160[m]	3.2	48.5[n]
Fertilisers, phosphate	42	162	510*[m]	14.5	25.8*[n]
Cement	4 474	9 002	17 873	7.2	12.1
Coke	704	1 815	2 300*[m]	9.9	4.9*[n]
Pig iron	1 783	4 296	8 200*	9.2	11.4*
Crude steel	2 260	5 390	9 090	9.1	9.1
Radio sets (000 units)	602[f]	809	640[m]	6.1[g]	−4.6[n]
Television sets (000 units)	309[f]	726	1 451[m]	18.6[g]	14.8[n]
Passenger cars (000 units)[h]	57	255	556	16.2	13.9
Commercial vehicles (000 units)[h]	76	161	416	7.8	17.1
Merchant vessels (000 grt)	—	100	407	na	26.4

Main products (000 t)	1960	1970	1976	Growth %pa	
				1960–70	1970–76
Construction					
Dwellings (000 units)[a]	na	188	200	na	1.0

[a]Permits issued, urban only [b]Consumption in Rio de Janeiro and São Paulo [c]Main establishments only [d]1962 [e]1962–70 [f]1965 [g]1965–70 [h]Including assembly [i]1971 [j]1960–71 [k]1977 [l]1971–77 [m]1975 [n]1970–75 [o]1973 [p]1970–73

Transport traffic	1960	1970	1976	Growth %pa	
				1960–70	1970–76
Passenger-kilometres (mn)					
Rail	15 395	12 351	10 649[a]	−2.2	−3.6[b]
Air	2 679	4 385	10 366	5.1	15.4
Cargo: tonne-kilometres (mn)					
Rail	12 688	17 531	55 220[a]	3.3	33.2[b]
Air	90	173	488	6.8	18.8
Sea: tonnes (mn)					
Goods loaded	11	40	90	13.8	14.4
Goods unloaded	16	28	61	5.8	13.9
Tourism					
Number of visitors (000)	na	194	518[c]	na	21.7[d]
Gross receipts ($ mn)	na	30	72[c]	na	19.1[d]

[a]1974 [b]1970–74 [c]1975 [d]1970–75

Finance and trade

Price indices	1960	1970	1976	Growth %pa	
(1970 = 100)				1960–70	1970–76
Consumer prices	2.5	100	368	44.5	24.3
Wholesale prices	3.0	100	392	42.0	25.6
Money stock					
(end-year, Cr mn)	690	34 740	245 100	48.0	38.5

Budget (1976) Revenue: Cr 166 220 mn = $ 15 434 mn = £ 8 546 mn
Expenditure: Cr 165 795 mn = $ 15 394 mn = £ 8 524 mn,
of which (1974), defence 14 %, roads and public works 11 %, education 5 %

Balance of payments ($ mn)	1972	1973	1974	1975	1976
Balance of goods (fob)	−252	−59	−4 748	−3 543	−2 297
Balance of services	−1 243	−1 728	−2 432	−3 212	−3 920
Balance of transfers	+5	+28	+2	—	+3
Current balance	−1 490	−1 759	−7 180	−6 754	−6 215
Long-term capital flow	+3 342	+3 699	+5 851	+5 274	+5 986
Reserves and debt (end-year, $ mn)					
International reserves	4 183	6 415	5 272	4 034	6 541
External public debt	8 002	9 557	12 022	14 121	na

External trade (1976) Imports: Cr 142 794 mn = $ 13 258 mn = £ 7 342 mn
Exports: Cr 107 106 mn = $ 9 945 mn = £ 5 507 mn

Main imports (1975)	% of total	Main exports (1975)	% of total
Machinery	29	Sugar	13
Crude oil	22	Coffee	11
Chemicals	13	Iron ore	11
Iron and steel	9	Soyabeans	8
Non-ferrous metals	4	Machinery	7
Transport equipment	4	Animal feeding stuffs	6
Wheat	3	Cocoa	3
Instruments	2	Motor vehicles	3
		Textile yarns and fabrics	3
Main sources (1976)		Main destinations (1976)	
United States	23	United States	18
Saudi Arabia	9	West Germany	9
Iraq	9	Netherlands	7
West Germany	8	Japan	6
Japan	7	Spain	4
Kuwait	4	Italy	4
Argentina	4	Soviet Union	4
Italy	3	United Kingdom	4
France	3	France	3

Special focus

Coffee prices (New York wholesale price, Brazilian coffee)

	$ per tonne	% change over previous year		$ per tonne	% change over previous year		$ per tonne	% change over previous year
1960	807	−1	1966	894	−7	1972	1 159	+18
1961	793	−2	1967	832	−7	1973	1 526	+32
1962	749	−6	1968	824	−1	1974	1 617	+6
1963	752	0	1969	902	+9	1975	1 821	+13
1964	1 029	+37	1970	1 230	+36	1976	3 295	+81
1965	965	−6	1971	986	−20			

Canada
Dominion of Canada

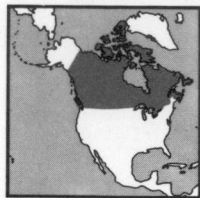

Location Northern half of North America
United States forms the southern border and
the land border to the north-west; the territory
stretches from the Atlantic Ocean to the Pacific
Land Area 9 976 139 km² = 3 851 809 mi²
Usage (1975): agricultural 688 670* km² (7* %),
of which, arable and cropland 437 670*
km² (4* %), pastures 251 000* km² (3* %);
forests 3 222 710 km² (32 %)
Climate Continental, with extremes of
temperature; very cold in the north, mild in south on Pacific coast
Weather at Ottawa, 103 m altitude
Temperature: hottest month July 14–27 °C,
coldest Jan minus 16–minus 6 °C
Rainfall (av monthly): driest month Feb 56 mm, wettest June 89 mm
Time Hours behind GMT (summer time[a] in brackets)

Newfoundland	3½ (2½)	Mountain Zone[e]	7	(6)
Atlantic Zone[b]	4 (3)	Pacific Zone[f]	8	(7)
Eastern Zone[c]	5 (4)	Yukon Territory	8	
Central Zone[d]	6 (5)			

[a]Summer time (daylight saving time) is at the discretion of individual provinces and
cities [b]Labrador, New Brunswick, Nova Scotia, Prince Edward Is, Quebec (east of
Pointe des Monts) [c]North-West Territory (east), Ottawa, Ontario, Quebec (west of
Pointe des Monts) [d]Manitoba, North-West Territory (central), Saskatchewan (east)
[e]Alberta, North-West Territory (mountain), Saskatchewan (west) [f]British Columbia
Measures UK (imperial) system, except 1 hundredweight = 100 pounds,
1 ton = 1 short ton = 2 000 lb. Conversion to the metric system is
taking place with a target date for completion at the end of 1980
Monetary unit Canadian dollar (C $) = 100 cents
Rate of exchange (1976 av): free C $ 0.9860 = $ 1, C $ 1.781 = £ 1

Summary

Political Parliamentary and federal monarchy; the provinces which first
formed the dominion became independent from the United Kingdom in
1867. There have been some calls for the separation of the mainly
French-speaking province of Quebec from the dominion. Member of
UN, Commonwealth, OECD, Nato and Colombo Plan
Economic One of the world's main industrial countries, but dependent on
the United States which accounts for two-thirds of trade. Motor vehicles
are the main export and other main exports are raw materials, notably
timber and its products, crude oil, natural gas, cereals and non-ferrous
metals. Motor vehicles are the largest item traded with the United States
(imports mainly parts and exports mainly completed vehicles)

People

Population 1960 17.91 mn, 1970 21.32 mn, 1976 23.14 mn
Growth: 1960–70 1.8 %pa, 1970–76 1.4 %pa
Density (1976): 2 people per km²
Vital statistics (rate per 1 000 people, 1976): births 15.8, deaths 7.2
Regions (province or territory, population in 000, 1976; total of 22.60 mn)

Newfoundland	549	Quebec	6 141	British	
Maritime provinces	*1 593*	Ontario	8 132	Columbia	2 406
Prince Edward Island	116	*Prairie provinces*	*3 713*	*Territories*	*64*
Nova Scotia	812	Manitoba	1 006	Yukon	21
New Brunswick	665	Saskatchewan	908	Northwest	42
		Alberta	1 800		

Cities (population in 000, 1976)

Ottawa (capital)	669	Hamilton	525	Victoria	212
Montreal	2 759	Calgary	458	Sudbury	155
Toronto	2 753	St Catherines	298	Regina	149
Vancouver	1 136	Kitchener	270	St John's	141
Winnipeg	571	London	265	Oshawa	134
Edmonton	543	Halifax	261	Saskatoon	132
Québec	534	Windsor	243	Thunder Bay	112*

Race (1971) United Kingdom and Ireland origins 45 %, French 29 %,
German 6 %, Italian 3 %, Ukrainian 3 %, Dutch 2 %, Indian and
Eskimo 1½ %
Language English and French
Mother tongue (1971): English 60 %, French 27 %
Religion (1971) Roman Catholic 46 %, United Church of Canada 17 %,
Anglican Church of Canada 12 %, Presbyterian 4 %, Lutheran 3 %
Education (1973/74) Pupils: primary 2 611 603, secondary, vocational
and teacher-training 2 719 051, higher 692 430. Teachers: primary and
secondary (1972/73) 300 000**, higher 46 825

Labour force (1977)	Number	% of total
Economic activity		
Agriculture	580 000	5
Mining and quarrying	151 000	1
Manufacturing	1 915 000	18
Construction	667 000	6
Distribution and hotels	1 684 000	16
Transport and communication	739 000	7
Finance and real estate	518 000	5
Other services	3 463 000	42
Unemployed	824 000	8
Total	10 645 000	100

Personnel Scientists and engineers (degree holders, 1971) 621 645
Physicians (1975): 39 104, 1 per 588 people
Standard of living
National income per person (1976): C $ 7 239 = $ 7 341 = £ 4 065
Consumption per person (1975): energy 9 880 kg coal equivalent,
electricity (production, 1976) 12 680 kW h, newsprint 27 kg, steel 577 kg
Newspapers (1973): number 121; circulation 5 207 000, 235 per
1 000 people
Telephones (Dec 1976): 13 785 600, 592 per 1 000 people

Resources and equipment

Livestock (000, 1976) Cattle 13 704, pigs 5 485, sheep 523, horses 345*,
chickens 82 683*
Mineral reserves Coal (1970–73) 97 041 mn tonnes
Lignite (1970) 11 736 mn tonnes
Crude oil (1975) 895 mn tonnes
Natural gas (1975) 1 613 bn cubic metres
Uranium (1974) 144 000 tonnes
Petroleum refinery capacity (1975) 104 mn tonnes
Electrical capacity (1975) 59 886 megawatts,
of which, hydro 37 253 megawatts, nuclear 2 666 megawatts
Hospital beds (1974) 206 763, 1 per 110 people
Roads (main, 1975) 834 152 km = 518 318 mi, density 0.084 km per km²
Railways (1975) 70 715 km = 43 940 mi, density 0.007 km per km²
Inland waterways (1976) St Lawrence system 3 670 km = 2 280 mi;
Mackenzie river 4 240 km = 2 635 mi;
Oil pipelines (1975) 31 716 km = 19 707 mi
Ships (registered, 1977) 1 283, total of 2 822 948 gross tons
Ports (goods traffic, 000 tonnes, 1976)

	International		Coastwise		Total
	loaded	unloaded	loaded	unloaded	
Vancouver	17 979	2 688	2 302	3 562	26 531
Sept-Iles	23 677	190	498	960	25 325
Cartier	17 814	1 614	5	2 253	21 686
Montreal	4 087	3 502	3 972	4 006	15 567

Other main ports are: Quebec, Hamilton, Halifax, and St John
Airports (passenger departures and arrivals, 000, 1976) Toronto 11 183,
Montreal 7 014, Vancouver 4 941; also 224 other airports with scheduled
flights

Durable equipment (at end-year)	000	no per 1 000 people	
Radio sets (1974)	20 252	894	no per
Television sets (1974)	8 232	363	km of road
Passenger cars (1975)	8 870	386	10.6
Commercial vehicles (1975)	2 158	94	2.6

Production

Gross domestic product
1976: C $ 191 882 mn = $ 194 606 mn = £ 107 738 mn
Growth in real terms: 1960–70 5.2 %pa, 1970–76 5.0 %pa
Structure of gross domestic product (1976)

By origin	C $ mn	% of gdp
Agriculture	7 092	4
Mining and quarrying	6 909	4
Manufacturing	35 985	19
Construction	12 639	7
Distribution	20 406	11
Transport and communication	12 986	7
Other	95 865	50
Total	191 882	100
By type of expenditure		
Government final consumption	38 548	20
Private final consumption	109 129	57
Stock investment	2 112	1
Gross fixed capital formation	44 309	23
Exports of goods and services	43 790	23
less Imports of goods and services	−45 137	−24
Total	191 882	100

Production indices (1970 = 100)	1960	1965	1970	1973	1974	1975	1976	Growth %pa 1960–70	1960–65	1965–70	1970–76
Agricultural	83	99	100	111	103	113	130	1.8	3.5	0.2	4.4
Industrial	55	78	100	123	127	121	127	6.2	7.4	5.0	4.1
of which, mining	56	76	100	124	123	114	115	6.0	6.3	5.6	2.4
Main products (000 t)											
Agriculture											
Wheat	14 108	17 674	9 024	16 159	13 295	17 078	23 587	−4.4	4.6	−12.6	17.3
Barley	4 212	4 753	8 889	10 223	8 802	9 520	10 513	7.8	2.4	13.3	2.8
Maize	663	1 511	2 634	2 803	2 577	3 645	3 770	14.8	17.9	11.8	6.2
Oats	6 146	6 169	5 445	5 041	3 929	4 467	4 831	−1.2	0.1	−2.5	−2.0
Potatoes	1 949	2 087	2 305	2 168	2 511	2 754	2 651	1.7	1.4	2.0	2.4
Apples	304	455	406	375	406	460	398	2.9	8.4	−2.3	−0.3
Rapeseed	252	513	1 637	1 207	1 163	1 749	837	20.6	15.3	26.1	−10.6
Linseed	571	741	1 218	493	351	445	277	7.9	5.4	10.5	−21.9
Tobacco	97	77	101	117	116	106	95*	0.4	−4.5	5.6	−1.0*
Milk	8 393	8 336	8 314	7 667	7 634	7 752	7 693	−0.1	−0.1	−0.1	−1.3
Eggs	307	293	334	314	313	305	297	0.8	−0.9	2.7	−1.9
Beef and veal	629	951	873	911	935	1 137	1 205*	3.3	8.6	−1.7	5.5*
Pigmeat	469	467	608	623	623	496	479*	2.6	−0.1	5.4	−3.9*
Poultrymeat	272	409	565	602	596	524	583	7.6	8.5	6.7	0.5
Fish catch	934	1 262	1 389	1 157	1 037	1 024	na	4.0	6.2	1.9	−5.9i
Softwood (000 m³)	88 000**	94 327	110 684	130 800	125 700	109 800	na	2.3**	1.4**	3.2	−0.2i
Hardwood (000 m³)	9 100**	9 343	10 747	12 900	12 200	11 300	na	1.7**	0.5**	2.8	1.0i
Energy											
Total energy (000 tce)	75 150	133 870	206 030	289 460	281 450	268 420	na	10.6	12.2	9.0	5.4i
Coal	8 020	8 641	11 598	16 818	17 784	21 710	20 798	3.8	1.5	6.1	10.2
Lignite	1 969	1 872	3 465	3 654	3 485	3 549	4 678	5.8	−1.0	13.1	5.1
Natural gasoline	na	2 807	4 331	6 269	5 803	5 444	na	na	na	9.1	4.7i
Crude oilg	25 630	39 457	61 868	88 007	82 531	69 898	64 100	9.2	9.0	9.4	0.6
Petroleum products	34 332	44 447	58 431	77 261	80 562	77 950	na	5.5	5.3	5.6	5.9i
Natural gas (mn m³)h	12 770*	36 000*	56 712	75 140	73 367	75 800*	74 470*	16.1*	23.0*	9.5*	4.6*
Manufactured gas (mn m³)	1 687	1 754	2 503	2 771	2 616	2 588	na	4.0	0.8	7.4	0.7i
Electricity (mn kW h)	114 457	144 274	204 723	263 335	280 256	272 624	293 410	6.0	4.7	7.2	6.2
of which, hydro (mn kW h)	105 883	117 064	156 709	192 843	210 937	202 404	na	4.0	2.0	6.0	5.2i
nuclear (mn kW h)	—	120	969	14 256	13 864	11 858	na	na	na	51.9	65.0i
Mining											
Iron ore (Fe content)b	11 140	21 822	29 187	29 211	28 772	27 569	34 167*	10.1	14.4	6.0	2.7*
Copper ore (Cu content)	398	461	610	824	821	724	723	4.4	3.0	5.8	2.9
Lead ore (Pb content)	192	275	358	338	294	338	247	6.4	7.4	5.4	−6.0
Molybdenum ore (Mo content)b	0.35	4.3	15.3	13.8	13.9	12.4	na	45.9	65.2	28.9	−4.1i
Nickel ore (Ni content)	195	235	277	249	269	245	na	3.6	3.8	3.3	−2.4i
Tungsten conc (oxide content)	—	1.73	1.69	2.10	1.61	1.35	na	na	na	−0.5	−4.4i
Uranium (U content)	na	3.42	3.23	3.71	3.42	3.60	na	na	na	−1.1	2.2i
Zinc ore (Zn content)	390	826	1 239	1 227	1 127	1 083	1 157	12.3	16.2	8.4	−1.1
Silver	1.06	1.00	1.38	1.48	1.33	1.22	1.27	2.7	−1.2	6.7	−1.4
Gold (000 kg)	144	112	75	61	53	50	52*	−6.3	−4.9	−7.7	−5.9*
Potash (oxide content)	—	1 353	3 103	4 453	5 776	4 850	na	na	na	18.1	9.3i
Salt	3 007	4 159	4 862	5 048	5 447	5 156	na	4.9	6.7	3.2	1.2i
Asbestos	1 015	1 259	1 508	1 690	1 644	1 037	na	4.0	4.4	3.7	−7.2t
Manufacturing											
Beer (000 hl)	11 489	13 584	16 862	19 006	19 384	20 402	20 100*	3.9	3.4	4.4	3.0*
Cigarettes (mn units)	34 699b	43 621b	50 170	56 034	59 603	58 258	61 558	3.8*	4.7b	2.8*	3.5
Cotton fabrics (mn m²)	230*	292	181	158	104	93	na	−1.9*	4.9*	−8.3	−12.5i
Man-made fibres	51	89	108	137	133	123	na	7.8	11.8	3.9	2.6i
Wood pulp	10 398	13 221	16 609	18 561	19 656	14 707	na	4.8	4.9	4.7	−2.4i
Newsprint	6 068	7 003	7 996	8 351	8 711	6 966	8 070	2.8	2.9	2.7	0.2
Other paper	2 513	2 755	3 257	4 227	4 507	3 160	na	2.6	1.9	3.4	−0.6i
Synthetic rubber	162	206	205	230	209	173	210	2.4	4.9	−0.1	0.4
Tyres (000 units)	8 458	13 424	19 109	19 926	na	na	na	8.5	9.7	7.3	1.4j
Sulphuric acid	1 518	1 964	2 475	2 980	2 821	2 723	2 840	5.0	5.3	4.7	2.3
Nitric acid	243	344	503	715	436	na	na	7.5	7.2	7.9	−3.5k
Caustic soda	338	583	860	1 029	1 030	755	922	9.8	11.5	8.1	1.2
Plastics and resinsc	114d	164	299	805	797	722	na	12.8e	12.9f	12.8	19.3i
Fertilisers, nitrogenousa	228	341	672	772	803	895*	916*	11.4	8.4	14.5	5.3*
Fertilisers, phosphatea	181	339	450	738	696	661	653*	9.5	13.4	5.8	6.4*
Fertilisers, potasha	—	1 067	3 566	3 820	5 073	5 623*	4 842*	na	na	27.3	5.2*
Cement	5 338	7 665	7 283	10 053	10 641	9 741	9 898	3.2	7.5	−1.0	5.2
Coke	3 293*	3 715	4 902	5 165	5 233	5 096	na	4.1*	2.4*	5.7	0.8i
Pig iron	4 025	6 582	8 424	9 737	10 416	9 872	10 026	7.7	10.3	5.1	3.0
Crude steel	5 270	9 132	11 198	13 386	13 623	13 025	13 135	7.8	11.6	4.2	2.7
Radio sets (000 units)b	649	1 117	2 004	2 652	2 524	1 009	753	11.9	11.5	12.4	−15.1
Television sets (000 units)b	342	556	543	811	672	510	440*	4.7	10.2	−0.5	−3.4*
Passenger cars (000 units)	326	707	923	1 227	1 166	1 045	1 137	11.0	16.7	5.5	3.5
Commercial vehicles (000 units)	72	140	236	347	359	379	503	12.6	14.2	11.0	13.4
Merchant vessels (000 grt)	116	183	33	234	142	206	244	−11.8	9.5	−29.0	40.0
Construction											
Dwellings (000 units)	124	153	176	247	257	217	236	3.6	4.3	2.8	5.0

aYears ending June 30th bShipments cCoverage is not complete d1962 e1962–70 f1962–65 gIncluding tar sands (synthetic crude oil) hExcluding shrinkage i1970–75
j1970–73 k1970–74

Canada

Transport traffic	1960	1970	1976	Growth %pa	
Passenger-kilometres (mn)				1960–70	1970–76
Rail	3 643	3 657	2 970*	0.0	−3.4*
Air	4 267	15 397	26 031	13.7	9.1
Cargo: tonne-kilometres (mn)					
Rail	95 548	160 749	199 000*	5.3	3.6*
Air	64	425	682	21.0	8.2
Sea[a]: tonnes (mn)					
Goods loaded	46	96	114	7.6	2.9
Goods unloaded	35	53	56	4.2	0.9
Tourism					
Number of visitors[d] (000)	29 655	37 688	35 910[b]	2.4	−1.0[c]
Gross receipts ($ mn)	433	1 182	1 775[b]	10.6	8.5[c]

[a]Including Great Lakes and St Lawrence international traffic   [b]1975   [c]1970–75
[d]Including excursionists (22 285 000 in 1975)

Finance and trade

Price and wage indices	1960	1970	1976	Growth %pa	
(1970 = 100)				1960–70	1970–76
Consumer prices	76.4	100.0	153.0	2.7	7.3
Wholesale prices	80.6	100.0	178.9	2.2	10.2
Wages (earnings)	59.1	100.0	191.4	5.4	11.4
Share prices	53.8	100.0	112.1	6.4	1.9
Money stock					
(end-year, C $ mn)	7 000**	15 270	25 150	8.1**	8.7

Budget (1976/77; year ending March 31st)
Revenue: C $ 39 990 mn = $ 40 200 mn = £ 23 172 mn
Expenditure: C $ 41 340 mn = $ 41 560 mn = £ 23 954 mn
of which (1975/76), social security 27 %, defence 9 %

Balance of payments ($ mn)	1972	1973	1974	1975	1976
Balance of goods (fob)	+2 082	+2 992	+1 939	−396	+1 387
Balance of services	−2 756	−3 229	−4 054	−4 702	−6 147
Balance of transfers	+286	+345	+576	+380	+530
Current balance	−388	+107	−1 540	−4 719	−4 230
Long-term capital flow	+1 604	+385	+1 120	+4 029	+7 650
Reserves and debt (end-year, $ mn)					
International reserves	6 050	5 768	5 825	5 326	5 843
Public debt: domestic	28 559	28 526	32 860	35 522	38 822
external	833	727	702	1 057	2 102

External trade (1976)
Imports: C $ 40 071 mn = $ 40 640 mn = £ 22 500 mn
Exports: C $ 39 672 mn = $ 40 235 mn = £ 22 275 mn

Main imports	% of total	Main exports	% of total
Motor vehicles	23	Motor vehicles	20
(of which, parts 12,		(of which, passenger	
passenger cars 8)		cars 10, parts 6)	
Machinery, non-electric	18	Machinery	10
Crude oil	9	Cereals	7
Food	7	Metal ores	7
Electrical machinery	6	Crude oil	6
Chemicals	5	Wood pulp	6
Textile yarns and fabrics	3	Newsprint	5
Instruments	2	Natural gas	5
		Non-ferrous metals	5
		Timber	5
		Chemicals	4
Main sources		Main destinations	
United States	69	United States	68
Japan	4	Japan	6
Venezuela	3	United Kingdom	5
United Kingdom	3	West Germany	2

Special focus

Trade with United States, 1976

Main imports from US	% of total	Main exports to US	% of total
Motor vehicles	30	Motor vehicles	27
(of which, parts 17,		(of which, passenger	
passenger cars 9)		cars 13, parts 8)	
Machinery, non-electric	22	Machinery, non-electric	9
(of which, agricultural 4,		(of which, power-	
construction and mining 3)		generating 4, agricultural 2)	
Electrical machinery	6	Crude oil	9
Chemicals	6	Natural gas	8
(of which, plastics 2)		Newsprint	6
Food	6	Timber	5
(of which, fruit and		Non-ferrous metals	5
vegetables 2)		Wood pulp	5
Small metal manufactures	3	Metal ores	4

Cayman Islands
Colony of Cayman Islands

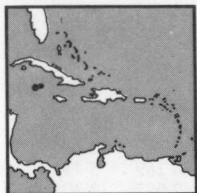

Location Caribbean Sea
The islands are about 300 km north-west of Jamaica, and about the same distance south of Cuba
Land Area 260 km² = 100 mi²
Climate Sub-tropical
Weather at George Town
Temperature: hottest months July, August 24–30 °C, coldest Jan, Feb 18–24 °C
Rainfall (av monthly): 119 mm

Time 5 hours behind GMT
Measures UK (imperial) system
Monetary unit Cayman Islands dollar (CI $) = 100 cents; local currency introduced May 1, 1972 to replace the Jamaican dollar at CI $ 1 = J $ 1
Rate of exchange (1976 av): par CI $ 1 = $ 1.20, free CI $ 1.51 = £ 1

Summary

Political UK colony; before the independence of Jamaica in 1962, the governor of Jamaica was also governor of the Cayman Islands. A revised constitution for the Cayman Islands came into operation in 1972, redefining the functions of the governor (from 1962 to 1971 the 'administrator') of Cayman Islands.
Economic The three main sources of income are tourism, financial activities as a tax-haven, and remittances from Cayman seamen employed with overseas shipping companies. The nature of the soil restricts agriculture, but there is some fishing and there are exports of turtle meat and products

People, resources and equipment

Population 1960 7 622, 1970 10 652, 1976 14 000
Growth: 1960–70 3.4 %pa, 1970–76 4.7 %pa
Density (1976): 54 people per km²
Vital statistics (rate per 1 000 people, 1976): births 20.1, deaths 5.8
Regions (main islands, population in 000, 1970)
Grand Cayman 9.2, Cayman Brac 1.3, Little Cayman 0.02
Town (population in 000, 1976) George Town 4*
Race (1970) Mixed 54%, African 26 %, European 19 %
Language English
Religion Mainly Protestant
Education (1974/75) Pupils 3 505, teachers 70*
Labour force (1970) 3 492
Personnel Physicians (1974): 13, 1 per 1 000* people
Seamen (1974): 700**
Standard of living National income per person (1976):
CI $ 2 500*** = $ 3 000*** = £ 1 600***
Consumption per person: energy (1975) 2 838 kg coal equivalent, electricity (production, 1975) 1 300* kW h
Telephones (Dec 1976): 4 700, 336 per 1 000 people
Electrical capacity (1972) 5.9 megawatts
Hospital beds (1973) 40, 1 per 300* people
Roads (1976) 180* km = 110* mi, density 0.69* km per km²
There are no railways
Ships (registered, 1977) 106, total of 123 787 gross tons
Port George Town
Airports Owen Roberts (Grand Cayman); also at Cayman Brac and Little Cayman
Durable equipment (at end-year)
Radio sets (1974): 3 500, 321 per 1 000 people
Motor vehicles (1976): 4 440*, 317* per 1 000 people, 25* per km of road

Production, finance and trade

Gross domestic product 1976 est: CI $ 40*** mn = $ 48*** mn = £ 26*** mn
Main products (1972) Turtle meat 14 t, turtle shells (exports) 490, turtle leather sets (exports) 12 000, electricity (1975) 17* mn kW h
Transport traffic (1974) *Sea* Number of calls by ships 164
Tourism Number of visitors (1975) 54 140; gross receipts (1973) $ 12* mn
Budget (1976) Revenue: CI $ 11.7 mn = $ 14 mn = £ 8 mn
Expenditure: CI $ 10.4 mn = $ 12 mn = £ 7 mn
External trade (1976) Imports: CI $ 29.8 mn = $ 36 mn = £ 20 mn
Exports: CI $ 0.5 mn = $ 0.6 mn = £ 0.3 mn
Main imports Food, textiles, building materials, motor vehicles, petroleum products
Main exports (1975) Turtle meat, leather and shell 95 %
Most trade is with United States and Jamaica

Chile

Republic of Chile
República de Chile

Location South-west of South America
With a coastline on the Pacific Ocean,
Argentina and Bolivia are to the east, and
Peru to the north. The Andes mountains run
the length of the country, forming the
border with Argentina; many islands are
included in the Republic, of which Easter Island
is about 3 000 km to the west in the Pacific
Ocean
Land Area 756 945 km² = 292 258 mi²
Climate Temperature, dry in the north, wet in the south
Weather at Santiago, 520 m altitude
Temperature: hottest month Jan 12–29 °C, coldest June 3–14 °C
Rainfall (av monthly): driest month Feb 2 mm, wettest June 84 mm
Time 3 hours behind GMT
Measures Metric system
Monetary unit Chilean peso (Ch $) = 100 centavos; the peso was
introduced from September 29, 1975 to replace the escudo at the rate
of 1 peso = 1 000 escudos
Rate of exchange (1976 av): free Ch $ 13.27 = $ 1, Ch $ 23.97 = £ 1

Summary

Political Republic with military government, which became independent
from Spain in 1818. A military junta took control from September 11,
1973. Member of UN, OAS, Sela and Lafta
Economic A mining economy, based on exports of copper and nitrates,
with some manufacturing developments; food accounts for about
one-quarter of imports. Inflation has been exceptionally high

People, resources and equipment

Population 1960 7.58 mn, 1970 9.37 mn, 1976 10.45 mn
Growth: 1960–70 2.1 %pa, 1970–76 1.8 %pa
Density (1976): 14 people per km²
Vital statistics (rate per 1 000 people, 1975): births 25.0, deaths 7.2
Cities (population in 000, 1975)
Santiago (capital) 3 263 Viña del Mar 229 Temuco 138
Valparaíso 592 Talcahuano 184 Talca 115
Concepción 500 Antofagasta 150 Arica 112
Race (1970) Mestizo 66** %, Spanish 25** %, Indian 5** %
Language Spanish
Religion (1970) Roman Catholic 89 %, Protestant 6 %
Education (1974) Pupils 2 900 000*, teachers 105 000*
Labour force (1970) 2 607 360; in agriculture 552 340 (21 %),
mining and quarrying 75 300 (3 %), manufacturing 415 440 (16 %),
construction 148 500 (6 %), transport and communication 155 520 (6 %)
Personnel Scientists and engineers engaged in research (1969): 4 904
Physicians (in government services, 1975): 4 414, 1 per 2 345 people
Standard of living
National income per person (1976): Ch $ 12 700* = $ 960* = £ 530*
Consumption per person (1975): energy 765 kg coal equivalent,
electricity (production, 1976) 903 kW h, newsprint 4 kg, steel 55 kg
Newspapers (1974): number 45
Telephones (Dec 1976): 473 400, 45 per 1 000 people
Livestock (000, 1976) Cattle 3 336, sheep 5 607, pigs 892,
goats 800*, horses 450*, chickens 19 800*
Mineral reserves Coal (1969–72) 3 945 mn tonnes
Lignite (1966) 5 365 mn tonnes
Crude oil (1975) 26 mn tonnes
Natural gas (1975) 66 000 mn cubic metres
Petroleum refinery capacity (1975) 6.9 mn tonnes
Electrical capacity (1975) 2 620 megawatts
Hospital beds (1975) 38 319, 1 per 270 people
Roads (1976) 75 197 km = 46 725 mi, density 0.099 km per km²
Railways (1974) 9 960* km = 6 190* mi, density 0.013⁹ km per km²
Inland waterways (1976) Rivers 2 200*km = 1 400*mi
Ships (registered, 1977) 143, total of 405 971 gross tons
Ports (goods traffic, 000 tonnes, 1976)

	loaded	unloaded		loaded	unloaded
Huasco	3 942	55	Chañaral	938	4
Caldera	1 222	6	Tocopilla	489	167
Valparaíso	481	545	Coquimbo	7	32

Airports Padahuel and Los Cerrillos (Santiago), Concepción, Arica,
Cerro Moreno (Antofagasta), Iquique, Puerto Montt, Calama,
Punta Arenas, Sombrero, El Salvador, Balmaceda

Durable equipment (Dec 1976)	000	no per 1 000 people	
Radio sets	3 300*	313*	no per km of road
Television sets	800*	76*	
Passenger cars	262*	25*	3.5*
Commercial vehicles	172*	16*	2.3*

Production

Gross domestic product 1976: Ch $ 152 061 mn = $ 11 459 mn = £ 6 344 mn
Growth in real terms: 1960–70 4.5 %pa, 1970–76 0.3 %pa
Structure of gross domestic product (1973) *By origin* Agriculture 7 %,
mining and quarrying 11 %, manufacturing 32 %, transport and
communication 4 %, other 46 %

Production indices (1970 = 100)	1960	1970	1976	Growth %pa 1960–70	1970–76
Agricultural	85	100	97	1.6	−0.6
Manufacturing	66	100	82	4.2	−3.3

Main products (000 t)

Agriculture	1960	1970	1976	1960–70	1970–76
Wheat	1 044	1 307	866	2.3	−6.6
Barley	88	97	89	1.0	−1.4
Maize	161	239	248	4.0	0.6
Potatoes	700	684	726	−0.2	1.0
Grapes	550**	598	880*	0.8**	6.7*
Wine	369	401	500*	0.8	3.7*
Sugar, raw value	77	228*	319	11.5*	5.7*
Apples	53*	82*	134	4.5*	8.5*
Peaches	40*	40*	68	0.0*	9.2*
Tobacco	6.6	8.5	8.0*	2.6	−1.0*
Milk	769	1 104	1 054	3.7	−0.8
Wool	11.1*	12.7*	9.6	1.4ᵃ	−4.6ᵃ
Beef and veal	139	149	179	0.7	3.1
Pigmeat	44	44	36	0.0	−3.3
Fish catch	340	1 181	1 128ᶠ	13.3	−0.9ᵍ
Timber (000 m³)	5 583	7 625	8 178ᶠ	3.2	1.4ᵍ
Energy					
Total energy (000 tce)	3 700*	5 200*	4 720*ᶠ	3.5*	−1.9*ᵍ
Coal	1 297	1 313	1 200	0.1	−1.5
Crude oil	943	1 601	1 092	5.4	−6.2
Petroleum products	1 276	3 483	3 770*ᶠ	10.6	1.6*ᵍ
Natural gas (mn m³)	30***	81***	88***	10.0***	1.4***
Manufactured gas (mn m³)	902*ᵇ	1 131	1 150*ᶠ	3.8*ᶜ	0.3*ᵍ
Electricity (mn kW h)	4 592	7 550	9 432	5.1	3.8
Mining					
Iron ore (Fe content)	3 804	6 940	6 300*	6.2	−1.6*
Copper ore (Cu content)	536	711	1 011	2.9	6.0
Manganese ore (Mn content)	19.8	11.1	7.7ᶠ	−5.6	−7.1ᵍ
Molybdenum ore (Mo content)	1.9	5.7	9.1ᶠ	11.9	9.8ᵍ
Vanadium ore (V content)	na	0.55*	0.54*ᶠ	na	−0.4ᵍ
Silver	0.045	0.076	0.192ᶠ	5.4	20.4ᵍ
Gold (000 kg)	1.69	1.62	4.28ᶠ	−0.4	21.4ᵍ
Salt	43	517	293ᶠ	28.2	−10.7ᵍ
Sulphur	31	106	21ᶠ	13.1	−27.7ᵍ
Manufacturing					
Fishmeal	42	191	169ᶠ	16.4	−2.4ᵍ
Beer (000 hl)	1 307	1 776	833ᶠ	3.1	−14.1ᵍ
Cigarettes (mn units)	6 140	6 590	8 149ᶠ	0.7	4.3ᵍ
Cotton yarn	21ᵈ	27	23ʰ	2.9ᵉ	−5.5ⁱ
Man-made fibres	3	9	8ᶠ	11.6	−1.8ᵍ
Wood pulp	102	356	405ᶠ	13.3	2.6ᵍ
Newsprint	52	124	135	9.1	1.6
Other paper	49**	138	146ᶠ	10.9**	1.1ᵍ
Tyres (000 units)	272	676	442	9.5	−6.8
Sulphuric acid	84	373	206ʲ	16.1	−13.8ᵏ
Fertilisers, nitrogenous	148	102	116ᶠ	−3.7	2.6ᵍ
Cement	835	1 349	964	4.9	−5.4
Coke	234*	303	310*ᶠ	2.6*	0.5*ᵍ
Pig-iron	266	481	417ᶠ	6.1	−2.8ᵍ
Crude steel	422	547	450	2.6	−3.2
Radio sets (000 units)	na	40	139ʲ	na	36.5ᵏ
Television sets (000 units)	na	80	223ʲ	na	29.2ᵏ
Passenger cars (000 units)ᵃ	6.6ᵇ	20.7	5.0	21.0ᶜ	−21.0
Commercial vehicles (000 units)ᵃ	1.2ᵇ	3.9	6.5	21.7ᶜ	9.0

ᵃAssembly only ᵇ1964 ᶜ1964–70 ᵈ1962 ᵉ1962–70 ᶠ1975 ᵍ1970–75 ʰ1973
ⁱ1970–73 ʲ1974 ᵏ1970–74

Chile

Transport traffic

Transport traffic	1960	1970	1976	Growth %pa	
Passenger-kilometres (mn)				1960–70	1970–76
Rail	1 906	2 338	2 460	2.1	0.9
Air	414	839	1 228	7.3	6.6
Cargo: tonne-kilometres (mn)					
Rail	1 953	2 533	2 160	2.6	−2.6
Air	10.6	42.2	75.7	14.8	10.2
Sea: tonnes (mn)					
Goods loaded	7.0	12.3	9.9[a]	5.8	−6.8[b]
Goods unloaded	3.2	5.4	6.2[c]	5.4	6.8[d]

[a]1973 [b]1970–73 [c]1972 [d]1970–72

Tourism Number of visitors (1975) 250 000, gross receipts (1973) $ 52 mn

Finance and trade

Price and wage indices	1960	1970	1976	Growth %pa	
(1970 = 100)				1960–70	1970–76
Consumer prices	9.5	100	85 835	26.5	208.3
Wholesale prices	9.0	100	256 661	27.3	270.0
Wages (earnings)	10.0[a]	100	90 555	38.9[b]	211.0
Money stock					
(end-year, Ch $ mn)	0.38	10.0	4 552[c]	38.6	204.3

[a]1963 [b]1963–70 [c]June

Budget (1977) Revenue: Ch $ 98 166 mn = $ 4 560* mn = £ 2 610* mn
Expenditure: Ch $ 102 389 mn = $ 4 755* mn = £ 2 720* mn
of which (1973), education 16 %, defence 15 %, social services 10 %

Balance of payments ($ mn)	1972	1973	1974	1975	1976
Balance of goods (fob)	−150	−10	+423	−7	+665
Balance of services	−329	−284	−616	−572	−644
Balance of transfers	+7	+15	+14	+14	+21
Current balance	−471	−279	−178	−565	+43
Long-term capital flow	+297	+278	+298	+106	+371
Reserves and debt (end-year, $ mn)					
International reserves	148	180	102	109	460
External public debt	3 224	3 340	4 312	4 287	4 171

External trade (1976) Imports: Ch $ 22 350* mn = $ 1 684* mn = £ 932* mn
Exports: Ch $ 27 640* mn = $ 2 083* mn = £ 1 153* mn

Main imports (1974)	% of total	Main exports (1974)	% of total
Food	30	Copper	72
(of which, wheat 13,		Iron ore	5
sugar 3, beef and veal 3)		Copper ore	5
Chemicals	12	Wood pulp	3
Machinery, non-electric	12	Chemicals	2
Crude oil	11	Fruit and vegetables	2
Motor vehicles	6		
Electrical machinery	4		
Iron and steel	3		
Textile fibres	3		
Main sources (1974)		*Main destinations* (1975)	
United States	22	West Germany	14
Argentina	17	Japan	11
West Germany	8	Argentina	10
Australia	5	United States	9
Saudi Arabia	5	United Kingdom	8
Brazil	4	Brazil	6
Ecuador	4	Netherlands	5
United Kingdom	4	Italy	5
France	3	France	4

Special focus

Rate of inflation (consumer price index; 1970 = 100)

	Index	% change over previous year
1966	39	+23
1967	46	+18
1968	58	+26
1969	75	+31
1970	100	+33
1971	119	+19
1972	211	+77
1973	959	+355
1974	5 797	+504
1975	27 518	+375
1976	85 835	+212

Colombia

Republic of Colombia
República de Colombia

Location North-west of South America
There are coastlines on the Pacific Ocean
and Caribbean Sea, with Venezuela to the
north-east, Brazil to the south-east,
Peru and Ecuador to the south-west and
Panama to the north-west, forming the link
with Central America
Land Area 1 138 914 km² = 439 737 mi²
Climate Tropical on coastland, temperate
on plateaux
Weather at Bogotá, 2 645 m altitude
Temperature: hottest month Feb 9–20 °C, coldest July 10–18 °C
Rainfall (av monthly): driest month July 51 mm, wettest Oct 160 mm
Time 5 hours behind GMT
Measures Metric system; also
length 100 vara = 1 cuadra = 80 metres = 87.5 yards
area 1 fanegada = 0.64 hectare = 1.58 acres
weight (*mass*) 16 onzas = 1 libra = 500 grams = 1.102 pounds
25 libras = 1 arroba = 12.5 kilograms = 27.56 pounds
4 arrobas = 1 quintal = 50 kilograms = 110.2 pounds
Monetary unit Colombian peso (Col $) = 100 centavos
Rate of exchange (1976 av): free Col $ 34.98 = $ 1, Col $ 63.17 = £ 1

Summary

Political Republic, which first became independent from Spain in 1819,
as part of the state of Greater Colombia (Gran Colombia) which
included the present states of Ecuador and Venezuela until 1830, and
Panama until 1903. Member of UN, OAS, Sela, Lafta and Andean Group
Economic Mainly an agricultural economy, with coffee accounting for
one-half of exports; mining is also important, especially of oil, gold and
platinum. Manufacturing makes up about one-quarter of domestic
product; textiles, chemicals and oil refining are significant industries

People, resources and equipment

Population 1960 15.42* mn, 1970 20.53* mn, 1976 24.37* mn
Growth: 1960–70 2.9* %pa, 1970–76 2.9* %pa
Density (1976): 21* people per km²
Vital statistics (rate per 1 000 people, 1976): births 30.0, deaths 9.0
Cities (population in 000, 1973)

Bogotá (capital)	2 855	Cartagena	293	Montería	170*
Medellín	1 417	Cúcuta	228	Ciénaga	168*
Cali	923	Pereira	212	Armenia	165
Barranquilla	727	Manizales	207	Valledupar	151*
Bucaramanga	341	Ibagué	176	Itagüi	144*

Race (1970) Mestizo 50 %, Mulatto 25 %, European 20 %, African 4 %,
American Indian 1 %
Language Spanish
Religion (1976) Roman Catholic 90* %, Protestant 4* % Jewish 1* %
Education (1974) Pupils 5 280 000*, teachers 203 000*
Labour force (1973) 5 975 000; in agriculture 1 546 000 (26 %),
mining 36 165 (1 %), manufacturing 678 322 (11 %),
construction 200 238 (3 %), distribution and hotels 575 609 (10 %),
transport and communication 167 019 (3 %), activities not adequately
described 963 687 (16 %), unemployed 856 517 (14 %)
Personnel (1973) Physicians: 10 625, 1 per 2 184 people
Standard of living
National income per person (1976): Col $ 19 000** = $ 540** = £ 300**
Consumption per person (1975): energy 671 kg coal equivalent,
electricity (production, 1976) 595* kW h, newsprint 2.1 kg, steel 27 kg
Newspapers (1974): number 36; circulation 1 449 000, 63 per 1 000
people
Telephones (Dec 1976): 1 295 900, 52 per 1 000 people
Livestock (000, Dec 1975) Cattle 23 860*, sheep 2 036*, pigs 1 900*,
goats 632*, horses 1 500*, chickens 47 000*
Mineral reserves Coal (1971) 4 100 mn tonnes
Lignite (1971) 1 230 mn tonnes
Crude oil (1975) 89 mn tonnes
Natural gas (1975) 128 000 mn cubic metres
Petroleum refinery capacity (1975) 8.7 mn tonnes
Electrical capacity (1975) 4 495* megawatts

Colombia

Hospital beds (1975) 44 642, 1 per 538 people
Roads (1976) 56 667 km = 35 211 mi, density 0.050 km per km²
Railways (1974) 3 431 km = 2 132 mi, density 0.003 km per km²
Inland waterways (1976) Magdalena river 1 450 km = 900 mi; also
Amazon, Orinoco and other rivers
Ships (registered, 1977) 52, total of 247 240 gross tons
Ports (goods traffic, 000 tonnes, 1976)

	loaded	unloaded		loaded	unloaded
Buenaventura	493	762	Cartagena	166	303
Barranquilla	462	438	Coveñas	19	8
Mamonal	261	257			

Airports (passenger departures and arrivals, 000, 1974)
Bogotá 2 087, Medellín 1 034, Cali 682, Barranquilla 635;
also (1977) 22 other airports with scheduled flights

Durable equipment	000	no per	
(Dec 1976)		1 000 people	
Radio sets	2 808	114	no per
Television sets	1 800*	73*	km of road
Passenger cars	356	14	6.3
Commercial vehicles	164	7	2.9

Production

Gross domestic product
1975: Col $ 419 010 mn = $ 13 429 mn = £ 6 044 mn
1976 est: Col $ 520 000** mn = $ 14 900** mn = £ 8 200** mn
Growth in real terms: 1960–70 5.2 %pa, 1970–75 6.2 %pa
Structure of gross domestic product (1974) *By origin* Agriculture 27 %,
mining and quarrying 1 %, manufacturing 22 %, construction 5 %,
transport and communication 6 %, other 39 %

Production indices	1960	1970	1976	Growth %pa	
(1970 = 100)				1960–70	1970–76
Agricultural	74	100	125	3.0	3.8
Industrial	58	100	133e	5.6	7.4f

Main products (000 t)
Agriculture

	1960	1970	1976	1960–70	1970–76
Rice	450	702	1 542	4.5	14.0
Maize	866	862	810	0.0	−1.0
Sorghum	—	165	374	na	14.6
Potatoes	653	1 110	1 400*	5.4	4.0*
Cassava	680	1 250	1 900*	6.3	7.2*
Sugar, raw value	329	676	935	7.5	5.6
Sugar, non-centrifugal	570	755	687*	2.9	−1.7*
Bananas	557	780	1 100*	3.4	5.9*
Coffee	480	507	522*	0.5	0.5*
Cocoa	13	19	26*	3.5	5.4*
Tobacco	25	42	54*	5.3	4.3*
Cotton	67	128	120	6.7	−1.1
Milk	1 753*	2 250	2 893*	2.5*	4.3*
Eggs	59*	92	104*	4.5*	2.1*
Beef and veal	340*	450**	598*	2.8**	4.9**
Pigmeat	50*	55*	78*	1.0*	6.0*
Fish catch	30	53*	67c	5.9*	4.8*d
Timber (000 m³)	25 000*	26 585*	24 940*c	0.6*	−1.3*d

Energy

	1960	1970	1976	1960–70	1970–76
Total energy (000 tce)	13 490	20 190	19 110c	4.1	−1.1d
Coal	2 600	2 750	3 200c	0.6	3.1d
Crude oil	7 584	11 327	7 547	4.1	−6.5
Petroleum products	3 289	6 647	7 400*c	7.3	2.2*d
Natural gas (mn m³)	404	1 464	1 625*c	13.7	2.1*d
Manufactured gas (mn m³)	106	150*	150*	3.5*	0.0*
Electricity (mn kW h)	3 750	8 750	14 500*	8.8	8.8*

Mining

	1960	1970	1976	1960–70	1970–76
Iron ore (Fe content)	178	454	537c	9.8	3.4d
Emeralds (000 CM)i	92	60	109g,j	−4.2	22.0*h
Platinum (000 kg)	na	0.81	0.66e	na	−5.0f
Silver (000 kg)	4.2	2.4	2.7c	−5.4	2.4d
Gold (000 kg)	13.5	6.3	11.9	−7.4	11.2
Salt	303	516	926c	5.5	12.4d

Manufacturing

	1960	1970	1976	1960–70	1970–76
Beer (000 hl)	6 425	7 058	7 649g	0.9	2.7h
Cigarettes (mn units)	15 353	19 080*	18 904*c	2.2*	−0.2*d
Man-made fibres	8	22*	29*c	11.1*	5.8*d
Paper	40	220	240c	18.6	1.8d
Caustic soda	19	42	60e	8.3	9.3f
Fertilisers, nitrogenous	—	61	100*c	na	10.3*d
Cement	1 589	2 757	3 612	5.7	4.6
Coke	420	498	520*c	1.7	0.9*d
Pig-iron	185	229*	300*c	2.2*	5.6*d

Main products (000 t)	1960	1970	1976	Growth %pa	
Manufacturing (cont)				1960–70	1970–76
Crude steel	157	239	255	4.3	1.1
Motor vehicles (000 units)a	—	17	30e	na	11.1d
Construction					
Dwellings (000 units)b	14*	23	32c	5.0*	6.6d

aAssembly only bPermits c1975 d1970–75 e1974 f1970–74 g1973 h1970–73
iGem stones only; may include gem stones other than emeralds jExports; emerald
mines were closed temporarily from July 1973

Transport traffic	1960	1970	1976	Growth %pa	
Passenger-kilometres (mn)				1960–70	1970–76
Roada	na	47 356b	60 042d	na	8.2c
Rail	598	249	511	−8.4	12.7
Air	777	2 063	2 976	10.3	6.3
Cargo: tonne-kilometres (mn)					
Road	na	11 924b	14 588d	na	7.0c
Rail	768	1 173	1 247	4.3	1.0
Air	42	79	154	6.5	11.8
Sea: tonnes (mn)					
Goods loaded	5.6	5.4	2.3	−0.5	−13.1
Goods unloaded	1.0	2.5	2.2	9.0	−2.0
Tourism					
Number of visitors (000)	na	162	443d	na	22.3c
Gross receipts ($ mn)	na	54	140d	na	21.0c

aPublic transport only b1972 c1972–75 d1975 e1970–75

Finance and trade

Price indices	1960	1970	1976	Growth %pa	
(1970 = 100)				1960–70	1970–76
Consumer prices	34.6	100.0	280.9	11.2	18.8
Wholesale prices	37.2	100.0	353.7	10.4	23.4
Share prices	66.2	100.0	84.0	4.2	−2.9
Money stock					
(end-year, Col $ mn)	3 947	22 397	79 383	19.0	23.5

Budget (1976) Revenue: Col $ 48 816 mn = $ 1 396 mn = £ 773 mn
Expenditure: Col $ 44 086 mn = $ 1 260 mn = £ 698 mn
of which, education 21 %, defence 9 %, health 10 %

Balance of payments ($ mn)	1972	1973	1974	1975	1976
Balance of goods (fob)	+130	+280	−16	+293	+702
Balance of services	−355	−370	−386	−445	−433
Balance of transfers	+35	+35	+51	+44	+72
Current balance	−190	−55	−350	−109	+341
Long-term capital flow	+264	+286	+229	+295	+115

Reserves and debt (end-year, $ mn)					
International reserves	325	534	449	521	1 158
External public debt	2 367	2 740	2 770	3 000	3 324

External trade (1976) Imports: Col $ 59 795 mn = $ 1 710 mn = £ 947 mn
Exports: Col $ 55 575 mn = $ 1 882 mna = £ 1 042 mn
aValued at the average rate of Col $ 29.54 = $ 1

Main imports (1975)	% of total	*Main exports* (1975)	% of total
Chemicals	22	Coffee	46
Machinery, non-electric	18	Beef and veal	16
Motor vehicles	12	Petroleum products	7
Iron and steel	8	Sugar	7
Electrical machinery	7	Cotton	5
Wheat	4	Textile yarns and	
Paper	3	fabrics	5
		Chemicals	4

Main sources (1975)		*Main destinations* (1975)	
United States	43	United States	32
West Germany	9	West Germany	15
Japan	9	Netherlands	6
France	4	Venezuela	6
United Kingdom	4	Sweden	3
Spain	3	United Kingdom	3
Canada	3	Italy	3

Special focus

Coffee and gold production ranking in central and south America

Rank		Coffee, 1977		Rank		Gold, 1976	
		000 tonnes	% of total			000 kilograms	% of total
1	Brazil	943*	36*	1	*Colombia*	11.9	26
2	*Colombia*	540*	21*	2	Brazil	9.3	20
3	Mexico	270*	10*	3	Mexico	4.8*	10*
4	El Salvador	180*	7*	4	Chile	4.6*	10*
5	Guatemala	147*	6*	5	Peru	3.0*	7*
	Totala	2 620*	100		Totala	46.0*	100

aFor central and south America only; includes others

Costa Rica

Republic of Costa Rica
República de Costa Rica

Location Central America
The Caribbean Sea is to the east and
Pacific Ocean to the west, with Nicaragua to
the north and Panama to the south-east
Land Area 50 700 km² = 19 600 mi²
Climate Tropical in lowlands, temperate on the
plateau
Weather at San José, 1 146 m altitude
Temperature: hottest month May 17–27 °C,
coldest Dec, Jan 14–24 °C
Rainfall (av monthly): driest month Feb 5 mm, wettest Sept 305 mm
Time 6 hours behind GMT
Measures Metric system; also
length 1 vara = 83.6 centimetres = 32.9 inches
weight (mass) 1 arroba = 25 libras = 11.5 kilograms = 25.35 pounds
1 quintal = 4 arrobas = 46 kilograms = 101.4 pounds
Monetary unit Costa Rican colón (CR ₡) = 100 céntimos
Rate of exchange (1976 av): par CR ₡ 8.57 = $ 1, free CR ₡ 15.48 = £ 1
The Central American peso of the CACM (CA $), equal to the
US dollar, is also used

Summary

Political Republic, which became independent from Spain in 1821
(from 1824 to 1838 as part of the Central American Federation)
Member of UN, OAS, Sela, Odeca and CACM
Economic An agricultural economy, with coffee and bananas accounting
for about one-half of exports. There are some mining prospects;
manufacturing accounts for 20 % of gross domestic product, and
includes especially chemicals and textiles

People, resources and equipment

Population 1960 1.25 mn, 1970 1.73 mn, 1976 2.02 mn
Growth: 1960–70 3.3 %pa, 1970–76 2.6 %pa
Density (1976): 40 people per km²
Vital statistics (rate per 1 000 people, 1976): births 29.3, deaths 4.8
Cities (population in 000, 1976)

San José (capital)	420*	Puntarenas	31	Cartago	23
Limón	44	Nicoya	30*	Turrialba	22*
Alajuela	35	Heredía	24	Liberia	18*

Race (1976) European (mainly Spanish) 80** %
Language Spanish
Religion (1976) Roman Catholic 98* %, Protestant 2* %
Education (1974) Pupils 545 556, teachers 23 500*
Labour force (1973) 585 313; in agriculture 213 226 (36 %),
manufacturing 69 917 (12 %), construction 39 078 (7 %)
Personnel (1974) Physicians: 1 293, 1 per 1 503 people
Standard of living
National income per person (1976): CR ₡ 9 116 = $ 1 064 = £ 589
Consumption per person (1975): energy 544 kg coal equivalent,
electricity (production, 1976) 812 kW h, newsprint 5 kg, steel 41 kg
Newspapers (1974): number 6; circulation 186 000, 97 per 1 000 people
Telephones (Dec 1976): 126 900, 62 per 1 000 people
Livestock (000, 1976) Cattle 1 894, pigs 230*, chickens 5 000*
Petroleum refinery capacity (1975) 470 000* tonnes
Electrical capacity (1975) 407* megawatts,
of which, hydro 241* megawatts
Hospital beds (1975) 7 549, 1 per 264 people
Roads (1975) 24 700* km = 15 300* mi, density 0.49* km per km²
Railways (1975) 1 300* km = 810* mi, density 0.026* km per km²
Ships (registered, 1977) 14, total of 6 811 gross tons
Ports (goods traffic, 000 tonnes, 1976)

	loaded	unloaded
Limón	862	923
Puntarenas	189	412
Golfito	359	65

Airports Santa María (San José), Limón, Golfito, Liberia, Palmar Sur,
Santa Cruz

Durable equipment	000	no per	
(at end-year)		1 000 people	
Radio sets (1976)	145	71	no per
Television sets (1976)	155*	76*	km of road
Passenger cars (1974)	55	28	2.2*
Commercial vehicles (1974)	37	19	1.5*

Production

Gross domestic product 1976: CR ₡ 20 100 mn = $ 2 345 mn = £ 1 298 mn
Growth in real terms: 1966–70 4.8 %pa, 1970–75 6.3 %pa
Structure of gross domestic product (1975) *By origin* Agriculture 20 %,
manufacturing, mining and quarrying 20 %, construction 5 %, transport
and communication 5 %, other 50 %

Production indices	1960	1970	1976	Growth %pa	
(1970 = 100)				1960–70	1970–76
Agricultural	59	100	129	5.5	4.4
Manufacturing	36	100	132f	10.8	9.7g
Main products (000 t)					
Agriculture					
Maize	80*	47	96*	−5.2*	12.6*
Rice	56*	80	92	3.6*	2.4
Sorghum	6*	12*	31	7.2*	17.1*
Sugar, raw value	52	153	173*	11.4	2.1*
Oranges	40**	59	70*	4.0**	2.9*
Bananas	487*	1 146*	1 240*	8.9*	1.3*
Palm oil	4	13	23*	13.7	10.0*
Coffee	54	73*	84	3.1*	2.4*
Cocoa	13.4	4.2	5.0*	−11.0	3.0*
Tobacco	0.8*	1.8	3.0*	8.4*	8.9*
Milk	123*	194	270*	4.7*	5.7*
Beef and veal	27*	59	67*	8.1*	2.1*
Timber (000 m³)	1 665	2 820	3 378d	5.4	3.7e
Other					
Total energy (000 tce)	50	118	164d	9.1	6.8e
Petroleum products	—	303	257d	na	−3.2e
Electricity (mn kW h)	438	1 028	1 640	8.9	8.1
of which, hydro (mn kW h)	393	940	1 306*d	9.1	6.8*e
Beer (000 hl)	89	109	112h	2.0	2.8i
Cigarettes (mn units)	1 280b	1 420	2 200*d	1.7c	9.1*e
Fertilisers, nitrogenousa	—	12	30	na	15.9
Cement	36b	187	330d	31.6c	12.0e

aYears ending June 30th b1964 c1964–70 d1975 e1970–75 f1973 g1970–73
h1971 i1970–71

Transport traffic
Passenger-kilometres Rail (1974) 81 mn, air (1976) 326 mn
Cargo: tonne-kilometres Rail (1974) 14 mn, air (1976) 13.5 mn
Sea (1975) Goods loaded 1.45 mn t, unloaded 1.22 mn t
Tourism (1975) Number of visitors 297 200, gross receipts $ 53 mn

Finance and trade

Consumer price index (1970 = 100) 1976 196.3; growth 1970–76 11.9 %pa
Money stock (end-year, CR ₡ mn) 1970 1 103, 1976 3 895;
growth 1970–76 23.4 %pa
Budget (1976) Revenue: CR ₡ 2 692 mn = $ 314 mn = £ 174 mn
Expenditure: CR ₡ 3 666 mn = $ 428 mn = £ 237 mn,
of which (1975), education 32 %, defence 4 %, public health 6 %

Balance of payments ($ mn)	1972	1973	1974	1975	1976
Balance of goods (fob)	−58	−67	−209	−134	−107
Balance of services	−48	−52	−67	−93	−110
Balance of transfers	+7	+7	+10	+10	+12
Current balance	−100	−112	−266	−218	−205
Long-term capital flow	+80	+92	+136	+238	+187
Reserves and debt (end-year, $ mn)					
International reserves	43	51	45	51	98
External public debt	293	337	477	698	933

External trade (1976) Imports: CR ₡ 6 632 mn = $ 774 mn = £ 428 mn
Exports: CR ₡ 4 777 mn = $ 557 mn = £ 309 mn

Main imports (1974)	% of total	*Main exports* (1974)	% of total
Chemicals	18	Coffee	28a
Food	10	Bananas	22b
Machinery, non-electric	9	Meat	8
Crude oil and products	9	Chemicals	8
Motor vehicles	7	Sugar	6
Electrical machinery	6	Textile yarns and fabrics	4
Paper	6		
Iron and steel	6		
Textile yarns and fabrics	3		
Main sources (1975)		*Main destinations* (1976)	
United States	37	United States	37
Japan	10	West Germany	10
Guatemala	6	Nicaragua	8
West Germany	6	Guatemala	7
Nicaragua	6	El Salvador	6
El Salvador	5	Finland	4
Venezuela	4	Netherlands	4
Mexico	4	Panama	3

a26 % in 1976 b25 % in 1976

Cuba

Republic of Cuba
República de Cuba

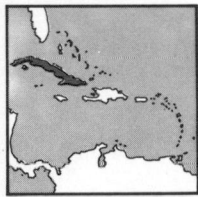

Location Caribbean Sea
The most westerly and largest of the Caribbean islands, about 150 km south of Florida in the United States and 160 km east of Mexico. Bahamas is to the north-east, Haiti to the east, and Cayman Islands and Jamaica are to the south
Land Area 114 524 km² = 44 218 mi²
Climate Sub-tropical
Weather at Havana, 24 m altitude
Temperature: hottest months July, Aug 24–32 °C, coldest Jan, Feb 18–26 °C
Rainfall (av monthly): driest months Feb, Mar 46 mm, wettest Oct 173 mm
Time 5 hours behind GMT (summer time, 4 hours behind)
Measures Metric system, also old Spanish measures, including:
area 1 caballería = 13.42 hectares = 33.16 acres
weight (mass) 1 arroba = 25 libras = 11.5023 kilograms = 25.36 pounds
Monetary unit Cuban peso (Cub $) = 100 centavos
Rate of exchange (1976 av): free Cub $ 0.827 = $ 1, Cub $ 1.49 = £1

Summary

Political Communist republic, which became independent from Spain in 1898; communist control began in 1959. Member of UN, OAS (suspended from 1962), Sela and Comecon
Economic The centrally planned economy is mainly dependent on exports of sugar, supplemented by nickel and copper. Nickel production is planned to increase from 37 000 tonnes pa to 100 000 tonnes pa by 1980. Fishing is being expanded, and industry developed; aid from the Soviet Union is substantial

People, resources and equipment

Population 1960 6.83 mn, 1970 8.55 mn, 1976 9.46 mn
Growth: 1960–70 2.3 %pa, 1970–76 1.7 %pa
Density (1976): 83 people per km²
Vital statistics (rate per 1 000 people, 1975): births 20.7, deaths 5.4
Regions (provinces[a], population in 000, 1970; total of 8.55 mn)

Camagüey	813	Las Villas	1 362	Oriente	2 999
Habana	2 335	Matanzas	501	Pinar del Río	542

[a]From December 2, 1976, 14 provinces were established
Cities (population in 000, 1970)

Habana[a] (capital)	1 751	Camagüey	198	Santa Clara	130
Santiago de Cuba	278	Holguín	132	Guantánamo	129

[a]Havana
Language Spanish
Religion Mainly Roman Catholic
Education (1975/76) Pupils 2 616 400, teachers 121 753
Labour force (1970) 2 633 309; in agriculture 790 356 (30 %), manufacturing 510 402 (19 %), construction 157 182 (6 %)
Personnel Scientists and engineers engaged in research (1969): 1 850
Physicians (1968): 7 000, 1 per 1 150 people
Standard of living
National income per person (1976): Cub $ 660*** = $ 800*** = £ 440***
Consumption per person (1975): energy 1 157 kg coal equivalent, electricity (production) 660* kW h, newsprint 2.5 kg, steel 85 kg
Newspapers (1972): number 10; circulation 834 000, 95 per 1 000 people
Telephones (Dec 1974): 289 000, 31 per 1 000 people
Livestock (000, 1976) Cattle 5 500*, pigs 1 460*, sheep 340*, horses 830*, chickens 16 750*
Petroleum refinery capacity (1975) 5.7* mn tonnes
Electrical capacity (1975) 1 700* megawatts
Hospital beds (1975) 39 863, 1 per 236 people
Roads (1974) 27 074 km = 16 823 mi, density 0.24 km per km²
Railways (1974) 14 872 km = 9 241 mi, density 0.13 km per km²
Ships (registered, 1977) 315, total of 667 518 gross tons
Ports Havana, Santiago de Cuba, Cienfuegos, Matanzas, Guayabal
Airports Havana, Camagüey, Santiago de Cuba, Santa Clara, Guantánamo, Holguín, Cienfuegos, Nueva Gerona, Manzanillo, Nicaro, Moa

Durable equipment	000	no per	
(at end-year)		1 000 people	
Radio sets (1974)	1 805	195	no per
Television sets (1974)	595	64	km of road
Passenger cars (1973)	70*	8*	2.6*
Commercial vehicles (1973)	33*	4*	1.2*

Production, finance and trade

Gross material product 1974: Cub $ 7 414 mn = $ 8 944 mn = £ 4 976 mn
Gross domestic product
1976 est: Cub $ 7 000*** mn = $ 8 500*** mn = £ 4 700*** mn

Production index (1970 = 100)	1960	1970	1976	Growth %pa	
				1960–70	1970–76
Agricultural	75	100	88	2.9	−2.1

Main products (000 t)	1960	1970	1976	1960–70	1970–76
Agriculture					
Rice	307	326*	420*	0.6*	4.3*
Maize	214*	115*	125*	−6.0*	1.4*
Sweet potatoes	272*	250*	250*	−0.8*	0.0*
Cassava	255*	220*	254*	−1.5*	2.4*
Sugar, raw value	5 862	7 559	6 279*	2.6	−3.0*
Oranges	79*	129	133*	5.0*	0.5*
Coffee	42*	29*	25*	−3.6*	−2.4*
Tobacco	52*	32*	53*	−4.7*	8.8*
Milk	330*	500*	625*	4.2*	3.8*
Beef and veal	193*	204*	195*	−0.6*	−0.8*
Fish catch	31	106	165*[f]	13.1	9.2*[g]
Timber (000 m³)	1 800*	1 839	1 885*[f]	0.2*	0.5*[g]
Energy					
Total energy (000 tce)	23**	190*	270[f]	23.5**	7.3*[g]
Crude oil	14*	159	150[f]	27.5*	−1.2*[g]
Petroleum products	3 150*	4 321	5 410*[f]	3.2*	4.6*[g]
Manufactured gas (mn m³)	75*	77	95*[f]	0.3*	4.3*[g]
Electricity (mn kW h)	2 981	4 888	6 150*[f]	5.1	4.7*[g]
Mining					
Chromium ore (oxide content)	10*	8	13[f]	−2.2*	9.5[g]
Copper ore (Cu content)	11.8*	3.8[j]	2.8[f]	−11.8*[k]	−5.0[l]
Manganese ore (Mn content)	20**	28[h]	na	4.3**[i]	na
Nickel ore (Ni content)	14**	37	37[f]	10.2**	0.3[g]
Salt	59*	89	157[f]	4.2*	12.0[g]
Manufacturing					
Beer (000 hl)	1 394[c]	1 002	2 111[f]	−3.6[d]	16.1[g]
Cigarettes (mn units)	13 611[c]	19 806	15 366[f]	4.3[d]	−4.9[g]
Cigars (mn units)	203[c]	364	383[f]	6.7[d]	1.0[g]
Paper	93[a]	80	120[f]	−2.1[b]	8.4[g]
Tyres (000 units)	343	202	368[f]	−5.2	12.7[g]
Ethyl alcohol (000 hl)	1 777[a]	1 405	na	−3.3[b]	na
Sulphuric acid	135[a]	322	418[f]	13.2[b]	5.4[g]
Fertilisers, nitrogenous[e]	—	4*	82*	na	65.5*
Cement	813	742	2 083[f]	−0.9	22.9[g]
Crude steel	63[a]	140	298[f]	12.1[b]	16.3[g]
Radio sets (000 units)	39[a]	19	113[f]	−9.8[b]	42.8[g]

[a]1963 [b]1963–70 [c]1961 [d]1961–70 [e]Years ending June 30th [f]1975 [g]1970–75 [h]1968 [i]1960–68 [j]1969 [k]1960–69 [l]1969–75

Transport traffic	1960	1970	1976	Growth %pa	
Passenger-kilometres (mn)				1960–70	1970–76
Rail	918[a]	1 130	695[e]	3.0[b]	−9.3[f]
Air	217	502	663	8.7	4.7
Cargo: tonne-kilometres (mn)					
Rail	1 064[a]	1 625	1 825[e]	6.2[b]	2.3[f]
Air	12.3	11.2	14.2	−0.9	4.0
Sea: tonnes (mn)					
Goods loaded	4.7[c]	7.8	5.9[e]	8.9[d]	−5.4[f]
Goods unloaded	8.8[c]	10.8	13.3[e]	3.5[d]	4.3[f]

[a]1963 [b]1963–70 [c]1964 [d]1964–70 [e]1975 [f]1970–75
Tourism (1975) Number of visitors 50 000*
Budget (1975)
Balanced at Cub $ 4 500** mn = $ 5 600** mn = £ 2 500** mn
Of expenditure: education 18**%, health and social security 22**%, defence 11**%
External trade (1975)
Imports: Cub $ 3 113 mn = $ 3 883 mn = £ 1 748 mn
Exports: Cub $ 2 947 mn = $ 3 680 mn = £ 1 654 mn

Main imports (1972)	% of total	Main exports (1974)	% of total
Machinery, non-electric	13	Sugar	86
Crude oil and products	12	Nickel and copper ores	6
Cereals	10	Tobacco products	2*
Iron and steel	5		
Chemicals	5		

Main sources (1974)		Main destinations (1974)	
Soviet Union	47	Soviet Union	36
Japan	8	Japan	17
United Kingdom	5	Spain	7
West Germany	4	East Germany	5

Dominica
State of Dominica

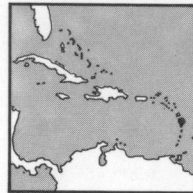

Location Eastern Caribbean Sea
Most northerly Windward Island; Guadeloupe is
to the north, Martinique to the south
Land Area 751 km² = 290 mi²
Climate Sub-tropical
Weather at Roseau, 18 m altitude
Temperature: hottest month June 23–32 °C,
coldest Jan 20–29 °C
Rainfall (av monthly): driest months
Feb, Mar 74 mm, wettest July 274 mm
Time 4 hours behind GMT
Measures UK (imperial) system
Monetary unit East Caribbean dollar (EC $) = 100 cents
Rate of exchange (1976 av): free EC $ 2.61 = $ 1, EC $ 4.71 = £ 1

Summary

Political An Associated State of the United Kingdom from March 1,
1967; formerly a UK colony. Full independence is planned. Member of
Caricom
Economic Bananas and citrus fruit are the main exports; light industry
is being developed to reduce agricultural dependence

People, resources and equipment

Population 1960 59 916, 1970 70 302, 1976 76 000*
Growth: 1960–70 1.6 %pa, 1970–76 1.3* %pa
Density (1976): 101* people per km²
Vital statistics (rate per 1 000 people, 1969): births 36.4, deaths 10.1
Towns (population in 000, 1976) Roseau (capital) 25*, Portsmouth,
Grand Bay
Race African, Mixed, European, Carib
Language English; also French patois
Education (1973/74) Pupils 23 536, teachers 714
Labour force (1970) 21 171
Personnel (1971) Physicians: 13, 1 per 5 480 people
Standard of living
National income per person (1976): EC $ 950** = $ 370** = £ 200**
Production per person (1975): electricity 187 kW h
Telephones (Dec 1976): 3 567, 47 per 1 000 people
Livestock (000, 1976) Cattle 4*, pigs 8*, goats 5*, chickens 102*
Electrical capacity (1975) 6 megawatts
Hospital beds (govt establishments only, 1973) 312, 1 per 234 people
Roads (1972) 1 208 km = 751 mi, density 1.6 km per km²
Ports Roseau, Portsmouth
Airport Melville Hall (55 km from Roseau)
Motor vehicles (1975) 4 000, 54 per 1 000 people, 3.3 per km of road

Production, finance and trade

Gross domestic product 1973: EC $ 51 mn = $ 26 mn = £ 11 mn
1976 est: EC $ 80**mn = $ 30**mn = £ 17**mn
Main products (000 t, 1976) Bananas 34*, limes 8*, grapefruit 3*,
coconuts 18*, copra 3*, electricity (mn kW h, 1975) 14
Transport traffic *Air* (1973) Passenger departures 23 600, arrivals 23 400
Sea (1970) Passenger departures 2 000, arrivals 1 900
Tourism (1973) Number of visitors 17 560
Consumer price index (1970 = 100) 1976 215; growth 1970–76 13.6 %pa
Budget (1974) Revenue: EC $ 12.6 mn = $ 6.1 mn = £ 2.6 mn
Expenditure: EC $ 15.9 mn = $ 7.7 mn = £ 3.3 mn
Budgetary aid is supplied by the United Kingdom
External trade (1975) Imports: EC $ 45 mn = $ 20.9 mn = £ 9.4 mn
Exports: EC $ 25 mn = $ 11.4 mn = £ 5.1 mn

Main imports (1969)	% of total	*Main exports* (1975)	% of total
Food	28	Bananas	58
(of which, meat 6,		Grapefruit	11
dairy produce 4)		Fruit juices	4
Chemicals	12	Essential oils	3
(of which, fertilisers 5)		Soap	2
Machinery	8		
Motor vehicles	8		
Main sources (1975)		*Main destination* (1975)	
United Kingdom	30	United Kingdom	78
United States	10		
Canada	10		

Dominican Republic
República Dominicana

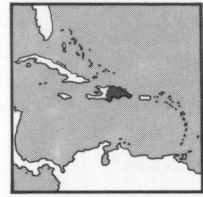

Location Caribbean Sea
Occupies the eastern two-thirds of the
Hispaniola island; Haiti is to the west,
occupying the rest of the island. Puerto
Rico is to the east of Hispaniola island
and Cuba to the west
Land Area 48 734 km² = 18 816 mi (64 % of
Hispaniola Island which totals 76 484 km²)
Climate Sub-tropical
Weather at Santo Domingo, 17 m altitude
Temperature: hottest month August 23–31 °C, coldest Jan 19–29 °C
Rainfall (av monthly): driest month Feb 36 mm, wettest Sept 185 mm
Time 4 hours behind GMT
Measures Metric system
Monetary unit Dominican Republic peso (DR $) = 100 centavos
Rate of exchange (1976 av): par DR $ 1 = $ 1, free DR $ 1.806 = £ 1

Summary

Political Republic, which became independent from Spain in 1821; it
was occupied by Haiti from 1822–44, and by US forces from 1916–24.
Member of UN, OAS and Sela
Economic Mainly an agricultural economy, with sugar accounting for
one-third to one-half of exports. Mining is of growing importance;
nickel has replaced bauxite as the main metal export. Light industry is
being encouraged, and tourism has expanded rapidly

People, resources and equipment

Population 1960 3.04*mn, 1970 4.06*mn, 1976 4.84*mn
Growth: 1960–70 2.9* %pa, 1970–76 3.0* %pa
Density (1976): 99* people per km²
Vital statistics (rate per 1 000 people, 1970–75): births 45.8*, deaths 11.0*
Cities (population in 000, 1975)

Santo Domingo[a]		San Pedro de Macorís	62	La Romana	47
(capital)	923	San Francisco de Macoris	58	Puerto Plata	42
Santiago	209	Barahona	51	San Juan	42

[a]Called Ciudad Trujillo from 1936–61
Race (1960) Mulatto 73 %, European 16 %, African 11 %
Language Spanish
Religion (1976) Roman Catholic 98** %, Protestant 1** %
Education (1972/73) Pupils 1 008 777, teachers 22 700*
Labour force (1970) 1 241 000; in agriculture 549 315 (44 %),
manufacturing 100 989 (8 %), construction 28 508 (2 %)
Personnel (1973) Physicians: 2 374, 1 per 1 866 people
Standard of living
National income per person (1976): DR $ 694 = $ 694 = £ 384
Consumption per person (1975): energy 458 kg coal equivalent,
electricity (production) 347 kW h, newsprint 0.5 kg, steel 29 kg
Newspapers (1974): number 10; circulation 197 000, 43 per 1 000 people
Telephones (Dec 1976): 127 300, 26 per 1 000 people
Livestock (000, 1976) Cattle 1 950*, pigs 820*, goats 350*, chickens 7 400*
Petroleum refinery capacity (1975) 1.5 mn tonnes
Electrical capacity (1975) 443* megawatts
Hospital beds (1973) 12 618, 1 per 356 people
Roads (1975) 11 844 km = 7 360 mi, density 0.24 km per km²
Railways (1975) 1 700** km = 1 000** mi, density 0.035** km per km²
Ships (registered, 1977) 20, total of 8 469 gross tons
Ports (total goods loaded and unloaded, 000 t, 1976) Haina 2 409,
Santo Domingo 1 102, Cabo Rojo 878, La Romana 492,
San Pedro de Macorís 84
Airports Santo Domingo, Santiago, Barahona, La Romana,
Puerto Plata, San Juan

Durable equipment	000	no per	
(at end-year)		1 000 people	
Radio sets (1974)	185	40	no per
Television sets (1974)	156	34	km of road
Passenger cars (1975)	66	14	5.6
Commercial vehicles (1975)	35	7	3.0

Production

Gross domestic product 1976: DR $ 3 915 mn = $ 3 915 mn = £ 2 168 mn
Growth in real terms: 1960–70 5.1 %pa, 1970–76 9.0 %pa
Structure of gross domestic product (1975) *By origin* Agriculture 21 %,
mining and quarrying 4 %, manufacturing 21 %, construction 7 %,
transport and communication 6 %, other 41 %

Dominican Republic

Production indices (1970 = 100)	1960	1970	1976	Growth %pa 1960–70	1970–76
Agricultural	95	100	118	0.5	2.8
Manufacturing	62	100	143c	4.9	9.4d

Main products (000 t)

Agriculture

	1960	1970	1976	1960–70	1970–76
Rice	114	210	286*	6.3	5.3*
Sweet potatoes	111	87	89*	−2.4	0.4*
Cassava	153	170	162*	1.1	−0.8*
Sugar, raw value	1 112	1 014	1 302*	−0.9	4.3*
Avocados	120**	124*	131*	0.3**	0.9*
Mangoes	190**	142*	165*	−2.9**	2.5*
Bananas	379	275	302ᵃ	−3.2	1.6ᵃ
Plantains	na	550*	550*	na	0.0*
Coffee	45	42	42*	−0.7	0.0*
Cocoa	40	38	32*	−0.5	−2.8*
Tobacco	27	22	31*	−2.0	5.9*
Milk	175*	283	293*	4.9*	0.6*
Beef and veal	22	32	40*	3.8	3.8*
Timber (000 m³)	1 940*	1 841	1 795ᵃ	−0.5*	−0.5ᵇ

Other

	1960	1970	1976	1960–70	1970–76
Petroleum products	—	—	1 159ᵃ	na	na
Electricity (mn kW h)	349	921	1 632*ᵃ	10.2	12.1*ᵇ
Bauxite	689	1 086	785ᵃ	4.7	−6.3ᵇ
Nickel ore (Ni content)	—	—	27ᵃ	na	na
Salt	86	37	40*ᵃ	−8.1	1.6*ᵇ
Beer (000 hl)	94	374	445ᵃ	14.8	3.5ᵇ
Cigarettes (mn units)	826	2 125	2 776ᶜ	9.9	6.9d
Ethyl alcohol (000 hl)	43ᵉ	60	75ᶜ	4.9f	5.7d
Cement	170	493	555ᵃ	11.2	2.4ᵇ
Cotton fabrics (mn m)	7	7	6ᶜ	0.0	−4.0d

ᵃ1975 ᵇ1970–75 ᶜ1974 ᵈ1970–74 ᵉ1963 ᶠ1963–70

Transport traffic *Air* (1976) Passenger-km 487 mn, cargo 1.8 mn t-km
Sea (1973) Goods loaded 2.6 mn t, unloaded 3.2 mn t
Tourism (1975) Number of visitors 232 900, gross receipts $ 61 mn

Finance and trade

Consumer price index (1970 = 100) 1976 180.8; growth 1970–76 10.4 %pa
Money stock (end-year, DR $ mn) 1976 403; growth 1970–76 14.6 % pa
Budget (1976) Revenue: DR $ 585 mn = $ 585 mn = £ 324 mn
Expenditure: DR $ 555 mn = $ 555 mn = £ 307 mn
International reserves (Dec 1976) $ 127 mn
External public debt (Dec 1975) $ 710 mn
External trade (1976) Imports: DR $ 878 mn = $ 878 mn = £ 486 mn
Exports: DR $ 716 mn = $ 716 mn = £ 397 mn

Main imports (1971)	% of total	Main exports (1976)	% of total
Machinery, non-electric	16	Sugar	38
Food	10	Ferro-nickel	15
Motor vehicles	10	Coffee	14
Chemicals	9	Cocoa	7
Crude oil and products	8	Tobacco	5
Electrical machinery	7	Bauxite	2
Main sources (1976)		**Main destinations (1976)**	
United States	68	United States	70
Japan	9	Netherlands	7
West Germany	4	Switzerland	7
Canada	4	Spain	3
United Kingdom	3	Belgium-Luxembourg	3

Ecuador

Republic of Ecuador
República del Ecuador

Location North-west of South America
With a coast line on the Pacific Ocean,
Colombia is to the north, and Peru to
the east and south. The Galápagos
Islands (Colón Archipelago), 1 000 km off
the coast, are part of Ecuador
Land Area 283 561 km² = 109 484 mi²
Estimates of area vary, depending on the figure
included for the eastern region
Climate Tropical, temperate in Central Valley

Weather at Quito, 2 879 m altitude
Temperature: hottest months Dec, Jan 8–22 °C, coldest April, May 8–21 °C
Rainfall (av monthly): driest month July 20 mm, wettest April 175 mm
Time 5 hours behind GMT

Measures Metric system; also local units, including:
length 1 vara = 84 centimetres = 33.1 inches
weight (mass) 1 libra = 0.46 kilogram = 1.014 pounds
1 arroba = 25 libras = 11.5 kilograms = 25.35 pounds
1 quintalᵃ = 4 arrobas = 100 libras = 46 kilograms = 101.4 pounds
1 fanegaᵇ = 8 arrobas = 200 libras = 92 kilograms = 202.8 pounds
ᵃOr media ᵇOr mula
Monetary unit Sucre (Su) = 100 centavos
Rate of exchange (1976 av): par Su 25.00 = $ 1, free Su 45.15 = £ 1

Summary

Political Republic with military government, which became independent
from Spain in 1821 and formed part of Greater Colombia from 1822 to
1830. The border with Peru has been subject to dispute. Member of UN,
OAS, Sela, Lafta, Andean Group and Opec
Economic Crude oil has improved the economy since the completion of
a pipeline from the oil fields in the eastern provinces across the Andes to
Esmeraldas. Traditional agricultural exports of bananas and coffee
remain important, making up one-third of exports. Oil revenues are
being used to improve agricultural growth and to help industrial
expansion, especially of petrochemicals, steel and cement

People, resources and equipment

Populationᵃ 1960 4.36* mn, 1970 5.96* mn, 1976 7.30* mn
ᵃExcluding nomadic Indian tribes
Growth: 1960–70 3.2* %pa, 1970–76 3.4* %pa
Density (1976): 26* people per km²
Vital statistics (rate per 1 000 people, 1974): births 35.2, deaths 9.2
Cities (population in 000, 1974)

Quito (capital)	557	Ambato	77	Portoviejo	59
Guayaquil	861	Machala	68	Riobamba	58
Cuenca	105	Esmeraldas	60		

Race (1965) American Indian 40 %, Mestizo 40 %, European 10 %, African 5 %
Language Spanish; Indian languages are also used
Religion (1969) Roman Catholic 94 %
Education (1975/76) Pupils 1 710 000*, teachers 59 500*
Labour force (1974) 1 940 628; in agriculture 896 897 (46 %),
manufacturing 226 265 (12 %), construction 86 192 (4 %)
Personnel (1973) Physicians: 3 109, 1 per 2 160 people
Standard of living
National income per person (1976): Su 15 520 = $ 621 = £ 344
Consumption per person, (1975): energy 442 kg coal equivalent,
electricity (production) 183* kW h, newsprint 2.2 kg, steel 27 kg
Newspapers (1974): number 22; circulation 285 000, 41 per 1 000 people
Telephones (Dec 1976): 174 000, 23 per 1 000 people
Livestock (000, 1976) Cattle 2 725*, sheep 2 150*, pigs 2 700*
Mineral reserves (1975) Crude oil 193 mn tonnes
Natural gas 44 000 mn cubic metres
Petroleum refinery capacity (1975) 2.2 mn tonnes
Electrical capacity (1975) 407* megawatts
Hospital beds (1972) 13 618, 1 per 478 people
Roads (1973) 21 300* km = 13 200* mi, density 0.075* km per km²
Railways (1974) 1 151 km = 715 mi, density 0.004 km per km²
Inland waterways
Esmeraldas, Guayas and Mira rivers: 200* km = 120* mi
Oil pipeline (1976) 504 km = 313 mi
Ships (registered, 1977) 55, total of 197 244 gross tons
Ports (goods traffic, 000 tonnes, 1975)

	loaded	unloaded		loaded	unloaded
Balaoᵃ	7 814	—	Puerto Bolívar	813	11
Salinas	43	1 920	Manta	90	284
Guayaquil	607	886	Esmeraldas	185	39

ᵃCrude oil port
Airports Mariscal Sucre (Quito), Simón Bolívar (Guayaquil), Esmeraldas,
Manta, Cuenca, Machala, Portoviejo, Macará, Loja, Tulcan

Durable equipment (at end-year)	000	no per 1 000 people	
Radio sets (1971)	1 700*	270*	no per km of road
Television sets (1974)	250	36	
Passenger cars (1974)	44	6	2.1*
Commercial vehicles (1974)	68	10	3.2*

Production

Gross domestic product 1976: Su 123 870 mn = $ 4 955 mn = £ 2 744 mn
Growth in real terms: 1970–76 9.3 %pa
Structure of gross domestic product (1975) *By origin* Agriculture 21 %,
mining and quarrying 5 %, manufacturing 15 %, construction 5 %,
transport and communication 5 %, other 49 %

Ecuador

Production indices (1970 = 100)	1960	1970	1976	Growth %pa 1960–70	1970–76
Agricultural	64	100	113	4.5	2.1
Manufacturing	35	100	193[e]	11.1	14.1[d]

Main products (000 t)

Agriculture

	1960	1970	1976	1960–70	1970–76
Rice	132	206	347	4.6	9.1
Maize	160*	270	250	5.4*	−1.3
Cassava	220*	410*	590*	6.4*	6.2*
Potatoes	203	542	550	10.3	0.2
Sugar, raw value	125	270	310*	8.0	2.3*
Bananas	2 224	2 700*	2 788	2.0*	0.5*
Oranges	140	173*	324	2.1*	11.0*
Coffee	33	60	82	6.2	5.3
Cocoa	41	54	65	2.8	3.1
Milk	403	700*	823	5.7*	2.7*
Fish catch	44	92*	223*[c]	7.7*	19.4*[d]
Timber (000 m³)	2 500**	2 778	3 466*[c]	1.1**	4.5*[d]

Other

	1960	1970	1976	1960–70	1970–76
Total energy (000 tce)	510	320	12 110[e]	−4.6	106.8[d]
Crude oil	360	193	9 480	−6.0	91.4
Petroleum products	560	1 149	1 977[e]	7.5	11.4[d]
Electricity (mn kW h)	387	949	1 290*[c]	9.4	6.3*[d]
Gold (000 kg)	0.47	0.26	0.24[c]	−5.6	−2.3[f]
Beer (000 hl)	406	647	889[e]	4.8	8.3[f]
Cigarettes (mn units)	850*	1 295	2 074[e]	4.3*	12.5[f]
Tyres (000 units)	52[a]	120	182[g]	12.7[b]	14.9[h]
Cement	201	458	604[c]	8.6	5.7[d]
Radio sets (000 units)	na	3	13[e]	na	44.3[f]

[a]1963 [b]1963–70 [c]1975 [d]1970–75 [e]1974 [f]1970–74 [g]1973 [h]1970–73

Transport traffic	1960	1970	1976	Growth %pa 1960–70	1970–76
Passenger-kilometres (mn)					
Rail	65	85	65[a]	2.7	−5.3[b]
Air	42*	256	318	19.8*	3.7
Cargo: tonne-kilometres (mn)					
Rail	121	56	46[a]	−7.4	−3.8[b]
Air	1.0*	9.6	7.3	25.4*	−4.5
Sea: tonnes (mn)					
Goods loaded	1.07	1.77	10.43[a]	5.1	42.6[b]
Goods unloaded	0.26	1.52	3.43[a]	19.5	17.6[b]

[a]1975 [b]1970–75

Tourism (1975) Number of visitors 172 900, gross receipts $ 20 mn

Finance and trade

Consumer price index (1970 = 100) 1976 208.1; growth 1970–76 13.0 %pa
Money stock (end-year, Su mn) 1976 24 757; growth 1970–76 26.7 %pa
Budget (1976) Revenue: Su 14 653 mn = $ 586 mn = £ 325 mn
Expenditure: Su 16 813 mn = $ 673 mn = £ 372 mn

Balance of payments ($ mn)	1972	1973	1974	1975	1976
Balance of goods (fob)	+39	+187	+350	+6	+235
Balance of services	−132	−207	−344	−259	−278
Balance of transfers	+15	+27	+31	+32	+41
Current balance	*−77*	*+ 7*	*+38*	*−220*	*−1*
Long-term capital flow	+159	+77	+105	+200	+295
Reserves and debt (end-year, $ mn)					
International reserves	143	241	350	286	515
External public debt	471	549	596	779	1 025

External trade (1976) Imports: Su 25 239 mn = $ 1 010 mn = £ 559 mn
Exports: Su 28 183 mn = $ 1 127 mn = £ 624 mn

Main imports (1975)	% of total	Main exports (1976)	% of total
Machinery	28	Crude oil	49
Motor vehicles	18	Coffee	18
Chemicals	15	Bananas	15
Iron and steel	11	Cocoa	3
Cereals	4		
Main sources (1975)		*Main destinations (1975)*	
United States	40	United States	47
Japan	13	Panama	15
West Germany	10	Chile	8
United Kingdom	4	Peru	7

Special focus

The impact of crude oil

	1971	1972	1973	1975
Crude oil production (000 t)	179	4 016	8 155	8 155
Sea traffic: all goods loaded (000 t)	1 746	5 187	11 880	10 430
Crude oil as % of exports (value)	½	17	51	57
Balance of payments in goods ($ mn)	−69	+39	+187	+6

El Salvador

Republic of El Salvador
República de El Salvador

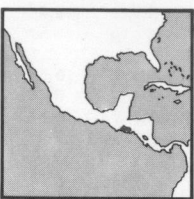

Location Central America
With a coastline on the Pacific Ocean, Guatemala is to the north-west and Honduras to the north and east
Land Area 21 041 km² = 8 124 mi²
Climate Tropical on coast, more temperate on inland heights
Weather at San Salvador, 682 m altitude
Temperature: hottest month May 19–33 °C, coldest Dec 16–32 °C
Rainfall (av monthly): driest month Feb 5 mm, wettest June 328 mm
Time 6 hours behind GMT
Measures Metric system; also old Spanish units, including:
length 1 vara = 83.59 centimetres = 32.9 inches
weight (mass) 1 libra = 16 onzas = 0.46 kilogram = 1.014 pounds
1 arroba = 25 libras = 11.5 kilograms = 25.35 pounds
1 quintal = 4 arrobas = 100 libras = 46 kilograms = 101.4 pounds
1 tonelada corta = 20 quintals = 0.92 tonne = 0.905 UK (long) ton
Monetary unit El Salvador colón (ES₵) = 100 centavos
Rate of exchange (1976 av): par ES₵ 2.50 = $ 1, free ES₵ 4.515 = £ 1
The Central American peso of the CACM (CA $), equal to the US dollar, is also used

Summary

Political Republic, which became independent from Spain in 1821 (from 1824 to 1839 as part of the Central American Federation). Honduras closed the border with El Salvador in 1969. Member of UN, OAS, Sela, Odeca and CACM
Economic Agriculture accounts for one-quarter of gross domestic product, with coffee, cotton and sugar major export earners. Manufacturing, accounting for one-sixth of gross domestic product, is comparatively well developed, especially for food, textiles and chemicals

People, resources and equipment

Population 1960 2.45* mn, 1970 3.44* mn, 1976 4.12* mn
Growth: 1960–70 3.5* %pa, 1970–76 3.1* %pa
Density (1976): 196* people per km²
Vital statistics (rate per 1 000 people, 1975): births 40.1, deaths 8.0
Cities (population in 000, 1973)

San Salvador (capital)	388	Zacatecoluca	60*	Sonsonate	52*
Santa Ana	180*	Santa Tecla	58*	San Vicente	49*
San Miguel	117*	Ahuachapán	56*	Usulután	45*

Race (1976) Mestizo 90** %, American Indian 9** %, European 1** %
Language Spanish
Religion (1976) Roman Catholic 90** %
Education (1974) Pupils 794 017, teachers 19 200*
Labour force (1971) 1 166 479; in agriculture 632 054 (54 %), manufacturing 113 983 (10 %), construction 32 555 (3 %)
Personnel (1973) Physicians: 950, 1 per 4 030 people
Standard of living
National income per person (1976): ES₵ 1 258 = $ 503 = £ 279
Consumption per person (1975): energy 248 kg coal equivalent, electricity (production, 1976) 290 kW h, newsprint 2.9 kg, steel 11 kg
Newspapers (1974): number 6; circulation 201 000, 51 per 1 000 people
Telephones (Dec 1976): 54 200, 13 per 1 000 people
Livestock (000, 1976) Cattle 1 109, pigs 425*, chickens 3 357
Petroleum refinery capacity (1975) 750 000 tonnes
Electrical capacity (1975) 306* megawatts
Hospital beds (1974) 6 931, 1 per 570 people
Roads (1974) 10 973 km = 6 818 mi, density 0.52 km per km²
Railways (1976) 602 km = 374 mi, density 0.029 km per km²
Ships (registered, 1977) 2, total of 1 987 gross tons
Ports (goods traffic, 000 tonnes, 1976)

	loaded	unloaded
Acajutla	467	1 458
La Unión	123	126
La Libertad	9	6

Airport Ilopango (San Salvador)

Durable equipment (Dec 1974)	000	no per 1 000 people	
Radio sets	940*	238*	no per km of road
Television sets	111	28	
Passenger cars	41	10	3.7
Commercial vehicles	19	5	1.7

El Salvador

Production

Gross domestic product 1976: ES₡ 5 464 mn = $ 2 186 mn = £ 1 210 mn
Growth in real terms: 1960–70 5.6 %pa, 1970–76 5.1 %pa
Structure of gross domestic product (1975) *By origin* Agriculture 24 %, manufacturing 17 %, transport and communication 5 %, other 54 %

Production indices (1970 = 100)	1960	1970	1976	Growth %pa 1960–70	1970–76
Agricultural	75	100	127	3.0	4.0
Manufacturing	36*	100	128[b]	10.8*	5.1[c]
Main products (000 t)					
Agriculture					
Maize	178	363	342	7.4	−1.0
Sorghum	82	147	156	6.0	1.0
Sugar, raw value	48	122	262	9.8	13.6
Oranges	22**	41	42*	6.4**	0.4*
Pineapples	18**	33	28*	6.2**	−2.7*
Coffee	99	129	159	2.7	3.5
Cotton	42	46	62	0.9	5.1
Milk	205*	181	186	−1.2*	0.5
Eggs	15**	23	27	4.4**	2.7
Beef and veal	16*	21	30*	2.8*	6.1*
Timber (000 m³)	2 960*	2 375*	3 928[a][b]	−2.2*	10.6[*c]
Other					
Petroleum products	—	172	630*[b]	na	29.6[*c]
Electricity (mn kW h)	256	671	1 193	10.1	10.0
Salt	14*	32*	23*[b]	8.6*	−6.4[*c]
Beer (000 hl)	145	184	386[b]	2.4	16.0[c]
Cigarettes (mn units)	854	1441	1 779[b]	5.4	4.3[c]
Cotton yarn	2.0	3.7	6.2[d]	6.3	13.8[e]
Cotton fabrics (mn m)	22	37	37[d]	5.3	0.0[e]
Rayon fabrics (mn m)	—	16.9	12.9[d]	na	−6.5[e]
Ethyl alcohol (000 hl)	29[b]	56[f]	na	11.6[g]	na
Sulphuric acid	na	11	12[b]	na	1.8[c]
Fertilisers, nitrogenous[a]	—	8.0*	5.3	na	−6.6[*]
Fertilisers, phosphate[a]	—	1.5*	2.4*	na	8.1[*]
Cement	86	167	340[b]	6.9	15.3[c]
Radio and television sets (000 units)	na	na	16[d]	na	na
Motor vehicles (000 units)[1]	na	na	1.0[b]	na	na

[a] Years ending June 30th [b] 1975 [c] 1970–75 [d] 1974 [e] 1970–74 [f] 1969 [g] 1963–69 [h] 1963 [1] Assembly of lorries only

Transport traffic (1976) *Air* Passenger-km 195 mn, cargo 14.0 mn t-km
Sea Goods loaded 0.60 mn t, unloaded 1.59 mn t
Tourism (1975) Number of visitors 266 000, gross receipts $ 18 mn

Finance and trade

Price indices (1970 = 100)	1960	1970	1976	Growth %pa 1960–70	1970–76
Consumer prices	93.8	100.0	161.8	0.6	8.4
Wholesale prices	89.3	100.0	208.3	1.1	13.0
Money stock (end-year, ES₡ mn)	193	317	1 001	5.1	21.1

Budget (1976) Revenue: ES₡ 805 mn = $ 322 mn = £ 178 mn
Expenditure: ES₡ 830 mn = $ 332 mn = £ 184 mn

Balance of payments ($ mn)	1972	1973	1974	1975	1976
Balance of goods (fob)	+52	+19	−58	−18	+105
Balance of services	−52	−76	−95	−103	−113
Balance of transfers	+12	+14	+18	+27	+30
Current balance	+12	−44	−134	−93	+22
Long-term capital flow	+16	+7	+144	+95	+48
Reserves and debt (end-year, $ mn)					
International reserves	82	62	98	127	205
External public debt	157	192	308	395	462

External trade (1976) Imports: ES₡ 1 762 mn = $ 705 mn = £ 390 mn
Exports: ES₡ 1 803 mn = $ 721 mn = £ 399 mn

Main imports	% of total	Main exports	% of total
Machinery	21	Coffee	53
Chemicals	20	Cotton	11
Food	9	Chemicals	6
Crude oil	7	Sugar	6
Motor vehicles	7	Textile yarns & fabrics	5
Textile yarns & fabrics	5	Clothing	2
Main sources		*Main destinations*	
United States	29	United States	33
Guatemala	15	West Germany	14
Japan	10	Guatemala	14
Venezuela	7	Netherlands	8
West Germany	6	Japan	8
Costa Rica	5	Costa Rica	6

Falkland Islands

Colony of Falkland Islands and Dependencies

Location South Atlantic
The Falkland Islands comprise some 200 islands about 770 km north-east of Cape Horn, with 2 main islands. The dependency of South Georgia is 1 300 km east-south-east of the Falklands, and the dependency of South Sandwich Islands is 800 km south-east of South Georgia
Land Area 16 260 km² = 6 280 mi², of which, South Georgia 3 750 km² = 1 450 mi², South Sandwich group 340 km² = 130 mi²
Climate Cool, with strong winds; rigorous in the dependencies
Weather at Stanley, 2 m altitude
Temperature: hottest month Jan 6–13 °C, coldest July minus 1–4 °C
Rainfall (av monthly): driest month Sept 38 mm, wettest Dec, Jan 71 mm
Time Stanley: 3 hours behind GMT; other: 4 hours behind GMT
Measures UK (imperial) system
Monetary unit Falkland Islands pound (FI £) = 100 new pence
Rate of exchange (1976 av): par FI £ = £ 1, free $ 1.806 = FI £ 1

Summary

Political UK Colony, to which, as the Islas Malvinas, Argentina makes a claim. South Orkney Islands, South Shetland Islands and part of the Antarctic Peninsula and continent, former dependencies of the Falklands, became the British Antarctic Territory from March 3, 1962
Economic Wool accounts for nearly all exports; there are reported to be oil reserves offshore, and the fishing potential is thought to be considerable

People, resources and equipment

Population 1962 2 190*, 1970 2 060*, 1976 1 920*
Growth: 1962–70 −0.8* %pa, 1970–76 −1.2* %pa
Density (1976): 0.1* people per km²
Vital statistics (rate per 1 000 people, 1975): births 16.8, deaths 11.5
Town (population, 1972) Stanley (capital) 1 081
Race (1972) British 97 %
Language English
Religion (1972) Christian 93 % (Anglican 54 %, Roman Catholic 11 %)
Education (1973) Pupils 280, teachers 21
Labour force (1972) 1 006
Personnel (1974) Physicians: 3, 1 per 606 people
Standard of living
National income per person (1976): FI £ 980* = $ 1 770* = £ 980*
Consumption per person (1975): energy 750* kg coal equivalent, electricity (production) 1 040 kW h
Telephones (Dec 1976): 500**, 260** per 1 000 people
Livestock (000, 1976) Sheep 645, cattle 9, horses 2, chickens 8*
Electrical capacity (1975) 1* megawatt
Hospital beds (1976) 27, 1 per 71 people
Roads (1976) 21 km = 13 mi, density 0.001 km per km²
Ships (registered, 1977) 5, total of 6 937 gross tons
Port Port Stanley
Airport Stanley
Radio sets (1975) 961, 500 per 1 000 people

Production, finance and trade

Gross domestic product
1976: FI £ 2.1*mn = $ 3.8*mn = £ 2.1*mn
Main products *Agriculture* (000 t, 1976) Milk 1*, mutton and goat meat 3*, wool 1.4*, sheepskins 0.6* *Other* (1975) electricity 2 mn kW h
Transport traffic (1974) *Sea* Goods loaded 2 000* t, unloaded 22 000* t
Budget (1975/76; year ending June 30th)
Revenue: FI £ 1.21 mn = $ 2.4 mn = £ 1.21 mn
Expenditure: FI £ 0.97 mn = $ 1.9 mn = £ 0.97 mn
External trade (1975) Imports: FI £ 1.53 mn = $ 3.39 mn = £ 1.53 mn
Exports: FI £ 1.17 mn = $ 2.61 mn = £ 1.17 mn

Main imports (1971)	% of total	Main export (1975)	% of total
Food	19	Wool	97
Electrical machinery	11		
Crude oil and products	9		
Alcoholic beverages	8		
Transport equipment	7		
Chemicals	6		

Main source (1971) United Kingdom 83 %
Main destination (1971) United Kingdom 93 %

French Guiana

Department of French Guiana
Département de la Guyane Française

Location North-east of South America
With a coastline on the Atlantic Ocean, Surinam
is to the west and Brazil to the south and east
Land Area 91 000 km² = 35 000 mi²
Climate Hot and humid
Weather at Cayenne, 6 m altitude
Temperature: hottest months Sept, Oct 23–33 °C
coldest Jan 23–29 °C
Rainfall (av monthly): driest month Sept 30 mm,
wettest May 551 mm
Time 3 hours behind GMT
Measures Metric system
Monetary unit French (metropolitan) franc (Fr) = 100 centimes; before
January 1, 1975 the local French Guiana franc, at par with the French
franc, was in use
Rate of exchange (1976 av): free Fr 4.780 = $ 1, Fr 8.633 = £1

Summary

Political French overseas department from 1946. Member of franc zone
Economic Main export is shrimps sold to the United States;
exploitation of the vast timber reserves has begun, and mineral reserves
are being developed. France supplies about two-thirds of imports and
subsidises the large trade deficit

People, resources and equipment

Population 1960 33 000*, 1970 51 000*, 1976 62 000*
Growth: 1960–70 4.4* %pa, 1970–76 3.3* %pa
Density (1976): 0.7* people per km²
Vital statistics (rate per 1 000 people, 1973): births 28.3, deaths 7.8
Towns (population in 000, 1975) Cayenne (capital) 30*, St Laurent 6*
Race (1974) Creole 66* %, French 12* %, American Indian 10* %
Language French
Religion (1976) Roman Catholic 92** %, Animist 6** %
Education (1973/74) Pupils 11 500*, teachers 620*
Labour force (1967) 17 012; in agriculture 3 132 (18 %)
Personnel (1975) Physicians: 37, 1 per 1 650 people
Standard of living
National income per person (1976): Fr 7 500** = $ 1 570** = £ 870**
Consumption per person (1975): energy 953 kg coal equivalent,
electricity (production) 930 kW h
Newspapers (1974): number 1; circulation 1 500, 26 per 1 000 people
Telephones (Dec 1976): 8 885, 140 per 1 000 people
Livestock (000, 1976) Pigs 5*, cattle 2*, chickens 46*, ducks 3*
Electrical capacity (1975) 18* megawatts
Hospital beds (1975) 973, 1 per 63 people
Roads (1976) 503 km = 313 mi, density 0.006 km per km²
Ports (1976) Cayenne: goods loaded 7 000 t, unloaded 120 000 t
St Laurent: goods loaded 5 000 t, unloaded 4 000 t
Airport Rochambeau (Cayenne)

Durable equipment (Dec 1974)	000	no per 1 000 people	no per km of road
Radio sets	2.8	48	
Television sets	3.0	52	
Passenger cars	7.7	128	15.3
Commercial vehicles	4.0	67	8.0

Production, finance and trade

Gross domestic product 1976 est: Fr 560** mn = $ 120** mn = £ 65** mn
Main products *Agriculture* (000 t, 1976) Cassava 4*, maize 1*, bananas 2*,
shrimps 2*. *Other* gold (1974) 35 kg, electricity (1975) 56 mn kW h
Transport traffic *Air* (1974) Passengers carried 56 200
Sea (1976) Goods loaded 15 000 t, unloaded 125 000 t
Consumer price index (1970 = 100) 1976 176.1; growth 1970–76 9.9 %pa
Budget (1976) Balanced at Fr 191 mn = $ 40 mn = £ 22 mn
External trade (1976) Imports: Fr 412 mn = $ 86 mn = £ 48 mn
Exports: Fr 19 mn = $ 4.0 mn = £ 2.2 mn

Main imports (1975)	% of total	*Main exports* (1975)	% of total
Food	24	Shrimps	44
Machinery	14	Timber	14
Petroleum products	8	Trailers	9
Main sources (1975)		*Main destinations* (1975)	
France	71	United States	46
Trinidad and Tobago	6	France	25
United States	4	Martinique	8

Grenada

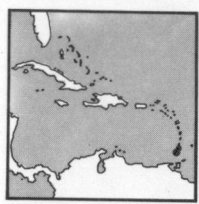

Location Eastern Caribbean Sea
Most southerly of the Windward Islands, with
Trinidad 150 km to the south, and St Vincent
110 km to the north. Some of the Grenadines,
islets lying between Grenada and St Vincent,
are included in the territory of Grenada
Land Area 344 km² = 133 mi²,
of which, Carriacou (Grenadines) 34 km²
Climate Semi-tropical
Weather at St George's, 1 m altitude
Temperature: hottest month Sept 25–31 °C, coldest Jan 23–29 °C
Rainfall (av monthly): 130 mm
Time 4 hours behind GMT
Measures UK (imperial) system
Monetary unit East Caribbean dollar (EC $) = 100 cents
Rate of exchange (1976 av): free EC $ 2.61 = $ 1, EC $ 4.71 = £ 1

Summary

Political Parliamentary monarchy, which became independent
February 7, 1974; formerly a UK associated state, and a UK colony
before March 3, 1967. Member of UN, OAS, Sela, Commonwealth,
Caricom and an EEC ACP state
Economic Tourism is an important earner of foreign exchange, with
gross receipts about equal to exports. Agriculture is diversified; main
exports are spices (nutmegs and mace), cocoa and bananas

People, resources and equipment

Population 1960 89 000, 1970 94 500*, 1976 96 000*
Growth: 1960–70 0.6* %pa, 1970–76 0.3* %pa
Density (1976): 279* people per km²
Vital statistics (rate per 1 000 people, 1975): births 27.4, deaths 5.9
Town (population in 000, 1975) St George's 6.6*
Race (1960) African 53 %, Mixed 42 %, European 1 %
Language English
Religion (1960) Christian 99.8 % (mainly Roman Catholic)
Education (1971/72) Pupils 34 863, teachers 1 082
Labour force (1970) 28 682; in agriculture 7 170* (25* %)
Personnel (1974) Physicians: 25, 1 per 4 000 people
Standard of living
National income per person (1976): EC $ 880** = $ 340** = £ 190**
Consumption per person (1975): energy 323 kg coal equivalent,
electricity (production) 240 kW h
Telephones (Dec 1976): 5 072, 52 per 1 000 people
Livestock (000, 1976) Cattle 5*, pigs 10*, sheep 7*, chickens 256*
Electrical capacity (1975) 7* megawatts
Hospital beds (1971) 692, 1 per 137 people
Roads (1976) 970* km = 600* mi, density 2.8*km per km²
Ships (registered, 1977) 2, total of 226 gross tons
Port St George's
Airports Pearls (29 km from St George's), Carriacou (Grenadines)
Durable equipment (at end-year)
Radio sets (1975): 22 000, 220 per 1 000 people
Motor vehicles (1974): 6 330, 60 per 1 000 people, 6.5* per km of road

Production, finance and trade

Gross domestic product 1973: EC $ 62 mn = $ 32 mn = £ 13 mn
1976 est: EC $ 100** mn = $ 38** mn = £ 21** mn
Main products *Agriculture* (000 t, 1976) Bananas 14*, lemons and
limes 1*, cocoa 3*, nutmegs 2.8, mace 0.4, fish catch (1975) 1.8*
Other Electricity (mn kW h, 1975) 25*
Transport traffic (1974) *Sea* Goods loaded 24 000* t, unloaded 104 000* t
Tourism (1976) Number of visitors 24 600, gross receipts $ 15* mn
Budget (1976) Balanced at EC $ 28 mn = $ 11 mn = £ 6 mn
External trade (1976) Imports: EC $ 66 mn = $ 25 mn = £ 14 mn
Exports: EC $ 34 mn = $ 13 mn = £ 7.2 mn

Main imports (1973)	% of total	*Main exports* (1976)	% of total
Food	37	Nutmeg and mace	44
Machinery and vehicles	14	Cocoa	30
Chemicals	8	Bananas	23
Main sources (1973)		*Main destinations* (1973)	
United Kingdom	27	United Kingdom	33
Trinidad and Tobago	20	West Germany	19

Guadeloupe

Department of Guadeloupe
Département de Guadeloupe

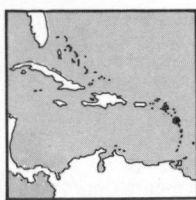

Location Eastern Caribbean Sea
Part of the Leeward Islands, with Antigua and
Montserrat to the north and Dominica to the
south. Dependencies include about two-thirds of
the island of St Martin
Land Area 1 780 km² = 687 mi²
Climate Sub-tropical
Weather at Pointe-à-Pitre, 533 m altitude
Temperature: hottest months Aug, Sept
21–28 °C, coldest Feb 17–24 °C
Rainfall (av monthly): driest month Feb 155 mm, wettest July 447 mm
Time 4 hours behind GMT
Measures Metric system
Monetary unit French (metropolitan) franc (Fr) = 100 centimes; before
January 1, 1975 the local Guadeloupe franc, at par with the French franc,
was in use
Rate of exchange (1976 av): free Fr 4.780 = $ 1, Fr 8.633 = £ 1

Summary

Political French overseas department from 1946; member of franc zone
Economic An agricultural economy with sugar and bananas making up
three-quarters of exports; trade is closely tied to France. Tourism and
light industry is being encouraged

People, resources and equipment

Population 1960 273 000*, 1970 327 000*, 1976 360 000*
Growth: 1960–70 1.8* %pa, 1970–76 1.6* %pa
Density (1976): 202* people per km²
Vital statistics (rate per 1 000 people, 1973): births 28.0, deaths 7.3
Regions (population in 000, 1972) Guadeloupe 306, Marie Galante 16,
La Désirade 2, Les Saintes 3, St Martin 5, St Barthélémy 2
Cities (population in 000, 1974) Basse-Terre (capital) 15*, Pointe-à-Pitre
(administrative capital) 30*, Grand-Bourg (Marie Galante) 13*
Race Creole, French, African, American Indian
Language French
Religion Mainly Roman Catholic
Education (1973/74) Pupils 103 015, teachers 4 160*
Labour force (1976) 125 000; in agriculture 26 000 (21 %)
Personnel (1974) Physicians: 217, 1 per 1 613 people
Standard of living
National income per person (1976): Fr 6 000** = $ 1 250** = £ 700**
Consumption per person (1975): energy 564 kg coal equivalent,
electricity (production) 480* kW h, newsprint 1.7 kg
Newspapers (1974): number 2; circulation 24 000, 70 per 1 000 people
Telephones (Dec 1976): 26 800, 74 per 1 000 people
Livestock (000, 1976) Cattle 86, goats 17*, pigs 26*, chickens 103*
Electrical capacity (1975) 40* megawatts
Hospital beds (1971) 3 566, 1 per 93 people
Hotel beds (1976) 3 067
Roads (1976) 1 975* km = 1 227* mi, density 1.11* km per km²
Ports Pointe-à-Pitre, Basse-Terre, Marigot (St Martin)
Airports Raizet (Pointe-à-Pitre), Basse-Terre, Marie Galante,
La Désirade, Terre-de-Haut (Les Saintes), L'Espérance (St Martin),
St Barthélémy

Durable equipment (at end-year)	000	no per 1 000 people	no per km of road
Radio sets (1974)	21	60	
Television sets (1974)	13	37	
Passenger cars (1972)	38	112	22
Commercial vehicles (1973)	16	47	9

Production, finance and trade

Gross domestic product 1970: Fr 1 326 mn = $ 239 mn = £ 99 mn
1976 est: Fr 2 600** mn = $ 550** mn = £ 300** mn
Main products (000 t) *Agriculture* (1976) Sweet potatoes 36*, sugar, raw
value 96, bananas 140*, milk 11* *Other* (1975) Electricity (mn kW h) 170*,
rum 10*
Transport traffic *Air* (Pointe-à-Pitre, 1975) Passenger departures and
arrivals 658 400 *Sea* (1973) Goods loaded 247 000 t, unloaded 176 000 t
Tourism (1972) Number of visitors 30 000**
Consumer price index (Basse-Terre, 1970 = 100) 1975 166.4;
growth 1970–75 10.7 %pa
Budget (1972) Balanced at Fr 583 mn = $ 116 mn = £ 46 mn
External trade (1976) Imports: Fr 1 515 mn = $ 317 mn = £ 175 mn
Exports: Fr 429 mn = $ 90 mn = £ 50 mn

Main imports (1975)	% of total		Main exports (1975)	% of total
Food	23		Sugar	41
Machinery	12		Bananas	37
Chemicals	9		Rum	7
Motor vehicles	7			
Petroleum products	5			
Main sources (1975)			Main destinations (1975)	
France	74		France	73
Martinique	5		United Kingdom	12
United States	5		Martinique	10

Guatemala

Republic of Guatemala
República de Guatemala

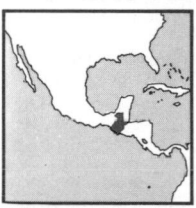

Location Central America
With the Caribbean Sea to the east and
Pacific ocean to the west, Mexico is to the north,
Belize to the east, and Honduras and
El Salvador to the south-east
Land Area 108 889 km² = 42 042 mi²
Climate Sub-tropical, temperate on highlands
Weather at Guatemala City, 1 480 m altitude
Temperature: hottest month May 16–29 °C,
coldest Jan 12–23 °C
Rainfall (av monthly): driest month Feb 3 mm, wettest June 274 mm
Time 6 hours behind GMT
Measures Metric system; also old Spanish units
Monetary unit Quetzal (Q) = 100 centavos
Rate of exchange (1976 av): par Q 1 = $ 1, free Q 1.806 = £ 1
The Central American peso of the CACM (CA $), equal to the US dollar,
is also used

Summary

Political Republic, which became independent from Spain in 1821
(from 1823 to 1839 as part of the Central American Federation).
Member of UN, OAS, Sela, Odeca and CACM
Economic Mainly agricultural, with coffee, cotton and sugar providing
over one-half of exports. The manufacturing sector is being developed,
and there has been substantial foreign investment in oil exploration and
nickel mining; petroleum refining is important. Tourism is to be expanded

People, resources and equipment

Population 1960 3.83* mn, 1970 5.27 mn, 1976 6.26 mn
Growth: 1960–70 3.2* %pa, 1970–76 2.9 %pa
Density (1976): 57 people per km²
Vital statistics (rate per 1 000 people, 1971): births 42.8, deaths 11.8
Cities (population in 000, 1973)

Guatemala City (capital)	707*	Jutiapa	50*
Escuintla	69*	Cobán	44*
Quezaltenango	66*	Jalapa	42*
Totonicapán	52*	Puerto Barrios	39*

Race (1973) American Indian 44 %, Ladino (Spanish/Indian) 52* %,
European 4* %
Language Spanish; also American Indian languages, especially Quiché,
Cakchiquel and Mam
Religion (1976) Roman Catholic 90** %
Education (1973) Pupils 697 680, teachers 23 260
Labour force (1973) 1 545 658; in agriculture 884 100 (57 %),
manufacturing 211 631 (14 %), construction 63 864 (4 %)
Personnel Scientists and engineers (1974): 5 551
Physicians (1971): 1 208, 1 per 4 340 people
Standard of living
National income per person (1976): Q 600* = $ 600* = £ 333*
Consumption per person (1975): energy 237 kg coal equivalent,
electricity (production) 181 kW h, newsprint 1.3 kg, steel 14 kg
Newspapers: number (1974) 11;
circulation (1972) 212 000, 39 per 1 000 people
Telephones (Dec 1973): 52 900, 10 per 1 000 people
Livestock (000, 1976) Cattle 2 270*, pigs 840*, sheep 520*
Mineral reserves Crude oil deposits were first discovered in 1974
Petroleum refinery capacity (1975) 1.25 mn tonnes
Electrical capacity (1975) 280* megawatts, of which, hydro 125* megawatts
Hospital beds (1973) 12 115, 1 per 480 people
Roads (1975) 13 450* km = 8 360* mi, density 0.12 km per km²
Railways (1976) 904* km = 560* mi, density 0.008 km per km²
Ships (registered, 1977) 8, total of 11 854 gross tons

Guatemala

Ports (goods traffic, 000 tonnes, 1974) Santo Tomás de Castilla:
loaded 225, unloaded 806; San José: loaded 163, unloaded 570;
Puerto Barrios: loaded 392, unloaded 121;
Champerico: loaded 101, unloaded 29
Airports (1976) La Aurora (Guatemala City): passenger departures and
arrivals 414 774; also Santa Elena (Flores) and Tikal

Durable equipment	000	no per	
(at end-year)		1 000 people	
Radio sets (1976)	262*	41*	no per
Television sets (1976)	110*	17*	km of road
Passenger cars (1974)	76	13	5.7
Commercial vehicles (1974)	40	7	3.0

Production

Gross domestic product 1976: Q 4 363 mn = $ 4 363 mn = £ 2 416 mn
Growth in real terms: 1960–70 5.5 %pa, 1970–76 5.9 %pa

Production indices	1960	1970	1976	Growth %pa	
(1970 = 100)				1960–70	1970–76
Agricultural	72	100	129	3.3	4.4
Industrial	58	100	124a	5.6	7.4b
Main products (000 t)					
Agriculture					
Maize	506	786	686	4.5	−2.2
Sugar, raw value	73	185	549*	9.7	19.9*
Bananas	244	487	550*	7.2	2.0*
Coffee	99	141	149*	3.6	0.9*
Cotton	21	70	99	12.8	6.0
Tobacco	1.4	3.2	8.0*	8.6	16.5*
Milk	144*	262	320*	6.2*	3.4*
Beef and veal	34*	55	78*	4.9*	6.0*
Timber (000 m³)	3 600**	4 580	5 665c	2.4**	4.3d
Other					
Petroleum products	—	724	955c	na	5.7d
Electricity (mn kW h)	281	780	1 100*c	10.7	7.1*d
Antimony ore (Sb content)	—	1.30	0.86c	na	−7.9d
Beer (000 hl)	166	299	526c	6.1	12.0d
Cigarettes (mn units)	1 889	2 986	2 737c	4.7	−1.7d
Cement	117	251	341c	7.9	6.3d

a1973 b1970–73 c1975 d1970–75

Transport traffic	1960	1970	1976	Growth %pa	
Passenger-kilometres (mn)				1960–70	1970–76
Air	30	104	132	13.2	4.1
Cargo: tonne-kilometres (mn)					
Rail	270	106	127a	−8.9	3.7b
Air	2.2	6.5	7.1	11.4	1.5
Sea: tonnes (mn)					
Goods loaded	0.32	0.53	0.81*c	5.0	11.2*d
Goods unloaded	0.59	1.30	1.30*c	8.2	0.0*d

a1975 b1970–75 c1974 d1970–74
Tourism (1976) Number of visitors 408 000, gross receipts $ 85 mn

Finance and trade

Consumer price index (1970 = 100) 1976 166.2; growth 1970–76 8.8 %pa
Money stock (end-year, Q mn) 1976 527; growth 1970–76 19.3 %pa
Budget (1976) Revenue: Q 428 mn = $ 428 mn = £ 237 mn
Expenditure: Q 525 mn = $ 525 mn = £ 291 mn

Balance of payments ($ mn)	1972	1973	1974	1975	1976
Balance of goods (fob)	+41	+51	−49	−31	−110
Balance of services	−83	−85	−109	−112	−101
Balance of transfers	+30	+43	+55	+78	+203
Current balance	−11	+8	−103	−66	−8
Long-term capital flow	+33	+60	+70	+169	+170
Reserves and debt (end-year, $ mn)					
International reserves	135	212	202	304	511
External public debt	183	196	198	281	551

External trade (1976) Imports: Q 982 mn = $ 982 mn = £ 544 mn
Exports: Q 782 mn = $ 782 mn = £ 433 mn

Main imports (1972)	% of total	Main exports (1976)	% of total
Chemicals	20	Coffee	32
Machinery	18	Sugar	14
Textile yarns and fabrics	9	Cotton	12
Food	7	Chemicals	8*
Main sources (1976)		*Main destinations* (1976)	
United States	36	United States	35
Japan	11	El Salvador	11
Venezuela	8	West Germany	11
El Salvador	7	Japan	8

Guyana

Location North-east of South America
With a coastline on the Atlantic Ocean, Brazil is
south, Venezuela west and Surinam east
Land Area 215 000 km² = 83 000 km²
Climate Tropical inland, sub-tropical on coast
Weather at Georgetown, 2 m altitude
Temperature: hottest months Sept, Oct
24–31 °C, coldest Jan, Feb 23–29 °C
Rainfall (av monthly): driest month Oct 76 mm,
wettest June 302 mm

Time 3 hours behind GMT
Measures UK (imperial) system; also metric system and
Rhynland acre = 0.426 hectare = 1.052 acres
Monetary unit Guyanese dollar (G $) = 100 cents
Rate of exchange (1976 av): par G $ 2.55 = $ 1, free G $ 4.606 = £ 1

Summary

Political Co-operative republic, which became independent May 26, 1966;
formerly the UK colony of British Guiana. Member of UN, Sela,
Commonwealth, Caricom and an EEC ACP state
Economic Agriculture and mining are the two main sectors. Sugar is
the largest export and, with rice, accounts for over 40 % of exports;
bauxite and alumina also make up over 40 % of exports. There are plans
to diversify the economy

People, resources and equipment

Population 1960 560 000*, 1970 714 000*, 1976 783 000*
Growth: 1960–70 2.5* %pa, 1970–76 1.5* %pa
Density (1976): 4* people per km²
Vital statistics (rate per 1 000 people, 1976): births 26.7, deaths 7.1
Cities (population in 000, 1970) Georgetown (capital) 182a, Linden 30*,
New Amsterdam 18*, Corriverton 17*, Rose Hall 8*
a1975
Race (1970) Asian Indian 51 %, African 31 %, Mixed 12 %, American
Indian 4 %, Portuguese 1 %, Chinese 1 %
Language English; Creole and American Indian languages are also used
Religion (1960) Hindu 33 %, Anglican 19 %, Roman Catholic 15 %,
Moslem 9 %, Methodist 4 %
Education (1973/74) Pupils 202 183, teachers 7 297
Labour force (1976) 253 000; in agriculture 61 000 (24 %)
Personnel (1974) Physicians: 237, 1 per 3 290 people
Standard of living
National income per person (1976): G $ 1 500** = $ 620** = £ 320**
Consumption per person (1975): energy 1 114 kg coal equivalent,
electricity (production, 1976) 511 kW h, newsprint 1.9 kg
Newspapers (1974): number 3; circulation 120 000, 155 per 1 000 people
Telephones (Dec 1976): 22 500, 29 per 1 000 people
Livestock (000, 1976) Cattle 280*, sheep 108*, pigs 125*, chickens
10 130*
Electrical capacity (public only, 1975) 95* megawatts
Hospital beds (1973) 3 987, 1 per 190 people
Roads (1975) 2 910* km = 1 810* mi, density 0.014* km per km²
Railways (industrial, 1975) 130* km = 80* mi, density 0.001* km per km²
Inland waterways (rivers, 1976) 900**km = 550**mi
Ships (registered, 1977) 70, total of 16 274 gross tons
Ports Georgetown, New Amsterdam
Airports Timehri (40 km from Georgetown); also 50* other airports
Durable equipment (at end-year)
Radio sets (1974): 268 000, 346 per 1 000 people
Passenger cars (1975): 25 500, 32 per 1 000 people, 9 per km of road
Commercial vehicles (1975): 12 200, 16 per 1 000 people, 4 per km of road

Production

Gross domestic product 1975: G $ 1 184 mn = $ 493 mn = £ 222 mn
1976 est: G $ 1 200** mn = $ 470** mn = £ 260** mn
Structure of gross domestic product (1973) *By origin* Agriculture 17 %,
mining and quarrying 13 %, manufacturing 10 %, other 60 %

Production index	1960	1970	1976	Growth %pa	
(1970 = 100)				1960–70	1970–76
Agricultural	90	100	113	1.1	2.1

Guyana

Main products (000 t)	1960	1970	1976	Growth %pa 1960–70	1970–76
Agriculture					
Rice	214	210	227	−0.2	1.3
Sugar, raw value	340	316	343	−0.7	1.4
Oranges	8**	8	11*	0.0**	5.4*
Milk	13	20	12*	4.4	−8.2*
Fish catch	6	17	20c	11.8	3.0d
Timber (000 m³)	282	254	266*c	−1.0	0.9*d
Other					
Electricity (mn kW h)	199a	344	400	11.6b	2.5
Bauxite	3 422	4 103	3 198c,g	1.8	−4.8d
Gold (000 kg)	0.07	0.14	0.56c	7.5	32.0d
Diamonds (000 CM)	101	61	50e	−4.9	−4.8f
Wheat flour	—	30	41c	na	6.4d
Beer (000 hl)	40	56	113c	3.4	15.1d
Cigarettes (mn units)	335	474	532c	3.5	2.3d

a1965 b1965–70 c1975 d1970–75 e1974 f1970–74 gProduction of alumina from bauxite: 218 thousand tonnes

Transport traffic (1974) *Rail* Passenger-km 6 mn
Sea Goods loaded 2.95* mn t, unloaded 1.07* mn t
Tourism (1975) Gross receipts $ 3 mn

Finance and trade

Consumer price index (1970 = 100) 1976 157.6; growth 1970–76 7.9 %pa
Money stock (end-year, G $ mn) 1976 214; growth 1970–76 23.9 %pa
Budget (1976) Revenue: G $ 446 mn = $ 175 mn = £ 97 mn
Expenditure: G $ 605 mn = $ 237 mn = £ 131 mn

Balance of payments ($ mn)	1972	1973	1974	1975	1976
Balance of goods (fob)	+15	−24	+40	+46	−58
Balance of services	−29	−39	−45	−63	−76
Balance of transfers	−1	−1	−3	−6	−5
Current balance	−15	−63	−9	−23	−140
Long-term capital flow	+15	+32	+43	+69	+37
Reserves and debt (end-year, $ mn)					
International reserves	37	14	63	100	27
External public debt	222	254	328	362	na

External trade (1976) Imports: G $ 913 mn = $ 358 mn = £ 198 mn
Exports: G $ 702 mn = $ 275 mn = £ 152 mn

Main imports	% of total	Main exports	% of total
Machinery	19	Sugar	34
Petroleum products	15	Bauxite	34
Food	12	Rice	10
Chemicals	10	Alumina	9
Motor vehicles	5	Rum	2
Textile yarns & fabrics	5		
Main sources		*Main destinations*	
United States	29	United Kingdom	27
United Kingdom	23	United States	20
Trinidad and Tobago	19	Algeria	8
Japan	4	Trinidad and Tobago	7
Canada	4	West Germany	6

Special focus

Main exports (% of total)	1965	1970	1971	1972	1973	1974	1975	1976
Sugar	26	28	27	34	30	49	50	34
Bauxite	21	34	32	34	37	25	24	34
Rice	13	7	7	8	9	8	10	10
Alumina	17	17	14	9	9	8	8	9

Haiti

Republic of Haiti
République d'Haïti

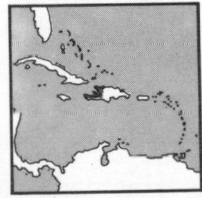

Location Caribbean Sea
Occupies the western part of the island of Hispaniola, with Dominican Republic to the east, occupying the rest of the land. The islands of Cuba and Jamaica are to the west
Land Area 27 750 km² = 10 714 mi²
Climate Tropical
Weather at Port-au-Prince, 37 m altitude
Temperature: hottest month July 23–34 °C, coldest Jan 20–31 °C
Rainfall (av monthly): driest months Dec, Jan 33 mm, wettest May 231 mm
Time 5 hours behind GMT

Measures Metric system; other units are also in use, including:
length 1 pouce = 27.1 millimetres = 1.07 inches
weight (*mass*) 1 livre française = 0.489 kilogram = 1.079 pounds
Monetary unit Gourde (Gde) = 100 centimes
Rate of exchange (1976 av): par Gde 5 = $ 1, Gde 9.031 = £ 1

Summary

Political One-party republic, which became independent from France in 1804 (from 1822–44 united with Dominican Republic, and from 1915–34 under United States occupation). Member of UN, OAS and Sela
Economic An agricultural economy, with coffee accounting for one-third of exports, and essential oils significant; bauxite is an important export. Light industry has been developed; tourism accounts for about one-fifth of foreign currency earnings

People, resources and equipment

Population 1960 3.62* mn, 1970 4.24* mn, 1976 4.67* mn
Growth: 1960–70 1.6* %pa, 1970–76 1.6* %pa
Density (1976): 168* people per km²
Vital statistics (rate per 1 000 people, 1970–75): births 35.8*, deaths 16.3*
Cities (population in 000, 1975) Port-au-Prince (capital) 459, Cap-Haïtien 30*, Les Cayes 14*, Gonaïves 14*, Jérémie 12*, Port-de-Paix 7*
Race (1976) African 95** %, Mulatto 5** %
Language French; also Creole
Religion (1966) Roman Catholic 66 %, Protestant 11 %, Voodoo
Education (1975/76) Pupils 547 000*, teachers 15 050*
Labour force (1976) 2 361 000; in agriculture 1 647 000 (70 %)
Personnel (1975) Physicians: 394*, 1 per 11 700* people
Standard of living
National income per person (1976): Gde 1 090** = $ 220** = £ 120**
Consumption per person (1975): energy 30 kg coal equivalent, electricity (production) 34 kW h, newsprint 0.2 kg, steel 5 kg
Newspapers (1974): number 7; circulation 93 000, 21 per 1 000 people
Telephones (Dec 1976): 17 800, 4 per 1 000 people
Livestock (000, 1976) Cattle 747*, pigs 1 770*, goats 1 384*
Electrical capacity (1975) 75* megawatts
Hospital beds (1973) 3 868, 1 per 1 160 people
Roads (1976) 4 000* km = 2 500* mi, density 0.14* km per km²
Railways (plantation, 1976) 240*km = 150*mi, density 0.009*km per km²
Ports (goods traffic, 000 tonnes, 1975)

	loaded	unloaded		loaded	unloaded
Miragoâne	523	3	Aux Cayes	15	3
Port-au-Prince	73	431			

Other ports include Cap-Haïtien and Fort-Liberté
Airports François Duvalier (Port-au-Prince), Cap-Haïtien

Durable equipment (at end-year)	000	no per 1 000 people	
Radio sets (1974)	91	20	no per km of road
Television sets (1974)	13	2.9	
Passenger cars (1973)	11.7	2.6	2.9
Commercial vehicles (1973)	1.3	0.3	0.3

Production

Gross domestic product 1975 (year ending Sept 30th):
Gde 4 612 mn = $ 922 mn = £ 402 mn
1976 est: Gde 5 200**mn = $ 1 040** mn = £ 580**mn
Growth in real terms: 1960–70 0.8 %pa, 1970–74 4.4 %pa

Production index (1970 = 100)	1960	1970	1976	Growth %pa 1960–70	1970–76
Agricultural	96	100	112	0.4	1.9
Main products (000 t)					
Rice	41*	80	131	6.9*	8.6
Maize	227	240	250*	0.6	0.7*
Sorghum	183	210	225*	1.4	1.2*
Cassava	110	130	147*	1.7	2.1*
Sugar, raw value	60*	64	60*	0.6	−1.1*
Bananas and plantains	280**	240*	250*	−1.5**	0.7*
Coffee	26*	33	36*	2.4*	1.5*
Tobacco	1.1*	2.2	3.0*	7.2*	5.3*
Sisal	27	15	19*	−5.7	4.0*
Timber (000 m³)	3 135**	3 802	3 989*b	1.9*	1.0*c
Electricity (mn kW h)	90**	118	158b	2.7**	6.0c
Bauxitea	346	673	733	6.9	1.4
Cigarettes (mn units)	362	421	674b	1.5	9.9c
Cotton fabrics (mn m)	4	3	3*b	−3.0	0.0*c
Cement	48	65	232	3.1	23.6

aExports b1975 c1970–75

Haiti

Transport traffic *Air* (1974) Passenger departures and arrivals 241 000
Sea (1975) Goods loaded 632 000 t, unloaded 461 000 t
Tourism Number of visitors (1976) 86 000 (excluding 202 000 cruise passengers), gross receipts (1975/76) $ 28 mn

Finance and trade

Price index (1970 = 100)	1960	1970	1976	Growth %pa	
				1960–70	1970–76
Consumer prices	76.0	100.0	199.4	2.8	12.2
Money stock (end-year, Gde mn)	104	191	535	6.2	18.8

Budget (1976) Revenue: Gde 666 mn = $ 133 mn = £ 74 mn
Expenditure: Gde 730 mn = $ 146 mn = £ 81 mn
International reserves (Dec 1976) $ 28 mn
External trade (1976) Imports: Gde 950* mn = $ 190* mn = £ 105* mn
Exports: Gde 622 mn = $ 124 mn = £ 69mn

Main imports (1974)	% of total	Main exports (1976)	% of total
Food	14	Coffee	37
(of which, cereals 6)		Bauxite	15
Petroleum products	11	Essential oils	9a
Machinery	10	Toys and sporting	
Chemicals	9	goods	8a
Vegetable oils and fats	7	Textile yarns and	
Motor vehicles	6	fabrics	8a
Main sources (1975)		Main destinations (1975)	
United States	54	United States	74
Netherlands Antilles	8	France	8
Japan	7	Belgium-Luxembourg	5
Canada	5	Italy	4
France	5	Canada	2
West Germany	3	Netherlands	2
Netherlands	3		
United Kingdom	3		

a1974

Honduras

Republic of Honduras
República de Honduras

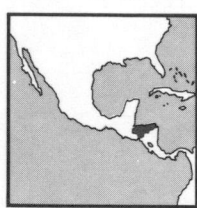

Location Central America
With coastlines on the Pacific Ocean and the Caribbean Sea, Guatemala is to the west, El Salvador to the south-west and Nicaragua to the south-east
Land Area 112 088 km² = 43 277 mi²
Climate Tropical on coast, moderate inland
Weather at Tegucigalpa, 1 004 m altitude
Temperature: hottest month May 12–33 °C, coldest Feb 4–27 °C
Rainfall: rain falls mainly May–November
Time 6 hours behind GMT
Measures Metric system; also old Spanish measures, including:
length 1 vara = 83.5 centimetres = 32.9 inches
weight (mass) 1 libra = 16 onzas = 0.46 kilogram = 1.014 pounds
1 arroba = 25 libras = 11.5 kilograms = 25.35 pounds
Monetary unit Lempira (La) = 100 centavos
Rate of exchange (1976 av): par La 2.00 = $ 1, free La 3.612 = £ 1
The Central American peso of the CACM (CA $), equal to the US dollar, is also used

Summary

Political Republic with military government, which became independent in 1838, after forming part of the Central American Federation. Member of UN, OAS, Sela, Odeca and CACM (suspended active participation since 1970)
Economic An agricultural economy, with bananas and coffee accounting for about one-half of exports; timber and mining are also important. There is some light industry. Hurricanes can affect the economy, and, in particular, one in September 1974 severely damaged crops and industries

People, resources and equipment

Population 1960 1.85**mn, 1970 2.37*mn, 1976 2.83*mn
Growth: 1960–70 2.5** %pa, 1970–76 3.0* %pa
Density (1976): 25* people per km²
Vital statistics (rate per 1 000 people, 1975): births 47.1, deaths 7.0
Cities (population in 000, 1976)

Tegucigalpa (capital)	300*	La Ceiba	55*	Puerto Cortés	44*
San Pedro Sula	200*	Choluteca	49*	El Progreso	28*

Race Mestizo, European, American Indian
Language Spanish
Religion (1964) Roman Catholic 71 %, Protestant 2 %
Education (1975) Pupils 538 113, teachers 16 894
Labour force (1976) 931 000; in agriculture 597 000 (64 %)
Personnel (1974) Physicians: 874, 1 per 3 100 people
Standard of living
National income per person (1976): La 788 = $ 394 = £ 218
Consumption per person (1975): energy 232 kg coal equivalent, electricity (production) 175 kW h, newsprint 0.7 kg, steel 11 kg
Newspapers (1974): number 7; circulation (4 only) 99 000, 37 per 1 000 people
Telephones (Dec 1976): 19 200, 7 per 1 000 people
Livestock (000, 1976) Cattle 1 800*, pigs 520*, chickens 7 800*
Petroleum refinery capacity (1975) 700 000 tonnes
Electrical capacity (1975) 144* megawatts
Hospital beds (1975) 4 602, 1 per 606 people
Roads (1976) 5 600* km = 3 500* mi, density 0.050 km per km²
Railways (1976) 1 780* km = 1 110* mi, density 0.016* km per km²
Ships (registered, 1977) 63, total of 104 903 gross tons
Ports Puerto Cortés (1976): goods loaded 636 000 t, unloaded 647 000 t; also La Ceiba, Tela and Amapala
Airports Toncontín (Tegucigalpa), Ramon Villeda (San Pedro Sula), Colozón (La Ceiba); also 10 other airports with scheduled flights

Durable equipment (Dec 1974)	000	no per 1 000 people	
Radio sets	158	58	no per
Television sets	46	17	km or road
Passenger cars	15	5	3*
Commercial vehicles	23	8	4*

Production

Gross domestic product 1976: La 2 402 mn = $ 1 201 mn = £ 665 mn
Growth in real terms: 1960–70 4.7 %pa, 1970–76 3.5 %pa
Structure of gross domestic product (1974) *By origin* Agriculture 29 %, mining and quarrying 3 %, manufacturing 14 %, construction 5 %, transport and communication 6 %, other 43 %

Production indices (1970 = 100)	1960	1970	1976	Growth %pa	
				1960–70	1970–76
Agricultural	68	100	124	4.0	3.7
Manufacturing	64	100	127a	4.6	8.3b
Main products (000 t)					
Agriculture					
Maize	262	274	289	0.4	0.9
Sorghum	53	47	47	−1.2	0.0
Cassava	14	35	11*	9.6	−17.5*
Sugar, raw value	20	81	88*	15.0	1.4*
Dry beans	36	60*	48	5.2*	−3.6*
Palm kernels	1.1*	2.3	2.5*	7.7*	1.4*
Palm oil	2.5*	5.9	6.0*	9.0*	0.3*
Oranges	35**	48	25*	3.2**	−10.3*
Bananas	799	1 200*	1 400*	4.2*	2.6*
Pineapples	2*	8*	20*	15.0*	16.5*
Coffee	23	36	57*	4.6	8.0*
Tobacco	3.6	6.2	6.5*	5.6	0.8*
Cotton	2.0	3.2	5.0*	4.8	7.7*
Milk	119	162	187*	3.1	2.4*
Beef and veal	20*	32	47*	4.8*	6.6*
Timber (000 m³)	3 390	4 100	3 868*c	1.9*	−1.2*d
Other					
Petroleum products	—	717	598c	na	−3.6d
Electricity (mn kW h)	97	310	480c	12.3	9.1d
Lead ore (Pb content)	4.0**	14.6	23.3c	13.8**	9.8d
Zinc ore (Zn content)	3.5**	16.3	30.3c	16.6**	13.2d
Silver	0.092	0.119	0.099c	2.6	−4.5f
Gold (000 kg)	0.075**	0.103	0.066e	3.2**	−10.5f
Beer (000 hl)	139	258	317e	6.4	5.3f
Cigarettes (mn units)	1 065	1 266	1 811e	1.7	9.4f
Cotton fabrics (mn m)	3	11	14e	13.9	6.2f
Cement	37	161	271c	15.8	11.0d

a1973 b1970–73 c1975 d1970–75 e1974 f1970–74

Honduras

Transport traffic *Air* (1976) Passenger-km 257 mn, cargo 3.6 mn t-km
Sea (1975) Goods loaded 1.26 mn t, unloaded 1.16 mn t
Tourism Number of visitors (1972) 140 000, gross receipts (1975) $ 8 mn

Finance and trade

Price index	1960	1970	1976	Growth %pa	
(1970 = 100)				1960–70	1970–76
Consumer prices	80.4	100.0	142.4	2.2	6.1

Money stock

(end-year, La mn)	64	166	388	9.9	15.2

Budget (1976) Revenue: La 353 mn = $ 176 mn = £ 98 mn
Expenditure: La 384 mn = $ 192 mn = £ 106 mn

Balance of payments ($ mn)	1972	1973	1974	1975	1976
Balance of goods (fob)	+36	+23	−89	−70	−24
Balance of services	−55	−65	−49	−68	−98
Balance of transfers	+7	+7	+33	+18	+13
Current balance	−13	−35	−106	−120	−109
Long-term capital flow	+23	+32	+63	+137	+108

Reserves and debt (end-year, $ mn)

International reserves	35	42	44	97	131
External public debt	169	215	276	451	581

External trade (1976) Imports: La 906 mn = $ 453 mn = £ 251 mn
Exports: La 784 mn = $ 392 mn = £ 217 mn

Main imports (1974)	% of total	Main exports (1976)	% of total
Chemicals	15	Bananas	27
Crude oil	14	Coffee	26
Machinery, non-electric	11	Timber	10
Food	8	Lead ore	8a
Paper	8	Meat	7
Motor vehicles	7	Silver	5a
Iron and steel	6		
Textile yarns and fabrics	6		
Electrical machinery	5		

Main sources (1976)		Main destinations (1976)	
United States	44	United States	56
Japan	9	West Germany	12
Trinidad and Tobago	7	Guatemala	4
Guatemala	6	Nicaragua	4
West Germany	5	Belgium-Luxembourg	3
Nicaragua	4	Japan	3
Costa Rica	3	Dominican Republic	2

a 1975

Special focus

Timber production (000 cubic metres)

	Hardwood	Softwood	Total
1970	1 690	2 410	4 100
1971	1 690	2 540	4 230
1972	1 659	2 775	4 434
1973	1 659*	2 775*	4 434*
1974	2 021	1 856	3 877
1975	2 043*	1 825*	3 868*
Growth			
1970–75	3.9* %pa	−5.4* %pa	−1.2* %pa

Jamaica

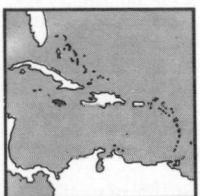

Location Caribbean Sea
The island of Jamaica has Cuba about 150 km
to the north and Haiti about 160 km to the east
Land Area 10 991 km² = 4 244 mi²
Climate Tropical, more temperate inland
Weather at Kingston, 34 m altitude
Temperature: hottest months July, Aug 23–32°C,
coldest Jan, Feb 19–30°C
Rainfall (av monthly): driest month Feb 15 mm,
wettest Oct 180 mm

Time 5 hours behind GMT (summer time, 4 hours behind)
Measures UK (imperial) system, converting to the metric system
Monetary unit Jamaican dollar (J $) = 100 cents; the dollar was
introduced as a decimal currency on September 8, 1969 to replace the
Jamaican pound (J £) at J $ 2 = J £ 1
Rate of exchange (1976 av): par $ 1.10 = J $ 1, free J $ 1.642 = £ 1

Summary

Political Parliamentary monarchy, which became fully independent on
August 6, 1962; formerly a UK colony (from 1958–61 a member of the
Federation of the West Indies). Member of UN, OAS, Sela,
Commonwealth, Caricom and an EEC ACP state
Economic Bauxite and alumina make up about two-thirds of exports and
mining accounts for 13 % of gross domestic product compared with 7 %
for agriculture and 16 % for manufacturing. Sugar is the other main
export, and there is a range of light industries. Tourism is important,
gross receipts being equal to one-seventh of total foreign currency
earnings

People, resources and equipment

Population 1960 1.63 mn, 1970 1.87 mn, 1976 2.06 mn
Growth: 1960–70 1.4 %pa, 1970–76 1.6 %pa
Density (1976): 187 people per km²
Vital statistics (rate per 1 000 people, 1976): births 30.0, deaths 7.1
Cities (population in 000, 1976)

Kingston (capital)	700*	May Pen	30*	Port Antonio	12*
Montego Bay	50*	Mandeville	16*	Ocho Rios	10*
Spanish Town	45*	Savannah la Mar	14*		

Race (1970) African 76* %, Mixed 15* %, European 1 %, Chinese 1 %
Language English; an English/African patois is also used
Religion (1960) Anglican 20 %, Baptist 19 %, Church of God 12 %,
Roman Catholic 7 %, Methodist 7 %, Presbyterian 5 %
Education (1974/75) Pupils 575 300*, teachers 16 500*
Labour force (1976) 895 000; in agriculture and mining 264 000 (29 %)
Personnel (1974) Physicians: 570, 1 per 3 509 people
Standard of living
National income per person (1976): J $ 1 190* = $ 1 310* = £ 725*
Consumption per person (1975): energy 1 427 kg coal equivalent,
electricity (production) 1 143 kW h, newsprint 4.3 kg, steel 43 kg
Newspapers (1974): number 3; circulation 180 000, 90 per 1 000 people
Telephones (Dec 1976): 108 500, 53 per 1 000 people
Livestock (000, 1976) Cattle 280*, goats 330*, pigs 235*, chickens 3 770*
Mineral reserves Bauxite (1956) 600* mn tonnes
Petroleum refinery capacity (1975) 2.0* mn tonnes
Electrical capacity (1975) 677* megawatts
Hospital beds (government establishments only, 1974) 7 780, 1 per
257 people
Roads (1976) 11 700* km = 7 300* mi, density 1.1* km per km²
Railways (1976) 373 km = 232 mi, density 0.034 km per km²
Ships (registered, 1977) 7, total of 7 075 gross tons
Ports Kingston (principal port); also Port Antonio, Savannah la Mar,
Port Morant, Portland Bight
Airports (passenger departures and arrivals, 000, 1976):
Norman Manley (Kingston) 965, Donald Sangster (Montego Bay) 623;
also Mandeville, Port Antonio, Ochos Rios, Negril

Durable equipment	000	no per	
(at end-year)		1 000 people	
Radio sets (1973)	633	320	no per
Television sets (1974)	97	49	km of road
Passenger cars (1972)	86	45	7.4*
Commercial vehicles (1972)	22	11	1.9*

Production

Gross domestic product 1976: J $ 2 768 mn = $ 3 045 mn = £ 1 686 mn
Growth in real terms: 1970–74 2.1 %pa
Structure of gross domestic product *By origin* (1974) Agriculture 7 %,
mining and quarrying 13 %, manufacturing 16 %, construction 12 %,
transport and communication 7 %, other 45 %. *By type* (1976) Gross
fixed capital formation 16 %, exports of goods and services 28 %,
government consumption 21 %

Production index	1960	1970	1976	Growth % pa	
(1970 = 100)				1960–70	1970–76
Agricultural	98	100	107	0.2	1.1
Main products (000 t)					
Agriculture					
Yams	70**	76	135*	0.8**	10.0
Copra	12	16	6*	2.5	−15.1*
Sugar, raw value	431	376	368*	−1.4	−0.3*
Oranges	58	87*	41*	4.1*	−11.8*
Grapefruit	24	33*	30*	3.2*	−1.6*
Lemons and limes	9	15*	20*	5.0*	5.0*
Bananas	254	185*	140*	−3.1*	−4.5*
Coffee	2.2	1.2	1.0*	−5.9	−3.0*
Cocoa	2.5	1.8	2.0*	−3.2	1.8*
Milk	39	50	54*	2.5	1.3*

Jamaica

Main products (000 t) Other	1960	1970	1976	Growth %pa 1960–70	1970–76
Petroleum products	—	1 496	1 480*b	na	−0.2*c
Electricity (mn kW h)	514	1 541	2 331b	11.6	8.6c
Bauxitea	5 838	12 106	10 310	7.6	−2.6
Aluminad	662	1 768	2 375b	10.3	6.1c
Beer (000 hl)	153	432	663b	10.9	8.9c
Cigarettes (mn units)	694	1 261	1 625b	6.2	5.2c
Cotton fabrics (mn m)	6	7	7b	1.6	0.0c
Tyres (000 units)	—	163	182	na	1.9
Sulphuric acid	—	9	19	na	13.2
Cement	212	457	365	8.0	−3.7
Radio sets (000 units)	—	10	7b	na	−7.0c
Television sets (000 units)	—	7	6b	na	−3.0c

aDried equivalent of crude ore mined b1975 c1970–75 dExports, manufactured from bauxite, with bauxite giving approximately 40 % alumina by weight; about one-half of bauxite is processed before export (27% in 1960)

Transport traffic	1960	1970	1976	Growth %pa 1960–70	1970–76
Passenger-kilometres (mn)					
Rail	46*	72	64a	4.6*	−3.9b
Air	na	335	1 379	na	26·6
Cargo: tonne-kilometres (mn)					
Rail	74*	127	156a	5.5*	7.1b
Air	na	2.5	11.5	na	29.0
Sea: tonnes (mn)					
Goods loaded	6.5	12.3	7.6	6.6	−7.7
Goods unloaded	1.6	4.6	2.6	11.1	−8·9
Tourism					
Number of visitors (000)	80	309	471	14.5	7.3
Gross receipts ($ mn)	na	96	106	na	1.7

a1973 b1970–73

Finance and trade

Price index (1970 = 100)	1960	1970	1976	Growth %pa 1960–70	1970–76
Consumer prices	66.8	100.0	209.7	4.1	13.1
Money stock (end-year, J $ mn)	57*	138	360	9.3*	17.3

Budget (1976) Revenue: J $ 616 mn = $ 678 mn = £ 375 mn
Expenditure: J $ 1 013 mn = $ 1 114 mn = £ 617 mn

Balance of payments ($ mn)	1972	1973	1974	1975	1976
Balance of goods (fob)	−152	−178	−59	−161	−132
Balance of services	−74	−97	−58	−150	−177
Balance of transfers	+29	+27	+25	+28	+6
Current balance	−197	−248	−92	−283	−303
Long-term capital flow	+127	+206	+226	+205	+88
Reserves and debt (end-year, $ mn)					
International reserves	160	127	190	126	32
External public debt	318	469	650	816	1 031

External trade (1976) Imports: J $ 830 mn = $ 913 mn = £ 505 mn
Exports: J $ 576 mn = $ 633 mn = £ 351 mn

Main imports (1975)	% of total	Main exports (1976)	% of total
Food	17	Alumina	43
(of which, cereals 8)		Bauxite	23
Machinery	14	Sugar	9
Crude oil	11	Bananas	2*
Chemicals	10		
Petroleum products	8		
Motor vehicles	5		
Textile yarns and fabrics	4		
Main sources (1976)		Main destinations (1976)	
United States	37	United States	43
Venezuela	14	United Kingdom	17
United Kingdom	11	Norway	12
Canada	6	Sweden	5
Netherlands Antilles	6	Canada	4

Special focus

Bauxite production in the Caribbean and nearby area, 1976

Rank in 1976		000 tonnes	
1	Jamaicaa	10 310 = 14* % of world production	
2	Surinam	4 751b	
3	Guyana	3 198b	
4	Dominican Rep	785b	} Hispaniola island = 1 310*b
5	Haiti	733c	

aThe only country in the world with larger production in 1976 was Australia with 23 540 thousand tonnes b1975 cExports

Martinique

Department of Martinique
Département de la Martinique

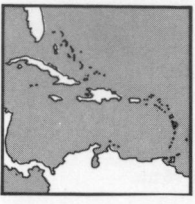

Location Eastern Caribbean Sea
One of the Windward Islands, with Dominica to the north and St Lucia to the south
Land Area 1 100 km² = 425 mi²
Climate Sub-tropical
Weather at Fort-de-France, 4 m altitude
Temperature: hottest month Sept 23–31 °C, coldest Jan 21–28 °C
Rainfall (av monthly): driest month March 74 mm, wettest Aug 262 mm

Time 4 hours behind GMT
Measures Metric system
Monetary unit French (metropolitan) franc (Fr) = 100 centimes; before January 1, 1975 the local Martinique franc, at par with the French franc, was in use
Rate of exchange (1976 av): free Fr 4.780 = $ 1, Fr 8.633 = £ 1

Summary

Political French overseas department from 1946; member of franc zone
Economic Mainly agricultural, with bananas the main export, and rum also important. There is petroleum refining using imported oil

People, resources and equipment

Population 1960 287 000*, 1970 338 000, 1976 369 000
Growth: 1960–70 1.6* %pa, 1970–76 1.5 %pa
Density (1976): 335 people per km²
Vital statistics (rate per 1 000 people, 1973): births 22.4, deaths 6.8
Cities (population in 000, 1974) Fort-de-France (capital) 99, Sainte Marie 20*
Race (1976) African or Mixed 90** %, Asian 5** %
Language French; also Creole
Religion (1976) Roman Catholic 90** %
Education (1973/74) Pupils 105 200*, teachers 4 750*
Labour force (1976) 123 000; in agriculture 23 000 (19 %)
Personnel (1975) Physicians: 236, 1 per 1 550 people
Standard of living
National income per person (1976): Fr 9 200** = $ 1 900** = £ 1 060 **
Consumption per person (1975): energy 1 062 kg coal equivalent, electricity (production) 501 kW h, newsprint 1.0* kg
Newspapers (1974): number 2; circulation 27 000, 83 per 1 000 people
Telephones (Dec 1976): 34 700, 93 per 1 000 people
Livestock (000, 1976) Cattle 47, sheep 35, pigs 40, chickens 1 030*
Petroleum refinery capacity (1975) 550 000 tonnes
Electrical capacity (1975) 44* megawatts
Hospital beds (1968) 3 281, 1 per 101 people
Roads (1976) 1 510* km = 940* mi, density 1.4* km per km²
Port Fort-de-France (1973): goods loaded 213 000 t, unloaded 569 000 t
Airport Lamentin (15 km from Fort-de-France)

Durable equipment (at end-year)	000	no per 1 000 people	no per
Radio sets (1974)	31	90	no per
Television sets (1974)	16	46	km of road
Motor vehicles (1972)	57	165	37*

Production, finance and trade

Gross domestic product 1971: Fr 1 873 mn = $ 337 mn = £ 139 mn
1976 est: Fr 4 000** mn = $ 840** mn = £ 460** mn
Main products *Agriculture* (000 t, 1976) Sweet potatoes 27*, pineapples 32*, bananas 196*, fish catch (1975) 3.4 *Other* (1975) Petroleum products 400**, electricity (mn kW h) 182, salt 163*, rum (000 hl, 1974) 220
Transport traffic (1974) *Sea* Goods loaded 397 000 t, unloaded 356 000 t
Tourism (1976) Number of visitors 120 500
Consumer price index (1970 = 100) 1976 184.3; growth 1970–76 10.7 %pa
Budget (1974) Balanced at Fr 291 mn = $ 60 mn = £ 26 mn
External trade (1976) Imports: Fr 1 827 mn = $ 382 mm = £ 212 mn
Exports: Fr 594 mn = $ 124 mn = £ 69 mn

Main imports (1975)	% of total	Main exports (1975)	% of total
Food	21	Bananas	48
Crude oil	12	Petroleum products	20
Machinery	11	Rum	13
Main sources (1975)		Main destinations (1975)	
France	65	France	69
Venezuela	6	Guadeloupe	23
Algeria	5	Italy	2

Mexico

United Mexican States
Estados Unidos Mexicanos

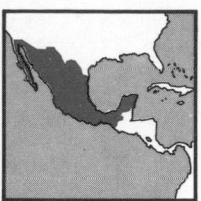

Location North America
With a coastline on the Pacific Ocean and to the east on the Gulf of Mexico, United States is to the north and Guatemala to the south; Cuba is 160 km to the east in the Caribbean Sea
Land Area 2 022 060 km² = 780 722 mi²
Usage (1975): agricultural 950 000* km² (47* %), of which, arable 262 200* km² (13* %), cropland 17 800* km² (1* %), pastures 670 000* km² (33* %); forests 716 000* km² (35* %)
Climate Tropical in the south, temperate in highlands
Weather at Mexico City, 2 309 m altitude
Temperature: hottest month May 12–26 °C, coldest Jan 6–19°C
Rainfall (av monthly): driest month Feb 5 mm, wettest July 170 mm
Time (hours behind GMT) Mexico City: 6
Baja California Sur, States of Sonora, Sinaloa and Nayarit: 7
Baja California Norte: 8 (summer time, 7)
Measures Metric system, also old Spanish measures including:
capacity 1 barril ≑ 76 litres = 16.72 gallons
weight (*mass*) 1 onza or marco = 28.775 grams = 1.015 ounces
1 libra = 16 onzas = 0.4604 kilogram = 1.015 pounds
1 arroba = 25 libras = 11.51 kilograms = 25.37 pounds
Monetary unit Mexican peso (Mex $) = 100 centavos
Rate of exchange (1976 av): free Mex $ 15.43 = $ 1, Mex $ 27.86 = £1

Summary

Political Republic, which became independent from Spain in 1821. Member of UN, OAS, Sela and Lafta
Economic A mixed economy, with a wide spread of exports, including agricultural products (especially coffee and cotton), metals, machinery, chemicals and textiles. Crude oil production increased rapidly in the 1970s, and oil became the largest single export in 1975. Growth has been high during the 1960s and early 1970s, and manufacturing then became the most important sector. This was helped by investment from the United States; inflows of capital are high

People, resources and equipment

Population 1960 36.05* mn, 1970 50.69* mn, 1976 62.33* mn
Growth: 1960–70 3.5* %pa, 1970–76 3.5* %pa
Density (1976): 31* people per km²
Vital statistics (rate per 1 000 people, 1976): births 34.6, deaths 6.5
Regions (population in 000, 1976; total of 62.33 mn)

Federal district	8 906	Hidalgo	1 409	Sinaloa	1 714
States		Jalisco	4 157	Sonora	1 414
Aguascalientes	430	México	6 245	Tabasco	1 054
Baja California	1 253	Michoacán	2 805	Tamaulipas	1 901
Campeche	337	Morelos	866	Tlaxcala	498
Chiapas	1 933	Nayarit	699	Veracruz	4 917
Chihuahua	2 000	Nuevo León	2 344	Yucatán	904
Coahuila	1 334	Oaxaca	2 337	Zacatecas	1 097
Colima	317	Puebla	3 055	*Territories*	
Durango	1 122	Querétaro	618	Baja California	
Guanajuato	2 811	San Luis		Sur	181
Guerrero	2 013	Potosí	1 527	Quintana Roo	131

Cities (population in 000, 1975)

Ciudad de México[a]		Chihuahua	346	Morelia	209
(capital)	11 340	Tampico	343	Reynosa	194
Guadalajara	1 963	Mexicali	331	Nuevo Laredo	193
Monterrey	1 638	San Luis Potosí	282	Durango	191
Ciudad Juárez	521	Cuernavaca	274	Matamoros	172
León	497	Veracruz Llave	266	Jalapa	172
Tijuana	496	Hermosillo	248	Poza Rica de	
Puebla		Culiacán	245	Hidalgo	161
de Zaragoza	482	Mérida	239	Mazatlán	154
Torreón	364	Aguascalientes	222	Ciudad Obregón	153
Acapulco	353	Saltillo	211	Querétaro	150

[a]Mexico City
Race (1970) Mestizo 55** %, American Indian 29** %, European 15** %
Language Spanish; also American Indian languages
Religion (1970) Roman Catholic 96 %, Protestant 2 %
Education (1975/76) Pupils: primary 11 570 900, secondary and vocational (1974/75) 2 388 134, teacher-training (1974/75) 89 868, higher (1974/75) 453 015. Teachers: primary 253 990, secondary and vocational (1974/75) 2 388 134, teacher-training (1974/75) 7 014, higher (1974/75) 34 869

Labour force (1975)	Number	% of total
Economic activity	000	
Agriculture	6 783	41
Mining and quarrying	241	1
Manufacturing	2 961	18
Electric power	71	—
Construction	756	5
Commerce	1 654	10
Transport and communication	490	3
Other	3 641	22
Total	16 597	100

Personnel Scientists and engineers (1970): 565 601
Physicians (1972): 38 000, 1 per 1 430 people
Standard of living
National income per person (1976): Mex $ 17 436 = $ 1 130 = £ 626
Consumption per person (1975): energy 1 221 kg coal equivalent, electricity (production, 1976) 742 kW h, newsprint 3.6 kg, steel 103 kg
Newspapers (1974): number 249
Telephones (Dec 1976): 3 308 800, 52 per 1 000 people
Livestock (000, 1976) Cattle 28 700, sheep 5 300, goats 8 800, pigs 12 100, horses 5 818*, mules 2 648*, asses 2 978*, chickens 163 000*
Mineral reserves (1975) Coal 12 000* mn tonnes
Crude oil 483 mn tonnes
Natural gas 338 000 mn cubic metres
Uranium 5 000 tonnes
Petroleum refinery capacity (1975) 38.7 mn tonnes
Electrical capacity (1975) 11 052 megawatts
Hospital beds (1974) 67 363, 1 per 863 people
Roads (1976) 193 390 km = 120 170 mi, density 0.096 km per km²
Railways (1974) 24 700* km = 15 350* mi, density 0.012* km per km²
Ships (registered, 1977) 311, total of 673 964 gross tons
Ports (goods traffic, 000 tonnes, 1975)

	International		Coastwise		Total
	loaded	unloaded	loaded	unloaded	
Coatzacoalcos[a]	7 290	1 968	12 506	202	21 966
Tampico[a]	1 795	1 706	1 705	4 365	9 571
Veracruz	635	1 535	92	2 753	5 015
Salina Cruz[a]	33	117	2 834	85	3 069
Guaymas	60	968	112	1 801	2 941

[a]Crude oil port

Airports (1973)	Passenger departures and arrivals 000	Cargo loaded and unloaded 000 tonnes
Mexico City	4 618	58.1
Acapulco	1 421	2.6
Guadalajara	840	5.8
Monterrey	388	3.3
Mérida	384	6.2

Also (1977) 30 other airports with scheduled flights

Durable equipment (at end-year)	000	no per 1 000 people	
Radio sets (1974)	17 514	301	no per
Television sets (1974)	4 885	84	km of road
Passenger cars (1976)	2 641	42	14
Commercial vehicles (1976)	1 033	16	5

Production

Gross domestic product
1976: Mex $ 1 220 800 mn = $ 79 119 mn = £ 43 819 mn
Growth in real terms: 1960–70 7.0 %pa, 1970–76 5.0 %pa
Structure of gross domestic product (1976)

By origin	Mex $ bn	% of gdp
Agriculture	110.3	9
Mining and quarrying	51.5	4
Manufacturing	298.6	24
Electricity, gas and water	14.9	1
Construction	74.9	6
Distribution and hotels	375.3	31
Transport and communication	36.5	3
Other	258.8	21
Total	1 220.8	100
By type of expenditure		
Government final consumption	158.7	13
Private final consumption	792.2	65
Stock investment	23.7	2
Gross fixed capital formation	266.9	22
Exports of goods and services	104.1	9
less Imports of goods and services	−124.8	−10
Total	1 220.8	100

Mexico

Production indices (1970 = 100)	1960	1970	1976	Growth %pa 1960–70	1970–76
Agricultural	66	100	115	4.3	2.4
Industrial	46.3	100.0	142.8	8.0	6.1

Main products (000 t)
Agriculture

	1960	1970	1976	1960–70	1970–76
Wheat	1 190	2 216	3 363	6.4	7.2
Maize	5 420	9 041	8 945	5.2	−0.2
Barley	184	238	460*	2.6	11.6*
Rice	328	402	400*	2.1	−0.1*
Sorghum	209	2 565	3 160*	28.5	3.5*
Potatoes	294	422	695*	3.7	8.7*
Sugar, raw value	1 628	2 365*	2 750*	3.8*	2.5*
Sugar, non-centrifugal	303	300*	219*	−0.1*	−5.1*
Dry beans	528	925	1 149	5.8	3.7
Soyabeans	5	280	319*	49.6	2.2*
Tomatoes	389	940	913*	9.2	−0.5*
Chillies and peppers	100**	265	380**	10.2**	6.2*
Grapes	71	144	245*	7.3	9.3*
Apples	68	163	255*	9.1	7.7*
Peaches	54	81	225*	4.1	18.5*
Oranges	766	1 555	2 300*	7.3	6.7*
Lemons	115	199	600*	5.6	20.2*
Avocadoes	119	168	293*	3.5	9.7*
Mangoes	165	209	425*	2.4	12.6*
Pineapples	181	308	300*	5.5	−0.4*
Bananas	614	1 136	1 340*	6.3	2.8*
Coffee	124	184	242	4.0	4.7
Cocoa	16	25	32*	4.6	4.2*
Tobacco	72	80	68	1.1	−2.7
Cotton	457	312	211	−3.7	−6.3
Milk	2 370*	3 053	3 960*	2.6*	4.4*
Eggs	138*	337	422*	9.3*	3.8*
Beef and veal	314	536	588*	5.5	1.5*
Pigmeat	126	341*	427*	10.5*	3.8*
Fish catch	198	387	499e	6.9	5.2f
Timber (000 m³)	11 000**	14 422	14 783*e	2.7**	0.5*f

Energy

	1960	1970	1976	1960–70	1970–76
Total energy (000 tce)	30 500*	56 000*	84 440e	6.3*	8.6*f
Coal	1 074	2 959	5 128e	10.7	11.6f
Crude oil	14 171	21 501	40 843	4.3	11.3
Petroleum products	13 430	22 053	30 780*e	5.1	6.9*f
Natural gas (mn m³)	6 260*	11 700ª	14 030*	6.5*	3.1*
Manufactured gas (mn m³)	na	192	200*e	na	0.8*f
Electricity (mn kW h)	10 813	28 608	46 238	10.2	8.3

Mining

	1960	1970	1976	1960–70	1970–76
Iron ore (Fe content)	521	2 612	3 645	17.5	5.7
Copper ore (Cu content)	60	61	87	0.2	6.1
Lead ore (Pb content)	191	177	166	−0.8	−1.0
Magnesite	—	8	40e	na	38.0f
Manganese ore (Mn content)	72	99	154e	3.2	9.4f
Zinc ore (Zn content)	262	266	259	0.2	−0.4
Silver	1.38	1.33	1.33*	−0.4	0.0*
Gold (000 kg)	9.14	6.17	4.47e	−3.9	−6.2f
Salt	994	4 063	3 803e	15.1	−1.3f
Sulphur	1 302	1 381	2 164e	0.6	9.4f

Manufacturing

	1960	1970	1976	1960–70	1970–76
Beer (000 hl)	8 521	14 321	19 684e	5.3	6.6f
Cigarettes (mn units)	36 453	40 633	44 342e	1.1	1.8f
Cotton yarn	104c	132	158e	3.5	3.7f
Cotton fabrics	99c	119	129	2.7d	1.4
Man-made fibres	32c	75	181e	12.9d	19.3f
Wood pulp	175	319	366e	6.2	2.8f
Paper	360**	897	1 184*	9.6**	5.7f
Sulphuric acid	249	1 235	1 805	17.4	6.5
Fertilisers, nitrogenousª	25*	362	581*	30.6*	8.2*
Fertilisers, phosphateª	33*	115	220*	13.3*	11.4*
Cement	3 089	7 267	12 477	8.9	9.4
Coke	426	1 300	2 088e	11.8	9.9f
Pig-iron	683	2 353	3 545	13.2	7.1
Crude steel	1 500	3 846	5 124	9.9	4.9
Radio sets (000 units)	676c	1 015	1 030e	6.0	0.3f
Television sets (000 units)	80	431	569e	18.3	5.7f
Passenger cars (000 units)b	25	136	232	18.5	9.3
Commercial vehicles (000 units)b	20	53	100e	10.2	13.5f

ªYears ending June 30th   bIncluding assembly   c1963   d1963–70   e1975   f1970–75

Transport traffic Passenger-kilometres (mn)	1960	1970	1976	Growth %pa 1960–70	1970–76
Railª	4 128	4 534	4 198b	0.9	−1.5c
Air	1 309	2 939	7 833	8.4	17.7
Cargo: tonne-kilometres (mn)					
Railª	14 004	22 863	32 542b	5.0	7.3c
Air	31.0	40.4	83.4	2.7	12.8
Sea: tonnes (mn)					
Goods loaded	4.20	9.70	14.28	8.7	6.6
Goods unloaded	1.11	3.38	7.17	11.8	13.4
Tourism					
Number of visitors (000)	690	2 250	3 218b	12.5	7.4c
Gross receipts ($ mn)	na	1 171	2 142b	na	12.8c

ªMain railways only   b1975   c1970–75

Finance and trade

Price and wage indices (1970 = 100)	1960	1970	1976	Growth %pa 1960–70	1970–76
Consumer prices	76.1	100.0	205.3	2.8	12.7
Wages (earnings)	50	100	251	7.2	16.6
Share prices	113.6	100.0	163.2	−1.3	8.5
Money stock (end-year, Mex $ bn)	17.3	53.8	158.0	12.0	19.7

Budget (1976)
Revenue: Mex $ 134 661 mn = $ 8 727 mn = £ 4 833 mn
Expenditure: Mex $ 191 954 mn = $ 12 440 mn = £ 6 890 mn

Balance of payments ($ mn)	1972	1973	1974	1975	1976
Balance of goods (fob)	−761	−1 237	−2 348	−2 817	−1 793
Balance of services	−219	−252	−642	−1 365	−1 677
Balance of transfers	+64	+74	+113	+140	+158
Current balance	−916	−1 415	−2 877	−4 042	−3 312
Long-term capital flow	+827	+1 820	+3 049	+4 667	+4 965
Reserves and debt (end-year, $ mn)					
International reserves	1 164	1 355	1 395	1 533	1 253
External public debt	4 745	7 209	10 183	13 456	17 517

External trade (1976)
Imports: Mex $ 90 989 mn = $ 5 897 mn = £ 3 266 mn
Exports: Mex $ 53 118 mn = $ 3 442 mn = £ 1 907 mn

Main imports (1975)	% of total	Main exports (1976)	% of total
Machinery	27	Crude oil	16
Chemicals	13	Coffee	10
Motor vehicles	10	Cotton	8
Cereals	9	Metals and ores	8
Iron and steel	7	Machinery	6
Petroleum products	5	Chemicals	5
Instruments	2	Shrimps	5
Paper	2	Textile yarns and fabrics	5
Main sources (1976)		*Main destinations (1976)*	
United States	63	United States	56
West Germany	7	Brazil	5
Japan	5	Japan	3
United Kingdom	4	West Germany	3
France	3	Israel	2
Canada	2	Venezuela	2

Special focus

Structure of gross domestic product (% of total)

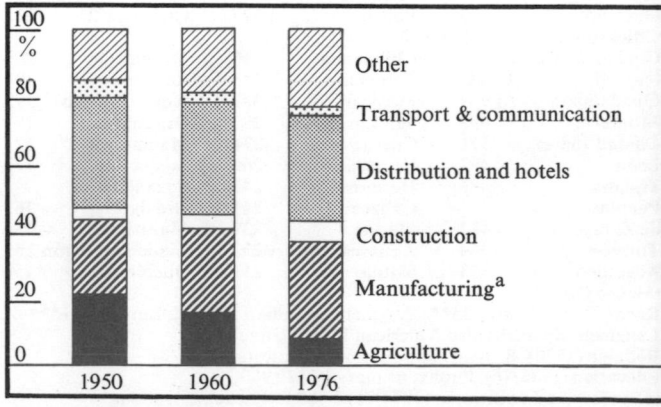

ª Including mining, electricity, gas and water

Montserrat

Colony of Montserrat

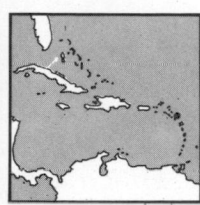

Location Eastern Caribbean Sea
Part of the Leeward Islands, with Nevis and
Antigua to the north, Guadeloupe to the south
Land Area 102 km² = 39 mi²
Climate Sub-tropical
Weather at Plymouth, 40 m altitude
Temperature: hottest month Sept 23–32 °C,
coldest Jan 21–28 °C
Rainfall (av monthly): driest month Feb 86 mm,
wettest Oct 196 mm

Time 4 hours behind GMT
Measures UK (imperial) system
Monetary unit East Caribbean dollar (EC $) = 100 cents
Rate of exchange (1976 av): free EC $ 2.61 = $ 1, EC $ 4.71 = £ 1

Summary

Political UK colony, which in January 1967 chose not to join other
UK Leeward Islands states in becoming an associated state. Member of
Caricom
Economic Cotton, limes and vegetables are the main products; there has
been some recovery in cotton production and market gardening and
livestock products are being further developed

People, resources and equipment

Population 1960 12 108, 1970 11 458, 1976 13 300*
Growth: 1960–70 −0.6 %pa, 1970–76 2.5* %pa
Density (1976): 130* people per km²
Vital statistics (rate per 1 000 people, 1974): births 24.3, deaths 10.5
Town (population in 000, 1976) Plymouth 3*
Race (1970) African and Mixed 96 %, European 3 %
Language English
Religion (1970) Christian 97 %
Education Pupils (1975/76) 3 220*, teachers (1973/74) 145
Labour force (1973) 4 651; in agriculture 3 048 (66 %),
construction 527 (11 %)
Personnel (1972) Physicians: 6, 1 per 2 000 people
Standard of living
National income per person (1976): EC $ 1 800** = $ 700** = £ 380**
Consumption per person: energy (1975) 716 kg coal equivalent,
electricity (production, 1976) 530* kW h
Telephones (Dec 1976): 1 930, 141 per 1 000 people
Livestock (000, 1976) Cattle 8*, sheep 3*, goats 3*, pigs 3*,
chickens 30*
Electrical capacity (1975) 4 megawatts
Hospital beds (1970) 86, 1 per 133 people
Roads (1975) 180* km = 110* mi, density 1.8* km per km²
Ships (registered, 1977) 3, total of 1 248 gross tons
Port (1975) Plymouth: goods loaded 950 t, unloaded 25 300 t
Airport Blackburne (14 km from Plymouth)
Radio There are 2 broadcasting stations
Television Programmes can be received from Antigua and St Kitts
Motor vehicles (1975) 1 300, 100 per 1 000 people, 7.2 per km of road

Production, finance and trade

Gross domestic product 1973: EC $ 17.3 mn = $ 8.8 mn = £ 3.6 mn
1976 est: EC $ 25** mn = $ 10** mn = £ 5** mn
Main products (1976) Cotton 29 t, tomatoes, limes, fish catch 120 t,
electricity 7*mn kW h
Transport traffic (1975) *Air* Passenger arrivals 16 030
Sea Goods loaded 951 t, unloaded 25 290 t
Tourism (1976) Number of visitors 11 211
Budget (total, 1976) Balanced at EC $ 8.5 mn = $ 3.2 mn = £ 1.8 mn
External trade (1976) Imports: EC $ 20.8 mn = $ 8.0 mn = £ 4.4mn
Exports: EC $ 1.12 mn = $ 0.43 mn = £ 0.24 mn
of which, domestic: EC $ 0.78 mn = $ 0.30 mn = £ 0.17 mn

Main imports	% of total	Main exports[a]	% of total
Food	25	Cotton	28
Electrical machinery	9	Potatoes	22
Petroleum products	7	Cattle	13
Chemicals	6	Lime juice	8
Motor vehicles	5*	Recapped tyres	5
		Tomatoes	5

Main sources		Main destinations[a]	
United Kingdom	31	Japan	28
United States	18	Trinidad and Tobago	14
Trinidad and Tobago	13	Guadeloupe and	
Canada	5	Martinique	14
West Germany	5	Dominica	11
		Antigua	9
[a]Domestic exports only		St Kitts-Nevis	6

Netherlands Antilles

De Nederlandse Antillen

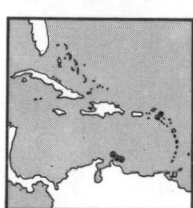

Location There are two main groups of islands,
about 800 km apart; the southern, about 100 km
from the Venezuelan coast, includes Curaçao,
Aruba and Bonaire, and the northern, part of the
Leeward Islands, includes Saba, St Eustatius
and the southern part of St Maarten island
Land Area 993 km² = 383 mi², of which,
Aruba 193 km², Bonaire 288 km²,
Curaçao 444 km²
Climate Sub-tropical

Weather at Willemstad, 23 m altitude
Temperature: hottest month Sept 26–32 °C, coldest Feb, Mar 23–29 °C
Rainfall (av monthly): driest month May 20 mm, wettest Nov 112 mm
Time 4 hours behind GMT
Measures Metric system
Monetary unit Netherlands Antillian guilder (NA Gld)
or florin (NA Fl) = 100 cents
Rate of exchange (1976 av):
par NA Gld 1.80 = $ 1, free NA Gld 3.251 = £ 1

Summary

Political Netherlands territory, with full control of internal affairs; there
are proposals for independence by 1980
Economic Main activity is refining of crude oil imported principally from
Venezuela. Tourism is important, and diversification of industry has
been planned

People, resources and equipment

Population 1960 192 000*, 1970 222 000*, 1976 241 000*
Growth: 1960–70 1.5* %pa, 1970–76 1.4* %pa
Density (1976): 243* people per km²
Vital statistics (rate per 1 000 people, 1973): births 20.0, deaths 4.8
Regions (population in 000, 1975; total of 240 000) *Southern* Curaçao 156,
Aruba 62, Bonaire 9, *Northern* St Maarten 10, St Eustatius 1, Saba 1
Town (population in 000, 1976) Willemstad (in Curaçao, capital) 65*
Race Dutch, Mixed, American Indian, Portuguese
Language Dutch; also used are Spanish, English and Papiamento
Religion (1976) Roman Catholic 80** %, Protestant 20** %
Education (1973/74) Pupils 50 500*, teachers 2 140*
Labour force (1972) 73 270; in agriculture 588 (1 %), manufacturing
10 549 (14 %), construction 5 705 (8 %)
Personnel (1968) Physicians: 120, 1 per 1 783 people
Standard of living National income per person (1976):
NA Gld 3 000** = $ 1 700** = £ 900**
Consumption per person (1975): energy 12 231 kg coal equivalent,
electricity (production) 5 800* kW h, newsprint 2.1 kg
Newspapers (1974): number 5; circulation 47 000, 197 per 1 000 people
Telephones (Dec 1976): 48 000, 200 per 1 000 people
Livestock (000, 1976): Cattle 8*, goats 21*, sheep 8*, chickens 100*
Petroleum refinery capacity (1975) 42.4 mn tonnes
Electrical capacity (1975) 290* megawatts
Hospital beds (1968) 1 969, 1 per 109 people
Roads (surfaced, 1975) 1 150* km = 715* mi, density 1.2* km per km²
Ports (goods traffic in crude oil ports, 000 tonnes, 1970) Aruba loaded
24 048, unloaded 27 192; Curaçao loaded 18 155, unloaded 22 101
Airports Plesman (16 km from Willemstad), Princess Beatrix (Aruba),
Bonaire, Juliana (St Maarten), St Eustatius, Saba

Durable equipment (at end-year)	000	no per 1 000 people	
Radio sets (1975)	132	545	no per
Television sets (1974)	34	143	km of road
Passenger cars (1973)	39*	167*	34*
Commercial vehicles (1973)	8*	34*	7*

Netherlands Antilles

Production, finance and trade

Gross domestic product 1975: NA Gld 700* mn = $ 390* mn = £ 170* mn
1976 est: NA Gld 900** mn = $ 500** mn = £ 280** mn
Main products (000 t) *Agriculture* (1976) Sorghum 7*, milk 4*, eggs 0.5*, meat 2*, cattle hides 0.2* *Other* (1975) Petroleum products 22 110*, electricity (mn kW h) 1 400*, phosphate rock 82, beer (000 hl, 1973) 58, sulphuric acid (1968) 132
Transport traffic *Air* (1975) Passenger departures 484 000, arrivals 487 000
Sea (1974) Goods loaded 37.3* mn t, unloaded 45.9* mn t
Tourism Number of visitors (1972) 553 000 (including 270 000 cruise tourists), gross receipts (1974) $ 150 mn
Consumer price index (1971 = 100) 1976 163.5; growth 1971–76 10.3 %pa
Budget (1972) Revenue: NA Gld 121.8 mn = $ 68 mn = £ 27 mn
Expenditure: NA Gld 121.2 mn = $ 67 mn = £ 27 mn
External trade (1976)
Imports: NA Gld 6 589 mn = $ 3 661 mn = £ 2 027 mn
Exports: NA Gld 4 535 mn = $ 2 519 mn = £ 1 395 mn

Main imports (1975)	% of total	Main exports (1975)	% of total
Crude oil	70	Petroleum products	91
Petroleum products	15	Crude oil	5
Food	3	Chemicals	2
Main sources (1975)		*Main destinations* (1975)	
Venezuela	57	United States	62
Saudi Arabia	17	Netherlands	5
United States	6	Ecuador	3
Nigeria	6	Mexico	3

Nicaragua

Republic of Nicaragua
República de Nicaragua

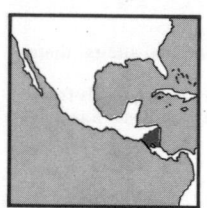

Location Central America
The Caribbean Sea is to the east and Pacific Ocean to the west, with Honduras to the north and Costa Rica to the south
Land Area 130 000 km² = 50 000 mi²
Climate Tropical, temperate in highlands
Weather at Managua, 55 m altitude
Temperature: hottest month May 23–34 °C, coldest Jan 21–31 °C
Rainfall: driest month June, wettest Jan

Time 6 hours behind GMT
Measures Metric system; also local measures, including:
length 1 vara = 84.1 centimetres = 33.1 inches
weight (mass) 1 arroba = 25 libras = 11.5 kilograms = 25.35 pounds
1 quintal = 4 arrobas = 46 kilograms = 101.4 pounds
Monetary unit Córdoba (C) = 100 centavos
Rate of exchange (1976 av): par C 7.00 = $ 1, free C 12.64 = £ 1
The Central American peso of the CACM (CA $), equal to the US dollar, is also used

Summary

Political One-party republic, which became independent in 1838 after forming part of the Central American Federation. Member of UN, OAS Sela, Odeca and CACM
Economic A broadly-based agricultural economy, with cotton and coffee the main exports. There is some mining, and natural gas was discovered offshore in 1974; industry is light.

People, resources and equipment

Population 1960 1.41 mn, 1970 1.83* mn, 1976 2.23* mn
Growth: 1960–70 2.6* %pa, 1970–76 3.3* %pa
Density (1976): 17* people per km²
Vital statistics (rate per 1 000 people, 1973): births 39.4*, deaths 6.6*
Cities (population in 000, 1976)

Managua (capital)	500**	Chinandega	54ᵃ	Estelí	27*
León	91ᵃ	Granada	51ᵃ	Diriamba	26*
Matagalpa	77ᵃ	Masaya	50ᵃ	Rivas	21*

ᵃ 1970
Race (1974) Mestizo/Mulatto/Zambo 69 %, European 19 %, African 9 %
Language Spanish; there is also some use of English
Religion (1964) Roman Catholic 83 %, Protestant 3 %
Education (1974) Pupils 418 994, teachers 11 367

Labour force 1971: 505 445; in agriculture 237 327 (47 %), manufacturing 62 509 (12 %), construction 20 252 (4 %)
1976: 714 000*
Personnel (1975) Physicians: 1 357*, 1 per 1 600* people
Standard of living
National income per person (1976): C 5 328 = $ 761 = £ 422
Consumption per person (1975): energy 479 kg coal equivalent, electricity (production) 387* kW h, newsprint 1.8 kg, steel 18 kg
Newspapers (1974): number 3; circulation 53 000, 26 per 1 000 people
Telephones (Dec 1976): 55 300, 25 per 1 000 people
Livestock (000, 1976) Cattle 2 600*, pigs 670*, horses 175*
Petroleum refinery capacity (1975) 750 000 tonnes
Electrical capacity (1975) 262* megawatts
Hospital beds (1975) 4 675, 1 per 461 people
Roads (1975) 12 500* km = 7 800* mi, density 0.096* km per km²
Railways (1975) 373 km = 232 mi, density 0.003 km per km²
Inland waterways Lake Nicaragua and rivers
Ships (registered, 1977) 30, total of 34 588 gross tons
Ports *Pacific* Corinto, Puerto Somoza, San Juan del Sur, Puerto Potosi
Caribbean Puerto Isabel, El Bluff (for Bluefields), Puerto Cabezas
Airports Las Mercedes (Managua), Puerto Cabezas, Bluefields, Siuna, Bonanza, Waspán, Is del Maíz (Corn Is)

Durable equipment (at end-year)	000	no per 1 000 people	
Radio sets (1974)	126	60	no per
Television sets (1974)	75	36	km of road
Passenger cars (1973)	32*	16*	2.6*
Commercial vehicles (1973)	20*	10*	1.6*

Production

Gross domestic product 1976: C 12 894 mn = $ 1 842 mn = £ 1 020 mn
Growth in real terms: 1960–70 6.9 %pa, 1970–76 5.6 %pa
Structure of gross domestic product (1975) *By origin* Agriculture 22 %, manufacturing 22 %, transport and communication 5 %, other 51 %

Production indices (1970 = 100)	1960	1970	1976	Growth %pa 1960–70	1970–76
Agricultural	60	100	127	5.2	4.1
Manufacturing	32	100	138ᵉ	12.1	11.3ᶠ
Main products (000 t)					
Agriculture					
Rice	34	81	80	9.1	−0.2
Maize	119	236	201	7.1	−2.6
Sugar, raw value	66	141	246	7.9	9.7
Bananas	220**	217	314*	−0.1**	6.3*
Cottonseed	59	111	199	6.5	10.2
Coffee	23	42	53	6.2	3.9
Tobacco	0.9	1.8	2.5*	7.2	5.6*
Cotton	33	68	113	7.5	8.8
Milk	140**	201	225*	3.7**	1.9*
Beef and veal	25*	61*	65*	9.3*	1.1*
Fish catch	3*	10	18	12.6*	10.5
Timber (000 m³)	2 270**	2 245	3 015ᶜ	−0.1**	6.1ᵈ
Other					
Petroleum products	—	438	626ᶜ	na	7.4ᵈ
Electricity (mn kW h)	183	615	835*ᶜ	12.9	6.3*ᵈ
Gold (000 kg)	6.4ᵃ	3.5	2.2ᶜ	−8.1ᵇ	−9.3ᵈ
Beer (000 hl)	41	137	144ᵍ	12.8	5.1ʰ
Cigarettes (mn units)	819	1 260	1 588ᶜ	4.4	4.7ᵈ
Cotton fabrics (mn m)	10	16	18ᵍ	4.8	12.0ʰ
Cement	98	98	193ᶜ	11.8	14.5ᵈ

ᵃ1963 ᵇ1963–70 ᶜ1975 ᵈ1970–75 ᵉ1973 ᶠ1970–73 ᵍ1971 ʰ1970–71
Transport traffic *Passenger-kilometres* Rail (1972) 28 mn, air (1976) 86 mn
Cargo: tonne-kilometres Rail (1972) 14 mn, air (1976) 2.1 mn
Sea (1974) Goods loaded 0.70 mn t, unloaded 1.30 mn t
Tourism Number of visitors (1976) 206 980, gross receipts (1975) $ 25 mn

Finance and trade

Money stock (end-year, C mn) 1976 1 727; growth 1970–76 18.3 %pa
Budget (1976) Revenue: C 1 514 mn = $ 216 mn = £ 120 mn
Expenditure: C 1 825 mn = $ 261 mn = £ 144 mn

Balance of payments ($ mn)	1972	1973	1974	1975	1976
Balance of goods (fob)	+44	−49	−161	−107	+44
Balance of services	−29	−74	−112	−94	−120
Balance of transfers	+7	+57	+15	+17	+10
Current balance	+22	−66	−257	−185	−66
Long-term capital flow	+50	+112	+173	+152	+65
Reserves and debt (end-year, $ mn)					
International reserves	80	117	105	122	147
External public debt	314	481	641	792	936

Nicaragua

External trade (1976) Imports: C 3 739 mn = $ 534 mn = £ 296 mn
Exports: C 3 807mn = $ 544 mn = £ 301 mn

Main imports	% of total	Main exports	% of total
Chemicals	20	Cotton	24
Machinery, non-electric	11	Coffee	22
Crude oil	11	Sugar	10
Transport equipment	9	Chemicals	8
Electrical machinery	7	Meat	7
Textiles	7	Fish	4
Main sources		Main destinations	
United States	31	United States	31
Venezuela	11	Japan	13
Costa Rica	9	West Germany	10
Japan	8	Costa Rica	8
Guatemala	8	Guatemala	6
El Salvador	7	El Salvador	5

Special focus

Main agricultural exports

% of total exports

	1960	1970	1971	1972	1973	1974	1975	1976
Cotton	23	19	22	25	23	36	25	24
Coffee	31	18	16	13	16	12	13	22
Sugar	5	6	6	6	5	3	11	10

Panama

Republic of Panama
República de Panamá

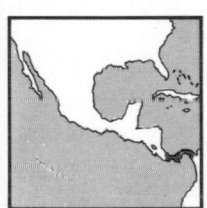

Location Central America
At the southern end of the Central American isthmus, the land runs from west to east, with the Caribbean Sea to the north and Pacific Ocean to the south, Costa Rica is to the north-west and Colombia to the east and south
Land Area 75 650 km² = 29 210 mi²
Climate Tropical
Weather at Balboa Heights, 36 m altitude
Temperature: hottest month April 23–32 °C, coldest Jan 22–31 °C
Rainfall (av monthly): driest month Feb 10 mm, wettest Nov 259 mm
Time 5 hours behind GMT
Measures Metric system; also
capacity 1 azumbre = 2.02 litres = 1.78 UK quarts
1 celemín = 4.625 litres = 1.02 UK gallons
Monetary unit Balboa (Ba) = 100 centésimos
Rate of exchange (1976 av): par Ba 1 = $ 1, free Ba 1.806 = £ 1

Summary

Political Republic with military government, which became independent in 1903, formerly being part of Colombia; the area of the Panama Canal Zone was ceded to the United States after independence, and is due to be returned to Panamanian control in the year 2000. Member of UN, OAS and Sela
Economic Exports, at about $ 220 mn in 1976, are exceeded in value by net external income from services ($ 320 mn in 1976), mainly receipts from the Canal Zone. The largest export in 1976 was petroleum products refined from imported oil; agriculture remains the main sector, with bananas and sugar significant exports. The Colón Free Zone is important and is being expanded. Copper deposits are being developed

People, resources and equipment

Population 1960 1.06* mn, 1970 1.43* mn, 1976 1.72* mn
Growth: 1960–70 3.0* %pa, 1970–76 3.1* %pa
Density (1976): 23* people per km²
Vital statistics (rate per 1 000 people, 1976): births 32.2, deaths 5.2
Cities (population in 000, 1976)
Panama City (capital) 456*, Colón 95*, David 40*, Santiago 29*
Race (1970) Mulatto or Mestizo 72** %, African 14** %, European 12** %
Language Spanish
Religion (including the Panama Canal Zone, 1968) Roman Catholic 92 %
Education Pupils (1975) 488 000*, teachers (1973) 18 115

Labour force 1970: 488 335; in agriculture 187 947 (38 %), manufacturing 38 847 (8 %), construction 27 946 (6 %)
1976: 580 000*
Personnel (1975) Physicians: 1 251, 1 per 1 333 people
Standard of living
National income per person (1976): Ba 1 055 = $ 1 055 = £ 584
Consumption per person (1975): energy 865 kg coal equivalent, electricity (production, 1976) 762* kW h, newsprint 3.3 kg, steel 33 kg
Newspapers (1973): number 9; circulation 145 000, 92 per 1 000 people
Telephones (Dec 1976): 140 000*, 80* per 1 000 people
Livestock (000, 1976) Cattle 1 361*, pigs 172*, horses 164*, chickens 3 776*
Petroleum refinery capacity (1975) 5.0 mn tonnes
Electrical capacity (1975) 275* megawatts
Hospital beds (1975) 5 880, 1 per 284 people
Roads (1975) 7 090* km = 4 400* mi, density 0.094* km per km²
Railways (1975) 350* km = 220* mi, density 0.005* km per km²
Inland waterway See Panama Canal Zone
Ships (registered, 1977) 3 267, total of 19 458 419 gross tons. Mainly owned by citizens of other countries
Ports (goods traffic, 000 tonnes, 1975) Bahía de las Minas (crude oil port), loaded 1 064, unloaded 4 268. The Panama Canal Zone ports of Cristóbal (for Colón) and Balboa (for Panama City) are also used
Airports (passenger departures and arrivals, 000, 1976) Tocumen (Panama City) 806; also Colón, Enrique Malek (David), Bocas del Toro, Changuinola

Durable equipment (at end-year)	000	no per 1 000 people	no per km of road
Radio sets (1975)	265	159	
Television sets (1975)	185	111	
Passenger cars (1974)	70	42	9.8
Commercial vehicles (1974)	21	13	3.0

Production

Gross domestic product 1976: Ba 2 028 mn = $ 2 028 mn = £ 1 123 mn
Growth in real terms: 1960–70 8.0 %pa, 1970–76 4.1 %pa
Structure of gross domestic product (1974) *By origin* Agriculture 16 %, manufacturing 13 %, electricity, gas and water 2 %, construction 8 %, transport and communication 6 %, other 55 %

Production indices (1970 = 100)	1960	1970	1976	Growth %pa 1960–70	Growth %pa 1970–76
Agricultural	66	100	116	4.2	2.5
Manufacturing	32	100	137[a]	12.1	6.5[b]
Main products (000 t)					
Agriculture					
Rice	96	131	144	3.2	1.6
Maize	63	56	55*	−1.2	−0.3*
Sugar, raw value	25	76	129	11.8	9.2
Tomatoes	10*	29	31*	11.2*	1.1*
Oranges	40	42	62	0.5	6.7
Bananas	536*	947	990*	5.9*	0.7*
Coffee	4.1	4.4	5.0*	0.7	2.2*
Milk	45*	73	74*	5.0*	0.2*
Beef and veal	20	34	46*	5.4	5.2*
Fish catch	11	62	80[a]	18.9	5.2[b]
Timber (000 m³)	1 744	1 439*	1 520*[a]	−1.9*	1.1*[b]
Other					
Petroleum products	—	3 402	3 770*[a]	na	2.1*[b]
Manufactured gas (mn m³)	17	19	2	1.1	−27.9
Electricity (mn kW h)	250*	956	1 310*	14.4*	5.4*
Beer (000 hl)	209	365	284[c]	5.7	−11.8[d]
Cigarettes (mn units)	658	996	1 045[a]	4.2	1.0[b]
Tyres (000 units)	29[g]	35	44[e]	2.7[h]	5.9[f]
Cement	109	181	277[a]	5.2	8.9[b]

[a] 1975 [b] 1970–75 [c] 1972 [d] 1970–72 [e] 1974 [f] 1970–74 [g] 1963 [h] 1963–70

Transport traffic *Rail* (1974) Passengers 0.52 mn
Air (1976) Passenger-km 437 mn, cargo 5.0 mn t-km
Sea (1975) Goods loaded 1.7 mn t, unloaded 4.7 mn t
Tourism (1976) Number of visitors 279 400, gross receipts $ 124 mn

Finance and trade

Price indices (1970 = 100)	1960	1970	1976	Growth %pa 1960–70	Growth %pa 1970–76
Consumer prices	88.1	100.0	144.8	1.3	6.4
Wholesale prices	na	100.0	202.2	na	12.5
Money stock[a]					
(end-year, Ba mn)	42	113	223[b]	10.3	18.6[c]

[a] Deposit money only [b] 1974 [c] 1970–74

Budget (1976) Revenue: Ba 282 mn = $ 282 mn = £ 156 mn
Expenditure: Ba 446 mn = $ 446 mn = £ 247 mn

Panama

Balance of payments ($ mn)	1972	1973	1974	1975	1976
Balance of goods (fob)	−263	−296	−510	−492	−517
Balance of services	+159	+189	+289	+328	+326
Balance of transfers	+5	−4	−4	−5	−4
Current balance	*−98*	*−111*	*−224*	*−168*	*−195*
Long-term capital flow	+128	+148	+113	+165	+718

Reserves and debt (end-year, $ mn)

International reserves	44	43	39	na	na
External public debt	467	698	746	1 086	1 408

External trade (excluding free zone of Colón, 1976)
Imports: Ba 838 mn = $ 838 mn = £ 464 mn
Exports: Ba 226 mn = $ 226 mn = £ 125 mn

Main imports (1975)	% of total	*Main exports* (1976)	% of total
Crude oil	37	Petroleum products	28
Machinery	14	Bananas	26
Chemicals	9	Shrimps	14
Food	6	Sugar	11
Motor vehicles	4*		
Textile yarns and fabrics	4*		

Main sources (1976)		*Main destinations* (1976)	
United States	32	United States	49
Ecuador	18	Panama Canal Zone	12
Venezuela	8	West Germany	9
Japan	6	Italy	7

Trade in the Colón Free Zone (1974)
Imports: Ba 462 mn = $ 462 mn = £ 198 mn
Re-exports: Ba 478 mn = $ 478 mn = £ 204 mn

Main products traded	% of total	
	Imports	Re-exports
Chemicals	21	29
Telecommunications equipment	14	11
Instruments	14	13
Clothing	8	6
Textile yarns and fabrics	7	7

Main sources	% of total	*Main destinations*	% of total
Japan	27	Panama	13
United States	21	Brazil	12
Hongkong	7	Netherlands Antilles	9
Taiwan	7	Colombia	7
Brazil	6	Mexico	6
Switzerland	5	Ecuador	5

Special focus

External trade: main products

% of imports	1970	1971	1972	1973	1974	1975	1976
Crude oil	17	16	15	18	33	37	28
% of exports							
Petroleum products	19	22	18	18	41	45	28
Bananas	56	54	53	46	23	21	26
Shrimps	9	10	12	12	.7	7	14

Panama Canal Zone

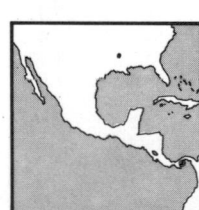

Location Central America
The strip of land 16 km wide which forms the zone is in the centre of Panama, running between the ports of Balboa on the Pacific Ocean and Cristóbal on the Caribbean Sea
Land Area 1 676 km² = 647 mi², of which, water 712 km² = 275 mi²
Climate Tropical
Weather at Balboa Heights: see Panama
Time 5 hours behind GMT
Measures US system
Monetary unit US dollar ($) = 100 cents
Rate of exchange (1976 av): free $ 1.806 = £ 1

Summary

Political United States territory, ceded by Panama in 1903; the canal was opened in 1914. US jurisdiction is due to finish in the year 2000, when control is to be returned to Panama.

Economic The canal is operated by the Panama Canal Company, which pays the costs of the Canal Zone Government from toll receipts

People, resources and equipment

Population 1960 42 000*, 1970 44 000*, 1976 44 000*
Growth: 1960–70 0.5* %pa, 1970–74 0.0* %pa
Density (1976): 26* people per km²
Vital statistics (rate per 1 000 people, 1975): births 13.2, deaths 1.9
Towns Balboa (capital), Cristóbal
Race (1976) United States citizens 89 %
Language English
Education (1976) Pupils 10 787, teachers 582
Labour force (1973) 18 500
Government and Panama Canal Company (1976) 13 984
Personnel (1975) Physicians: 151, 1 per 290 people
Standard of living
National income per person (1976): $ 5 000*** = £ 2 800***
Consumption per person (1975): energy 14 150 kg coal equivalent, electricity (production) 15 700 kW h
Telephones (Dec 1976): 15 000**, 340** per 1 000 people
Electrical capacity (1975) 178 megawatts, of which, nuclear 10 megawatts
Hospital beds (government establishments only, 1975) 590, 1 per 76 people
Roads (public, 1975) 232 km = 144 mi, density 0.14 km per km²
Railways (1975) 76* km = 47* mi, density 0.045* km per km²
Inland waterway Panama Canal 82 km = 51 mi
Ports (goods traffic, 000 tonnes, 1975) Cristóbal: loaded 2, unloaded 258; Balboa: loaded 9, unloaded 150
Airport Tocumen (Panama City) is used
Durable equipment (Dec 1974)
Passenger cars: 17 600, 400 per 1 000 people, 76 per km of road
Commercial vehicles: 500, 11 per 1 000 people, 2 per km of road

Production, finance and trade

Gross domestic product 1972: $ 190* mn = £ 76* mn
1976 est: $ 300*** mn = £ 170*** mn
Main product (1974/75) Electricity 692 mn kW h, of which, nuclear 25 mn kW h
Transport (1976) *Goods traffic through the Canal* (ocean-going commercial): Atlantic to Pacific 67 mn t, Pacific to Atlantic 54 mn t; toll receipts $ 135 mn
Number of vessels passing through (1975/76): 12 157, of which, Liberian 1 777 (15 %), United Kingdom 1 285 (11 %), United States 1 064 (9 %), Japanese 1 008 (8 %), Panamanian 930 (8 %), Greek 885 (7 %), Norwegian 685 (6 %)
Tourism (1976) Number of visitors 287 360
Budgets (1976)
Canal Zone Government Revenue: $ 74 mn = £ 41 mn
Expenditure: $ 72 mn = £ 40 mn
Panama Canal Company Revenue: $ 250 mn = £ 138 mn
Expenditure: $ 257 mn = £ 143 mn
External trade (1976) Imports: $ 150***mn = £ 80*** mn
Exports: $ 20*** mn = £ 11*** mn
Main import Petroleum products

Main sources	% of total	*Main destinations*	% of total
Venezuela	41***	Panama	40***
Panama	18***	United States	15***
United Kingdom	9***		
Netherlands Antilles	6***		

Paraguay

Republic of Paraguay
República del Paraguay

Location Central South America
Brazil is to the north and east, Argentina to the south, and Bolivia to the north and west. Land-locked but with river access to the Atlantic (River Plate)
Land Area 406 752 km² = 157 048 mi²
Climate Sub-tropical
Weather at Asunción, 139 m altitude
Temperature: hottest month January 22–35 ° C, coldest June 12–22 °C
Rainfall (av monthly): driest month Aug 38 mm, wettest Dec 157 mm
Time 4 hours behind GMT (summer time, 3 hours behind)

Paraguay

Measures Metric system; also local measures, including:
capacity 1 fanega = 12 almudes = 288 litres = 7.919 UK bushels
1 barril = 32 frascos = 96.93 litres = 21.32 UK gallons
weight (mass) 1 libra = 16 onzas = 0.459 kilogram = 1.012 pounds
1 arroba = 25 libras = 11.475 kilograms = 25.30 pounds
1 quintal = 4 arrobas = 45.90 kilograms = 101.2 pounds
1 tonelada = 20 quintales = 0.918 tonne = 0.904 UK (long) ton
Monetary unit Guaraní (G) = 100 céntimos
Rate of exchange (1976 av): par G 126 = $ 1, free G 227.6 = £ 1

Summary

Political Republic, which became independent from Spain in 1811;
during 1865–70 there was a devastating war with Argentina, Brazil
and Uruguay, and during 1932–35 a war against Bolivia. Member of
UN, OAS, Scla and Lafta
Economic An agricultural economy, with cattle ranching especially
important and a wide range of other products including cotton, oilseeds
and tobacco; there is some light industry. Hydro-electricity is important
and capacity is being further increased

People, resources and equipment

Population 1960 1.75*mn, 1970 2.30*mn, 1976 2.72*mn
Growth: 1960–70 2.8 %pa, 1970–76 2.8 %pa
Density (1976): 7* people per km²
Vital statistics (rate per 1 000 people, 1970–75): births 39.8*, deaths 8.9*
Cities (population in 000, 1972)

Asunción (capital)	473	Pedro Juan Caballero	49	Luque	40
Caaguazú	59	Concepción	45	Villarrica	34
Coronel Oviedo	55	Encarnación	41		

Race (1970) Mestizo (Spanish/Guaraní Indian) 75** %, European
20** %, American Indian 3** %, African 1** %
Language Spanish and Guaraní
Religion (1976) Roman Catholic 90* %
Education (1975) Pupils 544 808, teachers 24 340
Labour force (1972) 752 456; in agriculture 372 239 (49 %)
Personnel (1975) Physicians 2 230*, 1 per 1 190* people
Standard of living
National income per person (1976): G 72 280 = $ 574 = £ 318
Consumption per person (1975): energy 153 kg coal equivalent,
electricity (production) 193* kW h, newsprint 1.2 kg, steel 7 kg
Newspapers: number (1973) 11; circulation (5 newspapers only, 1972)
99 000, 41 per 1 000 people
Telephones (Dec 1976): 41 640, 15 per 1 000 people
Livestock (000, 1976) Cattle 5 049*, sheep 355*, pigs 800*,
horses 315*, chickens 8 520*
Petroleum refinery capacity (1975) 500 000 tonnes
Electrical capacity (Asunción only, 1975) 166* megawatts, of which,
hydro 97* megawatts
Hospital beds (1975) 3 816, 1 per 694 people
Roads (1973) 15 956 km = 9 915 mi, density 0.039 km per km²
Railways (1976) 498 km = 309 mi, density 0.001 km per km²
Inland waterways (including Paraguay-Paraná river system, 1976)
3 000* km = 2 000* mi
Ships (registered, 1977) 26, total of 21 930 gross tons
Port Asunción
Airport Presidente General Stroessner (15 km from Asunción)

Durable equipment	000	no per		
(Dec 1975)		1 000 people		
Radio sets	180	68	no per	
Television sets	54	20	km of road	
Passenger cars	22	8	1.4	
Commercial vehicles	20	7	1.2	

Production

Gross domestic product 1976: G 214 100 mn = $ 1 700 mn = £ 941 mn
Growth in real terms: 1960–70 4.5 %pa, 1970–75 6.1 %pa
Structure of gross domestic product (1975) *By origin* Agriculture 37 %,
manufacturing 16 %, construction 4 %, transport and communication 4 %,
other 39 %

Production indices	1960	1970	1976	Growth %pa	
(1970 = 100)				1960–70	1970–76
Agricultural	68	100	118	3.9	2.9
Industrial	67ᵃ	100	117ᶜ	5.9ᵇ	3.2ᵈ

Main products (000 t)	1960	1970	1976	Growth %pa	
Agriculture				1960–70	1970–76
Maize	143	220	351	4.4	8.1
Sweet potatoes	72	99	99*	3.2	0.0*
Cassava	979	1 782	1 450*	6.2	−3.4*
Sugar, raw value	30	52	58*	5.7	1.8*
Tomatoes	30**	45	53*	4.1**	2.7*
Soyabeans	2	40	253	34.9	36.0
Tung oil	5*	11	13ᶜ	8.2*	3.4ᵈ
Palm oil	4.0**	7.0	5.0*	5.8**	−5.5*
Oranges	120*	195*	128*	5.0*	−6.8*
Bananas	141	249	260*	5.9	0.7*
Coffee	1.5	4.3	9.0*	11.1	13.1*
Tobacco	10	18	39	6.6	13.8
Cotton	4	12	34	11.6	19.0
Milk	80*	88	122*	1.0*	5.6*
Beef and veal	107*	118*	90*	1.0*	−4.4*
Pigmeat	20*	42*	52*	7.7*	3.6*
Timber (000 m³)	1 734*	3 397	4 295ᶜ	7.0*	4.8ᵈ
Other					
Petroleum products	—	228	200*ᶜ	na	−2.6*ᵈ
Electricity (mn kW h)	96	218	510*ᶜ	8.5	18.5*ᵈ
of which, hydro (mn kW h)	—	154	450*ᶜ	na	24.0*ᵈ
Beer (000 hl)	51	175	253ᵉ	13.1	9.7ᶠ
Cigarettes (mn units)	528ᵃ	458	834ᶜ	−2.0ᵇ	12.7ᵈ
Cotton yarn	8*	12	24ᵉ	4.1*	20.1ᶠ
Cotton fabrics (mn m)	14	20	16ᶜ	3.6	−4.4ᵈ
Ethyl alcohol (000 hl)	34ᵃ	39	36ᶜ	2.0ᵇ	−1.6ᵈ
Cement	14	63	138ᶜ	16.2	17.0ᵈ

ᵃ1963 ᵇ1963–70 ᶜ1975 ᵈ1970–75 ᵉ1974 ᶠ1970–74
Transport traffic *Passenger-kilometres* Air (1976) 83 mn, rail (1973) 26 mn
Cargo: tonne-kilometres Air (1976) 1.0 mn, rail (1973) 30 mn
Tourism Number of visitors (1973) 95 100, gross receipts (1975) $ 12 mn

Finance and trade

Price indices	1960	1970	1976	Growth %pa	
(1970 = 100)				1960–70	1970–76
Consumer prices	72.7	100.0	180.5	3.2	10.3
Wholesale prices	75	100	280	2.9	18.7

Money stock					
(end-year, G mn)	2 674*	7 308	21 590	10.6*	19.8

Budget (1976) Revenue: G 19 384 mn = $ 154 mn = £ 85 mn
Expenditure: G 18 465 mn = $147 mn = £ 81 mn,
of which, education 16 %, defence 20 %, health 4 %

Balance of payments ($ mn)	1972	1973	1974	1975	1976
Balance of goods (fob)	+7	+1	−25	−51	−49
Balance of services	−19	−22	−32	−53	−50
Balance of transfers	+6	+6	+4	+14	+4
Current balance	−5	−16	−53	−89	−95
Long-term capital flow	+20	+30	+52	+118	+163
Reserves and debt (end-year, $ mn)					
International reserves	31	57	87	115	158
External public debt	179	216	307	369	386

External trade (1976) Imports: G 27 589 mn = $ 219 mn = £ 121 mn
Exports: G 22 423 mn = $ 178 mn = £ 99 mn

Main imports (1975)	% of total	Main exports (1976)	% of total
Crude oil and products	21	Cotton	19
Machinery	20	Meat	12
Motor vehicles	12	Vegetable oils	9
Iron and steel	8	Tobacco	8
Tobacco products	7	Timber	7
Chemicals	5		
Main sources (1976)		*Main destinations* (1976)	
Argentina	21	Netherlands	15
Brazil	17	United States	12
Algeria	13	West Germany	11
United States	10	Argentina	10
West Germany	8	Switzerland	10
United Kingdom	8	United Kingdom	6

Special focus

Beef prices (New York wholesale price)

	$ per tonne	% change over previous year
1970	911	+6
1971	937	+3
1972	1 097	+17
1973	1 404	+28
1974	1 173	−16
1975	992	−15
1976	1 155	+16

Peru

Republic of Peru
República del Perú

Location West of South America
With a coastline on the Pacific Ocean, Ecuador
and Colombia are to the north, Brazil and
Bolivia to the east and Chile to the south
Land Area 1 285 216 km² = 496 225 mi²
Climate Temperate on the coast, tropical in
jungles, cool in highlands
Weather at Lima, 120 m altitude
Temperature: hottest month Feb 19–28 °C,
coldest Aug 13–19 °C
Rainfall (av monthly): driest months Feb, Mar 1 mm, wettest Aug 8 mm
Time 5 hours behind GMT
Measures Metric system; also old Spanish measures
Monetary unit Sol (S) = 100 centavos
Rate of exchange (1976 av): free S 57.43 = $ 1, S 103.7 = £ 1

Summary

Political Republic with military government, which became fully
independent from Spain in 1824. Member of UN, OAS, Sela, Lafta and
Andean Group
Economic Mining products, especially copper, zinc and silver, are the
main exports. Fishmeal is also an important export and there is a wide
spread of agricultural production and a substantial manufacturing sector

People, resources and equipment

Population[a] 1960 10.02*mn, 1970 13.45*mn, 1976 16.09*mn
[a]Excluding Indian jungle population
Growth: 1960–70 3.0* %pa, 1970–76 3.0* %pa
Density (1976): 13* people per km²
Vital statistics (rate per 1 000 people, 1973): births 34.4, deaths 9.1
Cities (population in 000, 1972)

Lima (capital) 3 303   Trujillo   240   Huancayo   127
Arequipa   302   Chiclayo   188   Piura   126
Callao   297   Chimbote   159   Cuzco   121

Race (1970) Mestizo and American Indian 88** %, European 12** %
Language Spanish and Quechua; also Aymará
Religion (1966) Roman Catholic 99 %, Protestant 1 %
Education (1975) Pupils 4 159 732, teachers 130 000*
Labour force (1972) 3 871 613; in agriculture 1 581 846 (41 %),
mining and quarrying 53 134 (1 %), manufacturing 485 234 (13 %)
Personnel Scientists and engineers (1974): 84 923*
Physicians (1972): 8 023, 1 per 1 802 people
Standard of living
National income per person (1976): S 40 000** = $ 700** = £ 390**
Consumption per person (1975): energy 682 kg coal equivalent,
electricity (production) 530* kW h, newsprint 4.7 kg, steel 61 kg
Newspapers (1974): number 67; circulation (41 newspapers only)
1 436 000, 95 per 1 000 people
Telephones (Dec 1976): 295 200, 18 per 1 000 people
Livestock (000, 1976) Cattle 4 300*, sheep 17 300*, goats 1 970*,
pigs 1 950*, horses 713*, poultry 25 000*
Mineral reserves Coal (1973) 281 mn tonnes
Lignite (1966) 4 630 mn tonnes
Crude oil (1975) 100 mn tonnes
Natural gas (1975) 37 000 mn cubic metres
Petroleum refinery capacity (1975) 5.5 mn tonnes
Electrical capacity (1975) 2 354* megawatts,
of which, hydro 1 399* megawatts
Hospital beds (1972) 29 086, 1 per 497 people
Roads (1975) 56 416 km = 35 055 mi, density 0.044 km per km²
Railways (1974) 3 400* km = 2 110* mi, density 0.003* km per km²
Ships (registered, 1977) 681, total of 555 419 gross tons
Ports (goods traffic, 000 tonnes, 1976) Callao: loaded 2 355, unloaded
5 200; San Nicolás: loaded 4 600, unloaded nil; Chimbote: loaded 134,
unloaded 154; Pisco: loaded 248, unloaded 44
Airports (passenger departures and arrivals, 000, 1976) Jorge Chávez
(Lima) 2 424; also Chachani (Arequipa), Huanchaco (Trujillo), Piura,
Cuzco and (1977) 20 other airports with scheduled flights

Durable equipment (at end-year)	000	no per 1 000 people	
Radio sets (1975)	2 050	131	no per
Television sets (1975)	500	32	km of road
Passenger cars (1974)	267	17	4.9
Commercial vehicles (1974)	140	9	2.5

Production

Gross domestic product 1975: S 556 700 mn = $ 13 645 mn = £ 6 141 mn
1976 est: S 700 000**mn = $ 12 000**mn = £ 6 800**mn
Growth in real terms: 1963–70 4.8 %pa, 1970–75 5.5 %pa
Structure of gross domestic product *By origin* (1972) Agriculture 16 %,
mining and quarrying 7 %, manufacturing 23 %, construction 5 %,
transport and communication 5 %, other 44 %
By type (1975) Final consumption expenditure 89 % (of which,
government 13 %), stock investment 3 %, gross fixed capital
formation 18 %, exports of goods and services 12 %, less imports
of goods and services −22 %

Production indices (1970 = 100)	1960	1970	1976	Growth %pa 1960–70	1970–76
Agricultural	80	100	104	2.3	0.7
Manufacturing	46	100	145[d]	8.1	7.7[e]

Main products (000 t)
Agriculture

	1960	1970	1976	1960–70	1970–76
Rice	358	587	570	5.1	−0.5
Barley	180	170	170*	−0.6	0.0*
Maize	442	615	670*	3.4	1.4*
Potatoes	1 398	1 896	1 930*	3.1	0.3*
Cassava	414	498	475*	1.9	−0.8*
Sugar, raw value	821	773	950	−0.6	3.5
Onions	70	159	187*	8.6	2.7*
Oranges	130**	246	246*	6.6**	0.0*
Cottonseed	220	156	130*	−3.4	−3.0*
Coffee	33	65	60*	7.0	−1.3*
Tobacco	2.2	2.3	5.0*	0.4	13.8*
Cotton	133	92	65*	−3.6	−5.6*
Milk	419	825	825*	7.0	0.0*
Beef and veal	63	58	106*	−0.8	10.6*
Pigmeat	39	58	70*	4.0	3.2*
Fish catch	3 569	12 613	3 447[d]	13.5	−22.9[e]
Timber (000 m³)	4 649	5 735	6 562[d]	2.1	2.7[e]

Energy

	1960	1970	1976	1960–70	1970–76
Total energy (000 tce)	4 980	6 090	6 790[d]	2.0	2.2[e]
Coal	162	156	85*[d]	−0.4	−11.4*[e]
Crude oil	2 572	3 550	3 708	3.3	0.7
Petroleum products	2 175	4 070*	5 470*	6.5*	6.1*[e]
Natural gas (mn m³)	na	380*	450*	na	3.4*[e]
Electricity (mn kW h)	2 656	5 529	8 300*[d]	7.6	8.5*[e]
of which, hydro (mn kW h)	1 730	3 821	6 250*[d]	8.2	10.3*[e]

Mining

	1960	1970	1976	1960–70	1970–76
Iron ore (Fe content)	3 947	7 928	5 067*[d]	7.2	−8.6*[e]
Copper ore (Cu content)	209	206	176*[d]	−0.1	−3.1*[e]
Lead ore (Pb content)	166	157	166*	−0.6	1.2[e]
Tungsten conc (oxide content)	0.29	1.01*	0.73*[d]	13.3*	−6.3*[e]
Zinc ore (Zn content)	157	321	360*[d]	7.4	2.3*[e]
Silver	0.95	1.24*	1.24*	2.7*	0.0*
Gold (000 kg)	4.47	2.95	2.66[d]	−4.1	−2.0[e]
Salt	106	200*	350*[d]	6.6*	11.8*[e]

Manufacturing

	1960	1970	1976	1960–70	1970–76
Fishmeal	558	2 256	857	15.0	−14.9
Cigarettes (mn units)	2 132	2 904	3 740[d]	3.1	5.2[e]
Man-made fibres	1.4	5.8	21.0[d]	15.3	29.4[e]
Paper	47	124	177[d]	10.2	7.4[e]
Tyres (000 units)	268	609	801[d]	8.6	5.6[e]
Cement	600	1 144	1 936[d]	6.7	11.1[e]
Pig-iron	39	86	307[d]	8.2	29.0[e]
Crude steel	60	94	443[d]	4.6	36.3[e]
Radio sets (000 units)	19[b]	88	na	46.7[c]	na
Television sets (000 units)	22[b]	35	na	12.3[c]	na
Motor vehicles (000 units)[a]	0.2	14.5	34.0[d]	53.5	18.6[e]
Merchant vessels (000 grt)		35	20	na	−8.9

[a]Assembly only [b]1966 [c]1966–70 [d]1975 [e]1970–75

Transport traffic	1960	1970	1976	Growth %pa 1960–70	1970–76
Passenger-kilometres (mn)					
Rail	282	248	270[a]	−1.3	2.9[b]
Air	124*	789	1 367	20.3*	9.6
Cargo: tonne-kilometres (mn)					
Rail	506	610	735[a]	1.9	6.4[b]
Air	7.6*	24.6	25.9	12.5*	0.9
Sea: tonnes (mn)					
Goods loaded	8.1	14.3	9.0[c]	5.8	−8.9[d]
Goods unloaded	1.6	2.1	3.8*[e]	2.8	16.0*[f]
Tourism					
Number of visitors (000)	44	133	264	11.7	12.1
Gross receipts ($ mn)	na	52	118	na	14.6

[a]1973 [b]1970–73 [c]1975 [d]1970–75 [e]1974 [f]1970–74

Peru

Finance and trade

Price indices (1970 = 100)	1960	1970	1976	Growth %pa 1960–70	1970–76
Consumer prices	41.1	100.0	241.7	9.3	15.8
Share prices	119	100	98	−1.7	−0.3
Money stock (end-year, S mn)	7 000	39 130	148 970	18.8	25.0

Budget (1976) Revenue: S 111 305 mn = $ 1 938 mn = £ 1 073 mn
Expenditure: S 137 655 mn = $ 2 397 mn = £ 1 327 mn

Balance of payments ($ mn)	1972	1973	1974	1975	1976
Balance of goods (fob)	+133	+16	−403	−1 098	−740
Balance of services	−204	−320	−370	−492	−510
Balance of transfers	+40	+42	+48	+50	+58
Current balance	*−31*	*−261*	*−725*	*−1 541*	*−1 191*
Long-term capital flow	+106	+408	+720	+1 293	+844
Reserves and debt (end-year, $ mn)					
International reserves	484	568	968	467	330
External public debt	1 608	2 157	3 011	3 468	4 384

External trade (1976) Imports: S 118 362 mn = $ 2 061 mn = £ 1 141 mn
Exports: S 74 003 mn = $ 1 289 mn = £ 714 mn

Main imports (1974)	% of total	Main exports (1976)	% of total
Machinery	28	Copper	17
Chemicals	18	Fishmeal	13
Food	16	Zinc	11
Metals and manufactures	12	Silver	11
Mineral products	8	Coffee	9
Transport equipment	5	Sugar	7
		Iron ore	5
Main sources (1976)		*Main destinations* (1976)	
United States	32*	United States	25*
Japan	8*	Japan	14*
West Germany	7*	Soviet Union	8*
Venezuela	5*	Chile	6*
Brazil	4*	West Germany	4*

Special focus

Fishmeal and copper

	Production (000 tonnes)		Exports (% of total exports)	
	Fishmeal	Copper (Cu content)	Fishmeal	Copper
1960	558	209	9	22
1970	2 253	206	28	26
1971	1 935	207*	31	19
1972	897	219*	25	20
1973	423	215*	13	27
1974	880	213*	13	23
1975	689	176*	13	12
1976	857	300**	13	17

Puerto Rico

Commonwealth of Puerto Rico
Estado Libre Asociado de Puerto Rico

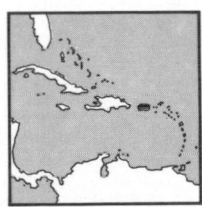

Location Caribbean Sea
The island of Puerto Rico is 80 km to the east of Dominican Republic (Hispaniola Island), and the US Virgin Islands are about 60 km to the east. The small islands of Vieques and Calebra, between the Puerto Rican island and the Virgin Islands, are part of Puerto Rico
Land Area 8 897 km² = 3 435 mi²
Climate Sub-tropical
Weather at San Juan, 25 m altitude
Temperature: hottest month Aug 24–29 °C, coldest Jan 21–27 °C
Rainfall (av monthly): driest month Feb 69 mm, wettest August 160 mm
Time 4 hours behind GMT
Measures UK (imperial) and US systems
Monetary unit US dollar ($) = 100 cents
Rate of exchange (1976 av): free $ 1.806 = £ 1

Summary

Political Self-governing Commonwealth associated with the United States; ceded originally by Spain in 1898. Consideration has been given to the alternative possibilities of complete independence or incorporation as a State of the United States

Economic Manufacturing accounts for 28 % of gross domestic product, with oil refining and petrochemicals very important, and clothing also a major industry. Industrial growth has been achieved by offering tax advantages, and by duty-free access to the US market

People, resources and equipment

Population 1960 2.36 mn, 1970 2.72 mn, 1976 3.21 mn
Growth: 1960–70 1.4 %pa, 1970–76 2.8 %pa
Density (1976): 361 people per km²
Vital statistics (rate per 1 000 people, 1975): births 22.3, deaths 6.1
Cities (population in 000, 1975)

San Juan (capital)	792	Ponce	176	Caguas	112
Bayamón	201	Carolina	143	Mayagüez	94

Language Spanish and English
Religion (1976) Roman Catholic 85* %
Education (1975/76) Pupils 917 754, teachers 34 600*
Labour force (1977) 915 800; in agriculture 61 200 (7 %), manufacturing 183 300 (20 %), construction 81 100 (9 %), transport and communication 40 200 (4 %)
Personnel (1973) Physicians: 3 479, 1 per 848 people
Standard of living
National income per person (1976): $ 2 200** = £ 1 200**
Consumption per person (1975): energy 3 203 kg coal equivalent, electricity (production) 5 190* kW h
Newspapers (1974): number 5; circulation 405 000, 134 per 1 000 people
Telephones (Dec 1976): 515 500, 158 per 1 000 people
Livestock (000, 1976) Cattle 562, pigs 269, chickens 4 969
Petroleum refinery capacity (1975) 17.1* mn tonnes
Electrical capacity (1975) 3 946 megawatts
Hospital beds (1973) 13 354, 1 per 220 people
Roads (1974) 16 827 km = 10 456 mi, density 1.9 km per km²
Railways (1976) 96 km = 60 mi, density 0.011 km per km²
Ports (goods traffic, 000 tonnes, 1972) Guayanilla (crude oil port): loaded 3 069, unloaded 8 989; San Juan: loaded 1 108, unloaded 7 105; also Yabucoa, Las Marcas, Lobos, Ponce, Mayaguez
Airports Isla Verde International (San Juan), Ponce, Borinquen (Aguadilla), Mayaguez, Vieques, Humacao, Culebra

Durable equipment (at end-year)	000	no per 1 000 people	
Radio sets (1975)	1 760	570	no per km of road
Television sets (1975)	630	204	
Passenger cars (1974)	608	198	36
Commercial vehicles (1974)	124	40	7

Production, finance and trade

Gross domestic product
1975/76 (year ending June 30th): $ 8 735 mn = £ 4 378 mn
1976 est: $ 9 000** mn = £ 5 000** mn
Growth in real terms: 1970–75 (years ending June 30th) 3.1 %pa
Structure of gross domestic product (1975/76) *By origin* Agriculture 3 %, manufacturing 28 %, electricity, gas and water 4 %, construction 5 %, distribution and hotels 20 %, transport and communication 6 %, other 34 % *By type* Final consumption expenditure 102 % (of which, government 21 %), stock investment 3 %, gross fixed capital formation 20 %, exports of goods and services 50 %, less imports of goods and services −75 %

Production index (1970 = 100)	1960	1970	1976	Growth %pa 1960–70	1970–76
Agricultural	124[e]	100	101	−2.4[f]	0.2
Main products (000 t)					
Sugar, raw value	1 007	407	279	−8.7	−6.1
Oranges	29	29	32	0.0	1.7
Pineapples	42	49	38	1.6	−4.1
Bananas	115	114	113*	−0.1	−0.1*
Coffee	12	15	12*	2.3	−3.6*
Tobacco	12	3	2	−13.3	−6.5
Milk	323	370	418	1.4	2.0
Eggs	10	14	20	3.7	6.1
Beef and veal	11	18	21	5.0	2.6
Pigmeat	9	13	19	3.7	6.5
Fish catch	13	46	81[a]	13.9	11.8[b]
Other					
Petroleum products	3 793	8 773	9 143[a]	8.7	0.8[b]
Electricity (mn kW h)	2 151	8 027	16 203*[a]	14.1	15.1*[b]
Salt	—	29	26*[c]	na	−2.7*[d]
Beer (000 hl)	688	937	594	3.1	−7.3
Cement	928	1 575	1 398	5.4	−2.0

[a]1975 [b]1970–75 [c]1974 [d]1970–74 [e]1961 [f]1961–70

Puerto Rico

Transport traffic (000, 1975/76) *Air* Passenger departures 2 416, arrivals 2 453 *Sea* Passenger departures 5, arrivals 5
Tourism (1975/76) Number of visitors 1 299 000, gross receipts $ 393 mn
Consumer price index (1970 = 100) 1960 76, 1976 153; growth: 1960–70 2.8 %pa, 1970–76 7.4 %pa
Budget (1975/76; year ending June 30th)
Revenue: $ 2 005 mn = £ 1 005 mn
Expenditure: $ 1 901 mn = £ 953 mn
External trade (1975/76; year ending June 30th)
Imports: $ 5 432 mn = £ 2 723 mn
Exports: $ 3 346 mn = £ 1 677 mn

Main imports (1974/75)	% of total	*Main exports* (1974/75)	% of total
Crude oil and products	27	Chemicals	25
Food	18	(of which, medicinal 9)	
(of which, meat 4)		Crude oil and products	14
Machinery	10	Clothing	11
Chemicals	8	Machinery	11
Textile yarns and fabrics	5	Fish	8
Motor vehicles	4*	Tobacco and products	4
Main sources (1975/76)		*Main destinations* (1975/76)	
United States	62	United States	84
Venezuela	10	Virgin Islands, US	3
Spain	1	Dominican Republic	1

Special focus

Structure of gross domestic product

	% of total					
	1950/51	1960/61	1970/71	1973/74	1974/75	1975/76
Agriculture	23	10	3	4	3	3
Manufacturing	16	23	23	27	26	28
Construction	5	6	8	7	6	5
Other	56	61	66	62	65	64

St Kitts-Nevis and Anguilla

State of Saint Christopher–Nevis and Colony of Anguilla

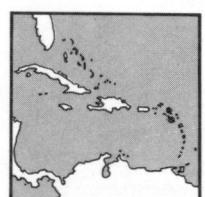

Location Eastern Caribbean Sea
Part of the Leeward Islands, St Kitts and Nevis are separated by a 3 km channel, with Antigua to the east and St Maarten (Netherlands Antilles) to the north; Anguilla is to the north of St Martin (French part of St Maarten island)
Land Area: St Kitts–Nevis 267 km² = 103 mi², Anguilla 91 km² = 35 mi²
Climate Sub-tropical
Weather at La Guérite, St Kitts, 48 m altitude
Temperature: hottest months July–Sept 24–30 °C, coldest Jan 22–27 °C
Rainfall (av monthly): driest month Feb 51 mm, wettest Nov 185 mm
Time 4 hours behind GMT
Measures UK (imperial) system
Monetary unit East Caribbean dollar (EC $) = 100 cents
Rate of exchange (1976 av): free EC $ 2.61 = $ 1, EC $ 4.71 = £ 1

Summary

Political St Kitts–Nevis is, since Feb 27, 1967, an Associated State of the United Kingdom, being formerly a UK colony. Anguilla was part of the original Associated State until March 19, 1969; after that date it came under direct UK administration until February 10, 1976 when Anguilla again became administered as a UK colony. Member of Caricom
Economic St Kitts–Nevis is agricultural with sugar the main crop; Anguilla depends mainly on salt production, livestock and fish. Tourism is important

People, resources and equipment

Population 1960 56 600, 1970 52 000*, 1976 54 600*
of which (1976), St Kitts–Nevis 48 000, Anguilla 6 600*
Growth: 1960–70 −0.8* %pa, 1970–76 0.8* %pa
Density (1976): 153* people per km²
Vital statistics (rate per 1 000 people, St Kitts–Nevis, 1975): births 22.9, deaths 8.9
Regions (population in 000, 1976) St Kitts 36, Nevis 12, Anguilla 6.6*

Towns (population in 000, 1976) Basseterre (on St Kitts, capital) 16, Charlestown (on Nevis) 1.5, The Valley (Anguilla)
Race Mainly African or Mixed
Language English
Education St Kitts–Nevis (1974/75) Pupils 15 000*, teachers 530*
Anguilla (1969/70) Pupils 1 837, teachers 77
Labour force (St Kitts-Nevis, 1970) 13 053
Personnel (St Kitts–Nevis, 1975) Physicians: 12, 1 per 4 000 people
Standard of living National income per person (1976):
St Kitts–Nevis EC $ 1 500** = $ 570** = £ 320**
Anguilla EC $ 1 100** = $ 420** = £ 230**
Production per person (1975): electricity 418 kW h
Newspapers (1974): number 1; circulation 1 500, 30 per 1 000 people
Telephones (St Kitts-Nevis, Dec 1976): 2 242, 45 per 1 000 people
Livestock (000, 1976) Cattle 8*, sheep 22*, pigs 18*, goats 14*
Electrical capacity (1975) 13 megawatts
Hospital beds (St Kitts–Nevis, 1975) 370*, 1 per 130* people
Roads (1976) St Kitts–Nevis 200* km = 124* mi, density 0.75* km per km²
Anguilla 56 km = 35 mi, density 0.62 km per km²
Railways (St Kitts–Nevis, 1976) 58 km = 36 mi, density 0.22 km per km²
Ship (St Kitts–Nevis, registered, 1977) 1, total of 256 gross tons
Ports Basseterre, Charlestown, Road Bay (Anguilla)
Airports Golden Rock (Basseterre, St Kitts), Newcastle (Charlestown, Nevis), Wall Blake (Anguilla)
Durable equipment (Dec 1974)
Passenger cars: 2 100, 38 per 1 000 people, 8.2 per km of road
Commercial vehicles: 300, 5 per 1 000 people, 1.2 per km of road

Production, finance and trade

Gross domestic product
St Kitts–Nevis 1973: EC $ 48 mn = $ 25 mn = £ 10 mn
1976 est: EC $ 80** mn = $ 30** mn = £ 17** mn
Anguilla 1970: EC $ 4 mn = $ 2 mn = £ 0.8 mn
1976 est: EC $ 7** mn = $ 3** mn = £ 1.5** mn
Main products (000 t, 1976) *Agriculture* Sugar, raw value 36*, eggs 0.3*, meat 1*, fish catch 1*
Other Electricity (1975) 23 mn kW h, salt (Anguilla, exports) 6
Transport traffic (1974) *Sea* Goods loaded 27 000* t, unloaded 44 000 t
Tourism (1975) Number of visitors: St Kitts-Nevis 11 697, Anguilla 1 652*
Budget (total, 1975) Balanced at EC $ 37 = $ 17 mn = £ 7.7 mn
Anguilla (1976): Revenue EC $ 3.2 mn = $ 1.2 mn = £ 0.68 mn
Expenditure EC $ 3.0 mn = $ 1.1 mn = £ 0.64 mn
External trade (1972) Imports: EC $ 31 mn = $ 16 mn = £ 6.5 mn
Exports: EC $ 11 mn = $ 5.7 mn = £ 2.3 mn

Main imports (1971)	% of total	*Main exports* (1971)	% of total
Machinery	27	Sugar	65
Food	21	Electrical equipment	24
Chemicals	6	Molasses	4
Petroleum products	5		
Main sources (1971)		*Main destinations* (1971)	
United Kingdom	35	United Kingdom	61
Puerto Rico	14	Puerto Rico	22
United States	10	United States	3
Trinidad and Tobago	9	Netherlands Antilles	3

Anguilla (1976) Exports: EC $ 1.0* mn = $ 0.4* mn = £ 0.2* mn
Main exports Salt 40* %, lobster 36* %, livestock 14* %
Main destinations Trinidad & Tobago 40* %, Puerto Rico 30* %, Guadeloupe 14* %, US Virgin Islands 10* %

St Lucia

State of Saint Lucia

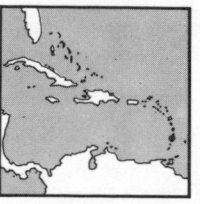

Location Eastern Caribbean Sea
One of the Windward Islands, with Martinique to the north and St Vincent to the south
Land Area 616 km² = 238 mi²
Climate Sub-tropical
Weather at Soufrière, 3 m altitude
Temperature: hottest month Aug 23–31 °C, coldest Jan 21–28 °C
Rainfall (av monthly): driest month April 86 mm, wettest Aug 269 mm
Time 4 hours behind GMT
Measures UK (imperial) system

St Lucia

Monetary unit East Caribbean dollar (EC $) = 100 cents
Rate of exchange (1976 av): free EC $ 2.61 = $ 1, EC $ 4.71 = £ 1

Summary

Political An Associated State of the United Kingdom from March 1, 1967; formerly a UK colony. Full independence has been proposed. Member of Caricom
Economic An agricultural economy, with bananas the principal crop and export, and a range of other produce including especially coconuts. There is some light industry, and further industrial development is planned. Tourism is important

People, resources and equipment

Population 1960 90 000**, 1970 101 000*, 1976 110 000*
Growth: 1960–70 1.2** %pa, 1970–76 1.4* %pa
Density (1976): 179* people per km²
Vital statistics (rate per 1 000 people, 1975): births 35.0, deaths 7.3
Cities (population in 000, 1976) Castries (capital) 47*, Soufrière 7*, Vieux Fort 7*
Race Mainly African
Language English, also a French patois
Religion Mainly Roman Catholic
Education (1973/74) Pupils 31 288, teachers 989
Labour force (1970) 28 988
Personnel (1974) Physicians: 24, 1 per 4 580 people
Standard of living
National income per person (1976): EC $ 1 200** = $ 460** = £ 260**
Production per person (1975): electricity 333 kW h
Telephones (Dec 1976): 6 630, 60 per 1 000 people
Livestock (000, 1976) Cattle 15*, pigs 29*, chickens 80*
Electrical capacity (1975) 14* megawatts
Hospital beds (1975) 545, 1 per 202 people
Roads (1976) 800* km = 500* mi, density 1.3* km per km²
Ships (registered, 1977) 3, total of 928 gross tons
Ports Castries, Vieux Fort, Soufrière
Airports Vigie (Castries), Hewanorra (Vieux Fort)

Durable equipment (at end-year)	000	no per 1 000 people	
Radio sets (1975)	82	740	no per
Television sets (1974)	1.7	16	km of road
Passenger cars (1975)	3.7	33	4.6
Commercial vehicles (1975)	1.8	16	2.2

Production, finance and trade

Gross domestic product 1973: EC $ 73 mn = $ 37 mn = £ 15 mn
1976 est: EC $ 140**mn = $ 54**mn = £ 30**mn
Structure of gross domestic product (1973) *By origin* Agriculture 18 %, construction 10 %, manufacturing 4 %, other 68 %
Main products *Agriculture* (000 t, 1976) Sweet potatoes 1*, yams 4*, bananas 71*, mangoes 41*, coconuts 40*, copra 7*, milk 1*, pigmeat 1*, fish catch 2*
Other (1975) Electricity 36 mn kW h
Transport traffic (1975) *Sea* Goods loaded 25 000 t, unloaded 125 000 t
Tourism (including cruise ship, 1973) Number of visitors 92 300
Consumer price index (1970 = 100) 1976 230; growth 1970–76 14.9 %pa
Budget (1974) Revenue: EC $ 29 mn = $ 14 mn = £ 6 mn
Expenditure: EC $ 50 mn = $ 24 mn = £ 10 mn
External trade (1974) Imports: EC $ 91 mn = $ 44 mn = £ 19 mn
Exports: EC $ 33 mn = $ 16 mn = £ 6.9 mn

Main imports (1973)	% of total	Main exports (1973)	% of total
Food	24	Bananas	64
Chemicals	9	Cardboard boxes	10
Machinery	9	Coconut oil	10
Paper and products	8		
Petroleum products	7		
Textile yarns and fabrics	4		
Main sources (1973)		Main destinations (1973)	
United Kingdom	30	United Kingdom	60
United States	16	Jamaica	10
Trinidad and Tobago	13	Barbados	8
Canada	5	United States	6
Barbados	4	Leeward and Windward Is	6

St Pierre and Miquelon

Department of Saint Pierre and Miquelon
Département des Iles Saint-Pierre-et-Miquelon

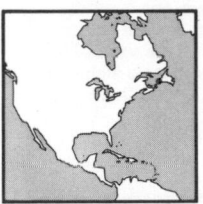

Location North-east of North America
The eight islands of St Pierre and Miquelon are 25 km off the Newfoundland (Canada) coast
Land Area 242 km = 93 mi²
Climate Cold in winter, mild in summer
Weather at St Pierre
Temperature: hottest month Aug 10*–17* °C, coldest Feb minus 8*–minus 1* °C
Rainfall (av monthly): driest month Sept 100* mm, wettest Dec 140* mm
Time 3 hours behind GMT
Measures Metric system
Monetary unit French (metropolitan) franc (Fr) = 100 centimes; before January 1, 1973 the CFA franc (CFA Fr) was in use (Fr 1 = CFA Fr 50)
Rate of exchange (1976 av): free Fr 4.780 = $ 1, Fr 8.633 = £ 1

Summary

Political French overseas department from July 1976; formerly a French overseas territory
Economic There is little agricultural production due to the rocky nature of the land; fishing is the main local industry. External income is principally from the supply of fuel and other ship's stores

People, resources and equipment

Population 1962 4 990, 1970 5 460*, 1976 6 000*
Growth: 1962–70 0.9* %pa, 1970–76 1.6* %pa
Density (1976): 25* people per km²
Vital statistics (rate per 1 000 people, 1974): births 16.6, deaths 9.1
Regions (population in 000, 1974) St Pierre 5 232, Miquelon 608
Town St Pierre (capital)
Race Mainly European (French)
Language French
Religion Mainly Roman Catholic
Education (1973/74) Pupils 1 675, teachers 104
Labour force (1971) 2 095
Personnel Scientists and engineers engaged in research (1972): 7
Physicians (1966): 5, 1 per 1 020 people
Standard of living National income per person (1976): Fr 24 000*** = $ 5 000*** = £ 2 900***
Consumption per person (1975): energy 4 122 kg coal equivalent
Telephones (Dec 1974): 1 430, 245 per 1 000 people
Hospital beds (1975) 78, 1 per 76 people
Port St Pierre
Airport St Pierre
Radio sets (Dec 1975) 2 100, 350 per 1 000 people
Television sets (Dec 1975) 1 700, 290 per 1 000 people
Motor vehicles (Dec 1975) 500*, 85* per 1 000 people

Production, finance and trade

Gross domestic product 1976 est: Fr 150***mn = $ 31***mn = £ 17***mn
Main products (1975) Fish catch 6 400 t, fishmeal (exports 1974) 350 t
Transport traffic (1974) *Sea* Goods loaded 2 000 t, unloaded 83 000 t
Tourism (1972) Number of visitors 11 270
Budget (1973) Balanced at Fr 3.4 mn = $ 0.8 mn = £0.3 mn
External trade (1974) Imports: Fr 126 mn = $ 26 mn = £ 11 mn
Exports: Fr 59 mn = $ 12 mn = £ 5 mn

Main imports	% of total	Main exports	% of total
Petroleum products	28	Petroleum products	53
Food	25	Cattle	30
(of which, livestock 8)		Fish	12
Ships and boats	8		
Machinery	7		
Chemicals	3		
Timber	3		
Main sources		Main destinations	
Canada	54	Ships' stores	57
France	38	Canada	30
West Germany	2	United States	11
		France	2

St Vincent
State of Saint Vincent

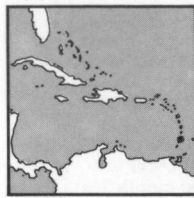

Location Eastern Caribbean Sea
Part of the Windward Islands with St Lucia to the north and Grenada to the south. The northern part of the Grenadines is included in the State of St Vincent, the southern being part of Grenada
Land Area 389 km² = 150 mi²,
of which, main island 344 km²
Climate Sub-tropical
Weather at Kingstown, 1 m altitude
Temperature: hottest month Sept 25–31 °C, coldest Jan 23–29 °C
Rainfall: driest period December–April
Time 4 hours behind GMT
Measures UK (imperial) system
Monetary unit East Caribbean dollar (EC $) = 100 cents
Rate of exchange (1976 av): free EC $2.61 = $ 1, EC $ 4.71 = £ 1

Summary

Political An Associated State of the United Kingdom from October 27, 1969; formerly a UK colony. Member of Caricom
Economic Agriculture and tourism are the main features of the economy; bananas account for one-half of exports, and arrowroot and vegetables are also important

People, resources and equipment

Population 1960 79 950, 1970 89 100, 1976 107 000**
Growth: 1960–70 1.1 %pa, 1970–76 3.1**%pa
Density (1976): 275** people per km²
Vital statistics (rate per 1 000 people, 1973): births 34.4, deaths 10.0
Towns (population in 000, 1974) Kingstown (capital) 24, Georgetown
Race (1972) African 65* %, Mixed 19* %, Indian 5* %, European 3* %, American Indian 2* %
Language English
Religion Methodist, Anglican, Roman Catholic
Education Pupils (1974/75) 32 850*, teachers (1971/72) 1 870*
Labour force (1976) 32 000*; in agriculture 16 000* (50 %)
Personnel (1974) Physicians: 19, 1 per 5 420 people
Standard of living
National income per person (1976): EC $ 900** = $ 340** = £ 190**
Consumption per person (1975): energy 152 kg coal equivalent, electricity (production) 154 kW h
Telephones (Dec 1976): 4 870, 45 per 1 000 people
Livestock (000, 1976) Cattle 7*, sheep 5*, pigs 4*, poultry 140*
Electrical capacity (1975) 9 megawatts
Hospital beds (1972) 529, 1 per 170 people
Roads (1976) 665 km = 413 mi, density 1.7 km per km²
Ships (registered, 1977) 25, total of 8 428 gross tons
Port Kingstown
Airport Arnos Vale (Kingstown)
Durable equipment (at end-year)
Radio sets (1974): 30 000, 290 per 1 000 people
Television sets (1974): 600, 6 per 1 000 people
Motor vehicles (1975): 4 200, 40 per 1 000 people, 6.3 per km of road

Production, finance and trade

Gross domestic product 1976 est: EC $ 100**mn = $ 38**mn = £ 21**mn
Main products *Agriculture* (000 t, 1976) Sweet potatoes 4*, cassava 2*, bananas 27*, coconuts 20*, copra 3*, coconut oil 1*, nutmegs 0.1*, arrowroot 1*, meat 1*, fish catch (1975) 0.3*
Other Electricity (1975) 16 mn kW h
Transport traffic (1974) *Sea* Goods loaded 37 000 t, unloaded 70 000 t
Tourism Number of visitors (1976) 17 953, gross receipts (1970) $ 2.7 mn
Budget (1976/77) Revenue: EC $ 23.1mn = $ 8.6 mn = £ 5.0 mn
Expenditure: EC $ 26.8 mn = $ 9.9 mn = £ 5.8 mn
External trade (1974) Imports: EC $ 52 mn = $ 25 mn = £ 11 mn
Exports: EC $ 14.7 mn = $ 7.2 mn = £ 3.1 mn

Main imports (1972)	% of total	Main exports (1974)	% of total
Food	29	Bananas	54
Machinery	11*	Arrowroot	6
Chemicals	9	Sweet potatoes	4
Main sources (1972)		Main destinations (1972)	
United Kingdom	28	United Kingdom	61
Trinidad and Tobago	17	Barbados	15
Canada	9	Trinidad and Tobago	11

Surinam
Suriname

Location North-east of South America
With a coastline on the Atlantic Ocean, Guyana (former British Guiana) is to the west, French Guiana to the east, and Brazil to the south
Land Area 163 265 km² = 63 037 mi²
Climate Tropical
Weather at Paramaribo, 4 m altitude
Temperature: hottest months September, October 23–33 °C, coldest February 22–29 °C
Rainfall (av monthly): driest month Oct 76 mm, wettest May 310 mm
Time 3½ hours behind GMT
Measures Metric system
Monetary unit Surinam guilder (S Gld) or florin (S Fl) = 100 cents
Rate of exchange (1976 av): par S Gld 1.785 = $ 1, free S Gld 3.224 = £ 1

Summary

Political Republic, which became independent on November 25, 1975; formerly a Netherlands dependency, with full autonomy from 1954. Formerly known as Dutch Guiana. Member of UN, OAS, and an EEC ACP state
Economic Bauxite and its products make up three-quarters of exports. There is some agricultural development, mainly rice production, with plans to diversify; timber is significant and there is some light industry. Mineral resources are considered to be great, including especially iron ore; tourism has been little developed. Netherlands, on independence, promised aid of about S Gld 3 500 mn over the period 1976 to 1990

People, resources and equipment

Population (1960) 285 000*, 1970 374 000*, 1976 435 000*
Growth: 1960–70 2.8* %pa, 1970–76 2.6* %pa
Density (1976): 3* people per km²
Vital statistics (rate per 1 000 people, 1966): births 36.9, deaths 7.2
Cities (population in 000, 1976) Paramaribo (capital) 150*, Nieuw Amsterdam, Moengo, Nieuw Nickerie, Albina
Race (1971) Indian 37 %, Creole 31 %, Indonesian 15 %, African 10 %, American Indian 3 %, Chinese 2 %, European 1 %
Language Dutch; also used are English, Hindi, Javanese, Chinese and local Indian and pidgin languages
Religion (1971) Hindu 29 %, Moslem 19 %, Roman Catholic 18 %, Moravian 13 %
Education (1974/75) Pupils 122 900*, teachers 4 850*
Labour force (1976) 108 000*; in agriculture 21 000* (20* %)
Personnel (1974) Physicians: 202, 1 per 2 030 people
Standard of living
National income per person (1976): S Gld 1 900** = $ 1 060** = £ 590**
Consumption per person (1975): energy 2 063 kg coal equivalent, electricity (production) 3 790 kW h, newsprint 1.4 kg
Newspapers (1974): number 6; circulation (5 only) 24 000, 59 per 1 000 people
Telephones (Dec 1976): 18 600, 42 per 1 000 people
Livestock (000, 1976) Cattle 28*, pigs 7*, chickens 910*
Electrical capacity (1975) 301* megawatts
Hospital beds (1974) 2 288, 1 per 180 people
Roads (1976) 2 000* km = 1 200* mi, density 0.012 km per km²
Railways (1976) 150* km = 100* mi, density 0.0009 km per km²
Inland waterways (rivers, 1976) 1 500* km = 1 000* mi
Ships (registered, 1977) 12, total of 7 277 gross tons
Ports Paramaribo, Nieuw Nickerie; also (inland river ports): Moengo, Paranam
Airports Zanderij (45 km from Paramaribo), Zorg en Hoop (Paramaribo), Nieuw Nickerie

Durable equipment (at end-year)	000	no per 1 000 people	
Radio sets (1975)	110	256	no per
Television sets (1975)	34	79	km of road
Passenger cars (1973)	21	54	11*
Commercial vehicles (1973)	6	14	3*

Production, finance and trade

Gross domestic product 1974: S Gld 870 mn = $ 487 mn = £ 270 mn
1976 est: S Gld 1 000** mn = $ 560** mn = £ 310** mn
Structure of gross domestic product (1972) *By origin* Agriculture 7 %, mining and quarrying 30 %, manufacturing 6 %, other 57 %

Surinam

Production index (1970 = 100)	1960	1970	1976	Growth %pa 1960–70	1970–76
Agricultural	44[d]	100	111	9.7[e]	1.8
Main products (000 t)					
Rice	79	145	173	6.3	3.0
Bananas	3	40*	46*	29.6*	2.4*
Timber (000 m³)	242*	212	298*[a]	−1.3*	7.0[b]
Electricity (mn kW h)	79	1 322	1 600*[a]	32.5	3.9*[b]
of which, hydro (mn kW h)	—	1 000	1 020*[a]	na	0.4*[b]
Bauxite	3 455	6 011	4 751[a,c]	5.7	−4.6[b]
Gold (000 kg)	0.153	0.035	0.004[a]	−13.7	−35.2[b]
Beer (000 hl)	22	86	82[a]	14.6	−0.9[b]
Cigarettes (mn units)	67	187	309[a]	10.8	10.6[b]
Ethyl alcohol (000 hl)	—	39	22[a]	na	−10.8[b]
Cement	—	47	35[a]	na	−5.7[b]
Aluminium	—	55	44	na	−3.6

[a]1975 [b]1970–75 [c]Alumina produced from bauxite: 1 186 000 t in 1974 [d]1961 [e]1961–70

Transport traffic *Air* (1976) Passenger-km 106 mn, cargo 1.1 mn t-km
Sea (1974) Goods loaded 4.9* mn t, unloaded 1.4* mn t
Tourism Number of visitors (1976) 64 700*, gross receipts (1974) $ 6 mn
Consumer price index (1970 = 100) 1976 163; growth 1970–76 8.5 % pa
Budget (1977) Revenue: S Gld 541 mn = $ 303 mn = £ 174 mn
Expenditure: S Gld 581 mn = $ 325 mn = £ 186 mn
External trade (1975) Imports: S Gld 467 mn = $ 262 mn = £ 145 mn
Exports: S Gld 495 mn = $ 277 mn = £ 154 mn

Main imports (1972)	% of total	Main exports (1975)	% of total
Machinery	18	Alumina	43
Chemicals	14	Bauxite	18
Food	12	Aluminium	8
Petroleum products	11		

Main sources (1975)		Main destinations (1975)	
United States	33*	United States	39*
Trinidad and Tobago	15*	Netherlands	13*
Netherlands	14*	West Germany	10*

Trinidad and Tobago

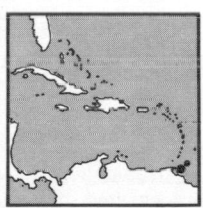

Location Eastern Caribbean Sea
Trinidad island is 16 km from the coast of
Venezuela; Tobago island is 34 km north-east of
Trinidad island. To the north is Grenada,
southernmost of the Windward Islands
Land Area 5 128 km² = 1 980 mi²,
of which, Trinidad 4 827 km² = 1 864 mi²,
Tobago 301 km² = 116 mi²
Climate Tropical
Weather at St Clair, 20 m altitude
Temperature: hottest month May 22–32 °C, coldest Jan 21–31 °C
Rainfall (av monthly): driest month Feb 41 mm, wettest Aug 246 mm
Time 4 hours behind GMT
Measures UK (imperial) system; conversion to the metric system is
taking place. Also, 1 fanega = 110 lb = 49.4 kg (for cocoa beans)
Monetary unit Trinidad and Tobago dollar (TT $) = 100 cents
Rate of exchange (1976 av): free TT $ 2.438 = $ 1, TT $ 4.403 = £ 1

Summary

Political Republic, which became independent on August 31, 1962.
Formerly a British colony. Member of UN, OAS, Sela, Caricom,
Commonwealth and an EEC ACP state
Economic Mainly based on petroleum and natural gas; refining is carried
out on home oil production and on crude oil imported from other
countries. Other industry is being developed

People, resources and equipment

Population 1960 0.92 mn, 1970 1.03 mn, 1976 1.07* mn
Growth: 1960–70 1.1 %pa, 1970–76 0.7* %pa
Density (1976): 210* people per km²
Vital statistics (rate per 1 000 people, 1975): births 23.0, deaths 6.5
Regions (population in 000, 1970): Trinidad 892, Tobago 39
Cities (population in 000, 1976) Port of Spain (capital) 103*,
San Fernando 37*, Arima 12*

Race (1973) African 43* %, Indian 40* %, Mixed 15* %, European 1* %
Language English
Religion (1976) Christian 57* %, Hindu 23* %, Moslem 6* %
Education (1972/73) Pupils 287 000*, teachers 9 500*
Labour force (1975) 390 950; in agriculture 51 900 (13 %)
Personnel (1975) Physicians: 550, 1 per 1 940 people
Standard of living
National income per person (1976): TT $ 5 100** = $ 2 090** = £ 1 160**
Consumption per person (1975): energy 3 132 kg coal equivalent,
electricity (production, 1976) 1 300* kW h, newsprint 6.9 kg, steel 144 kg
Newspapers (1974): number 4; circulation 98 000, 92 per 1 000 people
Telephones (Dec 1976): 70 400, 65 per 1 000 people
Livestock (000, 1976) Cattle 73*, goats 42*, pigs 55*, poultry 6 480*
Mineral reserves (1975) Crude oil 93 mn tonnes
Natural gas 113 bn cubic metres
Petroleum refinery capacity (1975) 23 mn tonnes
Electrical capacity (1975) 334* megawatts
Hospital beds (1975) 4 815, 1 per 222 people
Roads (1976) 6 500* km = 4 000* mi, density 1.27* km per km²
Ships (registered, 1977) 42, total of 17 192 gross tons
Ports (crude oil ports, goods traffic, 000 tonnes, 1976)

	loaded	unloaded		loaded	unloaded
Pointe-à-Pierre	11 436	11 654	Tembladora	1 843	18
Brighton	3 762	27	Port of Spain	365	1 085
Point Fortin	2 494	188	Chaguaramas	415	25

Also Scarborough (Tobago)
Airports Piarco (26 km from Port of Spain), Crown Point (Tobago)

Durable equipment (at end-year)	000	no per 1 000 people	
Radio sets (1974)	250	235	no per km of road
Television sets (1974)	100	94	
Passenger cars (1975)	101	94	16*
Commercial vehicles (1975)	26	24	4*

Production, finance and trade

Gross domestic product 1976: TT $ 6 544 mn = $ 2 684 mn = £ 1 486 mn
Growth in real terms: 1970–75 2.5 %pa
Structure of gross domestic product (1976) *By origin* Agriculture 3 %,
mining and quarrying 41 %, manufacturing 15 %, other 41 %

Production index (1970 = 100)	1960	1970	1976	Growth %pa 1960–70	1970–76
Agricultural	76[f]	100	111	3.1[g]	1.8
Main products (000 t)					
Rice	10	10	21	0.0	13.2
Sugar, raw value	223	222	205*	0.0	−1.3*
Grapefruit	39	21	11*	−6.0	−10.2*
Cocoa	6.2	6.2	5.0*	0.0	−3.5*
Coffee	1.8	2.7	3.0*	4.1	1.8*
Total energy (000 tce)	8 840	11 540	18 250[b]	2.7	9.6[c]
Crude oil	5 994	7 223	10 992	1.9	7.2
Petroleum products	10 614[a]	21 285	11 682[b]	6.4[a]	−11.3[c]
Natural gas (mn m³)	766	1 600	1 697	7.6	1.0
Electricity (mn kW h)	470	1 203	1 390*	9.6	2.4*
Beer (000 hl)	102	176	246[b]	5.6	6.9[c]
Cigarettes (mn units)	840	825	952[b]	−0.2	2.9[c]
Cement	177	271	241	4.4	1.9
Fertilisers, nitrogenous[d]	—	75*	57	na	−4.5*
Radio sets	na	12	16[b]	na	5.9[c]
Motor vehicles[e]	na	6.3	11.6	na	10.6

[a]Excluding jet fuel [b]1975 [c]1970–75 [d]Years ending June 30th [e]Assembly only [f]1961 [g]1961–70

Transport traffic (1976) *Air* Passenger-km 1 040 mn, cargo 24.2 mn t-km
Sea Goods loaded 20.3 mn t, unloaded 13.0 mn t
Tourism (1976) Number of visitors 156 680, gross receipts $ 87 mn
Consumer prices (1970 = 100) 1976 204.7; growth 1970–76 12.7 %pa
Money stock (end-year, TT $ mn) 1976 630; growth 1970–76 25.4 %pa
Budget (1976) Balanced at TT $ 2 024 mn = $ 830 mn = £ 460 mn
International reserves (Dec 1976) $ 1 013 mn
External public debt (Dec 1976) $ 150 mn
External trade (1976) Imports: TT $ 4 801 mn = $ 1 969 mn = £ 1 090 mn
Exports: TT $ 5 364 mn = $ 2 200 mn = £ 1 218 mn

Main imports	% of total	Main exports	% of total
Crude oil	57	Petroleum products	58
Machinery	10	Crude oil	34
Food	7	Sugar	2
Main sources		*Main destinations*	
Saudi Arabia	26	United States	
United States	20	and Puerto Rico	69
Indonesia	16	United Kingdom	5
Iran	11	Ship and aircraft bunkers	4

Trinidad and Tobago

Special focus

Crude oil and petroleum products 000 tonnes, 1976
Home crude oil production 10 992
plus imports   11 700* ⟶ of which from:
equals total available   22 690*   Saudi Arabia   5 540*
less exports   6 500*   Indonesia   3 070*
equals amount   Iran   2 280*
available for refining   16 200*   Ecuador   210*
                       Venezuela   150*
Petroleum products made   16 000*
of which:
residual fuel oil   10 239
motor spirit   2 388

Turks and Caicos Islands

Colony of Turks and Caicos Islands

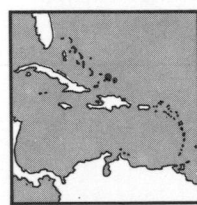

Location Western Atlantic Ocean
The two groups are at the south-east end of the
Bahamas chain of islands
Land Area 430 km² = 166 mi²
Climate Sub-tropical
Weather at Grand Turk, 3 m altitude
Temperature: hottest month Aug 26–32 °C,
coldest Jan, Feb 21–27 °C
Rainfall (av monthly): driest month March
29 mm, wettest Nov 114 mm
Time 5 hours behind GMT
Measures UK (imperial) system, converting to metric system
Monetary unit US dollar ($) = 100 cents (Jamaican currency until 1973)
Rate of exchange (1976 av): free $ 1.806 = £ 1

Summary

Political UK crown colony from 1962. Formerly a Jamaican dependency
Economic Main industries are fishing, tourism and salt

People, resources and equipment

Population 1960 5 716, 1970 5 675, 1976 6 000*
Growth: 1960–70 –0.1 %pa, 1970–76 0.9* %pa
Density (1976): 14* people per km²
Vital statistics (rate per 1 000 people, 1971): births 31.7, deaths 9.8
Towns (population, 1970) Cockburn Town (Grand Turk) 2 330,
Cockburn Harbour (South Caicos) 1 032
Race Mainly African
Language English
Religion Mainly Christian
Education Pupils (1975/76) 2 316, teachers (1973/74) 135*
Personnel (1973) Physicians: 3, 1 per 1 900 people
National income per person (1976) $ 1 200*** = £ 700***
Telephones (Dec 1976) 741, 124 per 1 000 people
Hospital beds (government establishments only, 1973) 20, 1 per 290 people
Roads (1976) 105 km = 65 mi, density 0.24 km per km²
Ships (registered, 1977) 8, total of 2 405 gross tons
Ports Grand Turk, Cockburn Harbour, Providenciales, Salt Cay
Airports Grand Turk, South Caicos, Providenciales, Salt Cay, North
Caicos, Middle Caicos, Pine Cay
Radio sets (Dec 1975) 3 000*, 500* per 1 000 people
Motor vehicles (1973) 704, 119 per 1 000 people, 7 per km of road

Production, finance and trade

Gross domestic product 1969: $ 1.9 mn = £ 0.8 mn
1976 est: $ 8*** mn = £ 4*** mn
Main products Fish catch (1975) 1 050 t, salt (exports, 1972) 2 271 t
Transport traffic *Sea* (1974) Goods loaded 4 000* t, unloaded 10 000* t
Tourism (1975) Number of visitors 8 781
Budget (1975) Revenue: $ 2.08 mn = £ 0.93 mn
Expenditure: $ 3.43 mn = £ 1.54 mn
External trade (1974) Imports: $ 6.0 mn = £ 2.7 mn
Exports: $ 0.5 mn = £ 0.2 mn

Main imports	% of total	*Main exports*	% of total
Manufactures	26	Crayfish	73
Food, drink and tobacco	24	Conch meat	25
Fuel and lubricants	10		

United States

United States of America

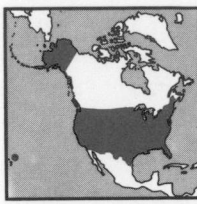

Location North America and Pacific Ocean
Continental United States has Canada on the
northern boundary and Mexico on the southern.
Alaska is bounded to the south-east by Canada
and to the west by the Soviet Union. Hawaii is in
the central Pacific 3 900 km (2 400 mi) to the west
of the mainland
Land Area 9 363 123 km² = 3 615 122 mi²
Usage (1975): agricultural 4 242 360* km²
(45* %), of which, arable 2 073 760* km² (22* %),
cropland 18 600* km² (0.2* %), pastures 2 150 000* km² (23* %)
forests 3 044 000* km² (33* %)
Climate Mainly temperate; sub-tropical in the south
Weather at Washington DC, 22 m altitude
Temperature: hottest month July 20–31 °C, coldest Jan minus 3–6 °C
Rainfall (av monthly): driest month Nov 66 mm, wettest July 112 mm
Time Hours behind GMT (summer time in brackets)

Eastern zone[a]	5	(4)	Alaska: Ketchikan to Skagway	8	(7)
Central zone[b]	6	(5)	Skagway to 141 °W	9	(8)
Mountain zone[c]	7	(6[e])	141 °W to 162 °W	10	(9)
Pacific zone[d]	8	(7)	162 °W to westernmost point	11	(10)
			Hawaii	10	

[a]Connecticut, Delaware, District of Columbia, Florida, Georgia, Maine, Maryland,
Massachusetts, Michigan, New Hampshire, New Jersey, New York, North Carolina,
Ohio, Pennsylvania, Rhode Island, South Carolina, Vermont, Virginia, West Virginia
[b]Alabama, Arkansas, Illinois, Indiana, Iowa, Kansas, Kentucky, Louisiana,
Minnesota, Mississippi, Missouri, Nebraska, North Dakota, Oklahoma,
South Dakota, Tennessee, Texas, Wisconsin   [c]Arizona, Colorado, Idaho, Montana,
New Mexico, Utah, Wyoming   [d]California, Nevada, Oregon, Washington State
[e]Arizona remains on 7 hours behind
Measures UK (imperial) system, except for the following main
differences: US liquid gallon = 0.833 UK gallon
US bushel = 0.969 UK bushel
US short ton = 2 000 lb
US short hundredweight = 100 lb
Metric units are being used to an increasing extent
Monetary unit Dollar ($) = 100 cents
Rate of exchange (1976 av): free $ 1.806 = £ 1
Gold convertibility of the dollar was suspended on August 15, 1971

Summary

Political Republic, which became independent from the United
Kingdom in 1776. There were 13 original states, the number now
being 50; the most recent admissions were Alaska in 1959 and
Hawaii in 1960. Member of UN, OAS, OECD, Nato, Colombo Plan,
South Pacific Commission and Anzus treaty
Economic The widespread nature of the economy is indicated by a
broad range of exports including machinery and chemicals as well
as cereals; manufacturing provides just under one-quarter of gross
domestic product (gdp) and absorbs the same proportion of the labour
force. The business cycle is relatively autonomous in the United States;
since exports account for only 8 % of gdp the economy is relatively
self-sufficient, although crude oil accounted for 22 % of total imports in
1976. The cycle reached a recent recession in 1974/75

People

Population 1960 180.68 mn, 1970 204.88 mn, 1976 215.12 mn
Growth: 1960–70 1.3 %pa, 1970–76 0.8 %pa
Density (1976): 23 people per km²
Vital statistics (rate per 1 000 people, 1976): births 14.7, deaths 8.9
Households (1976): 72.9 mn; average size 2.94 people

Age groups years	1976 population 000	%	1990 projections[a] population 000	%
under 5	15 339	7.1	19 437	8.0
5–13	32 955	15.3	32 568	13.4
14–17	16 897	7.9	12 771	5.2
18–21	16 771	7.8	14 507	6.0
22–44	66 515	30.9	88 320	36.3
45–64	43 707	20.3	46 087	18.9
65 and over	22 934	10.7	29 824	12.2
Total	215 118	100.0	243 513[b]	100.0

[a]Middle of three official projections   [b]Highest official projection is for 254 715
thousand and lowest 236 264 thousand

Regions

Regional division	State	Usual abbreviation	Capital	Area 1970 km²	Population[a], 1976 000	people per km²	Population growth % per annum 1960–70	1970–76	Vital statistics, 1975 rate per 1 000 people Births	Deaths
New England		*NE*		*172 514*	*12 221*	*71*	*1.2*	*0.5*	*12.1*	*9.1*
	Maine	Me	Augusta	86 026	1 070	12	0.2	1.2	14.4	9.7
	New Hampshire	NH	Concord	24 097	822	34	2.0	1.7	13.5	8.8
	Vermont	Vt	Montpelier	24 887	476	19	1.4	1.1	14.3	9.1
	Massachusetts	Mass	Boston	21 386	5 809	272	1.0	0.3	11.7	9.3
	Rhode Island	RI	Providence	3 144	927	295	1.1	−0.4	11.6	9.6
	Connecticut	Conn	Hartford	12 973	3 117	240	1.8	0.4	11.6	8.3
Middle Atlantic		*MA*		*266 108*	*37 282*	*140*	*0.8*	*0.0*	*12.8*	*9.5*
	New York	NY	Albany	128 401	18 084	141	0.8	−0.2	13.0	9.4
	New Jersey	NJ	Trenton	20 295	7 336	361	1.7	0.3	12.6	9.0
	Pennsylvania	Pa	Harrisburg	117 412	11 862	101	0.4	0.1	12.6	10.2
East North Central		*ENC*		*643 050*	*40 934*	*64*	*1.1*	*0.3*	*14.9*	*8.8*
	Ohio	Oh	Columbus	106 764	10 690	100	0.9	0.0	14.8	9.0
	Indiana	Ind	Indianapolis	93 993	5 302	56	1.1	0.2	15.5	8.8
	Illinois	Ill	Springfield	146 075	11 229	77	1.0	0.1	15.2	9.3
	Michigan	Mich	Lansing	150 779	9 104	60	1.3	0.4	14.6	8.1
	Wisconsin	Wis	Madison	145 438	4 609	32	1.1	0.7	14.1	8.7
West North Central		*WNC*		*1 339 664*	*16 805*	*13*	*0.6*	*0.4*	*14.7*	*9.5*
	Minnesota	Minn	St Paul	217 735	3 965	18	1.1	0.6	14.4	8.3
	Iowa	Ia	Des Moines	145 790	2 870	20	0.3	0.2	14.4	9.8
	Missouri	Mo	Jefferson City	180 486	4 778	26	0.8	0.3	14.4	10.3
	North Dakota	N Dak	Bismarck	183 022	643	4	−0.2	0.6	16.7	8.7
	South Dakota	S Dak	Pierre	199 551	686	3	−0.2	0.4	16.5	9.5
	Nebraska	Nebr	Lincoln	200 017	1 553	8	0.5	0.7	15.3	9.4
	Kansas	Kans	Topeka	213 063	2 310	11	0.3	0.4	15.0	9.6
South Atlantic		*SA*		*722 026*	*33 990*	*47*	*1.7*	*1.7*	*14.3*	*8.9*
	Delaware	Del	Dover	5 328	582	109	2.1	0.9	14.2	8.1
	Maryland	Md	Annapolis	27 394	4 144	151	2.4	0.9	12.9	7.8
	District of Columbia	DC	Washington	174	702	4 034	−0.1	−1.3	13.6	10.4
	Virginia	Va	Richmond	105 716	5 032	48	1.6	1.3	14.1	8.0
	West Virginia	W Va	Charleston	62 629	1 821	29	−0.6	0.7	15.6	10.8
	North Carolina	NC	Raleigh	136 197	5 469	40	1.1	1.2	14.8	8.4
	South Carolina	SC	Columbia	80 432	2 848	35	0.8	1.5	16.6	8.3
	Georgia	Ga	Atlanta	152 488	4 970	33	1.5	1.3	16.2	8.4
	Florida	Fla	Tallahassee	151 670	8 421	56	3.2	3.5	12.7	10.4
East South Central		*ESC*		*471 285*	*13 661*	*29*	*0.6*	*1.0*	*16.2*	*9.5*
	Kentucky	Ky	Frankfort	104 623	3 428	33	0.6	1.0	16.1	9.8
	Tennessee	Tenn	Nashville	109 411	4 214	39	1.0	1.1	14.9	9.2
	Alabama	Ala	Montgomery	133 667	3 665	27	0.5	1.0	16.1	9.3
	Mississippi	Miss	Jackson	123 584	2 354	19	0.2	1.0	18.7	9.7
West South Central		*WSC*		*1 136 704*	*21 204*	*19*	*1.3*	*1.5*	*17.3*	*8.7*
	Arkansas	Ark	Little Rock	137 539	2 109	15	0.8	1.5	16.3	10.3
	Louisiana	La	Baton Rouge	125 674	3 841	31	1.1	0.8	17.9	8.9
	Oklahoma	Okla	Oklahoma City	181 089	2 766	15	0.9	1.3	15.7	10.0
	Texas	Tex	Austin	692 402	12 487	18	1.6	1.8	17.6	8.0
Mountain		*Mt*		*2 237 457*	*9 833*	*4*	*1.9*	*2.8*	*18.3*	*7.3*
	Montana	Mont	Helena	381 086	753	2	0.3	1.2	16.1	8.7
	Idaho	Ida	Boise	216 412	831	4	0.7	2.5	19.8	7.9
	Wyoming	Wyo	Cheyenne	253 596	390	2	0.0	2.6	18.6	8.2
	Colorado	Colo	Denver	269 998	2 583	10	2.3	2.5	15.9	6.9
	New Mexico	N Mex	Santa Fe	315 113	1 168	4	0.7	2.2	18.3	7.0
	Arizona	Ariz	Phoenix	295 023	2 270	8	3.1	4.0	17.8	7.6
	Utah	Utah	Salt Lake City	219 931	1 228	6	1.7	2.4	26.3	6.2
	Nevada	Nev	Carson City	286 297	610	2	5.4	3.6	15.3	7.6
Pacific		*Pac*		*2 374 315*	*28 729*	*12*	*2.2*	*1.3*	*15.0*	*8.0*
	Washington	Wash	Olympia	176 616	3 612	20	1.8	0.9	14.3	8.4
	Oregon	Oreg	Salem	251 180	2 329	9	1.7	1.7	14.6	8.8
	California	Calif	Sacramento	411 013	21 520	52	2.3	1.2	15.0	8.1
	Alaska	Alaska	Juneau	1 518 800	382	0.3	2.9	3.9	21.2	4.3
	Hawaii	Hawaii	Honolulu	16 705	887	53	1.9	2.3	18.2	5.0
United States		**US**	**Washington DC**	**9 363 123**	**214 659**	**23**	**1.3**	**0.9**	**14.8**	**8.9**

[a]Resident population

United States

Causes of death (rate per 100 000 people)

	1960	1970	1975
Tuberculosis, all forms	6.1	2.6	1.6
Syphilis and its sequelae	1.6	0.2	0.1
Malignancies	149.2	162.8	171.7
Major cardiovascular diseases	515.1	496.0	455.8
of which, diseases of the heart	369.0	362.0	336.2
Diabetes mellitus	16.7	18.9	16.5
Influenza and pneumonia	37.3[a]	30.9	26.1
Cirrhosis of the liver	11.3	15.5	14.8
Motor vehicle accidents	21.3	26.9	21.5
All other accidents	31.0	29.5	26.9
Suicide	10.6	11.6	12.7

[a]Exclusive of new-born

Cities Populations are shown for a Standard Metropolitan Statistical Area (SMSA) and for the central city; an SMSA is in general an integrated area with a large population nucleus and with a central city. An SMSA may include populations of counties which are metropolitan in character; for New England, however, a unit is also used which includes cities and towns rather than counties and is referred to as a New England County Metropolitan Area (NECMA). An NECMA is not directly comparable with an SMSA

SMSA	SMSA rank in 1975	Population (000) Central cities 1970	SMSA 1970	1975
New York, NY/NJ	1	7 895	9 974	9 561
Chicago, Ill	2	3 367	6 978	7 015
Los Angeles–Long Beach, Calif	3	3 175	7 042	6 987
Philadelphia, Pa/NJ	4	1 949	4 824	4 807
Detroit, Mich	5	1 511	4 435	4 424
San Francisco–Oakland, Calif	6	1 077	3 109	3 140
Washington, DC/Md/Va	7	757	2 910	3 022
Boston, Mass	8	641	2 899	2 890
Nassau–Suffolk, NY	9	na	2 556	2 657
Dallas–Fort Worth, Tex	10	1 238	2 378	2 527
St Louis, Mo/Ill	11	622	2 411	2 367
Pittsburgh, Pa	12	520	2 401	2 322
Houston, Tex	13	1 233	1 999	2 286
Baltimore, Md	14	906	2 071	2 148
Minneapolis–St Paul, Minn/Wis	15	744	1 965	2 011
Newark, NJ	16	382	2 057	1 999
Cleveland, Ohio	17	751	2 064	1 967
Atlanta, Ga	18	497	1 596	1 790
Anaheim–Santa Ana–Garden Grove, Calif	19	446	1 421	1 700
San Diego, Calif	20	697	1 358	1 585
Miami, Fla	21	335	1 268	1 439
Denver–Boulder, Colo	22	582	1 239	1 413
Milwaukee, Wis	23	717	1 404	1 409
Seattle–Everett, Wash	24	584	1 425	1 407
Cincinnati, Ohio/Ky/Ind	25	453	1 385	1 381
Tampa–St Petersburg, Fla	26	494	1 089	1 348
Buffalo, NY	27	463	1 349	1 327
Kansas City, Mo/Kans	28	507	1 274	1 290
Riverside–San Bernardino–Ontario, Calif	29	308	1 141	1 226
Phoenix, Ariz	30	582	969	1 221
San Jose, Calif	31	446	1 065	1 174
Indianapolis, Ind	32	745	1 111	1 139
New Orleans, La	33	593	1 046	1 094
Portland, Oreg/Wash	34	383	1 007	1 083
Columbus, Ohio	35	540	1 018	1 069
Hartford–New Britain–Bristol, Conn (NECMA)	36	158	1 035	1 063
San Antonio, Tex	37	654	888	982
Rochester, NY	38	296	962	971
Providence–Warwick–Pawtucket, RI/Mass	39	340	909	904
Louisville, Ky/Ind	40	361	867	888
Sacramento, Calif	41	254	804	880
Memphis, Tenn/Ark/Miss	42	624	834	867

SMSA	SMSA rank in 1975	Population (000) Central cities 1970	SMSA 1970	1975
Fort Lauderdale–Hollywood, Fla	43	246	620	848
Dayton, Ohio	44	244	853	836
Bridgeport–Stamford–Norwalk–Danbury, Conn (NECMA)	45	157	793	799
Albany–Schenectady–Troy, NY	46	257	778	798
Birmingham, Ala	47	301	767	791
Salt Lake City, Utah	48	176	705	783
Toledo, Ohio/Mich	49	384	763	779
Norfolk–Virginia Beach–Portsmouth, Va/NC	50	591	733	773
Greensboro–Winston–Salem–High Point, NC	51	340	724	764
New Haven–West Haven–Waterbury–Meriden, Conn (NECMA)	52	191	745	760
Nashville–Davidson, Tenn	53	448	699	748
Oklahoma City, Okla	54	366	699	746
Honolulu, Hawaii	55	325	631	705
Jacksonville, Fla	56	529	622	693
Akron, Ohio	57	275	679	667
Worcester–Fitchburg–Leominster, Mass (NECMA)	58	177	637	648
Syracuse, NY	59	197	637	648
Gary–Hammond–East Chicago, Ind	60	330	633	643

Race (1970)

	Population (000)	% of total
European	177 749	87.5
African (Negro)	22 580	11.1
American Indian	793	0.4
Japanese	591	0.3
Chinese	435	0.2
Filipino	343	0.2
Other	721	0.4
Total	203 212	100.0

Language (1970)

Mother tongue	Population (000)	% of total
English	160 717	79.1
Spanish	7 824	3.9
German	6 093	3.0
Italian	4 144	2.0
French	2 598	1.3
Polish	2 438	1.2
Yiddish	1 594	0.8
Swedish	626	0.3
Norwegian	613	0.3
Slovak	510	0.3

Religion (Church membership, as % of total population, 1975)
Protestant 33 %, Roman Catholic 23 %, Jewish 3 %

Education (1974/75) Pupils: primary 27 141 000, secondary and vocational 19 981 000, higher 10 224 000. Teachers: primary 1 331 000, secondary and vocational 1 083 000, higher 633 000

Labour force (1976)

Economic activity	Number	% of total
Agriculture	3 616 000	4
Mining and quarrying	807 000	1
Manufacturing	21 732 000	22
Electricity, gas and water	1 289 000	1
Construction	5 904 000	6
Distribution and hotels	19 570 000	20
Transport and communication	4 638 000	5
Finance and real estate	5 002 000	5
Other services	31 329 000	32
Armed forces	2 144 000	2
Persons seeking first position	885 000	1
Total	96 917 000	100

Personnel (1973) Scientists and engineers: 1 614 000*
Physicians: 338 111, 1 per 622 people

Standard of living
National income per person (1976): $ 6 996 = £ 3 873
Consumption per person (1975): energy 10 999 kg coal equivalent, electricity (production, 1976) 9 844 kW h, newsprint 39 kg, steel 549 kg
Newspapers (1974): number 1 798; circulation 62.2 mn, 293 per 1 000 people
Telephones (Dec 1976): 154.58 mn, 718 per 1 000 people

United States

Personal income (total income by State, $ bn, 1976)
Total for United States: $ 1 382 bn

California	154.2	Washington	24.5	Nebraska	9.7
New York	128.4	Minnesota	24.4	Utah	6.7
Illinois	83.5	Connecticut	23.0	Dist Columbia	6.1
Texas	78.0	Tennessee	22.9	Hawaii	6.1
Pennsylvania	76.7	Louisiana	20.7	New Mexico	6.1
Ohio	68.8	Alabama	18.7	Rhode Island	6.0
Michigan	63.7	Kentucky	18.6	Maine	5.8
New Jersey	53.3	Iowa	18.5	New Hampshire	4.9
Florida	51.4	Colorado	16.8	Idaho	4.8
Massachusetts	38.3	Oklahoma	15.6	Nevada	4.5
Indiana	33.2	Kansas	15.0	Montana	4.2
Virginia	31.6	Oregon	14.7	Delaware	4.2
N Carolina	29.6	S Carolina	14.6	Alaska	3.9
Maryland	29.2	Arizona	13.2	North Dakota	3.5
Wisconsin	29.0	Mississippi	10.8	South Dakota	3.3
Missouri	28.7	Arkansas	10.7	Vermont	2.6
Georgia	27.7	West Virginia	9.8	Wyoming	2.6

Resources and equipment

Livestock (000, Dec 1975) Cattle 127 976, pigs 49 602, sheep 13 346, goats 1 120, horses 9 450*, chickens 379 192, turkeys 8 000*
Mineral reserves Coal (1972) 2 286 bn tonnes
Lignite (1972) 639 bn tonnes
Crude oil (1975) 4.42 bn tonnes
Natural gas (1975) 6 462 bn cubic metres
Uranium (1976) 523 000 tonnes
Petroleum refinery capacity (1975) 743 mn tonnes
Electrical capacity (1975) 524 270 megawatts,
of which, hydro 66 285 megawatts,
nuclear 38 943 megawatts, geothermal 559 megawatts
Hospital beds (1975) 1 401 624, 1 per 152 people
Roads (1975) 6 176 897 km = 3 838 146 mi, density 0.66 km per km²
Railways (1975) 331 311 km = 205 867 mi, density 0.035 km per km²
Inland waterways Great Lakes; Mississippi and other rivers
Ships (registered, 1977) 4 740, total of 15 299 681 gross tons;
of which, oil tankers 314, total of 5 976 499 gross tons
Ports (goods traffic, 000 tonnes, 1976)

	International		Coastwise		Total
	loaded	unloaded	loaded	unloaded	
New York	5 669	51 021	36 017	34 717	127 424
New Orleans	19 488	14 280	30 501	54 508	118 777
Baton Rouge	8 700	43 478	23 397	14 184	89 759
Houston	13 578	24 434	28 473	11 533	78 018
Philadelphia Harbour	4 442	29 522	5 867	10 140	49 971
Baltimore Harbour channels	13 557	17 827	3 258	10 568	45 210
Norfolk	28 142	5 911	5 367	4 499	43 919
Beaumont[a]	3 536	15 688	13 320	7 161	39 705
Corpus Christi[a]	3 905	17 451	14 735	1 432	37 523
Tampa Harbour	11 011	3 973	7 429	13 545	35 958
Mobile Harbour	5 227	7 455	10 285	7 266	30 233
Long Beach Harbour	5 058	13 562	2 991	6 504	28 115

[a]Crude oil port
Great Lakes

	International		Coastwise		Total
Duluth-Superior	3 312	359	23 820	2 127	29 618
Port of Detroit	409	4 530	360	18 202	23 501
Toledo Harbour	6 730	1 144	10 278	4 314	22 466
Port of Chicago	2 607	3 619	3 467	8 164	17 857

Airports (1976)

	Passengers (000)		Cargo (000 tonnes)	
	departures	arrivals	loaded	unloaded
Chicago	20 826	20 910	369	458
Los Angeles	12 993	12 987	382	312
Kennedy (New York)	10 450	10 583	553	509
San Francisco	8 692	8 872	213	184
Miami	6 427	6 457	258	180
Boston	5 702	5 694	102	86
Honolulu	5 642	5 724	67[a]	73[a]
Philadelphia	4 042	4 078	80	66
Newark (New York)	3 336	3 417	75	69
San Juan	2 450	2 499	46[a]	77[a]

[a]Excluding mail
Also 643 other main airports with scheduled flights

Durable equipment

	000 (Dec 1975)	no per 1 000 people	
Radio sets	402 000	1 875	
Television sets	126 300	589	
of which, colour	54 100	252	
Refrigerators	72 600	339	
Freezers	31 600	147	no per
Dishwashers	27 900	130	km of road
Passenger cars	106 713	498	17
Commercial vehicles	26 238	122	4.2
of which,			
buses and coaches	462	2	0.1
goods vehicles	25 776	120	4.2

Companies (industrial: sales, $ bn, year mainly in 1976)

Company	$ bn	Company	$ bn
Exxon Corp	51.6	Phillips Petroleum Co	5.7
General Motors Corp	47.2	The Dow Chemical Co	5.7
Ford Motor Co	28.8	Occidental Petroleum Corp	5.5
Texaco Inc	26.5	International Harvester Co	5.5
Mobil Oil Corp	23.1	Eastman Kodak Co	5.4
Standard Oil Co of California	20.2	Sun Co, Inc	5.4
Gulf Oil Corp	16.5	Union Oil Co of California	5.4
IBM Corp	16.3	RCA Corp	5.3
General Electric Co	15.7	Beatrice Foods Co	5.3
Chrysler Corp	15.5	Esmark, Inc	5.3
Sears, Roebuck and Co	15.0	Bethlehem Steel Corp	5.2
Standard Oil Co (Indiana)	12.5	Rockwell International Corp	5.2
Int Telephone & Tel Corp	11.8	United Technologies Corp	5.2
Safeway Stores, Inc	10.4	F W Woolworth Co	5.2
Shell Oil Co	10.0	Caterpillar Tractor Co	5.0
Atlantic Richfield Co	8.9	Kraft Inc	5.0
United States Steel Corp	8.6	LTV Corp	4.5
S S Kresge Co	8.4	Federated Dept. Stores, Inc	4.4
E I Du Pont de Nemours & Co	8.4	Xerox Corp	4.4
J C Penny Co, Inc	8.4	Ashland Oil, Inc	4.3
Continental Oil Co	8.2	Philip Morris Inc	4.3
Great Atlantic & Pacific Tea Co	7.2	Monsanto Co	4.3
Western Electric Co	6.9	American Brands, Inc	4.1
Procter & Gamble Co	6.5	General Foods Corp	4.0
Tenneco Inc	6.4	Cities Service Co	4.0
Union Carbide Corp	6.3	Firestone Tire & Rubber Co	3.9
Westinghouse Electric Corp	6.1	The Boeing Co	3.9
Kroger Co	6.1	Amerada Hess Corp	3.9
Goodyear Tire & Rubber Co	5.8	Greyhound Corp	3.7
R J Reynolds Industries, Inc	5.8	W R Grace & Co	3.6

Production

Gross domestic product 1976: $ 1 702 bn = £ 942 bn
Growth in real terms: 1960–70 3.9 %pa, 1970–76 2.9 %pa
Structure of gross domestic product

By origin (1975)	$ bn	% of gdp
Agriculture	52	3
Mining and quarrying	38	3
Manufacturing	349	23
Electricity, gas and water	37	2
Construction	67	4
Distribution and hotels	280	19
Transport and communication	96	6
Other	594	39
Total	1 513	100

By type of expenditure (1976)	$ bn	%
Government final consumption	318	19
Private final consumption	1 100	65
Stock investment	14	1
Gross fixed capital formation	276	16
Exports of goods and services	142	8
less Imports of goods and services	−148	−9
Total	1 702	100

Structure of national income (1976)	$ bn	% of national income
Compensation of employees	1 044	69
Operating surplus	296	20
Property and entrepreneurial income from the rest of the world	14	1
Indirect taxes net of subsidies	145	10
Total	1 505	100

United States

Production indices (1970 = 100)	1960	1965	1970	1973	1974	1975	1976	Growth %pa 1960–70	1960–65	1965–70	1970–76
Agricultural	88	95	100	110	112	118	122	1.3	1.6	1.0	3.4
Industrial	61.4	83.3	100.0	120.4	119.9	109.3	120.4	5.0	6.3	3.7	3.1
of which, mining	71.6	83.1	100.0	102.2	102.8	100.5	101.8	3.4	3.0	3.8	0.3
manufacturing	61.5	84.3	100.0	122.0	121.6	109.3	121.7	5.0	6.5	3.5	3.3
Main products (000 t)											
Agriculture											
Wheat	36 869	35 805	36 784	46 408	48 885	57 765	58 307	0.0	−0.6	0.5	8.0
Oats	16 740	13 493	13 313	9 680	8 909	9 319	7 930	−2.3	−4.2	−0.3	−8.3
Barley	9 340	8 559	9 061	9 180	6 622	8 153	8 111	−0.3	−1.7	1.1	−1.8
Maize (corn)	99 690*	104 217	105 463	143 435	118 461	148 062	159 173	0.6*	0.9*	0.2	7.1
Sorghum	15 745	17 087	17 363	23 623	15 983	19 307	18 382	1.0	1.6	0.3	1.0
Rice	2 476	3 460	3 801	4 208	5 098	5 826	5 246	4.4	6.9	1.9	5.5
Potatoes	11 662	13 211	14 781	13 586	15 520	14 623	16 228	2.4	2.5	2.3	1.6
Sugar beet	14 897	18 974	23 930	22 225	20 070	26 947	26 640	4.9	5.0	4.8	1.8
Sugar cane	15 179	20 881	21 769	23 430	22 509	25 875	26 097	3.7	6.6	0.8	3.1
Sugar, raw value	3 786	4 756	5 103	5 215	4 924	6 308	5 989	3.0	4.7	1.4	2.7
Tomatoes	4 530	4 921	5 416	6 270	7 274	8 666	6 857	1.8	1.7	1.9	4.0
Onions	1 201	1 274	1 385	1 345	1 500	1 423	1 597	1.4	1.2	1.7	2.4
Grapes	2 719	3 947	2 830	3 804	3 802	3 971	3 649	0.4	7.7	−6.4	4.3
Wine	1 033*	1 508*	1 167*	1 630*	1 534*	1 445*	1 463*	1.2*	7.9*	−5.0*	3.8*
Apples	2 363	2 940	2 838	2 824	2 941	3 215	2 826	1.8	4.5	−0.7	−0.1
Peaches	1 618	1 586	1 415	1 259	1 416	1 377	1 389	−1.3	−0.4	−2.3	−0.3
Oranges	4 150*	4 900*	7 491	8 834	8 515	9 294	9 506	6.1*	3.4*	8.9*	4.0
Grapefruit	1 538	1 718	1 984	2 428	2 442	2 271	2 585	2.6	2.2	2.9	4.5
Lemons and limes	526	625	562	805	652	1 053	679	0.7	3.5	−2.1	3.2
Pineapples	800**	852	813	735	635	617	626*	0.2**	1.3**	−0.9	−4.2*
Soyabeans	15 107	23 014	30 675	42 108	33 062	42 079	34 425	7.3	8.8	5.9	1.9
Groundnuts (peanuts)	779	1 084	1 351	1 576	1 664	1 750	1 701	5.7	6.8	4.5	3.9
Sunflowerseed	na	30*	85	353	291	357	389	na	na	23.2*	28.9
Linseed	772	899	751	409	344	381	187	−0.3	3.1	−3.5	−20.7
Tobacco	882	841	865	790	903	990	969	−0.2	−0.9	0.6	1.9
Cotton	3 107	3 252	2 219	2 825	2 513	1 807	2 304	−3.3	0.9	−7.4	0.6
Cottonseed	5 340	5 522	3 690	4 550	4 091	2 919	3 764	−3.6	0.3	−7.7	0.3
Milk	55 841	56 324	53 053	52 337	52 414	52 311	54 592	−0.5	0.2	−1.2	0.5
Cheese	950*	1 083	1 330	1 564	1 645	1 593	1 836	3.4*	2.7*	4.2	5.5
Butter	660*	611	518	417	436	446	444	−2.4*	−1.5*	−3.2	−2.5
Lard	1 165	928	868	561	620	451	485	−2.9	−4.4	−1.3	−9.2
Eggs	3 780	3 876	4 043	3 928	3 899	3 798	3 826	0.7	0.5	0.8	−0.9
Wool	64	51	40	34	31	28	25	−4.6	−4.4	−4.7	−7.5
Beef and veal	7 197*	8 962	10 006	9 715	10 696	11 253	12 129	3.4*	4.5*	2.2	3.3
Pigmeat	5 270*	5 055	6 086	5 779	6 244	5 215	5 630	1.5*	−0.8*	3.8	−1.3
Sheep and goat meat	349*	296	253	231	217	194	175	−3.2*	−3.2*	−3.1	−6.0
Poultry meat	3 801	4 858	6 203	6 380	6 471	6 324	7 096	5.0	5.0	5.0	2.2
Cattle hides	960**	960**	1 028	946*	1 053*	1 219*	1 267*	0.7**	0.0**	1.4**	3.5*
Fish catch	2 815	2 696	2 776	2 719	2 744	2 743	3 004	−0.1	−0.9	0.6	1.3
Hardwood (000 m³)	92 800	92 606	81 986	86 810	87 650	72 075	82 694	−1.2	0.0	−2.4	0.1
Softwood (000 m³)	216 100	229 958	245 959	249 558	246 525	229 958	258 703	1.3	1.3	1.4	0.8
Energy											
Total energy (000 tce)	1 366 450	1 633 260	2 062 530	2 141 610	2 078 080	2 036 670	2 050 000*	4.2	3.6	4.8	−0.1*
Coal	391 526	475 284	550 388	530 163	539 138	568 158	585 680	3.5	4.0	3.0	1.0
Lignite	2 491	2 761	5 409	12 848	14 058	18 174	23 260	8.1	2.1	14.4	27.5
Crude oil	347 975	384 946	475 289	454 190	432 794	413 090	401 594	3.2	2.0	4.3	−2.8
Petroleum products	369 331	441 269	530 603	607 763	591 151	601 209	645 000*	3.7	3.6	3.8	3.3*
Natural gas (mn m³)	344 800*	435 415*	595 057	615 357	586 531	546 580*	546 340*	5.6*	4.8*	6.4*	−1.4*
Manufactured gas (mn m³)	24 003	28 011	29 083	28 649	27 471	26 270*	26 750*	1.9	3.1	0.8	−1.4*
Electricity (mn kW h)	844 188	1 157 583	1 639 771	1 964 830	1 967 289	2 000 916	2 117 624	6.9	6.5	7.2	4.4
of which, hydro (mn kW h)	149 515	196 984	250 699	275 334	303 952	303 320	286 820*	5.3	5.7	4.9	2.2*
nuclear (mn kW h)	518	3 657	21 797	83 334	112 696	171 363	na	45.3	47.8	42.9	51.0ᵉ
Mining											
Iron ore (Fe content)	47 867	50 175	53 308	53 267	51 125	48 881	48 398*	1.1	0.9	1.2	−1.6*
Bauxite	2 448	2 022	2 562	2 324	2 408	2 199	2 434	0.5	−3.8	4.8	−0.9
Copper ore (Cu content)	980	1 226	1 560	1 558	1 449	1 282	1 461	4.8	4.6	4.9	−1.1
Lead ore (Pb content)	224	273	519	547	602	564	553	8.8	4.0	13.7	1.1
Molybdenum conc (Mo content)	31	35	51	53	51	48	na	5.1	2.5	7.8	−1.1ᵉ
Uranium (U content)	6.5*	8.0	9.9	10.2	8.9	8.9	9.8	4.3*	4.2*	4.4	−0.2
Zinc ore (Zn content)	395	554	485	434	453	426	433	2.1	7.0	−2.6	−1.9
Gold (000 kg)	52	53	56	37	35	33	33	0.7	0.4	1.1	−8.7
Phosphate rock	17 797	26 704	35 143	38 226	41 446	44 285	na	7.0	8.5	5.6	4.7ᵉ
Salt	23 114	31 467	41 636	39 834	42 217	37 222	na	6.1	6.4	5.8	−2.2ᵉ
Sulphur	5 118	6 214	7 196	7 727	8 028	7 326	na	3.5	4.0	3.0	0.4ᵉ
Manufacturing											
Beer (000 hl)ᵃ	110 938	126 739	157 996	167 824	179 604	185 257	188 800*	3.6	2.7	4.5	3.0*
Cigarettes (mn units)ᵃ	506 127	562 368	562 153	615 562	651 953	626 760	660 000*	1.1	2.1	0.0	2.7*
Cotton yarn	1 655	1 958	1 525	1 388	1 261	1 142	1 480	−0.8	3.4	−4.9	−0.5
Cotton fabrics (mn m)	8 564	8 447	5 711	4 650	4 297	3 745	4 158	−4.0	−0.3	−7.5	−5.2
Wool yarn	237	248	141	89	64	39	40	−5.1	0.9	−10.7	−18.9
Man-made fibres	774	1 500	2 250	3 478	3 368	3 010	3 380*	11.3	14.1	8.4	7.0*

United States

Main products (000 t) Manufacturing	1960	1965	1970	1973	1974	1975	1976	Growth %pa 1960–70	1960–65	1965–70	1970–76
Wood pulp, mechanical	3 000*	3 250*	3 719	4 271	4 274	4 004	3 900*	2.2*	1.6*	2.7*	0.8*
Wood pulp, chemical	18 887	26 085	33 599	36 927	37 174	32 812	38 000*	5.9	6.7	5.2	2.1*
Newsprint	1 818	1 978	3 035	3 095	3 080	3 120	2 846	5.3	1.7	8.9	−1.0
Other paper	27 900*	35 889	43 082	49 602	48 322	41 555	47 880*	4.4*	5.2*	3.7	1.8*
Synthetic rubber	1 460	1 842	2 232	2 607	2 419	1 940	2 313	4.3	4.8	3.9	0.6
Tyres (000 units)	119 824	167 854	190 251	223 418	211 390	186 708	187 953	4.7	7.0	2.5	−0.2
Sulphuric acid	16 223	22 540	26 784	28 613	29 982	29 376	29 954	5.1	6.8	3.5	1.9
Nitric acid	3 300*	4 870*	6 897	7 658	7 425	6 418	na	7.7*	8.1*	7.2*	−1.4e
Caustic soda	4 510	6 197	9 200	9 694	10 341	9 290	9 214	7.4	6.6	8.2	0.0
Plastics and resins	2 850	5 300	8 712	13 719	13 700*	11 000*	13 100**	11.8	13.2	10.5	7.0**
Fertilisers, nitrogenous[a,b]	2 544	4 465	7 562	8 433	9 158	8 474	9 262	11.5	11.9	11.1	3.4
Fertilisers, phosphate[a,b]	1 840*	3 440*	5 001	5 795	6 232	6 568	6 655	10.5*	13.3*	7.8*	4.9
Cement	56 063	65 078	67 682	75 796	72 109	61 816	64 898*	1.9	3.0	0.8	−0.7*
Coke[c]	55 500*	64 312	64 583	62 802	60 487	55 794	57 000*	1.5*	3.0*	0.1	−2.0*
Pig-iron	62 250	82 480	85 141	94 102	89 423	74 515	81 000*	3.2	5.8	0.6	−0.8*
Crude steel	90 067	119 260	119 309	136 804	132 196	105 817	116 311	2.9	5.8	0.0	−0.4
Aluminium	1 827	2 499	3 607	4 109	4 448	3 519	3 858	7.0	6.5	7.6	1.1
Radio sets (000 units)	10 695	14 082	8 261	9 500**	7 200**	5 500**	6 100**	−2.5	5.7	−10.1	−4.9**
Television sets (000 units)	5 828	11 028	9 483	12 300**	9 400**	6 400**	7 800**	5.0	13.6	−3.0	−3.2**
of which, black and white	5 708	8 382	4 851	5 000**	4 000**	2 600**	3 000**	−1.6	8.0	−10.4	−7.7**
colour	120	2 646	4 632	7 300**	5 200**	3 800**	4 800**	44.1	85.6	11.8	0.6**
Washing machines (000 units)[d]	3 364	4 430	4 094	5 504	4 948	4 228	4 492	2.0	5.7	−1.6	1.5
Dishwashers (000 units)[d]	555	1 260	2 116	3 702	3 320	2 702	3 140	14.3	17.8	10.9	6.8
Passenger cars (000 units)[d]	6 675	9 306	6 547	9 667	7 325	6 717	8 498	−0.2	6.9	−6.8	4.4
Commercial vehicles (000 units)[d]	1 194	1 752	1 692	3 014	2 748	2 270	2 978	3.5	8.0	−0.7	9.9
Merchant vessels (000 grt)	485	270	338	869	810	1 006	1 047	−3.5	−11.1	4.6	20.7
Construction											
Dwellings started (000 units)	1 296	1 510	1 469	2 058	1 352	1 171	1 549	1.3	3.1	−0.5	0.9

[a]Years ending June 30th   [b]Including production of Puerto Rico   [c]Including coke made in by-product coke ovens in gasworks   [d]Factory sales   [e]1970–75

Transport traffic Passenger-kilometres (mn)	1960	1965	1970	1973	1974	1975	1976	Growth %pa 1960–70	1960–65	1965–70	1970–76
Road	2 000 000*	2 400 000*	2 941 800	3 332 500	3 248 400	3 349 030	na	3.9*	3.7*	4.2*	2.6b
Rail (class 1 only)	34 211	27 985	17 284	14 957	16 635	15 715	15 680	−6.6	−3.9	−9.2	−1.6
Air	62 542	110 521	210 327	260 637	262 185	262 013	288 027	12.9	12.1	13.7	5.4
of which, domestic	49 175	93 639	171 820	211 378	216 937	218 609	240 011	13.3	13.7	12.9	5.7
Cargo: tonne-kilometres (mn)											
Road	430 000*	540 000*	617 989	736 800	722 200	713 810	na	3.7*	4.7*	2.7*	2.9b
Water[a]	540 000**	640 000**	780 000**	852 000	853 000	826 440	na	3.7**	3.5**	4.0**	1.2**b
Rail (class 1 only)	835 554	1 018 674	1 116 602	1 243 620	1 242 381	1 101 870	1 146 490	2.9	4.0	1.9	0.4
Air	1 455	3 440	7 440	8 827	8 955	8 620	9 084	17.7	18.8	16.7	3.4
of which, domestic	939	2 364	4 736	5 808	5 879	5 655	5 954	17.6	20.3	14.9	3.9
Sea: tonnes (mn)											
Goods loaded	116	158	218	250	246	246	258	6.5	6.4	6.6	2.8
Goods unloaded	192	245	293	422	425	409	448	4.2	5.0	3.6	7.4
Tourism											
Number of visitors (000)	6 453*	7 334	13 167	13 955	14 123	15 698	17 523	7.4*	2.6*	12.4	4.9
Gross receipts ($ mn)	919	1 380	2 331	3 413	4 033	4 839	5 806	9.8	8.5	11.1	16.4

[a]Includes Great Lakes traffic   [b]1970–75

Finance and trade

Price and wage indices (1970 = 100)	1960	1965	1970	1973	1974	1975	1976	Growth %pa 1960–70	1960–65	1965–70	1970–76
Consumer prices	76.3	81.2	100.0	114.4	127.0	138.6	146.6	2.7	1.3	4.3	6.6
Wholesale prices	86.0	87.5	100.0	122.0	145.0	158.4	165.8	1.5	0.3	2.7	8.8
Wages (earnings)	67.3	77.7	100.0	121.4	131.3	143.2	154.2	4.0	2.9	5.2	7.5
Share prices	65.1	102.4	100.0	131.9	101.8	105.7	125.3	4.4	9.5	−0.5	3.8
Money stock (end-year, $ bn)	146.5	175.9	225.5	286.9	278.2	302.3	319.2	4.4	3.7	5.1	6.0

Budget (1977/78; year ending September 30th—'fiscal 1978')
Revenue: $ 400 387 mn = £ 222 000*mn
Expenditure: $ 462 234 mn = £ 257 000*mn

Main revenue sources	$ mn	% of total revenue
Individual income taxes	178 828	45
Social insurance taxes and contributions	124 122	31
Corporation income taxes	58 949	15
Excise taxes	20 150	5

Main expenditures	$ mn	% of total expenditure
Income security	147 640	32
National defence	107 626	23
Health	44 261	10
Interest	43 841	9
Education and social services	27 471	6
Veterans' benefit	18 916	4
Transportation	16 318	4

Balance of payments ($ mn)	1972	1973	1974	1975	1976
Balance of goods (fob)	−6 418	+914	−5 363	+9 053	−9 321
Balance of services	+4 610	+10 085	+13 965	+13 551	+18 167
Balance of transfers	−4 016	−4 107	−7 430	−4 864	−5 345
Current balance	−5 824	+6 892	+1 171	+17 739	+3 501
Long-term capital flow	−5 384	−6 739	−6 702	−19 011	−14 168
Reserves and debt (end-year, $ mn)					
International reserves	13 150	14 380	16 060	15 880	18 320
Public debt: domestic	277 100	284 700	292 700	370 800	428 400
external	55 300	55 500	58 800	66 500	78 100

United States

External trade (including Puerto Rico, 1976)
Imports: $ 129 565 mn = £ 71 733 mn
Exports[a]: $ 114 997 mn = £ 63 668 mn
[a]Including military aid of $ 190 mn = £ 105 mn

Main imports	% of total	Main exports	% of total
Crude oil	22	Machinery, non-electric	19
Motor vehicles	12	(of which,	
(of which,		power-generating 3,	
passenger cars 8)		office machines 3)	
Food	9	Motor vehicles	10
(of which, coffee 2,		Cereals	10
fish 2, meat 1)		(of which, maize 5,	
Machinery, non-electric	6	wheat 3)	
(of which,		Chemicals	9
power-generating 2,		(of which, plastics and	
office machines 1)		resins 1)	
Electrical machinery	6	Electrical machinery	8
(of which,		(of which, switchgear 2,	
telecommunications 3)		telecommunications 2)	
Petroleum products	4	Aircraft	5
Chemicals	4	Soyabeans	3
Iron and steel	4	Instruments	3
Clothing	3	Coal	3
Non-ferrous metals	3	Metal small manufactures	2
Metal ores	2	Textile yarns and fabrics	2
Metal small		Iron and steel	2
manufactures	2	Paper	1
Paper	2	Fruit and vegetables	1
Textile yarns and fabrics	1	Tobacco	1

Main sources	% of total	Main destinations	% of total
Canada	21	Canada	21
Japan	13	Japan	9
West Germany	5	West Germany	5
Saudi Arabia	5	Mexico	4
Nigeria	4	United Kingdom	4
United Kingdom	4	Netherlands	4
Venezuela	3	France	3
Mexico	3	Italy	3
Taiwan	3	Belgium-Luxembourg	3
Indonesia	3	Brazil	2
Italy	2	Iran	2
France	2	Saudi Arabia	2
South Korea	2	Venezuela	2
Hongkong	2	Soviet Union	2
Libya	2	Australia	2
Algeria	2	Spain	2
		South Korea	2

Special focus

The business cycle in the United States (showing, for selected series, % change over one year earlier)

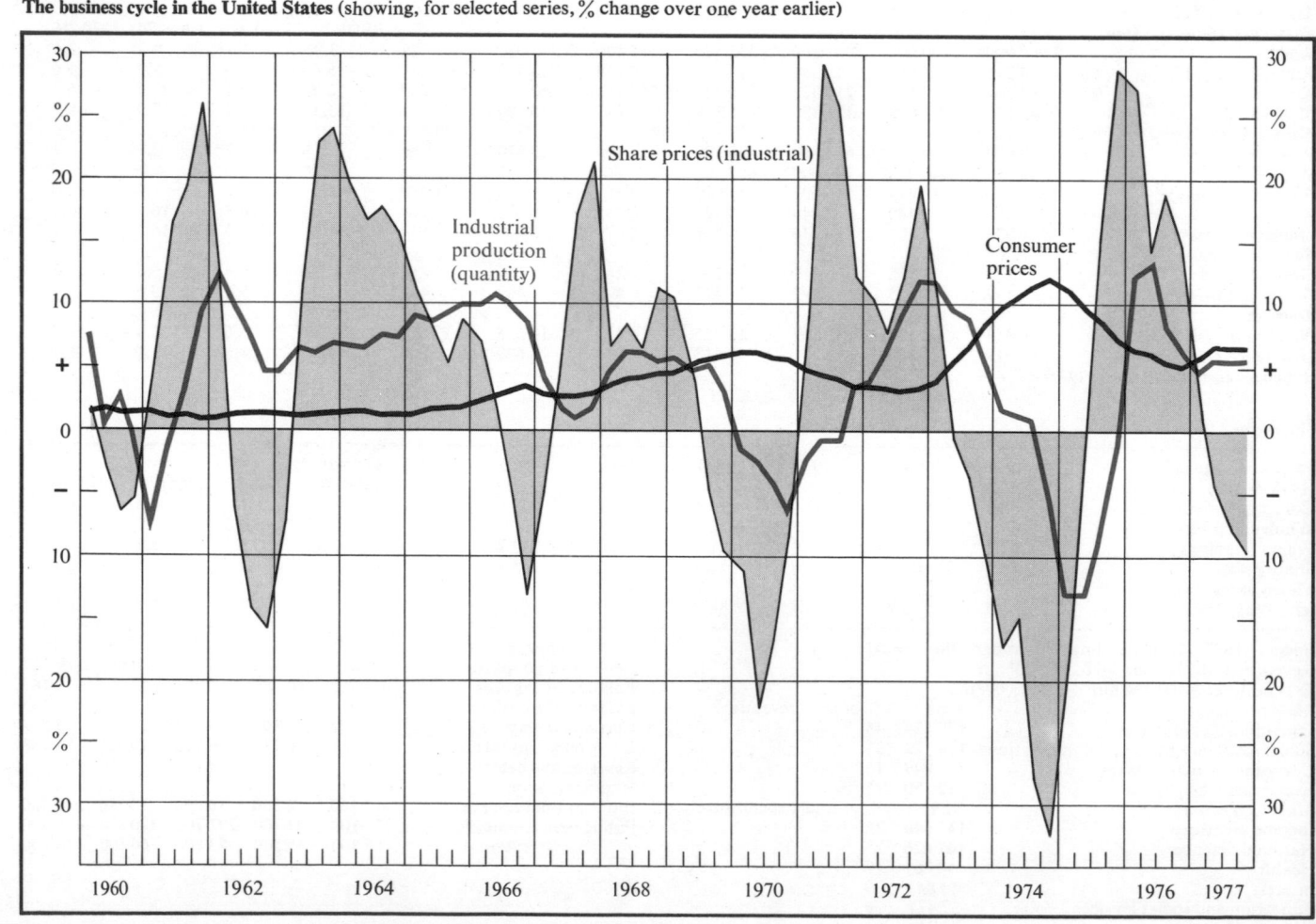

Uruguay

Oriental Republic of Uruguay
República Oriental del Uruguay

Location East central South America.
With a coastline on the Atlantic Ocean and on
the River Plate estuary, Brazil is to the north
and Argentina to the west. The Uruguay river
forms the frontier with Argentina
Land Area 177 508 km² = 68 536 mi²
Climate Temperate
Weather at Montevideo, 22 m altitude
Temperature: hottest month Jan 17–28°C,
coldest July 6–14 °C
Rainfall (av monthly): driest month Feb 66 mm, wettest April 99 mm
Time 3 hours behind GMT
Measures Metric system
Monetary unit Uruguayan new peso (Urug N$) = 100 centésimos;
the new peso replaced the old peso (Urug $) from July 1, 1975,
at the rate: 1 Urug N$ = 1 000 Urug $
Rate of exchange (1976 av): free Urug N$ 3.395 = $ 1,
Urug N$ 6.132 = £ 1

Summary

Political Republic, with military control, which became independent in
1825; formerly a province of Brazil and originally a Spanish territory.
Member of UN, OAS, Sela and Lafta
Economic The economy is based mainly on livestock, with meat, wool
and hides accounting for about one-half of exports; meat has replaced
wool as the main earner from 1971. Tourism is also important.
There has been a move to diversification, including development of
hydro-electricity, minerals and fishing. Inflation is high

People, resources and equipment

Population 1960 2.54*mn, 1970 2.70**mn, 1976 2.80*mn
Growth: 1960–70 0.6**%pa, 1970–76 0.6**%pa
Density (1976): 16* people per km²
Vital statistics (rate per 1 000 people, 1974): births 19.3, deaths 9.2
Cities (population in 000, 1975)

Montevideo (capital)	1 230	Mercedes	53	Minas	40
Salto	80	Las Piedras	42	Melo	38
Paysandú	80	Rivera	42		

Race Mainly European (of Spanish and Italian origin)
Language Spanish
Religion (1976) Roman Catholic 90**%, Jewish 2**%, Protestant 1**%
Education (1974) Pupils 564 000*, teachers 32 700*
Labour force (1975) 1 094 400; in agriculture 170 600 (16 %),
manufacturing 205 300 (19 %), construction 57 800 (5 %)
Personnel Scientists and engineers (1970): 20 069
Physicians (1972): 3 250, 1 per 911 people
Standard of living
National income per person (1976): Urug N $ 1 200* = $ 1 240* = £ 680*
Consumption per person (1975): energy 942 kg coal equivalent,
electricity (production) 934* kW h, newsprint 5.3 kg, steel 21 kg
Newspapers: number (1974) 30; circulation (25 only, 1972) 790 000,
267 per 1 000 people
Telephones (Dec 1976): 257 600, 92 per 1 000 people
Livestock (000, May 1976) Cattle 10 701, sheep 15 974, pigs 460*,
horses 418*, chickens 7 400*
Petroleum refinery capacity (1975) 2.45 mn tonnes
Electrical capacity (1975) 796* megawatts
Hospital beds (1971) 15 107, 1 per 193 people
Roads (1973) 49 634 km = 30 841 mi, density 0.28 km per km²
Railways (1976) 2 987 km = 1 856 mi, density 0.017 km per km²
Inland waterways (including Plate and Uruguay rivers, 1976)
1 300* km = 800* mi
Ships (registered, 1977) 45, total of 192 792 gross tons
Ports Montevideo; river ports: Paysandú, Salto
Airports (passenger departures and arrivals, 000, 1976): Carrasco
(Montevideo) 661; also Salto, Paysandú and 6 other airports with
scheduled flights

Durable equipment	000	no per	
(at end-year)		1 000 people	
Radio sets (1975)	1 500	538	no per
Television sets (1975)	351	126	km of road
Passenger cars (1974)	152*	55*	3.1*
Commercial vehicles (1974)	86*	31*	1.7*

Production

Gross domestic product
1976: Urug N$ 12 537 mn = $ 3 693 mn = £ 2 045 mn
Growth in real terms: 1960–70 1.6 %pa, 1970–76 1.0 %pa
Structure of gross domestic product (1976) *By origin* Agriculture 10 %,
manufacturing 25 %, transport and communication 7 %, other 58 %
By type Final consumption expenditure 87 % (of which, government
13 %), stock investment 0.3 %, gross fixed capital formation 13 %, exports
of goods and services 19 %, less imports of goods and services −19 %

Production indices	1960	1970	1976	Growth %pa	
(1970 = 100)				1960–70	1970–76
Agricultural	85	100	99	1.7	−0.1
Manufacturing	86	100	107c	1.5	1.4d

Main products (000 t)
Agriculture

	1960	1970	1976	1960–70	1970–76
Wheat	420	388	505	−0.8	4.5
Rice	53	142	213	10.4	7.0
Maize	78	139	210	5.9	7.1
Sorghum	14*	35	118	9.6*	22.5
Potatoes	59	118	166	7.2	5.9
Sugar, raw value	12	50	148	15.3	19.8
Grapes	120	134	155*	1.1	2.4*
Wine	81	86*	95*	0.6*	1.7*
Oranges	27*	59	42*	8.1*	−5.5*
Sunflowerseed	72	65	77	−1.0	2.9
Linseed	50	81	62	4.9	−4.3
Milk	773	763	750*	−0.1	−0.3*
Wool	50*	48	34	−0.4*	−5.6
Cattle hides	26	46*	58*	5.9*	4.0*
Beef and veal	249	318	381*	2.5	3.1*
Sheep and goat meat	56	49*	46*	−1.3*	−1.0*
Fish catch	11	13	34	1.9	16.9
Timber (000 m³)	1 260**	896	1 240*	−3.4**	5.6*

Other

	1960	1970	1976	1960–70	1970–76
Petroleum products	1 273	1 592*	1 786c	2.1*	2.3*d
Manufactured gas (mn m³)	30	26	24c	−1.4	−1.5d
Electricity (mn kW h)	1 310*	2 200	2 596*c	5.3*	3.4*d
Cigarettes (mn units)	2 214a	3 121	3 349c	3.9b	1.4d
Cement	415	497	632c	1.8	4.9d
Crude steel	10	16	16c	4.8	0.0d

a1961 b1961–70 c1975 d1970–75

Transport traffic	1960	1970	1976	Growth %pa	
Passenger-kilometres (mn)				1960–70	1970–76
Rail	603a	529	358e	−2.2b	−7.5f
Air	83	63	83	−2.7	4.7
Cargo: tonne-kilometres (mn)					
Rail	255a	250	281e	−0.3b	2.4f
Air	0.39	0.29	0.17	−2.9	−8.5
Sea: tonnes (mn)					
Goods loaded	1.24c	1.63	0.95e	7.1d	−10.3f
Goods unloaded	2.15c	2.41	2.60e	2.9d	1.5f
Tourism					
Number of visitors (000)	na	567	492g	na	−2.3
Gross receipts ($ mn)	na	43	57c	na	5.8f

a1964 b1964–70 c1966 d1966–70 e1975 f1970–75 gOf whom 299 thousand
(61%) from Argentina

Finance and trade

Price index	1960	1970	1976	Growth %pa	
(1970 = 100)				1960–70	1970–76
Consumer prices	2.62	100	2 086	43.9	65.9
Money stock					
(end-year, Urug N$ mn)	2.4	87.7	587.3a	43.4	60.9b

a1974 b1970–74

Budget (1976) Revenue: Urug N$ 1 722 mn = $ 507 mn = £ 281 mn
Expenditure: Urug N$ 2 036 mn = $ 600 mn = £ 332 mn

Balance of payments ($ mn)	1972	1973	1974	1975	1976
Balance of goods (fob)	+103	+79	−56	−111	+28
Balance of services	−55	−61	−98	−105	−110
Balance of transfers	+11	+19	+21	+13	+8
Current balance	+59	+37	−133	−203	−74
Long-term capital flow	+19	+15	+25	+122	+62
Reserves and debt (end-year, $ mn)					
International reserves	203	240	232	218	315
External public debt	398	466	732	1 028	1 005

Uruguay

External trade (1976)
Imports: Urug N$ 2 030* mn = $ 599* mn = £ 332* mn
Exports: Urug N$ 1 820* mn = $ 536* mn = £ 297* mn

Main imports (1972)	% of total	*Main exports* (1976)	% of total
Chemicals	19	Meat	21
Crude oil	14	Wool	18
Transport equipment	14	Hides	6
(of which, motor vehicles 7)			
Food	10		
(of which, sugar 4)			
Machinery	4		
Iron and steel	3		

Main sources (1975)		*Main destinations* (1975)	
Kuwait	16	Brazil	17
Brazil	13	West Germany	11
United States	10	Argentina	7
Argentina	9	Netherlands	7
West Germany	8	United States	6
United Kingdom	5	Italy	5
Iraq	3	United Kingdom	4
Italy	3	Israel	4
Nigeria	3	Soviet Union	4

Special focus

Main exports

	% of total exports								
	1960	1965	1970	1971	1972	1973	1974	1975	1976
Meat	24	32	35	32	46	38	36	20	21
Wool	52	47	31	31	25	30	23	23	18
Hides	12	8	10	10	11	8	6	4	6

Venezuela

Republic of Venezuela
República de Venezuela

Location North of South America
With a coastline on the Caribbean Sea,
Colombia is to the west and south, Guyana is
to the east and Brazil is to the east and
south
Land Area 912 050 km² = 352 144 mi²
Climate Tropical, cooler in highlands
Weather at Caracas, 1 042 m altitude
Temperature: hottest month May 17–27 °C,
coldest Jan 13–24 °C
Rainfall (av monthly): driest month Feb 10 mm, wettest July, Aug 109 mm
Time 4 hours behind GMT
Measures Metric system; a range of local units are used in agriculture,
including: 1 arroba = 11.5 kilograms = 25.35 pounds
Monetary unit Bolívar (B) = 100 céntimos
Rate of exchange (1976 av): par Bs 4.290 = $ 1, free Bs 7.748 = £ 1

Summary

Political Republic, which became independent from Spain in 1821
(forming part Greater Colombia until 1830). Member of UN, OAS, Sela,
Lafta, Andean group and Opec
Economic Oil, first discovered in 1917 has facilitated the development
of a mixed economy, but with oil accounting for virtually all exports.
Increased oil revenues after 1974 are being channeled into investment
projects to diversify further and expand non-petroleum sectors of the
economy, including especially increased production of steel and
aluminium, and an increase in hydro-electric capacity. Following a policy
of oil conservation, Venezuela fell from the position of second largest oil
producer in 1960 to be only fifth largest in 1976

People, resources and equipment

Population 1960 7.38*mn, 1970 10.43*mn, 1976 12.39*mn[a]
[a]Includes 32 000* tribal American Indians
Growth: 1960–70 3.5* %pa, 1970–76 2.9* %pa
Density (1976): 14* people per km²
Vital statistics (rate per 1 000 people, 1976): births 36.4, deaths 6.2

Regions (population in 000, 1976; total of 12.36 mn)

Federal district	2 177	Falcón	462	Táchira	580
States		Guárico	365	Trujillo	427
Anzoátegui	582	Lara	766	Yaracuy	254
Apure	188	Mérida	393	Zulia	1 504
Aragua	634	Miranda	1 005	*Federal territories*	
Barinas	269	Monagas	339	Amazonas	25
Bolívar	458	Nueva Esparta	135	Delta Amacuro	54
Carabobo	769	Portuguesa	343		
Cojedes	108	Sucre	525		

Cities (population in 000, 1974)

Caracas (capital)	2 500*	Maracay	260*	Cumaná	130*
Maracaibo	690*	Ciudad Guayana	250*	Maturin	130*
Valencia	380*	Cabimas	140*	Baruta	130*
Barquisimeto	340*	San Cristóbal	160*	Ciudad Bolívar	110*

Race (1976) Mestizo 70**%, European 20**%, African 8**%
Language Spanish
Religion (1970) Roman Catholic 93 %
Education (1974/75) Pupils: primary 1 990 123, secondary 583 163,
vocational 34 240, teacher-training 13 807, higher (1975/76) 213 542.
Teachers: primary 63 198, other 50 000*
Labour force (1975) 3 711 819; in agriculture 688 213 (19 %),
mining and quarrying 45 563 (1 %), manufacturing 573 197 (15 %),
construction 300 945 (8 %), transport and communication 207 393 (6 %)
Personnel Scientists and engineers engaged in research (1973): 2 720
Physicians (1975): 13 608, 1 per 897 people
Standard of living
National income per person (1976): Bs 8 900* = $ 2 070* = £ 1 150*
Consumption per person (1975): energy 2 639 kg coal equivalent,
electricity (production) 1 761 kW h, newsprint 8.2 kg, steel 194 kg
Newspapers (1974): number 47; circulation (31 only) 1 082 000, 93 per
1 000 people
Telephones (Dec 1976): 742 000, 59 per 1 000 people
Livestock (000, 1976) Cattle 9 404, goats 1 467*, pigs 1 880,
horses 457*, asses 557*, poultry 29 410
Mineral reserves Coal (1972) 845 mn tonnes
Lignite (1953) 26 mn tonnes
Crude oil (1975) 2 643 mn tonnes
Natural gas (1975) 1 189 bn cubic metres
Petroleum refinery capacity (1975) 77 mn tonnes
Electrical capacity (1975) 4 705 megawatts,
of which, hydro 1 568* megawatts
Hospital beds (1975) 35 867, 1 per 340 people
Roads (1974) 65 718 km = 40 835 mi, density 0·07 km per km²
Railways (1976) 419 km = 260 mi, density 0.0005 km per km². There are
plans to construct 4 000 km of railways over the period to 1990
Inland waterways Orinoco river: 1 120* km = 700* mi; also Lake
Maracaibo
Ships (registered, 1977) 179, total of 639 396 gross tons
Ports (crude oil ports, goods traffic, 000 tonnes, 1973)

	loaded	unloaded		loaded	unloaded
Bahía de			Guanta/Puerto		
Amuay	35 443	7	La Cruz	20 536	124
La Salina	32 932	—	Punta Cardón	19 761	11
Puerto			Puerto Ordaz	19 321	34
Miranda	29 439	—	Puerto Cabello	4 187	2 484
Maracaibo	21 735	433	Puerto Caripito	4 112	—
			San Félix	2 420	21

Also: La Guaira is the main port for Caracas. Margarita Island is a Free
Port
Airports (passenger departures and arrivals, 000, 1976): Simon Bolivar
(Caracas) 3 184, La Chinita International (Maracaibo) 518, Maturin 120;
also (1977) 34 other airports with scheduled flights

Durable equipment	000	no per		
(at end-year)		1 000 people		
Radio sets (1975)	2 050	168	no per	
Television sets (1975)	1 284	105	km of road	
Passenger cars (1973)	820	71	12.5	
Commercial vehicles (1973)	295	26	4.5	

Production

Gross domestic product 1976: Bs 133 071 mn = $ 31 019 mn = £ 17 175 mn
Growth in real terms: 1970–76 5.2 %pa
Structure of gross domestic product *By origin* (1975) Agriculture 6 %,
mining and quarrying 30 %, manufacturing 16 %, electricity, gas and
water 1 %, construction 5 %, distribution and hotels 9 %, transport and
communication 9 %, other 24 %
By type (1976) Final consumption expenditure 63 % (of which government
15 %), stock investment 1 %, gross fixed capital formation 31 %, exports
of goods and services 33 %, less imports of goods and services −27 %

Venezuela

Production indices (1970 = 100)	1960	1970	1976	Growth %pa 1960–70	1970–76
Agricultural	60	100	118	5.3	2.8
Manufacturing	55	100	118d	6.2	3.4e

Main products (000 t)

Agriculture

	1960	1970	1976	1960–70	1970–76
Rice	72	226	277	12.1	3.4
Maize	439	710	532*	4.9	−4.7*
Sorghum	1**	7	238*	21.5**	80.0*
Potatoes	134	125	135	−0.7	1.3
Cassava	340	317	353*	−0.7	1.8*
Sugar, raw value	190	435	478*	8.6	1.6*
Tomatoes	49	87	100*	5.9	2.4*
Oranges	37	184	250*	17.4	5.2*
Bananas	999	968	890	−0.3	−1.4
Sesameseed	16	126	62*	22.9	−11.1*
Cottonseed	18	23	36*	2.5	7.7*
Coffee	59	61	50*	0.3	−3.3*
Cocoa	18	19	16*	0.5	−2.8*
Tobacco	9	12	15*	2.7	3.8*
Cotton	10	13	23*	2.7	10.0*
Milk	434	830	1 193*	6.7	6.2*
Beef and veal	121	185*	239*	4.3*	4.3*
Pigmeat	25	45*	45*	6.1*	0.0*
Poultrymeat	22	86	157*	14.6	10.6*
Fish catch	85	126	146	4.0	2.4
Timber (000 m³)	5 100**	7 008	7 964*	3.2**	2.2*

Energy

	1960	1970	1976	1960–70	1970–76
Total energy (000 tce)	200 980	265 820	201 110d	2.8	−5.4e
Coal	35	40	86	1.3	13.6
Crude oil	149 372	194 306	119 756	2.7	−7.7
Petroleum products	43 761	66 700	43 570*d	4.3	−8.2*e
Natural gas (mn m³)	4 606	8 990	11 660	6.9	4.4
Electricity (mn kW h)	4 651	12 631	21 179d	10.5	10.9e

Mining

	1960	1970	1976	1960–70	1970–76
Iron ore (Fe content)	12 474	14 080	15 425d	1.2	1.8e
Gold (000 kg)	1.46	0.68	0.57d	−7.4	−3.5e
Diamonds (000 CM)	71	509	819d	21.8	10.0e

Manufacturing

	1960	1970	1976	1960–70	1970–76
Beer (000 hl)	2 411	4 954	6 135d	7.5	4.4e
Cigarettes (mn units)	6 840	10 463	14 300d	4.3	6.4e
Cotton yarn	8	13	16f	5.0	10.0g
Man-made fibres	3.3	11.8	18.9d	13.6	10.0e
Cotton fabrics (mn m)	36	80	94f	8.3	8.4g
Paper	103b	249	334d	13.4c	6.0e
Tyres (000 units)	753	1 577	2 123f	7.7	16.0g
Cement	1 501	2 318	3 455d	4.4	8.3e
Pig iron	—	510	538d	na	1.1e
Crude steel	47	927	882	34.7	−0.8
Aluminium	—	23	58d	na	20.8e
Radio sets (000 units)	na	71	74f	na	2.1g
Television sets (000 units)	na	98	86f	na	−6.3g
Passenger cars (000 units)a	6.5	48	56f	22.1	8.0g

a Assembly only b 1963 c 1963–70 d 1975 e 1970–75 f 1972 g 1970–72

Transport traffic	1960	1970	1976	Growth %pa 1960–70	1970–76
Passenger-kilometres (mn)					
Rail	25	36	42a	3.7	16.7b
Air	386	1 033	2 538	10.3	16.2
Cargo: tonne-kilometres (mn)					
Rail	20	13	15a	−4.2	15.4b
Air	12	58	77	17.1	4.6
Sea: tonnes (mn)					
Goods loaded	161	204	190e	2.4	−2.4d
Goods unloaded	2.2	4.3	5.5c	6.9	7.9d
Tourism					
Number of visitors (000)	na	117	426g	na	38.1h
Gross receipts ($ mn)	na	50	215e	na	33.9f

a 1971 b 1970–71 c 1973 d 1970–73 e 1975 f 1970–75 g 1974 h 1970–74

Finance and trade

Price indices (1970 = 100)	1960	1970	1976	Growth %pa 1960–70	1970–76
Consumer prices	90.8	100.0	142.0	1.0	6.0
Wholesale prices	78.6	100.0	162.5	2.4	8.4
Share prices	78.5	100.0	235.1	2.5	15.3
Money stock (end-year, Bs mn)	3 673*	7 148	31 090	6.9*	27.8

Budget (1976) Revenue: Bs 41 927 mn = $ 9 773 mn = £ 5 411 mn
Expenditure: Bs 43 888 mn = $ 10 230 mn = £ 5 664 mn
of which, education 13 %, public works 10 %, agriculture 7 %,
health and welfare 5 %, defence 4 %

Balance of payments ($ mn)	1972	1973	1974	1975	1976
Balance of goods (fob)	+931	+2 076	+6 985	+3 373	+2 875
Balance of services	−865	−1 176	−1 126	−988	−1 164
Balance of transfers	−94	−111	−194	−176	−253
Current balance	−29	+788	+5 666	+2 209	+1 458
Long term capital flow	−296	−105	−641	+44	−234

Reserves and debt (end-year, $ mn)					
International reserves	1 732	2 412	6 513	8 861	8 578
External public debt	1 795	1 984	1 901	1 404	3 078

External trade (1976) Imports: Bs 29 309 mn = $ 6 832 mn = £ 3 783 mn
Exports: Bs 39 250 mn = $ 9 149 mn = £ 5 066 mn

Main imports (1974)	% of total		Main exports (1976)	% of total
Machinery, non-electric	21		Crude oil	65
Chemicals	14		Petroleum products	29
Iron and steel	13		Iron ore	2*
Motor vehicles	11			
Electrical machinery	7			
Cereals	6			

Main sources (1975)			Main destinations (1975)	
United States	48		United States	33
West Germany	8		Netherlands Antilles	20*
Japan	8		Canada	13
Italy	6		United Kingdom	4
United Kingdom	4			
Canada	3			
Belgium-Luxembourg	3			

Special focus

Main world crude oil producers

Rank in 1976	1960 mn tonnes	% of total	1970 mn tonnes	% of total	1976 mn tonnes	% of total
1 Soviet Union	148	14	353	16	520	18
2 Saudi Arabia	62	6	177	8	424	15
3 United States	348	33	475	21	402	14
4 Iran	52	5	192	8	296	10
5 *Venezuela*	149	14	194	9	120	4
6 Iraq	47	4	76	3	112	4
World	1 054	100	2 267	100	2 860	100

Virgin Islands, British

Colony of British Virgin Islands (BVI)

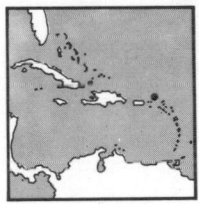

Location Eastern Caribbean Sea
US Virgin Islands are just to the west of this
group of islands and Puerto Rico is about 80 km
to the west; Anguilla is to the east
Land Area 153 km² = 59 mi²,
of which, Tortola 54 km², Anegada 39 km²,
Virgin Gorda 21 km²
Climate Sub-tropical
Weather at Road Town
Temperature: summer 26–31 °C, winter 22–28 °C
Rainfall (av monthly): 100 mm
Time 4 hours behind GMT
Measures UK (imperial) system
Monetary unit US dollar ($) = 100 cents
Rate of exchange (1976 av): free $ 1.806 = £ 1

Summary

Political UK Colony; from 1872–1956 part of the Colony of the Leeward
Islands
Economic Tourism is the major industry; domestic exports, mainly fish,
are negligible in relation to imports

People, resources and equipment

Population 1960 7 340, 1970 10 484, 1976 10 500*
Growth: 1960–70 3.6 %pa, 1970–76 0.0* %pa
Density (1976): 69* people per km²
Vital statistics (rate per 1 000 people, 1975): births 21.5, deaths 6.5

Virgin Islands, British

Regions (main islands, population in 000, 1970) Tortola 8.67,
Virgin Gorda 0.90, Anegada 0.27, Jost Van Dyke 0.12
Towns (population in 000, 1975) Road Town (on Tortola, capital) 3.5*,
East End-Long Look 2.0*
Race Mainly African
Language English
Religion Mainly Methodist
Education (1975/76) Pupils 2 686, teachers 145
Labour force (1970) 3 970; in agriculture 298 (8 %), construction 1 187
(30 %), finance, services and hotels 1 278 (32 %)
Personnel (1973) Physicians: 7, 1 per 1 500 people
Standard of living
National income per person (1976): $ 2 500** = £ 1 400**
Consumption per person (1975): energy 1 121 kg coal equivalent,
electricity (production) 1 140* kW h
Telephones (Dec 1976): 2 330, 220 per 1 000 people
Livestock (000, 1976) Cattle 2*, goats 11*, sheep 7*, pigs 2*
Electrical capacity (1975) 5* megawatts
Hospital beds (1976) 42, 1 per 250 people
Roads (1976) 80* km = 50* mi, density 0.52* km per km²
Ships (registered, 1977) 16, total of 4 057 gross tons
Ports Road Town, West End
Airports Beef Island (15 km from Road Town), Virgin Gorda, Anegada
Radio sets (Dec 1972) 9 000*, 750* per 1 000 people
Motor vehicles (Dec 1975) 1 927, 184 per 1 000 people, 24 per km of road

Production, finance and trade

Gross domestic product 1970: $ 11.3 mn = £ 4.7 mn
1976 est: $ 28** mn = £ 16** mn
Main products (1976) Fish catch 320* t, cattle hides 15* t,
electricity (1975) 12* mn kW h
Transport traffic *Sea* (1974) Goods loaded 6 000* t, unloaded 36 000 t
Tourism (1975) Number of visitors 64 568
Budget (1976) Revenue: $ 5.93 mn = £ 3.28 mn
Expenditure: $ 6.66 mn = £ 3.69 mn
External trade (1974) Imports: $ 11.6 mn = £ 5.0 mn
Exports: $ 0.42 mn = £ 0.18 mn

Main imports	% of total	Main exports	% of total
Food	30	Motor vehicles[a]	16
(of which, meat 8,		Timber[a]	14
fruit and vegetables 6)		Beverages[a]	10
Petroleum products	12	Fish	7
Machinery	8	Iron and steel[a]	6
Beverages	7	Machinery[a]	6
Chemicals	6		
Motor vehicles	5		
Main sources		*Main destinations*	
United States	26	US Virgin Islands	53
Puerto Rico	21	Anguilla	22
US Virgin Islands	17	St Martin (Guadeloupe)	9
United Kingdom	15	United Kingdom	5
Trinidad and Tobago	11	Antigua	4

[a]Re-exports

Virgin Islands, US
Virgin Islands of the United States (USVI)

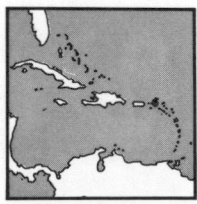

Location Eastern Caribbean Sea
Puerto Rico is about 60 km to the west, with
British Virgin Islands to the east
Land Area 344 km² = 133 mi²,
of which (land area), St Croix 207 km²,
St Thomas 83 km², St John 52 km²
Climate Sub-tropical
Weather at St Croix and St Thomas, 1 m altitude
Temperature: hottest month Aug 24–32 °C,
coldest Jan 21–29 °C
Rainfall (av monthly): 100* mm
Time 4 hours behind GMT
Measures US system
Monetary unit US dollar ($) = 100 cents
Rate of exchange (1976 av): free $ 1.806 = £ 1

Summary

Political US territory, purchased from Denmark in 1917; formerly
Danish West Indies
Economic Tourism and petroleum refining are the main industries; there
is some light manufacturing, including watches and textiles

People, resources and equipment

Population 1960 32 099, 1970 62 468, 1976 96 000*
Growth: 1960–70 6.9 %pa, 1970–76 7.4* %pa
Density (1976): 279* people per km²
Vital statistics (rate per 1 000 people, 1976): births 26.5, deaths 5.4
Regions (population in 000, 1970) St Croix 31.9, St Thomas 29.6,
St John 1.7
Towns (population in 000, 1970) Charlotte Amalie (on St Thomas,
capital) 12.2, Christiansted (St Croix) 3.0, Frederiksted (St Croix) 1.5
Race (1970) African 70** %, European 15** %
Language English; Spanish and Creole are also used
Religion Mainly Roman Catholic; also Protestant and Jewish
Education Pupils (1974/75) 25 100*, teachers (1971/72) 1 260
Labour force (1975) 40 000**
Personnel (1967) Physicians: 68, 1 per 824 people
Standard of living
National income per person (1976): $ 5 100** = £ 2 800**
Consumption per person (1975): 50 157 kg coal equivalent,
electricity (production) 8 350* kW h
Newspapers (1972): number 3; circulation 23 000, 300 per 1 000 people
Telephones (Dec 1976): 33 500, 340 per 1 000 people
Livestock (000, 1976) Cattle 7*, goats 5*, sheep 3*, pigs 4*,
chickens 62*
Petroleum refinery capacity (1975) 36 mn tonnes
Electrical capacity (1975) 224* megawatts
Hospital beds (1970) 248, 1 per 252 people
Roads (1976) 730* km = 450* mi, density 2.1* km per km²
Ports Charlotte Amalie (St Thomas), Frederiksted, Christiansted
Airports Alexander Hamilton (Christiansted), Harry S Truman
(Charlotte Amalie)

Durable equipment (Dec 1975)	000	no per 1 000 people	
Radio sets	75	815	no per
Television sets	30	326	km of road
Passenger cars	12	126	16
Commercial vehicles	2	22	3

Production, finance and trade

Gross domestic product 1976: $ 550** mn = £ 300** mn
Main products (000 t, 1976) Milk 3*, meat 1*, fish catch 1*,
phosphate rock (1974) 19, petroleum products (1975) 23 700*
electricity (1975) 768* mn kW h, watch assembly, woollen goods
processing, rum
Transport traffic *Passenger arrivals* (1972/73) Air 682 000, sea 491 000
Cargo (excluding traffic with the United States and Puerto Rico, 1974)
Sea: goods loaded 0.62* mn t, unloaded 25.6* mn t
Tourism (1976) Number of visitors 1.01 mn, gross receipts (1973) $ 100 mn
Budget (1975/76; year ending June 30th) Revenue: $ 121 mn = £ 61 mn
Expenditure: $ 113 mn = £ 56 mn
External trade (1976) Imports: $ 2 679 mn = £ 1 483 mn
Exports: $ 2 010 mn = £ 1 113 mn
Main imports Crude oil, food, machinery
Main exports Petroleum products, watch movements, woollen fabrics, rum

Main sources	% of total	Main destinations	% of total
Iran	43	United States	96
Nigeria	14	Netherlands	1
Qatar	10	Italy	1
Utd Arab Emirates	9		
United States	8		
Libya	8		

Asia

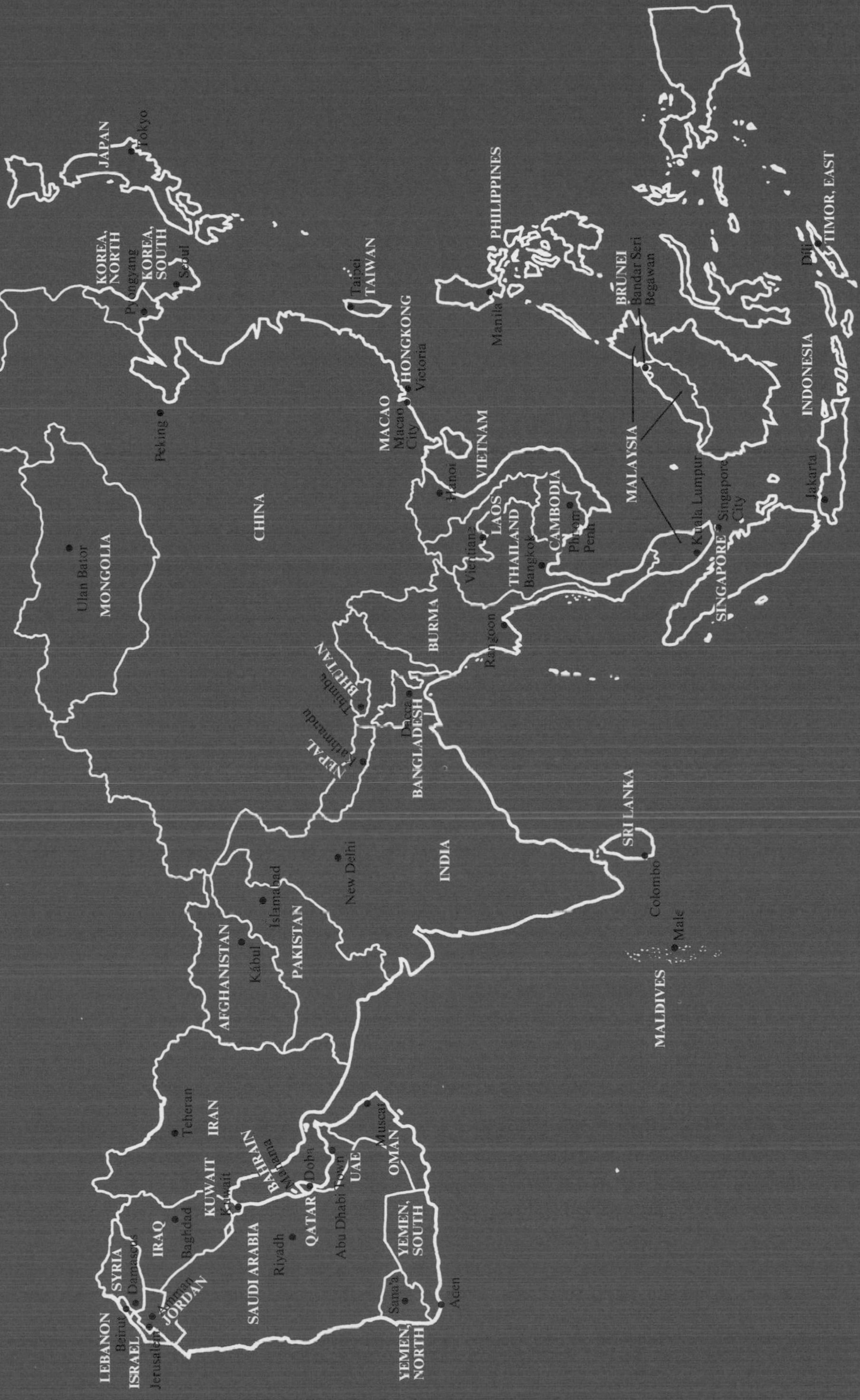

JAPAN
Tokyo

KOREA,
NORTH
Pyongyang

KOREA,
SOUTH
Seoul

Peking

TAIPEI
TAIWAN

PHILIPPINES

Manila

BRUNEI
Bandar Seri
Begawan

TIMOR, EAST

Dili

MONGOLIA
Ulan Bator

CHINA

MACAO
Macao
City

HONGKONG
Victoria

Hanoi

VIETNAM

LAOS

Vientiane

THAILAND
Bangkok

CAMBODIA
Phnom
Penh

MALAYSIA

Kuala Lumpur

SINGAPORE
Singapore
City

INDONESIA

Jakarta

BURMA

Rangoon

BHUTAN
Thimbu

NEPAL
Kathmandu

Dacca

BANGLADESH

SRI LANKA

Colombo

INDIA

New Delhi

Male

MALDIVES

AFGHANISTAN
Kabul

PAKISTAN
Islamabad

IRAN
Teheran

Muscat

OMAN

KUWAIT
Kuwait

BAHRAIN
Manama

QATAR
Doha

UAE
Abu Dhabi Town

IRAQ
Baghdad

SAUDI ARABIA
Riyadh

YEMEN,
SOUTH

SYRIA
Damascus

JORDAN
Amman

LEBANON
Beirut

ISRAEL
Jerusalem

Sana'a

YEMEN,
NORTH

Aden

Afghanistan
Democratic Republic of Afghanistan

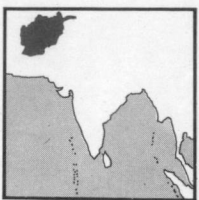

Location Central Asia
Pakistan is to the east and south, Iran to the
west, and Soviet Union to the north; there is
a small border with China to the north-east.
Land-locked
Land Area 647 500 km² = 250 000 mi²
Climate Continental (extremes of temperature)
Weather at Kábul, 1 815 m altitude
Temperature: hottest month July 16–33 °C,
coldest Jan minus 8–2 °C
Rainfall (av monthly): driest month Sept 1 mm, wettest April 102 mm
Time 4½ hours ahead of GMT
The Moslem solar year, beginning March 21st, is used; the year 1357
corresponds to the year ending March 20, 1979
Measures Metric system; also local measures, including:
length 1 gazi jerib = 0.7366 metre = 29 inches
area 1 jerib = 20 beswa = 0.195 hectare = 0.483 acre
weight (mass) 1 pow = 441.625 grams = 0.974 pound
1 charak = 4 pow = 1.7665 kilograms = 3.894 pounds
1 seer (Kabul) = 16 pow = 7.066 kilograms = 15.58 pounds
Monetary unit Afgháni (Af) = 100 puls
Rate of exchange (1976 av): par Af 45.00 = $ 1, free Af 81.28 = £ 1

Summary

Political One-party republic (before July 17, 1973, a constitutional
monarchy). There was a coup and change of government at end-April
1978. Member of UN and Colombo Plan
Economic An agricultural economy, but natural gas is important. Other
minerals are to be exploited. There is some light industry, mainly textiles

People, resources and equipment

Population 1960 13.80*mn, 1970 17.09*mn, 1976 19.80*mnª
ªOf whom 2.5**mn nomadic
Growth: 1960–70 2.2* %pa, 1970–76 2.5* %pa
Density (1976): 31* people per km²
Vital statistics (rate per 1 000 people, 1970–75): births 49.2, deaths 23.8
Cities (population in 000, 1975)
Kábul (capital) 749*   Herát   157*   Charekar   98*
Kandahár   209*   Kunduz 108*   Mazar-i-Sharif 97*
Race (excluding nomads, 1963) Pushtun or Pathan 59 %, Tadzhik 29 %,
Uzbek 5 %, Hazarah 3 %
Language Pushtu and Dari (Persian)
Religion (1976) Moslem (Sunni 90** %, Shiah 5** %)
Education (1975) Pupils 890 709, teachers 27 104
Labour force (1976) 6 728 000; in agriculture 5 344 000 (79 %)
Personnel Scientists and engineers (1966): 4 823
Physicians (1973): 701, 1 per 26 090 people
Standard of living
National income per person (1976): Af 4 800** = $ 110** = £ 60**
Consumption per person (1975): energy 52 kg coal equivalent,
electricity (production, 1975/76) 38* kW h, newsprint 0.06 kg, steel 1 kg
Newspapers (1974): number 15; circulation 499 000, 27 per 1 000 people
Telephones (Dec 1975): 25 000, 1 per 1 000 people
Livestock (000, 1976) Cattle 3 676*, sheep 18 000* (of which,
karakul 7 000**), goats 2 350*, buffaloes 37*, horses 370*, asses 1 250*,
camels 290*, chickens 10 000*
Mineral reserves Coal (1965) 85*mn tonnes
Natural gas (1976): 70 000**mn cubic metres
Electrical capacity (1975) 325* megawatts
Hospital beds (1974) 2 377, 1 per 7 910 people
Roads (1974) 17 973 km = 11 168 mi, density 0.03 km per km²
There are no railways, but 1 815 km have been proposed
Inland waterways Oxus river
Airports Kábul, Kandahár; also 13 other airports with scheduled flights
Durable equipment (at end-year)
Radio sets (1975): 113 000, 6 per 1 000 people
Passengers cars (1976): 20 300, 1.0 per 1 000 people, 1.1 per km of road
Commercial vehicles (1976): 8 500, 0.4 per 1 000 people, 0.5 per km of road

Production

Gross domestic product
1976 est: Af 100 000**mn = $ 2 200**mn = £ 1 200**mn

Production index (1970 = 100)	1960	1970	1976	Growth %pa 1960–70	Growth %pa 1970–76
Agricultural	85	100	126	1.6	3.9

Main products (000 t)

Agriculture

	1960	1970	1976	1960–70	1970–76
Wheat	2 279	2 081	2 930	−0.9	5.9
Maize	700	667	800	−0.5	3.1
Barley	378	370	400	−0.2	1.3
Rice	319	366	448	1.4	3.4
Apples	15*	15*	17*	0.0*	2.1*
Grapes	320**	350*	390*	0.9**	1.8*
Raisins	40*	34*	60*	−1.6*	9.9*
Sesameseed	30**	30*	40*	0.0**	4.9*
Linseed	15**	15*	16*	0.0**	1.1*
Cottonseed	36	49	111*	3.1	14.6*
Cotton	18	25	54*	3.3	13.7*
Milk, cow	259	310*	331*	1.8*	1.1*
Milk, sheep	178	180*	215*	0.1*	3.0*
Wool	11.4*	12.0*	14.0*	0.5*	2.6*
Hides and skins	22	25*	29*	1.3*	2.5*
Beef and buffalo meat	35**	39*	44*	1.1**	2.0*
Sheep and goat meat	85**	102*	115*	1.8**	2.0*
Timber (000 m³)	4 880**	6 416	6 961	2.8**	1.4

Other

	1960	1970	1976	1960–70	1970–76
Total energy (000 tce)	60	3 652	4 240ᵇ	50.8	3.0ᶜ
Coalª	48	164	189ᵇ	13.1	2.9ᶜ
Natural gas (mn m³)ª	—	2 583	2 959ᵇ	na	2.8ᶜ
Electricity (mn kW h)ª	119	396	748*ᵇ	12.8	13.5*ᶜ
of which, hydro (mn kW h)ª	113	379	730*ᵇ	12.9	14.0*ᶜ
Saltª	26	45	61ᵇ	5.6	6.3ᶜ
Cotton yarnª	0.1	0.7	1.1ᵈ	21.5	12.0ᵉ
Cotton fabrics (mn m)ª	24	57	60ᵇ	9.0	1.0ᶜ
Cementª	37	94	147ᵇ	9.1	9.3ᶜ

ªYears beginning March 21st   ᵇ1975   ᶜ1970–75   ᵈ1974   ᵉ1970–74
Transport traffic (1976) *Air* Passenger-kilometres 258 mn,
cargo: tonne-kilometres 13 mn
Tourism (1976) Number of visitors 90 980, gross receipts $ 30 mn

Finance and trade

Consumer price index (Kábul, 1970 = 100) 1976 122;
growth 1970–76 3.3 %pa
Money stock (Dec 21st, Af mn) 1970 7 341, 1976 15 910;
growth 1970–76 13.8 %pa
Budget (1975/76; year ending March 20th)
Revenue: Af 13 000 mn = $ 289 mn = £ 136 mn
Expenditure: Af 12 947 mn = $ 288 mn = £ 135 mn
International reserves (Dec 1976) $ 316 mn
External public debt (Dec 1976) $ 1 748 mn
External trade (1975/76; year ending March 20th)
Imports: Af 20 442 mn = $ 454 mn = £ 214 mn
Exports: Af 13 085 mn = $ 291 mn = £ 137 mn

Main imports (1974/75)	% of total	Main exports (1974/75)	% of total
Textile yarns and fabrics	25	Raisins	16
Sugar	11	Cotton	15
Chemicals	10	Natural gas	14
Petroleum products	9	Edible nuts	9
Tea	6	Carpets	9
Rubber manufactures	5	Grapes	8
Motor vehicles	4	Hides and skins	8
Machinery	4	Wool	3
Vegetable oils	4	Oilseeds and nuts	2
Main sources (1975/76)		Main destinations (1975/76)	
Soviet Union	24	Soviet Union	39
Japan	19	Pakistan	13
West Germany	12	India	12
India	12	West Germany	10
United States	7	United Kingdom	7
United Kingdom	4	United States	4
Pakistan	2	Switzerland	3

Special focus

Natural gasª

	Production (mn m³)	Exports as % of total exports
1966	—	—
1970	2 583	17
1974	2 946	14
1975	2 959	20

ªYears beginning March 21st

Bahrain

State of Bahrain
Dawlat al Bahrayn

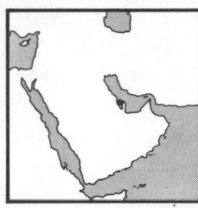

Location Middle East
An archipelago of 33 islands in the Gulf about 40 km from the eastern coast of Saudi Arabia; the Qatar peninsula is to the south-east. The two main islands are Bahrain and Muharraq
Land Area 662 km² = 256 mi²
Climate Hot, temperate Dec-March
Weather at Bahrain, 5 m altitude
Temperature: hottest month Aug 29–38 °C, coldest Jan 14–20 °C
Rainfall (av monthly): driest months June–Oct 0 mm, wettest Nov, Dec, Feb 18 mm
Time 3 hours ahead of GMT
Measures Metric system; also local measures, including:
length 1 dhara = 48.26 centimetres = 19 inches
weight (mass) 1 rafa = 10 maund = 254 kilograms = 560 pounds
Monetary unit Bahrain dinar (B D) = 1 000 fils
Rate of exchange (1976 av): par B D 0.394737 = $ 1 ($ 2.533 = B D 1), free B D 0.7130 = £ 1 (£ 1.403 = B D 1)

Summary

Political An independent shaikhdom from August 15, 1971; formerly a UK protected state. Member of UN, Arab League and Oapec
Economic Crude oil has been produced commercially since 1936; refined petroleum products, made from locally obtained oil and imports, are the main export. There has been some diversification, notably an aluminium smelter established in 1971 and shipbuilding and repairs (a large dry dock was brought into operation at end-1977). Banking is being developed. A causeway to link the main island with Saudi Arabia is planned

People, resources and equipment

Population 1960 147 000*, 1970 215 000*, 1976 259 000*
Growth: 1960–70 3.9* %pa, 1970–76 3.2* %pa
Density (1976): 391* people per km²
Vital statistics (rate per 1 000 people, 1968): births 30
Cities (population in 000, 1975) Al Manamah[a](capital) 95*, Al Muharraq 45*
[a]Manama
Race (1971) Bahraini 82 %, Omani 5 %, Indian 3 %, Pakistani 2 %, Iranian 2 %
Language Arabic
Religion (1971) Moslem 96 %, Christian 3 %
Education Pupils (1975/76) 59 497, teachers (1972/73) 2 800*
Labour force (1971) 60 301; in agriculture 3 990 (7 %), manufacturing 8 372 (14 %), construction 10 404 (17 %)
Personnel (1974) Physicians: 178, 1 per 1 350 people
Standard of living National income per person (1976): B D 800** = $ 2 030** = £ 1 100**
Consumption per person (1975): energy 12 079 kg coal equivalent, electricity (production) 1 590* kW h, steel 281 kg
Telephones (Dec 1976): 30 800, 116 per 1 000 people
Livestock (000, 1976) Cattle 5*, sheep 4*, goats 8*, chickens 172*
Mineral reserves (1975) Crude oil 43 mn tonnes
Natural gas 252 000 mn cubic metres
Petroleum refinery capacity (1975) 15.5 mn tonnes
Electrical capacity (1975) 110* megawatts
Hospital beds (1974) 981, 1 per 248 people
Roads (paved only, 1972) 30 km = 19 mi, density 0.05 km per km²
Ships (registered, 1977) 28, total of 6 409 gross tons
Ports (goods traffic, 000 tonnes, 1975)

	loaded	unloaded
Sitrah (crude oil port)	9 041	150
Sulman	7	554

Airport (passenger departures and arrivals, 1976) Bahrain 882 353

Durable equipment (Dec 1975)	000	no per 1 000 people	
Radio sets	100*	390*	no per
Television sets	30	117	km of paved road
Passenger cars	23	91	770*
Commercial vehicles	10	37	330*

Production, finance and trade

Gross domestic product 1970: B D 140 mn = $ 294 mn = £ 122 mn
1976 est: B D 250**mn = $ 630**mn = £ 350**mn

Main products (000 t)	1960	1970	1976	Growth %pa 1960–70	1970–76
Dates	15	15*	16*	0.0*	1.1*
Crude oil	2 256	3 847	2 916	5.5	−4.5
Petroleum products	9 840	12 222	10 490*[a]	2.2	−3.0*[b]
Natural gas (mn m³)	na	350	2 150*[a]	na	43.8[b]
Electricity (mn kWh)	49	243	400*[a]	17.4	10.5*[b]
Aluminium	—	—	116[a]	na	na

[a]1975 [b]1970–75

Transport traffic *Air* (1976) Passenger-kilometres 320 mn, cargo 7mn t-km
Sea (1975) Goods loaded 9.0 mn t, unloaded 0.7 mn t
Consumer price index (1970 = 100) 1976 211; growth 1970–76 13.3 %pa
Money stock (end-year, B D mn) 1970 37.95, 1976 127.94; growth 1970–76 22.5 %pa
Budget (1977) Revenue: B D 235 mn = $ 594 mn = £ 340 mn
of which, oil B D 150 mn = $ 379 mn = £ 217 mn
Expenditure: B D 250 mn = $ 632 mn = £ 362 mn
International reserves (Dec 1976) $ 442 mn
External trade (1976) Imports: B D 660 mn = $ 1 672 mn = £ 926 mn
Exports: B D 532 mn = $ 1 348 mn = £ 746 mn

Main imports (1975)	% of total	Main exports (1976)	% of total
Crude oil	50	Petroleum products	74
Machinery, non-electric	10	Aluminium	8
Electrical machinery	5	Machinery	4
Food	5		
Chemicals	4		
Iron and steel	3		
Main sources (1976)		Main destinations (1976)	
Saudi Arabia	42	Japan	17
United Kingdom	10	Saudi Arabia	15
Japan	8	Singapore	9
United States	4	United States	9

Bangladesh

People's Republic of Bangladesh
Gana Prajatantri Bangladesh

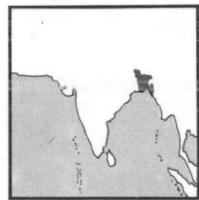

Location South Asia
With a coastline on the Bay of Bengal, mainly comprising the delta of the Ganges river, India borders the country to the west, north and east; Burma is to the south-east
Land Area 143 998 km² = 55 598 mi²
Climate Tropical monsoon
Weather at Chittagong, 27 m altitude
Temperature: hottest month June 25–31 °C, coldest Jan 13–26 °C
Rainfall (av monthly): driest month Jan 5 mm, wettest July 597 mm
Time 6 hours ahead of GMT
Measures UK (imperial) system; conversion to the metric system is planned. Local measures are also in use, including:
weight (mass) 1 tola = 4 sikis = 11.664 grams = 180 grains
1 seer = 16 chhataks = 80 tolas = 0.9333 kilogram = 2.057 pounds
1 maund = 8 punshuri = 40 seers = 37.32 kilograms = 82.29 pounds
Monetary unit Taka (Tk) = 100 paisa
Rate of exchange (1976 av): free Tk 15.35 = $ 1, free Tk 27.72 = £ 1

Summary

Political One-party republic, which became independent December 16, 1971; formerly East Pakistan, part of the Pakistan state formed from British India in 1947. Member of UN, Colombo Plan and Commonwealth
Economic An agricultural economy, with jute the main export and jute products accounting for 80 % of the total. There are plans to increase yields of jute, to release land for growing food grains. Natural gas fields are to be further exploited

People, resources and equipment

Population 1960 54.00**mn, 1970 68.12*mn, 1976 78.66*mn
Growth: 1960–70 2.4** %pa, 1970–76 2.4* %pa
Density (1976): 546* people per km²
Vital statistics (rate per 1 000 people, 1970–75): births 49.5, deaths 28.1

Bangladesh

Cities (population in 000, 1974) Dacca (capital) 1 730, Chittagong 890, Khulna 437, Narayanganj 187
Race Mainly Bengali
Language (1974) Bengali (85* %), Bihari, Hindi
Religion (1974) Moslem 85* %, Hindu, Buddhist, Christian
Education Pupils (1976) 10 000 000*, teachers (1973) 260 000*
Labour force (1976) 25 933 000; in agriculture 22 011 000 (85 %)
Personnel Scientists and engineers (1973/74): 23 500
Physicians (1973): 7 663, 1 per 9 554 people
Standard of living
National income per person (1976): Tk 1 300* = $ 85* = £ 47*
Consumption per person (1975): energy 28 kg coal equivalent, electricity (production) 18* kW h, newsprint 0.04 kg, steel 1 kg
Newspapers (1974): number 25
Telephones (Dec 1975): 80 000, 1 per 1 000 people
Livestock (000, 1976) Cattle 28 000*, buffaloes 719*, goats 12 772*, sheep 777*, chickens 31 100*, ducks 5 610*
Mineral reserves Coal (1966) 1 471 mn tonnes
Natural gas (1974) 231 000 mn cubic metres
Petroleum refinery capacity (1975) 1.68 mn tonnes
Electrical capacity (1975) 818* megawatts
Hospital beds (1975) 8 224, 1 per 9 420 people
Roads (1976) 6 300* km = 3 900* mi, density 0.04* km per km²
Railways (1976) 2 874 km = 1 786 mi, density 0.02 km per km²
Ships (registered, 1977) 133, total of 244 314 gross tons
Inland waterways (1976) 8 000* km = 5 000* mi
Ports (goods traffic, 000 tonnes, 1975/76)

	loaded	unloaded
Chittagong	553	3 161
Chalna	702	468

Airports Dacca, Chittagong

Durable equipment (at end-year)	000	no per 1 000 people	
Radio sets (1969)	531	8.0	no per
Television sets (1970)	10	0.2	km of road
Passenger cars (1976)	20	0.3	3.2
Commercial vehicles (1976)	15	0.2	2.4

Production

Gross domestic product 1975/76 (year ending June 30th):
Tk 102 785 mn = $ 6 923 mn = £ 3 469 mn
1976 est: Tk 115 000*mn = $ 7 500*mn = £ 4 100*mn
Structure of gross domestic product (1975/76) *By origin* Agriculture 56 %, manufacturing 5 %, construction 4 %, other 35 %

Production index (1970 = 100)	1960	1970	1976	Growth %pa 1960–70	1970–76
Agricultural	89ᶜ	100	104	1.7ᵈ	0.7
Main products (000 t)					
Agriculture					
Rice	14 500*	16 715	17 627	1.4*	0.9
Potatoes	332	865	903	10.0	0.7
Sweet potatoes	370ᶜ	852	790	12.7ᵈ	−1.3
Sugar cane	4 017	7 537	5 980	6.5	−3.8
Bananas	1 074	594	654*	−5.8	1.6*
Rapeseed	99	128	112	2.6	−2.2
Tea	19	32	33	5.4	0.5
Tobacco	26	41	45	4.7	1.5
Jute	809	1 248	863	4.4	−6.0
Milk, cow	610**	685	800*	1.2**	2.6*
Milk, goat	350**	502*	562*	3.7**	1.9*
Meat	180**	233*	253*	2.6**	1.4*
Fish catch	na	640*	640*	na	0.0*
Timber (000 m³)	7 000**	9 867	14 776	3.5**	7.0
Other					
Petroleum products	na	na	701ᵉ	na	na
Natural gas (mn m³)	na	na	510*	na	na
Electricity (mn kW h)	219	1 356	1 380*	20.0	0.3*
Jute textilesᵃ,ᵇ	260	597	478	8.7	−3.6
Cotton yarnᵃ	22	48	40	8.1	−3.0
Cotton fabrics (mn m)ᵃ	57	54	68	−0.5	3.9
Fertilisers, nitrogenousᵃ	32*ᶜ	43	131*	4.3*ᵈ	20.4*
Cementᵃ	96ᶜ	54	141	−7.9ᵈ	17.3
Crude steelᶠ	na	108**	76	na	−5.7**

ᵃYears ending June 30th ᵇHessian, sacking, carpet backing, packing, etc ᶜ1963
ᵈ1963–70 ᵉ1975 ᶠIngots only

Transport traffic *Passenger-kilometres* (mn)	1960	1970	1976	Growth %pa 1960–70	1970–76
Railᵃ	3 116ᵇ	3 317	3 331ᶠ	0.9ᶜ	0.1ᵍ
Air	na	na	426	na	na
Cargo: tonne-kilometres (mn)					
Railᵃ	1 736ᵇ	1 568	639ᶠ	−1.4ᶜ	−20.1ᵍ
Air	na	na	9.7	na	na
Sea: tonnes (mn)					
Goods loadedᵃ	1.16ᵈ	1.14	0.76ᶠ	−0.3ᵉ	−9.6ᵍ
Goods unloadedᵃ	3.65ᵈ	3.59	5.37ᶠ	−0.3ᵉ	10.5ᵍ

ᵃYears ending June 30th ᵇ1963 ᶜ1963–70 ᵈ1964 ᵉ1964–70 ᶠ1974 ᵍ1970–74
Tourism (1976) Number of visitors 60 680, gross receipts $ 2.1 mn

Finance and trade

Consumer price index (Dacca, government employees, 1972 = 100)
1976 252; growth 1972–76 26.0 %pa
Money stock (Dec 1976, Tk mn) 10 692
Budget (1976/77; year ending June 30th)
Revenue: Tk 9 664 mn = $ 625 mn = £ 365 mn
Expenditure: Tk 7 679 mn = $ 496 mn = £ 290 mn
(capital and development expenditure:
Tk 12 150 mn = $ 785 mn = £ 459 mn)

Balance of payments ($ mn)	1973	1974	1975	1976
Balance of goods (fob)	−432	−644	−836	−403
Balance of services	−78	−98	−146	−86
Balance of transfers	+274	+267	+412	+228
Current balance	−235	−475	−571	−260
Long-term capital flow	+115	+379	+573	+323
Reserves and debt (end-year, $ mn)				
International reserves	143	138	148	289
External public debt	830	1 844	2 538	2 884

External trade (1976)
Imports: Tk 13 192 mn = $ 859 mn = £ 476 mn
Exports: Tk 6 150 mn = $ 401 mn = £ 222 mn

Main imports (1975/76ᵃ) % of total		*Main exports* (1975/76ᵃ) % of total	
Wheat	16	Jute products	49
Chemicals	12	Jute	29
Rice	11	Leather	9
Machinery	10	Tea	4
Vegetable oils	10	Fish	3
Petroleum products	9		
Transport equipment	4		
Main sources (1976)		*Main destinations* (1976)	
United States	16	United States	16
Canada	7	United Kingdom	9
India	7	Italy	6
Japan	7	Mozambique	5
United Kingdom	6	Soviet Union	4
Iran	5	Egypt	4
Australia	4	Singapore	3

ᵃYear ending June 30th

Bhutan

Kingdom of Bhutan
Druk Yul

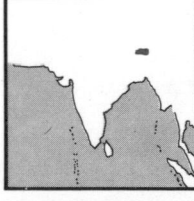

Location Central Asia
In the eastern Himalayas, with India to the south and Tibet to the north.
Land-locked
Land Area 47 000 km² = 18 000 mi², of which, forests 64 %
Climate Temperate, varying with altitude
Weather at Thimbu
Temperature (av monthly): hottest month July 17 °C, coldest Jan 4 °C
Rainfall (av monthly): 250 mm
Time 5½ hours ahead of GMT
Measures Metric system, as conversion from UK (imperial) system
Monetary unit Ngultrum (N) = 100 chetrums
The Indian rupee is also in use
Rate of exchange (1976 av): par N 1 = IR 1, free N 8.960 = $ 1, N 16.18 = £ 1

Bhutan

Summary

Political Constitutional monarchy, whose external relations were guided by the United Kingdom until 1949, and since then by India. Member of UN and Colombo Plan
Economic There has been little industrial development, and timber has been the main export. Transport is being improved, and hydro-electric stations have been set up. Tourism has become important

People, resources and equipment

Population 1960 0.84**mn, 1970 1.05*mn, 1976 1.20*mn
Growth: 1960–70 2.3**%pa, 1970–76 2.3*%pa
Density (1976): 26* people per km²
Vital statistics (rate per 1 000 people, 1970–75): births 43.6, deaths 20.5
City Thimbu (capital)
Race (1970) Tibetan 60**% (of whom, Bhotias 56 %), Gurungs 20**%, Assamese 12**%
Language Dzongkha (Tibetan/Burmese)
Religion (1970) Mahayana Buddhist 75**%, Hindu 25**%
Education (1976) Pupils 19 000, teachers 682
Labour force (1976) 588 000; in agriculture 551 000 (94 %)
Personnel (1971) Physicians: 22, 1 per 49 000 people
National income per person (1976) N 630** = $ 70** = £ 40**
Telephones (Dec 1976) 1 082, 1 per 1 000 people
Livestock (000, 1976) Cattle 198*, sheep 39*, goats 20*, pigs 56*
Electrical capacity (1974) 4 megawatts
Hospital beds (1971) 2 740, 1 per 390 people
Roads (1976) 1 500* km = 930* mi, density 0.03* km per km²
There are no railways
Airport Paro
Radio sets (Dec 1968) 3 000, 3 per 1 000 people
Motor vehicles (Dec 1971) 700, 0.6 per 1 000 people, 0.5 per km of road

Production, finance and trade

Gross domestic product 1976 est: N 800**mn = $ 90**mn = £ 50**mn
Main products *Agriculture* (000 t, 1976) Rice 280*, wheat 61*, maize 56*, potatoes 37*, jute 5*, milk 12* *Other* Fruit preserves, distillery products, textiles
Tourism Bhutan was opened-up for tourists from October 1974. Number of visitors (1976): 1 500*
Budget (1976/77; year ending March 31st)
Balanced at N 89.5 mn = $ 10 mn = £ 6 mn
External trade (1976) Imports: N 13***mn = $ 1.5***mn = £ 0.8***mn
Exports: N 9***mn = $ 1***mn = £ 0.6***mn
Main imports Textiles and light equipment
Main exports Timber, fruit and vegetables, distilled spirits, dolomite
Sources and destinations Mainly India; also Singapore

Brunei

State of Brunei
Negeri Brunei

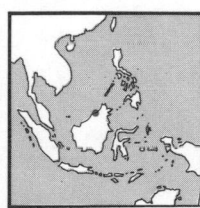

Location South-east Asia
On the north-west coast of the island of Borneo surrounded by Sarawak (Malaysia)
Land Area 5 765 km² = 2 226 mi²
Climate Tropical
Weather at Bandar Seri Begawan
Temperature: hottest month July 25*–31*°C, coldest Dec-Feb 24*–30*°C
Rainfall (av monthly): driest month Jan 110*mm, wettest Oct 465*mm
Time 8 hours ahead of GMT
Measures UK (imperial) system; also local measures, including:
weight (mass) 1 kati = 16 tahils = 0.605 kilogram = 1⅓ pounds
1 gantang = 6 katis = 3.63 kilograms = 8 pounds
1 pikul or picul = 100 katis = 60.48 kilograms = 133⅓ pounds
Monetary unit Brunei dollar (Br $) = 100 cents
Rate of exchange (1976 av): free Br $ 2.520 = $ 1, Br $ 4.550 = £ 1

Summary

Political Sultanate, with responsibility for internal affairs, but with the United Kingdom responsible for external affairs; full independence at end 1983 has been agreed. Brunei is the only Malay state which did not join in the formation of Malaysia in 1963

Economic The economy is dependent on crude oil, petroleum products and natural gas which account for virtually all exports. Transport facilities are being developed (notably a deep-water port at Muara), and new industry is to be encouraged

People, resources and equipment

Population 1960 84 000, 1970 130 000*, 1976 177 000*
Growth: 1960–70 4.5* %pa, 1970–76 5.3* %pa
Density (1976): 31* people per km²
Vital statistics (rate per 1 000 people, 1976): births 29.9, deaths 3.8
Cities (population in 000, 1976) Bandar Seri Begawanª (capital) 75*, Seria, Kuala Belait
ªFormerly called Brunei Town
Race (1976) Malay 59 %, Chinese 24 %, other indigenous 10 %
Language Malay; Chinese and English are also used
Religion (1971) Moslem 62 %, Buddhist 13 %, Christian 8 %
Education Pupils (1976) 46 770, teachers (1973) 2 371
Labour force (1971) 41 099; in agriculture 4 776 (12 %), mining and quarrying 2 915 (7 %), manufacturing 1 751 (4 %), construction 8 090 (20 %)
Personnel Scientists and engineers (1971): 589
Physicians (1974): 41, 1 per 3 660 people
Standard of living
National income per person (1976): Br $ 12 000** = $ 4 800** = £ 2 600**
Consumption per person (1975): energy 9 628 kg coal equivalent, electricity (production) 1 400 kW h, newsprint 1.4 kg
Newspapers (1972): number 1; circulation (1969) 7 000, 60 per 1 000 people
Telephones (Dec 1976): 11 070, 61 per 1 000 people
Livestock (000, 1976) Buffaloes 17*, cattle 3*, pigs 14*, chickens 894*
Mineral reserves (1975) Crude oil 254 mn tonnes
Natural gas 187 000 mn cubic metres
Petroleum refinery capacity (1975) 90 000* tonnes
Electrical capacity (1975) 81 megawatts
Hospital beds (1974) 490, 1 per 300 people
Roads (1976) 1 260* km = 780* mi, density 0.22* km per km²
Railways (1976) 10* km = 6* mi, density 0.002* km per km²
Ships (registered, 1977) 2, of 899 gross tons
Ports Muara, Bandar Seri Begawan, Kuala Belait
Airport Bandar Seri Begawan
Durable equipment (Dec 1975) Radio sets: 21 000, 125 per 1 000 people
Television sets: 10 000*, 60* per 1 000 people
Passenger cars: 22 300, 133 per 1 000 people, 17.7 per km of road
Commercial vehicles: 4 000, 24 per 1 000 people, 3.2 per km of road

Production, finance and trade

Gross domestic product
1976 est: Br $ 2 500**mn = $ 1 000**mn = £ 550**mn
Structure of gross domestic product (1971) *By origin* Agriculture 4 %, mining and quarrying 52 %, construction 10 %, other 34 %
Main products (000 t) *Agriculture* (1976) Rice 4*, cassava 3*, bananas 2*, meat 4*, fish catch 1.6, timber (000 m³) 227
Other (1975) Crude oil 8 639, petroleum products 54, natural gas (mn m³) 6 000, electricity (mn kW h) 230.
Growth of crude oil production: 1960–70 3.8 %pa, 1970–75 5.3 %pa
Transport traffic *Air* (1976) Passenger departures 83 186
Sea (1975) Goods loaded 16.2 mn t, unloaded 0.44 mn t
Budget (1976) Revenue: Br $ 1 185 mn = $ 470 mn = £ 260 mn
Expenditure: Br $ 617 mn = $ 245 mn = £ 136 mn
External trade (1976) Imports: Br $ 642 mn = $ 255 mn = £ 141 mn
Exports: Br $ 3 264 mn = $ 1 295 mn = £ 717 mn

Main imports (1974)	% of total	Main exports (1975)	% of total
Iron and steel	26	Crude oil	78
Machinery, non electric	14	Natural gas	17
Food	14	Petroleum products	4
(of which, cereals 5)			
Chemicals	8		
Metal small manufactures	7		
Motor vehicles	5		
Electrical machinery	4		
Main sources (1975)		*Main destinations* (1975)	
United States	23	Japan	78
Japan	22	South Africa	7
Singapore	17	United States	7
United Kingdom	12	Malaysia	5
Netherlands	5	Taiwan	2
Malaysia	5		
West Germany	3		

Burma

Socialist Republic of the Union of Burma
Pyidaungsu Socialist Thammada Myanma Nainggnan

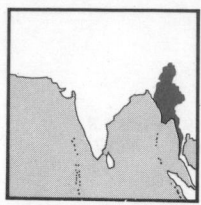

Location East Asia
With a coastline on the Bay of Bengal,
Bangladesh and India are to the west,
China to the north, and Laos and Thailand
to the east
Land Area 676 552 km² = 261 218 mi²
Climate Tropical
Weather at Rangoon, 5 m altitude
Temperature: hottest month April 24–36 °C,
coldest Jan 18–32 °C
Rainfall (av monthly): driest month Jan 3 mm, wettest July 582 mm
Time 6½ hours ahead of GMT
Measures UK (imperial) system; local measures are also in use, including:
length 1 taim or cubit = 0.457 metre = 1.5 feet
capacity 1 kwai = 4 zayoot = 4.040 litres = 0.888 UK gallon
weight (mass) 1 kati = 0.544 kilogram = 1.2 pounds
1 viss or peiktha = 3 kati = 1.633 kilograms = 3.6 pounds
Monetary unit Kyat (Kt) = 100 pyas
Rate of exchange (1976 av): par Kt 7.74289 = SDR 1,
free Kt 6.770 = $ 1, Kt 12.23 = £ 1

Summary

Political One-party socialist republic, which became independent from
the United Kingdom on January 4, 1948; a parliamentary democracy
until the 1962 revolution. Member of UN and Colombo Plan
Economic An agricultural economy with a wide spread of products, of
which the main are rice and timber. There is some mining and crude oil
production. Industry is limited. The economy has in general been closed
since 1962, but from 1977 there has been some encouragement of foreign
investment

People, resources and equipment

Population 1960 22.36*mn, 1970 27.03*mn, 1976 30.83*mn
Growth: 1960–70 1.9* %pa, 1970–76 2.2* %pa
Density (1976): 46* people per km²
Vital statistics (rate per 1 000 people, 1974): births 39.0*, deaths 12.0*
Cities (population in 000, 1973)

Rangoon (capital)	3 187*	Bassein	336*	Myingyan	220*
Mandalay	920*	Henzada	284*	Prome	148*
Moulmein	830*	Pegu	255*	Insein	144*

Race Tibeto-Burman
Language Burmese; tribal languages are also in use
Religion (1976) Buddhist 80** %; also Moslem, Hindu and Christian
Education (1976/77) Pupils 4 818 793, teachers 113 047
Labour force (1976) 11 933 000; in agriculture 8 238 000 (69 %)
Personnel (1975) Scientists and engineers: 18 500
Physicians: 5 550, 1 per 5 440 people
Standard of living
National income per person (1976): Kt 750* = $ 110* = £ 60*
Consumption per person (1975): energy 51 kg coal equivalent,
electricity (production, 1976) 32* kW h, newsprint 0.4 kg, steel 2 kg
Newspapers (1974): number 7; circulation 319 000, 11 per 1 000 people
Telephones (Dec 1976): 31 460, 1 per 1 000 people
Livestock (000, 1976) Cattle 7 300*, buffaloes 1 700*, goats 550*,
pigs 1 500*, chickens 16 000*, ducks 3 000*
Mineral reserves Coal (1960) 21 mn tonnes
Lignite (1951) 265 mn tonnes
Crude oil (1975) 10 mn tonnes
Natural gas (1975) 6 000 mn cubic metres
Petroleum refinery capacity (1975) 1.37* mn tonnes
Electrical capacity (1975) 263* megawatts
Hospital beds (1975) 20 122, 1 per 1 500 people
Roads (1975) 21 956 km = 13 643 mi, density 0.032 km per km²
Railways (1974) 4 328 km = 2 689 mi, density 0.006 km per km²
Inland waterways (1976) Irrawaddy delta 3 000* km = 2 000* mi
Ships (registered, 1977) 56, total of 67 502 gross tons
Port (goods traffic, 000 tonnes, 1976/77) Rangoon: loaded 690,
unloaded 465
Airports Mingaladon (Rangoon), Mandalay and 29 other airports with
scheduled flights

Durable equipment (Dec 1974)	000	no per 1 000 people	no per km of road
Radio sets	659	22	
Passenger cars	36	1	1.6
Commercial vehicles	39	1	1.8

Production

Gross domestic product 1975/76 (year ending September 30th):
Kt 23 519 mn = $ 3 488 mn = £ 1 832 mn
1976 est: Kt 24 800*mn = $ 3 700*mn = £2 000*mn
Growth in real terms: 1963–70 1.8 %pa, 1970–76 2.6 %pa
Structure of gross domestic product (1975/76) *By origin*
Agriculture 47 %, mining and quarrying 1 %, manufacturing 8 %,
transport and communication 3 %, other 41 %

Production index (1970 = 100)	1960	1970	1976	Growth %pa 1960–70	1970–76
Agricultural	80	100	113	2.3	2.0

Main products (000 t)
Agriculture

	1960	1970	1976	1960–70	1970–76
Rice	6 789	8 162	9 307*	1.9	2.2*
Maize	58	67	64*	1.5	−0.8*
Sugar, raw value	47	88*	75	6.5*	−2.6*
Sugar, non-centrifugal	142*	140*	135*	−0.1*	−0.6*
Dry beans	115**	140*	160*	2.0**	2.2*
Onions	40	100	103*	9.6	0.5*
Bananas	150**	176*	223*	1.6**	4.0*
Groundnuts	394	529	520*	3.0	−0.3*
Sesameseed	65	132	137	7.3	0.6
Tobacco	33	61	77*	6.3	4.0*
Jute	6	28	41	16.7	6.6
Rubber	9*	12*	16*	2.9*	4.9*
Cotton	12	15	16*	2.3	1.1*
Milk	260	325*	360*	2.1*	1.7*
Meat	95**	160*	170*	5.3**	1.0*
Fish catch	360	432	502	1.8	2.5
Timber (000 m³)	12 900**	15 797	21 655	2.0**	5.4
of which, teak (000 m³)	351	486	493*d	3.3	0.4*e
Energy					
Total energy (000 tce)	760	1 114	1 500b	3.9	6.1c
Crude oil	545	801	1 160	3.9	6.4
Petroleum products	528	877	980*b	5.2	2.2*c
Natural gas (mn m³)	21	9	15*b	−8.1	10.8*c
Electricity (mn kW h)	432	600	990*	3.3	8.7*
Mining					
Antimony (Sb content)	0.102	0.065	0.217b	−4.4	27.3c
Lead conc (Pb content)	18.0	4.5	10.5b	−13.0	18.5c
Tin conc (Sn content)	0.96	0.31	0.75b	−12.0	19.3c
Tungsten conc (oxide content)	0.58	0.28	0.44b	−7.0	9.7c
Silver	0.047	0.019	0.025b	−8.7	5.6c
Salt	148	160	98b	0.8	−9.3c
Manufacturing					
Cigarettes (mn units)	1 328	1 513	1 763b	1.3	3.1c
Cotton yarn	3.8	6.1	7.7b	4.8	4.8c
Fertilisers, nitrogenousa	—	—	47	na	na
Cement	45	156	184b	13.2	3.3c

aYears ending June 30th   b1975   c1970–75   d1974   e1970–74

Transport traffic	1960	1970	1976	Growth %pa 1960–70	1970–76
Passenger-kilometres (mn)					
Raila	1 528	2 370	2 912	4.5	3.5
Air	54	141	168	10.1	3.0
Cargo: tonne-kilometres (mn)					
Raila	781	763	382	−0.2	−10.9
Air	1.1	2.0	1.2	6.5	−8.2
Sea: tonnes (mn)					
Goods loaded	2.44	1.19	0.71	−6.9	−8.2
Goods unloaded	0.99	0.92	0.44	−0.7	−11.6

aYears ending September 30th
Tourism Number of visitors (1976) 18 280, gross receipts (1975) $ 3 mn

Finance and trade

Consumer price index (1970 = 100) 1976 285.4; growth 1970–76 19.1 %pa
Budget (1976/77; year ending March 31st)
Revenue: Kt 16 677 mn = $ 2 457 mn = £ 1 416 mn
Expenditure: Kt 17 318 mn = $ 2 552 mn = £ 1 470 mn

Balance of payments ($ mn)	1972	1973	1974	1975	1976
Balance of goods (fob)	−40	−55	−33	−87	−38
Balance of services	−24	−22	+11	−9	−9
Balance of transfers	+18	+18	+14	+17	+16
Current balance	−46	−60	−7	−79	−30
Long-term capital flow	+12	+64	+40	+40	+16

Reserves and debt (end-year, $ mn)

International reserves	52	100	191	141	126
External public debt	299	426	508	543	711

Burma

External trade (1976) Imports: Kt 1 311 mn = $ 194 mn = £ 107 mn
Exports: Kt 1 260 mn = $ 186 mn = £ 103 mn

Main imports (1974)	% of total	Main exports (1976)	% of total
Machinery	22	Rice	58
Crude oil and products	10	Teak	23
Textile yarns and fabrics	10	Fruit and vegetables	5[a]
Chemicals	10	Non-ferrous metals	5[a]
Railway vehicles	6	Jute	4[a]
Food	6	Fertilisers	4[a]
Oils and fats	5		
Motor vehicles	4		
Main sources (1975)		Main destinations (1975)	
Japan	33	Indonesia	14
China	9	Singapore	13
United Kingdom	8	Sri Lanka	11
United States	7	Japan	10
West Germany	7	Netherlands	6
South Korea	5	Mauritius	6

[a]1974

Special focus

The fluctuating importance of rice as an export earner

% of total exports	1960	1970	1971	1972	1973	1974	1975	1976
Rice	67	62	50	36	13[a]	42	47	58
Teak	8	13	23	26	34	26	24	23

[a]Affected by partial suspension of exports

Cambodia

Democratic Kampuchea

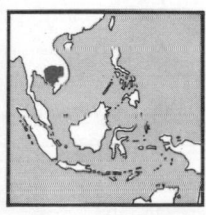

Location South-east Asia
With a coastline on the Gulf of Siam, Thailand is to the west and north, Laos to the north, and Vietnam to the east and south
Land Area 181 035 km² = 69 898 mi²
Climate Tropical
Weather at Phnom Penh
Temperature: average annual 27 °C
Rainfall: monsoon season is April to October
Time 7 hours ahead of GMT
Measures Metric system; also local measures, including: *weight (mass)*
1 damleng (taël) = 10 chi = 100 hun = 37.5 grams = 1.323 ounces
1 hap (picul) = 100 néal (livres) = 60 kilograms = 132.3 pounds
Monetary unit Riel (C Rl) = 100 sen
Rate of exchange (1976 av): nominal C Rls 1 140 = $ 1, C Rls 2 060 = £ 1
The general internal use of currency was suspended from 1975, a system of rationing being used

Summary

Political Communist republic from April 12, 1975, following a period of civil war; formerly called Khmer Republic. Before 1953 under French control. Member of UN
Economic An agricultural economy affected in the 1970s by the civil war; rubber, rice and timber have been main exports, but rice production fell markedly during the conflict. The new government is reported to be concentrating on production of rice for export, following a dispersal of city-dwellers to the country. Salt production is also important

People, resources and equipment

Population 1960 5.44*mn, 1970 7.06*mn, 1976 8.35*mn
Growth: 1960–70 2.6* %pa, 1970–74 2.8* %pa
Density (1976): 46* people per km²
Vital statistics (rate per 1 000 people, 1970–75): births 46.7*, deaths 19.0*
Cities (population in 000)
Phnom Penh (capital, 1977) 200*** (1962: 394, 1974: 2 000***)
Battambang 60**, Kompong Som (formerly called Sihanoukville), Kompong Chhnang
Race (1962) Khmer 93 %, Vietnamese 4 %, Chinese 3 %
Language Khmer
Religion (1970) Buddhist 85** %
Education (1972/73) Pupils 595 000*, teachers 24 500*

Labour force (1976) 3 280 000; in agriculture 2 483 000 (76 %)
Personnel (1971) Physicians: 438, 1 per 16 575 people
Standard of living National income per person (1976):
C Rls 100 000*** = $ 90*** = £50***
Consumption per person (1975): energy 16 kg coal equivalent, electricity (production) 18* kW h, steel 1 kg
Newspapers (1974): number 17
Telephones (Dec 1975): 71 000, 9 per 1 000 people
Livestock (000, 1976) Cattle 1 900**, buffaloes 870**, pigs 900**
Petroleum refinery capacity (1975) 600 000* tonnes
Electrical capacity (1975) 50* megawatts
Hospital beds (1971) 7 500, 1 per 968 people
Roads (1973) 15 029 km = 9 339 mi, density 0.08 km per km²
Railways (1977) 650** km = 400** mi, density 0.002** km per km²
Inland waterways (1977)
Mekong and Tonlé-Sap rivers 1 400* km = 900* mi
Ships (registered, 1977) 3, total of 3 558 gross tons
Port Kompong Som
Airport Pochentong (Phnom Penh)

Durable equipment (end-year)	000	no per 1 000 people	
Radio sets (1975)	110	14	no per km of road
Television sets (1974)	26	3	
Passenger cars (1972)	27	4	1.8
Commercial vehicles (1973)	11	1	0.7

Production

Gross domestic product 1969: C Rls 43 900 mn = $ 790 mn = £ 330 mn
1976 est: C Rls 900 000***mn = $ 800***mn = £440***mn

Production index (1970 = 100)	1960	1970	1976	Growth %pa 1960–70	1970–76
Agricultural	65	100	58	4.4	−8.6
Main products (000 t)					
Rice	2 335	3 814	1 800*	5.0	−11.8*
Maize	157	137	70*	−1.3	−10.6*
Bananas	130**	134	90*	0.3**	−6.4*
Tobacco	5.6	9.7	7.0*	5.7	−5.3*
Jute	1.0*	8.0*	3.0*	23.1*	−15.1*
Rubber	37	13	15*	−10.0	2.4*
Meat	40**	73*	73*	6.2**	0.0*
Fish catch	140**	52	85*	−9.4*	8.5*
Timber (000 m³)	3 300**	4 070	1 570*	2.1**	2.0*
Petroleum products	—	271	—	na	na
Electricity (mn kW h)	60	133	150*a	8.3	2.5*b
Salt	37	52	27*a	3.5	−12.3*b
Cigarettes (mn units)	1 272	3 874	2 622c	11.8	−12.2d
Cement	na	38	50*a	na	5.6*b

a1975 b1970–75 c1973 d1970–73
Transport traffic (1973) *Passenger-kilometres* Rail 54 mn, air 51 mn
Cargo: tonne-kilometres Rail 10 mn, air 0.5 mn
Sea Goods loaded 50 000 t, unloaded 583 000 t
Tourism (1973) Number of visitors 16 500

Finance and trade

Consumer price index (1970 = 100) 1973 554.2; growth 1970–73 77.0 %pa
Budget (1974) Revenue: C Rls 22 800 mn = $ 24 mn = £ 10 mn
Expenditure: C Rls 71 000 mn = $ 73 mn = £ 31 mn
External trade (1973) Imports: C Rls 14 200* mn = $ 70* mn = £ 28* mn
Exports: C Rls 2 733* mn = $ 13* mn = £ 5* mn

Main imports (1969)	% of total	Main exports (1973)	% of total
Machinery	19	Rubber	93
Chemicals (of which, medicinal 8)	17	Vegetables	5*
Motor vehicles	12		
Iron and steel	10		
Food	9		
Textile yarns and fabrics	6		
Main sources (1973)		Main destinations (1973)	
United States	69*	Hongkong	23*
Thailand	11*	Japan	22*
Singapore	5*	Malaysia	18*
Japan	5*	France	12*

Special focus

Rice production per person

	Kilograms		Kilograms		Kilograms
1965	407	1972	287	1975	185*
1970	544	1973	124	1976	216*
1971	376	1974	80		

China

People's Republic of China
Chung-hua Jen-min Kung-ho-kuo

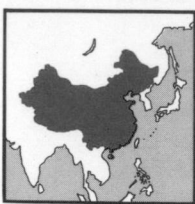

Location Asia
Comprises the main centre of the continent of Asia, with a coastline on the Yellow Sea, East China Sea and South China Sea; North Korea has a border in the north-east, Mongolia is to the north, Soviet Union is to the north and west, and to the south are Afghanistan, Pakistan, India, Nepal, Bhutan, Burma, Laos and Vietnam
Land Area 9 561 000 km² = 3 692 000 mi²
Usage (1975): pastures 2 130 000**km² (22** %), arable and cropland 1 280 000**km² (13** %), forests 1 500 000**km² (16** %)
Climate Continental, with extremes of temperature; sub-tropical in the south-east
Weather at Shanghai, 7 m altitude
Temperature: hottest months July, Aug 23–32 °C, coldest Jan 1–8 °C
Rainfall (av monthly): driest month Dec 36 mm, wettest June 180 mm
Time Zone I (Urumchi): 6 hours ahead of GMT
Zones II–IV (Chungking, Lanchow, Peking, Shanghai, Harbin): 8 hours ahead of GMT
Measures Metric system; also old Chinese measures, some of which were fixed in metric terms in 1959 (shown as 'new' below), including:
length old: 1 chih (Chinese foot) = 0.371 metre = 1.219 feet
new: 1 chih = ⅓ metre = 1.094 feet
new: 1 li = 1 500 chih = 0.5 kilometre = 0.31 mile
weight (mass) old: 1 chin (catty) = 0.605 kilogram = 1⅓ pounds
new: 1 shin chin (catty) = 0.5 kilogram = 1.102 pounds
Monetary unit Yuan (Y) = 10 chiao (jiao) = 100 fen;
the currency is called renminbi, and the renminbi yuan (RMB Y) is used for foreign currency
Rate of exchange (1976 av): free RMB Y 1.95 = $ 1, RMB Y 3.52 = £ 1

Summary

Political Communist republic, proclaimed as such in October 1949. China has claims to the territories of Hongkong, Macao and Taiwan. Here all figures refer to China 'mainland', with separate sections for Hongkong, Macao and Taiwan. Control was established over Tibet in 1951. Member of UN
Economic A broadly based economy with expanding mining and industrial sectors; crude oil has been an important growth area in the 1970s. Virtually all the information indicated is based on estimates; there are few official Chinese figures

People, resources and equipment

Population[a] 1960 650**mn, 1970 770**mn, 1976 850**mn
[a]Based on UN estimates; some sources indicate up to 1 000 mn for 1976. Also see special focus
Growth: 1960–70 1.7**%pa, 1970–76 1.7**%pa
Density (1976): 87** people per km²
Vital statistics (rate per 1 000 people, 1970–75): births 27**, deaths 10**
Regions (provinces unless otherwise shown, population in millions, 1977; total of 870** mn)

North-west	57**	*North-east*[c]	84**	*East*	258**
Ch'ing-hai	3**	Heilungkiang	28**	Anhwei	40**
Kansu	16**	Kirin	20**	Chekiang	36*[d]
Shensi	25**	Liaoning	36*[d]	Fukien	24*[d]
Ninghsia-Hui[a]	3**	*North*	92**	Kiangsi	26*[d]
Sinkiang-Uighur[a]	10**	Hopeh	45**	Kiangsu	52**
South-west	147**	Shansi	24*[d]	Shantung	70*[d]
Kweichow	25*[d]	Inner Mongolia[a]	8**	Shanghai[b]	10**
Szechwan	90*[d]	Peking[b]	8**	*Central south*	232**
Yunnan	30*[d]	Tientsin[b]	7*[d]	Honan	58**
Tibet[a]	2**			Hupeh	42*[d]
				Hunan	50*[d]
[a]Autonomous region	[b]Municipality			Kwangtung	50*[d]
[c]Also known as Manchuria				Kwangsi Chuang[a]	32*[d]
[d]Official figure announced 1976–78					

Cities (population in 000, 1976)

Peking (capital)	8 000**	Canton	3 000**[a]	Sian	1 500**[a]
Shanghai	10 000**	Wuhan	2 200**[a]	Nan-ch'ing	1 400**[a]
Tientsin	7 000**	Ch'ung-ch'ing	2 100**[a]	Ch'eng-tu	1 100**[a]
Shenyang	4 000**[a]	Ch'ang-ch'un	1 800**[a]	Ch'ing-tao	1 100**[a]
Lüta	3 600**[a]	Harbin	1 600**[a]	Fushun	1 000**[a]
[a]1965					

Race (1953) Han 94 %, Chuang 1 %, Uighur 1 %, Yi 1 %
Language Mainly Putonghua (also called Mandarin), based on northern Chinese (the Peking dialect); local dialects and languages are also used
Religion Pre-Communist estimates were: Confucian 40** %, Buddhist 20** %, Taoist 4** %, Moslem 3** %
Education Pupils (1975) 200 000 000*, teachers (1960) 3 000 000*
Labour force (1976) 400 000 000**; in agriculture 250 000 000** (63** %)
Personnel (1974) Physicians: 170 000**, 1 per 4 800** people
Standard of living
National income per person (1976): Y 750*** = $ 380*** = £ 210***
Consumption per person (1975): energy 660** kg coal equivalent, electricity 140** kW h, newsprint 1.2** kg, steel 42** kg
Newspapers (1955): number 392; circulation 12 mn, 19 per 1 000 people
Telephones (1976): 5 000 000**, 6** per 1 000 people
Livestock (000, 1976) Cattle 64 000**, buffaloes 30 000**, pigs 235 000**, sheep 74 000**, goats 60 500**, chickens 1 300 000**
Mineral reserves Coal 1 000 000**mn tonnes
Lignite 700**mn tonnes
Crude oil 2 360**mn tonnes
Petroleum refinery capacity (1974) 50**mn tonnes
Electrical capacity (1977) 40 000* megawatts
Hospital beds (1962) 660 000**, 1 per 1 020** people
Roads (1976) 808 000**km = 502 000**mi, density 0.08**km per km²
Railways (1976) 48 000**km = 30 000**mi, density 0.005**km per km²
Inland waterways (1976) Yangtze and other rivers and waterways: 160 000**km = 100 000**mi
Oil pipelines (1976) 3 500** km = 2 200** mi
Ships (registered, 1977) 622, total of 4 245 446 gross tons
Ports Shanghai, Tientsin, Ch'ing-tao, Talien (Dairen), Canton, Whampoa, Chan-chiang, Chinhuangtao, Lienyunkang
Airports Peking, Shanghai, Canton, Shenyang and 14 other airports with scheduled flights

Durable equipment (at end-year)	000	no per 1 000 people	
Radio sets (1970)	12 000**	16**	no per
Television sets (1976)	350**	0.4**	km of road
Passenger cars (1976)	50***	0.06**	0.06***
Commercial vehicles (1976)	1 500**	1.8**	1.9**

Production

Gross domestic product
1976 est: Y 660 000***mn = $ 340 000***mn = £ 190 000***mn
Growth in real terms: 1957–70 5.0** % pa, 1970–74 6.5** % pa
Structure of gross domestic product (1976) *By origin* Agriculture 25** %, manufacturing 50** %, other 25** %

Production index (1970 = 100)	1960	1970	1976	Growth %pa 1960–70	1970–76
Agricultural	74**	100	115**	3.1**	2.4**

Main products (000 t)
Agriculture

	1960	1970	1976	1960–70	1970–76
Wheat	22 200**	31 000**	43 000**	3.4**	5.6**
Rice	77 000**	102 000**	114 000**	2.8**	1.9**
Maize	19 000**	29 000**	34 000**	4.3**	2.7**
Barley	13 000**	19 000**	20 000**	3.9**	0.9**
Millet	15 000**	22 000**	20 000**	3.9**	−1.6**
Sweet potatoes	70 000**	103 000**	112 000**	4.0**	1.4**
Potatoes	23 500**	25 000**	41 000**	0.6**	8.6**
Sugar, raw value	1 260**	2 200**	4 000**	5.7**	10.5**
Soyabeans	10 200**	11 600**	12 000**	1.3**	0.6**
Groundnuts	1 900**	2 500**	2 800**	2.8**	1.9**
Cottonseed	2 700**	3 400**	4 800**	2.3**	6.0**
Tea	150**	170**	320**	1.2**	11.1**
Tobacco	520**	780**	950**	4.1**	3.3**
Jute	320**	490**	600**	4.3**	3.4**
Cotton	1 370**	1 700**	2 400**	2.2**	5.9**
Milk	2 600**	3 200**	3 700**	2.1**	2.4**
Eggs	2 500**	3 300**	3 700**	2.8**	1.9**
Beef and veal	1 800**	2 200**	2 000**	2.0**	−1.6**
Pigmeat	7 400**	8 400**	9 900**	1.3**	2.8**
Fish catch	5 800**	6 200**	6 800**	0.7**	1.5**
Timber (000 m³)	140 000**	170 000**	194 000**	2.0**	2.2**

Energy

	1960	1970	1976	1960–70	1970–76
Total energy (000 tce)	270 000**	380 000**	570 000**	3.5**	7.0**
Coal and lignite	250 000**	360 000**	450 000**	3.7**	3.8**
Crude oil	5 500**	20 000**	85 000**	13.8**	27.3**
Petroleum products	5 000**	20 000**	60 000**	15.0**	20.0**
Electricity (mn kW h)	58 000**	75 000**	120 000**	2.6**	8.1**

China

Main products (000 t)	1960	1970	1976	Growth %pa 1960–70	1970–76
Mining					
Iron ore					
(Fe content)	30 000**	24 000**	34 000**	−2.2**	6.0**
Bauxite	350**	500**	1 000**	3.6**	12.3**
Lead ore (Pb content)	80**	100**	100**	2.2**	0.0**
Magnesite	1 000**	1 000**	1 000**	0.0**	0.0**
Manganese ore					
(Mn content)	360**	300**	300**	−1.8**	0.0**
Tungsten conc					
(oxide content)	13.6**	7.6**	11.5**	−5.6**	7.2**
Zinc ore (Zn content)	80**	100**	100**	2.2**	0.0**
Salt	13 000**	15 000**	30 000**	1.4**	12.3**
Asbestos	80**	170**	150**	7.8**	−2.1**
Manufacturing					
Cotton yarn	1 683	1 500**	na	−0.8**	na
Wood pulp	600**	1 200**	1 600**	7.2**	4.9**
Newsprint	280**	550**	950**c	7.0**	11.5**d
Other paper	2 300**	3 400**	4 500**	4.0**	4.8**
Fertilisers, nitrogenous	300**	1 200**	3 000**	15.0**	16.5**
Fertilisers, phosphate	200**	700**	1 200**	13.4**	9.4**
Cement	12 270a	15 000**	30 000**	1.8**h	12.3**
Coke	25 000**	18 000**	30 000**c	−3.2**	10.8**d
Pig iron	27 000**	22 000**	25 000**c	−2.0**	2.6**d
Crude steel	18 000**	18 000**	22 000**	0.0**	3.4**

a1959 b1959–70 c1975 d1970–75

Transport traffic *Passenger-kilometres* Rail (1959) 45 670 mn, air 64 mn
Cargo: tonne-kilometres Rail (1971) 301 000*mn, air (1960) 2 mn

Finance and trade

Budget (1960; latest published)
Balanced at Y 70 020 mn = $ 35 010 mn = £ 12 500 mn
International reserves (1976) $ 2 500**mn
External trade (1976)
Imports: Y 11 900**mn = $ 6 100**mn = £ 3 400**mn
Exports: Y 14 000**mn = $ 7 200**mn = £ 4 000**mn

Main imports (1973)	% of total	Main exports (1976)	% of total
Machinery	17***	Food	30***
Iron and steel	17***	(of which, meat 10***)	
Cereals	17***	Textile yarns and fabrics	15***
Chemicals	10***	Crude oil	12***
(of which,		Clothing	6***
fertilisers 5***)		Textile fibres	6***
Non-ferrous metals	8***	Chemicals	5***
Main sources (1976)		*Main destinations* (1976)	
Japan	30*	Hongkong	20*
West Germany	11*	Japan	17*
France	6*	West Germany	3*
Australia	5*	Singapore	3*
Rumania	4*	Rumania	3*
Soviet Union	4*	North Korea	3***
North Korea	4***	Vietnam	3***
Canada	4*	United States	3*
United States	2*	Soviet Union	2*

Special focus

Population estimates
Starting point: 1953 census with reported total population 582.6 mna;
there are various estimates of the amount of under-enumeration at
the 1953 census, and the position is shown for 0 %, 5 % and 10 %
under-enumeration. The 1949 and 1953 censuses indicated a growth
rate for 1949–53 of 1.8 %pa

Assumed growth 1953–76 %pa	Resulting population 1976 (mn) by amount of under-statement at 1953 census		
	0 %	5 %	10 %
1.4	803	843	883
1.5	820	861	902
1.6	839	882	924
1.7	858	902	944
1.8	879	923	967
1.9	899	945h	990
2.0	921	967	1 012
2.1	939	987	1 034
2.2	962	1 010	1 058

aExcludes 7.6 mn for Taiwan bRoughly as indicated by US semi-official estimates
(950 mn)

Hongkong
Colony of Hongkong

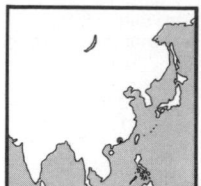

Location East Asia
The island of Hongkong is off the south-east
coast of China, about 150 km from Canton,
and the colony includes part of Kowloon
peninsula on the mainland, and other islands
Land Area 1 046 km² = 404 mi²
Climate Sub-tropical
Weather at Victoria, 33 m altitude
Temperature: hottest months July, Aug
26–31 °C, coldest Feb 13–17 °C
Rainfall (av monthly): driest month Dec 30 mm, wettest June 394 mm
Time 8 hours ahead of GMT
Measures UK (imperial) system, changing to metric system; local
measures are also used, including:
length 1 chih (Chinese foot) = 10 ts'un = 0.371 metre = 1.219 feet
1 li (Chinese mile) = 0.557 kilometre = 0.346 mile
weight (*mass*) 1 tael (leung) = 10 mace (tsin) = 37.8 grams = 1⅓ ounces
1 picul (tam) = 100 catty (kan) = 1 600 taels = 60.48 kilograms =
133⅓ pounds
Monetary unit Hongkong dollar (HK $) = 100 cents
Rate of exchange (1976 av): free HK $ 4.874 = $ 1, HK $ 8.803 = £ 1

Summary

Political UK colony from 1841, when the main island was ceded by
China; Kowloon and the New Territories were added by lease in 1860
and 1898 respectively, the latter on a 99 year lease
Economic An industrial economy, with manufacturing using nearly
one-half of the labour force, and supplying about one-quarter of gross
domestic product; manufactures are mainly from light industry, especially
textiles. Tourism and finance are important

People, resources and equipment

Population 1960 3.06*mn, 1970 3.96*mn, 1976 4.44*mn
Growth: 1960–70 2.6* %pa, 1970–76 1.9* %pa
Density (1976): 4 245* people per km²
Vital statistics (rate per 1 000 people, 1976): births 17.2, deaths 5.2
Regions (population in 000, 1976; total of 4.40 mn)
Hongkong island 1 027, Kowloon 750, New Kowloon 1 629,
New Territories 938, Marine 59
Cities (population in 000, 1976) Victoria (capital, on Hongkong island)
845, Kowloon (on mainland) 2 378a, Tsuen Wan 455, Yuen Long 40
aIncluding New Kowloon
Race Chinese 98** % (see also usage of language)
Language English and Chinese
Usage (1966): Cantonese (including Tanka) 81 %, Hokio 8 %,
Hakka 3 %, Sze Yap 3 %, Shanghai and Mandarin 3 %, English 1 %
Religion Mainly Buddhists; also Confucian, Taoist, Christian
Education (1973/74) Pupils 1 099 885, teachers 37 072
Labour force (1976) 1 952 000; in agriculture 49 040 (3 %),
manufacturing 867 310 (44 %), construction 110 150 (6 %), transport
and communication 142 000 (7 %)
Personnel (1975) Physicians: 2 880, 1 per 1 528 people
Standard of living
National income per person (1976): HK $ 10 200* = $ 2 090* = £ 1 160*
Consumption per person (1975): energy 1 119 kg coal equivalent,
electricity (production, 1976) 1 878 kW h, newsprint 13 kg, steel 138 kg
Newspapers (1973): number 113; circulation 1 362 000,
327 per 1 000 people
Telephones (Dec 1976): 1 132 400, 253 per 1 000 people
Livestock (000, 1976) Cattle 11, pigs 440, chickens 4 016, ducks 267
Electrical capacity (1975) 2 274 megawatts
Hospital beds (1976) 18 156, 1 per 245 people
Roads (1976) 1 085 km = 674 mi, density 1.04 km per km²
Railways (1976) 34 km = 21 mi, density 0.032 km per km²
Ships (registered, 1977) 113, total of 609 679 gross tons
Ports (goods traffic, 000 tonnes, 1976)
Hongkong: loaded 5 967, unloaded 17 374
Airport (1976) Kai Tak (Kowloon): passenger departures 2 267 000,
arrivals 2 149 000

Durable equipment (at end-year)	000	no per 1 000 people	
Radio sets (1975)	2 505	574	no per km of road
Television sets (1974)	785	185	
Passenger cars (1976)	121	27	112
Commercial vehicles (1976)	48	11	44

Hongkong

Production

Gross domestic product 1976: HK $ 47 403 mn = $ 9 726 mn = $ 5 385 mn
Growth in real terms: 1963–70 9.3 %pa, 1970–76 7.7 %pa
Structure of gross domestic product (1974) *By origin* Agriculture 2 %,
manufacturing 24 %, construction 4 %, transport and communication 6 %,
distribution and hotels 20 %, other 44 %

Main products (000 t)	1960	1970	1976	Growth %pa	
				1960–70	1970–76
Agriculture					
Rice	27	16	4*	−5.1	−20.6*
Cabbages	36**	30*	27*	−1.8**	−1.7*
Milk	6	7	5*	1.5	−5.5*
Pigmeat	24**	20**	21*	−1.8**	0.8**
Poultrymeat	4**	11*	15*	10.6**	5.3*
Fish catch	62	135	158	8.1	2.6
Other					
Manufactured gas (mn m³)	22	61	112	10.7	10.7
Electricity (mn kW h)	2 061ᵃ	5 097	8 340	13.8ᵇ	8.5
Iron ore (Fe content)	67	95	21	3.6	−22.2
Beer (000 hl)	119	238	238ᵉ	7.2	0.0ᶠ
Cigarettes (mn units)	4 015	6 402	4 550ᶜ	4.8	−6.6ᵈ
Cotton yarn	78	149	196	6.7	4.7
Cotton fabrics (mn m²)	387	646	823	5.3	4.1
Wool yarn	1.7	12.1	6.1ᶜ	21.7	−12.8ᵈ
Cement	150	430	764	11.1	10.0

ᵃ1963 ᵇ1963–70 ᶜ1975 ᵈ1970–75 ᵉ1971 ᶠ1970–71

Transport traffic	1960	1970	1976	Growth %pa	
Passenger-kilometres (mn)				1960–70	1970–76
Road	na	13 739	16 417	na	3.0
Rail	152	177	251	1.5	6.0
Air	194	999	2 632ᵃ	17.8	21.4ᵇ
Cargo: tonne-kilometres (mn)					
Rail	12	30	46	9.6	7.4
Air	4	16	78ᵃ	16.0	37.3ᵇ
Sea: tonnes (mn)					
Goods loaded	2.2	3.0	6.0	3.4	12.1
Goods unloaded	5.9	10.6	17.4	6.0	8.5
Tourism					
Number of visitors (000)	164	927	1 560	18.9	9.1
Gross receipts ($ mn)	na	293	740	na	16.7

ᵃ1975 ᵇ1970–75

Finance and trade

Consumer price index (1970 = 100)ᵃ 1976 158; growth 1970–76 7.9 %pa
ᵃBased on index with 1973/74 = 100, taking 1973/74 = 141.1 with base 1970 = 100
Money stock (end-year, HK $ mn) 1970 6 548, 1976 16 820;
growth 1970–76 17.0 %pa
Budget (1976/77; year ending March 31st)
Revenue: HK $ 6 431 mn = $ 1 341 mn = £ 773 mn
Expenditure: HK $ 5 289 mn = $ 1 103 mn = £ 636 mn
External public debt (Dec 1976) $ 413 mn
External trade (1976)
Imports: HK $ 43 292 mn = $ 8 882 mn = £ 4 918 mn
Exports: HK $ 41 557 mn = $ 8 526 mn = £ 4 721 mn

Main imports	% of total	Main exports	% of total
Food	15	Clothing	35
(of which, livestock 3)		Instruments	13
Textile yarns & fabrics	15	Electrical machinery	12
Electrical machinery	10	(of which, tele-	
Chemicals	8	communications 6)	
Machinery, non-electric	6	Textile yarns & fabrics	10
Petroleum products	6	Toys, games, etc.	5
Precious stones	5	Chemicals	3
Textile fibres	5	Precious stones	3
Main sources		*Main destinations*	
Japan	22	United States	29
China	18	West Germany	10
United States	12	United Kingdom	8
Taiwan	7	Japan	7
Singapore	6	Singapore	4

Special focus

Tourism, 1976

Origin of visitors	Number	%		Number	%
Japan	437 931	28	Taiwan	77 794	5
United States	238 605	15	Malaysia	63 395	4
Australia	157 360	10	United Kingdom	57 783	4
Thailand	88 240	6	Total	1 559 977	100

India

Bharat

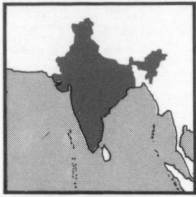

Location South Asia
The Indian peninsula has the Bay of
Bengal to the east and Arabian Sea to
the west (parts of the Indian Ocean);
Pakistan is to the west, China, Nepal
and Bhutan to the north, and Burma to
the east, with Bangladesh in the east
mainly surrounded by Indian territory
Land Area 3 287 590 km² = 1 269 346 mi²,
including 222 870 km² for Jammu and Kashmir.
Area excluding Pakistani held part of Jammu and Kashmir (83 807 km²):
3 203 783 km² = 1 236 988 mi²
Usage (1975): agricultural 1 797 000* km² (55* %), of which,
arable 1 625 000*km² (49* %), pastures 125 500*km² (4* %),
cropland 47 000*km² (1* %); forests 674 000*km² (21* %)
Climate Tropical; cool in highlands and mild in winter
Weather at New Delhi, 218 m altitude
Temperature: hottest month May 26–41 °C, coldest Jan 7–21 °C
Rainfall (av monthly): driest month Nov 3 mm, wettest July 180 mm
Time 5½ hours ahead of GMT
The Indian (Saka) year ends March 21st in the Gregorian calendar; the
Saka year 1900 corresponds to 1978/79
Measures Metric system
Monetary unit Indian rupee (I R) = 100 paisa
1 lakh = 1,00,000 = 100 000 = one hundred thousand
1 crore = 100 lakhs = 1,00,00,000 = 10 000 000 = ten million
Rate of exchange (1976 av): free I Rs 8.960 = $ 1, I Rs 16.18 = £ 1

Summary

Political Republic, formed as a Union of States, which became
independent from United Kingdom on August 15, 1947; Goa and other
Portuguese enclaves were absorbed in December 1961 and Sikkim was
absorbed in 1974 to become the 22nd state in 1975. There have been
conflicts with Pakistan, including a dispute over the status of
Jammu and Kashmir. Member of UN, Colombo Plan and Commonwealth
Economic Agriculture provides about 40 % of gross domestic
product, and food accounts for one-quarter of exports. Manufacturing is
also important, with textiles, iron and steel large exports. There are large
mineral reserves.

People

Populationᵃ 1960 429.18* mn, 1970 539.08*mn, 1976 610.08* mn
ᵃIncludes Indian-held part of Jammu and Kashmir
Growth: 1960–70 2.3* %pa, 1970–76 2.1* %pa
Density (1976): 186* people per km²
Vital statistics (rate per 1 000 people, 1973): births 34.6*, deaths 15.5*
Regions (population in 000, 1971; total of 548.2 mn, including Sikkim)

States			Maharashtra	50 412	Union territories	
Andhra Pradesh	43 503		Manipur	1 073	Andaman and	
Assam	14 625		Meghalaya	1 012	Nicobar Islands	115
Bihar	56 353		Nagaland	516	Arunachal Pradesh	468
Gujarat	26 697		Orissa	21 945	Chandigarh	257
Haryana	10 037		Punjab	13 551	Dadra and	
Himachal Pradesh	3 460		Rajasthan	25 766	Nagar Haveli	74
Jammu and			Sikkimᵇ	209	Delhi	4 066
Kashmirᵃ	4 617		Tamil Nadu	41 199	Goa, Daman & Diu	858
Karnatakaᶜ	29 299		Tripura	1 556	Lakshadweepᵈ	32
Kerala	21 347		Uttar Pradesh	88 341	Mizoramᵉ	332
Madhya Pradesh	41 654		West Bengal	44 312	Pondicherry	472

ᵃIndian-held part only ᵇWhich became a state in 1975 ᶜFormerly Mysore
ᵈFormerly Laccadive, Minicoy and Amindivi Islands ᵉFormerly part of Assam
Cities (population in 000, 1971)

New Delhi (capital)	302	Coimbatore	736	Baroda	467
Calcutta	7 031	Madurai	712	Tiruchirappalli	465
Bombay	5 971	Jaipur	637	Amritsar	458
Delhi	3 345	Agra	635	Jamshedpur	456
Madras	3 170	Varanasi	607	Trivandrum	455*
Hyderabad	1 796	Indore	561	Cochin	439
Ahmedabad	1 742	Jabalpur	535	Dhanbad	434
Bangalore	1 654	Allahabad	513	Salem	416
Kanpur	1 275	Ernakulam	495*	Gwalior	406
Poona	1 135	Surat	493	Ludhiana	401
Nagpur	930	Patna	491	Sholapur	398
Lucknow	814	Vadodara	467	Ulhasnagar	396

India

Race Mainly Indo-Aryan and Dravidian; also Mongoloid and Australoid; see also usage of language
Language Hindi and English; also other languages, as shown by usage
Usage (1961): Hindi 30 %, Telugu 9 %, Bengali 8 %, Marathi 8 %, Tamil 7 %, Urdu 5 %, Gujarati 5 %, Kannada 4 %, Malayalam 4 %, Bihari 4 %, Oriya 4 %, Rajasthani 3 %, Punjabi 2 %, Assamese 1 %
Religion (1971) Hindu 83 %, Moslem 11 %, Christian 3 %, Sikh 2 %, Buddhist 1 %, Jain 1 %
Education (1975/76) Pupils: primary 66 000 000*, secondary and vocational 24 900 000*, higher(1974/75) 2 230 225. Teachers: primary, secondary and vocational 2 560 000*, higher (1970/71) 119 000
Labour force (1971)

Economic activity	Number	% of total
Agriculture	129 963 000	72
Mining and quarrying	922 800	1
Manufacturing (incl repair services)	17 067 500	9
Electricity, gas and water	532 400	⅓
Construction	2 215 300	1
Distribution and hotels	8 748 300	5
Transport and communication	4 401 200	2
Finance and real estate	1 289 900	1
Other	15 233 000	9
Total	180 373 400	100

Personnel Scientists and engineers (1970): 1 187 500
Physicians (1973): 138 000, 1 per 4 162 people
Standard of living
National income per person (1976): I Rs 1 180* = $ 132* = £ 73*
Consumption per person (1975): energy 221 kg coal equivalent, electricity (production) 143* kW h, newsprint 0.3 kg, steel 14 kg
Newspapers (1974): number 822; circulation 9.2 mn, 16 per 1 000 people
Telephones (Dec 1976): number 2 095 960, 3 per 1 000 people

Resources and equipment

Livestock (000, 1976) Cattle 180 330*, buffaloes 61 090*, sheep 40 200*, pigs 7 060*, goats 70 400*, camels 1 180*, chickens 142 000*
Mineral reserves Coal (1972) 80 953 mn tonnes
Lignite (1972) 2 026 mn tonnes
Crude oil (1975) 141 mn tonnes
Natural gas (1975) 88 000 mn cubic metres
Uranium (1974) 3 400 tonnes
Petroleum refinery capacity (1975) 27.8 mn tonnes
Electrical capacity (1975) 22 172* megawatts
Hospital beds (1968) 325 500*, 1 per 1 571* people
Roads (1974) 1 232 300 km = 765 716 mi, density 0.37 km per km²
Railways (1975) 60 301 km = 37 469 mi, density 0.018 km per km²
Inland waterways (1976) 8 000* km = 5 000* mi
Ships (registered, 1977) 566, total of 5 482 176 gross tons
Ports (goods traffic, 000 tonnes, 1976/77)

	International loaded	International unloaded	Coastwise loaded	Coastwise unloaded	Total
Bombay	4 012	10 750	1 226	1 359	17 347
Mormugao	12 836	207	1	411	13 455
Vishakhapatnam	6 088	2 318	73	218	8 697
Calcutta	2 110	4 222	1 047	639	8 018
Madras	2 839	4 426	208	189	7 662
Cochin	419	3 331	768	231	4 749
Kandla	167	2 323	17	807	3 314
Paradip	3 149	147	—	2	3 298

Airports (passenger traffic, 000, 1975)

	departures	arrivals
Bombay	1 196	1 113
Delhi	831	807
Calcutta	482	475
Madras	270	263

Also (1977) 56 other airports with scheduled flights

Durable equipment (at end year)	000	no per 1 000 people	no per km of road
Radio sets (1975)	14 075	24	
Television sets (1974)	275	0.5	
Passenger cars (1976)	799	1.3	0.6
Commercial vehicles (1976)	537	0.9	0.4

Production

Gross domestic product 1975/76 (year ending March 31st):
I Rs 678.1 bn = $ 78 370 mn = £ 36 890 mn
1976 est: I Rs 760* bn = $ 85 000* mn = £ 47 000* mn
Growth in real terms: 1960–70 3.9 %pa, 1970–76 2.4 %pa

Structure of gross domestic product (1974/75) *By origin*
Agriculture 43 %, mining and quarrying 1 %, manufacturing 14 %, electricity, gas and water 1 %, construction 4 %, distribution and hotels 11 %, transport and communication 4 %, other 22 %
By type Government final consumption 8* %, private final consumption 74* %, stock investment 4* %, gross fixed capital formation 15* %, exports of goods and services 5* %, less imports of goods and services – 6* %

Production indices (1970 = 100)	1960	1970	1976	Growth %pa 1960–70	Growth %pa 1970–76
Agricultural	81	100	109	2.1	1.5
Industrial	55.3	100.0	131.7	6.1	4.7

Main products (000 t)

Agriculture	1960	1970	1976	1960–70	1970–76
Wheat	10 326	20 093	28 846	6.9	6.2
Rice	51 865	63 338	64 363	2.0	0.3
Maize	4 015	7 486	6 257	6.4	−2.9
Millet	6 886	12 176	9 544	5.9	−4.0
Sorghum	9 363	8 105	10 396	−1.4	4.2
Cassava	1 951	5 214	6 307	10.3	3.2
Groundnuts	4 812	6 111	5 262	2.4	−2.5
Sugar, raw value	2 814	4 634	4 630*	5.1	0.0*
Sugar, non-centrifugal[a]	4 255	7 100	7 000*	5.2	−0.2*
Mangoes	7 000**	8 300	8 847*	1.7**	1.1*
Coffee	50	64	84	2.5	4.6
Tea	326	418	511	2.5	3.4
Tobacco	292	337	347	1.4	0.5
Jute[b]	1 024	1 222	1 275	1.8	0.7
Cotton	1 008	954	1 019	−0.5	1.1
Rubber	24	82	148	13.1	10.3
Milk, cow	6 700**	7 630*	8 400*	1.3**	1.6*
Milk, buffalo	11 018*	13 800*	16 347*	2.3*	2.9*
Cheese	1 100**	1 408*	1 550*	2.5**	1.6*
Meat	550**	750*	829*	3.1**	1.7*
Fish catch	1 161	1 756	2 400	4.2	5.3
Timber (000 m³)	87 000**	110 770	130 947*	2.4**	2.8*
Energy					
Total energy (000 tce)	54 150	87 570	114 620[g]	4.9	5.5[h]
Coal	52 593	73 698	100 991	3.4	5.4
Lignite	47	3 545	3 895	54.0	1.6
Crude oil	454	6 809	8 623	31.1	4.0
Petroleum products	5 683	16 876	20 280*[g]	11.5	3.7*[h]
Natural gas (mn m³)	—	488	1 208	na	16.3
Electricity (mn kW h)[c]	17 794	56 543	85 613*	12.3	7.2*
of which,					
hydro (mn kW h)[c]	7 040	23 056	33 247	12.6	6.3
nuclear (mn kW h)[c]	—	1 339	2 627	na	11.9
Mining					
Iron ore (Fe content)	10 131	19 654	26 868*	6.8	5.3*
Bauxite	387	1 374	1 437	13.5	0.7
Chromium ore (oxide content)	50	135	242[g]	10.4	12.4[h]
Magnesite	156	354	314*	8.5	−2.4[h]
Manganese ore (Mn content)	544	632	575[g]	1.5	−1.9[h]
Salt	3 436	5 588	5 918[i]	5.0	1.4[j]
Manufacturing					
Beer (000 hl)	106[d]	311	583[g]	12.7[e]	13.4[h]
Cigarettes (mn units)	36 971	62 930	60 064[g]	5.5	−0.9[h]
Cotton yarn	788	965	1 028	2.0	1.1
Cotton fabrics (mn m)	6 629	7 849	8 034[g]	1.7	0.5[h]
Jute manufactures	1 085	954	1 195	−1.3	3.8
Man-made fibres	43	134*	153*[g]	12.0*	2.7*[h]
Newsprint	23	37	58	4.9	7.7
Other paper	400**	823	830[g]	7.5**	0.2[h]
Tyres (000 units)	1 264	3 058	3 640	9.2	3.0
Sulphuric acid	354	1 089	1 689	11.9	7.6
Plastics and resins	14[d]	95	88	23.7[e]	−1.3
Fertilisers, nitrogenous[f]	87	705	1 508	23.3	13.5
Fertilisers, phosphate[f]	53	225	320	15.6	6.0
Cement	7 845	13 956	18 499	5.9	4.8
Coke	4 778	8 679	8 846[g]	6.1	0.4[h]
Pig iron	4 275	7 118	10 000	5.2	5.8
Crude steel	3 200**	6 286	9 144	7.0**	6.4
Aluminium	18	161	210	2.5	4.5
Sewing machines (000 units)	297	246	361	−1.9	6.6
Radio sets (000 units)	268	1 771	1 677	20.8	−0.9
Motor vehicles (000 units)	52	86	81	5.1	−1.0
Merchant vessels (000 grt)	13	29	46	8.3	8.0

[a]Mainly gur [b]Including jute-like fibres [c]Years ending March 31st [d]1961 [e]1961–70 [f]Years ending June 30th [g]1975 [h]1970–75 [i]1974 [j]1970–74

India

Transport traffic

Transport traffic	1960	1970	1976	Growth %pa	
Passenger-kilometres (mn)				1960–70	1970–76
Road	na	105 150	130 000[d]	na	5.4[e]
Rail[a]	74 190	113 738	134 750	4.4	2.9
Air	1 115	3 555	7 196	12.3	12.5
Cargo: tonne-kilometres (mn)					
Road	na	45 300	65 000[d]	na	9.4[e]
Rail[a]	69 120	128 304	143 100	6.4	1.8
Air	46.5	115.2	279.3	9.5	15.9
Sea: tonnes (mn)[a]					
Goods loaded	6.7*	27.0	30.7[b]	15.0*	2.6[c]
Goods unloaded	18.0*	23.0	31.5[b]	2.5*	6.5[c]
Tourism					
Number of visitors (000)	123	281	545	8.6	11.7
Gross receipts ($ mn)	na	39	250	na	36.3

[a]Years ending March 31st [b]1975 [c]1970–75 [d]1974 [e]1970–74

Finance and trade

Price indices	1960	1970	1976	Growth %pa	
(1970 = 100)				1960–70	1970–76
Consumer prices	55	100	161	6.1	8.2
Wholesale prices	56	100	172	6.0	9.5
Share prices	89.5	100.0	100.1	1.1	0.0

Money stock
(end-year, I Rs bn) 27.4 67.7 147.2 9.5 13.8

Budget (central government, 1976/77; year ending March 31st)
Revenue: I Rs 81 320 mn = $ 9 098 mn = £ 5 243 mn
Expenditure: I Rs 102 920 mn = $ 11 515 mn = £ 6 636 mn
of which, defence 20 %

Balance of payments ($ mn)	1972	1973	1974	1975	1976
Balance of goods (fob)	+68	−191	−625	−288	na
Balance of services	−503	−607	−491	−468	na
Balance of transfers	+281	+268	+2 322	+609	na
Current balance	*−153*	*−529*	*+1 207*	*−148*	*na*
Long-term capital flow	+322	+467	−959	+928	na

Reserves and debt (end-year, $ mn)

International reserves	1 180	1 142	1 325	1 373	3 074
External public debt	12 007	13 076	14 405	14 580	na

External trade (1976/77; year ending March 31st)
Imports: I Rs 50 151 mn = $ 5 611 mn = £ 3 233 mn
Exports: I Rs 49 809 mn = $ 5 573 mn = £ 3 211 mn

Main imports	% of total	Main exports	% of total
Crude oil	23	Food	25
Wheat	16	(of which, tea 6,	
Machinery, non-electric	13	animal fodder 5)	
Chemicals	9	Textile yarns and fabrics	14
(of which, fertilisers 4)		(of which, cotton fabrics 7,	
Petroleum products	5	jute fabrics 4)	
Iron and steel	4	Iron and steel	8
Textile fibres	4	Clothing	7
Precious stones	4	Leather	5
		Precious stones	5
Main sources		*Main destinations*	
United States	21	United States	11
Iran	10	Japan	11
Saudi Arabia	7	United Kingdom	10
United Kingdom	6	Soviet Union	9
Soviet Union	6	West Germany	4
West Germany	6	Netherlands	4

Special focus

Population by main state, 1971

	mn	approximately equivalent country (by population)
Uttar Pradesh	88	Brazil
Bihar	56	United Kingdom
Maharashtra	50	France
West Bengal	44	
Andhra Pradesh	44	
Madhya Pradesh	42	Philippines or Thailand
Tamil Nadu	41	
Karnataka	29	Burma
Gujarat	27	Ethiopia
Rajasthan	26	Argentina
Orissa	22	Canada
Kerala	21	Rumania
Assam	15	Czechoslovakia
Punjab	14	Peru

Indonesia

Republic of Indonesia
Republik Indonesia

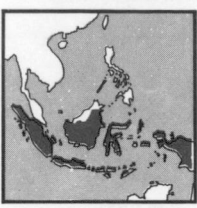

Location South-east Asia
A group of 13 667 islands off the coast of
mainland south-east Asia, of which 931
inhabited. Main islands or parts of islands are
Java, Sumatra, Sulawesi (Celebes), Maluku
(Moluccas), Kalimantan (Borneo) and
Irian Jaya (West Irian); in Borneo there is a
border to the north with Malaysia (Sabah and
Sarawak), in New Guinea a border to the east
with Papua New Guinea and in Timor a border
to the east with East Timor
Land Area 1 904 345 km² = 735 272 mi²,
of which, West Irian 412 781 km² = 159 376 mi²; excludes East Timor
Climate Tropical
Weather at Jakarta, 8 m altitude
Temperature: hottest months April, May 24–31 °C, coldest Jan, Feb
23–29 °C
Rainfall (av monthly): driest month Aug 43 mm, wettest Jan, Feb 300 mm
Time　hours ahead of GMT
Western zone[a]　7
Central zone[b]　8
Eastern zone[c]　9
[a]Java, Sumatra, Bali, Bangka, Billiton, Lombok, Madura　[b]Kalimantan, Sulawesi,
Flores, Soembawa, Soemba, Timor　[c]Maluku, Tanimbar, Kai, Aroe, Irian Jaya
Measures Metric system; local measures are also in use, including:
length 1 Java paal = 1.507 kilometres = 0.936 mile
area 1 bahu = 0.710 hectare = 1.754 acres
weight (mass) 1 picol = 100 catty = 61.76 kilograms = 136.2 pounds
Monetary unit Rupiah (Rp) = 100 sen
Rate of exchange (1976 av): par Rp 415 = $ 1, free Rp 749.6 = £ 1

Summary

Political Republic, which became independent from Netherlands on
December 27, 1949. West Irian was included in the Republic from
May 1, 1963; figures here include West Irian throughout unless
otherwise specified. Indonesia has included East Timor in its territory.
Member of UN, Colombo Plan, Opec and Asean
Economic An agricultural and oil economy; agriculture employs about
two-thirds of the labour force and provides about one-third of gross
domestic product; oil is the main export and makes up about one-fifth of
gross domestic product. Timber and rubber are also important exports.
Revenue from oil has helped expansion in the 1970s

People, resources and equipment

Population 1960 94.2*mn, 1970 119.47*mn, 1976 139.62*mn
Growth: 1960–70 2.4* %pa, 1970–76 2.6* %pa
Density (1976): 73* people per km²
Vital statistics (rate per 1 000 people, 1970–75): births 43*, deaths 17*
Regions (population in 000, 1974; total of 125.3 mn)

Bali	2 217	Maluku[c]	1 187	Sumatra	22 658
Java and Madura	79 004	Nusa Tenggara	4 718	Irian Jaya[d]	1 007
Kalimantan[a]	5 574	Sulawesi[b]	8 964		

[a]Borneo [b]Celebes [c]Moluccas [d]West Irian
Cities (population in 000, 1971)

Jakarta (capital)	4 576[c]	Jogyakarta	342	Telanaipura[b]	159
Surabaja	1 556	Banjarmasin	282	Pakanbaru	145
Bandung	1 202	Pontianak	218	Samarinda	138
Semarang	647	Teluk Betung	199	Balikpapan	137
Medan	636	Padang	196	Madiun	136
Palembang	583	Bogor	196	Pematang Siantar	129
Ujung Pandang[a]	435	Kediri	179	Pakalangan	112
Malang	422	Cirebon	179	Magelang	110
Surakarta	414	Manado	170	Tegal	106

[a]Formerly Makassar [b]Formerly Jambi [c]1976 estimate: 5 193 thousand
Race (1961) Javanese 45 %, Sundanese 14 %, Madurese 7 %,
Ruia 5 %, Minangkabus 4 %, Buginese 3 %, Batak 2 %, Chinese 2 %
Language Bahasa Indonesian; also many other languages and dialects
Religion (1971) Moslem 94 %, Christian 5 %, Hindu and other 1 %
Education (1975) Pupils 16 000 000*, teachers 525 000*
Labour force (1971) 41 261 216; in agriculture 26 473 477 (64 %),
mining and quarrying 85 828 (¼ %), manufacturing 2 681 952 (6 %),
construction 678 472 (2 %)
Personnel (1974) Physicians: 7 027*, 1 per 18 860* people

Indonesia

Standard of living

National income per person (1976): Rp 100 470 = $ 242 = £ 134
Consumption per person (1975): energy 178 kg coal equivalent,
electricity (production) 25 kW h, newsprint 0.4 kg, steel 10 kg
Newspapers (1974): number 170
Telephones (Dec 1976): 314 400, 2 per 1 000 people
Livestock (000, 1976) Cattle 6 765*, buffaloes 2 790*, goats 7 480*,
sheep 3 190*, pigs 4 360*, chickens 115 380*
Mineral reserves Coal (1972) 573*mn tonnes
Lignite (1950) 1 960*mn tonnes
Crude oil (1975) 1 614 mn tonnes
Natural gas (1975) 510 000 mn cubic metres
Petroleum refinery capacity (1975) 20.8 mn tonnes
Electrical capacity (1975) 1 100* megawatts
Hospital beds (1973) 88 086, 1 per 1 470 people
Roads (1974) 95 544 km = 59 368 mi, density 0.050 km per km^a
Railways (1976) 7 610 km = 4 729 mi, density 0.004 km per km^a
Ships (registered, 1977) 1 032, total of 1 163 173 gross tons
Ports (goods traffic, 000 tonnes, 1976)

	International		Coastwise		Total
	loaded	unloaded	loaded	unloaded	
Dumai^a	47 625	52	3 911	32	51 620
Tandjungpriok^b	124	4 282	403	3 939	8 748
Palembang^a	819	217	4 931	2 767	8 734
Surabaya	586	1 463	946	656	3 651
Belawan	878	876	199	1 116	3 069
Semarang	78	492	102	136	808

^{a}Crude oil port ^{b}1975

Airports (passengers, 1976) Jakarta: Halim Perdanak, departures 606 016,
arrivals 628 225; Kemajoran, departures 847 190, arrivals 839 012; also
Surabaja, Bandung, Semarang and 32 other airports with scheduled flights

Durable equipment (at end-year)	000	no per 1 000 people	
Radio sets (1975)	5 010*	37*	no per
Television sets (1975)	430*	3*	km of road
Passenger cars (1976)	420	3.0	4.4
Commercial vehicles (1976)	262	1.9	2.7

Production

Gross domestic product 1976: Rp 15 467 bn = $ 37 270 mn = £ 20 634 mn
Structure of gross domestic product (1975) By origin Agriculture 33 %,
mining and quarrying 20 %, manufacturing 9 %, construction 5 %,
transport and communication 4 %, other 29 %
By type Government final consumption 13 %, private final consumption
66 %, gross capital formation (including stocks) 20 %, exports of goods
and services 23 %, less imports of goods and services −22 %

Production index (1970 = 100)	1960	1970	1976	Growth %pa	
				1960–70	1970–76
Agricultural	84	100	120	1.7	3.1

Main products (000 t)	1960	1970	1976	1960–70	1970–76
Agriculture					
Rice	13 151	19 337	23 300	3.9	3.1
Maize	2 460	2 825	2 572	1.4	−1.6
Sweet potatoes	2 670	2 175	2 478*	−2.0	2.2*
Cassava	11 376	10 478	12 500*	−0.8	3.0*
Sugar, raw value	675	708	1 150*	0.5	8.4*
Bananas	1 200**	1 700**	3 100*	3.5**	10.5**
Soyabeans	443	498	482	1.2	−0.5
Groundnuts	427	468	550	0.9	2.7
Copra	636	748	949*	1.6	4.0*
Palm oil	141	217	439	4.4	12.5
Tea	79	65	73	−1.9	2.0
Coffee	96	185	179	6.8	−0.5
Tobacco	69	78	80*	1.2	0.4*
Rubber	619	809	848	2.7	0.8
Beef and buffalo meat	150**	180*	175*	1.8**	−0.5*
Fish catch	761	1 228	1 448	4.9	2.8
Timber (000 m³)	80 700	110 685	129 831	3.2	2.7
Energy					
Total energy (000 tce)	31 270	60 350	103 080^b	6.8	11.3^c
Coal	658	172	193	−12.6	1.9
Crude oil	20 844	42 103	74 029	7.3	9.8
Petroleum products	11 035	7 687*	12 750*^{b}	−3.5*	10.6*^{c}
Natural gas (mn m³)	2 431	3 072	8 840	2.4	19.3
Manufactured gas (mn m³)	63	33	43*^{b}	−6.3	5.4*^{c}
Electricity (mn kW h)	1 161	2 100	3 345*^{b}	6.1	9.7*^{c}

Main products (000 t)	1960	1970	1976	Growth %pa	
Mining				1960–70	1970–76
Bauxite	396	1 229	938	12.0	−4.4
Manganese ore (Mn content)	5.7*	4.6*	6.5*^{b}	−2.1*	7.1*^{c}
Nickel ore (Ni content)	0.4*	15.6*	21.0*^{b}	44.2*	6.1*^{c}
Tin conc (Sn content)	23.0	19.1	22.2	−1.8	2.5
Gold (000 kg)	0.18	0.24	0.32^b	3.0	6.3^c
Diamonds (000 CM)	na	20*	15*^{b}	na	−5.6*^{c}
Manufacturing					
Beer (000 hl)	176	189	492^b	0.7	21.1^c
Cigarettes (mn units)	21 198	32 530	41 866^d	4.4	8.8^e
Cotton yarn	8	31	54^d	14.5	20.2^e
Fertilisers, nitrogenous^a	—	39*	208*	na	32.2*
Cement	387	515	881^b	2.9	11.3^c

^{a}Years ending June 30th ^{b}1975 ^{c}1970–75 ^{d}1973 ^{e}1970–73

Transport traffic	1960	1970	1976	Growth %pa	
Passenger-kilometres (mn)				1960–70	1970–76
Road	na	39 124	60 009^a	na	23.8^b
Rail	7 255	3 378	3 258	−7.4	−0.6
Air	259	878	3 112	13.0	23.5
Cargo:tonne-kilometres (mn)					
Road	na	11 196	20 799^a	na	36.3^b
Rail	1 159	855	718	−3.0	−2.9
Air	5.5	15.6	47.6	11.0	20.4
Sea: tonnes (mn)					
Goods loaded	16.0	44.1	83.7	10.7	11.3
Goods unloaded	3.13	3.71	12.06	1.7	21.7

^{a}1972 ^{b}1970–72

Tourism Number of visitors (1974) 313 500, gross receipts (1975) $ 34 mn

Finance and trade

Consumer price index (1970 = 100) 1965 0.01, 1976 292;
growth 1965–70 144.5 %pa, 1970–76 19.6 %pa
Money stock (end-year, Rp bn) 1965 2.6, 1970 250.2, 1976 1 600.9;
growth 1965–70 149.6 %pa, 1970–76 36.3 %pa
Budget (1976/77; year ending March 31st)
Revenue: Rp 2 942 bn = $ 7 089 mn = £ 4 086 mn
Expenditure: Rp 3 178 bn = $ 7 658 mn = £ 4 414 mn

Balance of payments ($ mn)	1972	1973	1974	1975	1976
Balance of goods (fob)	+348	+552	+2 631	+1 419	+1 798
Balance of services	−732	−1 082	−2 083	−2 555	−2 721
Balance of transfers	+51	+55	+49	+27	+15
Current balance	−334	−476	+598	−1 109	−907
Long-term capital flow	+501	+521	+492	+2 244	+2 011
Reserves and debt (end-year, $ mn)					
International reserves	574	807	1 492	586	1 499
External public debt	5 098	6 780	9 195	11 764	14 555

External trade (1976) Imports: Rp 2 354 bn = $ 5 673 mn = £ 3 140 mn
Exports: Rp 3 547 bn = $ 8 547 mn = £ 4 732 mn

Main imports	% of total	Main exports	% of total
Machinery, non-electric	18	Crude oil	66
Food	14	Timber	9
(of which, rice 8)		Rubber	6
Electrical machinery	13	Petroleum products	4
Chemicals	10	Coffee	3
Iron and steel	8	Metal ores	3
Motor vehicles	8	Palm oil	2
Petroleum products	7		
Main sources		*Main destinations*	
Japan	26	Japan	42
United States	17	United States	29
Singapore	10	Singapore	8
West Germany	9	Trinidad & Tobago	7
Thailand	4	Netherlands	3

Special focus

Main exports (% of total exports)

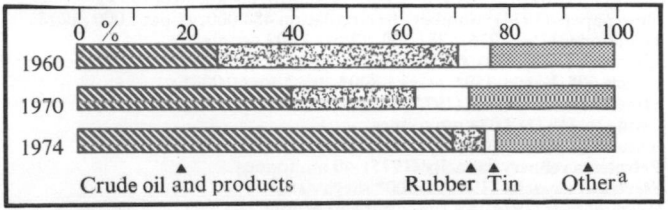

	0 %	20	40	60	80	100
1960						
1970						
1974						

Crude oil and products   Rubber Tin   Other^a

^{a}Includes timber

Iran

Empire of Iran
Keshuar-e Shahanshahi-ye Iran

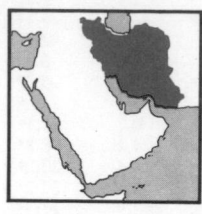

Location Western Asia
There are coastlines to the south on the Gulf and the Oman Sea and to the north on the Caspian Sea; Soviet Union is to the north, Afghanistan and Pakistan to the east, and Turkey and Iraq to the west
Land Area 1 648 000 km² = 636 000 mi²
Usage (1975): forests 180 000* km² (11* %), arable 159 000* km² (10* %), pastures 110 000* km² (7* %); the remainder is mainly desert
Climate Continental, with extremes of temperature
Weather at Teheran, 1 220 m altitude
Temperature: hottest month July 22–37 °C, coldest Jan minus 3–7 °C
Rainfall (av monthly): driest month July 3 mm, wettest Jan 46 mm
Time 4 hours ahead of GMT (summer time 5 hours ahead of GMT)
The Iranian year (Moslem solar) ends March 20th; the Iranian year 1354 corresponded to 1975/76; from March 21st, 1976, the calendar year was altered to date from the coronation of Cyrus the Great, so that the Gregorian year 1976/77 was the Iranian year 2535. The year was changed back to the Moslem year in August 1978
Measures Metric system, compulsory since 1933; some local measures are in unofficial use, including:
area 1 jerib = 0.108 hectare = 1 294 square yards
capacity 1 artaba = 50 chemicas = 0.66 hectolitre = 1.81 UK bushels
weight (mass) 1 rey = 11.88 kilograms = 26.19 pounds
Monetary unit Rial (Rl) = 100 dinars
Rate of exchange (1976 av): par Rls 82.2425 = SDR 1, free Rls 70.22 = $ 1, Rls 126.8 = £ 1

Summary

Political Constitutional monarchy led by the Shah. Formerly called Persia. The oil industry was nationalised in 1951. Member of UN, Opec, Cento and RCD
Economic An oil economy, with some diversification and plans to develop a strong industrial sector by using oil revenues. The revised Fifth development plan 1973–78 has provided total expenditure of some $ 119 000 mn, with some emphasis on defence

People, resources and equipment

Population 1960 21.52*mn, 1970 28.66*mn, 1976 33.59*mn
Growth: 1960–70 2.9 %pa, 1970–76 2.7* %pa
Density (1976): 20* people per km²
Vital statistics (rate per 1 000 people, 1976): births 42.5, deaths 16*
Cities (population in 000, 1976)

Teheran (capital)	4 496	Abadan	296	Ardabil	148
Esfahan	672	Kermanshah	291	Khorramshahr	147
Mashhad	670	Qom	247	Kerman	140
Tabriz	599	Rasht	187	Karaj	139
Shiraz	416	Rezaiyeh	164	Qazvin	139
Ahvaz	329	Hamadan	156	Yazd	136

Race (1970) Indo-European 66 %, Turk 25 %, Kurd 5 %, Semitic Arab 4 %
Language Farsi (Persian); Turkish and Kurdish are also in use
Religion (1966) Moslem 96 % (Shiah 93* %, Sunni 3* %), Christian 1 %
Education (1974/75) Pupils 6 244 078, teachers 216 745
Labour force (1972) 7 725 000; in agriculture 3 699 000 (48 %), mining and quarrying 15 000 (0.2 %), manufacturing 1 420 000 (18 %), construction 535 000 (7 %)
Personnel Scientists and engineers (1972): 127 793
Physicians (1973): 11 373, 1 per 2 752 people
Standard of living
National income per person (1976): Rls 119 200* = $ 1 700* = £ 940*
Consumption per person (1975): energy 1 353 kg coal equivalent, electricity (production) 450* kW h, newsprint 0.7 kg, steel 163 kg
Newspapers (1974): number 20; circulation 484 000, 15 per 1 000 people
Telephones (Dec 1976): 781 500, 23 per 1 000 people
Livestock (000, 1976) Cattle 6 650*, sheep 35 300*, goats 14 300*, camels 60*, horses 350*, asses 1 800*, chickens 60 000*
Mineral reserves Coal (1972) 385 mn tonnes
Crude oil (1975) 9 084 mn tonnes
Natural gas (1975) 10 613 bn cubic metres
Petroleum refinery capacity (1975) 40 mn tonnes
Electrical capacity (1975) 6 000* megawatts
Hospital beds (1974) 49 194, 1 per 650 people

Roads (1975) 52 000 km = 32 000 mi, density 0.032 km per km²
Railways (1976) 4 944 km = 3 072 mi, density 0.003 km per km²
Ships (registered, 1977) 193, total of 1 002 061 gross tons
Ports (goods traffic, 000 tonnes, 1976)

	loaded	unloaded
Kharg (crude oil terminal)	190 344	—
Bandar Khorramshahr	197	4 694

Also: Bandar Shahpur, Bushehr, Bandar Abbas, Bandar Pahlavi (Caspian Sea)
Airports (passengers, 000, 1976)

	departures	arrivals
Mehrabad (Teheran)	1 406	1 387
Abadan	250	253

Also (1977) 16 other airports with scheduled flights

Durable equipment (at end-year)	000	no per 1 000 people	
Radio sets (1974)	8 000	246	no per km of road
Television sets (1975)	1 700	51	
Passenger cars (1974)	589	18	11.3
Commercial vehicles (1974)	111	3	2.1

Production

Gross domestic product 1976/77 (year ending March 20th):
Rls 4 689 bn = $ 66 390 mn = £ 38 280 mn
1976 est: Rls 4 410* bn = $ 62 800* mn = £ 34 780* mn
Growth in real terms: 1960–70 9.2 %pa, 1970–75 9.2 %pa
Structure of gross domestic product (1975/76) *By origin* Agriculture 9 %, mining, quarrying and manufacturing 46 %, electricity, gas and water 1 %, construction 8 %, distribution and hotels 5 %, transport and communication 4 %, other 27 %

Production indices (1970 = 100)	1960	1970	1976	Growth %pa 1960–70	1970–76
Agricultural	75	100	135	2.9	5.1
Industrial[a]	30	100	142[o]	12.8	19.2[p]
Main products (000 t)					
Agriculture					
Wheat	2 923	4 262	6 044	3.8	6.0
Rice	709	1 350	1 566	6.7	2.5
Barley	808	1 083	1 487	3.0	5.4
Sugar, raw value	88*	624	734*	21.6*	2.7*
Watermelons	600**	700**	870*	1.5**	3.7**
Grapes	460**	620**	910*	3.0**	6.6*
Dates	141*	310*	310*	8.2*	0.0*
Cottonseed	179	288	296*	4.9	0.5*
Tea	7	20	24*	11.1	3.1*
Tobacco	11	17	20*	4.4	2.8*
Cotton	99	153	160*	4.4	0.7*
Milk, cow	1 200*	925*	1 250*	−2.6*	5.1*
Milk, sheep	540*	525*	617*	−0.3*	2.7*
Sheep and goat meat	200**	224*	219*	1.1**	−0.4*
Timber (000 m³)	6 200*	6 991	6 239*	1.2*	−1.9*
Energy					
Total energy (000 tce)	69 610	264 950	426 640[b]	14.3	10.0[c]
Coal[a]	230	530	1 150[b]	8.7	16.8[c]
Crude oil	52 392	191 740	296 500	13.8	7.5
Petroleum products	16 715	24 400*	33 380*[b]	3.8*	6.5*[c]
Natural gas (mn m³)	950	11 223	22 476	28.0	12.3
Electricity (mn kW h)[h]	2 446[h]	7 004	15 000*[b]	19.3[i]	16.4*[c]
Mining					
Iron ore (Fe content)[a]	—	3	610*[b]	na	189.5*[c]
Chromium ore (oxide content)[a]	33	86	84*[b]	10.0	−0.5*[c]
Zinc ore (Zn content)	9*	58	66[b]	20.5*	2.6[c]
Salt[a]	130	390	400[d]	11.6	0.6[c]
Manufacturing					
Beer (000 hl)	60	260	370[f]	15.8	19.3[e]
Cigarettes (mn units)	8 361	11 898	20 269[b]	3.6	11.2[c]
Cotton yarn	28	37	42[f]	2.8	7.1[e]
Cotton fabrics (mn m)	236	450	482[f]	6.7	3.5[e]
Wool yarn	7	25	29[f]	13.5	8.9[e]
Fertilisers, nitrogenous[l]	—	26	126	na	30.1
Fertilisers, phosphate[l]	—	—	83	na	na
Aluminium	—	—	50[b]	na	na
Cement	797	2 575	3 900*[b]	12.4	8.7*[c]
Radio sets (000 units)[j]	—	165	281[m]	na	19.4[n]
Television sets (000 units)	—	125	242[m]	na	24.6[n]
Motor vehicles (000 units)[k]	3	46	108[d]	31.8	23.6[c]

[a] Years beginning March 21st [b] 1975 [c] 1970–75 [d] 1974 [e] 1970–74 [f] 1972
[g] 1970–72 [h] 1964 [i] 1964–70 [j] Sales [k] Assembly only [l] Years ending June 30th
[m] 1973 [n] 1970–73 [o] 1972 [p] 1970–72

Iran

Transport traffic

	1960	1970	1976	Growth %pa	
				1960–70	1970–76
Passenger-kilometres (mn)					
Road	na	36 866	59 054[b]	na	9.9[c]
Rail	1 644	1 800	2 126[d]	0.9	4.2[e]
Air	76	684	3 059	24.6	28.4
Cargo: tonne-kilometres (mn)					
Road	na	43 200	69 030[b]	na	9.8[c]
Rail	2 150	2 720	4 917[d]	2.4	16.0[e]
Air	6.0	8.6	74.8	3.7	43.4
Sea: tonnes (mn)					
Goods loaded[a]	46	180*	191	14.6*	1.0*
Goods unloaded[a]	1.9	2.5	13.6	2.8	32.6
Tourism					
Number of visitors (000)	na	299	628	na	13.2
Gross receipts ($ mn)	na	45	142	na	21.1

[a]Years beginning March 21st [b]1975 [c]1970–75 [d]1974 [e]1970–74

Finance and trade

Price indices	1960	1970	1976	Growth %pa	
(1970 = 100)				1960–70	1970–76
Consumer prices	84.4	100.0	174.8	1.7	9.8
Wholesale prices	85.2	100.0	171.9	1.6	9.4
Money stock					
(end-year, Rls bn)	50.0[a]	128.3	668.0	12.5[b]	31.7

[a]1962 [b]1962–70

Budget (1976/77; year ending March 20th)
Revenue: Rls 1 802 bn = $ 25 515 mn = £ 14 710 mn
Expenditure: Rls 1 850 bn = $ 26 196 mn = £ 15 103 mn

Balance of payments ($ mn)

	1972	1973	1974	1975	1976
Balance of goods (fob)	+1 375	+2 137	+14 099	+7 300	+8 624
Balance of services	−1 768	−1 980	−1 799	−2 574	−3 540
Balance of transfers	+4	−2	−33	−18	−19
Current balance	−388	+154	+12 267	+4 707	+5 064
Long-term capital flow	+622	+448	−5 110	−3 281	−4 645
Reserves and debt (end-year, $ mn)					
International reserves	960	1 236	8 383	8 897	8 833
External public debt[a]	5 957	7 063	6 005	5 134	6 707

[a]At end-March of following year

External trade (1976)
Imports: Rls 905 200 mn = $ 12 890 mn = £ 7 139 mn
Exports: Rls 1 651 300 mn = $ 23 516 mn = £ 13 023 mn

Main imports (1975/76)[a]	% of total	*Main exports* (1975/76)[a]	% of total
Machinery, non-electric	22	Crude oil	91
Iron and steel	16	Petroleum products	5
Food	13	Cotton	1
(of which, cereals 5)		Carpets	1
Motor vehicles	12		
Chemicals	7		
Electrical machinery	7		
Textile yarns and fabrics	3		
Main sources (1976)		*Main destinations* (1976)	
West Germany	18	Japan	22
United States	17	West Germany	10
Japan	16	United Kingdom	9
United Kingdom	8	Netherlands	8
Italy	5	United States	8
France	5	France	7
Switzerland	3	Italy	6
Belgium-Luxembourg	3	US Virgin Islands	6

[a]Year ending March 20th

Special focus

Fifth development plan, 1972/73 to 1977/78 (1975 revision)
(converted at Rls 70 = $ 1)

Government finance
Revenue $ 119 bn, of which, 80 % from oil

Expenditure	$ bn	%
Current	48	41
of which: defence	28	24
social affairs	11	10
Fixed capital formation	40	34
Repayments of foreign debts	6	5
Foreign investments	11	9
Other	13	11
Total	119	100

Iraq

Republic of Iraq
Al Jumhuriyah al Iraqiyah

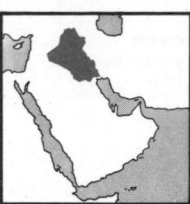

Location Middle East
With a short coastline on the Gulf between Iran and Kuwait, Iran is to the east, Kuwait and Saudi Arabia to the south, Jordan and Syria to the west and Turkey to the north
Land Area 434 924 km² = 167 925 mi²
Excludes one-half share (3 522 km²) of Iraq/Saudi Arabia neutral zone
Climate Very hot summers, cool winters
Weather at Baghdad, 34 m altitude
Temperature: hottest months July, Aug 24–43 °C, coldest Jan 4–16 °C
Rainfall (av monthly): driest months June–Sept 1 mm, wettest Mar 28 mm
Time 3 hours ahead of GMT
Measures Metric system; local measures are also used unofficially, including: *length* 1 dhirraa (Aleppo) = 68.5 centimetres = 27.0 inches
1 dhirraa (Baghdad) = 74.5 centimetres = 29.3 inches
1 dhirraa (Mosul) = 70 centimetres = 27.6 inches
area 1 féddan = 20 mishara (dönürm) = 5 hectares = 12.36 acres
weight (mass)
Baghdad: 1 mann = 6 hogga = 24 okiya = 25 kilograms = 55.11 pounds
1 tughar = 20 wazna = 80 mann = 2 tonnes = 1.968 UK (long) tons
Monetary unit Iraqi dinar (I D) = 5 riyals = 20 dirhams = 1 000 fils
Rate of exchange (1976 av): par I D 0.296053 = $ 1 ($ 3.378 = I D 1), free I D 0.5347 = £ 1 (£ 1.870 = I D 1)

Summary

Political Socialist republic, which before the July 1958 revolution was under the rule of King Faisal II. A Kurdish minority seeks independence and has been promised some autonomy. There has been a Neutral Zone between Iraq and Saudi Arabia, jointly administered; under a 1975 agreement, this is to be partitioned between the two countries. Member of UN, Arab League, Opec and Oapec
Economic An oil economy, but with nearly one-half of the people working in agriculture which accounts for under one-tenth of gross domestic product. Higher oil revenues are being used to increase industrial capacity. Increasing use is being made of Gulf oil ports and terminals for export; a new pipeline to Turkey (capacity 25 mn tonnes per year) was opened in January 1977. Syrian pipelines were closed in early 1976 following a dispute concerning charges

People, resources and equipment

Population 1960 6.89*mn, 1970 9.44*mn, 1976 11.51*mn
Growth: 1960–70 3.2* %pa, 1970–76 3.4* %pa
Density (1976): 26* people per km²
Vital statistics (rate per 1 000 people, 1970–75): births 48.1*, deaths 14.6*
Cities (population in 000, 1976)

Baghdad (capital)	2 800*	Al Basrah[b]	854*	An Najaf	136[a]
Al Mawsil[c]	857*	Kirkuk	560*	Al Hillah	111[a]

[a]1965 [b]Basra [c]Mosul
Race (1961) Arab 78 %, Kurd 18 %, Persian 1 %, Turk 1 %
Language Arabic; Kurdish and Turkish are also in use
Religion (1970) Moslem 96** %, Christian 3** %
Education Pupils (1974/75) 2 082 624, teachers 79 328
Labour force (1976) 2 868 000; in agriculture 1 227 000 (43 %)
Personnel Scientists and engineers (1972): 43 645
Physicians (1975): 4 500, 1 per 2 470 people
Standard of living
National income per person (1976): I D 380* = $ 1 280* = £ 710*
Consumption per person (1975): energy 713 kg coal equivalent, electricity (production) 306* kW h, newsprint 0.1 kg, steel 181 kg
Newspapers (1974): number 5
Telephones (Dec 1976): 319 600, 27 per 1 000 people
Livestock (000, 1976) Cattle 2 600*, buffaloes 220*, sheep 11 900*, goats 3 800*, camels 330*, asses 607*
Mineral reserves (1975) Crude oil 4 714 mn tonnes
Natural gas 769 000 mn cubic metres
Petroleum refinery capacity (1975) 9.2 mn tonnes
Hospital beds (1975) 22 942, 1 per 485 people
Roads (1975) 11 859 km = 7 369 mi, density 0.027 km per km²
Railways (1975) 1 955 km = 1 215 mi, density 0.004 km per km²
Inland waterways Euphrates and Tigris rivers
Ships (registered, 1977) 110, total of 1 135 245 gross tons
Ports Al Faw (Fao), Khor al Amaya and Mina al Bakr (crude oil ports and terminals); Basra, Umm Qasr

Iraq

Airports (1976) Baghdad: passenger departures 421 000, arrivals 402 000; also Basra and Mosul

Durable equipment (at end-year)	000	no per 1 000 people	
Radio sets (1975)	1 252	111	no per
Television sets (1973)	520	50	km of road
Passenger cars (1975)	118	10	10.0
Commercial vehicles (1975)	85	8	7.2

Production

Gross domestic product 1975: I D 4 022 mn = $ 13 587 mn = £ 6 115 mn
1976 est: I D 4 620* mn = $ 15 600* mn = £ 8 640* mn
Structure of gross domestic product (1975) *By origin* Agriculture 7 %, mining and quarrying 57 %, manufacturing 6 %, transport and communication 4 %, other 26 %

Production index (1970 = 100)	1960	1970	1976	Growth %pa 1960–70	1970–76
Agricultural	69	100	110	3.7	1.5
Main products (000 t)					
Agriculture					
Wheat	592	1 236	1 312	7.6	1.0
Barley	804	682	579	−1.6	−2.7
Tomatoes	140**	220	350*	4.6**	8.0*
Watermelons	250**	486	393*	6.9**	−3.5*
Dates	337*	350**	498*	0.4*	6.0**
Cotton	8	15	16*	6.5	1.1*
Milk, cow	186	216*	255*	1.5*	2.8*
Milk, sheep	258	236*	325*	−0.9*	5.5*
Beef and buffalo meat	40*	50*	51*	2.2*	0.3*
Sheep and goat meat	47*	77*	52*	5.0*	−6.3*
Other					
Total energy (000 tce)	62 510	100 430	165 620b	4.9	10.5c
Crude oil	47 467	76 457	112 415	4.9	6.6
Petroleum products	1 760	3 496	5 000*b	7.1	7.4*c
Natural gas (mn m³)	603	785	1 654b	2.7	16.1c
Electricity (mn kW h)	850*	1 909	3 400*b	8.4*	12.2*c
Salt	36	51	64*b	3.5	4.6*c
Sulphur	na	na	650b	na	na
Beer (000 hl)	50*	82	123d	5.1*	10.7e
Cigarettes (mn units)	4 918	6 624	7 000*b	3.0	1.1*c
Cotton fabrics (mn m)	20*	49	74d	9.4*	10.9e
Fertilisers, nitrogenousa	—	—	24d	na	na
Cement	813	1 542	1 800*b	6.6	3.1*c

aYears ending June 30th   b1975   c1970–75   d1974   e1970–74

Transport traffic	1960	1970	1976	Growth %pa 1960–70	1970–76
Passenger-kilometres (mn)					
Raila	656	369	645b	−5.6	11.8c
Air	52	202	863	14.5	27.4
Cargo: tonne-kilometres (mn)					
Raila	768	1 194	1 871b	4.5	9.4c
Air	1.1	2.0	15.7	6.2	41.0
Sea: tonnes (mn)					
Goods loaded	11.4	17.8	26.4d	4.5	10.4e
Goods unloaded	1.06	1.43	1.53f	3.0	2.3g
Tourism					
Number of visitors (000)	na	360	630	na	9.8
Gross receipts ($ mn)	na	42	78b	na	13.2c

aYears ending March 31st   b1975   c1970–75   d1974   e1970–74   f1973   g1970–73

Finance and trade

Price indices (1970 = 100)	1960	1970	1976	Growth %pa 1960–70	1970–76
Consumer prices	79.6	100.0	149.5	2.3	6.9
Wholesale prices	84.3	100.0	147.8	1.7	6.7
Money stock (end-year, I D mn)	103	218	755	7.7	23.0

Budget (1977) Balanced at I D 3 980 mn = $ 13 480 mn = £ 7 720 mn

Balance of payments ($ mn)	1972	1973	1974	1975	1976
Balance of goods (fob)	+697	+1 354	+4 226	+4 139	na
Balance of services	−154	−544	−1 373	−1 169	na
Balance of transfers	+3	−10	−235	−265	na
Current balance	*+546*	*+801*	*+2 619*	*+2 705*	*na*
Long-term capital flow	−345	+98	−532	−436	na
Reserves and debt (end-year, $ mn)					
International reserves	781	1 553	3 273	2 727	4 601
External public debt	667	741	670	1 132	na

External trade (1976) Imports: I D 1 025 mn = $ 3 461 mn = £ 1 916 mn
Exports: I D 2 738 mn = $ 9 248 mn = £ 5 120 mn

Main imports (1975)	% of total	Main exports (1975)	% of total
Motor vehicles	17	Crude oil	96
Machinery, non-electric	16	Petroleum products	2
Iron and steel	16	Dates	2
Food	15		
Electrical machinery	7		
Chemicals	6		
Textile yarns and fabrics	4		
Main sources (1976)		Main destinations (1976)	
West Germany	21	France	23
Japan	14	Italy	19
France	8	Turkey	9
United Kingdom	7	Japan	8
United States	5	United Kingdom	7

Special focus

Crude oil exports

	$ million		$ million
1970	1 030	1974	6 505
1971	1 073	1975	8 176
1972	1 022	1976	9 114*
1973	1 836	1977	9 506*

Israel

State of Israel
Medinat Yisra'el

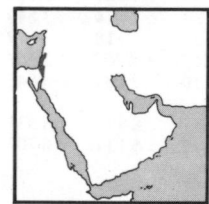

Location Middle East
With a coastline to the west on the Mediterranean Sea and an outlet to the Red Sea via Elat, Lebanon is to the north, Syria to the north-east, Jordan to the east and Egypt to the west
Land Area 20 700 km² = 8 000 mi² (area excluding Gaza Strip of 378 km², and other areas occupied after the June 1967 and October 1973 wars)
Climate Sub-tropical
Weather at Jerusalem, 757 m altitude
Temperature: hottest month August 18–31 °C, coldest Jan 5–13 °C
Rainfall (av monthly): driest months July, Aug 0 mm, wettest Jan, Feb 132 mm
Time 2 hours ahead of GMT
Measures Metric system; the metric dunam = 1 000 square metres is also in use
Monetary unit Israeli pound (Is £) = 100 agorot
Rate of exchange (1976 av): free Is £ 7.977 = $ 1, Is £ 14.41 = £ 1

Summary

Political Republic, which became independent on May 14, 1948; formerly under UK administration. There have since been two main conflicts with Arab countries: the 6 day war of June 1967 and the war of October 6–24, 1973. The Sinai oilfields, occupied by Israel in 1967, were returned to Egypt in October 1975. Member of UN
Economic The economy is diversified, with contributions to gross domestic product of 5 % from agriculture and one-fifth from manufacturing and mining; main exports are diamonds (which are imported, polished and traded), chemicals and clothing from the industrial sector, and citrus fruit and cotton from the agricultural sector. The economy is very dependent on inflows of capital and transfers from abroad

People, resources and equipment

Population 1960 2.11 mn, 1970 2.97 mn, 1976 3.53 mn
Growth: 1960–70 3.5 %pa, 1970–76 2.9 %pa
Density (1976): 171 people per km²
Vital statistics (rate per 1 000 people, 1976): births 27.6, deaths 6.8
Administered territories (population in 000, 1976)
Gaza Strip and Sinai 444, Judea and Samaria (West Bank of Jordan) 690
Cities (population in 000, 1975)

Jerusalem (capital)	350	Haifa	360	Bat Yam	118
Tel Aviv-Jaffa	354a	Ramat Gan	121	Holon	114

aThe population for Greater Tel Aviv (including, among other municipalities, Ramat Gan and Bat Yam was 1 181 thousand

Israel

Race (1976) Jewish 85 %, Arab 14* %
Language Hebrew and Arabic
Usage (1961): Hebrew 66 %, Arabic 16 %, Yiddish 5 %, Romanian 2 %
Religion (1970) Jewish 85 %, Moslem 11 %, Christian 2 %
Education Pupils (1974/75) 768 800*, teachers (1973/74) 66 000*
Labour force (1976) 1 169 700; in agriculture 72 800 (6 %), mining, quarrying and manufacturing 279 000 (24 %), construction 90 400 (8 %), transport and communication 80 500 (7 %)
Personnel Scientists and engineers (1970): 36 000
Physicians (registered, 1973) 9 143, 1 per 351 people
Standard of living
National income per person (1976): Is £ 26 230 = $ 3 288 = £ 1 820
Consumption per person (1975): energy 2 806 kg coal equivalent, electricity (production, 1976) 2 930 kW h, newsprint 13 kg, steel 221 kg
Newspapers (1974): number 23
Telephones (Dec 1976): 869 040, 245 per 1 000 people
Livestock (000, 1976) Cattle 323, sheep 202, goats 140, pigs 86*, chickens 11 500*, ducks 4 500
Mineral reserves (1975) Natural gas 1 000 mn cubic metres
Petroleum refinery capacity (1975) 11* mn tonnes
Electrical capacity (1975) 2 181 megawatts
Hospital beds (1975) 19 501, 1 per 177 people
Roads (1974) 10 657 km² = 6 622 mi², density 0.51 km per km²
Railways (1976) 902 km² = 560 mi², density 0.044 km per km²
Ships (registered, 1977) 58, total of 404 651 gross tons
Ports (goods traffic, crude oil ports, 000 tonnes, 1976)

	loaded	unloaded
Haifa	1 350	3 811
Ashdod	1 862	864
Eilat	418	401

Airports (1976) Ben Gurion (Tel Aviv): passenger departures 938 000, arrivals 933 000; also Jerusalem, Haifa, Eilat, Rosh Pinna, Beersheba, Mizpeh Ramon

Durable equipment (at end-year)	000	no per 1 000 people	
Radio sets (1972)	680	221	no per
Television sets (1974)	652	191	km of road
Passenger cars (1976)	297	84	28
Commercial vehicles (1976)	104	29	10

Production

Gross domestic product 1976: Is £ 108 807 mn = $ 13 640 mn = $ 7 551 mn
Growth in real terms: 1970–76 6.4 %pa
Structure of gross domestic product *By origin* (1975)
Agriculture 5 %, manufacturing, mining and quarrying 21 %, construction and utilities 9 %, other 65 %. *By type* (1976) Final consumption expenditure 94 % (of which, government 38 %), stock investment 3 %, gross fixed capital formation 22 %, exports of goods and services 31 %, less imports of goods and services −50 %

Production indices (1970 = 100)	1960	1970	1976	Growth %pa 1960–70	1970–76
Agricultural	56	100	139	5.9	5.6
Industrial	32.8	100.0	145.7	11.8	6.5
Main products (000 t)					
Agriculture					
Wheat	41	125	206	11.8	8.7
Potatoes	82	137	175	5.3	4.2
Tomatoes	110	159	237	3.7	6.9
Watermelons	60**	102	110*	5.4**	1.3*
Grapes	53	61	74	1.4	3.3
Oranges	425	938	1 200*	8.2	4.2*
Grapefruit	69	284	460*	15.2	8.4*
Olives	21	12	20	−5.4	8.9
Cotton	11	35	54	12.3	7.5
Milk	286	454	659	4.7	6.4
Eggs	61	74	96	1.9	4.4
Cheese	15	31	50	7.5	8.3,
Fish catch	14	24	26	5.8	1.1
Energy					
Crude oil	129	79	36	−4.8	−12.3
Petroleum products	1 344	4 984*	6 523*c	14.0*	5.5*d
Natural gas (mn m³)	—	134	58	na	−13.0
Electricity (mn kW h)	2 313	6 838	10 344	11.4	7.1
Mining					
Copper ore (Cu content)	5.8*	11.0	8.0c	6.6*	−6.2d
Phosphate rock	224	1 162	882e	17.9	−5.4d
Potash (oxide content)	82	530	570e	20.5	1.8f
Salt	37	66	117e	6.0	12.1d

Main products (000 t) *Manufacturing*	1960	1970	1976	Growth %pa 1960–70	1970–76
Beer (000 hl)	208	324	355c	4.5	1.8d
Cigarettes (mn units)	2 565	3 868	5 553e	4.2	7.5d
Cotton yarn	10.0	22.4	21.2	8.4	−0.9
Tyres (000 units)	407	1 392	1 680	13.1	3.2
Sulphuric acid	121	203	208	5.3	0.4
Fertilisers, nitrogenousa	14	28	45	7.2	8.5
Cement	806	1 384	2 042	5.6	6.7
Crude steel	40	120*	130*c	11.6*	1.6*d
Television sets (000 units)	—	68	61e	na	−2.1d
Motor vehiclesb (000 units)	3.2	9.6	7.0	11.6	−5.1

aYears ending June 30th bAssembly only c1975 d1970–75 e1974 f1970–74

Transport traffic *Passenger-kilometres* (mn)	1960	1970	1976	Growth %pa 1960–70	1970–76
Rail	350	358	280	0.2	−4.0
Air	439	2 426	4 368	18.6	10.3
Cargo: tonne-kilometres (mn)					
Rail	220	468	449	7.8	−0.7
Air	9.4	102.2	138.6	26.9	5.2
Sea: tonnes (mn)					
Goods loaded	1.18	3.34	3.67	11.0	1.6
Goods unloaded	2.03	4.27	5.12	7.7	3.1
Tourism					
Number of visitors (000)	114	419	733	13.9	9.8
Gross receipts ($ mn)	na	106	292	na	18.4

Finance and trade

Price and wage indices (1970 = 100)	1960	1970	1976	Growth %pa 1960–70	1970–76
Consumer prices	58.2	100.0	387.6	5.6	25.3
Wages (earnings)	40.4	100.0	397.8	9.5	25.9
Share prices	82.2	100.0	286.5	2.0	19.2
Money stock					
(end-year, Is £ mn)	880	3 386	13 486	14.4	25.9

Budget (total, 1977/78; year ending March 31st)
Balanced at: Is £ 122 500 mn = $ 10 030* mn = £ 5 600* mn; of expenditure, defence 34 %, development 22 %

Balance of payments ($ mn)	1972	1973	1974	1975	1976
Balance of goods (fob)	−1 081	−2 457	−3 021	−3 470	−2 740
Balance of services	−20	−184	−312	−581	−530
Balance of transfers	+1 054	+2 171	+1 737	+1 770	+2 239
Current balance	−48	−470	−1 596	−2 281	−1 030
Long-term capital flow	+688	+974	+669	+1 567	+875
International reserves					
(end-year, $ mn)	1 222	1 815	1 200	1 182	1 373

External trade (1976) Imports: Is £ 45 590 mn − $ 5 715 mn − £ 3 164 mn
Exports: Is £ 19 265 mn = $ 2 415 mn = £ 1 337 mn

Main importsa	% of total	Main exports	% of total
Diamonds	17	Diamonds	33
Crude oil	16	Chemicals	10
Machinery, non-electric	11	Citrus fruit	7
Food	10	(of which, oranges 5)	
(of which, cereals 6)		Machinery	7
Chemicals	7	Metal manufactures	6
Electrical machinery	5	Clothing	5
Iron and steel	5	Aircraft	3
Motor vehicles	3	Cotton	2
Soyabeans	3		
Main sourcesa		Main destinations	
United States	22	United States	18
United Kingdom	16	West Germany	8
West Germany	10	United Kingdom	8
Netherlands	6	Netherlands	7
Italy	4	Hongkong	6
Switzerland	4	France	6

aExcluding military goods

Special focus

Military expenditure

	$ mn	as % of gdp		$ mn	as % of gdp
1972	1 400**	18**	1975	3 500**	27**
1973	4 200**	42**	1976	4 200**	31**
1974	3 900**	29**			

Japan
Nihon or Nippon

Location North-east Asia
The chain of islands lies off the north-east coast of Asia; the Sea of Japan separates the islands from Soviet Union, China and South and North Korea, and the Pacific Ocean is to the east. The main island is Honshu, with Hokkaido island to the north, Shikoku and Kyushu islands to the south. Sakhalin island (Soviet Union) is to the north of Hokkaido
Land Area 372 313 km² = 143 751 mi²
Usage (1975): agricultural 60 030 km² (16 %), of which, arable 49 450 km² (13 %), cropland 6 280 km² (2 %), pastures 4 300* km² (1* %); forests 250 430* km² (67* %)
Climate Temperate
Weather at Tokyo, 6 m altitude
Temperature: hottest month Aug 22–30 °C, coldest Jan minus 2–8 °C
Rainfall (av monthly): driest month Jan 48 mm, wettest Sept 234 mm
Time 9 hours ahead of GMT
Measures Metric system; local units are also used, including:
length 1 shaku = 0.303 metre = 11.93 inches
1 chô = 60 ken = 360 shaku = 109.1 metres = 119.3 yards
area 1 tsubo = 3.31 square metres = 3.95 square yards
weight (mass) 1 kan = 6.25 kin = 3.75 kilograms = 8.27 pounds
Monetary unit Yen (Y) = 100 sen
Rate of exchange (1976 av): free ¥ 296.5 = $ 1, ¥ 535.6 = £ 1

Summary

Political Parliamentary monarchy; a revised constitution reducing the power of the Emperor came into effect in 1947. Okinawa (Ryukyu Islands) was returned to Japan by the United States from May 15, 1972. Member of UN, OECD and Colombo Plan
Economic An industrially developed economy with high growth; this has been helped by concentration on special industrial areas, notably steel, radio and television, motor cycles, motor cars and ships. Selling a comparatively limited range has been achieved with a wide spread of countries as export markets

People

Population 1960 94.10 mn, 1970 104.34 mn, 1976 112.77 mn
Growth: 1960–70 1.0 %pa, 1970–76 1.3 %pa
Density (1976): 303 people per km²
Vital statistics (rate per 1 000 people, 1976): births 16.3, deaths 6.3
Regions (population in 000, October 1970 census; total of 104.14 mn excluding Okinawa)

Hokkaido Island	5 184	Mie-ken	1 543	*Kyushu Island*	12 496
Hokkaido prefecture	5 184	Miyagi-ken	1 819	*Prefectures*	
Honshu Island	82 560	Nagano-ken	1 957	Fukuoka-ken	4 027
Prefectures		Nara-ken	930	Kagoshima-ken	1 729
Aichi-ken	5 386	Niigata-ken	2 361	Kumamoto-ken	1 700
Akita-ken	1 241	Okayama-ken	1 707	Miyazaki-ken	1 051
Aomori-ken	1 428	Osaka-fu	7 620	Nagasaki-ken	1 570
Chiba-ken	3 367	Saitama-ken	3 866	Oita-ken	1 156
Fukui-ken	744	Shiga-ken	890	Saga-ken	1 263
Fukushima-ken	1 946	Shimane-ken	774	*Shikoku Island*	3 904
Gifu-ken	1 759	Shizuoka-ken	3 090	*Prefectures*	
Gumma-ken	1 659	Tochigi-ken	1 580	Ehime-ken	1 418
Hiroshima-ken	2 436	Tokyo-to	11 408	Kagawa-ken	908
Hyogo-ken	4 668	Tottori-ken	569	Kochi-ken	787
Ibaraki-ken	2 144	Toyama-ken	1 030	Tokushima-ken	791
Ishikawa-ken	1 002	Wakayama-ken	1 043	*Okinawa*	
Iwate-ken	1 371	Yamagata-ken	1 226	*prefecture*	940[a]
Kanagawa-ken	5 472	Yamaguchi-ken	1 511		
Kyoto-fu	2 250	Yamanashi-ken	762		

[a] Population for Ryukyu Islands, under United States administration in 1970
Cities (population in 000, 1975)

Tokyo (capital)	11 282	Hiroshima	832	Shizuoka	449
Osaka	2 715	Sakai	740	Nagasaki	448
Yokohama	2 610	Chiba	654	Himeji	435
Nagoya	2 083	Sendai	586	Niigata	420
Kyoto	1 459	Amagasaki	537	Funabashi	416
Kobe	1 338	Okayama	517	Gifu	409
Sapporo	1 216	Higashi-Osaka	501	Kurashiki	397
Kita-Kyushu	1 061	Kumamoto	475	Wakayama	390
Kawasaki	989	Hamamatsu	472	Yokosuka	390
Fukuoka	965	Kagoshima	462	Kanazawa	389

Cities

Nishinomiya	386	Urawa	331	Naha	304
Toyonaka	384	Omiya	329	Takamatsu	297
Sagamihara	373	Takatsuki	327	Hirakata	297
Matsuyama	372	Asahikawa	322	Suita	293
Utsunomiya	343	Oita	316	Toyama	288
Matsudo	341	Hachioji	316	Toyohashi	284
Kawaguchi	341	Ichikawa	311	Kochi	279
Iwaki	336	Hakodate	306	Aomori	267
Fukuyama	332	Nagano	305	Fujisawa	266

Race (1976) Japanese 99.3 %, Korean 0.6 %
Language Japanese
Religion (1974) Shinto 79 %, Buddhist 77** %, Christian ¾ % (a majority adhere to both Shinto and Buddhist religions)
Education (1974/75) Pupils: primary 10 088 776, secondary 7 365 612, vocational 1 671 097, higher 2 155 893. Teachers: primary 406 347, secondary and vocational 503 608, higher 184 446
Labour force (1976)

Economic activity	Number	% of total
Agriculture	6 430 000	12
Mining and quarrying	180 000	—
Manufacturing	13 450 000	25
Electricity, gas and water	330 000	1
Construction	4 920 000	9
Distribution and hotels	11 510 000	21
Transport and communication	3 410 000	6
Finance and real estate	1 730 000	3
Other	10 760 000	20
Unemployed	1 080 000	2
Total	53 780 000	100

Personnel (1974) Scientists and engineers engaged in research: 375 379
Physicans: 126 822, 1 per 869 people
Standard of living
National income per person (1976): ¥ 1 323 960 = $ 4 465 = £ 2 472
Consumption per person (1975): energy 3 622 kg coal equivalent, electricity (production 1976) 4 538 kW h, newsprint 20 kg, steel 583 kg
Newspapers (1974): number 180; circulation (52 dailies only) 57 820 000, 526 per 1 000 people
Telephones (Mar 1977): 48 431 400, 427 per 1 000 people

Resources and equipment

Livestock (000, 1976) Cattle 3 500*, pigs 7 459, goats 94, sheep 10, chickens 245 000*
Mineral reserves Coal (1973) 7 443 mn tonnes
Lignite (1973) 1 185 mn tonnes
Crude oil (1975) 10 mn tonnes
Natural gas (1975) 36 000 mn cubic metres
Uranium (1975) 7 700 tonnes
Petroleum refinery capacity (1975) 277 mn tonnes
Electrical capacity (1975) 112 285 megawatts, of which, hydro 24 853 megawatts, nuclear 25 125* megawatts, geothermal 378* megawatts
Hospital beds (1975) 1 163 726, 1 per 96 people
Roads (1976) 1 078 357 km = 670 060 mi, density 2.9 km per km²
Railways (1974) 28 024 km = 17 413 mi, density 0.075 km per km²
Ships (registered, 1977) 9 642, total of 40.0 mn gross tons
Ports (goods traffic, 000 freight tonnes, 1976)

	International		Coastwise		Total
	loaded	unloaded	loaded	unloaded	
Kobe	19 992	20 190	44 383	51 303	135 868
Chiba	3 056	61 398	33 125	18 431	116 010
Yokohama	20 437	31 620	37 642	29 106	113 805
Nagoya	16 644	35 660	16 715	29 916	98 935
Osaka	5 866	10 261	21 929	40 763	78 819
Kawasaki	885	33 561	26 957	16 081	77 484
Tokyo	4 875	9 219	10 869	30 275	55 238
Hakodate	—	936	17 143	15 708	33 787

Airports (passenger traffic, 000, 1976)

	departures	arrivals
Tokyo	9 626	9 811
Osaka	5 706	5 722

Also 59 other airports with scheduled flights

Durable equipment (at end-year)	000	no per 1 000 people	
Radio sets (1975)	51 630	460	no per km of road
Television sets (1975)	26 030	232	
Passenger cars (1976)	18 476	163	17.1
Commercial vehicles (1976) of which,	11 610	102	10.8
buses and coaches (1976)	222	2	0.2
goods vehicles (1976)	11 387	100	10.6

Japan

Companies

Sales ($ bn, year ending March 31, 1977)

Mitsubishi Corporation	32.9	Toyo Menka Kaisha	8.6
Mitsui and Co	30.9	Nippon Steel Corp	8.6
Marubeni Corporation	22.0	Kanematsu-Gosho	8.0
C Itoh & Co	21.7	Toyota Motor Co	7.8[a]
Sumitomo Corporation	19.9	Nissan Motor Co	6.9
Nissho-Iwai Co	15.5	Nichimen	6.2

[a] Year ending June 30th

Production

Gross domestic product

1976: ¥ 164 604 bn = $ 555 157 mn = £ 307 326 mn
Growth in real terms (gross national product): 1963–70 10.8 %pa, 1970–76 5.5 %pa

Structure of gross domestic product	¥ bn	% of gdp
By origin (1974)		
Agriculture	6 996	7
Mining and quarrying	590	1
Manufacturing	29 800	28
Construction	9 355	9
Distribution and hotels	20 372	19
Transport, communication and utilities	6 987	7
Other services	32 616	31
Total	106 716	100
By type of expenditure (1976)		
Government final consumption	17 945	11
Private final consumption	93 449	57
Stock investment	3 093	2
Gross fixed capital formation	48 755	30
Exports of goods and services	22 656	14
less Imports of goods and services	−21 294	−13
Total	164 604	100

Production indices (1970 = 100)	1960	1965	1970	1973	1974	1975	1976	Growth %pa 1960–70	1960–65	1965–70	1970–76
Agricultural	70	84	100	104	105	112	108	3.6	3.6	3.6	1.2
Industrial	27.1	46.3	100.0	127.3	123.3	109.7	124.7	13.9	11.3	16.7	3.7
Main products (000 t)											
Agriculture											
Rice	16 114	16 126	16 493	15 778	15 964	17 101	15 292	0.2	0.0	0.5	−1.3
Wheat	1 531	1 287	474	202	232	241	222	−11.1	−3.4	−18.1	−11.9
Barley	2 301	1 234	573	216	233	221	210	−13.0	−11.7	−14.2	−15.4
Potatoes	3 594	4 060	3 617	3 418	2 949	3 261	3 200*	0.1	2.5	−2.3	−2.0*
Sweet potatoes	6 277	5 025	2 684	1 613	1 600	1 418	1 279	−8.1	−4.4	−11.8	−11.6
Taro	470**	478	542	385	431	500*	na	1.4**	0.3**	2.5	−1.6*[d]
Sugar, raw value	170	569	685	655	480	467	547	15.0	30.7	3.8	−3.7
Cabbages	1 684	3 173	3 694	3 678	3 650	3 700*	3 885*	8.2	13.5	3.1	0.8*
Tomatoes	242	534	792	867	820	1 110	1 176*	12.6	17.2	8.3	6.8*
Cucumbers	400**	774	967	1 005	965	1 000*	1 055*	9.2**	14.1**	4.6	1.5*
Aubergines	550**	624	722	714	663	650*	675*	2.7**	2.6**	3.0	−1.1*
Onions, green	410	568	615	567	557	610*	na	4.1	6.7	1.6	−0.2*[d]
Onions, dry	601	861	973	995	1 022	1 000*	1 050*	4.9	7.5	2.5	1.3*
Watermelons	741	742	1 007	1 209	1 085	1 200*	1 255*	3.1	0.0	6.3	3.7*
Grapes	155	225	234	271	295	311	303	4.2	7.7	0.8	4.4
Apples	876	1 132	1 021	963	868	894	900*	1.5	5.3	−2.0	−2.1*
Pears	250	360	464	495	524	499*	507	6.4	7.6	5.2	1.5
Peaches	170	229	279	281	259	289	304*	5.1	6.1	4.0	1.4*
Oranges	177*	237*	262	391	320	372	389*	4.0*	6.0*	2.0*	6.8*
Mandarines	894	1 331	2 552	3 508	3 553	3 665	3 873*	11.0	8.3	13.9	7.2*
Tea	78	77	91	101	95	105	100	1.6	−0.3	3.4	1.6
Tobacco	122	193	155	161	156	171	165*	2.4	9.6	−4.3	1.0*
Milk	1 887	3 224	4 769	4 908	4 868	4 963	5 265	9.7	11.3	8.1	1.7
Eggs	517*	1 007	1 788	1 799	1 799	1 786	1 815*	13.2*	14.3*	12.2	0.2*
Beef and veal	170*	250*	278	246	319	352	297	5.0*	8.0*	2.1*	1.1
Pigmeat	147	375	734	971	1 076	1 039	1 057	17.4	20.6	14.4	6.3
Poultrymeat	90**	170**	550*	634	676	752*	812*	19.8**	13.6**	26.5**	6.7*
Fish catch	6 193	6 908	9 366	10 748	10 805	10 524	10 620	4.2	2.2	6.3	2.1
Timber (000 m³)	60 400**	59 241	49 797	42 524	39 809	36 385*	38 134*	−1.9**	−0.4**	−3.4	−4.3*
Energy											
Total energy (000 tce)	60 630	62 120	54 820	37 490	38 190	37 440	na	−1.0	0.5	−2.5	−7.3[d]
Coal[a]	51 067	49 534	39 694	22 414	20 333	18 999	18 396	−2.5	−0.6	−4.3	−12.0
Lignite	1 409	573	197	86	75	60	52	−17.9	−16.5	−19.2	−19.9
Crude oil	526	671	770	700	672	606	580	3.9	5.0	2.8	−4.6
Petroleum products	25 744	70 796	163 255	224 000	217 514	205 300*	na	20.3	22.4	18.2	4.7*[d]
Natural gas (mn m³)	824	1 964	2 627	2 903	2 847	2 760	2 813	12.3	19.0	6.0	1.1
Manufactured gas (mn m³)	1 280*	1 926	3 180	4 260	4 950	5 404	6 500*	9.5*	8.5*	10.5	12.6*
Electricity (mn kW h)[b]	115 900*	193 000*	361 200*	470 287	459 041	475 794	511 780	12.0*	10.7*	13.4*	6.0*
of which, hydro (mn kW h)[b]	58 471	76 420	80 090	71 678	84 780	85 906	na	3.2	5.5	0.9	1.4[d]
nuclear (mn kW h)[b]	—	36	4 581	9 707	19 699	25 125	na	na	na	163.6	40.5[d]
geothermal (mn kW h)[b]	—	—	243	269	312	378	na	na	na	na	9.2[d]
Mining											
Iron ore (Fe content)	1 574	1 427	916	588	443	490*	400*	−5.3	−1.9	−8.5	−12.9*
Copper conc (Cu content)	89	107	119	91	82	85	82	2.9	3.8	2.1	−6.2
Lead conc (Pb content)	39	55	64	53	44	51	52	5.1	7.1	3.1	−3.6
Manganese conc (Mn content)	120	96	79	53	45	42	na	−4.1	−4.4	−3.8	−11.7[d]
Tungsten conc (oxide content)	0.59	0.43	0.85	1.10	1.02	0.97	na	3.7	−6.1	14.6	2.6[d]
Uranium (U content)	—	—	—	0.010	0.009	0.004	na	na	na	na	na
Zinc conc (Zn content)	157	221	280	264	241	254	260	6.0	7.1	4.9	−1.2
Silver	0.21	0.28	0.34	0.36	0.23	0.27	na	5.0	5.9	4.1	−4.5[d]
Gold (000 kg)[f]	10.5*	16.1	22.1	32.8	32.1	32.5	na	7.7*	8.9*	6.5	8.0[d]
Salt[b]	839	851	951	1 049	1 115	1 068	na	1.3	0.3	2.2	2.4[d]

For footnotes see page 178

Japan

Main products (000 t)	1960	1965	1970	1973	1974	1975	1976	Growth %pa 1960–70	1960–65	1965–70	1970–76
Manufacturing											
Beer (000 hl)[b]	9 287	20 152	30 536	38 362	36 435	38 966	37 236	12.6	16.8	8.7	3.4
Sake, refined (000 hl)	6 848	10 889	12 571	14 212	14 174	13 499	12 790	6.3	9.7	2.9	0.3
Cigarettes (mn units)[b]	125 048	181 034	221 957	269 930	292 371	287 000	287 340	6.0	7.7	4.3	4.4
Cotton yarn	536	537	495	525	480	432	467*	−0.8	0.0	−1.6	−1.0*
Man-made fibres	552	878	1 540	1 848	1 648	1 462	1 621	10.8	9.7	11.9	0.9
Cotton fabrics (mn m²)	3 220*	3 010*	2 616	2 380	2 163	2 124	2 237	−2.1*	−1.3*	−2.8*	−2.6
Wood pulp	3 524	5 160*	8 801	10 095	10 017	8 613	na	9.6	7.9*	11.3*	−0.4[d]
Newsprint	732	1 184	1 917	2 106	2 233	2 160	2 340	10.1	10.1	10.1	3.4
Other paper	4 300*	6 114	11 056	13 869	13 412	11 440	13 054	9.9*	7.3*	12.6	2.8
Synthetic rubber	23	161	697	967	858	789	941	40.6	47.6	34.1	5.1
Tyres (000 units)	10 100	28 432	66 556	87 746	70 778	74 580	85 280	20.7	23.0	18.5	4.2
Sulphuric acid	4 452	5 656	6 925	7 116	7 128	6 000	6 103	4.5	4.9	4.1	−2.1
Caustic soda	843	1 305	2 606	3 136	3 068	2 861	2 589	11.9	9.1	14.8	−0.1
Plastics and resins	556	1 609	5 128	6 392	6 722	4 373	4 954	24.9	23.7	26.2	−0.6
Fertilisers, nitrogenous[c]	922	1 394	2 131	2 199	2 138	2 341	1 557	8.7	8.6	8.9	−5.1
Fertilisers, phosphate[c]	800**	597	729	729	736	769	585	−0.9**	−5.7**	4.1	−3.6
Cement	22 537	32 486	57 189	78 120	73 108	65 517	68 712	9.8	7.6	12.0	3.1
Coke	8 038*	14 847	36 373	44 301	45 632	45 166	na	16.3*	13.1*	19.6	4.4[d]
Pig iron	12 341	28 160	69 714	92 043	92 704	89 016	88 612	18.9	17.9	19.9	4.1
Crude steel	22 138	41 161	93 322	119 322	117 131	102 313	107 399	15.5	13.2	17.8	2.4
Aluminium	133	294	733	1 103	1 124	1 016	919	18.6	17.2	20.0	3.8
Copper	248	366	705	951	996	819	864	11.0	8.1	14.0	3.4
Lead	69	103	195	219	218	184	177*	10.9	8.3	13.6	−1.6*
Zinc	175	353	659	834	837	702	715	14.2	15.1	13.3	1.4
Cameras (000 units)[e]	1 600**	3 916	5 813	5 685	6 644	7 281	na	13.8**	19.6**	8.2	4.6[d]
Radio sets (000 units)	12 851	22 937	32 618	24 484	18 027	14 283	16 770	9.7	12.3	7.3	−10.5
Television sets (000 units)	3 578	4 190	13 782	14 414	13 406	12 453	17 000*	14.4	3.2	26.9	3.6*
Motor cycles (000 units)	1 494	2 211	2 948	3 763	4 510	3 802	4 235	7.0	8.1	5.9	6.2
Passenger cars (000 units)	165	696	3 179	4 471	3 932	4 568	5 028	34.4	33.4	35.5	7.9
Commercial vehicles (000 units)	595	1 222	2 126	2 617	2 624	2 380	2 814	13.6	15.5	11.7	4.8
Merchant vessels (000 grt)	1 732	5 363	10 476	15 736	17 584	17 740	14 524	19.7	25.4	14.3	5.6
Construction											
Dwellings started (000 m²):											
Residential	31 371	58 418	111 295	161 066	118 502	124 912	138 471	13.5	13.2	13.8	3.7
Other	30 471	44 462	93 739	120 685	80 055	71 380	77 003	11.9	7.8	16.1	−3.2

[a]Including brown coal (about 7 % of total)   [b]Years beginning April 1   [c]Years ending June 30th   [d]1970–75   [e]Photographic   [f]Including gold refined

Transport traffic	1960	1970	1976	Growth %pa 1960–70	1970–76
Passenger-kilometres (mn)					
Road	na	284 200	360 900[a]	na	4.9[b]
Rail	180 893	288 133	321 100	4.8	1.8
Air	1 051	14 954	32 334	30.4	13.7
of which: international	482	6 587	13 974	29.9	13.4
Cargo: tonne-kilometres (mn)					
Road	na	135 900	129 700[a]	na	−0.9[b]
Rail	53 445	62 652	47 851	1.6	−4.4
Air	24	436	1 103	33.6	16.7
of which: international	19.6	357	952	33.7	17.8
Sea: tonnes (mn)					
Goods loaded	11.1	44.3	76.5	48.5	9.5
Goods unloaded	88	440	576	17.5	4.6
Tourism					
Number of visitors (000)	147	850	795	19.2	−1.1
Gross receipts ($ mn)	40	232	312	19.2	5.1

[a]1975   [b]1970–75

Finance and trade

Price and wage indices (1970 = 100)	1960	1970	1976	Growth %pa 1960–70	1970–76
Consumer prices	56.7	100.0	188.4	5.9	11.1
Wholesale prices	88.0	100.0	165.4	1.3	8.7
Wages (earnings)	34.3	100.0	259.1	11.3	17.2
Share prices	59.5	100.0	212.5	5.3	13.4
Money stock					
(end-year, ¥ bn)	4 146	21 359	56 179	17.8	17.5

Budget (1976/77; year ending March 31st)
Revenue: ¥ 26 026 bn = $ 89 039 mn = £ 51 313 mn
Expenditure: ¥ 33 505 bn = $ 114 625 mn = £ 66 059 mn
of which, social security 38 %, industrial development 12 %, land conservation and development 11 %, education and culture 9 %, defence 5 %

Balance of payments ($ mn)	1972	1973	1974	1975	1976
Balance of goods (fob)	+8 971	+3 686	+1 438	+5 030	+9 890
Balance of services	−1 883	−3 508	−5 843	−5 356	−5 870
Balance of transfers	−465	−314	−287	−355	−339
Current balance	*+6 624*	*−136*	*−4 693*	*−682*	*+3 680*
Long-term capital flow	−3 016	−8 453	−3 596	−82	−728
Reserves and debt (end-year, $ mn)					
International reserves	18 366	12 246	13 519	12 815	16 605
Public debt: domestic	21 250	26 900	32 039	49 966	71 762
external	256	150	126	115	106

External trade (1976) Imports: ¥ 19 229 bn = $ 64 853 mn = £ 35 902 mn
Exports: ¥ 19 930 bn = $ 67 218 mn = £ 37 211 mn

Main imports	% of total	Main exports	% of total
Crude oil	35	Motor vehicles	17
Food	14	Iron and steel	16
(of which, cereals 5)		Electrical machinery	14
Metal ores	7	(of which, tele-	
Timber	6	communications 8)	
Coal	6	Machinery, non-electric	11
Machinery	5	Ships	10
Chemicals	4	Chemicals	6
Textile fibres	3	Textile yarns and fabrics	5
Non-ferrous metals	2	Instruments	4
Natural gas	2	Metal manufactures	3
Main sources		*Main destinations*	
United States	18	United States	24
Saudi Arabia	12	South Korea	4
Australia	8	Liberia	4
Iran	7	Australia	3
Indonesia	6	Taiwan	3
Canada	4	Soviet Union	3
United Arab Emirates	4	West Germany	3
Kuwait	3	Saudi Arabia	3
South Korea	3	Hongkong	3
China	2	Iran	3
Malaysia	2	China	2
West Germany	2	Indonesia	2
Brunei	2	Canada	2
Taiwan	2	Singapore	2
Soviet Union	2	United Kingdom	2

Japan

Special focus

Growth in production (% change over previous year)

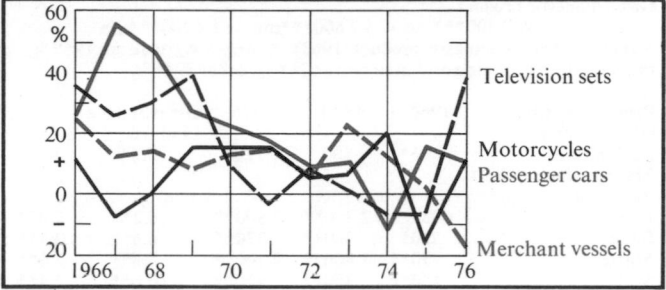

Jordan

Hashemite Kingdom of Jordan
Al Mamlakah al Urdunniyah al Hashimiyah

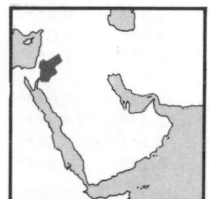

Location Middle East
Israel is to the west, between Jordan and the Mediterranean Sea, with Syria to the north, Iraq to the east and Saudi Arabia to the south. There is a short coastline with access to the sea at Aqaba
Land Area 97 740 km² = 37 740 mi² (includes 5 880 km² for West Bank of the River Jordan)
Climate Hot and dry summers, cool winters
Weather at Amman, 777 m altitude
Temperature: hottest month Aug 18–32 °C, coldest Jan 4–12 °C
Rainfall (av monthly): driest months June–Aug 0 mm, wettest Feb 74 mm
Time 2 hours ahead of GMT (summer time, 3 hours ahead)
Measures Metric system; also local measures including:
area 1 dunum = 0.1 hectare = 1 000 square metres = 1 196 square yards
Monetary unit Jordanian dinar (J D) = 1 000 fils
Rate of exchange (1976 av): par J D 0.387755 = SDR 1, free J D 0.3321 = $ 1 ($ 3.011 = J D 1), free £ 1.667 = J D 1

Summary

Political Constitutional monarchy, formed as an independent state in 1946; formerly under British mandate. Part of the west bank of the River Jordan was absorbed after the 1948–49 Israel/Arab conflict; since the 1967 conflict the West Bank area has been under Israeli occupation. Figures here in general exclude the West Bank. Member of UN and Arab League
Economic Phosphates form the basis of the economy, which has been helped by higher prices of recent years. Agriculture is important, but can be affected severely by weather conditions. The economy is dependent on foreign aid and is helped by remittances from Jordanian workers in other countries

People, resources and equipment

Population[a] 1960 1.69*mn, 1970 2.30*mn, 1976 2.78*mn[b]
[a]Including West Bank; includes registered Palestinian refugees, of whom there were 613 743 in 1960 and 722 687 at May 31, 1967 [b]Of whom, 0.76* mn in West Bank
Growth: 1960–70 3.1*%pa, 1970–76 3.2*%pa
Density (1976): 28* people per km²
Vital statistics (rate per 1 000 people, 1970–75): births 48*, deaths 15*
Regions (population in 000, 1976) East Jordan 2 018*, West Bank and East Jerusalem 761*
Cities (population in 000, 1975) Amman (capital) 634*, Zarqa 238*, Irbid 125*
Race Mainly Arab
Language Arabic
Religion (1961) Moslem 94 % (Sunni 80* %), Christian 6 %
Education (1974/75) Pupils 524 259, teachers 17 721
Labour force (1976) 670 000; in agriculture 194 000 (29 %)
Personnel Scientists and engineers (1973): 4 288
Physicians (1974): 763, 1 per 2 438 people
Standard of living
National income per person (1976): J D 260* = $ 780* = £ 430*
Consumption per person (1975): energy 408 kg coal equivalent, electricity (production) 164 kW h, newsprint 0.03 kg, steel 34 kg
Newspapers (1974): number 4; circulation 58 000, 22 per 1 000 people

Telephones (Dec 1976): 43 700, 16 per 1 000 people
Livestock (000, 1976) Cattle 35*, sheep 818*, goats 474*, camels 18*
Petroleum refinery capacity (1975) 1.1 mn tonnes
Electrical capacity (1973) 281 megawatts
Hospital beds (1974) 1 986, 1 per 937 people
Roads (1976) 4 152 km = 2 580 mi, density 0.045 km per km² (excludes 1 900* in West Bank)
Railways (1975) 371km = 231mi, density 0.004 km per km²
Oil pipeline Saudi Arabia (from Dhahran oil fields) to Lebanon (Sidon): on Jordanian territory 180 km = 110 mi. A transit Iraq/Israel pipeline is not in operation
Ships (registered, 1977) 2, of 696 gross tons
Port (goods traffic, 000 tonnes, 1976) Aqaba: loaded 1 632, unloaded 1 369
Airports (1976) Amman: passenger departures 386 000, arrivals 371 000; also Aqaba

Durable equipment (at end-year)	000	no per 1 000 people	
Radio sets (1975)	529	271	no per km of road
Television sets (1974)	85	45	
Passenger cars (1976)	40	20	9.6
Commercial vehicles (1976)	15	7	3.6

Production

Gross domestic product 1976: J D 400.4 mn = $ 1 206 mn = £ 667 mn
Structure of gross domestic product (1974) *By origin* Agriculture 16 %, mining and quarrying 3 %, manufacturing 11 %, construction 5 %, transport and communication 7 %, other 58 %
Industrial production index (1970 = 100) 1976 228.7; growth 1970–76 14.8 %
Main products (000 t) *Agriculture* (1976) Wheat 67, barley 13, olives 30*, tomatoes 145*, watermelons 44*, grapes 20*, oranges 6*, sheep milk 23*

Other	1960	1970	1976	Growth %pa 1960–70	1970–76
Petroleum products	—	440	756[a]	na	11.4[b]
Electricity (mn kW h)	na	165	443[a]	na	21.8[b]
Phosphate rock	362	913	1 353[a]	9.7	8.2[b]
Cigarettes (mn units)	946	1 610	1 998[a]	5.5	4.4[b]
Cement	165	378	533	8.6	5.9

[a]1975 [b]1970–75
Transport traffic (1976) *Passenger-kilometres* Road 15 558 mn, rail 18 mn, air 805 mn *Cargo; tonne-kilometres* Rail 6 mn, air 14 mn
Sea Goods loaded 1.63 mn t, unloaded 1.37 mn t
Tourism (1975) Number of visitors 707 600, gross receipts $ 101 mn

Finance and trade

Consumer price index (1970 = 100) 1976 192.4; growth 1970–76 11.5 %pa
Money stock (end-year, J D mn) 1970 105, 1976 264; growth 1970–76 16.5 %pa
Budget (1977) Revenue: J D 238 mn = $ 722 mn = £ 414 mn including foreign grants: J D 123 mn = $ 374 mn = £ 214 mn
Expenditure: J D 282 mn = $ 858 mn = £ 491 mn

Balance of payments ($ mn)	1972	1973	1974	1975	1976
Balance of goods (fob)	−190	−219	−277	−496	−704
Balance of services	+2	+34	+15	+123	+405
Balance of transfers	+191	+197	+271	+441	+381
Current balance	+4	+12	+9	+67	+82
Long-term capital flow	+18	+22	+33	+141	−73
Reserves and debt (end-year, $ mn)					
International reserves	271	304	347	492	491
External public debt	261	442	546	565	na

External trade (1976) Imports: J D 339.5 mn = $ 1 022 mn = £ 566 mn
Exports: J D 68.7 mn = $ 207 mn = £ 115 mn

Main imports (1975)	% of total	Main exports (1975)	% of total
Food	21	Phosphates	40
Crude oil	10	Oranges	11
Machinery, non-electric	10	Aircraft (re-exports)	8
Motor vehicles	9	Vegetables	6
Electrical machinery	6	Chemicals	4
Aircraft	6	Cement	3
Chemicals	5		
Textile yarns and fabrics	5		

Main sources (1976)		Main destinations (1976)[a]	
West Germany	17	Saudi Arabia	11
United States	9	Syria	9
United Kingdom	8	Iran	9
Japan	7	Kuwait	5
Italy	6	Bulgaria	4
India	4	Iraq	3

[a]The destination of 28 % of exports is not specified

Jordan

Special focus

Fluctuations in wheat production

	000 tonnes		000 tonnes		000 tonnes
1967	196	1971	168	1975	50
1968	95	1972	211	1976	67
1969	159	1973	50	1977	53*
1970	54	1974	245		

Korea, North

Democratic People's Republic of Korea
Choson Minchuchui Inmin Konghwaguk

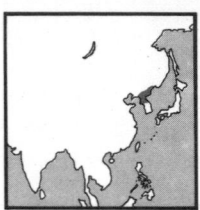

Location Eastern Asia
In the northern part of the Korean peninsula, with the Sea of Japan to the east and the Yellow Sea to the west; South Korea is to the south roughly below the 38th parallel, China to the north-west, and there is a short border with Soviet Union to the north-east
Land Area 120 538 km² = 46 540 mi² (excludes demilitarised zone of 1 262 km² = 487 mi²)
Climate Continental, with extremes of temperature
Weather at Wonsan, 37 m altitude
Temperature: hottest month Aug 20–27 °C, coldest Jan minus 8–1 °C
Rainfall (av monthly): driest month Jan 30 mm, wettest Aug 317 mm
Time 9 hours ahead of GMT
Measures Metric system
Monetary unit North Korean won (NK W) = 100 chon (jun)
Rate of exchange (1976 av):
Basic free NK W 0.96 = $ 1, NK W 1.73 = £ 1
Tourist official NK W 2.05 = $ 1, NK W 3.70 = £ 1

Summary

Political Communist republic, which was proclaimed as such September 9, 1948; formerly occupied by Soviet Union from the end of the 1939–45 world war. There was a conflict with South Korea during 1950–53 and a demilitarised zone was established in 1953 between the two Koreas. Member of Unesco, WHO, Unctad and other UN agencies
Economic Metals provide about one-half of exports, and agricultural produce is also important; industrialisation is increasing, helped since 1971 by increased trade with the west (following the Chinese pattern). There have been some external financing difficulties following the general fall in metal prices after 1974

People, resources and equipment

Population 1960 10.53*mn, 1970 13.89*mn, 1976 16.25*mn
Growth: 1960–70 2.8*%pa, 1970–76 2.6*%pa
Density (1976): 135* people per km²
Vital statistics (rate per 1 000 people, 1970–75): births 35.7*, deaths 9.4*
Cities (population in 000, 1976) Pyongyang (capital) 1 500**, Chongjin 300**, Hungnam 260**, Kaesong 240**, Wonsan 240**, Sinuiju 200**
Race Predominantly Korean
Language Korean
Religion Buddhist, Confucianist, Shamanist
Education (1973/74) Pupils 3 000 000**, teachers 110 000**
Labour force (1976) 7 168 000**; in agriculture 3 534 000* (49*%)
Personnel (1960) Physicians: 11 919, 1 per 889 people
Standard of living National income per person (1976):
NK W 430*** = $ 450*** = £ 250***
Consumption per person (1975): Energy 2 808** kg coal equivalent, electricity (production) 1 450** kW h, newsprint 0.1* kg, steel 251** kg
Newspapers (main, 1970): number 10
Livestock (000, 1976) Cattle 816**, sheep 268**, goats 200**, pigs 1 570**
Electrical capacity (1973) 7 000** megawatts
Hospital beds (1960) 55 000, 1 per 193 people
Roads (main, 1961) 5 600**km = 3 500**mi, density 0.05**km per km²
Railways (1976) 4 500**km = 2 800**mi, density 0.04**km per km²
Inland waterways Yalu river: 700* km = 430* mi
Also Aprok, Daidong, Dooman and Ryesung rivers
Ships (registered, 1977) 19, total of 89 482 gross tons
Ports Chongjin, Nampo
Airport Pyongyang

Radio sets (1970) 1 000 000**, 70** per 1 000 people
Television sets (1968) 2 500*, 0.2* per 1 000 people

Production

Gross domestic product
1976 est: NK W 7 300***mn = $ 7 600***mn = £ 4 200***mn
Structure of gross domestic product (1963) *By origin* Agriculture 19** %, manufacturing, mining and quarrying 62** %, other 19** %

Production index (1970 = 100)	1960	1970	1976	Growth %pa 1960–70	1970–76
Agricultural	93*	100	128*	2.4*	4.2*
Main products (000 t)					
Agriculture					
Rice	1 535	2 800**	3 800**	6.2**	5.2**
Barley	230*	360**	370**	4.6**	0.5**
Maize	950**	1 800**	2 100**	6.6**	2.6**
Millet	350**	350**	410**	0.0**	2.7**
Potatoes	650**	1 000**	1 300**	4.4**	4.5**
Sweet potatoes	201**	275**	330**	3.2**	3.1**
Apples	80**	115**	158**	3.7**	5.4**
Soyabeans	180**	228**	300**	2.4**	4.7**
Tobacco	17**	40**	41**	9.0**	0.4**
Cotton	3**	3**	3**	0.0**	0.0**
Eggs	34**	54**	72**	4.7**	4.9**
Pigmeat	50**	60**	63**	1.8**	0.8**
Fish catch	500**	800**	800**	4.8**	0.0**
Timber (000 m³)	3 800**	4 770**	5 240**	2.3**	1.6**
Energy					
Total energy (000 tce)	9 900**	26 340**	42 430***a	10.3**	10.0**b
Coal	6 800**	21 800**	35 000*a	12.4**	10.0**b
Lignite	3 800**	5 700**	9 000**b	4.1**	9.6**b
Electricity (mn kW h)	9 139	16 500**	23 000***a	6.1**	6.8**b
Mining					
Iron ore (Fe content)	1 550**	4 014**	3 760**	10.0**	−1.3**b
Copper ore (Cu content)	6**	13**	13**	8.0**	0.0**
Lead ore (Pb content)	50**	70**	100***a	3.4**	7.4**b
Magnesite	50**	1 600**	1 700**	40.0**	1.0**
Tungsten conc (oxide content)	3.0**	2.7**	2.7**	−1.0**	0.0**
Zinc ore (Zn content)	90**	130**	162**a	3.7**	4.5**b
Phosphate rock	100**	250**	450**a	9.6**	12.5**b
Salt	320**	540**	540**	5.4**	0.0**
Manufacturing					
Fertilisers, nitrogenous	100**	200**	270**	7.2**	5.1**
Fertilisers, phosphate	30**	90**	120**	11.6**	4.9**
Cement	2 285	4 000**	6 000**b	5.8**	8.5**b
Coke	800**	2 200**	3 000**a	10.6**	6.4**b
Pig iron	872	2 400**	2 900**a	10.7**	3.8**b
Crude steel	640**	2 200**	2 800**a	13.1**	4.9**b

a1975 b1970–75
Transport traffic *Rail* Passenger-kilometres (1960) 3 386 mn, cargo: tonne-kilometres (1964) 10 600 mn
Sea (1971) Goods loaded 1.0**mn t, unloaded 0.4**mn t

Finance and trade

Budget (1976) Balanced at NK W 12 513 mn = $ 13 000 mn = £ 7 200 mn
External trade (1976)
Imports: NK W 750***mn = $ 750***mn = £ 430***mn
Exports: NK W 800***mn = $ 800***mn = £ 460***mn

Main imports (1965)a	% of total	Main exports (1965)a	% of total
Machinery and transport equipment	23**	Metals and metal ores	50**
Petroleum products	11**	Raw materials (including coal, cement and tobacco)	18**
Food (of which, wheat 3**b)	9**		
Raw materials	8**	Food (including rice and apples)	16**
Chemicals	5**	Chemicals	2**
Main sources (1976)		*Main destinations* (1976)	
Soviet Union	32*	China	40***
China	30***	Soviet Union	20*
Japan	13*	Japan	9*
West Germany	6*	West Germany	6*

aLatest general information available b1973

Special focus

External debt, 1976 (end-year)	$ mn
Soviet Union and China	1 000***
Japan and other non-communist countries	1 500***

Korea, South

Republic of Korea
Taehan Minguk

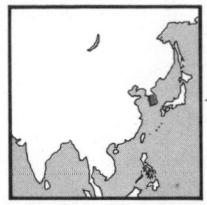

Location Eastern Asia
In the southern part of the Korean Peninsula, with the Sea of Japan to the east and the Yellow Sea to the west; North Korea is to the north, and Japan is about 160 km = 100 mi to the south-east across the Sea of Japan
Land Area 98 484 km² = 38 025 mi² (excludes demilitarised zone of 1 262 km² = 487 mi²)
Climate Continental, with extremes of temperature
Weather at Seoul, 87 m altitude
Temperature: hottest month Aug 22–31 °C, coldest Jan minus 9–0 °C
Rainfall (av monthly): driest month Feb 20 mm, wettest July 376 mm
Time 9 hours ahead of GMT
Measures Metric system; local units are also in use, including:
area 1 chungbo = 0.992 hectare = 2.45 acres
capacity 1 suk = 100 dai = 180.39 litres = 39.68 UK gallons
weight (mass) 1 kwan = 1 000 don = 3.75 kilograms = 8.27 pounds
Monetary unit South Korean won (SK W) = 100 chon (jun)
Rate of exchange (1976 av): par SK W 485 = £ 1, free SK W 876 = £ 1

Summary

Political One-party republic, which became independent in 1948; formerly under US occupation from the conclusion of the world war in 1945. There was a conflict with North Korea during 1950–53, and a demilitarised zone was established in 1953 between the two Koreas. Member of Unesco, WHO, Unctad and other UN agencies, and of Colombo Plan
Economic Agriculture has formed the main part of the economy but, with high growth of industry at around the 20 %pa level, manufacturing overtook agriculture in 1973 to become the largest sector. Textiles account for about one-third of exports, and electrical equipment is an important export. Steel and ship-building are also significant industries

People, resources and equipment

Population 1960 24.70* mn, 1970 32.24* mn, 1976 35.86* mn
Growth: 1960–70 2.7* %pa, 1970–76 1.8* %pa
Density (1976): 364* people per km²
Vital statistics (rate per 1 000 people, 1970–75): births 28.8*, deaths 8.9*
Regions (population in 000, 1975; census total of 34.71 mn)

Cities		Provinces			
Pusan	2 454	Cheju-do	412	Chungchong-pukto	1 522
Seoul	6 889	Cholla-namdo	3 985	Kangwon-do	1 862
		Cholla-pukto	2 456	Kyonggi-do	4 040
		Chungchong-namdo	2 949	Kyongsang-namdo	3 280
				Kyongsang-pukto	4 859

Cities (population in 000, 1975)

Seoul (capital)	6 889	Masan	372	Chungju	193
Pusan	2 454	Chonju	311	Jingu	155
Taegu	1 311	Seongnam	272	Gunsan	154
Inchon	800	Ulsan	253	Chunchon	141
Kwangju	607	Suwon	224	Jeju	135
Taejon	507	Mokpo	193	Yeosu	131

Race Predominantly Korean
Language Korean
Religion (1976) Buddhist 34 %, Confucian 13 %, Protestant 13 %, Roman Catholic 3 %, Chundo Kyo 2 %
Education (1975/76) Pupils 9 072 447, teachers 207 254
Labour force (1976) 13 061 000; in agriculture 5 601 000 (43 %), mining and quarrying 65 000 (½ %), manufacturing 2 678 000 (21 %), construction 529 000 (4 %), distribution and hotels 1 878 000 (14 %), transport and communication 390 000 (3 %), other 1 415 000 (11 %), unemployed 505 000 (4 %)
Personnel (1974) Physicians: 13 013, 1 per 2 670 people
Standard of living
National income per person (1976): SK W 310 760 = $ 641 = £ 355
Consumption per person (1975): energy 1 038 kg coal equivalent, electricity (production, 1976) 676* kW h, newsprint 4.4 kg, steel 85 kg
Newspapers (1974): number 44; circulation 5 867 000, 170 per 1 000 people
Telephones (Mar 1977): 1 681 000, 47 per 1 000 people
Livestock (000, 1976) Cattle 1 641, pigs 1 247, goats 250, chickens 20 140*
Mineral reserves Coal (1974) 1 450 mn tonnes
Petroleum refinery capacity (1975) 22 mn tonnes

Electrical capacity (1975) 5 135* megawatts
Hospital beds (1974) 22 089, 1 per 1 570 people
Roads (1976) 45 514 km = 28 281 mi, density 0.46 km per km²
Railways (1976) 5 653 km = 3 513 mi, density 0.057 km per km²
Ships (registered, 1977) 1 042, total of 2 494 724 gross tons
Ports (goods traffic, 000 tonnes, 1976)

	International		Coastwise		Total
	loaded	unloaded	loaded	unloaded	
Pusan	8 424	8 579	80	4 492	21 575
Ulsan[a]	1 444	12 044	3 868	622	17 978
Inchon	1 521	8 826	76	3 119	13 542
Ryosu	831	5 525	3 387	569	10 312
Pohang	470	4 250	48	1 278	6 046
Masan	155	468	15	939	1 577

[a] Crude oil port
Airports Seoul and 6 other airports with scheduled flights

Durable equipment (at end-year)	000	no per 1 000 people	
Radio sets (1974)	4 812	138	no per
Television sets (1974)	1 619	46	km of road
Passenger cars (1976)	96	2.7	2.1
Commercial vehicles (1976)	111	3.1	2.4

Production

Gross domestic product
1976: SK W 12 279 bn = $ 25 318 mn = £ 14 017 mn
Growth in real terms: 1960–70 8.7 %pa, 1970–76 11.0 %pa
Structure of gross domestic product

By origin (1975)	SK W bn	% of gdp
Agriculture	2 303	25
Mining and quarrying	117	1
Manufacturing	2 581	28
Electricity, gas and water	114	1
Construction	416	5
Distribution and hotels	1 609	17
Transport and communication	511	6
Other services	1 588	17
Total	9 239	100

By type of expenditure (1976)		
Government final consumption	1 488	12
Private final consumption	7 948	65
Stock investment	210	2
Gross fixed capital formation	2 829	23
Exports of goods and services	4 410	36
less Imports of goods and services	−4 595	−37
Total (including statistical discrepancy)	12 290	100

Production indices (1970 = 100)	1960	1970	1976	Growth %pa	
				1960–70	1970–76
Agricultural	58	100	124	5.7	3.7
Industrial	20	100	357	17.6	23.6

Main products (000 t)

Agriculture	1960	1970	1976	1960–70	1970–76
Rice	4 240*	5 476	7 250	2.6*	4.8
Barley	1 370	1 974	1 759	3.7	−1.9
Potatoes	421	605	591	3.7	−0.4
Sweet potatoes	780	2 136	1 950*	10.6	−1.5*
Cabbages	400**	847	900*	7.8**	1.0*
Apples	104	212	310*	7.4	6.5*
Soyabeans	130	232	295*	6.0	4.1*
Tobacco	27	56	111	7.6	12.1
Cotton	6.0*	4.5	3.0*	−2.8*	−6.5*
Milk	1	52	122*	48.5	15.3*
Eggs	34	130	168*	14.3	4.4*
Meat	100**	158	254	4.7**	8.2
Fish catch	455	843	2 407	6.4	19.1
Timber (000 m³)	8 700**	10 660	9 418*	2.0**	−2.0*
Energy					
Total energy (000 tce)	5 420	12 550	17 800[c]	8.8	7.2[d]
Coal	5 340*	12 394	16 428	8.8*	4.8
Petroleum products	—	9 804	16 100[c]	na	10.4[d]
Electricity (mn kW h)	1 758	9 597	24 240*	18.5	16.7*
Mining					
Iron ore (Fe content)	200	300	348	4.1	2.5
Tungsten conc (oxide content)	3.44	2.75	3.34[c]	−2.2	4.0[d]
Zinc ore (Zn content)	—	23.4	56.1	na	15.7
Silver	0.010	0.051	0.046[e]	17.7	−2.0[d]
Gold (000 kg)	2.05	1.60	0.37[c]	−2.5	−25.4[d]
Salt	399	405	665[c]	0.1	10.4[d]

Korea, South

Main products (000 t)	1960	1970	1976	Growth %pa	
Manufacturing				1960–70	1970–76
Beer (000 hl)	176	933	1 896	18.1	12.5
Cigarettes (mn units)	14 382	39 509	54 773	10.6	5.6
Cotton yarn	49	91	175	6.4	11.5
Cotton fabrics (mn m²)	188*	193	340	0.2*	9.9
Man-made fibres	0.09	50	338	88.1	37.5
Plywood (000 m²)	30	1 126	2 197	43.7	11.8
Newsprint	27	106	155	14.7	6.5
Other paper	30**	228	507c	22.5**	17.3d
Tyres (000 units)	175	923	3 491	18.1	24.8
Sulphuric acid	na	388	639	na	8.7
Fertilisers, nitrogenous	—	386	541c	na	7.0d
Fertilisers, phosphate	—	140	195c	na	6.8d
Cement	464	5 782	11 872	28.7	12.7
Pig-iron	14	41	1 938	11.3	90.2
Crude steelb	50	481	2 698	25.4	33.3
Aluminium	—	15	20*c	na	5.9*d
Radio sets (000 units)	na	1 088	6 578	na	35.0
Television sets (000 units)	—	114	2 291	na	64.9
Motor vehiclesa	—	28.5	49.7	na	9.7
Merchant vessels (000 grt)	—	—	814	na	na

a Assembly only   bIngots only   c1975   d1970–75

Transport traffic	1960	1970	1976	Growth %pa	
Passenger-kilometres (mn)				1960–70	1970–76
Rail	4 935	9 819	13 890	7.1	6.0
Air	22	445	4 519	34.9	47.2
Cargo: tonne-kilometres (mn)					
Rail	3 043	7 488	9 486	9.4	4.0
Air	0.17	6.0	355	42.7	97.4
Sea: tonnes (mn)					
Goods loaded	0.58	3.59	14.3	20.0	25.9
Goods unloaded	2.4	18.7	41.5	22.5	14.2
Tourism					
Number of visitors (000)	na	173	834	na	30.0
Gross receipts ($ mn)	na	19	275	na	56.1

Finance and trade

Price and wage indices	1960	1970	1976	Growth %pa	
(1970 = 100)				1960–70	1970–76
Consumer prices	29.1	100.0	233.1	13.1	15.1
Wholesale prices	31.0	100.0	267.0	12.4	17.8
Wages (earnings)	18.0	100.0	355.0	18.8	23.5

Money stock
(end-year, SK W bn) 23   308   1 544   29.7   30.9
Budget (1976) Revenue: SK W 2 752 bn = $ 5 674 mn = £ 3 142 mn
Expenditure: SK W 2 700 bn = $ 5 567 mn = £ 3 082 mn

Balance of payments ($ mn)	1972	1973	1974	1975	1976
Balance of goods (fob)	−575	−565	−1 938	−1 671	−589
Balance of services	+34	+68	−310	−441	−77
Balance of transfers	+169	+191	+221	+225	+348
Current balance	−370	−307	−2 027	−1 888	−318
Long-term capital flow	+494	+655	+1 079	+1 497	+1 432
Reserves and debt (end-year, $ mn)					
International reserves	740	1 094	1 056	1 550	2 960
External public debt	3 719	4 556	6 178	6 996	na

External trade (1976) Imports: SK W 4 246 bn = $ 8 755 mn = £ 4 847 mn
Exports: SK W 3 734 bn = $ 7 699 mn = £ 4 263 mn

Main imports	% of total	Main exports	% of total
Crude oil	18	Clothing	24
Machinery, non-electric	12	Textile yarns and fabrics	12
Chemicals	10	Electrical machinery	10
Electrical machinery	9	(of which,	
Cereals	5	telecommunications 4)	
Iron and steel	5	Footwear	5
Timber	5	Iron and steel	5
Ships	5	Plywood	4
Textile yarns and fabrics	4	Fish	4
Cotton	4	Ships	4
Main sources		*Main destinations*	
Japan	35	United States	32
United States	22	Japan	23
Saudi Arabia	8	West Germany	5
Kuwait	8	Hongkong	4
Indonesia	3	Canada	4
West Germany	3	United Kingdom	3

Structure of gross domestic product

	% of total					
	1960	1970	1972	1973	1974	1975
Agriculture	37	28	28	26	25	25
Manufacturing	14	22	23	26	28	28
Other	49	50	49	48	47	47

Kuwait

State of Kuwait
Dawlat al Kuwayt

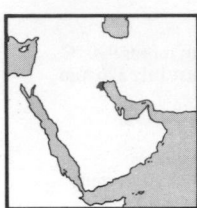

Location Middle East
With a coastline to the east on the Gulf, Iraq is to
the north and west, and Saudi Arabia to
the south. The Neutral Zone, partitioned and
shared with Saudi Arabia, is to the south
Land Area 17 818 km² = 6 880 mi² (including
partitioned neutral zone)
Climate Hot, mild in winter
Weather at Kuwait City, 5 m altitude
Temperature: hottest month Aug 30–40 °C,
coldest Jan 9–16 °C
Rainfall (av monthly): driest months June-Sept 0 mm, wettest Dec 28 mm
Time 3 hours ahead of GMT
Measures Metric system
Monetary unit Kuwaiti dinar (K D) = 10 dirhams = 1 000 fils
Rate of exchange (1976 av): free $ 3.420 = K D 1, £ 1.894 = K D 1

Summary

Political Parliamentary monarchy, which became fully independent in
June 1961; formerly under UK protection. The Neutral Zone, shared with
Saudi Arabia, was partitioned in 1966, but oil and other natural resources
of the zone continue to be shared. Member of UN, Arab League, Opec
and Oapec
Economic Crude oil and petroleum products account for over 90 % of
exports. Oil revenues have been used to operate a good social security
system, to increase overseas investments (especially through the Kuwait
Investment Company), to maintain a large shipping fleet and to develop
industry, especially petrochemicals. A policy of relative conservation of
oil reserves is observed

People, resources and equipment

Population 1960 0.28* mn, 1970 0.74* mn, 1976 1.06* mn
Growth: 1960–70 10.3* %pa, 1970–76 6.1* %pa
Density (1976): 59* people per km²
Vital statistics (rate per 1 000 people, 1975): births 43.4, deaths 4.8
Cities (population in 000, 1970) Al Kuwayta (capital) 218, Hawalli 107
aKuwait
Race (1976) Arab 89* %, (Kuwaiti 47 %), Asian 10* %, European 1* %
Language Arabic
Religion (1970) Moslem 95 %, Christian 5 %
Education (1974/75) Pupils 205 365, teachers 14 929
Labour force (1975) 304 582; in agriculture 7 514 (2 %), mining and
quarrying 4 859 (2 %), manufacturing 24 467 (8 %), construction
32 256 (11 %), transport and communication 15 685 (5 %), finance and
real estate 6 523 (2 %)
Personnel Scientists and engineers (1973): 10 754, of whom,
non-national 8 603
Physicians (1972): 1 050, 1 per 800 people
Standard of living
National income per person (1976): K D 3 300* = $ 11 300* = £ 6 200*
Consumption per person (1975): energy 8 718 kg coal equivalent,
electricity (production 1976) 4 910 kW h, newsprint 5.4 kg, steel 354 kg
Newspapers (1974): number 6; circulation 80 000, 86 per 1 000 people
Telephones (Dec 1976): 139 900, 128 per 1 000 people
Livestock (000, 1976) Cattle 9*, sheep 110*, goats 86*, camels 6*
Mineral reserves (1975) Crude oil 10 151 mn tonnes
Natural gas 1 070 bn cubic metres
Petroleum refinery capacity (1975) 30 mn tonnes
Hospital beds (1975) 4 255, 1 per 234 people
Roads (1974) 1 920* km = 1 190* mi, density 0.11 km per km²
There are no railways
Ships (registered, 1977) 226, total of 1 831 194 gross tons
Ports Shuwaykh, Shuaybah, Mina al Ahmadi and Mina Abd Allah
(oil ports), Mina Saud (oil port, Kuwait sector of partitioned Neutral
Zone)
Airport (1976) Kuwait: passenger departures 667 400, arrivals 682 400

Kuwait

Durable equipment

(at end-year)	000	no per 1 000 people	
Radio sets (1974)	215	231	no per
Television sets (1974)	182	188	km of road
Passenger cars (1975)	204	198	106
Commercial vehicles (1975)	68	66	35

Production, finance and trade

Gross domestic product 1975/76 (year ending March 31st):
K D 3 279 mn = $ 11 260 mn = £ 5 302 mn
1976 est: K D 3 500*mn = $ 12 000*mn = £ 6 600*mn
Structure of gross domestic product (1975/76) *By origin* Mining and quarrying 70 %, manufacturing 5 %, transport and communication 3 %, other 22 %. *By type* Final consumption expenditure 34 %, gross fixed capital formation 8 %, exports of goods and services 81 %, less imports of goods and services − 22 %
Main products (000 t) *Agriculture* (1976) Tomatoes 3*, watermelons 3*, eggs 2*, milk 11*, meat 19*

Other	1960	1970	1976	Growth %pa	
				1960–70	1970–76
Crude oil[a]	85 511	150 636	108 562	5.8	−5.3
Petroleum products[a]	10 390*	21 224	13 700*c	7.4*	−8.4*d
Natural gas (mn m³)[a]	941	4 041	5 581	15.7	5.5
Electricity (mn kW h)	249	2 213	5 210	24.4	15.4
Fertilisers, nitrogenous[b]	—	74	259	na	23.2

[a]Including one-half share of production in the Neutral Zone   [b]Years ending June 30th
[c]1975 · [d]1970–75

Transport traffic	1960	1970	1976	Growth %pa	
Air (mn)				1960–70	1970–76
Passenger-km	56*	557	1 135	25.8*	12.6
Cargo tonne-km	1.0*	14.1	27.1	31.0*	11.5
Sea: tonnes[a] (mn)					
Goods loaded[b]	83.5*	158.8	107.2c	6.6*	−7.6d
Goods unloaded	na	1.26	2.53c	na	14.9d

[a]Including Neutral Zone   [b]Crude oil only   [c]1975   [d]1970–75

Tourism (1970) Number of visitors 617 900
Consumer price index (1972 = 100) 1976 141.0; growth 1972–76 9.0 %pa
Money stock (end-year, K D mn) 1970 95, 1976 394; growth 1970–76 26.7 %pa
Budget (1975/76; year ending March 31st)
Revenue: K D 3 906 mn = $ 13 413 mn = £ 6 316 mn
Expenditure: K D 1 033 mn = $ 3 547 mn = £ 1 670 mn
International reserves (Dec 1976) $ 1 929 mn
External trade (1976) Imports: K D 970 mn = $ 3 318 mn = £ 1 838 mn
Exports: K D 2 873 mn = $ 9 827 mn = £ 5 442 mn

Main imports (1975)	% of total	Main exports (1975)	% of total
Motor vehicles	19	Crude oil	81
Food	15	Petroleum products	9
Electrical machinery	11	Chemicals	5
Machinery, non-electric	9	Natural gas	2
Textile yarns and fabrics	6		
Ships	5		
Clothing	4		
Main sources (1976)		*Main destinations* (1976)	
Japan	21	Japan	22
United States	15	Netherlands	10
West Germany	11	United Kingdom	8
United Kingdom	8	Brazil	6
France	5	France	4
Italy	5	Italy	4
India	4	Saudi Arabia	4

Special focus

Relative conservation of crude oil

Production of crude oil (including Neutral Zone)

	000 tonnes	% of world production
1960	85 511	8.1
1965	118 720	7.9
1970	150 636	6.6
1973	152 418	5.5
1974	128 609	4 6
1975	105 232	4.1
1976	108 562	3.8

Laos

Lao People's Democratic Republic
Pathet Sathalanalath Pasathipathay Pasason Lao

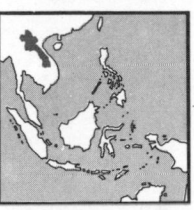

Location South-east Asia
China is to the north, Vietnam to the east, Cambodia to the south, with Thailand to the west and Burma to the north-west.
Land-locked
Land Area 236 800 km² = 91 400 mi²
Climate Tropical
Weather at Vientiane, 162 m altitude
Temperature: hottest month April 23–34 ° C, coldest Jan 14–28 °C
Rainfall (av monthly): driest month Dec 3 mm, wettest June 302 mm
Time 7 hours ahead of GMT
Measures Metric system; local measures are also in use, including:
length 1 va = 5 sok = 10 khup = 2 metres = 2.187 yards
1 sénh = 20 va = 40 metres = 43.74 yards
1 ngane = 100 talangua = 400 square metres = 478.4 square yards
area 1 rai = 4 ngane = 1 600 square metres = 1 914 square yards
weight (mass) 1 xang = 10 hoi = 1.2 kilograms = 2.646 pounds
1 mune = 10 xang = 12 kilograms = 26.46 pounds
1 sène = 2 picul = 10 mune = 120 kilograms = 264.6 pounds
Monetary unit New Kip (Kp) = 100 at; the new kip was introduced June 1976 to replace the old kip at 1 new kip = 20 old kips
Rate of exchange (1976): par new Kp 200 = $ 1, free new Kp 361.2 = £ 1
New rate, introduced mid-1976 is shown; in the first half of 1976 the equivalent rate was new Kp 37.5 = $ 1 (old Kp 750 = $ 1)

Summary

Political Communist republic, which became independent from France in 1949 and became communist controlled in 1975, following a cease-fire agreement in 1973. Member of UN and Colombo Plan
Economic Mainly an agricultural economy, with some tin production; timber and tin have been the main export products. Recovery is hoped for, after two decades of civil war and involvement in south-east Asian conflicts; exports have been very low, creating dependence on foreign aid from Soviet Union, Vietnam and China

People, resources and equipment

Population 1960 2.34*mn, 1970 2.96*mn, 1976 3.38*mn
Growth: 1960–70 2.4* %pa, 1970–76 2.2* %pa
Density (1976): 14* people per km²
Vital statistics (rate per 1 000 people, 1970–75): births 44.6*, deaths 22.8*
Cities (population in 000, 1973) Vientiane (capital) 177, Savannakhet 51, Pakse 45, Luang Prabang 44, Sayaboury 14
Race (1976) Lao-Lum 44* %, Lao-Theung 26* %, Thai 15* %, Lao-Soung (including Meo and Yao) and others 15* %
Language Lao; also tribal languages, including Meo
Religion Mainly Buddhist; also Christian and Animist
Education (1972/73) Pupils 295 500*, teachers 8 550*
Labour force (1976) 1 648 000; in agriculture 1 250 000 (76 %)
Personnel (1976) Physicians: 46, 1 per 73 500 people
Standard of living
National income per person (1976):
new Kp 17 000*** = $ 85*** = £ 47***
Consumption per person (1975): energy 63 kg coal equivalent, electricity (production) 77 kW h, newsprint 0.5 kg, steel 1 kg
Newspapers (1974): number 5
Telephones (Dec 1973): 5 000, 2 per 1 000 people
Livestock (000, 1976) Cattle 500*, buffaloes 1 100*, pigs 1 450*, chickens 15 800*
Electrical capacity (1975) 55* megawatts
Hospital beds (1975) 3 232, 1 per 1 021 people
Roads (1974) 7 395 km = 4 595 mi, density 0.031 km per km²
There are no railways
Inland waterway Mekong river
Airport Vientiane
Radio sets (Dec 1974) 125 000*, 38* per 1 000 people
Passenger cars (Dec 1974) 14 100, 4 per 1 000 people, 1.9 per km of road
Commercial vehicles (Dec 1974) 2 500, 1 per 1 000 people, 0.3 per km of road

Production

Gross domestic product
1976 est: new Kp 60 000*** mn = $ 300*** mn = £ 170*** mn

Laos

Production index (1970 = 100)	1960	1970	1976	Growth %pa	
				1960–70	1970–76
Agricultural	61	100	105	5.0	0.8

Main products (000 t)

Agriculture

	1960	1970	1976	1960–70	1970–76
Rice	500*	904*	850*	6.1*	−1.0*
Maize	12	28*	30*	8.8*	1.2*
Onions	18**	25*	30*	3.3**	3.1*
Pineapples	18**	25*	28*	3.3**	1.9*
Coffee	1.0*	3.2	2.0*	12.3*	−7.5*
Tobacco	1.5	3.8	4.0*	9.7	0.9*
Cotton	1.0*	3.0	2.0*	11.6*	−6.5*
Meat	33*	52*	54*	4.6*	0.6*
Timber (000 m³)	2 150**	2 836	3 154*	2.8**	1.8*

Other

	1960	1970	1976	1960–70	1970–76
Electricity[a] (mn kW h)	7	12	255*[b]	5.9	84.3*[c]
Tin conc (Sn content)	0.39	0.57	0.52[b]	4.0	−2.1[c]
Cigarettes (mn units)	na	361	628[d]	na	20.3[e]

[a] Public supply only [b] 1975 [c] 1970–75 [d] 1973 [e] 1970–73

Transport traffic (1976) *Air* Passenger-kilometres 22 mn, cargo: tonne-kilometres 0.5 mn
Tourism (1973) Number of visitors 23 100, gross receipts $ 3*mn

Finance and trade

Consumer price index (Vientiane, 1970 = 100) 1975 457.3; growth 1970–75 35.5 %pa
Budget (1973/74; year ending 30th June)
Revenue: new[a] Kp 1 115 mn = $ 37 mn = £ 16 mn
Expenditure: new[a] Kp 1 855 mn = $ 62 mn = £ 26 mn
of which, defence 49 %, education and culture 9 %
[a] Equivalent
External trade (1974) Imports: new[a] Kp 1 950 mn = $ 65 mn = £ 28 mn
Exports: new[a] Kp 330 mn = $ 11 mn = £ 4.8 mn
[a] Equivalent

Main imports	% of total	Main exports	% of total
Rice	20	Timber	81
Petroleum products	11	Tin	11
Electrical machinery	10	Plywood	3
Machinery, non-electric	8	Coffee	2
Motor vehicles	6		
Metal small manufactures	6		
Chemicals	6		
Iron and steel	5		
Main sources		Main destinations	
Thailand	49	Thailand	73
Japan	19	Malaysia	11
France	7	Hongkong	10
West Germany	7	Japan	4
United States	5	Singapore	2
Hongkong	4		

Lebanon

Republic of Lebanon
Al Jumhuriyah al Lubnaniyah

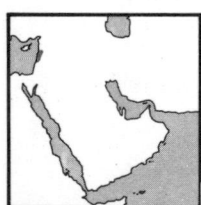

Location Middle East
With a coastline on the Mediterranean Sea, Syria is to the north and east, and Israel to the south
Land Area 10 400 km² = 4 000 mi²
Climate Sub-tropical, cool in highland
Weather at Beirut, 34 m altitude
Temperature: hottest month Aug 23–32 °C, coldest Jan 11–17 °C
Rainfall (av monthly): driest months July, Aug 1 mm, wettest Jan 190 mm
Time 2 hours ahead of GMT (summer time, 3 hours ahead)
Measures Metric system
Monetary unit Lebanese pound (L £) = 100 piastres
Rate of exchange (1976 av): free L £ 2.904 = $ 1, L £ 5.245 = £ 1

Summary

Political Republic, which was proclaimed independent on November 26, 1941; formerly a French mandated territory. There have been disputes with Israel concerning the Palestinian bases in Lebanon, and also civil war between Moslems and Christians. Israel occupied some border territory in March 1978. Member of UN and Arab League
Economic Trading, finance and tourism have been important to the economy, which also has a well-developed agricultural sector. Civil war has had severe effects on the economy.

People, resources and equipment

Population[a] 1960 2.11*mn, 1970 2.47*mn, 1976 2.96*mn
[a] UN estimates; official estimate for 1970, based on a sample survey: 2.13 mn. Excluding Palestinian refugees, of whom there were 187 529 registered in 1973
Growth: 1960–70 1.6* %pa, 1970–76 3.1* %pa
Density (1976): 285* people per km²
Vital statistics (rate per 1 000 people, 1973): births 24.5, deaths 4.3
Cities (population in 000, 1974) Beirut (capital) 700*, Tripoli 175*, Zahle 47*, Saïda[a] 25*, Tyr 14*
[a] Sidon
Race (1961) Arab 93* %, Armenian 5* %, Kurd 1* %
Language Arabic
Religion (1976) Moslem 50** %, Christian 50** % (of whom, Maronite 32** %)
Education (1972/73) Pupils 665 300*, teachers 33 240*
Labour force (1970) 571 755; in agriculture 101 760 (18 %), manufacturing 94 620 (17 %), construction 35 055 (6 %), transport and communication 38 235 (7 %), distribution and hotels 91 620 (16 %), finance and real estate 18 420 (3 %), other 192 045 (34 %)
Personnel Scientists and engineers (1969): 5 134
Physicians (1973): 2 300, 1 per 1 170 people
Standard of living
National income per person (1976): L £ 2 000*** = $ 700*** = £ 400***
Consumption per person (1975): energy 928 kg coal equivalent, electricity (production, 1976) 422 kW h, newsprint 2.8 kg, steel 101 kg
Newspapers (1974): number 37
Telephones (Dec 1972) 227 000, 87 per 1 000 people
Livestock (000, 1976) Cattle 84*, goats 330*, sheep 234*, pigs 23*
Petroleum refinery capacity (1975) 2.5* mn tonnes
Electrical capacity (1975) 608* megawatts
Hospital beds (1970) 10 727, 1 per 230 people
Roads (1976) 7 100* km = 4 400* mi, density 0.68* km per km²
Railways (1976) 425* km = 264* mi, density 0.041* km per km²
Oil pipelines Iraq–Syria to Tripoli; Saudi Arabia–Jordan to Sidon
Ships (registered, 1977) 163, total of 227 009 gross tons
Ports (goods traffic, 000 tonnes, 1974)

	loaded	unloaded
Tripoli[a]	14 918	595
Sidon[b]	10 168	—
Beirut	668	3 411

[a] Crude oil port; includes oil piped from Syria [b] Crude oil port; includes oil piped from Saudi Arabia
Airport (1974) Beirut: passenger departures and arrivals 2 807 000

Durable equipment (at end-year)	000	no per 1 000 people	
Radio sets (1975)	1 321	460	no per km of road
Television sets (1975)	410	143	
Passenger cars (1974)	214	76	30
Commercial vehicles (1974)	24	8	3.4

Production

Gross domestic product 1972: L £ 6 365 mn = $ 2 086 mn = £ 834 mn
1976 est: L £ 6 000*** mn = $ 2 000*** mn = £ 1 100*** mn
Structure of gross domestic product (1972) *By origin* Agriculture 10 %, manufacturing, mining and quarrying 14 %, construction 5 %, transport and communication 8 %, other 63 %

Production index (1970 = 100)	1960	1970	1976	Growth %pa	
				1960–70	1970–76
Agricultural	59	100	130	5.4	4.5

Lebanon

Main products (000 t)	1960	1970	1976	Growth %pa 1960–70	1970–76
Agriculture					
Wheat	40	43	30*	0.7	−5.8*
Potatoes	29	93	85*	12.4	−1.5*
Olives	30*	20*	46*	−4.0*	14.9*
Olive oil	6.0*	3.0*	11.0*	−6.7*	24.2*
Tomatoes	25	75	65*	11.6	−2.3*
Onions	23*	30*	32*	2.7*	1.1*
Grapes	70	90*	100*	2.5*	1.8*
Apples	53	119*	170*	8.4*	6.1*
Oranges	105*	175*	180*	5.2*	0.5*
Lemons	40*	60*	80*	4.1*	4.9*
Tobacco	3.5	7.2	11.0*	7.5	7.3*
Milk	60**	63*	68*	0.5**	1.3*
Eggs	3	27*	33*	24.0*	3.4*
Meat	50**	51*	45*	0.2**	−2.1*
Other					
Petroleum products	691	1932	1 980*a	10.8	0.5*b
Electricity (mn kW h)	422	1 230	1 250*	11.3	0.3*
Salt	12*	37*	35*a	11.9*	−1.1*b
Cigarettes (mn units)	1 230	1 281	883c	0.4	−8.9d
Cotton yarn	na	3.5	4.7e	na	10.3f
Fertilisers, phosphate	6**	39	67*a	20.0**	11.4*b
Cement	854	1 339	1 744c	4.6	6.8d

a1975 b1970–75 c1974 d1970–74 e1973 f1970–73

Transport traffic	1960	1970	1976	Growth %pa 1960–70	1970–76
Passenger-kilometres (mn)					
Rail	5.0*	7.4	1.9b	4.0*	−28.8c
Air	252	854	1 800	13.0	13.2
Cargo: tonne-kilometres (mn)					
Rail	36	20	42b	−5.7	20.4c
Air	18	145	525	22.9	23.9
Sea: tonnes (mn)					
Goods loaded a	21.7	32.6	25.8b	4.2	−5.7c
Goods unloaded	1.44	2.57	4.41b	6.0	14.5c

aIncluding Iraqi and Saudi Arabian crude oil loaded at Tripoli and Sidon b1974
c1970–74

Tourism Number of visitors (1975) 1 555 000, gross receipts (1974)
$ 415 mn

Finance and trade

Consumer price index (Beirut, 1970 = 100) 1974 125.5;
growth 1970–74 5.8 %pa
Money stock (end-year, L £ mn) 1960 1 115, 1970 1 677, 1976 4 906;
growth 1960–70 4.2 %pa, 1970–76 19.6 %pa
Budget (1978) Revenue L £ 1 403 mn = $ 470* mn = £ 260* mn
Expenditure L £ 2 083 mn = $ 690* mn = £ 390* mn
International reserves (Dec 1976) $ 1 677 mn
External trade (1976) Imports: L £ 2 200* mn = $ 750* mn = £ 410* mn
Exports: L £ 2 300* mn = $ 800* mn = £ 440* mn

Main imports (1973)	% of total	Main exports (1973)	% of total
Food	15	Fruit and vegetables	12
(of which, wheat 4)		Machinery, non-electric	10
Machinery, non-electric	10	Chemicals	8
Chemicals	9	Aircraft	6
Textile yarns and fabrics	9	Clothing	6
Motor vehicles	9	Textile yarns and fabrics	5
Iron and steel	6	Motor vehicles	5
Electrical machinery	6	Electrical machinery	4
Petroleum products	4	Iron and steel	3
Main sources (1973)		*Main destinations (1973)*	
United States	12	Saudi Arabia	15
West Germany	11	France	9
France	10	United Kingdom	8
Italy	10	Libya	7
United Kingdom	8	Kuwait	6
Switzerland	4	Syria	5

Special focus

Crude oil ports Goods loaded, 000 tonnes

	Sidon a	Tripoli b
1970	8 295	23 607
1971	16 222	19 572
1972	20 594	7 817
1973	19 413	21 337
1974	10 168	14 918

aPipeline terminal from Saudi Arabia bPipeline terminal from Iraq

Macao
Macau

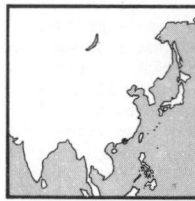

Location Eastern Asia
Occupies a peninsula in southern China, and the
offshore islands of Taipa and Colôane; the
territory is 60 km to the west of Hongkong
Land Area 16 km² = 6 mi²
Climate Sub-tropical
Time 8 hours ahead of GMT
Measures Metric system; local measures are
also in use, including:
length 1 côvado = 66 centimetres = 26 inches
1 vara = 1.1 metres = 43.4 inches
weight (mass) 1 arroba = 14.688 kilograms = 32.38 pounds
1 quintal = 58.75 kilograms = 129.5 pounds
Monetary unit Pataca (Pat) = 100 avos
Rate of exchange (1976 av): official Pat 1 = Portuguese Esc 5,
free Pat 6.045 = $ 1, Pat 10.92 = £ 1
From April 1977 the link with the Portuguese escudo was changed to
a link with the Hongkong dollar at Pat 1.075 = HK $ 1

Summary

Political Portuguese overseas province from 1951 to 1974; formerly
a trading post. After 1974 a Chinese territory under Portuguese
administration
Economic Textile yarns and fabrics are the main import and clothing the
main export. Some Hongkong exports are routed via Macao, partly
because of overseas import restrictions. Food is mainly imported from
China. Tourism, including especially gambling, is important.

People, resources and equipment

Population 1960 169 000*, 1970 249 000*, 1976 275 000*
Growth: 1960–70 4.0* %pa, 1970–76 1.7* %pa
Density (1976): 17 190* people per km²
City (population, 1976) Macao City (capital) 265 000*
Race (1974) Chinese 98** %
Language Chinese and Portuguese
Religion Mainly Buddhist and Roman Catholic
Education (1975/76) Pupils 32 619, teachers 1 310
Labour force (1976) 72 000**
Personnel (1974) Physicians: 181, 1 per 1 492 people
Standard of living
National income per person (1976): Pat 4 500*** = $ 750*** = £ 400***
Consumption per person (1975): energy 264 kg coal equivalent,
electricity (production) 530 kW h, newsprint 0.4 kg
Newspapers (1972): number 7
Telephones (Dec 1976): 11 760, 42 per 1 000 people
Livestock (000, 1976) Chickens 352*
Electrical capacity (1975) 70 megawatts
Hospital beds (1974) 1 275, 1 per 209 people
Roads (1974) 33 km = 21 mi, density 2.1 km per km²
Port Macao
Radio sets (Dec 1975) 61 000, 225 per 1 000 people
Motor vehicles (Dec 1974) 7 700, 29 per 1 000 people, 233 per km of road

Production, finance and trade

Gross domestic product
1976 est: Pat 1 300***mn = $ 210***mn = £ 120***mn
Main products *Agriculture* (000 t, 1976) Grapes 4*, wine 3*, meat 9*
Other (1975) Electricity 143 mn kW h, cigarettes 500 mn units
Transport traffic (1974) *Sea:* goods loaded 147 000 t, unloaded 299 000 t
Tourism (1976) Number of visitors 2 500 000*
Budget (1976) Balanced at Pat 133 mn = $ 22 mn = £ 12 mn
External trade (1976) Imports: Pat 977 mn = $ 162 mn = £ 89 mn
Exports: Pat 1 146 mn = $ 190 mn = £ 105 mn

Main imports (1975)	% of total	Main exports (1975)	% of total
Textile yarns and fabrics	41	Clothing	81
Food	20	Fish	5
Petroleum products	9	Textile yarns and fabrics	4
Chemicals	5		
Main sources (1976)		*Main destinations (1976)*	
Hongkong	68	West Germany	23
China	24	France	17
United States	3	United Kingdom	10
Japan	2	United States	10
		Hongkong	9

Malaysia

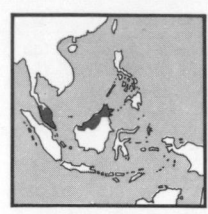

Location South-east Asia
Peninsular Malaysia occupies the southern part of the Kra peninsula, with Sumatra (Indonesia) to the west across the Strait of Malacca, and the South China Sea to the east; Thailand is to the north and Singapore island to the south. Sabah and Sarawak are adjoining territories on the north coast of Borneo; Indonesia occupies the larger southern part of Borneo, and Brunei forms an enclave within Sarawak

Land Area 329 749 km² = 127 317 mi²
of which, Peninsular Malaysia 131 587 km² = 50 806 mi²
Climate Tropical
Weather at Kuala Lumpur, 39 m altitude
Temperature: hottest months April, May 23–33 °C, coldest Dec 22–32 °C
Rainfall (av monthly): driest month July 99 mm, wettest April 292 mm
Time Peninsular (mainland): 7½ hours ahead of GMT
Sabah and Sarawak: 8 hours ahead of GMT
Measures Metric system, as a conversion from UK (imperial) system; local measures are also in use, including:
weight (mass) 1 pikul = 25 gantang = 60.48 kilograms = 133⅓ pounds
1 koyan = 40 pikul = 2.419 tonnes = 5 333⅓ pounds
Monetary unit Ringgit (Ma $) = 100 sen
Rate of exchange (1976 av): free Ma $ 2.542 = $ 1, Ma $ 4.591 = £ 1

Summary

Political Parliamentary monarchy, which became independent in August 1957 as the Federation of Malaya; formerly under UK protection. In 1963, Singapore, Sabah and Sarawak joined with the Federation to form Malaysia; Singapore withdrew in August 1965. Before 1973 Peninsular Malaysia was called West Malaysia. Member of UN, Colombo Plan, Commonwealth and Asean
Economic Agriculture makes up one-third of gross domestic product, with rubber the main export and timber and palm oil also important. Mining is substantial, with tin accounting for 11 % of exports and crude oil (from Sarawak) for 13 % in 1976. The industrial sector is developing, with a growth rate in the 1970s of about 9 %pa

People, resources and equipment

Population 1960 8.11*mn, 1970 10.39*mn 1976 12.30*mn
Growth: 1960–70 2.5* %pa, 1970–76 2.9* %pa
Density (1976): 37* people per km²
Vital statistics (rate per 1 000 people, 1974): births 32.1*, deaths 6.6*
Regions (population in 000, 1970; census total of 10.44 mn)

Peninsular Malaysia	8 810ᵃ	Malacca	404	Perlis	121
States:		Negeri Sembilan	482	Selangorᵇ	1 631
Johor	1 277	Pahang	505	Terengganu	405
Kedah	955	Pinang	776	*Sabah*	654ᵃ
Kelantan	685	Perak	1 569	*Sarawak*	976ᵃ

ᵃPopulation at end 1975 (000): Peninsular 10 115*, Sabah 838*, Sarawak 1 116*
ᵇKuala Lumpur, formerly part of this state, became a separate federal territory from February 1, 1974; population of Federal Territory, 1970: 648 thousand; 1977: 800* thousand
Cities (population in 000, 1970)

| Kuala Lumpur (capital) | 452 | Ipoh | 248 | Kelang | 114 |
| Pinang (George Town) | 270 | Johor Bahru | 136 | | |

Race (1975) Malay 47 %, Chinese 34 %, Indian and Pakistani 9 %, Ibans 3 %, Kadazans 2 %
Language Malay; also used are Chinese, English, Tamil, Iban Dusun, Bajan and other indigenous languages and dialects
Religion (1970) Moslem 50 %, Buddhist 26 %, Hindu 9 %, Christian 4 %
Education (1976) Pupils 2 930 000*, teachers 99 000*
Labour force (1976) 4 202 000; in agriculture 2 141 000 (51 %)
Personnel Scientists and engineers (1973): 30 000*
Physicians (1974): 1 556, 1 per 7 490 people
Standard of living
National income per person (1976): Ma $ 2 140* = $ 840* = £ 470*
Consumption per person (1975): energy 588 kg coal equivalent, electricity (production, 1976) 529* kW h, newsprint 3.3 kg, steel 46 kg
Newspapers (1974): number 31; circulation 1 038 000, 89 per 1 000 people
Telephones (Dec 1976): 329 600, 26 per 1 000 people
Livestock (000, 1976) Cattle 423*, buffaloes 298*, pigs 1 444*, goats 369*, chickens 44 930*

Mineral reserves (1975) Crude oil 195 mn tonnes
Natural gas 340 mn cubic metres
Petroleum refinery capacity (1975) 6.7* mn tonnes
Electrical capacity (1975) 1 177* megawatts
Hospital beds (1974) 40 124, 1 per 290 people
Roads (1976) 21 324 km = 13 250 mi, density 0.065 km per km²
Railways (1976) 2 290* km = 1 423* mi, density 0.007* km per km²
Ships (registered, 1977) 179, total of 563 666 gross tons
Ports (goods traffic, 000 tonnes, 1973)

	loaded	unloaded		loaded	unloaded
Miriᵃ	6 534	747	Tawau	2 378	314
Port Dickson	1 700	4 085	Labuan	2 294	155
Kelang	2 939	2 426	Kinabalu	73	581
Pinang	1 332	2 109	Kuching	59	346
Sandakan	2 899	415			

ᵃCrude oil port
Airports (passengers, 000, 1976) Kuala Lumpur: departures 781, arrivals 743; Pinang: departures 295, arrivals 294; also (1977) 34 other airports with scheduled flights

Durable equipment (at end-year)	000	no per 1 000 people	
Radio sets (1975)	450	37	no per
Television sets (1974)	390	33	km of road
Passenger cars (1974)	430	36	20
Commercial vehicles (1974)	140	12	6.6

Production

Gross domestic product 1976: Ma $ 27 944 mn = $ 10 993 mn = £ 6 087 mn
Growth in real terms: 1970–75 6.6 %pa
Structure of gross domestic product (1974) *By origin* Agriculture 32 %, mining and quarrying 6 %, manufacturing 17 %, electricity, gas and water 2 %, construction 5 %, other 38 %

Production indices (1970 = 100)	1960	1970	1976	Growth %pa 1960–70	1970–76
Agricultural: Peninsular	60	100	137	5.3	5.4
Agricultural: Sabah	68	100	159	3.9	8.1
Agricultural: Sarawak	108	100	140	−0.8	5.8
Industrial	na	100.0	164.3	na	8.6

Main products (000 t)
Agriculture

	1960	1970	1976	1960–70	1970–76
Rice	1 152	1 678*	1 908*	3.8*	2.2*
Cassava	170**	230*	337*	3.1**	6.6*
Bananas	300**	350*	454*	1.5**	4.4*
Copra	207	203*	166*	−0.2*	−3.3*
Palm kernels	24	92*	300*	14.4*	21.7*
Palm oil	92	431	1 390*	16.7	21.6*
Coffee	6.3*	6.3*	5.0*	0.0*	−3.8*
Cocoa	0.5*	5.5*	16.0*	27.1*	19.5*
Tea	2.5	3.4	3.4*	3.1	0.0*
Tobacco	2.5**	5.0**	10.0*	7.2**	12.3**
Pepper	na	24	35	na	6.5
Rubber	696	1 269	1 595*	6.2	3.9*
Pigmeat	25**	68*	50*	10.5**	−5.0*
Poultrymeat	35**	50**	73*	3.6**	6.5**
Fish catch	169	340	517	7.2	7.2
Timber (000 m³)	10 300**	24 135	36 361	8.9**	7.1

Energy

	1960	1970	1976	1960–70	1970–76
Total energy (000 tce)	110	1 270	7 000ᵉ	27.7	40.7ᶠ
Crude oilᵈ	60	859	8 024	30.5	45.1
Petroleum products	2 376	3 533	930*ᵉ	4.0	−23.4*ᶠ
Electricity (mn kW h)	1 242	3 545	6 510*	11.0	10.7*

Mining

	1960	1970	1976	1960–70	1970–76
Iron ore (Fe content)	3 209	2 516	172	−2.4	−36.1
Bauxite	748	1 139	660	4.3	−8.7
Tin conc (Sn content)	53	74	63	3.4	−2.5
Gold (000 kg)	0.75	0.15	0.11ᵉ	−14.7	−6.0ᶠ

Manufacturing

	1960	1970	1976	1960–70	1970–76
Cigarettes (mn units)ᵇ	3 600**	7 566	10 852ᵉ	7.7**	7.5ᶠ
Fertilisers, nitrogenousᵃ,ᵇ	—	31	34	na	1.5
Fertilisers, phosphateᵃ,ᵇ	—	—	21	na	na
Tyres (000 units)ᵇ	na	3 067	1 779	na	−8.7
Cement	286	1 030	1 733	13.7	9.1
Television sets (000 units)ᵇ	na	44	102ᵉ	na	18.3ᶠ
Passenger cars (000 units)ᵇ,ᶜ	—	21.2	44.2	na	13.0
Commercial vehicles (000 units)ᵇ,ᶜ	—	8.1	8.1	na	0.0

ᵃYears ending June 30th ᵇPeninsular Malaysia only ᶜAssembly only ᵈSarawak only ᵉ1975 ᶠ1970–75

Malaysia

Transport traffic	1960	1970	1976	Growth %pa	
Passenger-kilometres (mn)				1960–70	1970–76
Rail[a]	630	647	1 042[c]	0.3	10.0[d]
Air	39[b]	664[b]	1 814	32.8	18.2
Cargo: tonne-kilometres (mn)					
Rail[a]	710	1 208	829[c]	5.5	−7.3[d]
Air	1.1[b]	10.8[b]	36.5	25.3	22.5
Sea: tonnes (mn)					
Goods loaded	12.0	23.4	16.5[c]	6.9	−6.7[d]
Goods unloaded	4.0	9.6	10.9[c]	9.1	2.6[d]

[a]Including traffic of Singapore [b]Apportionment of traffic of Malayan Airways Ltd for 1960 and of Malaysia-Singapore Airlines Ltd for 1970 [c]1975 [d]1970–75
Tourism (1976) Number of visitors 1 224 800, gross receipts $ 25 mn

Finance and trade

Price index	1960	1970	1976	Growth %pa	
(1970 = 100)				1960–70	1970–76
Consumer prices	91.2	100.0	145.9	0.9	6.5
Money stock					
(end-year, Ma $ mn)	1 170	2 072	5 257	5.9	16.8

Budget (1976) Revenue: Ma $ 6 166 mn = $ 2 426 mn = £ 1 343 mn
Expenditure: Ma $ 7 414 mn = $ 2 917 mn = £ 1 615 mn

Balance of payments ($ mn)	1972	1973	1974	1975	1976
Balance of goods (fob)	+130	+652	+387	+441	na
Balance of services	−321	−485	−600	−575	na
Balance of transfers	−55	−62	−60	−52	na
Current balance	−248	+105	−273	−186	*na*
Long-term capital flow	+305	+232	+468	+560	na
Reserves and debt (end-year, $ mn)					
International reserves	970	1 345	1 618	1 524	2 472
External public debt	963	1 105	2 188	2 821	3 041

External trade (1976) Imports: Ma $ 9 967 mn = $ 3 921 mn = £ 2 171 mn
Exports: Ma $ 13 442 mn = $ 5 288 mn = £ 2 928 mn

Main imports (1975)	% of total	Main exports (1976)	% of total
Machinery	24	Rubber	23
Chemicals	8	Timber	18
Motor vehicles	8	Crude oil	13
Crude oil	8	Tin	11
Cereals	5	Palm oil	9
Iron and steel	5		
Petroleum products	4		
Main sources (1976)		*Main destinations* (1976)	
Japan	21	Japan	21
United States	13	Singapore	18
Singapore	9	United States	16
United Kingdom	7	Netherlands	7
Australia	7	United Kingdom	5
West Germany	6	West Germany	4
Kuwait	4	Australia	2
Thailand	4	Soviet Union	2

Special focus

Rubber prices (wholesale: at Singapore)

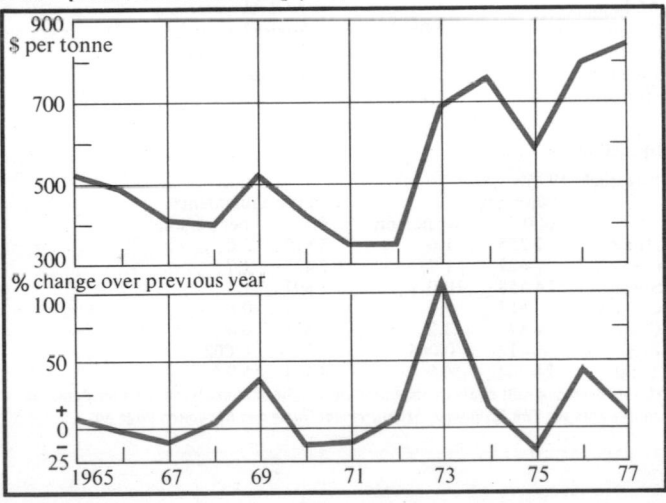

Maldives

Republic of Maldives
Divehi Raajje

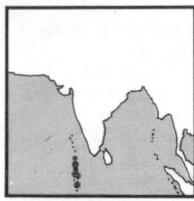

Location Northern Indian Ocean
A group of about 2 000 islands grouped into 12 clusters, situated about 700 km south-west of Sri Lanka
Land Area 298 km² = 115 mi²
Climate Tropical
Time 5 hours ahead of GMT
Measures UK (imperial) system, changing to metric system
Monetary unit Maldivian rupee (Mv R) = 100 laris
Rate of exchange (1976 av):
Official par Mv Rs 3.93 = $ 1, free Mv Rs 7.10 = £ 1
Non-official free Mv Rs 8.40 = $ 1, Mv Rs 15.20 = £ 1

Summary

Political Republic, which became independent in 1965; formerly under UK protection. Member of UN and Colombo Plan
Economic Fishing is the main industry, with copra also significant. In 1975 the United Kingdom decided to withdraw its RAF base from Gan, in the southern part of Maldives first established in 1956; this formerly provided $ 0.7 mn in earnings, and aid of £ 0.5 mn ($ 1.0 mn) was granted over 4 years to help counteract the effect on the economy. Shipping and tourism are important, and tourism has grown rapidly

People, resources and equipment

Population 1960 88 000*, 1970 110 000*, 1976 136 000*
Growth: 1960–70 2.3* %pa, 1970–76 3.6* %pa
Density (1976): 456* people per km²
Vital statistics (rate per 1 000 people, 1973): births 40*, deaths 17*
Cities (population in 000, 1976) Malé (capital) 17*
Race Mixed
Language Divehi
Religion Mainly Moslem (Sunni)
Education (1976) Pupils 3 894, teachers 138
Labour force (1974) 52 000*
Personnel (1973) Physicians: 5, 1 per 24 000 people
National income per person (1976): Mv Rs 600*** = $ 150*** = £ 80***
Newspapers (1974) Number 1
Telephones (Dec 1976) 480, 4 per 1 000 people
Hospital beds (1974) 45, 1 per 2 870 people
Ships (registered, 1977) 45, total of 110 681 gross tons
Port Malé
Airport Hulule Island (2 km from Malé)
Radio sets (Dec 1975) 2 500, 19 per 1 000 people

Production, finance and trade

Gross domestic product
1976 est: Mv Rs 85***mn = $ 20***mn = £ 12***mn
Main products (000 t, 1976) *Agriculture* Millet 2*, yams 1*, copra 1*, meat 1*, fish catch 32
Transport traffic *Sea* (000 t, 1974) Goods loaded 8, unloaded 11*
Tourism (1976) Number of visitors 9 200
Budget (1975) Revenue: Mv Rs 18.4 mn = $ 4.7 mn = £ 2.1 mn
Expenditure: Mv Rs 23.8 mn = $ 6.1 mn = £ 2.7 mn
External trade (1975) Imports: Mv Rs 43* mn = $ 10* mn = £ 5* mn
Exports: Mv Rs 16 mn = $ 4.2 mn = £ 1.8 mn
Main imports (1975) Food 30* % (rice 15* %), ships and boats 18 %, petroleum products 7* %, chemicals 5* %
Main exports (1974) Maldive (dried) fish 73 %, raw fish 17 %, shells 2 %
Main destinations (1974) Sri Lanka 75* %, Japan 10** %, Thailand 10** %

Mongolia

Mongolian People's Republic
Büged Nayramdah Mongol Arad Ulas

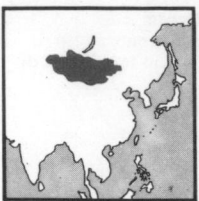

Location Central Asia
Soviet Union is to the north. and China to the
east, south and west. Land-locked
Land Area 1 565 000 km² = 604 000 mi²
Climate Dry and cold, extremely cold in winter
Weather at Ulan Bator, 1 325 m altitude
Temperature: hottest month July 11–22 °C,
coldest Jan minus 32–minus 19 °C
Rainfall (av monthly): driest months Jan, Feb
1 mm, wettest July 76 mm
Time 8 hours ahead of GMT
Measures Metric system; local units are also in unofficial use, including:
capacity 1 dan = 10 sulga = 65 litres = 14.30 UK gallons
weight (mass) 1 dzhin = 16 lan = 0.6 kilogram = 1.323 pounds
Monetary unit Tögrög or Tugrik (Tug) = 100 möngö
Rate of exchange (1976 av) *Basic* Tug 3.32 = $ 1, free Tug 6.00 = £ 1
Non-commercial Tug 5.04 = $ 1, free Tug 9.10 = £ 1

Summary

Political Communist republic, which became independent from China in
1921. Member of UN and Comecon
Economic An agricultural economy based mainly on livestock; livestock,
meat and wool are estimated to account for nearly two-thirds of exports.
Manufacturing is light, mainly food processing and textiles, but is growing
at nearly 10 %pa. Development of mining of non-ferrous metals,
especially copper and molybdenum, is taking place, with financing mainly
from Soviet Union

People, resources and equipment

Population 1960 0.95*mn, 1970 1.25*mn, 1976 1.49*mn
Many of the population are nomadic
Growth: 1960–70 2.8* %pa, 1970–76 3.0* %pa
Density (1976): 1* person per km²
Vital statistics (rate per 1 000 people, 1970–75): births 39*, deaths 9*
Cities (population in 000, 1976) Ulaanbaatar[a] (capital) 334*, Darhan 55*,
Erdenet 32*
[a]Ulan Bator
Race (1969) Khalkha 75 %, Kazakh 5 %, Durbet 3 %, Buryat 2 %,
Bayat 2 %, Russian 2 %
Language Mongolian
Religion Mainly Buddhist
Education Pupils (1976/77) 366 700, teachers (1975/76) 12 800
Labour force (1975) 550 000*; in agriculture 304 000* (55* %), industry
53 700* (10* %), construction 18 300* (3* %), transport and
communication 24 900* (5* %)
Personnel Scientists and engineers (1972): 1 908
Physicians (1974): 2 604, 1 per 538 people
Standard of living
National income per person (1976): Tug 2 500*** = $ 750*** = £ 400***
Consumption per person (1975): energy 1 091 kg coal equivalent,
electricity (production, 1976) 624 kW h, newsprint 1.7 kg
Newspapers (1974): number 1; circulation 112 000, 80 per 1 000 people
Telephones (Dec 1976): 37 800, 25 per 1 000 people
Livestock (000, 1975) Cattle 2 427, sheep 14 458, goats 4 595,
pigs 12, horses 2 255, camels 617
Mineral reserves Coal, lignite, uranium
Petroleum refinery capacity (1975) 70 000* tonnes
Electrical capacity (1975) 266 megawatts
Hospital beds (1974) 13 648, 1 per 103 people
Roads (main, 1970) 8 600 km = 5 300 mi, density 0.005 km per km²
Railways (1973) 1 425 km = 885 mi, density 0.001 km per km²
Inland waterways Selenge river 470 km = 290 mi; also Orhon river and
Lake Hövsgöl
Airport Ulan Bator
Radio sets (Dec 1975) 114 000, 79 per 1 000 people
Television sets (Dec 1975) 3 500, 2 per 1 000 people

Production

Gross domestic product
1976 est: Tug 4 000***mn = $ 1 200***mn = £ 670***mn
Structure of gross domestic product (1969) *By origin* Agriculture 22 %,
industry 20 %, construction 15 %, transport and communication 6 %,
other 37 %

Production indices (1970 = 100)	1960	1970	1976	Growth %pa 1960–70	1970–76
Agricultural	100**	100	120	0.0**	3.0
Industrial	39	100	155[a]	10.0	9.2[b]
Main products (000 t)					
Agriculture					
Wheat	216	250	280	1.5	1.9
Oats	30	24	39	−2.2	8.4
Barley	8	9	35*	1.0	25.4*
Potatoes	19	21	27*	1.0	4.3*
Milk, cow	120**	137*	145*	1.3**	1.0*
Milk, sheep and goat	80**	84*	87*	0.5**	0.6*
Wool	9.1*	11.4*	12.8*	2.3*	2.0*
Hides and skins	20*	23*	25*	1.4*	1.4*
Beef and veal	50**	55*	67*	1.0*	3.3*
Sheep and goat meat	100**	110*	128*	1.0**	2.6*
Horsemeat	25**	32	42*	2.5**	4.6*
Timber (000 m³)	1 000**	2 040*	2 390*	7.4**	2.7*
Other					
Total energy (000 tce)	200*	720	1 010[a]	13.7*	7.0[b]
Coal	—	85	171[a]	na	15.0[b]
Lignite	619	1 915	2 549[a]	12.0	5.9[b]
Electricity (mn kW h)	106	516	930	17.1	10.3
Salt	6	7	11*	1.5	7.8*
Beer (000 hl)	11	18	77	4.8	27.4
Vodka (000 hl)	9	27	39[a]	11.5	7.6[b]
Leather footwear (000 pairs)	904	1 600	1 999[a]	5.9	4.5[b]
Cement	—	96	159[a]	na	10.6[b]
Wool fabrics (mn m³)	na	0.90	1.20*	na	4.9*

[a]1975 [b]1970–75

Transport traffic	1960	1970	1976	Growth %pa 1960–70	1970–76
Passenger-kilometres (mn)					
Road	118	206	348*[a]	5.7	11.1*[b]
Rail	56	135	213[a]	9.1	9.5[b]
Air	23	106	142[a]	16.4	6.0[b]
Cargo: tonne-kilometres (mn)					
Road	201	624	1 054	12.0	9.1
Rail	3 036	1 528	2 701	−6.6	10.0
Air	0.8	1.5	3.8	6.5	16.8
Water	2.6	3.6	4.9[a]	3.3	6.4[b]

[a]1975 [b]1970–75

Finance and trade

Budget (1976) Revenue: Tug 3 176 mn = $ 957 mn = £ 529 mn
Expenditure: Tug 3 022 mn = $ 910 mn = £ 504 mn
External trade (1975) Imports: Tug 1 800*mn = $ 550*mn = £ 240*mn
Exports: Tug 730*mn = $ 220*mn = £ 100*mn

Main imports	% of total	Main exports	% of total
Machinery and		Livestock	27*
transport equipment	57*	Meat	19*
(of which, agricultural		Wool	16*
machinery 12*)		Fluorspar	2*
Clothing	5*	Clothing	2*
Crude oil and products	3*	Leather	1*
Timber	3*		
Main sources		*Main destinations*	
Soviet Union	90*	Soviet Union	79*
Czechoslovakia	2*	Czechoslovakia	6*
Bulgaria	2*	Hungary	3*
Hungary	2*		

Special focus

Livestock, 1975

	Number 000	per person	Cattle equivalent[a] 000	per person
Horses	2 255	1.6	2 819	2.0
Cattle	2 427	1.7	2 427	1.7
Sheep	14 458	10.0	1 807	1.3
Camels	617	0.4	864	0.6
Goats	4 595	3.2	574	0.4
Pigs	12	0.008	3	0.002
Total	24 364	16.9	8 494	5.9

[a]Using livestock unit equivalents, based on weight of animals; for further detail on
equivalents see *The Economist Measurement Guide and Reckoner*, page 86

Nepal

Kingdom of Nepal
Nepal Adhirajya

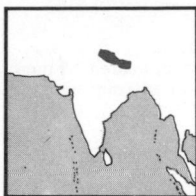

Location Central Asia
In the central Himalayas, with Tibet (Chinese autonomous region) to the north and India to the east (Sikkim state) and to the south. Land-locked
Land Area 140 797 km² = 54 362 mi²
Climate Temperate, varying with altitude
Weather at Kathmandu, 1 337 m altitude
Temperature: hottest month July 20–29 °C, coldest Jan 2–23 °C
Rainfall (av monthly): driest month Dec 3 mm, wettest July 373 mm
Time 5 hours 40 minutes ahead of GMT
Measures Metric system; local units are also in use, including:
weight (mass) 1 seer (terai) = 0.933 kilogram = 2.057 pounds
1 maund = 16 dharni = 40 seer = 37.32 kilograms = 82.29 pounds
Monetary unit Nepalese rupee (N R) = 2 mohur = 100 paisa (pice)
Rate of exchange (1976 av): par N Rs 12.50 = $ 1, free N Rs 22.58 = £ 1

Summary

Political Constitutional monarchy; member of UN and Colombo Plan
Economic Over 90 % of the labour force work in agriculture, which provides about 70 % of gross domestic product; rice and jute are main exports. Trading has always been mainly with India, which controls land outlets; since the absorption of Sikkim by India, trade with China has increased. Hydro-electric schemes are being developed. The economy is dependent on foreign aid

People, resources and equipment

Population 1960 9.24**mn, 1970 11.23*mn, 1976 12.86*mn
Growth: 1960–70 2.0**%pa, 1970–76 2.3*%pa
Density (1976): 91* people per km²
Vital statistics (rate per 1 000 people, 1975): births 42*, deaths 20*
Cities (population in 000, 1976) Kathmandu (capital) 415* (population including suburbs, Patan and Bhadgaon), Patan 135*, Bhadgaon 84*
Race (1961) Nepali 51 %, Tharu 14 %, Bihari 11 %, Tamang 6 %
Language Nepali; also Newari and Bhutia
Religion (1971) Hindu 89 %, Buddhist 7 %
Education (1974) Pupils 659 279, teachers 31 200*
Labour force (1976) 6 199 000; in agriculture 5 771 000 (93 %)
Personnel (1974) Physicians: 338, 1 per 36 450 people
Standard of living
National income per person (1976): N Rs 1 500** = $ 120** = £ 70**
Consumption per person: energy (1975) 10 kg coal equivalent, electricity (production, 1975) 10* kW h
Newspapers (1974): number 26; circulation (13 only) 93 000, 8 per 1 000 people
Telephones (Dec 1973): 8 000, 1 per 1 000 people
Livestock (000, 1976) Cattle 6 653*, buffaloes 3 930*, sheep 2 310*, goats 2 373*, chickens 20 532*
Electrical capacity (1975) 59* megawatts
Hospital beds (1974) 1 858, 1 per 6 630 people
Roads (1976) 3 000* km = 1 900* mi, density 0.021* km per km²
Railways (1976) 106 km = 66 mi, density 0.001 km per km²
Ropeway (1976) 42 km = 26 mi
Airports (passengers, 000, 1976) Tribhuvan (Kathmandu): departures 221, arrivals 214; also 3 other airports with scheduled flights
Durable equipment (Dec 1975)
Radio sets: 80 000, 6 per 1 000 people
Passenger cars: 6 000*, 0.5* per 1 000 people, 2* per km of road
Commercial vehicles: 4 000*, 0.3* per 1 000 people, 1* per km of road

Production

Gross domestic product 1974/75 (year ending July 15th):
N Rs 15 074 mn = $ 1 427 mn = £ 608 mn
1976 est: N Rs 20 000**mn = $ 1 600**mn = £ 900**mn
Growth in real terms: 1970–75 2.3 %pa
Structure of gross domestic product (1973/74) *By origin* Agriculture 69 %, manufacturing 10 %, construction 2 %, other 19 %

Production index (1970 = 100)	1960	1970	1976	Growth %pa 1960–70	1970–76
Agricultural	96	100	106	0.4	1.0
Main products (000 t)					
Agriculture					
Wheat	130	265	387*	7.4	6.5*
Rice	2 690	2 305	2 385	−1.5	0.6
Maize	840	833	787	−0.1	−0.9
Potatoes	190**	290*	312*	4.3**	1.2*
Mustardseed	70**	58	64*	−1.9**	1.6*
Tobacco	7.0**	6.9	9.0*	−0.1**	4.5*
Jute	35	40*	45	1.3*	2.0*
Milk, buffalo	370**	400*	458*	0.8*•	2.3*
Milk, cow	180**	204*	206*	1.2**	0.2*
Meat	40**	53*	65*	2.8**	3.5*
Timber (000 m³)	7 800**	8 931	9 260*	1.4**	0.6*
Other					
Electricity (mn kW h)	11	70	122*b	20.3	11.7*c
Cigarettes (mn units)a	na	1 125	2 265b	na	15.0c

a Years ending July 15th   b 1975   c 1970–75

Transport traffic *Rail* (1970) Passengers carried 556 000, cargo carried 66 200 t
Air (1976) Passenger-kilometres 66 mn, cargo: 0.7 mn t-km
Tourism (1976) Number of visitors 105 108, gross receipts (1974) $ 9.1 mn

Finance and trade

Price index (1970 = 100)	1960	1970	1976	Growth %pa 1960–70	1970–76
Consumer prices	68.3a	100.0	153.2	6.6b	7.4
Money stock (end-year, N Rs mn)	182*	728	1 702	14.9*	15.2

a 1964   b 1964–70

Budget (1976/77; year ending July 15th)
Revenuea: N Rs 1 321 mn = $ 106 mn = £ 62 mn
Expenditure: N Rs 2 372 mn = $ 190 mn = £ 111 mn
a Excludes foreign grants received: N Rs 386 mn = $31 mn = £18 mn

Reserves and debt (end year, $ mn)	1972	1973	1974	1975	1976
International reserves	103	121	132	110	140
External public debt	60	82	108	120	236

External trade (1976) Imports: N Rs 2 035 mn = $ 163 mn = £ 90 mn
Exports: N Rs 1 229 mn = $ 98 mn = £ 54 mn

Main imports (1975/76)	% of total	Main exports (1975/76)	% of total
Manufactures	26	Food (especially rice)	61
Machinery and transport equipment	21	Raw materials (especially jute)	18
Food	13	Manufactures (especially jute goods, curios, carpets)	9
Fuels and minerals	10		
Chemicals	9		
Raw materials	4		
Main sources (1976)		*Main destinations* (1976)	
India	70*	India	60*
Japan	10*	United States	8*
West Germany	4*	Japan	8*
United States	4*	West Germany	6*

Special focus

Tourism 1976

Origin of visitors	Number	%
India	19 339	18
United States	14 394	14
France	11 900	11
United Kingdom	9 210	9
West Germany	9 157	9
Japan	6 732	6
Australia	6 588	6
Italy	4 045	4
Total visitors	105 108	100

Oman

Sultanate of Oman
Sultanat Uman

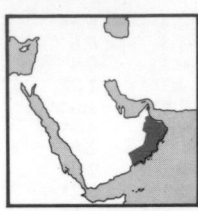

Location Middle East
At the south-east of the Arabian Peninsula, with a coastline to the south-east on the Arabian Sea and to the north-east on the Gulf of Oman; Saudi Arabia and United Arab Emirates are to the north-west and South Yemen to the south-west. The border with Saudi Arabia is not well defined. Territory includes the offshore Kuria Muria and Masirah islands
Land Area 212 457 km² = 82 030 mi²
An official estimate indicates 300 000 km²
Climate Hot in summer, mild in winter
Weather at Muscat, 5 m altitude
Temperature: hottest month June 31–38 °C, coldest Jan 19–25 °C
Rainfall (av monthly): driest months July, Aug 1 mm, wettest Jan 28 mm
Time 4 hours ahead of GMT
Measures Metric system; UK (imperial) system and local units are also in use, including:
length 1 dhara = 0.457 metre = 1½ feet
weight (mass) 1 Muscat maund = 24 kiyas = 4.04 kilograms = 8.9 pounds
1 bahár = 200 Muscat maunds = 0.808 tonne = 0.795 UK (long) ton
Monetary unit Rial Omani (O R) = 1 000 baizas
Rate of exchange (1976 av): par O R 0.345395 = $ 1 ($ 2.895 = O R 1), free £ 1.603 = O R 1

Summary

Political Constitutional monarchy, known as Muscat and Oman before 1970. The Kuria Muria islands were ceded to the United Kingdom from 1854 to 1967. Masirah island was used by the United Kingdom under a 1957 treaty until March 1977. After civil strife in Dhufar province a cease-fire was arranged on March 11, 1976. Member of UN and Arab League
Economic Crude oil has been the main feature since August 1967, when production began; other products are mainly agricultural, especially dates. Development expenditure in the early 1970s was mainly on transport, construction and social services, but emphasis is now being placed also on agriculture, mining, especially of copper, and on industry, especially cement. Oil reserves amount to 24 years production at the 1976 rate

People, resources and equipment

Population[a] 1960 490 000**, 1970 657 000**, 1976 791 000**
[a]Estimates vary; an official estimate is 1 500 000** for 1976
Growth: 1960–70 3.0**%pa, 1970–76 3.1**%pa
Density (1976): 4** people per km²
Cities (population in 000, 1976) Musqat[a] (capital) 50**[b], Salalah[c]
[a]Muscat [b]Including Matrah and Sib [c]Capital of Dhufar province
Race (1961) Arab 88 %, Baluchi 4 %, Persian 3 %, Indian 2 %
Language Arabic
Religion Mainly Moslem
Education Pupils (1976/77) 64 975, teachers (1974/75) 2 146
Labour force (1972) 150 000**; in agriculture 109 000** (73**%)
Personnel (1974) Physicians: 147, 1 per 5 050 people
Standard of living
National income per person (1976): O R 820* = $ 2 400* = £ 1 300*
Consumption per person (1975): energy 334 kg coal equivalent, electricity (production, 1976) 522 kW h
Telephones (Dec 1976): 7 300, 9 per 1 000 people
Livestock (000, 1976) Cattle 134, goats 165, sheep 57, camels 13
Mineral reserves (1975) Crude oil 433 mn tonnes
Natural gas 64 000 mn cubic metres
Electrical capacity (1975) 91 megawatts
Hospital beds (1974) 825*, 1 per 900* people
Roads (1976) 9 772 km = 6 072 mi, density 0.046 km per km²
There are no railways
Ships (registered, 1977) 10, total of 6 137 gross tons
Ports Mina al Fahal (crude oil port), Mina Qaboos (Matrah), Raysut
Airports (passengers, 000, 1976) Sib (Muscat): departures 152, arrivals 172; also Salalah
Durable equipment Radio sets (Dec 1970): 1 000**, 2** per 1 000 people
A television station was opened in November 1974
Passenger cars (Dec 1976): 13 300, 17 per 1 000 people, 1.4 per km of road
Commercial vehicles (Dec 1976): 20 400, 25 per 1 000 people, 2.1 per km of road

Production, finance and trade

Gross domestic product 1976: O R 839mn = $ 2 430 mn = £ 1 345mn
Main products (000 t, 1976) *Agriculture* Lucerne 120*, dates 50*, onions 7*, mangoes 2*, fish catch 198
Other Crude oil 18 293, electricity (mn kW h) 413, cement (1972) 626
Growth in crude oil production: 1970–76 1.6 %pa
Transport traffic *Air* (1976) Passenger-kilometres 320 mn, cargo: 7.0 mn t-km *Sea* (1974) Goods loaded 14* mn t, unloaded 0.9 mn t
Budget (1976) Revenue: O R 505 mn = $ 1 463 mn = £ 810 mn
Expenditure: O R 574 mn = $ 1 661 mn = £ 920 mn
External public debt (Dec 1976) $ 463 mn
External trade (1976) Imports: O R 250 mn = $ 725 mn = £ 401 mn
Exports: O R 544 mn = $ 1 575 mn = £ 872 mn

Main imports (1975)	% of total	Main export (1976)	% of total
Machinery, non-electric	16	Crude oil	99.7
Transport equipment	15		
Food	12		
(of which, cereals 4)			
Electrical machinery	11		
Iron and steel	6		
Petroleum products	5		
Chemicals	4		

Main sources (1975)		Main destinations (1975)	
United Kingdom	20	Japan	37
United Arab Emirates	18	Netherlands	20
West Germany	10	Trinidad & Tobago	7
United States	10	France	7
Japan	8	United Kingdom	7
Netherlands	5	Singapore	6
India	4	United States	6

Special focus

Crude oil production

	000 tonnes		000 tonnes		000 tonnes
1966	—	1970	16 583	1974	14 466
1967	3 076	1971	14 685	1975	17 016
1968	12 019	1972	14 056	1976	18 293
1969	16 367	1973	14 630		

Pakistan

Islamic Republic of Pakistan
Islami Jamhurija-e-Pakistan

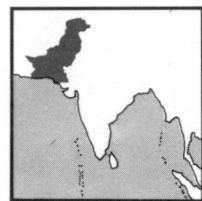

Location South-west Asia
With a coastline on the Arabian Sea to the south, Iran is to the west, Afghanistan to the north-west, China to the north-east, and India to the east, with Jammu and Kashmir, disputed territory, in the north-east
Land Area 803 940 km² = 310 403 mi², excluding Pakistani-held part of Jammu and Kashmir, the final status of which has not been determined (83 807 km² = 32 358 mi²)
Also excluding Junagadh, Manavadar, Gilgit, Baltistan and enclaves in India
Climate Sub-tropical, cold in highlands
Weather at Karachi, 4 m altitude
Temperature: hottest month June 28–34 °C, coldest Jan 13–25 °C
Rainfall (av monthly): driest month Oct 1 mm, wettest July 81 mm
Time 5 hours ahead of GMT
Measures Metric system, introduced in general from January 1, 1978 as conversion from the UK (imperial) system; local measures are also in use, including:
length 1 girah = 3 unglies = 2.25 inches = 5.715 centimetres
1 gaz = 2 haths = 0.9144 metre = 1 yard
1 coss = 1 000 dandas = 2 000 gaz = 1.8288 kilometres = 2 000 yards
weight (mass) 1 tola = 12 mashas = 11.664 grams = 180 grains
1 seer = 16 chattaks = 80 tolas = 0.933 kilogram = 2.057 pounds
1 maund = 40 seers = 37.32 kilograms = 82.29 pounds
Monetary unit Pakistan rupee (P R) = 100 paisa
1 lac = 1,00,000 = 100 000 = one hundred thousand
1 crore = 100 lacs = 1,00,00,000 = 10 000 000 = ten million
1 arab = 100 crores = 1,00,00,00,000 = 1 000 000 000 = one billion
Rate of exchange (1976 av): par P Rs 9.90 = $ 1, free P Rs 17.88 = £ 1

Pakistan

Summary

Political Republic, which became independent in August 1947; formerly part of British India. Originally the state was divided into West and East Pakistan, but East Pakistan became the separate state of Bangladesh in December 1971. Figures included here are for West Pakistan only before that date, unless otherwise mentioned. Member of UN, Colombo Plan, Cento and RCD

Economic Mainly an agricultural economy, with rice and cotton the largest exports; there is substantial industry, especially in textiles—cotton yarn, fabrics and carpets make up nearly one-third of exports. Growth has been low in the 1970s, partly because of the 1971 Bangladesh conflict

People, resources and equipment

Population[a] 1960 44.00**mn, 1970 60.61*mn, 1976 72.37*mn
Growth: 1960–70 3.1** %pa, 1970–76 3.3* %pa
Density (1976): 90* people per km²
Vital statistics (rate per 1 000 people, 1968): births 36*, deaths 12*
[a]Excludes population of Pakistani-held Jammu and Kashmir, estimated for 1976 at 1**mn; also excludes Junagadh, Manavadar, Gilgit, Baltistan and enclaves in India
Cities (population in 000, 1972)

Islamabad (capital)	77	Rawalpindi	615	Sargodha	201
Karachi	3 499[a]	Multan	542	Sukkur	159
Lahore	2 165	Gujranwala	360	Quetta	156
Lyallpur	822	Peshawar	268	Jhang	136
Hyderabad	628	Sialkot	204	Bahawalpur	134

[a]4 465* thousand in 1975
Race (1961) Punjabi 66* %, Sindhi 13* %, Pushtun 8* %
Language Urdu; English is also used, together with other languages and dialects
Religion (1976) Moslem 97** %, Hindu 2** %, Christian 1** %
Education (1974/75) Pupils: primary 5 165 771, secondary 1 751 681, vocational 26 850, teacher-training 16 883, higher (1973/74) 111 826. Teachers: primary 123 361, secondary 94 032, vocational 2 161, teacher-training 754, higher (1973/74) 5 054
Labour force (1976) 21 035 000; in agriculture 11 332 000 (54 %), manufacturing 2 819 000 (13 %), construction 866 000 (4 %), transport and communication 1 007 000 (5 %)
Personnel Scientists and engineers (1973/74): 111 000*
Physicians (registered, 1974): 17 929, 1 per 3 804 people
Standard of living
National income per person (1976): P Rs 1 800* = $ 180* = £ 100*
Consumption per person (1975): energy 183 kg coal equivalent, electricity (production) 125* kW h, newsprint 0.3 kg, steel 7 kg
Newspapers (1974): number 93; circulation (18 only) 358 000, 5 per 1 000 people
Telephones (Dec 1975): 240 000, 3 per 1 000 people
Livestock (000, 1976) Cattle 13 608*, buffaloes 10 795*, sheep 19 186*, goats 14 109*, camels 899*
Mineral reserves Coal (1972) 410 mn tonnes
Lignite (1977) 280*mn tonnes
Crude oil (1975) 5 mn tonnes
Natural gas (1975) 313 bn cubic metres
Petroleum refinery capacity (1975) 3.6 mn tonnes
Electrical capacity (1975) 1 911*megawatts
Hospital beds (1974) 36 417, 1 per 1 871 people
Roads (1976) 49 926km = 31 023 mi, density 0.062 km per km²
Railways (1976) 8 810 km = 5 474 mi, density 0.011 km per km²
Inland waterways Indus and other rivers and canals
Ships (registered, 1977) 84, total of 475 600 gross tons
Ports (goods traffic, 000 tonnes, 1975/76)
Karachi: loaded 2 376, unloaded 7 212
Airports (passenger traffic, 000, 1976)

	departures	arrivals
Karachi	943	847
Rawalpindi	363	321
Lahore	346	319

Also (1977) 16 other airports with scheduled flights

Durable equipment (at end-year)	000	no per 1 000 people	
Radio sets (1975)	1 100	16	no per
Television sets (1974)	125	1.8	km of road
Passenger cars (1973)	177	2.6	4.9
Commercial vehicles (1973)	79	1.2	2.2

Production

Gross domestic product 1976/77 (year ending June 30th):
P Rs 144 097 mn = $ 14 510 mn = £ 8 470 mn
1976 est: P Rs 137 000*mn = $ 13 840*mn = £ 7 660*mn
Growth in real terms: 1970–76 3.2 % pa
Structure of gross domestic product *By origin* (1974/75) Agriculture 31 %, mining and quarrying 1 %, manufacturing 15 %, construction 5 %, transport and communication 7 %, other 41 %
By type (1976/77) Final consumption expenditure 89 %, gross fixed capital formation 18 %, exports of goods and services 10 %, less imports of goods and services 17 %

Production indices (1970 = 100)	1960	1970	1976	Growth %pa 1960–70	1970–76
Agricultural	55*	100	111	6.2*	1.7
Industrial	31.2	100.0	108.4	12.4	1.4

Main products (000 t)

Agriculture					
Wheat	3 814	7 294	8 691	6.7	3.0
Rice	1 540*	3 298	4 106	7.9*	3.7
Maize	439	717	764	5.0	1.1
Sugar, raw value	330*	663	641*	7.2*	−0.5*
Chickpeas (gram)	610	500	601	−2.0	3.1
Cottonseed	608	1 114	1 029	6.2	−1.3
Mangoes	238	640*	600*	10.4*	−1.1*
Tobacco	60	116	61	6.8	−10.1
Cotton	301	542	515	6.0	−0.8
Milk, cow	1 500**	1 900**	2 038	2.4**	1.2**
Milk, buffalo	5 000**	6 200**	7 425	2.2**	3.1**
Butter	106**	175**	241**	5.1**	5.5**
Wool	9.6**	11.7**	12.7**	2.0**	1.4**
Hides and skins	80**	105**	123**	2.7**	2.7**
Beef and buffalo meat	138**	163**	191**	1.7**	2.7**
Fish catch	100**	177	206	5.9**	2.6
Timber (000 m³)	6 000**	7 963	8 963	2.9**	2.0
Energy					
Total energy (000 tce)	2 300**	6 626	8 200[b]	11.2**	4.3[c]
Coal and lignite[a]	831	1 315	1 100*	4.7	−2.9[a]
Crude oil	357	501	341	3.4	−6.2
Petroleum products	300**	3 800**	3 340**	29.0**	−2.5**[c]
Natural gas (mn m³)	816	3 485	5 140	15.6	6.7
Electricity (mn kW h)	1 000**	6 513	8 800*[b]	20.6**	6.2*[c]
of which, nuclear (mn kW h)	—	—	550*[b]	na	na
Mining					
Chromium ore (oxide content)[a]	8.0**	12.5	4.6[b]	4.6**	−18.1[c]
Salt	330*	562	543[b]	5.5*	−0.7[c]
Manufacturing					
Beer (000 hl)[a]	15	25	29[b]	5.2	3.0[c]
Cigarettes (mn units)[a]	8 172	22 369	26 804[b]	10.6	3.7[c]
Cotton yarn	161	273	325	5.4	3.0
Cotton fabrics (mn m)[a]	497	663	608[b]	2.9	−1.7[c]
Sulphuric acid	12	32	42	10.3	4.6
Fertilisers, nitrogenous[a]	—	129	316	na	16.1
Cement[a]	981	2 656	3 180	10.5	3.0

[a]Years ending June 30th [b]1975 [c]1970–75

Transport traffic	1960	1970	1976	Growth %pa 1960–70	1970–76
Passenger-kilometres (mn)					
Rail[a]	9 800*	9 566	12 360	−0.2*	4.4
Air	344	1 748	3 410	17.7	11.8
Cargo: tonne-kilometres (mn)					
Rail[a]	6 900*	7 644	8 300	1.0*	1.4
Air	18.8	77.6	148.3	15.2	11.4
Sea: tonnes (mn)					
Goods loaded[a]	0.71[b]	1.86	2.38	17.4[c]	4.2
Goods unloaded[a]	4.35[b]	5.70	7.21	4.6[c]	4.0

[a]Years ending June 30th [b]1964 [c]1964–70
Tourism (1976) Number of visitors 199 400, gross receipts $ 41 mn

Finance and trade

Price indices (1970 = 100)	1960	1970	1976	Growth %pa 1960–70	1970–76
Consumer prices	73.3[a]	100.0	228.8	3.2[a]	14.8
Share prices	63.5*	100.0	84.7	4.6*	−2.7
Money stock (end-year, P Rs mn)	5 000**	11 500**	34 044	8.7**	19.8**

[a]Including Bangladesh
Budget (1976/77; year ending June 30th)
Revenue: P Rs 20 478 mn = $ 2 068 mn = £ 1 207 mn
Expenditure: P Rs 35 854 mn = $ 3 622 mn = £ 2 114 mn, of which, defence 22 %

Pakistan

Balance of payments ($ mn)	1972	1973	1974	1975	1976
Balance of goods (fob)	−234	−103	−888	−1 160	−1057
Balance of services	−181	−156	−270	−295	−310
Balance of transfers	+174	+190	+242	+403	+594
Current balance	*−241*	*−71*	*−917*	*−1 052*	*−774*
Long-term capital flow	+201	+228	+700	+756	+730
Reserves and debt (end-year, $ mn)					
International reserves	281	480	461	406	532
External public debt	4 601	5 151	6 225	6 324	7 381

External trade (1976) Imports: P Rs 21 130 mn = $ 2 134 mn = £ 1 182 mn
Exports: P Rs 11 552 mn = $ 1 167 mn = £ 646 mn

Main imports (1975)	% of total	Main exports (1975)	% of total
Wheat	12	Rice	18
Crude oil	12	Cotton	15
Machinery, non-electric	12	Cotton fabrics	13
Iron and steel	10	Cotton yarn	11
Chemicals	9	Carpets	6
Electrical machinery	6	Leather	4
Petroleum products	6	Clothing	3
Main sources (1976)		*Main destinations* (1976)	
United States	18	Hong Kong	8
Japan	12	Japan	8
United Kingdom	8	United Kingdom	6
Saudi Arabia	7	United States	6
West Germany	7	West Germany	6
Kuwait	5	Saudi Arabia	6
United Arab Emirates	4	Iraq	5
China	3	Sri Lanka	4
Italy	3	Italy	4
France	3	United Arab Emirates	4

Special focus

External aid received
1973–75 annual average: $ 353 mn = $ 5.2 per person
= 4* % of national income
1976/77: $ 700mn = 5 % of national income
Planned aid requirement for 1978/79: $ 900 mn

Philippines
Republic of the Philippines
Republika ñg Pilipinas

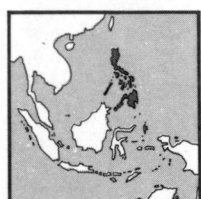

Location Western Pacific Ocean
A chain of some 7 000 islands to the north-east of Borneo, and north of the Celebes and Moluccas (Indonesia); main islands are Luzon and Mindanao
Land Area 300 000*km² = 116 000*mi², of which, Luzon 104 687 km², Mindanao 94 630 km²
Usage (1975): forests 123 000* km² (41* %), arable 51 250 km² (17 %), cropland 27 740 km² (9 %), pastures 6 560 km² (2 %)

Climate Tropical
Weather at Manila, 14 m altitude
Temperature: hottest month May 24–34 °C, coldest Jan 21–30 °C
Rainfall (av monthly): driest month Feb 13 mm, wettest July 432 mm
Time 8 hours ahead of GMT
Measures Metric system; some local units are also in use
Monetary unit Philippine peso (P P) = 100 centavos
Rate of exchange (1976 av): free P P 7.447 = $ 1, P P 13.45 = £ 1

Summary

Political One-party republic, which became fully independent July 4, 1946; a transitional period to independence from the United States began in 1935, and, before 1898, the islands had been a Spanish colony from 1565. Martial law was declared in September 1972. Member of UN, Colombo Plan and Asean
Economic A mainly agricultural economy, with sugar and coconut products making up about one-third of exports. Manufacturing is growing at about 6 %pa and contributes about 20 % to gross domestic product; mining, especially of copper, is also important. There is heavy dependence on the United States and Japan as export markets

People, resources and equipment

Population 1960 27.41*mn, 1970 36.85*mn, 1976 43.75*mn
Growth: 1960–70 3.0* %pa, 1970–76 2.9* %pa
Density (1976): 146* people per km²
Vital statistics (rate per 1 000 people, 1970–75): births 44*, deaths 11*
Cities (population in 000, 1975)

Manila (capital)[a]	1 438[b]	Zamboanga	240[c]	Cadiz	130[c]
Quezon City	995	Bacolod	196	Iligan	129[c]
Davao	516[c]	Angeles	176	San Pablo	126[c]
Cebu	419	Butuan	173[c]	Batangas	125[c]
Caloocan	364	Basilan	171[c]	Cabanatuan	118[c]
Iloilo	248	Cagayan de Oro	163[c]	Lipa	112
Pasay	241	Olongapo	134	General Santos	108

[a]The capital was changed back to Manila from Quezon City on May 31, 1976
[b]Population for metropolitan (greater) Manila was 4 500* thousand (including Quezon City, Caloocan and Pasay) [c]Excluding suburbs
Race (1960) Cebuano 24 %, Tagalog 21 %, Ilocano 12 %, Hiligaynon (Ilongo) 10 %, Bicolano 8 %, Pampanga 3 %
Language Pilipino (based on Tagalog), English and Spanish; many different local dialects are also in use
Religion (1970) Roman Catholic 85 %, Moslem 4 %, Aglipayan 4 %
Education (1972/73) Pupils: primary 7 622 424, secondary 1 631 363, vocational 159 813, higher 678 343. Teachers: primary 247 551, secondary 45 594, vocational 12 378, higher 32 651
Labour force (1975) 15 161 000; in agriculture 7 881 000 (52 %), manufacturing 1 720 000 (11 %), construction 503 000 (3 %)
Personnel Scientists and engineers (1967): 94 302
Physicians (1970): 14 000, 1 per 2 632 people
Standard of living
National income per person (1976): P P 2 720* = $ 360* = £ 200*
Consumption per person (1975): energy 326 kg coal equivalent, electricity (production) 291* kW h, newsprint 1.6 kg, steel 22 kg
Newspapers (1974): number 13; circulation 772 000, 18 per 1 000 people
Telephones (Dec 1976): 541 700, 12 per 1 000 people
Livestock (000, 1976) Cattle 2 323*, buffaloes 5 000*, pigs 9 700*, goats 1 370*, chickens 45 671
Mineral reserves Lignite (economic reserves, 1973) 91 mn tonnes
Uranium (1976) 300 tonnes
Petroleum refinery capacity (1975) 9.5* mn tonnes
Electrical capacity (1975) 3 020* megawatts
Hospital beds (1973) 62 939, 1 per 639 people
Roads (1974) 99 132 km = 61 598 mi, density 0.33 km per km²
Railways (1973) 1 150 km = 715 mi, density 0.004 km per km²
Ships (registered, 1977) 504, total of 1 146 529 gross tons
Ports (goods traffic, 000 tonnes, 1973)

	loaded	unloaded		loaded	unloaded
Manila and Bataan[a]	999	8 247	Iloilo	1 377	77
Batangas[a]	705	6 070	Masao and Bislig	970	20
Davao	1 544	128	Zamboanga	515	1

[a]Crude oil port
There is a total of 600* ports for inter-island traffic
Airports Luzon (Manila) and 45 other airports with scheduled flights

Durable equipment	000	no per		
(at end-year)		1 000 people		
Radio sets (1975)	1 850	43		
Television sets (1974)	711	17	no per	
Passenger cars (1974)	362	9	km of road	
Commercial vehicles (1974)	247	6	3.7	
			2.5	

Production

Gross domestic product 1976: P P 132 522 mn = $ 17 795 mn = £ 9 853 mn
Growth in real terms: 1960–70 5.1 %pa, 1970–76 6.2 %pa
Structure of gross domestic product

By origin (1975)	P P mn	% of gdp
Agriculture	31 482	28
Mining and quarrying	1 743	2
Manufacturing	21 984	19
Electricity, gas and water	583	1
Construction	2 728	2
Distribution	8 162	7
Transport and communication	2 405	2
Other	44 434	39
Total	113 521	100
By type of expenditure		
Government final consumption	13 227	10
Private final consumption	86 205	65
Stock investment	8 300	6
Gross fixed capital formation	32 755	25
Exports of goods and services	23 248	18
less Imports of goods and services	−31 841	−24
Total (incl statistical discrepancy)	131 894	100

Philippines

Production indices (1970 = 100)	1960	1970	1976	Growth %pa 1960–70	1970–76
Agricultural	73	100	142	3.2	6.0
Industrial	55	100	141	6.2	5.9

Main products (000 t)

Agriculture

	1960	1970	1976	1960–70	1970–76
Rice	3 705	5 343	6 490	3.7	3.3
Maize	1 210	2 005	2 767	5.2	5.5
Sweet potatoes	739	657	986*	−1.2	7.0*
Sugar, raw value	1 398	1 926	2 930	3.3	7.2
Copra	1 071	1 656	2 800*	4.4	9.1*
Pineapples	134	233	382*	5.7	8.6*
Bananas	349	1 311	1 150*	14.2	−2.2*
Coffee	32	49	80	4.3	8.5
Tobacco	60	61	60	0.2	−0.3
Rubber	3*	19	59	20.3*	20.8
Abaca (Manila hemp)	115	100*	137*	−1.4*	5.4*
Eggs	62	112	170*	6.1	7.2*
Beef and buffalomeat	50**	106	123*	7.8**	2.5*
Pigmeat	200**	306	373*	4.3**	3.3*
Poultrymeat	50**	95	136*	6.6**	6.2*
Fish catch	466	1 006	1 430	8.0	6.0
Timber (000 m³)	21 000**	33 727	33 527	4.9**	−0.1

Energy

	1960	1970	1976	1960–70	1970–76
Coal	148	42	149	−11.8	23.5
Petroleum products	1 278	8 209	8 790*b	20.4	1.4*c
Electricity (mn kW h)	2 731	8 666	12 360*b	12.2	7.3*c
Manufactured gas (mn m³)	13	22	15	5.4	−6.2

Mining

	1960	1970	1976	1960–70	1970–76
Iron ore (Fe content)	638	1 166	839b	6.2	−6.4c
Chromium ore (oxide content)	249	196	189b	−2.4	−0.8c
Copper ore (Cu content)	44	160	238*	13.8	6.8c
Nickel ore (Ni content)	—	0.10	9.50b	na	147.6c
Gold (000 kg)	12.8	18.7	15.6b	3.9	−3.6c
Salt	95	210	71*b	8.2	−19.5*c

Manufacturing

	1960	1970	1976	1960–70	1970–76
Cigarettes (mn units)	10 302	39 671	47 688b	14.4	3.7c
Cotton yarn	15	42	33	10.8	−4.0
Cotton fabrics (mn m)	148	195	205	2.8	0.8
Paper	40**	56*	149*	3.4**	17.7*
Tyres (000 units)	433	598	1 222	3.3	12.6
Sulphuric acid	—	196	280	na	6.1
Fertilisers, nitrogenous	7	48	57*b	20.5	3.5*c
Cement	795	2 447	4 220	11.9	9.5
Radio sets (000 units)	89	129	151b	3.8	3.2c
Television sets (000 units)	1	44	104b	46.0	18.8c
Motor vehicles (000 units)a	9.9	13.8	50.6	3.4	24.2

aAssembly only   b1975   c1970–75

Transport traffic	1960	1970	1976	Growth %pa 1960–70	1970–76
Passenger-kilometres (mn)					
Rail	865	752	768	−1.4	0.4
Air	286	1 456	3 050	17.7	13.1
Cargo: tonne-kilometres (mn)					
Rail	217	47	41	−14.2	−2.3
Air	3.8	24.9	111.9	20.7	28.5
Sea: tonnes (mn)					
Goods loaded	6.79	15.79	11.20	8.8	−5.6
Goods unloaded	4.16	12.54	16.43	11.7	4.6

Tourism (1976) Number of visitors 615 200, gross receipts $ 235 mn

Finance and trade

Price and wage indices (1970 = 100)	1960	1970	1976	Growth %pa 1960–70	1970–76
Consumer prices	59.4	100.0	216.0	5.3	13.7
Wholesale prices	56.1	100.0	270.4	6.0	18.0
Share pricesa	86.9	100.0	95.1	1.4	−0.8
Money stock (end-year, P P mn)	1 659	4 310	12 075	10.0	18.7

aCommercial; excludes mining and sugar

Budget (1975/76; year ending June 30th)
Revenue: P P 17 327 mn = $ 2 315 mn = £ 1 161 mn
Expenditure: P P 22 399 mn = $ 2 993 mn = £ 1 500 mn
of which, transport and communication 16 %, agriculture and natural resources 15 %, defence 12 %, education 10 %

Balance of payments ($ mn)	1972	1973	1974	1975	1976
Balance of goods (fob)	−125	+276	−450	−1 196	−1 116
Balance of services	−57	−31	−34	−45	−256
Balance of transfers	+188	+230	+277	+318	+268
Current balance	+7	+474	−207	−923	−1 105
Long-term capital flow	+115	+132	+227	+517	+1 137
Reserves and debt (end-year, $ mn)					
International reserves	551	1 038	1 504	1 358	1 640
External public debt	1 281	1 327	2 051	2 556	4 268

External trade (1976) Imports: P P 29 414 mn = $ 3 950 mn = £ 2 187 mn
Exports: P P 18 696 mn = $ 2 511 mn = £ 1 390 mn

Main imports	% of total	Main exports	% of total
Crude oil	21	Sugar	18
Machinery, non-electric	17	Coconut oil	12
Chemicals	10	Copper ores	10
Food	9	Timber	8
(of which, cereals 5)		Fruit and vegetables	7
Iron and steel	6	Copra	6
Electrical machinery	5	Clothing	3
Motor vehicles	5	Veneers and plywood	3
Petroleum products	2		
Main sources		Main destinations	
Japan	27	United States	36
United States	22	Japan	24
Saudi Arabia	8	Netherlands	10
Kuwait	6	West Germany	4
United Kingdom	3	Soviet Union	3
West Germany	3	United Kingdom	3
Australia	3	France	2

Special focus

Copra prices (Philippine wholesale price at European ports)

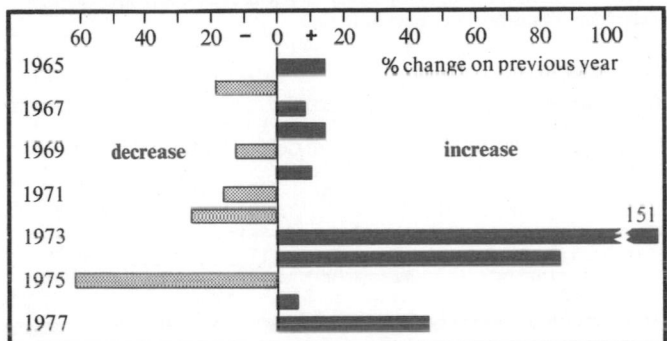

Qatar

State of Qatar
Dawlat Qatar

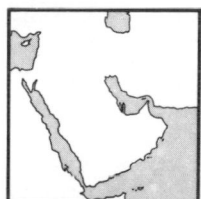

Location Middle East
Occupies mainly the Qatar peninsula which projects from the west coast of the Gulf, together with some offshore islands. Saudi Arabia is to the west, United Arab Emirates to the south-east, and Bahrain is off the coast about 60 km to the north-west
Land Area 11 000*km² = 4 000*mi²
Climate Hot, mild in winter
Time 3 hours ahead of GMT
Measures Metric system; the UK (imperial) system is also in use
Monetary unit Qatar riyal (Q R) = 100 dirhams
Rate of exchange (1976 av): par Q R 4.76190 = S D R 1, free Q R 3.962 = $ 1, Q R 7.157 = £ 1

Summary

Political Emirate, which became independent on September 1, 1971; formerly under UK protection. The status of the Huwar Islands is in dispute with Bahrain. Member of UN, Arab League, Opec and Oapec
Economic An oil economy, with some diversification in chemicals. Diversification is planned to increase with emphasis on transport and social services, and on industry, especially petrochemicals, steel and cement

Qatar

People, resources and equipment

Population[a] 1960 50 000**, 1970 110 000**, 1976 210 000**[b]
[a]Estimates including migrant workers [b]UN estimate indicates home population of 95 000**. Native Qataris: 50 000**
Growth: 1960–70 8.0** %pa, 1970–76 11.4** %pa
Density (1976): 19** people per km²
Cities (population in 000, 1976) Ad Dawhah[a] (capital) 150**, Dukhan, Umm Said, Al Wakrah, Al Khawr
[a]Doha
Race Arab
Language Arabic; English is also in use
Religion Moslem
Education (1974/75) Pupils 30 098, teachers 1 730
Labour force (1976) 86 700
Personnel Scientists and engineers (1974/75): 1 352 (of whom, 135* nationals). Physicians (1974): 96, 1 per 1 980 people
National income per person (1974)
Q R 35 000*** = $ 9 000*** = £ 5 000***
Consumption per person (1975) Energy 35 328 kg coal equivalent
Telephones (Dec 1976) 24 400, 113 per 1 000 people
Livestock (000, 1976) Cattle 6*, sheep 40*, goats 47*, camels 9*
Mineral reserves (1975) Crude oil 697 mn tonnes
Natural gas 1 354 bn cubic metres
Petroleum refinery capacity (1975) 370 000 tonnes
Hospital beds (1973) 661, 1 per 270 people
Roads (1976) 1 100*km = 700*mi, density 0.1*km per km²
There are no railways
Oil pipeline Dukhan to Umm Said
Ships (registered, 1977) 19, total of 84 710 gross tons
Ports Doha, Umm Said
Airport (passengers, 000, 1976) Doha: departures 209, arrivals 225
Television Transmission began in 1972

Production, finance and trade

Gross domestic product
1976 est: Q R 9 000**mn = $ 2 300**mn = £ 1 250** mn
Main products (000 t) *Agriculture* (1976) Cow milk 5*, goat milk 9*, meat 3*

Other	1960	1970	1976	Growth %pa	
				1960–70	1970–76
Crude oil	8 212	17 516	23 534	7.9	5.0
Petroleum products	28	31	171[a]	1.0	40.7[b]
Natural gas (mn m³)	na	1 100*	2 209	na	12.3*

[a]1975 [b]1970–75

Transport traffic *Air* (1976) Passenger-kilometres 320 mn, cargo: 7 mn t-km
Sea (1974) goods loaded 25.4* mn t, unloaded 0.58 mn t
Money stock (end-year, Q R mn) 1970 198, 1976 1 573; growth 1970–76 41.3 %pa
Budget (1976 fiscal-lunar-year ending December 21)
Revenue: Q R 8 811 mn = $ 2 224 mn = £ 1 231 mn
Expenditure: Q R 5 894 mn = $ 1 488 mn = £ 824 mn
External trade (1976) Imports: Q R 3 290 mn = $ 830 mn = £ 460 mn
Exports: Q R 8 754 mn = $ 2 209 mn = £ 1 223 mn

Main imports (1975)	% of total	*Main exports* (1975)	% of total
Electrical machinery	19	Crude oil	98
Machinery, non-electric	16	Chemicals	1
Motor vehicles	15		
Food	12		
(of which, fruit and			
vegetables 3, livestock 3)			
Iron and steel	8		
Textile yarns and fabrics	4		
Chemicals	4		

Main sources (1976)		*Main destinations* (1976)	
Japan	28	United Kingdom	21*
United Kingdom	17	France	15*
United States	8	US Virgin Islands	13*
West Germany	8	Netherlands	11*
United Arab Emirates	7*	Thailand	7*

Special focus

Main import suppliers

	% of total imports						
	1970	1971	1972	1973	1974	1975	1976
Japan	9	10	13	11	18	15	28
United Kingdom	24	37	26	27	14	21	17
United States	10	10	10	10	10	13	8

Saudi Arabia

Kingdom of Saudi Arabia
Al Mamlakah al Arabiyah as Sa'udiyah

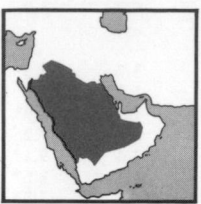

Location Middle East
Occupies the main centre of the Arabian peninsula, with the Red Sea to the west and the Gulf to the east; Jordan, Iraq and Kuwait are to the north, Bahrain, Qatar and United Arab Emirates to the east and Oman, South Yemen and North Yemen to the south. Some borders are ill-defined, especially those with Oman and the two Yemens
Land Area 2 150 000*km² = 830 000*mi²
Usage (1975): pastures 850 000* km² (40* %), forests 16 000* km² (1* %), arable and cropland 8 050* km² (0.4 * %), other 1 276 000* km² (59* %)
Climate Hot and dry, mild in winter
Weather at Riyadh, 591 m altitude
Temperature: hottest month July 26–42 °C, coldest Jan 8–21 °C
Rainfall (av monthly): driest months July, Sept, Oct 0 mm, wettest April 25 mm
Time 3 hours ahead of GMT
The Moslem lunar calendar is in use; this is 10 or 11 days earlier each year in terms of the Gregorian. The Moslem year 1398 is, in Gregorian time, December 12, 1977 to December 1, 1978 (approximate dates).
The year 1399 is 1978/79 Gregorian
Measures Metric system, made compulsory from December 3, 1964
Monetary unit Saudi riyal (SA R) = 20 quirsh = 100 hallalas
Rate of exchange (1976 av): par SA R 4.28255 = S DR 1, free SA R 3.530 = $ 1, SA R 6.376 = £ 1

Summary

Political Monarchy, with rule through a Council of Ministers. There has been a Neutral Zone between north-east Saudi Arabia and Iraq, established in 1922 and used by nomads; under a 1975 agreement, this is being partitioned between the two countries. There is also a Neutral Zone between Saudi Arabia and Kuwait, established in 1922 and partitioned in 1966; however, oil and other resources continue to be shared. Member of UN, Arab League, Opec and Oapec
Economic An oil economy, with crude oil making up 94 % of exports and petroleum products 6 %; agriculture employs over 60 % of the labour force, but contributes only 1 % to gross domestic product. Oil revenues are being used, under the second development plan, to expand social services and transport systems and to diversify the economy; industries being developed include petrochemicals and aluminium

People, resources and equipment

Population[a] 1960 5.98**mn, 1970 7.74**mn, 1976 9.24**mn
[a]Population estimates vary; shown here are UN figures
Growth: 1960–70 2.6** %pa, 1970–76 3.0** %pa
Density (1976): 4** people per km²
Vital statistics (rate per 1 000 people, 1970–75): births 50*, deaths 20*
Cities (population in 000, 1974)

Ar Riyad[a] (capital)	667	At Ta'if	205	Al Hufuf	101
Jiddah	561	Al Madinah[c]	198	Buraydah	70*
Makkah[b]	367	Ad Dammam	128	Jizan	60*

[a]Riyadh [b]Mecca [c]Medina
Race Arab
Language Arabic
Religion (1976) Moslem 99** % (mainly Sunni)
Education (1974/75) Pupils: primary 625 773, secondary 159 938, vocational 5 150, teacher-training 14 099, higher 19 773.
Teachers: primary 29 989, secondary 10 964, vocational 550, teacher-training 1 037, higher 1 818
Labour force (1976) 2 436 000; in agriculture 1 523 000 (63 %)
Personnel Scientists and engineers (1974/75): 33 376
Physicians (1973): 2 000*, 1 per 4 200* people
Standard of living
National income per person (1976): SA R 17 600* = $ 4 990* = £ 2 760*
Consumption per person (1975): energy 1 398 kg coal equivalent, electricity (production) 222 kW h, newsprint 0.9 kg, steel 159 kg
Newspapers (1974): number 11; circulation 96 000, 11 per 1 000 people
Telephones (Dec 1976): 160 000, 17 per 1 000 people
Livestock (000, 1976) Cattle 180*, sheep 1 379*, goats 779*, camels 614*
Mineral reserves (1975) Crude oil 15 168 mn tonnes
Natural gas 1 798 bn cubic metres
Petroleum refinery capacity (1975) 35 mn tonnes

Saudi Arabia

Electrical capacity (1975) 375* megawatts
Hospital beds (1975) 10 465, 1 per 857 people
Roads (1976) 26 267 km = 16 322 mi, density 0.012 km per km²
Railways (1976) 612 km = 380 mi, density 0.0003 km per km²
Ships (registered, 1977) 119, total of 1 018 713 gross tons
Ports (goods traffic, 000 tonnes, 1972)

	loaded	unloaded
Ras Tannurah[a,d]	321 190[b]	na
Jiddah[c]	na	1 022
Ad Dammam[d]	na	987
Yanbu al Bahr[c]	na	211

[a]Crude oil port [b]1973 [c]Red Sea port [d]Gulf port
Airports (passengers, 000, 1976) Jiddah: departures 1 622, arrivals 1 639; also Riyadh, Dhahran, Medina and 16 other airports with scheduled flights

Durable equipment	000	no per	
(at end-year)		1 000 people	
			no per
Radio sets (1975)	255	28	km of road
Television sets (1975)	124	14	
Passenger cars (1976)	226	24	8.6
Commercial vehicles (1976)	201	21	7.7

Production

Gross domestic product 1975/76 (year ending June 30th):
SA R 155 053 mn = $ 43 962 mn = £ 22 034 mn
1976 est: SA R 174 000*mn = $ 49 300*mn = £ 27 300*mn
Growth in real terms: 1970–76 12.3 %pa
Structure of gross domestic product (1975/76)

By origin	SA R mn	% of gdp
Agriculture	1 585	1
Mining and quarrying	110 136	71
Manufacturing	7 154	5
Electricity, gas and water	342	—
Construction	11 522	7
Distribution and hotels	4 642	3
Transport and communication	5 776	4
Other	13 896	9
Total	155 053	100
By type of expenditure		
Government final consumption	28 197	18
Private final consumption	18 923	12
Stock investment	2 091	1
Gross fixed capital formation	33 743	22
Exports of goods and services	20 284	78
less Imports of goods and services	−48 184	−31
Total	155 053	100

Production index	1960	1970	1976	Growth %pa	
(1970 = 100)				1960–70	1970–76
Agricultural	88[c]	100	119	3.4[d]	2.9
Main products (000 t)					
Agriculture					
Wheat	100**	120*	205*	1.8**	9.3*
Millet	70**	130*	150*	6.4**	2.4*
Sorghum	150**	190*	200*	2.4**	0.9*
Tomatoes	50**	100*	123*	7.2**	3.5*
Dates	220**	220*	262*	0.0**	3.0*
Oranges	5**	14**	16*	10.8**	2.2**
Meat	40**	70**	79*	5.7**	2.0**
Fish catch	15**	22	23	3.9**	1.2
Other					
Total energy (000 tce)[a]	85 220	250 150	530 100[e]	11.4	16.2[f]
Crude oil[a]	65 712	188 408	424 232	11.1	14.5
Petroleum products[a]	11 510	29 506	23 798*	9.9	−4.2[f]
Natural gas (mn m³)[a]	na	2 261	5 900*	na	17.3*
Electricity (mn kW h)	50**	724	1 988[e]	30.6**	22.4[f]
Fertilisers, nitrogenous[b]	—	—	100	na	na
Cement	60	675	1 056[g]	27.4	11.8[h]

[a]Including one-half share of production in the Neutral Zone [b]Year ending June 30th
[c]1966 [d]1966–70 [e]1975 [f]1970–75 [g]1974 [h]1970–74

Transport traffic	1960	1970	1976	Growth %pa	
Passenger-kilometres (mn)				1960–70	1970–76
Rail	24[a]	39	72[c]	7.2[b]	16.6[d]
Air	—	500	3 122	na	35.7
Cargo: tonne-kilometres (mn)					
Rail	77[a]	34	66[c]	−11.0[b]	18.0[d]
Air	—	5.2	85.2	na	59.4
Sea: tonnes (mn)					
Goods loaded	30.4	151.4	390.1*[c]	17.4	26.7*[d]
Goods unloaded	0.87	2.20	4.00*[c]	9.7	16.1*[d]

[a]1963 [b]1963–70 [c]1974 [d]1970–74
Tourism (1975) Gross receipts $ 509 mn
Pilgrims going to Mecca (1976; Moslem year 1396) 719 000

Finance and trade

Price index	1960	1970	1976	Growth %pa	
(1970 = 100)				1960–70	1970–76
Consumer prices	88.7[a]	100.0	273.1	1.7[b]	18.2
Money stock					
(end-year SA R mn)	921	2 404	24 280	10.1	47.0

[a]1963 [b]1963–70
Budget (1976/77, year ending June 30th; Moslem fiscal year 1396/97)
Balanced at SA R 110 935 mn = $ 31 426 mn = £ 18 336 mn

Balance of payments ($ mn)	1972	1973	1974	1975	1976
Balance of goods (fob)	+3 053	+5 428	+26 378	+21 034	+24 910
Balance of services	−1 155	−2 334	−1 838	−3 475	−6 963
Balance of transfers	−424	−890	−1 533	−3 679	−4 317
Current balance	+1 473	+2 203	+23 007	+13 880	+13 629
Long-term capital flow	+54	−700	−4 509	−4 416	−8 042
International reserves					
(end-year, $ mn)	2 500	3 877	14 285	23 319	27 025

External trade (1976)
Imports: SA R 30 690mn = $ 8 694mn = £ 4 813mn
Exports: SA R 127 520 mn = $ 36 125 mn = £ 20 000 mn

Main imports (1974)	% of total	*Main exports* (1976)	% of total
Food	17	Crude oil	94
(of which, cereals 6)		Petroleum products	6
Motor vehicles	15		
Machinery, non-electric	12		
Electrical machinery	9		
Textile yarns and fabrics	7		
Iron and steel	6		
Chemicals	5		
Main sources (1974)		*Main destinations* (1974)	
United States	17	Japan	16
Japan	16	France	12
Lebanon	15	Italy	10
West Germany	6	United Kingdom	9
United Kingdom	5	Spain	6

Special focus

Second development plan
Moslem 1395–1400, Gregorian 1975–80
Total planned expenditure:
SA R 498 000 mn at 1394/95 prices
= $ 141 000 mn = £ 60 000 mn at 1974/75 prices
Planned industrial centres (with projected population by late 1980s):
Jubail (Gulf) 170 000, Yanbu (Red Sea) 120 000

Singapore

Republic of Singapore
Republik Singapura
Hsin-chia-po Kung-ho-kuo
Singapore Kudiyarasu

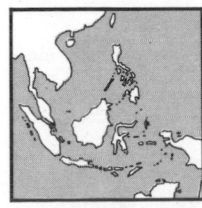

Location South-east Asia
The island is at the southern end of the Malay peninsula, with Malaysia to the north, separated by the Straits of Johore about 1 km wide; there is a causeway across these Straits linking the island with the peninsula. Islands which are part of Indonesia are several kilometres to the south
Land Area 581.5 km² = 224.5 mi²
Climate Tropical
Weather at Singapore, 10 m altitude
Temperature: hottest month May 24–32 °C, coldest Jan 23–30 °C
Rainfall (av monthly): driest month July 170 mm, wettest Dec 257 mm

Singapore

Time 7½ hours ahead of GMT
Measures Metric system, as a conversion from UK (imperial) system; local measures are also in use, including:
length 1 chhek = 10 chhuns = 0.375 metre = 14¾ inches
weight (mass) 1 kati = 16 tahils = 0.605 kilogram = 1⅓ pounds
1 pikul = 100 katis = 60.48 kilograms = 133⅓ pounds
Monetary unit Singapore dollar (S $) = 100 cents
Rate of exchange (1976 av): free S $ 2.471 = $ 1, S $ 4.463 = £ 1

Summary

Political Republic, which became independent on August 9, 1965, by separation from Malaysia; Singapore had been a separate UK colony from 1946 until it became a State of Malaysia on September 16th, 1963. Member of UN, Colombo Plan, Commonwealth and Asean
Economic Entrepôt trade is a major feature, especially for Malaysian and Indonesian produce, and there are significant rubber and tin markets. Refining of imported crude oil is very important, and petroleum products make up about one-quarter of exports; there is a high level of investment to develop industry, especially petrochemicals and electrical machinery. The port is one of the largest in the world and there are shipbuilding and repairing facilities. Tourism is developing

People, resources and equipment

Population 1960 1.63 mn, 1970 2.07 mn, 1976 2.28 mn
Growth: 1960–70 2.4 %pa, 1970–76 1.6 %pa
Density (1976): 3 921 people per km²
Vital statistics (rate per 1 000 people, 1976): births 18.5, deaths 5.1
City (1977) Singapore City (capital) 2 308 000 people
Race (1977) Chinese 76 %, Malay 15 %, Indian 7 %
Language Chinese, Malay, Tamil and English
Religion Buddhist, Taoist, Confucian, Moslem, Hindu, Christian
Education (1977) Pupils 516 129, teachers 20 550
Labour force (1976) 910 929; in agriculture 19 686 (2 %), manufacturing 233 954 (26 %), construction 42 026 (5 %), distribution and hotels 201 002 (22 %), transport and communication 101 615 (11 %)
Personnel Scientists and engineers (1974): 24 234
Physicians (1975): 1 622, 1 per 1 387 people
Standard of living
National income per person (1976): S $ 6 100* = $ 2 470* = £ 1 370
Consumption per person (1975): energy 2 151 kg coal equivalent, electricity (production, 1976) 2 019 kW h, newsprint 15 kg, steel 636 kg
Newspapers (1972): number 10; circulation 415 000, 190 per 1 000 people
Telephones (Dec 1976) 374 390, 163 per 1 000 people
Livestock (000, 1976) Cattle 9*, pigs 1 071, chickens 13 198*, ducks 1 953*
Petroleum refinery capacity (1975) 46 mn tonnes
Electrical capacity (1975) 1 115* megawatts
Hospital beds (govt establishments only, 1974) 8 178, 1 per 271 people
Roads (1976) 2 218 km = 1 378 mi, density 3.8 km per km²
Railways[a] (1976) 38 km = 24 mi, density 0.065 km per km²
[a]Linked to the Malaysian system
Ships (registered, 1977) 872, total of 6 791 398 gross tons
Ports (goods traffic, 000 tonnes, 1976)
Port of Singapore: loaded 20 273, unloaded 37 703
Airports (passengers, 000, 1976)

	departures	arrivals	
International	1 881	1 860	
Seletar	18	41	
Durable equipment (Dec 1975)	000	no per 1 000 people	
Radio sets	356	158	no per
Television sets	269	120	km of road
Passenger cars	149	66	69
Commercial vehicles	46	20	21

Production

Gross domestic product 1976: S $ 14 615 mn = $ 5 915 mn = £ 3 275 mn
Growth in real terms: 1960–70 9.2 %pa, 1970–76 9.1 %pa
Structure of gross domestic product (1976) *By origin* Agriculture 2 %, manufacturing 25 %, construction 8 %, distribution and hotels 26 %, transport and communication 12 %, other 27 %
Manufacturing production index (1970 = 100) 1976 184.8; growth 1970–76 10.8 %pa
Main products (000 t) *Agriculture* (1976) Sweet potatoes 1*, cassava 1*, coconuts 7*, bananas 1*, eggs 24*, fish catch 16

Other	1960	1970	1976	Growth %pa 1960–70	1970–76
Petroleum products	na	16 000**	21 000**	na	4.6**
Manufactured gas (mn m³)	17	51	91	11.6	10.1
Electricity (mn kW h)	659	2 205	4 604	12.8	13.1
Beer (000 hl)	348	413	462[a]	1.7	2.2[b]
Cigarettes (mn units)	1 526	2 787	3 241[a]	6.2	3.1[b]
Merchant vessels (000 grt)	—	6	143	na	69.6

[a]1975 [b]1970–75

Transport traffic[a]	1960	1970	1976	Growth %pa 1960–70	1970–76
Air (mn)					
Passenger-kilometres	190[b,c]	664[b]	6 362	36.7[d]	45.7
Cargo: tonne-kilometres	2.0[b,c]	10.8[b]	193.4	52.4[d]	61.7
Sea: tonnes (mn)					
Goods loaded	5.37	16.0	20.6	11.5	4.3
Goods unloaded	9.76	26.9	38.3	10.7	6.1

[a]For rail traffic see Malaysia [b]Apportionment of traffic of Malaysia-Singapore Airlines Ltd [c]1966 [d]1966–70
Tourism (1976) Number of visitors 1 320 600, gross receipts $ 283 mn

Finance and trade

Price index (1970 = 100)	1960	1970	1976	Growth %pa 1960–70	1970–76
Consumer prices	89.7	100.0	161.9	1.1	8.4
Money stock (end-year, S $ mn)	841[a]	1 631	4 000	9.9[b]	16.1

[a]1963 [b]1963–70
Budget (1976/77; year ending March 31st)
Revenue: S $ 3 108 mn = $ 1 265 mn = £ 729 mn
Expenditure: S $ 3 104 mn = $ 1 263 mn = £ 728 mn
of which, defence 28 %, education 13 %, social services 13 %

Balance of payments ($ mn)	1972	1973	1974	1975	1976
Balance of goods (fob)	−1 102	−1 339	−2 274	−2 481	−2 287
Balance of services	+569	+737	+1 193	+1 721	+1 460
Balance of transfers	+2	−3	−40	−38	−48
Current balance	−531	−606	−1 121	−796	−876
Long-term capital flow	+265	+358	+648	+726	+813

Reserves and debt (end-year, $ mn)

International reserves	1 748	2 286	2 812	3 007	3 364
External public debt	501	528	647	783	902

External trade (1976) Imports: S $ 22 404 mn = $ 9 067 mn = £ 5 020 mn
Exports: S $ 16 266 mn = $ 6 583 mn = £ 3 645 mn

Main imports	% of total	Main exports	% of total
Crude oil	21	Petroleum products	23
Electrical machinery	11	Electrical machinery	13
Machinery, non-electric	10	Rubber	12
Food	8	Ship and aircraft stores	7
Petroleum products	7	Machinery, non-electric	6
Rubber	6	Food	6
Transport equipment	5	Ships	4
Chemicals	5	Chemicals	4
Main sources		*Main destinations*	
Japan	16	Malaysia	15
Saudi Arabia	16	United States	15
Malaysia	14	Japan	10
United States	13	Hongkong	8
United Kingdom	4	Australia	5

Special focus

Registered shipping, 1977

Type of ship	Number	000 gross tons
Oil tankers	136	3 104
Liquefied gas carriers	3	3
Chemical tankers	2	3
Miscellaneous tankers	1	1
Bulk ore/oil carriers	4	275
Ore and bulk carriers	56	1 022
General cargo	439	1 938
Passenger/cargo ships	15	104
Container ships	18	162
Vehicle carriers	1	2
Livestock carriers	8	63
Fishing	9	5
Ferries/passenger vessels	9	38
Supply ships and tenders	26	17
Tugs	116	36
Research ships	4	1
Other	25	17
Total	872	6 791

Sri Lanka

Democratic Socialist Republic of Sri Lanka

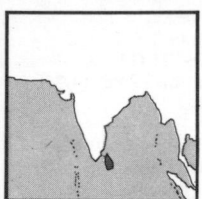

Location South Asia
The island of Sri Lanka is off the south-east coast
of India, separated by the Palk Strait
Land Area 65 610 km² = 25 332 mi²
Climate Tropical
Weather at Colombo, 7 m altitude
Temperature: hottest month May 26–31 °C,
coldest Dec 22–29 °C
Rainfall (av monthly): driest month Feb 69 mm,
wettest May 371 mm
Time 5½ hours ahead of GMT
Measures Metric system as a gradual conversion from the UK (imperial)
system; also, for copra: 1 candy = 254 kilograms = 560 pounds
Monetary unit Sri Lanka rupee (SL R) = 100 cents
Rate of exchange (1976 av): free SL Rs 8.458 = $ 1, SL Rs 15.28 = £ 1

Summary

Political Republic, which became independent from the United Kingdom
on February 4, 1948. Before May 22, 1972 called Ceylon. The sea border
with India was redefined on June 28, 1974 to include the island of
Kachchativu in Sri Lanka territory. Member of UN, Colombo Plan and
Commonwealth
Economic An agricultural economy, with tea, rubber and coconut
products making up nearly three-quarters of exports. Manufacturing
makes up one-eighth of gross domestic product; textiles, cement and
hydro-electric power are important

People, resources and equipment

Population 1960 9.90*mn, 1970 12.52*mn, 1976 13.73*mn
Growth: 1960–70 2.4* %pa, 1970–76 1.5* %pa
Density (1976): 209* people per km²
Vital statistics (rate per 1 000 people, 1972), births 29.9, deaths 7.8
Cities (population in 000, 1976) Colombo (capital) 607*,
Dehiwala-Mount Lavinia (1973) 136*, Jaffna 117*, Kandy 101*, Galle 78*
Race (1971) Lowland Sinhalese 43 %, Kandyan Sinhalese 29 %,
Sri Lanka Tamil 11 %, Indian Tamil 9 %, Sri Lanka Moor 6 %
Language Sinhala, Tamil
Religion (1971) Buddhist 67 %, Hindu 18 %, Christian 8 %, Moslem 7 %
Education (1974) Pupils 2 534 071, teachers 100 910
Labour force (1976) 4 869 000; in agriculture 2 635 000 (54 %)
Personnel Scientists and engineers (1972): 7 457
Physicians (1972): 3 251, 1 per 4 007 people
Standard of living
National income per person (1976): SL Rs 1 809 = $ 214 = £ 118
Consumption per person (1975): energy 127 kg coal equivalent,
electricity (production) 85* kW h, newsprint 0.5 kg, steel 4 kg
Newspapers: number (1974) 26; circulation (1971) 536 000, 42 per 1 000
people
Telephones (Dec 1975): 72 000, 5 per 1 000 people
Livestock (000, 1976) Cattle 1 744, buffaloes 854, goats 562
Petroleum refinery capacity (1975) 1.9 mn tonnes
Electrical capacity (1975) 281* megawatts, of which, hydro 195*
megawatts
Hospital beds (1973) 39 732, 1 per 333 people
Roads (1976) 31 150 km = 19 356 mi, density 0.47 km per km²
Railways (1976) 1 498 km = 931 mi, density 0.023 km per km²
Inland waterways (1976) 167*km = 104*mi
Ships (registered, 1977) 37, total of 92 581 gross tons
Ports (goods traffic, 000 tonnes, 1976)

	loaded	unloaded
Colombo	1 278	2 751
Trincomalee	62	67

Also Galle and Jaffna
Airports Bandaranaike (Katunayake, 34 km north of Colombo),
Ratmalana (14 km south of Colombo), Jaffna, Trincomalee

Durable equipment	000	no per	
(at end-year)		1 000 people	no per
Radio sets (1975)	530	38	km of road
Passenger cars (1976)	94	7	3.0
Commercial vehicles (1976)	52	4	1.7

Production

Gross domestic product
1976: SL Rs 26 488 mn = $ 3 132 mn = £ 1 734 mn
Growth in real terms: 1970–76 4.1 %pa
Structure of gross domestic product (1974) *By origin* Agriculture 35 %,
manufacturing 13 %, transport and communication 9 %, other 43 %

Production indices	1960	1970	1976	Growth %pa	
(1970 = 100)				1960–70	1970–76
Agricultural	77	100	97	2.7	−0.4
Manufacturing	56	100	108e	6.0	2.6f

Main products (000 t)
Agriculture

	1960	1970	1976	1960–70	1970–76
Rice	897	1 616	1 253	6.1	−0.4
Cassava	331	354	790*	0.7	14.3*
Copra	187	208	151	1.1	−5.2
Coffee	5.0**	7.3	9.0*	3.8**	3.6*
Tea	197	212	197	0.7	−1.2
Tobacco	4.1*	8.6*	7.0*	7.7*	−3.4*
Rubber	99	159	152	4.9	−0.8
Coir fibres	240**	210**	160*	−1.3**	−4.4**
Milk	84*	131	147*	4.5*	1.9*
Fish catch	58	98	136	5.4	5.6
Timber (000 m³)	3 900**	4 403	4 745	1.2**	1.2
Other					
Total energy (000 tce)	34	90	140a	10.3	9.2b
Petroleum products	—	1 713	1 560*a	na	−1.8*b
Manufactured gas (mn m³)	7	7	4a	0.0	−10.6b
Electricity (mn kW h)	357**	816	1149*a	8.6**	7.1*b
of which, hydro (mn kW h)	272	723	1002*a	10.3	6.7*b
Salt	57	65	119c	1.3	12.8b
Beer (000 hl)	57	89	32c	4.5	−22.6d
Cigarettes (mn units)	1 851	3 035	3 590c	5.1	4.3d
Cotton yarn	1.0	1.8	5.6c	6.0	32.8d
Cotton fabrics (mn m)	7	13	38c	6.4	30.8d
Cement	85	326	421	14.4	4.3
Tyres (000 units)	—	76	165c	na	21.4d
Radio sets (000 units)	—	45	117c	na	37.5f

a1975 b1970–75 c1974 d1970–74 e1973 f1970–73

Transport traffic	1960	1970	1976	Growth %pa	
Passenger-kilometres (mn)				1960–70	1970–76
Raila	1 714	2 938	2 900b	5.5	−0.3c
Air	48	96	305	7.2	21.2
Cargo: tonne-kilometres (mn)					
Raila	311	373	282	1.8	−4.6
Air	2.0	2.4	2.7	1.9	2.0
Sea: tonnes (mn)					
Goods loaded	0.73	0.96	1.09	2.8	2.1
Goods unloaded	4.02	2.20	1.38	−5.9	−7.5

aYears ending September 30th b1975 c1970–75
Tourism (1976) Number of visitors 126 530, gross receipts $ 28 mn

Finance and trade

Price index	1960	1970	1976	Growth %pa	
(1970 = 100)				1960–70	1970–76
Consumer prices	74.9	100.0	145.2	3.0	6.4
Money stock					
(end-year, SL Rs mn)	1 197	1 949	4 134	5.0	13.4

Budget (1976) Revenue: SL Rs 5 739 mn = $ 679 mn = £ 376 mn
Expenditure: SL Rs 8 653 mn = $ 1 023 mn = £ 566 mn

Balance of payments ($ mn)	1972	1973	1974	1975	1976
Balance of goods (fob)	−6	−5	−119	−127	−20
Balance of services	−39	−33	−58	−62	−51
Balance of transfers	+12	+13	+42	+79	+65
Current balance	−33	−25	−136	−109	−6
Long-term capital flow	+49	+52	+72	+89	+84
Reserves and debt (end-year, $ mn)					
International reserves	60	87	78	58	92
External public debt	608	673	868	997	1 082

External trade (1976) Imports: SL Rs 4 945 mn = $ 585 mn = £ 324 mn
Exports: SL Rs 4 840 mn = $ 572 mn = £ 317 mn

Main imports (1975)	% of total	*Main exports* (1976)	% of total
Food	50	Tea	44
(of which, cereals 42,		Rubber	18
sugar 5)		Copra and coconut oil	8
Crude oil	16	Nuts	5*
Chemicals	10	Industrial diamonds	5*
Machinery	7		
Iron and steel	2		
Textile yarns and fabrics	2		

Sri Lanka

External trade

Main sources (1976)	% of total	Main destinations (1976)	% of total
Saudi Arabia	13	China	11
Iran	11	United Kingdom	10
United States	8	United States	9
Japan	8	Pakistan	8
Pakistan	7	Japan	5
France	6	West Germany	4

Special focus

Tea prices (Sri Lanka/India wholesale price at New York)

	$ per tonne	% change over previous year		$ per tonne	% change over previous year
1970	1 010	+7	1974	1 367	+28
1971	1 074	+6	1975	1 517	+11
1972	1 118	+4	1976	1 638	+8
1973	1 065	−5	1977	3 082	+88

Syria

Syrian Arab Republic
El Jumhuriya el Arabiya es Suriya

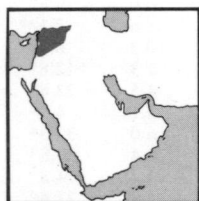

Location Middle East
With a coastline on the Mediterranean Sea, Turkey is to the north, Iraq to the east and south-east, Jordan to the south and Lebanon and Israel to the west at the southern part of the territory
Land Area 185 180 km² = 71 500 mi²
Climate Sub-tropical on coast, cold winters in highland
Weather at Damascus, 720 m altitude
Temperature: hottest month Aug 18–37 °C, coldest Jan 2–12 °C
Rainfall (av monthly): driest month Aug 0 mm, wettest Jan 43 mm
Time 2 hours ahead of GMT (summer time, 3 hours ahead)
Measures Metric system
Monetary unit Syrian pound (Sy £) = 100 piastres
Rate of exchange (1976 av): par Sy £ 3.862 = $ 1, free Sy £ 6.976 = £ 1
(Par value was changed from Sy £ 3.675 = $ 1 to Sy £ 3.925 = $ 1 from April 1976)

Summary

Political Socialist republic, which became independent from France in 1946. In February 1958 Syria joined with Egypt to form the United Arab Republic (also joined by North Yemen), but seceded from this union in September 1961. In September 1971 a federation was formed with Egypt and Libya, but this had little effect. There has been repeated conflict with Israel which occupied the Golan Heights after the war of June 1967 and enlarged its hold there in the war of October 1973. The Lebanese civil war of 1976 and subsequent difficulties there have involved Syria. Member of UN, Arab League and Oapec
Economic An agricultural and oil economy, with crude oil the main export, and cotton and tobacco significant exports. There is some industry, mainly petroleum refining and textiles. The Euphrates Dam is a major project (built with assistance from the Soviet Union) to provide better irrigation and hydro-electric power. Tourism and royalties from oil pipelines have been useful earners of foreign exchange

People, resources and equipment

Population[a] 1960 4.56*mn, 1970 6.26*mn, 1976 7.60*mn
[a]Including Palestinian refugees, of whom there were 173 936 registered on June 30th, 1973
Growth: 1960–70 3.2* %pa, 1970–76 3.3* %pa
Density (1976): 41* people per km²
Vital statistics (rate per 1 000 people, 1970–75): births 45*, deaths 15*
Regions (population in 000, 1970; total of 6.30 mn)

Damascus city	837	Deir es Zor	293	Idlib	384
Mohafazats (governorates)		Dira'a	232	El Ladhiqiya	390
Aleppo	1 317	Hama	515	Quneitra	16
Ar Rikqa	244	El Haseke	469	Es Suweidiya	140
Damascus	621	Homs	546	Tartus	302

Cities (population in 000, 1975) Damascus (capital) 1042[b], Aleppo 779, Homs 267, Hama 162, El Ladhiqiya[a] 157
[a]Latakia [b]Excluding suburbs

Race (1961) Arab 88 %, Kurdish 6 %
Language Arabic
Religion (1964) Moslem 88 % (of whom, Sunni 72 %, Alawi 11 %, Druzean 3 %), Christian 12 %
Education (1975/76) Pupils 1 793 000*, teachers 63 000*
Labour force (1975) 1 838 948; in agriculture 894 932 (49 %), mining and quarrying 11 450 (1 %), manufacturing 205 771 (11 %), construction 121 940 (7 %), transport and communication 75 862 (4 %)
Personnel (1975) Physicians: 2 400, 1 per 3 060 people
Standard of living
National income per person (1976): Sy £ 2 900* = $ 750* = £ 420*
Consumption per person (1975): energy 477 kg coal equivalent, electricity (production, 1976) 234 kW h, newsprint 0.1 kg, steel 64 kg
Newspapers (1974): number 6
Telephones (Dec 1976): 176 930, 23 per 1 000 people
Livestock (000, 1976) Cattle 555*, sheep 6 490, goats 956
Mineral reserves (1975) Crude oil 391 mn tonnes
Natural gas 76 bn cubic metres
Petroleum refinery capacity (1975) 3.0*mn tonnes
Electrical capacity (1975) 616* megawatts
Hospital beds (1975) 6 865, 1 per 1071 people
Roads (1973) 13 575 km = 8 435 mi, density 0.073km per km²
Railways (1976) 1 577 km = 980 mi, density 0.009 km per km²
Oil pipelines
Qarachuk via Homs to Baniyas and Tartus: 700**km = 450**mi
Iraq (Kirkuk) via Homs to Lebanon (for Tripoli)[a]: 500**km = 300**mi
Jordan (Saudi Arabian oil) to Lebanon (for Sidon): 160**km = 100**mi
[a]Also for Baniyas and Tartus; flow of oil was halted April 1976
Ships (registered, 1977) 32, total of 20 679 gross tons
Ports (goods traffic, 000 tonnes, 1976)

	loaded	unloaded
Tartus[a]	9 026	1 714
Baniyas[a]	7 698	2 860
Latakia	268	2 795

[a]Crude oil port
Airports Damascus, Aleppo, Latakia

Durable equipment (at end-year)	000	no per 1 000 people	
Radio sets (1972)	2 500	374	no per km of road
Television sets (1975)	224	30	
Passenger cars (1975)	50	7	3.7
Commercial vehicles (1975)	34	5	2.5

Production

Gross domestic product 1976: Sy £ 22 956 mn = $ 5 944 mn = £ 3 291 mn
Growth in real terms: 1960–70 6.6 % pa, 1970–76 10.3 %pa
Structure of gross domestic product (1976) *By origin* Agriculture 22 %, mining and quarrying 11 %, manufacturing 11 %, construction 7 %, transport and communication 5 %, other 44 % *By type* Gross fixed capital formation 35 %, exports of goods and services 23 %

Production indices (1970 = 100)	1960	1970	1976	Growth %pa 1960–70	1970–76
Agricultural	73	100	180	3.2	10.3
Industrial	46	100	173c	8.1	11.6d
Main products (000 t)					
Agriculture[a]					
Wheat	555	625	1 790	1.2	19.2
Barley	156	235	1 059	4.2	28.5
Potatoes	28	65	135*	8.8	13.0*
Tomatoes	75	216	517	11.1	15.7
Onions	32	50*	100	4.6*	12.3*
Watermelons	150**	129	500*	−1.5**	25.3*
Grapes	130**	206	280*	4.7**	5.2*
Cottonseed	184	234	254*	2.4	1.4*
Olives	84	83	233	−0.1	18.8
Tobacco	5.0**	6.7	14.0*	3.0**	13.1*
Cotton	111	149	155*	3.0	0.7*
Milk, cow	135**	198	270*	3.9**	5.3*
Milk, sheep	290**	196	250*	−3.8**	4.1*
Wool	4.0	7.0	7.0*	5.7	0.0*
Sheepskins	4.6**	6.6*	8.4*	3.7**	4.1*
Sheep and goat meat	26**	50*	60*	6.8**	3.1*
Energy					
Total energy (000 tce)	—	5 520	14 360c	na	21.1d
Crude oil	—	4 243	9 976	na	15.3
Petroleum products[b]	684	1 828	2 284c	10.3	4.5d
Natural gas (mn m³)	—	170*	210c	na	4.3*d
Electricity (mn kW h)	368	947	1 780	9.9	11.1
Mining					
Phosphate rock	—	—	860c	na	na
Salt	10	46	33c	16.5	−6.4d

Syria

Main products (000 t) Manufacturing	1960	1970	1976	Growth %pa 1960–70	1970–76
Beer (000 hl)	18	31	58e	5.6	17.0f
Cigarettes (mn units)	2 241	2 429	4 495c	0.8	13.1d
Cotton yarn	10	20	32c	7.6	9.8d
Cotton fabrics	25	27	36a	0.8	6.2d
Cement	489	964	1 110	7.0	2.4
Television sets (000 units)	—	8	39c	na	37.3d

aHarvest results can fluctuate markedly from year to year bExcluding petroleum coke c1975 d1970-75 e1974 f1970-74

Transport traffic Passenger-kilometres (mn)	1960	1970	1976	Growth %pa 1960–70	1970–76
Road	na	4 091	5 307**	na	4.4**
Rail	42	86	166	7.4	11.6
Air	36b	220	712	25.4c	21.6
Cargo: tonne-kilometres (mn)					
Road	na	3 326	3 751**	na	2.0**
Rail	107	102	305	−0.5	20.0
Air	1.2b	1.0	7.6	−2.3c	40.2
Sea: tonnes (mn)					
Goods loadeda	24.74b	34.54	16.99	4.3c	−11.2
Goods unloaded	1.23b	2.24	7.84	7.8c	23.2

aIncluding Iraqi crude oil loaded at Baniyas (22.55mn t or 83% in 1972)
b1962 c1962-70
Tourism Number of visitors (including excursionists, 1976) 1 392 569, gross receipts (1975) $ 93 mn

Finance and trade

Price indices (1970 = 100)	1960	1970	1976	Growth %pa 1960–70	1970–76
Consumer prices	83	100	195	1.9	11.8
Wholesale prices	87.8	100.0	189.5	1.3	11.2
Money stock					
(end-year, Sy £ mn)	825	2 341	8 599	11.0	24.2

Budget (total, 1977)
Balanced at Sy £ 17 050 mn = $ 4 344 mn = £ 2 489 mn

Balance of payments ($ mn)	1972	1973	1974	1975	1976
Balance of goods (fob)	−146	−213	−256	−495	−1 036
Balance of services	+92	+150	−37	−118	−191
Balance of transfers	+84	+401	+460	+706	+455
Current balance	+28	+339	+167	+93	−772
Long-term capital flow	+14	+25	0	−10	+270
Reserves and debt (end-year, $ mn)					
International reserves	135	413	500	735	361
External public debt	521	653	1 110	1 845	2 583

External trade (1976) Imports: Sy £ 9 203 mn = $ 2 383 mn = £ 1 319 mn
Exports: Sy £ 4 143 mn = $ 1 073 mn = £ 594 mn

Main importsa	% of total	Main exports	% of total
Gold	16	Crude oil	62
Machinery, non-electric	15	Cotton	15
Food	11	Textile yarns and fabrics	3
(of which, sugar 3,		Clothing	3
cereals 2)		Crude oil products	2
Iron and steel	9	Tobacco	2
Motor vehicles	9		
Crude oil	9		
Chemicals	6		
Electrical machinery	5		

Main sourcesa		Main destinations	
Switzerland	18	Italy	15
West Germany	12	West Germany	10
Italy	7	Soviet Union	9
Saudi Arabia	7	Jugoslavia	8
Japan	6	United Kingdom	7
France	6	Belgium-Luxembourg	7
United States	6	Netherlands	7
United Kingdom	4	Greece	5
Netherlands	2	Saudi Arabia	5

aIncludes gold, mainly from Switzerland

Special focus

Tourism (including excursionists), 1976

Origin of visitors	Number	%		Number	%
Lebanon	610 925	44	Libya	41 891	3
Jordan	219 324	16	Egypt	41 611	3
Turkey	150 955	11	Other	284 474	20
Iraq	43 389	3	Total visitors	1 392 569	100

Taiwan

Republic of China
Chung-hua Min-kuo

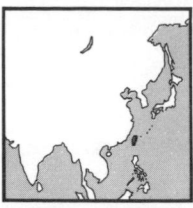

Location Eastern Asia
The island of Taiwan (Formosa) is about 300 km off the south-east coast of China, separated from the mainland by the Formosa Strait. Also included in the territory are the Pescadores islets, and Quemoy and Matsu islands which are close to mainland China
Land Area 35 981 km² = 13 893 mi²
Climate Sub-tropical
Weather at Taipei, 9 m altitude
Temperature: hottest month July 24–33 °C, coldest Feb 12–18 °C
Rainfall (av monthly): driest month Nov 66 mm, wettest Aug 305 mm
Time 8 hours ahead of GMT
Measures Metric system; some Chinese measures are also in use (see also Hongkong)
Monetary unit New Taiwan dollar (NT $) = 100 cents
Rate of exchange (1976 av): par NT $ 38 = $ 1, free NT $ 68.64 = £ 1

Summary

Political Republic, which became separated politically from mainland China in 1949, when a Communist government was installed in Peking. A security pact was signed with the United States in 1954 which pledged protection for Taiwan and the Pescadores, and in 1955 Quemoy and Matsu were added to the pact area; US military advisers were withdrawn from Quemoy and Matsu in June 1976. Taiwan lost membership of the UN in October 1971 when mainland China took over the Chinese seat there. Taiwan is regarded by China (mainland) as a province
Economic A mixed economy, with manufacturing making up about one-third of gross domestic product and agriculture one-sixth; main exports are textiles and clothing, electrical machinery (including radio and TV sets) and other machinery. Growth has been high, with industrial production increasing at 15 % pa

People, resources and equipment

Population 1960 10.79*mn, 1970 14.68 mn, 1976 16.33 mn
Growth: 1960–70 3.1* %pa, 1970–76 1.8 %pa
Density (1976): 454 people per km²
Vital statistics (rate per 1 000 people, 1976): births 25.9, deaths 4.7
Regions ((population in 000, 1976)

Taipei city	2 089	Kaohsiung	971	Taitung	290
Taiwan province	13 849	Miaoli	542	Taoyuan	896
Counties		Nantou	519	Yunlin	804
Chiayi	841	Penghu	114	*Cities*	
Changhua	1 118	Pingtung	867	Chilung	343
Hsinchu	624	Taichung	887	Kaohsiung	1 020
Hualien	349	Tainan	949	Taichung	561
Ilan	431	Taipei	1 757	Tainan	537

Cities (population in 000, 1976)

Taipei (capital)	2 089	Tainan	537	Pan-chiao	282
Kaohsiung	1 020	Chilunga	343	Chiayi	253
Taichung	561	Shan-chung	285	Hsinchu	230

aKeelung
Race (1976) Han Chinese 98* % (18* % from mainland China), aborigines (Indonesian) 2* %
Language Mandarin Chinese and dialects
Religion (1974) Buddhist 36 %, Taoist 21 %, Roman Catholic 2 %
Education (1976/77) Pupils 4 179 977, teachers 141 914
Labour force (1976) 6 836 000; in agriculture 2 366 000 (35 %), mining 57 000 (1 %), manufacturing 1 450 000 (21 %), construction 224 000 (3 %), transport 325 000 (5 %), other 2 414 000 (35 %)
Personnel (1976) Physicians: 16 982, 1 per 962 people
Standard of living
National income per person (1976): NT $ 36 843 = $ 970 = £ 537
Consumption per person (1970): energy 925 kg coal equivalent, electricity (1976) 1 646 kW h, newsprint 2* kg, steel 103 kg
Newspapers (1974): number 31; circulation 1 750 000**, 110** per 1 000 people
Telephones (Dec 1976): 1 396 000, 85 per 1 000 people
Livestock (000, Dec 1976) Cattle and buffaloes 253, pigs 3 676, sheep and goats 211, chickens 28 354, ducks 8 051
Mineral reserves (1976) Coal 216 mn tonnes
Crude oil 3 mn tonnes
Natural gas 30 856 mn cubic metres
Petroleum refinery capacity (1976) 13* mn tonnes
Electrical capacity (1976) 5 884 megawatts

Taiwan

Hospital beds (govt establishments only, 1976) 32 460*, 1 per 503* people
Roads (1976) 17 172 km = 10 670 mi, density 0.48 km per km²
Railways (1976) 4 200 km = 2 610 mi, density 0.12 km per km²
Ships (registered, 1977) 443, total of 1 559 000 gross tons
Ports (total goods traffic, 000 tonnes, 1976)

	loaded	unloaded
Kaohsiung	11 481	27 803
Chilung	7 807	12 565
Hualien	642	546

Airports Taipei, Kaohsiung, Hualien, Taichung, Chiayi, Taitung, Makung

Durable equipment (Dec 1976)	000	no per 1 000 people	no per km of road
Radio sets	1 493	90	15
Television sets	914	55	
Passenger cars	260	16	15
Commercial vehicles	55	3	3.2

Production

Gross domestic product
1976: NT $ 655 820 mn = $ 17 258 mn = £ 9 554 mn
Growth in real terms: 1960–70 9.1 %pa, 1970–76 8.3 % pa
Structure of domestic product (1976)

By origin (net)	NT $ mn	% of gdp
Agriculture	69 864	14
Mining and quarrying	5 310	1
Manufacturing	146 223	29
Electricity, gas and water	13 083	3
Construction	28 842	6
Commerce	59 250	12
Transport and communication	30 856	6
Other	153 372	30
Total	506 800	100

By type of expenditure		
Government final consumption	109 850	17
Private final consumption	346 600	53
Stock investment	−4 390	−1
Gross fixed capital formation	184 640	28
Export of goods and services	340 680	52
less Imports of goods and services	−321 550	−49
Total	655 820	100

Production indices (1970 = 100)	1960	1970	1976	Growth %pa 1960–70	1970–76
Agricultural	65	100	119	4.5	3.0
Industrial	22.0	100.0	233.1	16.4	15.1

Main products (000 t)	1960	1970	1976	Growth %pa 1960–70	1970–76
Agriculture					
Rice	1 912	2 463	2 713	2.6	1.6
Barley	315	851	664	10.4	−4.1
Maize	21	57	114	10.5	12.2
Sweet potatoes	2 979	3 441	1 851	1.5	−9.8
Cassava	159	308	294	6.8	−0.8
Sugar, raw value	808*	719*	779	−1.2*	1.3*
Groundnuts	102	122	89	1.8	−5.1
Soyabeans	53	65	53	2.1	−3.3
Bananas	114	462	213	15.0	−12.1
Pineapples	167	338	279	7.3	−3.1
Citrus fruit	53	209	384	14.7	10.7
Tea	16	28	25	5.7	−1.9
Tobacco	16	21	26	2.7	3.6
Pigmeat	181	393	503*	8.1	4.2*
Fish catch	259	613	811	9.0	4.8
Timber (000 m³)[b]	822	1 111	824	3.1	−4.9
Energy					
Total energy (000 tce)	4 260	6 210	na	3.8	na
Coal	3 962	4 473	3 236	1.2	−5.2
Crude oil	2.3	101	247	46.0	16.1
Petroleum products	1 000*	4 413	10 470	16.0*	15.5
Natural gas (mn m³)	25	947	1 836	43.8	11.7
Electricity (mn kW h)	3 628	13 213	26 877	13.8	12.6
Mining					
Salt	453	535	497	1.7	−1.2
Manufacturing					
Alcoholic beverages (000 hl)	981	1 785	2 838	6.2	8.0
Cigarettes (mn units)	11 851	16 175	21 911	3.2	5.2
Cotton yarn	40	105	147	10.1	5.7
Cotton fabrics (mn m)	176	528	811	11.6	7.4
Man-made fibres	1.8	88	341	47.5	25.3

Main products (000 t) Manufacturing	1960	1970	1976	Growth %pa 1960–70	1970–76
Paper[a]	97	320	500	12.7	7.7
Cement	1 183	4 541	8 749	14.4	11.5
Pig iron	24	96	105	14.9	1.5
Crude steel	201*	370	597	6.3	8.3
Aluminium	8	27	26	12.5	−0.6
Radio sets (000 units)	—	6 248	6 849	na	1.5
TV sets (000 units)	—	1 254	3 850	na	20.6
Merchant vessels (000 grt)	23	91	83	14.7	−1.5

[a]Excluding paper board [b]Excluding firewood

Transport traffic	1960	1970	1976	Growth %pa 1960–70	1970–76
Passenger-kilometres (mn)					
Road[a]	3 173	7 004	13 642	8.2	11.8
Rail	3 609	6 212	8 480	5.6	5.3
Air	64	954	2 908	31.0	20.4
Cargo: tonne-kilometres (mn)					
Road	317	1 364	4 965	15.7	24.0
Rail	2 071	2 631	2 886	2.4	1.6
Air	2.0*	24.6	181.6	28.5*	39.5
Sea: tonnes (mn)					
Goods loaded	3.02	9.00	19.93	11.5	14.2
Goods unloaded	3.87	16.02	40.95	15.3	16.9
Tourism					
Number of visitors (000)	24	472	1 008	34.7	13.5
Gross receipts ($ mn)	7.2[b]	90	454	43.5[c]	31.0

[a]Excluding city bus traffic [b]1963 [c]1963-70

Finance and trade

Price indices (1970 = 100)	1960	1970	1976	Growth %pa 1960–70	1970–76
Consumer prices	71.8	100.0	182.2	3.4	10.5
Wholesale prices	82.3	100.0	176.2	2.0	9.9
Money stock (end-year, NT $ mn)	6 110	34 510	130 560	18.9	24.8

Budget (total, 1976/77; year ending June 30th)
Revenue: NT $ 170 181 mn = $ 4 478 mn = £ 2 615 mn
Expenditure: NT $ 170 116 mn = $ 4 477 mn = £ 2 614 mn

Balance of payments ($ mn)	1972	1973	1974	1975	1976
Balance of goods (fob)	+647	+735	−831	−256	+777
Balance of services	−152	−169	−294	−337	−400
Balance of transfers	+18	0	+11	+7	+20
Current balance	+512	+565	−1 115	−587	+396
Long-term capital flow	+14	+279	+470	+364	+563
Reserves and debt (end-year, $ mn)					
International reserves	1 039	1 124	1 191	1 169	1 607
External public debt	1 596	1 801	2 612	3 103	3 521

External trade (1976)
Imports: NT $ 289 140 mn = $ 7 609 mn = £ 4 212 mn
Exports: NT $ 309 910 mn = $ 8 156 mn = £ 4 515 mn

Main imports	% of total	Main exports	% of total
Crude oil	14	Clothing	16
Machinery, non-electric	13	Electrical machinery	15
Chemicals	12	(of which, radio and	
Electrical machinery	11	TV sets 6)	
Cereals	6	Textile yarns and fabrics	12
Transport equipment	6	Footwear	7
Iron and steel	5	Fruit and vegetables	5
Textile fibres	5	Machinery, non-electric	4
Main sources		*Main destinations*	
Japan	32	United States	38
United States	24	Japan	14
Kuwait	9	Hongkong	7
Saudi Arabia	5	West Germany	5
West Germany	5	Canada	4

Special focus

Overseas Chinese, end 1976
Geographical distribution, excluding China and Taiwan

Place of residence	Chinese people (mn)
Asia	21.1[a]
America	1.1
Europe	0.2
Oceania	0.07
Africa	0.07
Total	22.6

Overseas Chinese visiting Taiwan (1976) 154 251

[a]Of whom, Thailand 5.0**mn, Hongkong 4.4**mn, Malaysia 4.0**mn, Indonesia 3.0**mn, Singapore 1.8*mn, Vietnam 1.5***mn

Thailand

Kingdom of Thailand
Muang Thai or Prathet Thai

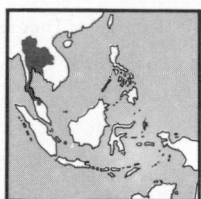

Location South-east Asia
With a coastline on the Gulf of Siam in the South China Sea, Laos and Cambodia are to the east, Burma to the north and west, and Malaysia to the south; Thailand includes the main part of the Isthmus and Peninsula of Kra north of Malaysia
Land Area 514 000 km² = 198 500 mi²
Climate Tropical
Weather at Bangkok, 2 m altitude
Temperature: hottest month April 25–35 °C, coldest Dec 20–31 °C
Rainfall (av monthly): driest month Dec 5 mm, wettest Sept 305 mm
Time 7 hours ahead of GMT
The Buddhist year 2521 corresponds to the Gregorian 1978
Measures Metric system; local units are also in use, including:
length 1 sen = 20 wa = 80 sok = 40 metres = 43.744 yards
area 1 ngan = 100 wa² = 400 square metres = 478.4 square yards
1 rai = 1 sen² = 4 ngan = 1 600 square metres = 1 914 square yards
capacity 1 sat or thang — 20 litres = 4.399 UK gallons
1 kwien = 2 ban = 100 sat = 2 000 litres = 439.9 UK gallons
weight (mass) 1 picul = 100 catty = 60 kilograms = 132.3 pounds
Monetary unit Baht (Bt) = 100 satangs
Rate of exchange (1976 av): par Bt 20 = $ 1, free Bt 36.12 = £ 1

Summary

Political Monarchy with military government, known as Siam before June 24, 1939. Member of UN, Colombo Plan and Asean
Economic A widely-based agricultural economy, with sugar, rice and maize the main exports. Tin is the main mineral, accounting for 5 % of exports, and there is a wide range of other minerals; natural gas reserves have been discovered off the coast. There is a range of manufacturing industry which provides 18 % of gross domestic product; petroleum refining is important

People, resources and equipment

Population 1960 26.39*mn, 1970 36.37*mn, 1976 42.96*mn
Growth: 1960–70 3.3* %pa, 1970–76 2.8* %pa
Density (1976): 84* people per km²
Vital statistics (rate per 1 000 people, 1970–75): births 43*, deaths 11*
Regions (population in 000, 1970; census total of 34.40 mn)
Central 10 612, North-east 12 026, North 7 489, South 4 272
Cities (population in 000, 1975)
Krung Thep[a] (capital) 4 340[b,c] Chiang Mai 93[d] Nakhon
Khon Kaen 200*[c] Songkhla 90*[c] Ratchasima 77[d]
[a]Bangkok [b]Includes Thon Buri [c]Agglomeration [d]1972
Race (1961) Thai 74 %, Chinese 18 %, Malay 3 %, Khmer 1 %
Language Thai
Religion (1973) Buddhist 94 %, Moslem 4 %, Christian 1 %
Education Pupils (1975/76) 7 870 000*, teachers (1973/74) 264 000*
Labour force (1976) 13 945 190; in agriculture 8 643 850 (62 %), mining and quarrying 49 950 (¼ %), manufacturing 1 514 050 (11 %), construction 358 870 (3 %), transport and communication 356 460 (3 %)
Personnel Scientists and engineers (1973): 18 827
Physicians (1975): 5 000*, 1 per 8 370* people
Standard of living
National income per person (1976): Bt 7 151 = $ 358 = £ 198
Consumption per person (1975): energy 284 kg coal equivalent, electricity (production) 189* kW h, newsprint 1.3 kg, steel 20 kg
Newspapers (1974): number 27
Telephones (Sept 1976): 333 800, 8 per 1 000 people
Livestock (000, 1976) Elephants 11**, cattle 4 311, buffaloes 5 438, pigs 4 300, chickens 53 860
Mineral reserves Lignite (known economic, 1967) 235 mn tonnes
Crude oil (1973) 0.2**mn tonnes
Petroleum refinery capacity (1975) 8.3 mn tonnes
Electrical capacity (1975) 2 500* megawatts
Hospital beds (1974) 51 215, 1 per 800 people
Roads (1976) 39 721 km = 24 681 mi, density 0.077 km per km²
Railways (1975) 3 765 km = 2 339 mi, density 0.007 km per km²
Ships (registered, 1977) 100, total of 260 664 gross tons
Ports (goods traffic, 000 tonnes, 1976) Bangkok: loaded 11 782, unloaded 12 942
Airports (passengers, 000, 1976) Don Muang (Bangkok): departures 1 374, arrivals 1 346. Also Chiang Mai, Haadyai and (1977) 14 other airports with scheduled flights

Durable equipment (at end-year)	000	no per 1 000 people	
Radio sets (1975)	5 500	131	no per
Television sets (1974)	715	17	km of road
Passenger cars (1976)	363	8	9.1
Commercial vehicles (1976)	294	7	7.4

Production

Gross domestic product 1976: Bt 332 177 mn = $ 16 609 mn = £ 9 196 mn
Growth in real terms: 1960–70 8.0 %pa, 1970–76 6.6 %pa
Structure of gross domestic product (1976) *By origin* Agriculture 31 %, mining and quarrying 1 %, manufacturing 18 %, construction 5 %, transport and communication 6 %, other 39 %
By type Final consumption expenditure 78 % (of which, government 11 %), stock investment 2 %, gross fixed capital formation 23 %, exports of goods and services 22 %, less imports of goods and services −24 %

Production index (1970 = 100)	1960	1970	1976	Growth %pa 1960–70	Growth %pa 1970–76
Agricultural	59	100	126	5.4	3.9
Main products (000 t)					
Agriculture					
Rice	7 834	13 270	15 800*	5.4	2.9*
Maize	544	1 938	2 675	13.5	5.5
Sorghum	10**	120	148	28.0**	3.6
Sweet potatoes	138	250*	347*	6.1*	5.6*
Cassava	1 222	3 431	7 850	10.9	14.8
Sugar, raw value	155	407	1 665	10.1	26.5
Beans, dry	60	220*	338*	13.9*	7.4*
Watermelons	121	270*	510*	8.4*	11.2*
Oranges	30**	44*	50*	3.9**	2.2*
Pineapples	256	300**	500*	1.6**	8.9**
Bananas	425	1 200*	1 464*	11.0*	3.4*
Soya beans	26	50	125	6.8	16.5
Groundnuts	152	125	169	−1.9	5.1
Copra	21*	28*	43*	2.9*	7.4*
Tobacco	74	93	73*	2.3	−3.9*
Kenaf[a]	181	381	233	7.7	−7.9
Cotton	45	9	12*	−15.0	4.9*
Rubber	172	287	392	5.2	5.3
Eggs	40**	105*	140*	10.1**	4.9*
Beef and buffalo meat	100**	142*	169*	3.6**	2.9*
Pigmeat	110**	170*	200*	4.4**	2.7*
Fish catch	221	1 448	1 640	20.7	2.1
Timber (000 m³)	14 500**	18 745	21 119	2.6**	2.0
Energy					
Total energy (000 tce)	40*	368	460[b]	24.8*	4.6[c]
Lignite	108	400	681	14.0	9.3
Crude oil	—	10	6[b]	na	−9.7[c]
Petroleum products	—	4 067	7 356[b]	na	12.6[c]
Electricity (mn kW h)	594	4 545	7 910*[b]	22.6	11.7*[c]
Mining					
Antimony ore (Sb content)	—	2.5	3.2[b]	na	5.3[c]
Manganese ore (Mn content)	0.3	8.3	8.7[b]	39.4	1.0[c]
Tin conc (Sn content)	12.3	21.8	27.9	5.9	4.2
Tungsten conc (oxide content)	0.26	0.90	2.0[b]	13.2	17.5[c]
Salt	335	200*	160*[b]	−5.0*	−4.4*[c]
Manufacturing					
Beer (000 hl)	86[f]	363	613[b]	22.8[g]	11.0[c]
Cigarettes (mn units)	10 861	15 291	22 618[b]	3.5	8.1[c]
Cotton yarn	13[f]	56	na	23.2[g]	na
Cotton fabrics (mn m²)	72[h]	306	443[b]	17.4[i]	9.7[e]
Man-made fibres	—	4.6	39.1[b]	na	53.4[c]
Cement	526	2 627	4 106	17.4	7.7
Crude steel	7	40	na	19.0	na

[a]Jute-like fibre [b]1975 [c]1970–75 [d]1974 [e]1970–74 [f]1963 [g]1963–70 [h]1961 [i]1961–70

Thailand

Transport traffic

	1960	1970	1976	Growth %pa	
				1960–70	1970–76
Passenger-kilometres (mn)					
Rail	2 353	4 113	5 530	5.7	5.1
Air	63	783	4 662	28.6	34.6
Cargo: tonne-kilometres (mn)					
Rail	1 147	2 209	2 630	6.8	3.0
Air	1.0	10.3	126.8	25.7	52.0
Sea: tonnes (mn)					
Goods loaded	2.87	5.96	12.66	7.6	13.4
Goods unloaded	2.08	8.86	13.09	15.6	6.7
Tourism					
Number of visitors (000)	98	629	1 098	20.4	9.7
Gross receipts ($ mn)	na	105	197	na	11.1

Finance and trade

Price indices (1970 = 100)

	1960	1970	1976	Growth %pa	
				1960–70	1970–76
Consumer prices	81.8	100.0	158.1	2.0	7.9
Wholesale prices	79.5	100.0	184.6	2.3	10.8
Money stock (end-year, Bt mn)	10 090	19 310	40 590	6.7	13.2

Budget (1976/77 ; year ending September 30th)
Revenue: Bt 48 082 mn = $ 2 404 mn = £ 1 406 mn
Expenditure: Bt 67 364 mn = $ 3 368 mn = £ 1 970 mn,
of which, education 22 %, defence 19 %, agriculture 9 %

Balance of payments ($ mn)

	1972	1973	1974	1975	1976
Balance of goods (fob)	−279	−320	−388	−662	−222
Balance of services	+168	+129	+59	−26	−269
Balance of transfers	+59	+144	+241	+80	+23
Current balance	*−51*	*−46*	*−87*	*−607*	*−469*
Long-term capital flow	+156	+80	+388	+255	+319
Reserves and debt (end-year, $ mn)					
International reserves	1 052	1 306	1 858	1 775	1 893
External public debt	653	721	1 122	1 273	1 619

External trade (1976) Imports: Bt 72 877 mn = $ 3 644 mn = £ 2 018 mn
Exports: Bt 60 797 mn = $ 3 040 mn = £ 1 683 mn

Main imports (1975)	% of total	*Main exports* (1975)	% of total
Crude oil	18	Sugar	13
Machinery, non-electric	18	Rice	12
Chemicals	13	Maize	12
Motor vehicles	8	Vegetables	10
Electrical machinery	6	Printed matter	8
Iron and steel	5	Rubber	7
Petroleum products	3	Tin	5
Textile yarns and fabrics	3	Fish	4
Cotton	3	Textile yarns and fabrics	3
Aircraft	2	Clothing	2
Main sources (1976)		*Main destinations* (1976)	
Japan	32	Japan	26
United States	13	Netherlands	13
Saudi Arabia	8	United States	10
Qatar	6	Singapore	7
West Germany	5	Indonesia	5
United Kingdom	4	Hongkong	5
Singapore	3	Malaysia	4
Taiwan	2	West Germany	3
Australia	2	Taiwan	3

Special focus

Rice prices (Bangkok wholesale price)

	$ per tonne	% change over previous year		$ per tonne	% change over previous year
1966	163	+20	1972	147	+14
1967	206	+26	1973	350	+138
1968	202	−2	1974	542	+55
1969	187	−7	1975	363	−33
1970	144	−23	1976	255	−30
1971	129	−10	1977	272	+7

Timor, East

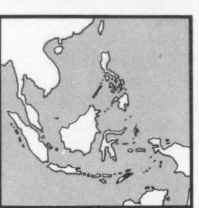

Location South-east Asia
Consists of the eastern part of the island of Timor;
the western section is part of Indonesia
Land Area 14 925 km² = 5 763 mi²
Climate Tropical
Weather at Díli, 0 m altitude
Temperature: hottest month Dec 26–32 °C,
coldest Aug 22–31 °C
Rainfall (av monthly): driest month August
5 mm, wettest Dec 140 mm
Time 8 hours ahead of GMT
Measures Metric system
Monetary unit Timor escudo (T Esc) = 100 centavos
Rate of exchange (1976 av): free T Esc 30.22 = $ 1, T Esc 54.59 = £ 1

Summary

Political Formerly a Portuguese overseas province, Indonesia claimed
East Timor as a province on July 17, 1976; there has been an independence
movement
Economic An agricultural economy, with coffee the main export

People, resources and equipment

Population 1960 517 000*, 1970 603 000*, 1976 688 000*
Growth: 1960–70 1.5* %pa, 1970–76 2.2* %pa
Density (1976): 46* people per km²
Vital statistics (rate per 1 000 people, 1970–75): births 44*, deaths 23*
City (population in 000, 1974) Díli (capital) 11*
Race Indonesian/Malay, Melanesian
Language Portuguese and local languages
Religion Animist, Moslem, Christian
Education (1971/72) Pupils 35 119, teachers 743
Labour force (1976) 212 000; in agriculture 130 000 (61 %)
Standard of living
National income per person (1976): T Esc 4 000*** = $ 130*** = £ 70***
Consumption per person (1975): energy 16 kg coal equivalent,
electricity 4* kW h, newsprint 0.1 kg
Telephones (Dec 1974): 966, 1.5 per 1 000 people
Livestock (000, 1976) Cattle 94*, buffaloes 130*, goats 192*, pigs 223*
Electrical capacity (1975) 3* megawatts
Roads (1972) 2 896 km = 1 800 mi, density 0.19 km per km²
There are no railways
Port Díli
Airport Díli
Radio sets (Dec 1974) 4 200, 6 per 1 000 people
Motor vehicles (Dec 1972) 1 300, 2 per 1 000 people, 0.4 per km of road

Production, finance and trade

Gross domestic product
1976 est: T Esc 3 000***mn = $ 100***mn = £ 55***mn
Main products (000 t) *Agriculture* (1976) Rice 22*, maize 15*,
sweet potatoes 16*, cassava 17*, copra 2*, coffee 4*, eggs 2*, pigmeat 5*
Other Electricity (mn kW h, 1975) 3*
Transport traffic *Air* (1972) Passenger departures 13 800, arrivals 14 500;
cargo loaded 87 t, unloaded 138 t
Sea (1973) Vessels entered 164 000 grt
Tourism (1972) Number of visitors 12 800
Budget (1974) Balanced at T Esc 216 mn = $ 8.5 mn = £ 3.6 mn
External trade (1974) Imports: T Esc 310 mn = $ 12 mn = £ 5.2 mn
Exports: T Esc 140 mn = $ 6 mn = £ 2.4 mn

Main imports	% of total	*Main exports*	% of total
Food	27*	Coffee	90*
Textiles	16	Copra	6*
Machinery	14		
Petroleum products	11		
Metals	9		
Chemicals	7		

Main sources		*Main destinations*	
Portugal	29	Singapore	44
Singapore	20	Portugal	15
Australia	12	United States	15
Macao	10	Belgium-Luxembourg	6
Japan	7	Denmark	6
Mozambique	5	Netherlands	5

United Arab Emirates

Ittihad al Imarat al Arabiyah

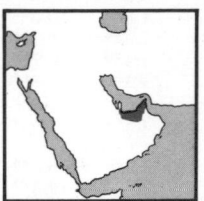

Location Middle East
With a coastline on the Gulf, known as the Trucial Coast, Oman is to the east, Saudi Arabia to the south and west and Qatar to the north-west. Some borders are ill-defined
Land Area 83 600 km² = 32 300 mi²
Climate Hot, mild in winter
Weather at Sharjah, 5 m altitude
Temperature: hottest month Aug 28–39 °C, coldest Jan 12–23 °C
Rainfall (av monthly): driest months May–Oct 0 mm, wettest Dec 36 mm
Time 4 hours ahead of GMT
Measures UK (imperial) and metric; local measures are also in use
Monetary unit UAE dirham (UAE Dh) = 100 fils
Rate of exchange (1976 av): par UAE Dh 3.94737 = $ 1, free UAE Dh 7.130 = £ 1

Summary

Political Federation of emirates, which was formed on December 2, 1971, by the federation of six formerly British-protected Trucial States: Abu Dhabi, Ajman, Dubai, Fujairah, Sharjah and Umm al Qaiwain; Ras al Khaimah joined in February 1972. There is a Federal Assembly, the seven rulers each retaining control over his own Shaikhdom. Member of UN, Arab League, Opec and Oapec
Economic The economies of Abu Dhabi, Dubai and Sharjah are based on oil, with Dubai also having an important entrepôt trade. Oil revenue has been used mainly for construction, transport and social services; the ports of Dubai, Abu Dhabi and Sharjah are being developed, and a large dry dock is being built for Dubai. Industry projects include an aluminium smelter at Dubai and petrochemicals at Abu Dhabi; industrial cities are being developed at Jebel Ali (near Dubai) and Rubais (in Abu Dhabi)

People, resources and equipment

Population[a] 1960 200 000**, 1970 300 000**, 1976 700 000**
Growth: 1960–70 4.1**%pa, 1970–76 15.2**%pa
Density (1976): 8** people per km²
*Total population; de jure (usually resident) population for 1976 229 000[a]
Regions (emirates, population in 000, 1975)
Abu Dhabi 236   Ras al Khaimah 57   Umm al Qaiwain 17
Dubai 207   Fujairah 26
Sharjah 88   Ajman 22
Cities (population in 000, 1975) Abu Dhabi Town (federal capital) 150**, Dubai Town 100**, Sharjah Town 50**, Al Ain 40**, Ras al Khaimah Town 30**
Race Arab
Language Arabic
Religion Mainly Moslem (Sunni and Shiah)
Education (1974/75) Pupils 64 752, teachers 2 988
Personnel (1973) Physicians: 211, 1 per 1 520 people
Standard of living National income per person (1976): UAE Dh 45 000*** = $ 11 000*** = £ 6 000***
Consumption per person (1975): energy 13 699 kg coal equivalent
Telephones (Dec 1976): 70 860, 90 per 1 000 people
Mineral reserves (1975) Crude oil 5 563 mn tonnes
Natural gas 937 bn cubic metres
Electrical capacity (1975) 175* megawatts
Roads (1974) 500*** km = 300*** mi, density 0.01*** km per km²
There are no railways
Ships (registered, 1977) 85, total of 152 100 gross tons
Ports Port Rashid (Dubai), Port Zayed (Abu Dhabi), Sharjah, Ras al Khaimah
Airports Dubai, Abu Dhabi, Sharjah, Ras al Khaimah
Durable equipment (Dec 1976) Radio sets: 150 000**, 195** per 1 000 people
Television sets: 80 000**, 104** per 1 000 people

Production, finance and trade

Gross domestic product
1976 est: UAE Dh 37 500**mn = $ 9 500**mn = £ 5 300**mn
Main products (1976) Fish catch 68 000* t, crude oil 93.3 mn t[a], natural gas (1975) 1 690*mn m³
Growth of crude oil production: 1970–76 16.3 %pa
[a]Of which: Abu Dhabi 76.8 mn t, Dubai 15.8 mn t, Sharjah 1.8 mn t

Transport traffic *Air* (1976) Passenger-kilometres 320 mn, cargo 7.0 mn t-km *Sea* (1974) Goods loaded 82.9* mn t, unloaded 1.0* mn t
Money stock (end-year, UAE Dh mn) 1973 970, 1976 4 725; growth 1973–1976 69.5 % pa
Budget (federal total, 1976)
Balanced at UAE Dh 4 152 mn = $ 1 052 mn = £ 582 mn
External trade (1976)
Imports: UAE Dh 13 150 mn = $ 3 331 mn = £ 1 844 mn
Exports: UAE Dh 33 770 mn = $ 8 555 mn = £ 4 736 mn

Main imports	% of total	Main exports	% of total
Machinery	30*	Crude oil	96
Iron and steel	11*		
Motor vehicles	10*		
Food	9		
Petroleum products	4*		
Metal manufactures	4*		
Chemicals	3		
Main sources		*Main destinations*	
Japan	17	Japan	28
United Kingdom	17	France	13
United States	13	United States	12
West Germany	7	Netherlands	10
India	5	United Kingdom	8
France	4	Spain	5
Netherlands	3	West Germany	4
Italy	3	Italy	3
Saudi Arabia	3	Netherlands Antilles	3
Iran	2		

Special focus

Abu Dhabi government finances, UAE Dh mn

	1971	1972	1973	1974	1975	1976
Revenue[a]	1 651	2 181	3 222	14 131	15 015	18 401
Expenditure of which:	1 104	1 736	3 391	6 923	11 457	18 205
current	712	1 274	2 513	4 677	6 506	12 796
development	369	371	524	1 010	2 250	4 463
Surplus or deficit	*+548*	*+445*	*−169*	*+7 208*	*+3 558*	*+196*

[a]Oil revenues are 95-98 % of the total

Vietnam

The Socialist Republic of Vietnam
Cong-Hoa Xa Hoi Chu Nghia Viet Nam

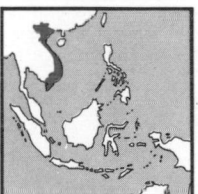

Location South-east Asia
With a coastline to the east and south on the South China Sea, China is to the north, Laos and Cambodia are to the west. Forms the south-east corner of mainland Asia
Land Area 329 556 km² = 127 242 mi²
Climate Tropical
Weather at Hanoi, 16 m altitude
Temperature: hottest month June 26 33 °C, coldest Jan 13–20 °C
Rainfall (av monthly): driest month Jan 18 mm, wettest Aug 343 mm
Time 7 hours ahead of GMT
Measures Metric system
Monetary unit New dong (D) = 10 chao = 100 sau
A new unified dong was introduced May 3, 1978 to replace the former separate currencies of the dong (North Vietnam) and new piastre (South Vietnam), on a one-for-one basis
Rate of exchange (1976 av): par D 2.13087 = SDR 1, free D 2.46 = $ 1, D 4.443 = £ 1

Summary

Political Communist republic, which declared its independence from France in 1945. There was conflict with French forces after declaration of independence; by the Geneva Agreements of July 21, 1954 the conflict was ended, and South Vietnam became a separate state (to the south of the 17th parallel) from December 1954. There was conflict between North and South Vietnam aimed at reunification; after years of war the United States withdrew from the fighting, and the North occupied the whole of South Vietnam in 1975. Complete reunification was announced officially on July 2, 1976, under the new name The Socialist Republic of Vietnam. Member of UN and Comecon

Vietnam

Economic Mainly agricultural but with significant coal reserves and a range of other minerals. Since the reunification of 1975 there has been concentration on the reconstruction of industry; plans include steel plant and hydro-electric schemes. Foreign investment is necessary for this development

People, resources and equipment

Population 1960 30.20*mn, 1970 39.19*mn, 1976 46.52*mn
Growth: 1960–70 2.6*%pa, 1970–76 2.9*%pa
Density (1976): 141* people per km²
Vital statistics (rate per 1 000 people, 1976): births 37*, deaths 15*
Regions (population in 000, 1976; total of 47.15 mn)

Cities					
Ho Chi Minh City	3 460	Ha Bac	1 466	Kien Giang	834
Hanoi	1 443	Thai Binh	1 416	Long An	829
Haiphong	1 191	Quang Nam-Da Nang	1 414	Bac Thai	753
Provinces				Quang Ninh	702
Nghe Tinh	2 705	An Giang	1 362	Ha Tuyen	686
Ha Nam Ninh	2 575	Cuu Long	1 319	Hoang Lien Son	677
Thanh Hoa	2 262	Dong Nai	1 260	Tay Ninh	626
Ha Son Binh	2 042	Tien Giang	1 137	Song Be	561
Hai Hung	1 930	Phu Khanh	1 066	Gia Lai-Kontum	465
Hau Giang	1 870	Dong Thap	991	Son La	410
Nghia Binh	1 789	Minh Hai	981	Dac Lac	373
Binh Tri Thien	1 752	Ben Tre	932	Lam Dong	343
Vinh Phu	1 580	Cao Lang	844	Lai Chau	266
		Thuan Hai	837		

Cities (population in 000, 1976)

Hanoi (capital)	1 443	Nha Trang	216[b]	My Tho	120[b]
Ho Chi Minh City[a]	3 460	Qui Nhon	214[b]	Cam Ranh	118[b]
Haiphong	1 191	Hue	209[b]	Vung Tou	108[c]
Da Nang	492[b]	Can Tho	182[b]	Da Lat	105[b]

[a]Formerly Saigon [b]1973 [c]1971
Race (1976) Vietnamese 95** %, Chinese 3** %
Language Vietnamese
Religion Mainly Buddhist; also pagan Taoist and Confucian
Education (1975/76) Pupils 10 200 000**, teachers 280 000**
Labour force (1976) 20 749 000; in agriculture 15 151 000 (73 %)
Personnel (1972) Physicians: 17 500**, 1 per 2 400** people
Standard of living
National income per person (1976): D 400*** = $ 160*** = £ 90***
Consumption per person (1973): energy 200** kg coal equivalent, electricity (production, 1976) 54**kW h, newsprint 0.5** kg, steel 4**kg
Livestock (000, 1976) Cattle 1 850*, buffaloes 2 260*, pigs 11 500*, chickens 56 000*, ducks 36 000*
Mineral reserves Coal (1952) 1 000 mn tonnes
Electrical capacity (1976) 1 100** megawatts
Hospital beds (1972) 70 000**, 1 per 600** people
Roads (1976) 60 000**km = 37 000**mi, density 0.18**km per km²
Railways (1977) 4 230*km = 2 630*mi, density 0.013*km per km²
Inland waterways (1976) 6 000**km = 3 700**mi
Ships (registered, 1977) 69, total of 128 525 gross tons
Ports Haiphong, Ho Chi Minh City, Da Nang, Hon Gai
Airports Hanoi, Ho Chi Minh City

Durable equipment (Dec 1976)	000	no per 1 000 people	
Radio sets	5 000**	100**	no per
Television sets	2 000**	40**	km of road
Passenger cars	100***	2***	2***
Commercial vehicles	200***	4***	3***

Production

Gross domestic product
1976 est: D 20 000*** mn = $ 8 000*** mn = £ 4 500*** mn

Production index (1970 = 100)	1960	1970	1976	Growth %pa 1960–70	1970–76
Agricultural	na	100	107	na	1.2

Main products (000 t)
Agriculture					
Rice	9 167	10 216	10 800**	1.1	0.9**
Maize	242	281**	320**	1.5**	2.2**
Sweet potatoes	711	1 120**	1 200**	4.6**	1.2**
Cassava	655	950**	1 150**	3.8**	3.2**
Dry beans	20*	26**	27**	4.5**	0.6**
Onions	80**	107**	110**	5.0**	0.5**
Soyabeans	15**	26**	33**	5.7**	4.0**
Groundnuts	50	78**	95**	4.5**	3.3**
Bananas	380**	400**	480**	0.5**	3.1**
Pineapples	60**	40**	34**	−4.0**	−2.7**

Main products (000 t)	1960	1970	1976	Growth %pa 1960–70	1970–76
Agriculture					
Copra	26	21**	23**	−2.1**	1.5**
Coffee	3**	6**	8**	7.2**	4.9**
Tea	8**	8**	9**	0.0**	2.0**
Tobacco	10**	12**	12*	1.8**	0.0**
Rubber	83**	37**	20**	−7.8**	−9.7**
Beef and buffalo meat	65**	91**	95**	3.4**	0.7**
Pigmeat	230**	350**	413**	4.3**	2.8**
Fish catch	472	817**	1 010**	5.6*	3.6**
Timber (000 m³)	13 800**	17 726**	18 832**	2.5**	1.0**
Other					
Coal	2 622	3 000**	4 250**	1.3**	6.0**
Electricity (mn kW h)	562	2 060**	2 500**	13.9**	3.3**
Phosphate rock	541	455**	1 400**[a]	−1.7**	25.2**[b]
Salt	261	320**	350**[a]	2.0**	1.8**[b]
Cement	408	800**	700**	7.0**	−2.2**

[a]1975 [b]1970-75

Transport traffic *Rail* (1960) Passenger-km 1 023 mn, cargo 809 mn t-km (of which, former South Vietnam: passenger-km 542 mn, cargo 141 mn t-km) *Air* (former South Vietnam only, 1975) passenger-km 120 mn, cargo 1.0 mn t-km *Sea* (1974) Goods loaded 0.91* mn t, unloaded 4.13* mn t (of which, former South Vietnam: goods loaded 0.16* mn t, unloaded 3.48* mn t)
Tourism (former South Vietnam only, 1973) Number of visitors 79 000, gross receipts $ 15 mn

Finance and trade

Budget (1977) Balanced at D 8 950mn = $ 3 440mn = £ 1 970mn
External trade (1976)
Imports: D 2 000**mn = $ 850**mn = £ 450**mn
Exports: D 600**mn = $ 250**mn = £ 130**mn

Main imports	% of total	Main exports	% of total
Machinery and transport equipment	30***	Coal	20***
Cereals	10***	Rubber	15***
Petroleum products	10***	Fish	10***
Main sources		Main destinations	
Soviet Union	35**	Soviet Union	35**
Japan	20**	Japan	20**
China	20***	China	20***
Australia	5**	Singapore	10**
Hongkong	4**	Hongkong	10**
France	4**	Hungary	5***
Hungary	4***		

Yemen, North

Yemen Arab Republic
Al Jumhuriyah al Arabiyah al Yamaniyah

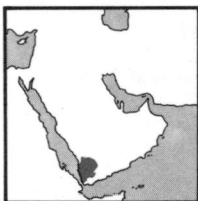

Location Middle East
In the south-west of the Arabian Peninsula, with a coastline on the Red Sea; Saudi Arabia is to the north and South Yemen to the south. The border is ill-defined to the north-east
Land Area 195 000*km² = 75 300* mi²
Climate Hot and humid
Time 3 hours ahead of GMT
Measures Local measures are mainly in use
Monetary unit Yemen rial (Y R) = 100 fils
A new unit, the fils, replaced the former buqsha from 1975 at 2.5 fils = 1 buqsha (formerly 1 rial = 40 buqshas)
Rate of exchange (1976 av): par Y R 4.5625 = $ 1, free Y R 8.241 = £ 1

Summary

Political Republic, which had been a monarchy until 1962. Also referred to as Yemen, Sana'a. After the independence of South Yemen in 1967, there was conflict between the North and South Yemens; an agreement was signed in 1972 for eventual unification, but had no immediate effect. Member of UN and Arab League
Economic An agricultural, especially pastoral economy, with cotton, coffee, and hides and skins the main exports. There is little industry. The economy is dependent on foreign aid and on remittances from Yemeni workers in other countries (especially Saudi Arabia); net inflow of transfers was $ 780 mn in 1976

Yemen, North

People, resources and equipment

Population 1960 4.43*mn, 1970 5.77*mn, 1976 6.87*mn
Growth: 1960–70 2.7*%pa, 1970–76 3.0*%pa
Density (1976): 35* people per km²
Vital statistics (rate per 1 000 people, 1970–75): births 50*, deaths 21*
Cities (population in 000, 1975) Sana'a 135*, Al Hudaydah[a] 148*, Ta'izz 80*
[a]Hodeida
Race (1976) Arab 99**%
Language Arabic
Religion (1976) Moslem 99**%
Education Pupils (1976/77) 248 508, teachers (1974/75) 6 850
Labour force (1976) 1 931 000; in agriculture 1 482 000 (77 %).
Yemenis working outside the country (1976): 800 000**
Personnel (1974) Physicians: 245, 1 per 26 450 people
Standard of living
National income per person (1976): Y R 900** = $ 200** = £ 110**
Consumption per person (1975): energy 49 kg coal equivalent,
electricity 7 kW h
Newspapers (1970): number 6; circulation 56 000, 10 per 1 000 people
Telephones (Dec 1971): 5 000, 1 per 1 000 people
Livestock (000, 1976) Cattle 1 000*, sheep 3 200*, goats 7 400*, camels 120*
Electrical capacity (1975) 14 megawatts
Hospital beds (1972) 4 200, 1 per 1 443 people
Roads (1975) 3 952 km = 2 456 mi, density 0.020 km per km²
There are no railways
Ships (registered, 1977) 4, total of 1 436 gross tons
Ports Hodeida, Al Mukha, Al Luhayyah
Airports (passengers, 000, 1976) Sana'a: departures 110, arrivals 141;
also Hodeida and Ta'izz
Durable equipment Radio sets (Dec 1975): 87 000, 13 per 1 000 people
Passenger cars (Dec 1973): 5 600, 0.9 per 1 000 people, 1.5 per km of road
Commercial vehicles (Dec 1973): 6 960, 1.1 per 1 000 people, 1.8 per km
of road

Production

Gross domestic product 1974: Y R 5 092 mn = $ 1 113 mn = £ 476 mn
1976 est: Y R 6 800**mn = $ 1 500**mn = £ 800**mn
Structure of gross domestic product *By origin* (1974) Agriculture 63 %,
manufacturing 2 %, construction 3 %, other 32 %
By type (1973) Final consumption expenditure 107 %,
stock investment 2 %, gross fixed capital formation 10 %,
exports of goods and services 3 %, less imports of goods and services − 22 %

Production index (1970 = 100)	1960	1970	1976	Growth %pa 1960–70	1970–76
Agricultural	95	100	142	0.5	6.1
Main products (000 t)					
Agriculture					
Wheat	20*	26	90	2.7*	23.0
Barley	140*	150*	80	0.7*	−9.9*
Maize	9*	70**	104	22.8**	6.8**
Sorghum	940*	980*	859	0.4*	−2.2*
Potatoes	18*	20*	86	1.1*	27.5*
Grapes	12*	15**	42	2.2**	18.7**
Cottonseed	3.8**	1.2	14.0*	−10.9**	50.6*
Dates	60*	60*	70	0.9*	2.6*
Coffee	5.4*	4.0*	5.0*	−3.0*	3.8*
Tobacco	2*	2	6*	0.0*	20.0*
Cotton	2.0*	0.7	10.0*	−10.0*	55.8*
Milk, cow	60**	49*	60*	−2.0**	3.4*
Milk, sheep	40**	37*	49*	−0.8**	4.8*
Milk, goat	120**	97*	128*	−2.1**	4.7*
Sheep and goat meat	45**	46**	49*	0.2*	1.1**
Sheep and goat skins	7.8**	7.8*	8.2*	0.0**	0.8**
Fish catch	7.5*	7.6*	9.7*	0.1*	4.1*
Other					
Electricity (mn kW h)	13**	18	49[a]	3.3**	22.2[b]
Salt	na	20**	na	na	na

[a]1975 [b]1970–75

Transport traffic *Air* (1976) Passenger-km 237 mn, cargo 0.4 mn t-km
Sea (1974) Goods loaded 23 000 t, unloaded 578 000 t

Finance and trade

Consumer price index (1972 = 100) 1976 261; growth 1972–76 27.1 %pa
Money stock (end-year, Y R mn) 1973 541, 1976 2 787;
growth 1973–76 72.7 % pa
Budget (1976/77; year ending June 30th)
Revenue: Y R 836 mn = $ 183 mn = £ 107 mn
Expenditure: Y R 1 198 mn = $ 263 mn = £ 153 mn

Balance of payments ($ mn)	1974	1975	1976
Balance of goods (fob)	−182	−231	−462
Balance of services	−12	−26	−21
Balance of transfers	+189	+386	+780
Current balance	−5	+129	+296
Long-term capital flow	+43	+18	+48
International reserves (end-year, $ mn)	199	338	720

External trade (1976) Imports: Y R 1 882 mn = $ 412 mn = £ 228 mn
Exports: Y R 35 mn = $ 8 mn = £ 4 mn

Main imports (1975)	% of total	*Main exports* (1975)	% of total
Food	43	Cotton	54
(of which, cereals 17,		Coffee	13
sugar 14)		Hides and skins	13
Machinery	9	Cottonseed	5
Textile yarns and fabrics	8	Cotton fabrics	4
Motor vehicles	7		
Chemicals	6		
Petroleum products	5		

Main sources (1976)		*Main destinations* (1976)	
Saudi Arabia	12	China	33
Japan	10	South Yemen	27
India	7	Italy	18
Australia	7	Saudi Arabia	16
United Kingdom	6		
China	6		
Netherlands	6		
Singapore	4		
Djibouti	4		
West Germany	4		

Yemen, South

People's Democratic Republic of Yemen
Jumhuriyat al Yaman ad Dimukratiyah ash Sha'biyah

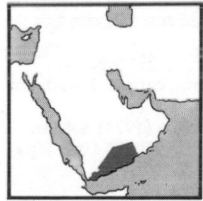

Location Middle East
At the south-west corner of the Arabian
Peninsula, with a coastline on the Arabian Sea;
Oman is to the east and North Yemen and
Saudi Arabia to the north. The borders with
North Yemen and Saudi Arabia are not clearly
defined. The islands of Kamaran and Perim are
part of the territory
Land Area 333 038km² = 128 587mi²
Climate Hot and dry
Weather at Khormaksar (Aden), 7 m altitude
Temperature: hottest month June 29–37 °C, coldest Jan 22–28 °C
Rainfall (av monthly): driest months April–June 1 mm, wettest Jan 5 mm
Time 3 hours ahead of GMT
Measures UK (imperial) system; local measures are also in use, including:
weight (mass) 1 seer = 80 tola = 0.933 kilogram = 2.057 pounds
1 local maund = 40 seer = 37.32 kilograms = 82.29 pounds
1 gadah = 90.72 kilograms = 200 pounds
Monetary unit Yemeni dinar (Y D) = 1 000 fils
Rate of exchange (1976 av): par Y D 0.345395 = $ 1 ($ 2.895 = Y D 1),
free £1.603 = Y D 1

Summary

Political People's republic; formerly the British Aden colony and
protectorate, it became independent on November 30, 1967. Also referred
to as Yemen, Aden. After independence, there was conflict with
North Yemen; an agreement was signed in 1972 for eventual unification,
but had no immediate effect. Member of UN and Arab League
Economic An agricultural economy, dependent on petroleum products for
exports, refined from imported crude oil. Fishing is significant. Tourism
has fallen since independence, and the port of Aden was affected by closure
of the Suez Canal; re-opening is expected to provide a recovery from 1975.
Aid from Saudi Arabia is expected to increase development plans, but
depends on the political situation. The Soviet Union has supplied oil for
refining; Aden is a suitable base for supplying Soviet fleets

Yemen, South

People, resources and equipment

Population 1960 1.00**mn, 1970 1.44*mn, 1976 1.75*mn
Growth: 1960–70 3.7** %pa, 1970–76 3.3* %pa
Density (1976): 5* people per km²
Vital statistics (rate per 1 000 people, 1970–75): births 50*, deaths 21*
Cities (population in 000, 1973) Aden (capital) 285, Shaykh Uthman 35*, Al Mukalla 25*
Race (1955) Arab 75 %, Indian 11 %, Somali 8 %
Language Arabic
Religion (1955) Moslem 91 %, Christian 4 %, Hindu 3 %
Education (1974/75) Pupils 237 096, teachers 8 500*
Labour force (1976) 450 000; in agriculture 275 000 (61 %)
Personnel (1969) Physicians: 103, 1 per 13 600 people
Standard of living
National income per person (1976): Y D 80*** = $ 230*** = £ 130***
Consumption per person (1975): energy 328 kg coal equivalent, electricity (production) 107* kW h, newsprint 0.1kg
Newspapers (1972): number 3; circulation 2 000, 1 per 1 000 people
Telephones (Dec 1973): number 10 000, 6 per 1 000 people
Livestock (000, 1976) Cattle 102*, sheep 930*, goats 1 230*, camels 40*
Petroleum refinery capacity (1975) 8.9* mn tonnes
Electrical capacity (Aden only, 1975) 70* megawatts
Hospital beds (1973) 2 340, 1 per 670 people
Roads (1976) 10 494 km = 6 521 mi, density 0.032 km per km²
There are no railways
Ships (registered, 1977) 16, total of 6 390 gross tons
Port Aden
Airport Khormaksar (Aden)

Durable equipment (at end-year)	000	no per 1 000 people	
Radio sets (1973)	520**	320**	no per
Television sets (1975)	31	18	km of road
Passenger cars (1973)	11	7	1.0
Commercial vehicles (1973)	8	5	0.8

Production, finance and trade

Gross domestic product 1970: Y D 59.2 mn = $ 142 mn = £ 59 mn
1976 est: Y D 150***mn = $ 430***mn = £ 240***mn
Agricultural production index (1970 = 100) 1960 101, 1976 119; growth 1960–70 −0.1 %pa, 1970–76 2.9 %pa
Main products *Agriculture* (000 t, 1976) Wheat 12*, millet and sorghum 65*, watermelons 38*, cottonseed 11*, dates 42*, coffee 2*, tobacco 1*, cotton 3*, goat milk 24*, fish catch 127*
Other (1975) Electricity 180* mn kW h, petroleum products 1.2* mn t, salt 75 000 t
Growth in production of petroleum products: 1970–75 −29* %pa
Transport traffic *Air* (1976) Passenger-kilometres 70 mn, cargo 1.3 mn t-km
Sea (1974) Goods loaded 2.31 mn t, unloaded 3.78 mn t
Tourism Number of visitors (1976) 17 996, gross receipts (1974) $ 4 mn
Consumer price index (1970 = 100) 1976 184.4; growth 1970–76 10.7 %pa
Money stock (end-year, Y D mn) 1970 29.03, 1976 81.51; growth 1970–76 18.8 %pa
Budget (1975/76; year ending March 31st)
Revenue: Y D 13.86 mn = $ 40 mn = £ 19 mn
Expenditure: Y D 25.55 mn = $ 74 mn = £ 35 mn
External trade (1976) Imports: Y D 143 mn = $ 414 mn = £ 229 mn
Exports: Y D 86 mn = $ 249 mn = £ 138 mn

Main imports (1975)	% of total	*Main exports* (1975)	% of total
Crude oil	41	Petroleum products	93
Food	20	Fish	3
(of which, cereals 8, sugar 5)			
Petroleum products	8		
Machinery	6		
Chemicals	3		
Textile yarns and fabrics	2		
Main sources (1976)		*Main destinations* (1976)	
Kuwait	13*	Canada	70*
Japan	10*	Japan	9*
United Kingdom	6*	Angola	4*
Netherlands	3*	North Yemen	4*
Italy	3*	Australia	3*
France	3*	China	2*

Europe

Albania

People's Socialist Republic of Albania
Republika Popullore Socialiste e Shqipërisë

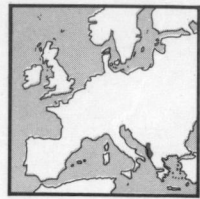

Location South-east Europe
With a coastline on the Adriatic Sea, Jugoslavia is to the north and east and Greece to the south
Land Area 28 748 km² = 11 100 mi²
Climate Mediterranean
Weather at Tirana, 89 m altitude
Temperature: hottest month August 17–31 °C, coldest Jan 2–12 °C
Rainfall (av monthly): driest months July, Aug 32 mm, wettest Nov 211 mm
Time 1 hour ahead of GMT (summer time, 2 hours ahead)
Measures Metric system
Monetary unit Lek (Lk) = 100 quintars (qindarka); a new lek replaced the old lek from August 1965 at 1 new lek = 10 old leks
Rate of exchange (1976 av) *Basic* par Lk 4.10 = $ 1, free Lk 7.40* = £ 1
Non-commercial par Lk 8.20 = $ 1, free Lk 14.80* = £ 1

Summary

Political Communist republic, which was proclaimed as a republic on January 11, 1946; independence from Turkey had been declared November 28, 1912, and liberation from occupation by Germany and Italy in the second world war was on November 29, 1944. Diplomatic relations with Soviet Union were broken off in 1961. Member of UN
Economic Mainly agricultural, but with important mining (especially crude oil and chromium) and manufacturing sectors; petroleum refining is significant. Chinese aid was provided for development in the late 1960's and up to 1977

People, resources and equipment

Population 1960 1.61mn, 1970 2.14 mn, 1976 2.55 mn
Growth: 1960–70 2.9 %pa, 1970–76 3.0 %pa
Density (1976): 89 people per km²
Vital statistics (rate per 1 000 people, 1971): births 33.3, deaths 8.1
Cities (population in 000, 1975)

Tiranë[a] (capital)	192	Vlorë	55	Berat	30
Shkodër	62	Elbasan	53	Fier	28
Durrës	60	Korcë	52	Gjirokastër	22

[a]Tirana

Race (1961) Albanian 95 %, Greek 2 %, Rumanian 1 %, Macedonian 1 %, Gypsy 1 %
Language Albanian
Religion Atheist; position at 1965 was: Moslem 69** %, Greek Orthodox 21** %, Roman Catholic 10** %
Education (1971/72) Pupils 632 111, teachers 24 738
Labour force (1976) 1 086 000*; in agriculture 681 000* (63* %)
Personnel (1972) Physicians: 14 371, 1 per 159 people
Standard of living
National income per person (1976): Lk 2 000*** = $ 490*** = £ 270***
Consumption per person (1975): energy 741 kg coal equivalent, electricity (production) 730* kW h, steel 48 kg
Newspapers (1974): number 2; circulation 115 000, 48 per 1 000 people
Telephones (1965): 13 991, 7 per 1 000 people
Livestock (000, 1976) Cattle 470*, sheep 1 163*, goats 674*, pigs 117*
Mineral reserves (1975) Crude oil 22 mn tonnes
Natural gas 12 bn cubic metres
Petroleum refinery capacity (1975) 2.1*mn tonnes
Electrical capacity (1973) 600** megawatts
Hospital beds (1969) 12 715, 1 per 164 people
Roads (1971) 5 000*km = 3 000*mi, density 0.17* km per km²
Railways (1974) 300* km = 186* mi, density 0.010* km per km²
Ships (registered, 1977) 20, total of 55 870 gross tons
Ports Durrës, Vlorë, Sarandë, Shëngjin
Airport Rinas (Tirana)

Durable equipment (at end-year)	000	no per 1 000 people	
Radio sets (1975)	175	71	no per
Television sets (1975)	4.5	2	km of road
Passenger cars (1970)	4*	2*	1*
Commercial vehicles (1970)	11*	5*	2*

Production

Gross domestic product
1976 est: Lk 5 300***mn = $ 1 300***mn = £ 720***mn

Production index (1970 = 100)	1960	1970	1976	Growth %pa 1960–70	1970–76
Agricultural	62[a]	100	127	4.9[a]	4.1

Main products (000 t)	1960	1970	1976	1960–70	1970–76
Agriculture					
Wheat	64	227	390*	13.5	9.4*
Maize	129	259	280*	7.2	1.3*
Potatoes	23	107	122*	16.6	2.2*
Grapes	34**	55*	69*	4.9**	3.8*
Olives	23**	26*	47*	1.2**	10.3*
Tobacco	7.1	12.2	14.0*	5.6	2.3*
Cotton	5	5*	7*	0.0*	5.7*
Milk	74	147	219*	7.1	6.9*
Meat	40**	54*	55*	3.0**	0.3*
Timber (000 m³)	1 500**	2 430*	2 330*	4.9**	−0.7*
Other					
Total energy (000 tce)	1 110	2 420	4 170[d]	8.1	11.5[e]
Lignite	291	606	850*[d]	7.6	7.0*[e]
Crude oil	728	1 487	2 300*[d]	7.4	9.1*[e]
Petroleum products	332	1 235	1 800*[d]	14.0	7.8*[e]
Natural gas (mn m³)	na	98	155*[d]	na	9.6*[e]
Electricity (mn kW h)	194	944	1 800*[d]	17.1	13.8*[e]
Chromium ore (oxide content)	116*	200*	320*[d]	5.6*	9.8*[e]
Copper ore (Cu content)	2.2*	5.6*	7.0*[d]	9.8*	4.6*[e]
Nickel ore (Ni content)	2.4*	5.4*	6.0*[d]	8.4*	2.1*[e]
Beer (000 hl)	69	116	144[f]	5.3	7.5[g]
Cigarettes (mn units)	4 222[b]	4 900*	5 700*[d]	2.1*[c]	3.1*[e]
Fertilisers, nitrogenous	—	28*	36*[d]	na	5.1*[e]
Fertilisers, phosphate	—	18*	25*[d]	na	6.8*[e]
Cement	73	350**	517*[d]	17.0**	8.1*[e]

[a]Agricultural production was some 10 % lower in 1960 than in the two years before and after; this is equivalent to higher growth of about 1 %pa over 1960–70 [b]1963 [c]1963–70 [d]1975 [e]1970–75 [f]1973 [g]1970–73
Transport traffic *Rail* (1971) Passenger-km 291 mn, cargo 188 mn t-km
Sea (1974) Goods loaded 2.7*mn t, unloaded 0.7*mn t

Finance and trade

Budget (1976) Revenue: Lk 7 300 mn = $ 1 780 mn = £ 986 mn
Expenditure: Lk 6 300 mn = $ 1 537 mn = £ 851 mn
External trade (1976)
Imports: Lk 1 000***mn = $ 250***mn = £ 130***mn
Exports: Lk 800***mn = $ 200***mn = £ 100***mn

Main imports (1964)[a]	% of total	Main exports (1964)[a]	% of total
Machinery and transport equipment	50	Crude oil, petroleum products and metals (especially chrome)	54
Metals and minerals	15	Food products	23
Industrial raw materials	12	Food raw materials	17
Food	9	Consumer goods (excluding food)	5
Chemicals	7		

Main sources (1964)[a]		Main destinations (1964)[a]	
China	63	China	40
Czechoslovakia	10	Czechoslovakia	19
Poland	8	East Germany	10
Italy	3	Poland	10
Rumania	2	Rumania	4
Hungary	2	Italy	3
Jugoslavia	1	France	3

[a]Latest detailed figures available

Andorra

The Valleys of Andorra
Les Valls d'Andorra

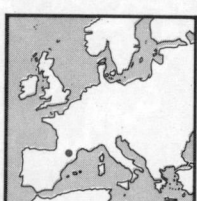

Location South-west Europe
In the eastern Pyrenees, between France and Spain. Land-locked
Land Area 453 km² = 175 mi²
Climate Temperate
Weather at Les Escaldes, 1 080 m altitude
Temperature: hottest month July 12–26 °C, coldest Jan minus 1–6 °C
Rainfall (av monthly): driest month Jan 34 mm. wettest May 105 mm
Time 1 hour ahead of GMT
Measures Metric system

Andorra

Monetary unit French franc and Spanish peseta
Rate of exchange (1976 av): free Fr 4.780 = $ 1, Fr 8.633 = £ 1
free Pa 66.90 = $ 1, Pa 120.8 = £ 1

Summary

Political Co-principality, self-governing under joint French and Spanish suzerainty
Economic Tourism (including skiing) is the main industry; duty free trading is important. There is some mining and timber industry, with agriculture in the valleys; tobacco growing and livestock raising are significant

People, resources and equipment

Population 1932 4 039, 1960 8 392, 1970 21 100*, 1976 28 300*
Growth: 1932–60 2.6 %pa, 1960–70 9.6* %pa, 1970–76 5.0* %pa
Density (1976): 62* people per km²
Vital statistics (rate per 1 000 people, 1976): births 16.5, deaths 5.0
Towns (population in 000, 1974) Andorra la Vella (capital) 9, Les Escaldes 7*
Race (1971; total of 23 092 people) Andorran citizens 6 829 (30 %), Spanish 14 035 (61 %), French 1 484 (6 %)
Language Catalan; also Spanish and French
Religion Mainly Roman Catholic
Education (1969/70) Pupils 2 654, teachers 111
National income per person (1974):
Fr 24 000*** = Pa 330 000*** = $ 5 000*** = £ 2 800***
Newspapers (1976): Number 1; circulation 4 000, 141 per 1 000 people
Telephones (Dec 1974) 3 356, 129 per 1 000 people
Livestock (000, 1976) Cattle 2*, sheep 20*ª, horses 1*
ªExcludes 15* thousand foreign sheep using Andorran pastures
Roads (1976) 100**km = 60**mi, density 0.2**km per km²
Airports None (nearest main airport: Barcelona in Spain)
Radio sets (Dec 1975) 6 600, 244 per 1 000 people
Television sets (Dec 1969) 1 700, 85 per 1 000 people
Motor vehicles (Dec 1976) 15 500**, 530** per 1 000 people, 150*** per km of road

Production, finance and trade

Gross domestic product
1976 est: Fr 720***mn = Pa 10 000***mn = $ 150***mn = £ 80***mn
Main products Tobacco, rye, wheat, potatoes, wool, timber, iron, lead
Tourism (1976) Number of visitors 6 450 000
Budget (1976) Revenue: Pa 454 mn = $ 6.8 mn = £ 3.8 mn
Expenditure: Pa 357 mn = $ 5.3 mn = £ 3.0 mn
External trade (1976) Imports: $ 156 mnª = £ 86 mnª
Exports: $ 6 mnª = £ 3 mnª
Import sources From France: Fr 473 mn = $ 99 mn = £ 55 mn
Spain: Pa 3 835 mn = $ 57 mn = £ 32 mn
Export destinations To France: Fr 18 mn = $ 4 mn = £ 2.1 mn
Spain: Pa 134 mn = $ 2 mn = £ 1.1 mn
ªTrade with France and Spain only

Austria

Republic of Austria
Republik Österreich

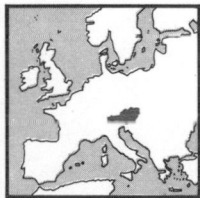

Location Central Europe
West Germany is to the north-west, Czechoslovakia to the north-east, Hungary to the east, Jugoslavia and Italy to the south and Switzerland to the west; Liechtenstein is between Austria and Switzerland. Land-locked but with access to the sea via the Danube river
Land Area 83 849 km² = 32 374 mi²
Usage (1975): arable, cropland and pastures 37 900 km² (45 %); forests 32 500 km² (39 %)

Climate Temperate
Weather at Vienna, 203 m altitude
Temperature: hottest month July 15–25 °C, coldest Jan minus 4–1 °C
Rainfall (av monthly): driest month Jan 39 mm, wettest July 84 mm
Time 1 hour ahead of GMT
Measures Metric system
Monetary unit Schilling (Sch) = 100 groschen
Rate of exchange (1976 av): free Sch 17.94 = $ 1, Sch 32.40 = £ 1

Summary

Political Republic, which was occupied by Germany from 1938 to 1945; full independence was obtained July 27, 1955, when the post-war occupation by United States, Soviet Union, United Kingdom and France was formally ended. Member of UN, OECD, Council of Europe and Efta
Economic Mainly industrial, with machinery, iron and steel and chemicals major exports; manufacturing makes up one-third of gross domestic product compared with 5 % from agriculture. Timber and paper are main products and account for about one-eighth of exports. Tourism is important, with gross receipts equal to about one-third of receipts from exports. West Germany is the main trading partner and two-thirds of tourists come from that country

People, resources and equipment

Population 1960 7.05 mn, 1970 7.43 mn, 1976 7.51 mn
Growth: 1960–70 0.5 %pa, 1970–76 0.2 %pa
Density (1976): 90 people per km²
Vital statistics (rate per 1 000 people, 1976): births 11.6, deaths 12.6
Households (1971) 2 535 916; population in households 7.36 mn (98.7 % of total)
Regions (provinces, population in 000, 1971; total of 7.46 mn)

Burgenland	272	Oberösterreich	1 223	Tirol	541
Kärnten	526	Salzburg	402	Vorarlberg	271
Niederösterreich	1 414	Steiermark	1 192	Wien	1 615

Cities (population in 000, 1971)

Wienª (capital)	1 615	Linz	203	Innsbruck	115
Graz	248	Salzburg	129	Klagenfurt	74

ªVienna
Race (1961) Austrian 99 %, German 1 %
Language German
Usage (1971): German 98 %, Serbo-Croatian ½ %
Religion (1971) Christian 94 % (of whom, Roman Catholic 88 %)
Education (1975/76) Pupils: primary 977 825, secondary 171 950, vocational 172 756, teacher-training 16 497, higher 82 600. Teachers: primary 54 922, secondary 12 650, vocational 3 704, teacher-training 1 544, higher 10 001
Labour force (1975)

Economic activity	Number	% of total
Agriculture	370 000	12
Mining and quarrying	24 000	1
Manufacturing	901 000	30
Electricity, gas and water	34 000	1
Construction	255 000	9
Distribution and hotels	476 000	16
Transport and communication	203 000	7
Finance and real estate	138 000	5
Other	568 000	19
Total	2 969 000	100

Personnel Scientists and engineers (1971): 118 294
Physicians (1975): 15 702, 1 per 479 people
Standard of living
National income per person (1976): Sch 86 525 = $ 4 823 = £ 2 671
Consumption per person (1975): energy 3 700 kg coal equivalent, electricity (production, 1976) 4 665 kW h, newsprint 16 kg, steel 286 kg
Newspapers (1974): number 30; circulation 2 316 000, 308 per 1 000 people
Telephones (Dec 1976): 2 281 250, 303 per 1 000 people
Livestock (000, Dec 1975) Cattle 2 500, pigs 3 683, sheep 169, goats 43, chickens 12 959
Mineral reserves (1975) Coal 10 mn tonnes
Lignite 365 mn tonnes
Crude oil 25 mn tonnes
Natural gas 16 bn cubic metres
Uranium (1976) 1 800 tonnes
Petroleum refinery capacity (1975) 11.5 mn tonnes
Electrical capacity (1975) 10 001 megawatts, of which, hydro 6 085 megawatts
Hospital beds (1975) 85 461, 1 per 88 people
Roads (1976) 102 858 km = 63 913 mi, density 1.2 km per km²
Railways (1975) 6 494 km = 4 035 mi, density 0.077 km per km²
Inland waterways (1975) 358 km = 222 mi, of which, Danube river: 351 km
Oil pipelines (1975) 597 km = 371 mi
Ships (registered, 1977) 11, total of 53 284 gross tons
Airports (passengers, 000, 1976) Vienna: departures 1 140, arrivals 1 139; also Graz, Linz, Salzburg, Innsbruck, Klagenfurt

Durable equipment (Dec 1976)	000	no per 1 000 people	
Radio sets	2 191	291	no per km of road
Television sets	1 974	263	
Passenger cars	1 828	243	18
Commercial vehicles	159	21	1.5

Austria

Production

Gross domestic product 1976: Sch 728.7 bn = $ 40 619 mn = £ 22 491 mn
Growth in real terms: 1960–70 4.7 %pa, 1970–76 4.1 %pa
Structure of gross domestic product (1976)

By origin	Sch bn	% of gdp
Agriculture	36.9	5
Manufacturing, mining and quarrying	230.1	32
Electricity, gas and water	22.5	3
Construction	68.8	9
Other	370.4	51
Total	728.7	100
By type of expenditure		
Government final consumption	120.0	16
Private final consumption	410.0	56
Stock investment	25.1	3
Gross fixed capital formation	189.3	26
Exports of goods and services	249.1	34
less Imports of goods and services	−264.7	−36
Total	728.7	100

Production indices (1970 = 100)	1960	1970	1976	Growth %pa 1960–70	1970–76
Agricultural	87	100	111	1.4	1.7
Industrial	59	100	125	5.4	3.8
Main products (000 t)					
Agriculture					
Wheat	702	810	1 234	1.4	7.3
Barley	589	913	1 287	4.5	5.9
Maize	213	612	936	11.1	7.3
Rye	353*	363	410	0.3*	2.0
Oats	343	272	283	−2.3	0.7
Potatoes	3 809	2 704	1 746	−3.4	−7.0
Sugar, raw value	323	331	416	0.2	3.9
Apples	702	309	304	−7.9	−0.3
Pears	433	160	138	−9.5	−2.4
Grapes	115*	440*	385*	14.4*	−2.2*
Wine	81	310	270*	14.4	−2.3*
Milk	2 842	3 229	3 192	1.3	−0.2
Beef and veal	124	182	183	3.9	0.1
Pigmeat	263	266	307	0.1	2.4
Timber (000 m³)	12 100	11 813	13 175	−0.2	1.8
Energy					
Total energy (000 tce)	9 740	10 660	10 910[b]	0.9	0.5[c]
Lignite	5 973	3 670	3 184	−4.7	−2.3
Crude oil	2 448	2 798	2 031	1.3	−5.2
Petroleum products	1 892	5 960	7 710*[b]	12.2	5.3*[c]
Natural gas (mn m³)	1 469	1 897	2 144	2.6	2.1
Manufactured gas (mn m³)	na	1 860	1 086	na	−8.6
Electricity (mn kW h)	15 965	30 036	35 037	6.5	2.6
of which,					
hydro (mn kW h)	11 882	21 240	23 745[b]	6.0	2.2[c]
Mining					
Iron ore (Fe content)	1 100	1 304	1 170	1.7	−1.8
Magnesite	1 625	1 609	1 266[b]	−0.1	−4.7[c]
Salt	354	563	527[b]	4.7	−1.3[c]
Manufacturing					
Beer (000 hl)	5 269	7 391	7 757[b]	3.4	1.0[c]
Cigarettes (mn units)	8 809	12 617	13 043[b]	3.7	0.7[c]
Cotton yarn	26.6	20.8	20.5	−2.4	−0.2
Cotton fabrics	18.4	17.9	17.2	−0.3	−0.7
Man-made fibres	58	86	101[b]	4.0	3.3[c]
Wood pulp	706	933	976[b]	2.8	0.9[c]
Newsprint	138	170	141	2.1	−3.0
Other paper	450**	847	1 107[b]	6.5**	5.5[c]
Sulphuric acid	140	289	303[d]	7.5	1.6[e]
Plastics and resins	26	187	404	21.8	13.7
Fertilisers, nitrogenous[a]	162	248	237	4.4	−0.8
Fertilisers, phosphate[a]	28*	91	99*	12.5*	1.4*
Coke	2 046	1 768	1 607[b]	−1.4	−1.9[c]
Cement	2 830	4 806	5 880	5.4	3.4
Pig iron	2 238	2 970	3 338	2.9	2.0
Crude steel	3 163	4 079	4 910	2.6	3.1
Aluminium	68	90	87	2.8	−0.5
Television sets (000 units)	112	369	404[b]	12.7	1.8[c]
Motor vehicles (000 units)	12.6	7.2	8.9	−5.4	3.6
Construction					
Dwellings (000 units)	40	43	46[b]	0.7	1.6[c]

[a]Years ending June 30th   [b]1975   [c]1970–75   [d]1973   [e]1970–73

Transport traffic	1960	1970	1976	Growth %pa 1960–70	1970–76
Passenger-kilometres (mn)					
Rail	6 840	6 478	6 504	−0.6	0.1
Air	66	452	823	21.3	10.5
Cargo: tonne-kilometres (mn)					
Rail	7 934	10 017	10 540	2.4	0.9
Water	995[a]	1 293	1 412[c]	3.8[b]	1.8[d]
Air	0.6	8.6	9.8	30.3	2.2
Tourism					
Number of visitors[e] (000)	4 595	8 867	11 598	6.8	4.6
Gross receipts ($ mn)	232	999	3 146	15.7	21.1

[a]1963   [b]1963–70   [c]1975   [d]1970–75   [e]Arrivals at accommodation

Finance and trade

Price and wage indices (1970 = 100)	1960	1970	1976	Growth %pa 1960–70	1970–76
Consumer prices	70.4	100.0	152.6	3.6	7.3
Wholesale prices	77.3	100.0	143.4	2.6	6.2
Wages (earnings)	46.4	100.0	204.7	8.0	12.7
Share prices	78.8	100	131.9	2.4	4.7
Money stock (end-year, Sch bn)	34.12	70.26	145.20	7.5	12.9

Budget (1976) Revenue: Sch 138 000 mn = $ 7 692 mn = £ 4 259 mn
Expenditure: Sch 171 670 mn = $ 9 569 mn = £ 5 298 mn

Balance of payments ($ mn)	1972	1973	1974	1975	1976
Balance of goods (fob)	−1 178	−1 519	−1 407	−1 418	−2 504
Balance of services	+1 026	+1 275	+1 111	+1 236	+1 124
Balance of transfers	−34	−83	−161	−153	−106
Current balance	−187	−329	−457	−335	−1 486
Long-term capital flow	+80	−222	+402	+1 062	−74
International reserves (end-year, $ mn)	2 718	2 874	3 430	4 439	4 410

External trade (1976) Imports: Sch 205 970 mn = $ 11 481 mn = £ 6 357 mn
Exports: Sch 152 110 mn = $ 8 479 mn = £ 4 695 mn

Main imports	% of total	Main exports	% of total
Machinery, non-electric	12	Machinery, non-electric	14
Motor vehicles	10	Iron and steel	10
Chemicals	9	Electrical machinery	9
Electrical machinery	8	Chemicals	8
Food	7	Textile yarns and fabrics	7
Crude oil	6	Paper and products	6
Textile yarns and fabrics	5	Timber	5
Clothing	4	Food	4
Iron and steel	3		
Petroleum products	3		
Main sources		*Main destinations*	
West Germany	41	West Germany	23
Italy	8	Italy	9
Switzerland	6	Switzerland	7
France	4	United Kingdom	5
United Kingdom	4	Poland	4
Soviet Union	4	Sweden	4
Netherlands	3	Jugoslavia	4
United States	3	Hungary	3

Special focus

Tourism, 1976
(arrivals at accommodation)

Origin of visitors	Number (000)	%
West Germany	7 369	64
Netherlands	857	7
United States	514	4
France	385	3
United Kingdom	384	3
Switzerland	336	3
Belgium-Luxembourg	310	3
Sweden	218	2
Italy	189	2
Denmark	148	1
Jugoslavia	104	1
Other	784	7
Total visitors	11 598	100

Belgium

Kingdom of Belgium
Koninkrijk België
Royaume de Belgique

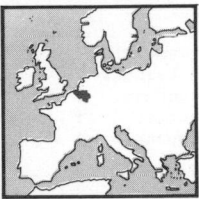

Location Western Europe
With a coastline on the North Sea, Netherlands is to the north-east, West Germany to the east, Luxembourg to the south-east and France to the south
Land Area 30 513 km² = 11 781 mi²
Usage (1973): arable 8 050 km² (26 %,)
pastures 7 330 km² (24 %), forests 6 010 km²
(20 %), cropland 300 km² (1 %)
Climate Temperate
Weather at Brussels, 100 m altitude
Temperature: hottest month July 12–23 °C, coldest Jan minus 1–4 °C
Rainfall (av monthly): driest month Mar 53 mm, wettest July 95 mm
Time 1 hour ahead of GMT (summer time, 2 hours ahead)
Measures Metric system
Monetary unit Belgian franc (B Fr) = 100 centimes
Rate of exchange (1976 av): free B Fr 38.60 = $ 1, B Fr 69.73 = £1,
par (central value within the European 'snake') B Fr 48.6573 = SDR 1

Summary

Political Parliamentary monarchy, established as an independent state in 1830; Flemings (Flemish speaking) and Walloons (French speaking) have officially recognised regional status. Member of UN, OECD, Council of Europe, Nato, EEC, WEU, Benelux and linked with Luxembourg since 1921 in the Belgium-Luxembourg Economic Union
Economic An industrial economy, with manufacturing making up 28 % of gross domestic product compared with only 3 % from agriculture. Iron and steel and chemicals are the main exports of the Belgium-Luxembourg Economic Union, with Belgian steel virtually all produced from imported ore; there are local coal resources. There is a large merchant fleet which contributes to the favourable balance of payments services account. Belgium provides an important assembly and manufacturing base centrally placed within the EEC; Brussels is the seat of the EEC Commission

People, resources and equipment

Population 1960 9.15 mn, 1970 9.66 mn, 1976 9.82 mn
Growth: 1960–70 0.5 %pa, 1970–76 0.3 %pa
Density (1976): 322 people per km²
Vital statistics (rate per 1 000 people, 1976): births 12.3, deaths 12.1
Households (1970) 3 234 228; population in households 9.53 mn (98.7 % of total)
Regions (provinces, population in 000, 1976; total of 9.82 mn)

Anvers	1 563	Liège	1 017	Namur	392
Brabant	2 223	Limbourg	692	Flandre orientale	1 325
Hainaut	1 318	Luxembourg	220	Flandre occidentale	1 073

Cities (population in 000, 1976)

Bruxelles[a] (capital)	1 042	Gent[d]	219[g]	La Louvière	113*[g]
Antwerpen[b]	662[g]	Charleroi	210*[g]	Deurne	81[g]
Liège[c]	433[g]	Bruges[e]	119	Oostende[f]	71

[a]Brussel or Brussels [b]Anvers or Antwerp [c]Luik [d]Gand or Ghent
[e]Brugge [f]Ostende or Ostend [g]1975
Race (1961) Belgian 95 %, Italian 2 %, French 1 %, Dutch 1 %
Language Flemish and French
Usage (by region, 1975): Flemish 57 %, French 32 %, German 1 %, Flemish and French 11 %
Religion (1976) Roman Catholic 90** %
Education Pupils (1975/76) 1 922 686, teachers (1967/68) 150 000*
Labour force (1976)

Economic activity	Number	% of total
Agriculture	127 853	3
Mining and quarrying	34 556	1
Manufacturing	1 082 582	27
Electricity, gas and water	33 314	1
Construction	298 599	7
Distribution and hotels	697 868	17
Transport and communication	265 074	7
Finance and real estate	226 900	6
Other services	1 005 758	25
Compulsory military service	31 110	1
Unemployed	224 277	6
Activities not known	3 592	—
Total	4 031 483	100

Personnel Scientists and engineers engaged in research (1969): 10 070
Physicians (1975): 18 506, 1 per 530 people

Standard of living

National income per person (1976): B Fr 245 900 = $ 6 371 = £ 3 527
Consumption per person (1975): energy 5 584 kg coal equivalent, electricity (production 1976) 4 822 kW h, newsprint 16 kg[a], steel 314 kg[a]
[a]Including Luxembourg
Newspapers: number (1974) 31; circulation 2 416 000, 247 per 1 000 people
Telephones (Dec 1976): 2 949 820, 300 per 1 000 people
Livestock (000, Dec 1976) Cattle 2 823, pigs 4 813, sheep 82, horses 50, chickens (1974) 31 790
Mineral reserves (1973) Coal 253 mn tonnes
Petroleum refinery capacity (1975) 48.2 mn tonnes
Electrical capacity (1975) 11 127 megawatts of which, nuclear 1 663 megawatts
Hospital beds (1974) 87 164, 1 per 112 people
Roads (1976) 114 814 km = 71 342 mi, density 3.8 km per km²
Railways (1975) 3 998 km = 2 484 mi, density 0.13 km per km²
Inland waterways (1975) 1 938 km = 1 204 mi
Oil pipelines (1975) 317 km = 197 mi
Ships (registered, 1977) 271, total of 1 595 489 gross tons
Ports (goods traffic, 000 tonnes, 1976)

	loaded	unloaded
Antwerp	26 933	39 113
Ghent	4 522	10 566
Zeebrugge	1 554	8 227

Airports (passengers, 000, 1976) Brussels: departures 2 209, arrivals 2 093; also: Deurne (Antwerp), Liège, Ostend, Charleroi

Durable equipment (Dec 1976)	000	no per 1 000 people	no per km of road
Radio sets	4 272	435	no per
Television sets	2 856	291	km of road
Passenger cars	2 700	275	24
Commercial vehicles	257	26	2.2

Production

Gross domestic product 1976: B Fr 2 604 bn = $67 460 mn = £37 340 mn
Growth in real terms: 1960–70 5.0 %pa, 1970–76 4.0 %pa
Structure of gross domestic product (1976)

By origin	B Fr bn	% of gdp
Agriculture	75	3
Mining and quarrying	18	1
Manufacturing	723	28
Electricity, gas and water	82	3
Construction	195	7
Distribution	309	12
Transport and communication	213	8
Other	989	38
Total	2 604	100
By type of expenditure		
Government final consumption	440	17
Private final consumption	1 619	62
Stock investment	11	—
Gross fixed capital formation	561	22
Exports of goods and services	1 256	48
less Imports of goods and services	—1 283	—49
Total	2 604	100

Production indices (1970 = 100)	1960	1970	1976	Growth %pa 1960–70	1970–76
Agricultural[c]	87	100	103	1.4	0.4
Industrial	61.5	100.0	118.4	5.0	2.9

Main products (000 t)

	1960	1970	1976	1960–70	1970–76
Agriculture					
Wheat	790	735	890*	—0.7	3.2*
Barley	382	527	610*	3.3	2.5*
Potatoes	1 894	1 597	715*	—1.7	—12.5*
Sugar, raw value	507	592	638*	1.5	1.3*
Apples	165	241	220*	3.8	—1.5*
Milk	3 904	3 745	3 650*	—0.4	—0.4*
Butter	89	92	90*	0.3	—0.4*
Eggs	157	236	206*	4.2	—2.2*
Beef and veal	204	244	280*	1.8	2.3*
Pigmeat	238	543	610*	8.6	2.0*
Poultrymeat	63	111	110*	5.8	—0.2*
Fish catch	64	53	44	—1.9	—3.0
Timber (000 m³)	2 411	2 404	2 500*	0.0	0.7*
Energy					
Total energy (000 tce)[c]	22 590	11 580	8 490[g]	—6.5	—6.0[h]
Coal	22 469	11 362	7 237	—6.6	—7.2
Petroleum products	6 328	26 145	27 401[g]	15.2	1.0[h]
Natural gas (mn m³)	70	55	34	—2.4	—7.7
Manufactured gas (mn m³)	3 504	3 280	2 668	—0.7	—3.4
Electricity (mn kW h)	15 152	30 523	47 350	7.2	7.6

Belgium

Main products (000 t)	1960	1970	1976	Growth %pa	
Mining				1960–70	1970–76
Iron ore (Fe content)	48	31	19	−4.3	−7.8
Manufacturing					
Beer (000 hl)	10 110	13 015	13 797[g]	2.6	1.2[h]
Cigarettes (mn units)	11 973	19 727	26 893[g]	5.1	6.4[h]
Cotton yarn	103	75	54	−3.2	−5.4
Cotton fabrics	84	70	63	−1.8	−1.8
Wool yarn	51	81	84	4.8	0.6
Wood pulp	145	340	306[g]	8.9	−2.1[h]
Newsprint	95	95	85	0.0	−1.9
Other paper	350**	677	599[g]	6.8**	−2.4[h]
Synthetic rubber	—	50	90	na	10.3
Sulphuric acid	1 423	1 794	1 890	2.4	0.9
Nitric acid	na	718	794[g]	na	2.0[h]
Fertilisers, nitrogenous[c, d]	307	494	610	4.9	3.6
Fertilisers, phosphate[c, d]	380**	767	516	7.3**	−6.4
Cement	4 388	6 729	7 506	4.4	1.8
Coke	7 525	7 119	5 728[g]	−0.6	−4.2[h]
Pig iron	6 553	10 821	9 868	5.1	−1.5
Crude steel	7 188	12 611	12 149	5.8	−0.6
Copper[e]	212	353	480	5.2	5.2
Lead	93	104	122	1.1	2.7
Zinc	248	235	240	−0.5	0.3
Radio sets (000 units)	1 011	1 943	1 796[g]	6.8	−1.6[h]
Television sets (000 units)	243	505	579[g]	7.6	2.8[h]
Passenger cars (000 units)[a]	196	734	1 047	14.1	6.1
Commercial vehicles (000 units)[a]	21	61	71	11.5	2.6
Merchant vessels (000 grt)	130	155	184	1.8	2.9
Construction					
Dwellings (000 units)[b]	48.9[f]	44.8	76.9	−1.0**	9.4

[a]Assembly only [b]Buildings started [c]Including Luxembourg [d]Years ending June 30th [e]Includes processing of refined copper [f]Buildings completed [g]1975 [h]1970–75

Transport traffic	1960	1970	1976	Growth % pa	
Passenger-kilometres (mn)				1960–70	1970–76
Road	na	46 089	56 535[d]	na	4.2[e]
Rail	8 578	8 260	8 203	−0.4	−0.1
Air	1 264	2 447	3 893	6.8	8.0
Cargo: tonne-kilometres (mn)					
Road	na	9 194	9 834[d]	na	1.4[e]
Rail	6 262	7 816	6 638	2.2	−2.7
Water	4 779[b]	6 734	5 124[d]	5.0[c]	−5.3[e]
Air	39.6	192.1	325.8	17.1	9.2
Sea: tonnes (mn)					
Goods loaded	17.0	29.3	32.9	5.6	2.0
Goods unloaded	24.7	71.1	57.5	11.1	−3.5
Tourism					
Number of visitors[f] (000)	na	4 171	8 040[d]	na	14.0[e]
Gross receipts[a] ($ mn)	110	348	959	12.2	18.4

[a]Including Luxembourg [b]1963 [c]1963–70 [d]1975 [e]1970–75 [f]Accomodation nights

Finance and trade

Price and wage indices	1960	1970	1976	Growth % pa	
(1970 = 100)				1960–70	1970–76
Consumer prices	74.3	100.0	163.2	3.0	8.5
Wholesale prices	82.1	100.0	147.1	2.0	6.6
Wages (earnings)	45.6	100.0	240.1	8.2	15.7
Share prices	108	100	109	−0.8	1.4
Money stock					
(end-year, B Fr mn)	220.4	418.5	684.0	6.6	8.5

Budget (total, 1976)
Revenue: B Fr 665 300 mn = $ 17 236 mn = £ 9 541 mn
Expenditure: B Fr 798 400 mn = $ 20 684 mn = £ 11 450 mn,
of which, education 22 %, defence 9 %, health and social services 25 %

Balance of payments[a] ($ mn)	1972	1973	1974	1975	1976
Balance of goods (fob)	+1 015	+1 224	+879	+548	−873
Balance of services	+324	+243	+410	+750	+1 070
Balance of transfers	−196	−313	−378	−592	−496
Current balance	+1 142	+1 153	+911	+706	−300
Long-term capital flow	−649		−378	+112	+342
Reserves and debt (end-year, $ mn)					
International reserves[a]	3 869	5 100	5 345	5 797	5 206
Public debt: domestic	15 628	18 773	22 567	23 383	29 331
external	219	161	149	104	86

[a]Belgium-Luxembourg

External trade (Belgium-Luxembourg, 1976)
Imports: B Fr 1 370 bn = $ 35 492 mn = £ 19 647 mn
Exports: B Fr 1 266 bn = $ 32 811 mn = £ 18 163 mn

Main imports	% of total	Main exports	% of total
Motor vehicles	11	Chemicals	12
Food	10	Iron and steel	12
Chemicals	9	Motor vehicles	11
Machinery, non-electric	8	Food	8
Crude oil	8	Machinery, non-electric	7
Electrical machinery	5	Textile yarns & fabrics	7
Non-ferrous metals	4	Electrical machinery	5
Textile yarns & fabrics	4	Petroleum products	5
Precious stones	3	Non-ferrous metals	4
Iron and steel	3		
Petroleum products	3		
Main sources[a]		*Main destinations[b]*	
West Germany	23	West Germany	23
Netherlands	17	France	21
France	16	Netherlands	17
United Kingdom	7	United Kingdom	6
United States	6	Italy	5
Saudi Arabia	4	United States	4
Italy	4	Sweden	2
Japan	2	Switzerland	2
Sweden	2		

[a]EEC 68 % [b]EEC 74 %

Special focus

Crude steel production, Belgium-Luxembourg (mn tonnes)

	1970	1972	1973	1974	1975	1976
Belgium	12.6	14.5	15.5	16.2	11.6	12.1
as % of Belgium-Luxembourg	70	73	75	72	71	73
Luxembourg	5.5	5.5	5.3	6.4	4.6	4.6
as % of Belgium-Luxembourg	30	27	25	28	29	27
Belgium-Luxembourg total	18.1	19.9	20.8	22.7	16.2	16.7

Bulgaria

People's Republic of Bulgaria
Narodna Republika Bŭlgariya

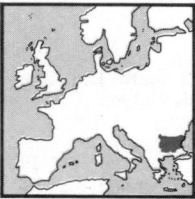

Location South-east Europe
In the eastern Balkans, with a coastline to the east on the Black Sea; Rumania is to the north, Jugoslavia to the west and Greece and Turkey are to the south
Land Area 110 912 km² = 42 823 mi²
Climate Continental
Weather at Sofia, 550 m altitude
Temperature: hottest month July 16–27 °C, coldest Jan minus 4–2 °C
Rainfall (av monthly): driest month Feb 28 mm, wettest May 87 mm
Time 2 hours ahead of GMT
Measures Metric system
Monetary unit Lev (Lv) = 100 stótinki
Rate of exchange (1976 av): par Lv 0.97 = $ 1, free Lv 1.752 = £ 1

Summary

Political Communist republic since 1946; won independence from Turkey in 1908. Involvement in the second world war ended in October 1944 by agreement with Soviet Union, United States and United Kingdom. Member of UN, Comecon and Warsaw Pact
Economic Industrialisation is increasing with a growth rate for industrial production of some 9 % pa, although agriculture remains significant – cigarettes and tobacco, fruit and vegetables, and wines and spirits are main exports. Industry includes especially electrical equipment, transport equipment, steel and chemicals. Tourism is important

People, resources and equipment

Population 1960 7.87 mn, 1970 8.49 mn, 1976 8.76 mn
Growth: 1960–70 0.8 % pa, 1970–76 0.5 % pa
Density (1976): 79 people per km²
Vital statistics (rate per 1 000 people, 1976): births 16.5, deaths 10.1

Bulgaria

Cities (population in 000, 1975)

Sofia (capital)	966	Stara Zagora	112	Tolbukhin	86
Plovdiv	309	Pleven	108	Shumen	84
Varna	252	Sliven	90	Yambol	76
Ruse	163	Gabrovo	90	Khaskovo	75
Burgas	144	Pernik	87	Pazardzhik	68

Race (1965) Bulgarian (including a small minority of Macedonian[a]) 88 %, Turkish 9 %, Gypsy 2 %
[a]Of whom there were 2½ % in 1956
Language Bulgarian
Religion (1962) Eastern Orthodox 27 %, Moslem 6 %
Education (1976/77) Pupils 1 502 346, teachers 90 527
Labour force (1975) 4 447 784; in agriculture 1 049 139 (24 %), manufacturing 1 438 242 (32 %), construction 350 356 (8 %), transport and communication 298 582 (7 %), other 1 311 465 (29 %)
Personnel Scientists and engineers (1973): 190 458
Physicians (1975): 18 770, 1 per 466 people
Standard of living
National income per person (1976): Lv 2 000*** = $ 2 100*** = £ 1 200***
Consumption per person (1975): energy 4 781 kg coal equivalent, electricity (production, 1976) 3 167 kW h, newsprint 5.6 kg, steel 252 kg
Newspapers (1974): number 13; circulation 1 971 000, 227 per 1 000 people
Telephones (Dec 1976): number 852 860, 97 per 1 000 people
Livestock (000, January 1977) Cattle 1 722, sheep 9 723, pigs 3 456, chickens 37 329
Mineral reserves Coal (1972) 34 mn tonnes
Lignite (1972) 5 198 mn tonnes
Crude oil (1975) 2 mn tonnes
Natural gas (1975) 13 bn cubic metres
Petroleum refinery capacity (1975) 11.5* mn tonnes
Electrical capacity (1975) 7 060 megawatts, of which, nuclear 880 megawatts
Hospital beds (1975) 75 037, 1 per 116 people
Roads (1976) 36 091 km = 22 426 mi, density 0.33 km per km²
Railways (1975) 4 290 km = 2 666 mi, density 0.039 km per km²
Inland waterways (including Danube river, 1973) 471 km = 293 mi
Ships (registered, 1977) 186, total of 964 156 gross tons
Ports Varna, Burgas
Airports Vrajdebna (Sofia), Plovdiv, Varna, Ruse, Burgas, Stara Zagora, Gorna Oryakhovitsa, Silistra, Tŭrgovishte, Vidin

Durable equipment (Dec 1975)	000	no per 1 000 people	
Radio sets	2 271	260	no per
Television sets	1 508	173	km of road
Passenger cars	198*	23*	5*
Commercial vehicles	43*	5*	1*

Production

Net material product 1976: Lv 15 145 mn = $ 15 613 mn = £ 8 644 mn
Growth in real terms: 1960–70 7.7 % pa, 1970–75 7.8 % pa
Gross domestic product
1976 est: Lv 19 500***mn = $ 20 000***mn = £ 11 100***mn
Structure of net material product (1975) *By origin* Agriculture 22 %, manufacturing, mining and electricity 51 %, construction 9 %, transport and communication 8 %, other 10 %

Production indices (1970 = 100)	1960	1970	1976	Growth % pa 1960–70	1970–76
Agricultural	70	100	113	3.6	2.0
Industrial	34	100	164	11.4	8.6

Main products (000 t)
Agriculture

	1960	1970	1976	1960–70	1970–76
Wheat	2 389	3 032	3 152	2.4	0.6
Barley	622	1 167	1 781	6.5	7.3
Maize	1 505	2 375	3 031	4.7	4.1
Potatoes	478	374	350	−2.4	−1.1
Tomatoes	634	685	785	0.8	2.3
Grapes	589	1 040	1 207	5.9	2.5
Wine	290*	375*	455*	2.6*	3.3*
Apples	261	363	369*	3.3	0.3*
Sunflowerseed	344	407	362	1.7	−1.9
Tobacco	62	122	165	7.0	5.1
Milk, cow	745	1 252	1 458	5.3	2.6
Milk, sheep	270	301	281	1.1	−1.1
Cheese	81	84	144	0.4	9.4
Wool	11.3	14.7	18.0*	2.7	3.4*
Meat	217	395	626	6.2	8.0
Fish catch	9	92	167	26.6	10.4
Timber (000 m³)	5 000**	5 050	4 415	0.1**	−2.2

Main products (000 t)	1960	1970	1976	Growth % pa 1960–70	1970–76
Energy					
Total energy (000 tce)	8 770	16 160	15 020[a]	6.3	−1.5[b]
Coal	571	397	288	−3.6	−5.2
Lignite	15 416	28 854	25 173	6.5	−2.3
Crude oil	200	334	120	5.3	−15.7
Petroleum products	124	5 971	10 220*[a]	47.3	11.3*[b]
Natural gas (mn m³)	—	474	111[a]	na	−25.2[b]
Manufactured gas (mn m³)	—	372	282[a]	na	−5.4[b]
Electricity (mn kW h)	4 657	19 513	27 741	15.4	6.0
Mining					
Iron ore (Fe content)	188	792	765	15.5	−0.6
Copper ore (Cu content)	12*	40*	55*[a]	12.8*	6.6*[b]
Lead ore (Pb content)	95*	98*	110*[a]	0.3*	2.3*[b]
Zinc ore (Zn content)	77*	76*	80*[a]	−0.1*	1.0*[b]
Manufacturing					
Beer (000 hl)	1 075	3 047	4 516[a]	11.0	8.2[b]
Cigarettes (mn units)	13 605	55 082	71 409[a]	15.0	5.3[b]
Cotton yarn	49	74	78	4.2	0.9
Cotton fabrics (mn m)	218	319	361	3.9	2.1
Man-made fibres	—	23	69*[a]	na	24.6*[b]
Wood pulp	38[c]	77	205[a]	10.6[d]	21.6[b]
Paper	80	232	341[a]	11.2	8.0[b]
Tyres (000 units)	172	546	1 226[a]	12.3	17.6[b]
Sulphuric acid	123	503	852	15.1	9.2
Nitric acid	195	613	840[a]	12.1	6.5[b]
Plastics and resins	7	89	156[a]	28.4	11.9[b]
Fertilisers, nitrogenous	84	602	672[a]	21.7	2.2[b]
Fertilisers, phosphate	41	148	246[a]	13.7	10.7[b]
Cement	1 586	3 668	4 362	8.7	2.9
Coke	20	837	1 364[a]	45.3	10.2[b]
Pig iron	192	1 252	1 581	20.6	4.0
Crude steel	253	1 800	2 459	21.7	5.3
Radio sets (000 units)	157	145	228[a]	−0.8	9.5[b]
Television sets (000 units)	0.4	193	124[a]	85.5	−8.5[b]
Motor vehicles (000 units)	—	11.0	20.9[a]	na	13.7[b]
Merchant vessels (000 grt)	na	na	133	na	na
Construction					
Dwellings (000 units)	49	46	57[a]	−0.6	4.4[b]

[a]1975 [b]1970–75 [c]1963 [d]1963–70

Transport traffic	1960	1970	1976	Growth % pa 1960–70	1970–76
Passenger-kilometres (mn)					
Rail	3 617	6 223	7 499	5.6	3.2
Air	na	305	500	na	8.6
Cargo: tonne-kilometres (mn)					
Road	na	7 902	10 480[c]	na	7.3[d]
Rail	6 981	13 858	17 055	7.1	3.5
Air	na	5.3	7.6	na	6.2
Sea: tonnes (mn)					
Goods loaded	0.9	2.3	2.7[a]	10.5	3.3[b]
Goods unloaded	1.4	13.8	20.0[a]	26.0	7.8[b]

[a]1975 [b]1970–75 [c]1974 [d]1970–74

Tourism Number of visitors (1976) 4 033 400, gross receipts (1975) $ 230 mn

Finance and trade

Price and wage indices (1970 = 100)	1960	1970	1976	Growth % pa 1960–70	1970–76
Consumer prices	91	100	101[a]	1.0	0.2[b]
Wages (earnings)	64	100	122	4.6	3.3

[a]1974 [b]1970–74

Budget (1976) Revenue: Lv 8 778 mn = $ 9 049 mn = £ 5 010 mn
Expenditure: Lv 8 758 mn = $ 9 029 mn = £ 4 999 mn
External trade (1976)
Imports (fob): Lv 5 436 mn = $ 5 604 mn = £ 3 103 mn
Exports: Lv 5 200 mn = $ 5 361 mn = £ 2 968 mn

Main imports (1974)	% of total	*Main exports* (1974)	% of total
Machinery	26	Machinery	30
Transport equipment	15	Cigarettes	11
Iron and steel	11	Transport equipment	9
Crude oil	7	Fruit and vegetables	8
Chemicals	4	Non-ferrous metals	6
Timber and paper	3	Chemicals	6
Coal	3	Iron and steel	6
Main sources (1976)		*Main destinations* (1976)	
Soviet Union	54	Soviet Union	54
East Germany	7	East Germany	8
West Germany	6	Czechoslovakia	5
Poland	5	Poland	5

Bulgaria

Special focus

Structure of net material product

	% of total		
	1960	1970	1975
Agriculture	32	23	22
Manufacturing, mining and electricity	46	49	51
Construction	7	9	9
Distribution and hotels	9	10	8
Transport and communication	4	7	8
Other	2	3	2

Cyprus

Republic of Cyprus
Kipriakí Demokratía
Kıbrıs Cumhuriyeti

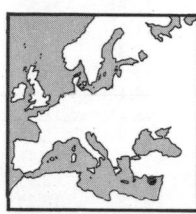

Location Eastern Mediterranean Sea
The island of Cyprus is about 70 km south of Turkey and 100 km west of Syria
Land Area 9 251 km² = 3 572 mi²
Climate Mediterranean
Weather at Nicosia, 175 m altitude
Temperature: hottest month July 21–37 °C, coldest Jan 5–15 °C
Rainfall (av monthly): driest month July 1 mm, wettest Jan 76 mm
Time 2 hours ahead of GMT (summer time, 3 hours ahead in North Cyprus)
Measures UK (imperial) and metric systems. The imperial system is gradually being replaced by the metric. Local measures are also in use, including: *length* 1 pic = 0.6096 metre = 2 feet
area 1 donum or scala = 1 337.8 square metres = 1 600 square yards
capacity 1 oke = 1.273 litres = 1.12 UK quarts
1 Cyprus litre = 3.182 litres = 2.8 UK quarts
1 kouza = 2 kartos = 8 okes = 10.229 litres = 9 UK quarts
weight (mass) 1 oke = 1.270 kilograms = 2.8 pounds
1 Cyprus litre = 2.286 kilograms = 5.04 pounds
1 cantar = 44 okes = 55.88 kilograms = 123.2 pounds
1 Aleppo cantar = 180 okes = 228.6 kilograms = 504 pounds
Monetary unit Cyprus pound (C £) = 1 000 mils
Rate of exchange (1976 av): free $ 2.437 = C £ 1, £ 1.349 = C £ 1
The Turkish lira is also in use in the northern sector; rate of exchange (1976 av): free T L 39.12 = C £ 1

Summary

Political Republic, which became independent from the United Kingdom in 1960. Between 1963 and 1974 there were a series of conflicts between Greek and Turkish Cypriots. After the 1974 violent conflict and Turkish invasion the island was effectively divided into North (Turkish sector) and South (Greek sector); the North sector is referred to internally as the 'Turkish Federated State of Cyprus'. Member of UN, Council of Europe and Commonwealth
Economic An agricultural economy, with potatoes, citrus fruit and wine main exports; mining is also significant, especially for copper and sulphur, recovered from pyrites. There is some industry, including especially petroleum refining and cement. The economy was affected severely by the conflict and partition. There has been some recovery in the southern sector; the northern sector is dependent on Turkish aid. Tourism has been important, but suffered from the 1974 conflict

People, resources and equipment

Population 1960 573 000*, 1970 600 000*, 1976 670 000**
Growth: 1960–70 0.5* %pa, 1970–76 1.9** %pa
Density (1976): 72** people per km²
Vital statistics (rate per 1 000 people, 1976): births 19.8*, deaths 9.7*
Regions (1976) North 150 000** (Turkish Cypriots), South 520 000**
Cities (population in 000, 1976) Nicosia (capital) 121*, Limassol 99*, Famagusta (1974) 39, Larnaca 38*, Paphos 12*, Kyrenia (1974) 4
Race (1976) Greek 73**%, Turkish 22**%, other 5**%
Language Greek and Turkish; English is also used
Religion (1973) Greek Orthodox 79 %, Moslem 18 %
Education (1976/77) Greek schools: pupils 106 008, teachers 4 766
Turkish schools: pupils 31 383, teachers 1 287

Labour force (South only, 1976) 204 353; in agriculture 51 324 (25 %), mining and quarrying 2 749 (1 %), manufacturing 32 668 (16 %), construction 22 961 (11 %), transport and communication 9 953 (5 %)
Personnel Scientists and engineers (1969): 4 650*
Physicians (South only, 1975): 547, 1 per 896 people
Standard of living
National income per person (1976): C £ 475 = $ 1 157 = £ 640
Consumption per person (1975): energy 1 278 kg coal equivalent, electricity (production, 1976) 1 197 kW h, newsprint 1.9 kg
Newspapers (1974): number 12; circulation (1972) 80 000, 124 per 1 000 people
Telephones (Dec 1976): 77 163, 113 per 1 000 people
Livestock (000, 1976) Cattle 33*, sheep 420*, goats 320*, pigs 115*
Petroleum refinery capacity (1975) 750 000 tonnes
Electrical capacity (1975) 239 megawatts
Hospital beds (South only, 1975) 3 286, 1 per 149 people
Roads (1976) 9 838 km = 6 113 mi, density 1.1 km per km²
There are no railways
Ships (registered, 1977) 800, total of 2 787 908 gross tons
Ports (goods traffic, 000 tonnes, 1974)

	loaded	unloaded
Larnaca	41ᵃ	622ᵇ
Famagusta	332	285
Limassol	249ᶜ	258ᵈ
Karavostassi	193	—

ᵃ1976: 683 thousand ᵇ1976: 639 thousand ᶜ1976: 881 thousand ᵈ1976: 726 thousand

Airports *North* Ercan (Tymbou) *South* Larnaca; Nicosia was the main airport until 1974

Durable equipment (at end-year)	000	no per 1 000 people	
Radio sets (1975)	206	320	no per
Television sets (1974)	85	133	km of road
Passenger cars (1976)	69	101	7.0
Commercial vehicles (1976)	17	25	1.7

Production

Gross domestic product 1976: C £ 319 mn = $ 778 mn = £ 430 mn
Growth in real terms: 1970–76 −1.0 %pa
Structure of gross domestic product (1976) *By origin* Agriculture 17 %, mining and quarrying 2 %, manufacturing 15 %, construction 6 %, transport and communication 10 %, other 50 %

Production indices (1970 = 100)	1960	1970	1976	Growth %pa	
				1960–70	1970–76
Agricultural	47	100	89	7.9	−1.9
Industrial	na	100	96	na	−0.7
Main products (000 t)					
Agriculture					
Wheat	46	43	66*	−0.7	7.4*
Barley	46	56	55*	2.0	−0.3*
Potatoes	81	208	135*	9.9	−7.0*
Olives	28	7	8*	−13.0	2.2*
Grapes	127	183	175*	3.7	−0.7*
Wine	30**	40	60*	2.9**	7.0*
Oranges	47*	92	60*	7.0*	−6.9*
Grapefruit	9*	37	57*	15.2*	7.5*
Milk, total	26	65	67*	9.6	0.5*
Timber (000 m³)	43	46	52*	0.7	2.1*
Other					
Petroleum products	—	—	315ᵇ	na	na
Electricity (mn kW h)	236	610	802	10.0	4.7
Chromium ore (oxide content)	6.5	15.0	13.6ᵇ	8.7	−1.9ᶜ
Copper ore (Cu content)ᵃ	35.5	17.3	9.9ᵇ	−6.9	−10.6ᶜ
Asbestos	21	26	32ᵇ	2.2	4.2ᶜ
Beer (000 hl)	46	93	112ᵇ	7.3	3.8ᶜ
Cigarettes (mn units)	423*	851	1 391ᵇ	7.2*	10.3ᶜ
Sulphurᵈ	363	442	96ᵇ	2.0	−26.3ᶜ
Cement	88	266	1 026	11.7	25.2

ᵃExports ᵇ1975 ᶜ1970–75 ᵈContent of pyrites

Transport traffic	1960	1970	1976	Growth %pa	
Air (mn)				1960–70	1970–76
Passenger–km	19	127	304	21.2	15.7
Cargo: tonne-kilometres	0.4	2.6	6.7	20.6	17.1
Sea: tonnes (mn)					
Goods loaded	1.57	1.55	1.81	−0.1	2.6
Goods unloaded	0.73	1.44	1.66	7.0	2.4

Tourism (1976) number of visitors 214 695, gross receipts $ 50 mn.
Excluding visitors to North Cyprus (mainly Turkish)

Cyprus

Finance and trade

Consumer price index (1970 = 100) 1976 148.6; growth 1970–76 6.8 %pa
Money stock (end-year, C £ mn) 1970 42.6, 1976 83.3; growth 1970–76 11.8 %pa
Budget (1976) *North* Balanced at T L 1 338 mn = $ 83 mn = £ 46 mn
South Revenue: C £ 77 mn = $ 188 mn = £ 104 mn
Expenditure: C £ 70 mn = $ 170 mn = £ 94 mn

Balance of payments ($ mn)	1972	1973	1974	1975	1976
Balance of goods (fob)	−160	−239	−237	−159	−157
Balance of services	+116	+145	+103	+49	+81
Balance of transfers	+19	+15	+61	+72	+64
Current balance	*−25*	*−78*	*−72*	*−37*	*−11*
Long-term capital flow	+36	+48	+46	+12	+59
Reserves and debt (end-year, $ mn)					
International reserves	320	307	268	215	292
External public debt	72	83	102	99	137

External trade (1976) Imports: C £ 207ᵃ mn = $ 505 mn = £ 279 mn
Exports: C £ 113ᵇ mn = $ 275 mn = £ 152 mn

Main importsᶜ	% of total	Main exportsᶜ	% of total
Food	18	Potatoes	17
(of which, cereals 7)		Cement	9
Textile yarns and fabrics	10	Clothing	8
Chemicals	9	Cigarettes	6
Crude oil	8	Wine	5
Machinery, non-electric	7	Citrus fruit	5
Petroleum products	6	Footwear	4
Iron and steel	5		
Electrical machinery	4		
Main sources		*Main destinations*	
United Kingdom	17	United Kingdom	28
Turkey	14*	Lebanon	16
Greece	8	Syria	6
Italy	8	Saudi Arabia	6
West Germany	6	Libya	5
United States	5	Soviet Union	4
France	5	Netherlands	3
Japan	4	Kuwait	2

ᵃOf which, South C £ 178 mn ᵇOf which, South C £ 106 mn ᶜSouth only

Special focus

The partitioned sectors

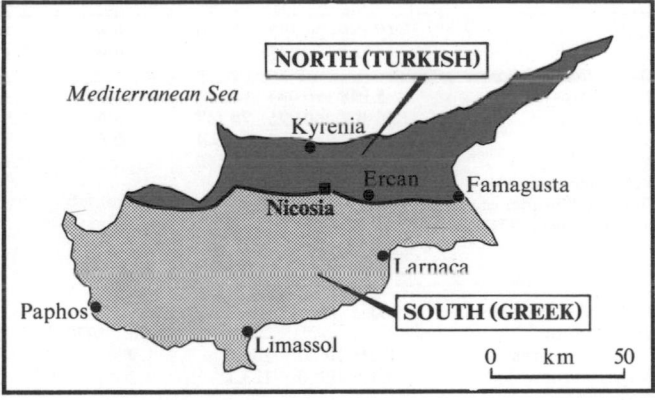

Mediterranean Sea

NORTH (TURKISH)

Kyrenia

Ercan Famagusta

Nicosia

Larnaca

Paphos

SOUTH (GREEK)

Limassol

0   km   50

Czechoslovakia

Czechoslovak Socialist Republic
Československá Socialistická Republika

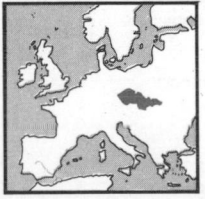

Location Eastern Europe
Poland and East Germany are to the north, West Germany to the west, Austria and Hungary to the south and Soviet Union to the east. Land-locked but with access to the sea via the Danube, Elbe and Oder rivers
Land Area 127 869 km² = 49 370 mi²
Usage (1975): agricultural 70 040 km² (55 %), of which, pastures 17 480 km² (14 %); forests 45 060 km² (35 %)
Climate Continental
Weather at Prague, 262 m altitude
Temperature: hottest month July 14*–23* °C, coldest Jan minus 4*–1* °C

Rainfall (av monthly): driest months Jan–Mar 18 mm, wettest July 68 mm
Time 1 hour ahead of GMT
Measures Metric system
Monetary unit Koruna or crown (Kčs) = 100 haler
Rate of exchange (1976 av) *Basic* par Kčs 5.97 = $ 1, free Kčs 10.78 = £ 1
Tourist par Kčs 10.45 = $ 1, free Kčs 18.87 = £ 1
Non-commercial par Kčs 11.94 = $ 1, free Kčs 21.57 = £ 1

Summary

Political Communist republic since 1948; became independent in 1918 on the break-up of Austria–Hungary. Occupied by Germany 1939–1945. A reform program introduced in 1968 led to occupation by Soviet Union forces on August 21, 1968. From January 1, 1969 a federal state was formed composed of two states, the Czech Socialist Republic and the Slovak Socialist Republic. Member of UN, Comecon and Warsaw Pact
Economic An industrial economy, with machinery accounting for about one-third of exports, iron and steel 11 %, and motor vehicles 8 %. There are large reserves of coal and lignite for smelting, but iron ore is mainly imported; there are few mineral resources other than coal. Other main industries include chemicals, textiles and footwear, and railway vehicles

People

Population 1960 13.65 mn, 1970 14.33 mn, 1976 14.92 mn
Growth: 1960 70 0.5 %pa, 1970 76 0.7 %pa
Density (1976): 117 people per km²
Vital statistics (rate per 1 000 people, 1976): births 19.2, deaths 11.4
Households (1970) 4 632 411; population in households 14.25 mn (99.3 % of total)
Regions (population in 000, December 31, 1976; total of 14.97 mn)

Czech	*10 158*
Praha	1 176
Středočeský (Central Bohemia)	1 139
Jihočeský (Southern Bohemia)	676
Západočeský (Western Bohemia)	880
Severočeský (Northern Bohemia)	1 149
Východočeský (Eastern Bohemia)	1 234
Jihomoravský (Southern Moravia)	2 005
Severomoravský (Northern Moravia)	1 899
Slovak	*4 815*
Bratislava	350
Západoslovenská (Western Slovakia)	1 648
Středoslovenská (Central Slovakia)	1 477
Východoslovenská (Eastern Slovakia)	1 341

Cities (population in 000, December 31, 1976)

Prahaᵃ (capital)	1 176	Košice	181	Hradec Králové	89
Brno	363	Plzeňᵇ	163	Pardubice	88
Bratislava	350	Olomouc	98	České Budějovice	85
Ostrava	317	Havířov	93	Liberec	83

ᵃPrague ᵇPilsen
Race (1976) Czech 64 %, Slovak 30 %, Magyar (Hungarian) 4 %
Language Czech and Slovak
Religion (1976) Roman Catholic 70** %, Protestant 15** %
Education (1975/76) Pupils: primary 1 881 414, secondary 128 545, vocational and teacher-training 293 718, higher 154 645. Teachers: primary 95 634, secondary 8 236, vocational and teacher-training 16 644, higher 17 009
Labour force (1976)

Economic activity	Number	% of total
Agriculture	1 125 000	15
Manufacturing, mining and utilities	2 887 000	39
Construction	714 000	10
Distribution and hotels	676 000	9
Transport and communication	494 000	7
Other	1 580 000	21
Total	7 476 000	100

Personnel Scientists and engineers (1973): 327 772
Physicians (1975): 35 383, 1 per 420 people
Standard of living National income per person (1976):
Kčs 29 000*** = $ 3 500***ᵃ = £ 2 000***ᵃ
ᵃConverted at average of basic and tourist rates
Consumption per person (1975): energy 7 151 kg coal equivalent, electricity (production, 1976) 4 198 kW h, newsprint 4.5 kg, steel 733 kg
Newspapers (1974): number 29; circulation 4 231 000, 288 per 1 000 people
Telephones (Dec 1976): 2 743 390, 183 per 1 000 people

Resources and equipment

Livestock (000, Dec 1976) Cattle 4 654, pigs 6 820, sheep 797, goats 98, chickens 42 559

Czechoslovakia

Mineral reserves Coal (1966) 11 573 mn tonnes
Lignite (1966) 9 857 mn tonnes
Crude oil (1975) 3 mn tonnes
Natural gas (1975) 14 bn cubic metres
Petroleum refinery capacity (1975) 16* mn tonnes
Electrical capacity (1975) 13 631 megawatts,
of which, nuclear 150 megawatts
Hospital beds (1975) 149 976, 1 per 99 people
Roads (1974) 145 455 km = 90 382 mi, density 1.1 km per km²
Railways (1975) 13 215 km = 8 211 mi, density 0.10 km per km²
Inland waterways (including Danube, Elbe, Vltava and Oder rivers, 1969)
483 km = 300 mi
Oil pipelines (1975) 1 454 km = 903 mi
Ships (registered, 1977) 14, total of 148 689 gross tons
Ports (river) *Danube* Bratislava, Komárno
Elbe and Vltava Děčín, Hřensko, Mělník, Prague, Ústí nad Labem
Oder Kozlí
Airports (passengers, 000, 1976)

	departures	arrivals
Prague	991	986
Bratislava	287	294

Also 10 other airports with scheduled flights

Durable equipment	000	no per	
(at end-year)		1 000 people	
Radio sets (1975)	3 916	265	no per
Television sets (1976)	3 793	253	km of road
Passenger cars (1976)	1 559	104	11
Commercial vehicles (1976)	301	20	2.1

Production

Net material product 1976: Kčs 413.4 bn = $ 50 350 mnᵃ = £ 27 890 mnᵃ
Growth in real terms: 1970–76 5.2 %pa
Gross domestic product
1976 est: Kčs 450***bn = $ 55 000***mnᵃ = £ 30 000***mnᵃ
ᵃConverted at average of basic and tourist rates
Structure of net material product (1976)

By origin	Kčs bn	% of gdp
Agriculture	34.6	8
Manufacturing, mining and utilities	278.9	67
Construction	52.5	13
Distribution and hotels	32.8	8
Transport and communication	12.5	3
Other	2.1	1
Total	413.4	100
By type of expenditure		
Government final consumption	29.0	7
Private final consumption	268.1	65
Stock investment	32.7	8
Gross fixed capital formation	88.0	21
Net external trade in goods and services	−4.4	−1
Total	413.4	100

Production indices	1960	1970	1976	Growth %pa	
(1970 = 100)				1960–70	1970–76
Agricultural	79	100	118	2.4	2.8
Industrial	56	100	146	6.0	6.5

Main products (000 t)	1960	1970	1976	Growth %pa	
Agriculture					
Wheat	1 503	3 174	4 807	7.8	7.1
Barley	1 745	2 280	2 901	2.7	4.1
Maize	572	513	514	−1.1	0.0
Rye	895	454	561	−6.6	3.6
Oats	1 020	776	379	−2.7	−11.3
Potatoes	5 093	4 793	4 214	−0.6	−2.1
Sugar, raw value	958*	763*	680*	−2.3*	−1.9*
Grapes	56	138	190*	9.4	5.5*
Wine	29	72	110*	9.5	7.3*
Apples	117ᵃ	232	185*	7.9ᵇ	−3.7*
Milk	3 830	4 794	5 400	2.2	2.0
Beef and veal	240*	366	402*	4.3*	1.6*
Pigmeat	483*	575	700*	1.8*	3.3*
Timber (000 m³)	12 623	14 524	16 891	1.4	2.5
Energy					
Total energy (000 tce)	63 470	79 460	81 930ᶜ	2.2	0.6ᵈ
Coal	26 400	28 195	28 266	0.7	0.1
Lignite	58 403	81 783	89 467	3.4	1.5
Crude oil	137	203	131	4.0	−7.0
Petroleum products	1 500*	9 896	15 223ᶜ	20.8*	9.0ᵈ

Main products (000 t)	1960	1970	1976	Growth %pa	
Energy				1960–70	1970–76
Natural gas (mn m³)	1 443	1 204	937	−1.8	−4.1
Manufactured gas (mn m³)	4 331	7 093	7 936	5.0	1.9
Electricity (mn kW h)	24 450	45 163	62 629	6.3	5.6
of which, nuclear (mn kW h)	—	—	187ᶜ	na	na
Mining					
Iron ore (Fe content)	948	447	496	−7.2	1.7
Antimony ore (Sb content)	1.60*	0.87*	0.28*	−5.9*	−17.2*
Magnesite	1 500**	2 928	2 885ᶜ	7.0**	−0.3ᵈ
Salt	168	213	244	2.4	2.3
Asbestos	2	28	43	28.4	7.4
Manufacturing					
Beer (000 hl)	14 093	21 178	22 629	4.2	1.1
Cigarettes (mn units)	18 620	20 472	23 232	1.0	2.1
Cotton yarn	102	114	125	1.1	1.5
Cotton fabrics (mn m)	460*	501	563	0.9*	2.0
Man-made fibres	62	101	147	5.0	6.5
Footwear (mn pairs)	96	117	128	2.0	1.4
Newsprint	34	81	78	9.1	−0.6
Other paper	579	765	967ᶜ	2.8	4.8ᵈ
Synthetic rubber	—	50	57	na	2.2
Tyres (000 units)	1 109	2 513	4 340	8.5	9.5
Ethyl alcohol (000 hl)	615	1 769	1 330ᶜ	11.1	−5.5ᵈ
Sulphuric acid	553	1 110	1 241	7.2	1.9
Plastics and resins	64	245	428ᶜ	14.4	11.8ᵈ
Fertilisers, nitrogenous	140	352	525ᶜ	9.6	8.3ᵈ
Fertilisers, phosphate	91	313	425ᶜ	13.1	6.3ᵈ
Cement	5 051	7 402	9 551	3.9	4.3
Coke	6 822	8 285	9 236ᶜ	2.0	2.2ᵈ
Pig iron	4 739	7 653	9 610	4.9	3.9
Crude steel	6 768	11 480	14 693	5.4	4.2
Aluminium	40	31	43ᶜ	−2.6	7.1ᵈ
Radio sets (000 units)	270	416	183ᶜ	4.4	−15.1ᵈ
Television sets (000 units)	263	383	456	3.8	3.0
Motor cycles (000 units)	160	108	128	−3.9	2.9
Passenger cars (000 units)	56	143	179	9.8	3.8
Commercial vehicles (000 units)	18	27	75	4.2	18.5
Construction					
Dwellings (000 units)	76	115	140*	4.2	3.3*

ᵃ1961 ᵇ1961–70 ᶜ1975 ᵈ1970–75

Transport traffic	1960	1970	1976	Growth %pa	
Passenger-kilometres (mn)				1960–70	1970–76
Roadᵃ	na	21 421	29 087	na	5.2
Rail	19 335	20 492	17 920	0.6	−2.2
Air	344	887	1 364	9.9	7.4
Cargo: tonnes-kilometres (mn)					
Road	5 108	10 093	16 292	7.0	8.3
Rail	44 407	60 995	70 747	3.2	2.5
Air	7.2	15.5	17.4	8.0	1.9
Water	na	2 434	2 568	na	0.9

ᵃPublic transport only
Tourism Number of visitors (1975) 14 078 500,
gross receipts (1971) $ 61*mn

Finance and trade

Price and wage indices	1960	1970	1976	Growth %pa	
(1970 = 100)				1960–70	1970–76
Consumer prices	91.0*ᵃ	100.0	101.6	1.9*ᵇ	0.3
Wages (earnings)	73.3	100.0	123.1	3.2	3.5

ᵃ1965 ᵇ1965–70
Budget (1976) Revenue: Kčs 292 165 mn = $ 35 590 mnᵃ = £ 19 710 mnᵃ
Expenditure: Kčs 290 071 mn = $ 35 330 mnᵃ = £ 19 570 mnᵃ
ᵃConverted at average of basic and tourist rates
External trade (1976)
Imports (fob): Kčs 55 996 mn = $ 9 380 mn = £ 5 194 mn
Exports: Kčs 52 137 mn = $ 8 733 mn = £ 4 836 mn

Main imports (1975)	% of total	Main exports (1975)	% of total
Machinery, non-electric	26	Machinery, non-electric	29
Food	8	Iron and steel	11
(of which, fruit and		Motor vehicles	8
vegetables 2, cereals 2)		Electrical machinery	6
Chemicals	8	Chemicals	5
Crude oil	8	Footwear	4
Electrical machinery	5	Coal	4
Non-ferrous metals	5	Textile yarns and fabrics	3
Metal ores	4	Railway vehicles	3
Motor vehicles	4	Clothing	3
Iron and steel	4		
Textile fibres	3		

Czechoslovakia

Main sources (1976)	% of total	Main destinations (1976)	% of total
Soviet Union	33	Soviet Union	34
East Germany	12	East Germany	13
Poland	9	Poland	9
Hungary	6	Hungary	6
West Germany	6	West Germany	5
Austria	3	Jugoslavia	4
Rumania	3	Rumania	3
Jugoslavia	3	Bulgaria	3
Bulgaria	3	Austria	2
United States	3	United Kingdom	2
Switzerland	2	Italy	1
United Kingdom	2		

Special focus

Energy, 1976 (figures in millions)

	Tonnes	Tonnes 'coal equivalent' (tce)	
Production: coal	28	24	
lignite	89	39	
Total production		63	*Consumption*
plus net imports		0	*as % of total available*
Total available		63	*100*
of which, consumed in:			
coke-oven plants		14	*22*
industry		14	*22*
transport		1	*2*
households and other		9	*14*
		38	*60*
net converted			
to other forms of energy		25	*40*
of which, electricity		23	*37*

Denmark

Kingdom of Denmark
Kongeriget Danmark

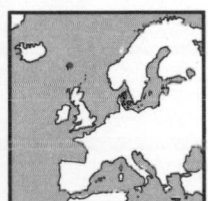

Location Northern Europe
Occupies the Jutland peninsula and some 500 islands, including the main islands of Zealand, Funen, and Lolland, between the North Sea to the west and the Baltic Sea to the east; West Germany is to the south and Sweden to the east across the Sound. The Kattegat is to the north-east and Skagerrak to the north
Land Area 43 069 km² = 16 629 mi²
Climate Temperate

Weather at Copenhagen, 9 m altitude
Temperature: hottest month July 14–22 °C, coldest Feb minus 3–2 °C
Rainfall (av monthly): driest month March 32 mm, wettest July 71 mm
Time 1 hour ahead of GMT
Measures Metric system
Monetary unit Danish krone (D Kr) = 100 øre
Rate of exchange (1976 av): free D Kr 6.045 = $1, D Kr 10.92 = £1
par (central value within the European 'snake') D Kr 7.644 = SDR 1.
The central par value was changed October 17, 1976 from
D Kr 7.57831 = SDR 1 to D Kr 7.89409 = SDR 1

Summary

Political Parliamentary monarchy; occupation by Germany during the second world war ended in 1945. Member of UN, OECD, Council of Europe, NATO, EEC (from January 1, 1973) and Nordic Council
Economic A mixed economy, with food accounting for about one-third of exports; meat, dairy produce and fish are main items. Manufacturing is important and makes up 26 % of gross domestic product compared with 5 % from agriculture. Main industries are food-processing, engineering, shipbuilding, petroleum refining and chemicals. Shipping and tourism are important

People, resources and equipment

Population 1960 4.58 mn, 1970 4.93 mn, 1976 5.07 mn
Growth: 1960–70 0.7 % pa, 1970–76 0.5 % pa
Density (1976): 118 people per km²
Vital statistics (rate per 1 000 people, 1976): births 12.9, deaths 10.7

Regions (population in 000, December 31, 1975; total of 5.07 mn)

København city	545	Vestsjællands	269	Vejle	317
Frederiksberg borough	94	Storstrøms	256	Ringkøbing	254
Counties		Bornholms	47	Aarhus	563
Københavns	631	Fyns	446	Viborg	227
Frederiksberg	306	Sønderjyllands	245	Nordjyllands	471
Roskilde	188	Ribe	206		

Cities (population in 000, December 31, 1976)

København[a] (capital)	1 251[b]	Aalborg	155	Helsingør	56
Aarhus	246	Esbjerg	79	Herning	55
Odense	168	Randers	64	Kolding	55

[a]Copenhagen [b]Includes Frederiksberg and Gentofte
Race Danish
Language Danish
Religion (1965) Evangelical Lutheran 94 %
Education (1975/76) Pupils 835 907, teachers 62 578
Labour force (1975) 2 485 619; in agriculture 230 021 (9 %), manufacturing 576 372 (23 %), construction 201 317 (8 %), transport and communication 162 250 (7 %), distribution and hotels 357 021 (14 %), finance and real estate 150 859 (6 %), other 807 779 (32 %)
Personnel Scientists and engineers engaged in research (1973): 4 717
Physicians (1972): 8 000, 1 per 624 people
Standard of living
National income per person (1976): D Kr 41 120 = $6 803 = £3 766
Consumption per person (1975): energy 5 268 kg coal equivalent, electricity (production) 3 469 kW h, newsprint 25 kg, steel 358kg
Newspapers (1974): number 51; circulation 1 792 000, 355 per 1 000 people
Telephones (Dec 1976): 2 528 585[a], 489[a] per 1 000 people
[a]Including Faroes and Greenland
Livestock (000, July 1976) Cattle 3 095, pigs 7 701, sheep 59, horses 56*, chickens 15 417*
Mineral reserves Lignite (1970) 20 mn tonnes
Crude oil (1975) 7 mn tonnes
Natural gas (1975) 50 bn cubic metres
Petroleum refinery capacity (1975) 11 mn tonnes
Electrical capacity (1975) 6 273 megawatts
Hospital beds (1970) 47 709, 1 per 103 people
Roads (1976) 66 515 km = 41 331 mi, density 1.5 km per km²
Railways (1975) 2 493 km = 1 549 mi, density 0.058 km per km²
Ships (registered, 1977) 1 407, total of 5 331 165 gross tons
Ports (goods traffic, 000 tonnes, 1976)

	International		Coastwise		Total
	loaded	unloaded	loaded	unloaded	
Copenhagen	896	6 423	473	2 253	10 045
Stigsnaes værket[a]	1 785	3 673	892	—	6 350
Esso-havnen[a]	460	2 575	1 185	1	4 220
Aarhus	319	2 357	63	949	3 688
Aalborg	238	2 374	107	405	3 124
Esbjerg	829	1 423	8	383	2 643
Shell-havnen[a]	59	1 927	408	225	2 619

[a]Crude oil port
Airports (passengers, 000, 1976) Kastrup (Copenhagen): departures 4 051, arrivals 4 002; also Aalborg, Tirstrup (Aarhus), Odense, Esbjerg, Billund, Skrydstrup, Stauning, Thisted, Karup, Sønderborg, Rønne

Durable equipment	000	no per		
(Dec 1976)		1 000 people		
Radio sets	1 753	345		
Television sets	1 634	322	no per	
of which, colour	511	101	km of road	
Passenger cars	1 340	264	20	
Commercial vehicles	255	50	3.8	

Production

Gross domestic product
1976: D Kr 232 894 mn = $38 527 mn = £21 327 mn
Growth in real terms: 1960-70 4.8 % pa, 1970–76 2.5 % pa
Structure of gross domestic product (1976)

By origin	D Kr mn	% of gdp
Agriculture	12 701	5
Mining and quarrying	59	—
Manufacturing	59 575	26
Electricity, gas and water	3 939	2
Construction	20 213	9
Distribution and hotels	33 375	14
Transport and communication	21 943	9
Other	81 089	35
Total	232 894	100

Denmark

Structure of gross domestic product

By type of expenditure	D Kr mn	% of gdp
Government final consumption	56 377	24
Private final consumption	135 975	58
Stock investment	929	—
Gross fixed capital formation	50 164	22
Exports of goods and services	77 835	33
less Imports of goods and services	−88 386	−38
Total	232 894	100

Production indices

(1970 = 100)	1960	1970	1976	Growth % pa 1960–70	1970–76
Agricultural	94	100	105	0.6	0.9
Industrial	58	100	114	5.7	2.2

Main products (000 t)

Agriculture	1960	1970	1976	1960–70	1970–76
Wheat	320	512	592	4.8	2.4
Barley	2 801	4 813	4 801	5.6	−0.1
Oats	681	631	263	−0.8	−13.6
Rye	454	134	213	−11.5	8.0
Potatoes	1 733	1 033	575	−5.0	−9.3
Sugar, raw value	342	291	415*	−1.6	6.1*
Apples	200**	121	129*	−5.0**	1.1*
Rapeseed	13	22	81	5.4	24.2
Milk	5 399	4 480	5 045	−1.8	2.0
Butter	167	131	139	−2.4	1.0
Cheese	113	111	157	−0.2	6.0
Eggs	138	86	71	−4.6	−3.2
Beef and veal	238	221	248	−0.7	1.9
Pigmeat	623	738	728	1.7	−0.2
Poultrymeat	56	79	97	3.5	3.5
Fish catch	581	1 226	1 912	7.7	7.7
Timber (000 m³)	1 852	2 293	1 582	2.2	−6.0
Energy					
Crude oil	—	—	167ᵈ	na	na
Petroleum products	41	9 750	7 662ᵈ	72.8	−4.7ᵉ
Manufactured gas (mn m³)	375	417	330	1.1	−3.8
Electricity (mn kW h)	5 179	18 864	19 700*	13.8	0.7*
Mining					
Salt	—	174	349	na	12.3
Manufacturing					
Fishmeal	51	245	339ᵈ	17.0	6.7ᵉ
Beer (000 hl)ᶜ	3 600**	7 122	8 881ᵈ	7.0**	4.5ᵉ
Cigarettes (mn units)	5 965	9 093	9 479ᵈ	4.3	0.8ᵉ
Man-made fibres	—	9.9	4.8ᵈ	na	−13.5ᵉ
Wood pulp	—	104	65ᵈ	na	−9.0ᵉ
Paper	195**	296	173ᵈ	4.3**	−10.2ᵉ
Fertilisers, nitrogenousᵇ	—	71	80	na	2.1
Fertilisers, phosphateᵇ	90**	95	78*	0.5**	−3.2*
Cement	1 442	2 604	2 355	6.1	−1.7
Pig iron	69	215	204ᶠ	12.0	−2.6ᵍ
Crude steel	317	473	722	4.1	7.3
Radio sets (000 units)	208	147	146ᵈ	−3.4	−0.1ᵉ
Television sets (000 units)	233	70	64ᵈ	−11.3	−1.8ᵉ
Motor vehicles (000 units)ᵃ	27.0	15.7	1.5ᵈ	−5.3	−37.5ᵉ
Merchant vessels (000 grt)	219	514	948	8.9	10.7
Construction					
Dwellings (000 units)	27.3	50.6	39.2	6.4	−4.2

ᵃAssembly only   ᵇYears ending July 31st   ᶜSales by breweries   ᵈ1975   ᵉ1970–75
ᶠ1972   ᵍ1970–72

Transport traffic

	1960	1970	1976	Growth % pa 1960–70	1970–76
Passenger-kilometres (mn)					
Railᵇ	3 303	3 477	3 420	0.5	−0.3
Airᵃ	602	1 552	2 603	9.9	9.0
Cargo: tonne-kilometres (mn)					
Railᵇ	1 417	1 729	1 800	2.0	0.7
Airᵃ	16	71	113	16.3	8.1
Sea: tonnes (mn)					
Goods loaded	3.66	6.77	7.16	6.3	0.9
Goods unloaded	17.8	31.5	30.9	5.9	−0.3

ᵃIncludes apportionment (two-sevenths) of international operations of Scandinavian Airlines System (SAS)   ᵇYears ending March 31st

Tourism (1976) Number of visitors 16 231 862, of whom, West German 12 350 005 (76 %); gross receipts $ 803 mn

Finance and trade

Price and wage indices

(1970 = 100)	1960	1970	1976	Growth % pa 1960–70	1970–76
Consumer prices	56.6	100.0	169.8	5.9	9.2
Wholesale prices	74	100	174	3.1	9.7
Wages (earnings)	36	100	237	10.8	15.5
Share prices	96	100	215	0.4	13.6

Money stock

	1960	1970	1976	1960–70	1970–76
(end-year, D Kr mn)	10 040	27 470	54 500	10.6	12.1

Budget (1976/77; year ending March 31st)
Revenue: D Kr 66 996 mn = $ 11 201 mn = £ 6 454 mn
Expenditure: D Kr 79 270 mn = $ 13 254 mn = £ 7 637 mn

Balance of payments ($ mn)

	1972	1973	1974	1975	1976
Balance of goods (fob)	−430	−1 188	−1 787	−1 304	−2 878
Balance of services	+457	+492	+649	+741	+748
Balance of transfers	−90	+232	+157	+74	+220
Current balance	−63	−464	−981	−488	−1 909
Long-term capital flow	+326	+481	+345	+136	+1 993

International reserves

	1972	1973	1974	1975	1976
(end-year, $ mn)	855	1 324	935	877	915

External trade (1976)
Imports: D Kr 75 011 mn = $ 12 409 mn = £ 6 869 mn
Exports: D Kr 55 034 mn = $ 9 104 mn = £ 5 040 mn

Main imports	% of total	Main exports	% of total
Crude oil and products	15	Food	31
(of which, crude oil 6)		(of which, meat 14,	
Machinery, non-electric	11	dairy produce 6,	
Food	9	fish 4)	
Chemicals	9	Machinery, non-electric	16
Motor vehicles	7	Chemicals	7
Electrical machinery	6	Electrical machinery	5
Iron and steel	5	Ships and boats	5
Textile yarns and fabrics	4	Petroleum products	3
Paper	3	Textile yarns and fabrics	3
Ships and boats	3		

Main sources		Main destinations	
West Germany	21	United Kingdom	17
Sweden	14	Sweden	16
United Kingdom	10	West Germany	14
Netherlands	5	Norway	7
United States	5	United States	6
Norway	5	Italy	5
Belgium-Luxembourg	4	France	4
France	4	Netherlands	3
Japan	3	Finland	2

Special focus

Fishing, 1976

	Landings 000 tonnes	D Kr mn
By Danish fishermen		
in Danish ports:		
Fish for reduction	1 535	623
Cod	158	421
Plaice	45	168
Other	138	435
Total in Danish ports	1 876	1 647
in foreign ports:		
Cod	1	3
Other	1	7
Total in foreign ports	2	10
By foreign vessels in Danish ports		
Herring	57	125
Fish for reduction	68	35
Other	33	66
Total by foreign vessels	158	226
Total in Danish ports	2 033	1 873
Total by Danish fishermen	1 878	1 657

Faroes

Faroe Islands
Føroyar
Færøerne

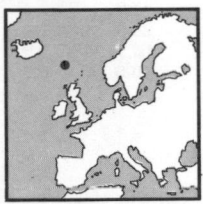

Location North Atlantic Ocean
A group of 18 islands, situated between Iceland to the north-west and Scotland to the south
Land Area 1 399 km² = 540 mi²
Climate Cool
Weather at Hoyvík, 20 m altitude
Temperature: hottest month Aug 9–14 °C, coldest Jan 1–6 °C
Rainfall (av monthly): driest month May 67 mm, wettest Dec 167 mm
Time GMT
Measures Metric system
Monetary unit Faroese krona (F Kr) = 100 øre, and Danish krone
Rate of exchange (1976 av): par F Kr 1 = D Kr 1, free F Kr 6.045 = $ 1, F Kr 10.92 = £ 1

Summary

Political Overseas part of the Kingdom of Denmark since 1380; there was a referendum in 1946 to consider the possibility of establishing a republic, and in 1948 a large degree of independence was obtained. Faroes elect two members of the Danish parliament. Member of Efta
Economic The economy is dependent on fish and fish products which make up virtually all exports. There is some agriculture

People, resources and equipment

Population 1960 34 600, 1970 38 600, 1976 41 200
Growth: 1960–70 1.1 % pa, 1970–76 1.1 % pa
Density (1976): 29 people per km²
Vital statistics (rate per 1 000 people, 1975): births 19.1, deaths 6.9
Town (population in 000, 1976) Thorshavn (capital) 11.5
Race Scandinavian
Language Faroese and Danish
Religion Mainly Lutheran
Education (1975/76) Pupils 9 750*, teachers 470**
Labour force (1970) 15 114; in fishing 3 207 (21 %), manufacturing 2 993 (20 %)
Personnel (1969) Physicians: 32, 1 per 1 200 people
National income per person (1976)
F Kr 28 000*** = $ 4 600*** = £ 2 600***
Consumption per person Energy (1975) 4 325 kg coal equivalent, electricity (1975/76) 2 260* kW h
Newspapers (1974) Number 1; circulation 3 700, 92 per 1 000 people
Livestock (000, Dec 1975) Sheep 69*, cattle 2*
Electrical capacity (March 1976) 28 megawatts
Hospital beds (1969) 240, 1 per 160 people
Ships (registered, 1977) 172, total of 57 110 gross tons
Ports Thorshavn, Fuglefjord, Klaksvig, Trangisvaag
Airport Vágar
Radio sets (1974) 23 000, 563 per 1 000 people

Production, finance and trade

Gross domestic product
1976 est: F Kr 1 300***mn = $ 215***mn = £ 120***mn
Main products (000t, 1976) Potatoes 1*, meat 1*, fish catch 342, electricity (1975/76; year ending March 31st) 93* mn kW h
Transport traffic (1974) *Sea* Goods loaded 105 000* t, unloaded 306 000 t, of which, petroleum products 99 000 t
Budget (1976/77; year ending March 31st)
Revenue: F Kr 315.7 mn = $ 53 mn = £ 30 mn
Expenditure: F Kr 315.5 mn = $ 53 mn = £ 30 mn
External trade (1976) Imports: F Kr 790 mn = $ 131 mn = £ 72 mn
Exports: F Kr 631 mn = $ 104 mn = £ 58 mn

Main imports	% of total	Main exports	% of total
Ships and boats	20	Fish	78
Petroleum products	14	(of which, salted	
Food	13	cod 28, fish fillets 20)	
Machinery	12	Fishmeal	10
Main sources		*Main destinations*	
Denmark	66	Denmark	22
Norway	19	United States	16
Poland	4	United Kingdom	15
Sweden	3	Italy	10

Finland

Republic of Finland
Suomen Tasavalta
Republiken Finland

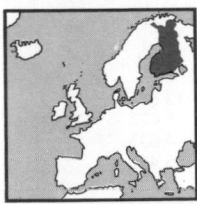

Location Northern Europe
With a coastline on the Baltic Sea, the Gulf of Bothnia is to the west and Gulf of Finland to the south; Soviet Union is to the east, Sweden to the west and there is a border with Norway to the north
Land Area 337 009 km² = 130 120 mi²
Usage (1975): forests 226 300* km² (66* %), arable, cropland and pastures 27 970 km² (8 %)
Climate Temperate, with very cold winters
Weather at Helsinki, 46 m altitude
Temperature: hottest month July 12–22 °C, coldest Feb minus 9–minus 4 °C
Rainfall (av monthly): driest month March 36 mm, wettest Oct 73 mm
Time 2 hours ahead of GMT
Measures Metric system
Monetary unit Markka (F Mk) = 100 penni
Rate of exchange (1976 av): free F Mk 3.864 = $ 1, F Mk 6.980 = £ 1

Summary

Political Republic, which declared independence from Soviet Union on December 6, 1917; involvement in the second world war ceased on September 19, 1944. A treaty of friendship with Soviet Union was first signed on April 6, 1948; the treaty is next due for review in 1990. Member of UN, OECD, Efta (as associate) and Nordic Council
Economic Forestry is the most important industry, providing nearly one-half of exports (of which paper accounts for one-quarter). Other main industries are machinery, shipbuilding, steel, clothing and chemicals.

People, resources and equipment

Population 1960 4.43 mn, 1970 4.61 mn, 1976 4.73 mn
Growth: 1960–70 0.4 %pa, 1970–76 0.4 %pa
Density (1976): 14 people per km²
Vital statistics (rate per 1 000 people, 1976): births 14.1, deaths 9.4
Regions (provinces, population in 000, December 31, 1976; total of 4.73 mn)

Ahvenanmaa (Åland)	22	Mikkeli (St Michel)	210
Häme (Tavastehus)	660	Oulu (Uleåborg)	407
Keski-Suomi (Mellersta		Pohjois-Karjala (Norra	
Finland)	241	Karelen)	177
Kuopio	251	Turku-Pori (Åbo-Björneborg)	699
Kymi (Kymmene)	346	Uusimaa (Nyland)	1 100
Lappi (Lapland)	196	Vaasa (Vasa)	425

Cities (population in 000, December 31, 1976)

Helsinki or Helsingfors (capital)	853[a]	Lahti	95
Tampere (Tammerfors)	235[a]	Oulu (Uleåborg)	92
Turku (Åbo)	231[a]	Pori (Björneborg)	80
Espoo (Esbo)	125	Kuopio	72
Vantaa (Vanda)	123	Jyväskylä	62

[a]1974
Race (1960) Finnish 92 %, Swedish 7 %
Language Finnish and Swedish
Usage (language mainly spoken, 1970): Finnish 93 %, Swedish 7 %
Religion (1974) Lutheran 92 %, Greek Orthodox 1 %
Education (1974/75) Pupils 987 829, teachers 62 028
Labour force (1976) 2 279 000; in agriculture 297 000 (13 %), manufacturing, mining and utilities 584 000 (26 %), construction 160 000 (7 %), distribution and hotels 319 000 (14 %), transport and communication 161 000 (7 %), finance and real estate 116 000 (5 %). unemployed 90 000 (4 %), other 552 000 (24 %)
Personnel Scientists and engineers (1973): 196 340
Physicians (1975): 6 701, 1 per 704 people
Standard of living
National income per person (1976): F Mk 20 680 = $ 5 351 = £ 2 962
Consumption per person (1975): energy 4 766 kg coal equivalent, electricity (production 1976) 6 199 kW h, newsprint 15* kg, steel 430 kg
Newspapers: number (1974) 62; circulation 2 058 000, 440 per 1 000 people
Telephones (Dec 1976): 1 935 680, 409 per 1 000 people
Livestock (000, June 1977) Cattle 1 762, pigs 1 145, sheep 55, reindeer (1975) 175, poultry 11 750
Mineral reserves (1976) Uranium 3 200 tonnes
Petroleum refinery capacity (1975) 10.2 mn tonnes
Electrical capacity (1975) 7 395 megawatts, of which, hydro 2 336 megawatts
Hospital beds (1975) 71 115, 1 per 66 people

Finland

Roads (1976) 73 763 km = 45 834 mi, density 0.22 km per km²
Railways (1975) 5 957 km = 3 702 mi, density 0.018 km per km²
Inland waterways (1975) 6 675 km = 4 148 mi
(excluding 40 000 km for timber floating)
Ships (registered, 1977) 337, total of 2 262 095 gross tons
Ports (goods traffic, 000 tonnes, 1976)

	International		Coastwise		Total
	loaded	unloaded	loaded	unloaded	
Skoldvik[a]	770	8 622	4 219	25	13 636
Helsinki	1 042	2 719	27	1 684	5 472
Naantali[a]	611	2 880	965	181	4 637
Kotka	2 108	874	—	449	3 431
Turku	518	1 137	40	857	2 552
Hamina	1 223	492	—	287	2 002

[a]Crude oil port

Airports (passengers, 000, 1976) Helsinki: departures 1 426, arrivals 1 414; also Tampere, Turku and 17 other airports with scheduled flights

Durable equipment (Dec 1976)	000	no per 1 000 people	
Radio sets	2 200	465	no per
Television sets	1 421	300	km of road
Passenger cars	1 033	218	14
Commercial vehicles	142	30	1.9

Production

Gross domestic product 1976: F Mk 108 754 = $ 28 145 mn = £ 15 580 mn
Growth in real terms: 1960–70 5.2 %pa, 1970–76 3.5 %pa
Structure of gross domestic product (1976) *By origin* Agriculture 10 %, mining and quarrying ½ %, manufacturing 27 %, electricity, gas and water 3 %, construction 9 %, distribution and hotels 11 %, transport and communication 6 %, other 34 %
By type Final consumption expenditure 72 % (of which, government 20 %), stock investment (including statistical discrepancy) 4 %, gross fixed capital formation 27 %, exports of goods and services 27 %, less imports of goods and services −30 %

Production indices (1970 = 100)	1960	1970	1976	Growth %pa 1960–70	1970–76
Agricultural	84	100	123	1.8	3.4
Industrial	49	100	124	7.4	3.7
Main products (000 t)					
Agriculture					
Wheat	368	409	654	1.1	8.1
Barley	440	933	1 553	7.8	8.9
Rye	186	131	178	−3.4	5.2
Oats	1 109	1 330	1 573	1.8	2.8
Potatoes	1 717	1 136	948	−4.0	−3.0
Milk	3 494	3 310	3 278	−0.5	−0.2
Butter	93	87	83	−0.7	−0.8
Cheese	31	41	57	2.8	5.6
Eggs	43	65	86	4.2	4.8
Beef and veal	72	106	112	4.0	0.9
Pigmeat	54	106	136	7.0	4.2
Fish catch	64	82	120	2.5	6.6
Timber (000 m³)	48 100	45 130	32 950	−0.6	−5.1
Energy					
Total energy (000 tce)	660	1 170	1 520[b]	5.9	5.4[c]
Petroleum products	1 090	7 742	7 763[b]	21.6	0.1[c]
Manufactured gas (mn m³)	68	59	27	−1.4	−12.2
Electricity (mn kW h)	8 628	21 158	29 319	9.4	5.6
of which, hydro (mn kW h)	5 269	9 353	12 189[b]	5.9	5.4[c]
Mining					
Iron ore (Fe content)	180	382	402[b]	7.8	1.0[c]
Chromium ore (oxide content)	—	50	113[b]	na	17.7[c]
Copper ore (Cu content)	30.4	33.7	41.7	1.0	3.6
Nickel ore (Ni content)	2.4	6.7	5.3[b]	11.0	−4.5[c]
Vanadium ore (V content)	0.57	1.20	1.28[b]	7.7	1.3[c]
Zinc ore (Zn content)	51	69	61	3.1	−2.0
Manufacturing					
Beer (000 hl)	917	2 424	2 706[d]	10.2	2.8[e]
Cigarettes (mn units)	6 090	6 476	7 902[b]	0.6	4.1[c]
Cotton yarn	18.9	17.0	11.8	−1.0	−5.9
Cotton fabrics	13.5	15.1	12.0	1.1	−3.7
Man-made fibres	16	40	32*[b]	9.6	−4.4*[c]
Wood pulp	3 699	6 222	5 174[b]	5.3	−3.6[c]
Newsprint	781	1 305	1 010	5.3	−4.2
Other paper	1 300**	2 955	3 002[b]	8.6**	0.3[c]
Tyres (000 units)	157	1 037	781*	20.8	−4.6*[c]

Main products (000 t) **Manufacturing**	1960	1970	1976	Growth %pa 1960–70	1970–76
Sulphuric acid	187	843	1 016	12.2	3.1
Nitric acid	6	261	398[b]	46.0	8.8[c]
Fertilisers, nitrogenous[a]	31	150	202	17.1	5.1
Fertilisers, phosphate[a]	96*	171	165	6.0*	−0.6
Cement	1 257	1 838	1 820	3.9	−0.2
Pig iron	137	1 164	1 321	23.8	2.1
Crude steel	254	1 167	1 643	16.5	5.9
Radio sets (000 units)	67	174	174[d]	10.0	0.0[e]
Merchant vessels (000 grt)	77	222	389	11.2	9.8
Construction					
Dwellings (000 units)	31.5	49.7	57.2	4.7	2.4

[a]Years ending June 30th [b]1975 [c]1970–75 [d]1974 [e]1970–74

Transport traffic	1960	1970	1976	Growth %pa 1960–70	1970–76
Passenger-kilometres (mn)					
Road	na	18 700*	23 100	na	3.6*
Rail	2 343	2 156	3 046	−0.8	5.9
Air	228	773	1 380	13.0	10.1
Cargo: tonne-kilometres (mn)					
Road	na	13 400	15 900[a]	na	3.5[b]
Rail	4 872	6 270	6 546	2.5	0.7
Water	na	4 400	4 700	na	1.1
Air	4.1	23.7	33.1	19.2	5.7
Sea: tonnes (mn)					
Goods loaded	10.4	12.4	12.0	1.8	−0.5
Goods unloaded	9.0	20.2	23.3	8.4	2.4

[a]1975 [b]1970–75

Tourism (1976) Number of visitors (excluding from Nordic countries) 264 155, gross receipts $ 323 mn

Finance and trade

Price and wage indices (1970 = 100)	1960	1970	1976	Growth %pa 1960–70	1970–76
Consumer prices	62	100	199	4.9	12.1
Wholesale prices	67	100	211	4.1	13.3
Wages (earnings)	43	100	260	8.8	17.3
Share prices	61	100	184	5.1	10.7
Money stock					
(end-year, F Mk mn)	1 495	3 445	9 601	8.7	18.6

Budget (1976) Revenue: F Mk 30 186 mn = $ 7 812 mn = £ 4 325 mn
Expenditure: F Mk 27 279 mn = $ 7 060 mn = £ 3 908 mn

Balance of payments ($ mn)	1972	1973	1974	1975	1976
Balance of goods (fob)	−56	−252	−889	−1 621	−582
Balance of services	−72	−123	−308	−531	−562
Balance of transfers	+10	−15	−18	−30	−28
Current balance	−117	−389	−1 215	−2 183	−1 171
Long-term capital flow	+319	+110	+267	+1 301	+954
Reserves and debt (end-year, $ mn)					
International reserves	727	618	634	470	498
Public debt: domestic	448	341	280	387	483
external	367	356	317	409	584

External trade (1976) Imports: F Mk 28 560 mn = $ 7 391 mn = £ 4 092 mn
Exports: F Mk 24 506 mn = $ 6 342 mn = £3 511 mn

Main imports	% of total	*Main exports*	% of total
Machinery, non-electric	16	Paper	25
Crude oil	14	Machinery, non-electric	11
Chemicals	9	Ships and boats	8
Electrical machinery	7	Timber	8
Food	7	Wood pulp	6
Motor vehicles	6	Clothing	5
Textile yarns and fabrics	5	Electrical machinery	4
Iron and steel	4	Chemicals	4
Petroleum products	4	Food	3
Main sources		*Main destinations*	
Soviet Union	18	Soviet Union	20
Sweden	16	Sweden	17
West Germany	15	United Kingdom	14
United Kingdom	8	West Germany	9
United States	5	Denmark	4

Special focus

Newsprint prices (unit value of Finnish exports)

	$ per tonne	% change over previous year		$ per tonne	% change over previous year
1970	126.6	+4	1974	259.6	+61
1971	131.9	+4	1975	347.9	+34
1972	137.7	+4	1976	325.2	−7
1973	161.3	+17	1977	355.1	+9

France

French Republic
République Française

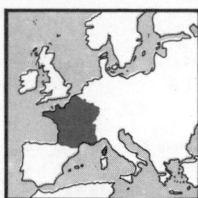

Location Western Europe
With coastlines on the North Sea, Atlantic Ocean and Mediterranean Sea, Belgium and Luxembourg are to the north, West Germany, Switzerland and Italy to the east and Spain to the south-west. Corsica is off the coast to the south-east in the Mediterranean Sea.
Land Area 547 026 km² = 211 208 mi²
Usage (1975): agricultural 322 600* km² (59* %), of which, arable 172 000* km² (31* %), pastures 134 500* km² (25* %), cropland 16 100* km² (3* %); forests 146 100* km² (27* %)
Climate Temperate, mild in the south
Weather at Paris, 75 m altitude
Temperature: hottest month July 14–25°C, coldest Jan 1–6°C
Rainfall (av monthly): driest month March 35 mm, wettest Aug 64 mm
Time 1 hour ahead of GMT (summer time, 2 hours ahead)
Measures Metric system
Monetary unit Franc (Fr) = 100 centimes
Rate of exchange: *free* 1976 av: Fr 4.780 = $ 1, Fr 8.633 = £ 1; 1977 av: Fr 4.913 = $ 1, Fr 8.576 = £ 1

Summary

Political Republic; the constitution of the Fifth Republic came into force on October 4, 1958. Member of UN, OECD, Council of Europe, franc zone, EEC, WEU, Nato (excluding the military side from which France withdrew in 1966) and South Pacific Commission
Economic An industrial economy with, however, food accounting for one-eighth of exports; agriculture absorbs 11 % of the labour force and contributes 5 % to gross domestic product. Main industrial exports are machinery, motor vehicles, chemicals, iron and steel, textiles and clothing. With small crude oil resources, nuclear power is important, and a nuclear defence capability has been developed

People

Population 1960 45.68 mn, 1970 50.77 mn, 1976 52.92 mn
Growth: 1960–70 1.1 %pa, 1970–76 0.7 %pa
Density (1976): 97 people per km²
Vital statistics (rate per 1 000 people, 1976): births 13.6, deaths 10.5
Households (1968) 15 762 508; population in households 48.24 mn (97.2 % of total)
Regions (régions and départements, population in 000, 1975; census total of 52.66 mn)

no	Région Parisienne		no	Bretagne	
	Région Parisienne	*9 879*		*Bretagne*	*2 595*
75	Paris	2 300	22	Côtes-du-Nord	526
77	Seine-et-Marne	756	29	Finistère	804
78	Yvelines	1 082	35	Ille-et-Vilaine	702
91	Essonne	923	56	Morbihan	564
92	Hauts-de-Seine	1 439		*Poitou-Charentes*	*1 528*
93	Seine-Saint-Denis	1 322	16	Charente	337
94	Val-de-Marne	1 216	17	Charente-Maritime	498
95	Val-d'Oise	841	79	Deux-Sèvres	336
	Champagne-Ardenne	*1 337*	86	Vienne	357
08	Ardennes	309		*Aquitaine*	*2 550*
10	Aube	285	24	Dordogne	373
51	Marne	530	33	Gironde	1 061
52	Haute-Marne	212	40	Landes	288
	Picardie	*1 679*	47	Lot-et-Garonne	293
02	Aisne	534	64	Pyrénées-Atlantiques	535
60	Oise	606		*Midi-Pyrénées*	*2 268*
80	Somme	538	09	Ariège	138
	Haute-Normandie	*1 596*	12	Aveyron	278
27	Eure	423	31	Haute-Garonne	777
76	Seine-Maritime	1 173	32	Gers	175
	Centre	*2 152*	46	Lot	151
18	Cher	316	65	Hautes-Pyrénées	227
28	Eure-et-Loir	335	81	Tarn	338
36	Indre	249	82	Tarn-et-Garonne	183
37	Indre-et-Loire	479		*Limousin*	*739*
41	Loir-et-Cher	284	19	Corrèze	240
45	Loiret	490	23	Creuse	146
	Basse-Normandie	*1 306*	87	Haute-Vienne	352
14	Calvados	561			
50	Manche	452			
61	Orne	294			

no	Bourgogne		no	Rhône-Alpes	
	Bourgogne	*1 571*		*Rhône-Alpes*	*4 781*
21	Côte-d'Or	456	01	Ain	376
58	Nièvre	245	07	Ardèche	257
71	Saône-et-Loire	570	26	Drôme	362
89	Yonne	300	38	Isère	860
	Nord	*3 914*	42	Loire	742
59	Nord	2 511	69	Rhône	1 430
62	Pas-de-Calais	1 403	73	Savoie	305
	Lorraine	*2 331*	74	Haute-Savoie	448
54	Meurthe-et-Moselle	723		*Auvergne*	*1 330*
55	Meuse	204	03	Allier	378
57	Moselle	1 006	15	Cantal	167
88	Vosges	398	43	Haute-Loire	205
	Alsace	*1 517*	63	Puy-de-Dôme	580
67	Bas-Rhin	882		*Languedoc-Roussillon*	*1 789*
68	Haut-Rhin	635	11	Aude	272
	Franche-Comté	*1 060*	30	Gard	495
25	Doubs	471	34	Hérault	648
39	Jura	239	48	Lozère	75
70	Haute-Saône	222	66	Pyrénées-Orientales	300
90	Territoire de Belfort	128		*Provence-Côte d'Azur*	*3 676*
	Pays de la Loire	*2 767*	04	Alpes-de-Haute-Provence	112
44	Loire-Atlantique	934	05	Hautes-Alpes	97
49	Maine-et-Loire	630	06	Alpes-Maritime	817
53	Mayenne	262	13	Bouches-du-Rhône	1 633
72	Sarthe	490	83	Var	626
85	Vendée	451	84	Vaucluse	390
	[a]Number of département		20	*Corse*	*290*

Cities (agglomerations, population in 000, 1975)

Paris (capital)	8 424	Tours	235	Thionville	142
Lyon[a]	1 153	Clermont-Ferrand	225	Hagondange-	
Marseille[b]	1 005	Valenciennes	224	Briey	133
Lille	929	Mulhouse	219	Montbéliard	130
Bordeaux	591	Montpellier	205	Nîmes	130
Toulouse	495	Orléans	205	Troyes	128
Nice	438	Douai	203	Denain	126
Nantes	438	Dijon	203	Pau	126
Rouen	389	Rheims	196	Besançon	124
Grenoble	389	Brest	186	Bayonne	120
Toulon	379	Le Mans	185	Saint-Nazaire	119
Strasbourg	355	Caen	183	Bruay-en-Artois	116
Saint-Etienne	335	Metz	181	Perpignan	114
Lens	313	Angers	181	Ville d'Aix-en-	
Nancy	279	Dunkerque[c]	165	Provence	111
Rennes	273	Limoges	165	Lorient	106
Le Havre	264	Avignon	154	Valence	104
Grass-Cannes-		Amiens	153	Annecy	101
Antibes	255	Béthune	145	Calais	100

[a]Lyons [b]Marseilles [c]Dunkirk

Race (1975) French 93½ % (of whom, French by birth 91 %), foreigners 6½ %. Also see special focus
Language French; also some Breton and Basque
Religion (1976) Roman Catholic 90* %, Moslem 4* %, Protestant 2* %, Jewish 2* %
Education (1976/77) Pupils: primary 4 978 000, secondary and vocational 5 262 000, teacher-training (1973/74) 20 647, higher (1974/75) 772 067. Teachers: primary (including pre-primary) 316 564, secondary and vocational 343 084, teacher-training (1973/74) 2 979, higher (1969/70) 31 039
Labour force (1975)

Economic activity	Number	% of total
Agriculture	2 350 600	11
Mining and quarrying	178 900	1
Manufacturing	5 784 800	26
Electricity, gas and water	178 400	1
Construction	1 880 900	9
Distribution and hotels	3 438 200	16
Transport and communication	1 171 000	5
Finance and real estate	1 194 600	5
Other services	4 861 700	22
Compulsory military service	293 800	1
Unemployed	501 100	2
Total	21 834 000	100

Personnel Scientists and engineers (1968): 992 000
Physicians (1974): 77 000*, 1 per 680* people
Standard of living
National income per person (1976): Fr 28 008 = $ 5 859 = £ 3 244
Consumption per person (1975): energy 3 944 kg coal equivalent, electricity (production, 1976) 3 613 kW h, newsprint 8 kg, steel 350 kg
Newspapers (1973): number 103: circulation 11 458 000, 220 per 1 000 people
Telephones (Dec 1976): 15 553 800, 293 per 1 000 people

France

Resources and equipment

Livestock (000, Dec 1976) Cattle 23 898, sheep 10 945, pigs 12 187*
goats 988*, horses 400*, poultry 210 000*
Mineral reserves Coal (known economic, 1973) 1 380 mn tonnes
Lignite (known economic, 1973) 27 mn tonnes
Crude oil (1975) 8 mn tonnes
Natural gas (1975) 127 bn cubic metres
Uranium (1976) 37 000 tonnes
Petroleum refinery capacity (1975) 169 mn tonnes
Electrical capacity (1975) 49 200 megawatts,
of which, hydro 17 574 megawatts, nuclear 3 098 megawatts
Hospital beds (1973) 534 023, 1 per 98 people
Roads (1976) 1 486 000* km = 923 000* mi, density 2.7* km per km²
(includes 690 000* km of local rural roads)
Railways (1975) 34 297 km = 21 311 mi, density 0.063 km per km²
Inland waterways (1975) 8 568 km = 5 324 mi,
of which, in regular use 7 080 km = 4 399 mi
Oil pipelines (1975) 5 231 km = 3 250 mi
Ships (registered, 1977) 1 327, total of 11 613 859 gross tons
Ports (goods traffic, including coastwise, 000 tonnes, 1976)

	loaded	unloaded	total
Marseilles[a,b]	15 908	88 071	103 979
Le Havre[b]	14 625	65 191	79 816
Dunkirk	5 640	27 874	33 514
Rouen[a]	6 551	9 035	15 586
Nantes-St Nazaire[a,b]	1 841	13 480	15 321
Bordeaux[a]	2 062	10 036	12 098

[a]Including subsidiary ports   [b]Crude oil port

Airports (1976)

	Passengers (000)		Cargo (000 tonnes)	
	departures	arrivals	loaded	unloaded
Paris, Orly	5 310	5 360	96	79
Paris, Ch. de Gaulle	3 746	3 771	129	132
Marseilles	1 184	1 207	15	15
Nice	1 110	1 099	10	7
Lyon	875	883	30	9
Paris, Le Bourget	760	775	10	6
Bordeaux	338	341	8	4

Also (1977) 70 other airports with scheduled flights

Durable equipment (at end-year)	000	no per 1 000 people	
Radio sets (1975)	18 197	346	no per
Television sets (1975)	14 197	268	km of road
Passenger cars (1976)	16 230	306	11
Commercial vehicles (1976)	2 195	41	1.5
of which:			
buses and coaches (1976)	50	1	0.03
goods vehicles (1976)	2 145	40	1.4

Companies (sales, $ bn, year ending in 1976)

Cie Française des Pétroles	10.1	Renault	5.4
SN Elf Aquitaine	8.0	Pechiney-Ugine Kuhlmann	4.7
PSA Peugeot-Citroën	7.3	Rhône-Poulenc SA	4.5
Cie Générale d'Electricité	6.7	Denain-Nord-Est-Longwy	4.1
Cie de St Gobain Pont-à-Mousson	6.0	Esso SAF	3.6

Production

Gross domestic product
1976: Fr 1 657.4 bn = $ 346 740 mn = £ 191 980 mn
Growth in real terms: 1960–70 5.8 %pa, 1970–76 4.1 %pa
Structure of gross domestic product (1976)

By origin	Fr bn	% of gdp
Agriculture	79.1	5
Manufacturing, mining, quarrying, electricity, gas and water	499.6	30
Construction	128.5	8
Distribution and hotels	203.4	12
Transport and communication	90.1	5
Other	656.7	40
Total	1 657.4	100
By type of expenditure		
Government final consumption	244.2	15
Private final consumption	1 031.6	62
Stock investment	18.2	1
Gross fixed capital formation	382.3	23
Exports of goods and services	336.9	20
less Imports of goods and services	−355.8	−21
Total	1 657.4	100

Production indices (1970 = 100)	1960	1965	1970	1973	1974	1975	1976	Growth %pa 1960–70	1960–65	1965–70	1970–76
Agricultural	81	93	100	111	112	109	107	2.1	2.6	1.5	1.1
Industrial	57	73	100	120	123	114	124	5.8	5.0	6.5	3.7
Main products (000 t)											
Agriculture											
Wheat	11 014	14 760	12 921	17 850	19 100	15 013	16 150	1.6	6.0	−2.6	3.8
Barley	5 716	7 378	8 126	10 948	9 972	9 336	8 319	3.6	5.2	2.0	0.4
Maize	2 854	3 468	7586	10 692	8 885	8 194	5 544	10.3	4.0	17.0	−5.1
Oats	2 735	2 509	2 103	2 208	2 059	1 898	1 402	−2.6	−1.7	−3.5	−6.5
Potatoes	14 894	11 223	8 904	7 340	7 490	7 228	4 673	−5.0	−5.5	−4.5	−10.2
Sugar, raw value	2 727	2 342	2 804	3 170	2 947	3 230*	2 957*	0.3	−3.0	3.7	0.9*
Grapes	9 677*	10 521*	11 445*	12 544	11 524	11 627*	10981	1.7*	1.7*	1.7*	−0.7*
Wine	6 311	6 842	7 437	8 243	7 548	7 641	7 181	1.6	1.6	1.7	−0.6
Apples	5 693	4 590	3 890*	3 476	3 162	3 285	2 736	−3.7*	−4.2	−3.3*	−5.7*
Rapeseed	83	334	592	653	685	487	510	21.7	32.1	12.1	−2.5
Milk	22 972	26 780	27 276	29 291	29 470	29 686	29 890	1.7	3.1	0.4	1.5
Butter	385	475	481	550	543	556	544	2.2	4.3	0.2	2.1
Cheese	398	579	781	881	916*	943	967	7.0	7.8	6.2	3.6
Eggs	537	582	658	720	735	761	755	2.0	1.6	2.5	2.3
Beef and veal	1 550**	1 636	1 624	1 547	1 886	1 868	1 951	0.5**	1.1**	−0.1	3.1
Pigmeat	1 110**	1 320	1 303*	1 486	1 510	1 545**	1 572	1.6**	3.5**	−0.2*	3.2**
Fish catch	734	768	782	823	808	806	806	0.6	0.9	0.4	0.5
Timber (000 m³)	33 000*	32 416	30 853	32 724	33 290	29 731	29 127	−0.7*	−0.4*	−1.0	−1.0
Energy											
Total energy (000 tce)	69 050	69 900	59 310	48 610	47 120	47 410	na	−1.5	0.2	−3.2	−4.4[ᵉ]
Coal	55 960	51 348	37 838	26 350	24 042	23 652	23 300	−3.8	−1.7	−5.9	−7.8
Lignite	2 276	2 689	2 785	2 764	2 790	3 186	3 135	2.0	3.4	0.7	2.0
Crude oil	1 983	2 988	2 309	1 254	1 080	1 024	1 052	1.5	8.5	−5.0	−12.3
Petroleum products	29 400**	55 100**	92 619	125 494	117 632	102 479	na	12.2**	13.4**	11.0**	2.0[ᵉ]
Natural gas (mn m³)	2 846	5 048	6 880	7 546	7 628	7 358	7 092	9.2	12.1	6.4	0.5
Manufactured gas (mn m³)	10 000**	9 519	8 847	7 238	7 098	6 550	6 097	−1.2**	−1.0**	−1.5	−6.0
Electricity (mn kW h)	72 118	101 442	140 708	174 480	180 402	178 514	191 199	6.9	7.0	6.8	5.2
of which, hydro (mn kW h)	40 344	46 429	56 612	47 543	56 830	59 892	na	3.4	2.8	4.0	1.1[ᵉ]
nuclear (mn kW h)	130	897	5 147	13 969	13 932	17 451	na	44.5	47.2	41.8	27.7[ᵉ]
Mining											
Iron ore (Fe content)	21 745	19 348	17 759	15 671	16 714	15 075	13 555	−2.0	−2.3	−1.7	−4.4
Bauxite	2 067	2 662	2 992	3 299	2 938	2 563	2 289	3.8	5.2	2.4	−4.4
Lead conc (Pb content)	19	18	29	25	23	22	28	4.3	−1.1	10.0	−0.6
Tungsten conc (oxide content)	—	—	0.09	0.95	0.90	0.78	na	na	na	na	54.8[ᵉ]

France

Main products (000 t)	1960	1965	1970	1973	1974	1975	1976	Growth %pa 1960–70	1960–65	1965–70	1970–76
Mining											
Uranium (U content)	na	1.09	1.14	1.62	1.67	1.74	2.06	na	na	0.9	10.3
Gold (000 kg)	1.23	1.78	1.76	1.72	1.26	1.46	na	3.7	7.7	−0.2	−3.6g
Potash (oxide content)	1 733	2 057	1 904	2 263	2 275	2 085	1 738	1.0	3.5	−1.5	−1.5
Salt	3 725*	4 370	5 664	6 117	5 995	5 347	5 575	4.3*	3.3*	5.3	−0.2
Manufacturing								1.9	2.8	1.1	1.6g
Beer (000 hl)	17 261	19 795	20 871	22 664	22 098	22 660	na	4.6*	4.3*	4.8	4.6g
Cigarettes (mn units)	44 650*	55 220	69 886	72 454	81 438	87 400	na	−1.5	−3.3	0.3	−1.6
Cotton yarn	314	266	270	281	279	232	245	−2.3	−3.3	−1.3	−1.1
Cotton fabrics	249	210	197	208	205	174	184	6.4	6.3	6.5	1.9
Man-made fibres	164	223	306	400	364	293	343	4.6	4.9	4.3	0.3
Wood pulp	1 144	1 451	1 787	1 935	1 977	1 753	1 824	−0.2d	0.8e	−0.9	−8.3*
Newsprint	436c	450	430	305	276	256	255*	5.4**d	4.3**e	6.4	2.7
Other paper	2 300**c	2 716	3 704	4 480	4 784	3 862	4 357	33.9	54.2	16.4	5.6
Synthetic rubber	17	148	316	458	463	350	437	6.0	7.3	4.8	1.2
Sulphuric acid	2 046	2 916	3 682	4 383	4 689	3 758	3 958	5.6	9.2	2.1	2.8
Nitric acid	1 490	2 319	2 575	3 414	3 754	3 287	3 035	16.1	14.9	17.4	6.0g
Plastics and resins	347	695	1 548	2 532	2 640	2 076	na	8.6	13.5	3.9	0.5
Fertilisers, nitrogenousa	574	1 082	1 313	1 476	1 642	1 694	1 354	2.9*	6.8*	−0.8*	−0.8
Fertilisers, phosphatea	995*	1 380*	1 324	1 611	1 693	1 709	1 259	7.3	9.3	5.3	0.3
Cement	14 349	22 423	29 009	30 713	32 469	29 708	29 500	0.4	−0.3	1.1	−3.7
Coke	13 607	13 382	14 152	11 880	12 282	11 445	11 313	3.1	2.2	4.1	−0.5
Pig iron	14 365	16 025	19 582	20 750	22 967	18 317	19 030	3.2	2.6	3.9	−0.4
Crude steel	17 281	19 604	23 773	25 264	27 023	21 528	23 235	4.8	7.4	2.3	0.2
Aluminium	238	340	381	359	393	383	385	4.2	5.2	3.1	1.2
Zincb	149	192	224	258	276	181	241	2.8	0.7	4.9	2.8
Radio sets (000 units)	2 214	2 298	2 921	3 017	3 374	3 051	3 458	8.7	13.8	3.9	2.7
Television sets (000 units)	655	1 250	1 511	1 695	1 694	1 606	1 774	8.0	3.9	12.3	5.5
Passenger cars (000 units)	1 136	1 374	2 458	3 202	3 046	2 951	3 388	2.2	0.7	3.8	8.3
Commercial vehicles (000 units)	234	242	292	394	418	346	471	4.9	−4.2	14.9	3.9
Merchant vessels (000 grt)	594	479	960	1 124	1 343	1 315	1 208				
Construction											
Dwellings (000 units)	280**f	450**f	471	517	516	516	530	5.3**	10.0**f	0.9**	2.0

aYears ending June 30th   bIncluding secondary zinc   c1961   d1961–70   e1961–65   fEstimate based on number of permits issued   g1970–75

Transport traffic

	1960	1970	1976	Growth %pa 1960–70	1970–76
Passenger-kilometres (mn)					
Road	na	293 000	415 000	na	6.0
Rail	32 000	41 080	51 170	2.5	3.7
Air	5 229	13 587	25 192	10.0	10.8
of which, international	2 912	10 988	18 506	14.2	9.1
Cargo: tonne-kilometres (mn)					
Road	37 087a	65 900	92 500	8.6b	5.8
Rail	56 930	70 403	68 518	2.1	−0.5
Water	11 358a	14 183	12 200	3.2b	−2.5
Air	152	542	1 380	13.5	16.9
of which, international	66	493	1 213	22.4	16.2
Sea: tonnes (mn)					
Goods loaded	19.1	25.2*	30.5c	2.8*	4.9*d
Goods unloaded	54.6	168.4*	190.7c	11.9*	3.2*d
Tourism					
Number of visitorse (000)	8 200**	13 700**	17 385	5.3**	4.1**
Gross receipts ($ mn)	525*	1 318	3 613	9.6*	18.3

a1963   b1963–70   c1974   d1970–74   eArrivals at all accommodation establishments

Finance and trade

Price and wages indices (1970 = 100)	1960	1970	1976	Growth %pa 1960–70	1970–76
Consumer prices	67.3	100.0	166.9	4.0	8.9
Wholesale prices	75.5	100.0	160.3	2.9	8.2
Wages (earnings)	44.4	100.0	237.7	8.5	15.5
Share prices	96.7	100.0	94.4	0.3	−1.0
Money stock					
(end-year, Fr bn)	90.0**	235.33	465.23	10.1**	12.0

Budget (1976) Revenue: Fr 350.7 bn = $ 73 377 mn = £ 40 628 mn
Expenditure: Fr 363.4 bn = $ 76 021 mn = £ 42 092 mn,
of which, education and culture 26* %, defence 18* %

Balance of payments ($ mn)	1972	1973	1974	1975	1976
Balance of goods (fob)	+1 299	+776	−3 862	+1 507	−4 671
Balance of services	+637	+573	+451	+1 085	+1 078
Balance of transfers	−1 639	−2 041	−2 532	−2 596	−2 439
Current balance	+297	−691	−5 942	−3	−6 033
Long-term capital flow	−648	−2 239	−182	−1 044	−1 636
Reserves and debt (end-year, $ mn)					
International reserves	10 015	8 529	8 852	9 728	10 194
Public debt: domestic	15 265	16 878	19 660	24 808	26 949
external	1 628	1 809	1 836	1 193	958

External trade (1976) Imports: Fr 308.12 bn = $ 64 460 mn = £ 35 691 mn
Exports: Fr 273.24 bn = $ 57 163 mn = £ 31 651 mn

Main imports	% of total	Main exports	% of total
Crude oil	18	Machinery, non-electric	14
Machinery, non-electric	11	Food	12
Food	10	(of which, cereals 4)	
Chemicals	8	Motor vehicles	12
Motor vehicles	5	Chemicals	10
Electrical machinery	5	Electrical machinery	7
Iron and steel	5	Iron and steel	7
Textile yarns and fabrics	3	Textile yarns and fabrics	3
Non-ferrous metals	3	Metal small manufactures	3
Instruments	2	Petroleum products	2
Metal small manufactures	2	Wines and spirits	2
Coal	2	Clothing	2
Main sourcesa		*Main destinationsb*	
West Germany	19	West Germany	17
Belgium-Luxembourg	10	Italy	11
Italy	9	Belgium-Luxembourg	10
United States	7	United Kingdom	6
Saudi Arabia	6	Netherlands	5
Netherlands	6	United States	5
United Kingdom	5	Switzerland	4
Iraq	2	Algeria	3
Switzerland	2	Spain	3

aEEC 50 %   bEEC 51 %

Special focus

Foreign residents in France, 1975

	Number (000)
Portuguese	759
Algerian	711
Spanish	497
Italian	463
Moroccan	260
Tunisian	140
Polish	94
Jugoslav	70
Total (including others)	3 442 = 6½ % of the total population

Germany, East

German Democratic Republic (GDR)
Deutsche Demokratische Republik (DDR)

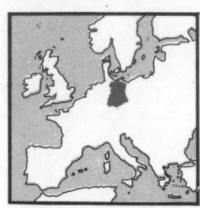

Location Northern Europe
With a northern coastline on the Baltic Sea, West Germany is to the west and south-west, Czechoslovakia to the south-east and Poland to the east across the Oder and Neisse rivers
Land Area 108 178 km² = 41 768 mi²
Usage (1975): agriculture 62 950 km² (58 %), of which, arable 46 990 km² (43 %), cropland 2 370 km² (2 %), pastures 13 590 km² (13 %); forests 29 520 km² (27 %)

Climate Temperate
Weather at East Berlin, 55 m altitude
Temperature: hottest month July 14–24 °C, coldest Jan minus 3–2 °C
Rainfall (av monthly): driest month March 33 mm, wettest July 73 mm
Time 1 hour ahead of GMT
Measures Metric system
Monetary unit Mark[a] (M) = 100 pfennig
Rate of exchange (1976 av): *Basic* free M 1.84 = $ 1, M 3.32 = £ 1
Tourist free M 2.50 = $ 1, M 4.52 = £ 1
Valuta or exchange mark (VM) par: VM 3.48 = $ 1, free VM 6.29 = £ 1
[a]DDR-Mark or 'Ostmark'

Summary

Political Communist republic, which was formed in 1949 from the zone of Germany occupied by Soviet Union in 1945 after the second world war. A treaty between East and West Germany was signed on December 21, 1972 to settle relationships between the two states. East Berlin is included here throughout as part of East Germany. Member of UN, Comecon and Warsaw Pact
Economic An industrial economy, with machinery the largest export; there is also a significant agricultural sector, providing about one-tenth of net material product. The industrial growth rate is high at some 6½ %pa; growth industries include electronics and machine tools

People, resources and equipment

Population 1960 17.24 mn, 1970 17.06 mn, 1976 16.79 mn
Growth: 1960–70 −0.1 % pa, 1970–76 −0.3 % pa
Density (1976): 155 people per km²
Vital statistics (rate per 1 000 people, 1976): births 11.6, deaths 14.0
Regions (districts, population in 000, Dec 31, 1976; total of 16.77 mn)

East Berlin	1 106	Gera	737	Potsdam	1 117
Cottbus	874	Halle	1 863	Rostock	871
Dresden	1 826	Karl-Marx-Stadt	1 962	Schwerin	589
Erfurt	1 238	Leipzig	1 435	Suhl	548
Frankfurt an der Oder	690	Magdeburg	1 283		
		Neubrandenburg	625		

Cities (population in 000, 1976)

East Berlin (capital)	1 106	Halle	236	Gera	116
Leipzig	565	Rostock	215	Schwerin	109
Dresden	509	Erfurt	204	Dessau	101
Karl-Marx-Stadt	306	Zwickau	122	Jena	101
Magdeburg	278	Potsdam	120	Brandenburg	95*

Race (1961) German 99 %, Wendish 1 %
Language German
Religion (1970) Protestant 80** %, Roman Catholic 10** %
Education (1975/76) Pupils: primary 2 578 782, secondary 47 854, vocational 412 785, higher (1974/75) 306 783. Teachers: primary and secondary 158 543, vocational and teacher-training (1969/70) 20 115, higher (1974/75) 33 570
Labour force (1971)

Economic activity	Number	% of total
Agriculture	959 979	12
Mining and quarrying	198 560	2
Manufacturing	3 093 689	38
Electricity, gas and water	91 150	1
Construction	614 004	7
Distribution and hotels	842 421	10
Transport and communication	555 241	7
Finance and real estate	87 691	1
Other	1 771 516	22
Total	8 214 251	100

Personnel (1974) Physicians: 30 798, 1 per 549 people

Standard of living National income per person (1976):
M 10 000*** = $ 4 000*** = £ 2 200***
Consumption per person (1975): energy 6 835 kg coal equivalent, electricity (production, 1976) 5 310 kW h, newsprint 8.2 kg, steel 566 kg
Newspapers (1974): number 40; circulation 7 753 000, 452 per 1 000 people
Telephones (Dec 1976): 2 750 600, 164 per 1 000 people
Livestock (000, Dec 1976) Cattle 5 471, pigs 11 291, sheep 1 870, goats 42, horses 68, poultry 48 444
Mineral reserves Coal (1956) 200 mn tonnes
Lignite (1966) 30 000 mn tonnes
Crude oil (1975) 3 mn tonnes
Natural gas (1975) 96 bn cubic metres
Petroleum refinery capacity (1975) 17.2* mn tonnes
Electrical capacity (1975) 16 928 megawatts, of which, nuclear 950 megawatts
Hospital beds (1974) 184 214, 1 per 92 people
Roads (1975) 126 933 km = 78 873 mi, density 1.2 km per km² (of which, 47 573 km non-urban)
Railways (1975) 14 298 km = 8 884 mi, density 0.13 km per km²
Inland waterways (1975) 2 538 km = 1 577 mi
Oil pipelines (1975) 951 km = 591 mi
Ships (registered, 1977) 447, total of 1 486 838 gross tons
Ports Rostock, Stralsund, Wismar
Airports Schönefeld (East Berlin), Schkeuditz (Leipzig), Dresden, Erfurt

Durable equipment (Dec 1975)	000	no per 1 000 people	
Radio sets	6 160*	366*	no per
Television sets	5 177	307	km of road
Passenger cars	1 880	112	15
Commercial vehicles	260	15	2.0

Production

Net material product 1976: M 148* bn = $ 59 000* mn[a] = £ 33 000* mn[a]
Growth in real terms: 1960–70 4.3 % pa, 1970– 76 5.1 % pa
Gross domestic product
1976 est: M 175*** bn = $ 70 000*** mn[a] = £ 39 000*** mn[a]
[a]Converted at tourist rate
Structure of net material product (1976, at 1975 prices)

By origin	M bn	% of gdp
Agriculture	14.2	9
Manufacturing, mining, quarrying and utilities	90.5	60
Construction	11.3	8
Distribution and hotels	21.9	15
Transport and communication	7.6	5
Other	4.4	3
Total (excluding statistical discrepancy)	147.5	100

Production indices (1970 = 100)	1960	1970	1976	Growth % pa 1960–70	1970–76
Agricultural	94	100	108	0.6	1.3
Industrial	55	100	145	6.2	6.4

Main products (000 t)

Agriculture	1960	1970	1976	1960–70	1970–76
Wheat	1 456	2 132	2 715	3.9	4.1
Barley	1 269	1 926	3 456	4.3	10.2
Rye	2 126	1 483	1 455	−3.5	−0.3
Oats	1 007	558	506	−5.7	−1.6
Potatoes	14 821	13 054	6 816	−1.3	−10.3
Sugar, raw value	880*	490*	560*	−5.7*	2.2*
Cabbages	310**	334	306*	0.7**	−1.4*
Apples	206[a]	198	469*	−0.6[b]	15.4*
Rapeseed	182	180	312*	−0.1	9.6*
Milk	5 730	7 091	8 092*	2.1	2.2*
Butter	175	216	278	2.1	4.3
Cheese	37	149	196*	15.0	4.7*
Eggs	193	244	287	2.4	2.7
Beef and veal	245*	363	421*	4.0*	2.5*
Pigmeat	687	815	1 123*	1.7	5.5*
Fish catch	114	322	375e	10.9	3.1f
Timber (000 m³)	8 400	7 372	8 706	−1.3	2.8
Energy					
Total energy (000 tce)	70 460	80 600	81 090e	1.3	0.1f
Coal	2 721	1 049	456	−9.1	−13.0
Lignite	225 465	261 482	246 889	1.5	−1.0
Crude oil	na	90*	80e	na	−2.3*
Petroleum products[e]	3 195	11 377	17 800*e	13.5	9.4*f
Natural gas (mn m³)	26	1 232	7 270e	47.1	42.6f

Germany, East

Main products (000 t)	1960	1970	1976	Growth % pa 1960–70	1970–76
Energy					
Manufactured gas (mn m³)	3 045	4 269	5 500	3.4	4.3
Electricity (mn kW h)	40 305	67 650	89 148	5.3	4.7
of which, nuclear (mn kW h)	—	464	2 740e	na	42.6f
Mining					
Iron ore (Fe content)	408	106	23e	−12.6	−26.3f
Potash (oxide content)	1 666	2 419	3 019e	3.8	4.5f
Salt	1 785	2 180	2 430e	2.0	2.2f
Manufacturing					
Margarine	181	188	179	0.4	−0.8
Beer (000 hl)	13 424	16 642	21 202	2.2	4.1
Cigarettes (mn units)	18 187	16 567	19 828	−0.9	3.0
Cotton yarn	73	68	59e	−0.6	−2.8f
Cotton fabrics (mn m²)	254	248	274	−0.2	1.7
Wool yarn	80	68	62e	−1.6	−1.8f
Man-made fibres	156	214	283e	3.2	5.7f
Wood pulp	640	842	864e	2.8	0.5f
Newsprint	89	97	105	0.9	1.3
Other paper	721	966	1 096	3.0	2.1
Synthetic rubber	87	118	145	3.1	3.5
Tyres (000 units)	2 714	4 692	6 432	5.6	5.4
Sulphuric acid	730	1 099	966	4.2	−2.1
Soda ash	594	676	829	1.3	3.5
Caustic soda	327	413	441	2.4	1.1
Plastics and resinsd	115	370	694	12.4	11.0
Fertilisers, nitrogenous	334	395	776	1.7	11.9
Fertilisers, phosphate	166	430	423	10.0	−0.3
Cement	5 032	7 984	11 345	4.7	6.0
Coke	3 206	2 572	1 693	−2.2	−6.7
Pig iron	1 995	1 994	2 529	0.0	4.0
Crude steel	3 750	5 053	6 739	3.0	4.9
Aluminium	40*	60*	60*e	4.0*	0.0*f
Copper	40*	50*	43*e	0.0*	−3.0*f
Radio sets (000 units)	810	807	1 068e	−0.1	5.7f
Television sets (000 units)	416	380	509e	−0.9	6.0f
Passenger cars (000 units)	64	127	164	7.0	4.3
Commercial vehicles (000 units)	12	27	36	8.1	4.9
Merchant vessels (000 grt)	na	334	355	na	1.0
Construction					
Dwellings (000 units)	80	76	151	−0.5	12.1

a1961–65 average   b1961–65 to 1970   cExcluding jet fuel, with production of 251 thousand tonnes in 1967 compared with a total for other petroleum products of 7 793 thousand tonnes   dIncomplete coverage   e1975   f1970–75

Transport traffic	1960	1970	1976	Growth % pa 1960–70	1970–76
Passenger-kilometres (mn)					
Road*	na	19 041	23 392e	na	4.2d
Rail	21 288	17 666	22 339	−1.8	4.0
Water	na	232	218	na	−1.0
Air	299a	947	1 448	15.5b	7.3
Cargo: tonne-kilometres (mn)					
Road	5 002	12 233	16 691c	9.3	6.4d
Rail	32 860	41 513	51 801	2.4	3.8
Water	2 252	2 358	1 947	0.5	−3.1
Air	8.3a	26.6	50.5	15.6b	11.3
Sea: tonnes (mn)					
Goods loaded	1.66	2.41	3.22	3.8	4.9
Goods unloaded	2.68	10.31	11.64	14.4	2.0

a1962   b1962–70   c1975   d1970–75   ePublic transport only

Tourism (1976) Number of visitors 17 312 700, of whom 40 % from Poland, 18 % from West Berlin

Finance and trade

Price and wage indices (1970 = 100)	1960	1970	1976	Growth % pa 1960–70	1970–76
Consumer prices	100.0	100.0	96.6	0.0	−0.6
Wages (earnings)	76	100	122	2.8	3.3

Budget (1976) Revenue: M 117 588 mn = $ 47 035 mna = £ 26 015 mna
Expenditure: M 117 128 mn = $ 46 851 mna = £ 25 913 mna
aConverted at tourist rate

External trade (1976)

Imports (fob): VM 45 921 mn = $ 13 196 mn = £ 7 301 mn
Exports: VM 39 536 mn = $ 11 361 mn = £ 6 286 mn

Main imports (1970)	% of total	*Main exports* (1975)	% of total
Machinery and metal products	36	Machinery, non-electric	25
Mineral products	32	Electrical machinery	7
Food and light industrial products	20	Motor vehicles	3
Agricultural and timber products	11	Ships	3
		Railway vehicles	2
		Furniture	2*

Main sources (1974)		*Main destinations* (1974)	
Soviet Union	30	Soviet Union	33
West Germany	9	Czechoslovakia	10
Czechoslovakia	7	West Germany	10
Poland	7	(of which,	
Hungary	5	West Berlin 3)	
Netherlands	3	Poland	9
Rumania	3	Hungary	6
Switzerland	3	Bulgaria	4
Bulgaria	3	Rumania	3
United Kingdom	3	United Kingdom	2
Austria	2	Jugoslavia	2

Special focus

East and West Germany, 1976

	East mn	%	West mn	%	Total mn	%
Area (km²)	0.11	30	0.25	70	0.36	100
Population	16.8	21	61.5	79	78.3	100
Production						
Wheat (tonnes)	2.7	29	6.7	71	9.4	100
Energy (tce, 1975)	81.1	33	165.9	67	247.0	100
Sulphuric acid (tonnes)	1.0	17	4.7	83	5.6	100
Crude steel (tonnes)	6.7	14	42.4	86	49.2	100

Germany, West

Federal Republic of Germany
Bundesrepublik Deutschland

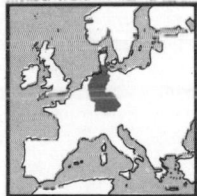

Location Northern central Europe
There are coastlines on the North and Baltic Seas, with Belgium, Luxembourg and Netherlands to the west, Denmark to the north, East Germany and Czechoslovakia to the east, Austria to the south-east, Switzerland to the south and France to the south-west. West Berlin lies within East Germany
Land Area 248 577 km² = 95 976 mi²
Usage (1975): agricultural 133 030 km² (54 %), of which, arable 75 380 km² (30 %), cropland 5 210 (2 %), pastures 52 440 km² (21 %); forests 71 620 (29 %)
Climate Temperate
Weather at Frankfurt, 103 m altitude
Temperature: hottest month July 14–25 °C, coldest Jan minus 1–3 °C
Rainfall (av monthly): driest month March 38 mm, wettest Aug 76 mm
Time 1 hour ahead of GMT
Measures Metric system
Monetary unit Deutsche mark (DM) = 100 pfennig
Rate of exchange: *free* 1976 av: DM 2.518 = $ 1, DM 4.548 = £ 1
1977 av: DM 2.322 = $ 1, DM 4.053 = £ 1
par (central value within the European 'snake')
1976 (to Oct 17): DM 3.21979 = SDR 1;
1976 (from Oct 18) and 1977: DM 3.15665 = SDR 1

Summary

Political Republic, which was formed from the British, French and US zones of Germany occupied in 1945 after the second world war; sovereignty was completed on May 5, 1955. The Saar was returned by France on January 1, 1957. A treaty between West and East Germany was signed on December 21, 1972, to settle relationships between the two states. West Berlin is included here throughout as part of West Germany. Member of UN, OECD, Council of Europe, Nato, EEC and WEU
Economic An industrial economy, with manufacturing absorbing about one-third of the labour force and providing about one-third of gross domestic product; exports are mainly machinery, motor vehicles, chemicals and iron and steel. Food and crude oil are the main imports

Germany, West

People

Population 1960 55.42 mn, 1970 60.71 mn, 1976 61.51 mn
Growth: 1960–70 0.9 %pa, 1970–76 0.2 %pa
Density (1976): 247 people per km²
Vital statistics (rate per 1 000 people, 1976): births 9.8, deaths 11.9
Households (1970) 21 990 469; population in households 60.18 mn
(97.5 % of census population)
Regions (population in 000, December 31, 1976; total of 61.44 mn)

West Berlin	1 951	Hessen	5 538
Länder		Niedersachsenᵇ	7 227
Baden-Württemberg	9 119	Nordrhein-Westfalenᶜ	17 073
Bayernᵃ	10 804	Rheinland-Pfalzᵈ	3 649
Bremen	710	Saarland	1 089
Hamburg	1 699	Schleswig-Holstein	2 583

ᵃBavaria ᵇLower Saxony ᶜNorth Rhine-Westphalia ᵈRhineland-Palatinate

Cities (population in 000, December 31, 1975)

Bonn (capital)ᵃ	284	Braunschweig	269	Leverkusen	166
West Berlin	1 985	Münster (Westfalen)	265	Neuss	148
Hamburg	1 717	Kiel	262	Bremerhaven	144
Münchenᵇ	1 315	Mönchengladbach	261	Darmstadt	137
Kölnᶜ	1 014	Wiesbaden	251	Oldeburg	135
Essen	678	Augsburg	250	Regensburg	134
Düsseldorfᵉ	664	Aachen	242	Remscheid	133
Frankfurt		Oberhausen	237	Heidelberg	129
am Mainᵈ	636	Lübeckʰ	232	Recklinghausen	126
Dortmund	630	Hagen	229	Wolfsburg	126
Stuttgart	600	Krefeld	228	Göttingen	124
Duisburg	592	Kassel	206	Koblenz	118
Bremen	573	Saarbrucken	205	Salzgitter	117
Hannoverᵍ	553	Herne	191	Siegen	117
Nürnbergᶠ	499	Mülheim an der Ruhr	189	Offenbach	
Bochum	415	Mainz	184	am Main	115
Wuppertal	405	Freiburg im Breisgau	175	Wurzburg	113
Gelsenkirchen	323	Hamm	172	Pforzheim	109
Bielefeld	316	Solingen	172	Hildesheim	105
Mannheim	314	Lüdwigshafen		Osnabrück	105
Karlsruhe	280	am Rhein	170	Wilhelmshaven	103

ᵃSeat of government and provisional capital; official capital Berlin   ᵇ Munich
ᶜCologne ᵈFrankfurt ᵉDusseldorf ᶠNuremberg ᵍHanover ʰLubeck
Race (1961) German 99 %, Italian ½ %
Language German
Religion (1970) Protestant 49 %, Roman Catholic 45 %
Education Pupils (1975/76): primary 6 425 217, secondary 3 176 508,
vocational 2 287 768, higher 840 757.
Teachers (1973/74): primary 269 310, secondary 187 478,
vocational 127 019, higher 93 841
Labour force (1975)

Economic activity	Number	% of total
Agriculture	1 732 000	6
Mining and quarrying	369 000	1
Manufacturing	9 347 000	35
Electricity, gas and water	247 000	1
Construction	2 073 000	8
Distribution and hotels	3 772 000	14
Transport and communication	1 537 000	6
Finance and real estate	1 360 000	5
Other	6 441 000	24
Total	26 878 000	100

of whom, foreign workers (gastarbeiter): 2 100 000* = 8* % of total
Personnel Scientists and engineers (1970): 1 083 000
Physicians (1974): 120 260, 1 per 516 people
Standard of living
National income per person (1976): DM 16 245 = $ 6 451 = £ 3 572
Consumption per person (1975): energy 5 345 kg coal equivalent,
electricity (production, 1976) 5 424 kW h, newsprint 17 kg, steel 490 kg
Newspapers (1974): number 320; circulation 17 872 000,
289 per 1 000 people
Telephones (Dec 1976): 21 161 800, 344 per 1 000 people

Resources and equipment

Livestock (000, Dec 1976) Cattle 14 496, pigs 20 589, sheep 1 091,
horses 355, chickens 88 085
Mineral reserves Coal (1971) 230 304 mn tonnes
Lignite (1972) 55 851 mn tonnes
Crude oil (1975) 71 mn tonnes
Natural gas (1975) 306 000 mn cubic metres
Uranium (1976) 1 500 tonnes

Petroleum refinery capacity (1975) 153.7 mn tonnes
Electrical capacity (1975) 74 356 megawatts,
of which, nuclear 3 504 megawatts
Hospital beds (1974) 716 530, 1 per 87 people
Roads (1976) 469 568 km = 291 776 mi, density 1.9 km per km²
Railways (1975) 32 006 km = 19 888 mi, density 0.13 km per km²
Inland waterways (1975) 5 892 km = 3 661 mi,
of which, in regular use 4 381 km = 2 722 mi
Oil pipelines (1975) 2 086 km = 1 296 mi
Ships (registered, 1977) 1 975, total of 9 592 314 gross tons
Ports (goods traffic, 000 tonnes, 1976)

	International		Coastwise		Total
	loaded	unloaded	loaded	unloaded	
Hamburg	12 208	37 317	1 677	337	51 539
Wilhelmshavenᵃ	650	27 619	1 574	355	30 198
Bremen	7 382	12 500	185	2 058	22 125
Emden	1 863	9 113	560	140	11 676
Nordenham	256	5 503	62	382	6 203
Lubeck	2 406	3 400	14	133	5 953
Brake	707	4 126	36	72	4 941
Puttgarden	1 421	982	—	—	2 403

ᵃCrude oil port
Airports (1976)

	Passengers (000)		Cargo (000 tonnes)	
	departures	arrivals	loaded	unloaded
Frankfurt	6 531	6 593	285	299
West Berlin	1 989	1 986	9	14
Dusseldorf	2 593	2 589	18	22
Munich	2 301	2 307	17	22
Hamburg	1 818	1 828	16	22
Stuttgart	1 188	1 226	13	13
Hanover	868	879	8	8
Cologne	867	901	15	20
Nuremberg	351	364	5	5
Bremen	289	290	2	2

Also (1977) 19 other airports with scheduled flights

Durable equipment (Dec 1976)	000	no per 1 000 people	
Radio sets	20 244	329	no per
Television sets	18 481	301	km of road
Passenger cars	19 180	312	41
Commercial vehicles	1 322	22	2.8
of which,			
buses and coaches	62	1	0.1
goods vehicles	1 260	21	2.7

Companies (sales, $ bn, year ending in 1976)

Veba AG	10.8	Ruhrkohle AG	4.9
Daimler-Benz AG	9.3	Mannesmann	4.7
Hoechst AG	9.3	RWE AG	4.5
Volkswagenwerk AG	8.5	Fried. Krupp Hütt. AG	3.9
BASF AG	8.3	Gutehoffnungshütte AV	3.5
Bayer AG	8.3	Adem Opel AG	3.5
Siemens AG	8.2	Ford-Werke AG	3.4
Thyssen AG	7.9	Salzgitter AG	3.4
AEG-Telefunken AG	5.3	Robert Bosch GmbH	3.3
Deutsche Shell AG	5.0	Flick Group	3.3
Esso AG	5.0	Deutsche BP	3.0

Production

Gross domestic product 1976: DM 1 122.8 bn = $ 445 910 mn = £ 246 880 mn
Growth in real terms: 1960–70 4.7 %pa, 1970–76 2.5 %pa
Structure of gross domestic product (1976)

By origin	DM bn	% of gdp
Agriculture	32.0	3
Mining, quarrying, electricity, gas and water	47.7	4
Manufacturing	417.6	37
Construction	79.3	7
Distribution	107.0	10
Transport and communication	65.7	6
Other	373.5	33
Total	1 122.8	100
By type of expenditure		
Government final consumption	228.6	20
Private final consumption	621.9	55
Stock investment	13.4	1
Gross fixed capital formation	232.9	21
Exports of goods and services	292.4	26
less Imports of goods and services	−266.4	−24
Total	1 122.8	100

Production indices (1970 = 100)	1960	1965	1970	1973	1974	1975	1976	Growth %pa 1960–70	1960–65	1965–70	1970–76
Agricultural	83	85	100	101	105	103	101	1.8	0.3	3.3	0.1
Industrial	57.3	75.7	100.0	112.6	109.9	103.2	112.0	5.7	5.7	5.7	1.9
of which, consumer goods	60.5	78.9	100.0	113.3	107.9	102.5	111.7	5.1	5.4	4.9	1.9
investment goods	55.5	74.2	100.0	109.6	105.8	101.5	110.6	6.1	6.0	6.1	1.7
Main products (000 t)											
Agriculture											
Wheat	4 965	4 348	5 662	7 134	7 761	7 014	6 702	1.3	−2.6	5.4	2.8
Barley	3 221	3 364	4 754	6 622	7 048	6 970	6 487	4.0	0.9	7.1	5.3
Rye	3 795	2 822	2 663	2 576	2 560	2 125	2 100	−3.5	−5.7	−1.1	−3.9
Oats	2 178	2 052	2 484	3 045	3 482	3 445	2 497	1.3	−1.2	3.9	0.1
Maize	20	96	507	573	521	531	480	38.2	36.8	39.5	−0.9
Potatoes	24 545	18 095	16 250	13 676	14 549	10 853	9 808	−4.0	−5.9	−2.1	−8.1
Sugar, raw value	1 956	1 562	2 056	2 453	2 439	2 534	2 736*	0.5	−4.4	5.7	4.9*
Cabbages	300**	421	586	589	528	465	423	7.0**	7.0**	6.8	−5.3
Grapes	952*	645*	1 266*	1 369*	871*	1 183*	1 108*	2.9*	−7.5*	14.4*	−2.2*
Wine	684	463	910	984	626*	850*	797*	2.9	−7.5	14.5	−2.2*
Apples	2 521	1 205	1 777	2 016	1 282	2 035	1 487	−3.4	−13.7	8.1	−2.9
Pears	646	286	550	414	332	387	388	−1.6	−15.0	14.0	−5.6
Plums	546	446	619	444	385	194	451	1.2	−4.0	6.8	−5.1
Rapeseed	68	107	185	222	301	199	222	10.5	9.5	11.6	3.1
Milk	19 250	21 183	21 856	21 265	21 508	21 604	22 165	1.3	1.9	0.6	0.2
Butter	431	501	505	512	511	521	545	1.6	3.1	0.2	1.3
Cheese	294	376	493	570	596	619	651	5.3	5.0	5.6	4.7
Lard	261	303	347	347	353*	373*	385*	2.9	3.0	2.8	1.7*
Eggs	450	680	900	925	890	893	883*	7.2	8.6	5.8	−0.3*
Beef and veal	896	959	1 292	1 193	1 366	1 337	1 405	3.7	1.4	6.1	1.4
Pigmeat	1 515	1 886	2 186	2 189	2 303	2 329	2 400	3.7	4.5	3.0	1.6
Poultrymeat	100	152	258	289	266	282	300*	10.0	8.7	11.2	2.5*
Cattle hides	82	102	162	155	176	173	178*	7.0	4.5	9.7	1.6*
Fish catch	674	633	613	478	526	442	454	−0.9	−1.3	−0.6	−4.9
Timber (000 m³)	25 300	26 336	28 196	31 941	32 869	27 503	30 025	1.1	0.8	1.4	1.1
Energy											
Total energy (000 tce)	183 320	183 890	174 330	170 280	170 540	165 870	na	−0.5	0.1	−1.0	−1.0c
Coal	143 255	135 464	116 341	102 994	100 893	96 755	95 890	−2.0	−1.1	−3.0	−3.2
Lignite	97 999	103 641	108 437	118 658	126 044	123 377	134 536	1.0	1.1	0.9	3.7
Crude oil	5 530	7 884	7 535	6 638	6 191	5 741	5 526	3.1	7.3	−0.9	−5.0
Petroleum products	26 892	63 710	100 984	115 096	105 630	91 784	na	14.2	18.8	9.6	−1.9c
Natural gas (mn m³)	950**	3 378	13 011	19 599	20 334	18 876	19 030	30.0**	29.0**	31.0	6.5
Manufactured gas (mn m³)	24 800**	23 200**	20 119	16 904	17 238	16 693	15 585	−2.1**	−1.3**	−2.8**	−4.2
Electricity (mn kW h)	118 986	172 340	242 612	298 995	311 710	301 802	333 652	7.4	7.7	7.1	5.5
of which, hydro (mn kW h)	15 409	15 365	17 758	15 516	17 876	17 111	na	1.4	−0.1	2.9	−0.7c
nuclear (mn kW h)	—	117	6 030	11 755	12 136	21 406	na	na	na	120.0	28.8c
Mining											
Iron ore (Fe content)	4 412	2 552	1 773	1 620	1 412	1 053	722	−8.7	−10.4	−7.0	−13.9
Lead conc (Pb content)	50	48	40	34	30	32	32	−2.1	−0.6	−3.5	−3.6
Zinc conc (Zn content)	114	109	127	123	116	116	115	1.1	−0.9	3.1	−1.6
Gold (000 kg)	2.6	2.5	3.2	9.3	9.7	10.9	na	1.9	−1.2	5.0	28.1c
Phosphate rock	na	43*	69*	93*	85*	75*	na	na	na	10.0*	1.7*c
Potash (oxide content)	2 316	2 740	2 645	2 975	3 090	2 607	na	1.3	3.4	−0.7	−0.3c
Salt	4 300**	6 863	10 511	9 677	11 497	9 500	na	9.3**	9.8**	8.9	−2.0c
Manufacturing											
Fishmeal	83	80	72	62	63	56	na	−1.3	−0.6	−2.1	−4.9c
Margarine	588	572	542	543	527	509	na	−0.8	−0.5	−1.1	−1.3c
Beer (000 hl)	47 324	67 439	81 609	87 450	87 688	88 426	na	5.6	7.3	3.9	1.6c
Cigarettes (mn units)	73 210	102 060	129 665	140 568	143 179	144 161	na	5.9	6.8	4.9	2.1c
Cotton yarn	360**	295	239	215	214	192	208	−4.0**	−3.9**	−4.1	−2.3
Cotton fabrics	250**	206	182	188	182	166	199	−3.1**	−3.8**	−2.5	1.5
Wool yarn	113	91	79	65	55	51	60	−3.5	−4.2	−2.8	−4.5
Man-made fibres	282	471	723	980	939	734	874	9.9	10.8	8.9	3.2
Wood pulp	1 451	1 390	1 732	1 760	1 849	1 531	na	1.8	−0.8	4.5	−2.4c
Newsprint	230	216	408	516	522	486	501	6.0	−1.3	13.5	3.5
Other paper	3 100**	4 020	5 108	5 839	6 000	4 801	na	5.1**	5.3**	4.9	−1.2c
Synthetic rubber	81	173	302	349	324	278	378	14.0	16.4	11.8	3.8
Tyres (000 units)	14 745	25 566	37 548	41 333	34 607	32 017	37 646	9.8	11.6	8.0	0.1
Ethylene	na	694	2 020	2 761	3 107	2 140	na	na	na	23.8	1.2c
Ethyl alcohol (000 hl)	na	na	2 218	2 137	2 432	1 829	na	na	na	na	−3.8c
Sulphuric acid	3 170	3 751	4 435	5 069	5 130	4 157	4 682	3.4	3.4	3.4	0.9
Hydrochloric acid	247	370	649	827	891	728	na	10.1	8.4	11.9	2.3c
Nitric acid	na	2 599	3 254	3 142	3 334	3 035	na	na	na	4.6	−1.4c
Caustic soda	776	1 178	1 682	2 513	2 819	2 489	3 088	8.0	8.7	7.4	10.7
Soda ash	1 117	1 165	1 334	1 422	1 456	1 249	na	1.8	0.9	2.7	−1.3a
Plastics and resins	982	1 999	4 321	6 434	6 271	5 046	6 443	16.0	15.3	16.7	6.9
Fertilisers, nitrogenousª	1 051	1 289	1 574	1 471	1 473	1 574	1 259	4.1	4.2	4.1	−3.7
Fertilisers, phosphateª	770**	890**	919	986	962	911	649	1.8**	3.0**	0.6**	−5.6
Fertilisers, potashª	2 000**	2 300**	2 293	2 498	2 539	2 659	1 848	1.4**	2.8**	−0.1**	−3.5

For footnotes see page 228

Germany, West

Main products (000 t)	1960	1965	1970	1973	1974	1975	1976	Growth %pa			
Manufacturing								1960–70	1960–65	1965–70	1970–76
Cement	24 905	34 133	38 325	41 011	35 977	33 516	34 097	4.4	6.5	2.3	−1.9
Coke	44 754	43 294	39 914	33 997	34 921	34 817	na	−1.1	−0.7	−1.6	−2.7c
Pig iron	25 890	27 186	33 897	37 093	40 504	30 330	32 170	2.7	1.0	4.5	−0.9
Crude steel	34 100	36 821	45 040	49 521	53 231	40 414	42 413	2.8	1.5	4.1	−1.0
Aluminium	169	234	309	533	689	678	697	6.2	6.7	5.7	14.5
Copperb	309	342	406	407	424	422	446	2.7	2.0	3.5	1.6
Lead	140	104	112	86	116	92	101	−2.2	−5.8	1.5	−1.7
Zinc	142	107	123	241	250	174	202	−1.4	−5.5	2.8	8.6
Radio sets (000 units)	4 313	4 071	6 729	5 953	5 340	4 415	na	4.5	−1.1	10.6	−8.1c
Television sets (000 units)	2 164	2 776	2 936	3 898	4 165	3 356	na	3.1	5.1	1.1	2.7c
Passenger cars (000 units)	1 817	2 734	3 528	3 643	2 840	2 905	3 546	6.8	8.5	5.2	0.1
Commercial vehicles (000 units)	239	243	318	306	265	286	330	2.9	0.3	5.5	0.6
Merchant vessels (000 grt)	1 092	1 023	1 687	1 804	2 109	2 545	1 786	4.4	−1.3	10.5	1.0
Construction											
Dwellings (000 units)	551*	574*	478	714	604	437	392	−1.4*	0.8*	−3.6*	−3.2

aYears ending June 30th   bOf which, secondary: 190 thousand tonnes in 1975   c1970–75

Transport traffic	1960	1970	1976	Growth %pa	
Passenger-kilometres (mn)				1960–70	1970–76
Road	na	na	529 700	na	na
Rail	39 300**	38 129	38 349	−0.3**	0.1
Air	1 284	8 255	14 982	20.5	10.4
of which, international	1 191	7 156	13 242	19.6	10.8
Cargo: tonne-kilometres (mn)					
Road	na	78 700	103 402	na	4.7
Rail	53 000**	71 287	59 202	3.0**	−3.0
Water	39 513a	48 800	45 805	3.1b	−1.1
Air	39	523	1 098	29.5	13.2
of which, international	39	502	1 068	29.1	13.4
Sea: tonnes (mn)					
Goods loaded	16.5	22.5	29.0	3.1	4.3
Goods unloaded	57.7	106.3	111.0	6.3	0.7
Tourism					
Number of visitorsc (000)	5 476	8 467	7 890	4.4	−1.2
Gross receipts ($ mn)	398	1 326	3 211	12.8	15.9

a1963 b1963–70 cArrivals at all accommodation establishments, including camping sites, etc

Finance and trade

Price and wage indices	1960	1970	1976	Growth %pa	
(1970 = 100)				1960–70	1970–76
Consumer prices	77.4	100.0	140.8	2.6	5.9
Wholesale pricesa	87.7	100.0	140.7	1.3	5.9
Wages (earnings)	44.3	100.0	168.8	8.5	9.1
Share prices	100.9	100.0	100.8	−0.1	0.1
Money stock					
(end-year, DM bn)	47.4	102.7	176.6	8.0	9.5

aIndustrial

Budget (Federal, 1977)
Revenue: DM 157.06 bn = $ 67 640 mn = £ 38 750 mn
Expenditure: DM 179.28 bn = $ 77 209 mn = £ 44 234 mn
of which, transport and communication 7 %, defence 21 %, social security 37 %

Balance of payments ($ mn)	1972	1973	1974	1975	1976
Balance of goods (fob)	+8 360	+15 467	+22 167	+17 672	+16 677
Balance of services	−3 227	−5 094	−6 102	−6 681	−6 221
Balance of transfers	−4 384	−6 001	−6 342	−7 095	−7 060
Current balance	+748	+4 372	+9 722	+3 896	+3 396
Long-term capital flow	+4 894	+4 841	−2 272	−6 870	−148
Reserves and debt (end-year, $ mn)					
International reserves	23 785	33 171	32 398	31 034	34 801
Total public debt	16 009	22 990	29 940	41 376	54 370

External trade (1976)
Imports: DM 224 433 mn = $ 89 131 mn = £ 49 348 mn
Exports: DM 260 574 mn = $ 103 485 mn = £ 57 294 mn

Main imports	% of total
Food	13
(of which, fruit and vegetables 4, meat 2, cereals 2)	
Crude oil	11
Chemicals	7
(of which, plastics 2)	
Machinery, non-electric	7
(of which, office machines 2)	
Petroleum products	5

Main exports	% of total
Machinery, non-electric	22
(of which, power 2, metal-working 2, textile 2, pumps 2)	
Motor vehicles	13
Chemicals	12
(of which, plastics 3, dyes, etc 1)	
Electrical machinery	9
(of which, switchgear 2, telecommunications 2)	

Main imports (cont)	% of total
Electrical machinery	5
Clothing	5
Motor vehicles	5
Iron and steel	4
Textile yarns and fabrics	4
Non-ferrous metals	3
(of which, copper 1)	
Metal ores	3
Paper	2
Instruments	2
Main sourcesa	
Netherlands	14
France	12
Belgium-Luxembourg	8
Italy	8
United States	8
United Kingdom	4
Switzerland	3
Japan	2
Libya	2
Iran	2
Austria	2
Sweden	2
Saudi Arabia	2
Soviet Union	2
East Germany	2

Main exports (cont)	% of total
Iron and steel	7
(of which, tubes and pipes 2)	
Textile yarns and fabrics	4
Food	4
Metal small manufactures	3
Instruments	3
Non-ferrous metals	2
Coal and coke	2
Main destinationsb	
France	13
Netherlands	10
Belgium-Luxembourg	8
Italy	7
United States	6
Austria	5
United Kingdom	5
Switzerland	4
Sweden	3
Soviet Union	3
Denmark	3
Iran	2
Spain	2
East Germany	2
Jugoslavia	2

aEEC 48 % bEEC 45 %

Special focus

Rhine traffic, 1976

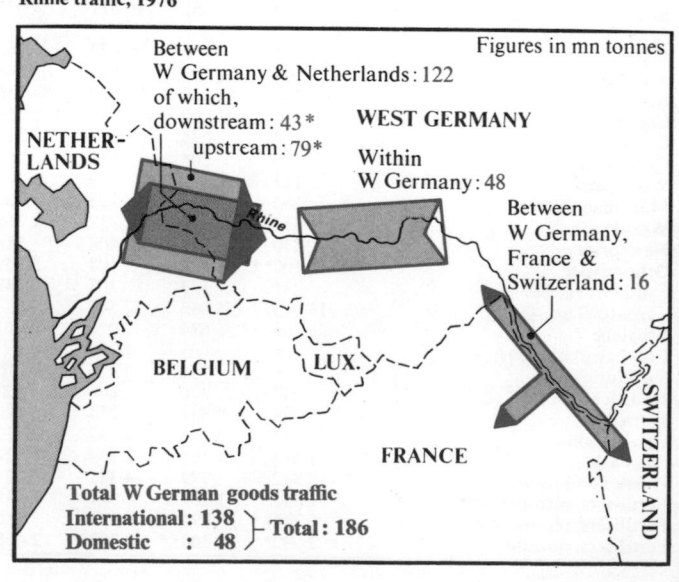

Figures in mn tonnes

Between
W Germany & Netherlands: 122
of which,
downstream: 43 *
upstream: 79 *

NETHER-LANDS

WEST GERMANY

Within
W Germany: 48

Between
W Germany,
France &
Switzerland: 16

BELGIUM

LUX.

FRANCE

SWITZERLAND

Total W German goods traffic
International: 138 ⎫
Domestic : 48 ⎬ Total: 186

Gibraltar
Colony of Gibraltar

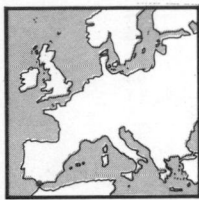

Location South-west Europe
Occupies a narrow peninsula on the southern coast of Spain; Morocco is 30 km to the south
Land Area 5.8 km² = 2.2 mi²
Climate Temperate, with mild winters
Weather at Gibraltar Town, 27 m altitude
Temperature: hottest month Aug 20–29 °C, coldest Feb 8–16 °C
Rainfall (av monthly): driest month July 1 mm, wettest Dec 151 mm
Time 1 hour ahead of GMT
Measures UK (imperial) system, changing to the metric system
Monetary unit Gibraltar pound (Gib £) = 100 new pence
Rate of exchange (1976 av): par Gib £ 1 = £ 1, free $ 1.806 = Gib £ 1

Summary

Political UK colony since 1713; Spain claims the territory, and closed the frontier in 1969. Within EEC as a UK territory, but with exemptions from some regulations
Economic A trading economy, dependent also on defence expenditure; tourism and ship bunkering are important and there are ship repair yards. United Kingdom provides development aid

People, resources and equipment

Population[a] 1960 24 000*, 1970 26 800*, 1976 30 117*
[a]Excluding armed forces
Growth: 1960–70 1.1* % pa, 1970–76 2.0* % pa
Density (1976): 5 190* people per km²
Vital statistics (rate per 1 000 people, 1975): births 17.7, deaths 8.8*
Town Gibraltar Town (capital)
Race (1970) Native-born 63 %, Spanish 14 %, British 10 %, Moroccan 9 %
Language English; Spanish is also used
Religion (1970) Roman Catholic 77 %, Moslem 8 %, Anglican 8 %
Education (1975/76) Pupils 4 641, teachers 289
Labour force (1975) 12 472
Personnel Scientists and engineers (1970): 41
Physicians (1971): 19, 1 per 1 450 people
Standard of living National income per person (1976):
Gib £ 1 500*** = $ 2 700*** = £ 1 500***
Consumption per person (1975): energy 1 267 kg coal equivalent, electricity 1 660 kW h
Newspapers (1974): number 1; circulation 3 000, 111 per 1 000 people
Telephones (Dec 1976): 8 148, 272 per 1 000 people
Electrical capacity (1975) 21 megawatts
Hospital beds (1971) 252, 1 per 110 people
Roads (1976) 50 km = 31 mi, density 8.6 km per km²
Ships (registered, 1977) 6, total of 10 549 gross tons
Port Gibraltar
Airport Gibraltar
Radio sets (Dec 1975) 9 900*, 330* per 1 000 people
Television sets (Dec 1976) 6 325, 210 per 1 000 people
Motor vehicles (Dec 1976) 6 147, 204 per 1 000 people, 123 per km of road

Production, finance and trade

Gross domestic product 1976: Gib £ 50** mn = $ 90** mn = £ 50** mn
Main products (1975) Electricity 50 mn kW h, soft drinks, canned meat
Transport traffic *Air* (1976) Passenger departures and arrivals 147 000
Sea (1975) Goods loaded 6 000 t, unloaded 345 000 t
Tourism Number of visitors (1976) 125 219, gross receipts (1974) $ 6 mn
Consumer price index (1970 = 100) 1976 211.1; growth 1970–76 13.3 % pa
Budget (1975/76; year ending March 31st)
Balanced at Gib £ 13.5 mn = $ 29 mn = £ 13.5 mn
including development aid: Gib £ 1.4 mn = $ 3.0 mn = £ 1.4 mn
External trade (1976) Imports: Gib £ 32 mn = $ 59 mn = £ 32 mn
Exports (re-exports only): Gib £ 14 mn = $ 25 mn = £ 14 mn

Main imports	% of total	Main exports	% of total
Manufactures	36	Petroleum products	87
Petroleum products	34	Tobacco and other	
Food	25	manufactures	10
Wines, spirits and tobacco	5	Wines and spirits	3
Main source		*Main destinations*	
United Kingdom	55*	Shipping, especially bunkers	

Greece
Hellenic Republic
Hellinik Thimokratia

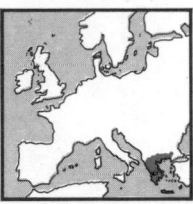

Location South-east Europe
The mainland has a long coastline on the Mediterranean Sea with the Ionian Sea to the west and Aegean Sea to the east; there are many islands, including Crete, around the coast, especially to the south. Albania, Jugoslavia and Bulgaria are to the north and Turkey to the east
Land Area 131 944 km² = 50 944 mi²
Climate Mediterranean (hot summer and mild winters)
Weather at Athens, 107 m altitude
Temperature: hottest month July 23–33 °C, coldest Jan 6–13 °C
Rainfall (av monthly): driest month July 6 mm, wettest Dec 71 mm
Time 2 hours ahead of GMT (summer time, 3 hours ahead)
Measures Metric system
Monetary unit Drachma (Dr) = 100 lepta
Rate of exchange (1976 av): free Dr 36.58 = $ 1, Dr 66.07 = £ 1

Summary

Political Republic, which secured independence from Turkey in 1830. After freedom from occupation by Germany in the second world war there was civil conflict. The King was exiled after a coup in April 1967 and a military junta ruled until July 1974; Greece was declared a republic in June 1973 and this was confirmed by a referendum in December 1974. A new republican constitution was introduced in June 1975. Member of UN, OECD, Council of Europe and Nato (excluding the military side from which Greece partially withdrew in 1974).
Economic A mixed economy with a large agricultural sector, absorbing 40 % of the labour force and contributing 17 % of gross domestic product. Manufacturing has developed rapidly with a growth rate of around 9 % pa, and textiles, clothing, iron and steel and petroleum products are major exports. Shipping and tourism are important earners of foreign exchange, but there remains a large balance of payments deficit on current account, balanced mainly by a large inflow of capital

People, resources and equipment

Population 1960 8.33 mn, 1970 8.79 mn, 1976 9.17 mn
Growth: 1960–70 0.5 % pa, 1970–76 0.7 % pa
Density (1976): 69 people per km²
Vital statistics (rate per 1 000 people, 1976): births 15.7, deaths 8.2
Regions (population in 000, 1971; total of 8.77 mn)
Mainland 7 469, Crete 457, other islands 843
Cities (population in 000, 1971)

Athinai[a] (capital)	2 540[b]	Vólos	88	Khaniá	53
Thessaloniki[c]	557	Iráklion	85	Kavalla	47
Patras	121	Lárissa	73	Serrai	41

[a]Athens [b]Of whom 187 thousand in Piraeus [c]Salonika
Race (1961) Greek 95 %, Macedonian 2 %, Turkish 1 %, Albanian 1 %
Language Greek
Religion (1961) Greek Orthodox 97 %, Moslem 1 %
Education (1974/75) Pupils 1 547 032, teachers 54 075
Labour force (1971) 3 234 996; in agriculture 1 312 600 (41 %), manufacturing 554 380 (17 %), construction 256 424 (8 %), transport and communication 211 672 (7 %), distribution and hotels 362 024 (11 %)
Personnel Scientists and engineers engaged in research (1969): 1 032
Physicians (1975): 18 421, 1 per 491 people
Standard of living
National income per person (1976): Dr 84 950 = $ 2 322 = £ 1 286
Consumption per person (1975): energy 2 090 kg coal equivalent, electricity (production) 1 674 kW h, newsprint 5.4 kg, steel 143 kg
Newspapers (1974): number 105; circulation 962 000, 107 per 1 000 people
Telephones (Dec 1976): 2 180 200, 238 per 1 000 people
Livestock (000, Dec. 1976) Cattle 1 200*, sheep 8 135*, goats 4 524*, pigs 710*, chickens 32 366*
Mineral reserves Lignite (1961) 1 575 mn tonnes
Crude oil (1975) 82 mn tonnes
Natural gas (1975) 11 bn cubic metres
Petroleum refinery capacity (1975) 12.9 mn tonnes
Electrical capacity (1975) 4 868 megawatts
Hospital beds (1975) 58 501, 1 per 155 people
Roads (1976) 36 574 km = 22 726 mi, density 0.28 km per km²
Railways (1975) 2 476 km = 1 539 mi, density 0.019 km per km²
Ships (registered, 1977) 3 344, total of 29 517 059 gross tons; of which, oil tankers 428, total of 9 725 491 gross tons

Greece

Ports (goods traffic, 000 tonnes, 1976)

	International loaded	unloaded	Coastwise loaded	unloaded	Total
Eleusis[a]	2 739	6 141	2 601	555	12 036
Salonika[a]	1 108	5 535	1 403	537	8 583
Piraeus	1 163	4 113	1 016	2 249	8 541
Megara[a]	4	3 258	—	183	3 445
Vólos	767	234	1 121	804	2 926
Antikyra	488	306	18	670	1 482

[a]Crude oil port

Airports (passengers, 000, 1976) Athens: departures 3 200, arrivals 3 179; also Salonika and (1977) 22 other airports with scheduled flights

Durable equipment (at end-year)	000	no per 1 000 people	
Radio sets (1974)	2 500*	279*	no per
Television sets (1975)	1 140	126	km of road
Passenger cars (1976)	496	54	14
Commercial vehicles (1976)	233	25	6.4

Production

Gross domestic product 1976: Dr 813.7 bn = $ 22 244 mn = £ 12 316 mn
Growth in real terms: 1960–70 7.6 % pa, 1970–76 5.2 % pa
Structure of gross domestic product (1976) *By origin* Agriculture 17 %, mining and quarrying 1 %, manufacturing 17 %, electricity, gas and water 1 %, construction 6 %, distribution 12 %, transport and communication 7 %, other 38 %
By type Final consumption expenditure 83 % (of which, government 15 %), stock investment 3 %, gross fixed capital formation 22 %, exports of goods and services 17 %, less imports of goods and services −26 %

Production indices (1970 = 100)	1960	1970	1976	Growth % pa 1960–70	1970–76
Agricultural	64	100	120	4.6	3.0
Manufacturing	43.6	100.0	166.4	8.6	8.9

Main products (000 t)
Agriculture

	1960	1970	1976	1960–70	1970–76
Wheat	1 692	1 931	2 351	1.3	3.3
Barley	232	737	957	12.3	4.4
Potatoes	425	756	991*	5.9	4.6*
Tomatoes	455	920	1 500*	7.3	8.5*
Watermelons	300**	583	550*	6.8**	−1.0*
Grapes	922	1 604	1 551*	5.7	−0.5*
Wine	290	483	453*	5.2	−1.0*
Apples	90	236	255*	10.1	1.3*
Peaches	61	197	333*	12.4	9.1*
Oranges	215	418	540*	6.9	4.3*
Lemons	80	137	180*	5.5	4.7*
Figs	26	130**	130*	17.5**	0.0**
Olives	382	850*	1 485*	8.3*	9.7*
Olive oil	89	198	285	8.3	6.3
Cottonseed	114	215	210*	6.5	−0.4*
Tobacco	61	95	127*	4.5	5.0*
Cotton	63	111	110*	5.8	−0.2*
Milk, total	1 012	1 357	1 704*	3.0	3.9*
Cheese	92	137*	178	4.1*	4.4*
Meat	170*	273*	415	4.9*	7.2*
Fish catch	87	98*	71*	1.2*	−5.2*
Timber (000 m³)	3 000**	3 130	2 953	0.4**	−1.0

Energy

	1960	1970	1976	1960–70	1970–76
Total energy (000 tce)	900	2 860	6 160[b]	12.3	16.6[c]
Lignite	2 550	7 680	22 241	11.7	19.4
Petroleum products	1 654	4 962	10 340*[b]	11.6	15.8*[c]
Manufactured gas (mn m³)	14	8	163[b]	−5.4	82.7[c]
Electricity (mn kW h)	2 277	9 399	15 151[b]	15.2	10.0[c]

Mining

	1960	1970	1976	1960–70	1970–76
Iron ore (Fe content)	161	380	950	9.0	16.5
Bauxite	884	2 283	2 455	10.0	1.2
Lead conc (Pb content)	9	10	14*[b]	0.4	8.0*[c]
Magnesite	187	755	1 426[b]	15.0	13.5[c]
Nickel ore (Ni content)	—	8.6	14.8[b]	na	11.4[c]

Manufacturing

	1960	1970	1976	1960–70	1970–76
Beer (000 hl)	435	777	1 286[b]	6.0	10.6[c]
Cigarettes (mn units)	12 359	17 029	21 592[b]	3.3	4.9[c]
Cotton yarn	24	42	80	5.7	11.3
Cotton fabrics	15	19	28[d]	2.4	10.2[e]
Man-made fibres	1.6	6.4	11.3[b]	14.9	12.0[c]
Paper	65*	160	234[b]	9.4*	7.9[c]
Sulphuric acid	135	623	911	16.5	6.5

Main products (000 t) Manufacturing	1960	1970	1976	Growth % pa 1960–70	1970–76
Nitric acid	na	70	159[d]	na	22.8[e]
Fertilisers, nitrogenous[a]	—	146	288	na	12.0
Fertilisers, phosphate[a]	60**	121	187	7.3**	7.5
Cement	1 649	4 848	8 754	11.4	10.3
Crude steel	125	435	612*[b]	13.3	7.1*[c]
Aluminium	—	91	133	na	6.5
Television sets (000 units)	—	134	196[d]	na	10.0[e]
Merchant vessels (000 grt)	17	73	76	15.7	0.7

[a]Years ending June 30th [b]1975 [c]1970–75 [d]1974 [e]1970–74

Transport traffic	1960	1970	1976	Growth % pa 1960–70	1970–76
Passenger-kilometres (mn)					
Rail	1 030	1 531	1 471	4.0	−0.7
Air	289	2 126	4 623	22.1	13.8
Cargo: tonne-kilometres (mn)					
Rail	363	688	845	6.6	3.5
Air	5.4	34.5	57.9	20.4	9.0
Sea: tonnes (mn)					
Goods loaded	2.31	4.16	13.25	6.0	21.3
Goods unloaded	4.52	13.39	25.95	11.5	11.7
Tourism					
Number of visitors (000)	344	1 407	3 845	15.1	18.2
Gross receipts ($ mn)	51	194	823	14.3	27.2

Finance and trade

Price and wage indices (1970 = 100)	1960	1970	1976	Growth % pa 1960–70	1970–76
Consumer prices	81.5	100.0	202.4	2.1	12.5
Wholesale prices	80.6	100.0	219.6	2.2	14.0
Wages (earnings)	47.8[a]	100.0	280.0	8.5[b]	18.7
Money stock (end-year, Dr bn)	15.60	54.60	159.58	13.3	19.6

[a]1961 [b]1961–70

Budget (1976) Revenue: Dr 171.15 bn = $ 4 679 mn = £ 2 590 mn
Expenditure: Dr 202.37 bn = $ 5 532 mn = £ 3 063 mn

Balance of payments ($ mn)	1972	1973	1974	1975	1976
Balance of goods (fob)	−1 326	−2 354	−2 352	−2 363	−2 692
Balance of services	+352	+431	+468	+619	+797
Balance of transfers	+570	+737	+644	+745	+811
Current balance	−404	−1 185	−1 241	−999	−1 084
Long-term capital flow	+686	+799	+759	+785	+547
Reserves and debt (end-year, $ mn)					
International reserves	1 032	1 047	936	931	925
External public debt	1 795	2 336	2 762	2 996	3 024

External trade (1976)
Imports: Dr 221 820 mn = $ 6 064 mn = £ 3 357 mn
Exports: Dr 93 810 mn = $ 2 565 mn = £ 1 420 mn

Main imports	% of total	Main exports	% of total
Ships	19	Fruit and vegetables	18
Crude oil	18	Textile yarns & fabrics	9
Machinery, non-electric	10	Clothing	8
Chemicals	8	Tobacco	7
Food	8	Iron and steel	6
(of which, meat 2)		Petroleum products	6
Motor vehicles	7	Chemicals	4
Electrical machinery	4	Aluminium	4
Iron and steel	4	Metal ores	3
Textile fibres	2	Minerals	3
Main sources		*Main destinations*	
West Germany	15	West Germany	21
Japan	12	Italy	9
Italy	8	France	7
Saudi Arabia	7	United States	6
United States	7	Netherlands	5
France	6	Libya	5
United Kingdom	4	Saudi Arabia	4
Sweden	4	United Kingdom	4

Special focus

Structure of gross domestic product

	% of total 1960	1970	1972	1974	1976
Agriculture	22	16	16	18	17
Manufacturing	15	17	17	20	17
Construction	6	7	9	7	6
Transport and communication	6	6	7	6	7
Other	51	54	51	49	52

Greenland
Grønland

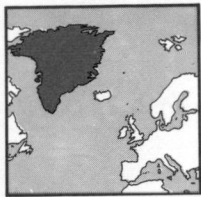

Location Arctic Ocean
Canada is to the south-west of the Greenland island across the Davis Strait, and Iceland to the south-east across Denmark Strait
Land Area 2 175 600 km² = 840 000 mi²
of which, ice-free part: 341 700 km²
Climate Very cold
Weather at Godthåb, 20 m altitude
Temperature: hottest month July 3–11 °C, coldest Jan minus 12– minus 7 °C
Rainfall (av monthly): driest month April 30 mm, wettest Sept 84 mm
Time Thule area: 4 hours behind GMT; other: 3 hours behind GMT
Measures Metric system
Monetary unit Danish krone (D Kr) = 100 øre
Rate of exchange (1976 av): free D Kr 6.045 = $ 1, D Kr 10.92 = £ 1

Summary

Political Overseas part of the Kingdom of Denmark since 1380 (as a colony before 1953). From 1953 there has been some independence, with Denmark controlling mainly external affairs; Greenland elects two members of the Danish parliament. There has been a proposal that home rule should be obtained by May 1979. Within EEC as a part of Denmark, although voting against EEC entry in a 1972 referendum
Economic Fishing and mining are the main industries; production of cryolite has fallen in recent years, but mining of zinc and lead began in 1973 and these metals make up about one-half of exports

People, resources and equipment

Population 1960 33 000, 1970 46 000, 1976 49 690
Growth: 1960–70 3.4 %pa, 1970–76 1.3 %pa
Density (1976): 0.1 people per km² (based on ice-free area)
Vital statistics (rate per 1 000 people, 1975): births 17.5, deaths 6.3
Cities (population in 000, 1977) Godthåb (capital) 9.0
Race (1976) Native-born Greenlanders 84 %, Danish 14* %
Language Danish and Eskimo
Religion (1965) Evangelical Lutheran 88 %
Education (1975/76) Pupils 12 750*, teachers 1 025*
Labour force (1970) 18 741; in agriculture (includes fishing) 3 493 (19 %), manufacturing and mining 2 525 (14 %), construction 2 753 (15 %)
Personnel (1970) Physicians: 42, 1 per 1 100 people
National income per person (1976)
D Kr 27 000*** = $ 4 500*** = £ 2 500***
Consumption per person (1975) Energy 5 465 kg coal equivalent, electricity 2 500* kW h
Livestock (000, 1975) Sheep 20, reindeer 2.5
Mineral reserves Coal (1967) 2 mn tonnes, uranium (1976) 5 800 tonnes
Electrical capacity (1975) 70* megawatts
Hospital beds (1970) 666, 1 per 70 people
Ports Godthåb, Faeringehavn, Frederikshab, Sukkertoppen
Airports Søndre Strømfjord, Godthåb, and (1977) 12 other airports with scheduled flights
Radio sets (1975) 13 000, 241 per 1 000 people
Motor vehicles (1975) 1 700, 34 per 1 000 people

Production, finance and trade

Gross domestic product
1976 est: D Kr 1 400** mn = $ 230** mn = £ 130** mn
Main products (000 t, 1976) Wool 0.006, fish catch 45, sheep skins (000) 5*, seal skins (000) 55*, electricity 124* mn kW h, cryolite (1974) 38, lead ore (Pb content, 1975) 24, zinc ore (Zn content, 1975) 85
Transport traffic (1974) *Sea* Goods loaded 80 000* t, unloaded 280 000* t
Budget (1975) Revenue: D Kr 73.6 mn = $ 12.8 mn = £ 5.8 mn
Expenditure: D Kr 77.2 mn = $ 13.4 mn = £ 6.0 mn
External trade (1976) Imports: D Kr 778 mn = $ 129 mn = £ 71 mn
Exports: D Kr 517 mn = $ 86 mn = £ 47 mn

Main imports	% of total	*Main exports*	% of total
Food	17	Fish and products	47
Machinery	17	Zinc ores	36
Petroleum products	16	Lead ores	10
Main sources		*Main destinations*	
Denmark	98	Denmark	44
United Kingdom	2	Finland	15
		United States	12

Guernsey
Bailiwick of Guernsey

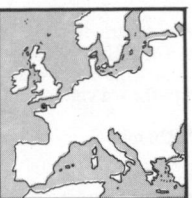

Location English Channel
Part of the Channel Islands, 40 km west of the Normandy peninsula (north-west France)
Land Area 78.4 km² = 30.3 mi²
of which, Guernsey 63.3 km², and dependencies Alderney 7.9 km², Great Sark 4.2 km², Little Sark 1.0 km², Herm 1.3 km², Brechou 0.3 km², Jethou 0.2 km², Lithou 0.2 km²
Climate See Jersey
Time GMT (summer time, 1 hour ahead of GMT)
Measures UK (imperial) system, converting to the metric system
Monetary unit UK pound and Guernsey pound (G £) = 100 new pence
The Guernsey pound is equal to the UK pound
Rate of exchange (1976 av): free $ 1.806 = G £ 1

Summary

Political Dependency of the Crown of England since the 12th century; autonomous state but with the United Kingdom responsible for defence and external relations. Main dependencies of Guernsey are Alderney, Sark and Herm; these are self-governing but with some control from the Bailiwick of Guernsey. Member of sterling area and within EEC via relationship with United Kingdom, but with special exemption from many regulations
Economic Horticulture and tourism are the main sectors of the economy, with tomatoes and flowers major exports. An offshore finance centre has been developed and there is some light manufacturing industry

People, resources and equipment

Population 1960 46 480*, 1970 53 030*, 1976 56 700*
Growth: 1960–70 1.3 %pa, 1970–76 1.1* %pa
Density (1976): 723* people per km²
Vital statistics (rate per 1 000 people, 1976): births 11.5, deaths 11.2
Regions (islands, population, 1976; census total of 53 637)
Guernsey 53 560, Alderney 1 800*, Sark 500*, Herm 69, Jethou 8, Brechou 6*, Lithou nil
Towns St Peter Port (Guernsey), St Anne (Alderney)
Race Norman/Breton/English
Language English and Norman–French patois
Religion Mainly Church of England, Methodist and Roman Catholic
Education (1977/78) Pupils 8 071, teachers 457
Labour force (1977) 26 903; in agriculture 17** %
Personnel (1977) Physicians: 59, 1 per 920* people
National income per person (1976)
G £ 2 400** = $ 4 300** = £ 2 400**
Telephones (Dec 1976) 30 600, 540 per 1 000 people
Livestock (000, 1977) Cattle 2.5, pigs 0.2
Hospital beds (1977) 540, 1 per 101 people
Ports St Peter Port, St Sampson, Braye Harbour (Alderney), Creux (Sark)
Airports La Villiaze (Guernsey), Alderney
Radio sets (Dec 1976) 18 000*, 316* per 1 000 people
Television sets (Dec 1976) 18 000*, 316* per 1 000 people
Passenger cars (Dec 1976) 21 100, 370 per 1 000 people[a]
Commercial vehicles (Dec 1976) 3 910, 69 per 1 000 people

Production, finance and trade

Gross domestic product 1974: G £ 90 mn = $ 210 mn = £ 90 mn
1976 est: G £ 130** mn = $ 230** mn = £ 130** mn
Main products (000 t, 1976) Tomatoes 47, milk 8, flowers
Transport traffic (1977) Passenger arrivals: air 181 800, sea 124 100
Tourism (1974) Number of visitors 280 000*, gross receipts $ 29* mn
Bank deposits (1977) G £ 650 mn = $ 1 200 mn
Budget (1977) Revenue: G £ 27.8 mn = $ 49 mn = £ 27.8 mn
Expenditure: G £ 21.5 mn = $ 38 mn = £ 21.5 mn
External trade (1975)[a]
Imports: G £ 70 mn = $ 126 mn = £ 70 mn
Exports: G £ 52 mn = $ 93 mn = £ 52 mn

Main imports	% of total	*Main exports*	% of total
Manufactured goods	60**	Manufactures	
Food	20**	(mainly electronic)	49**
Petroleum products	12**	Tomatoes	36**
		Flowers and fern	14**

Main source and destination United Kingdom
[a] Included within the United Kingdom area for external trade purposes

Hungary

Hungarian People's Republic
Magyar Népköztársaság

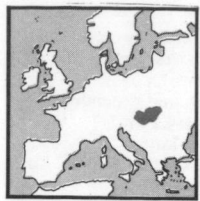

Location East Europe
Czechoslovakia is to the north, Austria to the
west, Jugoslavia to the south-west, Rumania to
the south-east and Soviet Union to the north-east.
Land-locked but with access to the sea via the
Danube
Land Area 93 030 km² = 35 920 mi²
Climate Continental
Weather at Budapest, 139 m altitude
Temperature: hottest month July 16–28 °C,
coldest Jan minus 4–1 °C
Rainfall (av monthly): driest month Sept 33 mm, wettest May 72 mm
Time 1 hour ahead of GMT
Measures Metric system
Monetary unit Forint (Ft) = 100 fillér
Rate of exchange (1976 av) *Non-commercial* free Ft 20.55 = $ 1,
Ft 37.12 = £ 1 *Commercial* par Ft 41.50 = $ 1, free Ft 75.00 = £ 1

Summary

Political Communist republic; freed by Soviet Union in 1945 from
German occupation (Hungary having been a German ally until 1944 and
occupied 1944/45). There was an unsuccessful anti-Soviet Union
revolution in 1956. Member of UN, Comecon and Warsaw Pact
Economic A mixed economy, with an important agricultural sector—food
makes up one-fifth of exports—and a large manufacturing sector;
machinery, motor vehicles, chemicals and steel are substantial export
products. Tourism is significant, with gross receipts equal to 5 % of exports

People, resources and equipment

Population 1960 9.98 mn, 1970 10.34 mn, 1976 10.60 mn
Growth: 1960–70 0.4 %pa, 1970–76 0.4 %pa
Density (1976): 114 people per km²
Vital statistics (rate per 1 000 people, 1976): births 17.5, deaths 12.5
Regions (population in 000, Dec 31, 1976; resident total of 10.63 mn)

Autonomous city		Heves	345
Budapest	2 082	Komárom	317
Counties (Megyék)		Nógrád	236
Baranya	433	Pest	959
Bács-Kiskun	570	Somogy	361
Békés	433	Szabolcs-Szatmár	573
Borsod-Abaúj-Zemplén	795	Szolnok	444
Csongrád	459	Tolna	258
Fejér	415	Vas	281
Györ-Sopron	426	Veszprém	428
Hajdú-Bihar	545	Zala	264

Cities (population in 000, Dec 31, 1976)

Budapest (capital)	2 082	Pécs	165	Kecskemét	93
Miskolc	203	Györ	122	Szombathely	79
Debrecen	192	Székesfehérvár	99	Szolnok	74
Szeged	173	Nyíregyháza	95	Tatabánya	73

Race (1960) Magyar (Hungarian) 98 %, German ½ %, Slovak ⅓ %,
Gypsy ⅓ %; also see language usage
Language Magyar (Hungarian)
Usage (1970): Magyar 98.5 %, German 0.4 %, Romany 0.3 %,
Slovak 0.2 %, Croatian 0.2 %, Rumanian 0.1 %
Religion (1965) Roman Catholic 55** %, Protestant 25** %, Jewish 1** %
Education (1976/77) Pupils: primary 1 072 423, secondary and vocational
373 372, higher 110 528. Teachers: primary 68 425, secondary and
vocational 14 454, higher 12 233
Labour force (1976) 5 093 200; in agriculture 1 136 500 (22 %),
manufacturing, mining and utilities 1 794 000 (35 %), construction 422 200
(8 %), transport and communication 402 900 (8 %), other 1 337 600 (26 %)
Personnel Scientists and engineers (including teachers, 1973): 336 143
Physicians (1975): 21 127, 1 per 500 people
Standard of living National income per person (1976):
Ft 44 000*** = $ 2 100*** = £ 1 200***
Consumption per person (1975): energy 3 624 kg coal equivalent,
electricity (production, 1976) 2 080 kW h, newsprint 5.2 kg, steel 361 kg
Newspapers (1974): number 27; circulation 2 431 000, 232 per 1 000 people
Telephones (Dec 1976): number 1 076 060, 101 per 1 000 people
Livestock (000, Dec 1976) Cattle 1 887, sheep 2 347, pigs 7 854,
horses 147, chickens 60 498
Mineral reserves Coal (1966) 714 mn tonnes
Lignite (1966) 5 679 mn tonnes
Crude oil (1975) 34 mn tonnes
Natural gas (1975) 103 bn cubic metres

Petroleum refinery capacity (1975) 9.7* mn tonnes
Electrical capacity (1975) 4 291 megawatts
Hospital beds (1975) 90 104, 1 per 117 people
Roads (1976) 99 595 km = 61 885 mi, density 1.1 km per km²
Railways (1975) 8 243 km = 5 122 mi, density 0.089 km per km²
Inland waterways (including Danube river and Lake Balaton, 1975)
1 688 km = 1 049 mi; of which, in regular use 1 302 km = 809 mi
Oil pipelines (1975) 1 236 km = 608 mi
Ships (registered, 1977) 19, total of 63 016 gross tons
Port (on Danube river) Budapest
Airport (passenger traffic, 000, 1976) Ferihegy (Budapest): departures 527,
arrivals 528

Durable equipment	000	no per	
(Dec 1976)		1 000 people	
Radio sets	2 559	241	no per
Television sets	2 477	233	km of road
Passenger cars	655	62	6.6
Commercial vehicles	143	13	1.4

Production

Net material product 1976: Ft 431.8 bn = $ 21 012 mn[a] = £ 11 633 mn[a]
Growth in real terms: 1960–70 5.5 %pa, 1970–76 5.8 %pa
Gross domestic product
1976 est: Ft 490*** bn = $ 24 000*** mn[a] = £ 13 000*** mn[a]
[a]Converted at non-commercial rate
Structure of net material product (1976) *By origin* Agriculture 16 %,
manufacturing, mining and utilities 47 %, construction 13 %, transport
and communication 6 %, distribution and hotels 13 %, other 6 %
By type Final consumption expenditure 74 % (of which, government
10 %), stock investment 10 %, gross fixed capital formation 21 %, exports
of goods and services 47 %, less imports of goods and services −52 %

Production indices	1960	1970	1976	Growth %pa	
(1970 = 100)				1960–70	1970–76
Agricultural	83	100	126	1.8	3.9
Industrial	51	100	141	7.0	5.9
Main products (000 t)					
Agriculture					
Wheat	1 768	2 723	5 148	4.4	11.2
Barley	986	553	747	−5.6	5.1
Maize	3 504	4 072	5 141	1.5	4.0
Rye	354	158	157	−7.7	−0.1
Potatoes	2 656	1 813	1 396	−3.7	−4.2
Sugar, raw value	477	304	404*	−4.4	4.9*
Cabbages	160**	185	203*	1.5**	1.5*
Tomatoes	202	293	395	3.8	5.1
Grapes	491	743	710	4.2	−0.8
Wine	296	438	480*	4.0	1.5*
Apples	290	661	930	8.6	5.9
Plums	128	227	230*	5.9	0.2*
Sunflowerseed	80**	96	188	1.8**	11.8
Tobacco	18.2	17.6	19.0*	−0.3	1.3*
Milk	1 955	1 863	2 084	−0.5	1.9
Eggs	103	182	222	5.9	3.4
Beef and veal	135	184	188*	3.1	0.4*
Pigmeat	480**	580*	724*	2.0**	3.7*
Poultry meat	122	222	302	6.2	5.3
Timber (000 m³)	3 500	5 034	5 488	3.7	1.4
Energy					
Total energy (000 tce)	16 730	23 210	23 710[a]	3.3	0.4[b]
Coal	2 847	4 151	2 934	3.8	−5.6
Lignite	23 676	23 679	22 323	0.0	−1.0
Crude oil	1 217	1 937	2 142	4.8	1.7
Petroleum products	2 575	5 713	9 579	8.3	9.0
Natural gas (mn m³)	342	3 469	6 082	26.1	9.8
Manufactured gas (mn m³)	544	941	1 029[a]	5.6	1.8[b]
Electricity (mn kW h)	7 617	14 542	22 049	6.7	7.2
Mining					
Iron ore (Fe content)	136	160	144	1.6	−1.7
Bauxite	1 190	2 022	2 919	5.4	6.3
Manganese ore					
(Mn content)	32.5	34.8	27.8[a]	0.7	−4.4[b]
Manufacturing					
Beer (000 hl)	3 555	5 006	6 619[a]	3.5	5.7[b]
Cigarettes (mn units)	15 782	22 050	24 549[a]	3.4	2.2[b]
Cotton yarn	47	57	58	1.9	0.5
Cotton fabrics (mn m²)	225	258	319	1.4	3.6
Man-made fibres	4.2	9.6	20.8	8.7	13.8
Wood pulp	47	82	67[a]	5.7	−4.0[b]
Paper	138	259	343[a]	6.5	5.8[b]
Tyres (000 units)	336	665	739	7.0	1.8

Hungary

Main products (000 t) Manufacturing	1960	1970	1976	Growth %pa 1960–70	1970–76
Sulphuric acid	178	471	652	10.2	5.6
Nitric acid	145	653	766[a]	16.2	3.3[b]
Plastics and resins	10	56	141	18.9	16.6
Fertilisers, nitrogenous	57	350	453[a]	19.9	5.3[b]
Fertilisers, phosphate	45	167	206[a]	14.1	4.3[b]
Cement	1 571	2 771	4 298	5.8	7.6
Coke	499	657	593[a]	2.8	−2.0[b]
Pig iron	1 259	1 837	2 227	3.8	3.3
Crude steel	1 887	3 108	3 652	5.1	2.7
Aluminium	49	66	70	2.9	1.1
Radio sets (000 units)	212	206	255[a]	−0.3	4.3[b]
Television sets (000 units)	139	364	412	10.1	2.1
Commercial vehicles (000 units)	4.4	10.2	12.9	8.8	4.0
Construction Dwellings (000 units)	58	78	98[a]	3.0	4.7[b]

[a] 1975 [b] 1970–75

Transport traffic Passenger-kilometres (mn)	1960	1970	1976	Growth %pa 1960–70	1970–76
Road	na	33 700**	61 080	na	10.4**
Rail	11 916	13 916	13 367	1.5	−0.7
Air	93[d]	334	510	17.3[e]	7.3
Cargo: tonne-kilometres (mn)					
Road	na	5 550**	8 681	na	7.7**
Rail	13 147	19 143	22 553	3.8	2.8
Water[a]	1 186[b]	1 596	1 468[f]	4.3[c]	−1.7[g]
Air	3.6[d]	6.7*	5.2	8.1*[e]	−4.1*

[a] National enterprises only [b] 1963 [c] 1963–70 [d] 1962 [e] 1962–70 [f] 1975 [g] 1970–75

Tourism (1976) Number of visitors 5 551 000, gross receipts $ 263 mn

Finance and trade

Consumer price index (1970 = 100) 1976 120.3; growth 1970–76 3.1 %pa
Wages (earnings) index (1970 = 100) 1976 140*; growth 1970–76 5.8* %pa
Budget (1976) Revenue: Ft 335 100 mn = $ 16 310 mn[a] = £9 030 mn[a]
Expenditure: Ft 338 100 mn = $ 16 450 mn[a] = £ 9 110 mn[a],
of which, education and culture 7 %, defence 4 %, public order 2 %,
social security 14 %, social welfare and health 4 %
[a] Converted at non-commercial rate
External trade (1976) Imports: 230 056 mn = $ 5 544 mn[a] = £ 3 067 mn[a]
Exports: Ft 204 834mn = $ 4 936 mn[a] = £ 2 731 mn[a]
[a] Converted at new 'commercial' rate: before 1976 trade was valued in 'exchange' forints

Main imports	% of total	Main exports	% of total
Machinery, non-electric	14	Machinery, non-electric	22
Chemicals	13	Food	20
Food	9	(of which, meat 5, fruit and	
Crude oil	8	vegetables 5, cereals 4,	
Motor vehicles	7	livestock 3)	
Textiles	6	Motor vehicles	10
Iron and steel	5	Chemicals	8
Non-ferrous metals	4	Iron and steel	6
Electrical machinery	3	Electrical machinery	4
Paper	2	Clothing	4
Main sources		Main destinations	
Soviet Union	27	Soviet Union	30
West Germany	10	East Germany	9
East Germany	9	West Germany	8
Czechoslovakia	6	Czechoslovakia	7
Austria	5	Poland	4
Poland	4	Italy	4
Italy	4	Austria	4

Special focus

Tourism 1976

Origin of visitors	Number (000)	%
Czechoslovakia	2 017	36
Poland	850	15
East Germany	655	12
Jugoslavia	639	12
West Germany	275	5
Austria	235	4
Soviet Union	216	4
Other	664	12
Total visitors	5 551	100

Iceland
Republic of Iceland
Lýðveldið Ísland

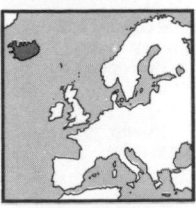

Location North Atlantic Ocean
The island is just south of the Arctic Circle
with Greenland to the west across the Denmark
Strait, Faroes and Scotland to the south-east
Land Area 103 000 km² = 39 800 mi²
Climate Temperate (warmed by the Gulf
Stream)
Weather at Reykjavík, 18 m altitude
Temperature: hottest month July 9–14 °C,
coldest Jan minus 2–2 °C
Rainfall (av monthly): driest months May, June 42 mm,
wettest Oct 94 mm
Time GMT
Measures Metric system
Monetary unit Icelandic króna (I Kr) = 100 aurar
Rate of exchange (1976 av): free I Kr 182.2 = $ 1, I Kr 329.0 = £ 1

Summary

Political Republic, which became independent from Denmark in 1944.
Fishing limits were extended to 50 nautical miles in 1972 and 200
nautical miles in October 1975. Member of UN, OECD, Council of
Europe, Nato, Efta and Nordic Council
Economic Mainly dependent on fishing which accounts for about
two-thirds of exports. The other main industry is production of
aluminium from imported alumina, making use of hydro-electricity;
the plant began operation in 1969. A ferro-silicon plant is being
constructed (to produce 50 000 tonnes by 1980) and further
power-intensive industries are planned

People, resources and equipment

Population 1960 176 000, 1970 204 000, 1976 220 000*
Growth: 1960–70 1.5 % pa, 1970–76 1.3* % pa
Density (1976): 2* people per km²
Vital statistics (rate per 1 000 people, 1976): births 19.4, deaths 6.2
Cities (population in 000, 1976)

Reykjavík (capital)	84	Akureyri	12	Keflavík	6
Kópavogur	13	Hafnarfjörður	12	Akranes	5

Race (1960) Native-born Icelanders 99 %
Language Icelandic
Religion (1960) Lutheran 98 %
Education (1975/76) Pupils 56 882, teachers 4 225
Labour force (1975) 93 170; in agriculture 13 000 (14 %)
Personnel Scientists and engineers (1970): 3 169
Physicians (1971): 321, 1 per 640 people
Standard of living National income per person (1976):
I Kr 1 002 000 = $ 5 500 = £ 3 046
Consumption per person (1975): energy 4 720 kg coal equivalent,
electricity (production, 1976) 11 030 kW h, newsprint 14 kg, steel 191 kg
Newspapers (1974): number 5; circulation 95 000, 436 per 1 000 people
Telephones (Dec 1976): 93 678, 424 per 1 000 people
Livestock (000, Dec 1976) Cattle 61, sheep 871, horses 48, poultry 296
Electrical capacity (1975) 514 megawatts,
of which, hydro 389 megawatts, geothermal 2 megawatts
Hospital beds (1970) 2 923, 1 per 70 people
Roads (excluding urban roads, 1976) 11 525 km = 7 161 mi,
density 0.11 km per km²
There are no railways
Ships (registered, 1977) 375, total of 166 702 gross tons
Ports (goods traffic, 000 tonnes, 1976)
Reykjavík *International* Loaded 64, unloaded 722
Coastwise Loaded 45, unloaded 91
Also: Akureyrí, Hafnarfjörður, Seyðisfjörður, Vestmannæyjar
Airports Keflavík, Reykjavík, Akureyrí, Vestmannæyjar, Hafnarfjörður

Durable equipment (at end-year)	000	no per 1 000 people	
Radio sets (1976)	63	285	no per km of road
Television sets (1976)	55	249	
Passenger cars (1975)	64	290	5.6
Commercial vehicles (1975)	7.5	34	0.7

Iceland

Production

Gross domestic product 1976: I Kr 264 973 mn = $ 1 454 mn = £ 805 mn
Growth in real terms: 1960–70 4.3 % pa, 1970–76 5.1 % pa
Structure of gross domestic product (1976) *By type* Final consumption expenditure 70 % (of which, government 10 %), stock investment −1 %, gross fixed capital formation 29 %, exports of goods and services 40 %, less imports of goods and services −38 %

Production index (1970 = 100)	1960	1970	1976	Growth % pa 1960–70	1970–76
Agricultural	89	100	121	1.2	3.2
Main products (000 t)					
Agriculture					
Potatoes	10.0*	5.4	6.9	−6.0*	4.2
Milk	104	117	126	1.2	1.2
Butter	1.0*	1.5	1.8	4.1*	3.1
Cheese	0.9	3.5	4.2*	14.5	3.1*
Eggs	1.0**	1.8	3.0	6.0**	8.9
Mutton and lamb	11	12	15	0.9	3.8
Fish catch	593	734	986	2.2	5.0
Other					
Electricity (mn kW h)	551	1 470	2 426	10.3	8.7
of which, hydro (mn kW h)	523	1 413	2 206a	10.5	9.3b
geothermal (mn kW h)	—	12	18a	na	8.4b
Fishmeal	55	67	107a	2.1	9.8b
Beer (000 hl)	14	27	26a	6.8	−0.8b
Cement	73	85	164a	1.5	14.0b
Aluminium	—	38	59a	na	9.3b
Dwellings (000 units)	1.48	1.33	2.07a	−1.1	9.2b

a 1975   b 1970–75

Transport traffic Air (mn)	1960	1970	1976	Growth % pa 1960–70	1970–76
Passenger-kilometres	261	1 746	1 914	20.9	1.5
Cargo: tonne-kilometres	6.1	13.0	30.6	7.9	15.3
Sea: tonnes (mn)					
Goods loaded	0.27	0.30	0.42	1.2	5.7
Goods unloaded	0.65	0.96	1.11	4.0	2.3

Tourism (1976) Number of visitors 70 180, gross receipts $ 12.3 mn

Finance and trade

Price and wage indices (1970 = 100)	1960	1970	1976	Growth % pa 1960–70	1970–76
Consumer prices	34.0	100.0	405.0	11.4	26.3
Wages (hourly rates)	na	100.0	448.1	na	28.4
Money stock					
(end-year, I Kr mn)	1 077	4 648	20 767	15.7	28.3

Budget (1976) Revenue: I Kr 60 342 mn = $ 331 mn = £ 183 mn
Expenditure: I Kr 58 857 mn = $ 323 mn = £ 179 mn

Balance of payments ($ mn)	1972	1973	1974	1975	1976
Balance of goods (fob)	−24	−35	−147	−286	−26
Balance of services	+4	+6	−8	−4	+1
Balance of transfers	—	+16	+1	−1	—
Current balance	−20	−13	−155	−144	−24
Long-term capital flow	+24	+23	+109	+135	+62
Reserves and debt (end-year, $ mn)					
International reserves	84	100	48	47	81
External public debt	170	252	352	462	493

External trade (1976) Imports: I Kr 85 660 mn = $ 470 mn = £ 260 mn
Exports: I Kr 73 500 mn = $ 403 mn = £ 223 mn

Main imports	% of total	Main exports	% of total
Petroleum products	12	Fish	57
Chemicals	12	Aluminium	17
(of which, alumina 4)		Fishmeal	4
Electrical machinery	11	Fish oil	3
Food	11	Mutton and lamb	2
Machinery, non-electric	10		
Motor vehicles	5		
Textile yarns and fabrics	4		
Iron and steel	3		
Main sources		*Main destinations*	
Soviet Union	12	United States	29
West Germany	11	United Kingdom	12
United States	11	West Germany	11
United Kingdom	10	Portugal	10
Denmark	10	Soviet Union	5
Norway	8	Switzerland	3
Sweden	6	Italy	3

Special focus

Aluminium, 1976

	quantity (tonnes)	value ($)	price (unit value) $ per tonne
Imports of alumina	130 966	20 546 000	157
Exports of aluminium	78 615	67 966 000	865

Ireland
Republic of Ireland
Éire

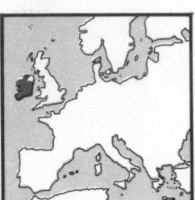

Location North-east Atlantic Ocean
The republic comprises the main part of the island of Ireland, which lies off the west coast of Europe, the rest being Northern Ireland, included in the United Kingdom. The island of Great Britain (rest of United Kingdom) is to the east
Land Area 70 283 km² = 27 136 mi²
Climate Temperate
Weather at Dublin, 47 m altitude
Temperature: hottest month July 11–20 °C, coldest Jan 1–8 °C
Rainfall (av monthly): driest month April 45 mm, wettest Aug 74 mm
Time GMT (summer time, 1 hour ahead of GMT)
Measures UK (imperial) system, converting to metric system
Monetary unit Irish pound (I £) = 100 new pence
Rate of exchange (1976 av): par I £ 1 = £ 1, free $ 1.806 = I £ 1

Summary

Political Republic since 1949; independent from the United Kingdom since 1921; the British Parliament, in 1920, provided for the 6 counties of the north-east (part of Ulster) to remain in the United Kingdom, the other 26 counties ('Southern Ireland') forming a separate state. There have been discussions with the United Kingdom concerning the possibility of combining with Northern Ireland in a Council of Ireland or other form of co-operation. Member of UN, OECD, Council of Europe, EEC (from January 1, 1973) and sterling area
Economic Mainly agricultural, and especially concerned with livestock and their products; in particular, meat and livestock account for about one-fifth of exports and dairy produce is important. Industrialisation is increasing, with main industries chemicals, textiles, clothing and machinery, and there is some mining of metals; natural gas production is planned to start in 1979, and further gas and oil exploration is being carried out. Tourism is important

People, resources and equipment

Population 1960 2.83 mn, 1970 2.94 mn, 1976 3.16 mn
Growth: 1960–70 0.4 % pa, 1970–76 1.2 % pa
Density (1976): 45 people per km²
Vital statistics (rate per 1 000 people, 1976): births 21.6, deaths 10.5
Regions (provinces, population in 000, 1971): total of 2.98 mn)
Connacht 391, Leinster 1 498, Munster 882, Ulster 207
Cities (population in 000, 1971)
Dublin (capital) 568   Limerick 63   Waterford 32
Cork 134   Dun Laoghaire 53   Galway 28
Race (1961) Irish 96 %, English and Welsh 2 %, Northern Irish 1 %
Language Irish and English
Religion (1971) Roman Catholic 94 %, Church of Ireland 3 %, Presbyterian 1 %
Education (1975/76) Pupils 824 352, teachers 34 985
Labour force 1971: 1 119 531; in agriculture 284 565 (25 %), mining and quarrying 11 523 (1 %), manufacturing 224 790 (20 %), electricity, gas and water 14 757 (1 %), construction 100 039 (9 %), distribution and hotels 179 825 (16 %), transport and communication 63 542 (6 %), finance and real estate 36 661 (3 %), other 203 829 (18 %)
1975: 1 140 000
Personnel Scientists and engineers (1971): 21 886
Physicians (1975): 3 772, 1 per 830 people
Standard of living
National income per person (1976): I £ 1 310 = $ 2 367 = £ 1 310
Consumption per person (1975): energy 3 097 kg coal equivalent, electricity (production, 1976) 2 722 kW h, newsprint 15 kg, steel 106 kg
Newspapers (1974): number 7; circulation 729 000, 236 per 1 000 people
Telephones (Dec 1976): 480 000, 151 per 1 000 people
Livestock (000, 1976) Cattle 6 954, sheep 3 526, pigs 925, chickens 8 776
Mineral reserves Coal (1967) 48 mn tonnes

Ireland

Petroleum refinery capacity (1975) 2.97 mn tonnes
Electrical capacity (1975) 1 986 megawatts
Hospital beds (1975) 33 772, 1 per 93 people
Roads (1975) 89 006 km = 55 306 mi, density 1.3 km per km²
Railways (1975) 2 006 km = 1 246 mi, density 0.029 km per km²
Ships (registered, 1977) 98, total of 211 872 gross tons
Ports Dublin, Dun Laoghaire, Cork, Waterford, Rosslare, Limerick
Airports Shannon, Dublin, Cork

Durable equipment (Dec 1976)	000	no per 1 000 people	
Radio sets	900*	280*	no per
Television sets	590	186	km of road
Passenger cars	551	173	6.2
Commercial vehicles	56	18	0.6

Production

Gross domestic product 1976: I £ 4 416 mn = $ 7 975 mn = £ 4 416 mn
Growth in real terms: 1960–70 4.2 % pa, 1970–76 3.1 % pa
Structure of gross domestic product (1974) *By origin* Agriculture 14 %, manufacturing, mining, utilities and construction 30 %, other 56 %
By type Final consumption expenditure 82 % (of which, government 19 %), stock investment 1 %, gross fixed capital formation 24 %, exports of goods and services 49 %, less imports of goods and services −56 %

Production indices (1970 = 100)	1960	1970	1976	Growth % pa 1960–70	1970–76
Agricultural	84	100	117	1.8	2.6
Industrial	52.3	100.0	123.9	6.7	3.6
Main products (000 t)					
Agriculture					
Wheat	469	381	200	−2.0	−10.2
Barley	442	782	922	5.9	2.8
Oats	426	207	130	−7.0	−7.5
Potatoes	1 829	1 468	1 179	−2.2	−3.6
Sugar, raw value	133	150*	189*	1.2*	3.9*
Cabbages	120*	130*	138*	0.8*	1.0*
Milk	2 666	3 629	4 550*	3.1	3.8*
Butter	58	73	98	2.3	5.0
Cheese	10*	29	48*	11.2*	8.8*
Beef and veal	252	298	450*	1.7	7.1*
Pigmeat	95	144	109*	4.3	−4.5*
Fish catch	43	79	94	6.3	3.0
Timber (000 m³)	350**	382	451	0.9**	2.8
Energy					
Total energy (000 tce)	2 060	2 690	2 330ᵉ	2.7	−2.8ᶠ
Coal	208	156	51	−2.8	−17.0
Petroleum products	1 355	2 573	2 340*ᵉ	6.6	−1.9*ᶠ
Manufactured gas (mn m³)	177	221	270	2.2	3.4
Electricity (mn kW h)ᶜ	2 180*	5 399	8 600	9.5*	8.1
Mining					
Copper conc (Cu content)	6.2	7.7	9.8ᵉ	2.2	4.9ᶠ
Lead conc (Pb content)	1.3	62	36*	47.3	−10.3ᶠ
Zinc ore (Zn content)	1.2	99	67*ᵉ	55.4	−7.5*ᶠ
Manufacturing					
Beer (000 hl)ᵃ	3 391	3 885	5 050*ᵍ	1.4	6.8*ʰ
Cigarettes (mn units)	5 122	5 550	8 315ᵉ	0.8	8.4ᶠ
Cotton yarn	5.1	4.9	3.6ᵉ	−0.4	−6.0ᶠ
Cotton fabrics (mn m²)	20	24	24ⁱ	1.8	0.0ʲ
Man-made fibres	—	—	11.5ᵉ	na	na
Wood pulp	22	35	31ᵉ	4.8	−2.4ᶠ
Paper	70**	89	102ᵉ	2.4**	2.7ᶠ
Fertilisers, nitrogenousᵇ	—	60*	100*	na	8.9*
Cement	745	859	1 570	1.4	10.6
Crude steel	40*	80*	82*ᵉ	7.2*	0.5*ᶠ
Passenger cars (000 units)ᵈ	33	47	43	3.7	−1.6
Commercial vehicles (000 units)ᵈ	4.0	4.3	4.2	0.7	−0.7
Merchant vessels (000 grt)	—	28	29	na	0.6
Construction					
Dwellings (000 units)	5.8	13.6	22.6	8.5	8.8

ᵃYears ending September 30th ᵇYears ending June 30th ᶜYears ending March 31st
ᵈAssembly only ᵉ1975 ᶠ1970–75 ᵍ1974 ʰ1970–74 ⁱ1973 ʲ1970–73

Transport traffic	1960	1970	1976	Growth % pa 1960–70	1970–76
Passenger-kilometres (mn)					
Rail	568	582	739	0.2	4.1
Air	404	1 777	1 528	16.0	−2.5
Cargo: tonne-kilometres (mn)					
Rail	345	499	523	3.7	0.8
Air	4.6	59.4	77.7	29.1	4.6
Sea: tonnes (mn)					
Goods loaded	1.00**	2.91*	3.39*ᵃ	11.3**	3.9*ᵇ
Goods unloaded	5.30**	9.90*	10.86*ᵃ	6.4**	2.3*ᵇ
Tourism					
Number of visitors (000)	1 450**	1 758	1 690	2.0**	−0.7
Gross receipts ($ mn)	111	178	253	4.8	6.0

ᵃ1974 ᵇ1970–74

Finance and trade

Price and wage indices (1970 = 100)	1960	1970	1976	Growth % pa 1960–70	1970–76
Consumer prices	62.8	100.0	219.9	4.8	14.0
Wholesale prices	68.5	100.0	231.0	3.9	15.0
Wages (earnings)	42.6	100.0	295.8	8.9	19.8
Share prices	40.4	100.0	116.8	9.5	2.6
Money stock					
(end-year, I £ mn)	203	435	875	7.9	12.4

Budget (1976) Revenue: I £ 1 551 mn = $ 2 801 mn = £ 1 551 mn
Expenditure: I £ 2 042 mn = $ 3 689 mn = £ 2 042 mn

Balance of payments ($ mn)	1972	1973	1974	1975	1976
Balance of goods (fob)	−378	−527	−1 073	−463	−608
Balance of services	+118	+81	+98	+47	+33
Balance of transfers	+125	+212	+303	+391	+314
Current balance	−135	−234	−671	−24	−262
Long-term capital flow	+20	+268	+713	+355	+544
Reserves and debt (end-year, $ mn)					
International reserves	1 126	1 025	1 267	1 532	1 837
External public debt	715	953	1 522	1 963	2 762

External trade (1976) Imports: I £ 2 332 mn = $ 4 212 mn = £ 2 332 mn
Exports: I £ 1 857 mn = $ 3 354 mn = £ 1 857 mn

Main imports	% of total	Main exports	% of total
Machinery, non-electric	13	Meat	13
Chemicals	11	(of which,	
Food	11	beef and veal 9)	
(of which, cereals 3)		Dairy produce	11
Petroleum products	9	(of which, milk 4)	
Electrical machinery	6	Chemicals	9
Textile yarns & fabrics	6	(of which, medicinal 3)	
Motor vehicles	6	Machinery, non-electric	8
Crude oil	4	Textile yarns & fabrics	6
Iron and steel	3	Livestock	6
Paper	3	Electrical machinery	4
		Clothing	3
		Instruments	3
Main sources		*Main destinations*	
United Kingdom	49	United Kingdom	49
United States	8	West Germany	9
West Germany	7	United States	7
France	5	Netherlands	6
Netherlands	3	France	5
Italy	3	Belgium-Luxembourg	4
Japan	2	Italy	2
Belgium-Luxembourg	2		
Sweden	2		

Special focus

Tourism, 1976

Origin of visitors	Number	%
United Kingdom	1 184 000ᵃ	70
United States	231 000	14
West Germany	69 000	4
France	51 000	3
Other	155 000	9
Total visitors	1 690 000	100

ᵃIncludes 399 000 visitors from Northern Ireland

Isle of Man

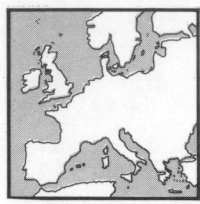

Location Irish Sea
The island is 130 km north-west of Liverpool (England); Northern Ireland is to the west
Land Area 588 km² = 227 mi²
Climate Temperate
Weather at Douglas, 87 m altitude
Temperature: hottest month Aug 12–17 °C, coldest Feb 2–7 °C
Rainfall (av monthly): driest month April 63 mm, wettest Dec 128 mm
Time GMT (summer time, 1 hour ahead of GMT)
Measures UK (imperial) system, converting to the metric system
Monetary unit UK pound and Isle of Man pound (IoM £) = 100 new pence. The Isle of Man pound is equal to the UK pound
Rate of exchange (1976 av): free $ 1.806 = IoM £ 1

Summary

Political Independent state for a thousand years; the millennium is in 1979. Attached to the Crown of England, the United Kingdom is responsible for defence and external relations. Member of sterling area and within EEC via relationship with United Kingdom, but with some special exemptions
Economic A mixed economy, with tourism, finance and manufacturing important; industry is being further encouraged by investment grants, and finance by tax advantages. Exports, mainly livestock, meat and fish, are small

People, resources and equipment

Population 1960 47 380*, 1970 55 410*, 1976 60 496
Growth: 1960–70 1.6 % pa, 1970–76 1.5* % pa
Density (1976): 103 people per km²
Vital statistics (rate per 1 000 people, 1977): births 11.1, deaths 16.7
Towns (population in 000, 1976) Douglas (capital)20, Onchan 6, Ramsey 5, Peel 3
Race Celtic; Manx-born (1976) 32 000* (53* %)
Language English; also Manx. Manx-speaking (1978): 284
Religion Christian, mainly Church of England
Education (1975/76) Pupils 12 060, teachers 500**
Labour force (1976) 23 278; in tourism 4 724 (20 %)
Personnel (1974) Physicians: 70, 1 per 840 people
National income per person (1976)
IoM £ 1 570* = $ 2 830* = £ 1 570*
Telephones (1976) 15 381, 254 per 1 000 people
Livestock (000, 1977) Cattle 39, sheep 99, pigs 4, poultry 108
Hospital beds (1977) 727, 1 per 83 people
Roads (1976) 800* km = 500* mi, density 1.4* km per km²
Railways (1977) 56 km = 35 mi, density 0.095 km per km²
Ports Douglas, Ramsey, Port St Mary, Peel
Airport Ronaldsway (Ballasalla)

Durable equipment (Dec 1976)	000	no per 1 000 people	
Radio sets	22.0*	360*	no per
Television sets	20.5*	338*	km of road
Motor vehicles	26.0*	430*	32*

Production, finance and trade

Gross domestic product
1976/77 (year ending March 31st): IoM £ 107 mn = $ 186 mn = £ 107 mn
1976 est: IoM £ 104* mn = $ 188* mn = £ 104* mn
Structure of gross domestic product (1976/77) *By origin* Agriculture 4** %, manufacturing 13 %, tourism 12 %, finance 26 %, other 45 %
Main products (000 t, 1976) Barley 8*, oats 4*, wheat 2*, potatoes 15*, milk 18, beef and veal 1*, mutton and lamb 1*, fish (landings) 10
Transport traffic *Passenger* (departures and arrivals, 000) Air (1976) 366, sea (1975/76) 857 *Cargo* Goods (000 tonnes, 1975/76) loaded 39, unloaded 300
Tourism (1977) Number of visitors 451 300, gross receipts $ 20** mn
Budget (1976/77; year ending March 31st)
Revenue: IoM £ 31.8 mn = $ 55.2 mn = £ 31.8 mn
Expenditure: IoM £ 30.8 mn = $ 53.4 mn = £ 30.8 mn
External trade (1976)[a]
Imports: IoM £ 40*** mn = $ 70*** mn = £ 40*** mn
Exports: beef and lamb IoM £ 0.8*** mn, fish IoM £ 0.3*** mn
[a] Included within the United Kingdom area for external trade purposes

Italy

Italian Republic
Repubblica Italiana

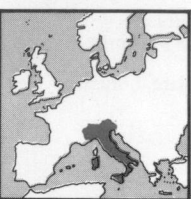

Location Southern Europe
The Italian peninsula projects into the Mediterranean Sea, with the Adriatic Sea to the east, Ionian Sea to the south-east, and Tyrrhenian Sea to the south-west; in the northern part, France is to the west, Switzerland and Austria to the north, and Jugoslavia to the east. Territory includes a number of islands, the main ones being Sicily and Sardinia
Land Area 301 225 km² = 116 304 mi²
Usage (1975): agricultural 175 170 km² (58 %), of which, arable 93 300 km² (31 %), cropland 29 830 km² (10 %), pastures 52 040 km² (17 %); forests 63 060 km² (21 %)
Climate Mediterranean (hot summers and mild winters)
Weather at Rome, 17 m altitude
Temperature: hottest month July 20–30 °C, coldest Jan 4–11 °C
Rainfall (av monthly): driest month July 15 mm, wettest Nov 129 mm
Time 1 hour ahead of GMT (summer time, 2 hours ahead)
Measures Metric system
Monetary unit Lira (L) = 100 centesimi
Rate of exchange: *free* 1976 av: L 832.3 = $ 1, L 1 503 = £ 1
1977 av: L 882.39 = $ 1, L 1 540 = £ 1

Summary

Political Republic from 1946; a peace treaty concerning the second world war was signed in 1947, and an agreement concerning the division of territory east of Trieste was made with Jugoslavia in 1975. Member of UN, OECD, Council of Europe, Nato, EEC and WEU
Economic An industrial economy, with manufacturing and mining providing about one-third of gross domestic product and absorbing about one-third of the labour force; 15 % of the labour force are in agriculture, but provide only 8 % of gross domestic product. Exports are mainly industrial, notably electrical and other machinery, motor vehicles, chemicals, textiles and clothing and iron and steel. Crude oil accounted for about one-fifth of imports in 1976, a large part being refined for export. Growth has slackened in the 1970s, and the south is still comparatively underdeveloped industrially. State-owned industry is important, and tourism is a useful earner of foreign exchange

People

Population 1960 49.64 mn, 1970 53.66 mn, 1976 56.17 mn
Growth: 1960–70 0.8 % pa, 1970–76 0.8 % pa
Density (1976): 186 people per km²
Vital statistics (rate per 1 000 people, 1976): births 14.0, deaths 9.7
Regions (population in 000, Dec 31, 1976; total of 56.32 mn)

Abruzzi	1 221	Liguria	1 865	Sicilia	4 902
Basilicata	617	Lombardia	8 866	Toscana	3 579
Calabria	2 049	Marche	1 397	Trentino-	
Campania	5 335	Molise	331	Alto Adige	869
Emilia-Romagna	3 947	Piemonte	4 543	Umbria	799
Friuli-Venezia Giulia	1 244	Puglia	3 819	Valle d'Aosta	114
Lazio	4 959	Sardegna	1 568	Veneto	4 301

Cities (population in 000, Dec 31, 1976)

Roma[a] (capital)	2 884	Cagliari	240	Rimini	126
Milano[b]	1 705	Brescia	215	La Spezia	121
Napoli[c]	1 224	Modena	179	Monza	121
Torino[d]	1 191	Parma	178	Siracusa[j]	121
Genova[e]	801	Reggio di Calabria	178	Vicenza	119
Palermo	673	Livorno[i]	178	Sassari	116
Bologna	486	Salerno	162	Terni	113
Firenze[f]	465	Ferrara	155	Forli	110
Catania	400	Prato	154	Piacenza	109
Bari	384	Foggia	153	Ancona	108
Venezia[g]	362	Ravenna	138	Bolzano	107
Verona	271	Perugia	137	Udine	104
Trieste	268	Pescara	135	Pisa	103
Messina	265	Reggio		Alessandria	103
Taranto	244	nell'Emilia	130	Cosenza	102
Padova[h]	242	Bergamo	128	Novara	102

[a] Rome [b] Milan [c] Naples [d] Turin [e] Genoa [f] Florence [g] Venice [h] Padua
[i] Leghorn [j] Syracuse

Italy

Race (1961) Italian 98 %, Austrian ½ %
Language Italian
Religion (1976) Roman Catholic 99** %
Education (1975/76) Pupils: primary 4 835 449, secondary (1973/74)
3 105 283, vocational (1973/74) 1 111 589, teacher-training (1973/74)
195 184, higher (1974/75) 929 300. Teachers: primary 252 736, secondary
(1972/73) 202 951, vocational (1972/73) 95 639, teacher-training (1972/73)
18 606, higher (1974/75) 42 639
Labour force (1976)

Economic activity	Number	% of total
Agriculture	3 084 000	15
Mining, quarrying, electricity, gas and water	343 000	2
Manufacturing	6 221 000	31
Construction	1 793 000	9
Distribution and hotels	2 698 000	13
Transport and communication	1 118 000	5
Other services	4 346 000	21
Persons seeking first position	516 000	3
Unemployed	260 000	1
Total	20 379 000	100

Personnel Scientists and engineers engaged in research (1972): 32 592
Physicians (1973): 109 166, 1 per 502 people
Standard of living
National income per person (1976): L 2 266 200 = $ 2 723 = £1 508
Consumption per person (1975): energy 3 012 kg coal equivalent,
electricity (production, 1976) 2 912 kW h, newsprint 4.4 kg, steel 319 kg
Newspapers (1974): number 79, circulation 6 963 000*,
126* per 1 000 people
Telephones (Dec 1976): 15 240 500, 271 per 1 000 people

Resources and equipment

Livestock (000, 1976) Cattle 8 446, sheep 8 152, pigs 8 814, goats 940,
buffaloes 82, horses 253, poultry 115 700*
Mineral reserves Coal (1973) 1 mn tonnes
Lignite (1972) 110 mn tonnes
Crude oil (1975) 92 mn tonnes
Natural gas (1975) 210 bn cubic metres
Uranium (1976) 1 200 tonnes
Petroleum refinery capacity (1975) 208* mn tonnes
Electrical capacity (1975) 43 305 megawatts,
of which, hydro 16 995 megawatts, nuclear 670 megawatts,
geothermal 421 megawatts
Hospital beds (1972) 575 162, 1 per 95 people
Roads (1975) 291 081 km = 180 869 mi, density 0.97 km per km²
Railways (1974) 20 171 km = 12 534 mi, density 0.067 km per km²
Inland waterways (1975) 2 237 km = 1 390 mi
Oil pipelines (1974) 2 516 km = 1 563 mi
Ships (registered, 1977) 1 690, total of 11 111 182 gross tons;
of which, oil tankers: 296, total of 4 684 889 gross tons;
bulk and ore carriers: 106, total of 2 075 962 gross tons;
oil/bulk carriers: 30, total of 1 911 445 gross tons

Ports (goods traffic, 000 tonnes, 1976)

	International loaded	International unloaded	Coastwise loaded	Coastwise unloaded	Total
Genoa	2 048	39 278	1 992	5 417	48 735
Trieste	1 306	32 212	1 065	878	35 461
Augusta[a]	4 910	17 007	9 242	2 281	33 440
Taranto	2 446	20 091	4 423	1 144	28 104
Venice	1 563	13 338	1 095	7 751	23 747
Porto Foxi[a]	3 031	9 595	4 606	102	17 334
Milazzo[a]	1 324	6 840	4 398	697	13 259
La Spezia	629	9 193	2 373	860	13 055
Naples	1 287	6 803	817	1 492	10 399

[a]Crude oil port
Airports (passenger departures and arrivals, 000, 1976)
Ciampino and Fiumicino (Rome) 9 369
Linate and Malpensa (Milan) 5 031
Also (1977) 28 other airports with scheduled flights

Durable equipment (Dec 1976)	000	no per 1 000 people	
Radio sets	13 024	231	no per
Television sets	12 377	220	km of road
Passenger cars	15 925	283	55
Commercial vehicles	1 249	22	4.3
of which, buses and coaches	47	1	0.2
goods vehicles	1 202	21	4.1

Industrial groups (sales, $ bn, year ending in 1976)

Istituto per la Ricostruzione Industriale (IRI)	14.6
Ente Nazionale Idrocarburi (ENI)	9.6
Montedison SpA	5.8
Societa Finanziaria Siderurgica SpA (Finsider)	4.7
Fiat SpA	4.6
Pirelli SpA	4.1[a]

[a]Figure for Pirelli-Dunlop

Production

Gross domestic product 1976: L 142 128 bn = $ 170 765 mn = £ 94 563 mn
Growth in real terms: 1960–70 5.5 % pa, 1970–76 2.9 % pa
Structure of gross domestic product (1976)

By origin	L bn	% of gdp
Agriculture	11 285	8
Manufacturing, mining, quarrying	47 890	34
Electricity, gas and water	2 528	2
Construction	10 930	8
Distribution and hotels	20 020	14
Transport and communication	7 880	6
Other	41 595	29
Total	142 128	100

By type of expenditure		
Government final consumption	19 436	14
Private final consumption	91 489	64
Stock investment	4 407	3
Gross fixed capital formation	28 810	20
Exports of goods and services	37 957	27
less Imports of goods and services	−39 971	−28
Total	142 128	100

Production indices (1970 = 100)	1960	1965	1970	1973	1974	1975	1976	Growth % pa 1960–70	1960–65	1965–70	1970–76
Agricultural	75	91	100	101	106	107	104	2.9	3.9	1.9	0.7
Industrial	50.0	70.3	100.0	114.4	119.5	108.5	122.0	7.2	7.0	7.3	3.4

Main products (000 t)

Agriculture	1960	1965	1970	1973	1974	1975	1976	1960–70	1960–65	1965–70	1970–76
Wheat	6 794	9 776	9 689	8 920	9 695	9 610	9 516	3.6	7.5	−0.2	−0.3
Rice	622	509	819	1 045	1 047	1 009	907	2.8	−3.9	10.0	1.7
Maize	3 813	3 317	4 754	5 089	5 043	5 326	5 321	2.2	−2.7	7.5	1.9
Potatoes	3 818	3 550	3 668	2 947	2 903	2 943	2 989	−0.4	−1.4	0.6	−3.3
Sugar, raw value	998	1 241	1 202	1 149	1 012	1 442	1 630	1.9	4.4	−0.6	5.2
Watermelons	401	602	749	817	706	767	822	6.4	8.5	4.5	1.5
Artichokes	402	542	671	662	688	680	677	5.2	6.1	4.3	0.1
Tomatoes	2 428	3 177	3 618	3 310	3 637	3 512	2 985	4.1	5.5	2.6	−3.2
Cauliflowers	650	693	745	615	601	584	570	1.4	1.3	1.5	−4.4
Chillies and peppers	225	316	424	477	481	483	495	6.5	7.0	6.0	2.6
Apples	1 834	2 185	2 062	2 050	1 886	2 127	2 048	1.2	3.6	−1.1	−0.1
Pears	622	962	1 906	1 570	1 507	1 453	1 480	11.8	9.1	14.7	−4.1
Peaches	836	1 300	1 127	1 176	1 252	1 139	1 419	3.0	9.2	−2.8	3.9
Oranges	862	1 175	1 325	1 566	1 770	1 580	1 624	4.4	6.4	2.4	3.4
Lemons	376	602	798	852	829	885	817	7.8	9.9	5.8	0.4
Grapes	8 643	10 675	10 724	11 842	11 809	10 756	10 250	2.2	4.3	0.1	−0.8
Wine	5 534	6 821	6 887	7 672	7 687	6 983	6 585	2.2	4.3	0.2	−0.7
Figs	295	247	198	138	123	123*	120*	−3.9	−3.5	−4.3	−8.0

Main products (000 t)	1960	1965	1970	1973	1974	1975	1976	Growth % pa 1960–70	1960–65	1965–70	1970–76
Agriculture											
Olives	2 106	2 232	2 124	2 696	2 237	3 228	1 820	0.1	1.2	−1.0	−2.5
Olive oil	424	456	464	595	475	695	328	0.9	1.5	0.3	−5.6
Tobacco	80	74	78	94	93	113	109	−0.2	−1.7	1.3	5.7
Milk	9 892	9 457	9 354	9 690	9 309	9 113	9 370*	−0.5	−0.9	−0.2	0.0*
Cheese	371	447	466	506	518	522	522*	2.3	3.8	0.8	1.9*
Eggs	354	458	577	609	631	671	775	5.0	5.3	4.7	5.0
Beef and veal	456	467	827	732	850	746	778*	6.1	0.5	12.1	−1.0*
Pigmeat	417	465	563	659	677	732	742*	3.0	2.2	3.9	4.7*
Poultrymeat	173	368	626	796	833	902	950*	13.7	16.3	11.2	7.2*
Fish catch	270*	354	401	399	431	417	420	4.0*	5.6*	2.5	0.8
Timber (000 m³)	12 000**	11 176	11 667	9 065	6 809	6 621	6 560	−0.3**	−1.4**	0.9	−9.2
Energy											
Total energy (000 tce)	18 420	20 340	26 390	27 650	27 610	27 030	na	3.7	2.0	5.3	0.5ᵉ
Lignite	794	1 011	1 393	1 301	1 176	1 213	1 222	5.8	4.9	6.6	−2.2
Crude oil	1 998	2 207	1 405	1 047	1 031	1 029	1 108	−3.5	2.0	−8.6	−3.9
Petroleum products	28 278	64 590	111 325	125 515	114 309	94 553	na	14.7	18.0	11.5	−3.2ᵉ
Natural gas (mn m³)	6 447	7 802	13 171	15 327	15 273	14 578	15 370	7.4	3.9	11.0	2.6
Manufactured gas (mn m³)	2 298	2 713	3 060	3 156	3 493	3 669	3 508	2.9	3.4	2.4	2.3
Electricity (mn kW h)	56 240	82 968	117 423	145 518	148 905	145 551	163 550	7.6	8.1	7.2	5.7
of which, hydro (mn kW h)	46 106	43 008	41 300	39 125	39 346	42 116	na	−1.1	−1.4	−0.8	0.4ᵉ
nuclear (mn kW h)	—	3 510	3 176	3 142	3 410	3 800	na	na	na	−2.0	3.7ᵉ
geothermal (mn kW h)	2 104	2 576	2 725	2 480	2 502	2 483	na	2.6	4.1	1.1	−1.8ᵉ
Mining											
Iron ore (Fe content)	619	420	318	220	289	277	231	−6.4	−7.5	−5.4	−5.2
Antimony ore (Sb content)	0.22	0.41	1.30	1.36	1.24	1.09	na	19.6	13.7	26.0	−3.5ᵉ
Bauxite	313	244	225	50	32	32	24	−3.2	−4.8	−1.6	−31.1
Mercury	1.90	1.98	1.53	1.13	0.90	1.09	na	−2.1	0.8	−5.0	−6.6ᵉ
Zinc concentrates (Zn content)	131	114	109	79	98	87	na	−1.8	−2.7	−0.9	−4.4ᵉ
Potash (oxide content)	49	228*	222	215	247	220	na	16.3	36.0*	−0.5*	−0.2ᵉ
Asbestos	55	72	119	149	148	147	na	8.0	5.5	10.6	4.3ᵉ
Manufacturing											
Beer (000 hl)	2 489	4 547	5 938	8 598	8 064	6 493	na	9.1	12.8	5.5	1.8ᵉ
Cigarettes (mn units)	50 013	58 798	71 618	66 751	66 359	67 085	na	3.7	3.3	4.0	−1.3ᵉ
Cotton yarn	196	158	176	149	150	140	164	−1.1	−4.2	2.2	−1.2
Cotton fabrics	139	103	129	128	140	126	na	−0.7	−5.8	4.6	−0.5ᵉ
Wool yarn	160	159	167	200	na	na	na	0.4	−0.1	1.0	6.2ᶠ
Man-made fibres	195	301	424	547	492	411	520*	8.1	9.1	7.1	3.5*
Wood pulp	465	669	925	916	921	758	na	7.1	7.5	6.7	−3.9ᵉ
Newsprint	259	379	311	263	273	243	250	1.8	7.9	−3.9	−3.6
Other paper	1 240**	1 829	3 238	3 837	3 927	3 253	na	10.1**	8.1**	12.1	0.1ᵉ
Synthetic rubber	67*	120*	155*	230*	240*	200*	240*	8.7*	12.4*	5.2*	7.5*
Tyres (000 units)	8 700**	14 900**	22 383	28 072	27 715	25 716	na	9.9**	11.3**	8.5**	2.8ᵉ
Ethylene	na	352	900	1 378	1 380	1 128	na	na	na	20.6	4.6ᵉ
Sulphurᵃ	785	726	747	858	542	572	na	0.5	1.6	0.6	−5.2ᵉ
Sulphuric acid	2 299	2 979	3 327	3 036	3 219	3 006	2 887	3.7	5.3	2.2	−2.3
Nitric acid	815	946	1 038	1 038	959	967	na	2.4	3.0	1.9	−1.4ᵉ
Caustic soda	426	691	999	1 179	1 138	1 002	1 134	8.9	10.1	7.6	2.1
Plastics and resins	346	931	1 567	2 479	2 468	2 053	2 650*	16.3	21.9	11.0	9.1*
Fertilisers, nitrogenousᵇ	593	836	960	1 045	1 129	1 132	1 000	4.9	7.1	2.8	0.7
Cement	16 014	20 695	33 076	36 312	36 309	34 235	36 323	7.5	5.3	9.8	1.6
Coke	4 140*	6 122	7 046	7 665	8 566	8 115	na	5.5*	8.1*	2.9	2.9ᵉ
Pig iron	2 824	5 631	8 529	10 271	11 935	11 591	11 889	11.7	14.8	8.7	5.7
Crude steel	8 229	12 681	17 277	20 995	23 803	21 568	23 450	7.7	9.0	6.4	5.2
Aluminium	84	124	146	184	212	190	213	5.7	8.1	3.3	6.5
Sewing machines (000 units)	493	477	1 005	922	884	832	826	7.4	−0.7	16.1	−3.2
Typewriters (000 units)	652ᶜ	765	552	594	647	577	633	−1.8ᵈ	4.1ᵍ	−6.3	2.3
Radio sets (000 units)	935*	1 195*	3 300*	2 150*	1 800*	na	na	13.4*	5.0*	22.5*	−14.1*ʰ
Television sets (000 units)	728*	1 042*	2 030*	2 200*	2 330*	na	na	10.8*	7.4*	14.3*	3.5*ʰ
Passenger cars (000 units)	596	1 104	1 720	1 825	1 631	1 349	1 469	11.2	13.1	9.3	−2.6
Commercial vehicles (000 units)	49	72	134	135	142	110	120	10.6	7.9	13.2	−1.8
Merchant vessels (000 grt)	434	442	598	781	868	843	664	3.3	0.4	6.2	1.8
Construction											
Dwellings (000 units)	263	355	377	197	181	220	184	3.7	6.2	1.2	−11.3

ᵃContent of iron and copper pyrites, and native sulphur   ᵇYears ending June 30th   ᶜ1961   ᵈ1961–70   ᵉ1970–75   ᶠ1970–73   ᵍ1961–65   ʰ1970–74

Transport traffic	1960	1970	1976	Growth % pa 1960–70	1970–76
Passenger-kilometres (mn)					
Road	87 401ᵃ	266 400*	348 644ᶜ	17.2*ᵇ	5.5*ᵈ
Rail	30 723	32 457	39 630	0.5	3.4
Air	1 339	8 395	10 780	20.0	4.3
of which, international	1 145	6 961	8 706	19.8	3.8
Cargo: tonne-kilometres (mn)					
Road	45 170ᵃ	58 658	62 795ᶜ	3.8ᵇ	1.4ᵈ
Rail	15 860	18 069	16 673	1.3	−1.3
Air	26	295	468	27.2	8.0
of which, international	24	282	451	27.8	8.1

Transport traffic	1960	1970	1976	Growth % pa 1960–70	1970–76
Sea: tonnes (mn)					
Goods loaded	11.6	34.8	30.9	11.6	−2.0
Goods unloaded	59.3	200.9	217.4	13.0	1.3
Tourism					
Number of visitorsᵉ (000)	7 300**	10 370	11 501	3.6**	1.7
Gross receipts ($ mn)	642	1 639	2 526	9.8	7.5

ᵃ1963   ᵇ1963–70   ᶜ1975   ᵈ1970–75   ᵉArrivals at hotels; arrivals at all forms of accommodation in 1976: 13 930 thousand

Italy

Finance and trade

Price and wage indices (1970 = 100)	1960	1970	1976	Growth % pa 1960–70	1970–76
Consumer prices	67.9	100.0	199.8	3.9	12.2
Wholesale prices	77.3	100.0	236.4	2.6	15.4
Wages[a]	46.4	100.0	279.2	8.0	18.7
Share prices	134.7	100.0	53.6	−2.9	−9.9
Money stock (end-year, L bn)	7 414	31 185	79 776	15.4	16.9

[a]Minimum contractual

Budget (1977) Revenue: L 43 100 bn = $ 48 845 mn = £ 27 983 mn
Expenditure: L 55 248 bn = $ 62 612 mn = £ 35 870 mn,
of which (1975), education 12 %, defence 7 %

Balance of payments ($ mn)	1972	1973	1974	1975	1976
Balance of goods (fob)	+818	−3 959	−8 511	−1 137	−4 041
Balance of services	+1 667	+1 991	+987	+1 235	+1 660
Balance of transfers	−218	−542	−516	−627	−477
Current balance	+2 266	−2 510	−8 039	−530	−2 856
Long-term capital flow	+307	+3 997	+3 342	+702	+60
Reserves and debt (end-year, $ mn)					
International reserves	6 085	6 436	6 941	4 774	6 654
Total public debt	57 037	67 670	77 145	97 456	92 942

External trade (1976) Imports: L 36 310 bn = $ 43 626 mn = £ 24 158 mn
Exports: L 30 903 bn = $ 37 130 mn = £ 20 561 mn

Main imports	% of total	Main exports	% of total
Crude oil	21	Machinery, non-electric	16
Food	15	Motor vehicles	9
(of which, meat 3, cereals 3)		Chemicals (of which, plastics 2)	8
Chemicals	8	Electrical machinery	7
Machinery, non-electric	8	(of which, domestic	
Motor vehicles	5	electrical equipment 2)	
Electrical machinery	5	Iron and steel	6
Iron and steel	4	Clothing	6
Textile yarns & fabrics	3	Textile yarns & fabrics	6
Non-ferrous metals	3	Petroleum products	5
Textile fibres	3	Footwear	4
Petroleum products	2	Fruit and vegetables	4
Timber	2		

Main sources[a]		Main destinations[b]	
West Germany	17	West Germany	19
France	14	France	15
United States	8	United States	6
Saudi Arabia	6	United Kingdom	5
Netherlands	5	Netherlands	4
Libya	4	Belgium-Luxembourg	4
Belgium-Luxembourg	4	Switzerland	4
United Kingdom	3	Libya	3
Soviet Union	3	Soviet Union	3
Iraq	3	Austria	2
Iran	3	Iran	2
Switzerland	2	Spain	2
Austria	2	Saudi Arabia	2

[a]EEC 44 % [b]EEC 48 %

Special focus

Rate of exchange for the lira

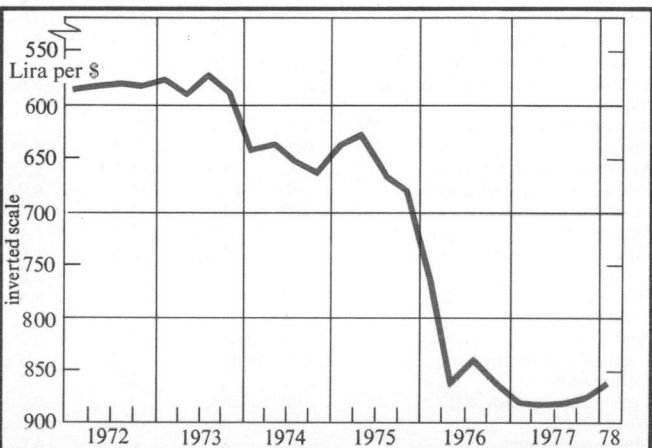

Jersey
Bailiwick of Jersey

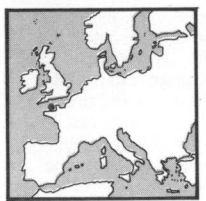

Location English Channel
Part of the Channel Islands, 20 km west of the
Normandy peninsula (north-west France)
Land Area 116.2 km² = 44.9 mi²
Climate Temperate
Weather at St Helier, 9 m altitude
Temperature: hottest month Aug 15–21 °C,
coldest Feb 4–8 °C
Rainfall (av monthly): driest month June 39 mm,
wettest Nov 101 mm
Time GMT (summer time, 1 hour ahead of GMT)
Measures UK (imperial) system, converting to the metric system
Monetary unit UK pound and Jersey pound (J £) = 100 new pence.
The Jersey pound is equal to the UK pound
Rate of exchange (1976 av): free $ 1.806 = J £ 1

Summary

Political Dependency of the Crown of England since the 11th century;
autonomous state but with the United Kingdom responsible for defence
and external relations. Member of sterling area and within EEC via
relationship with United Kingdom, but with special exemption from
many EEC regulations
Economic Main industries are tourism, banking and finance; agricultural
exports and light manufacturing industry are also important. There are no
wealth, inheritance, capital gains nor value added taxes

People, resources and equipment

Population 1960 61 300*, 1970 71 900*, 1976 74 500*
Growth: 1960–70 1.6* %pa, 1970–76 0.6* %pa
Density (1976): 641* people per km²
Vital statistics (rate per 1 000 people, 1976): births 10.6, deaths 11.4
Towns (population in 000, 1976) St Helier (capital) 25
Race Norman/Breton/English
Language English; also French and Norman-French patois
Religion Christian; mainly Church of England, Roman Catholic and
Methodist
Education (1977) Pupils 12 600, teachers 670
Labour force (1976) 39 389
Personnel (1974) Physicians 100*, 1 per 730* people
National income per person (1976) J £ 3 400** = $ 6 100** = £ 3 400**
Telephones (Dec 1977) 44 894, 600 per 1 000 people
Livestock (000, 1976) Cattle 7, pigs 2
Hospital beds (1974) 630, 1 per 117 people
Hotel beds (1977) 24 900
Ports St Helier, Gorey, St Aubin
Airport States of Jersey (St Peter)
Radio sets (Dec 1976) 26 000*, 348* per 1 000 people
Television sets (Dec 1976) 24 500, 328 per 1 000 people
Passenger cars (Dec 1976) 39 000, 522 per 1 000 people
Commercial vehicles (Dec 1976) 5 350, 72 per 1 000 people

Production, finance and trade

Gross domestic product 1975: J £ 156*mn = $ 347*mn = £ 156* mn
1976 est: J £ 185** mn = $ 330** mn = £ 185** mn
Structure of gross domestic product (1975) *By origin* Tourism 40 %,
agriculture 7 %, light industry 8 %, financial and other 45 %
Main products (000 t, 1977) Potatoes 40, cauliflowers 7, tomatoes 10,
milk 17, flowers 1.5
Transport traffic (1977) Passenger arrivals: air 708 227, sea 540 027
Tourism (1977) Number of visitors 1 125 000, gross receipts $ 120 mn
Bank deposits (1977) J £ 1 500 mn = $ 2 800 mn
Budget (1977) Revenue: J £ 62 mn = $ 108 mn = £ 62 mn
Expenditure: J £ 49 mn = $ 86 mn = £ 49 mn (excl capital expenditure)
External trade[a] (1976) Imports: J £ 123 mn = $ 222 mn = £ 123 mn
Exports: J £ 49 mn = $ 88 mn = £ 49 mn
[a]Included within the United Kingdom area for external trade purposes

Main imports	% of total	Main exports	% of total
Manufactured goods	46	Manufactures	33
Machinery and transport equipment	21	Potatoes	12
Food	19	Tomatoes	7
Petroleum products	8	Cauliflowers	3
Chemicals	6	Flowers	3

Main source and destination United Kingdom

Jugoslavia

Socialist Federal Republic of Jugoslavia
Socijalistička Federativna Republika Jugoslavija (SFRJ)

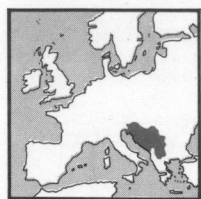

Location South-east Europe
With a western coastline on the Adriatic Sea,
Albania is to the south-west, Greece to the
south-east, Bulgaria and Rumania to the east,
Hungary and Austria to the north and there is
a short border with Italy to the north-west
Land Area 255 804 km² = 98 766 mi²
Climate Mediterranean on coast, continental
inland
Weather at Belgrade, 132 m altitude
Temperature: hottest month July 17–28 °C, coldest Jan minus 3–3 °C
Rainfall (av monthly): driest months Feb, Mar 46 mm, wettest June 96 mm
Time 1 hour ahead of GMT
Measures Metric system
Monetary unit Jugoslav dinar (Ju D) = 100 paras
Rate of exchange (1976 av): free Ju D 18.19 = $ 1, Ju D 32.86 = £ 1

Summary

Political Communist republic, set up as such in 1945 after the second
world war; there was a break in relations with Soviet Union in 1948,
since which time a policy of non-alignment has been followed. Member of
UN and has special relationships with OECD and Comecon
Economic A mixed economy, with agriculture absorbing nearly one-half
of the labour force and contributing one-sixth of gross material product.
There has been a good growth rate of about 7 %pa for industrial
production, but the economy has been dependent on a high inflow of
capital

People, resources and equipment

Population 1960 18.40 mn, 1970 20.37 mn, 1976 21.56 mn
Growth: 1960–70 1.0 %pa, 1970–76 1.0 %pa
Density (1976): 84 people per km²
Vital statistics (rate per 1 000 people, 1976): births 18.0, deaths 8.2
Regions (republics, population in 000, 1971; total of 20.52 mn)
Bosna i Hercegovina 3 746 Hrvatska[b] 4 426 Slovenija[d] 1 727
Crna Gora[a] 530 Makedonija[c] 1 647 Srbija[e] 8 447
[a]Montenegro [b]Croatia [c]Macedonia [d]Slovenia [e]Serbia; includes provinces of
Kosovo (1 244 thousand) and Vojvodina (1 953 thousand)
Cities (population in 000, 1971)
Beograd[a] 1 209 Ljubljana 258 Maribor 172
Zagreb 668 Novi Sad 214 Rijeka/Sušak 161
Skopje 389 Niš 194 Banja Luka 159
Sarajevo 292 Split 185 Priština 153
[a]Belgrade
Race (1971) Serbian 40 %, Croatian 22 %, Slovene 8 %, Albanian 6 %,
Macedonian 6 %, Montenegrin 2 %, Hungarian 2 %
Language Serbo-Croat, Slovene, Macedonian; Albanian and Hungarian
are also in use by minorities
Usage (1971): Serbo-Croat 74 %, Slovene 8 %, Albanian 6 %,
Macedonian 6 %, Hungarian 2 %
Religion (1970) Serbian Orthodox 50**%, Roman Catholic 30**%,
Moslem 9* %
Education (1975/76) Pupils 4 000 100*, teachers 173 000*
Labour force (1971) 8 889 816; in agriculture 3 965 027 (45 %),
manufacturing, mining and quarrying 1 574 512 (18 %), construction
397 863 (4 %), transport and communication 322 536 (4 %), at work in
other countries temporarily 589 168 (7 %)
Personnel Scientists and engineers (1972): 218 000*
Physicians (1974): 24 920, 1 per 849 people
Standard of living
National income per person (1976): Ju D 28 000** = $ 1 540** = £ 850**
Consumption per person (1975): energy 1 930 kg coal equivalent,
electricity (production, 1976) 2 021 kW h, newsprint 3.8 kg, steel 205 kg
Newspapers (1974): number 25; circulation 1 850 000, 87 per 1 000 people
Telephones (Dec 1976): 1 430 570, 66 per 1 000 people
Livestock (000, Jan 1977) Cattle 5 641, sheep 7 484, pigs 7 326,
chickens 53 779
Mineral reserves Coal (1971) 104 mn tonnes
Lignite (1971) 21 647 mn tonnes
Crude oil (1975) 50 mn tonnes
Natural gas (1975) 42 bn cubic metres
Uranium (1976) 4 500 tonnes
Petroleum refinery capacity (1975) 12* mn tonnes
Electrical capacity (1975) 9 043 megawatts,
of which, hydro 4 801 megawatts

Hospital beds (1975) 127 646, 1 per 167 people
Roads (1975) 112 000* km = 69 600* mi, density 0.44* km per km²
Railways (1975) 10 068 km = 6 256 mi, density 0.039 km per km²
Inland waterways (1975) 2 001 km = 1 243 mi
Oil pipelines (1975) 151 km = 94 mi
Ships (registered, 1977) 459, total of 2 284 526 gross tons
Ports (goods traffic, 000 tonnes, 1976) Rijeka *International* Loaded 1 725,
unloaded 9 858 *Coastwise* Loaded 1 784, unloaded 630
Also Split, Ploče, Dubrovnik, Koper and Bar
Airports (passengers, 000, 1976) Belgrade: departures 1 077, arrivals 970
Other main airports: Zagreb, Skopje, Sarajevo, Ljubljana, Split, Rijeka,
Dubrovnik, Priština, Maribor and Titograd

Durable equipment (at end-year)	000	no per 1 000 people	
Radio sets (1976)	4 526	209	no per
Television sets (1976)	3 463	160	km of road
Passenger cars (1975)	1 537	72	14
Commercial vehicles (1975)	179	8	1.6

Production

Gross material product 1976: Ju D 592.6 bn = $ 32 578 mn = £ 18 034 mn
Growth in real terms: 1960–70 6.7 %pa, 1970–76 5.7 % pa
Gross domestic product
1976 est: Ju D 640** bn = $ 35 000** mn = £ 19 500** mn
Structure of gross material product (1976) *By origin* Agriculture 17 %,
manufacturing, mining, quarrying and utilities 40 %, construction 10 %,
distribution and hotels 21 %, transport and communication 8 %, other 4 %
By type Final consumption expenditure 65 % (of which, government 5 %),
stock investment 4 %, gross fixed capital formation 34 %, exports of
goods and services 20 %, less imports of goods and services −25 %

Production indices (1970 = 100)	1960	1970	1976	Growth %pa 1960–70	1970–76
Agricultural	82	100	131	2.0	4.6
Industrial	45	100	153	8.3	7.3
Main products (000 t)					
Agriculture					
Wheat	3 574	3 792	5 979	0.6	7.9
Maize	6 160	6 933	9 106	1.2	4.6
Potatoes	3 270	2 964	2 928	−1.0	−0.2
Sugar, raw value	287	357	715	2.2	12.3
Tomatoes	271	313	433	1.5	5.6
Grapes	580	1 101	1 204	6.6	1.5
Wine	335	548	638	5.0	2.6
Apples	158	277	437*	5.8	7.9*
Plums	750**	896	569*	1.8**	−7.3*
Sunflowerseed	98	264	319	10.4	3.2
Tobacco	28	49	72	5.7	6.6
Milk	2 283	2 567	3 846	1.2	7.0
Eggs	81	139	184	5.5	4.8
Beef and veal	138	234	335	5.4	6.2
Pigmeat	291	338	360*	1.5	1.1*
Fish catch	31	46	59	4.1	4.1
Timber (000 m³)	17 469	16 957	14 036	−0.3	−3.1
Energy					
Total energy (000 tce)	14 050	21 420	28 210[e]	4.3	5.7[f]
Coal	1 283	643	587	−6.7	−1.5
Lignite	21 430	27 779	35 694	2.6	4.3
Crude oil	944	2 854	3 880	11.7	5.2
Petroleum products	1 258	6 615	10 068[e]	18.0	8.8[f]
Natural gas (mn m³)	53	977	1 730	33.8	10.0
Manufactured gas (mn m³)	561*[c]	616	740	1.6*[d]	3.1
Electricity (mn kW h)	8 928	26 024	43 574	11.3	9.0
of which, hydro (mn kW h)	5 984	14 741	19 317[e]	9.4	5.6[f]
Mining					
Iron ore (Fe content)	788	1 301	1 490	5.1	2.3
Bauxite	1 025	2 099	2 032	7.4	−0.5
Antimony ore (Sb content)	3.76	2.00	2.18[e]	−6.1	1.7[f]
Copper ore (Cu content)	33	91	120	10.6	4.7
Lead ore (Pb content)	91	127	123	3.4	−0.5
Magnesite	252	512	485[e]	7.3	−1.1[f]
Zinc ore (Zn content)	56	101	107	6.0	1.0
Gold (000 kg)	1.99	3.03	5.53[e]	4.3	12.8[f]
Salt	151	170	218[e]	1.2	5.1[f]
Manufacturing					
Beer (000 hl)	1 630	6 665	8 685	15.1	4.5
Cigarettes (mn units)	20 123	32 072	41 605[e]	4.8	5.3[f]
Cotton yarn	51	102	117	7.2	2.3
Cotton fabrics (mn m²)	257	390	384	4.3	−0.2
Wool yarn	21	38	44	6.3	2.5
Man-made fibres	21	41	82[e]	6.9	14.9[f]

Jugoslavia

Main products (000 t)	1960	1970	1976	Growth %pa	
Manufacturing				1960–70	1970–76
Wood pulp	209	547	595[e]	10.1	1.7[f]
Newsprint	28	75	90	10.3	3.1
Other paper	145**	510	660[e]	13.4**	5.3[f]
Tyres (000 units)	391	2 763	5 644	21.6	12.6
Sulphur[a]	167	160	156[e]	−0.4	−0.5[f]
Sulphuric acid	130	747	848[e]	19.1	2.1
Nitric acid	9	579	635[e]	51.0	1.9[f]
Plastics and resins	9	97	205	26.6	13.3
Fertilisers, nitrogenous	4	266	357[e]	52.0	6.1[f]
Fertilisers, phosphate	41	233	237[e]	19.0	−0.3[f]
Cement	2 398	4 399	7 620	6.2	9.6
Coke	1 083	1 309	1 344[e]	1.9	0.5[f]
Pig iron	1 019	1 377	2 121	3.1	7.5
Crude steel	1 442	2 228	2 698	4.4	3.2
Aluminium[g]	25	48	197	6.6	26.5
Radio sets (000 units)	244	277	109	1.3	−14.4
Television sets (000 units)	14	320	402	36.7	3.9
Passenger cars (000 units)[b]	10*	62	139	20.0*	14.4
Commercial vehicles (000 units)	5.5	28.8	20.8	18.0	−5.3
Merchant vessels (000 grt)	161	393	583	9.3	6.8
Construction					
Dwellings (000 units)	76	129	146[e]	5.5	2.5[f]

[a]Content of pyrites [b]Excludes assembly (48 000 in 1975) [c]1964 [d]1964–70
[e]1975 [f]1970–75 [g]Including secondary

Transport traffic	1960	1970	1976	Growth %pa	
Passenger-kilometres (mn)				1960–70	1970–76
Rail	10 261	10 939	9 884	0.6	−1.7
Air	103	774	2 150	22.3	18.6
Cargo: tonne-kilometres (mn)					
Rail	13 895	19 253	21 006	3.3	1.5
Air	1.2	6.5	20.6	18.0	21.2
Sea: tonnes (mn)					
Goods loaded	2.60	3.84	4.02	4.0	0.8
Goods unloaded	4.03	11.57	16.17	11.1	5.7
Tourism					
Number of visitors (000)	873	4 748	5 572	18.4	2.7
Gross receipts ($ mn)	na	276	802	na	19.5

Finance and trade

Price and wage indices (1970 = 100)	1960	1970	1976	Growth %pa	
				1960–70	1970–76
Consumer prices	31	100	269	12.4	17.9
Wholesale prices	61	100	244	5.1	16.0
Wages (nominal)	17	100	298	19.4	20.0
Money stock (end-year, Ju D bn)	7.5**	35.8	216.7	16.9**	35.0

Budget (1975) Revenue: Ju D 108 280 mn = $ 6 228 mn = £ 2 803 mn
Expenditure: Ju D 118 410 mn = $ 6 811 mn = £ 3 065 mn,
of which (1974), defence 34 %, culture and education 4 %, health
and social services 5 %

Balance of payments ($ mn)	1972	1973	1974	1975	1976
Balance of goods (fob)	−727	−1 284	−3 117	−2 988	−1 755
Balance of services	+983	+1 555	+1 659	+1 725	+1 639
Balance of transfers	+160	+215	+273	+228	+269
Current balance	+415	+485	−1 186	−1 036	+154
Long-term capital flow	+489	+607	+495	+951	na
Reserves and debt (end-year, $ mn)					
International reserves	731	1 338	1 147	871	2 049
External public debt	2 341	2 441	3 296	3 616	3 691

External trade (1976)
Imports: Ju D 134 050 mn = $ 7 369 mn = £ 4 079 mn
Exports: Ju D 88 770 mn = $ 4 880 mn = £ 2 701 mn

Main imports	% of total	Main exports	% of total
Machinery, non-electric	20	Transport equipment	12
Chemicals	11	(of which, ships and boats 6)	
Crude oil	11	Food	10
Food	8	Machinery, non-electric	8
(of which, wheat 2)		Electrical machinery	8
Transport equipment	7	Non-ferrous metals	8
(of which, motor vehicles 5)		Chemicals	7
Iron and steel	7	Timber	5
Electrical machinery	6	Clothing	5
Textile fibres	3	Textile yarns and fabrics	5
Textile yarns and fabrics	3	Footwear	5
Non-ferrous metals	2	Iron and steel	4

Main sources	% of total	Main destinations	% of total
West Germany	17	Soviet Union	23
Soviet Union	14	Italy	12
Italy	10	West Germany	9
Iraq	6	United States	7
United States	5	Czechoslovakia	5
Czechoslovakia	4	Poland	4
France	4	East Germany	4
United Kingdom	4	India	3
Austria	3	France	3
East Germany	3	Rumania	2

Special focus

Structure of gross material product	% of total		
	1960	1970	1976
Agriculture	26	19	17
Manufacturing, mining, quarrying and utilities	49	38	40
Construction	7	12	10
Distribution and hotels	11	22	21
Transport and communication	7	8	8
Other (incl statistical discrepancy)	0	1	4

Liechtenstein

Principality of Liechtenstein
Fürstentum Liechtenstein

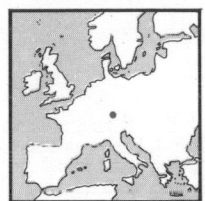

Location Central Europe
On the Upper Rhine, between Austria
(Vorarlberg province) to the east and
Switzerland to the west and south.
Land-locked
Land Area 160 km² = 62 mi²
Climate Temperate
Time 1 hour ahead of GMT
Measures Metric system
Monetary unit Swiss franc or
franken (S Fr) = 100 centimes or rappen
Rate of exchange (1976 av): free S Fr 2.500 = $1, S Fr 4.515 = £ 1

Summary

Political Principality, formed as such in 1719 from the County of Vaduz
and Lordship of Schellenberg; there is a customs and monetary union
with Switzerland, and Switzerland acts for diplomatic representation.
Member of the International Court of Justice, Unctad, ITU, UPU and
proposed (July 1978) as 21st member of Council of Europe
Economic A manufacturing economy, with a wide range of light industry,
most products being exported; there is some agriculture and tourism, and
there is financial activity resulting from use as a centre for holding
companies

People, resources and equipment

Population 1960 16 600, 1970 21 350, 1976 24 200*
Growth: 1960–70 2.5 %pa, 1970–76 2.1* %pa
Density (1976): 151* people per km²
Vital statistics (rate per 1 000 people, 1976): births 14.4, deaths 7.4
Towns (population in 000, 1976) Vaduz (capital) 4.6, Schaan 4.3,
Balzers 3.0, Triesen 2.9
Race Alemannic
Foreigners resident (1976): 8 472 (35 % of total)
Language German (Alemannish)
Religion (1976) Roman Catholic 84 %, Protestant 8 %
Education (1977/78) Pupils 3 807, teachers 176
Labour force (1976) 10 478; in agriculture 438 (4 %), industry, commerce
and construction 5 715 (55 %), services 4 305 (41 %)
Standard of living National income per person (1976):
S Fr 27 000* = $ 10 800* = £ 6 000*
Consumption per person (1976): electricity 1 983 kW h
Newspapers (1976): number 1; circulation 6 700, 280 per 1 000 people
Telephones (Dec 1976): 16 200*, 670* per 1 000 people
Livestock (000, 1976) Cattle 6, pigs 4, sheep 2, chickens 2
Roads (1976) 250* km = 150* mi, density 1.6* km per km²
Railways (1976)[a] 18.5 km = 11.5 mi, density 0.12 km per km²
[a]Administered by Austrian Federal Railways

Liechtenstein

Durable equipment 000 no per
(Dec 1976) 1 000 people

Radio sets	5.41	223	no per
Television sets	4.56	188	km of road
Passenger cars	9.89	409	39
Commercial vehicles	0.87	36	3.5

Companies registered (1976) number 15 000

Production, finance and trade

Gross domestic product
1976 est: S Fr 660* mn = $ 260* mn = £ 150* mn
Main products (000 tonnes, 1976) Potatoes 3, maize 19, milk 7, timber (000 m³) 9, electricity (mn kW h) 48; also a wide range of manufactures including instruments, ceramics, pharmaceuticals, light machinery and furniture
Tourism Number of visitors (1976) 74 462, gross receipts (1971) $ 4 mn
Budget (1977) Revenue: S Fr 190 mn = $ 79 mn = £ 45 mn
Expenditure: S Fr 189 mn = $ 79 mn = £ 45 mn
External trade (1976)
Imports: S Fr 598*** mn = $ 239*** mn = £ 132*** mn
Exports: S Fr 598 mn = $ 239 mn = £ 132 mn
Main exports Metal manufactures, furniture, chemicals, pottery
Main destinations (% of total) Switzerland 40 %, EEC 30 %, Efta (other than Switzerland) 8 %

Luxembourg

Grand Duchy of Luxembourg
Grand-Duché de Luxembourg
Grossherzogtum Luxemburg

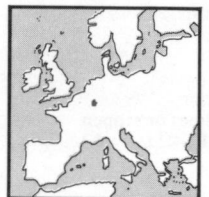

Location Western Europe
Belgium is to the north and west, France to the south, and West Germany to the east. Land-locked, but with access to the sea via the Moselle and Rhine
Land Area 2 586 km² = 998 mi²
Climate Temperate
Weather at Luxembourg City, 330 m altitude
Temperature: hottest month July 13–23 °C, coldest Jan minus 1–2 °C
Rainfall (av monthly): driest month Mar 42 mm, Aug 84 mm
Time 1 hour ahead of GMT (summer time, 2 hours ahead)
Measures Metric system
Monetary unit Luxembourg franc (L Fr) = 100 centimes; the Luxembourg franc is at par with the Belgian franc, which is also legal tender
Rate of exchange (1976 av): free L Fr 38.60 = $ 1, L Fr 69.73 = £ 1
par (central value within the European 'snake') L Fr 48.6573 = SDR 1

Summary

Political Parliamentary monarchy; an economic union was formed with Belgium in 1921, following a referendum on the future of the country in 1919; the Benelux economic union was agreed with Belgium and Netherlands in 1944 and came into force in 1960. Also a member of UN, OECD, Council of Europe, Nato, EEC and WEU
Economic Steel is the main industry, providing 16 % of gross domestic product, compared with 3 % from agriculture and about 20 % from other manufacturing. A policy of industrial diversification has been followed, especially since 1969. International finance is important

People, resources and equipment

Population 1960 309 000, 1970 339 000, 1976 358 000
Growth: 1960–70 0.9 %pa, 1970–76 0.9 %pa
Density (1976): 138 people per km²
Vital statistics (rate per 1 000 people, 1976): births 11.0, deaths 12.6
Regions (population in 000, 1966; total of 335 000)
Luxembourg-Ville 77;
districts: Diekirch 53, Grevenmaches 35, Luxembourg 170
Cities (population in 000, 1975) Luxembourg-Ville (capital) 78, Esch-sur-Alzette 28, Differdange 18, Dudelange 15, Petange 12
Race (1974) Luxembourgers 78* %, foreign workers 22* %, of whom, Italian 6* %, Portuguese 5* %
Language French, German and Letzeburgish

Religion (1976) Roman Catholic 97** %, Protestant 1** %
Education Pupils (1975/76) 58 300*, teachers (1973/74) 3 507
Labour force 1970: 129 255; in agriculture 9 641 (7 %), manufacturing and mining 43 526 (34 %), construction 11 770 (9 %), distribution and hotels 23 561 (18 %), transport and communication 7 743 (6 %), finance and real estate 6 085 (5 %)
1976: 147 700
Personnel (1974) Physicians: 368, 1 per 970 people
Standard of living National income per person (1976):
L Fr 224 000* = $ 5 800* = £ 3 200*
Consumption per person (1975): energy 15 504 kg coal equivalent, electricity (production, 1976) 4 307 kW h
Newspapers (1974): number 7; circulation (6 only) 161 000, 450 per 1 000 people
Telephones (Dec 1976): 157 830, 442 per 1 000 people
Livestock (000, May 1976) Cattle 214, pigs 83, sheep 4, chickens 188
Electrical capacity (1975) 1 157 megawatts, of which, hydro 932 megawatts
Hospital beds (1975) 3 848, 1 per 93 people
Roads (1975) 4 465 km = 2 774 mi, density 1.7 km per km²
Railways (1975) 275 km = 171 mi, density 0.11 km per km²
Inland waterways (1975) 37 km = 23 mi
Airport (passengers, 000, 1976)
Findel (Luxemburg City): departures 340, arrivals 345

Durable equipment 000 no per
(at end-year) 1 000 people

Radio sets (1975)	176	490	no per
Television sets (1974)	88	245	km of road
Passenger cars (1975)	117	326	26
Commercial vehicles (1975)	11.2	31	2.5

Company (main, sales, 1976)
Acieries Réunies de Burbach-Eich-Dudelange SA (Arbed) $ 920mn

Production

Gross domestic product 1975: L Fr 80 800 mn = $ 2 197 mn = £ 989 mn
1976 est: L Fr 90 000* mn = $ 2 300* mn = £ 1 290* mn
Growth in real terms: 1960–70 3.3 %pa, 1970–75 1.8 %pa
Structure of gross domestic product (1975) *By origin* Agriculture 3 %, manufacturing, mining and quarrying 35 % (of which, steel 16* %), electricity, gas and water 2 %, construction 11 %, distribution and hotels 13 %, transport and communication 4 %, other 31 %
By type Final consumption expenditure 75 % (of which, government 15 %), stock investment 3 %, gross fixed capital formation 29 %, exports of goods and services 81 %, less imports of goods and services −88 %

Production indices (1970 = 100)	1960	1970	1976	Growth %pa 1960–70	1970–76
Agricultural	103	100	111ᵍ	−0.3	2.5ʰ
Industrial	79	100	100	2.4	0.0

Main products (000 t)

Agriculture	1960	1970	1976	1960–70	1970–76
Barley	19	44	33	8.8	−4.7
Wheat	48	28	16	−5.2	−8.9
Oats	39	29	12	−2.9	−13.7
Potatoes	103	68	20	−4.1	−18.4
Grapes	17*	34*	20*	7.2*	−8.5*
Wine	12	24	13*	7.2	−9.7*
Milk	206	217	259ᵉ	0.5	3.6ᵈ
Timber (000 m³)	163	222	212ᵍ	3.1	−1.1ʰ
*Other*ᵃ					
Electricity (mn kW h)	1 538	2 148	1 542	3.4	−5.4
of which, hydro (mn kW h)	21	887	500ᶜ	45.4	−10.8ᵈ
Iron ore (Fe content)	1 926	1 571	603	−2.0	−14.7
Beer (000 hl)	427	551	803ᶜ	2.6	7.8ᵈ
Tyres	1 736ᵉ	2 974	2 269ᶜ	8.0ᶠ	−5.3ᵈ
Fertilisers, phosphateᵇ	115**	151	140ᶜ	2.8**	−1.5ᵈ
Cement	210	245	343ᶜ	1.5	7.0ᵈ
Pig iron	3 786	4 814	3 755	2.4	−4.1
Crude steel	4 084	5 462	4 565	2.9	−3.0

ᵃFor total energy production see Belgium ᵇYears ending June 30th ᶜ1975 ᵈ1970–75 ᵉ1963 ᶠ1963–70 ᵍ1974 ʰ1970–74

Transport traffic	1960	1970	1976	Growth %pa	
Passenger-kilometres (mn)				1960–70	1970–76
Rail	200	256	294	2.5	2.3
Air	—	77	165	na	13.5
Cargo: tonne-kilometres (mn)					
Rail	637	764	626	1.8	−3.3
Air	—	0.53	0.30	na	−9.0

Tourism (1976) Number of visitors 497 400

Luxembourg

Finance and trade

Consumer price index (1970 = 100) 1960 78, 1976 155.6;
growth 1960–70 2.5 %pa, 1970–76 7.6 %pa
Money stock The main part of money stock used is Belgian currency
which is also legal tender; Luxembourg notes and coins (1977): L Fr 600 mn
Budget (1977) Revenue: L Fr 33 406 mn = $ 932 mn = £ 534 mn
Expenditure: L Fr 34 338 mn = $ 958 mn = £ 549 mn
Balance of payments see Belgium
International reserves See Belgium; of total reserves held by
Belgium-Luxembourg, Luxembourg held (1976) $ 19 mn in SDRs and
reserve position in the International Monetary Fund
External trade (1975)[a]
Imports (fob)[b]: L Fr 73 460 mn = $ 1 997 mn = £ 899 mn
Exports[b]: L Fr 67 750 mn = $ 1 842 mn = £ 829 mn
Main export Steel 90* %
[a]Included with Belgium in the Belgium-Luxembourg customs union; see Belgium for
union products and direction   [b]Including services

Special focus

Steel prices (producers basic price, West Germany)

	$ per tonne	% change over previous year		$ per tonne	% change over previous year
1970	125	+25	1974	249	+21
1971	137	+10	1975	300	+20
1972	157	+15	1976	329	+10
1973	206	+31	1977	321	−2

Malta

Republic of Malta
Ir-Republika ta' Malta

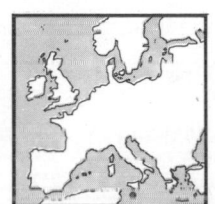

Location Mediterranean Sea
About halfway between Gibraltar and
Port Said, with Sicily 93 km to the north
and Tunisia 300 km to the west. The islands
include Malta, Gozo and Comino
Land Area 316 km² = 122 mi²,
of which, Malta 245 km², Gozo 67 km²
Climate Mediterranean (hot summers and
warm winters)
Weather at Valletta, 70 m altitude
Temperature: hottest month Aug 23–29 °C, coldest Jan 10–14 °C
Rainfall (av monthly): driest month July 0 mm, wettest Dec 110 mm
Time 1 hour ahead of GMT (summer time, 2 hours ahead)
Measures Metric and UK (imperial) systems, converting to a completely
metric system; local measures are also in use, including:
length 1 palmi = 10.29 inches = 26.13 centimetres
weight (*mass*) 1 rotolo or ratal = 0.794 kilogram = 1.75 pounds
1 wizna = 5 rotoli = 3.969 kilograms = 8.75 pounds
1 qantar = 20 wizna = 100 rotoli = 79.38 kilograms = 175 pounds
Monetary unit Maltese pound (M £) = 100 cents = 1 000 mils
Rate of exchange (1976 av): free $ 2.353 = M £ 1, £ 1.303 = M £ 1

Summary

Political Republic, which became independent from the United Kingdom
on September 21, 1964. The United Kingdom use of a defence base on the
island is planned to end in 1979. Member of UN, Council of Europe and
Commonwealth
Economic Tourism and light manufacturing industry have been developed
to replace a gradual loss of income from UK defence expenditure.
Receipts from tourism equal nearly one-third of exports; main
domestic products exported are clothing and textiles. Ship-repairing
is important

People, resources and equipment

Population 1960 329 000, 1970 326 000, 1976 330 000*
Growth: 1960–70 −0.1 %pa, 1970–76 0.2* %pa
Density (1976): 1 040* people per km²
Vital statistics (rate per 1 000 people, 1976): births 18.0, deaths 9.0
Regions (Maltese population in 000, 1976; total of 305 000)
Malta 283, Gozo and Comino 22
Towns (population in 000, 1976) *Malta* Valletta (capital) 14, Sliema 20*,
Birkirkara 17*, Qormi 15* *Gozo* Victoria (or Rabat) 5
Race (1961) Maltese 96 %, English 2 %, Italian 1½ %
Language Maltese and English

Religion (1970) Roman Catholic 98** %
Education (1975/76) Pupils 64 387, teachers 4 169
Labour force (1975) 112 596; in manufacturing 31 287 (28 %)
Personnel (1975) Physicians: 382, 1 per 850 people
Standard of living
National income per person (1976): M £ 652 = $ 1 533 = £ 849
Consumption per person (1975): energy 1 032 kg coal equivalent,
electricity (production) 1 020* kW h, newsprint 1.2 kg
Newspapers (1974): number 6
Telephones (Dec 1976): 62 324, 188 per 1 000 people
Livestock (000, 1976) Cattle 11*, pigs 25*, goats 11*, chickens 1 170*
Electrical capacity (1975) 115 megawatts
Hospital beds (1971) 3 431, 1 per 95 people
Roads (1974) 1 267 km = 787 mi, density 4.0 km per km²
Ships (registered, 1977) 44, total of 100 420 gross tons
Port Valletta
Airport Luqa (Valletta): passenger departures and arrivals (1976) 746 000

Durable equipment (at end-year)	000	no per 1 000 people	
Radio sets (1973)	129	401	no per
Television sets (1976)	63	190	km of road
Passenger cars (1976)	53	160	42
Commercial vehicles (1976)	13	40	10

Production

Gross domestic product 1976: M £ 203.7 mn = $ 479 mn = £ 265 mn
Growth in real terms: 1960–70 4.9 %pa, 1970–76 8.6* %pa
Agricultural production index (1970 = 100) 1960 63, 1976 95;
growth: 1960–70 4.8 %pa, 1970–76 −0.8 %pa
Main products *Agriculture* (000 t, 1976) Potatoes 20*, wheat 3*,
tomatoes 17*, cauliflowers 4*, onions 4*, grapes 4*, wine 2*, milk 25*,
pigmeat 4*, poultry meat 3*, fish catch 2 *Other* (1975) Manufactured
gas 5 mn m³, electricity (1975/76; year ending March 31st) 338 mn kW h,
cigarettes 740 mn units, salt 2

Transport traffic	1960	1970	1976	Growth %pa 1960–70	1970–76
Air (mn)					
Passenger-kilometres	—	202	340	na	9.1
Cargo: tonne-kilometres	—	2.7	3.8	na	5.9
Sea: tonnes (mn)					
Goods loaded	0.040	0.050	0.133	2.2	17.7
Goods unloaded	0.443	0.956	1.000	8.0	0.8
Tourism					
Number of visitors (000)	20	170	340	24.1	12.2
Gross receipts ($ mn)	na	24	67	na	18.7

Finance and trade

Price index (1970 = 100)	1960	1970	1976	Growth %pa 1960–70	1970–76
Consumer prices	84.0	100.0	133.7	1.8	5.0
Money stock					
(end-year, M £ mn)	27.6	58.6	144.6	7.8	16.2

Budget (total, 1976/77; year ending March 31st)
Revenue: M £ 100.8 mn = $ 234 mn = £ 135 mn,
of which, rent for defence facilities M £ 13.0 mn = $ 30 mn = £ 17 mn
Expenditure: M £ 96.1 mn = $ 223 mn = £ 129 mn

Balance of payments ($ mn)	1972	1973	1974	1975	1976
Balance of goods (fob)	−95	−105	−162	−158	−140
Balance of services	+88	+109	+130	+181	+129
Balance of transfers	+32	+32	+44	+41	+75
Current balance	*+25*	*+36*	*+13*	*+64*	*+63*
Long-term capital flow	+27	−12	+17	+32	+35
Reserves and debt (end-year, $ mn)					
International reserves	275	325	402	500	622
External public debt	24	28	36	48	58

External trade (1976) Imports: M £ 180 mn = $ 423 mn = £ 234 mn
Exports: M £ 97 mn = $ 229 mn = £ 127 mn

Main imports	% of total	*Main exports*	% of total
Food	20	Clothing	42
(of which, cereals 6)		Machinery	8
Textile yarns and fabrics	16	Petroleum products	7
Petroleum products	9	Food	6
Machinery, non-electric	8	Textile yarns and fabrics	5
Chemicals	6	Ships and boats	5
Electrical machinery	6	Tobacco	3
Main sources		*Main destinations*	
United Kingdom	24	West Germany	24
Italy	17	United Kingdom	18
West Germany	10	Libya	13
United States	10	Italy	5
France	5	Netherlands	4

Monaco

Principality of Monaco
Principauté de Monaco

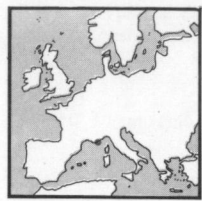

Location Southern Europe
Forms an enclave on the Mediterranean Sea
coast of France
Land Area 1.90 km² = 0.73 mi²
Climate Mediterranean
Weather at Monaco, 55 m altitude
Temperature: hottest month Aug 22–26 °C,
coldest Jan 8–12 °C
Rainfall (av monthly): driest month July 21 mm,
wettest Nov 123 mm
Time 1 hour ahead of GMT (summer time, 2 hours ahead)
Measures Metric system
Monetary unit French franc and Monégasque franc (Mn Fr) =
100 centimes. The Monégasque franc is equal to the French franc.
Rate of exchange (1976 av): free Mn Fr 4.780 = $ 1, Mn Fr 8.633 = £ 1

Summary

Political Principality, first formed in the tenth century. Relationships with
France are based on conventions, the latest being agreed in 1963. Member
of Unesco, WHO, UPU, Unctad and Wipo
Economic Tourism is the main industry, including casino operation;
income is also derived from sale of postage stamps and tobacco. There is
a range of light manufacturing industry

People, resources and equipment

Population 1960 21 380*, 1970 23 400*, 1976 25 300*
Growth: 1960–70 0.9*%pa, 1970–76 1.3*%pa
Density (1976): 13 300* people per km²
Vital statistics (rate per 1 000 people, 1974): births 8.2, deaths 12.3
Town (population in 000, 1976) Monte Carlo (capital) 10*
Race (1968) French 55 %, Monégasque 16 %, Italian 15 %, UK citizens
2½ %
Language French and Monégasque
Religion (1970) Roman Catholic 98 %
Education (1976/77) Pupils 4 966, teachers 357
Labour force 1968: 10 325; in agriculture 19 (0.2 %), manufacturing
1 470 (14 %), construction 596 (6 %), distribution and hotels 1 852 (18 %),
transport and communication 421 (4 %), other services 5 449 (53 %)
1975: 11 081
Personnel (1974) Physicians: 57, 1 per 430 people
National income per person (1976)
Mn Fr 40 000*** = $ 8 000*** = $ 4 600***
Telephones (Dec 1976): 23 740, 930 per 1 000 people
Hospital beds (1976) 318, 1 per 80 people
Roads (1976) 46 km = 29 mi, density 24 km per km²
Railways (1976)ᵃ 1.6 km = 1.0 mi, density 0.84 km per km²
ᵃOperated by SNCF (France)
Port Monaco
Airport Nearest main international airport is Nice (France)
Radio sets (1976) 8 800*, 346* per 1 000 people
Television sets (1976) 8 500*, 334* per 1 000 people
Motor vehicles (1970) 9 150, 390 per 1 000 people, 199 per km of road

Production, finance and trade

Gross domestic product
1976 est: Mn Fr 1 000***mn = $ 210***mn = £ 120*** mn
Tourism (1976) Number of visitors 181 023
Budget (1976) Revenue: Mn Fr 528 mn = $ 111 mn = £ 61 mn
Expenditure: Mn Fr 464 mn = $ 97 mn = £ 54 mn
External trade Included with France

Netherlands

Holland
Kingdom of the Netherlands
Koninkrijk der Nederlanden

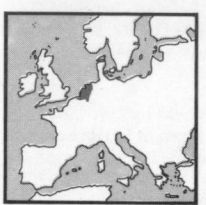

Location Western Europe
With a coastline on the North Sea, West
Germany is to the east and Belgium to the south
Land Area 41 160 km² = 15 892 mi²
Usage (1975): agricultural 20 820 km² (51 %),
of which, arable 8 040 km² (20 %), cropland
370 km² (1 %), pastures 12 410 km² (30 %);
forests 3 080 km² (7 %); water 7 600 km² (18 %)
Climate Temperate
Weather at Amsterdam, 3 m altitude
Temperature: hottest month July 13–22 °C, coldest Jan minus 1–4 °C
Rainfall (av monthly): driest month March 44 mm, wettest Aug 87 mm
Time 1 hour ahead of GMT (summer time, 2 hours ahead)
Measures Metric system
Monetary unit Guilder or florin (Gld or Fl) = 100 cents
Rate of exchange (1976 av): free Gld 2.644 = $ 1, Gld 4.776 = £ 1
par (central value within the European 'snake') Gld 3.35507 = SDR 1

Summary

Political Parliamentary monarchy; occupied by Germany during the
second world war and liberated in 1945. Joined with Belgium and
Luxembourg to form Benelux in 1960; member of UN, OECD, Council of
Europe, Nato, EEC and WEU
Economic An industrial economy with manufacturing absorbing about
one-quarter of the labour force and providing over one-quarter of gross
domestic product. Raw materials, especially crude oil, are mainly
imported, and chemicals and petroleum products are main exports;
agricultural produce remains a significant export, and electrical and other
machinery exports are also important. West Germany is the major trading
partner

People, resources and equipment

Population 1960 11.48 mn, 1970 13.03 mn, 1976 13.77 mn
Growth: 1960–70 1.3 %pa, 1970–76 0.9 %pa
Density (1976): 335 people per km²
Vital statistics (rate per 1 000 people, 1976): births 12.9, deaths 8.3
Regions (provinces, population in 000, 1975; total of 13.73 mn)

Groningen	540	Utrecht	868	Limburg	1 052
Friesland	561	Noord-Holland	2 296	Zuidelijke	
Drenthe	406	Zuid-Holland	3 049	IJsselmeerpoldersᵃ	19
Overijssel	986	Zeeland	332	Drontenᵃ	17
Gelderland	1 640	Noord-Brabant	1 967	No fixed residence	2

ᵃArea not included in a province
Cities (population in 000, 1975)

Amsterdam (capital)	989	Arnhem	280	Groningen	202
Rotterdam	1 032	Enschede	239	Dordrecht	185
's-Gravenhageᵃ	681	Haarlem	233	Breda	151
Utrecht	463	Nijmegen	213	Maastricht	146
Eindhoven	356	Tilburg	212	Zaanstad	137

ᵃDen Haag or The Hague; seat of government
Race (1968) Dutch 98½ %, German ¼ %, Belgian ¼ %, Moluccan ¼ **%
Language Dutch
Religion (1971) Roman Catholic 40 %, Dutch Reformed 24 %,
Reformed Church 7 %
Education (1974/75) Pupils: primary 1 448 177, secondary 740 280,
vocational 459 130, teacher-training 10 849, higher 264 297. Teachers:
primary 52 503, secondary 45 790, teacher-training (1969/70) 1 012,
higher (1971/72) 13 000*
Labour force (1976)

Economic activity	Numberᵃ	%
Agriculture	295 000	6
Mining and quarrying	8 000	—
Manufacturing	1 041 000	23
Electricity, gas and water	45 000	1
Construction	437 000	10
Distribution and hotels	815 000	18
Transport and communication	310 000	7
Finance and real estate	304 000	7
Other	1 287 000	28
Total	4 542 000	100

ᵃCivilian only; in terms of full-time equivalent for a year
Personnel Scientists and engineers (1971): 442 000*
Physicians (1975): 21 825, 1 per 625 people

Netherlands

Standard of living

National income per person (1976): Gld 15 577 = $ 5 892 = £ 3 262
Consumption per person (1975): energy 5 784 kg coal equivalent, electricity (production, 1976) 4 216 kW h, newsprint 27 kg, steel 330 kg
Newspapers (1973): number 93; circulation 4 175 000*, 311* per 1 000 people
Telephones (Dec 1976): 5 412 000, 392 per 1 000 people
Livestock (000, May 1977) Cattle 4 877, sheep 800, pigs 8 288, chickens 69 875
Mineral reserves Coal (1973) 3 705*mn tonnes
Crude oil (1975) 13 mn tonnes
Natural gas (1975) 1 848 bn cubic metres
Petroleum refinery capacity (1975) 103 mn tonnes
Electrical capacity (1975) 13 982* megawatts, of which, nuclear 524 megawatts
Hospital beds (1973) 136 216, 1 per 99 people
Roads (1976) 86 354 km = 53 658 mi, density 2.1 km per km²
Railways (1975) 2 832 km = 1 760 mi, density 0.069 km per km²
Inland waterways (1975) 4 360 km = 2 709 mi
Oil pipelines (1975) 613 km = 381 mi
Ships (registered, 1977) 1 254, total of 5 290 360 gross tons
Ports (goods traffic, 000 tonnes, 1976)

	loaded	unloaded		loaded	unloaded
Rotterdam	66 084	217 020	Terneuzen	1 936	3 974
Amsterdam	5 517	13 403	Vlissingen	1 881	3 165
IJmuiden	1 943	9 760	Vlaardingen	1 006	3 805

Airports (1976)

	Passengers (000) departures	arrivals	Cargo (000 tonnes) loaded	unloaded
Schipol (Amsterdam)	3 935	3 967	142	127
Rotterdam	153	160	4	6

Also: Eindhoven, Groningen, Maastricht, Enschede

Durable equipment (Dec 1976)	000	no per 1 000 people		
Radio sets	3 996	289	no per	
Television sets	3 774	273	km of road	
Passenger cars	3 760	272	44	
Commercial vehicles	326	24	3.8	

Production

Gross domestic product
1976: Gld 236 700 mn = $ 89 523 mn = £ 49 560 mn
Growth in real terms: 1963–70 6.0 %pa, 1970–76 3.6 %pa

Structure of gross domestic product

By origin (1972)	Gld mn	% of gdp
Agriculture	8 020	5
Mining and quarrying	470	—
Manufacturing	41 630	28
Electricity, gas and water	3 350	2
Construction	10 730	7
Distribution and hotels	20 630	14
Transport and communication	10 400	7
Other	51 500	35
Total	146 730	100

By type of expenditure (1976)		
Government final consumption	43 390	18
Private final consumption	135 400	57
Stock investment	3 400	1
Gross fixed capital formation	46 670	20
Exports of goods and services	128 520	54
less Imports of goods and services	−120 680	−51
Total	236 700	100

Production indices (1970 = 100)	1960	1970	1976	Growth %pa 1960–70	1970–76
Agricultural	82	100	118	2.0	2.8
Industrial	50	100	126	7.2	3.9

Main products (000 t)
Agriculture

	1960	1970	1976	1960–70	1970–76
Wheat	590	643	710	0.9	1.7
Barley	291	334	263	1.4	−3.9
Potatoes	4 173	5 648	4 783	3.1	−2.7
Sugar, raw value	669	713	945*	0.6	4.8*
Tomatoes	201	392	360	6.9	−1.4
Cucumbers	150**	311	323	7.6**	0.6
Onions	209	339	345	5.0	0.3
Apples	384	450	380	1.6	−2.8
Milk	6 838	8 238	10 538	1.9	4.2
Butter	99	121	216*	2.0	10.1*

Main products (000 t)	1960	1970	1976	Growth %pa 1960–70	1970–76
Agriculture					
Cheese	202	278	377	3.3	5.2
Eggs	335	279	334	−1.8	3.0
Beef and veal	231	349	425*	4.2	3.3*
Pigmeat	352	728	1 021	7.5	5.8
Poultrymeat	77	308	323*	1.5	0.8*
Fish catch	315	301	284	−0.5	−0.9
Timber (000 m³)	700	945	905	3.0	−0.7
Energy					
Total energy (000 tce)	15 450	49 020	111 790ᶜ	12.2	17.9ᵈ
Coal	12 498	4 334	—	−10.0	−90.0
Crude oil	1 918	1 919	1 371	0.0	−5.4
Petroleum products	33 196	57 245	52 249ᶜ	5.6	−1.8ᵈ
Natural gas (mn m³)	330	31 617	97 302	57.8	20.6
Manufactured gas (mn m³)	2 812	778	1 007	−12.1	4.4
Electricity (mn kW h)	16 516	40 859	58 059	9.5	6.0
of which, nuclear (mn kW h)	—	368	3 335ᶜ	na	55.4ᵈ
Mining					
Salt	1 096	2 871	3 387ᵉ	10.1	4.2ᶠ
Manufacturing					
Margarine	238	234	203	−0.2	−2.3
Beer (000 hl)	3 552	8 772	12 430ᶜ	9.5	7.2ᵈ
Cigarettes (mn units)	11 992	23 058	33 616ᶜ	6.8	7.8ᵈ
Cigars (mn units)	1 467	2 085	2 412ᶜ	3.6	3.0ᵈ
Cotton yarn	76	52	33	−3.7	−7.3
Cotton fabrics	66*	45	30ᶜ	−3.7*	−7.8ᵈ
Man-made fibres	58	124	149ᵍ	7.9	9.6ʰ
Wood pulp	127	188	174ᶜ	4.0	−1.5ᵈ
Newsprint	145	167	123	1.4	−5.0
Other paper	730**	1 429	1 505	7.0**	0.9
Synthetic rubber	12	200	247	32.5	3.6
Sulphuric acid	860	1 562	1 462	6.1	−1.1
Plastics and resins	115	795	1 722	21.3	13.7
Fertilisers, nitrogenousᵃ	406	893	1 153	8.2	4.3
Fertilisers, phosphateᵃ	207	284	179	3.2	−7.4
Cement	1 798	3 830	3 481	7.8	−1.6
Coke	4 517	1 997	2 813	−7.8	5.9
Pig iron	1 346	3 594	4 267	10.3	2.9
Crude steel	1 942	5 042	5 190	10.0	0.5
Aluminium	—	75	255	na	22.6
Zinc	36	46	126	2.5	18.3
Passenger cars (000 units)ᵇ	47.5	85.6	84.5	6.1	−0.2
of which, assembly only	32.3	18.3	12.3	−5.5	−6.4
Commercial vehicles (000 units)ᵇ	12.5	18.2	22.5	3.8	3.6
of which, assembly only	8.4	6.2	11.0	−3.0	10.0
Merchant vessels (000 grt)	567	461	578	−2.0	3.8
Construction					
Dwellings (000 units)	85	117	107	3.3	−1.5

ᵃYears ending June 30th ᵇIncluding assembly ᶜ1975 ᵈ1970–75 ᵉ1974 ᶠ1970–74 ᵍ1972 ʰ1970–72

Transport traffic	1960	1970	1976	Growth %pa 1960–70	1970–76
Passenger-kilometres (mn)					
Road	na	87 900	115 800	na	4.7
Rail	7 821	8 011	8 306	0.2	0.6
Air	2 672	5 769	10 613	8.0	10.7
Cargo: tonne-kilometres (mn)					
Road	7 652ᵇ	12 396	16 000	7.1ᶜ	4.3
Rail	3 409	3 532	2 695	0.3	−4.4
Water	20 201ᵇ	30 743	29 597ᵈ	6.2ᶜ	−0.8ᵉ
Air	117	393	663	12.9	9.1
Sea: tonnes (mn)					
Goods loaded	24.8	63.9	82.5	9.9	4.3
Goods unloaded	82.0	202.7	255.8	9.5	4.0
Tourism					
Number of visitorsᵃ (000)	1 477	2 235	2 910	4.2	4.5
Gross receipts ($ mn)	132	429	1 061	12.5	16.3

ᵃArrivals at hotels ᵇ1963 ᶜ1963–70 ᵈ1975 ᵉ1970–75

Finance and trade

Price and wage indices (1970 = 100)	1960	1970	1976	Growth %pa 1960–70	1970–76
Consumer prices	67.2	100.0	164.6	4.1	8.7
Wholesale prices	81	100	147	2.1	6.6
Wages (rates)	41	100	206	9.4	12.8
Share prices	70	100	92	3.6	−1.4
Money stock (end-year, Gld mn)	11 300	25 950	50 400	8.7	11.7

Netherlands

Budget (total, 1976) Revenue: Gld 76 961 mn = $ 29 108 mn = £ 16 114 mn
Expenditure: Gld 85 361 mn = $ 32 285 mn = £ 17 873 mn,
of which, education 26 %, defence 10 %, social services 26 %

Balance of payments ($ mn)	1972	1973	1974	1975	1976
Balance of goods (fob)	+437	+998	+595	+1 311	+1 776
Balance of services	+977	+1 402	+1 772	+1 322	+1 270
Balance of transfers	−125	−58	−304	−641	−357
Current balance	*+1 290*	*+2 342*	*+2 061*	*+1 992*	*+2 689*
Long-term capital flow	−879	−1 739	−1 147	−1 420	−3 233
Reserves and debt (end-year, $ mn)					
International reserves	4 785	6 546	6 957	7 109	7 387
Public debt: domestic	11 245	13 927	16 468	17 379	22 383
external	18	14	9	4	—

External trade (1976)
Imports: Gld 106 893 mn = $ 40 429 mn = £ 22 381 mn
Exports: Gld 106 017 mn = $ 40 097 mn = £ 22 198 mn

Main imports	% of total	Main exports	% of total
Crude oil	15	Food	19
Food	12	(of which, dairy produce 4,	
(of which, cereals 3)		meat 4,	
Chemicals	8	fruit and vegetables 4)	
Machinery, non-electric	8	Chemicals	15
Electrical machinery	7	Petroleum products	12
Motor vehicles	7	Electrical machinery	7
Clothing	4	Machinery, non-electric	6
Iron and steel	4	Transport equipment	6
Textile yarns and fabrics	4	(of which, ships and	
Petroleum products	3	boats 3)	
Non-ferrous metals	2	Natural gas	5
Paper	2	Textile yarns and fabrics	4
Instruments	2	Iron and steel	3
		Instruments	2

Main sources[a]		Main destinations[b]	
West Germany	24	West Germany	31
Belgium-Luxembourg	13	Belgium-Luxembourg	15
United States	9	France	11
France	7	United Kingdom	8
United Kingdom	6	Italy	5
Iran	4	United States	3
Saudi Arabia	4	Sweden	2
Italy	3	Denmark	2
Nigeria	3	Norway	2
Sweden	2		

[a]EEC 55 % [b]EEC 72 %

Special focus

Method of transportation, 1975

	Goods entering Netherlands 000 tonnes	%	Goods leaving Netherlands 000 tonnes	%
Road	*16 781*	*6*	*18 216*	*8*
Rail	*3 416*	*1*	*4 396*	*2*
Inland waterway	*35 188*	*12*	*77 515*	*36*
Iron ore and scrap iron	103	—	31 490	15
Other crude materials	19 496	7	3 778	2
Crude oil and products	1 066	—	17 815	8
Chemicals	1 592	1	1 553	1
Other	12 931	4	22 879	11
Oil pipeline	*—*	*—*	*37 861*	*18*
Sea	*234 893*	*81*	*78 013*	*36*
Iron ore and scrap iron	40 522	14	774	—
Other crude materials	6 647	2	1 851	1
Crude oil and products	130 199	45	43 054	20
Chemicals	5 118	2	5 572	3
Other	52 407	18	26 762	12
Air	*84*	*—*	*82*	*—*
Total	*290 473*	*100*	*216 083*	*100*
Iron ore and scrap iron	40 652	14	34 171	16
Other crude materials	30 387	10	6 865	3
Crude oil and products	131 408	45	99 340	46
Chemicals	8 924	3	11 141	5
Other	79 102	27	64 566	30

Norway

Kingdom of Norway
Kongeriket Norge

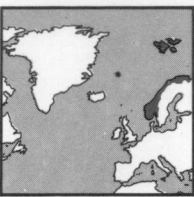

Location North-west Europe
With a long coastline on the North Sea, Atlantic and Arctic Oceans, Sweden is to the east and there are borders with Finland and Soviet Union in the north-east. The Svalbard archipelago is to the north in the Arctic Ocean and Jan Mayen Island is between Svalbard and Iceland
Land Area 324 219 km² = 125 182 mi², excluding Svalbard and Jan Mayen Islands, of 62 422 km²
Usage (1975): agricultural 8 980 km² (3 %), forests 83 300 (26 %), other 231 940 (71 %)
Climate Temperate, cold inland and arctic in the north
Weather at Oslo, 94 m altitude
Temperature: hottest month July 13–22 °C, coldest Jan minus 7–minus 2 °C
Rainfall (av monthly): driest month March 26 mm, wettest July 95 mm
Time 1 hour ahead of GMT
Measures Metric system
Monetary unit Norwegian krone (N Kr) = 100 øre
Rate of exchange (1976 av): free N Kr 5.456 = $ 1, N Kr 9.856 = £ 1
par (central value within the European 'snake', 1976)
to Oct 17, N Kr 6.87145 = S D R 1; from Oct 18, N Kr 6.94083 = S D R 1

Summary

Political Parliamentary monarchy, which became independent from Sweden in 1905. Member of UN, OECD, Council of Europe, Nato, Efta, and Nordic Council
Economic An industrial economy which has recently become a significant oil producer; crude oil accounted for 15 % of exports in 1976. There are also substantial exports of metals (especially aluminium and iron and steel), timber products and fish. Earnings from a large shipping fleet have been important, although fluctuating

People, resources and equipment

Population 1960 3.58 mn, 1970 3.88 mn, 1976 4.03 mn
Growth: 1960–70 0.8 % pa, 1970–76 0.6 % pa
Density (1976)[a]: 12 people per km²
[a]Based on area excluding Svalbard and Jan Mayen Islands
Vital statistics (rate per 1 000 people, 1976): births 13.3, deaths 9.9
Regions (fylker or county, population in 000, January 1, 1977; total of 4.04 mn)

Østfold	229	Telemark	160	Sør-Trøndelag	242
Akershus	358	Aust-Agder	87	Nord-Trøndelag	123
Oslo	462	Vest-Agder	133	Nordland	243
Hedmark	184	Rogaland	291	Troms	144
Oppland	179	Hordaland	388	Finnmark	79
Buskerud	211	Sogn og Fjordane	104		
Vestfold	183	Møre og Romsdal	233		

Cities (population in 000, January 1, 1977)

Oslo (capital)	462	Bærum	81	Tromsø	44
Bergen	213	Kristiansand	60	Ålesund	35
Trondheim	136	Drammen	51	Asker	35
Stavanger	87	Skien	47	Skedsmo	35

Race (1960) Norwegian 98 %, Swedish ½ %, Danish ¼ %
Language Bokmål (old Norwegian) and Landsmål or Nynorsk (new Norwegian)
Usage (1974): Bokmål 80* %, Landsmål 20* %
Religion (1976) Lutheran 96* %
Education (1975/76) Pupils 804 107, teachers (full time) 45 682
Labour force (1976) 1 821 000; in agriculture 168 000 (9 %), manufacturing, mining and utilities 445 000 (24 %), construction 148 000 (8 %), distribution and hotels 296 000 (16 %), transport and communication 161 000 (9 %), finance and real estate 82 000 (5 %)
Personnel Scientists and engineers (1973): 68 700
Physicians (1975): 6 886, 1 per 582 people
Standard of living
National income per person (1976): N Kr 35 000* = $ 6 400* = £ 3 500*
Consumption per person (1975): energy 4 607 kg coal equivalent, electricity (production, 1976) 20 400 kW h, newsprint 14 kg, steel 514 kg
Newspapers (1976): number 72; circulation 1 619 000, 402 per 1 000 people
Telephones (Dec 1976): 1 476 100, 366 per 1 000 people

Norway

Livestock (000, June 1976) Cattle 921, sheep 1 667, pigs 698, goats 68, chickens 6 594*

Mineral reserves
Coal (Svalbard, Norwegian-operated mines, 1972) 152 mn tonnes
Crude oil (1975) 806 mn tonnes
Natural gas (1975) 569 bn cubic metres
Petroleum refinery capacity (1975) 8.65* mn tonnes
Electrical capacity (1975) 17 090 megawatts,
of which, hydro 16 928 megawatts
Hospital beds (1975) 56 636, 1 per 71 people
Roads (1976) 78 116 km = 48 539 mi, density 0.24 km per km²
Railways (1975) 4 241 km = 2 635 mi, density 0.013 km per km²
Ships (registered, 1977) 2 738, total of 27 801 471 gross tons
Ports (goods traffic, 000 tonnes, 1976)

	loaded	unloaded		loaded	unloaded
Narvik	17 525ᵃ	25	Porsgrunn	2 061	1 587
Bergen	1 756	4 359	Stavanger	283	2 869
Tonsberg	1 291	3 401	Oslo	832	2 003

ᵃIncludes Swedish iron ore

Airports (passenger traffic, 000, 1976)

	departures	arrivals
Fornebu (Oslo)	1 455	1 470
Bergen	450	456
Stavanger	347	345
Gardermøn (Oslo)	167	169

Also (1977) 40 other airports with scheduled flights

Durable equipment (Dec 1976)

	000	no per 1 000 people	no per km of road
Radio sets	1 318	327	
Television sets	1 087	269	13
Passenger cars	1 023	254	
Commercial vehicles	148	37	1.9

Production

Gross domestic product
1976: N Kr 170 810 mn = $ 31 307 mn = £ 17 331 mn
Growth in real terms: 1970–76 4.8 % pa

Structure of gross domestic product

By origin (1975)	N Kr mn	% of gdp
Agriculture	7 824	5
Mining and quarrying	5 280	4
Manufacturing	31 851	22
Electricity, gas and water	5 333	4
Construction	11 729	8
Distribution and hotels	21 694	15
Transport and communication	17 231	12
Other	46 992	32
Total	147 934	100

By type of expenditure (1976)		
Government final consumption	29 140	17
Private final consumption	93 970	55
Stock investment	1 840	1
Gross fixed capital formation	62 030	36
Exports of goods and services	69 290	41
less Imports of goods and services	85 460	50
Total	170 810	100

Production indices (1970 = 100)	1960	1970	1976	Growth % pa 1960–70	1970–76
Agricultural	92	100	105	0.9	0.8
Industrial	60	100	119	5.2	2.9

Main products (000 t) Agriculture	1960	1970	1976	Growth % pa 1960–70	1970–76
Barley	400	581	486	3.8	−2.9
Oats	173	228	287	2.8	3.9
Potatoes	1 247	857	484	−3.7	−9.1
Apples	92	48	52*	−6.3	1.3*
Milk	1 614	1 704	1 864*	0.5	1.5*
Cheese	40	52	61	2.7	2.7
Beef and veal	50	56	62*	1.1	1.7*
Pigmeat	56	65	75*	1.5	2.4*
Fish catch	1 543	2 986	3 435	6.8	2.4
Timber (000 m³)	9 500**	8 542	8 968	−1.0**	0.8
Energy					
Total energy (000 tce)	4 270	7 640	23 560ᵈ	6.0	25.2ᵉ
Coalᵃ	404	484	545	1.8	2.0
Crude oil	—	—	13 691	na	na
Petroleum products	120*	5 581	6 600*ᵈ	46.8*	3.4*ᵉ

Main products (000 t) Energy	1960	1970	1976	Growth % pa 1960–70	1970–76
Manufactured gas (mn m²)	41	196	25	16.9	−29.0
Electricity (mn kW h)	31 121	57 606	82 198	6.4	6.1
of which, hydro (mn kW h)	30 915	57 260	82 100*	6.4	6.2*
Mining					
Iron ore (Fe content)	1 056	2 622	2 550	9.5	−0.5
Copper conc (Cu content)	15.4	19.8	31.5	2.5	8.0
Vanadium ore (V content)	0.60**	1.08*	1.03*ᵈ	6.0**	−0.9*ᵉ
Zinc conc (Zn content)	10.3	10.3	28.6	0.0	18.5
Manufacturing					
Fishmeal	141	351	337ᵈ	9.5	−0.8ᵉ
Beer (000 hl)	903	1 517	1 869ᶠ	5.3	5.4ᵍ
Man-made fibres	14.3	28.0	18.4*ᵈ	7.0	−8.1*ᵉ
Wood pulp	1 523	2 182	1 734ᵈ	3.7	−4.5ᵉ
Newsprint	226	554	470*	9.4	−2.7*
Other paper	560**	863	712ᵈ	4.4**	−3.8ᵉ
Sulphurᶜ	361	339	305ᶠ	−0.6	−2.6ᵉ
Sulphuric acid	87	312	401	13.6	4.3
Plastics and resins	30*	138	120ᵈ	16.5*	−2.7ᵉ
Fertilisers, nitrogenousᵇ	246	371	356	4.2	−0.7
Fertilisers, phosphateᵇ	43**	98	114	8.6**	2.5
Cement	1 151	2 635	2 680	8.6	0.3
Coke	na	311	260ᵈ	na	−3.5ᵉ
Pig iron	720	1 251	1 476	5.7	2.8
Crude steel	490	869	898	5.9	0.5
Aluminium	171	522	608	11.8	2.6
Magnesium	10.3	36.7	38.3ᵈ	13.5	0.9ᵉ
Radio sets (000 units)	129	144	111ʰ	1.1	−8.3ⁱ
Television sets (000 units)	64	103	108ᶠ	4.9	−1.2ᵍ
Merchant vessels (000 grt)	198	639	756	12.4	2.8
Construction					
Dwellings (000 units)	28.4	39.2	42.7	3.3	1.4

ᵃSpitsbergen (Svalbard), Norwegian-operated mines only; Soviet Union produces a similar amount at Spitsbergen, by arrangement with Norway ᵇYears ending June 30th ᶜContent of pyrites ᵈ1975 ᵉ1970–75 ᶠ1974 ᵍ1970–74 ʰ1973 ⁱ1970–73

Transport traffic Passenger-kilometres (mn)	1960	1970	1976	Growth % pa 1960–70	1970–76
Road	na	22 577	30 283ᶜ	na	7.6ᵈ
Rail	1 800**	1 573	1 990*	−1.3**	4.0*
Airᵃ	685	1 954	3 180	11.0	8.5
Cargo: tonne-kilometres (mn)					
Road	na	3 140*	4 371ᶜ	na	8.6*ᵈ
Rail	1 600**	2 731	2 774*	5.5**	0.3*
Airᵃ	16.8	72.7	118.9	15.8	8.5
Seaᵇ: tonnes (mn)					
Goods loaded	19.5	35.5	34.2	6.2	−0.6
Goods unloaded	9.3	20.8	22.0	8.3	0.9

ᵃIncludes apportionment (two-sevenths) of international operations of Scandinavian Airlines System (SAS) ᵇExcludes transit traffic, but includes Swedish iron ore, amounting to 18 mn tonnes loaded in 1976 ᶜ1974 ᵈ1970–74

Tourism (1976) Number of visitors (arrivals at hotels) 1 191 056, gross receipts $ 393 mn

Finance and trade

Price and wage indices (1970 = 100)	1960	1970	1976	Growth % pa 1960–70	1970–76
Consumer prices	64.4	100.0	163.2	4.5	8.5
Wholesale prices	78	100	163	2.5	8.5
Wages (earnings)	46	100	222	8.1	14.2
Share prices	93	100	102	0.7	0.3
Money stock (end-year, N Kr bn)	7.77	17.20	37.78	8.3	14.0

Budget (1977) Revenue: N Kr 46 372 mn = $ 8 711 mn = £ 4 990 mn
Expenditure: N Kr 60 874 mn = $ 11 435 mn = £ 6 551 mn

Balance of payments ($ mn)	1972	1973	1974	1975	1976
Balance of goods (fob)	−1 015	−1 533	−2 336	−2 911	−3 572
Balance of services	+977	+1 214	+1 321	+558	−18
Balance of transfers	−20	−47	−101	−160	−200
Current balance	−59	−365	−1 116	−2 515	−3 790
Long-term capital flow	+294	+833	+995	+2 587	+3 137
Reserves and debt (end-year, $ mn)					
International reserves	1 325	1 575	1 929	2 237	2 229
Public debt: domestic	4 226	4 955	6 302	6 373	7 855
external	281	228	219	983	1 844

Norway

External trade (1976)
Imports: N Kr 60 533 mn = $ 11 095 mn = £6 142 mn
Exports: N Kr 43 330 mn = $ 7 942 mn = £4 396 mn

Main imports	% of total	Main exports	% of total
Ships and boats	13	Crude oil	15
Machinery, non-electric	13	Ships and boats	11
Crude oil	7	Non-ferrous metals	11
Electrical machinery	6	(of which,	
Food	6	aluminium 7)	
Chemicals	6	Iron and steel	7
Motor vehicles	6	Machinery, non-electric	6
Iron and steel	5	Chemicals	6
Metal ores	4	Fish and products	6
Clothing	3	Paper	5
Textile yarns and fabrics	3	Electrical machinery	3
		Petroleum products	3
		Wood pulp	3

Main sources		Main destinations	
Sweden	18	United Kingdom	29
West Germany	15	Sweden	14
United Kingdom	10	West Germany	10
Japan	7	Denmark	7
United States	6	United States	5
Denmark	6	Liberia	4
Netherlands	5	Netherlands	4
France	4	France	3
Iran	3	Finland	2

Special focus

Fish utilisation 1976

	Fish landed	
	Tonnes	% of total
Fresh, chilled	69 840	2
Frozen	275 022	9
Cured (salted, smoked, dried)	239 600	8
Canned	18 900	1
Reduced for meal and/or oil	2 524 588	81
Bait	5 350	—
Total landed	3 133 300	100

Poland

Polish People's Republic
Polska Rzeczpospolita Ludowa

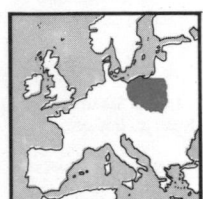

Location Eastern Europe
With a northern coastline on the Baltic Sea,
Soviet Union is to the east, Czechoslovakia to
the south, and East Germany to the west
Land Area 312 677 km² = 120 725 mi²
Usage (1975): agricultural 192 090 km² (61 %),
of which, arable 147 810 km² (47 %), cropland
3 030 (1 %), pastures 41 250 km² (13 %);
forests 86 080 km² (28 %)
Climate Temperate
Weather at Warsaw, 110 m altitude
Temperature: hottest month July 15–24 °C, coldest Jan minus 5–0 °C
Rainfall (av monthly): driest month Jan 27 mm, wettest July 96 mm
Time 1 hour ahead of GMT (summer time, 2 hours ahead)
Measures Metric system
Monetary unit Zloty (Zl) = 100 groszy
Rate of exchange (1976 av) *Basic* par Zl 3.32 = $ 1, free Zl 6.00 = £ 1
Non-commercial par Zl 19.92 = $ 1, free Zl 36 = £ 1
Tourist par Zl 33.20 = $ 1, free Zl 60 = £ 1

Summary

Political Communist republic, liberated from German occupation in the
second world war by Soviet Union in March 1945; the Oder-Neisse river
line was established as a new border with East Germany in 1945. Member
of UN, Comecon and Warsaw Pact
Economic An industrial economy, but with a large agricultural sector
which absorbs one-third of the labour force and provides about
one-sixth of net material product. Transport equipment (ships and motor
vehicles) and machinery are major exports, with chemicals and clothing
also important. Mining of coal is substantial and coal makes up about
one-sixth of exports. Industrial growth has been very high at 10½ %pa in
the 1970s, with a high level of investment

People, resources and equipment

Population 1960 29.56 mn, 1970 32.53 mn, 1976 34.36 mn
Growth: 1960–70 1.0 % pa, 1970–76 0.9 % pa
Density (1976): 110 people per km²
Vital statistics (rate per 1 000 people, 1976): births 19.5, deaths 8.8
Households (1970) 9 376 299; population in households 31.75 mn
(97.3 % of total)
Regions (district or voivodship, population in 000, December 31, 1976;
total of 34.53 mn)

Warszawskie	2 191	Krakowskie	1 137	Rzeszowskie	617
Białskopodlaskie	281	Krośnieńskie	426	Siedleckie	602
Białostockie	622	Legnickie	424	Sieradzkie	386
Bielskie	789	Leszczyńskie	345	Skierniewickie	390
Bydgoskie	1 006	Lubelskie	896	Słupskie	361
Chełmskie	222	Łomżyńskie	320	Suwalskie	416
Ciechanowskie	399	Łódzkie	1 093	Szczecińskie	867
Częstochowskie	730	Nowosądeckie	604	Tarnobrzeskie	538
Elbląskie	426	Olsztyńskie	671	Tarnowskie	585
Gdańskie	1 276	Opolskie	976	Toruńskie	594
Gorzowskie	439	Ostrołęckie	362	Wałbrzyskie	717
Jeleniogórskie	489	Pilskie	421	Włocławskie	405
Kaliskie	648	Piotrkowskie	584	Wrocławskie	1 038
Katowickie	3 535	Płockie	484	Zamojskie	471
Kieleckie	1 044	Poznańskie	1 187	Zielonogórskie	586
Konińskie	427	Przemyskie	375		
Koszalińskie	440	Radomskie	681		

Cities (population in 000, December 31, 1976)

Warszawa[a] (capital)	1 463	Lublin	282	Toruń	158
Łódź	810	Bytom	236	Kielce	157
Kraków[b]	701	Gdynia	225	Chorzów	156
Wrocław	584	Zabrze	204	Ruda Śląska	152
Poznań	527	Częstochowa	203	Tychy	140
Gdańsk	434	Białystok	201	Wałbrzych	129
Szczecin	376	Gliwice	200	Bielsko-Biała	124
Katowice	349	Sosnowiec	198	Olsztyn	122
Bydgoszcz	330	Radom	180	Opole	108

[a]Warsaw [b]Cracow

Race (1963) Polish 98½ %, Ukrainian ½ %, Byelorussian ½ %
Language Polish
Religion (1975) Roman Catholic 90* %, Polish Orthodox 1⅓ %,
Lutheran ¼ %
Education (1976/77) Pupils: primary 4 326 600, secondary 581 900,
vocational 2 066 100, higher 491 000. Teachers: primary 193 300,
secondary 25 600, vocational 76 100, higher 49 900
Labour force (1974)

Economic activity	Number	%
Agriculture	6 049 303	35
Manufacturing, mining, quarrying and utilities	5 295 361	30
Construction	1 235 459	7
Distribution and hotels	1 193 820	7
Transport and communication	1 008 239	6
Finance and real estate	94 912	1
Other	2 629 461	15
Total	17 506 555	100

Personnel Scientists and engineers (1973): 803 000
Physicians (1975): 58 226, 1 per 584 people
Standard of living National income per person (1976):
Zl 50 000*** = $ 2 500*** = £ 1 400***
Consumption per person (1975): energy 5 007 kg coal equivalent,
electricity (production, 1976) 3 030 kW h, newsprint 3.6 kg, steel 524 kg
Newspapers (1974): number 44; circulation 7 994 000, 237 per 1 000 people
Telephones (Dec 1976): 2 753 200, 80 per 1 000 people
Livestock (000, June 1977) Cattle 13 019, sheep 3 934, pigs 20 051,
horses 2 200*, chickens 200 000*
Mineral reserves Coal (1967) 45 741 mn tonnes
Lignite (1967) 14 862 mn tonnes
Crude oil (1975) 5 mn tonnes
Natural gas (1975) 114 bn cubic metres
Petroleum refinery capacity (1975) 14* mn tonnes
Electrical capacity (1975) 20 057 megawatts,
of which, hydro 827 megawatts
Hospital beds (1975) 264 103, 1 per 129 people
Roads (1976) 300 822 km = 186 922 mi, density 0.96 km per km²
Railways (1975) 23 766 km = 14 768 mi, density 0.076 km per km²
Inland waterways (1975) 3 759 km = 2 336 mi
Oil pipelines (1975) 1 851 km = 1 150 mi
Ships (registered, 1977) 773, total of 3 447 517 gross tons

Poland

Ports (goods traffic, 000 tonnes, 1976)

	loaded	unloaded
Szczecin	14 408	9 112
Gdańsk	15 740	6 864
Gdynia	5 540	7 415

Airports (passengers, 000 1976) Warsaw: departures 967, arrivals 869; also Cracow, Wrocław, Poznań, Gdańsk, Szczecin, Katowice, Bydgoszcz, Koszalin, Zielona Góra, Słupsk

Durable equipment (Dec 1976)	000	no per 1 000 people	
Radio sets	8 228	238	no per
Television sets	6 820	198	km of road
Passenger cars	1 290	37	4.3
Commercial vehicles[a]	524	15	1.7

[a]Including tractors

Production

Net material product 1976: Zl 1 595.9 bn = $ 80 115 mn[a] = £ 44 330 mn[a]
Growth in real terms: 1970–76 9.0 % pa

Gross domestic product
1976: Zl 1 900*** bn = $ 95 000*** mn[a] = £ 53 000*** mn[a]
[a]Converted at non-commercial rate

Structure of net material product (1976)

By origin	Zl bn	% of gdp
Agriculture	247.0	15
Manufacturing, mining, quarrying and utilities	826.6	52
Construction	200.4	13
Distribution and hotels	163.8	10
Transport and communication	123.7	8
Other	34.4	2
Total	1 595.9	100
By type of expenditure		
Government final consumption	179.4	11
Private final consumption	924.5	58
Stock investment	137.0	9
Gross fixed capital formation	460.8	29
Net external trade of goods and services	−105.8	−7
Total	1 595.9	100

Production indices (1970 = 100)	1960	1970	1976	Growth % pa 1960–70	1970–76
Agricultural	84	100	114	1.7	2.2
Industrial	44	100	181	8.6	10.4

Main products (000 t) *Agriculture*	1960	1970	1976	1960–70	1970–76
Wheat	2 303	4 608	5 745	7.2	3.7
Barley	1 310	2 149	3 617	5.1	9.1
Rye	7 878	5 433	6 922	−3.6	4.1
Oats	2 774	3 209	2 695	1.5	−2.9
Potatoes	37 855	50 301	49 951	2.9	−0.1
Sugar, raw value	1 500	1 505	1 801*	0.0	3.0*
Cabbages	1 400**	1 652	1 563*	1.7**	−0.9*
Tomatoes	164	355	380	8.0	1.1
Onions	182	365	332	7.2	−1.6
Cucumbers	190**	440	508*	8.8**	2.4*
Carrots	280**	490	444*	5.7**	−1.6*
Apples	627	691	840*	1.0	3.3*
Rapeseed	147	566	980	14.4	9.6
Linseed	56	65	49	1.5	−4.6
Tobacco	41	85	125	7.6	6.6
Flax	43	52	52*	2.0	0.0*
Milk	12 488	14 948	16 519	1.8	1.7
Butter	167	201	264	1.9	4.6
Cheese	142	246	353	5.7	6.2
Eggs	315	389	449	2.1	2.4
Beef and veal	333	544	858*	5.0	7.9*
Pigmeat	1 178	1 284	1 680*	0.9	4.6*
Fish catch	187	473	750	9.7	8.0
Timber (000 m³)	16 200*	18 473	21 596	1.3*	2.6
Energy					
Total energy (000 tce)	108 330	157 660	192 420[a]	3.8	4.1[b]
Coal	104 438	140 101	179 303	3.0	4.2
Lignite	9 327	32 766	39 302	13.4	3.1
Crude oil	194	424	460	8.1	1.4
Petroleum products	788	7 046	12 489[a]	24.5	12.1[b]

Main products (000 t) *Energy*	1960	1970	1976	Growth % pa 1960–70	1970–76
Natural gas (mn m³)	549	5 182	6 699	25.2	4.4
Manufactured gas (mn m³)	5 163	6 682	7 548	2.6	2.0
Electricity (mn kW h)	29 307	64 532	104 095	8.2	8.3
of which, hydro (mn kW h)	659	1 887	2 379[a]	11.1	4.7[b]
Mining					
Iron ore (Fe content)	624	707	210*	1.2	−18.3*
Copper ore (Cu content)	11	83	230*[a]	22.7	22.6*[b]
Lead ore (Pb content)	39	67	65*[a]	5.5	−0.6*[b]
Magnesite	22	39	27[a]	6.0	−7.1[b]
Nickel ore (Ni content)	1.3	1.5*	2.0*[a]	1.4*	5.9*[b]
Zinc ore (Zn content)	144	242	190*[a]	5.3	−4.7*[b]
Salt	1 946	2 904	3 818	4.1	4.7
Sulphur	25	2 683	4 891	59.6	10.5
Manufacturing					
Margarine	71	165	196	8.8	2.9
Beer (000 hl)	6 732	10 372	12 300	4.4	2.9
Cigarettes (mn units)	44 056	69 193	88 831	4.6	4.3
Cotton yarn	153	208	219	3.1	0.9
Cotton fabrics (mn m)	667	881	948	2.8	1.2
Wool yarn	58	84	106	3.7	3.9
Man-made fibres	78	138	228	5.9	8.7
Wood pulp	451	636	763	3.5	3.1
Newsprint	77	88	83	1.3	−1.0
Other paper	564	874	1 226	4.5	5.8
Synthetic rubber	20	62	117	11.8	11.2
Tyres (000 units)	1 233	2 995	4 770	9.3	8.1
Sulphuric acid	685	1 901	3 187	10.7	9.0
Nitric acid	602	1 526	2 185	9.7	6.2
Caustic soda	167	313	389	6.5	3.7
Soda ash	522	644	726	2.1	2.0
Plastics and resins	40	224	560	18.8	16.5
Fertilisers, nitrogenous	270	1 030	1 548	14.3	7.0
Fertilisers, phosphate	170*	599	928	13.4*	7.6
Cement	6 599	12 180	19 808	6.3	8.4
Coke	10 945	15 208	17 914	3.3	2.7
Pig iron	4 563	7 111	8 323	4.5	2.7
Crude steel	6 681	11 795	15 641	5.8	4.8
Aluminium[c]	26	99	103	14.3	0.7
Radio sets (000 units)	627	987	2 038	4.6	12.8
Television sets (000 units)	171	616	963	13.7	7.7
Passenger cars (000 units)	13	65	216	17.6	22.1
Commercial vehicles (000 units)	24	53	75*	8.2	6.0*
Merchant vessels (000 grt)	227	463	518	7.4	1.9
Construction					
Dwellings (000 units)	142*	193	273	3.1*	5.9

[a]1975 [b]1970–75 [c]Including secondary

Transport traffic *Passenger-kilometres* (mn)	1960	1970	1976	Growth % pa 1960–70	1970–76
Road[a]	na	29 140	48 274	na	8.8
Rail	30 942	36 891	42 800	1.8	2.5
Air	101	550	1 425	18.5	17.2
Cargo: tonne-kilometres (mn)					
Road[a]	na	15 761	36 779	na	15.2
Rail	66 547	99 262	130 956	4.1	4.7
Water	926[b]	2 295	1 950	13.8[c]	−2.7
Air	1.6	7.8	14.2	17.1	10.5
Sea: tonnes (mn)					
Goods loaded	11.4	24.6	35.8	8.0	6.5
Goods unloaded	10.2	11.2	23.6	0.9	13.2
Tourism					
Number of visitors (000)[d]	na	1 889	9 623	na	31.2
Gross receipts ($ mn)	na	32	157	na	30.4

[a]Public transport only [b]1963 [c]1963–70 [d]Including transit and excursionists (5 195 thousand in 1976)

Poland

Finance and trade

Price and wage indices (1970 = 100)	1960	1970	1976	Growth % pa 1960–70	1970–76
Consumer prices	87	100	118	1.4	2.8
Wages (earnings)	78	100	177	2.6	10.0

Budget (1976) Revenue: Zl 776.8 bn = $ 39 000 mn[a] = £ 21 600 mn[a]
Expenditure: Zl 754.7 bn = $ 37 900 mn[a] = £ 21 000 mn[a],
of which, education, science and culture 11 %, defence 7 %, health 8 %, social security 6 %
[a]Converted at non-commercial rate

External trade (1976)
Imports (fob): Zl 46 071 mn = $ 13 877 mn = £ 7 678 mn
Exports: Zl 36 710 mn = $ 11 057 mn = £ 6 118 mn

Main imports (1975)	% of total	Main exports (1975)	% of total
Machinery, non-electric	16	Coal	16
Iron and steel	13	Transport equipment	12
Food	7	(of which,	
(of which, cereals 4)		ships and boats 6,	
Electrical machinery	7	motor vehicles 4)	
Chemicals	6	Machinery, non-electric	11
Crude oil	6	Electrical machinery	8
Motor vehicles	4	Chemicals	6
Textile fibres	3	Food	6
Petroleum products	2	(of which, meat 3)	
Tools	2	Clothing	4
Iron ore	2	Iron and steel	4
Ships and boats	2		

Main sources (1976)		Main destinations (1976)	
Soviet Union	26	Soviet Union	30
West Germany	9	East Germany	10
East Germany	8	Czechoslovakia	8
United States	6	West Germany	6
Czechoslovakia	6	Hungary	3
France	6	France	3
United Kingdom	5	United Kingdom	3
Austria	4	Rumania	3
Switzerland	4	Itay	3

Special focus

Motor vehicle production

	Passenger cars 000	% change over previous year	Commercial vehicles 000	% change over previous year
1967	27.0	−2	40.2	+6
1968	39.0	+44	44.5	+11
1969	48.0	+23	52.0	+17
1970[a]	65.2	+36	52.8	+2
1971	86.1	+32	59.6	+13
1972	91.0	+6	67.0	+12
1973	115.0	+26	77.0	+15
1974	133.0	+16	80.1	+4
1975	164.0	+23	85.0	+6
1976	216.2	+32	75.1*	−12

[a]Year in which numbers of passenger cars produced first exceeded number of commercial vehicles

Portugal

Portuguese Republic
República Portuguesa

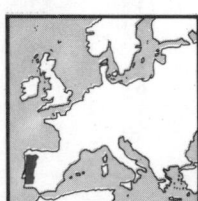

Location South-west Europe
Forms the western edge of the Iberian Peninsula, with a coastline on the Atlantic Ocean; Spain is to the east and north. The Azores and Madeira islands, in the Atlantic Ocean, form part of the territory
Land Area 92 082 km² = 35 553 mi²
of which, Azores 2 344 km², Madeira 797 km²
Climate Mediterranean in south, temperate in north

Weather at Lisbon, 77 m altitude
Temperature: hottest month Aug 17–28 °C, coldest Jan 8–14 °C
Rainfall (av monthly): driest month July 3 mm, wettest Jan 111 mm

Time GMT (summer time, 1 hour ahead of GMT)
Measures Metric system; some local measures are also in use, including:
1 arroba = 15 kilograms = 33.07 pounds
Monetary unit Escudo (Esc) = 100 centavos
Rate of exchange (1976 av): free Esc 30.22 = $ 1, Esc 54.59 = £ 1

Summary

Political Republic; a coup d'état on April 25th 1974 ended a one-party regime established since 1932. The Azores and Madeira Islands are included in the territory, and in the figures here. Most of the former Portuguese overseas territories were given their independence during 1974/75. Member of UN, OECD, Nato, Council of Europe and Efta
Economic A mixed economy, with a large agricultural labour force, but with manufacturing making up about one-third of gross domestic product; main exports are textiles and clothing, cork and products, wines and spirits, electrical machinery and chemicals. Tourism is important, but has been affected by political unrest. Many Portuguese have migrated to other European countries because of the low standard of living and lack of opportunity at home; their remittances home have helped the foreign exchange position

People, resources and equipment

Population[a] 1960 8.85 mn, 1970 9.01* mn, 1976 9.69* mn
[a]At end 1976 there were estimated to be 700 000** refugees from former overseas territories
Growth: 1960–70 0.2* %pa, 1970–76 1.2* %pa
Density (1976): 105* people per km²
Vital statistics (rate per 1 000 people, 1976): births 19.2, deaths 10.5
Regions (districts, population in 000, 1975; total of 9.45 mn)

Mainland	8 891	Leiria	400	Islands	558
Aveiro	608	Lisboa	1 870	Azores	292
Beja	194	Portalegre	142	Angra do Heroísmo	89
Braga	674	Porto	1 510	Horta	41
Bragança	174	Santarém	455	Ponta Delgada	163
Castelo Branco	251	Setúbal	583	Madeira	266
Coímbra	429	Viana do Castelo	262	Funchal	266
Evora	180	Vila Real	263		
Faro	277	Viseu	418		
Guarda	201				

Cities (population in 000, 1975)

Lisboa[a] (capital)	830[c]	Coímbra	56[d]	Setúbal	50[d]
Porto[b]	336[c]	Barreiro	54[d]	Braga	49[d]
Amadora	66[d]	Vila Nova de Gaia	51[d]	Almada	39[d]

[a]Lisbon [b]Oporto [c]Excluding suburbs [d]1970
Race (1970) Portuguese 99¾**%
Language Portuguese
Religion (1976) Roman Catholic 88 %
Education (1975/76) Pupils 1 750 932, teachers 97 874
Labour force (civilian, 1975) 2 997 000; in agriculture 847 000 (28 %), manufacturing 732 000 (24 %), construction 246 000 (8 %), distribution and hotels 367 000 (12 %), transport and communication 170 000 (6 %), finance and real estate 74 000 (2 %)
Personnel Scientists and engineers engaged in research (1972): 2 216
Physicians (1974): 10 312, 1 per 894 people
Standard of living
National income per person (1976): Esc 47 000** = $ 1 500** = £ 860**
Consumption per person (1975): energy 983 kg coal equivalent, electricity (production, 1976) 990 kW h, newsprint 3.1 kg, steel 118 kg
Newspapers (1974): number 32; circulation 799 000, 87 per 1 000 people
Telephones (Dec 1976): 1 118 970, 115 per 1 000 people
Livestock (000, 1976) Cattle 1 000*, sheep 3 800*, pigs 1 683*, goats 653*, chickens 16 200*
Mineral reserves Coal (known economic, 1972) 15 mn tonnes
Lignite (known economic, 1972) 27 mn tonnes
Uranium (1976) 6 800 tonnes
Petroleum refinery capacity (1975) 6.5 mn tonnes
Electrical capacity (1975) 3 149 megawatts,
of which, hydro 1 954 megawatts
Hospital beds (1975) 52 268, 1 per 184 people
Roads (1975) 46 241 km = 28 733 mi, density 0.50 km per km²
Railways (1975) 3 563 km = 2 214 mi, density 0.039 km per km²
Ships (registered, 1977) 350, total of 1 281 439 gross tons
Ports (goods traffic, 000 tonnes, 1976)

	International loaded	unloaded	Coastwise loaded	unloaded	Total
Lisbon	1 254	7 870	390	1 676	11 190
Leixoes	2 535	6 074	12	14	8 635
Setúbal	465	581	205	13	1 264
Douro	10	111	—	69	190

Portugal

Airports (passengers, 000, 1976)

	departures	arrivals
Lisbon	1 096	1 142
Funchal	281	281

Also Oporto and (1977) 11 other airports with scheduled flights

Durable equipment (at end-year)

	000	no per 1 000 people	
Radio sets (1975)	1 519	157	
Television sets (1976)	909	94	no per km of road
Passenger cars (1975)	873	90	19
Commercial vehicles (1975)	49	5	1.1

Production

Gross domestic product 1975: Esc 376.7 bn = $ 14 742 mn = £ 6 635 mn
1976 est: Esc 485**bn = $ 16 000**mn = £ 8 900**mn
Growth in real terms: 1960–70 6.4 %pa, 1970–75 4.3 %pa
Structure of gross domestic product (1975) *By origin* Agriculture 14 %, manufacturing 30 %, electricity, gas and water 2 %, construction 6 %, distribution and hotels 13 %, transport and communication 6 %
By type Final consumption expenditure 96 % (of which, government 15 %), stock investment −3 %, gross fixed capital formation 20 %, exports of goods and services 20 %, less imports of goods and services −32 %

Production indices (1970 = 100)	1960	1970	1976	Growth %pa 1960–70	1970–76
Agricultural	83	100	90	1.9	−1.7
Industrial	44	100	137	8.6	5.4

Main products (000 t)	1960	1970	1976	Growth %pa 1960–70	1970–76
Agriculture					
Wheat	492	564	694	1.4	3.5
Maize	466	628	429	3.0	−6.2
Rye	138	157	165	1.3	0.8
Rice	151	195	97	2.6	−11.0
Potatoes	1 041	1 285	1 003	2.1	−4.0
Cabbages	70**	120*	145*	5.5**	3.2*
Tomatoes	270**	740*	631	10.6**	−2.6*
Grapes	1 528	1 632*	1 250*	0.7*	−4.3*
Wine	1 146	1 162	813	0.1	−5.8
Oranges	99	90*	123*	−0.9*	5.3*
Figs	274	220*	190*	−2.2*	−2.4*
Olives	606	460*	257	−2.7*	−9.2*
Olive oil	86	67	43	−2.5	−7.1
Milk	400*	620*	685*	4.5*	1.7*
Meat	193	279	352	3.7	3.9
Fish catch	475	500	339	0.5	−6.3
Cork	179	154	108	−1.5	−5.7
Timber (000 m³)	5 760	6 370	7 887	1.0	1.6
Energy					
Total energy (000 tce)	900*	1 000*	1 010e	1.0*	0.2*f
Coal	434	271	193	−4.6	−5.5
Petroleum products	1 246	3 432	4 980*e	10.7	7.7*f
Manufactured gas (mn m³)d	77	118	220e	4.3	13.3f
Electricity (mn kW h)	3 300**	7 488	9 594	8.5**	4.2
of which, hydro (mn kW h)	3 110**	5 854	4 839	6.5**	−3.1
Mining					
Iron ore (Fe content)	151	58	24	−9.1	−13.7
Copper ore (Cu content)	3.4	4.7	5.1e	3.3	1.6f
Tin conc (Sn content)	0.654	0.435	0.350	−4.0	−3.5
Tungsten conc (oxide content)	1.75	1.86	1.77e	0.6	−0.9f
Uranium (U content)	na	0.080*	0.088	na	1.6*
Salt	284	401	533g	3.5	7.4h
Manufacturing					
Beer (000 hl)	379	1 319	2 934	13.3	14.3
Cigarettes (mn units)	5 560	8 264	12 582	4.0*	7.2
Cotton yarn	50	81	84*	4.9	0.6*
Cotton fabrics	40	45	61	1.2	5.2
Man-made fibres	2.1	6.7	12.3e	12.3	12.9f
Wood pulp	80	427	587	18.2	5.4
Paper	100**	220	381	8.2**	9.6
Cork products	na	348	264	na	−4.5
Tyres (000 units)	330	1 347	1 300	15.1	−0.6*
Sulphur[a]	301	213	202*e	−3.4	−1.0*f
Sulphuric acid	331	364	325	1.0	−1.9
Nitric acid	1	219	183e	73.0	−3.5f
Fertilisers, nitrogenous[b]	40*	119	203*	11.5*	9.4*
Fertilisers, phosphate[b]	75**	84	82*	1.1**	−0.3*
Cement	1 202	2 332	3 496	6.8	7.0
Pig iron	41	315	354	22.6	2.0

Main products (000 t) Manufacturing	1960	1970	1976	Growth %pa 1960–70	1970–76
Crude steel	20**	385	391	34.4**	0.2
Passenger cars (000 units)c	—	54.5	39.4	na	−5.3
Commercial vehicles (000 units)c	—	13.2	46.8	na	23.5
Merchant vessels (000 grt)	24	16	252	−4.0	58.3

[a]Content of pyrites [b]Years ending June 30th [c]Assembly only [d]Lisbon only [e]1975 [f]1970–75 [g]1974 [h]1970–74

Transport traffic	1960	1970	1976	Growth %pa 1960–70	1970–76
Passenger-kilometres (mn)					
Road[a]	na	3 519	4 500b	na	5.0c
Rail	2 156	3 546	4 856b	5.1	6.5c
Air	243	2 453	2 841	26.0	2.5
Cargo: tonne-kilometres (mn)					
Rail	762	776	856	0.2	1.6
Air	4.2	46.8	73.5	27.3	7.8
Sea: tonnes (mn)					
Goods loaded	2.53	4.06	4.26	4.8	0.8
Goods unloaded	3.85	9.24	14.64	9.1	8.0
Tourism					
Number of visitors (000)	353	3 343	2 175	25.2	−6.9
Gross receipts ($ mn)	26	237	321	24.7	5.2

[a]Public transport only [b]1975 [c]1970–75

Finance and trade

Price and wage indices (1970 = 100)	1960	1970	1976	Growth %pa 1960–70	1970–76
Consumer prices	64.6	100.0	244.5	4.5	16.1
Wholesale prices	79	100	208	2.4	13.0
Wages (earnings)	50.7a	100.0	240.4	7.8b	15.7
Money stock (end-year, Esc bn)	45.67	92.57	249.56	7.3	18.0

[a]1961 [b]1961–70

Budget (total, 1977)
Balanced at Esc 159 173 mn = $ 4 158 mn = £ 2 382 mn

Balance of payments ($ mn)	1972	1973	1974	1975	1976
Balance of goods (fob)	−734	−910	−1 995	−1 671	−2 095
Balance of services	+208	+154	+56	−194	−104
Balance of transfers	+881	+1 097	+1 110	+1 033	+973
Current balance	+354	+341	−830	−832	−1 226
Long-term capital flow	−96	−141	+272	−90	+12

Reserves and debt (end-year, $ mn)	1972	1973	1974	1975	1976
International reserves	2 312	2 839	2 354	1 534	1 302
External public debt	668	686	908	966	na

External trade (1976)
Imports: Esc 127 820 mn = $ 4 230 mn = £ 2 341 mn
Exports: Esc 54 680 mn = $ 1 809 mn = £ 1 002 mn

Main imports	% of total	Main exports	% of total
Food	17	Textile yarns and fabrics	15
(of which, cereals 6)		Clothing	11
Crude oil	13	Cork and products	7
Chemicals	12	Wines and spirits	7
Machinery, non-electric	11	Electrical machinery	7
Electrical machinery	6	Chemicals	5
Motor vehicles	6	Fruit and vegetables	5
Textile fibres	6	Fish and products	3
Iron and steel	5	Machinery, non-electric	3
Main sources		*Main destinations*	
West Germany	12	United Kingdom	18
United Kingdom	9	West Germany	11
United States	9	France	8
France	9	Sweden	8
Spain	5	United States	7
Italy	5	Italy	4

Special focus

Emigration from Portugal

Country of destination	1973 Number of people	%	1974 Number of people	%	1975 Number of people	%
France	20 692	26	10 568	24	2 866	12
West Germany	31 479	40	3 049	7	1 072	4
United States	8 160	10	9 540	22	8 975	36
Canada	7 403	9	11 650	27	5 857	24
Venezuela	4 294	5	2 550	6	1 903	8
Brazil	890	1	719	2	1 553	6
Other	6 599	8	5 321	12	2 585	10
Total	79 517	100	43 397	100	24 811	100

Rumania

The Socialist Republic of Rumania
Republica Socialistă România

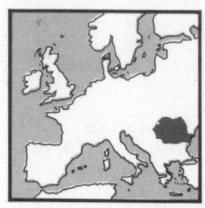

Location Eastern Europe
With a coastline to the east on the Black Sea, Soviet Union is to the north, Hungary to the west, Jugoslavia to the south-west and Bulgaria to the south
Land Area 237 500 km² = 91 700 mi²
Climate Continental
Weather at Bucharest, 92 m altitude
Temperature: hottest month July 16–30 °C, coldest Jan minus 7–1 °C
Rainfall (av monthly): driest month Feb 26 mm, wettest June 121 mm
Time 2 hours ahead of GMT
Measures Metric system
Monetary unit Leu = 100 bani; plural of leu is lei
Rate of exchange (1976 av) *Basic* par Lei 4.97 = $ 1, free Lei 8.98 = £ 1
Non-commercial par Lei 12.00 = $ 1, free Lei 21.67 = £ 1

Summary

Political Communist republic, which became a republic in 1947 on the abdication of King Michael. The formerly Rumanian territory of Bessarabia (Soviet Moldavia), a province on the north-east border, was annexed by Soviet Union in 1940; there has been a dispute with Hungary concerning the Hungarian minority in Transylvania (north-west Rumania) which was ceded to Rumania by Hungary after the first world war. Member of UN, Comecon and Warsaw Pact
Economic A mixed economy, with about one-third of the labour force in agriculture, and food a major export; crude oil and timber products are also important raw material exports, with machinery, chemicals and clothing significant industrial exports. Industrial production has increased at a high rate over the past and a continued high level is planned for the current plan period to 1980. There is a policy of energy conservation, especially for oil supplies, and increased use of hydro-electricity is planned

People, resources and equipment

Population 1960 18.41 mn, 1970 20.25 mn, 1976 21.45 mn
Growth: 1960–70 1.0 %pa, 1970–76 1.0 %pa
Density (1976): 90 people per km²
Vital statistics (rate per 1 000 people, 1975): births 19.7, deaths 9.3
Regions (districts, population in 000, 1975; total of 21.25 mn)

Municipiul Bucureşti	1 707	Covasna	195	Olt	521
Alba	404	Dîmboviţa	469	Prahova	794
Arad	497	Dolj	750	Satu Mare	392
Argeş	607	Galaţi	566	Sălaj	273
Bacău	686	Gorj	334	Sibiu	460
Bihor	628	Harghita	315	Suceava	652
Bistriţa-Năsăud	294	Hunedoara	519	Teleorman	543
Botoşani	492	Ialomiţa	396	Timiş	650
Braşov	505	Iaşi	736	Tulcea	263
Brăila	378	Ilfov	811	Vaslui	484
Buzău	524	Maramureş	492	Vîlcea	408
Caraş-Severin	374	Mehedinţi	330	Vrancea	388
Cluj	695	Mureş	617		
Constanţa	554	Neamţ	543		

Cities (population in 000, Jan 5, 1977)

Bucureşti[a] (capital)	1 934	Braşov	262	Arad	195
Constanţa	290	Ploieşti	255	Oradea	182
Iaşi	284	Craiova	249	Sibiu	170
Timişoara	283	Galaţi	247	Piteşti	165
Cluj-Napoca	262	Brăila	200	Tîrgu Mureş	153

[a]Bucharest
Race (1966) Rumanian 88 %, Hungarian 8 %, German 2 %
Language Rumanian
Religion (1976) Rumanian Orthodox 85** %, Roman Catholic 6** %, other Christian 4** %
Education (1976/77) Pupils 5 409 762, teachers 257 924
Labour force (civilian, 1976) 10 227 000; in agriculture 3 669 800 (36 %), manufacturing, mining, quarrying and utilities 3 267 900 (32 %), construction 848 100 (8 %), distribution and hotels 592 400 (6 %), transport and communication 505 600 (5 %)
Personnel Scientists and engineers (1968): 274 541
Physicians (1973): 25 870, 1 per 805 people

Standard of living National income per person (1976):
Lei 20 000*** = $ 1 600***[a] = £ 900***[a]
[a]Converted at non-commercial rate
Consumption per person (1975): energy 3 803 kg coal equivalent, electricity (production, 1976) 2 716 kW h, newsprint 2.1 kg, steel 464 kg
Newspapers (1974): number 20; circulation 2 716 000, 129 per 1 000 people
Telephones (Dec 1975): 1 196 000, 56 per 1 000 people
Livestock (000, Jan 1977) Cattle 6 351, sheep 14 331, pigs 10 193, poultry 91 503
Mineral reserves Coal (1966) 590 mn tonnes
Lignite (1966) 3 900 mn tonnes
Crude oil (1975) 174 mn tonnes
Natural gas (1975) 190 bn cubic metres
Petroleum refinery capacity (1975) 20* mn tonnes
Electrical capacity (1975) 11 577 megawatts, of which, hydro 2 632 megawatts
Hospital beds (1974) 191 910, 1 per 110 people
Roads (1975) 77 949 km = 48 435 mi, density 0.33 km per km²
Railways (1975) 11 039 km = 6 859 mi, density 0.046 km per km²
Inland waterways (1975) 1 659 km = 1 031 mi
Oil pipelines (1974) 2 314 km = 1 438 mi
Ships (registered, 1977) 207, total of 1 218 171 gross tons
Ports Constanţa, Galaţi, Brăila, Sulina, Tulcea
Airports Otopeni and Baneasa (Bucharest), Constanţa and 12 other airports with scheduled flights

Durable equipment (at end-year)	000	no per 1 000 people	
Radio sets (1976)	3 104	144	no per
Television sets (1976)	2 963	138	km of road
Passenger cars (1975)	138*	6*	2*
Commercial vehicles (1972)	50*	2*	1*

Production

Gross domestic product
1976 est: Lei 500*** bn = $ 40 000*** mn[a] = £ 23 000*** mn[a]
[a]Converted at non-commercial rate
Growth in real terms (net material product): 1960–70 8.4 %pa, 1970–74 11.7 %pa

Production indices (1970 = 100)	1960	1970	1976	Growth %pa 1960–70	1970–76
Agricultural	82	100	165	2.0	8.7
Industrial	30	100	184[b]	12.8	13.0[c]

Main products (000 t) *Agriculture*					
Wheat	3 450	3 356	6 724*	−0.3	12.3*
Barley	405	513	1 231	2.4	15.7
Maize	5 531	6 536	11 583	1.7	10.0
Potatoes	3 009	2 064	4 788	−3.7	15.0
Sugar, raw value	449	374	670*	−1.8	10.2*
Cabbages	450**	484	633*	0.7**	4.6*
Tomatoes	408	683	1 473*	5.3	13.7*
Grapes	980**	760	1 493	−2.5**	11.9
Wine	560*	432[a]	881*	−2.6*	12.6*
Apples	111	176	270*	4.7	7.4*
Plums	496	697	550*	3.5	−3.9*
Soyabeans	12	91	213	22.5	15.2
Sunflowerseed	522	770	799	4.0	0.6
Tobacco	15	22	45*	3.8	12.7*
Milk, cow	2 951	3 549	4 279*	1.9	3.1*
Milk, sheep	392	371	342*	−0.5	−1.3*
Eggs	107	161	270*	4.2	9.0*
Wool	13.1	17.8	19.5*	3.1	1.5*
Beef and veal	169	221	283*	2.7	4.2*
Pigmeat	276	450	760*	5.0	9.1*
Poultrymeat	61	122	303	7.2	16.4
Fish catch	18	59	127	12.4	13.8
Timber (000 m³)	19 030	22 286	20 587	1.6	−1.3
Energy					
Total energy (000 tce)	32 940	60 770	80 660[b]	6.3	5.8[c]
Coal	3 405	6 402	7 120*	6.5	1.8*
Lignite	3 363	14 129	18 730	15.4	4.8
Crude oil	11 500	13 377	14 700	1.5	1.6
Petroleum products	10 953	14 508	17 680*[b]	2.8	4.0*[c]
Natural gas (mn m³)	10 143	23 629	32 180	8.8	5.3
Manufactured gas (mn m³)	407	522	950*[b]	2.5	12.7*[c]
Electricity (mn kW h)	7 650	35 088	58 266	16.4	8.8
of which, hydro (mn kW h)	397	2 773	8 711[b]	21.4	25.7[c]
Mining					
Iron ore (Fe content)	467	881	737	6.6	−2.9
Bauxite	na	776*	779*[b]	na	0.1*[c]

Rumania

Main products (000 t) Mining	1960	1970	1976	Growth %pa 1960–70	1970–76
Lead ore (Pb content)	12*	40*	45*	12.8*	2.0*
Manganese ore (Mn content)	40	27	31*[b]	−3.9	2.8*[c]
Salt	1 045	2 862	4 210	10.6	6.6
Manufacturing					
Beer (000 hl)	1 633	4 375	7 449[b]	10.3	11.2[c]
Cigarettes (mn units)	20 622	26 491	29 000[b]	2.5	1.8[c]
Cotton yarn	52	109	145[b]	7.7	5.9[c]
Cotton fabrics (mn m²)	248	437	677	5.8	7.6
Wool yarn	19.4	35.7	50.6[b]	6.3	7.2[c]
Man-made fibres	4	77	148*[b]	34.0	14.0*[c]
Wood pulp	155	401	668[b]	10.0	10.7[c]
Paper	189	538	679[b]	11.0	4.8[c]
Synthetic rubber	—	61	147	na	15.8
Tyres (000 units)	337	2 457	4 120	22.0	9.0
Sulphur[a]	107	323	375*[b]	11.7	3.0*[c]
Sulphuric acid	226	994	1 555	16.0	7.7
Caustic soda	74	330	388	16.1	2.7
Soda ash	180	582	693[b]	12.5	3.6[c]
Plastics and resins	12	206	283[d]	32.4	8.3[c]
Fertilisers, nitrogenous	19	647	1 292[b]	42.3	14.8[c]
Fertilisers, phosphate	52	244	404[b]	16.7	10.6[c]
Cement	3 054	8 127	13 088	10.3	8.3
Coke	820	1 070	1 850*[b]	2.7	11.6*[c]
Pig iron	1 014	4 211	7 415	15.3	9.9
Crude steel	1 806	6 517	10 733	13.7	8.7
Aluminium	—	101	204[b]	na	15.1[c]
Radio sets (000 units)	167	455	712[b]	10.5	9.4[c]
Television sets (000 units)	—	280	548	na	11.8
Passenger cars (000 units)	1.2	23.6	71.2	34.7	20.2
Commercial vehicles (000 units)	9.3	37.1	39.2	14.8	0.9
Construction					
Dwellings (000 units)	253	159	165[b]	−4.5	0.7[c]

[a]Content of pyrites [b]1975 [c]1970–75 [d]1974 [e]1970–74

Transport traffic	1960	1970	1976	Growth %pa 1960–70	1970–76
Passenger-kilometres (mn)					
Road	na	7 858	18 668	na	15.5
Rail	10 737	17 793	23 077	5.2	4.4
Air	142[a]	354	769	14.0[b]	13.8
Cargo: tonne-kilometres (mn)					
Road	na	5 156	9 857	na	11.4
Rail	19 821	48 045	67 556	9.2	5.8
Water	na	na	1 859	na	na
Air	2.4[a]	7.1	10.6	16.8[b]	6.9
Sea: tonnes (mn)					
Goods loaded	1.37	4.26	7.00*[c]	12.0	13.2*[d]
Goods unloaded	0.81	4.86	12.00*[c]	19.6	25.4*[d]

[a]1963 [b]1963–70 [c]1974 [d]1970–74

Tourism (1976) Number of visitors 3 168 710, gross receipts $ 112 mn

Finance and trade

Consumer price index (1970 = 100) 1975 103; growth 1970–75 0.6 %pa
Wage index (1970 = 100) 1975 127; growth 1970–75 4.9 % pa
Budget (1976) Balanced at Lei 258 bn = $ 21 500 mn[a] = £ 11 906 mn[a]
[a]Converted at non-commercial rate
External trade (1976)
Imports (fob): Lei 30 294 mn = $ 6 095 mn = £ 3 373 mn
Exports: Lei 30 504 mn = $ 6 138 mn = £ 3 397 mn

Main imports (1973)	% of total	*Main exports* (1973)	% of total
Machinery, non-electric	15	Food	16
Iron and steel	9	(of which, meat 7,	
Electrical machinery	8	cereals 4)	
Food	8	Machinery	9
Textile yarns and fabrics	5	Crude oil and products	7
Chemicals	4	Chemicals	7
Coke	4	Clothing	6
Crude oil	3	Timber	5
Main sources (1976)		*Main destinations* (1976)	
Soviet Union	18	Soviet Union	18
East Germany	7	West Germany	9
West Germany	7	East Germany	7
Poland	5	Czechoslovakia	4
Iran	5	China	4
United States	5	Poland	4
Czechoslovakia	4	Italy	3
France	4	Hungary	3

Special focus

Five year plan, 1976–80

	Planned growth %pa
Industrial production	11½
Agricultural production	8
Investment	13
External trade turnover	16

San Marino

Republic of San Marino
Repubblica di San Marino

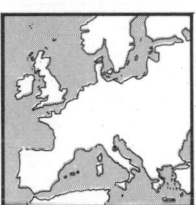

Location Central Italy
23 km inland from Rimini, on the Adriatic Sea;
completely enclosed by Italy. Land-locked
Land Area 61 km² = 24 mi²
Climate Mediterranean
Time 1 hour ahead of GMT (summer time,
2 hours ahead)
Measures Metric system
Monetary unit
San Marino lira (SM L) = 100 centesimi
The San Marino lira is at par with the Italian and Vatican liras, which are
also in use
Rate of exchange (1976 av): free SM L 832.3 = $ 1, SM L 1 503 = £ 1

Summary

Political Republic, founded in 301; there is a treaty of friendship with
Italy. Member of International Court of Justice, Unesco, Unctad and
UPU
Economic Tourism and agriculture are the main sectors; there is some
light industry, and there are proposals that this should be further
developed

People, resources and equipment

Population 1960 15 000*, 1970 18 000*, 1976 19 400*
Growth: 1960–70 1.8* %pa, 1970–76 1.3* %pa
Density (1976): 318* people per km²
Vital statistics (rate per 1 000 people, 1974): births 15.2, deaths 7.7
Cities (population in 000, 1977) San Marino (capital) 4.5
Race Mainly Italian
Language Italian
Religion Mainly Roman Catholic
Education (1974/75) Pupils 2 838, teachers 200
Labour force (1976) 9 596
Personnel (1973) Scientists and engineers: 228
National income per person (1976)
SM L 3.5***mn = $ 4 200*** = £ 2 300***
Telephones (Dec 1976) 5 696, 290 per 1 000 people
Roads (1970) 220 km = 137 mi, 3.6 km per km²
Radio sets (1974) 3 400, 181 per 1 000 people
Television sets (1974) 3 300, 174 per 1 000 people
Motor vehicles (1970) 9 584, 532 per 1 000 people, 44 per km of road

Production, finance and trade

Gross domestic product
1976 est: SM L 70*** bn = $ 85*** mn = £ 50*** mn
Main products Wheat, wine, textiles, cement, stone, paper, tiles, ceramics,
handicrafts
Tourism (1976) Number of visitors 2 435 474, of whom from:
Italy 1 950 823 (80 %), West Germany 197 110 (8 %), France 71 941 (3 %),
Belgium 43 132 (2 %), Switzerland 39 156 (2 %)
Budget (1976) Balanced at SM L 29 148 mn = $ 35 mn = £ 19 mn
External trade There is a customs union with Italy
Main exports Wine, textiles, stone, postage stamps, ceramics

Soviet Union

Union of Soviet Socialist Republics
Soyuz Sovyetskikh Sotsialisticheskikh Respublik

Location Eastern Europe and northern Asia
Occupies the main northern continental mass of
Europe/Asia, with coastlines on the Black Sea to
the south-west, Baltic Sea to the north-west,
Arctic Ocean to the north, and Sea of Okhotsk,
Bering Sea and Sea of Japan to the east.
Norway, Finland, Poland, Czechoslovakia and
Hungary are to the west, Rumania to the
south-west, and Turkey, Iran, Afghanistan,
China and Mongolia to the south; there is a
short border with North Korea in the east and Alaska (United States) is
to the east across the Bering Strait, and Japan to the south-east across
the Sea of Japan

Land Area 22 402 200 km² = 8 649 500 mi²
Usage (1975): agricultural 6 042 070 km² (27 %), of which, arable
2 273 000 km² (10 %), cropland 49 070 km² (0.2 %), pastures
3 720 000 km² (17 %); forests 9 200 000* km (41* %)
Climate Continental, arctic in the north
Weather at Moscow, 156 m altitude
Temperature: hottest month July 13–23 °C,
coldest Jan minus 16–minus 9 °C
Rainfall (av monthly): driest month Mar 36 mm, wettest July 88 mm

Time	Hours ahead of GMT
Chief towns	
Kiev, Leningrad, Moscow, Odessa	3
Archangel, Volgograd, Tbilisi	4
Ashkhabad, Sverdlovsk	5
Alma-Ata, Karaganda, Omsk	6
Novosibirsk, Krasnoyarsk	7
Irkutsk	8
Yakutsk	9
Khabarovsk, Vladivostok	10
Magadan, Yuzhno-Sakhalinsk	11
Petropavlovsk-Kamchatskiy	12
Anadyr	13

Measures Metric system
Monetary unit Rouble (Rub) = 100 kopecks
Rate of exchange (1976 av): free Rub 0.754 = $ 1, Rub 1.36 = £ 1

Summary

Political Communist republic, formed as such in 1917; comprises a union
of 15 republics. Member of UN, Comecon and Warsaw Pact. Two of the
republics, Byelorussian SSR and Ukrainian SSR have separate
membership of the UN
Economic An industrial economy with major agricultural and mining
sectors; crude oil and products were the largest export in 1976,
accounting for about one-quarter of total exports, and machinery
accounted for about one-tenth of exports. Timber, iron and steel and coal
were also major exports. Makes up 70 % of Warsaw Pact countries in
terms of population and gross domestic product, but over 90 % of defence
expenditure

People

Population 1960 214.33 mn, 1970 242.76 mn, 1976 256.67 mn
Growth: 1960–70 1.3 %pa, 1970–76 0.9 %pa
Density (1976): 11 people per km²
Vital statistics (rate per 1 000 people, 1976): births 18.5, deaths 9.5
Households (1970) 58.69 mn; population in households 217.44 mn
(90 % of total)

Regions (Dec 31, 1976)

Union republic	Capital	Area (000 km²)	Population (000)
Armenian SSR	Yerevan	30	2 894
Azerbaijan SSR	Baku	87	5 786
Byelorussian SSR[b]	Minsk	208	9 426
Estonian SSR	Tallin	45	1 447
Georgian SSR	Tbilisi	70	4 999
Kazakh SSR	Alma-Ata	2 717	14 527
Kirghiz SSR	Frunze	198	3 451
Latvian SSR	Riga	64	2 512
Lithuanian SSR	Vilnius	65	3 336
Moldavian SSR	Kishinev	34	3 896
Russian FSSR	Moscow	17 075	135 569
Tajik SSR	Dushanbe	143	3 589
Turkmen SSR	Ashkhabad	488	2 652
Ukrainian SSR	Kiev	604	49 343
Uzbek SSR	Tashkent	447	14 485
USSR	*Moscow*	*22 402*[a]	*257 912*

[a]Includes unallocated area   [b]White Russia

Cities (population in 000, Dec 31, 1975)

City	Pop	City	Pop	City	Pop
Moskva[a] (capital)	7 734	Ufa	923	Tula	506
Leningrad	4 372	Volgograd	918	Frunze	498
Kiev	2 013	Rostov-na-Donu	907	Kishinev	471
Tashkent	1 643	Alma-Ata	851	Zhdanov	467
Baku	1 406	Saratov	848	Tolyatti	463
Kharkov	1 385	Riga	806	Ivanovo	458
Gorky	1 305	Voronezh	764	Astrakhan	458
Novosibirsk	1 286	Zaporozhye	760	Dushanbe	448
Minsk	1 189	Krasnoyarsk	758	Vilnius	447
Kuibyshev	1 186	Krivoi Rog	634	Kemerovo	446
Sverdlovsk	1 171	Lvov	629	Voroshilovgrad	439
Tbilisi	1 030	Yaroslavl	577	Makeyevka	437
Odessa	1 023	Karaganda	570	Penza	436
Omsk	1 002	Krasnodar	543	Ulyanovsk	436
Chelyabinsk	989	Novokuznetsk	530	Nikolaev	436
Dnepropetrovsk	976	Vladivostok	526	Orenburg	435
Donetsk	967	Izhevsk	522	Ryazan	432
Kazan	958	Irkutsk	519	Tomsk	413
Perm	957	Barnaul	514	Tallin	408
Yerevan	928	Khabarovsk	513	Nizhny Tagil	396

[a]Moscow

Race (1970) Russian 53 %, Ukrainian 17 %, Uzbek 4 %,
Byelorussian 4 %, Tatar 2 %, Kazakh 2 %, Armenian 1 %,
Georgian 1 %, Moldavian 1 %, Lithuanian 1 %
Language Russian, Ukrainian and other local languages (see usage)
Usage (mother tongue, 1970): Russian 59 %, Ukrainian 15 %, Uzbek 4 %,
Byelorussian 3 %, Kazakh 2 %, Azerbaijani 2 %, Georgian 1 %,
Armenian 1 %, Lithuanian 1 %, Moldavian 1 %
Religion (1970) Russian Orthodox 13** %, Moslem 11** %,
Roman Catholic 1** %, Jewish 1** %, Russian Baptist Union 1** %
Education (1974/75) Pupils: primary 38 375 000, secondary 5 936 000,
vocational 4 105 600, teacher-training 372 200, higher 4 751 000.
Teachers: primary and secondary 2 415 000, vocational and
teacher-training 212 400, higher (1973/74) 302 000
Labour force 1970:

Economic activity	Number	%
Agriculture	30 761 092	26
Manufacturing, construction, transport and communication	52 771 253	45
Other productive	8 522 089	7
Non-productive (incl not known)	24 973 141	21
Total	117 027 575	100

1976 total: 129 154 000*
Personnel Scientists and engineers (1973): 8 384 000
Physicians (1975): 733 700, 1 per 347 people
Standard of living National income per person (1976):
Rub 2 000*** = $ 2 600*** = £ 1 500***
Consumption per person (1975): energy 5 546 kg coal equivalent,
electricity (production, 1976) 4 329 kW h, newsprint 4.3 kg, steel 554 kg
Newspapers (1974): number 675; circulation 97 664 000,
388 per 1 000 people
Telephones (Dec 1976): 18 000 000, 70 per 1 000 people

Soviet Union

Resources and equipment

Livestock (000, Dec 1976) Cattle 110 346, buffaloes 393, sheep 139 834, pigs 63 055, goats 5 539, horses 6 400*, camels 250*, chickens 747 744, turkeys (1975) 22 000*
Mineral reserves Coal (1971) 3 993 357 mn tonnes
Lignite (1971) 1 720 324 mn tonnes
Crude oil (1975) 8 149 mn tonnes
Natural gas (1975) 20 105 bn cubic metres
Petroleum refinery capacity (1976) 500***mn tonnes
Electrical capacity (1975) 217 484 megawatts,
of which, hydro 40 515 megawatts, nuclear 5 600* megawatts
Hospital beds (1975) 3 009 200, 1 per 85 people
Roads (1975) 1 403 000 km = 872 000 mi, density 0.063 km per km²
Railways (1974) 266 200 km² = 165 400 mi, density 0.012 km per km²
²Including 128 700 km of industrial railways
Inland waterways (1975) 145 400 km = 90 300 mi
Oil pipelines (1975) 56 500 km = 35 100 mi
Ships (registered, 1977)

Type	Number	Gross tons
General cargo	1 790	7 458 849
Oil tankers	497	4 385 489
Fishing	3 964	3 479 179
Fish factories	573	2 960 889
Other	1 343	3 153 885
Total	8 167	21 438 291

Ports *Barents Sea* Murmansk *White Sea* Arkhangelsk, Mezen, Onega *Baltic Sea* Klaipeda, Leningrad, Riga, Tallin, Ventspils, Vyborg *Danube* Izmail, Kilia, Reni *Black Sea* Batumi, Ilichevsk, Novorossijsk, Odessa, Poti, Sevastopol, Sochi, Sukhumi, Tuapse, Yalta
Sea of Azov Berdyansk, Zhdanov

Airports (passengers, 000, 1976) Sheremetyevo (Moscow): departures 1 996, arrivals 1 953; also Domodedovo and Vnukovo (Moscow), Leningrad, Kiev, Omsk, Odessa, Vladivostok and (1977) 40 other airports with scheduled flights

Durable equipment (at end-year)	000	no per 1 000 people	
Radio sets (1974)	116 100	461	no per
Television sets (1976)	57 200	222	km of road
Passenger cars (1975)	4 700**	18**	3.3**
Commercial vehicles (1975)	5 100**	20**	3.6**

Production

Net material product 1976: Rub 382.0 bn = $ 506.6 bn = £ 280.9 bn
Growth in real terms: 1960–70 7.2 %pa, 1970–76 5.6 %pa
Gross domestic product 1976: Rub 530***bn = $ 700***bn = £ 400***bn
Structure of net material product (1974)

By origin	Rub bn	% of gdp
Agriculture	63.1	17
Manufacturing, mining, quarrying, electricity, gas and water	200.8	53
Construction	42.7	11
Distribution and hotels	51.3	13
Transport and communication	24.1	6
Total	382.0	100
By type of expenditure		
Final consumption	279.8	73
Gross fixed capital formation²	100.1	26
Net exports of goods and services	2.1	1
Total	382.0	100

²Including stock investment

Production indices (1970 = 100)	1960	1965	1970	1973	1974	1975	1976	Growth %pa 1960–70	1960–65	1965–70	1970–76
Agricultural	72	80	100	113	106	101	110	3.3	2.0	4.6	1.6
Industrial	44	67	100	123	133	143	150	8.6	8.8	8.3	7.0
of which, mining	56	77	100	116	121	127	132	6.0	6.6	5.4	4.7
manufacturing	44	67	100	124	135	145	152	8.6	8.8	8.3	7.2
Main products (000 t)											
Agriculture											
Wheat	64 299	59 686	99 734	109 784	83 913	66 224	96 900²	4.5	−1.5	10.8	−0.5
Barley	16 021	20 304	38 172	55 044	54 208	35 808	69 539	9.1	4.9	13.4	10.5
Oats	11 999	6 186	14 203	17 516	15 302	12 495	18 113	1.7	−12.4	18.1	4.1
Rye	16 324	16 228	12 972	10 759	15 223	9 064	13 991	−2.3	−0.1	−4.4	1.3
Maize	9 823	8 030	9 428	13 216	12 104	7 328	10 138	−0.4	−4.0	3.3	1.2
Rice	190	583	1 279	1 765	1 913	2 009	2 001	21.0	25.1	17.0	7.7
Millet	3 230	2 205	2 100	4 416	2 907	1 125	3 198	−4.2	−7.4	−1.0	7.2
Potatoes	84 374	88 676	96 783	108 200	81 022	88 703	85 102	1.4	1.0	1.8	−2.1
Sugar, raw value	5 717*	9 196*	9 293*	9 538*	7 826	7 702*	7 350*	5.0*	10.0*	0.2*	−3.8*
Peas, dry	1 270	4 625	5 000**	6 066	6 510	4 036	6 886	14.7**	29.5	1.6**	5.5**
Vetches	1 000**	1 277	1 700**	1 574	1 527	1 100*	na	5.4**	5.0**	5.9**	−8.3**f
Lupins	200**	590	530**	513	441	350*	na	10.0**	24.0**	−2.1**	−8.0**f
Onions	200**	431	707	800*	910*	700*	1 135	13.5**	16.6**	10.4	8.2
Tomatoes	2 400**	2 473	3 064	3 500*	3 810*	3 590*	4 637	2.5**	0.6**	4.4	7.1
Watermelons	2 600**	2 700**	2 800	3 500*	3 100*	3 130*	3 160*	0.7**	0.8**	0.7**	2.0*
Grapes	1 871	3 723	4 011	4 583	4 608	5 400	5 400*	7.9	14.8	1.5	5.1*
Wine	777	1 339	2 684	2 070	2 680	2 965	2 965*	13.2	11.5	14.9	1.7*
Soyabeans	225*	421	595	424	360	780	480	10.2*	13.4*	7.2	−3.5
Sunflowerseed	3 967	5 449	6 144	7 385	6 784	4 993	5 277	4.5	6.6	2.4	−2.5
Linseed	400**	449	471*	407	520*	380*	337	1.6**	2.3**	1.0*	−5.4*
Cottonseed	2 804*	3 725*	4 416*	4 970*	5 400*	5 100*	5 383*	4.6*	5.9*	3.5*	3.3*
Tea	38	48	67	75	81	86	92	5.9	5.1	6.7	5.4
Tobacco	178	219	266	312	318	303	303	4.1	4.2	4.0	2.2
Flax	425	480	456	443	402	493	503*	0.7	2.5	−1.0	1.6*
Cotton	1 485*	1 937	2 129	2 424	2 661	2 649	2 590	3.7*	5.5*	1.9	3.3
Milk, cow	60 800*	71 763	82 400	87 800	91 300	90 300	88 658	3.1*	3.4*	2.8	1.2
Milk, goat	800**	700*	458	390*	350*	300*	300*	−5.4**	−2.6**	−8.1*	−6.8*
Butter	848	1 184	1 067	1 350	1 360	1 320	1 356	2.3	6.9	−2.0	4.1
Cheese	450**	690	1 051	1 211	1 292	1 251	1 363	8.8**	8.9**	8.8	4.4
Lard	480**	550*	620*	705*	755*	790*	600*	2.6*	2.7**	2.4**	−0.5*
Eggs	1 500*	1 600*	2 245	2 826	3 054	3 176	3 059	4.1*	1.3*	7.1*	5.3
Wool	214	214	251	258	277	280	258	1.6	0.0	3.3	0.5
Cattle hides	580**	600**	620**	664*	691*	695*	671*	0.7**	0.7**	0.7**	1.3**
Sheepskins	150**	140**	120**	114*	117*	128*	120*	−2.2**	−1.4**	−3.0**	0.0**
Pigskins	160**	150**	140**	180*	186*	186*	na	−1.3**	−1.3**	−1.3**	5.9**f
Beef and veal	3 252	3 917	5 393	5 873	6 384	6 473	6 120*	5.2	3.8	6.6	2.1*
Sheep and goat meat	1 019	1 013	1 002	954	974	975	950*	−0.2	−0.1	−0.2	−0.9*
Pigmeat	3 276	4 142	4 543	5 081	5 515	5 749*	4 502*	3.3	4.8	1.9	−0.2*

For footnotes see page 256

Soviet Union

Main products (000 t)	1960	1965	1970	1973	1974	1975	1976	Growth %pa 1960–70	1960–65	1965–70	1970–76
Agriculture											
Poultrymeat	766	696	1 071	1 295	1 420	1 525	1 414	3 4	−1.9	9.0	4.7
Fish catch	3 051	5 100	7 252	8 619	9 236	9 936	10 134	9.0	10.8	7.3	5.7
Timber (000 m³)	369 500	377 339	385 000	387 600	387 600*	395 054	384 534	0.4	0.4	0.4	0.0
Energy											
Total energy (000 tce)	685 000*	970 000*	1 250 000*	1 456 460	1 539 820	1 650 470	na	6.2*	7.2*	5.2*	5.7*
Coal	355 918	397 645	432 715	461 223	473 374	484 668	494 000	2.0	2.2	1.7	2.2
Lignite	134 206	147 444	144 745	153 467	157 179	160 216	160 000	0.8	1.9	−0.4	1.7
Crude oil	147 859	242 888	353 039	429 037	458 948	491 000	520 000	9.1	10.4	7.8	6.7
Natural gas (mn m³)	45 303	127 666	197 945	236 326	260 553	289 000	321 000	15.9	23.0	9.2	8.4
Manufactured gas (mn m³)	24 568	29 399	32 899	35 269	35 963	36 067	36 370	3.0	3.7	2.3	1.7
Electricity (mn kW h)	292 274	506 672	740 926	914 606	975 754	1 038 607	1 111 000	9.7	11.6	7.9	7.0
of which, hydro (mn kW h)	50 913	81 434	124 377	122 345	132 030	125 987	na	9.3	9.8	8.8	0.2ᶠ
nuclear (mn kW h)	na	1 647ᶜ	3 500	7 500*	9 000*	11 200*	na	na	na	20.7ᵈ	26.2*ᶠ
Mining											
Iron ore (Fe content)	54 075	80 996	106 058	118 151	123 155	127 483	130 700	7.0	8.4	5.5	3.5
Antimony ore (Sb content)	5.7**	6.2**	6.7**	7.1**	7.3**	7.5**	7.7**	1.6**	1.6**	1.6**	2.4**
Bauxite	3 500**	4 700**	4 300**	4 300**	4 300**	4 400**	4 500**	2.1**	6.1**	−1.8**	0.8**
Chromium ore (oxide content)	380**	600**	735**	800**	820**	870**	na	7.0**	9.6**	4.1**	3.4**ᶠ
Copper ore (Cu content)	500**	750**	925**	1 100**	1 100**	1 100**	na	6.3**	8.5**	4.3**	3.5**ᶠ
Lead ore (Pb content)	320**	350**	440**	470**	475**	480**	500**	3.2**	1.8**	4.7**	2.2**
Magnesite	na	na	1 500**	1 710**	1 730**	1 800**	na	na	na	na	3.7**ᶠ
Manganese ore (Mn content)	1 900**	2 484	2 446	2 839	2 847	2 951	na	2.6**	5.5**	−0.3	3.8ᶠ
Mercury	0.86**	1.38**	1.65**	1.79**	1.86**	1.90**	1.93**	6.7**	9.9**	3.6**	2.6**
Molybdenum ore (Mo content)	5.0**	6.2**	7.7**	8.5**	8.8**	9.1**	9.3**	4.4**	4.4**	4.4**	3.2**
Nickel ore (Ni content)	53**	80**	109**	135**	145**	152**	160**	7.5**	8.6**	6.4**	6.6**
Tungsten conc (oxide content)	5.7**	7.2**	8.5**	9.3**	9.6**	9.8**	10.0**	4.0**	4.8**	3.4**	2.7**
Vanadium ore (V content)	na	na	3.45**	3.85**	2.93**	3.20**	na	na	na	na	−1.5**ᶠ
Zinc ore (Zn content)	370**	470**	610**	670**	680**	690**	720**	5.1**	4.9**	5.3**	2.8**
Silver	0.80**	0.96**	1.18**	1.27**	1.31**	1.34**	1.37**	4.0**	3.7**	4.2**	2.5**
Gold (000 kg)	342**	156**	202**	220**	227**	233**	240**	−5.1**	−14.5**	5.3**	2.9**
Diamonds (000 CM)	1 000**	5 000**	7 850**	9 500**	9 500**	9 700**	9 900**	23.0**	38.0**	9.4**	4.0**
Phosphate rock	10 000**	13 470**	17 780**	21 250**	22 500**	24 120**	na	5.9**	6.1**	5.7**	6.3**
Potash (oxide content)	1 250**	2 368**	4 087**	5 900**	6 100**	6 050**	na	12.6**	13.6**	11.5**	8.2**ᶠ
Salt	6 703	9 485	12 428	12 860	13 356	14 300	na	6.4	7.2	5.6	2.8ᶠ
Asbestos	500**	750**	1 070**	1 280**	1 360**	1 900**	2 290**	7.9**	8.4**	7.4**	13.5**
Manufacturing											
Fishmeal	72	203	368	489	488	638	na	17.7	23.0	12.6	11.6ᶠ
Margarine	431	670	762	883	997	999	na	5.9	9.2	2.6	5.6ᶠ
Beer (000 hl)	24 979	31 690	41 857	50 809	54 003	57 050	na	5.3	4.9	5.7	6.4ᶠ
Cigarettes (mn units)	244 833	304 400	322 687	362 527	369 309	364 266	na	2.8	4.4	1.2	2.4ᶠ
Cotton yarn	1 169	1 292	1 435	1 535	1 557	1 573	na	2.1	2.0	2.1	1.9ᶠ
Cotton fabrics (mn m²)	5 390*	5 975	6 653	7 137	7 196	7 240	7 400	2.1*	2.1*	2.2	1.8
Flax yarn	201	209	252	267	267	260	na	2.3	0.8	3.8	0.6ᶠ
Wool yarn	221	236	350	393	408	417	na	4.7	1.3	8.2	3.6ᶠ
Man-made fibres	211	407	623	830	887	955	na	11.4	14.0	8.2	8.9ᶠ
Wood pulp	3 213	4 252	6 679	7 767	8 182	8 180*	na	7.6	5.7	9.5	4.1*ᶠ
Newsprint	434	744	1 101	1 298	1 333	1 361	1 400*	9.7	11.4	8.1	4.1*
Other paper	2 793	3 936	5 600	6 592	6 865	7 222	7 516*	7.2	7.1	7.3	5.0*
Tyres (000 units)	na	19 890	22 351	28 672	32 117	35 154	na	na	na	2.4	9.5ᶠ
Sulphurᵇ	3 700**	4 500**	5 000**	7 600**	7 800**	8 200**	na	3.1**	4.0**	2.1**	10.4**ᶠ
Sulphuric acid	5 398	8 518	12 059	14 855	16 663	18 645	20 014	8.4	9.5	7.2	8.8
Caustic soda	704	1 199	1 783	2 020	2 174	2 395	2 604	9.7	11.2	8.3	6.5
Soda ash	1 793	2 728	3 485	4 149	4 484	4 692	4 842	6.9	8.7	5.0	5.6
Plastics and resins	312	803	1 673	2 320	2 493	2 842	3 061	18.3	20.8	15.8	10.6
Fertilisers, nitrogenous	1 003	2 712	5 423	7 241	7 856	8 535	na	18.4	22.0	14.9	9.5ᶠ
Fertilisers, phosphate	810*	1 560*	2 449	2 982	3 504	4 103	na	11.9*	14.0*	9.9*	10.9ᶠ
Fertilisers, potash	1 084	2 368	4 087	5 918	6 586	7 944	8 500	14.2	16.9	11.5	13.0
Cement	45 520	72 388	95 248	109 521	115 145	122 057	124 000	7.7	9.7	5.6	4.5
Coke	56 233	67 462	75 404	81 401	82 641	83 543	na	3.0	3.7	2.2	2.1ᶠ
Pig iron	46 757	66 184	85 933	95 933	99 868	102 968	105 000	6.3	7.2	5.4	3.4
Crude steel	65 294	91 021	115 889	131 481	136 230	141 344	144 650	5.9	6.8	5.0	3.7
Aluminium	640**	840**	1 100**	1 360**	1 430**	1 530**	1 600**	5.6**	5.6**	5.6**	6.4**
Magnesium	25**	33**	50**	57**	60**	60**	63**	7.0**	5.7**	8.7**	3.9**
Watches and clocks (000 units)	26 000	30 600	40 171	47 453	50 585	55 065	57 900	4.4	3.3	5.6	6.3
Radio sets (000 units)	4 165	5 160	7 815	8 615	8 753	8 376	8 443	6.5	4.4	8.6	1.3
Television sets (000 units)	1 726	3 655	6 682	6 271	6 569	6 960	7 060	14.5	16.2	12.8	0.9
Washing machines (000 units)	895	3 430	5 243	2 987	3 075	3 286	3 509	19.3	30.8	9.0	−6.5
Refrigerators (000 units)	529	1 675	4 140	5 423	5 426	5 577	5 834	22.9	25.9	20.0	5.9
Passenger cars (000 units)	139	201	344	917	1 119	1 201	1 239	9.5	7.7	11.3	23.8
Commercial vehicles (000 units)	385	415	572	686	727	763	786	4.0	1.5	6.6	5.4
Merchant vessels (000 grt)ᵍ	na	na	na	419	394	396	616	na	na	na	na
Construction											
Dwellings (000 units)ᵉ	2 591	2 227	2 266	2 276	2 231	2 228	2 200	−1.3	−3.0	0.3	−0.5

ᵃProvisional estimate for 1977: 92 000* thousand tonnes　ᵇIncluding native sulphur, content of pyrites and sulphur obtained as a by-product　ᶜ1966　ᵈ1966–70
ᵉIncluding extensions, restorations and conversions　ᶠ1970–75　ᵍInformation is not complete

Soviet Union

Transport traffic	1960	1970	1976	Growth %pa	
				1960–70	1970–76
Passenger-kilometres (mn)					
Road[a]	60 962	202 482	303 568[f]	12.8	8.4[g]
Rail	170 759	265 406	312 517[f]	4.5	3.3[g]
Air	12 111	78 226	130 529	20.5	8.9
of which,					
international	na	2 721	6 828	na	16.6
Cargo: tonne-kilometres (mn)					
Road	119 723[b]	220 834	354 800	9.1[c]	8.2
Rail	1 504 400	2 494 721	3 295 400	5.2	4.7
Water	99 602	173 984	222 700	5.7	4.2
Air	563	1 877	2 698	12.8	6.2
of which,					
international	na	151[d]	241	na	16.9[e]
Sea: tonnes (mn)					
Goods loaded	38.8	107.0	119.8[f]	10.7	2.3[g]
Goods unloaded	5.9	14.4	35.5[f]	9.3	19.7[g]
Tourism					
Number of visitors (000)	712	2 059	3 879	11.2	11.1

[a]Excluding private car transport [b]1963 [c]1963–70 [d]1973 [e]1973–76 [f]1975 [g]1970–75

Finance and trade

Consumer price index (1970 = 100) 1976 99.7;
growth 1970–76 −0.1 %pa
Budget (1976) Revenue: Rub 225 400 mn = $ 298 900 mn = £ 165 700 mn
Expenditure: Rub 225 300 mn = $ 298 800 mn = £ 165 700 mn
of which, education, science & culture 15 %, defence 8 %,
social services 21 %
External trade (1976)
Imports (fob): Rub 28 731 mn = $ 38 105 mn = £ 21 126 mn
Exports: Rub 28 022 mn = $ 37 164 mn = £ 20 604 mn

Main imports (1975)	% of total	Main exports (1975)	% of total
Machinery	21	Crude oil	16
Iron and steel	9	Machinery	11
Cereals	8	Petroleum products	9
(of which, wheat 4)		Iron and steel	8
Sugar	6	Timber	4
Chemicals	6	Coal	4
Ships	4	Motor vehicles	4
Clothing	4	Non-ferrous metals	3
Textile yarns & fabrics	4	Wood pulp	3
Motor vehicles	3	Chemicals	2
Footwear	2	Iron ore	2
Fruit and vegetables	2	Cereals	2
Railway vehicles	2	Natural gas	2
Textile fibres	2		
Main sources (1976)[a]		*Main destinations* (1976)[b]	
East Germany	10	East Germany	11
Poland	9	Poland	10
Czechoslovakia	8	Czechoslovakia	8
Bulgaria	8	Bulgaria	8
United States	7	Hungary	6
West Germany	7	Cuba	5
Hungary	6	West Germany	4
Cuba	5	Italy	4
Japan	5	Finland	4
Finland	3	Jugoslavia	3
France	3	United Kingdom	3
Jugoslavia	3	France	3
Rumania	3	Rumania	3
Italy	2	Japan	3
Canada	2	Mongolia	2

[a]Eastern Europe 43 % [b]Eastern Europe 47 %

Special focus

Warsaw Pact and Nato, 1976

	Soviet Union	Warsaw Pact	Nato[a]	United States
Population (mn)	257	364	496	215
Armed forces (mn)	3.6	4.7	4.2	2.1
Defence expenditure ($ bn)	127***	136***	146*	103*
Gross domestic product ($ bn)	700***	1 004***	3 020*	1 702*
Production:				
crude steel (mn tonnes)	145	199	243	116
crude oil (mn tonnes)	520	538	502	402
electricity (bn kW h)	1 111	1 475	3 424	2 118

[a]Excluding France and Greece which have withdrawn from the military side of Nato

Spain
Kingdom of Spain

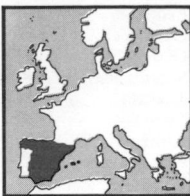

Location South-west Europe
Occupies the main part of the Iberian peninsula,
with coastlines to the west on the Atlantic Ocean,
north on the Bay of Biscay and south and east on
the Mediterranean Sea; France is to the north
and Portugal to the west. The state includes the
Balearic and Canary Islands, and Spanish areas
of North Africa: Ceuta, Melilla, Alhucemas,
Chafarinas Is and Peñón de Vélez (or Peñón de la
Gomera)
Land Area 504 782 km² = 194 897 mi²
Usage (1975): agricultural 319 210 (63 %), of which, arable 158 210 (31 %),
cropland 50 120 (10 %), pastures 110 880 (22 %); forests 149 440 (30 %)
Climate Mediterranean in east and south, temperate in north-west
Weather at Madrid, 660 m altitude
Temperature: hottest month July 17–31 °C, coldest Jan 1–8 °C
Rainfall (av monthly): driest month July 11 mm, wettest Oct 53 mm
Time 1 hour ahead of GMT (summer time, 2 hours ahead)
Measures Metric system; some local measures are also in use, including:
capacity 1 fanega = 55.5 litres = 1.53 bushels
weight (*mass*) 1 libra = 16 onzas = 460 grams = 1.0141 pounds
1 quintal = 4 arrobas = 100 libras = 46 kilograms = 101.41 pounds
Monetary unit Peseta (Pa) = 100 céntimos
Rate of exchange (1976 av): free Pa 66.90 = $ 1, Pa 120.8 = £ 1

Summary

Political Parliamentary monarchy; in 1936 a one-party state was formed
by General Franco, but a king was restored on November 22, 1975,
following the death of General Franco. There have been disputes with
Morocco concerning the Spanish areas of North Africa; these are included
in figures here as part of Spain, unless otherwise specified. There have also
been disputes with United Kingdom concerning Gibraltar; the former
Spanish overseas territory of Ifni was ceded to Morocco in 1969 and
Spanish Sahara was ceded to Morocco and Mauritania formally from
end-February 1976. There are internal tensions involving especially the
Basque (north) and also the Catalan (north-east) minorities; Catalonia
was given some regional autonomy from Sept 30, 1977, and a regional
Basque government is being established. Member of UN, OECD and
Council of Europe
Economic A mixed economy, with an important agricultural sector
making up one-tenth of gross domestic product, and a large industrial
sector which has developed following a period of very high growth.
Manufacturing absorbs 26 % of the labour force compared with 21 % for
agriculture. Exports account for only one-seventh of gross domestic
product and tourism is very important, with gross receipts at about
one-third the level of exports of goods

People

Population 1960 30.30*mn, 1970 33.78*mn, 1976 35.97*mn
Growth: 1960–70 1.1* %pa, 1970–76 1.1* %pa
Density (1976): 71* people per km²
Vital statistics (rate per 1 000 people, 1976): births 17.7*, deaths 8.0*
Regions (provinces, population in 000, Dec 1976; total of 36.45 mn)

Alava[d]	246	Granada	743	Las Palmas[b]	662
Albacete	332	Guadalajara	146	Pontevedra	857
Alicante	1 079	Guipúzcoa[d]	691	Salamanca	352
Almería	390	Huelva	403	Santa Cruz	
Avila	189	Huesca	212	de Tenerife[b]	680
Badajoz	639	Jaén	648	Santander	495
Baleares[a]	611	León	530	Segovia	149
Barcelona[c]	4 485	Lérida[c]	348	Sevilla	1 386
Burgos	349	Logroño	243	Soria	103
Cáceres	425	Lugo	409	Tarragona[c]	490
Cádiz	947	Madrid	4 576	Teruel	153
Castellón	415	Málaga	928	Toledo	466
Ciudad Real	479	Murcia	891	Valencia	1 969
Córdoba	715	Navarra[d]	491	Valladolid	458
Coruña, La	1 065	Orense	432	Vizcaya[d]	1 178
Cuenca	220	Oviedo	1 112	Zamora	230
Gerona[c]	448	Palencia	184	Zaragoza	801

[a]Balearic islands [b]Province of the Canary islands which have a total population of
1.34 mn [c]Part of Catalonia [d]Vizcaya, Guipúzcoa and Alava are the main Basque
areas, with Navarra also to some extent Basque country

Spain

Cities (population in 000, 1974)

Madrid (capital)	3 520	Alicante	213	Jérez de la	
Barcelona	1 810	Granada	203	Frontera	150[d]
Valencia	713	Vigo	197	Cartagena	147[d]
Sevilla[a]	589	La Coruña[c]	193	Cádiz	142
Zaragoza[b]	547	Gijón	188[d]	Salamanca	140
Bilbao	458	San Sebastián	178	Tarrasa	139
Málaga	403	Vitoria	169	Burgos	136
Las Palmas (Canarias)	328	Pamplona	169	Almería	126
Valladolid	275	Oviedo	164	Elche	123
Palma de Mallorca		Badalona	163[d]	León	119
(Baleares)	267	Santander	162	Baracaldo	109[d]
Córdoba	250	Sabadell	159[d]	Castellón	108
Hospitalet	242[d]	Santa Cruz		Santa Coloma	
Murcia	241	de Tenerife	158	de Grammanet	107[d]

[a]Seville [b]Saragossa [c]Corunna [d]1970

Race (1961) Spanish 73 %, Catalan 16 %, Galician 8 %, Basque 2 %
Language Spanish (Castilian), Catalan, Galician, Basque
Religion (1976) Roman Catholic 95* %
Education (1975/76) Pupils: primary 3 653 320, secondary (1974/75) 2 461 310, vocational (1974/75) 456 816, higher (1974/75) 453 816.
Teachers (1971/72): primary 142 987, secondary 60 794, vocational (1974/75) 29 739, higher (1974/75) 28 499
Labour force (1976)

Economic activity	Number	%
Agriculture	2 757 903	21
Mining and quarrying	102 264	1
Manufacturing	3 428 562	26
Electricity, gas and water	80 098	1
Construction	1 382 617	10
Transport and communication	665 806	5
Other	4 863 689	37
Total	13 280 939	100

Personnel Scientists and engineers (1967): 188 000*
Physicians (registered, 1974): 52 559, 1 per 670 people
Standard of living
National income per person (1976): Pa 178 150 = $ 2 663 = £ 1 475
Consumption per person (1975): energy 2 147 kg coal equivalent, electricity (production, 1976) 2 519 kW h, newsprint 5.8 kg, steel 284 kg
Newspapers (1974): number 115; circulation 3 396 000, 96 per 1 000 people
Telephones (Dec 1976): 8 597 800, 238 per 1 000 people

Resources and equipment

Livestock (000, 1976) Cattle 4 408, sheep 15 745, pigs 8 583, goats 2 339, horses 268, chickens 53 661
Mineral reserves Coal (1970) 2 370 mn tonnes
Lignite (1970) 1 192 mn tonnes
Crude oil (1975) 35 mn tonnes
Natural gas (1975) 11 bn cubic metres
Uranium (1976) 15 000 tonnes
Petroleum refinery capacity (1975) 69* mn tonnes
Electrical capacity (1975) 24 534* megawatts, of which, hydro 11 955* megawatts, nuclear 1 120 megawatts
Hospital beds (1973) 180 547, 1 per 193 people
Roads (1976) 145 328 km = 90 303 mi, density 0.29 km per km²
Railways (1975) 15 839 km = 9 842 mi, density 0.031 km per km²
Oil pipelines (1975) 1 102 km = 685 mi
Ships (registered, 1977) 2 726, total of 7 186 081 gross tons
Ports (goods traffic, 000 tonnes, 1976)

	International		Coastwise		Total
	loaded	unloaded	loaded	unloaded	
Bilbao	1 961	12 581	3 137	2 116	19 795
Barcelona	2 876	5 036	2 964	6 102	16 978
Cartagena	689	7 329	5 064	967	14 049
Santa Cruz de Tenerife[a]	1 693	6 498	3 917	1 524	13 632
Algeciras	582	5 920	5 117	985	12 604
Gijón Musel	698	8 094	1 134	2 625	12 551
Huelva	2 008	4 045	3 125	609	9 787
Corunna	464	4 524	2 038	348	7 374
Avilés	822	1 296	1 148	2 187	5 453
Las Palmas[a]	449	1 326	578	2 457	4 810

[a]Canary Islands

Airports (passengers, 000, 1976)

	departures	arrivals
Madrid	4 196	4 222
Palma de Mallorca	3 174	3 154
Barcelona	2 339	2 339
Las Palmas	1 906	1 908
Santa Cruz de Tenerife	1 407	1 399
Málaga	1 260	1 241
Alicante	907	904

Also (1977) 27 other airports with scheduled flights

Durable equipment (Dec 1976)	000	no per 1 000 people	
Radio sets	8 100*	224*	no per
Television sets	7 425	205	km of road
Passenger cars	5 351	148	37
Commercial vehicles	1 092	30	7.5

Production

Gross domestic product 1976: Pa 6 999 bn = $ 104 619 mn = £ 57 939 mn
Growth in real terms: 1960–70 7.5 %pa, 1970–76 4.9 %pa
Structure of gross domestic product

By origin (1975)

	Pa bn	% of gdp
Agriculture	547	9
Manufacturing, mining and utilities	1 769	30
Construction	492	8
Distribution and hotels	920	16
Transport and communication	351	6
Other	1 831	31
Total	5 910	100

By type of expenditure (1976)

Government final consumption	720	10
Private final consumption	4 869	70
Stock investment	102	1
Gross fixed capital formation	1 606	23
Exports of goods and services	976	14
less Imports of goods and services	−1 273	−18
Total	6 999	100

Production indices (1970 = 100)	1960	1970	1976	Growth %pa	
				1960–70	1970–76
Agricultural	72	100	120	3.3	3.2
Industrial	33	100	148	11.7	6.8

Main products (000 t)

Agriculture	1960	1970	1976	1960–70	1970–76
Wheat	3 528	4 127	4 436	1.6	1.2
Barley	1 562	3 103	5 473	7.1	9.9
Maize	1 012	1 848	1 545	6.2	−3.0
Oats	431	393	528	−0.9	5.0
Potatoes	4 620	5 301	5 633	1.4	1.0
Sugar, raw value	516	781	1 480	4.2	11.2
Tomatoes	1 148	1 809	2 078	4.6	2.3
Onions	731	890	887	2.0	−0.1
Chillies and peppers	330**	395	462	1.8**	2.6
Melons	530**	608	815	1.4**	5.0
Grapes	3 140	4 140	4 078	2.8	−0.2
Wine	2 126	2 560	2 475	1.9	−0.5
Apples	262	484	1 007	6.3	13.0
Pears	92	240	489	10.0	12.6
Peaches	93	229	447	9.4	11.8
Oranges	1 400**	1 950	1 751	3.4**	−1.8
Mandarins, etc.	129**	310	656	9.2**	13.3
Lemons	88	97	233	1.0	15.7
Bananas	303	421	327	3.3	−4.1
Almonds	140**	166	318	1.7**	11.4
Sunflowerseed	2	159	312	54.9	11.9
Olives	2 367	2 107	2 139	−1.2	0.2
Olive oil	464	480	442	0.3	−1.4
Tobacco	30	26	29	−1.3	1.8
Cotton	72	55	48	−2.7	−2.2
Milk	2 075	4 456	5 374	7.9	3.2
Eggs	162	464	617	11.1	4.9
Wool	14.5	13.5	11.4	−0.7	−2.8
Beef and veal	160	307	416*	6.7	5.2*
Pigmeat	258	492	649*	6.7	4.7*
Mutton and lamb	123	140	146	1.3	0.7
Poultrymeat	13	316	696	3.8	14.1
Fish catch	970	1 538	1 483*	4.7	−0.6*
Timber (000 m³)	12 490	13 653	12 122	0.9	−2.0

Spain

Main products (000 t)	1960	1970	1976	Growth %pa	
				1960–70	1970–76
Energy					
Total energy (000 tce)	16 700	15 980	20 370e	−0.4	5.0f
Coal	13 783	10 751	10 483	−2.5	−0.4
Lignite	1 762	2 831	4 140	4.9	6.5
Crude oil	64	151	1 982	9.0	53.6
Petroleum products	6 005	30 590	41 500*e	17.7	6.3*f
Manufactured gas (mn m³)	1 299c	2 439	2 656	9.4d	1.4
Electricity (mn kW h)	18 614	56 490	90 595	11.7	8.2
of which,					
hydro (mn kW h)	15 624	27 959	28 750*e	6.0	0.5*f
nuclear (mn kW h)	—	924	10 000*e	na	61.0*f
Mining					
Iron ore (Fe content)	2 798	3 514	3 800	2.3	1.3
Copper ore (Cu content)	8.2	9.5	19.4	1.5	12.6
Lead conc (Pb content)	73	73	62	0.0	−2.6
Magnesite	48	222	259e	16.5	3.1f
Mercury	1.50*	1.41	1.62e	−0.6*	2.8f
Tungsten conc (oxide content)	0.56	0.51	0.49e	−0.8	−1.0f
Uranium (U content)	0.060*c	0.051	0.170	−2.3*d	22.2
Zinc conc (Zn content)	86	98	81	1.3	−3.2
Potash (oxide content)	289	598	530e	7.5	−2.4f
Salt	1 391	2 080	2 257g	4.1	2.1h
Manufacturing					
Beer (000 hl)	3 433	12 307	16 620e	13.6	6.2f
Cigarettes (mn units)	26 334	50 494	54 510e	6.7	1.5f
Cotton yarn	111	96	71	−1.4	−4.9
Cotton fabrics	93	90	64	−0.3	−5.6
Man-made fabrics	60	118	201	7.0	9.3
Wood pulp	137	602	891e	16.0	8.1f
Newsprint	72	115	106	4.8	−1.3
Other paper	290**	1 166	1 864e	15.0**	9.8f
Footwear[i] (mn pairs)	31	86	166	10.7	11.6
Synthetic rubber	—	39	78	na	12.3
Tyres (000 units)	1 201	8 909	14 000*e	22.2	7.8*
Sulphur[a]	1 070	1 242	1 227e	1.5	−0.2f
Sulphuric acid	1 132	2 309	2 414	7.4	0.7
Nitric acid	103	539	899	18.0	8.9
Caustic soda	137	254	410	6.4	8.3
Soda ash	125	333	473e	10.3	7.3f
Plastics and resins	18	392	847	36.0	13.7
Fertilisers, nitrogenous	96*	592	825e	20.0*	6.8f
Fertilisers, phosphate	300*	487	458e	5.0*	−1.2f
Cement	5 234	16 702	25 291	12.3	7.1
Coke	2 574	4 029	4 777e	4.6	3.5f
Pig iron	1 924	4 278	6 952	8.3	8.4
Crude steel	1 919	7 394	10 910	14.4	6.7
Aluminium	29	120	210	15.4	9.8
Copper[b]	2**	86	150	46.0**	9.7
Zinc	72	88	158	2.1	10.2
Radio sets (000 units)	245	433	487e	5.9	2.4f
Television sets (000 units)	39	618	528e	31.8	−3.1f
Passenger cars (000 units)	42	455	755	26.8	8.8
Commercial vehicles (000 units)	17	77	110	16.2	6.1
Merchant vessels (000 grt)	161	926	1 624	19.1	9.8
Construction					
Dwellings (000 units)	128	298	370e	8.8	4.4f

[a]Content of pyrites [b]Primary only [c]1963 [d]1963–70 [e]1975 [f]1970–75 [g]1974
[h]1970–74 [i]Leather only

Transport traffic	1960	1970	1976	Growth %pa	
				1960–70	1970–76
Passenger-kilometres (mn)					
Road	26 780	84 257	128 706	12.1	7.3
Rail	7 341	13 293	16 684	6.1	3.9
Air	782	5 874	11 130	22.3	11.2
Cargo: tonne-kilometres (mn)					
Road	na	51 700	77 800	na	7.0
Rail	6 059	9 341	10 767	4.4	2.4
Air	8	107	290	29.0	18.0
Sea: tonnes (mn)					
Goods loaded	11.64	14.52	23.81	2.2	8.6
Goods unloaded	14.20	52.92	84.21	14.0	8.0
Tourism					
Number of visitors (000)	5 275	24 105	30 014	16.4	3.7
Gross receipts ($ mn)	297	1 681	3 083	18.9	10.6

Finance and trade

Price and wage indices	1960	1970	1976	Growth %pa	
(1970 = 100)				1960–70	1970–76
Consumer prices	54.9	100.0	207.9	6.2	13.0
Wholesale prices	71.5	100.0	190.8	3.4	11.4
Wages (earnings)	32.6a	100.0	284.7	15.0b	19.1
Share prices	37.3	100.0	126.2	10.4	4.0
Money stock (end-year, Pa bn)	205	739	2 390	13.7	21.6

[a]1962 [b]1962–70

Budget (1976) Revenue: Pa 884.0 bn = $ 13 214 mn = £ 7 318 mn
Expenditure: Pa 912.4 bn = $ 13 638 mn = £ 7 553 mn,
of which (1975), education 15 %, defence 15 %, pensions 8 %

Balance of payments ($ mn)	1972	1973	1974	1975	1976
Balance of goods (fob)	−2 316	−3 503	−7 047	−7 388	−7 305
Balance of services	+2 028	+2 680	+2 672	+2 757	+1 874
Balance of transfers	+868	+1 409	+1 143	+1 143	+1 144
Current balance	+581	+585	−3 233	−3 488	−4 287
Long-term capital flow	+795	+799	+2 712	+1 803	+2 008
Reserves and debt (end-year, $ mn)					
International reserves	5 014	6 772	6 485	6 090	5 284
External public debt	2 068	2 082	3 117	4 085	6 700

External trade (1976) Imports: Pa 1 170.4 bn = $ 17 495 mn = £ 9 689 mn
Exports: Pa 583.5 bn = $ 8 722 mn = £ 4 830 mn

Main imports	% of total	Main exports	% of total
Crude oil	26	Fruit and vegetables	12
Machinery, non-electric	13	Machinery, non-electric	9
Chemicals	9	Iron and steel	7
Food	9	Motor vehicles	6
(of which, maize 3)		Chemicals	6
Iron and steel	5	Footwear	6
Electrical machinery	5	Ships	5
Metal ores	4	Electrical machinery	4
Oilseeds and nuts	3	Petroleum products	4
Instruments	3	Textile yarns and fabrics	3
Textile fibres	2	Fish	3
Motor vehicles	2	Rubber and products	2
		Clothing	2
Main sources[a]		*Main destinations*[b]	
United States	12	France	16
West Germany	10	West Germany	11
Saudi Arabia	9	United States	10
France	8	United Kingdom	6
Iran	7	Italy	5
United Kingdom	5	Netherlands	4
Italy	5	Morocco	3
Brazil	3	Belgium-Luxembourg	3
Japan	3	Algeria	3
Iraq	3	Venezuela	2
Libya	3	Portugal	2
Netherlands	3	Libya	2
Utd Arab Emirates	2		

[a]EEC 33 % [b]EEC 46 %

Special focus

Growth of industrial production (% change over previous year)
based on index of industrial production

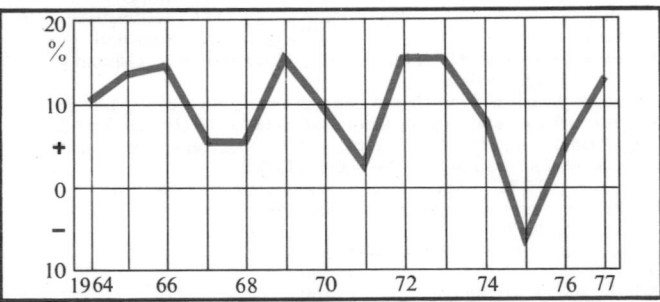

Sweden

Kingdom of Sweden
Konungariket Sverige

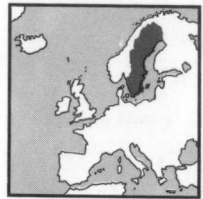

Location Northern Europe
Occupies the main part of the Scandinavian peninsula, with coastlines in the north-east on the Gulf of Bothnia, south-east on the Baltic Sea, and west on the Kattegat; Norway is to the west, Finland to the north-east, and Denmark across the Sound to the south-west
Land Area 449 964 km² = 173 732 mi²
Usage (1975): agricultural 37 150 km² (8 %); forests 264 240 km² (59 %)
Climate Temperate, very cold in north
Weather at Stockholm, 44 m altitude
Temperature: hottest month July 14–22 °C, coldest Feb minus 5–minus 1 °C
Rainfall (av monthly): driest month Mar 25 mm, wettest Aug 76 mm
Time 1 hour ahead of GMT
Measures Metric system
Monetary unit Swedish krona (S Kr) = 100 öre
Rate of exchange (1976 av): free S Kr 4.356 = $ 1, S Kr 7.868 = £ 1
par (central rate within the European 'snake', 1976): to Oct 17, S Kr 5.50094 = S DR 1; from Oct 18, S Kr 5.55651 = S DR 1.
Sweden withdrew from the 'snake' August 28 1977

Summary

Political Parliamentary monarchy; neutral in first and second world wars. Member of UN, OECD, Council of Europe, Efta and Nordic Council
Economic An industrial economy, with major exports of machinery, motor vehicles and iron and steel; forestry is also important, with timber, wood pulp and paper making up about one-fifth of exports. There is a large merchant fleet. With a high level of social welfare, government current expenditure makes up one-quarter of gross domestic product, and is about equal to one-half of private consumer expenditure

People

Population 1960 7.48 mn, 1970 8.04 mn, 1976 8.22 mn
Growth: 1960–70 0.7 %pa, 1970–76 0.4 %pa
Density (1976): 18 people per km²
Vital statistics (rate per 1 000 people, 1976): births 11.9, deaths 11.0
Households (1970) 3.05 mn; people in households 7.92 mn (98.0 % of total)
Regions (counties, population in 000, December 31, 1976; total of 8.24 mn)

Stockholm	1 501	Blekinge	155	Örebro	274
Uppsala	233	Kristianstad	274	Västmanland	260
Södermanland	252	Malmöhus	740	Kopparberg	283
Östergötland	389	Halland	223	Gälveborg	295
Jönköping	302	Göteborg och Bohus	714	Västernorrland	268
Kronoberg	170	Älvsborg	420	Jämtland	134
Kalmar	241	Skaraborg	264	Västerbotten	238
Gotland	55	Värmland	285	Norrbotten	266

Cities (population in 000, December 31, 1976)

Stockholm (capital)	1 364	Norrköping	120	Jönköping	108
Göteborg (Gothenburg)	692	Västerås	118	Borås	104
Malmö	453	Örebro	117	Helsingborg	101
Uppsala	140	Linköping	110	Sundsvall	94

Race (1970) Swedish 93 %, Finnish 3 %, Norwegian ½ %, German ½ %, Dutch ½ %, Jugoslavian ½ %
Language Swedish; Finnish and Lappish are also in use by minorities
Religion (1976) Swedish State Church (Evangelical Lutheran) 95** %
Educational Pupils (1976/77): primary 709 000, secondary and vocational 545 000, teacher-training 12 600, higher 113 000.
Teachers (1975/76): primary, secondary and vocational 98 000, teacher-training 1 100, higher (1973/74) 6 000
Labour force (1976)

Economic activity	Number	%
Agriculture	254 200	6
Mining and quarrying	21 100	1
Manufacturing	1 100 400	26
Electricity, gas and water	33 400	1
Construction	294 500	7
Distribution and hotels	591 900	14
Transport and communication	275 400	7
Finance and real estate	240 700	6
Other services	1 276 200	31
Unemployed	66 000	2
Total	4 154 300	100

Personnel Scientists and engineers engaged in research (1973): 11 762
Physicians (1974): 13 260, 1 per 615 people
Standard of living
National income per person (1976): S Kr 35 040 = $ 8 043 = £ 4 453
Consumption per person (1975): energy 6 178 kg coal equivalent, electricity (production, 1976) 10 513 kW h, newsprint 32 kg, steel 772 kg
Newspapers (1974): number 111; circulation 4 362 000, 536 per 1 000 people
Telephones (Dec 1976): 5 673 400, 689 per 1 000 people

Resources and equipment

Livestock (000, 1976) Cattle 1 876, pigs 2 485, sheep 389, chickens 11 692*
Mineral reserves Coal (1967) 90 mn tonnes
Uranium (1976) 4 000 tonnes
Petroleum refinery capacity (1975) 14 mn tonnes
Electrical capacity (1975) 23 135 megawatts, of which, hydro 12 716 megawatts, nuclear 2 522 megawatts
Hospital beds (1974) 124 350, 1 per 66 people
Roads (1975) 179 654 kmª = 111 632 mi, density 0.40 km per km²
ªIncludes 65 121 km of subsidised private roads, open to public traffic
Railways (1975) 12 070 km = 7 500 mi, density 0.027 km per km²
Inland waterways (1975) 736 km = 457 mi
Ships (registered, 1977) 728, total of 7 429 394 gross tons
Ports (goods traffic, 000 tonnes, 1976)

	International		Coastwise		Total
	loaded	unloaded	loaded	unloaded	
Gothenburg	3 218	11 626	4 809	311	19 964
Helsingborg	2 809	3 819	205	867	7 700
Luleå	4 809	1 883	185	618	7 495
Stockholm	612	2 729	211	2 199	5 751
Malmö	517	2 044	113	997	3 671
Norrköping	684	1 991	23	955	3 653
Oxelösund	824	1 855	715	167	3 561

Airports (passengers, 000, 1976)

	departures	arrivals
Arlanda (Stockholm)	1 913	1 931
Torslanda (Gothenburg)	617	637
Sturup (Malmö)	283	289

Also (1977) 29 other airports with scheduled flights

Durable equipment (Dec 1976)

	000	no per 1 000 people	
Radio sets	3 203	389	no per
Television sets	2 988	363	km of road
Passenger cars	2 881	350	16
Commercial vehicles	178	22	1.0

Production

Gross domestic product
1976: S Kr 323 278 mn = $ 74 214 mn = £ 41 088 mn
Growth in real terms: 1960–70 8.5 %pa, 1970–76 2.2 %pa
Structure of gross domestic product

By origin (1975)	S Kr mn	% of gdp
Agriculture	12 913	4
Mining and quarrying	2 566	1
Manufacturing	83 148	29
Electricity, gas and water	6 176	2
Construction	17 763	6
Distribution and hotels	27 242	9
Transport and communication	13 617	5
Other	124 586	43
Total	288 011	100

By type of expenditure (1976)		
Government final consumption	82 851	26
Private final consumption	172 893	53
Stock investment	7 514	2
Gross fixed capital formation	66 599	21
Exports of goods and services	90 760	28
less Imports of goods and services	−97 339	−30
Total	323 278	100

Production indices (1970 = 100)	1960	1970	1976	Growth %pa 1960–70	1970–76
Agricultural	94	100	112	0.6	1.8
Industrial	55	100	114	6.2	2.2

Main products (000 t)
Agriculture

	1960	1970	1976	1960–70	1970–76
Wheat	824	962	1 763	1.5	10.6
Barley	847	1 904	1 825	8.4	−0.7
Rye	230	228	427	−0.1	11.0

Sweden

Main products (000 t)	1960	1970	1976	Growth %pa 1960–70	1970–76
Agriculture					
Oats	1 176	1 686	1 251	3.7	−4.8
Potatoes	1 753	1 490	1 058	−1.6	−5.5
Sugar, raw value	361	220	302*	−4.8	5.4*
Apples	291	146	96*	−6.7	−6.7*
Rapeseed	61	192	279	12.1	6.4
Milk	3 926	2 932	3 247	−2.9	1.7
Butter	84	42	61	−6.7	6.4
Cheese	54	63	87	1.5	5.5
Eggs	97	99	105	0.2	1.0
Beef and veal	129	164	149*	2.4	−1.6*
Pigmeat	215	229	292*	0.6	4.1*
Fish catch	255	295	209	1.5	−5.6
Timber (000 m³)	44 900*	59 967	55 660	2.9*	−1.2
hardwood (000 m³)	4 700*	6 697	6 835ᶜ	3.6*	0.4ᵈ
softwood (000 m³)	40 200*	53 270	45 580ᶜ	2.8*	−3.1ᵈ
Energy					
Total energy (000 tce)	4 190	5 210	8 580ᶜ	2.2	10.5ᵈ
Petroleum products	2 907	11 981	11 980*ᶜ	15.2	0.0*ᵈ
Manufactured gas (mn m³)	357	511	587ᶜ	3.7	2.8ᵈ
Electricity (mn kW h)	34 718	60 645	86 416	5.7	6.1
of which,					
hydro (mn kW h)	31 066	41 538	57 669ᶜ	2.9	6.8ᵈ
nuclear (mn kW h)	—	56	11 969ᶜ	na	192.4ᵈ
Mining					
Iron ore (Fe content)	13 014	19 804	18 370	4.3	−1.2
Copper conc (Cu content)	17.5	26.3	40.6ᶜ	4.2	9.1ᵈ
Lead conc (Pb content)	55.3	78.3	70.4ᶜ	3.5	−2.1ᵈ
Tungsten conc (oxide content)	—	—	0.14ᶜ	na	na
Zinc conc (Zn content)	74	93	111ᶜ	2.3	3.6ᵈ
Gold (000 kg)	2.93	3.90	2.41ᶜ	2.9	−9.2ᵈ
Manufacturing					
Margarine	118	137	148ᵉ	1.5	1.9ᶠ
Beer (000 hl)	2 795	4 374	4 386ᵉ	4.6	0.1ᶠ
Cigarettes (mn units)	6 200**	8 975	10 319ᶜ	3.8**	2.8ᵈ
Cotton yarn	25.4	11.6	5.7	−7.5	−11.2
Man-made fibres	28.5	36.1	20.5ᵉ	2.4	−10.7ᵈ
Wood pulp	4 949	8 142	8 344ᶜ	5.1	0.5ᵈ
Newsprint	582	1 030	1 136	5.9	1.6
Other paper	1 700**	3 329	3 810	7.0**	2.3
Tyres	2 203	5 349	2 960	9.3	−9.4
Sulphurᵃ	209	289	211ᶜ	3.3	−6.1ᵈ
Sulphuric acid	410	709	793ᶜ	5.6	2.2ᵈ
Nitric acid	118	270	344ᵉ	8.6	6.2ᶠ
Caustic soda	183	377	470ᵉ	7.5	5.7ᶠ
Plastics and resins	75	304	442ᶜ	15.0	7.8ᵈ
Fertilisers, nitrogenousᵇ	43	144	169	12.8	2.7
Fertilisers, phosphateᵇ	100**	143	145	3.6**	0.3
Cement	2 919	4 061	2 798	3.3	−6.0
Coke	134	530	820ᶜ	14.7	9.1ᵈ
Pig iron	1 626	2 842	3 140	5.7	1.7
Crude steel	3 218	5 494	5 140	5.5	−1.1
Aluminium	16.0	64.1	83.3	14.9	4.5
Radio sets (000 units)	261	170	194ᵉ	−4.2	−3.4ᶠ
Television sets (000 units)	256	178	370ᵉ	−3.6	20.1ᶠ
Passenger cars (000 units)	109	272	336ᶜ	9.6	4.3ᵈ
Commercial vehicles (000 units)	20.3	32.0	49.2ᶜ	4.7	9.0ᵈ
Merchant vessels (000 grt)	711	1 711	2 367	9.2	5.6
Construction					
Dwellings (000 units)	68	110	56	4.9	−10.6

ᵃSulphur content of pyrites ᵇYears ending May 31st ᶜ1975 ᵈ1970–75
ᵉ1974 ᶠ1970–74

Transport traffic	1960	1970	1976	Growth %pa 1960–70	1970–76
Passenger-kilometres (mn)					
Rail	5 180	4 693	5 363	−1.0	2.2
Airᵃ	1 012	2 449	4 041	9.2	8.7
Cargo: tonne-kilometres (mn)					
Rail	10 949	17 311	15 458	4.7	−1.9
Airᵃ	26.9	111.0	172.5	15.2	7.6
Sea: tonnes (mn)					
Goods loadedᵇ	19.5	27.0	32.4	3.3	3.1
Goods unloaded	25.3	46.7	55.1	6.3	2.8

ᵃIncluding apportionment (three-sevenths) of traffic of Scandinavian Airlines System
ᵇExcludes Swedish iron ore loaded at Narvik (Norway)

Tourism (1976) Number of visitors (arrivals at all forms of accommodation) 1 780 591, gross receipts $ 303 mn

Finance and trade

Price and wage indices (1970 = 100)	1960	1970	1976	Growth %pa 1960–70	1970–76
Consumer prices	67.0	100.0	161.7	4.1	8.3
Wholesale prices	74	100	173	3.1	9.6
Wages (earnings)	43.6	100.0	199.3	8.7	12.2
Share prices	76	100	168	2.8	9.0
Money stock					
(end-year, S Kr mn)	8 500**	17 760	33 480	7.6**	11.1

Budget (1976/77; year ending June 30th)
Revenue: S Kr 102 066 mn = $ 23 720 mn = £ 13 845 mn
Expenditure: S Kr 108 247 mn = $ 25 156 mn = £ 14 684 mn,
of which, education 13 %, defence 10 %, health 5 %,
other social security 30 %

Balance of payments ($ mn)	1972	1973	1974	1975	1976
Balance of goods (fob)	+1 211	+2 273	+607	+695	+441
Balance of services	−690	−747	−1 120	−1 665	−2 085
Balance of transfers	−251	−312	−419	−654	−745
Current balance	+269	+1 215	−933	−1 624	−2 389
Long-term capital flow	+208	+187	+277	+1 361	+193
Reserves and debt (end-year, $ mn)					
International reserves	1 575	2 529	1 736	3 077	2 491
Public debt: domestic	8 995	10 805	14 631	na	na
external	383	443	566	na	na
total	*9 378*	*11 248*	*15 197*	*16 752*	*19 486*

External trade (1976)
Imports: S Kr 84 000 mn = $ 19 284 mn = £ 10 676 mn
Exports: S Kr 80 217 mn = $ 18 415 mn = £ 10 195 mn

Main imports	% of total	Main exports	% of total
Crude oil and products	16	Machinery, non-electric	18
(of which, crude oil 7)		Motor vehicles	10
Machinery, non-electric	14	Paper	9
Chemicals	8	Electrical machinery	8
Motor vehicles	8	Ships and boats	7
Food	7	Wood pulp	7
Electrical machinery	7	Iron and steel	7
Iron and steel	5	Timber	5
Clothing	4	Chemicals	5
Textile yarns and fabrics	4	Metal small manufactures	4
Instruments	3	Metal ores	3
Non-ferrous metals	3	Food	2
Ships and boats	2	Non-ferrous metals	2
Main sources		*Main destinations*	
West Germany	19	United Kingdom	11
United Kingdom	11	Norway	11
Denmark	7	West Germany	10
United States	7	Denmark	10
Norway	6	Finland	6
Finland	6	France	5
Netherlands	4	United States	5
France	4	Netherlands	4
Belgium-Luxembourg	3	Belgium-Luxembourg	3
Japan	3	Italy	3
Italy	3	Poland	2
Soviet Union	2	Switzerland	2
Switzerland	2	Soviet Union	2
Austria	2		

Special focus

Structure of gross domestic product

By type of expenditure	% of total 1960	1970	1972	1974	1976
Government final consumption	16	21	23	24	26
Private final consumption	61	55	54	53	53
Stock investment	3	3	—	2	2
Gross fixed capital formation	22	22	22	22	21
Exports of goods and services	23	24	24	33	28
less Imports of goods and services	−24	−25	−23	−33	−30
Total	100	100	100	100	100

Switzerland

Swiss Confederation
Schweizerische Eidgenossenschaft
Confédération Suisse
Confederazione Svizzera

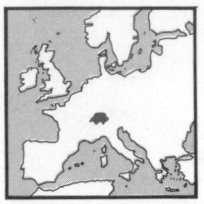

Location Central Europe
West Germany is to the north, France to the
west, Italy to the south and Austria to the east
Land Area 41 288 km² = 15 941 mi²
Usage (1975): agricultural 20 190 km² (49 %),
forests 10 520 km² (25 %)
Climate Temperate
Weather at Zurich, 493 m altitude
Temperature: hottest month July 13–24 °C,
coldest Jan minus 3–2 °C
Rainfall (av monthly): driest month Dec 64 mm, wettest July 136 mm
Time 1 hour ahead of GMT
Measures Metric system
Monetary unit Swiss franc or franken (S Fr) = 100 centimes or rappen
Rate of exchange: 1976 av (free) S Fr 2.500 = $ 1, S Fr 4.515 = £ 1
1977 av (free) S Fr 2.403 = $ 1, S Fr 4.195 = £ 1

Summary

Political Republic, with neutrality guaranteed by major countries in 1815.
Member of a number of UN agencies (although not a member of UN)
including FAO, ILO and Unesco; also of OECD, Council of Europe and
Efta
Economic An industrial economy, with machinery, chemicals, watches and
textiles major exports. There is a significant agricultural sector but little
mining; tourism and international finance are large earners of foreign
exchange. With a reputation as a safe haven for currency, there are large
inflows of cash

People, resources and equipment

Population 1960 5.36 mn, 1970 6.19 mn, 1976 6.35 mn
Growth: 1960–70 1.4 %pa, 1970–76 0.4 %pa
Density (1976): 154 people per km²
Vital statistics (rate per 1 000 people, 1976): births 12.0, deaths 8.8
Households (1970) 2.05 mn; population in households 6.01 mn
(95.9 % of total)
Regions (cantons and demi-cantons, population in 000, January 1, 1977;
total of 6.30 mn)

Zürich[a]	1 112	Fribourg[b]	182	Graubünden[a,d]	162
Bern[a]	985[e]	Solothurn[a]	222	Aargau[a]	441
Luzern[a]	291	Basel	430	Thurgau[a]	183
Uri[a]	33	*Basel-Stadt[a]*	221	Ticino[c]	261
Schwyz[a]	92	*Basel-Landschaft[a]*	219	Vaud[b]	521
Unterwalden	51	Schaffhausen[a]	69	Valais[b]	211
Obwalden[a]	25	Appenzell	60	Neuchâtel[b]	163
Nidwalden[a]	26	*Ausser-Rhoden[a]*	47	Genève[b]	336
Glarus[a]	36	*Inner-Rhoden[a]*	13		
Zug[a]	73	St Gallen[a]	383		

[a]Mainly German speaking   [b]Mainly French speaking   [c]Mainly Italian speaking
[d]Romansch is also a main language   [e]Includes the mainly French speaking Jura
districts, which are being formed into a new Jura canton
Cities (population in 000, January 1, 1977)

Bern[a] (capital)	283	Lausanne	227	St Gallen[f]	87
Zürich[b]	708	Luzern[e]	156	Thun	65
Basel[c]	369	Winterthur	107	Neuchâtel	60
Genève[d]	323	Biel	89	Fribourg	54

[a]Berne [b]Zurich [c]Bâle or Basle [d]Genf or Geneva [e]Lucerne [f]St-Gall
Race (1976) Swiss 85 %, foreigners 15 % (of whom, Italian 7** %,
Spanish 2** %, West German 2** %, French 1** %)
Language German, French, Italian, Romansch
Usage (mother tongue, 1970): German 65 %, French 18 %, Italian 12 %,
Spanish 2 %, Romansch 1 %. Also see regions for areas mainly speaking
various languages
Religion (1970) Roman Catholic 49 %, Protestant 48 %
Education Pupils (1974/75): primary 561 645, secondary 323 743,
vocational 164 463, teacher-training 11 251, higher 62 584.
Teachers (1969/70): primary 14 672, secondary and vocational 10 920,
higher (1975/76) 5 414

Labour force 1970:

Economic activity	Number	% of total
Agriculture	230 664	8
Mining and quarrying	6 800	—
Manufacturing	1 129 763	38
Electricity, gas and water	23 447	1
Construction	285 151	10
Distribution and hotels	524 187	17
Transport and communication	169 910	6
Finance and real estate	159 053	5
Other	466 802	16
Total	2 995 777	100

1976 total: 3 128 000*
Personnel Scientists and engineers engaged in research (1971): 14 396
Physicians (1974): 10 904, 1 per 591 people
Standard of living
National income per person (1976): S Fr 20 614 = $ 8 246 = £ 4 566
Consumption per person (1975): energy 3 642 kg coal equivalent,
electricity (production, 1976) 5 490 kW h, newsprint 22 kg, steel 232 kg
Newspapers (1974): number 92; circulation 2 535 000, 391 per 1 000 people
Telephones (Dec 1976): 4 016 320, 638 per 1 000 people
Livestock (000, April 1976) Cattle 2 005, pigs 2 006, sheep 377, goats 70*,
chickens 6 138
Petroleum refinery capacity (1975) 6.85 mn tonnes
Electrical capacity (1975) 12 816* megawatts,
of which, hydro 9 800* megawatts, nuclear 2 422* megawatts
Hospital beds (1970) 71 276, 1 per 87 people
Roads (1975) 62 158 km = 38 623 mi, density 1.5 km per km²
Railways (1974) 4 969 km = 3 088 mi, density 0.12 km per km²
Inland waterways (1975) 21 km = 13 mi
Oil pipelines (1975) 239 km = 149 mi
Ships (registered, 1977) 28, total of 252 746 gross tons
Port Basle (Rhine port)
Airports (passengers, 000, 1976)

	departures	arrivals
Zurich	3 216	3 218
Geneva	1 730	1 719
Basle/Mulhouse[a]	342	353
Also Berne and St Moritz		

[a]Situated on French territory

Durable equipment	000	no per	
(Dec 1976)		1 000 people	
Radio sets	2 108	335	no per
Television sets	1 809	287	km of road
Passenger cars	1 864	296	30
Commercial vehicles	178	28	2.9

Production

Gross domestic product
1976: S Fr 140 710 mn = $ 56 284 mn = £ 31 165 mn
Growth in real terms: 1960–70 4.7 % pa, 1970–76 0.3 % pa
Structure of gross domestic product

By origin (1967)	S Fr mn	% of gdp
Agriculture	4 080	6
Manufacturing	23 370	37
Electricity, gas, water,		
mining and quarrying	1 910	3
Construction	6 200	10
Transport and communication	4 410	7
Other	23 485	37
Total	63 455	100

By type of expenditure (1976)		
Government final consumption	18 810	12
Private final consumption	88 745	62
Stock investment	−1 050	—
Gross fixed capital formation	29 175	27
Exports of goods and services	48 050	35
less Imports of goods and services	−43 020	−37
Total	140 710	100

Production indices	1960	1970	1976	Growth %pa	
(1970 = 100)				1960–70	1970–76
Agricultural	87	100	114	1.4	2.2
Industrial	59	100	96	5.4	−0.7

Main products (000 t)

Agriculture					
Wheat	378	346	408	−0.8	2.8
Barley	76	142	183	6.4	4.3
Potatoes	1 291	977	829	−2.7	−2.7
Grapes	130	165	110*	2.4	−6.5*
Wine	90	118	122*	2.7	0.5*

Switzerland

Main products (000 t)

	1960	1970	1976	Growth %pa 1960–70	1970–76
Agriculture					
Apples	465	280	200*	−4.9	−5.5*
Milk	3 084	3 183	3 452	0.3	1.4
Butter	35	29	35	−1.9	3.2
Cheese	67	85	112	2.4	4.7
Beef and veal	99	135	149	3.1	1.6
Pigmeat	132	196	239*	4.0	3.4*
Timber (000 m³)	3 650**	4 190	3 510	1.4**	−2.9
Energy					
Total energy (000 tce)	2 350	3 970	4 970e	5.4	4.6f
Petroleum products	—	5 229	4 440*e	na	−3.2*f
Manufactured gas (mn m³)	362	396	94	0.9	−21.3
Electricity (mn kW h)b	19 072	33 173	42 282e	5.7	5.0f
of which,					
hydro (mn kW h)b	18 826	29 330	33 069e	4.5	2.4f
nuclear (mn kW h)b	—	2 450	7 373e	na	24.7f
Mining					
Salt	149	333	237e	8.4	−6.6f
Manufacturing					
Chocolate	45	61	61	6.1	0.0
Beer (000 hl)	3 127	4 733	4 325e	4.2	−1.8f
Cigarettes (mn units)	10 909	29 229	27 788	10.3	−0.8
Cotton yarnd	na	39.0*	39.2g	na	0.2*h
Cotton fabrics (mn m)	157	161	146g	0.2	−3.2h
Wool yarnd	na	16.0*	16.4i	na	0.6*j
Man-made fibres	29.0	60.4	75.9e	7.6	4.7f
Wood pulp	238	268	223e	1.2	−3.6f
Newsprint	90	143	143e	4.7	0.0f
Other paper	400**	588	457e	3.9**	−4.9f
Sulphuric acid	160	168	165k	0.5	−0.9l
Plastics and resins	30*	60*	na	7.2*	na
Fertilisers, nitrogenousa	19.0	29.0	29.8*	4.3	0.5*
Cement	3 036	4 797	3 546	4.7	−4.9
Pig iron	50*	28*	35*e	−5.6*	4.6*f
Crude steel	275*	524*	420*e	6.7*	−4.3*f
Aluminium	40	91	78	8.7	−2.5
Watches (000 units)	30 195	52 607	47 191e	5.7	2.1f
Passenger carsc (000 units)	15.4	19.0	17.0g	2.1	−3.6h
Construction					
Dwellings (000 units)	na	63.6	54.9e	na	−2.9f

aYears ending June 30th bYears ending September 30th eAssembly only
dIncluding mixed yarn e1975 f1970–75 g1973 h1970–73 i1974 j1970–74
k1972 l1970–72

Transport traffic

	1960	1970	1976	Growth %pa 1960–70	1970–76
Passenger-kilometres (mn)					
Rail	7 974	9 339	8 130	1.6	−2.3
Air	1 138	4 420	8 493	14.5	11.5
Cargo: tonne-kilometres (mn)					
Rail	4 346	7 035	5 659	4.9	−3.6
Air	36	188	346	18.1	10.7
Tourism					
Number of visitorsa (000)	4 949	6 840	5 879	3.3	−2.5
Gross receiptsb ($ mn)	365	905	2 191	9.5	15.9

aArrivals at hotels bIncluding international fare receipts

Finance and trade

Price and wage indices (1970 = 100)	1960	1970	1976	Growth %pa 1960–70	1970–76
Consumer prices	72.0	100.0	147.3	3.3	6.7
Wholesale prices	83.2	100.0	132.2	1.9	4.8
Wages (earnings)	59.7	100.0	156.0	5.3	7.7
Share prices	90.4	100.0	77.1	1.0	−4.2

| Money stock (end-year, S Fr mn) | 19 340 | 45 690 | 62 910 | 9.0 | 5.5 |

Budget (1976) Revenue: S Fr 13 781 mn = $ 5 512 mn = £ 3 052 mn
Expenditure: S Fr 15 185 mn = $ 6 074 mn = £ 3 363 mn,
of which, education and research 10 %, defence 20 %,
social security 18 %

Balance of payments ($ mn)	1972	1973	1974	1975	1976
Balance of goods (fob)	−1 269	−1 685	−2 109	+98	+433
Balance of services	+1 995	+2 636	+2 985	+3 182	+3 665
Balance of transfers	504	−670	−705	−694	−597
Current balance	+220	+280	+171	+2 587	+3 501
Long-term capital flow	−2 887	−2 756	−1 584	−4 464	−7 239
Reserves and debt (end-year, $ mn)					
International reserves	7 557	8 520	9 011	10 428	12 993
Public debt: total	1 610	2 002	3 196	na	na

External trade (1976)

Imports: S Fr 36 874 mn = $ 14 750 mn = £ 8 167 mn
Exports: S Fr 37 015 mn = $ 14 806 mn = £ 8 198 mn

Main imports	% of total	Main exports	% of total
Chemicals	11	Machinery, non-electric	23
Food	10	Chemicals	21
Machinery, non-electric	8	Electrical machinery	9
Petroleum products	8	Watches	8
Electrical machinery	7	Textile yarns and fabrics	6
Motor vehicles	7	Precious metals, jewellery	5
Precious metals, jewellery	5	Metal small manufactures	3
Clothing	5	Instruments (excl watches)	3
Iron and steel	4	Food	3
Textile yarns and fabrics	4		

Main sources	% of total	Main destinations	% of total
West Germany	28	West Germany	16
France	13	France	9
Italy	10	Italy	7
United States	7	United States	7
United Kingdom	7	United Kingdom	6
Austria	4	Austria	5
Netherlands	4	Sweden	3
Belgium-Luxembourg	4	Japan	3
Japan	3	Belgium-Luxembourg	3

Special focus

Tourism 1976

Origin of visitorsa	Number	%		Number	%
West Germany	1 589 874	27	Japan	183 246	3
United States	845 068	14	Austria	148 688	3
France	644 404	11	Spain	123 463	2
Italy	352 745	6	Sweden	98 607	2
United Kingdom	341 178	6	Other	982 548	17
Belgium	287 923	5	Total	5 878 899	100
Netherlands	281 155	5			

aArrivals at hotels

Turkey

Republic of Turkey
Türkiye Cumhuriyeti

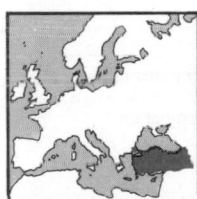

Location South-east Europe
There are coastlines to the north on the Black Sea, south on the Mediterranean Sea and west on the Aegean Sea; Soviet Union is to the north-east, Iran to the east, Iraq and Syria to the south-east, Greece to the west and Bulgaria to the north-west. The Sea of Marmara is within the north-west part of the territory, having an outlet through the Bosphorus to the Black Sea, and through the Dardanelles to the Mediterranean Sea. The area to the east of the Sea of Marmara (Anatolia) is generally regarded as forming part of 'Asia' and the area to the west (Thrace) as forming part of 'Europe'
Land Area 780 576 km² = 301 382 mi²
Climate Mediterranean on the south coast, continental inland
Weather at Ankara, 861 m altitude
Temperature: hottest month Aug 15–31 °C, coldest Jan minus 4–4 °C
Rainfall (av monthly): driest month Aug 10 mm, wettest Dec 48 mm
Time 2 hours ahead of GMT (summer time, 3 hours ahead)
Measures Metric system. Old Turkish units are not generally in use; these included especially:
area 1 dönüm = 919 sq metres = 1 099 square yards
Monetary unit Turkish lira (T L) = 100 kuruş or piastres
Rate of exchange (1976 av): free T L 16.05 = $ 1, T L 29.00 = £ 1

Summary

Political Republic since 1923; formerly the dominant part of the Ottoman Empire. Remained neutral in the second world war. Member of UN, OECD, Council of Europe, Nato, Cento and RCD
Economic An agricultural economy, with fruit and vegetables and cotton together making up about one-half of exports; about two-thirds of the labour force are in agriculture, which provides about one-quarter of gross domestic product. Industrialisation is increasing, especially in textiles and clothing, which make up one-sixth of exports, and chemicals. There is some mining, including crude oil and chromium, etc. Tourism is expanding, and there are substantial receipts of foreign exchange from remittances sent home by workers in other countries.

Turkey

People, resources and equipment

Population 1960 27.51*mn, 1970 34.85*mn, 1976 41.09*mn
Growth: 1960–70 2.4* %pa, 1970–76 2.8* %pa
Density (1976): 53* people per km²
Vital statistics (rate per 1 000 people, 1967): births 39.6, deaths 14.6
Regions (population in 000, 1976; total of 41.1*mn)
Anatolia 37 500**, Thrace 3 600**
Cities (population in 000, 1973)

Ankara (capital)	1 701ᵃ	Gaziantep	353	Samsun	242
Istanbul	3 135	Konya	324	Maraş	237
Izmir	819	Eskişehir	303	Malatya	234
Adana	475ᵃ	Kayseri	297	Izmit	233
Bursa	427	Diyarbakır	251	Erzurum	226

ᵃ1975; excluding suburbs
Race (1965) Turk 90 %, Kurd 7 %, Arab 1 %
Language Turkish; Kurdish and Arabic are also in use
Religion (1965) Moslem 99 %, Christian ½ %
Education (1975/76) Pupils: primary 5 512 000, secondary 1 363 188, vocational 312 522, teacher-training and higher 209 052.
Teachers: primary 172 488, secondary 37 899, vocational 12 240, teacher-training and higher 24 816
Labour force (1975) 16 349 380; in agriculture 10 482 966 (64 %), mining and quarrying 108 506 (1 %), manufacturing 1 243 567 (8 %), construction 447 324 (3 %), transport and communication 512 327 (3 %). Workers in other countries (1976) 713 000*
Personnel Scientists and engineers engaged in research (1967): 5 856
Physicians (1975): 21 714, 1 per 1 858 people
Standard of living
National income per person (1976): T L 15 700* = $ 980* = £ 540*
Consumption per person (1975): energy 630 kg coal equivalent, electricity (production, 1976) 445 kW h, newsprint 2.3 kg, steel 69 kg
Newspapers: number (1973) 450; circulation (1961) 1 299 000, 45 per 1 000 people
Telephones (Dec 1976): 1 130 980, 27 per 1 000 people
Livestock (000, Dec 1976) Cattle 14 102, buffaloes 1 056, sheep 41 504, goats 18 508, horses 853, chickens 45 711
Mineral reserves Coal (1972) 1 291 mn tonnes
Lignite (1972) 5 991 mn tonnes
Crude oil (1975) 14 mn tonnes
Natural gas (1975) 1 bn cubic metres
Uranium (1976) 4 100 tonnes
Petroleum refinery capacity (1975) 16.5* mn tonnes
Electrical capacity (1975) 4 165 megawatts, of which, hydro 1 770 megawatts
Hospital beds (1975) 85 872, 1 per 470 people
Roads (1976) 195 982 kmᵃ = 121 778 mi, density 0.25 km per km²
ᵃIncludes village roads
Railways (1975) 8 138 km = 5 057 mi, density 0.010 km per km²
Oil pipelines (1975) 495 km = 308 mi
Ships (registered, 1977) 448, total of 1 288 282 gross tons
Ports (goods traffic, 000 tonnes, 1975)

	loaded	unloaded		loaded	unloaded
Mersin	495	2 670	Izmir	516	503
Istanbul	144	2 090	Samsun	76ᵃ	717ᵃ
Iskenderun	486	776	Izmit	166	275

ᵃ1974
Airports (passengers, 000, 1975)

	Departures	Arrivals
Yeşilköy (Istanbul)	1 438	1 369
Esenboğa (Ankara)	450	453

Also Giğli (Izmir) and (1977) 15 other airports with scheduled flights

Durable equipment (Dec 1976)	000	no per 1 000 people	
Radio sets	4 228	102	no per
Television sets	1 769	43	km of road
Passenger cars	512	12	2.6
Commercial vehicles	262	6	1.3

Production

Gross domestic product 1976: T L 653.3 bn = $ 40 703 mn = £ 22 527 mn
Growth in real terms: 1970–75 7.5 %pa
Structure of gross domestic product By origin (1975) Agriculture 26 %, mining and quarrying 1 %, manufacturing 19 %, electricity, gas and water 2 %, construction 5 %, distribution 13 %, transport and communication 8 %, other 26 %
By type (1973) Final consumption expenditure 85 % (of which, government 14 %), stock investment 1 %, gross fixed capital formation 18 %, exports of goods and services 8 %, less imports of goods and services −12 %

Production index (1970 = 100)	1960	1970	1976	Growth %pa 1960–70	1970–76
Agricultural	79	100	133	2.4	4.9
Main products (000 t)					
Agriculture					
Wheat	8 590	10 081	16 578	1.6	8.6
Barley	3 700*	3 250	4 900	−1.3*	7.1
Maize	1 090	1 040	1 310	−0.5	3.9
Potatoes	1 400	1 915	2 850	3.2	6.8
Sugar, raw value	700	645	1 284*	−0.8	12.1*
Tomatoes	1 200**	1 810	2 750	4.2**	7.2
Onions	400	680	760	5.4	1.9
Aubergines	400**	480	496*	1.8**	0.5*
Chillies and peppers	260**	307	378*	1.7**	3.5*
Watermelons	3 000**	3 490	4 087*	1.5**	2.7*
Grapes	3 000**	3 850	3 347*	2.5**	−2.3*
Raisins	226	310*	335*	3.2*	1.3*
Apples	208	748	1 000*	13.7	5.0*
Oranges	243	523	561*	8.0	1.2*
Lemons	64	126	280*	7.0	14.2*
Figs	145	214	188	4.0	−2.1
Hazelnuts	100**	255	260*	9.8**	0.3*
Sunflowerseed	123	375	550	11.8	6.6
Cottonseed	306	640	760	7.6	2.9
Olives	426	681	1 097	4.8	8.3
Olive oil	79	118	201	4.1	9.3
Tea	6	34	60	19.0	9.9
Tobacco	139	150	314	0.8	13.1
Cotton	176	400	475	8.6	2.9
Milk, cow	2 241	2 551	3 100	1.3	3.3
Milk, buffalo	258	279	278	0.8	−0.1
Milk, sheep	882	860	1 004	−0.2	2.6
Milk, goat	811	603	624	−2.9	0.6
Butter	100**	113	120*	1.2**	1.0*
Cheese	70**	94*	114*	3.0**	3.3*
Eggs	66	96	155	3.8	8.3
Wool	26.1	26.0	29.7*	0.0	2.2*
Beef and buffalo meat	170**	213*	232*	2.3**	1.4*
Sheep and goat meat	280**	349*	386*	2.2**	1.7*
Fish catch	89	184	155*	7.5	−2.8*
Timber (000 m³)	9 700**	16 809	16 844	5.7**	0.0
Energy					
Total energy (000 tce)	4 900	11 020	12 430ᶜ	8.4	2.4ᵈ
Coal	3 653	4 574	4 639	2.3	0.2
Lignite	1 911	4 437	8 250	8.8	10.9
Crude oil	375	3 542	2 568	25.2	−5.2
Petroleum products	340*	7 052	12 110*ᶜ	35.4*	11.4*ᵈ
Manufactured gas (mn m³)	321	713	688ᶜ	8.3	−0.7ᵈ
Electricity (mn kW h)	2 815	8 623	18 270	11.8	13.3
of which, hydro (mn kW h)	1 002	3 033	8 373	11.7	18.4
Mining					
Iron ore (Fe content)	494	1 663	1 840*	12.9	1.7*
Antimony ore (Sb content)	1.01	2.38	3.42*ᶜ	8.9	7.5*ᵈ
Bauxite	—	52	570ᶜ	na	61.4ᵈ
Chromium ore (oxide content)	221	302	363ᶜ	3.2	3.8ᵈ
Copper ore (Cu content)	28.6	19.9	43.5*ᶜ	−3.6	16.9*ᵈ
Magnesite	—	300	459ᶜ	na	8.9ᵈ
Mercury	0.074	0.324	0.187ᶜ	15.9	−10.4ᵈ
Zinc ore (Zn content)	4.1	24.4	27.1ᶜ	19.5	2.1ᵈ
Salt	445	648	740*ᶜ	3.8	2.7*ᵈ
Manufacturing					
Beer (000 hl)	600**ᶜ	1 000*ᶜ	1 527ᶜ	5.2**ᶜ	8.8*ᵈ
Cigarettes (mn units)	26 256	37 253	53 401ᶜ	3.6	7.5*ᵈ
Cotton yarn	69	185	127ᶜ	10.3	−7.2ᵈ
Cotton fabrics (mn m)	384	820*ᶜ	730*ᶜ	7.9**	−1.9*ᵈ
Man-made fibres	1.2	13.0*	52.7*ᶜ	27.0*	32.3*ᵈ
Wood pulp	37	101	263ᶜ	10.6	21.1ᵈ
Newsprint	10	12	80	1.4	37.2
Other paper	36**	154	332ᶜ	15.6**	16.6ᵈ
Tyres (000 units)	—	1 139	2 410	na	13.3
Sulphurᵃ	36	70	54*ᶜ	6.9	−5.0*ᵈ
Sulphuric acid	24	28	190	1.5	37.6
Fertilisers, nitrogenous	1	82	172ᶜ	55.4	16.0ᵈ
Fertilisers, phosphate	10**	61	322ᶜ	19.8**	39.6ᵈ
Cement	2 038	6 374	12 392	12.1	11.7
Coke	613	1 341	1 260*ᶜ	8.1	−1.2*ᵈ
Pig iron	248	1 034	1 516	15.3	6.6
Crude steel	266	1 312	1 456	17.3	1.7

Turkey

Main products (000 t) Manufacturing	1960	1970	1976	Growth %pa 1960–70	1970–76
Radio sets (000 units)	na	198	272[f]	na	8.3[g]
Television sets (000 units)	—	4	571[c]	na	169.7[d]
Passenger cars (000 units)[b]	—	5.0	74.6[c]	na	71.7[d]
Commercial vehicles (000 units)[b]	—	11.8	22.4	na	11.3
Merchant vessels (000 grt)	—	11	23	na	13.1

[a]Includes content of pyrites and mineral sulphur [b]Assembly only [d]1970–75
[c]Estimate to include production other than by government sector [f]1974 [g]1970–74

Transport traffic	1960	1970	1976	Growth %pa 1960–70	1970–76
Passenger-kilometres (mn)					
Rail	4 396	5 561	4 660	2.4	−2.9
Air	141	640	2 019	16.3	21.1
Cargo: tonne-kilometres (mn)					
Road	na	17 500*	25 230	na	6.3*
Rail	4 327	5 556	7 288	2.5	4.6
Air	2.1	6.0	17.3	11.3	19.3
Sea: tonnes (mn)					
Goods loaded	2.22	3.43	3.78[a]	4.4	2.0[b]
Goods unloaded	3.29	8.28	17.75[a]	9.7	16.5[b]
Tourism					
Number of visitors (000)	na	926[c]	1 676	na	12.6[d]
Gross receipts ($ mn)	6	51	180	24.9	23.4

[a]1975 [b]1970–75 [c]1971 [d]1971–76

Finance and trade

Price indices (1970 = 100)	1960	1970	1976	Growth %pa 1960–70	1970–76
Consumer prices	56.6	100.0	276.1	5.9	18.4
Wholesale prices	61.0	100.0	272.2	5.1	18.2
Money stock					
(end-year, T L mn)	9 090	34 580	152 500	14.3	28.1

Budget (1976/77; year ending February 28th)
Revenue: T L 141 924 mn = $ 8 718 mn = £ 5 024 mn
Expenditure: T L 153 637 mn = $ 9 437 mn = £ 5 438 mn,
or which, defence 22 %, education 14 %, health and social welfare 3 %

Balance of payments ($ mn)	1972	1973	1974	1975	1976
Balance of goods (fob)	−522	−560	−1 831	−2 834	−2 605
Balance of services	−151	−83	−302	−438	−465
Balance of transfers	+797	+1 259	+1 498	+1 424	+1 106[a]
Current balance	+124	+615	−634	−1 848	−1 964
Long-term capital flow	+695	+361	+404	+1 315	+2 334
Reserves and debt (end-year, $ mn)					
International reserves	1 401	2 120	1 861	1 064	1 123
External public debt	3 277	3 739	4 224	4 565	na

[a]Receipt of workers remittances from abroad: $ 983 mn

External trade (1976) Imports: T L 82 941 mn = $ 5 168 mn = £ 2 860 mn
Exports: T L 30 768 mn = $ 1 917 mn = £ 1 061 mn

Main imports	% of total	Main exports	% of total
Machinery, non-electric	25	Cotton	22
(of which, textile 3)		Fruit and vegetables	22
Crude oil	20	(of which, hazelnuts 11,	
Chemicals	15	raisins 3)	
Iron and steel	9	Tobacco	13
Electrical machinery	6	Textile yarns and fabrics	11
Motor vehicles	6	Clothing	5
Non-ferrous metals	2	Non-ferrous metal ores	3
Instruments	1	Fertilisers	2
Main sources		*Main destinations*	
West Germany	17	West Germany	19
Iraq	13	United States	10
United States	9	Switzerland	9
United Kingdom	8	Italy	9
Italy	8	United Kingdom	7
France	6	France	6
Switzerland	6	Belgium-Luxembourg	4
Libya	5	Soviet Union	4
Japan	5	Netherlands	3

Special focus

Turkish workers in other countries, 000, 1976

Country	Number	%		Number	%
West Germany	526	74	Australia	21	3
France	40	6	Other	61	8
Netherlands	39	5	Total	713	100
Austria	26	4			

United Kingdom

United Kingdom of Great Britain and Northern Ireland

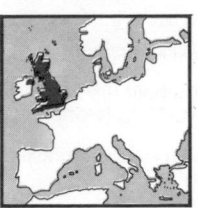

Location North-west Europe
Comprises the island of Great Britain (which includes England, Wales and Scotland), 35 km off the north-west coast of France, and Northern Ireland (the north-east part of the island of Ireland) which is 30 km west of Scotland across the Irish Sea. The islands of Great Britain and of Ireland (including the territory of the Republic of Ireland), together with the Isle of Man, are known as the British Isles
Land Area 244 103 km² = 94 249 mi², of which Great Britain 229 983 km² (England 130 441 km², Wales 20 768 km², Scotland 78 775 km²), Northern Ireland 14 120 km²
Usage (1975): agricultural 186 230 km² (76 %), of which, arable and cropland 69 810 km² (29 %), pastures 116 420 km² (48 %); forests 20 200 km² (8 %)
Climate Temperate
Weather at London (Kew), 5m altitude
Temperature: hottest month July 13–22 °C, coldest Jan 2–6 °C
Rainfall (av monthly): driest months Mar, Apr 37 mm, wettest Nov 64 mm
Time GMT (summer time, 1 hour ahead of GMT)
Measures UK (imperial) system, gradually converting to the metric system
Monetary unit Pound (£) = 100 new pence (decimal currency introduced February 15, 1971; formerly £ 1 = 20 shillings, 1 shilling = 12 pence)
Rate of exchange: 1976 av (free) $ 1.806 = £ 1
1977 av (free) $ 1.745 = £ 1

Summary

Political Parliamentary monarchy; the United Kingdom was formed in 1801 by a union of Great Britain and Ireland, and the Republic of Ireland withdrew in 1921. Also referred to as Britain. Northern Ireland had a separate parliament until 1973; Northern Ireland is sometimes referred to as Ulster, although it includes only a part of Ulster, the remainder being in Ireland. Member of UN, OECD, Council of Europe, Nato, EEC (from January 1, 1973), WEU, Commonwealth, Cento, Colombo Plan and South Pacific Commission
Economic An industrialised economy, with machinery, chemicals and motor vehicles accounting for one-half of exports, and a wide range of other manufactures making up the balance. Crude oil and products accounted for one-seventh of imports in 1976, but production of North Sea oil is becoming significant. Manufacturing occupies about one-third of the labour force and provides about one-quarter of gross domestic product. There is a substantial government sector which absorbs about 30 % of the employed labour force

People

Population 1960 52.37 mn, 1970 55.42 mn, 1976 55.89 mn
Growth: 1960–70 0.6 % pa, 1970–76 0.1 % pa
Density (1976): 229 people per km²
Vital statistics (rate per 1 000 people, 1976): births 12.1, deaths 12.2
Households (1976) 19.9* mn
Regions (standard regions, population in 000, 1976; total of 55.93 mn)

England	*49 184*	South West	4 254
North	3 122	West Midlands	5 165
Yorkshire and Humberside	4 892	North West	6 554
East Midlands	3 733	Wales	2 767
East Anglia	1 803	Scotland	5 205
South East	16 894	Northern Ireland	1 538
of which, Greater London	7 028		

Cities[g] (population in 000, 1976)

London (capital)	7 028	Edinburgh	467	Sandwell	313[a]
Birmingham	1 059[a]	Bradford	459[b]	Wigan	311[e]
Glasgow	856	Bristol	416	Sefton	306[d]
Leeds	744[b]	Kirklees	372[b]	Wakefield	306[b]
Sheffield	558[c]	Belfast	368	Newcastle-upon-Tyne	296[f]
Liverpool	540[d]	Wirral	348[d]	Leicester	289
Manchester	490[e]	Coventry	337[a]	Cardiff	281

[a]Part of West Midlands metropolitan county which has a total of 2 743 thousand
[b]Part of West Yorkshire metropolitan county which has a total of 2 072 thousand
[c]Part of South Yorkshire metropolitan county which has a total of 1 318 thousand
[d]Part of Merseyside metropolitan county which has a total of 1 578 thousand
[e]Part of Greater Manchester metropolitan county which has a total of 2 684 thousand
[f]Part of Tyne and Wear metropolitan county which has a total of 1 183 thousand
[g]‘Metropolitan districts’, introduced when the local authority areas were changed in April 1974

United Kingdom

Race 1971 : English 80** %, Scottish 9** %, Irish 2** % (including
1* % in Great Britain), Welsh 2** %, Northern Irish 2** %, West Indian
1* %, Indian 1* %, Pakistani 1** %, other 2** %
1976: Pakistani and 'new' Commonwealth origin 3.3 %
Language English; there are also Welsh and Gaelic speaking minorities
Usage (1961): English 99 %, Welsh 1 %
Religion (1970) Church of England 50** %, Roman Catholic 10** %,
Moslem 3** %, Church of Scotland 2** %, Methodist 1** %, Jewish
1** %, Baptist 1** %
Education (1974/75) Pupils: primary 6 250 000, secondary and vocational
4 699 000, higher 650 600. Teachers (1973/74): primary 227 535,
secondary and vocational 240 134, higher 36 950
Labour force (excluding Northern Ireland, 1971)

Economic activity	Number	% of total
Agriculture	634 750	3
Mining and quarrying	391 460	2
Manufacturing	8 135 790	33
Electricity, gas and water	362 300	1
Construction	1 669 130	7
Distribution and hotels	3 704 510	15
Transport and communication	1 563 920	6
Finance and real estate	1 383 660	6
Other	5 887 090	24
Unemployed	1 288 470	5
Total	25 021 080	100

Total labour force (including Northern Ireland, Dec 1976) 26 258 000
Personnel
Scientists and engineers (excluding Northern Ireland, 1968): 43 588
Physicians (1974): 75 379, 1 per 742 people
Standard of living
National income per person (1976): £ 1 965 = $ 3 550
Consumption per person (1975): energy 5 265 kg coal equivalent,
electricity (production, 1976) 4 956 kW h, newsprint 22 kg, steel 385 kg
Newspapers (1974): number 109; circulation 24 800 000*,
443* per 1 000 people
Telephones (March 1977): 22 012 300, 394 per 1 000 people

Resources and equipment

Livestock (000, June 1977) Cattle 13 899, sheep 28 030, pigs 7 665,
chickens 127 851, turkeys 5 108
Mineral reserves Coal (1977) 190 000* mn tonnes
Crude oil (1976) 3 200 mn tonnes
Natural gas (1976) 1 443 bn cubic metres
Uranium (potential, 1976) 7 400 tonnes
Petroleum refinery capacity (1975) 147 mn tonnes
Electrical capacity (1975) 78 911 megawatts,
of which, hydro 2 456 megawatts, nuclear 5 734 megawatts
Hospital beds (government establishments only, 1974) 499 958,
1 per 112 people
Roads (1976) 368 370 km = 228 895 mi, density 1.5 km per km²
Railways (1976) 18 354 km = 11 405 mi, density 0.075 km per km²
Inland waterways (1974) 1 147 km = 713 mi,
of which, in regular use: 569 km = 354 mi
Oil pipelines (1975) 2 658 km = 1 652 mi
Ships (registered, 1977)

Type	Number	Gross tons
Oil tankers	438	14 834 078
Ore and bulk carriers	228	5 346 365
General cargo	985	4 223 161
Bulk/oil carriers	38	2 913 602
Container ships	89	1 470 518
Other	1 654	2 858 627
Total	3 432	31 646 351

Ports (goods traffic, 000 tonnes, 1976)

	International loaded	International unloaded	Coastwise loaded	Coastwise unloaded	Total
London[a]	5 864	28 183	3 335	6 782	44 164
Milford Haven[b]	5 787	25 586	9 636	2 160	43 169
Tees (Middlesbrough) and Hartlepool	11 852	12 311	4 691	1 608	30 462
Southampton[b]	3 396	16 763	5 001	1 392	26 552
Immingham[b]	3 304	15 187	3 248	641	22 380
Liverpool	3 162	17 575	995	498	22 230
Manchester	3 188	5 670	2 512	3 508	14 878
Clyde (Glasgow)	859	7 113	395	1 566	9 933

[a]Including Medway ports [b]Mainly crude oil port

Airports (passengers, 000, 1976)

	Departures	Arrivals
Heathrow (London)	11 622	11 621
Gatwick (London)	2 847	2 867
Manchester	1 379	1 381
Luton	912	895

Also (1977) 56 other airports with scheduled flights

Durable equipment (Dec 1976)	000	no per 1 000 people	
Radio sets	43 000*	770*	
Television sets	17 995	322	no per
of which, colour	9 569	171	km of road
Passenger cars	14 355	257	39
Commercial vehicles	1 909	34	5.2
of which, buses and coaches	114	2	0.3
goods vehicles	1 795	32	4.9

Industrial groups (sales, $ bn, year ending in 1976)

British Petroleum Co Ltd	23.2	Unilever Ltd	6.8
Shell Transport and Trading Co Ltd	18.0	British Steel Corporation	5.5
BAT Industries Ltd	10.7	Imperial Group Ltd	5.5
Electricity Council and Boards	7.5	National Coal Board	4.4
		BL (British Leyland)	4.2
Imperial Chemical Industries Ltd	7.5	General Electric Co Ltd	3.7
		Esso Petroleum Co Ltd	3.5
Post Office	6.9	British Gas Corporation	3.5
		Rio Tinto-Zinc Corp Ltd	3.0

Production

Gross domestic product 1976: £ 121 966 mn = $ 220 295 mn
Growth in real terms: 1960–70 2.8 %pa, 1970–76 2.0 %pa
Structure of gross domestic product (1976)

By origin	£ mn	% of gdp
Agriculture	3 116	3
Mining and quarrying	2 458	2
Manufacturing	30 464	25
Electricity, gas and water	3 905	3
Construction	7 793	6
Distribution	10 379	9
Transport and communication	10 315	8
Other	53 536	44
Total	121 966	100
By type of expenditure		
Government final consumption	26 573	22
Private final consumption	73 102	60
Stock investment	372	–
Gross fixed capital formation	23 414	19
Exports of goods and services	35 446	29
less Imports of goods and services	−36 941	−30
Total	121 966	100

Production indices (1970 = 100)	1960	1965	1970	1973	1974	1975	1976	Growth % pa 1960–70	1960–65	1965–70	1970–76
Agricultural	82	97	100	108	112	107	103	2.0	3.4	0.5	0.4
Industrial[s]	76.0	89.1	100.0	110.0	106.3	100.6	101.3	2.8	3.2	2.3	0.2
of which, manufacturing	75.0	87.6	100.0	110.5	108.9	102.2	103.2	2.9	3.1	2.7	0.5
construction	76.8	95.1	100.0	107.9	93.8	86.1	84.5	2.7	4.4	1.0	−2.8
Main products (000 t)											
Agriculture											
Wheat	3 040	4 171	4 236	5 002	6 130	4 488	4 740	3.4	6.5	0.3	1.9
Barley	4 309	8 191	7 529	9 007	9 133	8 513	7 648	5.7	13.7	−1.7	0.2
Oats	2 091	1 232	1 217	1 080	955	795	764	−5.3	−10.0	−0.2	−7.5
Potatoes	7 273	7 578	7 482	6 845	6 791	4 551	4 789	0.3	0.8	−0.2	−7.2
Sugar, raw value	982	936	984	1 046	622	697*	760*	0.0	−1.0	1.0	−4.2*

Main products (000 t)	1960	1965	1970	1973	1974	1975	1976	Growth % pa 1960–70	1960–65	1965–70	1970–76
Agriculture											
Broad beans, dry	80	80*	160	188	200	91	110*	7.2	0.0*	14.9*	−6.0*
Peas, green	400**	413	549	684	777	692	715*	3.2**	0.6**	5.9	4.5*
Cabbages	750**	797	959	745	824	733	674*	2.5**	1.2**	3.8	−5.7*
Cauliflowers	250**	344	296	346	343	293	300*	1.7**	6.6**	−3.0	0.2*
Carrots	350**	450**	549	606	495	573	555	4.6**	5.1**	4.1**	0.2
Onions	42	85	144	223	231	201	175	13.1	15.1	11.1	3.3
Tomatoes	90	81	108	120	119	181	179	1.8	−2.1	5.9	8.8
Apples	698	616	596	490	397	386	389*	−1.6	−2.5	−0.7	−6.9
Strawberries	30**	34	46	59	61	45	47*	4.3**	2.6**	6.0	0.4*
Hops	12.0*	13.2	12.2	10.4	10.2	8.3	8.0	0.2*	1.9*	−1.6	−6.8
Rapeseed	—	3*	8*	30	56	61	111	na	na	22.0*	55.0*
Milk	12 080	12 862	12 971	14 402	13 993	13 937	14 420	0.7	1.2	0.2	1.8
Butter	45	41	64	97	54	48	90	3.6	−1.8	9.3	5.9
Cheese	112	115	134	182	218	235	204	1.8	0.5	3.1	7.2
Eggs	756	835	892	826	814	799	805	1.7	2.0	1.3	−1.7
Wool	36.6	38.7	30.6	32.0	32.8	32.5	31.4	−1.8	1.1	−4.6	0.4
Cattle hides	79	87*	116	98	132	153*	126*	3.9	2.0*	5.9*	1.4*
Beef and veal	780	881	869	810	998	1 150	1 035*	1.1	2.5	−0.3	2.9*
Sheep and goat meat	231	247	231	237*	252	261	254*	0.0	1.3	−1.3	1.6*
Pigmeat	664	906	925	979	983	815	816*	3.4	6.4	0.4	−2.1*
Poultrymeat	290	391	579	663	652	626	663	7.1	6.2	8.2	2.3
Fish catch	924	1 047	1 099	1 136	1 089	980	1 051	1.7	2.5	1.0	−0.7
Timber (000 m³)	3 264	3 502	3 421	3 364	3 337	2 998	3 343	0.5	1.4	−0.5	−0.4
Energy											
Total energy (000 tce)	197 640	193 420	163 730	175 000	162 500	184 000	na	−1.9	−0.4	−3.3	2.4ᶠ
Coal	196 711	190 499	147 109	131 981	110 451	128 682	123 809	−2.9	−0.6	−5.0	−2.8
Crude oil	148	84	83	88	410	1 564	12 036	−5.6	−10.7	−0.2	129.2
Petroleum products	42 651	64 366	99 942	105 950	103 060	86 647	90 284	8.9	8.6	9.2	−1.7
Natural gas (mn m³)ᵈ	1	13	11 270	29 590	35 730	37 230	39 410	154.0	67.0	287.0	23.2
Manufactured gas (mn m³)	22 727	23 288	20 580	10 788	7 806	5 440*	3 680*	−1.0	0.5	−2.4	−24.9*
Electricity (mn kW h)	137 300*	196 495	249 193	282 048	273 319	272 219	276 970	6.1*	7.4*	4.9	1.8
of which, nuclear (mn kW h)	2 079	16 338	26 012	27 997	33 617	30 338	37 000*	28.7	51.0	9.7	6.0*
Mining											
Iron ore (Fe content)	4 688	4 229	3 365	1 926	936	1 143	1 190	−3.3	−2.0	−4.5	−15.9
Tin conc (Sn content)	1.2	1.3	1.7	3.6	3.2	3.3	3.3	3.5	1.8	5.2	12.2
Salt	5 860	7 000	9 188	8 374	8 283	7 630	na	4.6	3.6	5.6	−3.6ᶠ
Manufacturing											
Margarine	374	319	315	341	299	296	343	−1.7	−3.1	−0.2	1.4
Beer (000 hl)	43 369	48 447	55 165	60 582	63 057	64 565	65 500	2.4	2.2	2.6	2.9
Cigarettes (mn units)	122 248	112 796	145 308	127 266	125 333	121 329	na	1.7	−1.6	5.2	−3.5ᶠ
Cotton yarn	270	220	159	115	101	95	107	−5.2	−4.0	−6.3	−6.4
Cotton fabrics (mn m)	1 183	920	620	454	409	406	374	−6.1	−4.7	−7.5	−8.3
Jute yarn	144	132	88	68	57	51	62	−4.8	−1.7	−7.8	−5.6
Wool yarn	248	249	227	235	210	187	190	−0.9	0.1	−1.8	−2.9
Man-made fibres	269	391	599	731	628	562	618	8.3	7.8	8.9	0.5
Wood pulp	265*	285*	432	356	342	319	na	5.0*	1.5*	8.7*	−5.9ᶠ
Newsprint	753	780	757	439	383	315	321	0.1	0.7	−0.6	−13.3
Other paper	3 386	3 834	4 184	4 265	4 245	3 343	3 856	2.1	2.5	1.8	−1.3
Synthetic rubber	92	174	316	364	336	261	320	13.1	13.6	12.7	0.3
Tyres (000 units)	15 904	23 392	30 545	30 287	27 499	26 008	na	6.7	8.0	5.5	−3.2ᶠ
Ethylene	na	538	998	1 247	1 275	959	na	na	na	13.2	−0.8ᶠ
Sulphuric acid	2 745	3 358	3 352	3 886	3 855	3 166	3 272	2.0	4.1	−0.1	−0.4
Plastics and resins	600*	1 020*	1 560*	2 108	2 108	1 804	2 510	10.0*	11.2*	8.9*	8.2*
Fertilisers, nitrogenousᵃ	401	598	710	751	755	997	1 055	5.9	8.3	3.5	6.8
Fertilisers, phosphateᵃ	400*	433*	434	444	417	440	464	0.8*	1.6*	0.0*	1.1
Cement	13 501	16 977	17 171	19 986	17 781	16 891	15 781	2.4	4.7	0.2	−1.4
Coke	20 600**	18 800	18 141	17 864	13 580	15 859	na	−1.3**	−1.8**	−0.7	−2.7ᶠ
Pig iron	16 016	17 740	17 672	16 838	13 903	12 131	13 835	1.0	2.1	−0.1	−4.0
Crude steel	24 695	27 440	28 315	26 649	22 426	20 198	22 274	1.4	2.1	0.6	−3.9
Aluminium	29	36	40	252	293	308	334	3.3	4.4	2.1	42.7
Copperᵉ	219	228	206	171	160	151	137	−0.6	0.8	−2.0	−6.6
Lead	52	45	140	120	137	105	na	10.4	−2.9	25.5	−5.6ᶠ
Tin	26.8	16.8	22.0	20.4	12.1	11.6	na	−2.0	−8.9	5.5	−12.0ᶠ
Zinc	75	107	147	84	84	53	42	7.0	7.4	6.6	−18.9
Radio sets (000 units)	2 504	1 912	1 313	1 376	990	696	na	−6.2	−5.2	−7.2	−11.9ᶠ
Television sets (000 units)	2 141	1 591	2 214	3 130	2 640	2 120	2 100	0.3	−5.8	6.8	−0.9
of which, monochrome (000 units)	2 141	1 591	1 725	1 010	670	540	560	−2.1	−5.8	1.6	−17.1
colour (000 units)	—	—	489	2 120	1 950	1 580	1 540	na	na	na	21.1
Washing machines (000 units)ᵇ	1 019	891	956	1 320	990	1 250	1 240	−0.6	−2.7	1.4	4.4
Domestic refrigerators (000 units)ᵇ,ᶜ	1 047	1 074	1 118	1 240	1 177	1 380	1 630	0.6	0.5	0.8	6.5
Passenger cars (000 units)	1 353	1 722	1 641	1 747	1 534	1 268	1 333	2.0	4.9	−1.0	−3.4
Commercial vehicles (000 units)	458	455	458	417	403	381	372	0.0	−0.1	0.1	−3.4
Merchant vessels (000 grt)	1 331	1 073	1 237	1 010	1 262	1 294	1 347	−0.7	−4.2	2.9	1.4
Construction											
Dwellings (000 units)	304	391	362	304	280	322	325	1.8	5.2	−1.5	−1.8

ᵃYears ending May 31st   ᵇManufacturers' sales   ᶜIncluding deep-freeze units   ᵈGas and petroleum fields only   ᵉOf which, secondary: 86 thousand tonnes in 1976   ᶠ1970–75
ᵍIncluding construction

United Kingdom

Transport traffic	1960	1970	1976	Growth %pa	
Passenger-kilometres (mn)				1960–70	1970–76
Road[a]	215 000	362 000	411 000[b]	5.3	2.6[c]
Rail[a]	34 676	35 708	36 840[b]	0.3	0.6[c]
Air	6 372	17 432	30 948	10.6	10.0
of which, international	5 603	15 440	28 598	10.7	10.8
Cargo: tonne-kilometres (mn)					
Road[a]	52 000	87 800	91 800[b]	5.4	0.9[c]
Rail[a]	30 500*	26 807	20 448	−1.3*	−4.4
Air	181	604	914	12.9	7.1
of which, international	175	568	898	12.5	7.9
Sea: tonnes (mn)					
Goods loaded	33.0	50.3	55.6	4.3	1.7
Goods unloaded	128.0*	200.0	181.5	4.6*	−1.6
Tourism					
Number of visitors (000)	na	6 730	10 089	na	7.0
Gross receipts ($ mn)	473	1 037	2 839	8.2	18.3

[a]Excluding Northern Ireland [b]1975 [c]1970–75

Finance and trade

Price and wage indices (1970 = 100)	1960	1970	1976	Growth %pa	
				1960–70	1970–76
Consumer prices	67.9	100.0	215.0	3.9	13.6
Wholesale prices	73.8	100.0	219.6	3.1	14.0
Wages (earnings)	54	100	245	6.4	16.1
Share prices	68.8	100.0	114.6	3.8	2.3
Money stock					
(end-year, £ mn)	6 975*	10 140*	19 467	3.8*	11.5*

Budget (1977/78; year ending March 31st)
Revenue: £ 38 773 mn = $ 69 752 mn
Expenditure: £ 43 989 mn = $ 79 135 mn,
of which, education 4 %, defence 15 %, social services and health 26 %

Balance of payments ($ mn)	1972	1973	1974	1975	1976
Balance of goods (fob)	−1 491	−5 441	−11 745	−6 547	−5 972
Balance of services	+2 418	+4 286	+4 356	+3 821	+4 989
Balance of transfers	−670	−1 103	−949	−1 033	−1 541
Current balance	+257	−2 258	−8 338	−3 759	−2 524
Long-term capital flow	−2 262	−610	+4 388	+988	+956
Reserves and debt (end-year, $ mn)					
International reserves	5 647	6 476	6 939	5 459	4 230
Public debt[a]: domestic	88 492	87 384	92 365	95 806	101 787
external	4 896	4 678	4 467	6 190	6 611

[a]At end-March; nominal amounts

External trade (1977) Imports: £ 36 996 mn = $ 64 577 mn
Exports: £ 33 308 mn = $ 58 139 mn

Main imports	% of total	Main exports	% of total
Food	15	Machinery, non-electric	18
(of which,		Chemicals	12
fruit and vegetables 3,		(of which, plastics 2)	
meat 3, cereals 2)		Motor vehicles	8
Crude oil	11	Electrical machinery	8
Machinery, non-electric	11	Diamonds	6
Chemicals	7	Crude oil and products	6
Motor vehicles	6	Food	4
Diamonds	5	Textile yarns & fabrics	3
Electrical machinery	5	Iron and steel	3
Non-ferrous metals	3	Metal small	
Textile yarns & fabrics	3	manufactures	3
Iron and steel	3	Non-ferrous metals	3
Paper	3	Instruments	2
Petroleum products	3	Clothing	2
Clothing	2	Alcoholic spirits	2
Instruments	2		
Metal ores	2		
Timber	2		
Main sources[a]		*Main destinations*[b]	
United States	10	United States	9
West Germany	10	West Germany	8
France	7	France	7
Netherlands	7	Netherlands	6
Belgium-Luxembourg	5	Belgium-Luxembourg	6
Italy	4	Ireland	5
Switzerland	4	Switzerland	4
Ireland	4	Sweden	4
Sweden	3	Italy	3
Canada	3	Nigeria	3
Saudi Arabia	3	Denmark	2
Japan	3	Norway	2
South Africa	2	Australia	2

Main sources (cont)	% of total	Main destinations (cont)	% of total
Norway	2	Canada	2
Denmark	2	Iran	2
Iran	2	South Africa	2
Soviet Union	2	Saudi Arabia	2
Finland	2	Japan	1

[a]EEC 38 % [b]EEC 36 %

Special focus

Employment by sector

	Employed labour force, mid-year				
Number (000)	1972	1973	1974	1975	1976
Private sector	17 711	18 192	18 148	17 662	17 447
Government sector	6 679	6 778	6 912	7 267	7 314
Public Corporations	1 929	1 890	1 962	2 012	1 951
Local Authorities	2 771	2 890	2 844	2 993	3 012
Central Government	1 979	1 998	2 106	2 262	2 342
Total	24 390	24 970	25 060	24 929	24 761
Percentage					
Private sector	72.6	72.9	72.4	70.8	70.5
Government sector	27.4	27.1	27.6	29.2	29.5
Public Corporations	7.9	7.6	7.8	8.1	7.9
Local Authorities	11.4	11.6	11.3	12.0	12.2
Central Government	8.1	8.0	8.4	9.1	9.5
Total	100.0	100.0	100.0	100.0	100.0

Vatican

State of the Vatican City
Stato della Città del Vaticano

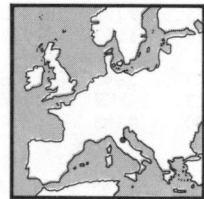

Location Southern Europe
The city state is within the city of Rome in Italy; there are also areas outside the main city state
Land Area 0.44 km² = 0.17 mi²
Climate Mediterranean (also see Italy)
Time 1 hour ahead of GMT (summer time, 2 hours ahead)
Measures Metric system
Monetary unit
Vatican City lira (V L) = 100 centesimi
At par with the Italian lira which is also in use
Rate of exchange (1976 av): free V L 832.3 = $ 1, V L 1 503 = £ 1

Summary

Political Ecclesiastical state, seat of the Holy See, the government of the Roman Catholic Church
Economic Supported by contributions from adherents to the Roman Catholic Church. Operates a radio station

People, resources and equipment

Population 1960 1 000*, 1970 1 000*, 1976 1 000*
Growth: 1960–70 0.0* % pa, 1970–76 0.0* % pa
Density (1976): 2 300* people per km²
Vital statistics (rate per 1 000 people, 1977): births 0, deaths 3
Language Italian
Religion Roman Catholic
Newspapers (1977): number 1 (L'Osservatore Romano)
Railways A railway station is operated
Airport A heliport is operated
Radio A radio station is operated (Radio Vatican)
Passenger cars (1976) 200*, 200* per 1 000 people

Oceania

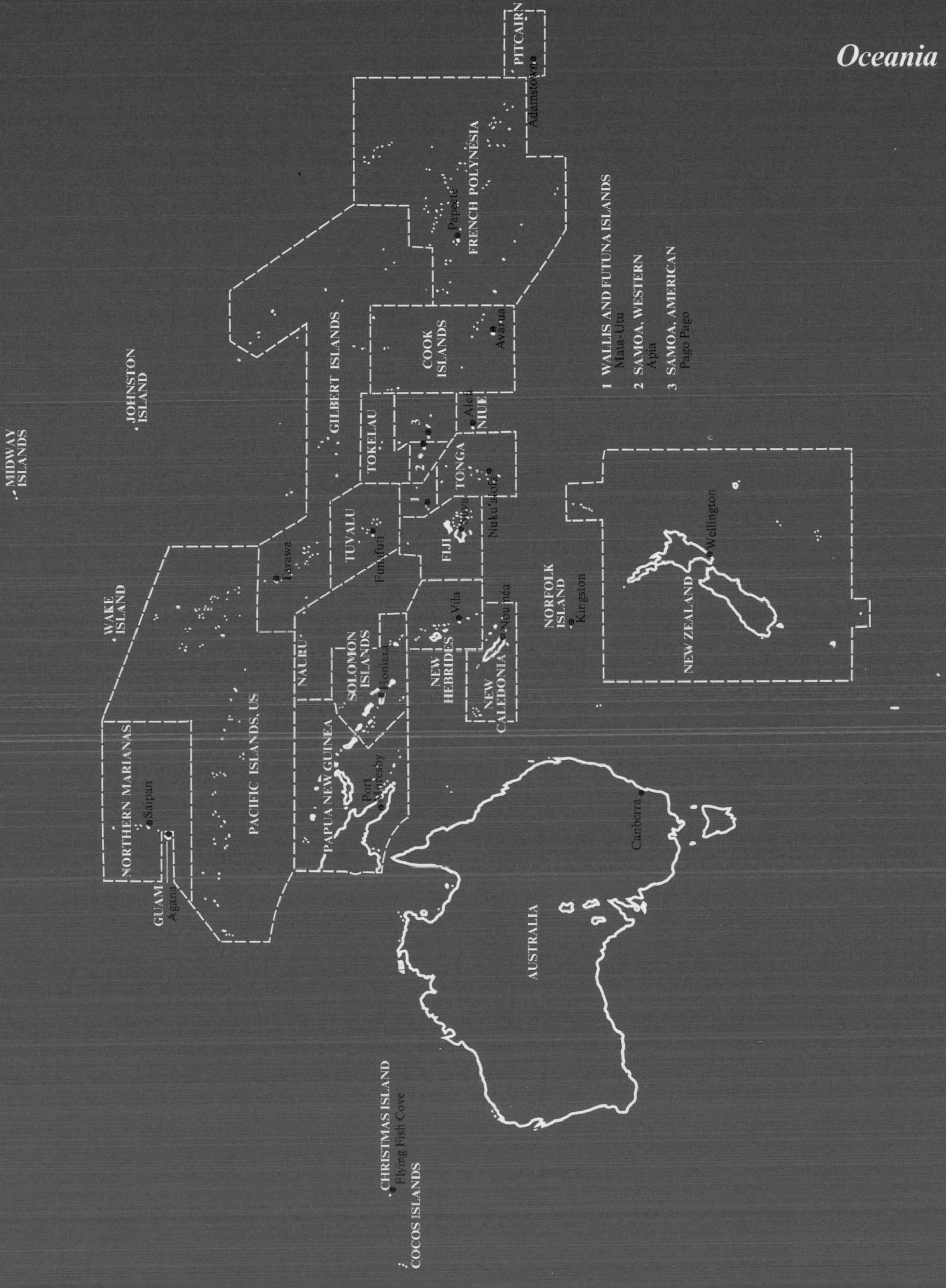

PITCAIRN

Adamstown

FRENCH POLYNESIA

Papeete

WALLIS AND FUTUNA ISLANDS
1 WALLIS AND FUTUNA ISLANDS
   Mata-Utu
2 SAMOA, WESTERN
   Apia
3 SAMOA, AMERICAN
   Pago Pago

JOHNSTON ISLAND

COOK ISLANDS

Avarua

MIDWAY ISLANDS

GILBERT ISLANDS

TOKELAU

Alofi
NIUE

TONGA

Nuku'alofa

Wellington

NEW ZEALAND

NORFOLK ISLAND

Kingston

WAKE ISLAND

TUVALU

Funafuti

Tarawa

FIJI

NAURU

NEW HEBRIDES

Vila

NEW CALEDONIA

Nouméa

SOLOMON ISLANDS

Honiara

NORTHERN MARIANAS

Saipan

PACIFIC ISLANDS, US

PAPUA NEW GUINEA

Port Moresby

GUAM

Agana

AUSTRALIA

Canberra

CHRISTMAS ISLAND

Flying Fish Cove

COCOS ISLANDS

Australia
Commonwealth of Australia

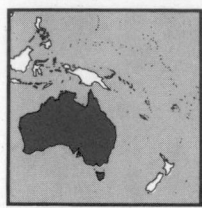

Location South-west Pacific Ocean
Comprises the main island continent of Australia with the Coral and Tasman Seas (Pacific Ocean) to the east and Indian Ocean to the west, and the offshore island of Tasmania. Indonesia, East Timor and Papua New Guinea are to the north of the Torres Strait and across the Timor and Arafura Seas, and New Zealand is 1 800 km to the south-east across the Tasman Sea
Land Area 7 686 848 km² = 2 967 909 mi²
Usage (1975): agricultural 5 008 740* km² (65* %), of which arable 457 000* km² (6* %), cropland 1 740 km² (0.02 %), pastures 4 550 000* km² (59* %); forests 1 377 000* km² (18* %)
Climate Temperate in south, sub-tropical or tropical in north, hot and dry inland
Weather at Canberra, 560 m altitude
Temperature: hottest months Jan, Feb 13–28 °C, coldest July 1–11 °C
Rainfall (av monthly): driest month April 41 mm, wettest Oct 56 mm
Time Hours ahead of GMT (summer time in brackets)

New South Wales, Tasmania, Victoria	10	(11)
Queensland	10	
Northern Territory	9½	
South Australia	9½	(10½)
Western Australia	8	

Measures Metric system, as conversion from UK (imperial) system; there is a general target date for completion of conversion by 1980
Monetary unit Australian dollar (A $) = 100 cents; decimal currency, introduced on February 14, 1966, to replace the Australian pound (A £) at A $ 2 = A £ 1
Rate of exchange (1976 av): free A $ 0.8162 = $ 1 ($1.225 = A $ 1), A $ 1.474 = £ 1

Summary

Political Parliamentary monarchy, which became independent from the United Kingdom on January 1, 1901. Member of UN, OECD, Commonwealth, South Pacific Commission, Colombo Plan and Anzus Treaty. There is a free trade agreement (Nafta) with New Zealand
Economic A mixed industrial economy, with important agricultural and mining sectors, and with a manufacturing sector contributing one-fifth of gross domestic product; main exports are metal ores, wool, coal, meat and wheat, with a wide range of other manufactures and agricultural produce. Australia has become a major raw material supplier to Japan

People

Population 1960 10.28 mn, 1970 12.51 mn, 1976 13.92 mn
Growth: 1960–70 2.0 % pa, 1970–76 1.8 % pa
Density (1976): 2 people per km²
Vital statistics (rate per 1 000 people, 1976): births 16.7, deaths 8.3
Households (1971) 3.67 mn; population in households 12.16 mn (95 % of total)
Regions (population in 000, mid-1976; total of 13.92 mn)

States		Territories	
New South Wales	4 914	Victoria 3 746 Northern Territory[a]	101
Queensland	2 112	Western Australian Capital	
South Australia	1 262	Australia 1 170 Territory	203
Tasmania	407		

[a]Self-governing from July 1, 1978; to become the seventh state
Cities (population in 000, 1976)

Canberra (capital)	215	Perth	805	Townsville	82[a]
Sydney	2 936	Newcastle	363	Gold Coast	80[a]
Melbourne	2 604	Wollongong	211	Toowoomba	63[a]
Brisbane	958	Hobart	162	Rockhampton	51[a]
Adelaide	900	Geelong	132	Ballarat	38[a]

[a]1975
Race (1966) English, Scottish and Irish 96 %, Italian 1 %, Greek 1 %. There were 106 208 aboriginals in 1971 (0.8 % of total population)
Language English
Religion (1971) Church of England 31 %, Roman Catholic 27 %, Methodist 9 %, Presbyterian 8 %, Greek Orthodox 3 %, other Christian 9 %
Education (1973) Pupils: primary 1 811 027, secondary, vocational and teacher-training 1 042 384, higher (1975) 288 571.
Teachers: primary 70 911, secondary, vocational and teacher-training 65 211, higher (1975) 20 951

Labour force (1973)

Economic activity	Number	% of total
Agriculture	407 000	7
Mining and quarrying	68 000	1
Manufacturing, electricity, gas and water	1 442 000	25
Construction	491 000	8
Distribution and hotels	1 166 000	20
Transport and communication	427 000	7
Finance and real estate	398 000	7
Other services	1 241 000	21
Armed forces	74 000	1
Unemployed	108 000	2
Total	5 822 000	100

Personnel (1972) Physicians: 17 972, 1 per 721 people
Standard of living
National income per person (1976): A $ 5 133 = $ 6 288 = £ 3 482
Consumption per person (1975): energy 6 485 kg coal equivalent, electricity (production, 1975/76) 5 532 kW h, newsprint 38 kg, steel 467 kg
Newspapers (1973): 58; circulation 5 126 000, 386 per 1 000 people
Telephones (June 1976): number 5 501 510, 397 per 1 000 people

Resources and equipment

Livestock (000, 1976) Cattle 33 434, sheep 148 643, pigs 2 173*, camels 2*, horses 446*, chickens 50 608, turkeys 503*
Mineral reserves Coal (1972) 111 865 mn tonnes
Lignite (1973) 86 702 mn tonnes
Crude oil (1975) 316 mn tonnes
Natural gas (1975) 855 bn cubic metres
Uranium (1976) 289 000 tonnes
Iron ore (metal content, 1973) 5 600 mn tonnes
Bauxite (1973) 5 600 mn tonnes
Petroleum refinery capacity (1975) 36* mn tonnes
Electrical capacity (public only, 1975) 19 506 megawatts, of which, hydro 5 535 megawatts
Hospital beds (1972) 160 552, 1 per 81 people
Roads (1974) 837 866 km = 520 626 mi, density 0.11 km per km²
Railways (government, 1976) 40 753 km = 25 323 mi, density 0.005 km per km²
Ships (registered, 1977) 424, total of 1 374 197 gross tons
Ports (goods traffic, 000 freight tons, 1975/76)

	International		Coastwise		Total
	loaded	unloaded	loaded	unloaded	
Kembla	5 377	661	1 976	8 137	16 151
Fremantle	6 392	4 204	2 341	2 106	15 043
Sydney	6 003	5 002	489	2 168	13 662
Melbourne	2 829	5 079	2 243	2 585	12 736
Newcastle	9 417	1 192	1 015	4 345	11 624

Airports (passengers, 000, 1976)

	Departures	Arrivals
Sydney	3 227	3 264
Melbourne	2 371	2 393
Brisbane	1 162	1 179
Perth	421	431
Cairns	145	143
Darwin	113	113

Also (1977) 261 other airports with scheduled flights

Durable equipment (at end-year)	000	no per 1 000 people	
Radio sets (1974)	2 851	208	no per
Television sets (1974)	3 022	221	km of road
Passenger cars (1976)	5 284	378	6.3
Commercial vehicles (1976)	1 291	92	1.5

Production

Gross domestic product
1976/77 (year ending June 30th):
A $ 82 224 mn = $ 94 722 mn = £ 55 295 mn
1976: A $ 77 170 mn = $ 94 533 mn = £ 52 354 mn
Growth in real terms (years ending June 30th):
1960–70 5.2 % pa, 1970–76 3.9 % pa

Australia

Structure of gross domestic product

By origin (1975/76)	A $ mn	% of gdp
Agriculture	3 516	5
Mining and quarrying	2 574	4
Manufacturing	13 545	19
Electricity, gas and water	1 982	3
Construction	5 151	7
Distribution	9 479	13
Transport and communication	4 807	7
Other	30 224	42
Total	71 278	100
By type of expenditure (1976/77)		
Government final consumption	13 249	16
Private final consumption	47 814	58
Stock investment	1 010	1
Gross fixed capital formation	19 169	23
Exports of goods and services	13 067	16
less Imports of goods and services	−13 252	−16
Total (incl statistical discrepancies)	82 224	100

Production indices (1970 = 100)	1960	1970	1976	Growth % pa 1960–70	1970–76
Agricultural	75	100	115	3.0	2.4
Manufacturing	61	100	116	5.0	2.5

Main products (000 t)	1960	1970	1976	Growth % pa 1960–70	1970–76
Agriculture					
Wheat	5 402	7 890	11 825	3.8	7.0
Barley	775	2 351	2 850	11.7	3.3
Oats	849	1 613	1 073	6.6	−6.6
Rice	128	247	417	6.8	9.1
Maize	171	192	131	1.2	−6.2
Sorghum	163	547	1 124*	12.9	12.8*
Potatoes	589	762	697	2.6	−1.5
Sugar, raw value	1 308	2 514	3 296	6.8	4.6
Tomatoes	128	176	165*	3.2	−1.1*
Onions	58	93	106*	4.8	2.2*
Peas, green	90**	140*	131*	4.5**	−1.1*
Apples	268	425	430	4.7	0.2
Pears	108	190	164	5.8	−2.4
Grapes	452	758	731	5.3	−0.6
Wine	132	288	363	8.1	3.9
Raisins	71	94	65	2.9	−6.0
Peaches	62	115	85*	6.4	−4.9*
Oranges	146	322	384*	8.2	3.0*
Pineapples	90	141	125*	5.2	−2.0*
Bananas	125	132	119*	0.9	−1.7*
Sunflowerseed	3	13	80	16.0	35.4
Tobacco	9.7	18.9	15.0*	6.9	−3.8*
Cotton	2	29	25	31.0	−2.4
Milk	6 592	7 522	6 442	1.3	−2.5
Butter	201	203	148	0.1	−5.1
Cheese	45	77	113	5.5	6.6
Tallow	130*	130*	250*	0.0*	11.5*
Eggs	144	185	183	2.5	−0.2
Wool	419	497	452	1.7	−1.6
Cattle hides	63*	118*	203*	6.5*	9.5*
Sheepskins	40*	113*	98*	11.0*	−2.3*
Beef and veal	643	1 055	1 840	5.1	9.7
Mutton and lamb	583	813	578	3.4	−5.5
Pigmeat	109	182	174	5.3	−0.7
Poultrymeat	46	116	204	9.7	9.9
Fish catch	61	101	114	5.4	2.0
Timber (000 m³)	14 500**	13 462	13 450	−0.7**	0.0
Energy					
Total energy (000 tce)	28 530	67 620	111 050ᵍ	9.0	10.4ʰ
Coal	20 400*	45 214	74 854	8.3*	8.8
Lignite	14 200*	24 175	31 240	5.5*	4.4
Crude oilᵃ	—	4 125	20 515	na	30.6
Petroleum products	11 313	22 987	24 604ᵍ	7.3	1.4ʰ
Natural gas (mn m³)ᵃ	—	781	5 906	na	40.1
Manufactured gas (mn m³)ᵃ	2 512	4 130*	7 100**ᵍ	5.1*	11.4**ʰ
Electricity (mn kW h)ᵃ	22 300*	53 892	76 597	9.2*	6.0
of which, hydro (mn kW h)ᵃ	4 036	9 161	15 217ᵍ	9.7	10.7ʰ
Mining					
Iron ore (Fe content)ᵃ	2 800*	28 676	59 300	26.0*	12.9
Antimony ore (Sb content)ᵃ	1.05*	0.97	1.54ᵍ	−0.8*	9.7ʰ
Bauxiteᵃ	43*	8 294	21 080	69.0*	16.8

Main products (000 t)	1960	1970	1976	Growth % pa 1960–70	1970–76
Mining					
Copper conc (Cu content)ᵃ	104*	143	217	3.2*	7.2
Lead conc (Pb content)ᵃ	317*	459	391	3.8*	−2.6
Magnesiteᵃ	62.4*	23.5	36.0ᵍ	−9.3*	8.9ʰ
Manganese ore (Mn content)ᵃ	37*	396	673ᵍ	26.7*	11.2ʰ
Nickel conc (Ni content)ᵃ	—	18.0	49.1ᵍ	na	22.2ʰ
Tin conc (Sn content)	2.21	8.83	9.08	14.9	0.5
Tungsten conc (oxide content)ᵃ	0.90*	1.76	1.58ᵍ	6.9*	−2.1ʰ
Uranium (U content)	0.285ⁱ	0.254	0.360	−1.9ʲ	6.0
Zinc conc (Zn content)ᵃ	301*	502	468	5.2*	−1.2
Silver	0.47	0.80*	0.76	5.5*	−0.9*
Gold (000 kg)	33.80	19.42	15.50	−5.4	−3.7
Saltᵃ	473*	2 054	5 057ᵍ	15.8*	19.8ʰ
Manufacturing					
Margarineᵃ	43	63	87ᵍ	3.9	6.5ʰ
Beer (000 hl)ᵃ	10 532	15 533	19 573ᵍ	4.0	4.7ʰ
Cigarettes (mn units) ᵃ,ᵉ	17 135	27 174	29 972ᵏ	4.7	2.5ˡ
Cotton yarnᵃ	21.1*	29.3	25.8	3.3*	−2.1
Cotton fabrics (mn m²)ᵃ	38	58	55	4.3	−0.9
Wool yarnᵃ	23.0	23.8	22.6	0.3	−0.8
Man-made fibresᵃ	10*	31*	24*ᵍ	12.0*	−5.0*ʰ
Wood pulpᵃ	370*	859	967ᵍ	8.8*	2.4ʰ
Newsprint	90	170	205	6.6	3.1
Other paperᵃ	370**	879	946ᵍ	9.0**	1.5ʰ
Synthetic rubber	—	33.0	42.2	na	4.2
Tyres (000 units)ᵃ	4 223	7 227	8 690	5.5	3.1
Sulphur ᵇ,ᵈ	116	103	108ᵍ	−1.2	1.0ʰ
Sulphuric acidᵃ	1 088	1 762	1 300	4.9	−4.9
Nitric acidᵃ	17	126	148	22.1	2.7
Caustic sodaᵃ	47	112	137	9.1	3.4
Plastics and resinsᵃ	59	225	407	14.3	10.4
Fertilisers, nitrogenousᵃ	24	160*	180*	20.6*	2.0*
Fertilisers, phosphateᵃ	530*	804	474	4.3*	−8.4
Cement	2 799	4 499	5 039	4.9	1.9
Coke	3 000*	4 977	5 239ᵍ	5.2*	1.0ʰ
Pig ironᵃ	2 600**	5 769	7 420	8.3**	4.3
Crude steelᵃ	3 600*	6 874	7 760	6.7*	2.0
Aluminiumᵃ	12	168	232	30.2	5.5
Radio sets (000 units)ᵃ	403	729	941ᵏ	6.1	6.6ˡ
Television sets (000 units)ᵃ	435	320	533	−2.3	8.9
Passenger cars (000 units)ᵃ,ᶜ	220**	390	370	6.0**	−0.9
Commercial vehicles (000 units)ᵃ,ᶜ	70**	88	86	2.3**	−0.4
Merchant vessels (000 grt)	28	54	46	6.8	−2.6
Construction					
Dwellings (000 units)ᶠ	95*	143	142	4.2*	−0.1

ᵃYears ending June 30th ᵇContent of pyrites ᶜIncluding assembly ᵈShipments ᵉIncluding cigars ᶠIncluding extensions, restorations and conversions ᵍ1975 ʰ1970–75 ⁱ1964 ʲ1964–70 ᵏ1974 ˡ1970–74

Transport traffic	1960	1970	1976	Growth % pa 1960–70	1970–76
Passenger-kilometres (mn)					
Road	na	126 900ᵈ	157 200ᵉ	na	7.4ᶠ
Air	3 008	9 268	19 384	11.9	13.1
Cargo: tonne-kilometres (mn)					
Road	na	24 752	33 000ᵇ	na	5.9ᶜ
Rail	13 090	25 403	31 509	6.9	3.7
Air	105	267	399	9.8	6.9
Sea: tonnes (mn)					
Goods loadedᵃ	11.3	79.7	158.6	21.6	12.2
Goods unloadedᵃ	17.9	32.5	26.9	6.1	−3.1
Tourism					
Number of visitors (000)	85	338	532	14.8	7.9
Gross receipts ($ mn)	na	147	306	na	13.0

ᵃYears ending June 30th ᵇ1975 ᶜ1970–75 ᵈ1971 ᵉ1974 ᶠ1971–74

Finance and trade

Price and wage indices (1970 = 100)	1960	1970	1976	Growth % pa 1960–70	1970–76
Consumer prices	78.4	100.0	184.9	2.5	10.8
Wholesale prices	90.9	100.0	176.3	1.0	9.9
Wages (earnings)	57.6	100.0	226.6	5.7	14.6
Share prices	56.5	100.0	80.9	5.9	−3.5
Money stock (end-year, A $ mn)	3 551	5 446	10 646	4.4	11.8

Australia

Budget (Central government, 1975/76; year ending June 30th)
Revenue: A $ 17 901 mn = $ 22 571 mn = £ 11 315 mn
Expenditure: A $ 21 448 mn = $ 27 044 mn = £ 13 558 mn,
of which, defence 7 %, social security and welfare 22 %,
health 8 %

Balance of payments ($ mn)	1972	1973	1974	1975	1976
Balance of goods (fob)	+2 104	+2 785	+101	+2 217	+2 025
Balance of services	−1 348	−1 994	−2 292	−2 474	−3 017
Balance of transfers	−199	−322	−426	−334	−427
Current balance	*+557*	*+469*	*−2 617*	*−591*	*−1 419*
Long-term capital flow	+1 663	−668	+808	+592	+1 458
Reserves and debt (end-year, $ mn)					
International reserves	6 141	5 697	4 269	3 256	3 170
Public debt[a]: domestic	14 997	19 100	21 231	21 992	23 492
external	1 717	1 793	1 535	1 567	1 637

[a]Mid-year

External trade (1976) Imports: A $ 10 184 mn = $ 12 476 mn = £ 6 909 mn
Exports: A $ 10 743 mn = $ 13 160 mn = £ 7 288 mn

Main imports (1976/77[a])	% of total	*Main exports* (1976/77[a])	% of total
Machinery, non-electric	16	Metal ores	14
Motor vehicles	10	(of which, iron ore 8)	
Electrical machinery	10	Wool	13
Crude oil and products	10	Coal	11
Chemicals	9	Meat	8
Textile yarns & fabrics	6	Wheat	7
Food	4	Sugar	6
Instruments	3	Non-ferrous metals	5
Paper	3	Alumina	5
Metal small manufactures	3	Iron and steel	4
Clothing	2	Machinery	3
Main sources (1976)		*Main destinations* (1976)	
Japan	21	Japan	34
United States	21	United States	9
United Kingdom	12	New Zealand	5
West Germany	6	United Kingdom	4
New Zealand	3	Soviet Union	4
Canada	3	West Germany	3
Saudi Arabia	3	Italy	3
Hong Kong	3	Canada	3
Italy	2	France	2
Kuwait	2	China	2
Singapore	2	Malaysia	2
Sweden	2	Indonesia	2

[a]Year ending June 30th

Special focus

Wool prices (UK wholesale price, Australia/New Zealand 64s)

	£ per tonne	$ per tonne	% change over previous year ($ basis)
1970	787	1 890	−13
1971	708	1 722	−9
1972	1 174	2 937	+71
1973	2 749	6 741	+129
1974	2 117	4 952	−27
1975	1 736	3 858	−22
1976	2 192	3 960	+3
1977	2 454	4 283	+8

Christmas Island

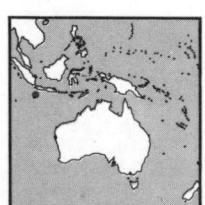

Location East Indian Ocean
Australia is 1 410 km to the south-east of the
island, Java (Indonesia) 360 km to the north,
and Cocos Islands 417 km to the west
Land Area 135 km² = 52 mi²
Climate Sub-tropical
Temperature (annual av): 27 °C (with little
variation)
Rainfall (av monthly): 144 mm; wet season
Nov–April
Time 7 hours ahead of GMT
Measures UK (imperial) system, changing to metric system
Monetary unit Australian dollar (A $) = 100 cents
Rate of exchange (1976 av): free $ 1.225 = A $ 1, A $ 1.474 = £ 1

Summary

Political Australian external territory from October 1, 1958
Economic Phosphates are the main product; the British Phosphate
Commission acts as managing agents for the Christmas Island Phosphate
Commission (controlled by Australia and New Zealand)

People, resources, equipment, production and trade

Population[a] 1960 3 140*, 1970 3 000*, 1976 3 260*
[a]Mainly employees of British Phosphate Commission
Growth: 1960–70 −0.5* %pa, 1970–76 1.4* %pa
Density (1976): 24* people per km²
Vital statistics (rate per 1 000 people, 1973): births 10, deaths 2
Town Flying Fish Cove
Race (1976) Chinese 56 %, Malay 29 %, European 12 %. There is no
indigenous people
Language English
Education (1976) Pupils: primary and secondary 625, vocational 500
Labour force (1971) 1 377
National income per person (1976) A $ 2 500*** = $ 3 000*** = £ 1 700***
Consumption per person (1975) Energy 20 537 kg coal equivalent,
electricity (production) 9 235 kW h
Telephones (Dec 1976) 350, 107 per 1 000 people
Electrical capacity (1975) 9 megawatts
Port (anchorage) Flying Fish Cove
Airport Christmas Island
Gross domestic product
1976 est: A $ 50***mn = $ 60***mn = £34***mn
Main products (1975) Phosphate rock 1 487 000 t, electricity 28 mn kW h
Transport traffic (1974) Sea: loaded 1.69 mn t, unloaded 0.05 mn t
External trade *Main exports* (1976/77; year ending June 30th)
Phosphate rock 996 000 t, phosphate dust 123 000 t
Export destinations Mainly Australia, New Zealand, south-east Asia

Cocos Islands
or Keeling Islands

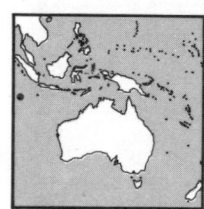

Location East Indian Ocean
Includes 27 islands, of which the main islands are
Home and West, with Perth (Australia) 2 770 km
to the south-east, Colombo (Sri Lanka)
2 300 km to the north-west, and Christmas Island
417 km to the east
Land Area 14 km² = 5.5 mi²
Climate Sub-tropical
Temperature (annual range): 21–32 °C
Rainfall (av monthly): 142 mm
Time 6½ hours ahead of GMT
Measures UK (imperial) system and metric system
Monetary unit Australian dollar (A $) = 100 cents
Rate of exchange (1976 av): free $ 1.225 = A $ 1, A $ 1.474 = £ 1

Summary

Political Australian external territory from November 23, 1955; internal
control was exercised by the Clunies-Ross family from 1831. Australia
announced an agreement in July 1978 to buy the territory from
Mr J Clunies-Ross for A $ 6¼ mn
Economic Coconuts, copra and coconut oil are the main products, and
aviation services are important

People, resources, equipment, production and trade

Population 1960 605*, 1970 624*, 1976 500*
Growth: 1960–70 0.3* %pa, 1970–76 −3.6* %pa
Density (1976): 36* people per km²
Vital statistics (rate per 1 000 people, 1973): births 23, deaths 6
Race (1976) Malay 75** %, European (mainly Scottish) 25** %
Language English
National income per person (1976) A $ 1 000*** = $ 1 200*** = £ 700***
Telephones (Dec 1976) 83, 180 per 1 000 people
Airport West Island
Radio sets (1975) 650, 1 076 per 1 000 people
Gross domestic product
1976 est: A $ 0.6***mn = $ 0.7***mn = £ 0.4***mn
Main products (000 t, 1976) Coconuts 3*, copra 0.24*
External trade *Main export* (1975/76) Copra 423 t

Cook Islands

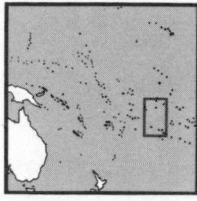

Location South central Pacific Ocean
Comprises 15 main islands spread over about
2 mn km² of the Pacific Ocean, roughly
3 000 km north-east of New Zealand. There are
two main groups, Northern and Southern
Land Area 241 km² = 93 mi²
Climate Sub-tropical (warm and humid)
Weather at Rarotonga, 5 m altitude
Temperature: hottest month Jan 23–29 °C,
coldest July 18–25 °C
Rainfall (av monthly): driest month July 112 mm, wettest March 284 mm
Time 10½ hours behind GMT
Measures UK (imperial) system, converting to metric system
Monetary unit Cook Islands dollar (CK $) = 100 cents
The Cook Islands dollar is equal to the New Zealand dollar, also in use
Rate of exchange (1976 av): free CK $ 1.004 = $ 1, CK $ 1.813 = £ 1

Summary

Political New Zealand associated territory, having internal self-government and with New Zealand responsible for external affairs; the Cook Islands can proclaim full independence should they so wish. Territorial member of South Pacific Commission
Economic An agricultural economy, with fruit products as the main exports; there is some light industry. Tourism is being developed, aided by a new international airport opened at Rarotonga in 1973. Remittances sent home by workers in New Zealand substantially help the economy

People, resources and equipment

Population 1960 18 100*, 1970 20 900*, 1976 18 100
Growth: 1960–70 1.4* %pa, 1970–76 — 2.4* %pa
Density (1976): 75 people per km²
Vital statistics (rate per 1 000 people, 1976): births 25.2, deaths 5.9
Regions (population in 000, 1976) main islands, Rarotonga 9.8, Aitutaki 2.4, Atiu 1.5, Mangaia 1.3
Town Avarua (capital), on Rarotonga
Race Mainly Polynesian
Language English and Polynesian
Religion (1966) Christian 95 %
Education (1975) Pupils 6 615, teachers 360
Labour force (1966) 5 768
Personnel Scientists and engineers (1970): 164
Physicians (1972): 22, 1 per 909 people
Standard of living
National income per person (1976): CK $ 800** = $ 800** = £ 450**
Consumption per person (1975): electricity 443 kW h
Newspapers (1972): number 1; circulation 1 000, 48 per 1 000 people
Telephones (Dec 1976): 956, 53 per 1 000 people
Livestock (000, 1976) Pigs 10*, goats 2*, horses 2, chickens 63*
Electrical capacity (1975) 3* megawatts
Hospital beds (1974) 179, 1 per 111 people
Ports Avatiu, Avarua, Mangaia
Airports Rarotonga, Aitutaki
Radio sets (Dec 1975) 7 000, 390 per 1 000 people

Production, finance and trade

Gross domestic product 1972: CK $ 8.1mn = $ 9.7mn = £ 3.9mn
1976 est: CK $ 18**mn = $ 18**mn = £ 10**mn
Main products *Agriculture* (000 t, 1976) Cassava 3*, sweet potatoes 8*, coconuts 11*, copra 1*, oranges 2*, mandarines 5*, grapefruit 1*, lemons 1*, bananas 1*, pineapples 3, mangoes 2*, avocados 1* fish catch 1* *Other* (1975) Electricity 8 mn kW h
Budget (1976/77; year ending March 31st)
Balanced at CK $ 10.98 mn = $ 10.7 mn = £ 6.5 mn; includes NZ assistance of CK $ 3.70 mn = $ 3.6 mn = £ 2.2 mn
External trade Imports (1973): CK $ 4.9 mn = $ 6.7 mn = £ 2.8 mn
Exports (1975): CK $ 2.0 mn = $ 2.4 mn = £ 1.1 mn

Main imports (1973)	% of total	*Main exports* (1975)	% of total
Food	22	Preserved fruit	70
Machinery	11	Clothing	13
Textile yarns and fabrics	10	Oilseeds and nuts	8
Chemicals	7	Fresh fruit	3
Main sources (1973)		*Main destinations* (1975)	
New Zealand	83	New Zealand	98
Japan	5	Japan	1

Fiji

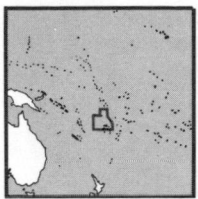

Location South central Pacific Ocean
Includes main islands Viti Levu and Vanua Levu
and 840 islands and islets, of which about 100 are
permanently inhabited; New Zealand is about
2 000 km to the south. The island of Rotuma is
part of the territory
Land Area 18 272 km² = 7 055 mi²
of which, Viti Levu 10 390 km², Vanua Levu
5 535 km²
Climate Tropical
Weather at Suva, 6 m altitude
Temperature: hottest months Jan–Mar 23–30 °C, coldest July–Aug 20–26 °C
Rainfall (av monthly): driest month July 124 mm, wettest Mar 368 mm
Time 12 hours ahead of GMT
Measures Metric system, as conversion from UK (imperial) system; all customs work metric from January 1, 1977
Monetary unit Fiji dollar (F $) = 100 cents; decimal currency, introduced from January 13, 1969, to replace the Fiji pound (F £) at F $ 2 = F £ 1
Rate of exchange (1976 av): free $ 1.111 = F $ 1, F $ 1.626 = £ 1

Summary

Political Parliamentary monarchy, which became independent from the United Kingdom on October 10, 1970. Member of UN, Commonwealth, Colombo Plan, South Pacific Commission and an EEC ACP state
Economic An agricultural economy dependent mainly on exports of sugar. Tourism is also important, with gross receipts amounting to about one-half of exports. Light manufacturing industry is being encouraged with tax concessions

People, resources and equipment

Population 1960 390 000*, 1970 520 000*, 1976 588 000
Growth: 1960–70 2.9* %pa, 1970–76 2.1 %pa
Density (1976): 32 people per km²
Vital statistics (rate per 1 000 people, 1975): births 29.0, deaths 6.9
Towns (population in 000, 1976) Suva (capital) 64, Lautoka 29*, Nadi 13*, Labasa 13*, Nausori 13*, Ba 9*, Vatukoula 6*, Sigatoka 4*, Levuka 3*
Race (1976) Indian 50 %, Fijian 44 %, European (including mixed) 2.6 %, Rotuman 1.2 %, Chinese 0.8 %
Language English, Fijian and Hindustani
Religion (1966) Hindu 40 %, Methodist 38 %, Roman Catholic 8 %, Moslem 8 %, Church of England 1 %
Education Pupils (1976) 167 882, teachers (1974) 5 770*
Labour force (1976) 176 322; in agriculture 77 009 (44 %)
Personnel Scientists and engineers (1969): 315
Physicians (1971): 256, 1 per 2 070 people
Standard of living
National income per person (1976): F $ 1 005* = $ 1 117* = £ 618*
Consumption per person (1975): energy 582 kg coal equivalent, electricity (production) 410 kW h, newsprint 1.9 kg
Newspapers (1974): number 1; circulation 20 000, 36 per 1 000 people
Telephones (Dec 1976): 30 759, 52 per 1 000 people
Livestock (000, 1976) Cattle 156*, goats 55*, pigs 31*, horses 35*, chickens 785*
Electrical capacity (1975) 83 megawatts
Hospital beds (1971) 1 513, 1 per 350 people
Roads (1975) 2 960 km = 1 839 mi, density 0.16 km per km²
Railways[a] (1976) 644 km = 400 mi, density 0.035 km per km²
[a] For sugar estates; permanent only. In addition 225 km of portable tracks
Ships (registered, 1977) 33, total of 10 879 gross tons
Ports (goods traffic, 000 tonnes, 1976)

	loaded	unloaded
Suva	78	428
Lautoka	251	100
Levuka	—	10

Airports Nadi (217 km from Suva), Nausori (24 km from Suva), and 10 other airports with scheduled flights

Durable equipment	000	no per	
(Dec 1975)		1 000 people	no per
Radio sets	300	524	km of road
Passenger cars	21	37	7.1
Commercial vehicles	11	19	3.7

Fiji

Production

Gross domestic product 1976: F $ 596.3 mn = $ 662 mn = £ 367 mn
Growth in real terms: 1970–74 7.1 %pa
Structure of gross domestic product *By origin* (1973) Agriculture 21 %, mining and quarrying 2 %, manufacturing 12 %, construction 7 %, transport and communication 6 %, other 52 %
By type (1976) Final consumption expenditure 87 % (of which, government 14 %), stock investment 2 %, gross fixed capital formation 16 %, exports of goods and services 39 %, less imports of goods and services −44 %

Main products (000 t)	1960	1970	1976	Growth %pa 1960–70	1970–76
Agriculture					
Rice	30	20	24*	−4.0	3.1*
Sweet potatoes	10**	10**	8*	0.0**	−3.6**
Cassava	80**	86*	91*	0.7**	0.9*
Sugar, raw value	287	361	298*	2.3	−3.2*
Copra	30	28	29*	−0.7	0.6*
Milk	11	41	45*	14.0	1.5*
Fish catch	na	4.1	5.5	na	5.0
Timber (000 m³)	80**	123	163	4.6**	4.8
Other					
Electricity (mn kW h)	55	158	241a	11.1	8.8b
Gold (000 kg)	2.25	3.23	1.90a	3.7	−10.1b
Beer (000 hl)	20	72	157a	13.7	16.9b
Cigarettes (mn units)	231	389	473a	5.3	4.0b
Cement	—	60	73a	na	4.0b

a1975 b1970–75

Transport traffic (1976) *Air* Passenger-kilometres 108 mn, cargo 1.2 mn t-km *Sea* goods loaded 0.44 mn t, unloaded 0.72 mn t
Tourism (1976) Number of visitors 168 665, gross receipts $ 76 mn

Finance and trade

Consumer price index (1970 = 100) 1976 186.2; growth 1970–76 10.9 %pa
Money stock (end-year, F $ mn) 1976 83.7; growth 1970–76 13.7 %pa
Budget (1976) Revenue: F $ 128.8 mn = $ 143 mn = £ 79 mn
Expenditure: F $ 129.7 mn = $ 144 mn = £ 80 mn

Balance of payments ($ mn)	1972	1973	1974	1975	1976
Balance of goods (fob)	−68	−112	−92	−73	−107
Balance of services	+28	+49	+61	+49	+58
Balance of transfers	+9	+6	+2	−1	−4
Current balance	−32	−57	−29	−25	−52
Long-term capital flow	+27	+50	+50	+47	+19
Reserves and debt (end-year, $ mn)					
International reserves	69	74	109	149	116
External public debt	46	66	69	80	87

External trade (1976) Imports: F $ 237 mn = $ 263 mn = £ 146 mn
Exports: F $ 125 mn = $ 139 mn = £ 77 mn

Main imports (1975)	% of total	*Main exports* (1975)	% of total
Food	17	Sugar	72
(of which, cereals 6)		Petroleum products	12
Petroleum products	17	Gold	6
Chemicals	8	Coconut (copra) oil	4
Machinery, non-electric	8		
Electrical machinery	7		
Textile yarns and fabrics	5		
Instruments	4		
Motor vehicles	4		
Main sources (1976)		*Main destinations* (1976)	
Australia	29	United Kingdom	40
Japan	18	Australia	18
New Zealand	14	New Zealand	10
United Kingdom	11	Singapore	5
Singapore	9	United States	4
United States	4	Western Samoa	3
Hongkong	3	Malaysia	3

Special focus

Central position in the south-west Pacific

Distance from Nadi Airport to:	kilometres	nautical miles
Auckland (New Zealand)	2 156	1 164
Sydney (Australia)	3 167	1 710
Port Moresby (Papua New Guinea)	3 390	1 831
Manila (Philippines)	7 146	3 859
Vancouver (Canada)	9 451	5 103
San Francisco (United States)	8 790	4 746
Hawaii (United States)	5 104	2 756

French Polynesia

Territory of French Polynesia
Territoire de la Polynésie Française

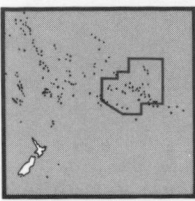

Location South central Pacific Ocean
Comprises about 130 islands scattered over a large area; the main island is Tahiti (Windward Islands). Hawaii is 4 500 km to the north
Land Area 4 000 km² = 1 500 mi², of which, Tahiti 1 042 km²
Climate Tropical
Weather at Papeete, 92 m altitude
Temperature: hottest months Jan-April 22–32 °C coldest July, Aug 20–30 °C
Rainfall (av monthly): driest month Aug 43 mm, wettest Jan 251 mm
Time 10 hours behind GMT
Measures Metric system
Monetary unit CFP franc (CFP Fr) = 100 centimes
Rate of exchange (1976 av):
par CFP Fr 1 = Fr 0.055 (CFP Fr 18.18 = Fr 1),
free CFP Fr 86.90 = $ 1, CFP Fr 157.0 = £ 1

Summary

Political French overseas territory, which chose to become a territory in November 1958. Member of franc zone and territorial member of South Pacific Commission
Economic An agricultural economy, with copra the main product (phosphate deposits were exhausted in 1966); tourism is very important. France provides substantial aid; this territory includes the French nuclear test centre on Mururoa Atoll

People, resources and equipment

Population 1960 80 000**, 1970 115 000*, 1976 134 000*
Growth: 1960–70 3.7** %pa, 1970–76 2.6* %pa
Density (1976): 33* people per km²
Vital statistics (rate per 1 000 people, 1972): births 33.7, deaths 7.2
Regions (circonscriptions, population in 000, 1971; total of 119 000)

Archipel de la Société a	100	Archipel des Tuamotu-Gambier	8
Iles du Vent b	85	Iles Australes	5
Iles sous le Vent c	16	Iles Marquises d	6

aSociety Archipelago bWindward Islands cLeeward Islands dMarquesa Islands
Cities (population in 000, 1977) Papeete a (capital) 63
aOn Tahiti (Windward Islands, Society Archipelago)
Race (1970) Polynesian 85** %, Chinese 11** %, European 3** %
Language French; Polynesian languages, including Tahitian, are also used
Religion (1970) Protestant 60** %, Roman Catholic 30** %
Education (1973/74) Pupils 42 770, teachers 1 771
Labour force (1971) 34 906
Personnel Scientists and engineers (1973): 95
Physicians (1972): 62, 1 per 1 900 people
Standard of living National income per person (1976):
CFP Fr 400 000** = $ 4 600** = £ 2 500**
Consumption per person (1975): energy 877 kg coal equivalent, electricity (production) 725 kW h
Newspapers (1972): number 3; circulation 11 000, 94 per 1 000 people
Telephones (Dec 1976): number 14 679, 108 per 1 000 people
Livestock (000, 1976) Cattle 13*, pigs 16*, sheep 3*, goats 4*, chickens 175*
Electrical capacity (1975) 29* megawatts
Hospital beds (1972) 961, 1 per 121 people
Roads (1976) 583 km = 362 mi, density 0.14 km per km²
Port (goods traffic, 000 tonnes, 1973) Papeete: loaded 34, unloaded 442
Airports Faa (Papeete), and 17 other airports with scheduled flights

Durable equipment at end-year	000	no per 1 000 people	
Radio sets (1974)	70	560	no per
Television sets (1974)	13	105	km of road
Passenger cars (1972)	17	148	30
Commercial vehicles (1972)	6	50	10

Production, finance and trade

Gross domestic product 1975: CFP Fr 45 000 mn = $ 577 mn = £ 260 mn
1976 est: CFP Fr 55 000** mn = $ 630** mn = £ 350** mn
Main products (000 t, 1976) Cassava 6*, sweet potatoes 2*, potatoes 1*, sugar cane 2*, bananas 1*, coconuts 165*, copra 23*, fish catch 2.8, electricity (mn kW h, 1975) 95, beer (000 hl) 64*, mother-of-pearl 0.1*
Transport traffic (1976) *Air* (Papeete only, 000) Passenger departures 119, arrivals 117 *Sea* see port

French Polynesia

Tourism (1976) Number of visitors 117 246, gross receipts $ 49 mn
Consumer price index (Papeete only, 1970 = 100) 1976 171.5;
growth 1970–76 9.4 %pa
Budget (1976) Balanced at CFP Fr 14 000 mn = $ 161 mn = £ 89 mn
External trade (1976)
Imports: CFP Fr 25 700 mn = $ 296 mn = £ 164 mn
Exports: CFP Fr 1 911 mn = $ 22 mn = £ 12 mn
Main imports (1974) Food 20 %, petroleum products 7 %,
machinery, motor vehicles, cement, textiles
Main exports (1974) Re-exports (mainly nuclear material) 81 %,
cocnut oil 15 %, vanilla 1 %
Main sources France 53 %, United States 19 %
Main destination France 82 %

Gilbert Islands

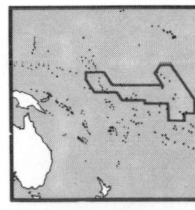

Location South-west Pacific Ocean
Comprises 28 main islands in 4 sections: the
main Gilbert Island group of 16 main islands,
Phoenix Islands of 8 main islands, Line Islands
of 3 main islands, and Ocean Island (Banaba).
The international date line divides the Gilbert
and Ocean islands to the west from the
Phoenix and Line Islands to the east
Land Area 860km² = 330mi²,
of which, Ocean Island 6 km²

Climate Tropical
Weather at Tarawa, 3 m altitude
Temperature: hottest month Oct 26–32 °C, coldest July 25–31 °C
Rainfall (av monthly): driest month Nov 58 mm, wettest Dec 317 mm

Time	Hours ahead of GMT (+)		Hours behind GMT (−)
Ocean Island	11¼	Phoenix Islands	11
Gilbert Islands	12	Line Islands	10

Measures UK (imperial) system, converting to the metric system
Monetary unit Australian dollar (A $) = 100 cents
Rate of exchange (1976 av): free $ 1.225 = A $ 1, A $ 1.474 = £ 1

Summary

Political UK colony, formerly the Gilbert and Ellice Islands (the
Ellice Islands became a separate colony from October 1, 1975 as Tuvalu);
internal self-government was obtained January 1, 1977 and full
independence is planned. The Banabans, people of Ocean Island, wish to
form an independent country associated with Fiji; they have been living on
Rambi (Rabi) in the Fiji group since the end of the second world war. The
Gilbert Islands are also known as Kingsmill Islands. Territorial member
of the South Pacific Commission
Economic The main product has been phosphates obtained by the
British Phosphate Commission from Ocean Island; the Gilbert Islands
receive a royalty from these, as do the original inhabitants of Ocean
Island (Banabans). Revenue from the phosphates is expected to end in
1979; the Banabans wish to restore some form of agriculture and a
fishing industry to Ocean Island. Other main products in the Gilbert
groups are coconuts, copra, bananas and fish

People, resources and equipment

Population 1960 41 000**, 1970 49 300*, 1976 59 000*
Growth: 1960–70 1.9* %pa, 1970–76 3.0* %pa
Density (1976): 69* people per km²
Vital statistics (including Tuvalu, rate per 1 000 people, 1971):
births 22.3, deaths 6.5
Regions (population in 000, 1968; total of 47 578)
Gilbert Islands (includes Tarawa) 44.2, Phoenix Islands nil,
Line Islands (includes Christmas Island) 1.2, Ocean Island 2.2
Town (population in 000, 1973) Tarawa (capital) 17
Race (1968) Micronesian 94** %, Polynesian 3** %, European 1** %,
Mixed 1** %
Language Gilbertese and English
Religion (1976) Roman Catholic 50** %, Protestant 50** %
Education (1973) Pupils 18 500**, teachers 420**
Labour force (including Tuvalu, 1968) 13 279; seamen (1975) 1 000*,
phosphate workers (1975) 1 150*

Personnel Scientists and engineers (including Tuvalu, 1971): 112
Physicians (1975): 26, 1 per 2 270 people
Standard of living
National income per person (1976): A $ 800** = $ 1 000** = £ 500**
Consumption per person (1975): energy 346 kg coal equivalent,
electricity (production) 85* kW h
Telephones (incl Tuvalu, Dec 1974): number 485, 8 per 1 000 people
Livestock (000, 1976) Pigs 10*, chickens 154*
Electrical capacity (1975) 2* megawatts
Hospital beds (including Tuvalu, 1974) 634, 1 per 99 people
Roads (1976) 640*km = 400*mi, density 0.74*km per km²
Ships (registered, 1977) 2, total of 1 333 gross tons
Ports Betio Harbour (Tarawa), Ocean Island (moorings)
Airports Bonriki (Tarawa), Abemama, Tabiteuea, Butaritari, Marakei,
Nonouti, Beru
Radio sets (Dec 1975) 8 200, 140 per 1 000 people

Production, finance and trade

Gross domestic product
1976 est: A $ 50**mn = $ 60**mn = £ 34**mn
Main products (000 t) *Agriculture* (1976) Coconuts 74*, copra 11*,
bananas 3*, fish catch 0.7* *Other* (1975) Electricity (mn kW h) 5*,
phosphate rock 529
Transport traffic (including Tuvalu, 1975) *Sea* Goods loaded 535 000 t,
unloaded 20 000 t
Budget (1977) Revenue: A $ 11.77 mn = $ 13.1 mn = £ 7.5 mn
Expenditure: A $ 9.72 mn = $ 10.8 mn = £ 6.2 mn
External trade (1976)
Imports: A $ 10.1 mn = $ 12.3 mn = £ 6.8 mn
Exports: A $ 18.1 mn = $ 22.2 mn = £ 12.3 mn
Main imports Food 29 %, petroleum products, textiles, metal manufactures
Main exports Phosphates 95 %, copra 5 %

Main sources[a] (1974)	% of total	*Main destinations* (1974)	% of total
Australia	60	Australia	44
United Kingdom	11	New Zealand	42
New Zealand	3	United Kingdom	14

[a]Including trade of Tuvalu

Guam

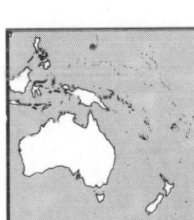

Location West Pacific Ocean
Southernmost of the Mariana Islands; Hawaii
(United States) is about 6 000 km to the east
Land Area 549 km² = 212 mi²
Climate Tropical
Weather at Sumay, 20 m altitude
Temperature: hottest months May, June
25–31 °C, coldest Feb 23–29 °C
Rainfall (av monthly): driest month Apr 51 mm,
wettest Aug 376 mm

Time 10 hours ahead of GMT (summer time, 11 hours ahead)
Measures US system
Monetary unit US dollar ($) = 100 cents
Rate of exchange (1976 av): $ 1.806 = £ 1

Summary

Political US territory, self-governing; the territory includes a major
US naval and air base. Territorial member of South Pacific Commission
Economic Tourism is important, and reduces dependence on military
expenditure. Operation as a free port has encouraged some light industry,
including watch assembly; crude oil refining began in 1970. The main
agricultural export product is copra

People, resources and equipment

Population 1960 67 004, 1970 84 996, 1976 102 000*
Growth: 1960–70 2.4 % pa, 1970–76 3.1* % pa
Density (1976): 186* people per km²
Vital statistics (rate per 1 000 people, 1975): births 30.4, deaths 4.2
Town (population in 000, 1970) Agana (capital) 5 (excluding armed forces)
Race (excluding armed forces) Chamorros (Indonesian/Filipino/Spanish)
and Micronesian

Guam

Language English and Chamorro
Religion (1976) Roman Catholic 96* %
Education (1974/75) Pupils 31 300*, teachers 1 600*
Labour force (1970) 32 699; in armed forces 9 997 (31 %)
Personnel (1975) Physicians: 66, 1 per 1 575 people
Standard of living
National income per person (1976): $ 6 000*** = £ 3 300***
Consumption per person (1975): energy 6 097 kg coal equivalent,
electricity (production) 12 500* kW h
Newspapers (1974): number 2; circulation 22 000, 210 per 1 000 people
Telephones (Dec 1976): 39 060, 370 per 1 000 people
Livestock (000, 1976) Cattle 2*, pigs 8*, chickens 105*
Petroleum refinery capacity (1975) 1.7* mn tonnes
Electrical capacity (1975) 310* megawatts
Hospital beds (government establishments only, 1973) 250, 1 per 376 people
Roads (1976) 370* km = 230* mi, density 0.67* km per km²
Port Apra Harbour
Airport Guam International (Agana)

Durable equipment (at end-year)	000	no per 1 000 people	
Radio sets (1975)	85	817	no per km of road
Television sets (1974)	46	442	
Passenger cars (1975)	48	468	130
Commercial vehicles (1975)	10	98	27

Production, finance and trade

Gross domestic product 1976 est: $ 650** mn = £ 360** mn
Main products (000 t) *Agriculture* (1976) Coconuts 27*, copra 1*,
eggs 1.1*, fish 0.1*
Other (1975) Petroleum products 1 100*, electricity (mn kW h) 1 300*
Transport traffic (1974/75) *Sea* Goods loaded 0.10 mn t,
unloaded 0.48 mn t
Tourism Number of visitors (1976) 205 000, gross receipts (1973) $ 90 mn
Budget (1975) Revenue: $ 117 mn = £ 53 mn
(of which, US federal grant-in-aid $ 15 mn = £ 7 mn)
Expenditure: $ 119 mn = £ 54 mn
External trade (1974) Imports: $ 259 mn = £ 111 mn
Exports: $ 20 mn = £ 9 mn
Main imports Crude oil, machinery, food, textiles
Main exports Petroleum products, copra, watches, scrap metal
Main sources United States, Hongkong, Taiwan, Australia
Main destinations United States, Japan, Taiwan, Hongkong

Johnston Island

Johnston and Sand Islands

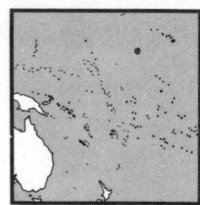

Location North Pacific Ocean
Two islands, Johnston Island and Sand Island,
comprise the territory, with Hawaii about
1 100 km to the north-east
Land Area 1.0* km² = 0.4* mi²
Climate Tropical
Time 11 hours behind GMT
Measures US system
Monetary unit US dollar ($) = 100 cents
Rate of exchange (1976 av): $ 1.806 = £ 1

Summary

Political US territory; a US air base
Economic Mainly a US military operation; the landing field takes up
virtually all of the islet

People and resources

Population 1960 156, 1970 1 007, 1976 1 400***
Growth: 1960–70 20.5 % pa, 1970–76 5.6*** % pa
Density (1976): 1 400*** people per km²
Regions Sand Island is generally uninhabited
Language English
Airport Johnston Island

Midway Islands

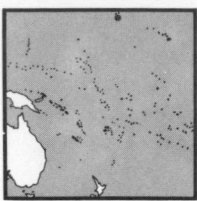

Location North Pacific Ocean
A coral atoll and two islands with Honolulu
(Hawaii) about 2 000 km to the south-east
Land Area 5 km² = 2 mi²
Climate Sub-tropical
Time 11 hours behind GMT
Measures US system
Monetary unit US dollar ($) = 100 cents
Rate of exchange (1976 av): $ 1.806 = £ 1

Summary

Political US territory; an air and naval station is operated on a
caretaker basis
Economic The airport takes up the main part of the island. There is some
agriculture

People, equipment and production

Population 1960 2 356, 1970 2 220, 1976 2 300*
Growth: 1960–70 −0.6 % pa, 1970–76 0.6* % pa
Density (1976): 460* people per km²
Language English
National income per person (1976) $ 6 000*** = £ 3 300***
Telephones (Dec 1976) 1 345, 580 per 1 000 people
Gross domestic product 1976 est: $ 14*** mn = £ 8*** mn

Nauru

Republic of Nauru
Naoero

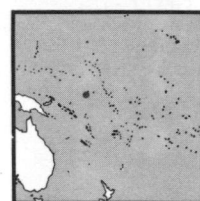

Location West Pacific Ocean
Australia is 2 000 km to the south-west, with
Ocean Island (Gilbert Islands) 300 km to the east
Land Area 21 km² = 8 mi²
Climate Tropical
Weather at Nauru, 27 m altitude
Temperature: hottest months Apr, May
24–32 °C, coldest Jan 23–31 °C
Rainfall (av monthly): driest month May 53 mm,
wettest Jan 315 mm
Time 11½ hours ahead of GMT
Measures UK (imperial) system, converting to the metric system
Monetary unit Australian dollar (A $) = 100 cents
Rate of exchange (1976 av): free $ 1.225 = A $ 1, A $ 1.474 = £ 1

Summary

Political Republic, which became independent on January 31, 1968;
formerly administered under a UN trusteeship agreement by Australia,
New Zealand and the United Kingdom. Member of ICAO, ITU and UPU
(UN agencies), Commonwealth (associate member) and South Pacific
Commission
Economic Phosphate mining is the only industry, and reserves are
expected to last until about 2000; Nauruans bought the phosphate rights
from the former owners, British Phosphate Commissioners, in 1967, and
payment of the A $ 21 mn agreed was made by April 1969; control was
passed over July 1, 1970. Income from phosphates is being used to
develop shipping and air services, and there are plans to develop the
island as a tax haven and tourist centre. There is some agriculture

People, resources and equipment

Population 1960 4 400*, 1970 6 650*, 1976 7 500*
Growth: 1960–70 4.2* % pa, 1970–76 2.0* % pa
Density (1976): 357* people per km²
Vital statistics (rate per 1 000 people, 1972): births 33.5, deaths 5.5
Administrative centre Domaneab
Race (1976) Nauruan 59 %, other Pacific Islanders[a] 25 %, Chinese[a] 9 %,
European[a] 7 %
[a]Mainly temporary immigrants working on phosphate deposits
Language Nauruan and English
Religion (1976) Nauruan Protestant 25* %
Education (1975) Pupils 1 973, teachers 130*

Nauru

Labour force (phosphate workers, 1974) 3 346
Personnel (1971) Physicians: 10, 1 per 700 people
Standard of living National income per person (1976):
A $ 10 000*** = $ 12 000*** = £ 7 000***
Consumption per person (1975): energy 6 626 kg coal equivalent, electricity (production) 3 610 kW h
Telephones (Dec 1976): 1 450, 190 per 1 000 people
Livestock (000, 1976) Pigs 2*, chickens 4*
Electrical capacity (1975) 9* megawatts
Hospital beds (1971) 207, 1 per 33 people
Roads (1976) 24** km = 15** mi, density 1.1** km per km²
Railways (1976) 5.2 km =3.2 mi, density 0.25 km per km²
Ships (registered, 1977) 5, total of 48 353 gross tons
Port There is anchorage offshore
Airport Nauru
Radio sets (Dec 1975) 3 600, 490 per 1 000 people

Production, finance and trade

Gross domestic product
1976 est: A $ 80***mn = $ 100***mn = £ 55***mn
Main products (000 t) Coconuts (1976) 2*, phosphate rock (1974/75) 1 535, electricity (mn kW h, 1975) 26
Transport traffic *Air* (1976) Passenger-kilometres 34 mn, cargo 0.2 mn t-km
Sea (1971/72) Goods loaded 2.2* mn t, unloaded 0.04* mn t
Budget (1975/76; year ending June 30th)
Revenue: A $ 45.5 mn = $ 57 mn = £ 29 mn
Expenditure: A $ 26.7 mn = $ 34 mn = £ 17 mn
External trade Imports (1974): A $ 10* mn = $ 14* mn = £ 6* mn
Exports (1976): A $ 70** mn = $ 90** mn = £ 50** mn
Main imports Machinery, transport equipment, food
Main exports Phosphates 100* %

Main sources (1974)	% of total	Main destinations (1974)	% of total
Australia	58*	Australia	57*
Netherlands	30*	Japan	23*
United Kingdom	6*	New Zealand	18*
Japan	5*		

New Caledonia

Territory of New Caledonia
Territoire de la Nouvelle-Calédonie

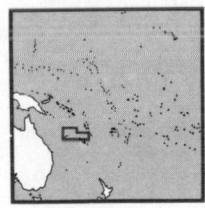

Location South-west Pacific Ocean
The territory comprises the main island of New Caledonia, and the dependencies of Loyalty Islands, 100 km to the east, Isle of Pines, 50 km to the south-east, Walpole Island, 150 km east of the Isle of Pines, Huon Islands 270 km to the north-west, Belep Archipelago 11 km to the north-east and Chesterfield Islands 550 km west of the northern headland. Queensland (Australia) is about 1 200 km to the west, and New Hebrides 400 km to the north-east
Land Area 19 058 km² = 7 358 mi²
Climate Sub-tropical
Weather at Nouméa, 9 m altitude
Temperature: hottest month Jan 22–30 °C, coldest Aug 16–24 °C
Rainfall (av monthly): driest month Oct 51 mm, wettest Mar 144 mm
Time 11 hours ahead of GMT
Measures Metric system
Monetary unit CFP franc (CFP Fr) = 100 centimes
Rate of exchange (1976 av):
par CFP Fr 1 = Fr 0.055 (CFP Fr 18.18 = Fr 1),
free CFP Fr 86.90 = $ 1, CFP Fr 157.0 = £ 1

Summary

Political French overseas territory. A governor, assisted by a local parliament, was first installed in 1885. Member of franc zone and territorial member of South Pacific Commission
Economic An agricultural and mining economy, with nickel accounting for virtually all exports; there are other mineral reserves in the territory. Some light industry is being developed, and there are plans to diversify by expansion of coffee growing and exploitation of timber resources

People, resources and equipment

Population 1960 77 000*, 1970 109 000*, 1976 135 000*
Growth: 1960–70 3.5* % pa, 1970–76 3.6* % pa
Density (1976): 7* people per km²
Vital statistics (rate per 1 000 people, 1975): births 30.5, deaths 7.0
City (population in 000, 1976) Nouméa (capital) 74
Race (1976) Melanesian 42 %, European (mainly French) 38 %, Wallisians 7 %, Polynesians 5 %
Language French; many languages and dialects are also in use
Religion (1976) Roman Catholic 65** %
Education (1974) Pupils 38 500*, teachers 1 860*
Labour force (1976) 50 469; in agriculture 13 564 (27 %), mining and quarrying 2 110 (4 %), manufacturing 5 469 (11 %), construction 4 475 (9 %), transport and communication 2 632 (5 %)
Personnel Scientists and engineers (1971): 69
Physicians (1975): 125, 1 per 1 040 people
Standard of living National income per person (1976):
CFP Fr 400 000*** = $ 4 600*** = £ 2 500***
Consumption per person (1975): energy 9 933 kg coal equivalent, electricity (production) 13 780 kW h
Newspapers (1974): number 2; circulation 8 000, 61 per 1 000 people
Telephones (Dec 1976): 20 612, 149 per 1 000 people
Livestock (000, 1976) Cattle 92*, pigs 30*, goats 10*, sheep 5*, poultry 166*
Electrical capacity (1975) 323 megawatts
Hospital beds (1975) 1 501, 1 per 87 people
Roads (1974) 5 214 km = 3 240 mi, density 0.27 km per km²
Port Nouméa
Airports Tontouta (53 km from Nouméa), Magenta (Nouméa), and 12 other airports with scheduled flights (including 4 in Loyalty Islands)

Durable equipment (at end-year)	000	no per 1 000 people	no per
Radio sets (1975)	60	450	no per
Television sets (1975)	15	114	km of road
Passenger cars (1971)	34	285	6.3
Commercial vehicles (1971)	14	116	2.6

Production, finance and trade

Gross domestic product 1970: CFP Fr 36 000 mn = $ 356 mn = £ 149 mn
1976 est: CFP Fr 55 000** mn = $ 630** mn = £ 350** mn
Main products (000 t)
Agriculture (1976) Sweet potatoes 2*, potatoes 1*, maize 1*, cassava 1*, taro 3*, yams 9*, bananas 4*, plantains 4*, coconuts 19*, copra 1*, coffee 2*, milk 8*, beef and veal 3*, fish catch 1.0, timber (000 m³) 15*

Other	1960	1970	1975	Growth % pa 1960–70	1970–75
Electricity (mn kW h)	475	873	1 791	6.3	15.5
Nickel ore (Ni content)	53	138	133	10.0	−0.7
Cement	—	—	58	na	na

Transport traffic *Air* (1976) Passenger departures 89 000, arrivals 88 000
Sea (1974) Goods loaded 3.6 mn t, unloaded 1.3 mn t
Tourism (1976) Number of visitors 34 983
Consumer price index (Nouméa, 1970 = 100) 1960 73, 1976 166; growth: 1960–70 3.2 % pa, 1970–76 8.8 % pa
Budget (1977) Balanced at CFP Fr 11 887 mn = $ 133 mn = £ 74 mn
External trade (1976)
Imports: CFP Fr 24 074 mn = $ 277 mn = £ 153 mn
Exports: CFP Fr 26 195 mn = $ 301 mn = £ 167 mn

Main imports (1973)	% of total	Main exports (1976)	% of total
Food	24	Ferro-nickel	44
Machinery	13	Nickel matte	27
		Nickel	24
Main sources (1976)		Main destinations (1976)	
France	39	France	51
Australia	10	Japan	29
Singapore	9	United States	10*
Bahrain	8		

Special focus

Nickel prices (wholesale price, Canadian ports)

	$ per tonne	% change over previous year		$ per tonne	% change over previous year
1970	2 846	+20	1974	3 825	+13
1971	2 932	+3	1975	4 538	+19
1972	3 079	+5	1976	4 973	+10
1973	3 373	+10			

New Hebrides

Nouvelles-Hébrides
Anglo-French Condominium of the New Hebrides

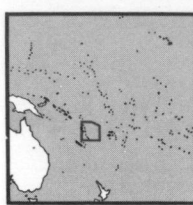

Location South-west Pacific Ocean
Comprises 12 main and about 50 small islands, with Fiji 800 km to the east and New Caledonia 400 km to the south-west. Main islands are Efate, Espíritu Santo, Malekula and Erromanga
Land Area 14 800 km² = 5 700 mi²
Climate Tropical in north, sub-tropical in south
Weather at Vila, 57 m altitude
Temperature: hottest month Feb 24–31 °C, coldest July, Aug 19–26 °C
Rainfall (av monthly): driest month Aug 89 mm, wettest Mar 297 mm
Time 11 hours ahead of GMT
Measures Metric and UK (imperial) systems
Monetary unit New Hebridean franc (NH Fr) = 100 centimes, and Australian dollar (A $) = 100 cents
Rate of exchange (1976 av):
par NH Fr 1 = Fr 0.061875 (NH Fr 16.16 = Fr 1),
NH Fr 1 = CFP Fr 1.125, free NH Fr 77.25 = $ 1, NH Fr 139.5 = £ 1;
free NH Fr 94.64 = A $ 1, $ 1.225 = A $ 1, A $ 1.474 = £ 1

Summary

Political Condominium, governed jointly by France and United Kingdom since 1906. A form of self-government was introduced early in 1978 and full independence by 1980 has been proposed. Member of franc zone and territorial member of South Pacific Commission
Economic An agricultural economy with copra, fish and meat main exports; a freezing plant at Santo processes fish, mainly landed by foreign vessels, for re-export. There is mining of manganese and tourism is to be developed

People, resources and equipment

Population 1960 65 000**, 1970 83 000*, 1976 97 000*
Growth: 1960–70 2.5** % pa, 1970–76 2.6* % pa
Density (1976): 7* people per km²
Vital statistics (rate per 1 000 people, 1966): births 45*, deaths 20*
Cities (population in 000, 1976) Vila[a] (capital) 17, Santo[b] 5*
[a]On Efate [b]On Espíritu Santo; also known as Luganville
Race (1967) Melanesian 93 %, European 2 %, French Polynesian 2 %, Mixed European 2 %, Mixed other 1 %
Language English, French and Bislama; many Melanesian languages and dialects are also in use
Religion (1967) Christian 84 %
Education Pupils (1977) 22 733, teachers (1974) 910
Labour force (1967) 35 133
Personnel Scientists and engineers (1972): 161
Physicians (1969) 25, 1 per 3 200 people
Standard of living National income per person (1976):
NH Fr 40 000*** = $500*** = £ 300***
Consumption per person (1975): energy 561 kg coal equivalent, electricity (production) 158 kW h
Telephones (Dec 1976): 2 289, 23 per 1 000 people
Livestock (000, 1976) Cattle 110*, pigs 64*, goats 7*, chickens 131*
Electrical capacity (1975) 8* megawatts
Hospital beds (1975) 924, 1 per 102 people
Roads (1976) 640* km = 400* mi, density 0.043* km per km²
Ships (registered, 1977) 10, total of 12 189 gross tons
Ports Vila, Santo
Airports Bauer Field (Vila), Pekoa (Santo) and 13 other airports with scheduled flights
Durable equipment Radio sets (Dec 1975): 11 000, 116 per 1 000 people
Passenger cars (Dec 1974): 2 600, 28 per 1 000 people, 4.1 per km of road
Commercial vehicles (Dec 1974): 800, 9 per 1 000 people, 1.2 per km of road

Production, finance and trade

Gross domestic product
1976 est: NH Fr 4000*** mn = $ 50***mn = £ 30***mn
Main products (000 t) *Agriculture* (1976) Maize 1*, bananas 1*, coconuts 264*, copra 37*, cocoa 1*, milk 2*, beef and veal 3*, pigmeat 1*, fish catch 8*, timber (000 m³) 10* *Other* (1975) Manganese ore (Mn content, exports) 19.1, electricity (mn kW h) 15

Transport traffic *Air* (1974) Passenger departures 18 400, arrivals 19 000
Sea (1974) Goods loaded 104 000 t, unloaded 144 000 t
Tourism (1976) Number of visitors 17 929 (excluding 48 742 cruise ship passengers)
Budget
Condominium (1976): revenue NH Fr 872 mn = $ 11.3 mn = £ 6.3 mn
expenditure NH Fr 878 mn = $ 11.4 mn = £ 6.3 mn
British (1974/75): revenue A $ 4.5 mn = $ 5.5 mn = £ 3.0 mn
expenditure A $ 4.2 mn = $ 5.2 mn = £ 2.9 mn
French (1976): revenue NH Fr 1 496 mn = $ 19.4 mn = £ 10.7 mn
expenditure NH Fr 1 497 mn = $ 19.4 mn = £ 10.7 mn
External trade (1976) Imports: NH Fr 2 628 mn = $ 34 mn = £ 19 mn
Exports: NH Fr 1 292 mn = $ 17 mn = £9 mn

Main imports (1972)	% of total	Main exports (1975)	% of total
Food	21	Copra	43
(of which, cereals 8,		Fish	33
fish 4, meat 3)		Manganese	9
Motor vehicles	8	Beef and veal	8
Machinery, non-electric	7	Cocoa	5
Wood products	7		
Petroleum products	6		
Chemicals	6		
Ships and boats	5		
Electrical machinery	5		
Beer, wines and spirits	4		
Metal tanks and boxes	3		

Main sources (1975)		Main destinations (1975)	
Australia	30	France	43
France	25	United States	28
Japan	8	Japan	15
New Caledonia	7	New Caledonia	8
United Kingdom	5		
Singapore	4		
New Zealand	4		
West Germany	3		

New Zealand

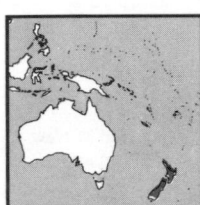

Location South-west Pacific Ocean
Comprises two main islands, North and South, with other small islands, including especially Stewart Island off the south coast of South Island; Australia is about 1 800 km to the west
Land Area 268 704 km² = 103 747 mi²
Climate Temperate
Weather at Wellington, 126 m altitude
Temperature: hottest months Jan, Feb 13–21 °C, coldest July 6–12 °C
Rainfall (av monthly): driest months Jan-Mar 81 mm, wettest July 137 mm
Time 12 hours ahead of GMT (summer time, 13 hours ahead)
Measures Metric system, as conversion from UK (imperial) system, with substantial completion by end 1976
Monetary unit New Zealand dollar (NZ $) = 100 cents; decimal currency, introduced from July 10, 1967, to replace the New Zealand pound (NZ £) at NZ $ 2 = NZ £ 1
Rate of exchange (1976 av): free NZ $ 1.004 = $ 1, NZ $ 1.813 = £ 1

Summary

Political Parliamentary monarchy, which became independent from the United Kingdom on September 26, 1907. Member of UN, OECD, Commonwealth, South Pacific Commission, Colombo Plan and Anzus Treaty. There is a free trade agreement (Nafta) with Australia
Economic Mainly agricultural, with main exports meat, wool and butter, accounting for one-half of exports. Timber and paper are important and export of aluminium from imported bauxite, using hydro-electricity resources, is becoming significant. Tourism is growing at a high rate. There is some light industry, and export markets are being diversified to counteract the effect of UK entry into the EEC which has severely affected New Zealand's traditional market

People, resources and equipment

Population 1960 2.37 mn, 1970 2.81 mn, 1976 3.13 mn
Growth: 1960–70 1.7 %pa, 1970–76 1.8 %pa
Density (1976): 11 people per km²
Vital statistics (rate per 1 000 people, 1975): births 18.5, deaths 8.2

New Zealand

Regions (population in 000, March 23, 1976; total of 3.13 mn)

North Island	*2 268*	*South Island*	*861*
Northland	107	Marlborough	35
Central Auckland	797	Nelson	76
South Auckland	472	Westland	24
East Coast	48	Canterbury	429
Hawke's Bay	145	Otago	189
Taranaki	107	Southland	109
Wellington	592		

Cities (population in 000, March 23, 1976)

Wellington (capital)	350	Napier/Hastings	109	New Plymouth	44
Auckland	797	Palmerston North	89	Nelson	42
Christchurch	326	Invercargill	54	Wanganui	40
Hamilton	155	Tauranga	48	Whangarei	39
Dunedin	120	Rotorua	47	Gisborne	32

Race (1971) European 89 %, Maori 8 %, Samoan 0.8 %, Cook Island Maori 0.5 %, Chinese 0.4 %, Indian 0.3 %, Niuean and Tokelauan 0.2 %

Language English; Maori is also in use

Religion (1971) Church of England 31 %, Presbyterian 20 %, Roman Catholic 16 %, Methodist 6 %, Baptist 2 %

Education (1976) Pupils 931 729, teachers 45 000*

Labour force (1976) 1 276 120; in agriculture 129 310 (10 %), manufacturing 308 490 (24 %), construction 114 520 (9 %), transport and communication 110 750 (9 %), finance and real estate 81 110 (6 %)

Personnel (1975) Physicians: 4 110, 1 per 754 people

Standard of living National income per person (1976):
NZ $ 3 678* = $ 3 663* = £ 2 029*

Consumption per person (1975): energy 3 111 kg coal equivalent, electricity (production, 1975/76) 6 457 kW h, newsprint 34 kg, steel 300 kg

Newspapers (1974): number 40; circulation (1970) 1 058 000, 376 per 1 000 people

Telephones (March 1977): 1 632 478, 517 per 1 000 people

Livestock (000, 1976) Cattle 9 777, sheep 56 400, pigs 505, goats 42*, horses 74*, chickens 6 100*

Mineral reserves Coal (1969) 678 mn tonnes

Lignite (1969) 396 mn tonnes

Crude oil (1975) 16 mn tonnes

Natural gas (1975) 180 bn cubic metres

Petroleum refinery capacity (1975) 3.5* mn tonnes

Electrical capacity (public supply, 1976) 4 901 megawatts, of which, hydro 3 471 megawatts, geothermal 192 megawatts

Hospital beds (1973) 31 959, 1 per 93 people

Roads (1975) 95 026 km = 59 046 mi, density 0.35 km per km²

Railways (1976) 4 797 km = 2 981 mi, density 0.018 km per km²

Ships (registered, 1977) 102, total of 199 462 gross tons

Ports (goods traffic, 000 manifest tons, 1976)

	International loaded	unloaded	Coastwise loaded	unloaded	Total
Whangarei[a]	49	3 439	3 457	688	7 633
Auckland	1 061	2 762	203	1 578	5 604
Wellington	559	1 179	1 512	1 803	5 053
Tauranga	1 497	417	23	720	2 657
Lyttleton	311	762	282	740	2 095

[a]Crude oil port

Airports (passengers, 000, 1975)

	Departures	Arrivals
Auckland	893	920
Wellington	681	680
Christchurch	595	586

Also (1977) 30 other airports with scheduled flights

Durable equipment

(at end-year)	000	no per 1 000 people	
Radio sets (1975)	2 704	876	no per
Television sets (1976)	821	261	km of road
Passenger cars (1976)	1 194	379	13
Commercial vehicles (1976)	230	73	2.4

Production

Gross domestic product 1976/77 (year ending March 31st):
NZ $ 13 189 mn = $ 12 855 mn = £ 7 800 mn
1976 est: NZ $ 12 680*mn = $ 12 630*mn = £ 6 994*mn
Growth in real terms: 1960–70 4.1 %pa, 1970–76 3.4 %pa

Structure of gross domestic product (1975/76) *By type*
Final consumption expenditure 78 % (of which, government 18 %), stock investment 3 %, gross fixed capital formation 26 %, exports of goods and services 24 %, less imports of goods and services −31 %

Production indices (1970 = 100)	1960	1970	1976	Growth %pa 1960–70	1970–76
Agricultural	78	100	111	2.5	1.8
Industrial[b]	52	100	152	6.8	7.2

Main products (000 t)

Agriculture	1960	1970	1976	1960–70	1970–76
Wheat	237	287	427	1.9	6.8
Barley	71	174	344	9.4	12.0
Oats	33	58	60	5.8	0.6
Maize	18	59	232	12.6	25.6
Potatoes	196	250	250*	2.5	0.0*
Peas, dry	25	50	59*	7.2	2.8*
Tomatoes	42	51	69*	2.0	5.2*
Onions	15	32	47*	7.9	6.6*
Apples	74	132	159*	6.0	3.1*
Milk	5 335	5 960	6 538	1.1	1.5
Dry skim milk	66	136	245	7.5	10.3
Butter	211	237	261	1.2	1.6
Cheese	95	100	103	0.5	0.5
Tallow	73	100	104	3.2	0.6
Eggs	41	52	53	2.4	0.3
Wool	190	238	223	2.3	−1.1
Cattle hides	23	51	62*	8.3	3.3*
Sheepskins	34	114	108*	12.8	−0.9*
Beef and veal	240	393	613	5.0	7.7
Mutton and lamb	448	563	510	2.3	−1.6
Fish catch	44.3	59.3	70.4	3.0	2.9
Timber (000 m³)	5 287	8 706	10 019	5.1	2.4
Energy					
Total energy (000 tce)	2 310	3 510	4 670[e]	4.3	5.9[f]
Coal	813	450	460	−5.7	0.4
Lignite	2 247	1 916	1 986	−1.6	0.6
Crude oil	1	58	136[e]	50.0	18.6[f]
Petroleum products	—	2 921	2 970*[e]	na	0.3*[f]
Natural gas (mn m³)	—	107	876	na	42.0
Manufactured gas (mn m³)	161	179	77[e]	1.1	−15.5[f]
Electricity (mn kW h)[b,d]	6 361	12 926	20 064	7.3	7.6
of which,					
hydro (mn kW h)[b,d]	5 483	10 190	16 868	6.4	8.8
geothermal (mn kW h)[b,d]	384	1 243	1 272	12.5	0.4
Mining					
Magnesite	0.8	0.5	1.0*	−4.6	12.3*
Gold (000 kg)	1.04	0.35	0.08[e]	−10.3	−24.7[f]
Salt	17	53	40[e]	12.0	−5.5[f]
Manufacturing					
Beer (000 hl)	2 450	3 353	4 096	3.2	3.4
Cigarettes (mn units)	3 127	5 364	6 535	5.5	3.3
Wood pulp[b]	251	562	913	8.4	8.4
Newsprint[b]	77	207	219	10.4	1.0
Other paper[b]	110**	239	336	8.0**	5.9
Tyres (000 units)	740	1 433	1 673	6.8	2.6
Fertilisers, phosphate[a]	220**	330	350*	4.1**	1.0*
Cement	617	829	999	3.0	3.1
Aluminium	—	—	110*[e]	na	na
Radio sets (000 units)	148	119	158	−2.2	4.8
Television sets (000 units)	8	45	147[e]	18.9	2.7[f]
Passenger cars (000 units)[c]	29.0*	55.3	69.5	6.7*	3.9
Commercial vehicles (000 units)[c]	7.7*	16.5	11.5	7.9*	−5.8

[a]Years ending June 30th [b]Years ending March 31st [c]Assembly only
[d]Public supply [e]1975 [f]1970–75

Transport traffic *Passenger-kilometres* (mn)	1960	1970	1976	Growth %pa 1960–70	1970–76
Road	na	24 349	27 940[b]	na	4.7[c]
Rail[a]	720	560	589	−2.5	0.8
Air	528	1 683	4 324	12.3	17.0
Cargo: tonne-kilometres (mn)					
Road	na	5 536	6 168[b]	na	3.7[c]
Rail[a]	1 914	2 741	3 648	3.7	4.9
Air	14.9	39.3	143.7	10.2	24.1
Sea: manifest tons (mn)					
Goods loaded	1.96*	4.77	8.22	9.3*	9.5
Goods unloaded	5.13*	8.32	10.04	5.0*	3.2
Tourism					
Number of visitors (000)	na	164*	384	na	15.2*
Gross receipts ($ mn)[a]	na	29	161	na	33.1

[a]Years ending March 31st [b]1973 [c]1970–73

New Zealand

Finance and trade

Price and wage indices (1970 = 100)

	1960	1970	1976	Growth %pa 1960–70	1970–76
Consumer prices	69.0	100.0	190.2	3.8	11.3
Wholesale prices	75.0	100.0	193.8	2.9	11.7
Wages (earnings)	62	100	212	5.0	13.3
Share prices	57	100	95	5.8	−0.9

Money stock

(end-year, NZ $ mn)	748.8	862.0	1 837.1	1.4	13.4

Budget (1976/77; year ending March 31st)
Revenue: NZ $ 4 255 mn = $ 4 147 mn = £ 2 516 mn
Expenditure: NZ $ 4 491 mn = $ 4 377 mn = £ 2 656 mn

Balance of payments ($ mn)

	1972	1973	1974	1975	1976
Balance of goods (fob)	+520	+571	−487	−764	−46
Balance of services	−324	−443	−703	−702	−753
Balance of transfers	+52	+84	+63	+67	+49
Current balance	*+249*	*+212*	*−1 128*	*−1 400*	*−751*
Long-term capital flow	+73	−77	+390	+981	+743

International reserves

(end-year, $ mn)	833	893	640	428	492

External trade (1976)
Imports: NZ $ 3 271 mn = $ 3 258 mn = £ 1 804 mn
Exports: NZ $ 2 815 mn = $ 2 804 mn = £ 1 553 mn

Main imports (1976/77[a])	% of total	Main exports (1976/77[a])	% of total
Machinery, non-electric	14	Meat	25
Chemicals	13	(of which,	
Crude oil	9	lamb and mutton 12,	
Motor vehicles	8	beef and veal 10)	
Iron and steel	7	Wool	21
Textile yarns and fabrics	7	Butter	8
Electrical machinery	6	Wood pulp and paper	5
Petroleum products	5	Hides and skins	4
Food	5	Aluminium	4
Metal small		Milk, dried	3
manufactures	3	Cheese	3
Instruments	3	Machinery	2
Ships and boats	2	Fruit and vegetables	2
		Casein	2

Main sources (1976)		Main destinations (1976)	
Australia	20	United Kingdom	19
United Kingdom	17	Japan	14
Japan	15	Australia	13
United States	14	United States	12
Iran	5	Soviet Union	4
West Germany	4	West Germany	3
Singapore	3	France	3
Canada	2	Netherlands	3
Hongkong	2	Canada	2

[a]Year ending June 30th

Special focus

Main export destinations (% of total exports)

	1960	1965	1970	1972	1974	1976
United Kingdom	53	48	34	30	20	19
Japan	3	5	10	11	13	14
Australia	4	4	9	8	11	13
United States	13	13	17	15	14	12
Soviet Union	1	—	2	1	3	4
West Germany	4	4	2	3	2	3
France	7	5	2	3	3	3
Netherlands	1	1	1	2	3	3
Canada	1	1	3	2	3	2
Other	13	19	20	25	28	27

Niue

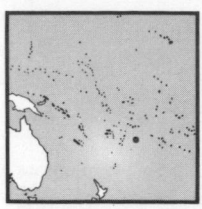

Location South-west Pacific Ocean
Niue island has Rarotonga (Cook Islands)
930 km to the east, with New Zealand 2 000 km
to the south-west, Tonga 500 km to the west,
and Western Samoa 600 km to the north-west
Land Area 260 km² = 100 mi²
Climate Sub-tropical
Weather at Niue, 20 m altitude
Temperature: hottest month Feb 23–31 °C,
coldest July, Aug 19–27 °C
Rainfall (av monthly): driest month June 76 mm, wettest Mar 312 mm
Time 11 hours behind GMT
Measures UK (imperial) system, converting to the metric system
Monetary unit New Zealand dollar (NZ $) = 100 cents
Rate of exchange (1976 av): free NZ $ 1.004 = $ 1, NZ $ 1.813 = £ 1

Summary

Political New Zealand associated territory, having internal
self-government from October 1974, and with New Zealand responsible
for external affairs. Territorial member of South Pacific Commission
Economic An agricultural economy, with main exports of fruit and copra,
most of which go to New Zealand. There are some mining prospects,
including the possibility of uranium production. New Zealand provides
a large subsidy. There is some migration of Niueans to New Zealand

People, resources and equipment

Population 1960 4 850*, 1970 4 980*, 1976 3 840**
Growth: 1960–70 0.2* %pa, 1970–74 −4.2** %pa
Density (1976): 15 people per km²
Vital statistics (rate per 1 000 people, 1975): births 41.3, deaths 6.0
Town (population in 000, 1971) Alofi (capital) 1.0
Race Mainly Niuean (Polynesian; there are close links with Samoans and
Tongans)
Language Niue dialect (Polynesian)
Education (1974) Pupils 1 393, teachers 84
Labour force (1971) 1 422; most Niueans work on family plantations
Personnel Scientists and engineers (1971): 2
Physicians (1975): 5, 1 per 780 people
National income per person (1976): NZ $ 550*** = $ 550*** = £ 300***
Telephones (Dec 1976) 231, 58 per 1 000 people
Livestock (000, 1976) Cattle 1*, pigs 1*, chickens 26*
Hospital beds (1975) 30, 1 per 131 people
Roads (1976) 229 km = 142 mi, density 0.88 km per km²
Port Alofi
Airport Niue Island International
Radio sets (Dec 1975) 800, 160 per 1 000 people
Motor vehicles (March 1976) 242, 63 per 1 000 people, 1.1 per km of road

Production, finance and trade

Gross domestic product
1976 est: NZ $ 2.5***mn = $ 2.5***mn = £1.4***mn
Main products (000 t, 1976) Taro 1*, coconuts 4*, copra 0.5*,
passion fruit 0.1*, honey 0.04*, meat 1*
Transport traffic (1973) *Sea* Cargo loaded and unloaded 5 166 t
Budget (1975/76; year ending March 31st)
Revenue: NZ $ 3.1 mn = $ 3.5 mn = £ 1.7 mn
(of which, New Zealand subsidy: NZ $ 2.5 mn = $ 2.9 mn = £ 1.4 mn)
Expenditure: NZ $ 3.7 mn = $ 4.2 mn = £ 2.0 mn
External trade (1975) Imports: NZ $ 2.09 mn = $ 2.5 mn = £ 1.1 mn
Exports: NZ $ 0.20 mn = $ 0.24 mn = £ 0.11 mn

Main imports	% of total	Main exports	% of total
Machinery	22	Fruit	45
Food	21	Copra	37
(of which, meat 7,		Plaited ware	11
cereals 3)		Honey	6
Transport equipment	11		
Metal small			
manufactures	10		
Petroleum products	8		
Chemicals	4		
Beer, wine and spirits	4		
Main source		*Main destination*	
New Zealand	79	New Zealand	73

Norfolk Island

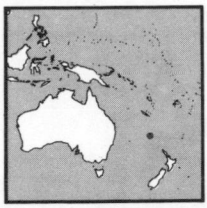

Location South-west Pacific Ocean
The island is in the north Tasman Sea, with
Sydney (Australia) 1 500 km to the south-west
Land Area 36 km² = 14 mi²
Climate Sub-tropical
Weather at Norfolk Island, 99 m altitude
Temperature: hottest month Feb 20–25 °C,
coldest Aug 13–18 °C
Rainfall (av monthly): driest month Nov 66 mm,
wettest July 155 mm
Time 11½ hours ahead of GMT
Measures UK (imperial) system, converting to the metric system
Monetary unit Australian dollar (A $) = 100 cents
Rate of exchange (1976 av): free $ 1.225 = A $ 1, A $ 1.474 = £ 1

Summary

Political Australian external territory; settled in 1856 by 194 people
from Pitcairn Island. There have been proposals for a limited form of
self-government. Territorial member of South Pacific Commission
Economic There is some agriculture, and forestry is being developed.
Tourism is the main foreign exchange earner

People, resources and equipment

Population 1960 790*, 1970 1 560*, 1976 1 850*
Growth: 1960–70 7.0* %pa, 1970–76 2.9* %pa
Density (1976): 51* people per km²
Vital statistics (rate per 1 000 people, 1975): births 10.2, deaths 6.8
Town Kingston (capital)
Race European, Polynesian and European/Polynesian
Language English   **Labour force** (1971) 860
National income per person (1976)
A $ 3 500*** = $ 4 300*** = £ 2 400***
Telephones (Dec 1976) 400, 220 per 1 000 people
Livestock (000, 1976) Cattle 2*
Roads (1976) 80* km = 50* mi, density 2.2* km per km²
Port Kingston   **Airport** Norfolk Island
Radio sets (Dec 1975) 1 150, 615 per 1 000 people

Production, finance and trade

Gross domestic product 1976 est: A $ 7***mn = $ 8.5***mn = £ 5***mn
Main products Cereals, fruit, vegetables, beef and veal, palm, timber
Tourism (1976/77) Number of visitors 18 844
Budget (1976/77; year ending June 30th)
Revenue: A $ 1.68 mn = $ 1.93 mn = £ 1.13 mn
Expenditure: A $ 1.58 mn = $ 1.82 mn = £ 1.07 mn
External trade (1976/77; year ending June 30th)
Imports: A $ 6.47 mn = $ 7.45 mn = £ 4.35 mn
Exports: A $ 0.80 mn = $ 0.92 mn = £ 0.54 mn
Main imports Food, building materials, petroleum products, radios
Main exports Bean seed, palm seed, fish
Main sources (1974/75) Australia 49 %, New Zealand 12 %
Main destinations (1974/75) Australia 66 %, New Zealand 23 %

Pacific Islands, US

US Trust Territory of the Pacific Islands and
Commonwealth of Northern Marianas

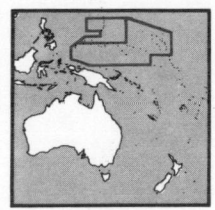

Location West Pacific Ocean
Includes over 2 000 islands in Micronesia of
which about 100 are inhabited; Caroline,
and Marshall islands are the main groups;
Northern Marianas consists of the Marianas
excluding Guam. Philippines is to the west and
Papua New Guinea to the south
Land Area (land) *Northern Marianas*
479 km² = 185 mi²
Trust Territory 1 378 km² = 532 mi²

Climate Tropical
Weather at Saipan, Northern Marianas, 206 m altitude
Temperature: hottest month June 24–29 °C, coldest Jan, Feb 22–27 °C
Rainfall (av monthly): driest month Jan 69 mm, wettest Sept 338 mm

Time Northern Marianas and western Caroline Islands (west of 160° E):
10 hours ahead of GMT; Truk island: 11 hours ahead of GMT
Other Caroline islands and Marshall islands: 12 hours ahead of GMT
Measures US system
Monetary unit United States dollar ($) = 100 cents
Rate of exchange (1976 av): $ 1.806 = £ 1

Summary

Political Territory administered by the United States under UN
trusteeship since July 18, 1947. In 1975 the Northern Marianas voted to
join the United States as a commonwealth, and the former district
(Marianas excluding Guam) became a US commonwealth territory from
January 1, 1978; figures here in general include those for Northern
Marianas. The other trust territories are considering proposals for some
form of independence by 1981 when the Trusteeship Agreement ends.
Territorial member of the South Pacific Commission
Economic An agricultural economy with fish and copra the main exports.
Tourism is being developed

People, resources and equipment

Population 1960 74 000*, 1970 90 940, 1976 123 000*
of which (1976), Northern Marianas 15 000*, Trust Territory 108 000*
Growth: 1960–70 2.1* %pa, 1970–76 5.2* %pa
Density (1976): *Northern Marianas* 31* people per km²
Trust Territory 78* people per km²
Vital statistics (rate per 1 000 people, 1976): births 28.7*, deaths 3.0*
Regions (population in 000, 1973) *Northern Marianas* 14*
Trust Territory Marshall Islands 25, Caroline Islands 75
(Truk 32, Panape 18, Palau 13, Yap 8, Kosrae 5)
Towns (population in 000, 1973)
Northern Marianas Saipan (capital) 10.8
*Trust Territory*ᵃ Moen 8.8, Majuro 7.5, Koror 7.2
ᵃAdministered from Saipan
Race Mainly Micronesian and Chamorro; also some Polynesian
Language English, Chamorro, Yapese, Trukese and other Micronesian
Religion (1970) Roman Catholic 50** %, Protestant 40*** %
Education Pupils (1975/76) 39 169, teachers (1973/74) 1 922
Labour force (1975) 14 493
Personnel (1975) Physicians: 55, 1 per 2 180 people
Standard of living National income per person (1976):
Northern Marianas $ 1 200*** = £ 660***
Trust Territory $ 900*** = £ 500***
Consumption per person (1975): energy 909 kg coal equivalent,
electricity (production) 960* kW h
Telephones (Dec 1974): 7 075, 60 per 1 000 people
Livestock (000, 1976) Cattle 17*, pigs 18*, goats 7*, chickens 170*
Electrical capacity (1975) 25* megawatts
Hospital beds (1975) 538, 1 per 220 people
Roads (1975) 1 035 km = 643 mi, density 0.56 km per km²
Ports *Northern Marianas* Saipan, Tinian
Trust Territory Koror, Takatik, Colonia, Moen, Majuro, Kusaie
Airports *Northern Marianas* Isley (Saipan), Tinian, Rota *Carolines* Palau,
Ponape, Truk, Yap *Marshalls* Majuro, Kwajalein

Durable equipment (Dec 1976)	000	no per 1 000 people	no per km of road
Radio sets	72	660	
Television sets	3.0	28	
Passenger cars	7.2	67	7.0
Commercial vehicles	1.0	9	1.0

Production, finance and trade

Gross domestic product 1976 est: *Northern Marianas* $ 20***mn =
£ 11***mn *Trust Territory* $ 100***mn = £ 55***mn
Main products (000 t, 1976) Cassava 5*, sweet potatoes 3*,
bananas 2*, coconuts 84*, copra 8*, meat 1*, fish catch 6.1,
electricity (mn kW h, 1975) 115*
Transport traffic *Air* (1976) Passenger-km 140 mn, cargo 4 mn t-km
Sea (1972) Goods loaded 10 000* t, unloaded 80 000* t
Budget (1975/76) Revenue: $ 72 mn = £ 38 mn
Expenditure: $ 62 mn = £ 33 mn
External trade (1976) Imports: $ 38* mn = £ 21* mn
Exports: $ 4.8 mn = £ 2.7 mn

Main imports (1976)	% of total	Main exports (1976)	% of total
Food	38*	Fish	62
Petroleum products	20**	Copra	33
Beverages	14*	Handicrafts	2
Main sources (1972)		*Main destination* (1972)	
United States	50*	Japan	54
Japan	27*		

Papua New Guinea

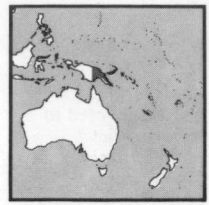

Location West Pacific Ocean
Includes the eastern side of the island of
New Guinea, the Bismarck archipelago (New
Britain, New Ireland and Admiralty Islands),
Northern Solomon Islands (including
Bougainville and Buka), islands at the eastern
tip of New Guinea (D'entre Casteaux,
Trobriand, Woodlark islands and Louisiade
archipelago) and about 600 smaller islands.
West Irian (Indonesia) is to the west,
comprising the western part of the island of New Guinea, Australia
to the south, and Solomon Islands to the east
Land Area 461 691 km² = 178 260 mi²
Climate Tropical
Weather at Port Moresby, 38 m altitude
Temperature: hottest month Dec 24–32 °C, coldest Aug 23–28 °C
Rainfall (av monthly): driest month Aug 18 mm, wettest Feb 193 mm
Time 10 hours ahead of GMT
Measures Metric system, as conversion from the UK (imperial) system
Monetary unit Kina (Ka) = 100 toea; the kina replaced the Australian
dollar (A $) from April 19, 1975 at Ka 1 = A $ 1
Rate of exchange (1976 av): free $ 1.265 = Ka 1, Ka 1.427 = £ 1

Summary

Political Parliamentary monarchy, which became independent on
September 16, 1975, having become self-governing on December 1, 1973;
formerly Papua was an Australian territory, and New Guinea was a UN
trust territory, administered by Australia. Member of UN, Commonwealth, Colombo Plan, South Pacific Commission and an EEC ACP state
Economic An agricultural and mining economy, where development of
the extensive mineral reserves is increasing markedly; copper from
Bougainville was first exported in 1972 and is now the major export,
reducing dependence on coffee, cocoa and copra. Foreign investment is
high, and the copper mining operation is largely owned overseas.
Mineral exploration and expansion of hydro-electricity and of forestry
are important developments

People, resources and equipment

Population 1960 1.90* mn, 1970 2.49* mn, 1976 2.83* mn
Growth: 1960–70 2.7* %pa, 1970–76 2.2* %pa
Density (1976): 6* people per km²
Vital statistics (rate per 1 000 people, 1970–75): births 41*, deaths 17*
Regions (provinces[c], population in 000, 1971; total of 2.49 mn)

Bougainville[a]	96	Madang	171	Northern	67
Central	194	Milne Bay	109	Sepik, East	182
Chimbu	160	Manus[b]	25	Sepik, West	94
Gulf	59	Morobe	249	Western	71
Highlands, Eastern	240	New Britain, East[b]	114		
Highlands, Southern	193	New Britain, West[b]	62		
Highlands, Western	346	New Ireland[b]	60		

[a]Northern Solomons [b]Bismarck archipelago [c]A new province, Enga, has been
created

Cities (population in 000, 1971) Port Moresby (capital) 77, Lae 39,
Rabaul 27, Madang 17
Race (1971) Papuan and Melanesian 98 %, European 2 %
Language English, Pidgin and Hiri Motu; 700 other languages and
dialects are also in use
Religion (1966) Christian 92 %; also Pantheistic
Education (1975) Pupils 285 000*, teachers 9 900*
Labour force (1976) 1 410 000; in agriculture 1 182 000 (84 %)
Personnel Scientists and engineers (1973): 2 646,
of whom, non-nationals 2 501
Physicians (1971): 233, 1 per 10 800 people
Standard of living
National income per person (1976): Ka 358* = $ 453* = £ 251*
Consumption per person (1975): energy 278 kg coal equivalent,
electricity (production, 1975/76) 375 kW h, newsprint (1972) 0.3 kg
Newspapers (1974): 1; circulation 18 000, 7 per 1 000 people
Telephones (June 1977): 37 531, 13 per 1 000 people
Livestock (000, 1976) Cattle 155*, pigs 1 173*, goats 15*, chickens 1 085*
Electrical capacity (1975) 250* megawatts, of which, hydro 45* megawatts
Hospital beds (1972) 15 255, 1 per 169 people
Roads (1975) 18 200* km = 11 300* mi, density 0.039* km per km²
There are no railways
Ships (registered, 1977) 64, total of 16 217 gross tons

Ports Port Moresby, Lae, Madang, Samarai, Wewak,
Kieta (Bougainville), Rabaul (New Britain), Kavieng (New Ireland)
Airports Jackson (Port Moresby), Lae, Nadzab, Lakunai (Rabaul); also
144 other airports with scheduled flights
Radio sets (Dec 1975) 110 000*, 39* per 1 000 people
Passenger cars (Dec 1976) 17 700, 6 per 1 000 people, 1.0 per km of road
Commercial vehicles (Dec 1976) 19 200, 7 per 1 000 people, 1.1 per km of
road

Production

Gross domestic product
1976/77 (year ending June 30th): Ka 1202 mn = $ 1 527 mn = £ 892 mn
1976 est: Ka 1 130* mn = $ 1 430* mn = £ 790* mn
Growth in real terms: 1970–74 8.4 %pa
Structure of gross domestic product *By origin* (1974/75) Agriculture 25 %,
mining and quarrying 11 %, manufacturing 6 %, construction 7 %,
transport and communication 6 %, other 45 %
By type (1976/77) Final consumption expenditure 80 % (of which,
government 30 %), stock investment 2 %, gross fixed capital
formation 16 %, exports of goods and services 45 %, less imports of
goods and services −44 %

Production index (1970 = 100)	1960	1970	1976	Growth %pa 1960–70	1970–76
Agricultural	61*	100	114	5.0*	2.2
Main products (000 t)					
Agriculture					
Sweet potatoes	200**	360**	411*	6.0**	2.2**
Cassava	50**	70**	86*	3.4**	3.5**
Taro	150**	200**	220*	3.0**	1.6**
Yams	110**	145**	170*	2.8**	2.7**
Coconuts	520**	684	744*	2.8**	1.4*
Copra	104	131	132*	2.3	0.1*
Bananas	500**	800**	840*	5.0**	0.8*
Coffee	7.5**	27.5	41.0*	14.0**	6.9*
Cocoa	7.6	22.9	32.0*	11.7	5.7*
Tea	—	2.4*	6.0*	na	16.5*
Rubber	4*	5*	6*	2.0*	3.1*
Pigmeat	12**	16*	19*	3.0**	2.9*
Fish catch	10.0**	23.4*	63.0	6.5**	18.0*
Timber (000 m³)	3 500**	4 756	5 892	3.1**	3.6
Other					
Electricity (mn kW h)[a]	57	160	1 048	10.9	36.8
Copper ore (Cu content)	—	—	172[b]	na	na
Silver	0.001	0.001	0.039[d]	0.0	150.0[e]
Gold (000 kg)	1.40	0.74	19.57[b]	−6.2	92.5[c]

[a]Years ending June 30th [b]1975 [c]1970–75 [d]1974 [e]1970–74
Transport traffic *Air* (1976) Passenger–km 220 mn, cargo 5.5 mn t-km
Sea (1972/73) Goods loaded 1.2 mn t, unloaded 1.1 mn t

Finance and trade

Consumer price index (1971 = 100) 1976 168.4;
growth 1971–76 11.0 %pa
Budget (1976/77; year ending June 30th[a])
Revenue: Ka 432 mn = $ 549 mn = £ 321 mn
Expenditure: Ka 436 mn = $ 554 mn = £ 324 mn
[a]The financial year has changed to the calendar year, with the first full financial
calendar year being 1978
External public debt (Dec 1976) $ 336 mn
External trade (1976/77; year ending June 30th)
Imports (fob): Ka 393 mn = $ 499 mn = £ 292 mn
Exports: Ka 517 mn = $ 657 mn = £ 384 mn

Main imports (1973/74)	% of total	Main exports (1976/77)	% of total
Food	25	Copper ore	37
(of which, cereals 6,		Coffee	26
meat 6)		Cocoa	11
Machinery, non-electric	15	Copra	4
Petroleum products	9	Timber	4
Chemicals	6	Tuna	3
Electrical machinery	5	Copra oil	2
Motor vehicles	5		
Metal small manufactures	4		
Textile yarns and fabrics	3		
Clothing	3		

Main sources (1975/76)		Main destinations (1975/76)	
Australia	47	Japan	29
Japan	14	West Germany	25
Singapore	12	Australia	15
United States	7	United States	8
United Kingdom	5	Spain	6

Papua New Guinea

Special focus

Main exports (% of total exports)ᵃ

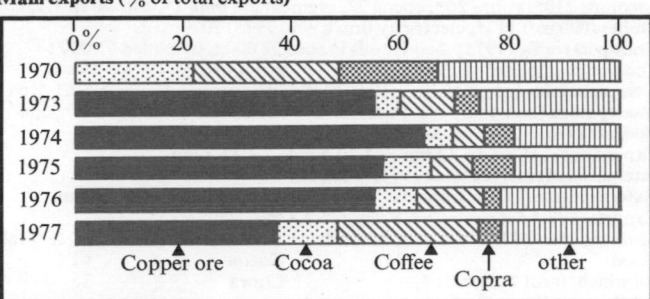

ᵃYears ending June 30th

Pitcairn

Colony of Pitcairn Islands

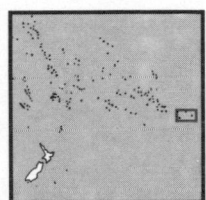

Location South central Pacific Ocean
The territory comprises 4 islands, Pitcairn,
Ducie, Henderson and Oeno; the group is about
half-way between New Zealand and Panama,
with Tahiti 2 200 km to the north-west
Land Area 5 km² = 2 mi²
Climate Sub-tropical
Weather at Pitcairn Island, 73 m altitude
Temperature: hottest month Feb 23–28 °C,
coldest July 18–26 °C
Rainfall (av monthly): driest month Apr 107 mm, wettest Oct 307 mm
Time 9 hours behind GMT
Measures UK (imperial) system
Monetary unit New Zealand dollar (NZ $) = 100 cents
Rate of exchange (1976 av): free NZ $ 1.004 = $ 1, NZ $ 1.813 = £ 1

Summary

Political UK colony, founded originally in 1790 by mutineers from
HMS Bounty together with some Tahitians; the population of 194 moved
to Norfolk Island in 1856, but 43 returned to Pitcairn by 1864, resettling
the territory. Territorial member of South Pacific Commission
Economic An agricultural economy, with fishing important; there is
some trading with passing ships, and sale of curios and postage stamps.
There has been steady emigration to New Zealand over the period from
1950

People, resources, equipment, production, finance and trade

Population 1960 103*, 1970 94*, 1976 74
Growth: 1960–70 −0.9* %pa, 1970–76 −3.9* %pa
Density (1976): 15 people per km²
Town Adamstown (capital)
Race European, Polynesian and European/Polynesian
Language English
Religion Mainly Seventh-day Adventist
Education (1976) Teachers 1
National income per person (1976)
NZ $ 2 700*** = $ 2 700*** = £ 1 500***
Telephones (Dec 1976) 29, 390 per 1 000 people
Port (anchorage) Bounty Bay
Gross domestic product
1976 est: NZ $ 0.20***mn = $ 0.20***mn = £ 0.11***mn
Main products Fish, sweet potatoes, oranges, bananas, coffee, curios
Transport traffic Number of ships calling: 1972 52, 1974 34
Budget (1974/75) Revenue: NZ $ 134 269 = $ 0.19 mn = £ 0.08 mn
Expenditure: NZ $ 92 262 = $ 0.13 mn = £ 0.05 mn
External trade *Main imports* Food (especially cereals and sugar),
tractors, telecommunications equipment, construction materials
Main exports Postage stamps, fruit and vegetables, curios

Samoa, American

American Samoa

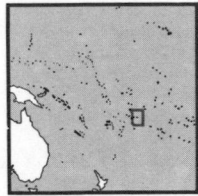

Location South central Pacific Ocean
Comprises six main islands, with New Zealand
about 2 900 km to the south-west and Hawaii
about 4 000 km to the north-east. Three of the
islands are part of the Samoan group and are to
the east of 171° west longitude, Western Samoa
being to the west; the other three main islands
are part of the Manu'a group
Land Area 197 km² = 76 mi²
Climate Tropical
Time 11 hours behind GMT
Measures US system
Monetary unit United States dollar ($) = 100 cents
Rate of exchange (1976 av): $ 1.806 = £ 1

Summary

Political US territory from 1900 (with Swain's Island included from 1925
and Tutuila and Aunu'u from 1929). Territorial member of South Pacific
Commission
Economic An agricultural economy, with fishing and fish canning
important. There is some diversification into light manufacturing
industry and tourism is being developed

People, resources and equipment

Population 1960 20 051, 1970 27 159, 1976 31 000*
Growth: 1960–70 3.1 %pa, 1970–76 2.2* %pa
Density (1976): 157* people per km²
Vital statistics (rate per 1 000 people, 1976): births 36.5, deaths 4.4
Regions (population in 000, 1970) Eastern Tutuila (including Aunu'u) 16,
Western Tutuila 9, Swain's Island 0.07, Manu'a 2
Towns (population in 000, 1970) Pago Pago (capital) 2.5,
Fagatogo (seat of government)
Race Mainly Samoan (Polynesian)
Language Samoan and English
Religion Mainly Christian
Education Pupils (1973/74) 12 505, teachers (1972/73) 630*
Labour force (1974) 9 700
Personnel Scientists and engineers (1973): 327
Physicians (1974): 26, 1 per 1 120 people
Standard of living
National income per person (1976): $ 3 000*** = £ 1 700***
Consumption per person: energy (1975) 1 682 kg coal equivalent,
electricity (production, 1976) 2 065 kW h
Newspapers (1974): 1; circulation 3 500, 113 per 1 000 people
Telephones (June 1977): 4 400, 138 per 1 000 people
Livestock (000, 1976) Pigs 8*, goats 8*, chickens 38*
Electrical capacity (1975) 13* megawatts
Hospital beds (1975) 181, 1 per 166 people
Roads (1976) 150* km = 93* mi, density 0.76* km per km²
Port Pago Pago
Airport (passenger departures and arrivals, 000, 1976)
Tafuna (Pago Pago) 138

Durable equipment	000	no per	
(at end-year)		1 000 people	
Radio sets (1974)	4.2	140	no per
Television sets (1975)	5.0	160	km of road
Passenger cars (1974)	2.4	81	16
Commercial vehicles (1974)	0.3	10	2

Production, finance and trade

Gross domestic product 1976 est: $ 120***mn = £ 70***mn
Main products (000 t, 1976) Taro 9*, bananas 3*, coconuts 10*, copra 1*,
fish catch 0.1, electricity (mn kW h) 64, canned fish (1975) $ 64 mn
Transport traffic *Air* See airport
Sea (1974) Goods loaded 65 000 t, unloaded 319 000 t
Tourism (1973/74) Number of visitors 35 422
Budget (1976)
Balanced at $ 44 mn = £ 24 mn (of which, US grants: $ 40 mn = £ 22 mn)
External trade (1976) Importsᵃ: $ 51 mn = £ 28 mn
Exports: $ 65 mn = £ 36 mn
Main imports Food, petroleum products, beverages, fish (for canneries)
Main exports Canned tuna 89 %, watches 6 %, pet food 4 %
Main sources (1973/74) United States 74 %, Japan 10 %
Main destination (1970) United States 95 %
ᵃExcluding fish for canneries

Samoa, Western

Western Samoa
Samoa i Sisifo

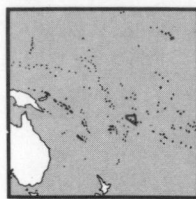

Location South central Pacific Ocean
Comprises nine islands in the Samoan group,
with New Zealand about 2 900 km to the
south-west and Hawaii about 4 000 km to the
north-east; Western Samoa is west of 171 ° west
longitude, American Samoa being to the east
Land Area 2 842 km² = 1 097 mi²,
of which, Savai'i 1 715 km², Upolu 1 121 km²
Climate Tropical
Weather at Apia, 2 m altitude
Temperature: hottest month Jan 24–30 °C, coldest July 23–29 °C
Rainfall (av monthly): driest month July 81 mm, wettest Jan 455 mm
Time 11 hours behind GMT
Measures UK (imperial) system, converting to the metric system
Monetary unit Western Samoan tala (WS $) = 100 sene; the tala was
introduced from July 10, 1967, to replace the pound (WS £) at
WS $ 2 = WS £ 1
Rate of exchange (1976 av): free $ 1.255 = WS $ 1, WS $ 1.440 = £ 1

Summary

Political Parliamentary monarchy, which became independent from New
Zealand on January 1, 1962. Member of UN, Commonwealth, South
Pacific Commission and an EEC ACP state
Economic An agricultural economy, with cocoa and copra the main
exports. An industrial free zone has been established at Vaitele, near Apia,
to encourage industry and forest resources are to be exploited. Tourism
earnings and remittances from Samoans working overseas make a
significant contribution to foreign exchange receipts

People, resources and equipment

Population 1960 112 000*, 1970 143 000*, 1976 151 000*
Growth: 1960–70 2.5* %pa, 1970–76 0.9* %pa
Density (1976): 53* people per km²
Vital statistics (rate per 1 000 people, 1976): births 36.9, deaths 6.7
Regions (population in 000, 1976; Nov census total of 152 000)
Savai'i 42, Upolu 110
Town (population in 000, 1974) Apiaª (capital) 33
ªOn Upolu
Race (1966) Samoan (Polynesian) 89 %, part Samoan (Polynesian) 10 %,
European 1 %
Language Samoan and English
Usage (1966): Samoan and English 52 %, Samoan 48 %
Religion (1966) Congregationalist 52 %, Roman Catholic 21 %,
Methodist 15 %
Education (1975) Pupils 48 800*, teachers 1 760*
Labour force (1971) 37 901; in agriculture 25 403 (67 %), manufacturing,
mining and quarrying 819 (2 %), construction 1 617 (4 %), transport and
communication 1 248 (3 %)
Personnel (1975) Physicians: 55, 1 per 2 700 people
Standard of living
National income per person (1976): WS $ 250** = $ 320** = £ 170**
Consumption per person (1975): energy 160 kg coal equivalent,
electricity (production) 133 kW h, newsprint 1.2 kg
Telephones (Dec 1976): 3 300, 22 per 1 000 people
Livestock (000, 1976) Cattle 20*, pigs 30*, chickens 490*
Electrical capacity (1975) 7 megawatts
Hospital beds (1975) 655, 1 per 230 people
Roads (1976) 930* km = 580* mi, density 0.33* km per km²
Ports Apia (Upolo), Asau (Savai'i)
Airports Faleolo (37 km from Apia), Fagali'i (Apia), Asau

Durable equipment	000	no per	
(at end-year)		1 000 people	
Radio sets (1973)	50	329	no per
Television sets (1972)	0.1	0.3	km of road
Passenger cars (1975)	2.0	13	2.2
Commercial vehicles (1975)	2.0	13	2.2

Production, finance and trade

Gross domestic product 1976 est: WS $ 40**mn = $ 50**mn = £ 28**mn
Main products (000 t, 1976) Taro 14*, bananas 36*, papayas 11*,
coconuts 210*, copra 20*, cocoa 2*, pigmeat 2*, milk 1*, fish catch 1.1,
timber (000 m³) 117*, electricity (mn kW h, 1975) 20
Transport traffic (1975) *Sea* Goods loaded 22 000 t, unloaded 51 000 t
Tourism Number of visitors (1976) 33 776, gross receipts (1974) $5 mn
Consumer price index (1970 = 100) 1976 179.7; growth 1970–76 10.3 %pa
Money stock (end-year, WS $ mn) 1976 4.02; growth 1970–76 15.0* %pa
Budget (1976) Revenue: WS $ 16.25 mn = $ 20.4 mn = £ 11.3 mn
Expenditure: WS $ 16.32 mn = $ 20.5 mn = £ 11.3 mn
International reserves (Dec 1976) $ 5.2 mn
External trade (1976) Imports: WS $ 23.6 mn = $ 30 mn = £ 16 mn
Exports: WS $ 5.4 mn = $ 6.8 mn = £ 3.8 mn

Main imports (1973)	% of total	*Main exports* (1976)	% of total
Food	25	Cocoa	41
(of which, meat 7, cereals 4,		Copra	35
fish 4, sugar 3)			
Machinery, non-electric	9		
Electrical machinery	8		
Motor vehicles	6		
Metal small manufactures	6		
Chemicals	5		
Iron and steel	5		
Textile yarns and fabrics	4		
Petroleum products	4		
Beer, wines and spirits	3		

Main sources (1976)		*Main destinations* (1976)	
New Zealand	28	New Zealand	36
Australia	20	West Germany	35
Japan	15	Japan	8
United States	8	Netherlands	6
Fiji	7	American Samoa	5
Singapore	5	Australia	4
American Samoa	4	United States	4

Special focus

Main exports

	% of total exports			
	1970	1972	1974	1976
Cocoa	31	26	24	41
Copra	40	41	61	35
Timber and veneers	—	10	5	1

Solomon Islands

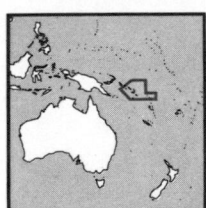

Location South-west Pacific Ocean
Comprises 15 main islands (including
Guadalcanal, Santa Isabel, Malaita, San
Cristobal) and numerous small islands of the
Southern Solomons; Bougainville (Northern
Solomons, Papua New Guinea) is to the north-
west, and New Hebrides to the south-east
Land Area 29 800* km² = 11 500* mi²
Climate Tropical
Weather at Tulagi, 2 m altitude
Temperature: hottest month Dec 24–32 °C, coldest June–Sept 24–30°C
Rainfall (av monthly): driest month June 173 mm, wettest Feb 401 mm
Time 11 hours ahead of GMT
Measures UK (imperial) system, converting to the metric system
Monetary unit Solomon Islands dollar (SI $) = 100 cents; the Solomon
Islands dollar replaced the Australian dollar in 1977 at SI $ 1 = A $ 1
Rate of exchange (1976 av): free $ 1.225 = SI $ 1, SI $ 1.474 = £ 1

Summary

Political Parliamentary monarchy, which became independent from
United Kingdom on July 7, 1978; formerly a UK protectorate, with
self-government from January 1, 1976. Before June 15, 1975, known as
British Solomon Islands. Member of South Pacific Commission
Economic An agricultural economy, with copra, timber and fish the main
exports. Fishing has expanded rapidly and is being further developed.
There is some light manufacturing, and mineral resources are being
explored

Solomon Islands

People, resources and equipment

Population 1960 125 000**, 1970 163 000*, 1976 200 000*
Growth: 1960–70 2.7**%pa, 1970–76 3.5*%pa
Density (1976): 7* people per km²
Vital statistics (rate per 1 000 people, 1969): births 36.1, deaths 13.0
Cities (population in 000, 1976) Honiaraª (capital) 15*
ªOn Guadalcanal
Race (1970) Melanesian 93 %, Polynesian 4 %, Micronesian 1 %, European 1 %
Language English; Pidgin and many local languages and dialects are also in use
Religion (1970) Christian 94 %, Animist 5 %
Education (1973) Pupils 28 039, teachers 1 266
Labour force (wage-earning, 1974) 13 384; in agriculture 4 088 (31 %), construction 985 (7 %), transport and communication 1 143 (9 %)
Personnel Scientists and engineers (1971/72): 129
Physicians (1975): 37, 1 per 4 470 people
Standard of living
National income per person (1976): SI $ 240** = $ 300** = £ 160**
Consumption per person (1975): energy 241 kg coal equivalent, electricity 80* kW h
Telephones (Dec 1976): 1 838, 9 per 1 000 people
Livestock (000, 1976) Cattle 23*, pigs 34*, chickens 133*
Electrical capacity (1975) 8* megawatts
Hospital beds (1971) 1 413, 1 per 120 people
Roads (1975) 1 220 km = 760 mi, density 0.041 km per km²
Ships (registered, 1977) 9, total of 1 746 gross tons
Ports Honiara, Gizo, Yandina
Airports Henderson (16 km from Honiara), Kukum (Honiara), and 21 other airports with scheduled flights
Radio sets (Dec 1975) 10 000, 53 per 1 000 people
Passenger cars (Dec 1974) 936, 5 per 1 000 people, 0.8 per km of road
Commercial vehicles (Dec 1974) 607, 3 per 1 000 people, 0.5 per km of road

Production, finance and trade

Gross domestic product 1975: SI $ 47.3 mn = $ 62 mn = £ 28 mn
1976 est: SI $ 50** mn = $ 60** mn = £ 34** mn
Main products (000 t, 1976) Sweet potatoes 48*, taro 15*, yams 13*, coconuts 183*, copra 24*, pigmeat 1*, fish catch 19, timber (000 m³) 421, electricity (public supply only, consumption, mn kW h, 1975) 16*
Transport traffic (1973) Sea Goods loaded 154 000 t, unloaded 62 000 t
Consumer price index (Honiara, 1971 = 100) 1976 150.6; growth 1971–76 8.5 %pa
Budget (1976) Revenue: SI $ 10.3 mn = $ 12.6 mn = £ 7.0 mn
Expenditure: SI $ 8.1 mn = $ 9.9 mn = £ 5.5 mn
External trade (1976) Imports: SI $ 21.9 mn = $ 27 mn = £ 15 mn
Exports: SI $ 19.3 mn = $ 24 mn = £ 13 mn

Main imports (1975)	% of total	Main exports (1975)	% of total
Machinery, non-electric	17	Copra	39
Food	14	Timber	27
(of which, rice 4)		Fish	24
Metal small manufactures	12		
Petroleum products	10		
Chemicals	8		
Electrical machinery	5		
Motor vehicles	4		
Textile yarns and fabrics	3		
Iron and steel	3		
Ships and boats	3		

Main sources (1975)		Main destinations (1975)	
Australia	35	Japan	29
United Kingdom	14	United Kingdom	11
Japan	13	France	8
Singapore	10	Denmark	8
United States	4	West Germany	7
West Germany	3	Netherlands	6
Malaysia	3	Norway	6
New Zealand	3	Sweden	5

Tokelau

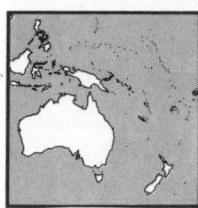

Location South-west Pacific Ocean
Includes 3 islands; Western Samoa is about 430 km to the south
Land Area 10 km² = 4 mi²
Climate Tropical
Weather at Atafu, 3 m altitude
Temperature: hottest month May 27–31 °C, coldest Feb 26–30 °C
Rainfall (av monthly): driest May 175 mm, wettest Feb 409 mm
Time 11 hours behind GMT
Measures UK (imperial) system, converting to the metric system
Monetary unit New Zealand dollar (NZ $) = 100 cents
Rate of exchange (1976 av): free NZ $ 1.004 = $ 1, NZ $ 1.813 = £ 1

Summary

Political New Zealand overseas territory from January 1, 1949. Territorial member of South Pacific Commission
Economic An agricultural economy with copra the main export. Due to the restricted potential of the islands, the islanders began in 1965 a policy of resettlement in New Zealand, but this was suspended in 1975 when over-population became less important

People, resources, production, finance and trade

Population 1960 1 900*, 1970 1 640*, 1976 1 575
Growth: 1960–70 −1.5* %pa, 1970–76 −0.7* %pa
Density (1976): 157 people per km²
Vital statistics (rate per 1 000 people, 1972): births 24.5, deaths 5.6
Regions (population in 000, 1976; total of 1 575)
Atafu 546, Nukunonu 363, Fakaofo 666
Race Tokelauan (Polynesian; akin to Samoan)
Language Tokelauan and English
Religion Mainly Christian
Education (1976) Pupils 560**, teachers 51
Labour force (1972) 359
National income per person (1976) NZ $ 500*** = $ 500*** = £ 280***
Livestock (000, 1976) Pigs 1*, chickens 3*
Gross domestic product
1976 est: NZ $ 0.8*** mn = $ 0.8*** mn = £ 0.4*** mn
Main products (000 t, 1976) Coconuts 2*, copra 0.16*, eggs 0.006*
Budget (1976/77; year ending March 31st)
Revenue: NZ $ 111 000 = $ 108 000 = £ 66 000
Expenditure: NZ $ 957 000 = $ 933 000 = £ 566 000
The deficit is made up by grants from New Zealand
External trade (1976/77)
Imports: NZ $ 1.0*** mn = $ 1.0*** mn = £ 0.6*** mn
Exports: NZ $ 0.013mn = $ 0.013mn = £ 0.009mn
Main imports Food, petroleum products, building material
Main export Copra 100 %
Main source and destination New Zealand

Tonga

Kingdom of Tonga
Pule'anga Tonga

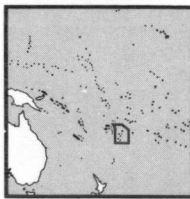

Location South-west Pacific Ocean
Comprises an archipelago of about 150 islands; Fiji is about 600 km to the west and Sydney (Australia) about 3 000 km to the south-west. There are 3 main groups, Tongatapu, Vava'u and Ha'apai
Land Area 700 km² = 270 mi²
Climate Sub-tropical (warm and humid during January–March)
Weather at Nuku'alofa, 3 m altitude
Temperature: hottest month Feb 23–29 °C, coldest July 18–25 °C
Rainfall (av monthly): driest month Oct 99 mm, wettest Mar 218 mm
Time 13 hours ahead of GMT
Measures UK (imperial) system, converting to the metric system

Tonga

Monetary unit Tongan pa'anga (T $) = 100 seniti, which replaced the Australian dollar (A $) from April 3, 1967 at T $ 1 = A $ 1.
Rate of exchange (1976 av): free $ 1.304 = T $ 1, T $ 1.385 = £ 1

Summary

Political Monarchy, which became independent on June 4th, 1970; formerly a UK protectorate. Also known as the Friendly Islands. Member of Gatt, ITU, UPU and WHO (UN agencies), Commonwealth, South Pacific Commission (as associate) and an EEC ACP state
Economic An agricultural economy, with copra the major export, and bananas also important. Tourism is being developed and the transport and communications system improved. Oil has been found and further exploration is taking place

People, resources and equipment

Population 1960 63 000*, 1970 86 000*, 1976 90 000*
Growth: 1960–70 3.1* % pa, 1970–76 0.8* % pa
Density (1976): 129* people per km²
Vital statistics (rate per 1 000 people, 1976): births 13.0, deaths 1.9
Regions (main groups, population in 000, 1976)
Tongatapu 54.4　　Ha'apai 10.8　　Niuas 2.3
Vava'u 15.1　　'Eua 4.5
Town (population in 000, 1976) Nuku'alofaª (capital) 18
ªOn Tongatapu
Race (1966) Tongan (Polynesian) 98 %, European and part European 1 %
Language Tongan and English
Religion (1966) Christian 99 % (Wesleyan 77%)
Education (1974) Pupils 27 800*, teachers 1 160*
Labour force (1966) 18 998; in agriculture 14 064 (74 %), manufacturing, mining and quarrying 502 (3 %), transport and communication 372 (2 %)
Personnel (1974) Physicians: 30, 1 per 3 240 people
Standard of living
National income per person (1976): T $ 330** = $ 430** = £ 240**
Consumption per person (1975): electricity 54 kW h
Telephones (Dec 1976): 552, 6 per 1 000 people
Livestock (000, 1976) Cattle 4*, pigs 48*, goats 5*, poultry 147*
Electrical capacity (1975) 3 megawatts
Hospital beds (government establishments, 1974) 330, 1 per 297 people
Roads (1975) 433 km = 269 mi, density 0.62 km per km²
Ships (registered, 1977) 12, total of 14 180 gross tons
Ports Nuku'alofa (Tongatapu), Neiafu (Vava'u)
Airports Fua'amotu (22 km from Nuku'alofa), Vava'u, Ha'apai, 'Eua

Durable equipment (Dec 1975)	000	no per 1 000 people	no per km of road
Radio sets	11	108	2.3*
Passenger cars	1.0*	11*	0.9*
Commercial vehicles	0.4*	4*	

Production, finance and trade

Gross domestic product
1975/76 (year ending June 30th): T $ 28.3 mn = $ 38 mn = £ 19 mn
1976 est: T $ 30**mn = $ 40**mn = £ 22**mn
Structure of gross domestic product (1970/71) *By origin* Agriculture 47 %, construction 5 %, transport and communication 5 %, other 43 %
Main products (000 t, 1976) Sweet potatoes 78*, cassava 12*, tomatoes 2*, oranges 2*, lemons 2*, bananas 4*, coconuts 125*, copra 17*, pigmeat 1*, fish catch 1.0, electricity (mn kW h, 1975) 5
Transport traffic *Air* (1974) Aircraft arrivals 399
Sea (1974) Goods loaded 21 000 t, goods unloaded 46 000 t
Tourism (1975) Number of visitors 70 000*, gross receipts $ 3* mn
Consumer price index (1970 = 100) 1976 177.6; growth 1970–76 10.0 %pa
Budget (1975/76; year ending June 30th)
Revenue: T $ 5.3 mn = $ 7.1 mn = £ 3.6 mn
Expenditure: T $ 5.9 mn = $ 7.9 mn = £ 4.0 mn
External trade (1976) Imports: T $ 11.7 mn = $ 15 mn = £ 8.4 mn
Exports: T $ 3.3 mn = $ 4.4 mn = £ 2.4 mn

Main imports (1975)	% of total	*Main exports* (1976)	% of total
Food	30	Copra	54
(of which, meat 12,		Desiccated coconut	11
cereals 7)		Bananas	8
Timber	7	Kava	6
Chemicals	6	Watermelons	5
Machinery	6		
Textile yarns and fabrics	5		
Metal small manufactures	5		
Petroleum products	5		
Motor vehicles	5		
Iron and steel	4		
Tobacco	4		
Main sources (1976)		*Main destinations* (1976)	
New Zealand	40	Netherlands	30
Australia	22	New Zealand	29
United Kingdom	11	West Germany	21
Japan	6	United Kingdom	6
Fiji	5	Fiji	5

Tuvalu

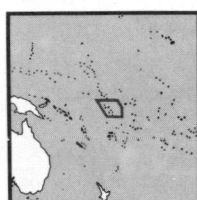

Location South-west Pacific Ocean
Comprises a group of atolls including nine main islands; Gilbert Islands are to the north, Fiji is to the south, and Australia is 4 000 km to the south-west. Funafuti is the main island
Land Area (inhabited) 26 km² = 10 mi²
Climate Tropical
Weather at Funafuti, 2 m altitude
Temperature: hottest month Nov 26–33 °C, coldest July 24–31 °C
Rainfall (av monthly): driest month May 185 mm, wettest Feb 450 mm
Time 12 hours ahead of GMT
Measures UK (imperial) system, converting to metric system
Monetary unit Australian dollar (A $) = 100 cents; Tuvaluan coinage was introduced from August 1, 1977
Rate of exchange (1976 av): free $ 1.225 = A $ 1, A $ 1.474 = £ 1

Summary

Political A UK colony from October 1, 1975, with separate administration from January 1, 1976 when it was separated from the former colony of Gilbert and Ellice Islands; formerly the Ellice Islands. Full independence is planned for October 1, 1978. Territorial member of the South Pacific Commission
Economic An agricultural economy, with copra the main export crop; fishing is important. Remittances home from 1 500* Tuvaluans working in Nauru or in ships are significant

People, resources, equipment, production and trade

Population 1960 7 000**, 1970 7 800**, 1976 9 000*
Growth: 1960–70 1.1** % pa, 1970–76 2.4** % pa
Density (1976): 346* people per km²
Vital statistics (including Gilbert Islands, rate per 1 000 people, 1971): births 22.3, deaths 6.5
Town (population in 000, 1976) Funafuti (capital) 1***
Race Mainly Polynesian
Language Tuvaluan (Samoan) and English
Religion Mainly Protestant
Education (1975) Pupils 1 958, teachers 50**
Personnel (1975) Physicians: 1, 1 per 9 000 people
National income per person (1976) A $ 270*** = $ 330*** = £ 180***
Livestock (000, 1976) Pigs 1**, chickens 20**
Port Funafuti
Airport Funafuti
Gross domestic product
1976 est: A $ 2.5***mn = $ 3.0***mn = £ 1.7***mn
Main products (000 t, 1976) Coconuts 10**, copra 1**, taro 1**, fish
Budget (1977) Balanced at A $ 1.56 mn = $ 1.74 mn = £ 0.99 mn; includes UK grants of A $ 0.68 mn = $ 0.75 mn = £ 0.43 mn
External trade (1975)
Main imports Food, textiles, metal manufactures, petroleum products
Main export Copra: A $ 23 184 = $ 30 380 = £ 13 670
Main sources Australia, New Zealand
Main destinations United Kingdom, New Zealand, Australia

Wake Island

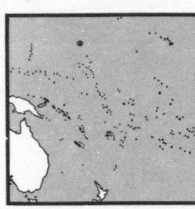

Location North Pacific Ocean
An atoll of three islands, Wake, Wilkes and
Peale; Hawaii is about 3 200 km to the east, and
Guam 2 100 km to the west
Land Area 8 km² = 3 mi²
Climate Sub-tropical
Time 12 hours ahead of GMT
Measures US system
Monetary unit
United States dollar (US $) = 100 cents
Rate of exchange (1976 av): $ 1.806 = £ 1

Summary

Political US territory; administered by the US Air Force
Economic The airport occupies much of the island; used as a missile
testing base

People, equipment and production

Population 1960 1 097, 1970 1 647, 1976 2 000**
Growth: 1960–70 4.1 % pa, 1970–76 3.3** % pa
Density (1976): 250** people per km²
Language English
National income per person (1976) $ 6 000*** = £ 3 300***
Consumption per person (1975) Energy 24 025 kg coal equivalent
Telephones (Dec 1976) 121, 60 per 1 000 people
Airport Wake Island
Gross domestic product 1976 est: $ 13*** mn = £ 7*** mn

National income per person (1976)
CFP Fr 35 000*** = $ 400*** = £ 220***
Telephones (Dec 1976) 85, 9 per 1 000 people
Livestock (000, 1976) Cattle 0.1, pigs 5, horses 0.4
Hospital beds (1970) 108, 1 per 80 people
Ports Mata-Utu, Singave
Airport Hihifo (Uvéa)
Gross domestic product
1976 est: CFP Fr 300***mn = $ 3.5***mn = £ 2.0***mn
Main products (000 t, 1976) Coconuts 10*, copra 1*; also yams, taro,
bananas, fish, timber
Budget (1977) Balanced at CFP Fr 131 mn = $ 1.5 mn = £ 0.8 mn
External trade Imports (1976): CFP Fr 217 mn = $ 2.5 mn = £ 1.4 mn
Exports (1974): CFP Fr 1.5**mn = $ 0.02**mn = £ 0.009**mn
Main exports Trochus shell, handicrafts

Wallis and Futuna Islands

Territory of Wallis and Futuna
Territoire de Wallis et Futuna

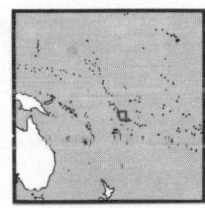

Location South-west Pacific Ocean
Includes Wallis, Futuna and Alofi islands, with
Western Samoa about 300 km to the east. Uvéa
is the main island of the Wallis archipelago;
Futuna and Alofi islands (Hoorn group) are
200 km to the south of the Wallis group
Land Area 275 km² = 106 mi²,
of which, Wallis islands 159 km²
Climate Tropical
Time 12 hours ahead of GMT
Measures Metric system
Monetary unit CFP franc (CFP Fr) = 100 centimes
Rate of exchange (1976 av):
par CFP Fr 1 = Fr 0.055 (CFP Fr 18.18 = Fr 1),
free CFP Fr 86.90 = $ 1, CFP Fr 157.0 = £ 1

Summary

Political French overseas territory from July 29th, 1961; formerly
dependencies of New Caledonia. Member of franc zone and territorial
member of South Pacific Commission
Economic An agricultural economy, with copra and timber the main
export products. Remittances sent home from workers in New Caledonia
provide an important source of income

People, resources, equipment, production, finance and trade

Population 1960 7 500***, 1970 8 600*, 1976 9 190*ª
ªExcludes 11 000** who live in New Caledonia and New Hebrides
Growth: 1960–70 1.4*** % pa, 1970–76 1.1* % pa
Density (1976): 33* people per km²
Vital statistics (rate per 1 000 people, 1970): births 43.3, deaths 10.6
Regions (population in 000, 1976) Wallis Island 6.0, Futuna Island 3.2
Towns (population in 000, 1969) Mata-Utuª (capital) 0.6, Singaveᵇ
ªOn Uvéa ᵇOn Futuna Island
Race Mainly Polynesian
Language French; local languages are also in use
Religion Mainly Roman Catholic
Education (1973) Pupils 2 700
Personnel (1974) Physicians: 3, 1 per 3 300 people

Main sources

General (applying to all sections)

Statistical Yearbook, United Nations (annual)
Monthly Bulletin of Statistics, United Nations (monthly)
Main Economic Indicators, Organisation for Economic Co-operation and
  Development (monthly)
Statistical Yearbook, Unesco (annual)
African Statistical Yearbook, United Nations
Statistical Yearbook for Asia and the Pacific, United Nations (annual)
Individual countries: statistical yearbooks and monthly statistical bulletins

Climate, time and measures

Tables of Temperature, Relative Humidity, Precipitation
  and Sunshine for the World, UK Meteorological Office
The Economist Measurement Guide and Reckoner

People

Demographic Yearbook, United Nations (annual)
Population and Vital Statistics Report, United Nations, Series A
Year Book of Labour Statistics, International Labour Office (annual)
Bulletin of Labour Statistics, International Labour Office (quarterly)
Labour force statistics, Organisation for Economic Co-operation and
  Development (annual)

Resources and equipment

World Road Statistics, International Road Federation (annual)
Annual Bulletin of Transport Statistics for Europe, United Nations (annual)
Statistical Tables, Lloyd's Register of Shipping (annual)

Production

Yearbook of National Accounts Statistics, United Nations (annual)
National Accounts of OECD countries, Organisation for Economic
  Co-operation and Development (annual)
World Bank Atlas, World Bank (annual)
FAO Production Yearbook, Food and Agriculture Organisation of the
  United Nations (annual)
FAO Trade Yearbook, Food and Agriculture Organisation of the
  United Nations (annual)
Monthly Bulletin of Agricultural Economics and Statistics, Food and
  Agriculture Organisation of the United Nations (monthly)
Yearbook of Fishery Statistics, Food and Agriculture Organisation of the
  United Nations (annual)
Review of Fisheries in OECD Member Countries, Organisation for
  Economic Co-operation and Development (annual)
Yearbook of Forest Products, Food and Agriculture Organisation of the
  United Nations (annual)
Annual Bulletin of General Energy Statistics for Europe, United Nations
  (annual)

Transport and tourism

Digests of Statistics, International Civil Aviation Organisation (annual
  volumes)
International Tourism and Tourism Policy in OECD Member Countries,
  Organisation for Economic Co-operation and Development (annual)
World Travel Statistics, World Tourism Organisation (annual)

Finance and trade

International Financial Statistics, International Monetary Fund (monthly
  and annual supplement)
IMF Survey, International Monetary Fund (twice each month, except one
  issue in December)
Balance of Payments Yearbook, International Monetary Fund (monthly
  and annual)
World Debt Tables, World Bank (annual and supplements)
Direction of Trade, International Monetary Fund (monthly and annual)
Statistics of Foreign Trade, Series A, B and C, Organisation for Economic
  Co-operation and Development (quarterly and annual)
Yearbook of International Trade Statistics, United Nations (annual)
Individual countries: external trade yearbooks and monthly bulletins

Glossary

Abbreviations: organisations and states

ACP state	African, Caribbean or Pacific state having special arrangements with the EEC under the Lomé Convention
Asean	Association of South East Asian Nations
Benelux	Benelux Economic Union (Belgium, Luxembourg and Netherlands)
BLS	Botswana, Lesotho and Swaziland
BIOT	British Indian Ocean Territory
BVI	British Virgin Islands
CACM	Central American Common Market
Caricom	Caribbean Community
CEAO	Communauté Economique de l'Afrique de l'Ouest (West African Economic Community)
Cento	Central Treaty Organisation
CFA	Communauté Financière Africaine (African Financial Community)
CFP	Communauté Française du Pacifique (French Pacific Community)
Comecon	Council for Mutual Economic Assistance (also CMEA)
Ecowas	Economic Community of West African States
EEC	European Economic Community
EEC ACP	see ACP
Efta	European Free Trade Association
FAO	Food and Agriculture Organisation of the United Nations
Icao	International Civil Aviation Organisation
ILO	International Labour Organisation
IMF	International Monetary Fund
ITU	International Telecommunications Union
Lafta	Latin American Free Trade Association (also Alalc)
Nafta	New Zealand-Australia Free Trade Agreement
Nato	North Atlantic Treaty Organisation
Oapec	Organisation of Arab Petroleum Exporting Countries
OAS	Organisation of American States
OAU	Organisation of African Unity
Ocam	Organisation Commune Africaine et Mauricienne (African and Mauritian Communal Organisation)
Odeca	Organización de Estados Centroamericanos (Organisation of Central American States)
OECD	Organisation for Economic Co-operation and Development
Opec	Organisation of the Petroleum Exporting Countries
RCD	Regional Co-operation for Development
Sela	Sistema Económica Latino Americana
UAE	United Arab Emirates
Udeac	Union Douanière et Economique de l'Afrique Centrale (Customs and Economic Union of Central Africa)
UK	United Kingdom
UN	United Nations
Unctad	United Nations Conference on Trade and Development
Unesco	United Nations Educational, Scientific and Cultural Organisation
UPU	Universal Postal Union
US	United States
USVI	United States Virgin Islands
WEU	Western European Union

Symbols and abbreviations

General

*	provisional or estimated figure
**	rough estimate
***	very rough estimate ('guesstimate')
av	arithmetic average
conc	concentrates
excl	excluding
gdp	gross domestic product
GMT	Greenwich mean time
govt	government
I or Is	island or islands
incl	including
max	maximum
mfrs	manufactures
min	minimum
na	not available or not applicable
no	number
%pa	percentage per annum (compound growth rate per annum)
Rep	Republic
Terr	Territory
Utd	United

Numbers

—	nil or negligible (less than one-half of the relevant unit)
000	thousand
mn	million = 1 000 000 = 10^6
bn	billion = 1 000 000 000 = 10^9

Measures

°C	degrees Celsius (centigrade)
CM	metric carat
grt	gross registered ton (shipping)
hl	hectolitre = 100 litres
kg	kilogram
km	kilometre
km²	square kilometre
kW h	kilowatt-hour
m	metre
m²	square metre
m³	cubic metre
mi	mile
mi²	square mile
mm	millimetre
MW	megawatt = 1 000 kilowatts = 1 000 000 watts
nrt	net registered ton (shipping)
t	tonne or metric ton = 1 000 kilograms
tce	tonnes of coal equivalent

Monetary (see also individual countries)

Esc	Portuguese escudo
Fr	French franc
Gld or Fl	Netherlands guilder or florin
L	Italian lira
£	UK (sterling) pound
Pa	Spanish peseta
R	South African rand
$	US dollar
SDR	IMF special drawing right

Conversion factors

Length
1 inch = 25.4 millimetres, 1 millimetre = 0.03937 inch
1 yard = 0.9144 metre, 1 metre = 1.094 yards
1 mile = 1.609 kilometres, 1 kilometre = 0.621 mile

Area
1 square yard = 0.8361 square metre, 1 square metre = 1.196 square yards
1 square mile = 2.590 square kilometres,
1 square kilometre = 0.3861 square mile

Capacity
1 cubic yard = 0.7646 cubic metre, 1 cubic metre = 1.308 cubic yards
1 UK gallon = 0.04546 hectolitre, 1 hectolitre = 22.00 UK gallons
1 US liquid gallon = 0.03785 hectolitre,
1 hectolitre = 26.42 US liquid gallons
1 barrel crude oil = 42 US gallons = 159 litres

Weight (mass)
1 ounce troy = 0.0311 kilogram, 1 kilogram = 32.15 ounces troy
1 pound = 0.4536 kilogram, 1 kilogram = 2.205 pounds
1 UK (long) ton = 1.016 tonnes, 1 tonne = 0.9842 UK (long) ton
1 short ton = 0.9072 tonne, 1 tonne = 1.102 short tons
1 tonne = 2 204.62 pounds

Temperature
Degrees Fahrenheit = degrees Celsius multiplied by 9, divided by 5, plus 32
Degrees Celsius = degrees Fahrenheit, minus 32, multiplied by 5, divided by 9

Energy (approximate equivalents)
1 tonne crude oil = 1.5 tonnes of coal equivalent
1 million cubic metres of natural gas = 1 330 tonnes of coal equivalent
1 million kilowatt-hours of electricity = 125 tonnes of coal equivalent
8 kilowatt-hours of electricity = 1 kilogram of coal equivalent
1 cubic metre of natural gas = 9 320 kilocalories = 39 megajoules

Notes

General

1. Years refer to calendar years, unless otherwise specified; for figures measured at a point of time, the year refers to mid-year or annual average unless otherwise specified. Reference to a month generally means end-month, except for climate where the monthly average is referred to

A period expressed as 1975/76 means a year ending at a date in 1976; a period expressed as 1975–76 means a period from the calendar year 1975 to the calendar year 1976 or an average of the two years where an average is specified

The Moslem year 1397 is December 23 1976 to December 11 1977 Gregorian time; the Moslem year 1398 is 1977/78 Gregorian. The Buddhist year 2520 is 1977 Gregorian

For any section, the year stated at the beginning of the section applies throughout the section, unless otherwise mentioned. Points for specific items are:
(a) Population figures are for mid-year unless otherwise specified.
(b) Crop production figures, unless otherwise specified, are for harvest years referred to calendar years in which the entire harvest or the bulk of the harvest took place; years for livestock, unless otherwise stated, are for enumeration in years ending September of the year referred to.
(c) Under resources and equipment, stock items (road length, etc) refer in general to end-year or mid-year

2. Figures do not necessarily add to totals because of rounding

3. Monetary figures, where converted from national currencies to the $ or £, have been converted at the average rate for the year or period concerned. The monetary rate of exchange has been used unless otherwise specified; this rate is shown for 1976 in the monetary unit section

4. The term 'agriculture' is used in a general way to include fishing and forestry when referring to sectors of the economy; the agricultural production index, however, excludes fishing and forestry

5. Main items only are shown for most sections

6. Where countries are ranked for any year or period, a country for which the information is not available for the correct time period is included with the latest available information. Only countries with available information or estimates are included

7. Definitions for many items may vary from country to country

8. Ratios comparing items to population have not been marked as estimates where the population is an estimate; they have only been shown as estimates where the item concerned is itself an estimate

9. Figures for latest years are in many cases provisional

10. Territories with no permanent inhabitants are not included; for example, the Antarctic territories

11. Many of the world pages indicate countries ranked in order of importance; if the rough position for any particular country is not known, finding that country can usually be made easier by checking the figure concerned in the individual country section before referring to the ranked list. This is especially useful for tables in which some countries may not be included. Where countries are ranked equally, they are shown in alphabetical order within that equal rank

Land and climate

Area includes water recognised as part of the territory concerned
Arable refers to land under temporary crops, including under market gardens, temporary pastures and land temporarily fallow; *cropland* refers to land cultivated with permanent crops which need not be replanted after each harvest, including cocoa, coffee, rubber etc; *pastures* refers to land used permanently for forage crops; *forests* refers to both natural and planted stands

Temperatures shown for a month are the average daily lowest and highest for the month

Monetary unit

Par refers to a rate which is fixed in terms of one of the main two currencies or other currency specified; 'free' refers to any rate not so fixed, which may vary according to a free market rate or according to the rates varied officially from time to time (within the period of one year). There may be several rates, especially for communist countries, and where conversions are made other than by a commercial rate, the rate used is specified

The SDR (special drawing right) is the currency unit of the International Monetary Fund

There are alternative methods of conversion, including notably a 'purchasing power' rate of exchange; the latter has not been used here as there is not yet comprehensive information, estimated relevant rates having become available only for a comparatively few countries

Political

Main terms used are:
Independent Controlling both internal and external affairs
Republic An independent state having an elected form of government, generally with a president as the head of state
One-party republic A republic ruled in general by one person or party
Military government Control by the armed forces
Monarchy An independent state ruled by a hereditary head
Parliamentary monarchy An independent state having a hereditary head, but controlled by an elected form of government
Communist A state controlling most or all of the means of production and distribution
Territory A state under the control of another
Protectorate A state controlled by another in some of its internal affairs, and all external affairs
Colony A state controlled by another with no or negligible representation
Associated state of the United Kingdom A state fully self-governing in internal affairs, with the United Kingdom responsible for defence and external affairs
French overseas department A subdivision of the French Republic, similar to a department of metropolitan France
French overseas territory A subdivision of the French Republic, having elected representation to the French parliament

People

Population refers to home population.
Regions Total population estimates usually include a figure allowing for under or over enumeration for a census; hence the total of figures for regions, usually based on a census, will not always agree with the estimated total population figure
City populations in general include suburbs
Capital refers in general to the city which is the administrative centre; this can differ from the seat of government or religious centre
Race and language Some general meanings are:
Mestizo Mixed European/American Indian
Mulatto Mixed European/African
Zambo or Sambo Mixed American Indian/African
Creole For race, originally persons of European origin born outside of Europe, the term can now mean those of mixed origin
For language, a composite or pidgin form of language, usually in relation to French
Standard of living Items shown are a rough guide only; for example, energy and steel consumption depend on the type of industry in the country, and energy consumption depends also on climate
National income Refers to total income for the country, including property income from other countries. National income equals gross domestic product (see later, under production), plus net property income from other countries, less depreciation (capital consumption)
Energy consumption Refers to all forms of energy, including consumption by industry
Newspapers Refers to daily newspapers only

Resources and equipment

Mineral reserves In general total resources are shown including probable and possible reserves in addition to 'known economic' reserves. For crude oil and uranium, reserves are those reasonably assured
Roads Generally all-weather, excluding tracks
Railways Refers to route-length (which is equal to or shorter than track-length)
Ships Refers to registered vessels of 100 gross tons and over at mid-year. Registration does not always imply ownership in the country concerned
Airports Passenger and cargo traffic refers to total domestic and international, excluding direct transit
Ports Goods traffic is international only, unless otherwise specified
Companies Includes state-owned or partly owned organisations

Production

Gross domestic product Total product for the country, which excludes property income from other countries. For each country, the latest official figure is shown where available, together with an estimate for 1976 if there is no official figure for that year
Growth in real terms Refers to growth of quantity, as measured by gross domestic product at constant prices
Gross material product Used by communist countries, this differs from gross domestic product of other countries primarily in excluding the value of non-material services, such as public administration and defence, personal and professional services and similar activities
Net material product is gross material product less depreciation (capital consumption)
Production indices Agricultural excludes fishing and forestry Industrial includes mining, manufacturing, utilities (electricity, gas and water), and in general excludes construction

Main products

Agriculture Rice refers to paddy rice, onions to dry onions, cocoa to cocoa beans, coffee to green coffee, cotton to lint cotton, wool to scoured wool, milk to cow milk, cheese to all kinds of cheese and eggs to hen eggs
Energy Total energy refers to indigenous production only, from coal, lignite, crude oil, natural gas, and hydro, geothermal or nuclear electricity. . Lignite includes brown coal. Natural gas excludes shrinkage
Mining Gold refers to refined gold, silver to refined silver, mercury to Hg content of ores, sulphur to elemental sulphur unless otherwise specified
Manufacturing Cotton yarn refers to pure yarn (but including mixed in some cases), paper to paper and paperboard, tyres to motor vehicle tyres, sulphuric, hydrochloric and nitric acids to pure acid (in terms of 100 % acid), nitrogenous, phosphate and potash fertilisers to nutrient content, coke to coke-oven coke, pig iron to pig iron and ferro-alloys, aluminium to primary aluminium, copper to primary and secondary copper, lead to primary lead, zinc to primary zinc and merchant vessels to merchant vessels launched

Transport traffic

Water transport Refers to inland waterways only, excluding coastwise shipping traffic
Cargo Includes freight and mail for air, freight only for rail
Sea Refers to international traffic only
Tourists 'Visitors' refers to tourist arrivals at frontiers, excluding excursionists and cruise passengers unless otherwise stated

Finance

Consumer price index Growth rates indicate the amount of 'inflation'
Share price index Refers to industrial share prices
Budget Refers to current revenue and expenditure (excluding capital expenditures), unless otherwise specified; are generally estimates
Balance of payments Balance of goods (fob) differs from the crude balance of trade (where imports are shown cif); transfers include private and government transfers.
The current balance is the balance of goods, services and transfers; in general the sum of this and long-term capital movements (items often of opposite sign) is balanced by monetary movements.
For each item a — sign means a net outflow of funds and a + sign a net inflow

International reserves Includes gold, SDRs, foreign exchange holdings and the reserve position in the International Monetary Fund
Public debt Total outstanding, including undisbursed. 'Domestic' refers to debt in local currency

External trade

Imports are shown cif (cost, insurance, freight included)
Exports are shown fob (free on board)
Unless otherwise specified, exports means exports of goods, and imports means imports of goods
Main imports and exports Food includes livestock, electrical machinery includes electrical equipment, motor vehicles includes spare parts and bunkers includes supplies to ships and aircraft. Classification and meaning of terms is based in general on the Standard International Trade Classification (SITC)
Main sources and destinations The ultimate source or destination can differ from the country indicated; for imports, sources are usually the last country of shipment, but this can differ from the country of production because goods may have been trans-shipped, smuggled or brought in on false papers
Ranking of main imports and exports, and sources and destinations, has been based on actual figures before rounding
Customs areas There are some cases of customs unions where information is not available separately for the individual countries making up the customs union. It may be noted that extracting figures for countries which form part of a customs union usually results in an increase in the value of external trade; for example, if Puerto Rico is to be separated from the United States/Puerto Rico total to give a United States only figure, imports from Puerto Rico are added to the United States total – while the only imports to be deducted from the United States total are imports into Puerto Rico from countries other than the United States. If there were a United States of Europe the total of world external trade would fall markedly (internal trade increasing), and if trade between the States of the United States of America were regarded as external, the total of world external trade would be greatly enlarged

Country name index

Special focus index